MATHEMATICS
for
CDS
ENTRANCE EXAMINATION
(Fully Solved)

Dr R S AGGARWAL

S Chand And Company Limited
(ISO 9001 Certified Company)

S Chand And Company Limited

(ISO 9001 Certified Company)

Head Office: D-92, Sector–2, Noida 201301, U.P. (India), Ph. 91-120-4682700

Registered Office: A-27, 2nd Floor, Mohan Co-operative Industrial Estate, New Delhi – 110 044 Phone: 011-49731800

www.schandpublishing.com; e-mail: info@schandpublishing.com

Branches

Chennai	:	Ph: 044-2363 2120, chennai@schandpublishing.com
Guwahati	:	Ph: 0361-4066 369, guwahati@schandpublishing.com
Hyderabad	:	Ph: 040-4018 6018, hyderabad@schandpublishing.com
Jalandhar	:	Ph: 0181-4645 630, jalandhar@schandpublishing.com
Kolkata	:	Ph: 033-2335 7458, 2335 3914, kolkata@schandpublishing.com
Lucknow	:	Ph: 0522-4003 633, lucknow@schandpublishing.com
Mumbai	:	Ph: 022-2500 0297, mumbai@schandpublishing.com
Patna	:	Ph: 0612-4011400, patna@schandpublishing.com

First Edition 1990

Subsequent Editions and Reprints 1993, 94, 95, 96, 98, 99, 2000, 2001, 2003, 2004, 2007, 2008, 2009, 2010, 2011, 2012, 2013 (Twice), 2014, 2015

Revised Edition 2018; Reprints 2018, 2019, 2021; 2022 (Fourth)

Revised Edition 2024

Reprint 2026

ISBN : 978-93-5870-666-6 **Product Code:** H1MCD61MATH40ENZX24O

PRINTED IN INDIA

By Shri Shyamlal Printing Press Private Limited , Plot 20/4, Site-IV, Industrial Area Sahibabad, Ghaziabad – 201 010 and Published by S Chand And Company Limited, A-27, 2nd Floor, Mohan Co-operative Industrial Estate, New Delhi – 110 044

PREFACE TO THE REVISED EDITION

Mathematics for CDS Entrance Examination has long been recognised for its comprehensive collection of objective type questions paired with detailed solutions. This revised edition builds upon this strong foundation by introducing new questions specifically designed to align with the latest examination patterns and trends. These updates ensure that students are well-prepared for the evolving challenges of the CDS examination.

The fully solved latest paper CDS-II, 2025 (held on 14 September, 2025) has been added.

Each chapter now includes latest questions and their solutions, reflecting the most recent examinations up to CDS-I, 2025 (April, 2025). These carefully selected questions provide candidates with a clear understanding of the current format and types of questions they can expect. This targeted guide is intended to significantly enhance their readiness for the test and instil greater confidence as they approach the exam.

As always, suggestions and feedback from the readers are welcomed greatly.

Dr R S AGGARWAL

PREFACE TO THE FIRST EDITION

This book on Mathematics is really an asset to those who plan to appear in competitive examinations for CDS.

The market is flooded with different types of guide books for the above examinations. Most of these guides carry a small portion of problems on mathematics. These problems have been given along with answers. But, how to get these answers is not given therein. This book contains a huge accumulation of objective type questions with their solutions by short-cut methods. The references of questions asked in various examinations have been given on memory basis.

It is very much hoped that the subject matter will create a confidence among the candidates and the book will help them like an ideal teacher.

I convey my gratitude to Shri T.N. Goel, Manager and Shri R.K. Gupta, M.D., S. Chand and Company Ltd., New Delhi for taking all pains, and interest in the publication of the book.

Constructive suggestions will be highly appreciated and welcome.

Dr R S AGGARWAL

What is CDS?

CDS Exam or Combined Defence Services Examination is conducted by the Union Public Service Commission (UPSC) for recruitment of Commissioned Officers in the Indian Military Academy, Officers Training Academy (OTA), Indian Naval Academy and Indian Air Force Academy.

The Notification for the examination is usually released in the months of December and May and the examinations are conducted in April and September respectively.

Only unmarried graduates are eligible to sit for the exam. Examination is conducted twice a year. Successful candidates are admitted into the respective Academies after an interview conducted by the Services Selection Board (SSB). This exam has 2 steps. First, there is a written test and after passing it there is the interview.

Age Limit

Indian Military Academy	19-24 years
Air Force Academy	20-24 years
Indian Naval Academy	19-24 years
Officers Training Academy	19-25 years

Educational Qualifications

CDS Exam Educational Qualifications is given below.

(i) For I.M.A. and Officers' Training Academy, Chennai — Degree of a recognised University or equivalent.

(ii) For Indian Naval Academy— Degree in Engineering from a recognised University/ Institution

(iii) For Air Force Academy—Degree of a recognised University (with Physics and Mathematics at 10+2 level) or Bachelor of Engineering.

CDS Exam Pattern

CDS Exam Pattern for IMA, INA, AFA

Subject	Code	Duration	M.M.
English	11	2 hours	100
General Knowledge	12	2 hours	100
Elementary Mathematics	13	2 hours	100

CDS Exam Pattern for OTA

Subject	Code	Duration	M.M.
English	11	2 hours	100
General Knowledge	12	2 hours	100

DETAILED SYLLABUS OF CDS

ELEMENTARY MATHEMATICS

1. Arithmetic

(*i*) Number System—Natural numbers, Integers, Rational and Real numbers. Fundamental operations, addition, substraction, multiplication, division, Square roots, Decimal fractions. Unitary method, time and distance, time and work, percentages, applications to simple and compound interest, profit and loss, ratio and proportion, variation.

(*ii*) Elementary Number Theory—Division algorithm. Prime and composite numbers. Tests of divisibility by 2, 3, 4, 5, 9 and 11. Multiples and factors. Factorisation Theorem. H.C.F. and L.C.M. Euclidean algorithm. Logarithms to base 10, laws of logarithms, use of logarithmic tables.

2. Algebra

Basic Operations, simple factors, Remainder Theorem, H.C.F., L.C.M., Theory of polynomials, solutions of quadratic equations, relation between its roots and coefficients (Only real roots to be considered). Simultaneous linear equations in two unknowns—analytical and graphical solutions. Simultaneous linear inequations in two variables and their solutions. Practical problems leading to two simultaneous linear equations or inequations in two variables or quadratic equations in one variable & their solutions. Set language and set notation, Rational expressions and conditional identities, Laws of indices.

3. Trigonometry

Sine x, cosine x, Tangent x when $0° \leq x \leq 90°$ Values of sin x, cos x and tan x, for $x = 0°, 30°, 45°, 60°$ and $90°$

Simple trigonometric identities.

Use of trigonometric tables.

Simple cases of heights and distances.

4. Geometry

Lines and angles, Plane and plane figures, Theorems on (i) Properties of angles at a point, (ii) Parallel lines, (iii) Sides and angles of a triangle, (iv) Congruency of triangles, (v) Similar triangles, (vi) Concurrence of medians and altitudes, (vii) Properties of angles, sides and diagonals of a parallelogram, rectangle and square, (viii) Circles and its properties including tangents and normals, (ix) Loci.

5. Mensuration

Areas of squares, rectangles, parallelograms, triangle and circle. Areas of figures which can be split up into these figures (Field Book), Surface area and volume of cuboids, lateral surface and volume of right circular cones and cylinders, surface area and volume of spheres.

6. Statistics

Collection and tabulation of statistical data, Graphical representation frequency polygons, histograms, bar charts, pie charts etc. Measures of central tendency.

CDS Mathematics : Trend Analysis

S.No.	Chapter Names	2025		2024		2023		2022		2021	
		II	I	II	I	II	I	II	I	II	I
1	Numbers	16	17	20	15	5	8	2	6	10	1
2	H.C.F. and L.C.M. of Numbers		1	2	1	4	5	1	1	2	2
3	Decimal Fractions								1		
4	Square Roots		1	1		1	1	1	4	4	
5	Percentage	2	1	3						2	1
6	Average		1	2					1	4	2
7	Ratio and Proportion	2	2	3	4	3	1	4		2	2
8	Partnership				1				2	4	3
9	Profit and Loss	1	1		2	1	1		1	1	3
10	Unitary Method			2					1	1	1
11	Time and Work			1	4		2	3		2	2
12	Time and Distance	2	2	3	5		2	9	4	2	2
13	Problems on Ages					1			2	2	2
14	Simple Interest			1	1		1	1	1	1	2
15	Compound Interest	1	2	1	2	2				1	2
16	Logarithms	3	2			1	1	1		1	2
17	Area	5	3	7	19	15	12	9	7	5	3
18	Volume and Surface Areas	5	8	4	3	3	7	8	8	10	10
19	Basic Operations and Factorization	9	3	3	4	7	3	5	4	4	5
20	H.C.F. and L.C.M. of Polynomials	3	2	3	2	2	1	2	1	2	2

S.No.	Chapter Names	2025		2024		2023		2022		2021	
		II	I	II	I	II	I	II	I	II	I
21	Linear Equations					5	3	1		1	
22	Quadratic Equations	3	5	4	2	8	3	3	3	3	
23	Rational Expressions	4			1				2	1	1
24	Surds and Indices			2	1		1		4	1	3
25	Set Theory	1		2			2	4	1		
26	Identities and T-Ratios	15	13	14	8	11	9	8	15	10	15
27	Height and Distance	1	2	2	3	7	2	1	2	2	1
28	Lines and Angles	2				6	17	12	1	2	
29	Polygons	2		1	1	1			3		5
30	Triangles	7	10	9	2	6	3	8	6	6	6
31	Quadrilaterals and Parallelograms	3	1		8		3	5	2	1	1
32	Circles	3	8	6	7	6	7	4	3	3	4
33	Loci								1	2	2
34	Coordinate Geometry								2		1
35	Statistical Data (Frequency distribution, Mean, Mode, Median and Standard Deviation)	10	9	4	1	2	5	8	6	8	9
36	Data Interpretation		6		3	3			5		5
	Total	**100**	**100**	**100**	**100**	**100**	**100**	**100**	**100**	**100**	**100**

CONTENTS

UNIT-1 ARITHMETIC

UNIT-2 MENSURATION

UNIT-3 ALGEBRA

UNIT-4 TRIGONOMETRY

UNIT-5 GEOMETRY

UNIT-6 COORDINATE GEOMETRY

UNIT-7 STATISTICS

UNIT 1

ARITHMETIC

Number System: Natural numbers, Integers, Rational and irrational numbers. Fundamental operations 1 Addition, Subtraction, Multiplication & Division : Square Roots and Decimal Fractions.

Elementary Number Theory: Division Algorithm; Prime & Composite numbers. Tests of divisibility by 2, 3, 4, 5, 9, & 11. Factors & Multiples. Factorisation Theorem, H.C.F. & L.C.M.; Euclidean — Algorithm.

Unitary Method: Time & Distance; Time & Work; Percentages; Application to Simple & Compound Interest, Profit & Loss; Ratio & Proportion; Variation.

Logarithms: Logarithm to base 10, Laws of logarithms, Use of logarithmic tables.

Numbers

CHAPTER

> **Highlights:** Number System: Natural numbers, Integers, Rational & real numbers. Fundamental operations: Addition, Subtraction, Multiplication & Division. Elementary Number Theory: Division Algorithm, Prime & Composite numbers, Test of divisibility by 2, 3, 4, 5, 8, 9, 11. Multiples & Factors, Factorization theorem & Division Algorithm.

Numbers: In Hindu Arabic system, we use ten symbols 0, 1, 2, 3, 4, 5, 6, 7, 8, 9 called digits, to represent any number.

A group of figures, denoting a number is called a numeral. For a given numeral, we start from extreme right as, 'unit's place; ten's place; hundred's place' and so on.

Ex. ***We represent a number, '793802465' as shown below:***

Ten Crores 10^8	Crores 10^7	Ten Lacs (millions) 10^6	Lacs 10^5	Ten Thousands 10^4	Thousands 10^3	Hundreds 10^2	Tens 10^1	Units 10^0
7	9	3	8	0	2	4	6	5

We read it as:

'Seventy-nine crores, thirty eight lacs, two thousand, four hundred and sixty five'

In the above numeral:

The place value of 5 is $(5 \times 1) = 5$;

The place value of 6 is $(6 \times 10) = 60$;

The place value of 4 is $(4 \times 100) = 400$ and so on.

The place value of 7 is $7 \times 10^8 = 700000000$.

The face value of a digit in a number is the value of digit itself wherever it may be.

Thus, the face value of 7 in the above numeral is 7; the face value of 9 is 9 and so on.

Various Types of Numbers

Natural Numbers: *Counting numbers are called natural numbers.*

Thus, $N = \{1, 2, 3, 4, 5, ...\}$ is the set of all natural numbers.

Whole Numbers: *All counting numbers together with zero form the set of all whole numbers.*

Thus, $W = \{0, 1, 2, 3, 4, 5, ...\}$ is the set of all whole numbers.

Clearly, every natural number is a whole number.

But, 0 is a whole number which is not a natural number.

Integers: *The set I of all natural numbers, 0 and negatives of counting numbers, is the set of all integers.*

Thus, $I = \{ ...., -3, -2, -1, 0, 1, 2, 3, ...\}$ is the set of all integers.

Postitive Integers: *The set $I^+ = \{1, 2, 3, 4, ...\}$ is the set of all positive integers.* Clearly, positive integers and natural numbers are synonyms.

Negative Integers: *The set $I^- = \{-1, -2, -3, ...\}$ is the set of all negative integers.*

Note that 0 is neither positive nor negative.

Non-negative Integers: *The set* $\{0, 1, 2, 3, ...\}$ *is the set of non-negative integers.*

Non-positive Integers: *The set* $\{0, -1, -2, -3, ...\}$ *is the set of all non-positive integers.*

Rational Numbers: *The numbers of the form $\frac{p}{q}$, where p and q are integers and $q \neq 0$, are known as rational numbers. e.g.* $\frac{3}{5}, \frac{9}{7}, \frac{-2}{3}, \frac{0}{1}, \frac{-143}{272}$ etc.

Thus, $Q = \left[\frac{p}{q} : p \;\&\; q \text{ are integers } \&\; q \neq 0\right]$ is the set of all rational numbers.

Since every natural number a can be written as $\frac{a}{1}$, so every natural number is a rational number.

Since 0 can be written as $\frac{0}{1}$ and every non-zero integer a can be written as $\frac{a}{1}$, so *every integer is a rational number.*

Terminating & Repeating Decimals: *Every rational number has a peculiar characteristic that it when expressed in decimal form is expressible either in terminating decimals or in repeating decimals.*

e.g. $\frac{1}{2} = 0.5$; $\frac{1}{3} = 0.333 ... = 0.\overline{3}$, $\frac{22}{7} = 3.\overline{142857}$, $\frac{8}{45} = 0.1\overline{7}$, $\frac{69}{550} = 0.12\overline{54}$ etc.

Irrational Numbers: All numbers which when expressed in decimal form are in non-terminating and non-repeating form, are known as irrational numbers. *e.g.* $\sqrt{2}, \sqrt{3}, \sqrt{5}, \sqrt{7}, \neq$ etc.

Note that the exact value of π is not $\frac{22}{7}$, as $\frac{22}{7}$ is rational while π *is irrational.*

$\frac{22}{7}$ is the approximate value of π. Similarly, 3.14 is not an exact value of π.

Also, 0.101001000100001 ... is a non-terminating and non-repeating decimal, so it is an example of an irrational number.

Real Numbers: *The totality of all rational and all irrational numbers forms the set R of all real numbers.*

Thus, every natural number, every whole number, every integer, every rational number and every irrational number is a real number.

Note that:

(*i*) The sum (or difference) of a rational and an irrational number is irrational.

e.g. $(2+\sqrt{3}), (3-\sqrt{5}), \left(\frac{5}{2}-\sqrt{2}\right), (-1+\sqrt{7})$ etc. are all irrational.

(*ii*) The Product of a rational and an irrational number is irrational, *e.g.* $2\sqrt{3}, -3\sqrt{5}$ etc. are all irrational.

More Discussion About Numbers

Even & Odd Numbers: Integers divisible by 2 are known as even integers, while those which are not divisible by 2 are known as odd integers.

Thus, ..., – 6, – 4, – 2, 0, 2, 4, 6, ... etc. are even integers.

And, ..., – 5, – 3, – 1, 1, 3, 5, etc. are odd integers.

Prime Numbers: *A number greater than 1 is called a prime number if it has exactly two factors, namely 1 and itself.*

Ex. 1. ***Which of the following numbers are prime?***

(*i*) 165 **(*ii*) 247** **(*iii*) 571.**

Sol. (*i*) The nearest value of $\sqrt{165}$ is 13, *i.e.* $13 > \sqrt{165}$.

So, we test the divisibility of 165 by each of the prime numbers less that 13.

We find that 165 is divisible by 11.

So, 165 is not a prime number.

(*ii*) Clearly, $16 > \sqrt{247}$.

So, we divide 247 by each prime number less than 16 which are 2, 3, 5, 7, 9, 11, & 13.

We find that 247 is divisible by 13.

Hence, 247 is not a prime number.

(*iii*) Clearly, $24 > \sqrt{571}$.

So, we divide 571 by each prime number less than 24 which are 2, 3, 5, 7, 9, 11, 13, 17, 19 and 23.

We find that 571 is not divisible by any of them.

So, 571 is a prime number.

Composite Numbers: *Numbers greater than 1 which are not prime, are known as composite numbers. e.g. 4, 6, 8, 9, 10, 12 etc. are all composite numbers.*

Remarks: (*i*) 1 is neither prime nor composite.

(*ii*) 2 is the only even number which is prime.

(*iii*) There are 25 prime numbers between 1 & 100.

Division on Numbers

Division Algorithm: *Let a and b be two given integers such that $b \neq 0$. On dividing a by b, let q be the quotient and r the remainder, then $a = bq + r$.*

Clearly, $0 \le r < b$.

In general, we have:

$$\begin{array}{r} 16\,)\,123\,(\,7 \\ 112 \\ \hline 11 \end{array}$$

Dividend = (Divisor × Quotient) + Remainder

Ex. 1. ***On dividing 16890 by a certain number, the quotient is 123 and the remainder is 39. Find the divisor.***

Sol. $\text{Divisor} = \dfrac{(\text{Dividend} - \text{Remainder})}{\text{Quotient}} = \left(\dfrac{16890 - 39}{123}\right) = 137.$

Ex. 2. ***What least number must be added to 5204 to get a number exactly divisible by 180?***

Sol. On dividing 5204 by 180, the remainder is 164.

∴ Required number to be added = (180 – 164) = 16.

Ex. 3. ***Find the number nearest to 5028, which is exactly divisible by 64.***

Sol. On dividing 5028 by 64, we get 36 as remainder, which is greater than half of the divisor.

∴ Required number = 5028 + (64 – 36) = 5056.

Tests of Divisibility

(*i*) **Divisibility by 2:** *A given number is divisible by 2, if its unit digit is any one of 0, 2, 4, 6 & 8.*

Ex. ***The numbers 16970, 217352, 67904, 81376 & 34918 end in 0, 2, 4, 6, & 8 respectively. So, they are all divisible by 2.***

(*ii*) **Divisibility by 5:** *A given number is divisible by 5, if its unit digit is either 0 or 5.*

Ex. ***The numbers 63180 & 75635 end in 0 and 5 respectively.***

So, both of them are divisible by 5.

(*iii*) **Divisibility by 3:** *A given number is divisible by 3, if the sum of its digits is divisible by 3.*

Ex. 1. ***Show that 6903258 is divisible by 3.***

Sol. Sum of its digits is 33, which is divisible by 3.

So, the given number is divisible by 3.

Ex. 2. ***Show that 734689 is not divisible by 3.***

Sol. Sum of its digits is 37, which is not divisible by 3.

So, the given number is not divisible by 3.

(*iv*) **Divisibility by 9:** *A given number is divisible by 9, if the sum of its digits is divisible by 9.*

Ex. 1. ***Show that 7864506 is divisible by 9.***

Sol. Sum of its digits is 36, which is divisible by 9.

So, the given number is divisible by 9.

Ex. 2. ***Show that 65879413 is not divisible by 9.***

Sol. Sum of its digits is 43, which is not divisible by 9.

So, the given number is not divisible by 9.

(*v*) **Divisibility by 4:** A given number is divisible by 4, if the number formed by its last two digits is divisible by 4.

Ex. 1. ***Show that 738924 is divisible by 4.***

Sol. Clearly, the number formed by the last two digits of the given number is 24, which is divisible by 4.

Hence, the given number is divisible by 4.

Ex. 2. ***Show that 8844634 is not divisible by 4.***

Sol. Clearly, the number formed by the last two digits of the given number is 34, which is not divisible by 4.

So, the given number is not divisible by 4.

(*vi*) **Divisibility by 8:** *A given number is divisible by 8, if the number formed by its last three digits is divisible by 8.*

Ex. 1. ***Show that 769573512 is divisible by 8.***

Sol. Clearly, the number formed by the last three digits of the given number is 512, which is divisible by 8.

So, the given number is divisible by 8.

Ex. 2. ***Show that 4862412 is not divisible by 8.***

Sol. Clearly, the number formed by the last three digits of the given number is 412, which is not divisible by 8.

So, the given number is not divisible by 8.

(*vii*) **Divisibility by 11:** *A given number is divisible by 11, if the difference of the sum of its digits at odd places and the sum of its digits at even places, is either 0 or a number divisible by 11.*

Ex. 1. ***Show that 684387 is divisible by 11.***

Sol. Sum of digits at odd places = $(7 + 3 + 8) = 18$.

Sum of digits at even places = $(8 + 4 + 6) = 18$.

Difference of sums = (18 − 18) = 0.

Hence, the given number is divisible by 11.

Ex. 2. ***Show that 915436708 is not divisible by 11.***

Sol. Sum of digits at odd places = (8 + 7 + 3 + 5 + 9) = 32.

Sum of digits at even places = (0 + 6 + 4 + 1) = 11.

∴ Difference of sums = (32 − 11) = 21, which is not divisible by 11.

Hence, the given number is not divisible by 11.

Co-primes: *Two numbers a and b are said to be co-primes, if their H.C.F. is 1.*

Ex. ***Each one of the pairs (4, 7), (5, 9), (6, 7) etc. is a pair of co-primes.***

An Important Result: If a number is divisible by p as well as by q, where p and q are co-primes, then the given number is divisible by pq.

If p and q are not co-primes, then the given number need not be divisible by pq even when it is divisible by both p and q.

Ex. 1. ***Show that 569172 is divisible by both 2 and 4, but not divisible by 8. Why so?***

Sol. Given number ends in 2. So, it is divisible by 2.

Number formed by last 2 digits of the given number is 72, which is divisible by 4. So, the given number is divisible by 4.

But, the number formed by the last 3 digits of the given number is 172, which is not divisible by 8.

So, the given number is not divisible by 8.

Clearly, 2 and 4 are not co-primes. So, the given number need not be divisible by their product, namely 8.

Ex. 2. ***Show that 2784936 is divisible by 88.***

Sol. The number formed by the last 3 digits of the given number is 936, which is divisible by 8. So, the given number is divisible by 8.

In the given number, we have:

(Sum of digits at odd places) − (Sum of digits at even places)

= [(6 + 9 + 8 + 2) − (3 + 4 + 7)] = 11, which is divisible by 11.

So, the given number is divisible by 11.

Thus, the given number is divisible by both 8 and 11, where 8 and 11 are co-primes.

Hence, the given number is divisible by 88.

Multiplication By Short Cut-Methods:

(i) **Multiplication of a Number By 99, 999, 9999 etc.**

Rule: *Place as many zeros to the right of the multiplicand as is the number of nines and from the number so formed, subtract the multiple and to get the answer.*

Ex. 1. ***Multiply 567958 by 99999.***

Sol. Using the above rule, we have:

567958 × 99999 = (56795800000 − 567958) = 56795232042.

(*ii*) ***Multiplication of a Number By 5^n.***

Rule: Put n zeros to the right of the multiplicand and divide the number so formed by 2^n.

Ex. 2. ***Multiply 975436 by 625.***

Sol. We may write, $625 = 5^4$.

$\therefore\ 975436 \times 625 = 975436 \times 5^4 = \dfrac{9754360000}{16} = 609647500.$

(*iii*) **Multiplication By Distributive Law:**

We use the laws:

I. $a \times b + a \times c = a \times (b + c)$ II. $a \times b - a \times c = a \times (b - c)$

Ex. 3. ***Evaluate: 978 × 184 + 978 × 816.***

Sol. $978 \times 184 + 978 \times 816 = 978 \times (184 + 816) = 978 \times 1000 = 978000.$

Ex. 4. ***Evaluate: 857 × 318 – 857 × 218 .***

Sol. $857 \times 318 - 857 \times 218 = 857 \times (318 - 218) = 857 \times 100 = 85700.$

(*iv*) **Multiplication By Using Formulae:**

Multiplication can be performed quickly by using the formulae:

(*i*) $(a + b)^2 = a^2 + b^2 + 2ab$ (*ii*) $(a - b)^2 = a^2 + b^2 - 2ab$

(*iii*) $(a + b)^2 - (a - b)^2 = 4ab$ (*iv*) $(a + b)^2 + (a - b)^2 = 2(a^2 + b^2)$

(*v*) $(a^2 - b^2) = (a + b)(a - b)$ (*vi*) $(a^3 - b^3) = (a - b)(a^2 + ab + b^2)$

(*vii*) $(a^3 + b^3) = (a + b)(a^2 - ab + b^2)$

Ex. 5 ***Evaluate: [(1786)² – (214)²].***

Sol. $[(1786)^2 - (214)^2] = (1786 + 214)(1786 - 214) = (2000 \times 1572) = 3144000$

Ex. 6. ***Find the value of:*** $\dfrac{(997+496)^2-(997-496)^2}{997\times 496}$

Sol. Given Exp. $\dfrac{(a+b)^2-(a-b)^2}{ab}$, where $a = 997$ & $b = 496$

$= \dfrac{4\,ab}{ab} = 4$ $[\because (a + b)^2 - (a - b)^2 = 4ab]$

Ex. 7. ***Evaluate:*** $\left(\dfrac{657\times 657\times 657-368\times 368\times 368}{657\times 657+657\times 368+368\times 368}\right)$.

Sol. Given Exp. $= \left(\dfrac{a^3-b^3}{a^2+ab+b^2}\right)$, where $a = 657$ & $b = 368$

$= (a - b) = (657 - 368) = 289.$

Ex. 8. ***Evaluate*: (287 × 287 + 213 × 213 + 2 × 287 × 213).**

Sol. Given Exp. $= (a^2 + b^2 + 2ab)$, where $a = 287$ & $b = 213$

$= (a + b)^2 = (287 + 213)^2 = (500)^2 = 250000.$

Ex. 9. ***Evaluate:*** **(*i*) (1307)²** **(*ii*) (1498)²**

Sol. (*i*) $(1307)^2 = (1300 + 7)^2 = (1300)^2 + (7)^2 + 2 \times 1300 \times 7$ $[\because (a+b)^2 = a^2 + b^2 + 2ab]$

$= (1690000 + 49 + 18200) = 1708249.$

(*ii*) $(1498)^2 = (1500 - 2)^2 = (1500)^2 + 2^2 - 2 \times 1500 \times 2$ $[\because (a-b)^2 = a^2 + b^2 - 2ab]$

$= (2250000 + 4 - 6000) = 2244004.$

Ex. 10. ***Find the total number of prime factors in*** $(6^{41} \times 3^{56} \times 11^7 \times 65^9)$.

Sol. Given Exp. $= (2 \times 3)^{41} \times 3^{56} \times 11^7 \times (5 \times 13)^9 = (2^{41} \times 3^{97} \times 5^9 \times 11^7 \times 13^9)$

$\therefore$ Total number of prime factors $= (41 + 97 + 9 + 7 + 9) = 163$.

OPERATIONS ON REAL NUMBERS

Let R be the set of all real numbers.

Properties of Addition on R:

1. **Closure Property:** *The sum of two real numbers is always a real number.*
   i.e. $a \in R, b \in R \Rightarrow a + b \in R$ for all $a, b \in R$.
   We say that R is ***closed for addition.***
2. **Associative Law:** $(a + b) + c = a + (b + c)$ for all $a, b, c \in R$.
3. **Commutative Law:** $a + b = b + a$ for all $a, b \in R$.
4. **Additive Identity:** *Zero is a real number such that* $0 + a = a + 0 = a$ for all $a \in R$.
   0 is called the additive identity in R.
5. **Additive Inverse:** *For each* $a \in R$., *there exists* $-a \in R$ *such that* $a + (-a) = (-a) + a = 0$.
   The real number – a is called the additive inverse of *a*.

Properties of Multiplication on R:

1. *R is closed for multiplication*
   i.e. $a \in R, b \in R \Rightarrow ab \in$ R for all $a, b \in R$.
2. $(ab)\,c = a\,(bc)$ for all $a, b, c \in R$ [Associative Law]
3. $ab = ba$ for all $a, b \in R$ [Commutative Law]
4. $a(b + c) = ab + ac$ [Distributive Law]
5. $1 \in R$ such that $1.\ a = a\ .\ 1 = a$ for all $a \in R$.
   1 is called the *multiplicative identity* in *R*.
6. For each non-zero real number $a \in R$ there exists real number $\frac{1}{a}$ such that

   $a\ .\ \frac{1}{a} = \frac{1}{a}\ .\ a = 1$. The number $\frac{1}{a}$ is called the *multiplicative inverse* or *reciprocal of a.*

   Note that 0 has no reciprocal.

Properties of Subtraction & Division on R:

(*i*) **R** is closed for subtraction.

(*ii*) Subtraction on **R** does not satisfy the commutative and associative laws

(*iii*) **R** *is not closed for division, since*

$$2 \in \mathbf{R},\ 0 \in \mathbf{R} \text{ but } \frac{2}{0} \notin \mathbf{R}.$$

Properties of Addition & Subtraction on I:

(*i*) *I is closed for addition.*
i.e. $a \in I, b \in I \Rightarrow a + b \in I$ for all $a, b \in I$.

(*ii*) $(a + b) + c = a + (b + c)$ for all $a, b, c \in I$ [Associative Law]

(*iii*) $a + b = b + a$ for all $a, b \in I$ [Commutative Law]

(*iv*) *0 is the additive identity in I, since* $0 + a = a + 0 = a$ for all $a \in I$.

(*v*) *For each* $a \in I$ there exists $-a \in I$ *such that* $a + (-a) = (-a) + a = 0$.
Integer – *a* is called the additive inverse of *a*.

(*vi*) *I is closed for subtraction, since* $a \in I, b \in I \Rightarrow a - b \in I$ for all $a, b \in I$.

(*vii*) In general, $a - b \neq b - a$, as $2 - 3 \neq 3 - 2$.
(*viii*) In general, $(a - b) - c \neq a - (b - c)$ as $(2 - 3) - 4 \neq 2 - (3 - 4)$.

Properties of Multiplication & Division on I:

(*i*) *I is closed for multiplication.*
(*ii*) *Multiplication on I is commutative as well as associative.*
(*iii*) *Multiplication distributes addition on I.*
(*iv*) *1 is the multiplicative identity.*

Factors in I: *Let a, b $\in I$. We say that a is a factor of b, if there exists an integer p such that b = ap. In this case, we write, a/b.*

If a is a factor of b, then b is called a multiple of a.

Ordering in I: *For any two integers a, b we say that $a > b$, if $a = b + c$ for some integer c.* If $a > b$, then we write, $b < a$.

Trichotomy Law: *For any two integers a and b one and only one of the following holds:*

(*i*) $a > b$ (*ii*) $a < b$ (*iii*) $a = b$

Properties of Addition on Q:

(*i*) *Q is closed for addition.*
(*ii*) *Addition on Q is commutative as well as associative.*
(*iii*) $0 \in Q$ *is the additive identity.*
(*iv*) *For each* $\frac{a}{b} \in Q$ *there exists* $\frac{-a}{b} \in Q$ *such that*

$$\frac{a}{b} + \left(\frac{-a}{b}\right) = \left(\frac{-a}{b}\right) + \frac{a}{b} = 0.$$ *We call* $-\frac{a}{b}$ *the additive inverse of* $\frac{a}{b}$.

(*v*) *Q is closed for subtraction.*

Properties of Multiplication on Q:

(*i*) *1 is closed for multiplication.*
(*ii*) *Multiplication on Q is commutative as well as associative.*
(*iii*) *1 is the multiplicative identity in Q.*
(*iv*) *Every non-zero rational number (a/b) has its multiplicative inverse or reciprocal (b/a). Note that the rational number 0 has no reciprocal.*

An Important Property: *Between any two rational numbers, there are an infinite number of rational numbers.*

Note: *If x and y are any two rational numbers such that $x < y$,* then $\frac{1}{2}(x + y)$ *is a rational number lying between x and y.*

Ex. ***Write three rational numbers between*** $\frac{1}{2}$ ***and*** $\frac{3}{4}$.

Sol. Let $a = \frac{1}{2}\left(\frac{1}{2} + \frac{3}{4}\right) = \frac{5}{8}$.

$\therefore \frac{1}{2}, \frac{5}{8}, \frac{3}{4}$ are rationals in ascending order.

Let $b = \frac{1}{2}\left(\frac{1}{2} + \frac{5}{8}\right) = \frac{9}{16}$ and $c = \frac{1}{2}\left(\frac{5}{8} + \frac{3}{4}\right) = \frac{11}{16}$.

$\therefore \quad \frac{1}{2}, \frac{9}{16}, \frac{5}{8}, \frac{11}{16}, \frac{3}{4}$ are rationals in ascending order

Properties of Irrationals:

(*i*) *The set P of all irrationals is not closed for addition*, since the sum of two irrationals need not be irrational.

e.g. $(2+\sqrt{3}) \in P, (2-\sqrt{3}) \in P$ but $(2+\sqrt{3})+(2-\sqrt{3}) = 4 \in P$.

(*ii*) *The set P of all irrationals is not closed for multiplication,* since the product of two irrationals need not be irrational.

e.g. $\sqrt{2} \in P, -\sqrt{3} \in P$ but $\sqrt{2}\times(-\sqrt{2}) = 2 \notin P$.

Some More Properties of Real Numbers:

1. **Factors & Multiples:** *For real numbers a and b, we say that a is a factor of b, if b = ac for some real number c and we write, a/b.*

   If a/b then b is called a multiple of a.

   (*i*) $a/b, b/c \Rightarrow a/c$ (Transitivity)

   (*ii*) a/a for all $a \in R$ (Reflexivity)

   (*iii*) a/b and $a/c \Rightarrow a/(b+c)$ and $a/(b-c)$.

2. **Ordering:** *For real numbers a and b, we say that a is greater than b, if a = b + c for some $c \in R$, and write, $a > b$.*

   Clearly, $a > b \Leftrightarrow b < a$.

I. **Trichotomy Law:** *If a and b are any two real numbers, then one and only one of the following conditions holds.*

   (*i*) $a > b$ (*ii*) $a < b$ (*iii*) $a = b$

II. *If a and b are positive real numbers, then*

   (*i*) $a > b \Rightarrow \frac{1}{a} < \frac{1}{b}$ (*ii*) $a > b \Rightarrow -a < -b$

   (*iii*) $a > b \Rightarrow a + c > b + c$ (*iv*) $a > b \Rightarrow ac > bc$, when $c > 0$

3. $ab = 0 \Rightarrow a = 0$ or $b = 0$ for all $a, b \in R$.

4. **Absolute value of a Real Number:** *The numerical value of a real number x is called its absolute value, to be denoted by $|x|$.*

   Thus, $|2| = 2$ and $|-2| = 2$.

   If x is any real number, then we define:

$$|x| = \begin{cases} x \text{ when } x \geq 0 \\ -x \text{ when } x < 0 \end{cases}$$

Ex. 1. ***Find the value of $|2x - 5| + |3 - 5x|$, when $x = 2$.***

Sol. $|2x - 5| + |3 - 5x| = |2 \times 2 - 5| + |3 - 5 \times 2|$, when $x = 2 = |-1| + |-7| = (1 + 7) = 8$.

Ex. 2. ***Find all real values of x for which $|x| < 2$.***

Sol. Note that $|x| < 2$ means $x < 2$ and $-x < 2$.

$\therefore \quad |x| < 2 \Rightarrow x < 2$ and $-x < 2 \Leftrightarrow x < 2$ and $x > -2 \Leftrightarrow -2 < x < 2$.

Simplification:

In simplifying an expression, we must remove the brackets strictly in the order ( ), { } and [ ] and then we must apply the operations:

***Of, Division, Multiplication, Addition and Subtraction** strictly in this order.*

Remember: 'BODMAS', where B stands for bracket, O for of; D for division; M for multiplication; A for addition and S for subtraction.

Ex. 1. ***Simplify:***

(i) 153 ÷ 17 × 9 ***(ii) 5005 – 5000 ÷ 10***

Sol. (*i*) $153 \div 17 \times 9 = 9 \times 9 = 81$ [*First apply division*]

(*ii*) $5005 - 5000 \div 10 = 5005 - 500 = 4505.$ [*First apply division*]

Ex. 2. ***Simplify:***

$$\cfrac{1}{2+\cfrac{1}{2+\cfrac{1}{5-\cfrac{1}{5}}}}$$

Sol. Given Exp. $= \cfrac{1}{2+\cfrac{1}{2+\cfrac{1}{\frac{24}{5}}}} = \cfrac{1}{2+\cfrac{1}{2+\frac{5}{24}}} = \cfrac{1}{2+\frac{24}{53}} = \cfrac{1}{\frac{130}{53}} = \frac{53}{130}.$

Surds:

Surd: *If a is rational, n is a positive integer and* $\sqrt[n]{a} = a^{1/n}$ *is irrational, then* $\sqrt[n]{a}$ *is called a surd of order n.*

Ex. $\sqrt{5}$ ***is a surd of order 2 and*** $\sqrt[3]{4}$ ***is a surd of order 3.***

Comparison of Surds: *If two surds are of the same order, then the one whose radicand is larger, is the larger of the two.*

For Example: $\sqrt[3]{19} > \sqrt[3]{15}.$

Ex. 1. ***Which is larger*** $\sqrt{2}$ ***or*** $\sqrt[3]{3}$ **.**

Sol. Given surds are of order 2 and 3 respectively whose L.C.M. is 6.

Convert each into a surd of order 6, as shown below.

$$\sqrt{2} = 2^{1/2} = 2^{\frac{1}{2}\times\frac{3}{3}} = 2^{3/6} = (2^3)^{1/6} = (8)^{1/6} = \sqrt[6]{8}$$

$$\sqrt[3]{3} = 3^{1/3} = 3^{\frac{1}{3}\times\frac{2}{2}} = 3^{2/6} = (3^2)^{1/6} = 9^{1/6} = \sqrt[6]{9}$$

Clearly, $\sqrt[6]{9} + \sqrt[6]{8}$. So $\sqrt[3]{3} > \sqrt{2}$.

Ex. 2. ***Simplify:*** $\sqrt[4]{81} + \sqrt[3]{216} + \sqrt[5]{32}$.

Sol. $\sqrt[4]{81} + \sqrt[3]{216} + \sqrt[5]{32} = \sqrt[4]{3\times3\times3\times3} + \sqrt[3]{6\times6\times6} + \sqrt[5]{2\times2\times2\times2\times2} = (3+6+2) = 11.$

PROGRESSIONS

Sequence: *A succession of numbers formed and arranged in a definite order according to certain definite rule, is called a sequence.*

The number occurring at the *n*-th place of a sequence is called its *n*-th term.

Progression: *If the terms of a sequence increase or decrease continuously, then such a sequence is called a progression.*

Arithmetical Progression (A.P.): *If each term of a progression differs from its preceding term by a constant, then such a progression is called an arithmetical progression. This constant difference is known as common difference of the A.P.*

An A.P. with first term = a & common difference = d is given by:

$$a, (a+d), (a+2d), (a+3d),$$

The n-th term of the A.P. with 1st term = a and common difference = d is given by: $T_n = a + (n-1)d$.

The sum of the n terms of an A.P. is given by:

$$S_n = \frac{n}{2}[2a+(n-1)d] = \frac{n}{2}\text{ (first term + last term)}.$$

Ex. 1. ***How many numbers between 16 and 200 are divisible by 7?***

Sol. The required numbers are 21, 28, 35, ..., 196.

This is clearly an A.P. in which $a = 21$ & $d = 7$.

Let the number of terms of this A.P. be n.

Then, $T_n = 196 \Rightarrow a + (n-1)d = 196$, *i.e.* $21 + (n-1) \times 7 = 196 \Rightarrow n = 26$.

$\therefore$ Required number of numbers = 26.

Ex. 2. ***Find the sum of first 100 natural numbers.***

Sol. We have to find: $1 + 2 + 3 + ... + 100$.

This is an A.P. in which $a = 1$, $d = 1$ and $n = 100$.

$$\therefore \text{Required Sum} = \frac{n}{2} \times [\text{first term + last term}] = \frac{100}{2} \times (1 + 100) = (50 \times 101) = 5050.$$

Remark: If a. b, c are in A.P., we say that b is the arithmetic mean between a and c and, we have: $b = \frac{1}{2}(a+c)$.

Geometrical Progressions (G.P.): *A progression of numbers in which every term bears a constant ratio with its preceding term, is called a geometrical progression.*

The constant ratio is called the common ratio of the G.P., A G.P. with lst term = a & common ratio = r is given by: $a, ar, ar^2, ar^3, ...$

In this G.P., we have: $T_n = ar^{n-1}$.

Sum of n terms of a G.P. is, $S_n = \dfrac{a(1-r^n)}{1-r}$.

Remark: If a, b, c are in G.P., then $b^2 = ac$.

Harmonical Progression (H.P.): Numbers $a_1, a_2, ..., a_n$ *are said to be in H.P.* if, $\frac{1}{a_1}, \frac{1}{a_2}, ..., \frac{1}{a_n}$ are in A.P.

Remark: a, b, c are in H.P. if $\frac{1}{a}, \frac{1}{b}, \frac{1}{c}$ *are in A.P. and so in this case,* $b = \left(\dfrac{2ac}{a+c}\right)$

Some More Results:

(i) $(1 + 2 + 3 + ... + n) = \dfrac{n(n+1)}{2}$

(ii) $(1^2 + 2^2 + 3^2 + ... + n^2) = \dfrac{n(n+1)(2n+1)}{6}$

(iii) $(1^3 + 2^3 + 3^3 + ... + n^3) = \dfrac{n^2(n+1)^2}{4}$.

EXERCISE 1

1. What is the remainder when 2^{101} is divided by 101? **[CDS (I) Exam, 2025]**
(*a*) 1 (*b*) 2 (*c*) 5 (*d*) 7

2. What is the largest power of 10 that divides the product $29 \times 28 \times 27 \times ... 2 \times 1$? **[CDS (I) Exam, 2023]**
(*a*) 4 (*b*) 5 (*c*) 6 (*d*) 7

3. What is the remainder if we divide 3^{10} by 7? **[CDS (II) Exam, 2024]**
(*a*) 0 (*b*) 1 (*c*) 2 (*d*) 4

4. Consider the following numbers: **[CDS (I) Exam, 2023]**
1. 437
2. 797
3. 1073

How many of the above numbers are prime ?
(*a*) Only one (*b*) Only two (*c*) All three (*d*) None

5. If $43^x \times 47^y = (2021)^2$, $x \neq 0, y \neq 0$ then what is the value of the following?
$$\frac{4xy + x + y}{2xy - x - y}$$
[CDS (I) Exam, 2022]
(*a*) 5 (*b*) 15 (*c*) 25 (*d*) 45

6. Let *a*, *b*, *c*, *d*, be positive integers. If $\dfrac{1}{a+\dfrac{1}{b+\dfrac{1}{c+\dfrac{1}{d}}}} = \dfrac{17}{60}$, then what is the product of *a*, *b*, *c*, *d* ? **[CDS (I) Exam, 2022]**
(*a*) 24 (*b*) 51 (*c*) 68 (*d*) 102

7. Let $p = 2^{2n+2} + m$ and $q = 2^{4n} - m$ (where *n* is even natural number). What should be the least value of *m* such that *p* as well as *q* is divisible by 5? **[CDS (I) Exam, 2022]**
(*a*) –1 (*b*) 1 (*c*) 4 (*d*) 6

8. How many pairs of (x, y) can be chosen from the set {2, 3, 6, 8, 9} such that $\frac{x}{y} + \frac{y}{x} = 2$, where $x \neq y$? **[CDS (I) Exam, 2021]**
(*a*) Zero (*b*) One
(*c*) Two (*d*) Three

9. How many terms are there in the following product? **[CDS (I) Exam, 2021]**
$(a_1 + a_2 + a_3)(b_1 + b_2 + b_3 + b_4)(c_1 + c_2 + c_3 + c_4 + c_5)$
(*a*) 15 (*b*) 30 (*c*) 45 (*d*) 60

10. What is the remainder when $2^p - 1$ is divided by *p*, where $p > 5$ is a prime number? **[CDS (I) Exam, 2025]**
(*a*) 1 (*b*) 2 (*c*) 3 (*d*) 4

11. How many zeros are there in the product $1^{50} \times 2^{49} \times 3^{48} \times .. \times 50^1$? **[CDS (II) Exam, 2021]**
(*a*) 262 (*b*) 261 (*c*) 246 (*d*) 235

12. Let p be the remainder when 7^{84} is divided by 3^{42} and q be the remainder when 7^{84} is divided by 344. What is $(p - q)$ equal to? **[CDS (I) Exam, 2025]**

(*a*) 0 (*b*) 1 (*c*) 2 (*d*) 6

13. The sum of digits of a 2-digit number is 12. When the digits are reversed, the number becomes greater by eighteen. What is the difference between the digits in the number ?

(*a*) 1 (*b*) 2 (*c*) 3 (*d*) 4

[CDS (I) Exam, 2023]

14. If the points P and Q represent real numbers $0.7\overline{3}$ and $0.5\overline{6}$ on the number line, then what is the distance between P and Q? **[CDS (I) Exam, 2021]**

(*a*) $\frac{1}{6}$ (*b*) $\frac{1}{5}$ (*c*) $\frac{16}{45}$ (*d*) $\frac{11}{90}$

15. Let *XYZ* be a 3-digit number. Let $S = XYZ + YZX + XSY$. Which of the following statements is/are correct? **[CDS (I) Exam, 2021]**

1. S is always divisible by 3 and $(X + Y + Z)$
2. S is always divisible by 9
3. S is always divisible by 37

Select the correct answer using the code given below:

(*a*) 1 only (*b*) 2 only (*c*) 1 and 2 (*d*) 1 and 3

16. What is the remainder when $111^{222} + 222^{333} + 333^{444}$ is divided by 5?

[CDS (II) Exam, 2020]

(*a*) 1 (*b*) 2 (*c*) 3 (*d*) 4

17. If $I = a^2 + b^2 + c^3$, where *a* and *b* are consecutive integers and $c = ab$, then I is

(*a*) an even number and it is not a square of an integer
(*b*) an odd number and it is not a square of an integer
(*c*) square of an even integer
(d) square of an odd integer **[CDS (II) Exam, 2020]**

18. What is the remainder when 3^{255} is divided by 28? **[CDS (II) Exam, 2024]**

(*a*) 1 (*b*) 11 (*c*) 24 (*d*) 27

19. A 60-page book has *n* lines per page. If the number of lines where reduced by 3 in each page, the number of pages would have to be increased by 10 to give the same writing space. What is the value of *n*? **[CDS (II) Exam, 2020]**

(*a*) 18 (*b*) 21 (*c*) 24 (*d*) 30

20. What is the remainder when $(17^{29} + 19^{29})$ is divided by 18? **[CDS (I) Exam, 2019]**

(*a*) 6 (*b*) 2 (*c*) 1 (*d*) 0

21. What is the largest value of *n* such that 10^n divides the product? **[CDS (I) Exam, 2019]**

$2^5 \times 3^3 \times 4^8 \times 5^3 \times 6^7 \times 7^6 \times 8^{12} \times 9^9 \times 10^6 \times 15^{12} \times 20^{14} \times 22^{11} \times 25^{15}$?

(*a*) 65 (*b*) 55 (*c*) 50 (*d*) 45

22. Consider the multiplication $999 \times abc =$ def123 in decimal notation, where *a*, *b*, *c*, *d*, *e* and f are digits. What are the values of *a*, *b*, *c*, *d*, *e* and *f* respectively? **[CDS (I) Exam, 2019]**

(*a*) 6, 6, 8, 6, 8, 7 (*b*) 8, 6, 8, 6, 7, 8
(*c*) 6, 8, 8, 7, 8, 6 (*d*) 8, 6, 8, 8, 6, 7

23. A person carries ₹500 and wants to buy apples and oranges out of it. If the coast of one apple is ₹5 and the cost of one orange is ₹7, then what is the number of ways in which a person can buy both apples and oranges using total amount? **[CDS (II) Exam, 2019]**

(*a*) 10 (*b*) 14 (*c*) 15 (*d*) 17

24. Let x be the smallest positive integer such that when 14 divides x, the remainder is 7; and when 15 divides x, the remainder is 5. Which one of the following is correct?

[CDS (II) Exam, 2019]

(*a*) $20 < x < 30$ (*b*) $30 < x < 40$ (*c*) $40 < x < 50$ (*d*) $x > 50$

25. If $\frac{36}{11} = 3 + \cfrac{1}{x + \cfrac{1}{y + \cfrac{1}{z}}}$ when x, y and z are natural numbers, then what is $(x + y + z)$ equal to?

[CDS (II) Exam, 2019]

(*a*) 6 (*b*) 7 (*c*) 8 (*d*) 9

26. Which one of the following decimal numbers is a rational number with denominator 37 ?

[CDS (I) Exam, 2018]

(*a*) 0.459459459... (*b*) 0.459459459 (*c*) 0.0459459459... (*d*) 0.00459459...

27. If $x = \cfrac{1}{2 + \cfrac{3}{4 + \cfrac{5}{6 + \cfrac{7}{8 + \cfrac{9}{10}}}}}$ then which one of the following is correct ?

[CDS (I) Exam, 2024]

(*a*) $0 < x < 0.5$ (*b*) $x = 0.5$ (*c*) $0.5 < x < 1.0$ (*d*) $x > 1.0$

28. There are two natural numbers m and n $(m > n)$. When m is divided by 12, it leaves a remainder 4. When n is divided by 12, it leaves a remainder 6. Which of the following statements is/are correct?

I. The remainder when $(m + n)$ is divided by 12 is 10.

II. The remainder when $(m - n)$ is divided by 12 is 10.

Select the correct answer using the code given below:

(*a*) I only (*b*) II only (*c*) Both I and II (*d*) Neither I nor II

[CDS (I) Exam, 2025]

29. If $\frac{37}{13} = 2 + \cfrac{1}{x + \cfrac{1}{y + \cfrac{1}{z}}}$ where x, y, z are natural numbers, then what is z equal to?

(*a*) 1 (*b*) 2 (*c*) 3

(*d*) Cannot be determined due to insufficient data **[CDS (II) Exam, 2024]**

30. If n is a natural number, then what is the sum of all distinct remainders of $4^n + 6^n + 9^n + 11^n$ when divided by 10 for various values of n ? **[CDS (I) Exam, 2024]**

(*a*) 3 (*b*) 4 (*c*) 6 (*d*) 7

31. When a ball bounces, it rises to $\frac{2}{3}$ of the height from which it fell. If the ball is dropped from a height of 36 m, how high will it rise at the third bounce? **[CDS (I) Exam, 2015]**

(*a*) $10\frac{1}{3}$ m (*b*) $10\frac{2}{3}$ m (*c*) $12\frac{1}{3}$ m (*d*) $12\frac{2}{3}$ m

32. What is the digit at the 100th place of number $(225)^{40}$?

(*a*) 6 (*b*) 5 (*c*) 4 (*d*) 2

[CDS (II) Exam, 2024]

33. The number of values of *x* satisfying $x+\frac{100}{x}>50$, where *x* is a natural number less than or equal to 100 is **[CDS (II) Exam, 2015]**

(*a*) 51 (*b*) 53 (*c*) 55 (*d*) 57

34. If *a, b, c, d* are natural numbers, then how many possible remainders are there when $1^a + 2^b + 3^c + 4^d$ is divided by 10? **[CDS (II) Exam, 2024]**

(*a*) 3 (*b*) 4 (*c*) 5 (*d*) 6

35. The largest natural number which divides every natural number of the form $(n^3-1)(n-2)$, where *n* is a natural number greater than 2 is **[CDS (II) Exam, 2015]**

(*a*) 6 (*b*) 12 (*c*) 24 (*d*) 48

36. The average of *m* number is n^4 and the average of *n* number is m^4. The average of $(m+n)$ numbers is

(*a*) mn (*b*) m^2+n^2

(*c*) $mn(m^2+n^2)$ (*d*) $mn(m^2+n^2-mn)$ **[CDS (II) Exam, 2015]**

37. The digit in the units place of the resulting number of the expression $(234)^{100}+(234)^{101}$ is

(*a*) 6 (*b*) 4 (*c*) 2 (*d*) 0

[CDS (II) Exam, 2015]

38. The seven digit number 876*p*37*q* is divisible by 225. The value of *p* and *q* can be respectively

(*a*) 9, 0 (*b*) 0, 0 (*c*) 0, 5 (*d*) 9, 5

[CDS (II) Exam, 2015]

39. What is the unit digit of 7^{139}?

(*a*) 9 (*b*) 7 (*c*) 6 (*d*) 3

[CDS (II) Exam, 2016]

40. Let *S* be a set of first ten natural numbers. What is the possible number of pairs (*a, b*) where *a, b* ∈ S and *a* ≠ *b* such that the product *ab* (> 12) leaves remainder 4 when divided by 12?

(*a*) 4 (*b*) 6 (*c*) 8 (*d*) 10

[CDS (II) Exam, 2016]

41. What is the remainder when $13^5 + 14^5 + 15^5 + 16^5$ is divided by 29?

(*a*) 8 (*b*) 5 (*c*) 3 (*d*) 0

[CDS (II) Exam, 2016]

42. What is the difference between the sum of the cubes and that of squares of first ten natural numbers?

(*a*) 2280 (*b*) 2640 (*c*) 3820 (*d*) 4130

[CDS (II) Exam, 2016]

43. What is the remainder when the number $(4444)^{4444}$ is divided by 9?

(*a*) 4 (*b*) 6 (*c*) 7 (*d*) 8

[CDS (I) Exam, 2017]

44. What is the value of $\frac{(443+547)^2+(443-547)^2}{443\times443+547\times547}$? **[CDS (I) Exam, 2017]**

(*a*) 0 (*b*) 1 (*c*) 2 (*d*) 3

45. The number of prime number which are less than 100 is

(*a*) 24 (*b*) 25 (*c*) 26 (*d*) 27

[CDS (I) Exam, 2017]

46. What number must be subtracted from both the numerator and the denominator of the fraction $\frac{27}{35}$ so that it becomes $\frac{2}{3}$? **[CDS (I) Exam, 2017]**

(*a*) 6 (*b*) 8 (*c*) 9 (*d*) 11

47. Which one among the following is the largest?

(*a*) $\frac{7}{9}$ (*b*) $\frac{11}{14}$ (*c*) $\frac{3}{4}$ (*d*) $\frac{10}{13}$

[CDS (I) Exam, 2017]

48. A real number M is squared to give the value N. What is the minimum value of $(M+N)$?

(*a*) – 0.25 (*b*) – 0.50 (*c*) 3 (*d*) 5

[CDS (I) Exam, 2025]

49. What is the sum of all 3-digit numbers that give a remainder of 5 when they are divided by 50?

(*a*) 9005 (*b*) 9540 (*c*) 9600 (*d*) 9640

[CDS (I) Exam, 2025]

50. N is the smallest 5-digit number which when divided by $2, 2^2, 2^3\ 2^4, ..., 2^n$ leaves a remainder 1. What is the value of n?

(*a*) 12 (*b*) 13 (*c*) 14 (*d*) 15

[CDS (I) Exam, 2025]

51. In a division sum, the divisor is 12 times the quotient and 5 times the remainder. If the remainder is 48, what is the dividend?

(*a*) 240 (*b*) 576 (*c*) 4800 (*d*) 4848

52. What is the smallest natural number n such that $(n+1) \times n \times (n-1) \times (n-2) \times ... 3 \times 2 \times 1$ is divisible by 910 ?

(*a*) 91 (*b*) 90 (*c*) 13 (*d*) 12

[CDS (I) Exam, 2024]

53. If x and y are positive real numbers, then:

(*a*) $x > y \Rightarrow -x > -y$ (*b*) $x > y \Rightarrow -x < -y$ (*c*) $x > y \Rightarrow \frac{1}{x} > \frac{1}{y}$ (*d*) $x > y \Rightarrow -\frac{1}{x} < -\frac{1}{y}$

54. $\sqrt{.0001}$ is:

(*a*) an integer (*b*) a rational (*c*) an irrational (*d*) imaginary

55. When the number (12345678910111213 ... 99100) is divided by 16, what will be the remainder ?

(*a*) 15 (*b*) 12 (*c*) 4 (*d*) 3

[CDS (I) Exam, 2024]

56. The expression $555^{777} + 777^{555}$ is divisible by which of the following ?

1. 2 2. 3 3. 37

Select the correct answer using the code given below :

(*a*) 1 and 2 only (*b*) 2 and 3 only

(*c*) 1 and 3 only (*d*) 1, 2, and 3

[CDS (I) Exam, 2024]

57. If x is a negative real number, then:

(*a*) $|x| = x$ (*b*) $|x| = -x$ (*c*) $|x| = \frac{1}{x}$ (*d*) $|x| = -\frac{1}{x}$

58. $1399 \times 1399 = ?$

(*a*) 1687401 (*b*) 1943211 (*c*) 1957201 (*d*) 1901541

59. $1916 \times 1916 = ?$

(*a*) 3671056 (*b*) 3556996 (*c*) 3748096 (*d*) None

60. $\frac{885 \times 885 \times 885 + 115 \times 115 \times 115}{885 \times 885 + 115 \times 115 - 885 \times 115} = ?$

(*a*) 1000 (*b*) 770 (*c*) 885 (*d*) 115

61. What is the minimum value of p for which $\frac{1}{532900} + \frac{p^2}{266450} + \frac{p^4}{523900}$ is an integer?

(*a*) 729 (*b*) 243 (*c*) 27 (*d*) 1

[CDS (I) Exam, 2025]

62. Given that $\frac{100 \times 99 \times 98 \times \ldots \times 3 \times 2 \times 1}{100^n}$ is an integer. What is the largest value of n for which this is true?

(*a*) 20 (*b*) 21 (*c*) 24 (*d*) None of the above

[CDS (I) Exam, 2025]

63. A fraction becomes 4 when 1 is added to both the numerator and the denominator and it becomes 7 when 1 is subtracted from both the numerator and the denominator. The numerator of the fraction is:

(*a*) 2 (*b*) 3 (*c*) 7 (*d*) 15

64. A real number x is such that the sum of the number and four times its square is the least. What is that number ?

(*a*) –0.625 (*b*) –0.125 (*c*) 0.125 (*d*) 1

[CDS (I) Exam, 2025]

65. The difference of the square of two natural numbers m and n $(m > n)$ is 72. How many pairs of natural numbers will satisfy ?

(*a*) 3 (*b*) 4 (*c*) 5 (*d*) 6

[CDS (I) Exam, 2025]

66. The sum of two numbers is 37 and their product is 342. The numbers are:

(*a*) 19, 18 (*b*) 23, 14

(*c*) 24, 13 (*d*) 28, 9

67. A number exceeds its one-seventh by 84. The number is:

(*a*) 12 (*b*) 14 (*c*) 98 (*d*) 196

68. $\frac{4}{5}$ of a number exceeds its $\frac{2}{3}$ by 8. The number is:

(*a*) 30 (*b*) 60 (*c*) 75 (*d*) 90

69. A number is as much greater than 21 as is less than 71. The number is:

(*a*) 39 (*b*) 41 (*c*) 46 (*d*) 49

70. Consider the following statements:

1. If $(3m^3 + 2m^2 + 5m + n)/m$ is not an integer, where *m* and *n* are integers, then *n* is not divisible by m.
2. $5(8^m) + 2^{3m}$ is divisible by 48 for all whole numbers m.

Which of the statements given above is/are correct ?

(*a*) 1 only (*b*) 2 only (*c*) Both 1 and 2 (*d*) Neither 1 nor 2

[CDS (I) Exam, 2024]

71. The sum of three numbers is 102. If the ratio between first and second be 2 : 3 and that between second and third be 5 : 3, then the second number is:

(*a*) 30 (*b*) 45 (*c*) 27 (*d*) 48

72. Of the three numbers, the first is twice the second and is half of the third. If the average of these numbers be 56, the numbers in order are:

(*a*) 48, 96, 24 (*b*) 48, 24, 96 (*c*) 96, 24, 48 (*d*) 96, 48, 24

73. A 4-digit number *N* has exactly 15 distinct divisors. What is the total number of distinct divisors of N^2 ?

(*a*) 16 (*b*) 30 (*c*) 45 (*d*) 225

[CDS (II) Exam, 2024]

74. What is the sum of the largest and the smallest 4-digit numbers made by using single digit prime numbers (without repetition)?

(*a*) 7887 (*b*) 7997 (*c*) 8998 (*d*) 9889

[CDS (II) Exam, 2024]

75. What is the value of x $(0 \le x \le 8)$ if $(100^{97} + 100^{54} + x + 1)$ leaves a remainder 0 when divided by 9?

(*a*) 8 (*b*) 6 (*c*) 4 (*d*) 1

[CDS (II) Exam, 2024]

76. If $11x + 5y$ is a prime number where *x*, *y* are natural numbers then what is the minimum value of $(x + y)$?

(*a*) 3 (*b*) 4 (*c*) 5 (*d*) 6

[CDS (II) Exam, 2024]

77. How many two digit numbers are there which are divisible by 6?

(*a*) 16 (*b*) 15 (*c*) 17 (*d*) 18

78. If $(1^2 + 2^2 + 3^2 + ... + 10^2) = 385$, then the value of $(2^2 + 4^2 + 6^2 + ... + 20^2)$ is:

(*a*) 770 (b) 1540 (c) 1155 (*d*) (385×385)

79. If *a*, *b*, *c* are in H.P., then *b* is equal to:

(*a*) $\dfrac{a+c}{2a}$ (*b*) $\dfrac{2\,ac}{a+c}$ (*c*) $\dfrac{a+c}{2\,ac}$ (*d*) None

80. If $|2x - 3| = 7$, then the value of *x* are:

(*a*) 5, – 5 (*b*) 2, – 5 (*c*) – 2, 5 (*d*) 2, – 2

81. The value of $1+\dfrac{1}{1+\dfrac{1}{1+\dfrac{1}{9}}}$ is:

(*a*) $\frac{29}{19}$ (*b*) $\frac{10}{19}$ (*c*) $\frac{29}{10}$ (*d*) $\frac{10}{9}$

82. The value of $\left(\frac{2^n + 2^{n-1}}{2^{n+1} - 2^n}\right)$ is:

(*a*) $\frac{1}{2}$ (*b*) $\frac{3}{2}$ (*c*) $2^{\frac{n-1}{n+1}}$ (*d*) None

83. Which is largest amongst $\sqrt[4]{4}, \sqrt[3]{3}$ and $\sqrt{2}$?

(*a*) $\sqrt{2}$ (*b*) $\sqrt[3]{3}$

(*c*) $\sqrt[4]{4}$ (*d*) cannot be determined

84. The value of $\left[\frac{1}{(216)^{-2/3}} + \frac{1}{(256)^{-3/4}} + \frac{1}{(243)^{-1/5}}\right]$ is:

(*a*) 107 (*b*) 105 (*c*) 103 (*d*) None

85. If $a^x = b, b^y = c$ and $c^z = a$, then xyz equals:

(*a*) abc (*b*) $\frac{1}{abc}$ (*c*) 1 (*d*) None

86. When every even power of every odd integer (greater than 1) is divided by 8, what is the remainder? **[CDS (I) Exam, 2023]**

(*a*) 3 (*b*) 2

(*c*) 1 (*d*) The remainder is not necessarily 1.

87. Consider the following statements : **[CDS (I) Exam, 2023]**

1. If n is a natural number, then the number $\frac{n(n^2+2)}{3}$ is also natural number.

2. If m is an odd integer, then the number $\frac{m^4+4m^2+11}{16}$ is an integer.

Which of the statements given above is/are correct?

(*a*) 1 only (*b*) 2 only (*c*) Both 1 and 2 (*d*) Neither 1 nor 2

88. It is given that 5 does not divide $n - 1$, n and $n + 1$, where n is a positive integer. Which one of the following is correct ? **[CDS (I) Exam, 2023]**

(*a*) 5 divides $(n^2 + 1)$

(*b*) 5 divides $(n^2 - 1)$

(*c*) 5 divides $(n^2 + n)$

(*d*) 5 divides $(n^2 - n)$

89. What is the largest 5-digit number, which leaves remainder 7, when divided by 18 as well as by 11 ? **[CDS (I) Exam, 2023]**

(*a*) 99981 (*b*) 99988 (*c*) 99997 (*d*) 99999

90. Consider the following statements in respect of all factors of 360 : **[CDS (I) Exam, 2023]**

1. The number of factors is 24.

2. The sum of all factors is 1170.

Which of the above statements is/are correct ?

(*a*) 1 only (*b*) 2 only (*c*) Both 1 and 2 (*d*) Neither 1 nor 2

91. Consider a 6-digit number of the form XYXYXY. The number is divisible by :

(*a*) 3 and 7 only (*b*) 7 and 13 only
(*c*) 3, 13 and 37 only (*d*) 3, 7, 13 and 37

[CDS (I) Exam, 2023]

92. If $x = \sqrt{4\sqrt{4\sqrt{4\sqrt{4...}}}}$, then what is the value of x? **[CDS (I) Exam, 2023]**

(*a*) 2 (*b*) 4 (*c*) 8 (*d*) 16

93. Let m and n be natural numbers. What is the minimum value of $(m + n)$ such that $33m + 22n$ is divisible by 121 ? **[CDS (I) Exam, 2023]**

(*a*) 3 (*b*) 4 (*c*) 5 (*d*) 10

94. Let a, b, c and d be four positive integers such that $a + b + c + d = 200$. If $S = (-1)^a + (-1)^b + (-1)^c + (-1)^d$, then what is the number of possible values of S? **[CDS (I) Exam, 2023]**

(*a*) One (*b*) Two (*c*) Three (*d*) Four

95. The number $97^{30} - 14^{30}$ is divisible by : **[CDS (I) Exam, 2023]**

(*a*) 37 but not 83 (*b*) 83 but not 37 (*c*) Both 37 and 83 (*d*) Neither 37 nor 83

96. Consider the following statements : **[CDS (I) Exam, 2023]**

1. $n^3 - n$ is divisible by 6.
2. $n^5 - n$ is divisible by 5.
3. $n^5 - 5n^3 + 4n$ is divisible by 120.

Which of the statements given above are correct?

(*a*) 1 and 2 only (*b*) 2 and 3 only
(*c*) 1 and 3 only (*d*) 1, 2 and 3

97. What is the last digit of the sum $S = 9^{27} + 27^9$? **[CDS (I) Exam, 2023]**

(*a*) 3 (*b*) 6 (*c*) 7 (*d*) 9

Each item contains a question followed by two Statements. Answer each item using the following instructions :

(a) Choose this option if the Question can be answered by one of the Statements alone but not by the other.
(b) Choose this option if the Question can be answered by either Statement alone.
(c) Choose this option if the Question can be answered by using both the Statements together, but cannot be answered by using either Statement alone
(d) Choose this option if theQuestion cannot be answered even by using both Statements together

98. A number 277XY5 (where X, Y are digits) is divisible by 25. **[CDS (I) Exam, 2023]**

Question: What is the value of X?

Statement I : The given number is divisible by 9.

Statement II: $X > 5$.

99. Question: Is $m > n$, where m, n are non-zero numbers?

Statement I : $\frac{m}{n} > 1$.

Statement II : $m > 2n$.

100. Question : Is $p^2 + q^2 + q$ odd, where p, q are positive integers ? **[CDS (I) Exam, 2023]**

Statement I : $2p + q$ is odd.

Statement II : $q - 2p$ is odd.

101. A 2-digit number is such that the sum of the number and the number obtained by reversing the order of the digits of the number is 55. Further, the difference of the given number and the number obtained by reversing the order of the digits of the number is 45. What is the product of the digits ? **[CDS (II) Exam, 2023]**

(*a*) 5 (*b*) 2 (*c*) 1 (*d*) O

102. What is number of digits in the expansion of 125^{100}? (Given $\log_{10}2 = 0{\cdot}301$)

(*a*) 69 (*b*) 70 (*c*) 209 (*d*) 210

Direction: Consider the following for the next three (3) items that follow :

Mark option (a) if the question can be answered by using one of the statements alone, but cannot be answered using the other statement alone.

Mark option (b) if the question can be answered by using either statement alone.

Mark option (c) if the question can be answered by using both the statements together, but cannot be answered using either statement alone.

Mark option (d) if the question cannot be answered even by using both the statements together.

103. Question: What is the 3-digit number which is divisible by 10 ? **[CDS (II) Exam, 2023]**

Statement-I: If the digits in hundred's place and ten's place of the number are interchanged, the resulting number is diminished by 180.

Statement-II: If the digit in hundred's place is halved and digit in ten's place and unit place of the number are interchanged, the resulting number is diminished by 336.

104. Question : Is $(a - b + c) > (a + b - c)$, where a, b and care real numbers?

Statement-I: b is negative.

Statement-II: c is negative.

105. Question: What is the cost of 15 pens, 21 pencils and 18 note books? **[CDS (II) Exam, 2023]**

Statement-I: The cost of 7 pens, 6 pencils and 5 note books is 200.

Statement-II: The cost of 3 pens, 8 pencils and 7 note books is 210.

106. A question is given followed by two statements I and II. Consider the Question and the Statements and mark the correct option.

Question : The largest of five different integers is 8 and least is 2. What is the average of these integers ?

Statement-I : The sum of all the 5 integers is a multiple of 5

Statement-II : The number of odd integers is odd **[CDS (II) Exam, 2024]**

(*a*) If the Question can be answered by one of the Statements alone, but not by the other.

(*b*) If the Question can be answered by either Statement alone.

(*c*) If the Question can be answered by using both the Statements together, but cannot be answered by using either Statement alone.

(*d*) If the Question cannot be answered even by using both Statements together.

107. The sum of first 47 terms of the series $\frac{1}{4} + \frac{1}{5} - \frac{1}{6} - \frac{1}{4} - \frac{1}{5} + \frac{1}{6} + \frac{1}{4} + \frac{1}{5} - \frac{1}{6}$...... is

(*a*) 0 (*b*) $-\frac{1}{6}$ (*c*) $\frac{1}{6}$ (*d*) $\frac{9}{20}$

[CDS (II) Exam, 2015]

108. A number consists of two digits, whose sum is 7. If the digits are reversed, the number is increased by 27. The product of digits of the number is **[CDS (II) Exam, 2015]**

(*a*) 6 (*b*) 8 (*c*) 10 (*d*) 12

109. Consider all positive two digit numbers each of which when divided by 7 leaves a remainder 3. What is their sum? **[CDS (II) Exam, 2015]**

(*a*) 661 (*b*) 666 (*c*) 676 (*d*) 777

110. Consider the following statements in respect of two different non-zero integers p and q:

1. For $(p + q)$ to be less than $(p - q)$, q must be negative.

2. For $(p + q)$ to be greater than $(p - q)$, both p and q must be positive.

Which of the above statements is/are correct?

(*a*) 1 only (*b*) 2 only (*c*) Both 1 and 2 (*d*) Neither 1 nor 2

[CDS (I) Exam, 2016]

111. Let a two digit number be k times the sum of its digits. If the number formed by interchanging the digits is m times the sum of the digits, then the value of m is

(*a*) $9 - k$ (*b*) $10 - k$ (*c*) $11 - k$ (*d*) $k - 1$

[CDS (I) Exam, 2016]

112. Let S be a set of first fourteen natural numbers. The possible number of pairs (*a*, b), where a, b$\in$S and a $\neq$ b such that ab leaves remainder 1 when divided by 15, is

(*a*) 3 (*b*) 5

(c) 6 (*d*) None of the above **[CDS (I) Exam, 2016]**

113. What is the maximum value of m, if the number N = 90 × 42 × 324 × 55 is divisible by 3^m?

(*a*) 8 (*b*) 7 (*c*) 6 (*d*) 5

[CDS (I) Exam, 2016]

114. Consider the following statements:

1. Every natural number is a real number. **2.** Every real number is a rational number.

3. Every integer is a real number. **4.** Every rational number is a real number.

Which of the above statements are correct?

(*a*) 1, 2 and 3 (*b*) 1, 3 and 4

(c) 2 and 3 only (*d*) 3 and 4 only **[CDS (I) Exam, 2016]**

115. If m and n are distinct natural numbers, then which of the following is/are integer/integers?

1. $\frac{m}{n}+\frac{n}{m}$ **2.** $mn\left(\frac{m}{n}+\frac{n}{m}\right)\left(m^2+n^2\right)^{-1}$ **3.** $\frac{mn}{m^2+n^2}$

Select the correct answer using the code given below. **[CDS (I) Exam, 2016]**

(*a*) 1 and 2 (*b*) 2 only (*c*) 2 and 3 (*d*) 3 only

116. If a and b are negative real numbers and c is a positive real number, then which of the following is/are correct?

1. $a - b < a - c$ **2.** If $a < b$, then $\frac{a}{c} < \frac{b}{c}$ **3.** $\frac{1}{b} < \frac{1}{c}$

Select the correct answer using the code given below. **[CDS (I) Exam, 2016]**

(*a*) 1 (*b*) 2 only (*c*) 3 only (*d*) 2 and 3

117. In an examination, a student was asked to divide a certain number by 8. By mistake he multiplied it by 8 and got the answer 2016 more than the correct answer. What was the number?

(*a*) 252 (*b*) 256 (*c*) 258 (*d*) 260

[CDS (II) Exam, 2016]

118. Which one of the following is correct in respect of the number 1729?

(*a*) It cannot be written as the sum of the cubes of two positive integers

(*b*) It can be written as the sum of the cubes of two positive integers in one way only

(*c*) It can be written as the sum of the cubes of two positive integers in two ways only

(*d*) It can be written as the sum of the cubes of two positive integers in three ways only

[CDS (II) Exam, 2016]

119. Sunil wants to spend ₹200 on two types of sweets, costing ₹7 and ₹10 respectively. What is the maximum number of sweets he can get so that no money is left over?

[CDS (I) Exam, 2017]

(*a*) 25 (*b*) 26 (*c*) 27 (*d*) 28

120. Consider the following statements: **[CDS (I) Exam, 2017]**

1. Of two consecutive integers, one is even.
2. Square of an odd integer is of the form 8n + 1.

What is the above statements is/are correct?

(*a*) 1 only (*b*) 2 only (*c*) Both 1 and 2 (*d*) Neither 1 nor 2

121. $(N^{p-1} - 1)$ is a multiple of p, if N is prime to p and p is a **[CDS (I) Exam, 2017]**

(*a*) Prime number (*b*) Rational number

(*c*) Real number (*d*) Composite number

122. The ratio of two numbers is 1 : 5 and their product is 320. What is the difference between the squares of these two numbers? **[CDS (I) Exam, 2017]**

(*a*) 1024 (*b*) 1256 (*c*) 1536 (*d*) 1640

123. Let *a*, *b*, *c* and d be positive integers.

Question : Which one of *a*, *b*, *c*, closest to the product *abcd* ?

Statement I : $a > b > c$

Statement II : *c* is not the smallest

(*a*) If the Question can be answered by one of the Statements alone, but not by the other.

(*b*) If the Question can be answered by either Statement alone.

(*c*) If the Question can be answered by using both the Statements together, but cannot be answered by using either Statement alone.

(*d*) If the Question cannot be answered even by using both Statements together.

[CDS (I) Exam, 2024]

124. Let $mn = k$, where m and n are prime numbers and *k* is an even number.

Question : What is the value of $mn - n + 1$?

Statement I : $m > n$

Statement II : One of the numbers is 2.

(*a*) If the Question can be answered by one of the Statements alone, but not by the other.

(*b*) If the Question can be answered by either Statement alone.

(*c*) If the Question can be answered by using both the Statements together, but cannot be answered by using either Statement alone.

(*d*) If the Question cannot be answered even by using both Statements together.

[CDS (I) Exam, 2024]

125. **Question :** If *p* is a positive integer, then what is the remainder when p^n is divided by $p + 1$?

Statement I : *n* is even.

Statement II : *p* is even.

(*a*) If the Question can be answered by one of the Statements alone, but not by the other.

(*b*) If the Question can be answered by either Statement alone.

(*c*) If the Question can be answered by using both the Statements together, but cannot be answered by using either Statement alone.

(*d*) If the Question cannot be answered even by using both Statements together.

[CDS (I) Exam, 2024]

126. Let x, y, z be variables such that $(x + y + z) = k$, where k is a constant. If $(x + z - y) \times (x - z + y)$ is proportional to yz, then $(y + z - x)$ is proportional to:

(*a*) x (*b*) y (*c*) yz (*d*) xz

[CDS (I) Exam, 2025]

127. Which is the maximum value of the sum of the numbers 36, 33, 30, 27, 24, ...?

(*a*) 240 (*b*) 237 (*c*) 234 (*d*) 231

[CDS (I) Exam, 2025]

128. Which one of the following is a true statement:

(*a*) The difference of two natural numbers is always a natural number

(*b*) The difference of two integers is always an integer

(*c*) The quotient of two rational numbers is always rational

(*d*) None of these

129. Of the following statements for natural numbers a, b and c

I. If a is divisible by b and b is divisible by c, then a must be divisible by c.

II. If a is a factor of both b and c, then a must be a factor of $b + c$.

II. If a is a factor of both b and c, then a must be a factor of $b - c$,

the true ones are:

(*a*) I and II (*b*) I and III (*c*) II and III (*d*) I, II and III

130. p varies directly as $(x^2 + y^2 + z^2)$. When $x = 1, y = 2, z = 3$, then $p = 70$. What is the value of p when $x = -1, y = 1, z = 5$?

(*a*) 100 (*b*) 125 (*c*) 135 (*d*) 140

131. A number is always divisible by 25 only if

(*a*) the last digit of the number is zero.

(*b*) the last digit of the number is 5.

(*c*) the last two digits of the number is divisible by 5.

(*d*) the last two digits of the number is divisible by 25.

132. Which of the following statements is correct?

(*a*) If n is odd, $(10)^n - 1$ is divisible by 11 (*b*) If n is even, $(10)^n$ - 1 is divisible by 11

(*c*) If n is odd, $(5)^n - 1$ is divisible by 5 (*d*) If n is even, $(5)^n - 1$ is divisible by 5

133. Consider the following statements:

Assertion (A): For any two integers a and b with $b \neq 0$, there exist unique integers q and r, such that $a = bq + r, 0 \leq r < | b |$.

Reason (R): An integer a which is not exactly divisible by 3 can be written in one of the forms: $a = 3n + 1$ or $a = 3n + 2$, where n is an integer.

Of these statements,

(*a*) both A and R are true and R is the correct explanation of A.

(*b*) both A and R are true but R is not the correct explanation of A.

(*c*) A is true and R is false.

(*d*) A is false and R is true.

134. Consider the following statements:

Assertion (A): Every non-zero real number has an inverse with respect to multiplication of reals.

Reason (R): The product of every real number with zero is zero.

of these statements,

(*a*) both *A* and *R* are true and *R* is the correct explanation of *A*.

(*b*) both *A* and *R* are true but *R* is not the correct explanation of *A*.

(*c*) *A* is true and *R* is false.

(*d*) A is false and R is true.

135. Which one of the following is a true statement?

(*a*) 0 is the additive indentity for natural numbers

(*b*) Every rational number has its reciprocal

(*c*) Integers are closed for subtraction

(*d*) Real numbers are closed for division

136. If the numerator and denominator of a pure fraction be increased by the same non-zero positive quantity, the resulting fraction is always:

(*a*) equal to original fraction

(*b*) less than the original fraction

(*c*) greater than the original fraction

(*d*) sometimes less than and sometimes greater than the original fraction

137. Which one of the following is a true statement?

(*a*) Two co-primes are always prime (*b*) No even number is prime

(*c*) Every odd number is prime (*d*) A prime number may be even or odd

138. Four-fifth of a number is more than three-fourth of the number by 4. The number is:

(*a*) 64 (*b*) 72 (*c*) 80 (*d*) 84

139. When simplified, the product $\left(1-\frac{1}{3}\right)\left(1-\frac{1}{4}\right)\left(1-\frac{1}{5}\right)\ldots\left(1-\frac{1}{n}\right)$ equals:

(*a*) $\frac{1}{n}$ (*b*) $\frac{2}{n}$ (*c*) $\frac{2(n-1)}{n}$ (*d*) $\frac{2}{n(n+1)}$

140. When simplified, the product $\left(2-\frac{1}{3}\right)\left(2-\frac{3}{5}\right)\left(2-\frac{5}{7}\right)\ldots\left(2-\frac{997}{999}\right)$ is:

(*a*) $\frac{5}{999}$ (*b*) $\frac{1001}{999}$ (*c*) $\frac{1001}{3}$ (*d*) $\frac{3}{1001}$

ANSWERS

1. (*b*)	**2.** (*c*)	**3.** (*d*)	**4.** (*b*)	**5.** (*a*)	**6.** (*a*)	**7.** (*b*)	**8.** (*a*)	**9.** (*d*)
10. (*a*)	**11.** (*a*)	**12.** (*a*)	**13.** (*b*)	**14.** (*a*)	**15.** (*d*)	**16.** (*b*)	**17.** (*d*)	**18.** (*d*)
19. (*b*)	**20.** (*d*)	**21.** (*a*)	**22.** (*d*)	**23.** (*b*)	**24.** (*b*)	**25.** (*a*)	**26.** (*a*)	**27.** (*a*)
28. (*c*)	**29.** (*b*)	**30.** (*b*)	**31.** (*b*)	**32.** (*a*)	**33.** (*c*)	**34.** (*c*)	**35.** (*d*)	**36.** (*d*)
37. (*d*)	**38.** (*a*)	**39.** (*d*)	**40.** (*d*)	**41.** (*d*)	**42.** (*b*)	**43.** (*c*)	**44.** (*c*)	**45.** (*b*)
46. (*d*)	**47.** (*b*)	**48.** (*a*)	**49.** (*b*)	**50.** (*c*)	**51.** (*d*)	**52.** (*d*)	**53.** (*b*)	**54.** (*b*)
55. (*b*)	**56.** (*d*)	**57.** (*b*)	**58.** (*c*)	**59.** (*a*)	**60.** (*a*)	**61.** (*c*)	**62.** (*d*)	**63.** (*d*)

64. (*b*)	**65.** (*a*)	**66.** (*a*)	**67.** (*c*)	**68.** (*b*)	**69.** (*c*)	**70.** (*d*)	**71.** (*b*)	**72.** (*b*)
73. (*c*)	**74.** (*d*)	**75.** (*b*)	**76.** (*c*)	**77.** (*b*)	**78.** (*b*)	**79.** (*b*)	**80.** (*c*)	**81.** (*a*)
82. (*b*)	**83.** (*b*)	**84.** (*c*)	**85.** (*c*)	**86.** (*c*)	**87.** (*c*)	**88.** (*a*)	**89.** (*c*)	**90.** (*c*)
91. (*d*)	**92.** (*b*)	**93.** (*b*)	**94.** (*c*)	**95.** (*c*)	**96.** (*d*)	**97.** (*b*)	**98.** (*c*)	**99.** (*b*)
100. (*d*)	**101.** (*d*)	**102.** (*d*)	**103.** (*c*)	**104.** (*d*)	**105.** (*c*)	**106.** (*c*)	**107.** (*b*)	**108.** (*c*)
109. (*c*)	**110.** (*c*)	**111.** (*c*)	**112.** (*d*)	**113.** (*b*)	**114.** (*b*)	**115.** (*b*)	**116.** (*d*)	**117.** (*b*)
118. (*c*)	**119.** (*b*)	**120.** (*c*)	**121.** (*a*)	**122.** (*c*)	**123.** (*c*)	**124.** (*d*)	**125.** (*a*)	**126.** (*c*)
127. (*c*)	**128.** (*b*)	**129.** (*d*)	**130.** (*c*)	**131.** (*d*)	**132.** (*b*)	**133.** (*b*)	**134.** (*b*)	**135.** (*c*)
136. (*c*)	**137.** (*d*)	**138.** (*c*)	**139.** (*b*)	**140.** (*c*)				

SOLUTIONS

1. If p is a prime and a is not divisible by p, then:

$a^{p-1} \equiv 1 \pmod p$

$2^{101} = 2 \times 2^{100} \Rightarrow 2^{101} \equiv 2 \times 1 \equiv 2 \pmod{101}$

$\therefore$ The remainder is 2.

2. We have $29 \times 28 \times 27 \times ... 2 \times 1$?

In this product, there are 25, 20, 15, 10 and 5.

Therefore total number of factors of 5 = 6

$\therefore$ Largest power = 6

3. Pattern in $3k$ mod 7 for small powers of k.

$3^1 = 3 \Rightarrow 3 \bmod 7 = 4$

$3^2 = 9 \Rightarrow 9 \bmod 7 = 2$

$3^3 = 27 \Rightarrow 27 \bmod 7 = 6$

$3^4 = 81 \Rightarrow 81 \bmod 7 = 4$

$3^5 = 243 \Rightarrow 243 \bmod 7 = 5$

$3^6 = 729 \Rightarrow 729 \bmod 7 = 1$

So, $3^6 \equiv 1 \bmod 7$

$3^{10} = 3^6 \times 3^4 \equiv 1 \times 4 = 4 \bmod 7$

$\therefore$ The remainder when 3^{10} is divided by 7 is 4.

4. $437 = 19 \times 23$ (not prime number)

$1072 = 29 \times 37$ (not prime number)

$\therefore \sqrt{797} = 28.2 > 2, 3, 5, 7, 11, 13, 17, 19, 23$

These numbers are not a factor of 797. $\therefore$ 797 is prime.

5. $43^x \times 47^y = (2021)^2 \Rightarrow 43^x \times 47^y = (43 \times 47)^2$

$\Rightarrow 43^x \times 47^y = 43^2 \times 47^2$

Now, we have, $x = 2$ and $y = 2$

Now, $\dfrac{4xy + x + y}{2xy - x - y} = \dfrac{4 \times 2 \times 2 + 2 + 2}{2 \times 2 \times 2 - 2 - 2} = \dfrac{16 + 4}{8 - 4} = \dfrac{20}{4} = 5$

6. From R.H.S. we have,

$$\frac{17}{60} \Rightarrow \frac{1}{\frac{60}{17}} = \frac{1}{3+\frac{9}{17}} = \frac{1}{3+\frac{1}{\frac{17}{9}}} = \frac{1}{3+\frac{1}{1+\frac{8}{9}}} = \frac{1}{3+\frac{1}{1+\frac{1}{\frac{9}{8}}}} = \frac{1}{3+\frac{1}{1+\frac{1}{1+\frac{1}{8}}}}$$

∴ We have

$$\frac{1}{a+\frac{1}{b+\frac{1}{c+\frac{1}{d}}}} = \frac{1}{3+\frac{1}{1+\frac{1}{1+\frac{1}{8}}}}$$

∵ We have

$$\frac{1}{a+\frac{1}{b+\frac{1}{c+\frac{1}{d}}}} = \frac{1}{3+\frac{1}{1+\frac{1}{1+\frac{1}{8}}}}$$

Now, we get, $a = 3, b = 1, c = 1$ and $d = 8$

Hence, $a \times b \times c \times d = 3 \times 1 \times 1 \times 8 = 24.$

7. ∵ p is divisible by 5

$$\frac{p}{5} = \frac{2^{n+2}+m}{5} = \frac{\left(\left(2^2\right)\right)^n \times 2^2 + m}{5} = \frac{4^n \times 4 + m}{5}$$

$$\Rightarrow \frac{-1^n \times 4 + m}{5}$$

$$\Rightarrow \frac{1 \times 4 + m}{5} [\because \text{n is even natural number}]$$

$$\Rightarrow \frac{4+m}{5}$$

$$\frac{4+m}{5} = \frac{5}{5} = 5$$

⇒ The least value of $m = 1$

Now, given that $q = 2^{4n} - m$

$(2^4)^n - m = 16^n - m$

If $m = 1, \Rightarrow q = 16 - 1 = 15$

∵ 15 is divisible by 5

∵ $m = 1$

Therefore, the least value of m such that p as well as q, is divisible by 5 is 1.

8. We have,

$\frac{x}{y} + \frac{y}{x} = 2$ Therefore, the distance between P and Q is $\frac{1}{6}$

$\Rightarrow \quad \dfrac{x^2+y^2}{xy}=2$

$\Rightarrow \quad x^2+y^2=2xy$

$\Rightarrow \quad (x-y)^2=0$

9. We have, $N = 3 \times 4 \times 5 = 60$

10. Since p is a prime number greater than 5

putting value $p = 7$

Expression $= 2^7 - 1 = 128 - 1 = 127 = 2^7 - 1 = 128 - 1 = 127$

Divide 127 by $p = 7$

$$\frac{127}{7} = 7 \times 18 + 1$$

The remainder when $2^7 - 1$ is divided by 7 is 1.

11. We have, $1^{50} \times 2^{49} \times 3^{48} \times ... \times 50^1$

In this product, it contains 10^{105}, 5^{46}, 15^{36}, 25^{26}, 35^{16}, 45^6, 5 and 2^{49}, 4^{47}, 12^{39}, 22^{39}, 32^{19}, 42^9

$\Rightarrow$ Total number of zeros $= 10^{105} \times 10^{157} = 10^{262.}$

12. $342 = 7^3 - 1$ (since $7^3 = 343$).

We know that $a^n - b^n$ is divisible by $a - b$.

Consider $7^3 = 1 \pmod{342}$ because $7^3 - 1 = 343 - 1 = 342$, which is divisible by 342.

Now, we have $7^{84} = (7^3)^{28}$

$$7^{84} = 1^{28} \pmod{342} = 1$$

So, $p = 1$.

$$344 = 7^3 + 1 \text{ (since } 7^3 = 343).$$

$a^n + b^n$ is divisible by $a + b$ when n is odd.

$a^n - b^n$ is divisible by $a + b$ when n is even.

Consider $7^3 = -1 \pmod{344}$ because $7^3 + 1 = 343 + 1 = 344$ which is divisible by 344.

Now, we have $7^{84} = (7^3)^{28}$

$$7^{84} = (-1)^{28} \pmod{344}$$

So, $q = 1$

$p - q = 1 - 1 = 0$

13. Let units digit and tens digit of a 2-digit number is x and y respectively.

Number $= 10y + x$

New number $= 10x + y$

Now, according to the question,

$\Rightarrow 9x - 9y = 18$

$\Rightarrow x - y = 2$

14. Let $a = 0.73333$

$\Rightarrow \quad 100a = 73.3333$ ... (i)

$\Rightarrow \quad 10\,a = 7.3333$ ... (ii)

Subtracting (ii) from (i) we get,

$90a = 66$

$\Rightarrow \quad a = \frac{11}{15}$

Similarly,

Let b = 0.56 = 0.56666...

$\Rightarrow \quad b = \frac{51}{90} = \frac{17}{30} \quad \Rightarrow \; a - b = \frac{11}{15} - \frac{17}{30} = \frac{22-17}{30}$

$= \frac{1}{6}$

15. We have,

S = XYZ + YZX + ZXY

= 100X + 10Y + Z + 100Y + 10Z + X + 100Z + 10X + Y

= 111X + 111Y + 111Z

$\because$ S = 111 (X + Y + Z)

= 3 × 337 (X + Y + Z)

Thus s is divisible by 3, 37 and $(x + y + z)$. Hence option (d) is correct.

16. 111^{222}:

$222 \div 4$, we get remainder = 2,

So unit place digit $1^2 = 1$

222^{333} :

$333 \div 4$, we get remainder 1,

So unit place digit $2^1 = 2$

333^{444}:

$444 \div 4$, we get remainder = 0 or say 4,

So unit place digit $3^4 = 1$

By summing all the unit place digits = 1 + 2 + 1 = 4

When 4 is divided by 5 we finally get the remainder = 4

17. Let $a = 1$, $b = 2$ [$\therefore$ a and b are consecutive integers]

So, $C = ab = 1 \times 2 = 2$

Now, we have,

$I = 1^2 + 2^2 + 2^2 = 9$

Now, Let $a = 2$, $b = 3$

$\Rightarrow \quad C = ab = 2 \times 3 = 6$

Now,

$I = 2^2 + 3^2 + 6^2 = 49$

$\therefore \quad$ I = Square of an odd integer.

18. $\therefore 3^{255}$ can be written as $(3^3)^{85}$

$$= \frac{(3^3)^{85}}{28} = \frac{(27)^{85}}{28} = \frac{(28-1)^{85}}{28} = \frac{(-1)^{85}}{28} = \frac{-1}{28}$$

Remainder can be calculated as (–1) or (28 – 1) = 27

19. Total no. of lines = 60 × n = 60n

Now, according to question

$70 \times (n-3) = 70n - 210$

We have,

$$60n = 70n - 210$$
$$10n = 210$$
$$n = 21$$

20. We have,

$N = 17^{29} + 19^{29}$

$\Rightarrow \quad N = (17^2)^{14} \times 17 + 19^{29} = 289^{14} \times 17 + 19^{29}$

When 289, is divided by 18, the remainder is 1, and when 19 is divided by 18, by the remainder is also 1.

So, we have,

Remainder of $289^{14} \times 17 + 19^{29} = 1 \times 17 + 1 = 18$

Which is completely divisible by 18.

Hence the overall remainder is 0.

21. We have,

$2^5 \times 3^3 \times 4^8 \times 5^3 \times 6^7 \times 7^6 \times 8^{12} \times 9^9 \times 10^6 \times 15^{12} \times 20^{14} \times 22^{11} \times 25^{15}$

$\Rightarrow 2^5 \times 3^3 \times 2^{16} \times 5^3 \times 2^7 \times 3^7 \times 7^6 \times 2^{36} \times 3^{27} \times 2^6 \times 5^6 \times 5^{12} \times 3^{12} \times 5^{14} \times 2^{28} \times 11^{11} \times 2^{11} \times 5^{30}$

$\therefore$ We have,

$= 2^{109} \times 5^{65}$ [Ignore rest of the factors.]

$= (2 \times 5)^{65} \times 2^{14}$

$\therefore \quad 10^n = 10^{65} \Rightarrow n = 65$

22. We have, $999 \times abc$ = def 132

$(1000 - 1) \times abc$ = def 132

abc 000 + abc = def 132

$\Rightarrow \quad ab\,(c-1)\,(9-a)\,(9-b)\,(10-c)$ = def 132.

By comparing, we have,

$10 - c = 2$

$\Rightarrow \quad c = 8$

$9 - b = 3$

$\Rightarrow \quad b = 6$

$9 - a = 1$

$\Rightarrow \quad a = 8$

$c - 1 = f$

$\Rightarrow f = 8 - 1 = 7$

$b = e = 6$

$a = d = 8$

$\therefore$ We have,

$999 \times 868 = 867132.$

23. Let he buys a apples and b oranges.

$5a + 7b = 500$

Possible non-zero values of (a, b) are (2, 70), (9, 65), (16, 60), (23, 55), (30, 50), (37, 45), (44, 40), (51, 35), (58, 30), (65, 25), (72, 20), (79, 15), (86, 10), (93, 5)

$\therefore$ Total no. of ways = 14 .

24. We have,

$$x = 14p + 7$$

And $$x = 15p + 5$$

If we put $p = 2$, then we get $x = 35$ in both the cases.

$\therefore$ We have

$30 < x < 40$

25. We have,

$$\frac{36}{11} = 3 + \cfrac{1}{x + \cfrac{1}{y + \cfrac{1}{z}}}$$

$$\Rightarrow \quad 3 + \frac{3}{11} = 3 + \cfrac{1}{x + \cfrac{1}{y + \cfrac{1}{z}}} \Rightarrow \quad \frac{3}{11} = \cfrac{1}{x + \cfrac{1}{y + \cfrac{1}{z}}}$$

$$\Rightarrow \quad \cfrac{1}{\cfrac{11}{3}} = \cfrac{1}{x + \cfrac{1}{y + \cfrac{1}{z}}} \quad \Rightarrow \quad \cfrac{1}{3 + \cfrac{2}{3}} = \cfrac{1}{x + \cfrac{1}{y + \cfrac{1}{z}}}$$

$$\Rightarrow \quad \cfrac{1}{3 + \cfrac{1}{\cfrac{3}{2}}} = \cfrac{1}{x + \cfrac{1}{y + \cfrac{1}{z}}} \quad \Rightarrow \quad \cfrac{1}{3 + \cfrac{1}{1 + \cfrac{1}{2}}} = \cfrac{1}{x + \cfrac{1}{y + \cfrac{1}{z}}}$$

By comparing, we have,

$x = 3, y = 1$ and $z = 2$

$$x + y + z = 3 + 1 + 2 = 6.$$

26. We have,

$459 = 17 \times 27$

option (b) is terminating decimal,

Now, Let $x = 0.459459459 \ldots$

$\Rightarrow \quad 1000\,x = 459.459459 \ldots$

$\Rightarrow \quad 1000\,x = 459 + 0.459459 \ldots$

$\Rightarrow \quad 1000\,x = 459 + x$

$$\Rightarrow \quad 999x = 459$$

$$\Rightarrow \quad x = \frac{459}{999} = \frac{17\times 27}{37\times 27} = \frac{17}{37}$$

27. We have,

$$x = \cfrac{1}{2+\cfrac{3}{4+\cfrac{5}{6+\cfrac{7}{8+\cfrac{9}{10}}}}}$$

$$x = \cfrac{1}{2+\cfrac{3}{4+\cfrac{5}{6+\cfrac{7}{\frac{89}{10}}}}} = \cfrac{1}{2+\cfrac{3}{4+\cfrac{5}{6+\frac{70}{89}}}} = \cfrac{1}{2+\cfrac{3}{4+\cfrac{5}{\frac{604}{89}}}} = \cfrac{1}{2+\cfrac{3}{4+\frac{445}{604}}} = \cfrac{1}{2+\cfrac{3}{\frac{2861}{604}}}$$

$$= \cfrac{1}{2+\frac{1812}{2861}} = \cfrac{1}{\frac{7534}{2861}} = \frac{2861}{7534} = 0.38$$

$$\therefore \quad 0 < x < 0.5$$

28. Statement I: $(m+n)$ mod 12

$\Rightarrow m+n \equiv 4+6 \equiv 10 \pmod{12}$

The statement I is correct.

Statement II: $(m-n)$ mod 12 $\Rightarrow m-n \equiv 4-6 \equiv -2 \equiv 10 \pmod{12}$

The statement II is correct.

$\therefore$ The correct answer is: Both I and II

29. $\frac{37}{13} = 2+\cfrac{1}{x+\cfrac{1}{y+\frac{1}{z}}}$; $2+\frac{11}{13} = 2+\cfrac{1}{x+\cfrac{1}{y+\frac{1}{z}}}$; $\frac{11}{13} = \cfrac{1}{x+\cfrac{1}{y+\frac{1}{z}}}$

Taking reciprocals on both sides, $\frac{13}{11} = x+\cfrac{1}{y+\frac{1}{z}}$ *i.e.*, $1+\frac{2}{11} = x+\cfrac{1}{y+\frac{1}{z}}$ ...(1)

x, *y*, *z* are natural numbers. As y and z are natural numbers, $y+\frac{1}{z}$ is greater than 1.

The integer and fractional parts on both sides of (1) must be equal (as both fractions are proper fractions).

$\therefore x = 1$ and $\cfrac{1}{y+\frac{1}{z}} = \frac{2}{11}$; *i.e.*, $x = 1$ and $y+\frac{1}{z} = \frac{11}{2} = 5\frac{1}{2}$

The integer and fractional parts of $y+\frac{1}{z}$ are *y* and $\frac{1}{z}$ respectively.

$\therefore y = 5$ and $\frac{1}{z} = \frac{1}{x} \therefore z = 2.$

Alternative Sol:

We notice that on the RHS, at each stage, the numerator is 1.

We can proceed as shown below.

$$\frac{37}{13} = 2 + \frac{11}{13} = 2 + \frac{1}{\frac{13}{11}}; \frac{13}{11} = 1 + \frac{2}{11} = 1 + \frac{1}{\frac{11}{2}}; \frac{11}{2} = 5 + \frac{1}{2}$$

$$\therefore \frac{37}{13} = 2 + \frac{1}{1 + \frac{1}{5 + \frac{1}{2}}} = 2 + \frac{1}{x + \frac{1}{y + \frac{1}{z}}}$$

By inspection, $1 = x$, $5 = y$, $2 = z$.

30. When n is even

$$\text{Remainder} = \frac{6+6+1+1}{10} = \frac{14}{10} = 4$$

When n is odd

$$\text{Remainder} = \frac{4+6+9+1}{100} = \frac{20}{10} = 0$$

$\therefore$ required sum $= 4 + 0 = 4$

31. The ball is dropped from a height of 36 m. After each bounce, the ball rises to $\frac{2}{3}$ rds of the height from which it fell.

The ball will rise to a height of $36 \times \frac{2}{3} = 24$ m after the first bounce. It will then fall 24 m

The ball will rise to a height of $24 \times \frac{2}{3} = 16$ m after the second bounce and to a height of

$16 \times \frac{2}{3} = 10\frac{2}{3}$ m m after the third bounce.

32. We have,

$$R = \frac{(225)^{40}}{1000} = \frac{(25 \times 9)^{40}}{1000} = \frac{25^{40} \times 9^{40}}{1000} = \frac{5^{80} \times 9^{40}}{1000}$$

$$= \frac{5^3 \times 5^{77} \times 9^{40}}{1000} = \frac{5^{77}}{8} = \frac{(5^2)^{38} \times 5}{8}$$

$$= \frac{25^{38} \times 5}{8} = \frac{5}{8} = 5$$

Last 3 digit $= 5 \times 5^3 = 625$

$\therefore$ Digit at 100th place = 6

33. $x + \frac{100}{x} > 50$ and x ! {1, 2, 3,…….99, 100}

For x = 1, 2 the equation is satisfied. It is also satisfied for x = 50, 51, ……100.

It is also satisfied for x = 49, 48.
It is not satisfied for x = 3, 4,……..,47 (i.e. for 45 values).
There are 55 positive integral values (less than or equal to 100) which satisfy the given inequality.

34. $$\frac{1^a+2^b+3^c+4^d}{10} = \frac{(1^a+3^c)+(2^b+4^d)}{10} = \frac{\text{odd}_1+\text{odd}_2+(\text{even}_1+\text{even}_2)}{10}$$
$$= \frac{\text{even}+\text{even}}{10} = \frac{\text{even}}{10}$$

∴ Number of possible remainders = 5 (0, 2, 4, 6, 8)

35. Let E = $(n^3 - n)(n - 2) = (n - 2)(n - 1)\,n(n+1)$
For n = 3, E = 1(2)(3)(4) = 24
For n = 4, E = 2(3)(4)(5) = 120
For n = 5, E = (3)(4)(5)(6) =360
The product of any 4 consecutive numbers is divisible by 4!, viz. 24.
(If there were some other conditions imposed on n, we should consider whether 48 would also be a factor of all corresponding values of E.)

36. The average of m numbers is n^4. The average of n numbers is m^4.
∴ The average of these m + n numbers is
$$\bar{x} = \frac{mn^4+nm^4}{m+n} = \frac{mn(n^3+m^3)}{m+n} = mn(m^2 - mn + n^2)$$

37. For all numbers ending in 4, the last digits of successive powers are 4, 6, 4, 6…… ∴ $(234)^{100}$ ends in 6 and 234^{101} ends in 4. ∴ The units digit of $234^{100} + 234^{101}$ is 6 + 4, viz 0.

38. The 7-digit number 876p37q is divisible by 225, i.e, by 25 and 9.
∴ q has to be 5 and the sum of the 7 digits, 36 + p has to be a multiple of 9. ∴ p = 0 or 9.

39. Successive power of 7 have the units digit of 7, 9, 3, 1;
7, 9, 3, 1; …….As 139 = 4(34) + 3, the units digits of 7^{139} is 3.

40. Number leaving a remainder of 4 when divided by 12 are
4, 16, 28, 40, 52, 64 etc.
If a = 1 b can be 4
If a = 2 b can be 2 or 8
If a = 3, b has no possible value.
If a = 4, b can be 1, 4, 7, 10
If a = 5, b can be 8
If a = 6, b has no possible value.
If a = 7, b can be 4
If a = 8, b can be 2, 5, 8.
If a = 9, b has no possible value
As $a \neq b$ we leave out (4, 4) and (8, 8)
There are 10 possible values for (*a*, *b*).

41. $13^5 + 14^5 + 15^5 + 16^5 \Rightarrow (135 + 165 + (14^5 + 15^5)$
$(13^5 + 16^5)$ will be divisible by 29, here $n = 5$ (odd)
[∵ $(a^n + b^n)$ is divisible by $(a + b)$ if n is odd]
Similarly, $(14^5 + 15^5)$ is divisible by 29.

$\therefore$ The given expression is a multiple of 29 or Rem $\frac{E}{29} = 0$.

43. $1^3 + 2^3 + ... + 10^3 = \left[\frac{10(11)}{2}\right]^2 = 55^2 = 3025$

$$1^2 + 2^2 + ... + 10^2 = \frac{10(11)(21)}{6} = 5(11)(7) = 385$$

The required difference = 2640

43. $\frac{(4444)^{4444}}{9}$

Divide 4444 by 9, we get remainder 7

$\therefore \frac{(7)^{4444}}{9}$

$\Rightarrow \frac{(-2)^{4444}}{9}$ $\quad [\because 7-9=-2]$

$\Rightarrow \frac{(2)^{4444}}{9}$ $\quad \Rightarrow \frac{(2^3)^{1481}}{9} \times \frac{2}{9}$ $\quad \left[\because \frac{4444}{3} = 1481\right]$

$\Rightarrow \frac{(8)^{1481}}{9} \times \frac{2}{9}$ $\quad \Rightarrow \frac{(-1)^{1481}}{9} \times \frac{2}{9}$ $\quad [\because 8-9=-1]$

$\Rightarrow \frac{8}{9} \times \frac{2}{9} = \frac{16}{9}$ So, the remainder of 16 divided by 9 is 7

44. Put $a = 443$, $b = 547$

$$\therefore \frac{(a+b)^2 + (a-b)^2}{a^2 + b^2} = \frac{2(a^2 + b^2)}{a^2 + b^2} = 2$$

45. The number of prime numbers which are less than 100 is 25

46. Let x be the number which is to be subtracted from both numerator and denominator of the given fraction

According to the question, $\frac{27-x}{35-x} = \frac{2}{3} \Rightarrow 81 - 3x = 70 - 2x \Rightarrow x = 11$

47. Comparing $\frac{7}{9}$ and $\frac{3}{4}$, we get

$\frac{7}{9} \times \frac{4}{4} = \frac{28}{36}$ and $\frac{3}{4} \times \frac{9}{9} = \frac{27}{36}$ $\quad [\because$ Taking LCM of 9 and 4 = 36]

So, $\frac{7}{9}$ is greater ... (*i*)

Comparing $\frac{11}{14}$ and $\frac{10}{13}$ we get $\frac{11 \times 13}{14 \times 13} = \frac{143}{182}$ and $\frac{10}{13} \times \frac{14}{14} = \frac{140}{182}$

So, $\frac{11}{14}$ is greater ... (*i*)

Now comparing (*i*) and (ii) we get $\frac{7}{9}\times\frac{14}{14}=\frac{98}{126}$ and $\frac{11\times 9}{14\times 9}=\frac{99}{126}$

So, $\frac{11}{14}$ is greater

48. Let $f(M) = M + M^2$ ...(i)

To find the minimum value, take the derivative and set it to zero:

$$F(M) = 1 + 2M$$

$$1 + 2M = 0$$

$$M = -0.5$$

From eq (i)

$$f(M) = M + M^2$$

$$f(-0.5) = (-0.5) + (-0.5)^2 = -0.5 + 0.25 = -0.25$$

49. The 3-digit numbers that give a remainder of 5 when divided by 50 are: 105, 155, 205, ..., 955.

$a = 105$, $l = 955$ and $d = 50$

$$n=\frac{(l-a)}{d}+1$$

$$n=\frac{955-105}{50}+1$$

$$n=\frac{850}{50}+1$$

$$n = 17 + 1 \Rightarrow n = 18$$

$$S_{18}=\frac{n}{2}(a+l)=\frac{18}{2}(105+955)$$

$$S_{18}=9\times 1060=9540$$

50. $N - 1$ = multiple of 2^n

$$N = 2^n \times k + 1$$

$N - 1$ = multiple of 2^n

For smallest $N \geq 10000$

Let, $n = 13 \Rightarrow 2^{13} = 8192 \Rightarrow N = 8192 \times 2 + 1 = 16385$

$n = 14 \Rightarrow 2^{14} = 16384 \Rightarrow N = 16384 \times 1 + 1 = 16385$

$n = 14$ follows the condition.

$\Rightarrow$ The value of n is 14

51. Divisor = (5 × 48) = 240, Quotient = $\left(\frac{240}{12}\right)$ = 20.

∴ Dividend = (Divisor × Quotient) + Remainder = (240 × 20 + 48) = 4848.

52. 910 = 2 × 5 × 7 × 13

If $n + 1 = 13 \Rightarrow n = 12$

53. When each side of an inequality is multiplied by a negative real number, then the inequality is reversed. So, $x > y \Rightarrow -x < -y$.

54. $\sqrt{.0001} = \sqrt{\dfrac{1}{10000}} = \dfrac{1}{100}$, which is rational.

55. Remainder = $\dfrac{9100}{16} = 12$

56. (1) $\dfrac{1+1}{2} = \dfrac{2}{2} = 1$

(2) $555^{777} + 777^{555}$ is divisible by 3.

(3) $555^{777} + 777^{555}$ is divisible by 37.

57. When x is a negative real number, then $|x| = -x$.

58. $1399 \times 1399 = (1400-1)^2 = (1400)^2 + 1^2 - 2 \times 1400 \times 1 = (1960000 + 1 - 2800) = 1957201.$

59. $1916 \times 1916 = (1900 + 16)^2 = (1900)^2 + 16^2 + 2 \times 1900 \times 16 = 3671056.$

60. Given Exp. = $\left(\dfrac{a^3+b^3}{a^2+b^2-ab}\right)$, where $a = 885$ & $b = 115 = (a+b) = (885+115) = 1000.$

61. $\dfrac{1}{532900} + \dfrac{p^2}{266450} + \dfrac{p^4}{532900}$

$\Rightarrow \dfrac{1}{532900} + \dfrac{p^2}{266450} + \dfrac{p^4}{532900} = \dfrac{1}{532900} + \dfrac{2p^2}{532900} + \dfrac{p^4}{532900}$

$\Rightarrow \dfrac{1+2p^2+p^4}{532900} \Rightarrow \dfrac{(p^2+1)^2}{(730)^2}$

Since both numerator and denominator both are perfect square and the result is an integer, minimum possible value of expression is 1

So. $p^2 + 1 = 730$

$p^2 = 730 - 1 = 729$

$p = 27$

The minimum value of p for which the expression is an integer is 27.

62. Power of 2 in 100!

$\dfrac{100}{2} + \dfrac{100}{4} + \dfrac{100}{8} + \dfrac{100}{16} + \dfrac{100}{32} + \dfrac{100}{64} = 50+25+12+6+3+1 = 97$

Power of 5 in 100!

$\dfrac{100}{5} + \dfrac{100}{25} + \dfrac{100}{125} = 20+4+0 = 24$

2^{2n} must $\le 2^{97} \Rightarrow 2n \le 97 \Rightarrow n \le 48.5$

And 5^{2n} must $\le 5^{24} \Rightarrow 2n \le 24 \Rightarrow n \le 12$

So, the limiting factor is power of 5 $= n \le 12$

The largest value of $n = 12$

63. Let the fraction be $\dfrac{x}{y}$. Then,

$\dfrac{x+1}{y+1} = 4$ and $\dfrac{x-1}{y-1} = 7$. So, $x - 4y = 3$ and $7y - x = 6$.

On solving $x - 4y = 3$ and $7y - x = 6$, we get $x = 15$.

64. According to the question, $\Rightarrow f(x) = x + 4 \times x^2$

For the critical point which is $x, f'(x) = 0$

Differentiating we get,

$\Rightarrow f'(x) = 1 + 4 \times (2x) = 0 \Rightarrow f'(x) = 1 + 8x = 0 \Rightarrow 8x = -1 \Rightarrow x = -\frac{1}{8} = -0.125.$

$\therefore$ The real number is calculated as –0.125.

65. According to the question,

$$m^2 - n^2 = 72$$

However, there are only three pairs of natural numbers (m, n) that satisfy the equation $m^2 - n^2 = 72$, and $m > n$.

(19, 17), (11, 7), and (9, 3).

The number of pairs that satisfy the above conditions is 3.

66. $x + y = 37$ and $xy = 342$.

$$(x - y) = \sqrt{(x + y)^2 - 4xy} = 1.$$

Solving $x + y = 37$ and $x - y = 1$, we get $x = 19$ and $y = 18$.

67. $x - \frac{1}{7}x = 84 \Rightarrow 6x = 84 \times 7 \Rightarrow x = 98.$

68. $\frac{4}{5}x - \frac{2}{3}x = 8 \Rightarrow 12x - 10x = 120 \Rightarrow x = 60$

69. $x - 21 = 71 - x \Rightarrow 2x = 92 \Rightarrow x = 46$

70. (1) $\frac{3m^3 + 2m^2 + 5m + n}{m} = 3m^2 + 2m + 5 + \frac{n}{m}$

(2) $\frac{8(8^m) + 2^{3m}}{48} = \frac{5 \times 8^m + 8^m}{48} = \frac{8^m(6)}{48}$

If $m = 0 = \frac{6}{48}$

$\therefore$ The statement is not correct.

71. $x : y = 2 : 3$ and $y : z = 5 : 3 = \frac{3}{5} \times 5 : \frac{3}{5} \times 3 = 3 : \frac{9}{5}$

$\therefore\ x : y : z = 2 : 3 : \frac{9}{5} = 10 : 15 : 9.$

Let the numbers be $10x$, $15x$ and $9x$. Then,

$\therefore\ 10x + 15x + 9x = 102 \Rightarrow x = 3.$

$\therefore$ Second number = $15x = 45$.

72. Let the numbers be $2x$, x and $4x$.

Then, $2x + x + 4x = 3 \times 56 \Rightarrow x = 24$.

So, the numbers are 48, 24 and 96.

73. According to the question,

Let the 4-digit number $N = (p^a)(q^b)$

Factors = $(a + 1)(b + 1)$

As per the question, $(a + 1)(b + 1) = 15$

$\Rightarrow$ 15 can be written as (3×5) or (1×15)

Case 1 : (3×5)	Case 2 : (1×15)
$(a+1)(b+1)=3\times 5$ Then, on comparing, we get $(a=2$ and $b=4)$ $\Rightarrow$ Number $N=(p^2)(q^4)$ $\Rightarrow N^2=(p^4)(q^8)$ $\Rightarrow$ Number of factors of $N^2=(4+1)(8+1)=5\times 9=45$	$(a+1)(b+1)=1\times 15$ Then, on comparing we get $(a=0\ \&\ b=14)$ $\Rightarrow$ Number $N=(p^0)(q^{14})=q^{14}$ $\Rightarrow N^2=q^{28}$ $\Rightarrow$ Number of factors of $N^2=(0+1)(28+1)=1\times 29=29$

But, $N=q^{14}$ cannot satisfy the 4-digit number condition.

Number of factors = 45

74. All single-digit prime no. = 2, 3, 5, 7

Largest no. = 7532

Smallest no. = 2357

Sum= 7532 + 2357 = 9889

The sum of the largest and the smallest 4-digit numbers made by using single digit prime numbers is 9889.

75. If $10^4 = 10000$, the sum of all digits is 1

$10^6 = 1000000$ also the sum of all digits is 1

sum of all digits of $(100^{97} + 100^{54} + x + 1)\ x + 3$

According to the divisibility rule for 9,

If we add 6 to $x + 3$ so that their sum is 9, then $x + 3$ will be divisible by 9 with 0 remainder.

The value of x is 6.

76. Let's test the values of x and y to find the minimum sum $(x+y)$

Taking $x=1$ and checking the values of y starting from 1

For $y=1$:

$$11+5\times 1=11+5=16 \text{ (not a prime number)}$$

For $y=2$:

$$11+5\times 2=11+10=21 \text{ (not a prime number)}$$

For $y=3$:

$$11+5\times 3=11+15=26 \text{ (not a prime number)}$$

For $y=4$:

$$11+5\times 4=11+20=31 \text{ (prime number)}$$

Thus, the minimum value of $(x+y)$ when $x=1$ and $y=4$

$$x+y=1+4=5$$

The minimum value of $(x+y)$ is 5.

77. Required numbers are: 12, 18, 24, ..., 96.

This is an A.P. in which $a = 12$ & $d = 6$.

Now, $T_n = 96 \Rightarrow a + (n-1)d = 96$, *i.e.* $12 + (n-1)\times 6 = 96 \Rightarrow n = 15$.

78. $2^2 + 4^2 + ... + 20^2 = (1\times 2)^2 + (2\times 2)^2 + (2\times 3)^2 + ... + (2\times 10)^2$

$= 2^2\times 1^2 + 2^2\times 2^2 + 2^2\times 3^2 + ... + 2^2\times 10^2$

$= 2^2\times [1^2 + 2^2 + 3^2 + ... + 10^2] = 4\times 385 = 1540.$

79. a, b, c are in H.P. $\Rightarrow \frac{1}{a}, \frac{1}{b}, \frac{1}{c}$ are in A.P. $\Rightarrow \frac{1}{b} = \frac{1}{2}\left(\frac{1}{a} + \frac{1}{c}\right)$ *i.e.* $b = \frac{2ac}{a+c}$.

80. $|2x - 3| = 7 \Leftrightarrow 2x - 3 = 7$ or $-(2x - 3) = 7 \Leftrightarrow 2x = 10$ or $-2x = 4 \Leftrightarrow x = 5$ or $x = -2$.

81. Given Exp. $= 1 + \dfrac{1}{1 + \frac{9}{10}} = 1 + \frac{10}{19} = \frac{29}{19}$.

82. Given Exp. $= \dfrac{2^{n-1}(2+1)}{2^n(2-1)} = \frac{3}{2}$.

83. Given surds are of order 4, 3 & 2 whose L.C.M. is 12.

$\sqrt[4]{4} = 4^{1/4} = 4^{\frac{1}{4}\times\frac{3}{3}} = 4^{3/12} = (4^3)^{1/12} = (64)^{1/12} = \sqrt[12]{64};$

$\sqrt[3]{3} = 3^{1/3} = 3^{\frac{1}{3}\times\frac{4}{4}} = 3^{4/12} = (3^4)^{1/12} = (81)^{1/12} = \sqrt[12]{81};$

$\sqrt{2} = 2^{1/2} = 2^{\frac{1}{2}\times\frac{6}{6}} = 2^{6/12} = (2^6)^{1/12} = (64)^{1/12} = \sqrt[12]{64};$

Clearly, $\sqrt[12]{81}$ is largest, *i.e.* $\sqrt[3]{3}$ is largest

84. Given Exp. $= (216)^{2/3} + (256)^{3/4} + (243)^{1/5} = (6^3)^{2/3} + (4^4)^{3/4} + (3^5)^{1/5}$

$= 6^{\left(3\times\frac{2}{3}\right)} + 4^{\left(4\times\frac{3}{4}\right)} + 3^{\left(5\times\frac{1}{5}\right)} = (6^2 + 4^3 + 3^1) = (36 + 64 + 3) = 103$

85. $a = c^z = (b^y)^z = b^{yz} = (a^x)^{yz} = a^{xyz}$. So, $xyz = 1$.

86. Let odd number = 3

Now, according to the question

$$9 = 8 \times 1 + 1$$

$\therefore$ We can check any odd number greater than 1.

Remainder is 1.

87. (1) Let $n = 1$

$\therefore \dfrac{n(n^2 + 2)}{3} = \dfrac{1\times 3}{3} = 1$

So, it is correct statement.

(2) Let $m = 3$

$\therefore \dfrac{m^4 + 4m^2 + 11}{16} = \dfrac{3^4 + 4(3)^2 + 11}{16} = \dfrac{81 + 36 + 11}{16} = 8$

$\therefore$ it is also a correct statement.

88. Let $n - 1$, and $n + 1$ be 7, 8, and 9

Now, $n^2 + 1 = 8^2 + 1 = 65$

$n^2 - 1 = 63$

$n^2 + n = 64 + 8 = 72$

$n^2 - n = 64 - 8 = 56$

$\therefore$ 5 divides $n^2 + 1$

89. Let required number = n

Now, by option

(a) 99981 − 7 = 99974 is not divisible by 11.

(b) 99988 – 7 = 99981 is not divisible by 11.
(c) 99997 – 7 = 99990 is divisible by 11.
(d) 99999 –7 = 99992 is not divisible by 11.

90. We have,
$360 = 2^3 \times 3^2 \times 5$
Total no. of factors = $(3 + 1) \times (2 + 1) \times (1 + 1) = 24$
Factors of 360 = 1, 2, 3, 4, 5, 6, 8, 9, 10, 12, 15, 18, 20, 24, 30, 36, 40, 45, 60, 72, 90, 120, 180, 360
∴ Sum of all factors = 1170

91. $xy\,xy\,xy = (100000x + 1000x + 10x) + (10000y + 100y + y)$
$= 101010x + 1010y = 10101\,(10x + y)$
$= 3 \times 7 \times 13 \times 37\,(10x + y)$

92. We have,
$x = \sqrt{\sqrt{4\sqrt{4\sqrt{4\sqrt{4}}}}}$
$x = \sqrt{4x}$
Squaring both sides, we get
$x^2 = 4x$
$\Rightarrow x^2 = -4x = 0 \quad \Rightarrow x\,(x - 4) = 0 \quad \Rightarrow x = 0, 4$

93. We have,
$33m + 22n = 11\,(3m + 2n)$
Now,
$$\frac{11(3m+2n)}{121} = \frac{3m+2n}{11}$$
Let m = 3 and $n = 1$
$3m + 2n = 11$
∴ $m = 3$ and $n = 1$
$m + n = 3 + 1 = 4$

94. We have, $a + b + c + d = 200$
Case 1: When all four numbers be even
S = 1 + 1 + 1 + 1 = 4
Case 2: When all four numbers be odd
S = – 1 – 1 – 1 – 1 – 1 = – 4
Case 3: When two number be even and other two be odd.
S = 1 + 1 – 1 – 1 = 0
∴ Number of possible value of S = 3

95. ∴ We know that $a^b - b^n$ is always divisible by $(a - b)$ and $(a + b)$
$97^{30} - 14^{30}$ is divisible by 97 – 14 = 83
and 97 + 14 = 111 = 37 × 3
∴ $97^{30} - 14^{30}$ is divisible by both 37 and 83.

96. (1) $n^3 - n = n(n^2 - 1) = n(n - 1)\,(n + 1)$
Product of three consecutive number is always divisible by 6
(2) $n^5 - n = n(n^4 - 1) = n(n^2 - 1)\,(n^2 + 1)$
$= n(n^2 - 1)\,(n^2 - 4) + 5)$

$= n(n^2 - 1)(n^2 - 4) + 5n(n^2 - 1)$
$= n(n - 1)(n + 1)(n + 2)(n - 2) + 5n(n - 1)(n + 1)$
$= [(n - 2)(n - 1)n(n + 1)(n + 2)] + 5n(n - 1)(n + 1)$

Product of 5 numbers is always divisible by 5.

(3) $n^5 - 5n^3 + 4n$

$n(n^4 - 5n^2 + 4) = n(n^2 - 4)(n^2 - 1)$
$= n(n - 2)(n + 2)(n + 1)(n - 1) = (n - 2)(n - 1)n(n + 1)(n + 2)$

97. For 9^n

If n is odd, the unit digit is 9 and if n is even, the unit digit is 1.

$\therefore$ the last digit of 9^{27} is 9 [$\because$ 27 is odd]

the last digit of 27^9 is 7

Now, $9 + 7 = 16$

$\therefore$ last digit of S = 6

98. Statement I:

Sum of digits $= 21 + x + y$

When $y = 2$

$\therefore$ Sum $= 23 + x$

When $x = 4$, then it is divisible by 9

Now, when $y = 7$

$\therefore$ Sum $= 28 + x$

Now, it is divisible by 9, when $x = 8$

$\therefore$ Number of possible values of x = two

Statement II:

In $277 \times y5$,

Whey $y = 2$ or 7

277×25 and 277×75 are divisible by 25, which is independent of x.

From statement I and II, we get

$x = 8$

$\therefore$ option (c) is correct

99. Statement I : $\frac{m}{n} > 1 \quad \Rightarrow m > n$

Statement II: $m > 2n$

let $n = 1$

$m > 2 \times 1$ when $m > 1$

$\therefore$ option (b) is correct.

100. Statement I:

For any value of p, $2p$ will be even.

$\therefore$ q is odd [$\therefore$ $2p + q$ is odd]

Statement II:

$\therefore$ $q - 2p$ is odd

$\therefore$ q must be odd.

$\therefore$ $p^2 + q^2 + q$ is even

101. Let the unit digit and tens digit of a two digit number be x and y respectively.

$\therefore$ number $= 10y + x$

Now, according to the question,

$10y + x + 10x + y = 55$

$\Rightarrow x + y = 5$ ...(i)

and $(10y + x) - (10x + y) = 45$

$\Rightarrow y - x = 5$ ...(ii)

by solving (i) and (ii), we get

$y = 5$ and $x = 0$

$\therefore$ Number $= 10 \times 5 + -0 = 50$

$\therefore$ required products $= 5 \times 0 = 0$

102. Let $A = 125^{100}$

By applying log on both sides, we get

$\log_{10} A = 100 \log_{10} 125$

$300 \log_{10} 5 = 300 (\log_{10} 10 - \log_{10} 2) = 300 (1 - 0.301) = 300 \times 0.699 = 209.7$

$\therefore$ required number of digits = 210

103. Let three-digit number is *pqr*

$\therefore$ Number $= 100p + 10q + r$

(1) New number $= 100q + 10p + r$

Now,

$100p + 10q + r - 100q - 10p - r = 180$

$\Rightarrow p - q = 2$ ...(i)

(2) New number $= 100 \times \frac{p}{2} + 10r + q$

Now,

$100p + 10q + r - 100\left(\frac{p}{2}\right) - 10r - q = 336$

$50p + 9q - 9r = 336$

$r = 0$ [$\because$ Number is divisible by 10]

$50p + 9q = 336$ ...(ii)

Multiplying (i) by 9 and adding (i) & (ii)

$p = 6$ and $q = 4$

$\therefore$ (c) is correct

104. $a - b + c > a + b - c$ and $a, b, c \in R$

(1) $b < 0$

(2) $c < 0$

$\therefore$ (d) is correct

105. Let cost of 1 pen $= a$

Cost of 1 pencil = b

Cost of 1 notebook = c

$\therefore 15a + 21y + 18z = ?$

(1) $7a + 6b + 5c = 200$ ...(i)

(2) $3a + 8b + 7c = 210$ ...(ii)

Let $15a + 21b + 18c = \alpha(7a + 6b + 5c) + \beta(3a + 8b + 7c)$

$(15 = 7\alpha + 3\beta)$ ... (i)

$(21 = 6\alpha + 8\beta)$ ... (ii)

$18 = 5\alpha + 7\beta)$

Multiplying (i) by 8 and (ii) by 3 and substracting (i) and (ii), we get $57 = 38\alpha$

$$\Rightarrow \alpha = \frac{57}{38} = \frac{3}{2}, \quad \beta = \frac{21 - 6 \times \frac{3}{2}}{8} = \frac{12}{8} = \frac{3}{2}$$

$$\therefore\ 5\alpha + 7\beta = 5 \times \frac{3}{2} + 7 \times \frac{3}{2} = \frac{15+21}{2} = 18$$

$$\therefore\ \alpha = \beta = \frac{3}{2} \text{ is sufficient}$$

$$\therefore\ \text{Required cost} = \frac{3}{2} \times 200 + \frac{3}{2} \times 210 = ₹65$$

∴ Option (c) is correct.

106. Let the five different integers be a, b, c, d, and e where $2 \le a < b < c < d < 8$.

Statement-I:

The sum of the five integers must end in 0 or 5 to be a multiple of 5.

Let's consider possible sets of integers:

(2, 3, 4, 5, 6), (2, 3, 5, 7, 8), (2, 4, 5, 6, 8), (3, 4, 5, 6, 7) and (4, 5, 6, 7, 8)

Statement II: If the number of odd integers is odd, we can have 1, 3, or 5 odd numbers,

Possible sets satisfying both conditions could be:

Set 1: (2, 3, 5, 7, 8) (Odd count = 3)

Set 2: (2, 4, 5, 6, 8) (Odd count = 1)

Set 3: (2, 3, 4, 6, 8) (Odd count = 1) and so on.....

Both statements together:

If the number of odd integers is odd, and the sum of all the 5 integers is a multiple of 5.

we can have 1, 3, or 5 odd numbers,

Possible sets satisfying both conditions could be:

Set 1: (2, 3, 5, 7, 8) (Odd count = 3)

Set 2: (2, 4, 5, 6, 8) (Odd count = 1)

Set 1: Sum = 2 + 3 + 5 + 7 + 8 = 25 (divisible by 5)

Set 2: Sum = 2 + 4 + 5 + 6 + 8 = 25 (divisible by 5)

$$\text{Average for both sets} = \frac{\text{Sum of integers}}{5} = \frac{25}{5} = 5$$

107. We can see that each 6 consecutive terms, when taken together form a group whose sum is zero.

$$\underbrace{\left(\frac{1}{4} + \frac{1}{5} - \frac{1}{6} - \frac{1}{4} - \frac{1}{5} + \frac{1}{6}\right)}_{0} + \underbrace{\left(\frac{1}{4} + \frac{1}{5} - \frac{1}{6} - \frac{1}{4} - \frac{1}{5} + \frac{1}{6}\right)}_{0} \ldots\ldots\ldots\ldots$$

In 47 terms, there are 7 groups of 6 terms each (7 × 6 = 42 terms) and the remaining five terms will be $\frac{1}{4} + \frac{1}{5} - \frac{1}{6} - \frac{1}{4} - \frac{1}{5}$, whose sum is $\frac{-1}{6}$.

108. Let the number be '*ab*'.

$a + b = 7$-------(1) and '*ba*' = '*ab*' + 27

$\Rightarrow 10b + a = 10a + b + 27 \Rightarrow b - a = 3$ ... (2)

$\therefore$ $b = 5, a = 2$. The product of the digits is 10.

109. The numbers to be considered are 10, 17, 24, ... 94, i.e. 7(1) +3, ----- –7(13) + 3. There are 13 numbers Their average is $\frac{10 + 94}{2}$, viz. 52. $\therefore$ Their sum is 676.

110. Clearly both the statements are true for any nonzero integers p, q

111. Let the two digit number be $10x + y$

$10x + y = k(x + y)$ ... (1)

also $10y + x = m(x + y)$ ...(2)

(1) + (2) = $11x + 11y = (x + y)(k + m)$

$11(x + y) = (x + y)(k + m)$

$\because$ $x + y \neq 0 \Rightarrow k + m = 11$

$m = 11 - k$

112. S = {1, 2, 3, ..., 14}

a, b, ∈ s such that all leaves remainder 1 when divided by 15

possible values of above

16, 31, 46, 61, 76, 91, 106, 121, 136, 151, 166, 181, 196

Only possible ordered pairs are

(2, 8), (7, 13) *i.e.,* two only

113. $N = 90 \times 42 \times 324 \times 55 = 3^2 \times 10 \times 3 \times 14 \times 3^4 \times 4 \times 55 = 3^7 \times 10 \times 14 \times 4 \times 55$

$\therefore$ The maximum Value of m for which N is divisible by 3^m is 7

114. (1) Every natural number is a real number (true)

(2) Every real number need not be a rational number (false)

(3) Every integer is a real number (True)

(4) Every rational number is a real number (true)

$\therefore$ Statements 1, 3, 4 are true

115. Statement (1) When m, n are natural number $\frac{m}{n} + \frac{n}{m}$ need not be an integer.

Statement (1) need not be true

Statement (2) $mn\left(\frac{m}{n} + \frac{n}{m}\right)\left(m^2 + n^2\right)^{-1} = m\left(\frac{m^2 + n^2}{mn}\right)\left(m^2 + n^2\right)^{-1} = \frac{1}{n}$

Which is always an integer

$\therefore$ Statement (2) is true.

Statement (3) $\frac{mn}{m^2 + n^2} \Rightarrow \frac{1}{\frac{m^2 + n^2}{mn}} = \frac{1}{\frac{1}{m} + \frac{1}{n}}$ need not be an integer.

Statement (3) is not true.

116. $a < 0, b < 0$ while $0 < c$

(1) $a - b < a - c$ The RHS is negative The LHS could be positive False

(2) If $a < b$, then $\frac{a}{b} < \frac{b}{c}$ True

(3) $\frac{1}{l} < \frac{1}{c}$ True $\left(\because \frac{1}{b} < 0 < \frac{1}{c}\right)$

(2) and (3) are true.

117. Let the number be 8*x*.

$64x - x = 2016 \Rightarrow x = \frac{2016}{63} = \frac{224}{7} = 32$

$\therefore$ The number in 8(32), ie 256

118. 1729, called Remazajuns Member, is smallest positive number that can be expressed as the sum of two 'positive' cubes in 2 different ways.

$1729 = 1728 + 1 = 12^3 + 1^3 = 1000 + 729 = 10^3 + 9^3$

If the condition 'positive' is removed, we can consider 728.

$728 = 729 + (-1) = 9^3 + (-1)^3 = 512 + 216 = 8^3 + 6^3$

119. Let cost of two types of sweets be 7*x* and 10*y*

$\therefore$ According to the question $7x + 10y = 200$

$\Rightarrow y = \frac{200 - 7x}{10} \Rightarrow y = 20 - \frac{7}{10}x$ ... (ii)

Let $x > y$ and assume $7x = 140$

$7x = 140 \Rightarrow x = 20$

Put $x = 20$ in (*i*)

$y = 20 - \frac{7 \times 20}{10} = 20 - 14 = 6$

So, Maximum number of sweets $= x + y = 20 + 6 = 26$

120. Statement 1: Of two consecutive integers, one is even e.g., 2 3, 4 5, 6 7

So, it is correct

Statement 2 : $3^2 = 9 = 8 \times 1 + 1$, $5^2 = 25 = 8 \times 3 + 1$, $9^2 = 81 = 8 \times 10 + 1$, $13^2 = 169 = 8 \times 21 + 1$

They are of the form 8n + 1

So, it is also correct.

Hence, both statements 1 and 2 are correct.

121. N is prime to *p* means N and p are co-prime numbers

According to the question, $N^{p-1} - 1 = Kp, K = \frac{N^{P-1} - 1}{p}$

So, here *p* will be prime number.

122. Let the numbers be *x*, 5*x*

According to the question, $5x^2 = 320$, $x^2 = 64$

Difference between squares of two numbers $= (5x)^2 - x^2 = 25x^2 - x^2 = 24x^2 = 24 \times 64 = 1536$

123. By combining I and II, we get $a > b > c > d$

124. by combining I and II, we get

Let $n = 2$

From I

$n = 2$

Question cannot be answered from both I and II, since we cannot determine the exact value of m.

125. I. $\frac{P^n}{p+1} = \frac{P^{\text{even}}}{(p+1)} = \frac{+1}{p+1}$ (R = 1)

II. $\frac{P^n}{P+1}$ (not find remainder)

126. Given that $(x + z - y) \times (x - z + y) \propto yz$

$$(k - 2y) \times (k - 2z) = Cyz$$
$$k^2 - 2kz - 2ky + 4yz = Cyz$$
$$k^2 - 2k(y + z) = Cyz - 4yz$$
$$k^2 - 2k(y + z) = (C - 4) \times yz$$

$\Rightarrow$ $(x + y + z) = k$. So, $y + z = k - x$

Substitute $(y + z)$ in the equation $k^2 - 2k(y + z) = (C-4)yz$:

$\Rightarrow$ $k^2 - 2k(k - x) = (C - 4) \times yz$

$\Rightarrow$ $-k^2 + 2kx = (C - 4) \times yz$

$\Rightarrow$ $k(2x - k) = (C - 4) \times yz$

We know $(y + z - x) = (k - x) - x = k - 2x$

From $k(2x - k) = (C - 4)yz$,

$\Rightarrow -k(k - 2x) = (C - 4)yz \quad \Rightarrow k - 2x = \frac{(C-4)yz}{-k} \quad \Rightarrow k - 2x = \frac{(4-C)yz}{k}$

Since k, C are constants, $\frac{(4-C)}{k}$ is also a constant.

Therefore, $(k - 2x)$ is proportional to (yz).

Since $(y + z - x) = k - 2x$

$\therefore$ $(y + z - x)$ is proportional to yz.

127. Last term $(l) = a + (n - 1) \times d$

For the last term to be ≥ 0:

$$36 + (n - 1)(-3) \geq 0$$
$$36 - 3(n - 1) \geq 0$$
$$3n \leq 39$$
$$n \leq 13$$

Thus, the sequence has a maximum of $n = 13$ terms.

$S_{13} = \frac{13}{2}[2(36) + (13-1)(-3)] \Rightarrow S_{13} = \frac{13}{2}[72 + 12(-3)] \Rightarrow S_{13} = \frac{13}{2}[72 - 36]$

$\Rightarrow S_{13} = \frac{13}{2}[36] \Rightarrow S_{13} = 13 \times 18 \Rightarrow S_{13} = 234$

128. $2 \in N$, $3 \in N$ but $2 - 3 = -1 \notin N$.

$5 \in Q$, $0 \in Q$ but $\frac{5}{0} \notin Q$.

$a \in I$, $b \in I \Rightarrow a - b \in I$ is always true.

129. I. Let a/b and b/c.

Then, $b = am$ and $c = bn$ for some $m, n \in N$.

$\therefore$ $c = bn = (am)\,n = a\,(mn)$. So, a/c.

II. Let a/b and a/c. Then, $b = am$ & $c = an$ for $m, n \in N$.

$\therefore$ $b + c = a(m+n)$. So, $a/(b+c)$.

III. Let a/b and a/c. Then, $b = am$ & $c = an$ for $m, n \in N$.

$\therefore$ $(b - c) = a\,(m - n)$. So, $a/(b - c)$.

130. When $x = 1, y = 2, z = 3$ then

$(x^2 + y^2 + z^2) = 1^2 + 2^2 + 3^2 = 14 \Rightarrow p \propto (x^2 + y^2 + z^2) \Rightarrow 70 \propto 14 \Rightarrow 70 = k(14) \Rightarrow k = 5$

When $x = -1, y = 1, z = 5$, then

$$(x^2 + y^2 + z^2) = (1 + 1 + 25) = 27$$

$$p = k(x^2 + y^2 + z^2) \Rightarrow p = 5 \times (27) = 135$$

131. Since 30 ends in 0 but 30 is not divisible by 25. So, (*a*) is incorrect.

Since 35 ends in 5 but 35 is not divisible by 25. So, (*b*) is incorrect.

Since 115 has last 2 digits 15, which is divisible by 5. But, 115 is not divisible by 25.

So, (*c*) is incorrect. Clearly, (*d*) is true.

132. (*a*) When $n = 1$, $10^n - 1 = 9$, which is not divisible by 11.

(*b*) For even values of n, the number $(10^n - 1)$ consists of even number of nines and hence it will be divisible by 11.

(*c*) $5^1 - 1 = 4$, which is not divisible by 11.

(*d*) $5^2 - 1 = 24$, which is not divisible by 11.

So, (*b*) is correct.

133. *A* is a restatement of division algorithm theorem. Also, every integer a is either a multiple of 3 or when divided by 3, it will give 1 or 2 as remainder.

So, $a = 3n + 1$ or $a = 3n + 2$

Thus, both (*A*) and (*R*) are true statements.

But, *R* is not a correct explanation of *A*.

134. Clearly, both *A* and *R* are true but *R* is not a correct explanation of *A*.

135. Integers are closed for subtraction is a true statement.

Note that $0 \notin N$ and $0 \in Q$ but 0 has no reciprocal.

$5 \in R, 0 \in R$ but $\frac{5}{0} \notin R$.

136. A pure fraction has its numerator less than its denominator. Let $\frac{a}{b}$ be a pure fraction. Then, $a < b$.

$$b > a \Rightarrow bp > ap \Rightarrow ab + bp > ab + ap.$$

So, $\frac{a+p}{b+p} > \frac{a}{b}$.

137. Clearly, (*d*) is true.

138. $\frac{4}{5}x - \frac{3}{4}x = 4 \Rightarrow 16x - 15x = 80 \Rightarrow x = 80$.

139. Given Exp. $= \frac{2}{3} \times \frac{3}{4} \times \frac{4}{5} \times \frac{5}{6} \times \ldots \times \frac{(n-1)}{n} = \frac{2}{n}$.

140. Given Exp. $= \frac{5}{3} \times \frac{7}{5} \times \frac{9}{7} \times \ldots \times \frac{1001}{999} = \frac{1001}{3}$.

H.C.F. and L.C.M. of Numbers

Factors and Multiples: *If a divides b exactly, then we say that a is a factor of b and we write, a/b. In this case, b is called a multiple of a.*

Highest Common Factor (H.C.F or G.C.D. or G.C.M.): *The H.C.F. of two or more than two numbers is the greatest number that divides each one of them exactly. This is also known as greatest common divisor (G.C.D.) or greatest common measure (G.C.M.).*

To Find H.C.F. of Given Numbers:

(*i*) **H.C.F. By Factorization:** *Express each number as the product of primes. Now, take the product of least powers of common factors to get the H.C.F.*

Ex. 1. ***Find the H.C.F. of 96, 528 and 792.***

Sol. We have:

$96 = 2^5 \times 3$, $528 = 2^4 \times 3 \times 11$ and $792 = 2^3 \times 3^2 \times 11$.

Taking the product of least powers of common factors, we get:

H.C.F. $= (2^3 \times 3) = 24$.

(*ii*) **H.C.F. By Division:** *Divide larger number by smaller one. Now, divide the divisor by the remainder. Repeat the process of dividing the preceding divisor by the remainder last obtained, till the remainder 0 is obtained. The last divisor is the required H.C.F.*

Ex. 2. ***Find the H.C.F. of 370 and 592.***

Sol.

```
370 ) 592 ( 1
      370
      ---------
      222 ) 370 ( 1
            222
            ---------
            148 ) 222 ( 1
                  148
                  ---------
                   74 ) 148 ( 2
                        148
                        ---
                         ×
```

∴ H.C.F. of given numbers = 74.

Suppose we have to find the H.C.F. of three numbers. In this condition, the H.C.F. of the H.C.F. of any two numbers and the third number will be the H.C.F. of given three numbers.

Ex. 3. ***Find the H.C.F. of 312, 351 and 650.***

Sol. By division method, we can find that:

H.C.F. of 312 and 351 = 39.

H.C.F. of 39 and 650 = 13.

∴ H.C.F. of 312, 351 and 650 = 13.

Ex. 4. ***Reduce*** $\frac{777}{1147}$ ***to lowest terms.***

Sol. H.C.F. of 777 & 1147 is 37.

On dividing Nr & Dr of the given fraction by 37, we get:

$\frac{777}{1147} = \frac{21}{31}$.

Lowest Common Multiple (*L.C.M.*): *The least number which is exactly divisible by each one of the given numbers, is called their L.C.M.*

L.C.M. By Factorization (*Rule*): *Resolve each one of the given numbers into prime factors. Then, the product of highest powers of all the factors, gives the L.C.M.*

Ex. 5. ***Find the L.C.M. of 96, 108 and 280.***

Sol. $96 = 2^5 \times 3$, $108 = 2^2 \times 3^3$ and $280 = 2^3 \times 5 \times 7$.

$\therefore$ $L.C.M. = (2^5 \times 3^3 \times 5 \times 7) = 30240$.

Two Important Rules:

(*i*) ***Product of Two Numbers = (Their H.C.F.) × (Their L.C.M.)***

(*ii*) ***H.C.F. of given numbers always divides their L.C.M.***

H.C.F. and L.C.M. of Fractions

(*i*) *H.C.F. of given fractions* $\frac{H.C.F.\ of\ Numerators}{L.C.M.\ of\ Denominators}$

(*ii*) *L.C.M. of given fractions* = $\frac{L.C.M.\ of\ Numerators}{H.C.F.\ of\ Denominators}$

Ex. 6. ***Find the H.C.F. and L.C.M. of*** $\frac{8}{9}, \frac{32}{81}$ ***and*** $\frac{10}{27}$.

Sol. $\text{H.C.F.} = \frac{\text{H.C.F. of 8, 32, 10}}{\text{L.C.M. of 9, 81, 27}} = \frac{2}{81}$, $\text{L.C.M.} = \frac{\text{L.C.M. of 8, 32, 10}}{\text{H.C.F. of 9, 81, 27}} = \frac{160}{9}$.

Ex. 7. ***The H.C.F. of two numbers is 14 and their L.C.M. is 11592. If one of the numbers is 504, find the other.***

Sol. $\text{Other number} = \frac{\text{H.C.F.} \times \text{L.C.M.}}{\text{Given Number}} = \frac{14 \times 11592}{504} = 322$.

Ex. 8. ***Find the least number which when divided by 6, 7, 8, 9 and 12 leaves the same remainder 2 in each case.***

Sol. Required number = (L.C.M. of 6, 7, 8, 9, 12) + 2 = 506.

Euclidean Algorithm: *If d is the H.C.F. of two positive integers a and b, then there exist unique integers m and n such that:* $d = am + bn$.

An Important Result: *If p is prime and a, b are any integers, then* $p/ab \Rightarrow p/a$ *or* p/b,

EXERCISE 2

1. What is the HCF of $3^{29} - 9$ and $3^{38} - 9$? **[CDS (I) Exam, 2023]**

(*a*) $3^9 - 1$ (*b*) $3^{11} - 1$ (*c*) $3^{11} - 3$ (*d*) $3^{11} - 9$

2. What is the HCF of $2^{36} - 1$ and $2^{45} - 1$? **[CDS (II) Exam, 2024]**

(*a*) 1023 (*b*) 512 (*c*) 511 (*d*) 255

3. What is the HCF of $x^3 - 19x + 30$ and $x^2 - 5x + 6$? **[CDS (I) Exam, 2021]**
(*a*) $(x + 2)\ (x - 3)$ (*b*) $(x - 2)\ (x + 3)$ (*c*) $(x + 2)\ (x - 1)$ (*d*) $(x - 3)\ (x - 2)$

4. The sum of LCM and HCF of two numbers is 536 and the difference between LCM and HCF is 296. If one of the numbers is 104, then what is the other number? **[CDS (I) Exam, 2021**
(*a*) 420 (*b*) 480 (*c*) 484 (*d*) 506

5. The LCM of two prime numbers *p* and *q* is 2231, where $p > q$. What is the value of $p-q$? **[CDS (I) Exam, 2021**
(*a*) 67 (*b*) 70 (*c*) 74 (*d*) 81

6. Three runners are running in a circular track, and they complete one round in 20, 30 and 35 minutes respectively. When will they next meet at the starting point? **[CDS (I) Exam, 2021]**
(*a*) After 3 hours 30 minutes (*b*) After 4 hours 30 minutes
(*c*) After 3 hours (*d*) After 7 hours

7. LCM of two numbers is 28 times their HCF. The sum of the HCF and the LCM is 1740. If one of these numbers is 240, then what is the other number ? **[CDS (I) Exam, 2020]**
(*a*) 420 (*b*) 640 (*c*) 820 (*d*) 1040

8. HCF of two numbers is 12. Which one of the following can never be their LCM ? **[CDS (I) Exam, 2019]**
(*a*) 80 (*b*) 60 (*c*) 36 (*d*) 24

9. X, Y and Z start at same point and same time in the same direction to run around a circular stadium. X completes a round is 252 seconds, Y in 308 seconds and Z in 198 seconds. After what time will they meet again at the starting point? **[CDS (I) Exam, 2019]**
(*a*) 26 minutes 18 seconds
(*b*) 42 minutes 35 seconds
(*c*) 45 minutes
(*d*) 46 minutes 12 seconds

10 .What is the LCM of $\frac{1}{3}, \frac{5}{6}, \frac{2}{9}, \frac{4}{27}$? **[CDS (I) Exam, 2019]**
(*a*) $\frac{5}{18}$ (*b*) $\frac{1}{27}$ (*c*) $\frac{10}{27}$ (*d*) $\frac{20}{3}$

11. There are two numbers which are greater than 21 and their LCM and HCF are 3003 and 21 respectively. What is the sum of these numbers? **[CDS (I) Exam, 2018]**
(*a*) 504 (*b*) 508 (*c*) 514 (*d*) 528

12. The product of two integers *p* and *q*, where $p > 60$ and $q > 60$, is 7168 and their HCF is 16. The sum of these two integers is **[CDS (I) Exam, 2018]**
(*a*) 256 (*b*) 184 (*c*) 176 (*d*) 164

13. What is the remainder obtained when 1421 × 1423 × 1425 is divided by 12?
(*a*) 1 (*b*) 2 (*c*) 3 (*d*) 4
[CDS (I) Exam, 2015]

14. A student was asked to multiply a number by 25. He instead multiplied the number by 52 and got the answer 324 more than the correct answer. The number to be multiplied was
(*a*) 12 (*b*) 15 (*c*) 25 (*d*) 32
[CDS (I) Exam, 2015]

15. What is the number of possible pairs of (P, Q) if the number 357P25Q is divisible by both 3 and 5?
(*a*) 7 (*b*) 6 (*c*) 5 (*d*) None of the above
[CDS (I) Exam, 2015]

16. The sides of a triangular field are 41m, 40m and 9m. The number of rose beds that can be prepared in the field if each rose bed, on an average, needs 900 square cm space, is
(*a*) 2000 (*b*) 1800 (*c*) 900 (*d*) 800
[CDS (II) Exam, 2015]

17. The HCF of two expressions p and q is 1. What is the reciprocal of their LCM?
(*a*) p + q (*b*) p − q (*c*) pq (*d*) $(pq)^{-1}$
[CDS (I) Exam, 2017]

18. The product of two numbers is 2160 and their HCF is 12. If the sum of the squares of the two numbers is 4896, then what is the mean of the two numbers? **[CDS (I) Exam, 2023]**
(*a*) 24 (*b*) 36 (*c*) 48 (*d*) 96

19. H.C.F. of $2^3 \times 3^2 \times 5$, $2^2 \times 3^3 \times 5^2$ and $2^4 \times 3 \times 5^3 \times 7$ is:
(*a*) 30 (*b*) 48 (*c*) 60 (*d*) 105

20. G.C.D. of $\frac{1}{2}, \frac{2}{3}, \frac{3}{4}$ & $\frac{4}{5}$ is:
(*a*) 1 (*b*) 12 (*c*) $\frac{4}{5}$ (*d*) $\frac{1}{60}$

21. Which of the following is a pair of co-primes?
(*a*) (14, 35) (*b*) (18, 25) (*c*) (31, 93) (*d*) (32, 62)

22. $\frac{1095}{1168}$ in simplest form is:
(*a*) $\frac{3}{16}$ (*b*) $\frac{15}{16}$ (*c*) $\frac{17}{26}$ (*d*) $\frac{25}{26}$

23. L.C.M. of 22, 54, 108, 135 and 198 is:
(*a*) 330 (*b*) 1980 (*c*) 5940 (*d*) 11880

24. L.C.M. of 87 and 145 is:
(*a*) 870 (*b*) 1305 (*c*) 435 (*d*) 1740

25. L.C.M. of $\frac{1}{3}, \frac{5}{6}, \frac{5}{9}, \frac{10}{27}$ is:
(*a*) $\frac{5}{54}$ (*b*) $\frac{5}{27}$ (*c*) $\frac{10}{3}$ (*d*) $\frac{5}{3}$

26. L.C.M. of 3^3, 5×2^2, $2^3 \times 5^2 \times 3$ and $2^2 \times 3^2 \times 5 \times 11$ is:
(*a*) 660 (*b*) 5400 (*c*) 10800 (*d*) 59400

27. L.C.M. of 9, 12, 15, 18, 24 and 56 is:
(*a*) 2630 (*b*) 2520 (*c*) 13 (*d*) 2525

28. The G.C.M. of two numbers is 38 and their L.C.M. is 98154. If one of the numbers is 1558, the other number is:
(*a*) 1260 (*b*) 3450 (*c*) 2394 (*d*) 2395

29. The least common multiple of 15, 24, 30, 40 is:
(*a*) 120 (*b*) 240 (*c*) 360 (*d*) 480

30. The largest number which exactly divides 210, 315, 147 and 168 is:
(*a*) 3 (*b*) 7 (*c*) 21 (*d*) 4410

31. The greatest possible length which can be used to measure exactly the lengths 7m, 3m 85 cm and 12 m 95 cm, is:
(*a*) 15 cm (*b*) 25 cm (*c*) 35 cm (*d*) 42 cm

32. The largest number which divides 77, 147 and 252 to leave the same remainder in each case, is:

(*a*) 9 (*b*) 15 (*c*) 25 (*d*) 35

33. There are two numbers *p* and *q* such that their HCF is 1. Which of the following statement are correct? **[CDS (II) Exam, 2016]**

1. Both p and q may be prime.

2. One number may be prime and the other composite.

3. Both the numbers may be composite.

Select the correct answer using the code given below.

(*a*) 1 and 2 only (*b*) 2 and 3 only (*c*) 1 and 3 only (*d*) 1, 2 and 3

34. Let L be the LCM and H be the HCF of two given numbers. L and H are in the ratio 3 : 2. If the sum of the two numbers is 45, then what is the product of the numbers?

[CDS (II) Exam, 2023]

(a) 243 (b) 486

(c) 504 (d) Cannot be determineddue to insufficient data

Mark option (a) if the question can be answered by using one of the statements alone, but cannot be answered using the other statement alone.

Mark option (b) if the question can be answered by using either statement alone.

Mark option (c) if the question can be answered by using both the statements together, but cannot be answered using either statement alone.

Mark option (d) if the question cannot be answered even by using both the statements together.

35. Question: What is the sum of two natural numbers? **[CDS (II) Exam, 2023]**

Statement-I: LCM of the two numbers is 144.

Statement-II: One of the numbers is 72.

36. Suppose p and q are the LCM and HCF respectively of two positive numbers. If $p : q = 14 : 1$ and $pq = 1134$, then what is the difference between the two numbers? **[CDS (I) Exam, 2022]**

(*a*) 27 (*b*) 35

(*c*) 45 (*d*) Cannot be determined due to insufficient data

37. If $(x - k)$ is the HCF of $x^2 + ax + b$ and $x^2 + cx + d$, then what is the value of k?

[CDS (I) Exam, 2021]

(*a*) $\frac{d-b}{c-a}$ (*b*) $\frac{d-b}{a-c}$ (*c*) $\frac{d+b}{c+a}$ (*d*) $\frac{d-b}{c+a}$

38. Consider the following statements in respect of prime numbers p and q :

I. Their LCM is always an odd number.

II. Sum of their LCM and HCF is always an even number.

Which of the statements given above is/are correct ?

(*a*) I only (*b*) II only (*c*) Both I and II (*d*) Neither I nor II

[CDS (I) Exam, 2024]

39. A number when divided by 7 leaves a remainder 3 and the resulting quotient when divided by 11 leaves a remainder 6. If the same number when divided by 11 leaves a remainder m and the resulting quotient when divided by 7 leaves a remainder n. What are the values of m and n respectively?

(a) 1 and 4 (b) 4 and 1 (c) 3 and 6 (d) 6 and 3

[CDS (II) Exam, 2015]

40. There are 437 fruit plants in an orchard planted in rows. The distance between any two adjacent rows is 2m and the distance between any two adjacent plants is 2m. Each row has the same number of plants. There is 1m clearance on all sides of the orchard. What is the cost of fencing the area at the rate of ₹100 per metre?

(a) ₹15,600 (b) ₹16,800
(c) ₹18,200 (d) More Information is required

[CDS (II) Exam, 2015]

41. The LCM of two numbers is 12 times their HCF. The sum of HCF and LCM is 403. If one of the numbers is 93, then the other number is

(a) 124 (b) 128 (c) 134 (d) 138

[CDS (II) Exam, 2015]

42. A is a set of positive integers such that when divided by 2, 3, 4, 5 and 6 leaves the remainder 1, 2, 3, 4 and 5 respectively. How many integers between 0 and 100 belong to the set A?

(a) No integer (b) One (c) Two (d) Three

[CDS (II) Exam, 2016]

43. The sum of the squares of two positive integers is 208. If the square of the larger number is 18 times the smaller number, then what is the difference of the larger and smaller numbers?

(a) 2 (b) 3 (c) 4 (d) 6

[CDS (II) Exam, 2016]

44. A boy saves ₹4.65 daily. What is the least number of days in which he will be able to save an exact number of rupees? **[CDS (II) Exam, 2016]**

(a) 10 (b) 20 (c) 21 (d) 25

ANSWERS

1. (d)	**2.** (c)	**3.** (d)	**4.** (b)	**5.** (c)	**6.** (d)	**7.** (a)	**8.** (a)	**9.** (d)
10. (d)	**11.** (a)	**12.** (c)	**13.** (c)	**14.** (a)	**15.** (a)	**16.** (a)	**17.** (d)	**18.** (c)
19. (c)	**20.** (d)	**21.** (b)	**22.** (b)	**23.** (c)	**24.** (c)	**25.** (c)	**26.** (d)	**27.** (b)
28. (c)	**29.** (a)	**30.** (c)	**31.** (c)	**32.** (d)	**33.** (d)	**34.** (b)	**35.** (c)	**36.** (c)
37. (b)	**38.** (d)	**39.** (a)	**40.** (b)	**41.** (a)	**42.** (b)	**43.** (c)	**44.** (b)	

SOLUTIONS

1. We have

$3^{29} - 9 = 3^2 (3^{27} - 1)$

$3^{38} - 9 = 3^2 (3^{36} - 1)$

$\therefore$ We know that H.C.F. $[(a^{m-1})(a^{n-1})] = [a^{\text{H.C.F. of }(m,n)} - 1]$

$\therefore$ HCF $= 3^2 [3^{\text{HCF }(27, 36)} - 1] = 3^2 [3^9 - 1] = 3^{11} - 9$

2. HCF of $(a^m - 1)$ and $(a^n - 1)$ is $(a^{\text{HCF}(m,n)} - 1)$.

HCF of 36 and 45 = 9

HCF of $2^{36} - 1$ and $2^{45} - 1 = 2^9 - 1 = 512 - 1 = 511$

3. We have,

$x^3 - 19x + 30$

$\Rightarrow (x - 3)(x^3 + 3x - 10)$

$\Rightarrow (x-3)(x^3+5x-2x-10)$

$\Rightarrow (x-3)(x+5)(x-2)$

Now, $x^2-5x+6 = x^2-3x-2x+6$

$= (x-3)(x-2)$

$\therefore$ HCF $= (x-3)(x-2)$

4. We have,

LCM + HCF = 536 ... (i)

LCM – HCF = 296 ... (ii)

from (i) and (ii),

2 LCM = 832

$\Rightarrow$ LCM = 416

$\Rightarrow$ HCF = 536 – 416 = 120

Let the other number be n

$\Rightarrow 104 \times n = \text{LCM} \times \text{HCF}$

$$\Rightarrow n = \frac{416\times120}{104} = 480$$

5. We have, HCF of p & q is 1 [$\therefore p$ and q are prime numbers]

$p \times q = 2231$

Now, we get 1 at unit place

$7 \times 3, 1 \times 1, 9 \times 9$

$\therefore$ 2231 is nearest to 2300

$\therefore 97 \times 23 = 2231$

Now, we have,

$p = 97$ and $q = 23$ $p - q = 97 - 23 = 74$

6. LCM of 20, 30 and 35 minutes = 420 minutes

$\therefore$ 420 minutes = 7 hrs.

7. Let LCM and HCF of two numbers be a and b respectively.

$\therefore$ We have,

$a = 28b$ and $a + b = 1740$

$\Rightarrow 28b + b = 1740$

$$\Rightarrow b = \frac{1740}{29} = 60$$

$\Rightarrow a = 28b = 28 \times 60 = 1680$

Now,

One number × other number = $a \times b$

$$\text{Other number} = \frac{60\times1680}{240} = 420$$

8. Let the two numbers be $12p$ and $12q$ respectively, where p and q are coprime integers.

LCM = 12 pq which must be divisible by 12

In the given options, only 80 is not divisible by 12

$\therefore$ 80 can never be their LCM.

9. Time after which they meet again at the starting point = LCM of 252, 308 and 198 seconds = 2772 seconds = 46 minutes 12 seconds.

10. Required LCM = $\frac{\text{LCM of }(1,5,2\text{ and }4)}{\text{HCF of }(3,6,9,27)} = \frac{20}{3}$

11. Let the two numbers be 21*a* and 21*b* respectively.

∴ We have,

$3003 \times 21 = 21a \times 21b$

$\Rightarrow 3003 = 21ab$

$\Rightarrow ab = 143 = 13 \times 11$ or 143×1

$a = 13, b = 11$ [∴ both numbers are greater than 21]

∴ First number = $21\,a = 21 \times 13 = 273$

Second number = $216 = 21 \times 11 = 231$

∴ Required number = $273 + 231 = 504$

12. Let the two numbers be 16*a* and 16*b* respectively.

Now, we have,

$16a \times 16b = 7168$

$\Rightarrow ab = 28$

$\Rightarrow ab = 2 \times 2 \times 7$

Neither *p* nor *q* is 2

($\because p > 60$ and $q > 60$)

Hence, $p = 16a = 16 \times 4 = 64$

$q = 16b = 16 \times 7 = 112$

Or $p = 112$ and $q = 64$

∴ Required sum = $64 + 112 = 176$

13. 1421 leaves a remainder of 5 when divided by 12.

∴ $1421 = 12k + 5$

$1423 = 12k + 7$ and $1425 = 12k + 9$

$(1421)(1423)(1425) = (12k + 5)(12k + 7)(12k + 9)$

$= 12k_1 + (5)(7)(9)$

The remainder of 1421(1423) (1425) divided by 12 is that of (5) (7) (9) divided by 12 i.e., 315 divided by 12 i.e., 3.

14. Let the number to be multiplied be N.

The student multiplied N by 52 instead of 25. Hence he got a product which was 324 more than the correct product.

$52\text{N} - 25\text{N} = 324 \Rightarrow \text{N} = \frac{18(18)}{9(3)} = 12.$

15. Sum of the digits of the number = 22 + P + Q

This must be divisible by 3.

As the number is divisible by 5, Q must be 0 or 5.

If Q = 0, 22 + P must be divisible by 3,

∴ P can be 2, 5 or 8.

If Q = 5, 22 + P + Q i.e., 27 + P must be divisible by 3

∴ P = 0, 3, 6 or 9

(P, Q) has 7 values.

16. $40^2 + 9^2 = 41^2$. ∴ The triangle is right angled.

Area of the field $= \frac{1}{2}(40)(9)\ m^2 = 180\ m^2$.

Number of rose beds that can be prepared = Area of the field divided by the Average space needed for each rose bed.

$= \frac{180}{900(10^{-4})} = 2000$

17. Given HCF of p and q is 1

So, LCM of p and q is pq

So, its Reciprocal $= \frac{1}{pq} = (pq)^{-1}$

18. Let two numbers are $12p$ and $12q$

Now, according to the question

$12\,p \times 12\,q = 2160 \quad \Rightarrow pq = 15$

$(12a)^2 + (12b)^2 = 4896$

$\Rightarrow \quad 144\,(a^2 + b^2) = 4896 \quad \Rightarrow a^2 + b^2 = 34$

$(a + b)^2 = a^2 + b^2 + 2ab = 34 + 2 \times 15 = 64$

$\Rightarrow \quad a + b = 8$

∴ Required average $= \frac{12a + 12b}{2} = \frac{12(a+b)}{2} = 6 \times 8 = 48$

19. Required H.C.F. $= (2^2 \times 3 \times 5) = 60$.

20. G.C.D. $= \frac{\text{H.C.F. of 1, 2, 3, 4}}{\text{L.C.M. of 2, 3, 4, 5}} = \frac{1}{60}$.

21. H.C.F. of 14 & 35 = 7; H.C.F of 18 & 25 = 1;

H.C.F. of 31 & 93 = 31 & H.C.F. of 32 & 62 = 2.

∴ Pair of co-primes is (18, 25).

22. H.C.F. of 1095 and 1168 is 73

Dividing Nr and Dr by 73, we get: $\frac{1095}{1168} = \frac{15}{16}$.

23.

2	22	–	54	–	108	–	135	–	198
11	11	–	27	–	54	–	135	–	99
3	1	–	27	–	54	–	135	–	9
3	1	–	9	–	18	–	45	–	3
3	1	–	3	–	6	–	15	–	1
	1	–	1	–	2	–	5	–	1

∴ L.C.M. $= (2 \times 11 \times 3 \times 3 \times 3 \times 2 \times 5) = 5940$.

24. H.C.F. of 87 and 145 is 29.

∴ L.C.M. $= \left(\frac{87 \times 145}{29}\right) = 435$.

25. L.C.M. $= \frac{\text{L.C.M. of 1, 5, 5, 10}}{\text{H.C.F. of 3, 6, 9, 27}} = \frac{10}{3}$.

26. L.C.M. = Product of largest powers of all prime factors. = $(3^3 \times 5^2 \times 2^3 \times 11) = 59400$.

27.

3	9	–	12	–	15	–	18	–	24	–	56
2	3	–	4	–	5	–	6	–	8	–	56
2	3	–	2	–	5	–	3	–	4	–	28
2	3	–	1	–	5	–	3	–	2	–	14
3	3	–	1	–	5	–	3	–	1	–	7
	1	–	1	–	5	–	1	–	1	–	7

$\therefore$ L.C.M. = $(3 \times 2 \times 2 \times 2 \times 3 \times 5 \times 7) = 2520$.

28. Other Number = $\dfrac{\text{H.C.F.} \times \text{L.C.M.}}{\text{Given Number}} = \dfrac{38 \times 98154}{1558} = 2394$.

29.

3	15	–	24	–	30	–	40
2	5	–	8	–	10	–	40
5	5	–	4	–	5	–	20
4	1	–	4	–	1	–	4
	1	–	1	–	1	–	1

$\therefore$ L.C.M. = $(3 \times 2 \times 5 \times 4) = 120$.

30. Required number = H.C.F. of 210, 315, 147 & 168 = 21.

31. Required length = H.C.F. of 700 cm, 385 cm, 1295 cm = 35 cm.

32. Required number = H.C.F. of (147 – 77), (252 – 147) & (252 – 77) = H.C.F. of 70, 105 & 175 = 35

33. The HCF of two number is 1.

1. Both could be prime For example HCF (2, 3) = 1.

2. One may be prime, the other composite. For example, HCF (2, 15) = 1

3. Both may be composite. For example HCF (6, 35) = 1

All 3 statements are correct.

34. Let two numbers are p and q.

p and q are $3x$ and $2x$ respectively

H.C.F. $(p, q) = x$

L.C.M. (p,q) = $6x$

Now, $p + q = 3x + 2x = 5x = 45 \Rightarrow x = 9$

$\therefore$ number = 27, 18

$\therefore$ required product = $27 \times 18 = 486$

35. (1) LCM = 144

(2) 1st number × 2nd Number = LCM × HCF

$\therefore$ 2nd number = $\dfrac{144}{72} = 2$

36. Let p and q be $14y$ and y respectively,

Now, we have,

$$pq = 1134$$

$$\Rightarrow \quad 14\,y \times y = 1134$$

$\Rightarrow \quad y^2 = \dfrac{1134}{14} = 81$

$\Rightarrow \quad y = 9$

$\therefore \quad p = 14y = 14 \times 9 = 126$

$q = y = 9$

Let the two numbers be $9a$ and $9b$.

$\therefore$ We have,

$126 \times 9 = 9a \times 9b \Rightarrow ab = 14$

If we put $a = 7$ and $b = 2$, we have first number $= 9a = 63$

Second number $= 9b = 18$

Required difference $= 63 - 18 = 45$

37. Put $x = k$ in $x^2 + ax + b = 0$ [$\because x - k$ is a factor of each expression]

We have,

$(k)^2 + ka + b = 0$ ... (i)

Also, put $x - k$ in $x^2 + cx + d = 0$,

We have,

$k^2 + ck + d = 0$ ... (ii)

From equation (i) and (ii), we get $k^2 + ak + b = k^2 + ck + d$

$k(a - c) = d - b$

$\Rightarrow k = \dfrac{d - b}{a - c}$

38. LIf $p = 2, q = 3 \Rightarrow$ LCM = 6 (even)

So, LCM is not always odd

Statement I is incorrect

Statement II: Sum of their LCM and HCF is always even

HCF of two distinct primes = 1

LCM of p and $q = p \times q$

Sum $= p \times q + 1$

$p = 2, q = 3 \Rightarrow \text{LCM} = 6,\ \text{HCF} = 1 \Rightarrow \text{Sum} = 6 + 1 = 7$ odd

$p = 3, q = 5 \Rightarrow \text{LCM} = 15,\ \text{HCF} = 1 \Rightarrow \text{Sum} = 15 + 1 = 16$ even

Statement II is incorrect

Thus neither statement I nor II is correct.

39. This is an example of successive division. Let the number be N. The number and successive quotients, the successive divisors and the corresponding remainders are tabulated below.

Quotients	N	P_1
Divisors	7	11
Remainder	3	6

One value of N is 6(7) + 3 =45. In general, N = 77k + 45

$\therefore$ N = 11(7k + 4) + 1 i.e. m = 1.

and $q_1 = 7k + 4$, $q_2 = k$ and $n = 4$.

$\therefore$ (m, n) = (1, 4)

40. There are 437 fruit plants in several rows, with each row having the same number of plants. As 437 = 19 (23), there are 19 rows with 23 trees in each row. (This can also be seen as 23 rows with 19 trees in each row.)
The distance between any two adjacent plants is 2 m and the distance between any two adjacent rows is 2 m. Therefore, the dimensions of the orchard are:
Length: [1 + 22 (2) + 1] m = 46 m,
Breadth: [1 + 18 (2) + 1] m = 38 m,
Perimeter = 2(84) m = 168 m.
Cost of fencing = ₹16800.

41. Let the HCF of the two numbers be h. The LCM is 12h, h + 12h = 403
$\Rightarrow$ h = 31. The LCM is 12(31).
One number is 3(31).
$\therefore$ The other number is 4(31) viz 124.

42. The divisors and remainder are tabulated below

Divisor	2 3 4 5 6
Remainder	1 2 3 4 5

$\therefore$ The required number is of the form n LCM (2, 3, 4, 5, 6) – 1,
i.e. $60n$- 1. From 0 to 100, the only number of the form is 59.
There is only one such integer in set A.

43. Let the smaller and larger numbers be a and b respectively
$a^2 + b^2 = 208$ and $b^2 = 18a$

$\therefore \quad a^2 + 18a - 208 = 0$ ... (1)
The two parts are 26 and 8.
(1) $(a + 26)(a - 8) = 0 \Rightarrow a = -26$ or 8. As $a > 0$, $a = 8$

and $b = \sqrt{18a} = \sqrt{18(8)} = 12$ and $b - a = 4$

44. $S = 4.65 = \frac{465}{100} = \frac{93}{20}$

The least number with which S has to be multiplied to get an integer is 20.

CHAPTER 3

Decimal Fractions

Decimal Fractions: *Fractions in which the denominators are powers of 10 are called decimal fractions.*

$\frac{1}{10}, \frac{1}{100}, \frac{1}{1000}$ etc. are respectively the tenth, hundredth and thousandth part of 1.

$\frac{7}{10}$ is 7 tenths, written as .7 (called decimal seven)

$\frac{13}{100}$ is 13 hundredths written as .13 (called decimal one three)

$\frac{9}{100}$ is 9 hundredths written as .09 (called decimal zero nine)

Converting a Decimal into a Vulgar Fraction:

Rule: *In the denominator, put 1 under the decimal point and annex with it as many zeros as is the number of digits after the decimal point. Now, remove the decimal point and reduce the fraction ta lowest terms.*

Ex. 1. ***Convert (i) 0.85 and (ii) 2.004 into vulgar fractions***

Sol. (*i*) $0.85 = \frac{85}{100} = \frac{17}{20}$. (*ii*) $2.004 = \frac{2004}{1000} = \frac{501}{250} = 2\frac{1}{250}$.

Remark: *Annexing zeros to the extreme right of a decimal fraction does not change its value.* Thus, 0.7 = 0.70 = 0.700 etc.

Addition & Subtraction of Decimal Fractions:

Rule: *The given numbers are so placed under each other that the decimal point lie in column. Now, the numbers can be added or subtracted as usual, putting the decimal point under the decimal points.*

Ex. 2. ***(i) 51.3 + 6.079 + .38 + .5 = ?*** ***(ii) 5.063 – 3.87 = ?***

Sol. (*i*)

```
   51.3
    6.079
     .38
   + .5
  ------
   58.259
  ------
```

(*ii*)

```
   5.063
  –3.87
  ------
   1.193
  ------
```

Multiplication of Two or More Decimal Fractions:

Rule: *Multiply the given numbers considering them without the decimal point. In the product, the decimal point is marked off to obtain as many places of decimal as is the sum of the number of decimal places in the given numbers.*

Ex. 3. ***Find the products:***

(i) $\mathbf{2.5143 \times 1.1}$ ***(ii)*** $\mathbf{3.71 \times 1.21}$ ***(iii)*** $\mathbf{0.7 \times 0.2 \times 0.02}$ ***(iv)*** $\mathbf{2 \times 0.2 \times 0.02 \times 0.0002}$

Sol. (*i*) $25143 \times 11 = 276573.$

Sum of decimal places = (4 + 1) = 5.

$\therefore$ $2.5143 \times 1.1 = 2.76573.$

(*ii*) $371 \times 121 = 44891$

Sum of decimal places = (2 + 2) = 4.

$\therefore$ $3.71 \times 1.21 = 4.4891.$

(*iii*) $7 \times 2 \times 2 = 24$. Sum of decimal places = (1 + 1 + 2) = 4.

$\therefore$ $0.7 \times 0.2 \times 0.02 = 0.0028.$

(*iv*) $2 \times 2 \times 2 \times 2 = 16$. Sum of decimal places = 6.

$\therefore$ $2 \times 0.2 \times 0.02 \times 0.002 = 0.000016.$

Dividing a Decimal Fraction by a Decimal Fraction:

Rule: *Multiply both the dividend and the divisor by a suitable power of I0 to make the divisor a whole number. Now proceed as above.*

Ex. 4. ***Evaluate: (i) 36 + 0.06 (ii) 0.00066 + 0.11***

Sol. (*i*) $\dfrac{36}{.06} = \dfrac{36 \times 100}{.06 \times 100} = \dfrac{3600}{6} = 600$ (*ii*) $\dfrac{0.00066}{0.11} = \dfrac{0.00066 \times 100}{0.11 \times 100} = \dfrac{0.066}{11} = 0.006$

H.C.F. & L.C.M. of Decimal Fractions:

Rule: *In given decimal numerals, make same number of decimal places by annexing zeros if needed. Considering these numbers without decimal places, find H.C.F. or L.C.M. as the case may be. Now, in the result, mark off as many decimal places as are there in each of the given numbers.*

Ex. 5. ***Find the H.C.F. and L.C.M. of 1.75, 5.6 and 7.***

Sol. We may write the given numbers as 1.75, 5.60 & 7.00. Without decimal places, these numbers are 175, 560 & 700.

H.C.F. of 175, 560 and 700 is 35

$\therefore$ H.C.F. of 1.75, 5.60 & 7.00 is .35.

L.C.M. of 175, 560 and 700 is 2800.

$\therefore$ L.C.M. of 1.75, 5.60 & 7.00 is 28.00, *i.e.* 28.

Comparison of Fractions Using Decimals:

Rule: *Convert each one of the given fractions in decimal form. Now, arrange them in ascending or descending order as per requirement.*

Ex. 6. ***Arrange*** $\mathbf{\dfrac{2}{3}, \dfrac{7}{12}, \dfrac{3}{8}}$ ***and*** $\mathbf{\dfrac{16}{25}}$ ***in ascending order.***

Sol. We have:

$\dfrac{2}{3} = 0.666;\ \dfrac{7}{12} = 0.583;\ \dfrac{3}{8} = 0.375$ and $\dfrac{16}{25} = 0.640$

Clearly, $0.375 < 0.583 < 0.640 < 0.666.$

Hence, $\dfrac{3}{8} < \dfrac{7}{12} < \dfrac{16}{25} < \dfrac{2}{3}$. Thus, $\dfrac{3}{8}, \dfrac{7}{12}, \dfrac{16}{25}, \dfrac{2}{3}$ are in ascending order.

Recurring Decimals: *If in a decimal fraction, a figure or a set of figures is repeated, then such a number is known as a recurring decimal.*

Ex. ***(i)*** $\mathbf{\dfrac{1}{3} = 0.333.... = 0.\overline{3}}$ (*Put a bar over repeated figure*) (*ii*) $\frac{1}{7} = 0.142857\,142857 = 0.\overline{142857}$. *(Put a bar over repeated figures)*

Pure Recurring Decimal: *A decimal fraction in which all the figures after the decimal point are repeated, is called a pure recurring decimal. e.g.* $0.5353\ldots = 0.\overline{53}$

Mixed Recurring: *A decimal fraction in which some figures after the decimal point do not repeat uni some figures repeat, is called a mixed recurring decimal. e.g.* $0.1\overline{7}$, $0.5\overline{36}$ *etc.*

To Convert a Pure Recurring Decimal into Vulgar Fraction:

Rule: *Write the repeated figure or figures only once in the numerator without decimal point and take as many nines in the denominator as is the number of repeating figures.*

Ex. 7. ***Express the following as vulgar fractions:***

(*i*) $0.\overline{7}$ (*ii*) $0.\overline{41}$ (*iii*) $0.\overline{057}$

Sol. Using the above law, we have:

(*i*) $0.\overline{7} = \frac{7}{9}$ (*ii*) $0.\overline{41} = \frac{41}{99}$ (*iii*) $0.\overline{057} = \frac{57}{999}$.

To Convert a Mixed Recurring Decimal to Vulgar Fraction:

Rule: *In the numerator take the dijference between the number formed by all the digits after decimal point (taking repeated digits only once) and that formed by the digits which are not repeated. In the denominator, take the number formed by as many nines as there are repeating digits followed by many zeros as is the number of non-repeating digits.*

Ex. 8. ***Express the following as vulgar fractions:***

(*i*) $0.1\overline{7}$ (*ii*) $0.12\overline{54}$ (*iii*) $2.63\overline{4}$

Sol. Using the above rule, we have:

(*i*) $0.1\overline{7} = \frac{17-1}{90} = \frac{16}{90} = \frac{8}{45}$.

(*ii*) $0.12\overline{54} = \frac{1254-12}{9900} = \frac{1242}{9900} = \frac{69}{550}$

(*iii*) $2.63\overline{4} = 2 + 0.63\overline{4} = 2 + \left(\frac{634-63}{900}\right) = 2 + \frac{571}{900} = \frac{2371}{900}$.

EXERCISE 3

1. What is the value of following? **[CDS (I) Exam, 2022]**

$$\frac{(5.4)^3 - 0.064}{(5.4)^2 + 2.16 + 0.16}$$

(*a*) 4 (*b*) 4.4 (*c*) 5 (*d*) 5.4

2. Students of a class are made to sit in rows of equal number of chairs. If number of students is increased by 2 in each row, then the number of rows decreases by 3. If number of students is increased by 4 in each row, then the number of row decreases by 5. What is the number of students in the class? **[CDS (I) Exam, 2022]**

(*a*) 100 (*b*) 105 (*c*) 110 (*d*) 120

3. What is $0.\overline{53} + 0.\overline{53}$ equal to ? **[CDS (I) Exam, 2019]**

(*a*) $1.0\overline{68}$ (*b*) $1.06\overline{8}$ (*c*) $1.\overline{068}$ (*d*) 1.068

4. If $\frac{a}{b+c}=\frac{b}{c+a}=\frac{c}{a+b}$, then which one of the following statements is correct?

[CDS (I) Exam, 2019]

(*a*) Each fraction is equal to 1 or –1. (*b*) Each fraction to $\frac{1}{2}$ or 1.

(*c*) Each fraction is equal to $\frac{1}{2}$ or 1. (d) Each fraction is equal to $\frac{1}{2}$ only.

5. What is the difference between 0.9 and $0.\overline{9}$ **[CDS (I) Exam, 2018]**

(*a*) 0 (*b*) 0.009 (*c*) 0.1 (d) 0.09

6. The value of $\left(0.\overline{63}+0.\overline{37}\right)$ is **[CDS (II) Exam, 2015]**

(*a*) 1 (*b*) $\frac{100}{91}$ (*c*) $\frac{100}{99}$ (*d*) $\frac{1000}{999}$

7. Which one of the following rational numbers has non-terminating and repeating decimal expansion? **[CDS (II) Exam, 2016]**

(*a*) $\frac{15}{1600}$ (*b*) $\frac{23}{8}$ (*c*) $\frac{35}{50}$ (*d*) $\frac{17}{6}$

8. 0.8697 + 0.249 + 0.09 + 0.9 = ?

(*a*) 2.1087 (*b*) 0.8964 (*c*) 0.89559 (*d*) None

9. 1000.01 – 99.987 = ?

(*a*) 0.14 (*b*) 900.023 (*c*) 901.023 (*d*) 901.977

10. 1.086 – 0.3983 – 0.669 = ?

(*a*) 0.6208 (*b*) 1.3948 (*c*) 0.0048 (*d*) 0.0187

11. 16.7 + 12.38 – ? = 10.09

(*a*) 16.98 (*b*) 17.89 (*c*) 18.99 (*d*) 20.09

12. 0.000009875 when expressed in scientific notation, is:

(*a*) 9875×10^{-5} (*b*) 98.75×10^{-7} (*c*) 9.875×10^{-6} (*d*) 987.5×10^{-4}

13. 3 × 0.3 × 0.03 × 0.003 × 30 = ?

(*a*) 0.0243 (*b*) 0.000243 (*c*) 0.00243 (*d*) 0000243

14. 25 % 0.0005 = ?

(*a*) 50 (*b*) 500 (*c*) 5000 (*d*) None

15. 0.000033 % 0.11 = ?

(*a*) 0.003 (*b*) 0.03 (*c*) 0.0003 (*d*) 0.3

16. If 2805 % 2.55 = 1100, then 280.5 % 25.5 is equal to:

(*a*) 1.01 (*b*) 1.1 (*c*) 0.11 (*d*) 11

17. If $\frac{1}{3.718}$ = 0.2689, then the value of $\frac{1}{0.0003718}$ is.

(*a*) 2689 (*b*) 2.689 (*c*) 26890 (*d*) .2689

18. If 168 × 54 = 9072, then 0.9072 + 16.8 is equal to:

(*a*) 5.4 (*b*) 0.54 (*c*) 0.054 (*d*) 0.0054

19. 0.777777 ÷ 0.011 = ?

(*a*) 70.707 (*b*) 77.07 (*c*) 0.70707 (*d*) None

20. 0.863 × 0.637 + 0.863 × 0.363 = ?

(*a*) 1 (*b*) 0.863 (*c*) 8.63 (*d*) 1.226

21. $\dfrac{0.625 \times 0.0729 \times 28.9}{0.0081 \times 0.025 \times 1.7} = ?$

(*a*) 0.3825 (*b*) 3.825 (*c*) 38.25 (*d*) 3825

22. The value of: $\left[\dfrac{(2.3)^3 - 0.027}{(2.3)^2 + 0.69 + 0.09}\right]$ is:

(*a*) 2.6 (*b*) 2 (*c*) 2.33 (*d*) 2.27

23. The value of: $\left[\dfrac{(0.05)^2 + (0.41)^2 + (0.073)^2}{(0.005)^2 + (0.041)^2 + (0.0073)^2}\right]$ is:

(*a*) 0.1 (*b*) 10 (*c*) 100 (*d*) 1000

24. The value of: $\dfrac{1}{6}$ is:

(*a*) 1.28 (*b*) 0.976 (*c*) 9.76 (*d*) 12.8

25. The value of: $\left(\dfrac{6.5 \times 4.7 + 6.5 \times 5.3}{1.3 \times 7.9 - 1.3 \times 6.9}\right)$ is:

(*a*) 3.945 (*b*) 39 (*c*) 34.45 (*d*) 50

26. The value of: $\left(\dfrac{0.943 \times 0.943 - 0.943 \times 0.057 + 0.057 \times 0.057}{0.943 \times 0.943 \times 0.943 + 0.057 \times 0.057 \times 0.057}\right)$ is:

(*a*) 0.886 (*b*) 1.1286 (*c*) 0.32 (*d*) None

27. $(26.872)^2 - (16.872)^2 = ?$

(*a*) 43.744 (*b*) 437.44 (*c*) 52.744 (*d*) 527.44

28. The value of: $\dfrac{(3.537 - 0.948)^2 + (3.537 + 0.948)^2}{(3.537 \times 3.537 + 0.948 \times 0.948)}$ is:

(*a*) 4.485 (*b*) 2.589 (*c*) 4 (*d*) 2

29. $(4.7 \times 13.23 + 4.7 \times 9.43 + 4.7 \times 77.34)$ equals:

(*a*) 773 (*b*) 943 (*c*) 470 (*d*) 132

30. The H.C.F. of 0.54, 1.8 and 7.2 is:

(*a*) 1.8 (*b*) 0.18 (*c*) 0.018 (*d*) 18

31. The G.C.D. of 1.08, 0.36 and 0.9 is:

(*a*) 0.03 (*b*) 0.9 (*c*) 0.18 (*d*) 0.108

32. The L.C.M. of 3, 2.7 and 0.09 is:

(*a*) 2.7 (*b*) 0.27 (*c*) 0.027 (*d*) 27

33. $1.\overline{6}$ expressed as a fraction is:

(*a*) $\dfrac{16}{10}$ (*b*) $\dfrac{16}{9}$ (*c*) $\dfrac{5}{3}$ (*d*) None

34. The value of $0.1\overline{36}$ is:

(*a*) $\dfrac{136}{1000}$ (*b*) $\dfrac{136}{999}$ (*c*) $\dfrac{136}{990}$ (*d*) $\dfrac{3}{22}$

35. The value of $0.53\overline{6}$ is:

(*a*) $\dfrac{536}{1000}$ (*b*) $\dfrac{536}{999}$ (*c*) $\dfrac{536}{990}$ (*d*) $\dfrac{161}{300}$

36. The value of $0.\overline{63} + 0.\overline{37}$ is:

(*a*) 1 (*b*) $\dfrac{100}{99}$ (*c*) $\dfrac{100}{90}$ (*d*) None of these

37. Which part contains the fractions in ascending order?

(*a*) $\frac{11}{14}, \frac{16}{19}, \frac{16}{21}$ (*b*) $\frac{16}{19}, \frac{11}{14}, \frac{16}{21}$ (*c*) $\frac{16}{21}, \frac{11}{14}, \frac{16}{19}$ (*d*) $\frac{16}{19}, \frac{16}{21}, \frac{11}{14}$

38. Which pair of rational numbers lies between $\frac{1}{4}$ & $\frac{3}{4}$?

(*a*) $\frac{262}{1000}, \frac{752}{1000}$ (*b*) $\frac{252}{1000}, \frac{748}{1000}$ (*c*) $\frac{24}{100}, \frac{74}{100}$ (*d*) $\frac{9}{40}, \frac{31}{40}$

ANSWERS

1. (*c*)	**2.** (*d*)	**3.** (*a*)	**4.** (*d*)	**5.** (*c*)	**6.** (*c*)	**7.** (*d*)	**8.** (*a*)	**9.** (*b*)
10. (*d*)	**11.** (*c*)	**12.** (*c*)	**13.** (*c*)	**14.** (*c*)	**15.** (*c*)	**16.** (*d*)	**17.** (*a*)	**18.** (*c*)
19. (*a*)	**20.** (*b*)	**21.** (*d*)	**22.** (*b*)	**23.** (*c*)	**24.** (*a*)	**25.** (*d*)	**26.** (*d*)	**27.** (*b*)
28. (*d*)	**29.** (*c*)	**30.** (*b*)	**31.** (*c*)	**32.** (*d*)	**33.** (*c*)	**34.** (*d*)	**35.** (*d*)	**36.** (*b*)
37. (*c*)	**38.** (*b*)							

SOLUTIONS

1. $\frac{(5.4)^3 - 0.064}{(5.4)^2 + 2.16 + 0.16}$

Here $a = 5.4$ and $b = 0.4$

We know that,

$a^3 - b^3 = (a - b)(a^2 + ab + b^2)$

$$\frac{(5.4)^3 - (0.4)^3}{(5.4)^2 + 5.4 \times 0.4 + (0.4)^2} \Rightarrow \frac{a^3 - b^3}{a^2 + ab + b^2} = a - b = 5.4 - 0.4 = 5$$

2. Let total number of rows is x and number of students in each row is y.

Total number of students = xy

Now, we have,

$(x - 3) \times (y + 2) = xy$

$\Rightarrow 2x - 3y = 6$ ... (i)

And $(x - 5) \times (y + 4) = xy$

$\Rightarrow 4x - 5y = 20$ ... (ii)

By multiplying equation (i) by 2 and subtracting equation (ii) from it:

$\Rightarrow q = 8$

From equation (i), we have,

$2p = 6 + 3q$

$\Rightarrow 2p = 6 + 24$

$\Rightarrow p = 15$

$\therefore$ total number of students = $p \times q = 15 \times 8 = 120$

3. $0.\overline{53} + 0.5\overline{3}$

0.52352353 ... + 0.533333

= 1.0686868 ...

= $1.0\overline{68}$

4. Let, $\frac{a}{b+c} = \frac{b}{c+a} = \frac{c}{a+b} = \alpha$

$\Rightarrow b+c = \dfrac{a}{\alpha}$ ... (i) $c+a = \dfrac{b}{\alpha}$ ... (ii) $a+b = \dfrac{c}{\alpha}$ ... (iii)

By adding (i), (ii) and (iii), we get

$2(a+b+c) = \dfrac{1}{\alpha}(a+b+c) \quad \Rightarrow \alpha = \dfrac{1}{2}$

If we put

$$\frac{a}{b+c} = \frac{b}{c+a} = \frac{c}{a+b} = 1$$

$\Rightarrow a = (b+c)$
$b = (c+a)$
$c = (a+b)$

We have,

$a+b+c = 2(a+b+c)$

$\Rightarrow 1 = 2$ (which is not possible.)

If we put

$$\frac{a}{a+c} = \frac{b}{c+a} = \frac{c}{a+b} = -1$$

$\Rightarrow a = -(b+c)$
$b = -(c+a)$
$c = -(a+b)$

$\Rightarrow (a+b+c) = -2(a+b+c) \Rightarrow 1 = -2$ (which is not possible)

5. $0.0\overline{9} - 0.9$

$\Rightarrow \dfrac{9}{9} - \dfrac{9}{10} = 1 - \dfrac{9}{10} = \dfrac{1}{10} = 0.1$

6. 0.6363.........
0.3737.........

1.0101.........

Let $x = 1.0101$.........

$100x = 101.0101$........

And $99x = 100 \Rightarrow x = \dfrac{100}{99}$.

7. In the simplified form $\dfrac{N}{D}$, if the only factors of D are 2'a and 5'a N is a terminating decimal.

If D has any factor other then 2a 5, the fraction is non-terminating. Among the given fraction, only 17/6 is non terminating.

8.
0.8697
0.249
0.09
+0.9
2.1087

9.
1000.010
−99.987
900.023

10. Given Exp. = 1.086 – (.3983 + .669) = (1.086 – 1.0673) = .0187.

11. Let 16.7 + 12.38 – x = 10.09.

Then, x = 29.08 – 10.09 = 18.99

12. $0.000009875 = \frac{0.000009875 \times 10^6}{10^6} = 9.875 \times 10^{-6}$.

13. 3 × 3 × 3 × 3 × 30 = 2430. Sum of decimal places = 6.

∴ Given Product = 0.002430 = 0.00243.

14. $\frac{25}{0.0005} = \frac{25000}{5} = 5000$.

15. $\frac{0.000033}{0.11} = \frac{0.000033 \times 100}{10} = \frac{0.0033}{11} = 0.0003$.

16. Given: $\frac{2805}{2.55} = 1100$.

$\therefore \frac{280.5}{25.5} = \frac{28.05}{2.55} = \frac{2805}{2.55} \times \frac{1}{100} = 1100 \times \frac{1}{100} = 11$.

17. $\frac{1}{0.0003718} = \frac{10000}{3.718} = \left(10000 \times \frac{1}{3.718}\right) = 10000 \times 0.2689 = 2689$.

18. Given: $\frac{9072}{168} = 54$.

$\therefore \frac{0.9072}{16.8} = \frac{9.072}{168} = \frac{9072}{168} \times \frac{1}{1000} = \frac{54}{1000} = 0.054$.

19. $\frac{0.777777}{0.011} = \frac{777.777}{11} = 70.707$.

20. Given Exp. = 0.863 × (0.637 + 0.363) = 0.863 × 1 = 0.863.

21. Since the sum of decimal places in Nr is the same as the sum of decimal places in Dr, so we may remove the decimal points.

∴ Given Exp. = $\frac{625 \times 729 \times 289}{81 \times 25 \times 17} = 3825$.

22. Given Exp. = $\frac{(2.3)^3 - (0.3)^3}{(2.3)^2 + 2.3 \times 0.3 + (0.3)^2} = \frac{a^3 - b^3}{a^2 + ab + b^2}$, where a = 2.3 & b = 0.3

= $(a - b)$ = (2.3 – 0.3) = 2.

23. Given Exp. = $\frac{a^2 + b^2 + c^2}{\left(\frac{a}{10}\right)^2 + \left(\frac{b}{10}\right)^2 + \left(\frac{c}{10}\right)^2}$, where a = 0.05, b = 0.41 and c = 0.073

$= \frac{100(a^2 + b^2 + c^2)}{(a^2 + b^2 + c^2)} = 100$.

24. Given Exp. = $\frac{0.896 \times (0.752 + 0.248)}{0.7 \times (0.034 + 0.966)} = \frac{0.896}{0.700} = \frac{128}{100} = 1.28$.

25. Given Exp. = $\frac{0.65 \times (4.7 + 5.3)}{1.3 \times (7.9 - 6.9)} = \frac{6.5 \times 10}{1.3 \times 1} = 50$

26. Given Exp. $= \left(\dfrac{a^2-ab+b^2}{a^3+b^3}\right)$, where $a = 0.943$ & $b = .057$

$= \dfrac{1}{(a-b)} = \left(\dfrac{1}{0.943+0.057}\right) = 1.$

27. Given Exp. = (26.872 + 16.872) (26.872 – 16.872) = (43.744 × 10) = 437.44.

28. Given Exp. $= \dfrac{(a-b)^2+(a+b)^2}{a^2+b^2}$, where $a = 3.537$ & $b = 0.948$

$= \dfrac{2(a^2+b^2)}{(a^2+b^2)} = 2.$

29. Given Exp. = 4.7 × (13.23 + 9.43 + 77.34) = 4.7 × 100 = 470.

30. With equal number of decimal places, the given numbers are 0.54, 1.80 and 7.20.
H.C.F. of 54, 180 and 720 is 18.
∴ H.C.F. of given numbers = 0.18.

31. Given numbers are: 1.08, 0.36 and 0.90.
G.C.D. of 108, 36 and 90 is 18.
∴ G.C.D. of given numbers = 0.l8.

32. With equal number of decimal places, the given numbers are 3.00, 2.70 and 0.09.
L.C.M. of 300, 270 and 9 is 2700.
∴ L.C.M. of given numbers = 27.00.

33. $1.\bar{6} = 1 + 0.\bar{6} = 1 + \dfrac{6}{9} = \dfrac{15}{9} = \dfrac{5}{3}$

34. $0.1\overline{36} = \dfrac{136-1}{990} = \dfrac{135}{990} = \dfrac{3}{22}.$

35. $0.53\bar{6} = \dfrac{536-53}{900} = \dfrac{483}{900} = \dfrac{161}{300}.$

36. $0.\overline{63} + 0.\overline{37} = \dfrac{63}{99} + \dfrac{37}{99} = \dfrac{100}{99}$

37. $\dfrac{11}{14} = 0.785, \dfrac{16}{19} = 0.842$ & $\dfrac{16}{21} = 0.762$. Since 0.762 < 0.785 < 0.842, we have $\dfrac{16}{21} < \dfrac{11}{14} < \dfrac{16}{19}$.

38. $\dfrac{1}{4} = 0.25, \dfrac{3}{4} = 0.75, \dfrac{262}{1000} = 0.262, \dfrac{752}{1000} = 0.752,$

$\dfrac{252}{1000} = 0.252, \dfrac{748}{1000} = 0.748, \dfrac{24}{100} = 0.24, \dfrac{74}{100} = 0.74,$

$\dfrac{9}{40} = 0.225$ & $\dfrac{31}{40} = 0.775$. Clearly 0.25 < .252 < 0.748 < 0.75. So, (*b*) is true

4

CHAPTER

Square Roots

Square Root: *If $x^2 = y$, we say that square root of y is x and we write, $\sqrt{y} = x$.*

Thus, $\sqrt{4} = 2, \sqrt{225} = 15, \sqrt{.16} = .4$ etc.

Square Root by Factorization:

Rule: *When a given number is a perfect square, we resolve it into the product of prime factors and take the product of prime factors choosing one out of every pair of the same primes.*

Ex. 1. ***Evaluate:*** $\sqrt{4624}$

Sol. Resolving 4624 into prime factors, we get:

$4624 = 2^2 \times 2^2 \times 17^2.$

$\therefore \sqrt{4624} = (2 \times 2 \times 17) = 68.$

2	4624
2	2312
2	1156
2	578
17	289
	17

General Method: *First we extract the square root of a given number and then we shall explain to make the ideas clear.*

Ex. 2. ***Find the square root of 1734489.***

Sol.

1	$\overline{1}\ \overline{73}\ \overline{44}\ \overline{89}$ (1317
	1
23	73
	69
261	444
	261
2627	18389
	18389
	×

$\therefore \sqrt{1734489} = 1317.$

Explanation. In given number, mark off the digits in pairs. Each pair and remaining one digit is called a period.

Now, $1^2 = 1$. On subtracting, we get 0 as remainder.

Now, bring down the next period *i.e.* 73.

Now, trial divisor is $1 \times 2 = 2$ and trial dividend is 73.

The quotient is 3, we place 3 as the second digit in the root and we annex 3 to the trial divisor.

Now, $23 \times 3 = 69$. The remainder is 4.

Bring down the next period, *i.e.* 44.

Now, trial divisor is 26 and trial dividend is 444.

The quotient is 1. But, 261 × 1 = 261.

Now, the remainder is 183. Bring down the last period *i.e.* 89.

Now, the trial divisor is 131 × 2 = 262 and the trial dividend is 18389. Take 7 as the quotient and 2627 as the divisor.

The remainder is then zero. $\therefore \quad \sqrt{1734489} = 1317.$

Square Root of Decimal Fractions:

Rule: *We make even number of decimal places by affixing a zero, if necessary. Now, we mark off" periods and extract the square roots as shown in the following examples.*

Ex. 3. ***Evaluate*** $\sqrt{\mathbf{176.252176}}$.

Sol. Here the number of decimal places is already even. So, we mark off periods and proceed.

Divisor	Dividend	Quotient
1	$\overline{1}\ \overline{76}\ \overline{25}\ \overline{21}\ \overline{76}$	(13.276
	1	
23	76	
	69	
262	725	
	524	
2647	20121	
	18529	
26546	159276	
	159276	
	×	

$\therefore \quad \sqrt{176.252176} = 13.276$

Ex. 4. ***Evaluate*** $\sqrt{\mathbf{0.4}}$ ***upto 3 places of decimal.***

Sol. 0.4 = 0.40 (*Making even decimal places*)

Divisor	Dividend	Quotient
6	$0.\overline{15}$	(.632
	36	
123	400	
	369	
1262	3100	
	2524	
	×	

$\therefore \quad \sqrt{0.4} = 0.632.$

An important property: (*i*) $\sqrt{ab} = \sqrt{a} \times \sqrt{b}$ (*ii*) $\sqrt{\dfrac{a}{b}} = \dfrac{\sqrt{a}}{\sqrt{b}}$

Ex. 5. ***If*** $\sqrt{\mathbf{21}} = \mathbf{4.582}$, ***find the value of*** $\sqrt{\dfrac{3}{7}}$

Sol. $\sqrt{\dfrac{3}{7}} = \sqrt{\dfrac{3}{7} \times \dfrac{7}{7}} = \dfrac{\sqrt{21}}{7} = \dfrac{4.582}{7} = 0.6545.$

Ex. 6. ***If*** $\sqrt{\mathbf{15}}$ ***= 3.8729, find the value of*** $\left(\dfrac{\sqrt{5}+\sqrt{3}}{\sqrt{5}-\sqrt{3}}\right)$.

Sol. $\dfrac{\sqrt{5}+\sqrt{3}}{\sqrt{5}-\sqrt{3}} = \dfrac{\sqrt{5}+\sqrt{3}}{\sqrt{5}-\sqrt{3}} \times \dfrac{\sqrt{5}+\sqrt{3}}{\sqrt{5}+\sqrt{3}} = \dfrac{(\sqrt{5}+\sqrt{3})^2}{(5-3)} = \dfrac{5+3+2\sqrt{5}\times\sqrt{3}}{2} = \left(\dfrac{8+2\sqrt{15}}{2}\right)$

$= (4 + \sqrt{15}) = (4 + 3.8729) = 7.8729.$

Ex. 7. ***If*** $\sqrt{1+\frac{25}{144}}=\left(1+\frac{x}{12}\right)$, ***find the value of x.***

Sol. The given equation is: $\left(1+\frac{x}{12}\right)=\sqrt{\frac{169}{144}}$

or $\left(1+\frac{x}{12}\right)=\frac{13}{12}$ or $\frac{x}{12}=\left(\frac{13}{12}-1\right)$ or $\frac{x}{12}=\frac{1}{12}$. So, $x=1$.

EXERCISE 4

1. What is the square root of $15-4\sqrt{14}$? **[CDS (I) Exam, 2022]**
   (*a*) $2\sqrt{2}-\sqrt{7}$ (*b*) $3\sqrt{2}-2\sqrt{7}$ (*c*) $\sqrt{15}-\sqrt{7}$ (*d*) $\sqrt{5}-\sqrt{3}$
2. What is the square root of $23-4\sqrt{15}$? **[CDS (II) Exam, 2021]**
   (*a*) $\sqrt{6}-3\sqrt{2}$ (*b*) $7-3\sqrt{5}$ (*c*) $\sqrt{3}-2\sqrt{5}$ (*d*) $\sqrt{5}-4\sqrt{3}$
3. Which one of the following is a set of solutions of the equation $x^{\sqrt{x}}=\sqrt[n]{x^x}$, if *n* is a positive integer? **[CDS (I) Exam, 2020]**
   (*a*) $\{1, n^2\}$ (*b*) $\{1, \sqrt{n}\}$ (*c*) $\{1, n\}$ (*d*) $\{n, n^2\}$
4. What is the least perfect square which is divisible by 3, 4, 5, 6 and 7? **[CDS (I) Exam, 2020]**
   (*a*) 1764 (*b*) 17640 (*c*) 44100 (*d*) 176400
5. The smallest integer with 4 digits which is a perfect square is **[CDS (I) Exam, 2019]**
   (*a*) 1000 (*b*) 1024 (*c*) 1089 (*d*) None of the above
6. What is the square root of the $16+6\sqrt{7}$? **[CDS (II) Exam, 2019]**
   (*a*) $4+\sqrt{7}$ (*b*) $4-\sqrt{7}$ (*c*) $3+\sqrt{7}$ (*d*) $3-\sqrt{7}$
7. What is $\sqrt{17-4\sqrt{15}}+\sqrt{8-2\sqrt{15}}$ equal to? **[CDS (I) Exam, 2025]**
   (*a*) $\sqrt{3}$ (*b*) $2\sqrt{3}$ (*c*) $2(\sqrt{5}-\sqrt{3})$ (*d*) $2(\sqrt{5}+\sqrt{3})$
8. What is the value of $\frac{\sqrt{5}-\sqrt{3}}{\sqrt{5}+\sqrt{3}}-\frac{\sqrt{5}+\sqrt{3}}{\sqrt{5}-\sqrt{3}}$? **[CDS (I) Exam, 2018]**
   (*a*) $-2\sqrt{15}$ (*b*) $2\sqrt{15}$ (*c*) $\sqrt{15}$ (*d*) $-\sqrt{15}$
9. The square root of $\frac{(0.75)^2}{1-0.75}+(0.75+0.75)^2+1$ is **[CDS (I) Exam, 2015]**
   (*a*) 1 (*b*) 2 (*c*) 3 (*d*) 4
10. The difference between the squares of two consecutive odd integers is always divisible by
    (*a*) 3 (*b*) 7 (*c*) 8 (*d*) 16
    [CDS (I) Exam, 2015]
11. Which one of the following is correct?
    (*a*) $\sqrt{2}<\sqrt[4]{6}<\sqrt[3]{4}$ (*b*) $\sqrt{2}<\sqrt[4]{6}>\sqrt[3]{4}$
    (*c*) $\sqrt[4]{6}<\sqrt{2}<\sqrt[3]{4}$ (*d*) $\sqrt[4]{6}>\sqrt{2}>\sqrt[3]{4}$ **[CDS (II) Exam, 2015]**

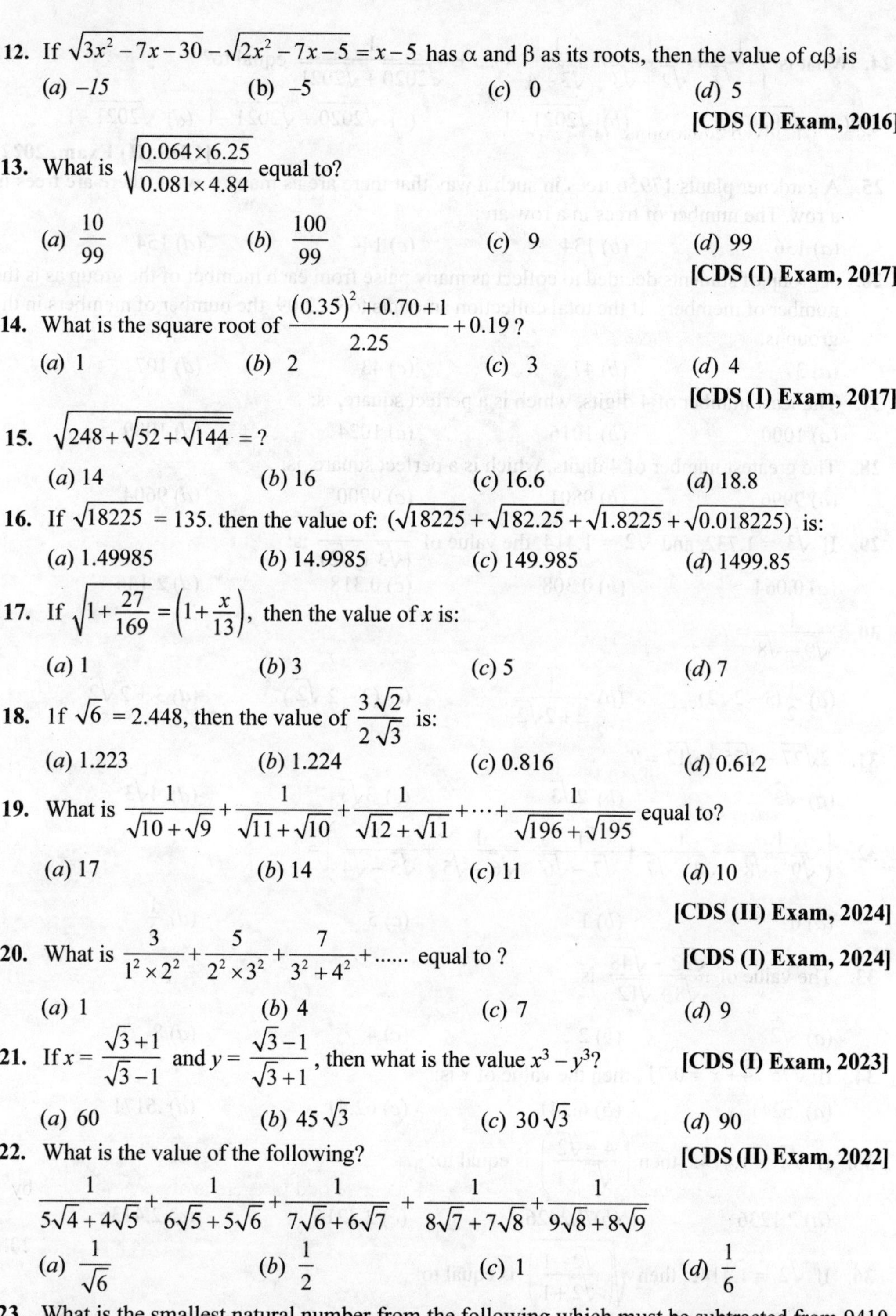

12. If $\sqrt{3x^2-7x-30}-\sqrt{2x^2-7x-5}=x-5$ has α and β as its roots, then the value of $\alpha\beta$ is

(*a*) –15 (b) –5 (*c*) 0 (*d*) 5

[CDS (I) Exam, 2016]

13. What is $\sqrt{\dfrac{0.064\times6.25}{0.081\times4.84}}$ equal to?

(*a*) $\dfrac{10}{99}$ (*b*) $\dfrac{100}{99}$ (*c*) 9 (*d*) 99

[CDS (I) Exam, 2017]

14. What is the square root of $\dfrac{(0.35)^2+0.70+1}{2.25}+0.19$?

(*a*) 1 (*b*) 2 (*c*) 3 (*d*) 4

[CDS (I) Exam, 2017]

15. $\sqrt{248+\sqrt{52+\sqrt{144}}}=?$

(*a*) 14 (*b*) 16 (*c*) 16.6 (*d*) 18.8

16. If $\sqrt{18225}=135$. then the value of: $(\sqrt{18225}+\sqrt{182.25}+\sqrt{1.8225}+\sqrt{0.018225})$ is:

(*a*) 1.49985 (*b*) 14.9985 (*c*) 149.985 (*d*) 1499.85

17. If $\sqrt{1+\dfrac{27}{169}}=\left(1+\dfrac{x}{13}\right)$, then the value of x is:

(*a*) 1 (*b*) 3 (*c*) 5 (*d*) 7

18. If $\sqrt{6}=2.448$, then the value of $\dfrac{3\sqrt{2}}{2\sqrt{3}}$ is:

(*a*) 1.223 (*b*) 1.224 (*c*) 0.816 (*d*) 0.612

19. What is $\dfrac{1}{\sqrt{10}+\sqrt{9}}+\dfrac{1}{\sqrt{11}+\sqrt{10}}+\dfrac{1}{\sqrt{12}+\sqrt{11}}+\cdots+\dfrac{1}{\sqrt{196}+\sqrt{195}}$ equal to?

(*a*) 17 (*b*) 14 (*c*) 11 (*d*) 10

[CDS (II) Exam, 2024]

20. What is $\dfrac{3}{1^2\times2^2}+\dfrac{5}{2^2\times3^2}+\dfrac{7}{3^2+4^2}+\ldots\ldots$ equal to ? **[CDS (I) Exam, 2024]**

(*a*) 1 (*b*) 4 (*c*) 7 (*d*) 9

21. If $x=\dfrac{\sqrt{3}+1}{\sqrt{3}-1}$ and $y=\dfrac{\sqrt{3}-1}{\sqrt{3}+1}$, then what is the value x^3-y^3? **[CDS (I) Exam, 2023]**

(*a*) 60 (*b*) $45\sqrt{3}$ (*c*) $30\sqrt{3}$ (*d*) 90

22. What is the value of the following? **[CDS (II) Exam, 2022]**

$$\frac{1}{5\sqrt{4}+4\sqrt{5}}+\frac{1}{6\sqrt{5}+5\sqrt{6}}+\frac{1}{7\sqrt{6}+6\sqrt{7}}+\frac{1}{8\sqrt{7}+7\sqrt{8}}+\frac{1}{9\sqrt{8}+8\sqrt{9}}$$

(*a*) $\dfrac{1}{\sqrt{6}}$ (*b*) $\dfrac{1}{2}$ (*c*) 1 (*d*) $\dfrac{1}{6}$

23. What is the smallest natural number from the following which must be subtracted from 9410 to make the remaining number a perfect square? **[CDS (II) Exam, 2022]**

(*a*) 4 (*b*) 3 (*c*) 2 (*d*) 1

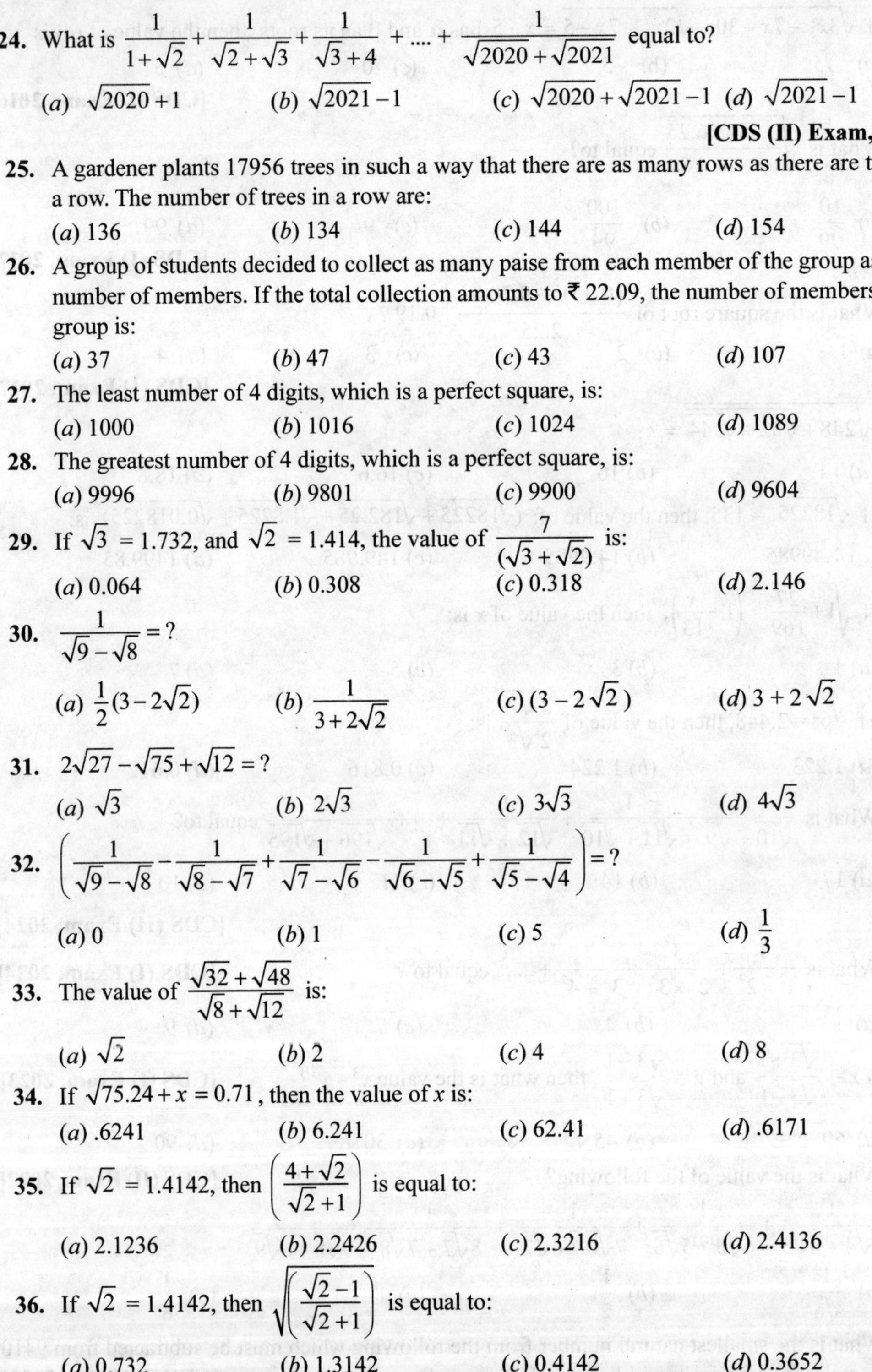

24. What is $\frac{1}{1+\sqrt{2}}+\frac{1}{\sqrt{2}+\sqrt{3}}+\frac{1}{\sqrt{3}+4}+....+\frac{1}{\sqrt{2020}+\sqrt{2021}}$ equal to?

(*a*) $\sqrt{2020}+1$ (*b*) $\sqrt{2021}-1$ (*c*) $\sqrt{2020}+\sqrt{2021}-1$ (*d*) $\sqrt{2021}-1$

[CDS (II) Exam, 2022]

25. A gardener plants 17956 trees in such a way that there are as many rows as there are trees in a row. The number of trees in a row are:

(*a*) 136 (*b*) 134 (*c*) 144 (*d*) 154

26. A group of students decided to collect as many paise from each member of the group as is the number of members. If the total collection amounts to ₹ 22.09, the number of members in the group is:

(*a*) 37 (*b*) 47 (*c*) 43 (*d*) 107

27. The least number of 4 digits, which is a perfect square, is:

(*a*) 1000 (*b*) 1016 (*c*) 1024 (*d*) 1089

28. The greatest number of 4 digits, which is a perfect square, is:

(*a*) 9996 (*b*) 9801 (*c*) 9900 (*d*) 9604

29. If $\sqrt{3}$ = 1.732, and $\sqrt{2}$ = 1.414, the value of $\frac{7}{(\sqrt{3}+\sqrt{2})}$ is:

(*a*) 0.064 (*b*) 0.308 (*c*) 0.318 (*d*) 2.146

30. $\frac{1}{\sqrt{9}-\sqrt{8}}=?$

(*a*) $\frac{1}{2}(3-2\sqrt{2})$ (*b*) $\frac{1}{3+2\sqrt{2}}$ (*c*) $(3-2\sqrt{2})$ (*d*) $3+2\sqrt{2}$

31. $2\sqrt{27}-\sqrt{75}+\sqrt{12}=?$

(*a*) $\sqrt{3}$ (*b*) $2\sqrt{3}$ (*c*) $3\sqrt{3}$ (*d*) $4\sqrt{3}$

32. $\left(\frac{1}{\sqrt{9}-\sqrt{8}}-\frac{1}{\sqrt{8}-\sqrt{7}}+\frac{1}{\sqrt{7}-\sqrt{6}}-\frac{1}{\sqrt{6}-\sqrt{5}}+\frac{1}{\sqrt{5}-\sqrt{4}}\right)=?$

(*a*) 0 (*b*) 1 (*c*) 5 (*d*) $\frac{1}{3}$

33. The value of $\frac{\sqrt{32}+\sqrt{48}}{\sqrt{8}+\sqrt{12}}$ is:

(*a*) $\sqrt{2}$ (*b*) 2 (*c*) 4 (*d*) 8

34. If $\sqrt{75.24+x}=0.71$, then the value of x is:

(*a*) .6241 (*b*) 6.241 (*c*) 62.41 (*d*) .6171

35. If $\sqrt{2}$ = 1.4142, then $\left(\frac{4+\sqrt{2}}{\sqrt{2}+1}\right)$ is equal to:

(*a*) 2.1236 (*b*) 2.2426 (*c*) 2.3216 (*d*) 2.4136

36. If $\sqrt{2}$ = 1.4142, then $\sqrt{\left(\frac{\sqrt{2}-1}{\sqrt{2}+1}\right)}$ is equal to:

(*a*) 0.732 (*b*) 1.3142 (*c*) 0.4142 (*d*) 0.3652

37. If $\frac{1872}{\sqrt{x}} = 234$, the value of x is:

(*a*) 324 (*b*) 64 (*c*) 8 (*d*) 256

ANSWERS

1. (*a*)	**2.** (*c*)	**3.** (*a*)	**4.** (*c*)	**5.** (*b*)	**6.** (*c*)	**7.** (*a*)	**8.** (*a*)	**9.** (*d*)
10. (*c*)	**11.** (*a*)	**12.** (*a*)	**13.** (*b*)	**14.** (*a*)	**15.** (*b*)	**16.** (*c*)	**17.** (*a*)	**18.** (*b*)
19. (*c*)	**20.** (*a*)	**21.** (*c*)	**22.** (*d*)	**23.** (*d*)	**24.** (*b*)	**25.** (*b*)	**26.** (*b*)	**27.** (*c*)
28. (*b*)	**29.** (*c*)	**30.** (*d*)	**31.** (*c*)	**32.** (*c*)	**33.** (*b*)	**34.** (*b*)	**35.** (*b*)	**36.** (*c*)
37. (*b*)								

SOLUTIONS

1. We have,

$x = 15 - 4\sqrt{14}$

$8 + 7 - 2 \times 2 \times \sqrt{14}$

$= 8+7-\left[2\times\sqrt{4\times14}\right] = \left(\sqrt{8}\right)^2 + \left(\sqrt{7}\right)^2 - 2\times\sqrt{56} = \left(\sqrt{8}-\sqrt{7}\right)^2$

$\Rightarrow \sqrt{x} = \sqrt{\left(\sqrt{8}-\sqrt{7}\right)^2} = \sqrt{8}-\sqrt{7} = 2\sqrt{2}-\sqrt{7}$

2. Let $x = 23 - 4\sqrt{15}$

$= 3 + 20 - 4\sqrt{15}$

$= \left(\sqrt{3}\right)^2 + (20)^2 - 2\times\sqrt{3}\times2\sqrt{5} = \left(\sqrt{3}\right)^2 + \left(2\sqrt{5}\right)^2 - 2\times\sqrt{3}\times2\sqrt{5} = \left(\sqrt{3}-2\sqrt{5}\right)^2$

$\Rightarrow \sqrt{x} = \sqrt{(\sqrt{3}-2\sqrt{5})^2} = \sqrt{3}-2\sqrt{5}$

3. We have,

$x^{\sqrt{x}} = \sqrt{x^x}$ by taking log on both sides, we get $\log x^{\sqrt{x}} = \log\sqrt{x^x}$ $[\because \log a^b = b \log a]$

$\Rightarrow \sqrt{x}\log x = \frac{x}{n}\log x \Rightarrow n\sqrt{x}\log x - x\log x = 0$

$\Rightarrow \sqrt{x}\log x(n-\sqrt{x}) = 0$

$\Rightarrow \sqrt{x}\log x = 0$ or $n-\sqrt{x} = 0$

$\Rightarrow \log x = \log 1$ or $n = \sqrt{x}$

$\Rightarrow x = 1 \quad \Rightarrow n^2 = x$

$\therefore$ set $= \{1, n^2\}$

4. LCM of 3, 4, 5, 6, and 7 = 2 × 2 × 3 × 5 × 7 = 420

$\therefore$ Least perfect square = 3 × 5 × 7 × 420 = 44100

5. $\because$ $30^2 = 900$

$31^2 = 961$

$32^2 = 1024$

$\therefore$ Required integer = 1024

6. Let $x = 16 + 6\sqrt{7}$

$$= 7 + 9 + 6\sqrt{7} = \left(\sqrt{7}\right)^2 + (3)^2 + 2\times 3\times\sqrt{7} = \left(3+\sqrt{7}\right)^2$$

$\Rightarrow \sqrt{x} = (3+\sqrt{7}) = 3+\sqrt{7}$

7. $\sqrt{17-4\sqrt{15}} = \sqrt{(2\sqrt{3})^2 + (\sqrt{5})^2 - 2\times 2\sqrt{3}\times\sqrt{5}} = \sqrt{(2\sqrt{3}-\sqrt{5})^2} = 2\sqrt{3}-\sqrt{5}$

$\sqrt{8-2\sqrt{15}} = \sqrt{(\sqrt{3})^2 + (\sqrt{5})^2 - 2\times\sqrt{3}\times\sqrt{5}} = \sqrt{(\sqrt{5}-\sqrt{3})^2} = \sqrt{5}-\sqrt{3}$

Now, adding both terms:

$$2\sqrt{3}-\sqrt{5}+\sqrt{5}-\sqrt{3} = \sqrt{3}$$

8. $\dfrac{\sqrt{5}-\sqrt{3}}{\sqrt{5}+\sqrt{3}} - \dfrac{\sqrt{5}+\sqrt{3}}{\sqrt{5}-\sqrt{3}} = \dfrac{\left(\sqrt{5}-\sqrt{3}\right)^2 - \left(5+\sqrt{3}\right)^2}{\left(\sqrt{5}+\sqrt{3}\right)\left(\sqrt{5}-\sqrt{3}\right)}$

$$= \frac{\left(8-2\sqrt{15}\right)-\left(8+2\sqrt{15}\right)}{5-3} = \frac{8-2\sqrt{5}-8-2\sqrt{15}}{2} = \frac{-4\sqrt{15}}{2} = -2\sqrt{15}$$

9. $\dfrac{(0.75)^3}{1-0.75} + [0.75 + (0.75)^2 + 1]$

$$= \frac{[(0.75)^3 + (1-0.75)[0.75 + (0.75)^2 + 1]}{1-0.75} = \frac{(0.75)^3 + 1 - (0.75)^3}{1-0.75}$$

$[\because (a-b)(a^2+ab+b^2) = a^3 - b^3] = \dfrac{1}{0.25} = 4.$

10. Let the consecutive odd integers be 2k– 1 and 2k + 1.
$(2k+1)^2 - (2k-1)^2 = 2(4k) = 8k$. This is always a multiple of 8.
Note: The odd integers need not be consecutive. If our interest is the remainder when the number is divided by 8, then we should construct the following table.

N	1	3	5	7
N^2	1	1	1	1

All odd numbers can be expressed as 8k + 1, 8k + 3, 8k + 5 or 8k + 7. The squares, in each of the 4 cases are of the form 8p + 1. (We can construct similar tables for any divisor.)

11. All the choices involve the same 3 surds. We can compare them 2 at a time.

Let $a = 2^{\frac{1}{2}}$ $b = 6^{\frac{1}{4}}$.

$\therefore a^4 = 4^{\frac{1}{3}}$ and $b^4 = 6$ *i.e.* $a < b$.

Let $c = 4\frac{1}{3}$ $b^{12} = 6^3 = 216$ and $c^{12} = 4^4 = 256$ *i.e.* $b < c$

$\therefore\ a < b < c.$

12. $\sqrt{3x^2-7x-30} - \sqrt{2x^2-7x-5} = x-5$

By clearing the root express. We get a fourth degree equation. The roots turns out to be 6, 5, 5, $-\dfrac{5}{2}$

The product of two of these could be 30, –15, 25 or $\frac{-25}{2}$
Among these, only –15 is there.

13. $\sqrt{\frac{0.064\times 6.25}{0.081\times 4.84}} \Rightarrow \frac{8\times 25}{9\times 22}=\frac{100}{99}$

14. We know,
$(a+b)^2 = a^2+b^2+2ab$
So, $a = 0.35$, $b = 1$

Given, $\frac{(0.35)^2+0.70+1}{2.25}+0.19=?$

$$\Rightarrow \frac{(0.35+1)^2}{2.25}+0.19 \qquad \Rightarrow \frac{(1.35)^2}{(1.50)^2}+0.19$$

$$\Rightarrow \frac{(9)^2}{(10)^2}+0.19 \qquad \Rightarrow \frac{81}{100}+0.19$$

$$\Rightarrow \frac{100(a+b)}{x-y} \qquad \Rightarrow \frac{100(a-b)}{x+y}$$

Square root of 1 is 1.

15. $\sqrt{248+\sqrt{52+\sqrt{144}}}=\sqrt{248+\sqrt{52+12}}=\sqrt{248+\sqrt{64}}=\sqrt{248+8}=\sqrt{256}=16$

16. $\sqrt{18225}+\sqrt{182.25}+\sqrt{1.8225}+\sqrt{0.018225} = \sqrt{18225}+\sqrt{\frac{18225}{100}}+\sqrt{\frac{18225}{10000}}+\sqrt{\frac{18225}{1000000}}$

$$=\sqrt{18225}+\frac{\sqrt{18225}}{10}+\frac{\sqrt{18225}}{100}+\frac{\sqrt{18225}}{1000} = \left(135+\frac{135}{10}+\frac{135}{100}+\frac{135}{1000}\right)$$

$= (135 + 13.5 + 1.35 + 0.135) = 149.985.$

17. $\left(1+\frac{x}{13}\right)=\sqrt{\frac{169+27}{169}}=\sqrt{\frac{196}{169}}=\frac{14}{13}.$

$$\therefore \frac{x}{13}=\left(\frac{14}{13}-1\right) \Leftrightarrow \frac{x}{13}=\frac{1}{3} \Leftrightarrow x=1.$$

18. $\frac{3\sqrt{2}}{2\sqrt{3}}=\frac{3\sqrt{2}}{2\sqrt{3}}\times\frac{\sqrt{3}}{\sqrt{3}}=\frac{\sqrt{6}}{2}=\frac{2.448}{2}=1.224.$

19. $\frac{1}{(\sqrt{10}+\sqrt{9})}\times\frac{(\sqrt{10}-\sqrt{9})}{(\sqrt{10}-\sqrt{9})}+\frac{1}{\sqrt{11}+\sqrt{10}}\times\frac{(\sqrt{11}-\sqrt{10})}{(\sqrt{11}-\sqrt{10})}+\frac{1}{\sqrt{12}+\sqrt{11}}\times\frac{(\sqrt{12}-\sqrt{11})}{(\sqrt{12}-\sqrt{11})}+\ldots$

$$+\frac{1}{\sqrt{196}+\sqrt{195}}\times\frac{(\sqrt{196}-\sqrt{195})}{(\sqrt{196}-\sqrt{195})}$$

$$=\sqrt{10}-\sqrt{9}+\sqrt{11}-\sqrt{10}+\sqrt{12}-\sqrt{11}.....+\sqrt{196}+\sqrt{195}$$

$$=-\sqrt{9}+\sqrt{196}=-3+14=11$$

20. We have,

$$\frac{3}{1^2\times 2^2}+\frac{5}{2^2\times 3^2}+\frac{7}{3^2+4^2}\ldots\ldots$$

$$=\frac{2^2-1^2}{1^2\times 2^2}+\frac{3^2-2^2}{2^2\times 3^2}+\frac{4^2-3^2}{3^2\times 4^2}+\ldots..=\frac{1}{1^2}-\frac{1}{2^2}+\frac{1}{2^2}-\frac{1}{3^2}+\frac{1}{3^2}-\frac{1}{4^2}+\ldots\ldots-\frac{1}{\infty^2}=1-0=1$$

21. $x=\dfrac{\sqrt{3}+1}{\sqrt{3}-1}\times\dfrac{\sqrt{3}+1}{\sqrt{3}+1}=\dfrac{4+2\sqrt{3}}{2}=2+\sqrt{3}$

$$y=\frac{\sqrt{3}-1}{\sqrt{3}+1}\times\frac{\sqrt{3}-1}{\sqrt{3}-1}=2-\sqrt{3}$$

$$x-y=2+\sqrt{3}-(2-\sqrt{3})=2\sqrt{3}$$

$$x^3-y^3=(x-y)^3+3xy\,(x-y)=\left(2\sqrt{3}\right)^3+3\times1\times2\sqrt{3}\,)=24\sqrt{3}+6\sqrt{3}=30\sqrt{3}$$

22. $\dfrac{1}{5\sqrt{4}+4\sqrt{5}}+\dfrac{1}{6\sqrt{5}+5\sqrt{6}}+\dfrac{1}{7\sqrt{6}+6\sqrt{7}}+\dfrac{1}{8\sqrt{7}+7\sqrt{8}}+\dfrac{1}{9\sqrt{8}+8\sqrt{9}}$

$$\Rightarrow\frac{1}{\sqrt{20}\sqrt{5}+\sqrt{4}}+\frac{1}{\sqrt{30}\left(\sqrt{6}+\sqrt{5}\right)}+\frac{1}{\sqrt{42}\left(\sqrt{7}+\sqrt{6}\right)}+\frac{1}{\sqrt{56}((\sqrt{8}+\sqrt{7})}+\frac{1}{\sqrt{72}\left(\sqrt{9}+\sqrt{8}\right)}$$

$$\Rightarrow\frac{\sqrt{5}-\sqrt{4}}{\sqrt{20}}+\frac{\sqrt{6}-\sqrt{5}}{\sqrt{30}}+\frac{\sqrt{7}-\sqrt{6}}{\sqrt{42}}+\frac{\sqrt{8}-\sqrt{7}}{\sqrt{56}}+\frac{\sqrt{9}-\sqrt{8}}{\sqrt{72}}$$

$$\Rightarrow\frac{1}{\sqrt{4}}-\frac{1}{\sqrt{5}}+\frac{1}{\sqrt{5}}-\frac{1}{\sqrt{6}}+\frac{1}{\sqrt{6}}-\frac{1}{\sqrt{7}}+\frac{1}{\sqrt{7}}-\frac{1}{\sqrt{8}}+\frac{1}{\sqrt{8}}-\frac{1}{\sqrt{9}}$$

$$\Rightarrow\frac{1}{2}-\frac{1}{3}=\frac{1}{6}$$

23.

9	$\overline{94}\ \overline{10}$	97
+9	81	
178	1310	
	1309	
	1	

$\because\ 97^2=9409$

$\because$ Required no. $=9410-9409=1$

24. $\dfrac{1}{1+\sqrt{2}}+\dfrac{1}{\sqrt{2}+\sqrt{3}}+\dfrac{1}{\sqrt{3}+\sqrt{4}}+\ldots.+\dfrac{1}{\sqrt{2020}-\sqrt{2021}}$

$$=\sqrt{2}-1+\sqrt{3}-\sqrt{2}+\sqrt{4}-\sqrt{3}+\ \ldots+\sqrt{2021}-\sqrt{2020}$$

$$=\sqrt{2021}-1$$

25. Number of trees in each row = $\sqrt{17956}$ = 134.

26. Number of members = $\sqrt{2209}$ = 47.

27. Least number of 4 digits = 1000.

In finding the square root, we find that:

$(31)^2 < 1000 < (32)^2$.

$\therefore$ Required number = $(32)^2 = 1024$.

28. Greatest number of 4 digits = 9999.

In finding the square root, we find that:

$9999 = (99)^2 + 198$. So, $(99)^2 = 9999 - 198 = 9801$.

$\therefore$ Required number = 9801

29. $\frac{7}{\sqrt{3}+\sqrt{2}} = \frac{7}{\sqrt{3}+\sqrt{2}} \times \frac{\sqrt{3}-\sqrt{2}}{\sqrt{3}-\sqrt{2}} = \frac{\sqrt{3}-\sqrt{2}}{(\sqrt{3})^2-(\sqrt{2})^2} = (\sqrt{3}-\sqrt{2}) = (1.732 - 1.414) = 0.318.$

30. $\frac{1}{\sqrt{9}-\sqrt{8}} = \frac{1}{\sqrt{9}-\sqrt{8}} \times \frac{\sqrt{9}+\sqrt{8}}{\sqrt{9}+\sqrt{8}} = \frac{3+2\sqrt{2}}{9-8} = 3+2\sqrt{2}.$

31. $2\sqrt{27} - \sqrt{75} + \sqrt{12} = 2\sqrt{3\times3\times3} - \sqrt{5\times5\times3} + \sqrt{2\times2\times3} = 6\sqrt{3} - 5\sqrt{3} + 2\sqrt{3} = 3\sqrt{3}.$

32. $\frac{1}{(\sqrt{9}-\sqrt{8})} = \frac{1}{(\sqrt{9}-\sqrt{8})} \times \frac{(\sqrt{9}+\sqrt{8})}{(\sqrt{9}+\sqrt{8})} = \frac{\sqrt{9}+\sqrt{8}}{1} = (\sqrt{9}+\sqrt{8}).$

Similarly, $\frac{1}{(\sqrt{8}-\sqrt{7})} = (\sqrt{8}+\sqrt{7})$; $\frac{1}{(\sqrt{7}-\sqrt{6})} = (\sqrt{7}+\sqrt{6})$;

$\frac{1}{(\sqrt{6}-\sqrt{5})} = (\sqrt{6}+\sqrt{5})$ and $\frac{1}{(\sqrt{5}-\sqrt{4})} = (\sqrt{5}+\sqrt{4})$.

Given Expression = $(\sqrt{9}+\sqrt{8}) - (\sqrt{8}+\sqrt{7}) + (\sqrt{7}+\sqrt{6}) - (\sqrt{6}+\sqrt{5}) + (\sqrt{5}+\sqrt{4})$

$= (\sqrt{9}+\sqrt{4}) = (3+2) = 5$

33. $\sqrt{32} = \sqrt{4\times4\times2} = 4\sqrt{2}$; $\sqrt{48} = \sqrt{4\times4\times3} = 4\sqrt{3}$;

$\sqrt{8} = \sqrt{2\times2\times2} = 2\sqrt{2}$ and $\sqrt{12} = \sqrt{2\times2\times3} = 2\sqrt{3}$

$\therefore \frac{\sqrt{32}+\sqrt{48}}{\sqrt{8}+\sqrt{12}} = \frac{4\sqrt{2}+4\sqrt{3}}{2\sqrt{2}+2\sqrt{3}} = \frac{4(\sqrt{2}+\sqrt{3})}{2(\sqrt{2}+\sqrt{3})} = 2.$

34. $\sqrt{75.24+x} = 8.71 \Rightarrow 75.24 + x = 8.71\times8.71$

$\therefore x = (8.71)^2 - (75.24) = (72.8641 - 75.24) = 0.6241.$

35. $\frac{4+\sqrt{2}}{\sqrt{2}+1} = \frac{(4+\sqrt{2})}{(\sqrt{2}+1)} \times \frac{(\sqrt{2}-1)}{(\sqrt{2}-1)} = (3\sqrt{2}-2) = (3\times1.4142-2) = 2.2426.$

36. $\frac{\sqrt{2}-1}{\sqrt{2}+1} = \frac{(\sqrt{2}-1)}{(\sqrt{2}+1)} \times \frac{(\sqrt{2}-1)}{(\sqrt{2}-1)} = (\sqrt{2}-1)^2.$

$\therefore \sqrt{\left(\frac{\sqrt{2}-1}{\sqrt{2}+1}\right)} = (\sqrt{2}-1) = (1.4142-1) = 0.4142.$

37. $\frac{1872}{\sqrt{x}} = 234 \Rightarrow \sqrt{x} = \frac{1872}{234} = 8 \Rightarrow x = 64.$

CHAPTER 5

Percentage

Percent: *By a certain percent, we mean that many hundredths.*
Thus, x percent means x hundredths, written as x %.

To express x% as a fraction: $x\% = \frac{x}{100}$.

To express a fraction as a percent: $\frac{a}{b} = \left(\frac{a}{b} \times 100\right)\%$

Thus, $15\% = \frac{15}{100} = \frac{3}{20}$. And, $\frac{4}{5} = \left(\frac{4}{5} \times 100\right)\% = 80\%$

SOLVED PROBLEMS

Ex. 1. ***Express each one of the following as a decimal:***

***(i)* 34 %** ***(ii)* 0.5 %** ***(iii)* 0.03 %**

Sol. (*i*) $34\% = \frac{34}{100} = 0.34$. (*ii*) $0.5\% = \frac{0.5}{100} = 0.005$. (*iii*) $0.03\% = \frac{0.03}{100} = 0.0003$.

Ex. 2. ***Find: (i) 70 % of 28 (ii) 56% of 5.***

Sol. (*i*) 70 % of 28 $= \left(\frac{70}{100} \times 28\right)\% = 19.6$ (*ii*) 56% of 5 $= \left(\frac{56}{100} \times 5\right) = 2.8$

Ex. 3. ***What percent of 96 is 8?***

Sol. Required percentage $= \left(\frac{8}{96} \times 100\right)\% = 8\frac{1}{3}\%$

Two Important Results (Short Cut Method)

(*i*) If A's income is r% more than that of B, then:

B's income is less than that of A by: $\left\{\frac{r}{(100+r)} \times 100\right\}\%$.

(*ii*) If A's income is r% less than that of B, then:

B's income is more than that of A by: $\left\{\frac{r}{(100-r)} \times 100\right\}\%$

Ex. 4. ***If A's monthly income is 40 % more than that of B, how much percent is B's income less than that of A?***

Sol. Required percentage $= \left[\frac{40}{(100+40)} \times 100\right]\% = 28\frac{4}{7}\%$.

Ex. 5. ***If A's height is 25 % less than that of B, how much percent is B's height more than that of A?***

Sol. Required percentage = $\left[\frac{25}{(100-25)}\times100\right]\% = 33\frac{1}{3}\%$

Two Important Rules (Short Cut Method)

(*i*) If the price of a commodity increases by r%, then reduction in consumption, so as not to increase the expenditure, is:

$$\left\{\frac{r}{(100+r)}\times100\right\}\%$$

(*ii*) If the price of a commodity decreases by r%, then increase in consumption, so as not to decrease the expenditure; is:

$$\left\{\frac{r}{(100-r)}\times100\right\}\%.$$

Ex. 6. ***If the price of sugar is increased by 15 %, find by how much percent a householder must reduce her consumption of sugar so as not to increase the expenditure.***

Sol. Reduction in consumption

$$=\left[\frac{15}{(100+15)}\times100\right]\% = \left(\frac{15}{115}\times100\right)\% = 13\frac{1}{23}\%$$

Results on Population:

Let the percent population of a town be P and let there be an increase of R % per annum. Then,

(*i*) Population after n years = $P\left(1+\frac{R}{100}\right)^n$. (*ii*) Population n years ago = $\frac{P}{\left(1+\frac{R}{100}\right)^n}$.

Ex. 7. ***The population of a town is 176400. If it increases annually at 5 %, what will be its population 2 years hence? What was it 2 years ago?***

Sol. Population after 2 years = $176400\times\left(1+\frac{5}{100}\right)^2 = \left(176400\times\frac{21}{20}\times\frac{21}{20}\right) = 194481.$

Population 2 years ago = $\frac{176400}{\left(1+\frac{5}{100}\right)^2} = \left(176400\times\frac{20}{21}\times\frac{20}{21}\right) = 160000.$

Results on Depreciation:

Let the present value of a machine be P and let it depreciate at R% per annum. Then:

(*i*) Value of machine after n years = $P\left(1-\frac{R}{100}\right)^n$

(*ii*) Value of machine n years ago = $\frac{P}{\left(1-\frac{R}{100}\right)^n}$.

Ex. 8. ***The value of a machine depreciates at the rate of 10 % per annum. If its present value is ₹81000, what will be its worth after 2 years? What was the value of the machine 2 years ago?***

Sol. Value of the machine after 2 years = ₹ $\left[81000\times\left(1-\frac{10}{100}\right)^2\right]$

$$= ₹\left(81000\times\frac{9}{10}\times\frac{9}{10}\right) = \text{Rs. } 65610.$$

Value of the machine 2 years ago $= ₹\ \dfrac{81000}{\left(1-\dfrac{10}{100}\right)^2} = ₹\left(81000\times\dfrac{10}{9}\times\dfrac{10}{9}\right) = \text{Rs. } 100000.$

EXERCISE 5

1. The fraction equivalent to $\frac{2}{5}$% is:

(*a*) $\frac{1}{40}$ (b) $\frac{1}{125}$ (*c*) $\frac{1}{250}$ (*d*) $\frac{1}{500}$

2. 0.025 in terms of rate percent is:

(*a*) 25% (*b*) 2.5% (*c*) .25% (*d*) $37\frac{1}{2}$%

3. $\left(0.756\times\frac{3}{4}\right)$ is equivalent to:

(*a*) 18.9% (*b*) 37.8% (*c*) 56.7% (*d*) 75%

4. What percent of $\frac{2}{7}$ is $\frac{1}{35}$?

(*a*) 25% (*b*) 2.5% (*c*) 10% (*d*) 1000%

5. What will be 80% of a number whose 200% is 90?

(*a*) 144 (*b*) 72 (*c*) 36 (*d*) 54

6. A number decreased by $27\frac{1}{2}$% gives 87. The number is:

(*a*) 120 (*b*) 110 (*c*) 135 (*d*) 58

7. A number increased by $37\frac{1}{2}$% gives 33. The number is:

(*a*) 27 (*b*) 25 (*c*) 22 (*d*) 24

8. If $37\frac{1}{2}$% of a number is 900, then $62\frac{1}{2}$% of the number will be:

(*a*) 1200 (*b*) 1350
(*c*) 1500 (*d*) 540

9. 100 quintals is what percent of 10 metric tonnes? **[CDS (I) Exam, 2025]**

(*a*) 1% (*b*) 10% (*c*) 100% (*d*) 1000%

10. What is the square root of 64%?

(*a*) 0·08% (*b*) 0·8% (*c*) 8% (*d*) 80%

[CDS (II) Exam, 2024]

Direction : Consider the following for the four (04) items that follow:
500 candidates appeared in an examination comparing tests in English, Hindi and Mathematics. 30 candidates failed in English only; 85 failed in Hindi only; 50 failed in Mathematics only; 15 failed in both English and Hindi; 17 failed in both Mathematics and English; 5 failed in all three tests.

11. What is the percentage of candidates who failed in at least two subjects? **[CDS Exam, 2022]**

(*a*) 5.4% (*b*) 6.4% (*c*) 6.8% (*d*) 7.8%

12. What is the percentage of candidates who failed in only one subject? **[CDS Exam, 2022]**
(*a*) 28% (*b*) 31% (*c*) 35.8% (*d*) 38.8%

13. What is the percentage of candidates who failed in at least one subject? **[CDS Exam, 2022]**
(*a*) 31% (*b*) 35.4% (*c*) 38.8% (*d*) 41.5%

14. How many candidates passed in two or more subjects? **[CDS Exam, 2022]**
(*a*) 461 (*b*) 405 (*c*) 345 (*d*) 306

15. If *p* varies directly as *q* and inversely as square of *r*, what is the percentage increase in *p* due to an increase in *q* by 20% and a increase in r by 20%? **[CDS (II) Exam, 2021]**
(*a*) 87.5% (*b*) 85% (*c*) 82.5% (*d*) 80%

16. The price of an article X increases by 20% every year and price of article Y increases by 10% every year. In the year 2010, the price of article X was ₹5000 and price of article Y was ₹2000. In which year the difference in their prices exceeded ₹5000 for the first time? **[CDS (I) Exam, 2020]**
(*a*) 2012 (*b*) 2013 (*c*) 2014 (*d*) 2015

17. If the annual income of *x* is 20% more than that of *y*, then the income of *y* is less than that of *x* by *p*%. What is the value of *p*? **[CDS (I) Exam, 2020]**
(*a*) 10 (*b*) $16\frac{2}{3}$ (*c*) $17\frac{1}{3}$ (*d*) 20

18. The number of items in a booklet is N. In the first year there is an increase of *x*% in this number and in the subsequent year there is a decrease of *x*%. At the end of the two years, what will be the number of items in the booklet? **[CDS (II) Exam, 2020]**
(*a*) Less than N
(*b*) Equal to N
(*c*) More than N
(*d*) It depends on the value of N

19. In a hostel the rent per room is increased by 20%. If number of rooms in the hostel is also increased by 20% and the hostel is always full, then what is the percentage change in the total collection at the cash counter? **[CDS (I) Exam, 2019]**
(*a*) 30% (*b*) 40% (*c*) 44% (*d*) 48%

20. *x*, *y* and *z* are three numbers such that *x* is 30% of *z* and *y* is 40% of *z*. If *x* is *p*% of *y*, then what is the value of *p*? **[CDS (I) Exam, 2019]**
(*a*) 45 (*b*) 55 (*c*) 64 (*d*) 75

21. If the price of wheat rises by 25%, then by how much percent must a man reduce his consumption in order to keep his budget the same as before? **[CDS (I) Exam, 2019]**
(*a*) 15% (*b*) 20% (*c*) 25% (*d*) 30%

22. The train fare and bus fare between two stations is in the ratio 3 : 4. If the train fare increases by 20% and bus fare increases by 30%, then what is the ratio between revised train fare and revised bus fare? **[CDS (II) Exam, 2019]**
(*a*) $\frac{9}{13}$ (*b*) $\frac{17}{12}$ (*c*) $\frac{32}{43}$ (*d*) $\frac{19}{21}$

23. In an examination, 50% of the candidates failed in English, 40% failed in Hindi and 15% failed in both the subjects. The percentage of candidates who passed in both English and Hindi is **[CDS (II) Exam, 2018]**
(*a*) 20% (*b*) 25% (*c*) 60% (*d*) 75%

24. A milkman claims to sell milk at its cost price only, but he is making a profit of 20% since he has mixed some amount of water in the milk. What is the percentage of milk in the mixture?

(*a*) 80% (*b*) $\frac{250}{3}$% (*c*) 75% (*d*) $\frac{200}{3}$%

[CDS (I) Exam, 2015]

25. A person could save 10% of his income. But 2 years later, when his income increased by 20%, he could save the same amount only as before. By how much percentage has his expenditure increased? **[CDS (I) Exam, 2015]**

(*a*) $22\frac{2}{9}$% (*b*) $23\frac{1}{3}$% (*c*) $24\frac{2}{9}$% (*d*) $25\frac{2}{9}$%

26. The value of a single discount on some amount which is equivalent to a series of discounts of 10%, 20% and 40% on the same amount, is equal to **[CDS (II) Exam, 2015]**

(*a*) 43.2% (*b*) 50% (*c*) 56.8% (*d*) 70%

27. 20% of a number when added to 20 becomes the number itself, then the number is

(*a*) 20 (*b*) 25 (*c*) 50 (*d*) 80

[CDS (II) Exam, 2015]

28. *A*'s salary was increased by 40% and then decreased by 20%. On the whole *A*'s salary is increased by **[CDS (II) Exam, 2015]**

(*a*) 60% (*b*) 40% (*c*) 20% (*d*) 12%

29. In an election 10% of the voters on the voter list did not cast their vote and 60 voters cast their ballot papers blank. There were only two candidates. The winner was supported by 47% of total voters in the voter list and he got 308 votes more than his rival. The number of voters on the voter lists is **[CDS (II) Exam, 2015]**

(*a*) 3600 (*b*) 6200 (*c*) 6028 (*d*) 6400

30. The salary of a person is increased by 10% of his original salary. But he received the same amount even after increment. What is the percentage of his salary he did not receive?

(*a*) 11% (*b*) 10% (*c*) (100/11)% (*d*) (90/11)%

[CDS (I) Exam, 2016]

31. The expenditure of a household for a certain month is ₹20,000, out of which ₹8,000 is spent on education, ₹5,900 on food, ₹2,800 on shopping and the rest on personal care. What percentage of expenditure is spent on personal care? **[CDS (I) Exam, 2016]**

(*a*) 12% (*b*) 16·5% (*c*) 18% (*d*) 21·8%

32. A candidate scoring *x*% marks in an examination fails by a marks, while another candidate who scores y% marks gets b marks more than the minimum required pass marks. What is the maximum marks for the examination? **[CDS (II) Exam, 2016]**

(*a*) $\frac{100(a+b)}{x-y}$ (*b*) $\frac{100(a-b)}{x+y}$ (*c*) $\frac{100(a+b)}{y-x}$ (*d*) $\frac{100(a-b)}{x-y}$

33. If the mean age of combined group of boys and girls is 18 years and the mean of age of boys is 20 and that girls is 16, then what is the percentage of boys in the group?

(*a*) 60 (*b*) 50 (*c*) 45 (*d*) 40

[CDS (II) Exam, 2016]

34. If a% of a + b% of b = 2% of ab, then what percent of a is b?

(*a*) 50% (*b*) 75%

(*c*) 100% (*d*) Cannot be determined **[CDS (I) Exam, 2017]**

35. $\frac{5}{9}^{th}$ **part of the population in a village are males. If 30% of the males are married, the percentage of unmarried females in the total population is** **[CDS (I) Exam, 2017]**

(*a*) $20\frac{2}{9}$ (*b*) $27\frac{2}{9}$ (*c*) $27\frac{7}{9}$ (*d*) $29\frac{2}{9}$

36. After deducting a commission of 5%, a T.V. costs ₹ 15200. Its gross value (in rupees) is:

(*a*) 15860 (*b*) 16000 (*c*) 16400 (*d*) 14440

37. 96% of the population of a town is 23040. The total population of the town is:

(*a*) 32256 (*b*) 24000 (*c*) 24936 (*d*) 25640

38. If 75% of the students in a school are boys and the number of girls is 420, the number of boys is:

(*a*) 1176 (*b*) 1350 (*c*) 1260 (*d*) 1125

39. A man spends ₹ 3500 and saves $12\frac{1}{2}$**% of his income. His monthly income (in rupees) is:**

(*a*) 3937.50 (*b*) 4000 (*c*) 4250 (*d*) 4160

40. In an examination, 65% of the total examinees passed. If the number of failures is 168, the number of those who passed is:

(*a*) 480 (*b*) 312 (*c*) 648 (*d*) 218

41. An ore contains 14% copper. How many kilograms of the ore are required to get 77 kg of copper?

(*a*) 87.78 kg (*b*) 66.22 kg (*c*) 1078 kg (*d*) 550 kg

42. The population of a town increases by 5% annually. If it is 15435 now, its population 2 years ago was:

(*a*) 14000 (*b*) 15000 (*c*) 13700 (*d*) 14800

43. The population of a town is 8000. It increases by 10% during first year and by 20% during second year. The population after 2 years will be:

(*a*) 10560 (*b*) 10400 (*c*) 10620 (*d*) 8960

44. The value of a machine depreciates at the rate of 16% per annum. If the price of a new machine is ₹ 62500, its value after 3 years (in rupees) will be:

(*a*) 32500 (*b*) 36054 (*c*) 37044 (*d*) 38400

45. A man spends 75% of his income. His income is increased by 20% and he increased his expenditure by 10%. His savings are increased by:

(*a*) 10% (*b*) 25% (*c*) 37.5% (*d*) 50%

46. 5% income of *A* is equal to 15% income of *B* and 10% income of *B* is equal to 20% income of *C*. If income of *C* is ₹ 2000, then total income of *A*, *B* and *C* (in rupees) is:

(*a*) 6000 (*b*) 9000 (*c*) 12000 (*d*) 18000

47. If *A's* income is 25% more than *B's* and *B's* income is 20% more than *C's*, by what percent is *A's* income more than *C's*?

(*a*) 15% (*b*) 25% (*c*) $33\frac{1}{2}$% (*d*) 50%

48. The ratio of the number of boys and girls in a school is 3: 2. If 20% of the boys and 30% of the girls are scholarship holders, the percentage of the students who are not scholarship holders is:

(*a*) 50 (*b*) 72 (*c*) 75 (*d*) 76

49. On decreasing the price of fans by 30%, the sale is increased by 20%. What is the effect on money receipt by the shopkeeper?

(*a*) 10% increase (*b*) 10% decrease (*c*) 16% increase (*d*) 16% decrease

50. Water tax is increased by 20% and its consumption is decreased by 20%, The change in the expenditure is:

(*a*) nil (*b*) 5% decrease

(*c*) 4% increase (*d*) 4% decrease

51. A man received 15% increase in his salary. If his new salary is ₹ 1932, his original salary (in rupees) was:

(*a*) 1642.20 (*b*) 1820 (*c*) 1680 (*d*) 1720.60

52. A man's wages were decreased by 50%. Again the reduced wages were increased by 50%. He has a loss of:

(*a*) 0% (*b*) 2.5% (*c*) .25% (*d*) 25%

53. A mixture of 40 litres of milk and water contains 10% water. How much water should be added to it so that water may be 20% in the new mixture?

(*a*) 5 litres (*b*) 4 litres (*c*) 6.5 litres (*d*) 7.5 litres

54. The total population of an area is 10,000 out of which males and females are equal in number. Out of the total population 30% are Newspaper readers. Out of the total newspaper readers, one-third read English Newspaper. Out of the total English Newspaper readers, 20% are females. What is the number of males who do not read English Newspaper?

(*a*) 800

(*b*) 2100

(*c*) 4200

(*d*) Cannot be determined due to insufficient data

[CDS (II) Exam, 2024]

55. A question is given followed by two statements I and II. Consider the Question and the Statements and mark the correct option. **[CDS (II) Exam, 2024]**

Question : If the price of petrol goes up by 20%, by what percentage should the consumption be reduced so that the expenditure remains the same?

Statement-I : Price of petrol per litre was Rs. 90

Statement-II : Consumption was 24 litre before price hike

Which one of the following is correct in respect of the above Question and the Statements?

(*a*) The Question can be answered by using one of the Statements alone, but cannot be answered using the other Statement alone

(*b*) The Question can be answered by using either Statement alone

(*c*) The Question can be answered by using both the Statements together, but cannot be answered using either Statement alone

(*d*) The Question can be answered even without using any of the Statements

56. Twelve percent of bananas bought by a fruit vendor got lost during transportation. On selling the remaining bananas, the vendor's overall profit turned out to be 4%. If the vendor had not lost any bananas and had sold them at the price of the remaining bananas, what would have been his profit percentage? **[CDS (I) Exam, 2023]**

(*a*) $8\frac{1}{9}\%$ (*b*) $9\frac{1}{11}\%$ (*c*) $18\frac{2}{11}\%$ (*d*) None of the above

57. A person bought an article and sold it at a profit of 20%. Had he bought it at 20% less, what would have been the profit percentage if the selling price had been the same ?

[CDS (II) Exam, 2023]

(*a*) 25% (*b*) 40% (*c*) 50% (*d*) 60%

ANSWERS

1. (*c*)	**2.** (*b*)	**3.** (*c*)	**4.** (*c*)	**5.** (*c*)	**6.** (*a*)	**7.** (*d*)	**8.** (*c*)	**9.** (*c*)
10. (*d*)	**11.** (*d*)	**12.** (*b*)	**13.** (*c*)	**14.** (*a*)	**15.** (*a*)	**16.** (*b*)	**17.** (*b*)	**18.** (*a*)
19. (*c*)	**20.** (*d*)	**21.** (*b*)	**22.** (*a*)	**23.** (*b*)	**24.** (*b*)	**25.** (*a*)	**26.** (*c*)	**27.** (*b*)
28. (*d*)	**29.** (*b*)	**30.** (*c*)	**31.** (*b*)	**32.** (*c*)	**33.** (*b*)	**34.** (*c*)	**35.** (*c*)	**36.** (*b*)
37. (*b*)	**38.** (*c*)	**39.** (*b*)	**40.** (*b*)	**41.** (*d*)	**42.** (*a*)	**43.** (*a*)	**44.** (*c*)	**45.** (*d*)
46. (*d*)	**47.** (*d*)	**48.** (*d*)	**49.** (*d*)	**50.** (*d*)	**51.** (*c*)	**52.** (*d*)	**53.** (*a*)	**54.** (*c*)
55. (*d*)	**56.** (*d*)	**57.** (*c*)						

SOLUTIONS

1. $\frac{2}{5}\% = \left(\frac{2/5}{100}\right) = \frac{2}{5\times100} = \frac{1}{250}.$

2. $0.025 = \left(\frac{25}{1000}\times100\right)\% = 2.5\%$

3. $0.756\times\frac{3}{4} = \left(\frac{736}{1000}\times\frac{3}{4}\times100\right)\% = 56.7\%.$

4. Required percent $= \left[\frac{(1/35)}{(2/7)}\times100\right]\% = \left(\frac{1}{35}\times\frac{7}{2}\times100\right)\% = 10\%.$

5. Let 200% of $x = 90$. Then, $\frac{200}{100}\times x = 90$ or $x = 90\times\frac{1}{2} = 45.$

$\therefore$ Number = 45. Now, 80% of 45 = $\left(\frac{80}{100}\times45\right) = 36.$

6. $72\frac{1}{2}\%$ of $x = 87 \Leftrightarrow \frac{145}{2\times100}\times x = 87 \Leftrightarrow x = \frac{87\times2\times100}{145} = 120.$

7. $137\frac{1}{2}\%$ of $x = 33 \Leftrightarrow \frac{275}{2\times100}\times x = 33 \Leftrightarrow x = \frac{33\times2\times100}{275} = 24.$

8. $\frac{75}{2\times100}\times x = 900 \Rightarrow x = \frac{900\times2\times100}{75} = 2400.$

$\therefore$ Number = 2400.

So, $62\frac{1}{2}\%$ of $2400 = \frac{125}{2\times100}\times2400 = 1500.$

9. 1 quintal = 100 kg

1 metric tonne = 1000 kg

So, 100 quintals = 100 × 100 = 10,000 kg

10 metric tonnes = 10 × 1000 = 10,000 kg

Required percentage = $\left(\frac{10000}{10000}\right)\times100 = 100\%$

10. Square root of 64% = $\sqrt{\frac{64}{100}} \times 100 = \frac{8}{10} \times 100 = 80$

11. We have,

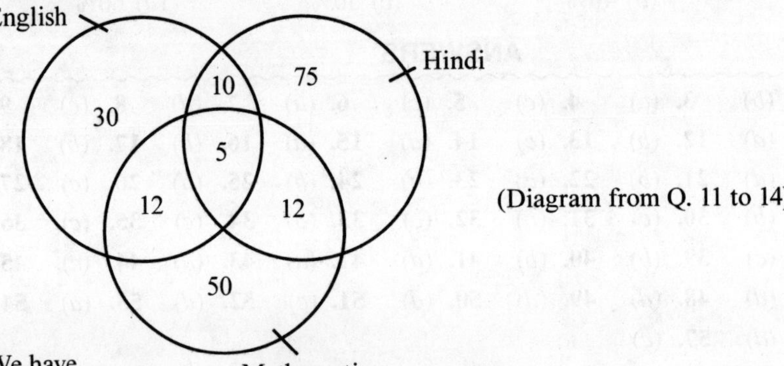

(Diagram from Q. 11 to 14)

We have,

Candidates failed in only English and Mathematics = 12

Candidates failed in only English and Hindi = 10

Candidates failed in only Hindi and Mathematics = 12

Candidates failed in English, Hindi and Mathematics = 5

Now,

Candidates failed in at least two subjects = 12 + 10 + 12 + 5 = 39

$\therefore$ Required percentage = $\frac{39}{500} \times 100 = 7.8\%$

12. From diagram, we have,

Candidates failed in only one subject = 30 + 75 + 50 = 155

$\therefore$ Required percentage = $\frac{155}{500} \times 100 = 31\%$

13. From diagram, we have,

Candidates failed in at least one subject = 30 + 75 + 50 + 10 + 12 + 12 + 5 = 194

$\therefore$ Required percentage = $\frac{194}{500} \times 100 = 38.8\%$

14. From diagram,

Candidates failed in at least one subject – 10 + 12 + 12 + 5 = 39

Now,

Candidates passed in two or more subjects = 500 – 39 = 461

15. We have,

$p \propto \frac{q}{r^2} \Rightarrow p = \frac{kq}{r^2}$, where k is constant

Now, according to the question,

$$p = k \times \frac{120}{100} q \times \left(\frac{80}{100}\right)^2 r^2 = k \times 1.2q \times \frac{0.64}{r^2} = 1.875\ kq/r^2$$

$$\therefore \text{ Required percentage} = \frac{1.875kq/r^2 - kq/r^2}{kq/r^2} \times 100$$

$$\Rightarrow \frac{0.875kq/r^2}{kq/r^2} \times 100 = 87.5\%$$

16. Price of X in 2011 $= 5000 \times \frac{120}{100} =$ ₹6000

Price of Y in 2011 $= 2000 \times \frac{110}{100} =$ ₹2200

Difference = 6000 – 2200 = ₹3800 (less than ₹5000)

Now,

Price of X in 2012 $= 6000 \times \frac{120}{100} =$ ₹7200

Price of Y in 2012 $= 2200 \times \frac{110}{100} =$ ₹2420

Difference = 7200 – 2420 = ₹4780 (less than ₹5000)

Now,

Price of X in 2013 $= 7200 \times \frac{120}{100} =$ ₹8640

Price of Y in 2013 $= 2420 \times \frac{110}{100} =$ ₹2662

Difference = 8640 – 2662 = ₹5978 (more than ₹5000)

17. Let annual income of $y = y$

$\therefore$ Annual income of $x = y \times \frac{120}{100} = 1.2y$

Now, according to the question,

$$p = \frac{1.2y - y}{1.2} \times 1000 = \frac{0.2y}{1.2} \times 100 = 16\frac{2}{3}\%$$

18. Let N = 100

$x = 10$

According to the question,

$\therefore$ Final number of items $= 100 \times \frac{110}{100} \times \frac{90}{100} = 99$

19. Let number of rooms in hosted be a and rent per room be b.

$\therefore$ Total collection of cash $= ab$

Now, according to the question,

Total number of rooms $= a \times \frac{120}{100} = 1.2a$

Rent per person $= b \times \frac{120}{100} = 1.2b$

$\therefore$ Required percentage $= \frac{1.2a \times 1.2b - ab}{ab} = \frac{1.44ab - ab}{ab} \times 100 = \frac{0.44ab}{ab} \times 100 = 44\%$

20. According to the question,

$$x = z \times \frac{30}{100} = 0.32 \quad \text{... } (i)$$

$$y = z \times \frac{40}{100} = 0.4z \quad ...(ii)$$

and $$x = \frac{py}{100} \quad ...(iii)$$

Put the value of x and y in equation (iii), we get

$$0.3z = \frac{p}{100} \times 0.4z$$

$$\Rightarrow \quad p = 25 \times 3 = 75$$

21. Let the initial price of wheat be a and consumption be b respectively.

Let the new consumption be b'

$\because$ Expenditure = price × consumption

$$\therefore \quad ab = ab' \times \frac{125}{100}$$

$$ab = 1.25\ ab'$$

$$b' = \frac{b}{1.25} = 0.8b$$

$$\therefore \text{ Required percentage} = \frac{b - 0.8b}{b} \times 100$$

$$= \frac{0.2b}{b} \times 100 = 20\%$$

22. Let the train fare and bus fare between two stations be $3x$ and $4x$ respectively.

After increasing,

$$\text{Train fare} = 3x \times \frac{120}{100} = 3.6x$$

$$\text{But fare} = 4x \times \frac{130}{100} = 5.2x$$

$$\therefore \text{ Required ratio} = 3.6x : 5.2x = 9 : 13$$

23. Let the total number of students, who are appearing in the examination be a

No. of candidates that failed in English = $\frac{50}{100} \times a = 0.5a$

No. of candidates that failed in Hindi = $\frac{40}{100} \times a = 0.4a$

No. of candidates that failed in both Hindi and English = $a \times \frac{15}{100} = 0.15\ a$

Now, no. of candidates that either failed in Hindi or English or both

$$= 0.5a + 0.4a - 0.15a = 0.75a$$

$\Rightarrow$ No. of candidates that passed in both Hindi and English = $a - 0.75a = 0.25a$

$$\therefore \text{ Required percentage} = \frac{0.25a}{a} \times 100 = 25\%$$

24. The milkman buys milk at (say) ₹40 per litre, adds water and sells the mixture at ₹40 per litre. The profit comes not from the price that he charges, but from the increase in quantity. To obtain 20% profit, he has to increase the quantity by 20%. i e., to 10 litres of milk he adds 2 litres of water, to get 12 litres of mixture. The percentage of milk in the mixture is $\frac{10}{12}$ (100%) = $\frac{250}{3}$%

25. Let the original income of the person be ₹100.
Original savings = ₹10.
Original expenditure = 100 – 10 = 90.
2 years later, the income of the person was ₹120.
Savings then were (still) ₹10. Expenditure then was ₹110.
Expenditure of the person increased by $\frac{20}{90}$(100%), v/z $22\frac{2}{9}$%.

26. Consider the series of 3 discounts of 10%, 20%, and 40%.
Let the article be marked at M.
The final selling price is (60%) of (80%) of (90%) of (M) = (0.6) (0.8) (0.9) M = 0.432 M = 43.2% of M.
This is equivalent to a single discount of 56.8%.

27. Let the number be x.

$$20 + 0.2x = x \text{ (given) } 0.8x = 20$$

$$x = 25.$$

28. Let A's salary initially be 1. Finally it is (1.4) (0.8) = 1.12
∴ It increases by 0.12 or by 12%.

29. Let the number of voters be 100n.
$10n$ voters did not cast their vote and 60 voters cast a blank vote.
The other $90n – 60$ voters cast their votes for one of the two candidates.
The winner won by 308 votes.
∴ The winning candidate got $\frac{90n - 60 - 308}{2}$ + 308 votes and the losing candidate got $\frac{90n - 60 - 308}{2}$ votes.

Given $\frac{90n - 60 - 308}{2} + 308 = 47n$.

∴ $90n – 368 + 616 = 94n$

$\Rightarrow 248 = 4n \Rightarrow n = 62$

The number of voters is 6200.

30. Let the salary of the person be ₹100 Salary, after 10% increment = ₹110
But he received a ₹ 100 ie he received ₹10 less

∴ % of the salary not received = $\frac{10}{110} \times 100 = \frac{100}{11}$%

31. Of the total expenditure of 20,000. Money was spent on education, food, shopping (8000 + 5900 + 2800) i.e., 16700. The remaining 3300 is on personal care.

The required percentage $= \frac{3300}{20000} = \frac{33}{200} = \frac{16.5}{100} = 16.5\%$

32. Let the pass marks be P and the maximum Marks be M.

$$\frac{xM}{100} = P - a \text{ and } \frac{yM}{100} = P + b.$$

$$\frac{(y-x)M}{100} = a + b \text{ or } M = \frac{100(a+b)}{y-x}$$

33. The data is tabulated below. G denote the group of girls and B denotes the group of boys.

Average	16	18	20
Group	G	GUB	B

The ratio of the number of boys to the number of girls = $\frac{18-16}{20-18} = \frac{1}{1}$.

$\therefore$ The percentage of boys in the group is 50.

34. $\frac{a}{100} \times a + \frac{b}{100} \times b = \frac{2}{100} \times ab$

$\Rightarrow a^2 + b^2 = 2ab \Rightarrow a^2 + b^2 - 2ab = 0 \Rightarrow (a-b)^2 = 0 \Rightarrow a - b = 0$

$a = b$

$\therefore$ Answer is 100% as a and b are equal

35. Let total population be 90

Number of males = $\frac{5}{9} \times 90 = 50$

$\therefore$ Number of Females = 40

Number of married males = $\frac{30}{100} \times 50 = 15$

Number of unmarried females = 40 − 15 = 25

$\therefore$ Percentage of unmarried females in the total population = $\frac{25}{90} \times 100 = \frac{250}{9} = 27\frac{7}{9}\%$

36. 95% of $x = 15200 \Rightarrow x = \left(15200 \times \frac{100}{95}\right) = 16000.$

37. 96% of $x = 23040 \Rightarrow x = \left(23040 \times \frac{100}{96}\right) = 24000.$

38. Let total number of students be x.

Then, 25 % of $x = 420 \Rightarrow x = \left(420 + \frac{100}{25}\right) = 1680.$

$\therefore$ Number of boys = 75% of 1680 = 1260.

39. Let total income = x. Then,

$87\frac{1}{2}\%$ of $x = 3500 \Leftrightarrow \frac{175}{2 \times 100} x = 3500.$

$\therefore x = \left(\frac{3500 \times 2 \times 100}{175}\right) = 4000.$

40. Let the number of all examinees be x.

Then, 35% of $x = 168 \Rightarrow x = \left(168 \times \frac{100}{35}\right) = 480.$

$\therefore$ Number of those who passed = (480 – 168) = 312.

41. To obtain 14 kg copper, ore needed = 100 kg.

$\therefore$ 14: 100:: 77: $x \Leftrightarrow x = \left(\frac{100 \times 77}{14}\right) = 550$ kg.

42. Population 2 years ago $= \dfrac{15435}{\left(1+\frac{5}{100}\right)^2} = \left(15435 \times \frac{20}{21} \times \frac{20}{21}\right) = 14000.$

43. Population after 2 years $= 8000 \times \left(1+\frac{10}{100}\right) \times \left(1+\frac{20}{100}\right) = \left(8000 \times \frac{11}{10} \times \frac{6}{5}\right) = 10560.$

44. Value of machine after 3 years $= 62500 \times \left(1-\frac{16}{100}\right)^3 = \left(62500 \times \frac{21}{25} \times \frac{21}{25} \times \frac{21}{25}\right) = 37044.$

45. Let income = 100. Then, expenditure = 75 & saving = 25.

New income = 120, New expenditure = 110 % of 75 = 82.5.

New saving = (120 – 82.5) = 37.5.

Increase in saving $= \left(\frac{12.5}{25} \times 100\right)\% = 50\ \%$

46. $\frac{5}{100} A = \frac{15}{100} B$ and $\frac{10}{100} B = \frac{20}{100} C.$

$\therefore\ A = \left(\frac{15}{100} \times \frac{100}{5}\right) B = 3B$ and $B = \left(\frac{20}{100} \times \frac{100}{10}\right) C = 2C.$

Now, $C = 2000 \Rightarrow B = (2 \times 2000) = 4000$ & $A = (3 \times 4000) = 12000.$

$\therefore\ A + B + C = (12000 + 4000 + 2000) = 18000.$

47. Let *C*'s income = 100. Then, *B*'s income = 120 .

And, *A*'s income = 125 % of 120 = 150.

$\therefore$ *A*'s income is 50% more than *C*'s.

48. Let the number of boys & girls be $3x$ and $2x$.

Number of those who are not scholarship holders

$= (80\% \text{ of } 3x + 70\% \text{ of } 2x) = \left(\frac{12x}{5} + \frac{7x}{5}\right) = \frac{19x}{5}$

$\therefore$ Required percentage $= \left(\frac{19x}{5} \times \frac{1}{5x} \times 100\right)\% = 76\%$

49. Let cost of each fan = ₹ 100 & Sale = 100 fans.

Money receipt = ₹ (100 × 100) = ₹. 10000

New cost per fan = ₹ 70 & New Sale = 120 fans.

$\therefore$ New Money Receipt = ₹ (70 × 120) = ₹ 8400.

Decrease in money receipt $= \left(\frac{1600}{10000} \times 100\right)\% = 16\%$

50. Let Tax = ₹ 100 per unit & Consumption = 100 units

Original Expenditure = ₹ (100 × 100) = ₹. 10000.

New Tax = ₹ 120 per unit & New consumption = 80 units.

New Expenditure = ₹ (120 × 80) = ₹ 9600.

$\therefore$ Decrease in Expenditure = $\left(\frac{400}{10000}\times 100\right)\% = 4\%$

51. 115% of x = 1932 $\Rightarrow x = \left(1932\times\frac{100}{115}\right) = 1680.$

52. Let original wages = ₹ 100, Reduced wages = ₹ 50.
Increased wages = 150% of ₹ 50 = ₹ 75.
Loss = (100 – 75)% = 25%.

53. Water in original mix. = $\left(\frac{10}{100}\times 40\right)$ litres = 4 litres.

Let x litres of water be added to this mix.

Then, $\frac{4+x}{40+x}\times 100 = 20 \Leftrightarrow x = 5$ *litres.*

54. Total number of newspaper reader = $\frac{10000\times 30}{100} = 3000$

Total number of English newspaper readers = $\frac{1}{3}\times 3000 = 1000$

Total number of Female who read English newspapers = $\frac{1000\times 20}{100} = 200$

Hence, number of males who read English newspapers = 1000 – 200 = 800
Total Males = 5,000
$\therefore$ Total number of males who do not read English newspapers = 5000 – 800 = 4200.

55. According to question price of petrol increased by 20%. So,
Percentage decrease in the consumption of petrol = $\frac{20}{100+20}\times 100 = \frac{50}{3} = 16\frac{2}{3}\%$

Thus, the Question can be answered even without using any of the statements

56. Let CP of 100 bananas = ₹100
Number of banana lost during transportation = 12
Remaining bananas = 100 – 12 = 88
SP of 88 bananas = $100\times\frac{104}{100}$ = ₹104

SP of 1 bananas = $\frac{104}{88}$ = ₹$\frac{13}{11}$

SP of 100 bananas = $100\times\frac{13}{11}$ = ₹118.18

$\therefore$ Profit % = $\frac{118.18-100}{100}\times 100 = 18.18\% = 18\frac{2}{11}\%$

57. Let C.P. = ₹x
S.P. = $1.2x$
Now, when buying at 20 × less C.P. = $0.8x$

$\therefore$ profit % = $\frac{1.2x-0.8x}{0.8x}\times 100 = 50\%$

6 CHAPTER

Average

Mean or Average = $\dfrac{\textbf{\textit{Sum of observations}}}{\textbf{\textit{Number of observations}}}$

Rule 1: *If each one of some given numbers is increased by k, then their average is increased by k.*

Rule 2: *If each one of some given numbers is multiplied by k, then their average is multiplied by k.*

Rule 3: *If a man covers some journey from A to B at u km/hr and returns back to A at a uniform speed of v km/hr, then the average speed during the whole journey is* $\left(\dfrac{2uv}{u+v}\right)$ *km/hr.*

SOLVED PROBLEMS

Ex. 1. ***Find the average of first 18 multiples of 6.***

Sol. Required average $= \dfrac{6(1+2+3+...+18)}{18} = \dfrac{6}{18} \times \dfrac{18 \times (18+1)}{2} = 57 \left[\because\ 1+2+...+n = \dfrac{n(n+1)}{2}\right]$

Ex. 2. ***A batsman makes a score of 87 runs in the 17th inning and thus increased his average by x^3. Find his average after 17th inning.***

Sol. Let the average of 17 innings be x.

Then, average of 16 innings $= (x-3)$.

$\therefore\ 17x - 16(x-3) = 87$ or $x = 39$.

Hence, the average after 17th inning = 39.

Ex. 3. ***A man covers a distance from a town A to a town B on scooter at 30 km per hour and rides back from B to A at 20 km per hour. Find his average speed during the whole journey.***

Sol. Average speed $= \left(\dfrac{2uv}{u+v}\right)$ km/hr. $= \dfrac{2 \times 30 \times 20}{(30+20)}$ km/hr = 24 km/hr.

Ex. 4. ***The average temperature for Monday, Tuesday and Wednesday was 40°C. The average for Tuesday, Wednesday and Thursday was 41°C. If the temperature on Thursday be 42° C, find the temperature on Monday.***

Sol. $(M + T + W) = (3 \times 40) = 120$, $(T + W + Th) = (3 \times 41) = 123$.

On subtraction, we get: $(Th - M) = 3$.

$\therefore\ M = (Th - 3) = (42 - 3) = 39.$

Hence, the temperature on Monday was 39° C.

Ex. 5. ***The average of 11 results is 50. If the average of first six results is 49 and that of last six is 52, find the sixth result.***

Sol. Sum of 11 results $= (11 \times 50) = 550$.

Sum of first 6 results $= (6 \times 49) = 294$

Sum of last 6 results $= (6 \times 52) = 312$.

$\therefore$ 6th result $= (294 + 312 - 550) = 56$.

Ex. 6. *If a, b, c, d, e are five consecutive odd integers, then what is their average?*

(a) $a + 4$ (b) $\frac{abcde}{5}$ (c) $5(a + b + e + d + e)$ (d) None of these

Sol. Clearly, $b = a + 2$; $c = a + 4$, $d = a + 6$ & $e = a + 8$.

$$\therefore \text{Average} = \frac{a+(a+2)+(a+4)+(a+6)+(a+8)}{5} = (a+4).$$

EXERCISE 6

1. In a class of 100 students, the average weight is 30 kg. If the average weight of the girls is 24 kg and that of the boys is 32 kg, then what is the number of girls in the class? **[CDS (I) Exam, 2019]**

   (a) 25 (b) 26 (c) 27 (d) 28

2. The average of 50 consecutive natural numbers is x. What will be the new average when the next four natural numbers are also included? **[CDS (I) Exam, 2019]**

   (a) $x + 1$ (b) $x + 2$ (c) $x + 4$ (d) $x + \frac{x}{54}$

3. Consider two-digit numbers which remain the same when the digits interchange their positions. What is the average of such two-digit numbers? **[CDS (I) Exam, 2019]**

   (a) 33 (b) 44 (c) 55 (d) 66

4. A library has an average number of 510 visitors on Sunday and 240 on other days. What is the average number of visitors per day in a month of 30 days beginning with Saturday? **[CDS (II) Exam, 2019]**

   (a) 276 (b) 282 (c) 285 (d) 375

5. At present the average of the ages of a father and a son is 25 years. After seven years, the son will be 17 years old. What will be the age of the father after 10 years. **[CDS (I) Exam, 2018]**

   (a) 44 years (b) 45 years (c) 50 years (d) 52 years

6. If the average of 64, 69, 72, 75, x lies between 62 and 76 (excluding 62 and 76), then what is the number of possible integer values of x? **[CDS (I) Exam, 2025]**

   (a) 68 (b) 69 (c) 70 (d) 71

7. A man walks at an average speed of 3 km/hr from his residence and reaches office 40 minutes early. If he walks at an average speed of 2 km/hr, he reaches 40 minutes late. What is the distance between his residence and office? **[CDS (I) Exam, 2022]**

   (a) 6 km (b) 8 km (c) 10 km (d) 12 km

8. The average of the temperatures recorded at noontime from Monday to Sunday is 31°C. If the lowest temperature recorded is 30°C, then what is the maximum of temperature that is possible to record at noontime on any one of the days? **[CDS (II) Exam, 2024]**

   (a) 34°C (b) 35°C (c) 36°C (d) 37°C

9. What is the difference between the average of first 50 even natural numbers and the average of first 50 odd natural numbers? **[CDS (II) Exam, 2024]**

   (a) 0 (b) 0.5 (c) 1 (d) 2

10. Average marks in Mathematics of Section A comprising 30 students is 65 and that of Section B comprising 35 students is 70. What are the average marks (approximately) of both the sections if it was detected later that an entry of 47 marks was wrongly made as 74 ?

(*a*) 67·28 (*b*) 67·58 (*c*) 68·11 (*d*) 68·63

[CDS (I) Exam, 2023]

Consider the following for the next (05) items that follow:

A, B, C, D, E, F and G are cousins. D is thrice as old as A. Further, C is as many years younger to B, as G to E and E to D. The average age of D and G is 16 years; the average age of A and E is 11 years; the average age of B and C is also 11 years. B and C have equal weight. A's weight is 10 kg less than that of B; D is 4 kg heavier than E; E is 4 kg heavier than F; F is 4 kg heavier than G. Further, D has age-weight ratio of 9 : 20, where age is in years and weight in kg; A has age-weight ratio of 2 : 5. Moreover, none of them is more than 40 kg.

11. What is D's age (in years) ? **[CDS (II) Exam, 2023]**

(*a*) 15 (*b*) 16 (*c*) 17 (*d*) 18

12. What is the average age (in years) of B, C, D, E and G? **[CDS (II) Exam, 2023]**

(*a*) 12 (*b*) 13 (*c*) 14 (*d*) 15

13. What is the difference between the weights (in kg) of G and C? **[CDS (II) Exam, 2023]**

(*a*) 4 (*b*) 3 (*c*) 2 (*d*) 1

14. What is the average weight (in kg) of A, B, C, D, E,F and G? **[CDS (II) Exam, 2023]**

(*a*) $\frac{201}{7}$ (*b*) $\frac{197}{7}$ (*c*) 30 (*d*) 32

15. Consider the following statements: **[CDS (II) Exam, 2023]**

1. The age of F cannot be determined due to insufficient data.
2. The average weight of D and F is equal to weight of E.
3. The weight difference is maximum for D and A.

Which of thestatements given above are correct?

(*a*) 1 and 2 only (*b*) 2 and 3 only (*c*) 1 and 3 only (*d*) 1, 2 and 3

Mark option (a) if the question can be answered by using one of the statements alone, but cannot be answered using the other statement alone.

Mark option (b) if the question can be answered by using either statement alone.

Mark option (c) if the question can be answered by using both the statements together, but cannot be answered using either statement alone.

Mark option (d) if the question cannot be answered even by using both the statements together.

16. Question: Is average of the largest and the smallest of 4 given numbers greater than the average of the 4 numbers ? **[CDS (II) Exam, 2023]**

Statement-I: The difference between the largest and the second largest numbers is less than the difference between the second smallest and the smallest of the numbers.

Statement-II: The difference between the largest and the smallest numbers is greater than the difference between the second largest and the second smallest of the numbers.

17. A motorist travels to a place 150 km away at an average speed of 50 km/hour and returns at 30 km/hour. What is the average speed for the whole journey? **[CDS (II) Exam, 2016]**

(*a*) 35 km/hour (*b*) 37 km/hour (*c*) 37.5 km/hour (*d*) 40 km/hour

18. The mean of 5 numbers is 15. If one more number is included, the mean of the 6 numbers becomes 17. What is the included number? **[CDS (I) Exam, 2017]**

(*a*) 24 (*b*) 25 (*c*) 26 (*d*) 27

19. The mean marks obtained by 300 students in a subject are 60. The mean of top 100 students was found to be 80 and the mean of last 100 students was found to be 50. The mean marks of the remaining 100 students are **[CDS (I) Exam, 2017]**

(*a*) 70 (*b*) 65 (*c*) 60 (*d*) 50

20. Let a, b, c, d, e, f, g be consecutive even numbers and j, k, l, m, n be consecutive odd numbers. What is the average of all the numbers? **[CDS (I) Exam, 2017]**

(*a*) $\frac{3(a+n)}{2}$ (*b*) $\frac{(5l+7d)}{4}$ (*c*) $\frac{(a+b+m+n)}{4}$ (*d*) None of the above

21. *A* and *B* are two towns. Mr. Kishore covers the distance from *A* to *B* at an average speed of 57 km/hr. However, he covers the distance from *B* to *A* at an average speed of 38 km/hr. His average speed during the whole journey in km/hr is:

(*a*) 47.5 (*b*) 45.6 (*c*) 46.35 (*d*) 44.8

22. With an average speed of 40 km/hr, a train reaches its destination in time. If it goes with an average speed of 35 km/hr, it is late by 15 minutes. The total journey is:

(*a*) 80 km (*b*) 70 km (*c*) 60 km (*d*) 40 km

23. On a journey across Delhi, a taxi averages 30 km/hr for 60% of the distance, 20 km/hr for 20% of it and 10 km/hr for the remainder. The average speed for the whole journey (in km/hr) is:

(*a*) 20 (*b*) 22.5 (*c*) 25 (*d*) 24.625

24. Average daily wages for 250 workers in a factory is ₹30 and that of 200 workers in another factory is ₹35. The average daily wages (in rupees) of the workers of the two factories combined is:

(*a*) ₹36.05 (*b*) ₹32.22 (*c*) ₹29.85 (*d*) ₹31.72

25. The average weight of *A*, *B*, *C* is 45 kg. If the average weight of *A* and *B* be 40 kg and that of *B* and *C* be 43 kg, then the weight of *B* is:

(*a*) 17 kg (*b*) 20 kg (*c*) 26 kg (*d*) 31 kg

26. The average age of three boys is 16 years. If their ages are in the ratio 4 : 5 : 7, the age of the youngest boy is:

(*a*) 8 years (*b*) 9 years (*c*) 12 years (*d*) 16 years

27. Average monthly income of a family of four earning members was ₹2940. One of the earning members died and therefore the average income came down to ₹2600. The income of the deceased was:

(*a*) ₹3280 (*b*) ₹3960 (*c*) ₹2770 (*d*) ₹5540

28. The average weight of 19 students was 25 kg. By the admission of a new student the average weight is reduced to 24.8 kg. The weight of the new student is:

(*a*) 24.8 kg (*b*) 20.8 kg (*c*) 20.6 kg (*d*) 21 kg

29. The average age of a committee of 8 members is 40 years. A member aged 55 years retired and his place was taken by another member aged 39 years. The average age of the present committee is:

(*a*) 35 years (*b*) 36 years (*c*) 38 years (*d*) 39 years

30. The average age of 11 players of a cricket team is decreased by 2 months when two of them aged 17 years and 20 years are replaced by two players. The average age of the players is:

(*a*) 18 years 3 months (*b*) 17 years 1 month

(*c*) 17 years 7 months (*d*) 17 years 11 months

31. The average of the ages of 15 students in a class is 19 years. When 5 new students are admitted to the class, the average age of the class becomes 18.5 years. What is the average age of the 5 newly admitted students? **[CDS (I) Exam, 2018]**
(*a*) 17 years (*b*) 17.5 years (*c*) 18 years (*d*) 18.5 years

32. Let the average score of a class of boys and girls in an examination be p. The ratio of boys and girls in the class is 3 : 1. If the average score of the boys is $(p + 1)$, then what is the average score of the girls? **[CDS (I) Exam, 2018]**
(*a*) $(p - 1)$ (*b*) $(p - 2)$ (*c*) $(p - 3)$ (*d*) p

33. The average marks of section A are 65 and that of section B are 70. If the average marks of both the sections combined are 67, then the ratio of number of students of section A to that of section B is: **[CDS (II) Exam, 2018]**
(*a*) 3 : 2 (*b*) 1 : 3 (*c*) 3 : 1 (*d*) 2 : 3

34. The average weight of students in a class is 43 kg. Four new students are admitted to the class whose weights are 42 kg, 36.5 kg, 39 kg and 42.5 kg respectively. Now the average weight of the students of the class is 42.5 kg. The number of students in the beginning was:
(*a*) 10 (*b*) 15 (*c*) 20 (*d*) 25
[CDS (I) Exam, 2015]

35. The average score of class X is 83.
The average score of class Y is 76.
The average score of class Z is 85.
The average score of class X and Y is 79.
and average score of class Y and Z is 81.
What is the average score of X, Y and Z?
(*a*) 81.5 (*b*) 80.5 (*c*) 79.0 (*d*) 78.0
[CDS (II) Exam, 2016]

36. A cricketer has a certain average of 10 innings. In the eleventh inning he scored 108 runs, thereby increasing his average by 6 runs. What is his new average?
(*a*) 42 (*b*) 47 (*c*) 48 (*d*) 60
[CDS (II) Exam, 2016]

37. In a class of 100 students, are 70 boys whose average marks in a subject are 75. If the average marks of the complete class is 72, then what is the average marks of the girls?
(*a*) 64 (*b*) 65 (*c*) 68 (*d*) 74
[CDS (II) Exam, 2016]

ANSWERS

1. (*a*)	**2.** (*b*)	**3.** (*c*)	**4.** (*c*)	**5.** (*c*)	**6.** (*c*)	**7.** (*b*)	**8.** (*d*)	**9.** (*c*)
10. (*a*)	**11.** (*d*)	**12.** (*c*)	**13.** (*b*)	**14.** (*a*)	**15.** (*d*)	**16.** (*a*)	**17.** (*c*)	**18.** (*d*)
19. (*d*)	**20.** (*d*)	**21.** (*b*)	**22.** (*b*)	**23.** (*a*)	**24.** (*b*)	**25.** (*d*)	**26.** (*c*)	**27.** (*b*)
28. (*d*)	**29.** (*c*)	**30.** (*c*)	**31.** (*a*)	**32.** (*c*)	**33.** (*a*)	**34.** (*c*)	**35.** (*a*)	**36.** (*c*)
37. (*b*)								

SOLUTIONS

1. Let the number of girls be a
Number of boys in the class be $100 - a$

Now, we have

$24a + (100 - a) \times 32 = 3000$

$32a - 24a = 3200 - 3000$

$\Rightarrow 8a = 200 \Rightarrow a = 25$

2. Let the 50 consecutive numbers be a, $a + 1$, $a + a$, ... $a + 49$ and their average is x

$$\Rightarrow \frac{x + x + 1 + x + 2 + ... + x + 49}{50} = a$$

Now, sum is

$$\frac{50}{2}(0 + 49) = 49 \times 25$$

Now, after including next 4 numbers

$$\frac{a + a + 1 + a + 2 + ... + a + 53}{54} \Rightarrow \frac{54a + 27 \times 53}{54}$$

Now, put the value of x in eqn (i)

$$\Rightarrow \frac{54 \times \left[\frac{50a - 49 \times 25}{50}\right] + 27 \times 53}{54} \Rightarrow x - \frac{49}{2} + \frac{53}{2} = x + 2$$

3. We have,

The two digit numbers which remain the same when the digits interchange their positions are:
11, 22, 33, 44, 55, 66, 77, 88, 99

$$\text{Average} = \frac{11 + 22 + 33 + 44 + 55 + 66 + 77 + 88 + 99}{9} = \frac{\frac{9}{2}[11 + 99]}{9} = \frac{110}{2} = 55$$

4. We have,

$$\text{Required average} = \frac{510 \times 5 + 240 \times 25}{30} = \frac{2550 + 6000}{30} = \frac{8550}{30} = 285$$

5. Average of the ages of a father and a son after 10 years = 25 + 10 = 35

Son's age after 10 years = 25 + 10 = 35

Son's age after 10 years = 17 + 3 = 20 years

Let the father's age after 10 years be 'a'

$$\frac{a + 20}{2} = 35$$

$\Rightarrow a + 20 = 70$

$\Rightarrow a = 50$

6. Let the sum of the given numbers $= 64 + 69 + 72 + 75 + x = 280 + x$

$$\text{The average} = \frac{280 + x}{5}$$

Since the average lies between 62 and 76:

$$62 < \frac{280 + x}{5} < 76 \Rightarrow 310 < 280 + x < 380 \Rightarrow 30 < x < 100$$

The possible integer values of x are from 31 to 99 (both inclusive).

Total number of integers $= 99 - 31 + 1 = 69$

7. Let the distance between his residence and office = d km.

The usual speed is 5 km/hr and usual time is t hr.

Now, we have,

$$d = 3\left(t - \frac{40}{60}\right) \quad \text{... (i)}$$

and
$$d = 2\left(t + \frac{40}{60}\right) \quad \text{... (ii)}$$

From (i) and (ii)

$$3\left(t - \frac{40}{60}\right) = 2\left(t + \frac{40}{60}\right)$$

$$\Rightarrow \quad 3t - 3 \times \frac{2}{3} = 2t + \frac{4}{3} \Rightarrow t = \frac{4}{3} + 2 = \frac{10}{3}\text{ hr}$$

By putting $t = \frac{10}{3}$ hr in (*ii*), we get,

$$d = 2\left(\frac{10}{3} + \frac{2}{3}\right) = 2 \times \frac{12}{3} = 8 \text{ km}$$

8. Total temperature of seven days = 31° × 7° = 217°

The lowest temperature recorded is 30°C

Let all six days temperature be 30°C

Total temperature of 6 days = 6 × 30° = 180°

Remaining day temperature = 217°– 180° = 37°

The maximum temperature is 37°

9. Average of the first 50 even natural numbers:

$(2+4+6+\ldots+100)$

Where, $(a = 2,\ l = 100,\ n = 50)$

$$S_n = \frac{50}{2} \times [2+100]$$

$$S_n = 25 \times 102 = 2550$$

Its average = $\frac{2550}{50} = 51$

Average of the first 50 odd natural numbers: $(1+3+5+\ldots+99)$

Where, (a = 1, l = 99, n = 50)

$$S_n = \frac{50}{2} \times [1+99]$$

$$S_n = 25 \times 100 = 2500$$

Its average = $\frac{2500}{50} = 50$

Difference of both averages $= 51 - 50 = 1$

10. Average = $\frac{30\times 67\times 35\times 70+47+74}{30+35}=67.28$

11. Ages of D = 18 years

12. Required average age = $\frac{a+r+s+t+v}{5}=\frac{12+10+18+16+14}{5}=\frac{70}{5}=14$

13. Required difference = $v_{wt}-r_{wt}=28-25=3$ kg

14. Required average weight = $\frac{15+25+25+40+36+32+28}{7}=\frac{201}{7}$

15. (1) is correct

(2) Average weight of D and F = $\frac{40+32}{2}=36$ = weight of E

(3) Difference of weight of D and A = 40 – 15 = 25 (maximum)

16. Let p, q, r and s are four numbers.

(1) $p>q<r<s$

$p-q>r>-s \Rightarrow p+s>r+q$ ...(i)

(2) $p-s>q-r$

$\Rightarrow p+r>q+s$ ...(ii)

Now, average of largest and smallest numbers

$\frac{p+s}{2}=a$ and average of a 4 numbers = $\frac{p+q+r+s}{4}=b \Rightarrow \frac{p+s}{2}>\frac{p+q+r+s}{4}$

$\Rightarrow 4p+4s>2p+2q+2r+2s$

$p+s>q+r$

$\therefore$ (1) is sufficient

17. The average speed = $\frac{\textit{Total}\text{ distance}}{\textit{Total time}}=\frac{300}{\frac{150}{50}+\frac{150}{30}}$ km/hr $\frac{300}{8}$ km/hr $=\frac{75}{2}$ hr/hr = 37.5 hr/hr

18. Sum of 5 numbers = 15 × 5 = 75

Sum of 6 numbers = 17 × 6 = 102

The included number = 102 – 75 = 27

19. No. of students × Average = Sum of marks

According to the question,

$300\times 60=18000$ ... *(i)*

$100\times 80=8000$ ... *(ii)*

$100\times 50=5000$ ... *(iii)*

$\therefore\ 300-200=18000-[8000+5000]$

$\Rightarrow\ 100=18000-13000 \Rightarrow 100=5000$

$\therefore\ \textit{Mean}=\frac{5000}{100}=50$

20. Let a, b, c, d, e, f, g be 2, 4, 6, 8, 10, 12, 14

And j, k, l, m, n be 1, 3, 5, 7, 9

Average $=\frac{55+26}{12}=\frac{81}{12}=\frac{27}{4}$

Through option, we can check the average

(*a*) $\frac{3(a+n)}{2} = \frac{3(2+9)}{2} = \frac{33}{2}$ (*b*) $\frac{5l+7d}{4} = \frac{5\times5+7\times8}{4} = \frac{25+56}{4} = \frac{81}{4}$

(*c*) $\frac{a+b+m+n}{4} = \frac{2+4+7+9}{4} = \frac{22}{4}$

In the above options, average is not $\frac{27}{4}$. So, the answer is option (*d*).

21. Average speed = $\frac{2uv}{u+v} = \frac{2\times57\times38}{(57+38)}$ km/hr = 45.6 km/hr

22. Let total journey be x km. Then,

$$\frac{x}{35} - \frac{x}{40} = \frac{15}{60} \Leftrightarrow x = 70 \text{ km.}$$

23. Let total journey be x km.

Time taken = $\left(\frac{60x}{100\times30} + \frac{20x}{100\times20} + \frac{20x}{100\times10}\right)$ hrs = $\frac{x}{20}$ hrs.

$\therefore$ Average speed = $\left(x\times\frac{20}{x}\right)$ km/hr = 20 km/hr.

24. Total daily wages of 450 workers = ₹(250 × 30 + 200 × 35) = ₹14500.

$\therefore$ Required average = ₹$\left(\frac{14500}{450}\right)$ ₹32.22.

25. $A + B + C = (45 \times 3) = 135$, $A + B = (40 \times 2) = 80$ & $B + C = (43 \times 2) = 86$.

$\therefore$ $B = (A + B) + (B + C) - (A + B + C) = (80 + 86 - 135) = 31$ kg.

26. Let their ages be $4x$, $5x$ and $7x$.

Their total age = (16 × 3) = 48.

$\therefore$ $4x + 5x + 7x = 48$ or $x = 3$.

So, the age of youngest = $4x$ = 12 years.

27. Income of deceased = (4 × 2940 – 3 × 2600) = ₹3960.

28. Weight of new student = (20 × 24.8 – 19 × 25) kg = 21 kg.

29. Total age of committee now = (8 × 40 – 55 + 39) = 304.

$\therefore$ Average present age = $\left(\frac{304}{8}\right)$ years = 38 years.

30. Decrease in total age = (11 × 2) months = 1 year 10 months

Total age of reserves = [(17 + 20) years] – (1 year 10 months) = 35 years 2 months.

$\therefore$ Average age of reserves = 17 years 7 months.

31. Let the average age of the 5 newly admitted students be a

Now, we have

$15 \times 9 + 5 \times a = 20 \times 18.5 \Rightarrow 285 + 5a = 370 \Rightarrow 5a = 85 \Rightarrow a = 17$ years

32. Let the average number of boys and girls be 3x and x respectively

Let the average score of the girls be q

Now, we have

$3x(p + 1) + qx = (3x + x) \times p = 4xp \Rightarrow 3px + 3x + qx = 4xp \Rightarrow 3(x + q) = xp$

$\Rightarrow 3 + q = p \Rightarrow q = p - 3 \Rightarrow$ Average score of girls = $p - 3$

33. Let the number of students in section A and B x and y respectively.
Now, according to the question,

$$65x + 70y = 67\,(x + y) \Rightarrow 65x + 70y = 67\,x + 67y \Rightarrow 2x = 3y \Rightarrow \frac{x}{y} = \frac{3}{2}$$

34. Let the number of students initially be n.
Total initial weight of students = $43n$.

Given, $\frac{43n + 42 + 36.5 + 39 + 42.5}{n + 4} = 42.5$

Upon solving, $n = 20$.

35. The data is tabulated below

Average	76	79	83
Class	y	xy	x

76	81	85
y	yz	z

Let the number of students in X, Y, Z be x, y, z respectively

$$\frac{x}{y} = \frac{79-76}{83-79} = \frac{3}{4}, \frac{z}{y} = \frac{81-76}{85-81} = \frac{5}{4}$$

$x : y : z = 3 : 4 : 5$

The relevant data is tabulated below.

Average	76	83	85
Class	y	x	z
Number	4	3	5

The deviation about 76 in 3 (83 – 76) + 5 (85 – 76) = 21 + 45 = 66
The average deviation is 66/12 =5.5
The average = 76 + 5.5 = 81.5

36. The data is tabulated below.

Average	x	$x + 6$	108
No. of innings	10	11	1

The negative deviation of the first 10 innings from the averages of 2 + 6 is 60.
The positive desertion is 60, is $(x + 6) + 60 = 108$ or $x = 42$. The new average = $x + 6 = 48$

37. The data is tabulated below

	Girls	Class	Boys
Average		72	75
No	30		70

The positive deviation for the boys is 70(3). The negative deviation for the girls has to be 70 (3). The average deviation.

$= \frac{70(3)}{30} = 7$ the girls average = 72 – 7 = 65

CHAPTER

Ratio and Proportion

Highlights: Ratio, Proportion, Variation.

1. **Ratio:** *The ratio of two quantities in the some units is the fraction that one quantity is of the other.*

   Thus, the ratio a to b is the fraction $\frac{a}{b}$, written as $a : b$.

   In the ratio $a : b$, we call a the *first term* or *antecedent* and b is called the *second term* or *consequent*.

   Rule: *The value of a ratio remains unchanged, if each one of its terms is multiplied or divided by a same non-zero number.*

2. (*i*) If $a > b$, then $a : b$ is called a ratio of greater inequality.

   (*ii*) If $a < b$, then $a : b$ is called a ratio of less inequality.

   Rule 1: *If $a > b$ and same positive number is added to each term of $a : b$, then the ratio is diminished.*

   Rule 2: *If $a < b$ and some positive number is added to each term of $a : b$, then the ratio is increased.*

3. (*i*) $a^2 : b^2$ is called *duplicate ratio of $a : b$.* (*ii*) $a^3 : b^3$ is called *triplicate ratio of $a : b$.*

   (*iii*) $\sqrt{a} : \sqrt{b}$ is called *sub-duplicate ratio of $a : b$.*

   (*iv*) $a^{1/3} : b^{1/3}$ is called *sub-triplicate ratio of $a : b$.*

   (*v*) If $a : b$ and $c : d$ are two ratios, then $ac : bd$ *is called the ratio compounded of the given ratio.*

4. **Proportion:** *The equality of two ratios is called proportion.*

   If $a : b = c : d$, we say that a, b, c, d are proportional and, we write, $a : b :: c : d$.

   Here a and d are known as extremes and b, c are known as means.

   We always have:

   Product of Means = Product of Extremes

5. (*i*) In $a : b :: c : d$, we say that d is the fourth proportional to a, b & c.

   (*ii*) If x is the third proportional to a, b then $a : b :: b : x$.

   (*iii*) Mean proportional between a and b is $\sqrt{ab}$.

   (*iv*) If $\frac{a}{b} = \frac{c}{d}$ then $\frac{a+b}{a-b} = \frac{c+d}{c-d}$ & $\frac{a-b}{a+b} = \frac{c-d}{c+d}$.

6. **Variation:** We say that x is directly proportional to y, if $x = ky$ for some constant k and we write, $x \propto y$.

   Also, we say that x is inversely proportional to y, if $x = \frac{k}{y}$ for some constant k and we write,

$$x \propto \frac{1}{y}$$

SOLVED PROBLEMS

Ex. 1. ***Find: (i) Fourth proportional to 3, 7, 15. (ii) Third proportional to 16 and 36. (iii) Mean proportion between 8 and 32.***

Sol. (*i*) Let it be x. Then, $3 : 7 :: 15 : x$.

$\therefore\ 3x = 7 \times 15$ or $x = \left(\frac{7\times15}{3}\right) = 35.$

$\therefore$ Fourth proportional to 3, 7, 15 is 35.

(*ii*) Let it be x. Then, $16 : 36 :: 36 : x$

$\therefore\ 16x = 36 \times 36$ or $x = \frac{36\times36}{16} = 81.$

(*iii*) Mean proportion $= \sqrt{8\times32} = \sqrt{256} = 16.$

Ex. 2. ***Find: (i) The duplicate ratio of 3 : 5. (ii) The triplicate ratio of 2 : 3. (iii) The sub-duplicate ratio of 16 : 25. (iv) The sub-triplicate ratio of 27 : 8 (v) The compounded ratio of (2 : 5) and (3 : 4).***

Sol. (*i*) Duplicate ratio of 3 : 5 is $3^2 : 5^2$ *i.e.* 9 : 25.

(*ii*) Triplicate ratio of 2 : 3 is $2^3 = 3^3$ *i.e.* 8 : 27.

(*iii*) Sub-duplicate ratio of 16 : 25 is $\sqrt{16} : \sqrt{25}$ *i.e.* 4 : 5.

(*iv*) Sub-triplicate ratio of 27 : 8 is $(27)^{1/3} : 8^{1/3} = 3 : 2$.

(*v*) Compounded ratio of (2 : 5) & (3 : 4) is $(2 \times 3 : 5 \times 4) = 6 : 20 = 3 : 10$.

Ex. 3. ***Divide ₹1024 in the ratio 9 : 7.***

Sol. Sum of terms = (9 + 7) = 16.

First part = ₹ $\left(1024\times\frac{9}{16}\right)$ = ₹576.

Second part = ₹ $\left(1024\times\frac{7}{16}\right)$ = ₹448.

Ex. 4. ***Divide 581 among A, B, C so that 4A = 5B = 7C.***

Sol. Let $4A = 5B = 7C = k$. Then, $A = \frac{k}{4}$, $B = \frac{k}{5}$ and $C = \frac{k}{7}$..

$\therefore\ A : B : C = \frac{k}{4} : \frac{k}{5} : \frac{k}{7} = 35 : 28 : 20$

$\therefore$ *A's* share $= \left(581\times\frac{35}{83}\right) = 245;$ *B's* share $= \left(581\times\frac{28}{83}\right) = 196;$

C's share $= \left(581\times\frac{20}{83}\right) = 140.$

Ex. 5. ***A bag contains rupee, 50-paise and 25-paise coins in the ratio 5 : 6 : 7. If the total amount is ₹390, find the number of coins of each kind.***

Sol. Ratio of values of the coins $= \frac{5}{1} : \frac{6}{2} : \frac{7}{4} = 20 : 12 : 7.$

Value of 1 rupee coins = ₹ $\left(390\times\frac{20}{39}\right)$ = ₹200.

Value of 50-paise coins = ₹ $\left(390\times\frac{12}{39}\right)$ = ₹120.

Value of 25-paise coins = ₹$\left(390 \times \frac{7}{39}\right)$ = ₹70.

∴ Number of 1-rupee coins = 200;

Number of 50-paise coins = (120 × 2) = 240;

Number of 25-paise coins = (70 × 4) = 280.

Ex. 6. ***In a mixture of 28 litres, the ratio of milk and water is 5 : 2. If 2 litres of water is added to the mixture, find the ratio of milk and water in the new mirture.***

Sol. Milk in given mix. = $\left(28 \times \frac{5}{7}\right)$ = 20 litres, water in this mix. = 8 litres. After adding 2 litres of water to this mixture, it contains 20 litres of milk and 10 litres of water.

So, ratio of milk and water in new mix. = 20 : 10 = 2 : 1.

Ex. 7. ***A and B are two alloys of gold and copper prepared by mixing metals in the ratio 7 : 2 and 7 : 11 respectively. If equal quantities of alloys are melted to form a third alloy C, find the ratio of gold and copper in C.***

Sol. Let 1 kg of each one of A and B be taken to form 2 kg of C.

Gold in 1 kg of $A = \frac{7}{9}$, copper in 1 kg of $A = \frac{2}{9}$;

Gold in 1 kg of $B = \frac{7}{18}$; copper in 1 kg of $B = \frac{11}{18}$.

Gold in 2 kg of $C = \left(\frac{7}{9} + \frac{7}{18}\right) = \frac{21}{18} = \frac{7}{6}$.

Copper in 2 kg of $C = \left(\frac{2}{9} + \frac{11}{18}\right) = \frac{15}{18} = \frac{5}{6}$.

Ratio of gold and copper in $C = \frac{7}{6} : \frac{5}{6} = 7 : 5$.

Ex. 8. ***Find three numbers in the ratio 2 : 3 : 5, the sum of whose squares is 608.***

Sol. Let the numbers be $2x$, $3x$ and $5x$. Then,

$4x^2 + 9x^2 + 25x^2 = 608$ or $38x^2 = 608$ or $x^2 = 16$ or $x = 4$.

∴ The numbers are 8, 12 and 20.

Ex. 9. ***A mixture contains alcohol and water in the ratio 4 : 3. If 7 litres of water is added to it, the ratio of alcohol and water becomes 3 : 4. Find the quantity of alcohol in the mixture.***

Sol. Let alcohol = $4x$ and water = $3x$.

Then $\frac{4x}{3x+7} = \frac{3}{4} \Leftrightarrow 16x = 3(3x+7) \Leftrightarrow x = 3$.

∴ Alcohol = $4x$ = 12 litres.

EXERCISE 7

1. If $\frac{a}{b} = \frac{1}{3}$, $\frac{b}{c} = 2$, $\frac{c}{d} = \frac{1}{3}$, $\frac{d}{e} = 3$ and $\frac{e}{f} = \frac{1}{4}$, then what is the value of $\frac{abc}{def}$?

[CDS (I) Exam, 2020]

(*a*) $\frac{1}{4}$ (*b*) $\frac{3}{4}$ (*c*) $\frac{3}{8}$ (*d*) $\frac{27}{4}$

2. If $m : n = 1 : 2$ and $p : q = 3 : 4$, then what is $(2m + 4p) : (n + 3)$ equal to ?

[CDS (I) Exam, 2024]

(*a*) 1 : 1 (*b*) 1 : 3 (*c*) 2 : 1 (*d*) 2 : 3

3. If a : b = c : d = 1 : 6, then what is the value of $\frac{a^2+c^2}{b^2+d^2}$? **[CDS (I) Exam, 2019]**

(*a*) $\frac{1}{600}$ (*b*) $\frac{1}{60}$ (*c*) $\frac{1}{36}$ (*d*) $\frac{1}{6}$

4. If A : B = 1 : 2, B : C = 3 : 4, C : D = 2 : 3 and D : E = 3 : 4, then what is B : E equal to ?

(*a*) 3 : 2 (*b*) 1 : 8 (*c*) 3 : 8 (*d*) 4 : 1

[CDS (I) Exam, 2018]

5. If $\sqrt{2} : (1+\sqrt{3}) :: \sqrt{6} : x$, then x is equal to:

(*a*) $\sqrt{3}-3$ (*b*) $1+\sqrt{3}$ (*c*) $1-\sqrt{3}$ (*d*) $\sqrt{3}+3$

6. The fourth proportional to 3, 5 and 21 is:

(*a*) 35 (*b*) $\frac{7}{5}$ (*c*) $\frac{5}{7}$ (*d*) 12.6

7. Third proportional to 9 and 12 is:

(*a*) $6\sqrt{3}$ (*b*) 10.5 (*c*) 16 (*d*) None

8. What least number must be added to each one of 6, 14, 18, 38 to make them in proportion?

(*a*) 1 (*b*) 2 (*c*) 3 (*d*) 4

9. If *a, b, c, d, e* and *f* satisfy $2a = 3b = 6c = 9d = 12e = 18f$, then what is the value of $(a + b)/(c + d + e + f)$? **[CDS (I) Exam, 2023]**

(*a*) 4/7 (*b*) 2 (*c*) 5/2 (*d*) 9/2

10. If $\frac{2a}{3} = \frac{4b}{5} = \frac{3c}{4}$, then what is the value of $\frac{18}{a}\sqrt{a^2+c^2-b^2}$? **[CDS (II) Exam, 2023]**

(*a*) $3\sqrt{5}$ (*b*) $\sqrt{355}$ (*c*) $\sqrt{375}$ (*d*) $3\sqrt{15}$

11. If (3a + 6b + c + 2d) × (3a – 6b – c + 2d) = (3a – 6b + c – 2d) × (3a + 6b – c – 2d), then which one of the following is correct ? **[CDS (II) Exam, 2023]**

(*a*) $ab = cd$ (*b*) $ac = bd$ (*c*) $ad = bc$ (*d*) $ad + bc = 0$

12. The ratio of sum of two numbers to their difference is 5 : 1. What is the ratio of the sum of their squares to the difference of their squares? **[CDS (I) Exam, 2025]**

(*a*) 13 : 5 (*b*) 25 : 1 (*c*) 9 : 4 (*d*) 16 : 1

13. In a class room the ratio of number of girls to that of boys is 3 : 4. The average height of students in the class is 4.6 feet. If the average height of the boys in the class in 4.8 feet, then what is the average height of the girls in the class? **[CDS (I) Exam, 2020]**

(*a*) Less than 4.2 feet (*b*) More than 4.2 feet but less han 4.3 feet

(*c*) More than 4.3 feet but less than 4.4 feet (*d*) More than 4.4 feet but less than 4.5 feet

14. A mixture of 100 L contains kerosene and turpentine oil in the ratio 3 : 2. What is the minimum quantity of kerosene in litres (whole number) that should be mixed in the mixture so that the resulting mixture has 20% of kerosene? **[CDS (I) Exam, 2025]**

(*a*) 10 L (*b*) 20 L (*c*) 25 L (*d*) Not possible

15. When a ball is allowed to fall, the time it takes to fall any distance varies as the square root of the distance and it takes 4 seconds to fall 78.40 m. How long would it takes to fall 122.50 m? **[CDS (II) Exam, 2020]**

(*a*) 5 seconds (*b*) 5.5 seconds (*c*) 6 seconds (*d*) 6.5 seconds

16. The mean weight of 150 students in a class is 60 kg. The mean weight of boys in the class is 70 kg and that of girls is 55 kg. What is the ratio of number of boys to number of girls? **[CDS (II) Exam, 2024]**

(*a*) 1 : 2 (*b*) 1 : 1 (*c*) 2 : 1 (*d*) 2 : 3

17. The monthly incomes of *A* and *B* are in the ratio 4 : 3. Each saves ₹600. If their expenditures are in the ratio 3 : 2, then what is the monthly income of *A*? **[CDS (II) Exam, 2019]**

(*a*) ₹1800 (*b*) ₹2000 (*c*) ₹2400 (*d*) ₹3600

18. In the following table of inverse variation, what are the values of A, B and C respectively? **[CDS (I) Exam, 2018]**

M	15	–6	2	c
N	–4	A	B	60

(*a*) 10, – 30, –1 (*b*) 10, –1, 30 (*c*) –30, 10, –1 (*d*) –1, –30, 10

19. 16 litres of a mixture contains milk and water in the ratio 5: 3. If 4 litres of milk is added to this mixture, the ratio of milk to water in the new mixture would be **[CDS (I) Exam, 2015]**

(*a*) 2: 1 (*b*) 7: 3 (*c*) 4: 3 (*d*) 8: 3

20. The efficiency of P is twice that of Q, whereas the efficiency of P and Q together is three times that of R. If P, Q, and R work together on a job, in what ratio should they share their earnings? **[CDS (I) Exam, 2015]**

(*a*) 2: 1: 1 (*b*) 4: 2: 1 (*c*) 4: 3: 2 (*d*) 4: 2: 3

21. If $a : b = 3 : 5$ and $b : c = 7 : 8$, then $2a : 3b : 7c$ is equal to **[CDS (II) Exam, 2015]**

(*a*) 42 : 105 : 320 (*b*) 15 : 21 : 35 (*c*) 6 : 15 : 40 (*d*) 30 : 21 : 350

22. The speeds of three buses are in the ratio 2 : 3 : 4. The time taken by these buses to travel the same distance will be in the ratio **[CDS (II) Exam, 2015]**

(*a*) 2 : 3 : 4 (*b*) 4 : 3 : 2 (*c*) 4 : 3 : 6 (*d*) 6 : 4 : 3

23. In a mixture of milk and water of volume 30 litre, the ratio of milk and water is 7 : 3. The quantity of water to be added to the mixture to make the ratio of milk and water 1 : 2 is **[CDS (II) Exam, 2015]**

(*a*) 30 (*b*) 32 (*c*) 33 (*d*) 35

24. There are two containers A and B. In container A, the ratio of milk and water is 1 : 3 and in container B, the ratio of milk and water is (*m* : *n*). If the mixture in the containers A and B are mixed in the ratio 2 : 3 to get 20 litres of a mixture having milk and water in the ratio 3 : 7, then what is the value of $\frac{m}{n}$? **[CDS (II) Exam, 2024]**

(*a*) $\frac{1}{2}$ (*b*) $\frac{2}{3}$ (*c*) $\frac{3}{4}$ (*d*) $\frac{4}{5}$

25. In an office, one-third of the workers are women, half of the women are married and one-third of the married women have children. If three-fourth of the men are married and one-third of the married men have children, then what is the ratio of married women to married men? **[CDS (II) Exam, 2016]**

(*a*) 1 : 2 (*b*) 2 : 1 (*c*) 3 : 1 (*d*) 1 : 3

26. There are twelve friends A, B, C, D, E, F, G, H, I, J, K and L who invested money in some business in the ratio of 1 : 2 : 3 : 4 : 5 : 6: 7 : 8 : 9 : 10 : 11 : 12 and the duration for which they invested the money is in the ratio of 12 : 11 : 10 : 9 : 8 : 7 : 6 : 5 : 4 : 3 : 2 : 1 respectively. Who will get the maximum profit at the end of the year? **[CDS (II) Exam, 2016]**

(*a*) F only (*b*) G only (*c*) Both F and G (*d*) Neither F nor G

27. Incomes of Mahesh and Kamal are in the ratio 1 : 2 and their expenses are in the ratio 1 : 3. Which one of the following statements is correct? **[CDS (II) Exam, 2016]**

(*a*) Mahesh saves more than what Kamal saves

(*b*) Savings of both of them are equal

(*c*) Kamal saves more than what Mahesh saves

(*d*) It is not possible to determine who saves more

28. The speeds of three cars are in the ratio 2 : 3 : 4. What is the ratio between the times taken by these cars to travel the same distance? **[CDS (II) Exam, 2016]**

(*a*) 4 : 3 : 2 (*b*) 2 : 3 : 4 (*c*) 4 : 3 : 6 (*d*) 6 : 4 : 3

29. If A : B = 3 : 4, then what is the value of the expression $\left(\frac{3A^2+4B}{3A-4B^2}\right)$?

(*a*) $\frac{43}{55}$ (*b*) $-\frac{43}{55}$

(*c*) $\frac{47}{55}$ (*d*) Cannot be determined **[CDS (I) Exam, 2017]**

30. The cost of diamond varies directly as the square of its weight. A diamond broke into four pieces with their weights in the ratio of 1 : 2 : 3 : 4. If the loss in total value of the diamond was ₹70,000, what was the price of the original diamond? **[CDS (I) Exam, 2017]**

(*a*) ₹1,00,000 (*b*) ₹1,40,000 (*c*) ₹1,50,000 (*d*) ₹1,75,000

31. A bottle contains spirit and water in the ratio 1 : 4 and another identical bottle contains spirit and water in the ratio 4 : 1. In what ratio should the mixtures in the two bottles be mixed to get a new mixture in which the ratio of spirit to water is 1 : 3? **[CDS (I) Exam, 2024]**

(*a*) 5 : 1 (*b*) 6 : 1 (*c*) 10 : 1 (*d*) 11 : 1

32. 25 kg of alloy X is mixed with 125 kg of alloy Y. If the amount of lead and tin in the alloy X is in the ratio 1 : 2 and the amount of lead and tin in the alloy Y is in the ratio 2 : 3, then what is the ratio of lead to tin in the mixture? **[CDS (I) Exam, 2017]**

(*a*) 1 : 2 (*b*) 2 : 3 (*c*) 3 : 5 (*d*) 7 : 11

33. What least number must be subtracted from each of the numbers 14, 17, 34, 42 so that the remainders may be proportional?

(4) 0 (*b*) 1 (*c*) 2 (*d*) 7

34. A fraction bears the same ratio to $\frac{1}{27}$ as $\frac{3}{7}$ does to $\frac{5}{9}$. The fraction is :

(*a*) $\frac{7}{45}$ (*b*) $\frac{1}{35}$ (*c*) $\frac{45}{7}$ (*d*) $\frac{5}{21}$

35. ₹1980 are divided among *A*, *B*, *C* so that half of *A's* part, one-third of *B's* part and one-sixth of *C's* part are equal. Then, *B's* part is:

(*a*) ₹660 (*b*) ₹360 (*c*) ₹1080 (*d*) ₹540

36. ₹1360 have been divided among *A*, *B*, *C* such that *A* gets $\frac{2}{3}$ of what *B* gets and *B* gets $\frac{1}{4}$ of what *C* gets. Then, *B's* share is:

(*a*) ₹120 (*b*) ₹160 (*c*) ₹240 (*d*) ₹320

37. ₹4850 have been divided among *A*, *B*, *C* such that if their shares be diminished by ₹15, ₹10 and ₹25 respectively, the remainders are in the ratio 3 : 4 : 5. Then, *B's* share is?

(*a*) ₹1595 (*b*) ₹1610 (*c*) ₹1626.66 (*d*) ₹1600

38. A sum of ₹53 is divided among *A*, *B*, *C* in such a way that *A* gets ₹7 more than what *B* gets and *B* gets ₹8 more than what *C* gets. The ratio of their shares is:

(*a*) 16 : 9 : 18 (*b*) 25 : 18 : 10 (*c*) 18 : 25 : 10 (*d*) 15 : 8 : 30

39. A bag contains ₹600 in the form of one-rupee, 50-paise and 25-paise coins in the ratio 3 : 4 : 12. The number of 25-paise coins is:

(*a*) 600 (*b*) 900 (*c*) 1200 (*d*) 1376

40. A sum of money is divided among *A*, *B*, *C* such that to each rupee *A* gets, *B* gets 65 paise and *C* gets 35 paise. If *C's* share is ₹560, the sum is:

(*a*) ₹2400 (*b*) ₹2800 (*c*) ₹3200 (*d*) ₹3600

41. ₹5625 are divided among *A*, *B* , *C* so that *A* may receive one-half as much as *B* and *C* together receive and *B* receives one-fourth of what *A* and *C* together receive. The share of *A* is more than that of B by?

(*a*) ₹750 (*b*) ₹775 (*c*) ₹1500 (*d*) ₹1600

42. A certain amount was divided between *X* and *Y* in the ratio 4 : 3. If *B's* share was ₹4800, the total amount was:

(*a*) ₹11200 (*b*) ₹6400 (*c*) ₹19200 (*d*) ₹39200

43. The ratio of number of boys and girls in a school of 720 students is 7 : 5. How many more girls should be admitted to make the ratio 1 : 1?

(*a*) 90 (*b*) 120 (*c*) 220 (*d*) 240

44. A boy 1.4 m tall casts a shadow 1.2 m long at the time when a building casts a shadow 5.4 m long. The height of the building is:

(*a*) 4.63 m (*b*) 3.21 m (*c*) 6.3 m (*d*) 5.6 in

45. The incomes of *A* and *B* are in the ratio 3 : 2 and their expenditures in the ratio 5 : 3. If each saves ₹1500 then *B's* income is:

(*a*) ₹6000 (*b*) ₹4500 (*c*) ₹3000 (*d*) ₹7500

46. Consider the following statements: **[CDS (I) Exam, 2021]**

1. If *x* is directly proportional to *z* and *y* is directly proportional to *z*, then $(x^2 - y^2)$ is directly proportional to z^2.
2. If x is inversely proportional to *z* and *y* is inversely proportional to *z*, then (*xy*) is inversely proportional to z^2.

Which of the above statements is/are correct?

(*a*) 1 only (*b*) 2 only (*c*) Both 1 and 2 (*d*) Neither 1 nor 2

47. The incomes of A, B and C are the ratio 7 : 9 : 12 and their expenditures are in the ratio 8 : 9 : 15. If A's saving one-fourth of his income, then the ratio of savings of A, B and C is **[CDS (I) Exam, 2021]**

(*a*) 56 : 99 : 69 (*b*) 99 : 56 : 69 (*c*) 69 : 56 : 99 (*d*) 99 : 69 : 56

48. A question is given followed by two statements I and II. Consider the Question and the Statements and mark the correct option.

Question: The ratio of P's salary to Q's salary is 6 : 5. How much is P's expenditure?

Statement-I: The ratio of P's saving to Q's saving is 3 : 2.

Statement-II: The ratio of P's expenditure to Q's expenditure is 1 : 1.

Which one of the following is correct in respect of the above Question and the Statements? **[CDS (II) Exam, 2024]**

(*a*) The Question can be answered by using one of the Statements alone, but cannot be answered using the other Statement alone

(*b*) The Question can be answered by using either Statement alone

(*c*) The Question can be answered by using both the Statements together, but cannot be answered using either Statement alone

(*d*) The Question cannot be answered even by using both the Statements together

49. The population of two villages is 1525 and 2600 respectively. If the ratio of male to female population in the first village is 27 : 34 and the ratio of the male to female population in the second village is 6 : 7, then what is the ratio of male to female population of these two villages taken together? **[CDS (I) Exam, 2020]**

(*a*) $\frac{33}{41}$ (*b*) $\frac{85}{82}$ (*c*) $\frac{71}{90}$ (*d*) $\frac{5}{6}$

Each item contains aQuestion followed by twoStatements. Answer each item using the following instructions :

(a) Choose this option if the Question can be answered by one of the Statements alone but not by the other.

(b) Choose this option if the Question can be answered by either Statement alone.

(c) Choose this option if the Question can be answered by using both the Statements together, but cannot be answered by using either Statement alone

(d) Choose this option if theQuestion cannot be answered even by using both Statements together

50. Question : What are the unique values of a, b and c if 2 is a root of the equation $ax^2 + bx + c = 0$? **[CDS (I) Exam, 2023]**

Statement I : Ratio of c to a is 1.

Statement II : Ratio of b to a is $(-\frac{5}{2})$.

51. Question: What is the ratio $x : y : z$ equal to if $x, y, z \neq 0$?

Statement-I: $\frac{x+z}{y} = \frac{z}{x}$

Statement-II: $\frac{x-y}{x} = \frac{x}{z}$

52. Consider the following statements : **[CDS (II) Exam, 2023]**

1. If $(a + b)$ is directly proportional to $(a - b)$, then $(a^2 + b^2)$ is directly proportional to ab.
2. If a is directly proportional to b, then $(a^2 - b^2)$ is directly proportional to ab.

Which of the statements given above is/are correct?

(*a*) 1 only (*b*) 2 only (*c*) Both 1 and 2 (*d*) Neither 1 nor 2

ANSWERS

1. (*c*)	**2.** (*a*)	**3.** (*d*)	**4.** (*c*)	**5.** (*d*)	**6.** (*a*)	**7.** (*c*)	**8.** (*b*)	**9.** (*b*)
10. (*b*)	**11.** (*c*)	**12.** (*a*)	**13.** (*c*)	**14.** (*d*)	**15.** (*a*)	**16.** (*b*)	**17.** (*c*)	**18.** (*a*)
19. (*b*)	**20.** (*a*)	**21.** (*c*)	**22.** (*d*)	**23.** (*c*)	**24.** (*a*)	**25.** (*d*)	**26.** (*c*)	**27.** (*d*)
28. (*d*)	**29.** (*d*)	**30.** (*a*)	**31.** (*b*)	**32.** (*d*)	**33.** (*c*)	**34.** (*b*)	**35.** (*d*)	**36.** (*c*)
37. (*b*)	**38.** (*b*)	**39.** (*b*)	**40.** (*c*)	**41.** (*a*)	**42.** (*a*)	**43.** (*b*)	**44.** (*c*)	**45.** (*a*)
46. (*c*)	**47.** (*a*)	**48.** (*d*)	**49.** (*d*)	**50.** (*d*)	**51.** (*c*)	**52.** (*a*)		

SOLUTIONS

1. We have,

$$\frac{a}{b}=\frac{1}{3} \Rightarrow a=\frac{b}{3}\ \ \frac{b}{c}=2 \Rightarrow c=\frac{b}{2}\ \ \frac{c}{d}=\frac{1}{2} \Rightarrow d=2c \Rightarrow d=2\times\frac{b}{2}=b$$

$$\frac{d}{e}=3 \Rightarrow e=\frac{d}{3} \Rightarrow e=\frac{b}{3}$$

$$\frac{e}{f}=\frac{1}{4} \Rightarrow f=4e \Rightarrow f=\frac{4b}{3}$$

Now,

$$\frac{abc}{def}=\frac{\frac{b}{3}\times b\times\frac{b}{2}}{b\times\frac{b}{3}\times\frac{4b}{3}}=\frac{\frac{1}{6}}{\frac{4}{9}}=\frac{1\times 9}{6\times 4}=\frac{3}{8}$$

2. Let $m = a$, $n = 2$ and $p = 3b$, $q = 4b$

$$\frac{2m+4p}{n+3q}=\frac{2\times a+4\times 3b}{2a+3\times 4b}=1:1$$

3. $\dfrac{a}{b}=\dfrac{c}{d}=\dfrac{1}{6} \Rightarrow a=c=x$ and $b=d=6x$

Now, $\dfrac{a^2+c^2}{b^2+d^2}=\dfrac{x^2+x^2}{(6x)^2+(6x)^2}=\dfrac{2x^2}{36x^2+36x^2}=\dfrac{2x^2}{72x^2}=\dfrac{1}{36}$

4. A : B = 1 : 2, and B : C = 3 : 4

C : D = 2 : 3 and D : E = 3 : 4 $\Rightarrow$ A : B : C = 3 : 6 : 8

A : B : C : D = 6 : 12 : 16 : 24

A : B : C : D : E = 18 : 36 : 48 : 72 : 96

Now, B : C = 36 : 96 = 3 : 8

5. Clearly, $\dfrac{\sqrt{2}}{1+\sqrt{3}}=\dfrac{\sqrt{6}}{x} \Rightarrow x=\dfrac{\sqrt{6}(1+\sqrt{3})}{\sqrt{2}}=\sqrt{3}(1+\sqrt{3})=(\sqrt{3}+3).$

6. Let it be x. Then, 3 : 5 : : 21 : x

$\therefore\ 3\times x=5\times 21$ or $x=\left(\dfrac{5\times 21}{3}\right)=35.$

7. Let it be x. Then, it is 4th proportion between 9, 12, 12. So, 9 : 12 :: 12 : x or $9x = 12\times 12$ or $x = 16$.

8. Let x must be added. Then,

$$\frac{6+x}{14+x}=\frac{18+x}{38+x} \Leftrightarrow (6+x)(38+x)=(18+x)(14+x)$$

$\therefore\ x^2+44x+228=x^2+32x+252 \Leftrightarrow 12x=24 \Leftrightarrow x=2.$

9. $2a=3b=6c=9d=12e=18f=p$

$$\Rightarrow a=\frac{p}{2},b=\frac{p}{3},c=\frac{p}{6},d=\frac{p}{9},e=\frac{p}{12},f=\frac{p}{18}$$

LCM of (2, 3, 6, 9, and 18) = 36

$\Rightarrow a=18p,\ b=12p,\ c=6p,\ d=4p,\ e=3p, f=2p$

Now, $$\frac{a+b}{c+d+e+f}=\frac{30p}{15p}=2$$

10. We have,

Let $62<\frac{280+x}{5}<76 \quad \Rightarrow a=\ 310<280+x<380$

Now, $30<x<100==99-31+1=69$

$$=(2+4+6+\ldots+100)=\frac{x}{(1/27)}=\frac{(3/7)}{(5/9)} \Leftrightarrow 27x=\frac{3}{7}\times\frac{9}{5} \Leftrightarrow x=\frac{1}{35}..$$

11. $(3a+6b+c+2d)\times(3a-6b-c+2d)=(3a-6b+c-2d)\times(3a+6b-c-2d)$

$\Rightarrow \{(3a+2d)+(6b+c)\}\{(2a+3d)-(6b+c)\}$

$=\{(3a-2d)+(c+6b)\}\times\{(3a-2d)-(c-6b)\}$

Using $(a+b)(a-b)=(a^2-b^2)$

$\Rightarrow (3a+2d)^2-(6b+c)^2=(3a-2d)^2-(c-6b)^2$

$\Rightarrow 9a^2+4d^2+12\,ad-36\,b^2-c^2-12\,bc=9a^2+4d^2-12\,ad-c^2+36\,b^2+12\,bc$

$\Rightarrow 24bc=24\,ad \Rightarrow bc=ad$

12. Let the both number be x and y.

According to question,

$$\frac{(x+y)}{(x-y)}=\frac{5}{1} \Rightarrow x+y=5x-5y \Rightarrow 4x=6y \Rightarrow \frac{x}{y}=\frac{6}{4}=\frac{3}{2}=\frac{3k}{2k} \text{ (Let)}$$

Sum of square of both number $=(3k)^2+(2k)^2=13k^2$

Difference of square of both number $=(3k)^2-(2k)^2=5k^2$

Required ratio $=13k^2:5k^2=13:5$

13. Let the no. of girls and boys in the classroom be $3x$ and $4x$ respectively.

Let the average height of girls be A feet.

$3x+\text{A}+4x+4.8=4.6\,(3x+4x)$

$$3\text{A}+19.2=4.6\times7 \Rightarrow 3\text{A}=32.2-19.2 \Rightarrow 3\text{A}=13 \Rightarrow \text{A}=\frac{13}{3}=4.33$$

14. Total mixture = 100 Lit

Quantity of kerosene $=\frac{100}{(3+2)}\times3=\frac{100}{5}\times3=60$ Lit

Quantity of turpentine oil = (100 – 60) = 40 Lit

There is 60 litre kerosene in mixture.

So when we add more kerosene the percentage of kerosene in mixture increase.

Thus, it is not possible to get 20% of kerosene in resulting mixture.

15. We have,

$t_1 = 4$ seconds, $d_1 = 78.40$ m, $d_2 = 122.5$ m

Now,

$$\frac{4^2}{t_2^2} = \frac{78.4}{122.5} \quad \left[\because \frac{t_1^2}{t_2^2} = \frac{d_1^2}{d_2}, \text{ where, } t_1 \text{ and } t_2 \text{ is the time taken and } d_1 \text{ and } d_2 \text{ is distance}\right]$$

$$\Rightarrow t_2^2 = \frac{1225 \times 16}{784} \Rightarrow t_2 = \frac{35 \times 4}{28} = 5 \text{ seconds}$$

16. Let the number of boys $= x$

Then, number of girls $= 150 - x$

According to question,

$$x + 70 + 50\,(150 - x) = 150 \times 60$$

$$\Rightarrow \quad 70x + 7500 - 50x = 9000 \Rightarrow 20x = 1500$$

$$x = \frac{1500}{20} = 75$$

So, number of girls = Number of boys = 75

Required ratio = 75 : 75 = 1 : 1

17. Let the monthly incomes of *A* and *B* are 4*x* and 3*x* respectively and their expenditures are 3*y* and 2*y* respectively,

Now, we have,

$$4x - 3y = 600 \quad \text{... }(i)$$

$$3x - 2y = 600 \quad \text{... }(ii)$$

From (i) and (ii)

We have, $4x - 3y = 3x - 2y \Rightarrow x = y$

Put $x = y$ in eqn (i), we get,

$$4x - 3y = 600$$

$$4x - 3x = 600 \Rightarrow x = 600$$

∴ Income of A = $4x = 4 \times 600 =$ ₹2400

18. We have,

$$M \propto \frac{1}{N} \Rightarrow M = \frac{K}{N} \text{ (where } k \text{ is constant)} \Rightarrow MN = k \Rightarrow k = 15 \times -4 = -60$$

$$\Rightarrow \quad -6 \times A = -60 \Rightarrow A = 10 \text{ and } 2 \times B = -60 \Rightarrow B = -30$$

and $\quad C \times 60 = -60 \Rightarrow C = -1$

19. 16 litres of the mixture contains milk and water in the ratio 5 : 3.

∴ It contains 10 litres of milk and 6 litres of water.

4 litres of milk is added to the mixture.

∴ The resulting mixture would contain 14 litres of milk and 6 litres of water.

Ratio of milk to water in the new mixture ₹ = 7 : 3.

20. The word 'efficiency' has been used to refer to the rate of doing work. Let the rates of work of P, Q, R be *p*, *q*, *r* respectively.

$p = 2q$ and $p + q = 3r$

$2q + q = 3r$ *i.e.,* $q = r$

Ratio of the rates of P, Q, and R = 2 : 1 : 1.

If P, Q, and R work together on the job, each person works for the same time.

∴ The earnings of P, Q and R, will then be in the same ratio as the ratio of the rates of P, Q and R.

∴ Ratio of the earnings of P, Q and R = 2 : 1 : 1.

21. a : b = (3: 5) × 7 ⇒ a : b = 21 : 35

b : c = (7 : 8) × 5 ⇒ b : c = 35 : 40

a : b : c = 21 : 35 : 40

2a : 3b : 7c = 42 : 105 : 280 = 6 : 15 : 40.

22. The speeds of the buses are in the ratio 2 : 3 : 4. Therefore, the time taken by the 3 buses to cover a certain distance would be in the ratio 6 : 4 : 3. (The product of corresponding terms is 12 in each case.)

23. 30 litres of mixture contains milk and water in ratio 7 : 3

∴ It contains 21 litres of milk and 9 litres of water.

According to question, *n* litres of water is to be added,

$$\frac{21}{9+x} = \frac{1}{2} \quad \Rightarrow \quad 42 = 9 + x \quad \Rightarrow \quad x = 33$$

24. Total mixture = 20 litres

A		B
Milk : Water		Milk : Water
1 : 3		$m : n$
	20 litre	
A	:	B
2		3
20/5×2 = 8 litre		20/5×3 = 12 litre

Volume of container A $= \frac{20}{(2+3)} \times 2 = 8$ litres

Volume of container B $= \frac{20}{(2+3)} \times 3 = 12$ litres

Milk from A $= \frac{8}{4} \times 1 = 2$ litres

Water from A = (8 – 2) litres = 6 litres

Milk from B $= \frac{m}{m+n} \times 12 = \frac{12m}{m+n}$

Water from B $= \frac{n}{m+n} \times 12 = \frac{12n}{m+n}$

According to question

$$\frac{\frac{12m}{m+n}+2}{\frac{12n}{m+n}+6} = \frac{3}{7} \Rightarrow \frac{12m+2(m+n)}{12n+6(m+n)} = \frac{3}{7} \Rightarrow \frac{14m+2n}{6m+18n} = \frac{3}{7}$$

$$\Rightarrow \quad 80m = 40n \Rightarrow \frac{m}{n} = \frac{40}{80} = \frac{1}{2}$$

25. The data about children is not needed far the question

$$\frac{\textit{Married women}}{\text{Worker}} = \frac{1}{3}\left(\frac{1}{2}\right)$$

$$\frac{\textit{Married Men}}{\text{Worker}} = \frac{2}{3}\left(\frac{3}{4}\right)$$

$\therefore$ The required rates in $\frac{1}{6} : \frac{1}{2}$ or 1 : 3.

26. The data is tabulated below.

Person	A	B	C	D	E	F	G	H	I	J	K	L
Capital	1	2	3	4	5	6	7	8	9	10	11	12
Period	12	11	10	9	8	7	6	5	4	3	2	1

If the sum of two numbers is given, the product is maximum
When they are equal (or close to each other as possible).

$$1\ (12) < 2(11) < 3(10) < 4\ (9) < 5\ (8) < 6\ (7)$$

$\therefore$ F and G would get the maximum profit

27. The data is tabulated below.

	Mahesh	**Kamal**
Saving		
Expense	1	3
Income	1	2

M	**K**
1	0
2	0
3	0

	Mahesh	**Kamal**
Saving	***3***	***6***
Expense	***0***	***0***
Income	***3***	***6***

If Kamal spends his entire income, Mahesh's savings are more (Sec-I)
If Mahesh doesn't spend anything, Kamal's savings are more (Sec –II)
We cannot determine who saves more.

Alternative Sol.

If the ratio of two people's expenses is $\frac{E_1}{E_2}$ and the ratio of their saving is $\frac{S_1}{S_2}$, then the ratio of the incomes lies between these values.

The ratio of Mahesh's expenses to Kamal's expanses is $\frac{1}{3}$.

The ratio of Mahesh's income to Kamal's income is $\frac{1}{2}$

The ratio of Mahesh's savings to Kamal's savings has to the more then $\frac{1}{2}$.

It could be less than 1 or more than 1.

28. The ratio of the speed of the 3 cars in 2 : 3 : 4.
We assume a distance of 12 units (the LCM). The ratio of the times taken in 6 : 4 : 3.

29. Put, $A = 3x$

$$B = 4x\,,\ \frac{3A^2 + 4B}{3A - 4B\ 2} \Rightarrow \frac{3(3x)^2 + 4(4x)}{3(3x) - 64x^2} = \frac{x(27x+16)}{x(9-64x)}$$

In the above equation x can have different values, so the value of the expression is 'cannot be determined.

30. Let the weights of four pieces of diamond be x, $2x$, $3x$, $4x$

So, Total weight of diamond = $10x$

$\therefore$ Total cost of original diamond = $100x^2$ ... (*i*)

and

Total cost of 4 pieces is $x^2 + 4x^2 + 9x^2 + 16x^2 = 30x^2$

Loss in value of diamond $= 100x^2 - 30x^2\ \frac{(x+y)}{(x-y)} = \frac{5}{1}$

$\therefore$ According to the question, $x + y = 5x - 5y$, $4x = 6y$

Putting x^2 in (*i*), we get

Total cost of original

diamond $= (3k)^2 + (2k)^2 = 13k^2 = (3k)^2 - (2k)^2 = 5k^2$

31. By mixture and alligation method,

$\frac{1}{5} \times 20 = 4$ $\qquad$ $\frac{4}{5} \times 20 = 16$

$\frac{1}{4} \times 20 = 5$

11 $\qquad$ 1

$\therefore$ New mixture = 11 : 1

32. X Y

25kg 125kg

Ratio of lead and tin in alloy X = 1 : 2

Amount of lead in alloy X = $\frac{1}{3} \times 25 = \frac{25}{3} kg$

Amount of tin in alloy X = $\frac{2}{3} \times 25 = \frac{50}{3} kg$

Similarly, ratio of lead and tin in alloy Y = 2 : 3

Amount of lead in alloy Y = $\frac{2}{5} \times 125 = 50kg$

Amount of tin in alloy Y = $\frac{3}{5} \times 125 = 75kg$

Amount of lead in the mixture = $\frac{25}{3} + 50 = \frac{175}{3}$

Amount of tin in the mixture = $\frac{50}{3} + 75 = \frac{275}{3}$

Their ratio in the mixture = $\frac{175}{3} : \frac{275}{3} = 7 : 11$

33. Let x must be subtracted. Then,

$$\frac{14-x}{17-x}=\frac{34-x}{42-x} \Leftrightarrow (14-x)(42-x)=(34-x)(17-x)$$

$$\Leftrightarrow x^2-56x+588=x^2-51x+578$$

$$\Leftrightarrow 5x=10 \Leftrightarrow x=2.$$

34. Let the fraction be x. Then,

$$\frac{x}{(1/27)}=\frac{(3/7)}{(5/9)} \Leftrightarrow 27x=\frac{3}{7}\times\frac{9}{5} \Leftrightarrow x=\frac{1}{35}.$$

35. $\frac{1}{2}A=\frac{1}{3}B=\frac{1}{6}C=k(\text{say})$. Then, $A=2k$, $B=3k$ and $C=6k$.

$\therefore A:B:C=2k:3k:6k=2:3:6$

$\therefore B\text{'s part}=₹\left(1980\times\frac{3}{11}\right)=₹540.$

36. $A=\frac{2}{3}B$ and $B=\frac{1}{4}C$.

So, $A:B=2:3$ and $B:C=1:4=3:12$

$\therefore A:B:C=2:3:12$

$\therefore B\text{'s share}=₹\left(1360\times\frac{3}{17}\right)=₹240.$

37. Divide ₹[4850 – (15 + 10 + 25)] in the ratio 3 : 4 : 5.

Then, B's part $=\left(4800\times\frac{4}{12}+10\right)=₹1610.$

38. Suppose C gets x. Then, B gets $(x+8)$ and A gets $(x+15)$.

$\therefore x+(x+8)+(x+15)=53$ or $x=10$.

$\therefore A:B:C=(10+15):(10+8):10=25:18:10.$

39. Ratio of values of coins $=\frac{3}{1}:\frac{4}{2}:\frac{12}{4}=3:2:3$

Values of 25-paise coins $=₹\left(600\times\frac{3}{8}\right)=₹225.$

$\therefore$ Number of these coins $=(225\times 4)=900.$

40. $A:B:C=100:65:35=20:13:7.$

If C's share is ₹7, then total sum $=(20+13+7)=40.$

$\therefore 7:40::560:x$ or $x=\frac{40\times 560}{7}=3200.$

41. $A+B+C=5625,\ B=\frac{1}{4}(A+C)\Rightarrow A+C=4B$

$\therefore B+4B=5625 \Leftrightarrow 5B=5625 \Leftrightarrow B=1125.$

$A+B+C=5625\ \&\ A=\frac{1}{2}(B+C)\Leftrightarrow B+C=2A$

$\therefore\ A+2A=5625 \Leftrightarrow 3A=5625 \Leftrightarrow A=1875$

$\therefore\ A-B=(1875-1125)=750.$

42. Let their amounts be $4x$ & $3x$

Then, $3x = 4800 \Rightarrow x = 1600$.

$\therefore$ Total amount = $7x$ = ₹11200.

43. Number of boys = $\left(720 \times \frac{7}{12}\right) = 420$.

Number of girls = (720 – 420) = 300.

$\therefore$ Number of girls to be admitted = (420 – 300) = 120.

44. 1.2 : 1.4 : : 5.4 : x. So, $x = \frac{1.4 \times 5.4}{1.2} = 6.3$m.

45. Let their incomes be $3x$ and $2x$ and their expenditure be $5y$ and $3y$.

Then, $3x - 5y = 1500$ & $2x - 3y = 1500$.

On solving, we get $x = 3000$. So, B's income = $2x = 6000$.

46. Statement-1

Given that

$x \propto z$ and y and $y \propto z$

$\Rightarrow x = zk$ ... (i) and $y = zt$ ... (ii) [Where k and t are constants]

Squaring eqn. (i) and (ii), we get,

$$x^2 = z^2k^2 \text{ and } y^2 = z^2t^2$$

By subtracting we get,

$$x^2 - y^2 = z^2 (k^2 - t^2) \Rightarrow x^2 - y^2 \propto z^2$$

Hence, statement –1 is correct.

Statement - 2

Given that

$x \propto \frac{1}{z}$ and $y \propto \frac{1}{z}$

$\Rightarrow xz = k$ ... (i) and $zy = t$ ... (ii) [where k and t are constants]

multiplying eqn (i) and (ii), we get,

$xz \,.\, yz = kt \Rightarrow xyz^2 = kt \Rightarrow xy = \frac{kt}{z^2} \Rightarrow xy \propto \frac{1}{z^2}$

Hence, statement –2 is correct.

47. Let the incomes of A, B and C be $7x$, $9x$ and $8x$ respectively and their expenditures are $8y$, $9y$ and $15y$ respectively.

Now, we have

$7x - 8y = \frac{1}{4} \times 7x \Rightarrow 7x - \frac{7x}{4} = 8y \Rightarrow \frac{21x}{4} = 8y \Rightarrow \frac{x}{y} = \frac{32}{21}$

Now,

Savings of A = 7 × 32 – 8 × 21 = 56

Savings of B = 9 × 32 – 9 × 21 = 99

Savings of C = 12 × 32 – 15 × 21 = 69

$\therefore$ Ratio of savings of A : B : C = 56 : 99 : 69.

48. P and Q's salary is neither mention in question nor in both statement. So, the question cannot be answered even by using both the statements together.

49. We have,

Number of males in first village = $1525 \times \dfrac{27}{61} = 675$

Number of females in first village = 1525 – 675 = 850

In the second village

Number of males = $2600 \times \dfrac{6}{13} = 1200$

Number of females = 2600 – 1200 = 1400

Hence, required ration = $\dfrac{675+1200}{850+1400} = \dfrac{1875}{2250} = 5:6$

50. We have,

$ax^2 + bx + c$ ...(i)

and Let $\alpha = 2$

Statement I:

$\therefore \alpha\beta = \dfrac{c}{a} = 1 \quad \Rightarrow 2\beta = 1 \quad \Rightarrow \beta = \dfrac{1}{2}$

Now, quadratic equation is :

$k(x^2 - x(\alpha + \beta) + \alpha\beta] = 0$

$\Rightarrow k\left(x^2 - \dfrac{5x}{2} + 1\right) = 0$

$\Rightarrow k(2x^2 - 5x + 2) = 0$ ...(ii)

By comparing (i) and (ii), we get

$a = 2k, b = -5k, c = 2k$

$\therefore$ There is no unique value of a, b and c.

Statement II We have,

$\alpha + \beta = -\dfrac{5}{2} \quad \Rightarrow \beta = \dfrac{5}{2} - 2 = \dfrac{1}{2}$

Again, no, unique value of a, b and c.

Hence, option (d) is correct

51. (1) $\dfrac{x+z}{y} = \dfrac{z}{x} = p$ (let)

$x + y = py$ and $z = px$

$\therefore x + px = py$

$\Rightarrow x(p+1) = py \quad \Rightarrow y = \left(\dfrac{p+1}{p}\right)x$

$\therefore x : y : z = p : p+1 : p$

(2) $\dfrac{z-y}{x} = \dfrac{x}{z} = p_1$ (let)

$\Rightarrow z - y = p_1 x$ and $x = p_1 z$

$\therefore y = z - p_1 x$

$\Rightarrow y = z - p_1^2 z = z(1 - p_1^2)$

$\therefore\ x:y:z = p_1 z : z(1-p_1^2) : z = p_1 : (1-p_1^2) : 1$

Using both, we get

$p = \frac{1}{p_1} = 1 \Rightarrow p_1 = \frac{1}{p}$

$\therefore\ \frac{1}{p} : 1 - \frac{1}{p^2} : 1$

$p : p^2 - 1 : p^2$

Now, $\frac{p}{p} = \frac{p+1}{p^2-1} = \frac{p}{p^2}$

$\Rightarrow p^2 - 1 = p + 1$ and $p^2 = p \quad \Rightarrow p = -1, 2,\ p = 0, 1$

$\therefore$ It is not possible

$\therefore$ (c) is correct

52. Statement I

$(a+b) \propto (a-b)$

$\Rightarrow a + b = \alpha(a-b)$

$\therefore\ a^2 + b^2 = \frac{(a+b)^2 + (a-b)^2}{2} - \frac{a^2(a-b)^2 + (a-b)^2}{2} = (a-b)^2\left(\frac{1+a^2}{2}\right)$

$\Rightarrow a^2 + b^2 = (a^2 + b^2 - 2ab)\left(\frac{1+\alpha^2}{2}\right) \quad \Rightarrow ab\,(1+\alpha^2) = (a^2+b^2)\left(\frac{1+\alpha^2}{2} - 1\right)$

$\Rightarrow a^2 + b^2 = \frac{2(\alpha^2+1)}{(\alpha^2-1)} ab \quad \Rightarrow (a^2+b^2) \propto ab$

Statement II

$a \propto b \Rightarrow a = \alpha\, b$

$\therefore\ a^2 - b^2 = (a-b)(a+b) = (\alpha b - b)(\alpha b + b) = b^2(\alpha^2 - 1)$

Hence, statement II is false.

8 CHAPTER Partnership

Partnership: *When two or more than two persons run a business jointly, they are called partners and the deal is known as partnership.*

Rule I: *When investments of all the partners are for the same time, the profit or loss is divided among them in the ratio of their investments.*

Rule II: *When investments are for different times, then equivalent capitals are calculated for a unit of time by multiplying the capital with the number of units of time.*

Gain or loss is now divided in the ratio of these capitals.

Remark: *A partner who manages the business is called a working partner and the one who simply invests the money is called a sleeping partner.*

SOLVED PROBLEMS

Ex. 1. ***A and B started a business investing ₹25000 and ₹20000 respectively. After 5 months, A left and C joined by investing ₹15000. At the end of the year, there was a profit of ₹14100. Find the share of each.***

Sol. Money invested by A = ₹25000 for 5 months.

Money invested by B = ₹20000 for 12 months.

Money invested by C= ₹15000 for 7 months.

$\therefore A : B : C = (25000 \times 5) : (20000 \times 12) : (15000 \times 7) = 25 : 48 : 21$

$\therefore$ A's share = ₹ $\left(14100 \times \frac{25}{94}\right)$ = ₹3750.

$\therefore$ B's share = ₹ $\left(14100 \times \frac{48}{94}\right)$ = ₹7200.

$\therefore$ C's share = ₹ $\left(14100 \times \frac{21}{94}\right)$ = ₹3150.

Ex. 2. ***Suresh started a business by investing ₹65000. After 4 months Anil joined him with some investment. At the end of the year, the total profit was divided between them in the ratio 2 : 3. How much was invested by Anil?***

Sol. Suppose Anil invested ₹x. Then,

$$\frac{65000 \times 12}{x \times 8} = \frac{2}{3} \Leftrightarrow x = \left(\frac{65000 \times 12 \times 3}{8 \times 2}\right) = 146250.$$

So, Anil invested ₹146250.

Ex. 3. ***A invested ₹76000 in a business. After a few months B joined him with an investment of ₹57000. At the end of the years, the total profit was divided between them in the ratio 2 : 1. After how many months did B join?***

Sol. Suppose B joined after x months.

Then, $\frac{76000\times 12}{57000\times(12-x)}=\frac{2}{1}\Leftrightarrow 3(12-x)=24\Leftrightarrow x=4.$

So, B joined after 4 months.

Ex. 4. ***A, B, C started a business with their investments in the ratio 1 : 3 : 5. After four months A invested the same amount as before and B as well as C with drew half of their investments. Find the ratio of their profits at the end of the year.***

Sol. Suppose their initial investments were x, $3x$ & $5x$ respectively.

Ratio of their investments during the year

$$= (x\times 4+2x\times 8):\left(3x\times 4+\frac{3}{2}x\times 8\right):\left(5x\times 4+\frac{5}{2}x\times 8\right)$$

$$= 20x : 24x : 40x = 5 : 6 : 10.$$

$\therefore$ Ratio of their profits = 5 : 6 : 10.

EXERCISE 8

1. ₹120 is distributed among A, B and C so that A's share is ₹20 more than B's and ₹20 less than C's. What is B's share? **[CDS (I) Exam, 2018]**

(*a*) ₹10 (*b*) ₹15 (*c*) ₹20 (*d*) ₹25

2. A merchant commences with a certain capital and gains annually at the rate of 25%. At the end of 3 years, he has ₹10,000. What is the original amount that the merchant invested? **[CDS (I) Exam, 2018]**

(*a*) ₹5,120 (*b*) ₹5,210 (*c*) ₹5350 (*d*) ₹5,500

3. ₹9400 is distributed among P, Q, R in such a way that if ₹93, ₹24, ₹55 are deducted from their respective shares, then they have money in the ratio 3 : 4 : 5. What is the share of P? **[CDS (I) Exam, 2024]**

(*a*) ₹2307 (*b*) ₹2376 (*c*) ₹2508 (*d*) ₹2896

4. According to Mr. Sharma's will, half of his property goes to his wife and the rest is equally divided between his two sons, Ravi and Raj. Some years later, Ravi dies and leaves half of his property to his widow and rest to his brother Raj. When Raj dies he leaves half of his property to his widow and remaining to his mother, who is still alive. The mother now owns ₹88,000 worth of the property. The total worth of the property of Mr. Sharma was **[CDS (II) Exam, 2018]**

(*a*) ₹1,00,000 (*b*) ₹1,24,000 (*c*) ₹1,28,000 (*d*) ₹1,32,000

5. Three persons start a business with capitals in the ration $\frac{1}{3}:\frac{1}{4}:\frac{1}{5}$. The first person withdraws half his capital after 4 months. What is his share of profit if the business fetches an annual profit of ₹96,800? **[CDS (I) Exam, 2020]**

(*a*) ₹32,000 (*b*) ₹34,500 (*c*) ₹36,000 (*d*) ₹36,800

6. A man who recently died left a sum of ₹3,90,000 to be divided among his wife, five sons and four daughters. He directed that each son should receive 3 times as much as each daughter receives and that each daughter should receive twice as much as their mother receives. What was the wife's share? **[CDS (I) Exam, 2019]**

(*a*) ₹14,000 (*b*) ₹12,000 (*c*) ₹10,000 (*d*) ₹9,000

7. X and Y entered into partnership with ₹700 and ₹600 respectively. After 3 months X withdrew $\frac{2}{7}$ of his stock but after 3 months, he puts back $\frac{3}{5}$ of what he had withdrawn. The profit at the end of the year is ₹726. How much of this should X receive? **[CDS (II) Exam, 2016]**
(*a*) ₹336 (*b*) ₹366 (*c*) ₹633 (*d*) ₹663

8. Three partners *A, B, C* invest ₹34000, ₹26000 and ₹10000 respectively, in a business. Out of a total profit of ₹17500, *A's*: share (in rupees) is:
(*a*) 8750 (*b*) 8500 (*c*) 7500 (*d*) 3750

9. Arun, Maya and Satya started a shop by investing ₹27000, ₹81000 and ₹72000 respectively. At the end of the year the profit was distributed in the ratio of their investments. If Maya's share of profit be ₹36000, the total profit was:
(*a*) ₹63000 (*b*) ₹80000 (*c*) ₹108000 (*d*) ₹116000

10. *A* and *B* entered into partnership investing ₹12000 and ₹16000 respectively. After 3 months, *B* withdrew ₹5000 while *A* invested ₹5000 more. Out of a total annual profit of ₹16000, the share of A exceeds that of *B* by:
(*a*) ₹1000 (*b*) ₹1500 (*c*) ₹ 2000 (*d*) ₹2500

11. Kamal started a business with ₹21000 and is joined afterwards by Vinod with ₹36000. After how many months did Vinod join, if the profits at the end of the year are divided equally?
(*a*) 3 months (*b*) 4 months (*c*) 5 months (*d*) 8 months

12. *A* and *B* enter into partnership. *A* invests ₹16000 for 8 months and *B* remains in the business for 4 months. Out of a total annual profit, *B* claims $\frac{2}{7}$ of the profit. The contribution of *B* (in rupees) was :
(*a*) 10500 (*b*) 11900 (*c*) 13600 (*d*) 12800

13. *A* and *B* invest in a business in the ratio 3 : 2. If 5% of the total profit goes to charity and *A*'s share is ₹855, the total profit (in rupees) is:
(*a*) 1425 (*b*) 1500 (*c*) 1537.50 (*d*) 1576

14. If *A*'s capital is equal to twice *B*'s capital and *B*'s capital is three times *C*'s capital, then the ratio of their capitals is:
(*a*) 2:1:3 (*b*) 1:2:6 (*c*) 6:3:1 (*d*) 1:3:6

15. *A, B, C* are three partners in a business. If twice the investment of *A* is equal to thrice the capital of *B* and the capital of *B* is four times the capital of *C*, out of a total profit of ₹5940, the share of *C* is:
(*a*) ₹540 (*b*) ₹700 (*c*) ₹740 (*d*) ₹900

16. *A, B, C* subscribe ₹47000 for a business. If *A* subscribes ₹7000 more than *B* and *B* ₹5000 more than *C*, then out of a total profit of ₹9400, *B* receives:
(*a*) ₹1737.90 (*b*) ₹2000 (*c*) ₹3000 (*d*) ₹4400

17. *A* and *B* entered into partnership with capitals in the ratio 4 and 5. After 3 months, *A* withdrew $\frac{1}{4}$ of his capital and *B* withdrew $\frac{1}{5}$ of his capital. The gain at the end of 10 months was ₹760. The share of *A* in this profit is:
(*a*) ₹360 (*b*) ₹330 (*c*) ₹430 (*d*) ₹380

18. *A, B, C* hired a car for ₹520 and used it for 7, 8 and 11 hours respectively. The amount of hire charges paid by *B* was:
(*a*) ₹140 (*b*) ₹160 (*c*) ₹180 (*d*) ₹220

19. Four milkmen rented a pasture. *A* grazed l8 cows for 4 months, *B* 25 cows for 2 months, *C* 28 cows for 5 months and *D* 21 cows for 3 months. If *A*'s share of rent in ₹360, the total rent of the field (in rupees) is:

(*a*) 1500 (*b*) 1600 (*c*) 1625 (*d*) 1650

20. *A*, *B*, *C* start a business with capitals in the ratio $\frac{1}{3}:\frac{1}{4}:\frac{1}{5}$. *A* withdraws half his capital at the end of 4 months. Out of a total annual profit of ₹847, *A*'s share is:

(*a*) ₹252 (*b*) ₹280 (*c*) ₹315 (*d*) ₹412

ANSWERS

1. (*c*)	**2.** (*a*)	**3.** (*a*)	**4.** (*c*)	**5.** (*a*)	**6.** (*c*)	**7.** (*b*)	**8.** (*b*)	**9.** (*b*)
10. (*c*)	**11.** (*c*)	**12.** (*d*)	**13.** (*b*)	**14.** (*c*)	**15.** (*a*)	**16.** (*c*)	**17.** (*b*)	**18.** (*b*)
19. (*c*)	**20.** (*b*)							

SOLUTIONS

1. Let B's share = ₹x

A's share = ₹$(x + 20)$

C's share ₹$(x + 40)$

Now, we have,

$x + x \times 20 + x \times 40 = 120$

$\Rightarrow x = 60 = 120$

$\Rightarrow x = 120$

$\therefore$ B's share = ₹20

2. Let the original amount that the merchant invested be ₹x

Now, we have,

$$10000 = x \times \left(1 + \frac{25}{100}\right)^3$$

$$\Rightarrow 10000 = x \times \left(\frac{5}{4}\right)^3 \Rightarrow 10000 = \frac{125x}{64} \Rightarrow x = 80 \times 64 = ₹5120$$

3. Amount after deduction = ₹(9400 – 172) = ₹9228

$\therefore$ P's share = ₹$(9229 \times \frac{3}{12})$ = ₹2307

4. Let the total worth of the property of MN Sharma = ₹x

His wife's share = $\frac{x}{2}$

Ravi's share = Raj share = $\frac{x}{2} \times \frac{1}{2} = \frac{x}{4}$

After death of Ravi

Raj share = $\frac{x}{4} + \frac{1}{2} \times \frac{x}{4} = \frac{x}{4} + \frac{x}{8} = ₹\frac{3x}{8}$

After death of Raj

mother's share = $\frac{x}{2}+\frac{1}{2}+\frac{3x}{8}=\frac{x}{2}+\frac{3x}{16}=\frac{11x}{16}$

Now, according to the question,

$\frac{11x}{16}=88{,}000 \Rightarrow \frac{x}{16}=8{,}000 \Rightarrow x = 128000$

5. Let the investments of three persons A, B and C are $20x$, $15x$ and $12x$ respectively.
Now, we have,
A's investment = $20x \times 4 + 10x \times 8 = 80x + 80x = 160x$
B's investment = $15x \times 12 = 180x$
C's investment = $12x \times 12 = 144x$
$\therefore$ Total investments = $160x + 180x + 144x + = 484x$
Now,
$484x = 96{,}800$

$\Rightarrow x = \frac{96800}{484} = 200$

$\therefore$ A's annual profit = $160x = 160 \times 200 =$ ₹32000

6. Let his wife' share = ₹x
Each daughter's share = ₹$2x$
Each son's share = ₹$6x$
Now, we have,
$x + 5 \times 6x + 4x \times 2x = 390000$
$\Rightarrow x + 30x + 48x = 390000$
$\Rightarrow 39x = 390000 \Rightarrow x =$ ₹10000

7. The data is tabulated below.

Period	X	Y	Period
3	700	600	12
3	500		
6	620		

The ratio of their profits is

$$\frac{3(7)+3(5)+6(6.2)}{6(12)}=\frac{7+5+12.4}{24}=\frac{24.4}{24}=\frac{244}{240}=\frac{61}{60}$$

$\therefore$ X's share = $\frac{61}{121}(726) = \frac{61}{11}(66) = 366$

8. Ratio of investments of A, B, C = 34000 : 26000 : 10000 = 17 : 13 : 5

$\therefore$ A's share = Rs.$\left(17500\times\frac{17}{35}\right)$ = ₹8500.

9. Ratio of shares = 27000 : 81000 : 72000 = 3 : 9 : 8.
If Maya's share is ₹9, total profit = ₹(3 + 9 + 8) = 20

$\therefore 9 : 20 :: 36000 : x$ or $x = \dfrac{20 \times 36000}{9} = 80000.$

10. Ratio of shares of A and $B = (12000 \times 3 + 17000 \times 9 : 16000 \times 3 + 11000 \times 9)$
$= 189000 : 147000 = 9 : 7$

A's share $= ₹\left(16000 \times \dfrac{9}{16}\right) = ₹9000.$

B's share $= ₹(16000 - 9000) = ₹7000.$

$\therefore$ (A's share) – (B's share) $= ₹(9000 - 7000) = ₹2000.$

11. Suppose Vinod joined after x months.
Then, he invested the money for $(12 - x)$ months.
$\therefore 21000 \times 12 = 36000 \times (12 - x)$ or $x = 5.$

12. Suppose B contributed ₹x.
$\therefore A : B = (16000 \times 8 : x \times 4) = 32000 : x.$

Ratio of profits of A and $B = \dfrac{5}{7} : \dfrac{2}{7} = 5 : 2$ (Note)

$\therefore \quad \dfrac{32000}{x} = \dfrac{5}{2} \Leftrightarrow 5x = 64000 \Leftrightarrow x = 12800.$

13. Let total profit = ₹x.

Profit to be distributed = 95 % of $x = \dfrac{19x}{20}.$

$\therefore A$'s share $= ₹\left(\dfrac{19x}{20} \times \dfrac{3}{5}\right) = ₹\dfrac{57x}{100}.$

$\therefore \quad \dfrac{57x}{100} = 855$ or $x = \left(\dfrac{855 \times 100}{57}\right) = ₹1500.$

14. Let C's capital $= x$. Then, B's capital $= 3x$ & A's capital $= 6x$.
$\therefore \; A : B : C = 6x : 3x : x = 6 : 3 : 1.$

15. $2A = 3B$ and $B = 4C$
Let $C = x$. Then, $B = 4x$ & $2A = 3 \times 4x = 12x \Rightarrow A = 6x.$
$\therefore A : B : C = 6 : 4 : 1.$

$\therefore$ C's share $= ₹\left(5940 \times \dfrac{1}{11}\right) = ₹540.$

16. Suppose C subscribes ₹x. Then,
B's contribution $= (x + 5000)$ & A's contribution $= (x + 12000).$
$\therefore (x + 12000) + (x + 5000) + x = 47000 \Leftrightarrow x = 10000.$
$\therefore A : B : C = 22000 : 15000 : 10000 = 22 : 15 : 10.$

B' s share $= ₹\left(9400 \times \dfrac{15}{47}\right) = ₹3000.$

17. Let their initial capitals be $4x$ and $5x$.

After 3 months A withdrew $\dfrac{1}{4} \times 4x = x$ & B withdrew $\dfrac{1}{5} \times 5x = x.$

$\therefore$ Ratio of capitals $= (4x \times 3 + 3x \times 7 : 5x \times 3 + 4x \times 7) = 33x : 43x = 33 : 43.$

$\therefore$ *A*'s share = ₹$\left(760 \times \frac{33}{76}\right)$ = ₹330.

18. Ratio of hire charges = 7 : 8 : 11.

$\therefore$ Charges paid by *B* = ₹$\left(520 \times \frac{8}{26}\right)$ = ₹160.

19. $A : B : C : D = (18 \times 4 : 25 \times 2 : 28 \times 5 : 21 \times 3) = 72 : 50 : 140 : 63$

Let total rent be ₹x. Then,

A's share = $\left(x \times \frac{72}{325}\right)$. So, $\frac{72x}{325} = 360$ or $x = 1625$.

20. Ratio of capitals in the beginning = $\frac{1}{3} : \frac{1}{4} : \frac{1}{5} = 20 : 15 : 12$.

Suppose they originally invest ₹20, ₹15 & ₹12.

Ratio of their investments = $(20 \times 4 + 10 \times 8) : (15 \times 12) : (12 \times 12) = 40 : 45 : 36$.

$\therefore$ *A's* share = ₹$\left(847 \times \frac{40}{121}\right)$ = ₹280.

9 CHAPTER Profit and Loss

Definitions and Review of Concepts

(*i*) **Cost Price:** *The price at which an article is bought is called its cost price, abbreviated as C.P.*

(*ii*) **Selling Price:** *The price at which an article is sold is called its selling price, abbreviated as S.P.*

(*iii*) **Profit** or **Gain** = (S.P.) – (C.P.) (*iv*) **Loss** = (C.P.) – (S.P.)

(*v*) Loss or Gain is always reckoned on C.P.

(*vi*) Gain % = $\left(21200 \times \frac{36}{53}\right)$ (*vii*) Loss % = $\left(\frac{\text{Loss} \times 100}{\text{C.P.}}\right)$

(*viii*) S.P. = $\left(\frac{100 + \text{Gain\%}}{100} \times \text{C.P.}\right)$ (*ix*) S.P. = $\left(\frac{100 - \text{Loss\%}}{100} \times \text{C.P.}\right)$

(*x*) C.P. = $\left(\frac{100}{100 + \text{Gain\%}} \times \text{S.P.}\right)$ (*xi*) C.P. = $\left(\frac{100}{100 - \text{Loss\%}} \times \text{S.P.}\right)$

(*xii*) If an article is sold at a gain of 27%, then S.P. = 127% of C.P.

(*xiii*) If an article is sold at a loss of 18%, then S.P. = (l00 – 18)% of C.P. = 82% of C.P.

SOLVED PROBLEMS

Ex. 1. ***Find C.P., when: (i) S.P. = ₹77, Gain = 10% (ii) S.P. = ₹94.30, Loss = 8%***

Sol. (*i*) C.P. = $\frac{100}{(100+10)} \times 77 = \text{Rs.}\left(\frac{100}{110} \times 77\right) = ₹70.$

(*ii*) C.P. = $\frac{100}{(100-8)} \times 94.30 = ₹\left(\frac{100}{92} \times 94.30\right) = ₹102.50.$

Ex. 2. ***By selling an article for ₹110, a man loses 12%. For how much should he sell it to gain 8%?***

Sol. S.P. = ₹110, Loss = 12%. So, C.P. = $₹\left(\frac{100}{88} \times 110\right) = ₹125.$

Now, C.P. = ₹125, Gain required = 8%

$\therefore$ S.P. = $₹\left(\frac{108}{100} \times 125\right) = ₹135.$

Ex. 3. ***A man sells two houses for ₹536850 each. On one he gains 5% and on the other he loses 5%. Find his gain or loss per cent on the whole transaction.***

Sol. In such a problem selling price is immaterial. There is always a loss given by:

loss % = $\left(\frac{\text{common gain or loss \%}}{10}\right)^2 = \left(\frac{5}{20}\right)^2 = \frac{1}{4} = 0.25\%.$

Ex. 4. ***By selling 100 pencils, a shopkeeper gains the selling price of 20 pencils. Find his gain percent.***

Sol. (S.P. of 100 pencils) – (C.P. of 100 pencils) = S.P. of 20 pencils.

∴ S.P. of 80 pencils = C.P. of 100 pencils.

Let C.P. of each pencil be Re 1.

Then, C.P. of 80 pencils = ₹80.

S.P. of 80 pencils = ₹100

$$\therefore \quad \text{Gain} = \left(\frac{20}{80}\times 100\right)\% = 25\%.$$

Ex. 5. ***A dishonest dealer professes to sell his goods at cost price but uses a weight of 960 gms for 1 kg. Find his gain per cent.***

Sol. Formula: $\text{Gain}\ \% = \left(\frac{\text{Error}}{\text{True value} - \text{Error}}\times 100\right)\% = \left(\frac{40}{960}\times 100\right)\% = 4\frac{1}{6}\%.$

Problems on Discount

Remarks: (*i*) *List price is called the marked price of the article.*

(*ii*) *Discount is reckoned on the marked price.*

Ex. 6. ***A cycle dealer marks his goods 25% above his cost price and allows a discount of 8% on it. Find his gain per cent.***

Sol. Let C.P. = ₹100. Then, marked price = ₹125.

S.P. = 92% of ₹125 = ₹115. ∴ Gain = 15%.

Ex. 7. ***Successive discounts of 20% and 10% is equivalent to a single discount of how many per cent?***

Sol. Let marked price be ₹100.

Final S.P. after two discounts = 90% of 80% of ₹100 = ₹72.

This is equivalent to a single discount of (100 – 72)% = 28%.

Ex. 8. ***If a commission of 10% is given on the list price, the gain is 20%. Find the gain per cent, if the commission is increased to 20%.***

Sol. Let list price = ₹100. Commission = 10%.

∴ S.P. = ₹90, Gain = 20%.

$$\therefore \quad \text{C.P.} = ₹\left(\frac{100}{120}\times 90\right) = ₹75.$$

New S.P. = ₹80.

$$\text{New gain}\ \% = \left(\frac{5}{75}\times 100\right)\% = 6\frac{2}{3}\%.$$

EXERCISE 9

1. A shopkeeper marks the price of an article at ₹200. After allowing a discount of 10%, he still gains 20% on the cost price. What is the cost price of the article? **[CDS (I) Exam, 2022]**

(*a*) ₹170 (*b*) ₹160 (*c*) ₹150 (*d*) ₹120

2. A trader gives successive discounts of 20%, 10% and 5% respectively. What is the overall discount? **[CDS (I) Exam, 2021]**

(*a*) 30% (*b*) 31.6% (*c*) 32.8% (*d*) 35%

3. The cost price of 100 mangoes is equal to the selling price of 80 mangoes. What is the profit percentage? **[CDS (I) Exam, 2021]**

(*a*) 16% (*b*) 20% (*c*) 24% (*d*) 25%

4. X sells his goods 25% cheaper than Y and 25% dearer than Z. How much percentage is Z's goods cheaper than Y? **[CDS (I) Exam, 2021]**

(*a*) $\frac{100}{3}\%$ (*b*) 40% (*c*) 50% (*d*) $\frac{200}{3}\%$

5. A person bought a book at $\frac{3}{4}^{\text{th}}$ of its listed price and sold it at 50% more than its listed price. What is the percentage of gain in the transaction ? **[CDS (I) Exam, 2024]**

(*a*) 20% (*b*) 40% (*c*) 75% (*d*) 100%

6. A shopkeeper sells his articles at their cost price but uses a faulty balance which reads 1000 gm for 800 gm. What is the actual profit percentage? **[CDS (I) Exam, 2020]**

(*a*) 20% (*b*) 25% (*c*) 30% (*d*) 40%

7. If a television set is sold at ₹x, a loss of 28% would be incurred. If it is sold at ₹y, a profit of 12% would be incurred. What is the ratio of y to x? **[CDS (I) Exam, 2020]**

(*a*) 41 : 9 (*b*) 31 : 9 (*c*) 23 : 9 (*d*) 14 : 9

8. A person sold an article for ₹75 which cost him ₹x. he finds that he realised x% profit on his outlay. What is x equal to ? **[CDS (I) Exam, 2020]**

(*a*) 20% (*b*) 25% (*c*) 50% (*d*) 100%

9. The cost price of y articles is equal to selling price of z articles, If $y : z = 5 : 4$, what is the profit percentage? **[CDS (I) Exam, 2024]**

(*a*) 20% (*b*) 25% (*c*) 30% (*d*) 40%

10. A trader sells two computers at the same price, making a profit of 30% on one and a loss of 30% on the other. What is the net loss or profit percentage on the transaction? **[CDS (II) Exam, 2019]**

(*a*) 6% loss (*b*) 6% gain (*c*) 9% loss (*d*) 9% gain

11. A person sells article X for ₹34,500 and makes a profit of 15%. He sells article Y at a loss of 10%. He neither loses nor gains on the whole because of these two transactions. What is the selling price of article Y? **[CDS (I) Exam, 2025]**

(*a*) ₹40,000 (*b*) ₹40,500 (*c*) ₹41,000 (*d*) ₹51,500

12. A person bought two articles X and Y from a departmental store. The sum of prices before sales tax was ₹130. There was no sales tax on the article X and 9% sales tax on the article. Y, The total amount the person paid, including the sales tax was ₹136.75. What was the price of the article Y before sales tax? **[CDS (II) Exam, 2018]**

(*a*) ₹75 (*b*) ₹85 (*c*) ₹122 (*d*) ₹125

13. X bought 4 bottles of lemon juice and Y bought one bottle of orange juice. Orange juice per bottle costs twice the cost of lemon juice per bottle. Z bought nothing but contributed ₹50 for his share of the drink which they mixed together and shared the cost equally. If Z's ₹50 is covered from his share, then what is the cost of one bottle of orange juice? **[CDS (II) Exam, 2018]**

(*a*) ₹75 (*b*) ₹50 (*c*) ₹46 (*d*) ₹30

14. A milk vendor bought 28 litres of milk at the rate of ₹8.50 per litre. After adding some water he sold the mixture at the same price. If his gain is 12.5%, how much water did he add? **[CDS (II) Exam, 2018]**

(*a*) 4.5 litres (*b*) 4 litres (*c*) 3.5 litres (*d*) 3 litres

15. A cloth merchant buys cloth from a weaver and cheats him by using a scale which is 10 cm longer than a normal metre scale. He claims to sell cloth at the cost price to his customers, but while selling uses a scale which is 10 cm shorter than a normal metre scale. What is his gain?

(*a*) 20% (*b*) 21% (*c*) $22\frac{2}{9}\%$ (*d*) $23\frac{1}{3}\%$

[CDS (I) Exam, 2016]

16. A shopkeeper increases the cost price of an item by 20% and offers a discount of 10% on this marked price. What is his percentage gain?

(*a*) 15% (*b*) 12% (*c*) 10% (*d*) 8%

[CDS (II) Exam, 2016]

17. Rajendra bought a mobile with 25% discount on the selling price. If the mobile cost him ₹4,785, what is the original selling price of the mobile?

(*a*) ₹6300 (*b*) ₹6400 (*c*) ₹6500 (*d*) ₹6600

[CDS (I) Exam, 2017]

18. The per cent profit made when an article is sold for ₹78 is twice as when it is sold for ₹69. The cost price of the anicle is:

(*a*) ₹51 (*b*) ₹60 (*c*) ₹57 (*d*) ₹49

19. A retailer buys a radio for ₹225. How overhead expenses are ₹15. He sells the radio for ₹300. The profit per cent of the retailer is:

(*a*) 10% (*b*) 25% (*c*) 50% (*d*) 52%

20. A man gains 10% by selling an article for a certain price. If he sells it at double the price, the profit made is:

(*a*) 20% (*b*) 60% (*c*) 100% (*d*) 120%

21. If an article is sold at a gain of 6% instead of at a loss of 6%, then the seller gets ₹6 more. The cost price of the article is:

(*a*) ₹50 (*b*) ₹94 (*c*) ₹100 (*d*) ₹106

22. A radio is sold at a gain of 16%. If it had been sold for ₹20 more, 20% would have been gained. The cost price of the radio is:

(*a*) ₹350 (*b*) ₹400 (*c*) ₹500 (*d*) ₹600

23. Profit after selling an article for ₹425 is the same as loss after selling it for ₹355. The cost of the article is:

(*a*) ₹385 (*b*) ₹390 (*c*) ₹395 (*d*) ₹400

24. A merchant sold his goods for ₹75 at a profit percent equal to its cost price. The cost price was:

(*a*) ₹40 (*b*) ₹42 (*c*) ₹37.50 (*d*) ₹50

25. A sold a watch to *B* at a gain of 5% and *B* sold it to *C* at a gain of 4%. If *C* paid ₹1092 for it, the price paid by *A* is:

(*a*) ₹993.72 (*b*) ₹996 (*c*) ₹1000 (*d*) ₹995.90

26. An article is sold at a certain price. By selling it at $\frac{2}{3}$ of that price one loses 10%. The gain percent at original price is:

(*a*) $33\frac{1}{3}\%$ (*b*) 35% (*c*) 40% (*d*) 20%

27. A man sells an article at a gain of 10%. If he had bought it at 10% less and sold it for ₹132 less, he would have still gained 10%. The cost price of the article is:
(*a*) ₹1200 (*b*) ₹1188 (*c*) ₹1320 (*d*) ₹1336.40

28. A dealer sold a radio at a loss of 2.5%. Had he sold it for ₹100 more, he would have gained $7\frac{1}{2}$%. In order to gain $12\frac{1}{2}$%, he should sell it for:
(*a*) ₹850 (*b*) ₹925 (*c*) ₹1080 (*d*) ₹1125

29. Akbar sells a chair at a gain of $7\frac{1}{2}$%. If he had bought it at $12\frac{1}{2}$% less and sold it for ₹5 more, he would have gained 30%. The cost price of the chair is:
(*a*) ₹72 (*b*) ₹80 (*c*) ₹88 (*d*) ₹96

30. By selling sugar at ₹11.16 per kg, a man loses 7%. To gain 7%, it must be sold (per kg) at:
(*a*) ₹11.24 (*b*) ₹12.84 (*c*) ₹14.64 (4) ₹13.24

31. Due to an increase of 30% in the price of eggs, 3 eggs less are available for ₹9.10. The present rate per egg is:
(*a*) 91 paise (*b*) 78 paise (*c*) 64 paise (*d*) 61 paise

32. A man purchased sugar worth ₹400. He sold $\frac{3}{4}$ th at a loss of 10% and the remainder at a gain of 10%. On the whole, he gets:
(*a*) a loss of 5% (*b*) a gain of $5\frac{1}{2}$% (*c*) a loss of $5\frac{1}{19}$% (*d*) a loss of $5\frac{5}{19}$%

33. 6% more is gained by selling a radio for ₹1425 than by selling it for ₹1353. The cost price of the radio is:
(*a*) ₹1200 (*b*) ₹1302 (*c*) ₹1339.50 (*d*) ₹1444.28

34. If the selling price of 12 articles is equal to the cost price of 15 articles, the gain per cent is:
(*a*) 25% (*b*) 20% (*c*) $6\frac{2}{3}$% (*d*) 75%

35. By selling 8 dozens of pencils, a shopkeeper gains the selling price of 1 dozen pencils. His gain per cent is:
(*a*) 12.5% (*b*) 87.5% (*c*) $14\frac{2}{7}$% (*d*) $13\frac{1}{7}$%

36. By selling 36 bananas, a vendor loses the selling price of 4 bananas. His loss per cent is:
(*a*) $11\frac{1}{9}$% (*b*) 10% (*c*) $12\frac{1}{2}$% (*d*) $10\frac{2}{3}$%

37. A fruit seller buys lemons at 2 for a rupee and sells them at 5 for three rupees. His gain per cent is:
(*a*) 10% (*b*) 15% (*c*) 20% (*d*) $66\frac{2}{3}$%

38. Toffees are bought at the rate of 8 for a rupee. To gain 60%, they must be sold at:
(*a*) 6 for a rupee (*b*) 5 for a rupee (*c*) 9 for two rupees (*d*) 24 for five rupees

39. By selling 45 oranges for ₹80, a man loses 20%. How many should he sell for ₹48 so as to gain 20% in the transaction?
(*a*) 25 (*b*) 18 (*c*) 15 (*d*) 20

40. A tradesman by means of false balance defrauds to the extent of 8% in buying goods and also defrauds to 8% in selling. His gain per cent is:

(*a*) 16% (*b*) 15.48% (*c*) 16.64% (*d*) 36%

41. A dishonest dealer sells his goods at the cost price and still eams a profit of 25% by underweighing. What weight does he use for a kg?

(*a*) 750 gm (*b*) 800 gm (*c*) 825 gm (*d*) 850 gm

42. A trader lists his articles 20% above cost price and allows a discount of 10% on cash payment. His gain per cent is:

(*a*) 5% (*b*) 6% (*c*) 8% (*d*) 10%

43. At what per cent above the cost price must an article be marked so as to gain 22.5% after allowing a discount of 2%?

(*a*) 20% (*b*) 24% (*c*) 25% (J) 25.5%

44. Mohan buys an article with 25% discount on its marked price. He makes a profit of 10% by selling it at ₹660. The marked price is:

(*a*) ₹600 (*b*) ₹700 (*c*) ₹800 (*d*) ₹685

ANSWERS

1. (*c*)	**2.** (*b*)	**3.** (*d*)	**4.** (*b*)	**5.** (*d*)	**6.** (*b*)	**7.** (*d*)	**8.** (*c*)	**9.** (*b*)
10. (*c*)	**11.** (*b*)	**12.** (*a*)	**13.** (*b*)	**14.** (*c*)	**15.** (*c*)	**16.** (*d*)	**17.** (*c*)	**18.** (*b*)
19. (*b*)	**20.** (*d*)	**21.** (*a*)	**22.** (*c*)	**23.** (*b*)	**24.** (*d*)	**25.** (*c*)	**26.** (*b*)	**27.** (*a*)
28. (*d*)	**29.** (*b*)	**30.** (*b*)	**31.** (*a*)	**32.** (*a*)	**33.** (*a*)	**34.** (*a*)	**35.** (*c*)	**36.** (*b*)
37. (*c*)	**38.** (*b*)	**39.** (*b*)	**40.** (*c*)	**41.** (*b*)	**42.** (*c*)	**43.** (*c*)	**44.** (*c*)	

SOLUTIONS

1. M.P. = ₹200

Discount = 10%

$$\text{S.P.} = \frac{\text{M.P.} \times (100 - \text{Discount})}{100} = \frac{200 \times 90}{100} = ₹180$$

$$\text{C.P.} = \frac{180}{120} \times 100 = ₹150$$

2. Given that

Successive discounts are 20%, 10% and 5%

$$\therefore \text{Net discount} = 100 \times \frac{100-20}{100} \times \frac{100-10}{100} \times \left(\frac{100-5}{100}\right) = 100 \times \frac{80}{100} \times \frac{90}{100} \times \frac{95}{100} = 68.4\%$$

$\therefore$ Overall discount = 100 – 68.4 = 31.6%

3. Let S.P. = C.P.

Now, we have,

100 × C.P. = 80 × S.P.

$$\Rightarrow \frac{\text{C.P.}}{\text{S.P.}} = \frac{4}{5}$$

$$\text{Profit \%} = \frac{5-4}{4} \times 100 = 25\%$$

4. Let S.P. of y = ₹x

S.P. of $x = x + \frac{75}{100} = \frac{3x}{4}$

S.P. of $z = \frac{3x}{4} \times \frac{100}{125} = \frac{3x}{5}$

Now, required percentage $= \frac{x - \frac{3x}{5}}{x} \times 100 = \frac{2x}{5x} \times 100 = 40\%$

5. Let listed price = ₹x

C.P. = ₹$\frac{3x}{4}$;   S.P. $= \frac{150x}{100} =$ ₹$\frac{3x}{2}$

$\therefore$ gain % $= \frac{\frac{3x}{2} - \frac{3x}{4}}{\frac{3x}{4}} \times 100 = \frac{\frac{6x - 3x}{4}}{\frac{3x}{4}} \times 100 = 100\%$

6. Actual profit % $= \frac{1000 - 800}{800} \times 100 = \frac{200}{800} \times 100 = 25\%$

7. Let C.P. = ₹100

S.P. $= \frac{100 \times 72}{100} = 72 \Rightarrow$ $x =$ ₹72

Now, S.P. of television = ₹y

Profit 12

S.P. = C.P. + profit = 100 + 12

$y = 112$

$\frac{y}{x} = \frac{112}{72} = 14:9$

8. C.P. = ₹x

S.P. = ₹75

Profit % $= \frac{75 - x}{x} \times 100 \quad \Rightarrow \quad x = \frac{75 - x}{x} \times 100 \quad \Rightarrow \quad \frac{x}{100} = \frac{75 - x}{x}$

$\Rightarrow x^2 = (75 - x) \times 100 \quad \Rightarrow \quad x^2 = 7500 - 100x$

$\Rightarrow x^2 + 100x - 7500 = 0 \quad \Rightarrow \quad x^2 + 150x - 50x - 7500 = 0$

$x(x + 150) - 50(x + 150) = 0$

$\Rightarrow (x - 50)(x + 150) = 0 \quad \Rightarrow \quad x = 50$ or

$x = -150$ (not possible)

9. According to the question, we have,

C.P. (y) = SP (z)

$\Rightarrow \quad \frac{CP}{SP} = \frac{z}{y} = \frac{4}{5}$

Now, profit % $= \frac{1}{4} \times 100 = 25\%$

10. Let S.P. of two computers = ₹x

C.P. of first computer = $\frac{x}{130}\times 100 = \frac{10x}{13}$

C.P. of 2nd computer = $\frac{x}{70}\times 100 = \frac{10x}{7}$

Total C.P. = $\frac{10x}{13}+\frac{10x}{7} = \frac{70x+130x}{91} = \frac{200x}{91}$

Total S.P. = ₹$2x$

$$\therefore \text{Loss\%} = \frac{\frac{200x}{91}-2x}{\frac{200x}{91}}\times 100 = \frac{18x}{200x}\times 100 = 9\%$$

11. Given,

Selling price of article X = ₹34500

Let the cost price of article Y = ₹Y

Cost price of article X = ₹$\left(34500\times\frac{100}{115}\right)$ = ₹30000

Selling price of article $Y = \left(y\times\frac{90}{100}\right) =$ ₹$\frac{9y}{10}$

According to question

$$34500+\frac{9y}{10} = y+30000$$

$$\Rightarrow \quad y-\frac{9y}{10} = 34500-30000 \Rightarrow \frac{y}{10} = 4500 \Rightarrow y = 45000$$

Selling price of article $Y = 45000\times\frac{9}{100}$ = ₹ 40500

12. We have,

Price of x + price of y = ₹130 ...(*i*)

And price of x + price of y + 9% of price of y = ₹136.75 ...(*ii*)

Subtracting equation (i) from (ii), we get,

$\frac{9}{100}\times$ price of $y = 136.75 - 130 = 6.75$

$\Rightarrow$ Price of $y = 6.75\times\frac{100}{9} = 75$

13. Let C.P. of one bottle of lemon juice = ₹x

C.P. of one bottle of orange juice = ₹$2x$

Total C.P. = $4x + 2x$ = ₹$6x$

Cost of each of them = $\frac{6x}{3}$ = ₹$2x$ [$\because$ the cost was shared equally]

$2x = 50$ [$\because$ z contributed ₹50]

$\therefore$ Cost of one bottle of orange juice = ₹50

14. Total cost price of milk = 28 × 8.5 = ₹238

$$\text{S.P.} = \frac{238 \times 112.5}{100} = ₹267.75$$

Let milk vendor added x litres of water to the milk.

∵ We have, 8.5 (28 + x) = 267.75

$$\Rightarrow \quad 28 + x = \frac{267.75}{8.5} \Rightarrow 28 + x = 31.5 \Rightarrow x = 31.5 - 28 = 3.5 \text{ litres}$$

15. Let the merchant – purchases 100m at ₹c
But he receives 110 m at ₹c
He sells 90m for c

$$\text{S.P of } 110 = \frac{110}{90} \times c$$

$$\text{Profit} = \frac{11c}{9} - c = \frac{2c}{9}$$

$$\text{Profit \%} = \frac{\frac{2c}{9}}{c} \times 100 = \frac{200}{9} = 22\frac{2}{9}\%$$

16. If the cost price is 100, the marked price is 120. The offer price (or selling price) is 108. There is a profit of 8%

17. Original SP = ₹M. R. P
According to the question
75% of MRP = 4875

$$\Rightarrow \frac{75}{100} \times MRP = 4875 \quad \Rightarrow MRP = \frac{487500}{75} = ₹\,6500$$

18. Let C.P. be ₹x. Then,

$$\frac{3d}{\frac{3d + 4d + 2d}{120}} = \frac{2(69 - x)}{x} \times 100 \Leftrightarrow 78 - x = 2(69 - x) \Leftrightarrow x = 60.$$

19. C.P. = ₹(225 + 15) = ₹240, S.P. = ₹300.

$$\therefore \text{Profit \%} = \left(\frac{60}{240} \times 100\right)\% = 25\%.$$

20. Let C.P. = ₹x. Then,

$$\text{1st S.P.} = 110\% \text{ of } x = \left(\frac{11x}{10}\right). \text{ Second S.P.} = ₹\frac{11x}{5}.$$

$$\text{Profit now} = \left[\frac{\left(\frac{11}{5}x - x\right)}{x} \times 100\right]\% = 120\%.$$

21. Let C.P. = ₹x. Then,

$$106\% \text{ of } x - 94\% \text{ of } x = 6 \Leftrightarrow 12\% \text{ of } x = 6 \Leftrightarrow x = \left(\frac{6 \times 100}{12}\right) = 50.$$

22. Let C.P. be ₹x. Then,

$$120\% \text{ of } x - 116\% \text{ of } x = 20 \Leftrightarrow 4\ \% \text{ of } x = 20 \Leftrightarrow x = \left(\frac{20 \times 100}{4}\right) = 500.$$

23. Let C.P. be ₹x. Then, $425 - x = x - 355 \Leftrightarrow 2x = 780 \Leftrightarrow x = 390.$

24. Let C.P. = ₹x. Then, $x + x\%$ of $x = 75$

$$\therefore x + \frac{x^2}{100} = 75 \text{ or } x^2 + 100x - 7500 = 0 \Leftrightarrow (x + 150)(x - 50) = 0.$$

$\therefore x = 50.$

25. Let C.P. be ₹x. Then

$$104\% \text{ of } (105\% \text{ of } x) = 1092 \Leftrightarrow \frac{104}{100} \times \frac{105}{100} x = 1092.$$

$$\therefore x = \left(\frac{1092 \times 100 \times 100}{104 \times 105}\right) = 1000.$$

26. Let C.P. be ₹100.

Then, S.P. at 10% loss = ₹90.

$$\frac{2}{3} \text{ of actual S.P.} = ₹90 \Leftrightarrow \text{Actual S.P.} = \left(90 \times \frac{3}{2}\right) = ₹135.$$

$\therefore$ Gain = 35%.

27. Let C.P. be ₹x. Then, S.P. = 110% of $x = \frac{11x}{10}$.

$$\text{New C.P.} = \left(\frac{90}{100} \times x\right) = \frac{9x}{10}.$$

$$\text{New S.P.} = \left(110\% \text{ of } \frac{9x}{10}\right) = \frac{99x}{100}.$$

$$\therefore \frac{11x}{10} - \frac{99x}{100} = 132. \text{ So, } x = 1200.$$

28. Let C.P. be ₹x. Then,

$$107\frac{1}{2}\% \text{ of } x - 97\frac{1}{2}\% \text{ of } x = 100 \Leftrightarrow 10\% \text{ of } x = 100 \Leftrightarrow x = \frac{100 \times 100}{10} = 1000.$$

Now, C.P. = ₹1000 & Gain = $12\frac{1}{2}\%$.

$$\therefore \text{S.P.} = \left(112\frac{1}{2}\% \text{ of } 1000\right) = \left(\frac{225}{200} \times 1000\right) = \text{Rs.}1125.$$

29. Let C.P. = ₹100. Then, first S.P. = ₹107.50.

New C.P. = ₹87.50 and gain on it = 30%.

$\therefore$ Second S.P. = 130% of 87.50 = ₹113.75.

Difference in two selling prices = ₹(113.75 – 107.50) = ₹6.25

$$\therefore \text{Actual S.P.} = ₹\left(\frac{100}{6.25} \times 5\right) = ₹80.$$

30. Let C.P. per kg = ₹x.

$\therefore$ 93% of $x = 11.16 \Leftrightarrow x = \left(\frac{11.16 \times 100}{93}\right) = 12.$

So, C.P. = ₹12 per kg.

$\therefore$ S.P. = 107% of ₹12 = $\left(\frac{107}{100} \times 12\right) = 12.84.$

31. Let original rate per egg be x paise. New rate = $\frac{130x}{100}$ paise.

$\therefore$ Number of eggs originally bought for ₹9.10 = $\frac{910}{x}$..

Number of eggs bought now = $\frac{910 \times 10}{13x} = \frac{9100}{13x} = \frac{700}{x}.$

$\therefore \quad \frac{910}{x} - \frac{700}{x} = 3 \Leftrightarrow 3x = 210 \Leftrightarrow x = 70$

So, present rate per egg = 130% of 70 = 91 paise.

32. C.P. of $\frac{3}{4}$th = $\left(\frac{3}{4} \times 400\right)$ = ₹300, C.P. of $\frac{1}{4}$th = ₹100.

S.P. of $\frac{3}{4}$ th = (90% of 300) = ₹270, S.P. of $\frac{1}{4}$th = 110% of 100 = ₹110.

$\therefore$ Total S.P. = ₹(270 + 110) = ₹380

Loss % = $\left(\frac{20}{400} \times 100\right)\% = 5\%.$

33. Difference between two selling prices = ₹72.

$\therefore$ 6% of C.P. = ₹24. So, S.P. = $72 \times \frac{100}{6} = 1200.$

34. Let C.P. of each article be Re. 1.
C.P. of 12 articles = ₹12, S.P. of 12 articles = ₹15.

$\therefore$ Gain = $\left(\frac{3}{12} \times 100\right)\% = 25\%.$

35. (S.P. of 8 dozen) – (C.P. of 8 dozen) = S.P. of 1 dozen
$\therefore$ S.P. of 7 dozen = C.P. of 8 dozen.
Let C.P. be Re. l, per dozen.
Then, C.P. of 7 dozen = ₹7, S.P. of 7 dozen = ₹8.

$\therefore$ Gain = $\left(\frac{1}{7} \times 100\right)\% = 14\frac{2}{7}\%.$

36. (C.P. of 36 bananas) – (S.P. of 36 bananas) = S.P. of 4 bananas
(S.P. of 40 bananas) = (C.P. of 36 bananas)
Let C.P. of each be Re. 1. Then,
C.P. of 40 bananas = ₹40 & S.P. of of 40 bananas = ₹36.

$\therefore$ Loss = $\left(\frac{4}{40} \times 100\right)\% = 10\%$

37. Suppose he buys $(2 \times 5 \times 3) = 30$ lemons.

Then, C.P. $= \left(\frac{1}{2} \times 30\right) = ₹15$, S.P. $= \left(\frac{3}{5} \times 30\right) = ₹18$.

Gain $= \left(\frac{3}{15} \times 100\right)\% = 20\%$.

38. Suppose he buys 8 toffees. Then, C.P. = Re 1, S.P. = 160% of Re 1 $= ₹\frac{8}{5}$.

For $₹\frac{8}{5}$, toffees sold = 8

For Re 1, toffees sold $= \frac{8 \times 5}{8} = 5$.

39. S.P. of 45 oranges = ₹80, Loss = 20%.

$\therefore$ C.P. of 45 oranges $= \left(\frac{100}{80} \times 80\right) = ₹100$.

Now, C.P. = 100, Gain = 20%. S.P. = ₹120.

$120 : 45 :: 48 : x \Leftrightarrow x = \frac{45 \times 48}{120} = 18$.

40. In such problems, we use the rule:

Gain % $= \left[\frac{(100 + \text{common gain\%})^2}{100} - 100\right] = \left[\frac{(108)^2}{100} - 100\right] = 16.64\%$.

41. Formula: Gain % $= \left[\frac{\text{Error}}{(\text{True value}) - (\text{Error})} \times 100\right]$

$\therefore 25 = \frac{x}{(1000 - x)} \times 100 \Leftrightarrow x = 200$.

Thus, he uses a weight of 800 gm instead of 1 kg.

42. Let C.P. = ₹100. Then, marked price = ₹120.

$\therefore$ S.P. = 90% of ₹120 = ₹108.

$\therefore$ Gain % = 8%.

43. Let C.P. = ₹100. Then, S.P. = ₹122.50.

Marked price $= \left(\frac{100}{98} \times 122.50\right) = ₹125$.

$\therefore$ Marked price is 25% above cost price.

44. Let the marked price be ₹x.

Then, C.P. paid by Mohan = 75% of $x = \frac{3x}{4}$.

$\therefore$ 110% of $\frac{3x}{4} = 660 \Leftrightarrow x = \left(660 \times \frac{100}{110} \times \frac{4}{3}\right) = 800$.

Unitary Method

Direct Proportion: *Two quantities are said to be directly proportional if on the increase or decrease of the one, the other increases or decreases proportionately.*

Ex. 1. ***More is the work, more men would finish it in a specified time.***

$\therefore$ Work done is directly proportional to the number of men.

Indirect Proportion: *Two quantities are said to be indirectly proportional, if on the increase of the one, the other decreases proportionately.*

Ex. 2. ***More is the speed, less is the time taken to cover a specified distance.***

So, time taken is inversely proportional to the speed.

Ex. 3. ***Less is the number of days required to finish a work, more persons are to be employed.***

So, time taken is inversely proportional to number of men.

Chain Rule: *The method of finding the fourth proportional when the other three are given is called rule of three.*

Repeated use of rule of three is called chain rule.

SOLVED PROBLEMS

Ex. 1. ***If 25 men can do a piece of work in 36 days, in how many days will 15 men do it?***

Sol. Less men, more days (*indirect*)

$\therefore$ Inverse ratio of men = Ratio of time taken.

Let the required number of days be x.

Then, $15 : 25 :: 36 : x \Leftrightarrow x = \left(\frac{25 \times 36}{15}\right) = 60.$

$\therefore$ Required number of days $= 60$.

Ex. 2. ***If 15 men can build a wall 108 m long in 6 days, what length of similar wall can be built by 25 men in 3 days?***

Sol. Since we have to find the length, we shall compare each item with the length as shown below.

More men, more length built (*Direct*)

Less days, less length built (*Direct*)

$$\left.\begin{array}{l}\text{Men } 15 : 25 \\ \text{Days } 6 : 3\end{array}\right\} :: 108 : x$$

$\therefore 15 \times 6 \times x = 25 \times 3 \times 108$ or $x = \left(\frac{25 \times 3 \times 108}{15 \times 6}\right) = 90.$

EXERCISE 10

1. If 22.5 metres of a uniform rod weighs 85.5 kg, what will be the weight of 9 metres of the same rod?

(*a*) 34.2 kg (*b*) 38.4 kg (*c*) 42 kg (*d*) None of these

2. On a scale of a map, 0.6 cm represents 6.6 km. If the distance between two points on the map is 80.5 cm, the actual distance between these points is:

(*a*) 9 km (*b*) 72.5 km (*c*) 190.75 km (*d*) 885.5 km

3. If 15 men can reap a field in 35 days, in how many days will 21 men reap the field?

(*a*) 49 days (*b*) 9 days (*c*) 25 days (*d*) $29\frac{1}{6}$ days

4. Ten pipes of the same type fill a tank in 24 minutes. If two pipes go out of order, how long will the remaining pipes take to fill the tank?

(*a*) 40 min. (*b*) 45 min. (*c*) $19\frac{1}{5}$ min. (*d*) 30 rnin.

5. A rope makes 140 rounds of a cylinder with base radius l4 cm. How many times can it go round a cylinder with base radius 20 cm?

(*a*) 98 (*b*) 17 (*c*) 200 (*d*) 120

6. A person purchases one kg of tea powder from each of the four places A, B, C, D at the rate of ₹1000 per 1 kg, 2 kg, 4 kg, 5 kg. If on an average he purchased kg of tea powder per ₹1000, then what is the approximate value of ? **[CDS (II) Exam, 2024]**

(*a*) 1.95 (*b*) 2.00 (*c*) 2.05 (*d*) 2.10

7. In a mixture of 80 litres of a liquid and water, 25% of the mixture is the liquid. How much water should be added to the mixture so that the liquid becomes 20% of the mixture?

[CDS Exam, 2021]

(*a*) 15 litres (*b*) 20 litres (*c*) 24 litres (*d*) 25 litres

8. In a mixture of 80 litres of a liquid and water, 25% of the mixture is the liquid. How much water should be added to the mixture so that the liquid becomes 20% of the mixture.

[CDS Exam, 2022]

(*a*) 15 litres (*b*) 20 litres (*c*) 24 litres (*d*) 25 litres

9. An alloy A contains two elements, copper and tin in the ratio of 2 : 3, whereas an alloy B contains the same elements in the ratio of 3 : 4. If 20 kg of ally A, 28 kg of alloy B and some more pure copper are mixture to from a third alloy C which now contains copper and tin the ration of 6 : 7, then what is the quantity of pure copper mixed in the alloy C? **[CDS Exam, 2015]**

(*a*) 3 kg (*b*) 4 kg (*c*) 5 kg (*d*) 7 kg

10. The cost of 2.5 kg rice is ₹125. The cost of 9 kg rice is equal to that of 4 kg pulses. The cost of 14 kg pulses is equal to that of 1.5 kg tea. The cost of 2kg tea is equal to that of 5 kg nuts. What is the cost of 11 kg nuts? **[CDS (II) Exam, 2016]**

(*a*) ₹2310 (*b*) ₹3190 (*c*) ₹4070 (*d*) ₹4620

11. Three taps A, B and C together can fill a tank in 6 hours. Tap C alone can fill the tank in 12 hours. To fill the tank, when it is empty, all the three taps are started together. After working *t* hours, tap C is closed and the tank is filled in 8 more hours. What is *t* equal to ?

[CDS (I) Exam, 2023]

(*a*) 1 (*b*) 2 (*c*) 4 (*d*) 6

12. Three glasses P, Q and R have capacities in the ratio 1 : 2 : 3. All these glasses are completely filled with mixtures of milk and water. The ratio of milk to water in P is 1 : 2, in Q it is 2: 3 and in R it is 3: 1. If the content of all three glasses are put into a bigger container, what will be the ratio of milk to water in the container ? **[CDS (II) Exam, 2023]**

(*a*) 203 : 117 (*b*) 203 : 157 (*c*) 172 : 91 (*d*) 165 : 88

13. If 18 binders bind 900 books in 10 days, how many binders will be required to bind 660 books in 12 days?

(*a*) 55 (*b*) 14 (*c*) 13 (*d*) 11

14. If 40 persons consume 240 kg of rice in 15 days, in how many days will 30 persons consume 48 kg of rice?

(*a*) $6\frac{1}{4}$ days (*b*) $3\frac{3}{4}$ days (*c*) 4 days (*d*) 9 days

15. If 5 men working 6 hours a day can reap a field in 20 days, in how many days will 15 men reap the field, working 8 hours a day?

(a) 5 days (*b*) 6 days (*c*) $7\frac{1}{2}$ days (*d*) 9 days

16. If 20 men working 7 hours a day can do a piece of work in 10 days, in how many days will 15 men working for 8 hours a day do the same piece of work?

(*a*) $15\frac{5}{21}$ days (*b*) $11\frac{2}{3}$ days (*c*) $6\frac{9}{16}$ days (*d*) $4\frac{1}{5}$ days

17. If x men working x hours a day can do x units of a work in x days, then y men working y hours per day would be able to complete, how many units of work?

(a) $\frac{x^2}{y^3}$ (*b*) $\frac{x^3}{y^2}$ (*c*) $\frac{y^2}{x^3}$ (*d*) $\frac{y^3}{x^2}$

18. If the work done by x men in $(x + 1)$ days is equal to the work done by $(x + 5)$ men in $(x - 2)$ days, then what is the value of x ? **[CDS (I) Exam, 2024]**

(*a*) 5 (*b*) 6 (*c*) 7 (*d*) 8

19. If $\frac{4}{5}$th of a cistern is filled in l minute, how much more time will be required to fill the rest of it?

(*a*) 10 sec (*b*) 12 sec (*c*) 15 sec (*d*) 20 sec

20. A garrison had provisions for a certain number of days. After 10 days, $\frac{1}{5}$th of men desert and it is found that the provisions will now last just as long as before. How long was that?

(*a*) 15 days (*b*) 25 days (*c*) 35 days (*d*) 50 days

21. If 4 examiners can examine a certain number of answer books in 8 days by working 5 hours a day, for how many hours a day would 2 examiners have to work in order to examine twice the number of answer books in 20 days?

(*a*) 6 hours (*b*) $7\frac{1}{2}$ hours (*c*) 8 hours (*d*) 9 hours

22. If 5 engines consume 6 metric tonnes of coal when each is running 9 hours a day, how much coal (in metric tonnes) will be needed for 8 engines, each running 10 hours a day, it being given that 3 engines of the former type consume as much as 4 engines of latter type?

(*a*) 8 (*b*) $3\frac{1}{8}$ (*c*) 6.48 (*d*) $8\frac{5}{9}$

23. If 9 men working $7\frac{1}{2}$ hours a day can finish a work in 20 days, then how many days will be taken by 12 men working 6 hours a day to finish the work, it being given that 3 men of latter type work as much as 2 men of the former type in the same time?

(*a*) $12\frac{1}{2}$ (*b*) 13 (*c*) $9\frac{1}{2}$ (*d*) 11

24. 15 men take 21 days of 8 hours each to do a piece of work. How many days of 6 hours each would 21 women take, if 3 women do as much work as 2 men?

(*a*) 20 (*b*) 18 (*c*) 25 (*d*) 30

ANSWERS

1. (*a*)	**2.** (*d*)	**3.** (*c*)	**4.** (*d*)	**5.** (*a*)	**6.** (*c*)	**7.** (*b*)	**8.** (*b*)	**9.** (*b*)
10. (*d*)	**11.** (*b*)	**12.** (*a*)	**13.** (*d*)	**14.** (*c*)	**15.** (*a*)	**16.** (*b*)	**17.** (*d*)	**18.** (*a*)
19. (*c*)	**20.** (*d*)	**21.** (*c*)	**22.** (*a*)	**23.** (*a*)	**24.** (*d*)			

SOLUTIONS

1. Less length, less weight (*Direct*)

$\therefore 22.5 : 9 :: 85.5 : x \Leftrightarrow x = \left(\frac{9\times 85.5}{22.5}\right) = 34.2$

So, required weight = 34.2 kg.

2. More distance on the map, more actual distance (*Direct*)

$0.6 : 80.5 :: 6.6 : x \Leftrightarrow x = \frac{80.5\times 6.6}{0.6} = 885.5$ km.

3. More men, less days (*indirect*)

$\therefore$ Inverse ratio of men = Ratio of days.

$21 : 15 :: 35 : x \Leftrightarrow x = \frac{15\times 35}{21} = 25$ days.

4. Less pipes, more time (*indirect*)

So, inverse ratio of number of pipes = Ratio of time.

$\therefore 8 : 10 :: 24 : x \Leftrightarrow x = \left(\frac{10\times 24}{8}\right) = 30$ min.

5. Larger radius, less number of rounds.

$\therefore$ Inverse ratio of radii = Ratio of number of rounds.

$20 : 14 :: 140 : x \Leftrightarrow x = \left(\frac{14\times 140}{20}\right) = 98.$

6. Total purchase = 4 kg (A + B + C + D)

cost of 1 kg tea powder in A's place = ₹1000

cost of 1 kg tea powder in B's place = $\frac{1000}{2}$ = ₹500

cost of 1 kg tea powder in C's place = $\frac{1000}{4}$ = ₹250

cost of 1 kg tea powder in D's place = $\frac{1000}{5}$ = ₹200

Total cost = 1000 + 500 + 250 + 200 = ₹1950

According to the above calculation,

At ₹1950 = 4 kg

At ₹1000 = $\frac{4}{1950}$ ×1000 = 2.05 kg

On average, he purchases 2.05 kg of sugar for every ₹1000 spent.

7. We have,

Amount of liquid = $80 \times \frac{25}{100} = 20$ litres

Let x be the amount of water be added

We have, $20 = \frac{20}{100} \times (80 + x) \Rightarrow 100 = 80 + x \Rightarrow x = 100 - 80 = 20$ litres

8. We have, Amount of liquid = $80 \times \frac{25}{100} = 20$ litres

Let x be the amount of water be added we have, $20 = \frac{20}{100} \times (80 + x)$

$\Rightarrow 100 = 80 + x \qquad \Rightarrow \qquad x = 100 - 80 = 20$ litres

9. Let quantity of pure copper added by 'x' kg

Amount of copper in 20 kg of alloy A = $\frac{2}{5} \times 20 = 8$ kg

Amount of copper in 28 kg of alloy B = $\frac{3}{7} \times 28 = 12$ kg

Total amount of copper in alloy C = $8 + 12 + x = 20 + c$

Total amount of alloy = $20 + 28 + x = 48 + xv$

$20 + x = \frac{6}{13} \times (48 + x) \Rightarrow 260 + 13x \;\; 288 + 6x \Rightarrow 13x - 6x = 288 \;\; 260$

$\Rightarrow 7x = 28 \Rightarrow x = \frac{28}{7} = 4$ kg

10. The data is tabulated below.

Cost	Rice	Pulses	Tea	Nuts
125	2.5			
	9	4		
		14	1.5	
			2	5
				11

The required cost is

$$125\left(\frac{9}{2.5}\right)\left(\frac{14}{4}\right)\left(\frac{2}{1.5}\right)\left(\frac{11}{5}\right) = 125\left(\frac{18}{5}\right)\left(\frac{7}{2}\right)\left(\frac{4}{3}\right)\left(\frac{11}{5}\right) = 5(3)\,(7)(4)\,(11)$$

60 (77) = 4620.

11. Tap (A + B + C) = $\frac{1}{6}$

Tap C = $\frac{1}{12}$

Tap (A + B) = $\frac{1}{6} - \frac{1}{12} = \frac{1}{12}$

Now, according to the question,

Part of tank filled by (A + B) in 8 hours = $\frac{8}{12} = \frac{2}{3}$

Before tap C was closed,

Part of tank filled by A, B, and C together $= 1 - \frac{2}{3} = \frac{1}{3} \Rightarrow t = \frac{1/3}{1/6} = 2$ hours

12. Let capacities of glasses *P*, *Q* and *R* are *x*, 2*x* and 3*x* respectively

Now, according to the question,

$$\text{Quantity of water} = \frac{\frac{2}{3}\times x + \frac{3}{5}\times 2x + \frac{1}{4}\times 3x}{6x} = \frac{\frac{2}{3}+\frac{6}{5}+\frac{3}{4}}{6} = \frac{157}{360}$$

$$\text{Quantity of milk} = \frac{\frac{1}{3}\times x + \frac{2}{5}\times 2x + \frac{3}{4}\times 3x}{6x} = \frac{\frac{1}{3}+\frac{4}{5}+\frac{9}{4}}{6} = \frac{20+48+135}{360} = \frac{203}{360}$$

∴ required ratio = 203 : 157

13. Less books, less binders (*Direct*)

More days, less binders (*Indirect*)

$$\left.\begin{array}{ll}\text{Books} & 900 : 660\\ \text{Days} & 12 : 10\end{array}\right\} :: 18 : x$$

$900 \times 12 \times x = 660 \times 10 \times 18$ or $x = \left(\frac{660\times 10\times 18}{900\times 12}\right) = 11$ binders.

14. Less persons, more days (*Indirect*)

Less rice, less days (*Direct*)

$$\left.\begin{array}{ll}\text{Persons} & 30 : 40\\ \text{Weight} & 240 : 48\end{array}\right\} :: 15 : x$$

$30 \times 240 \times x = 40 \times 48 \times 15$ or $x = \left(\frac{40\times 48\times 15}{30\times 240}\right) = 4$ days.

15. More men, less days (*indirect*)

More working hours, less days (*indirect*)

$$\left.\begin{array}{ll}\text{Men} & 15 : 5\\ \text{Working hrs.} & 8 : 6\end{array}\right\} :: 20 : x$$

$(15 \times 8 \times x) = (5 \times 6 \times 20)$ or $x = \left(\frac{5\times 6\times 20}{15\times 8}\right) = 5$ days.

16. Less men, more days (*indirect*)

More working hours, less days (*indirect*)

$$\left.\begin{array}{ll}\text{Men} & 15:20\\ \text{Working hrs.} & 8:7\end{array}\right\}::10:x$$

$15 \times 8 \times x = 20 \times 7 \times 10$ or $x = \left(\dfrac{20\times7\times10}{15\times8}\right) = 11\dfrac{2}{3}$ days.

17. More men, more units (*Direct*)

More working hours, more units (*Direct*)

More days, more units (*Direct*)

$$\left.\begin{array}{ll}\text{Men} & x:y\\ \text{Working hrs} & x:y\\ \text{Days} & x:y\end{array}\right\}::x:z$$

$\therefore x \times x \times x \times z = y \times y \times y \times x$. So, $z = \left(\dfrac{y^3}{x^2}\right)$ units.

18. $\therefore \quad M_1 \times D_1 = M_2 \times D_2$

$\Rightarrow \quad x \times (x + 1) = (x + 5) \times (x - 2) \Rightarrow x^2 + x = (x^2 + 3x - 10)$

$\Rightarrow \quad 2x = 10 \Rightarrow x = 5$

19. Part filled = $\dfrac{4}{5}$, Remaining part = $\dfrac{1}{5}$.

Less part, less time (*Direct*)

$$\frac{4}{5}:\frac{1}{5}::60:x \Leftrightarrow x = \left(\frac{1}{5}\times60\times\frac{5}{4}\right) = 15\,\text{sec}.$$

20. Initially, let there be x men having provisions for y days. After 10 days, x men had provisions for $(y - 10)$ days.

Thus, $\left(x - \dfrac{x}{5}\right)$ men had provisions for y days.

$\therefore x(y - 10) = \left(\dfrac{4x}{5}\right)y$ or $xy - 50x = 0$. So, $y = 50$.

21. Less examiners, more hours per day (*Indirect*)

More days, less hours per day (*Indirect*)

More answer books, more hours per day (*Direct*)

$$\left.\begin{array}{ll}\text{Examiners} & 2:4\\ \text{Days} & 20:8\\ \text{Answer books} & 1:2\end{array}\right\}::5:x$$

$$\therefore \quad x = \left(\frac{4\times8\times2\times5}{2\times20\times1}\right) = 8 \text{ hours per day.}$$

22. More engines, more coal (*Direct*)

More hours a day, more coal (*Direct*)

More rate of consumption, more coal (*Direct*)

$$\left.\begin{array}{ll}\text{Engines} & 5:8 \\ \text{Hrs/Day} & 9:10 \\ \text{Rate of consumption} & \frac{1}{3}:\frac{1}{4}\end{array}\right\} :: 6 : x$$

$$\therefore x = \left(8 \times 10 \times \frac{1}{4} \times 6 \times \frac{1}{5} \times \frac{1}{9} \times 3\right) = 8 \text{ metric tonnes.}$$

23. More men, less days (*Direct*)
Less hours a day, more days (*Indirect*)
More speed, less days (*Indirect*)

$$\left.\begin{array}{ll}\text{Men} & 12:9 \\ \text{Hrs/day} & 6:7\frac{1}{2} \\ \text{Speed} & \frac{1}{2}:\frac{1}{3}\end{array}\right\} :: 20 : x$$

$$\therefore \quad x = \left(9 \times \frac{15}{2} \times \frac{1}{3} \times 20 \times \frac{1}{12} \times \frac{1}{6} \times 2\right) = 12\frac{1}{2} \text{ days.}$$

24. 3 women ≡ 2 men. So, 21 women ≡ 14 men.
Less men, more days (*Indirect*)
Less hours, more days (*Indirect*)

$$\left.\begin{array}{ll}\text{Men} & 14:15 \\ \text{Working hrs} & 6:8\end{array}\right\} :: 21 : x$$

$$\therefore \quad x = \left(\frac{15 \times 8 \times 21}{14 \times 6}\right) = 30 \text{ days.}$$

CHAPTER 11

Time and Work

Highlights: Problems on time and work, pipes and cisterns.

General Rules (*Time* and *Work*):

(*i*) If A can do a piece of work in n days, then the part of the work done by him in 1 day $= \frac{1}{n}$.

(*ii*) If A's 1 day's work $= \frac{1}{n}$, then the time taken by A to finish the work $= n$ days.

General Rules (*Pipes* and *Cisterns*):

(*i*) If an inlet pipe can fill a cistern in x hours, then the part filled in 1 hour $= \frac{1}{x}$.

(*ii*) If an inlet pipe can fill a tank in x hours and an outlet empties the full tank in y hours, then the net part filled in 1 hour when both the pipes are opened $= \left(\frac{1}{x} - \frac{1}{y}\right)$.

SOLVED PROBLEMS

Ex. 1. ***A can build a wall in 30 days, which B alone can build in 40 days. If they build it together and get a payment of ₹385, what is B' share?***

Sol. Ratio of their shares = Ratio of their 1 day's work $= \frac{1}{30} : \frac{1}{40} = 4 : 3$.

$\therefore$ B's share $= ₹\left(385 \times \frac{3}{7}\right) = ₹165$.

Ex. 2. ***A is thrice as good a workman as B and is therefore able to finish a piece of work in 60 days less than B. Find the time in which they can do it, working together.***

Sol. Ratio of work done by A and B in same time = 3 : 1.

Ratio of times taken to finish the same work = 1 : 3.

Suppose B takes x days and A takes $(x - 60)$ days to finish a work.

Then,

$$\frac{x-60}{x} = \frac{1}{3} \Leftrightarrow 3(x-60) = x \Leftrightarrow x = 90.$$

$\therefore$ Time taken by B to finish the work = 90 days.

Time taken by A to finish the work = (90 – 60) = 30 days.

$\therefore$ $(A + B)$'s 1 day's work $= \left(\frac{1}{90} + \frac{1}{30}\right) = \frac{4}{90} = \frac{2}{45}$.

Hence, both together can finish the work in $\frac{45}{2} = 22\frac{1}{2}$ days.

Ex. 3. ***A & B can do a piece of work in 12 days; B & C in 15 days; C& A in 20 days. In how many days will they finish it together and separately?***

Sol. $(A+B)$'s 1 day's work $= \frac{1}{12}$; $(B+C)$'s 1 day's work $= \frac{1}{15}$;

$(C+A)$'s 1 day's work $= \frac{1}{20}$.

Adding: $2(A+B+C)$'s 1 day's work $= \left(\frac{1}{12}+\frac{1}{15}+\frac{1}{20}\right) = \frac{12}{60} = \frac{1}{5}$.

$\therefore$ $(A+B+C)$'s 1 day's work $= \frac{1}{10}$.

So, they together can finish the work in 10 days.

A's 1 day's work = [$(A+B+C)$'s 1 day's work] – [$(B+C)$'s 1 day's work]

$= \left(\frac{1}{10}-\frac{1}{15}\right) = \frac{1}{30}$.

$\therefore$ A alone can finish the work in 30 days.

Similarly, B's 1 day's work $= \left(\frac{1}{10}-\frac{1}{20}\right) = \frac{1}{20}$.

$\therefore$ B alone can finish it in 20 days.

And, C's 1 day's work $= \left(\frac{1}{10}-\frac{1}{12}\right) = \frac{1}{60}$.

$\therefore$ C alone can finish the work in 60 days.

Problems on Pipes & Cisterns

Ex. 4. ***Pipe A can fill a tank in 20 hours while pipe B alone can fill it in 30 hours and pipe C can empty the full tank in 40 hours. If all the pipes are opened together, how much time will be needed to make the tank full?***

Sol. Net part filled in 1 hour $= \left(\frac{1}{20}+\frac{1}{30}-\frac{1}{40}\right) = \frac{7}{120}$.

$\therefore$ The tank will be full in $\frac{120}{7}$ hrs $= 17\frac{1}{7}$ hours.

Ex. 5. ***If 3 men or 4 women can build a wall in 43 days, then in how many days can 7 men and 5 women build this wall?***

Sol. 3 men $\equiv$ 4 women.

$\therefore$ 7 men + 5 women $\equiv \left(\frac{4}{3}\times 7+5\right)$ women $= \frac{43}{3}$ women.

Let the required number of days be x.

Less women, more days (*indirect*)

$\frac{43}{13} : 4 :: 43 : x$. So, $x = \left(4\times 43\times\frac{3}{43}\right) = 12$ days.

EXERCISE 11

1. A, B and C can complete a work in x, $1\cdot5x$ and $2x$ days respectively. If they complete the work together, in what ratio should they be paid? **[CDS (I) Exam, 2023]**

(*a*) 2 : 3 : 4 (*b*) 6 : 4 : 3 (*c*) 3 : 2 : 1 (*d*) 4 : 3 : 2

2. If A and B can finish a work in 10 days, B and C can finish the same work in 12 days, C and A can finish the same work in 15 days; then in how many days can A, B and C together finish half of the work ? **[CDS (II) Exam, 2023]**

(*a*) 8 days (*b*) 5 days (*c*) 4 days (*d*) 3 days

3. Let the work done by $(x-1)$ men in $(x+1)$ days be y. Let the work done by $(x+2)$ men in $(x-1)$ days be z. If $y : z = 9 : 10$, then what is the value of x? **[CDS (I) Exam, 2021]**

(*a*) 8 (*b*) 9 (*c*) 10 (*d*) 12

4. 20 men are supposed to complete a work in 10 days. After working for 5 days, they realise that only one-fourth of the work is done. How many more men they need to employ to finish the work in time? **[CDS (II) Exam, 2021]**

(*a*) 40 (*b*) 30 (*c*) 20 (*d*) 15

5. Let the work done by $(3n-1)$ men in $(2n+1)$ days be x and work done by $(3n+1)$ men in $(4n-3)$ days be y. If $y : x = 6 : 11$, then what is the value of n? **[CDS (II) Exam, 2021]**

(*a*) 6 (*b*) 7 (*c*) 8 (*d*) 9

6. Sheela can stitch a suit in 2 days, while Meena can stitch a suit in $1\frac{1}{2}$ days. How many days will both take in stitching 30 suits? **[CDS (II) Exam, 2021]**

(*a*) 32 days (*b*) 33 days (*c*) 30 days (*d*) 40 days

7. A can do a certain work at twice the speed of B. Further, B can do the same work at 1·5 times the speed of C. All of them together can finish the work in 12 days. In how many days can C alone finish the work ? **[CDS (I) Exam, 2023]**

(a) 36 days (b) 45 days (c) 60 days (d) 66 days

8. Three persons A, B and C together can do a piece of work in 36 days. A and B together can do five times as much work as C alone; B and C together can do as much work as A alone. If A and C together can do n times as much work as B alone, then what is the value of n?

[CDS (II) Exam, 2023]

(a) 1·5 (b) 2 (c) 2·5 (d) 3

9. If x men working x hours per day can do x units of work in x days, then y men working y hours per day in y days would be able to do k units of work. What is the value of k?

[CDS (II) Exam, 2020]

(*a*) $x^2 y^{-3}$ (*b*) $x^3 y^{-2}$ (*c*) $y^2 x^{-3}$ (*d*) $y^3 x^{-2}$

10. A, B, C, D can complete a work in 3, 6, 9, 12 hours respectively. Further, only one person can work at a time in each hour and nobody can work for two consecutive hours. It is not necessary to engage all. What is the minimum number of hours that they will take to finish the work? **[CDS (I) Exam, 2024]**

(*a*) $\frac{36}{25}$ (*b*) $\frac{12}{5}$ (*c*) 4 (*d*) 2

11. 50 men can complete a work in 40 days. They begin the work together but a batch of 5 men left after each period of 10 days. What is the time to complete the work ?

[CDS (I) Exam, 2024]

(*a*) 45 days (*b*) 50 days (*c*) 55 days (*d*) 60 days

12. If the ratio of the work done by $(x+2)$ workers in $(x-3)$ days to the work done by $(x+4)$ workers in $(x-2)$ days is 3 : 4, then what is the value of x? **[CDS (II) Exam, 2019]**

(*a*) 8 (*b*) 10 (*c*) 12 (*d*) 15

13. A work when done by 10 women is completed in 12 days. The same work can be completed in 8 days when done by 5 men. How many days will it takes to complete when 6 women and 3 men are employed to perform the same job? **[CDS (I) Exam, 2018]**

(*a*) 12 (*b*) 10 (*c*) 8 (*d*) 5

14. A man undertakes to do a certain work in 150 days. He employes 200 men. He finds that only a quarter of the work is done in 50 days. How many additional men should he employ so that the whole work is finished in time? **[CDS (I) Exam, 2018]**

(*a*) 75 (*b*) 85 (*c*) 100 (*d*) 120

15. If 5 tractors can plough 5 hectares of land in 5 days, then what is the number of tractors required to plough 100 hectares in 50 days? **[CDS (I) Exam, 2018]**

(*a*) 100 (*b*) 20 (*c*) 10 (*d*) 5

16. Two workers A and B working together completed a job in 5 days. Had A worked twice as efficiently as he actually did and B worked one-third as efficiently as the actually did, the work would have completed in 3 days. In how many days could A alone complete the job? **[CDS (II) Exam, 2018]**

(*a*) $3\frac{1}{2}$ days (*b*) $4\frac{1}{6}$ days (*c*) $5\frac{1}{2}$ days (*d*) $6\frac{1}{4}$ days

17. A is thrice as good a workman as B and takes 10 days less to do a piece of work than B takes. The number of days taken by B alone to finish the work is **[CDS (II) Exam, 2018]**

(*a*) 12 (*b*) 15 (*c*) 20 (*d*) 30

18. Twelve (12) men work 8 hours per day and require 10 days to build a wall. If 8 men are available, how many hours per day must they work to finish the work in 8 days? **[CDS (II) Exam, 2018]**

(*a*) 10 hours (*b*) 12 hours (*c*) 15 hours (*d*) 18 hours

19. X and Y together can finish a job in 6 days. X can alone do the same job in 12 days. How long will Y alone take to do the same job? **[CDS (II) Exam, 2018]**

(*a*) 16 days (*b*) 12 days (*c*) 10 days (*d*) 8 days

20. Twelve (12) persons can paint 10 identical rooms in 16 days. In how many days can 8 persons paint 20 such rooms? **[CDS (II) Exam, 2018]**

(*a*) 12 (*b*) 24 (*c*) 36 (*d*) 48

21. Water flows through a cylindrical pipe of internal diameter 7 cm at the rate of 5 m/s. The time, in minutes, the pipe would take to fill an empty rectangular tank 4m × 3m × 2.31m is

(*a*) 28 (*b*) 24 (*c*) 20 (*d*) 12

[CDS (I) Exam, 2015]

22. *A* and *B* are two taps which can fill a tank individually in 10 minutes and 20 minutes respectively. However there is a leakage at the bottom, which can empty a filled tank in 40 minutes. If the tank is empty initially, how much time will both the taps take to fill the tank with leakage? **[CDS (II) Exam, 2015]**

(*a*) 2 minutes (*b*) 4 minutes (*c*) 5 minutes (*d*) 8 minutes

23. If 4 men working 4 hours per day for 4 days complete 4 units of work, then how many units of work will be completed by 2 men working for 2 hours per day in 2 days?

(*a*) 2 (*b*) 1 (*c*) $\frac{1}{2}$ (*d*) $\frac{1}{8}$

[CDS (II) Exam, 2015]

24. If m persons can paint a house in d days, how many days will it take for $(m + 2)$ persons to paint the same house? **[CDS (II) Exam, 2015]**

(*a*) $md + 2$ (*b*) $md - 2$ (*c*) $\frac{m+2}{md}$ (*d*) $\frac{md}{m+2}$

25. A river 2.5 m deep and 45m wide is flowing at the speed of 3.6 km/hour. The amount of water that runs into the sea per minute is **[CDS (II) Exam, 2015]**

(*a*) 6650 m^3 (*b*) 6750 m^3 (*c*) 6850 m^3 (*d*) 6950 m^3

26. Two pipes A and B can fill a tank in 60 minutes and 75 minutes respectively. There is also an outlet *C*. If A, B and C are opened together, the tank is full in 50 minutes. How much time will be taken by C to empty the full tank? **[CDS (I) Exam, 2016]**

(*a*) 100 minutes (*b*) 110 minutes
(*c*) 120 minutes (*d*) 125 minutes

27. A can do 50% more work than *B* in the same time. B alone can do a piece of work in 30 hours. B starts working and had already worked for 12 hours when A joins him. How many hours should B and A work together to complete the remaining work? **[CDS (II) Exam, 2016]**

(*a*) 6 hours (*b*) 12 hours (*c*) 4.8 hours (*d*) 7.2 hours

28. A tank can be filled by pipe *X* in 2 hours and pipe *Y* in 6 hours. At 10 a.m. pipe *X* was opened. At what time will the tank be filled if pipe *Y* is opened at 11 a.m.? **[CDS (II) Exam, 2016]**

(*a*) 12: 45 hours (*b*) 5:00 p.m. (*c*) 11:45 a.m. (*d*) 11:50 a.m.

29. If 15 men take 21 days of 8 hours each to do a piece of work, then what is the number of days of 6 hours each that 21 women would take, if 3 women would do as much work as 2 men?

(*a*) 18 (*b*) 20 (*c*) 25 (*d*) 30

[CDS (I) Exam, 2017]

30. A and B working together can finish a piece of work in 12 days while B alone can finish it in 30 days. In how many days can A alone finish the work? **[CDS (I) Exam, 2017]**

(*a*) 18 days (*b*) 20 days (*c*) 24 days (*d*) 25 days

31. If 5 men can do a piece of work in 10 days and 12 women can do the same work in 15 days, the number of days required to complete the work by 5 men and 6 women is

(*a*) $7\frac{1}{2}$ days (*b*) 8 days (*c*) $9\frac{1}{2}$ days (*d*) 12 days

[CDS (I) Exam, 2017]

32. *A* and *B* can do a piece of work in 18 days; *B* and *C* in 24 days; *C* and *A* in 36 days. In how many days can they do it all working together?

(*a*) 12 (*b*) 13 (*c*) 16 (*d*) 26

33. *A* and *B* can do a piece of work in 12 days; *B* and *C* in 15 days; *C* and *A* in 20 days. In how many days can *A* alone do it?

(*a*) $15\frac{2}{3}$ (*b*) 24 (*c*) 30 (*d*) 40

34. *A* can do a piece of work in 24 days while *B* alone can do it in 16 days. With the help of *C*, they finish the work in 8 days. In how many days can *C* alone do the work?

(*a*) 32 (*b*) 36 (*c*) 40 (*d*) 48

35. A can do a piece of work in 20 days and *B* can do it in 12 days. *B* worked at it for 9 days. In how many days *A* alone can finish the remaining work?

(*a*) 3 (*b*) 5 (*c*) 7 (*d*) 11

36. *A* can complete a job in 9 days, *B* in 10 days and *C* in 15 days. *B* and *C* start the work and are forced to leave after 2 days. The number of days taken by *A* to finish the remaining work is:
(*a*) 8 (*b*) 9 (*c*) 6 (*d*) 14

37. *A* is twice as good a workman as *B* and together they finish a piece of work in 14 days. In how many days can *A* alone finish the work?
(*a*) 11 (*b*) 21 (*c*) 28 (*d*) 42

38. *A* is thrice as good a workman as *B* and takes 10 days less to do a piece of work than *B* takes. The number of days taken by *B* to finish the work is:
(*a*) 12 (*b*) 15 (*c*) 20 (*d*) 30

39. *A* does half as much work as *B* in three-fourth of the time. If together they take 18 days to complete the work, how many days will *B* take to do it?
(*a*) 30 (*b*) 35 (*c*) 40 (*d*) 45

40. A works twice as fast as *B*. If *B* can complete a piece of work independently in 12 days, then the number of days taken by *A* and *B* together to finish the work is:
(*a*) 4 (b) 6 (*c*) 8 (d) 18

41. *A* alone can finish a work in 10 days and *B* alone can do it in 15 days. If they work together and finish it, then out of a total wages of ₹225, the amount (in rupees) that *A* will get, is:
(*a*) 90 (*b*) 112.50 (*c*) 135 (*d*) 150

42. *A* can do a piece of work in 14 days which *B* can do in 21 days. They begin together but 3 days before the completion of the work, *A* leaves off. The total number of days to complete the work is:
(*a*) $6\frac{3}{5}$ (*b*) $8\frac{1}{2}$ (*c*) $10\frac{1}{5}$ (*d*) $13\frac{1}{2}$

43. *A* and *B* working separately can do a piece of work in 9 and 12 days respectively. If they work for a day alternately, *A* beginning, in how many days, the work will be completed?
(a) $10\frac{1}{2}$ (*b*) $10\frac{1}{4}$ (*c*) $10\frac{2}{3}$ (*d*) $10\frac{1}{3}$

44. *A* and *B* can do a piece of work in 45 and 40 days respectively. They began the work together but *A* leaves after some days and B finished the remaining work in 23 days. After how many days did *A* leave?
(a) 6 (*b*) 3 (*c*) 9 (*d*) 12

45. *A* can do a piece of work in 80 days. He works at it for 10 days and then *B* alone finishes the work in 42 days. The number of days taken by the two together to finish the work is:
(*a*) 24 (*b*) 25 (*c*) 30 (*d*) 35

46. A sum of money is sufficient to pay *A*'s wages for 21 days and *B's* wages for 28 days. The money is sufficient to pay the wages of both for:
(*a*) 12 days (*b*) 14 days (*c*) $12\frac{1}{4}$ days (*d*) $24\frac{1}{2}$ days

47. If 12 men or 18 women can reap a field in 14 days, then the number of days that 8 men and 16 women will take to reap the same field is:
(*a*) 5 (*b*) 7 (*c*) 8 (*d*) 9

48. If 1 man or 2 women or 3 boys can do a piece of work in 44 days, then the same piece of work will be done by 1 man, 1 woman and 1 boy in:
(*a*) 21 days (*b*) 24 days (*c*) 26 days (*d*) 33 days

49. A certain number of men complete a piece of work in 60 days. If there were 8 men more, the work could be finished in 10 days less. How many men were originally there?
(*a*) 30 (*b*) 32 (*c*) 36 (*d*) 40

50. A question is given followed by two statements I and II. Consider the question and the Statements and mark the correct option.

Question : In how many days can A, B and C together finish the work?

Statement-I : A and B together can finish the work in 24 days.

Statement-II : B and C together can finish the work in 36 days. **[CDS (II) Exam, 2024]**

Which one of the following is correct in respect of the above Question and the Statements?
(*a*) The Question can be answered by using one of the Statements alone, but cannot be answered using the other Statement alone
(*b*) The Question can be answered by using either Statement alone
(*c*) The Question can be answered by using both the Staetments together, but cannot be answered using either Statement alone
(*d*) The Question cannot be answered even by using both the Statements together

51. 8 men or 12 women can do a piece of work in 24 days. In how many days can the work be done by 8 men and 12 women? **[CDS (I) Exam, 2022]**
(*a*) 12 days (*b*) 18 days
(*c*) 24 days (*d*) Cannot be determined due to insufficient data

52. If 20 persons can clean 20 floors in 20 days, then in how many days can 16 persons clean 16 floors? **[CDS (I) Exam, 2021]**
(*a*) 25 days (*b*) 24 days (*c*) 20 days (*d*) 16 days

53. A person agrees to work for 30 days, on a condition that for every day's work he should receive ₹500, and that for every day's absence from work he should forfeit ₹100. At the end of the time he received ₹11,400. How many days did he work? **[CDS (II) Exam, 2021]**
(*a*) 20 (*b*) 21 (*c*) 24 (*d*) 25

54. If 6 men and 8 women can do a piece of work in 10 days; and 13 men and 24 women can do the same work in 4 days, then what is the ratio of daily work done by a man to that of a woman? **[CDS (I) Exam, 2020]**
(*a*) 2 : 1 (*b*) 1 : 2 (*c*) 4 : 3 (*d*) 3 : 4

55. If 17 women and 24 men can do a piece of work in 5 days and 12 women and 23 men can do it in 6 days, then which one of the following is correct? **[CDS (I) Exam, 2023]**
(a) Efficiency of 13 women = Efficiency of 18 men
(b) Efficiency of 11 women= Efficiency of 16 men
(c) Efficiency of 13 women = Efficiency of 17 men
(d) Efficiency of 11 women = Efficiency of 15 men

ANSWERS

1. (*b*)	**2.** (*c*)	**3.** (*a*)	**4.** (*a*)	**5.** (*b*)	**6.** (*c*)	**7.** (*d*)	**8.** (*b*)	**9.** (*d*)	**10.** (*c*)
11. (*b*)	**12.** (*b*)	**13.** (*c*)	**14.** (*c*)	**15.** (*c*)	**16.** (*d*)	**17.** (*b*)	**18.** (*c*)	**19.** (*b*)	**20.** (*d*)
21. (*b*)	**22.** (*d*)	**23.** (*c*)	**24.** (*d*)	**25.** (*b*)	**26.** (*a*)	**27.** (*d*)	**28.** (*c*)	**29.** (*d*)	**30.** (*b*)
31. (*a*)	**32.** (*c*)	**33.** (*c*)	**34.** (*d*)	**35.** (*b*)	**36.** (*c*)	**37.** (*b*)	**38.** (*b*)	**39.** (*a*)	**40.** (*a*)
41. (*c*)	**42.** (*c*)	**43.** (*b*)	**44.** (*c*)	**45.** (*c*)	**46.** (*a*)	**47.** (*d*)	**48.** (*b*)	**49.** (*d*)	**50.** (*d*)
51. (*d*)	**52.** (*c*)	**53.** (*c*)	**54.** (*a*)	**55.** (*a*)					

SOLUTIONS

1. Ratio of time taken by A, B and C = $x : 1.5 : 2x$
$= 2 : 3 : 4$

Now, ratio of money = $\frac{1}{2}:\frac{1}{3}:\frac{1}{4} = 6 : 4 : 3$

2. Number of days to complete the work taken by A = x days
Number of days to complete the work taken by B = y days
Number of days to complete the taken by C = z days
Now, according to the question,

$$\frac{1}{x}+\frac{1}{y}=\frac{1}{10} \quad \text{...(i)}$$

$$\frac{1}{y}+\frac{1}{z}=\frac{1}{12} \quad \text{...(ii)}$$

and $$\frac{1}{z}+\frac{1}{x}=\frac{1}{15} \quad \text{...(iii)}$$

On adding (i), (ii) and (iii), we get,

$$2\left(\frac{1}{x}+\frac{1}{y}+\frac{1}{z}\right)=\frac{15}{60} \quad \Rightarrow \frac{1}{x}+\frac{1}{y}+\frac{1}{z}=\frac{1}{8}$$

$\Rightarrow$ Number of days top complete half the work = 4 days

3. We have according to the question,

$$\frac{(x-1)(x+1)}{y}=\frac{(x+2)(x-1)}{z} \Rightarrow \frac{x+1}{x+2}=\frac{y}{z} \Rightarrow \frac{x+1}{x+2}=\frac{9}{10} \Rightarrow 10x + 10 = 9x + 18 \Rightarrow x = 8$$

4. We have,

$M_1 = 20, D_1 = 5, W_1 = \frac{1}{4}$

Let $M_2 = x,\ D_2 = 5, W_2 = \frac{3}{4}$

$$\therefore \frac{20\times 5}{\frac{1}{4}}=\frac{x\times 5}{\frac{3}{4}} \Rightarrow 20=\frac{x}{3} \Rightarrow x = 60$$

$\therefore$ Required men to be added $= 60 - 20 = 40$

5. Let x and y be $6k$ and $11k$ respectively .
Now, we have,

$$\frac{(3n-1)\times(2n+1)}{6k}=\frac{(3n+1)\,(4n-3)}{11k}$$

$\Rightarrow\ 11\,(6n^2 + n - 1) = 6 \times (12\,n^2 - 5n - 3)$
$\Rightarrow\ 66n^2 + 4n - 11 = 72\,n^2 - 30n - 18$
$\Rightarrow\ 6n^2 - 41n - 7 = 0 \Rightarrow 6n^2 - 42n + n - 7 = 0$
$\Rightarrow\ 6n\,(n - 7) + 1\,(n - 7) = 0 \Rightarrow (6n + 1)\,(n - 7) = 0$
$\Rightarrow\ n = -\frac{1}{7}$ (not possible) $\Rightarrow n = 7$

6. We have,

Combined efficiency of Sheela and Meena = $\frac{1}{2}+\frac{2}{3}=\frac{7}{6}$

$\therefore$ Number of days to stitch 35 suits = $\frac{6}{7}\times 35 = 30$ days

7. Number of days taken by C to complete the work = x days

No. of days taken by B to complete the work = $\frac{x}{15}=\frac{2x}{3}$ days

No. of days taken by A to complete the work = $\frac{x}{3}$ days

Now, according to the question

$$\frac{3}{x}+\frac{3}{2x}+\frac{1}{x}=\frac{1}{12} \Rightarrow \frac{6+3+2}{x}=\frac{1}{6} \Rightarrow x = 66 \text{ days}$$

8. Number of days taken by A to complete the work = x days
Number of days taken by B to complete the work = y days
Number of days taken by C to complete the work = z days
Now, according to the question,

$$\frac{1}{x}+\frac{1}{y}+\frac{1}{z}=\frac{1}{36} \quad \text{...(i)}$$

$$\frac{1}{x}+\frac{1}{y}=\frac{5}{z} \quad \text{...(ii)}$$

$$\frac{1}{y}+\frac{1}{z}=\frac{1}{x} \quad \text{...(iii)}$$

$$\frac{1}{x}+\frac{1}{z}=\frac{n}{y} \quad \text{...(iv)}$$

From (i) and (ii), we get

$$\frac{5}{z}+\frac{1}{z}=\frac{1}{36} \Rightarrow \frac{6}{z}=\frac{1}{36} \Rightarrow z = 216$$

From (i) and (iii), we get

$$\frac{2}{x}=\frac{1}{36} \Rightarrow x = 72$$

From (ii)

$$\frac{1}{72}+\frac{1}{y}=\frac{5}{216} \Rightarrow \frac{1}{y}=\frac{2}{216} \Rightarrow y = 108$$

From (iv), we get

$$\frac{1}{x}+\frac{1}{z}=\frac{n}{y} \Rightarrow \frac{1}{72}+\frac{1}{216}=\frac{n}{108} \Rightarrow \frac{4}{216}=\frac{n}{108} \Rightarrow n=\frac{108\times 4}{216}=2$$

9. Let the work done by each person = p units
Now, we have,

Case (i) $x = x \times x \times x \times p \Rightarrow p = \frac{1}{x^2}$

Case (ii) we have,

Total work $= k = y \times y \times y \times p \Rightarrow k = y \times y \times y \times \dfrac{1}{x^2} \Rightarrow k = \dfrac{y^3}{x^2}$

10. 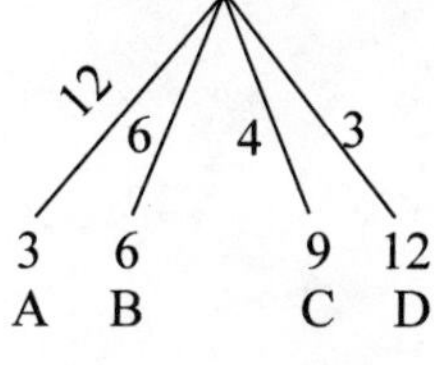

1 hr.	2nd hr	3rd hr.	4th hr. = 4 hrs.
A	B	A	B
12	6	12	6 = 36

$\therefore$ Minimum number of hours = 4

11. According to the question,

Total work = 50 × 40 = 2000

1st 10 days work = 50 × 10 = 500

Next 10 days work = 45 × 10 = 450

Next 10 days work = 40 × 10 = 400

Next 10 days work = 35 × 10 = 350

Next 10 days work = 30 × 10 = 300

= 2000

$\therefore$ Number of days to complete the work = 10 × 5 = 50 days

12. We have, according to the question,

$$\frac{(x+2)(x-3)}{(x+4)(x-2)} = \frac{3}{4}$$

$$\Rightarrow \frac{x^2 - x - 6}{x^2 + 2x - 8} = \frac{3}{4} \Rightarrow 4x^2 - 4x - 24 = 3x^2 + 6x - 24$$

$$\Rightarrow x^2 = 10x \Rightarrow x = 10$$

13. 5 women complete work in $\dfrac{12 \times 10}{5} = 24$ days

5 men complete work in 8 days

$$\therefore \quad 1\,M = 3\,W \Rightarrow 6\,W + 3\,M = \frac{6}{3}\,M + 3\,W = 5\,\text{M}$$

Now, no. of required days = 8

14. We have, no. of days to complete the remaining $\dfrac{3}{4}$th of the work = 150 days

Now, we have, no. of additional men = x

$$\therefore \quad \frac{150}{100} = \frac{(200+x)}{200}$$

$$\Rightarrow 300 = 200 + x \Rightarrow x = 100$$

15. Let no. of tractors required to plough 100 hectares = n

Now, according to the question,

$$\frac{5\times 5}{5}=\frac{50\times n}{100} \Rightarrow 10n = 100 \Rightarrow n = 10$$

16. We have,

$$\frac{1}{A}+\frac{1}{B}=\frac{1}{5} \quad ...(i)$$

And $\frac{2}{A}+\frac{1}{3B}=\frac{1}{B}$ ...(*ii*)

Multiplying eqn. (i) by 2 and subtracting (*i*) and (ii), we get,

$$\frac{2}{B}-\frac{1}{3B}=\frac{2}{5}-\frac{1}{3}$$

$$\Rightarrow \frac{5}{3B}=\frac{6-5}{15} \Rightarrow \frac{5}{3B}=\frac{1}{15} \Rightarrow B = 25 \text{ days}$$

Now, from (i) we have

$$\frac{1}{A}+\frac{1}{25}=\frac{1}{5}$$

$$\Rightarrow \frac{1}{A}=\frac{1}{5}-\frac{1}{25} \Rightarrow \frac{1}{A}=\frac{4}{25} \Rightarrow A=\frac{25}{4}=6\frac{1}{4} \text{ days}$$

17. Ratio of and B's efficiency = 3 : 1.
Let A takes x days to complete the work
B takes $(x + 10)$ days to complete the work

$$\frac{x}{x+10}=\frac{1}{3}$$

$\Rightarrow 3x = x + 10 \Rightarrow 2x = 10 \Rightarrow x = 5$ days

18. Total number of hours 1 man took = 12 × 8 × 10 = 960 hours

Total number of hours 8 man took = $\frac{960}{8} = 120$ hours.

$\therefore$ Required number of hours = $\frac{120}{8}$ = 15 hours

19. $(x + y)$ 1 day's work = $\frac{1}{6} \Rightarrow x$, 1 day's work = $\frac{1}{12}$

$\therefore$ Required number of days = 12

20. We have,

$M_1 = 12, D_1 = 16, W_1 = 10, M_2 = 8, D_2 = x, W_2 = 20$

$$\therefore \frac{12\times 16}{10}=\frac{8\times x}{20}$$

$\therefore$ $x = 48$ days

21. Time (in minutes) = $\frac{\text{Volume of tank}}{\text{Volume flowing perminute}}$.

Volume flowing per second = $\pi\times\left(\frac{7}{200}\right)^2 \times 5m^3$ m^3.

Volume flowing per minute = $\pi \times \left(\frac{7}{200}\right)^2 (5)(60)\,m^2$ m³.

$\therefore$ Time needed = = $\dfrac{4(3)(231)}{100 \times \pi \times \left(\frac{7}{200}\right)^2 \times 5 \times 60}$ minutes = 24 minutes.

22. The data (the time taken) and the corresponding flow rates are tabulated below.

	A	**B**	**Leak**	**A, B, leak**
Flow rate	4	2	1	5
Time	10	20	40	

If the tank capacity is taken as 40 units, the flow rates of A, B, and the leak are 4, 2, 1 respectively. The net flow rate (if all 3 are operating) is 5. It would take $\frac{40}{5}$, viz. 8 minutes to fill the tank.

23. $\dfrac{M_1 D_1 H_1}{W_1} = \dfrac{M_2 D_2 H_2}{W_2}$

M → Number of men D → Number of days

H → Number of hours/day W → Amount of work

$\dfrac{4(4)(4)}{4} = \dfrac{2(2)(2)}{W} \Rightarrow W = \dfrac{1}{2}$ unit.

24. $M_1D_1 = M_2D_2$; (m)(d) = (m + 2)D_2

$D_2 = \dfrac{md}{m+2}$.

25. The depth of the river (d) = 2.5 m.

The width of the river (w) = 45 m.

The speed of the water (v) = $3.6 \dfrac{km}{hr} = \dfrac{3600}{60}\dfrac{m}{min} = 60 \dfrac{m}{min}$.

Volume of water flowing past any point (or flowing into the sea) every minute = dwv = (2.5)(45)(60) m³ = 6750 m³.

26. Let the part of the tank emptied by C in 1 minute be $\frac{1}{x}$

$\Rightarrow \dfrac{1}{60} + \dfrac{1}{75} - \dfrac{1}{x} = \dfrac{1}{50} \Rightarrow \dfrac{1}{x} = \dfrac{1}{60} + \dfrac{1}{75} - \dfrac{1}{50} \Rightarrow \dfrac{1}{x} = \dfrac{1}{100}$

⇒ C can empty full tank in 100 minutes.

27. The rate is tabulated below. As A can do 50% more work then B, we assume B', rate as 2 units/hour. Therefore A's rate is 3 units/hour

The total work is 60 units.

	A	B
Rate	3	2
Time		30
Work		60

In 12 hours B does 24 units. The remaining 36units can be completed by A and B is 36/5, ie 7.2 hours.

28. Part of tank filled by pipe X in 1 hr = $\frac{1}{2}$

Part of tank filled by pipe

Pipe Y in 1 hr = $\frac{1}{6}$

Part of tank filled by both the pipes in 1 hr = $\frac{1}{2}+\frac{1}{6}=\frac{2}{3}$

Time taken by both pipes to fill $\frac{2}{3}$ part of tank = 60 minutes

Time taken to fill $\frac{1}{2}$ tank = $60 \times \frac{3}{2} \times \frac{1}{2}$ = 45 minutes.

∴ The tank will be be filled at 11 : 45 a.m.

29. 3 women work = 2 men work

1 women work $\frac{2}{3}$ men work

21 women work = $21 \times \frac{2}{3}$ men = 14 men work

Also , $M_1 D_1 T_1 = M_2 D_2 T_2$

$$\Rightarrow \quad D_2 = \frac{M_1 D_1 T_1}{M_2 T_2} = \frac{15 \times 21 \times 8}{14 \times 6} = 30 \text{ days}$$

30. (A + B) complete a work in 12 days

B' complete a work in 30 days

Total work = 60 unit [∵ LCM of 12 and 30 is 60]

(A + B) 1 day work = $\frac{60}{12}$ = 5 unit

B's 1 day work = $\frac{60}{30}$ = 2 unit

∴ A's 1 day work = (5 – 2) = 3 unit

∴ A complete 60 unit of work in $\left(\frac{60}{3}\right)$ = 20 days

31. Work done by men in 10 days = 5 M × 10

Work done by women 15 in days = 12 W × 15

According to question, $5M \times 10 = 12W \times 15$

⇒ 50 M = 180 W ⇒ 5M = 18 W

5 Men + 6 women = 5 M + 6 W = 18 W + 6 W = 24 W

∴ $M_1 D_1 = M_2 D_2$

$12 \times 15 = 24 \times D_2$

$$\Rightarrow \quad D_2 = \frac{12 \times 15}{24} = 7\frac{1}{2} \text{ days}$$

32. $(A + B)$'s 1 day's work $= \frac{1}{18}$;

$(B + C)$'s 1 day's work $= \frac{1}{24}$ & $(C + A)$'s 1 day's work $= \frac{1}{36}$.

Adding: $2(A + B + C)$'s 1 day's work $= \left(\frac{1}{18} + \frac{1}{24} + \frac{1}{36}\right) = \frac{9}{72} = \frac{1}{8}$.

$\therefore (A + B + C)$'s 1 day's work $= \frac{1}{16}$.

Hence, all of them can finish the work in 16 days.

33. $2(A + B + C)$'s : 1 day's work $= \left(\frac{1}{12} + \frac{1}{15} + \frac{1}{20}\right) = \frac{12}{60} = \frac{1}{5}$.

$\therefore (A + B + C)$'s 1 day's work $= \frac{1}{10}$

$\therefore A$'s 1 day's work = [$(A + B + C)$'s 1 day's work] – [$(B + C)$'s 1 day's work]

$= \left(\frac{1}{10} - \frac{1}{15}\right) = \frac{1}{30}$.

$\therefore A$ alone can do the work in 30 days.

34. C's 1 day's work $= \frac{1}{8} - \left(\frac{1}{24} + \frac{1}{16}\right) = \left(\frac{1}{8} - \frac{5}{48}\right) = \frac{1}{48}$.

$\therefore C$ alone can finish the work in 48 days.

35. B's 9 day's work $= \left(9 \times \frac{1}{12}\right) = \frac{3}{4}$. Remaining work $= \frac{1}{4}$.

A's 1 day's work $= \frac{1}{20}$.

$\therefore \frac{1}{20} : \frac{1}{4} :: 1 : x \Leftrightarrow x = \left(\frac{1}{4} \times 1 \times 20\right) = 5$ days.

36. $(B + C)$'s 2 day's work $= 2\left(\frac{1}{10} + \frac{1}{15}\right) = \frac{1}{3}$.

Remaining work $= \left(1 - \frac{1}{3}\right) = \frac{2}{3}$.

A's 1 day's work $= \frac{1}{9}$.

$\therefore \frac{1}{9} : \frac{2}{3} = 1 : x$ or $x = \left(\frac{2}{3} \times 1 \times \frac{9}{1}\right) = 6$ days.

37. Ratio of 1 day's work of A and $B = 2 : 1$.

$(A + B)$'s 1 day's work $= \frac{1}{14}$.

Dividing $\frac{1}{14}$ in the ratio 2 : 1, we get:

A's 1 day's work $= \left[\frac{1}{14} \times \frac{2}{3}\right] = \frac{1}{21}$.

$\therefore A$ alone can finish the work in 21 days.

38. Ratio of the time taken by A and $B = 1:3$.

If diff. of time is 2 days, B takes 3 days.

If diff. of time is 10 days, B takes $= \left(\frac{3}{2}\times 10\right) = 15$ days.

39. Suppose B takes x days.

Then, A takes $\left(2\times\frac{3}{4}x\right) = \frac{3x}{2}$ days.

Now, $(A+B)$'s 1 day's work $= \frac{1}{18}$.

$\therefore\ \frac{1}{x}+\frac{2}{3x} = \frac{1}{18}$ or $x = 30$.

40. B's 1 day's work $= \frac{1}{12}$. So, A's 1 day's work $= \frac{1}{6}$.

$(A+B)$'s 1 day's work $= \left[\frac{1}{12}+\frac{1}{6}\right] = \frac{3}{12} = \frac{1}{4}$.

So, both together can finish the work in 4 days.

41. Ratio of their 1 day's work $= \frac{1}{10}:\frac{1}{15} = 3:2$.

A's share $= ₹\left(225\times\frac{3}{5}\right) = ₹135$.

42. B's 3 day's work $= \left(3\times\frac{1}{21}\right) = \frac{1}{7}$. Remaining work $= \left(1-\frac{1}{7}\right) = \frac{6}{7}$.

$(A+B)$'s 1 day's work $= \left(\frac{1}{14}+\frac{1}{21}\right) = \frac{5}{42}$.

$\frac{5}{42}:\frac{6}{7}::1:x \Leftrightarrow x = \frac{6}{7}\times 1\times\frac{42}{5} = 7\frac{1}{5}$ days.

$\therefore$ Total number of days taken $= \left(3+7\frac{1}{5}\right) = 10\frac{1}{5}$ days.

43. $(A+B)$'s 2 day's work $= \left(\frac{1}{9}+\frac{1}{12}\right) = \frac{7}{36}$.

Work done in 5 pairs of days $= \left(5\times\frac{7}{36}\right) = \frac{35}{36}$.

Remaining work $= \left(1-\frac{35}{36}\right) = \frac{1}{36}$.

On 11th day it is A's turn.

$\frac{1}{9}:\frac{1}{36}::1:x \Leftrightarrow x = \left(\frac{1}{36}\times 1\times 9\right) = \frac{1}{4}$ day.

$\therefore$ Total time taken $= 10\frac{1}{4}$ days.

44. B's 23 day's work $= \left(23 \times \frac{1}{40}\right) = \frac{23}{40}$. Remaining work $= \left(1 - \frac{23}{40}\right) = \frac{17}{40}$.

$(A + B)$'s 1 day's work $= \left(\frac{1}{45} + \frac{1}{40}\right) = \frac{17}{360}$.

$\frac{17}{360}$ work is done by A & B in 1 day

$\therefore$ $\frac{17}{40}$ work is done by them in $\left(\frac{360}{17} \times \frac{17}{40}\right) = 9$ days.

Hence, A left after 9 days.

45. A's 10 day's work $= \frac{10}{80} = \frac{1}{8}$. Remaining work $= \left(1 - \frac{1}{8}\right) = \frac{7}{8}$.

$\frac{7}{8}$ work is done by B is 42 days

$\therefore$ Whole work will be done by B in $\left(42 \times \frac{8}{7}\right) = 48$ days.

$\therefore$ $(A + B)$'s 1 day's work $= \left(\frac{1}{80} + \frac{1}{48}\right) = \frac{1}{30}$.

Hence, both together can finish the work in 30 days.

46. Let total money be x.

A's 1 day's wages $\frac{x}{21}$ & B's 1 day's wages $= \frac{x}{28}$.

$\therefore$ $(A + B)$'s 1 day's wages $= \left(\frac{x}{21} + \frac{x}{28}\right) = \frac{x}{12}$.

$\therefore$ The money is sufficient to pay 12 days' wages of both.

47. 12 men $\equiv$ 18 women $\Leftrightarrow$ 8 men $\equiv \left(\frac{18}{12} \times 8\right)$ women = 12 women.

$\therefore$ 8 men + 16 women = 28 women.

Now, more women, less days.

$\therefore$ $28 : 18 :: 14 : x$. So, $x = \frac{18 \times 14}{28} = 9$ days.

48. 1 man $\equiv$ 3 boys and 1 woman $= \frac{3}{2}$ boys.

$\therefore$ (1 man + 1 woman + 1 boy) $\equiv \left(3 + \frac{3}{2} + 1\right)$ boys $= \frac{11}{2}$ boys.

Now, more boys, less number of days.

$\therefore$ $\frac{11}{2} : 3 :: 44 : x \Leftrightarrow x = \left(3 \times 44 \times \frac{2}{11}\right) = 24$ days.

49. Let there be originally x men.

Then, $(x + 8)$ men can finish the job in 50 days.

So, x men can finish it in $\left[\frac{(x+8)\times 50}{x}\right]$ days.

$$\therefore \quad \frac{(x+8)\times 50}{x} = 60 \Leftrightarrow 60x = 50x + 400 \Leftrightarrow x = 40.$$

50. From statement I $\frac{1}{A}+\frac{1}{B}=\frac{1}{24}$

From statement II $\frac{1}{B}+\frac{1}{C}=\frac{1}{36}$

From both statement $\frac{1}{A}+\frac{1}{B}+\frac{1}{B}+\frac{1}{C}=\frac{1}{24}+\frac{1}{36}$

$$\frac{1}{A}+\frac{2}{B}+\frac{1}{C}=\frac{5}{72}$$

Thus, the question cannot be answered even by using both the statements together.

51. We have,
8 men = 12 women $\Rightarrow$ 2 men = 3 women
$\Rightarrow$

$\therefore$ Required days to complete the work $= \frac{8\times 24\times 3}{8\times 3+12\times 2} = \frac{8\times 24\times 3}{48} = 12$ days

52. We have,
$M_1 = 20, D_1 = 20, W_1 = 20$
$M_2 = 16, W_2 = 16, D_2 = ?$
We know that

$$\frac{M_1\times D_1}{W_1} = \frac{M_2\times D_2}{W_2} \Rightarrow \frac{20\times 20}{20} = \frac{16\times D_2}{16} \Rightarrow D_2 = 20 \text{ days}$$

53. Let number of days he worked = x day
Now, we have,
$500x - 100(30 - x) = 11{,}400 \Rightarrow 600x = 14400 \Rightarrow x = 24$

54. Let the efficiencies of 1 man and 1 woman be x and y respectively.
Now, we have,
$6x + 8y) \times 10 = (13x + 24y) \times 4$

$$\Rightarrow 30x + 40y = 26x + 48y \Rightarrow 4x = 8y \Rightarrow x = 2y \Rightarrow \frac{x}{y} = \frac{2}{1} - 2:1$$

55. Let capacity of a man = M
Let capacity of a woman = W
$5(17W + 24M) = 6(12W + 23M) \Rightarrow 85W + 120M = 72W + 138M$
$\Rightarrow 13W = 18M$

12 CHAPTER

Time and Distance

Highlights: Problems on distance, time and speed.
Problems on Trains, Boats and Streams.

FORMULAE (Problems on Time & Distance)

(*i*) Speed = $\frac{\text{Distance}}{\text{Time}}$ (*ii*) Time = $\frac{\text{Distance}}{\text{Speed}}$

(*iii*) Distance = (Speed × Time) (*iv*) x km/hr = $\left(x \times \frac{5}{18}\right)$ m/sec

(v) y m/sec = $\left(y \times \frac{18}{5}\right)$ km/hr

(*vi*) If a certain distance is covered at x km/hr and the same distance is covered at y km/hr, then the average speed during whole journey = $\frac{2xy}{(x+y)}$ km/hr.

(*vii*) If a man changes his speed in the ratio $m : n$, then the ratio of times taken becomes $n : m$.

PROBLEMS ON TIME & DISTANCE

Ex. 1. ***A man covers a certain distance at 75 km/hr and returns back to the starting point at 50 km/hr. Find his average speed during the whole journey.***

Sol. Average speed = $\frac{2xy}{(x+y)} = \left(\frac{2 \times 75 \times 50}{75+50}\right)$ km/hr = 60 km/hr.

Ex. 2. ***A distance is covered in 3 hours 48 min. at 5 kmph. How much time will be taken to cover it at 28.5 kmph?***

Sol. Distance = (Speed × Time) = $\left(5 \times \frac{19}{5}\right)$ km = 19 km.

Now, distance = 19 km and speed = 28.5 km/hr.

$\therefore$ Time Taken = $\frac{\text{Distance}}{\text{Speed}} = \frac{19}{28.5}\text{ hrs.} = \left(\frac{19}{28.5} \times 60\right)\text{min.} = 40\text{ min.}$

Ex. 3. ***Ritu goes to school at 10 kmph and reaches the school 6 minutes late. Next day, she covers this distance at 12 km per hour and reaches the school 9 minutes earlier than the scheduled time. What is the distance of her school from her house?***

Sol. Let the required distance be x km.

With the two speeds, the difference of time taken in 15 min.

$\therefore \frac{x}{10} - \frac{x}{12} = \frac{15}{60} \Leftrightarrow 6x - 5x = 15 \Leftrightarrow x = 15$

So, the required distance is 15 km.

Ex. 4. ***A and B are two stations 330 km apart. A scooterist starts from A at 6.30 a.m. and travels towards B at 60 kmph. Another scooterist starts from B at 7.30 a.m. and travels towards A at 75 kmph. At what time do they meet?***

Sol. Suppose they meet x hours after 6.30 a.m. Then,

(Distance moved by 1st in x hours) + [Distance moved by 2nd in $(x - 1)$ hours] = 330

$\therefore\ 60x + 75(x - 1) = 330 \Leftrightarrow 135x = 405 \Leftrightarrow x = 3.$

So, they will meet at 9.30 a.m.

Ex. 5. ***A goods train leaves a station at a certain time at a fixed speed. After 6 hours, an express train leaves the same station and moves in the same direction at a uniform speed of 90 kmph. This train catches the goods train in 4 hours. Find the speed of the goods train.***

Sol. Let the speed of the goods train be x kmph.

Distance covered by goods train in 10 hours = Distance covered by express train in 4 hours

$\therefore\quad 10x = 4 \times 90$ or $x = 36.$

Hence, the speed of the goods train = 36 kmph.

FORMULAE (Problems on Trains)

(*i*) *Time taken by a train x metres long in passing a signal post or a pole or a standing man is the same as the time taken by the train to cover x metres with its own speed.*

(*ii*) *Time taken by a train x metres long in passing a stationary object of length y metres is the same as the time taken by the train to cover (x + y) metres with its own speed.*

(*iii*) Suppose two trains or bodies are moving in the same direction at u km/hr and v km/hr such that $u > v$, then their *relative speed* = $(u - v)$ km/hr. *If two trains of lengths x km and y km move in the same direction at u km/hr and v km/hr (where u > v), then time taken to cross each other is* $\left(\dfrac{x+y}{u-y}\right)$ *hrs.*

(*iv*) Suppose two trains or bodies are moving in the opposite directions at u km/hr and v km/hr. Then, relative speed = $(u + v)$ km/hr.

If two trains of lengths x km and y km move in the opposite directions at u kmph and v kmph, then time taken to cross each other = $\left(\dfrac{x+y}{u+v}\right)$ *hrs.*

(*v*) If two trains start at the same time from A and B towards each other and after crossing, they take a & b hours in reaching B and a respectively. Then,

A's speed: B's speed = $(\sqrt{b} : \sqrt{a})$.

SOLVED PROBLEMS ON TRAINS

Ex. 1. ***A train 180 m long is running with a speed of 54 kmph. In what time will it pass a tunnel 540 m long?***

Sol. Speed of the train = $\left(54 \times \dfrac{5}{18}\right)$ m/sec = 15 m/sec.

In passing a tunnel it will have to cover (180 + 540) m.

$\therefore$ Required time = $\left(\dfrac{720}{15}\right)$ sec = 48 sec.

Ex. 2. ***A train 100 m long is running with a speed of 70 kmph. In what time will it pass a man who is running at 10 kmph in the same direction in which the train is going?***

Sol. Speed of train relative to man = (70 – 10) km/hr = $\left(60\times\frac{5}{18}\right)$ m/sec = $\left(\frac{50}{3}\right)$ m/sec.

Time taken by the train to cross the man = Time taken by it to cover 100 m at $\left(\frac{50}{3}\right)$ m/sec

$= \left(100\times\frac{3}{50}\right)$ sec = 6 sec.

Ex. 3. ***A train 110 m long is running with a speed of 60 kmph. In what time will it pass a man who is running at 6 kmph in the direction opposite to that in which the train is going?***

Sol. Speed of the train relative to man = (60 + 6) kmph = $\left(66\times\frac{5}{18}\right)$ m/sec = $\left(\frac{55}{3}\right)$ m/sec.

Time taken by the train to cross the man = Time taken by it to cover 110 m at $\left(\frac{55}{3}\right)$ m/sec

$= \left(110\times\frac{3}{55}\right)$ sec = 6 sec.

Ex. 4. ***Two trains 132 metres and 108 metres in length are running towards each other on parallel lines, one at the rate of 32 km. per hour and another at 40 km. per hour. In what time will they be clear of each other from the moment they meet?***

Sol. Relative speed of the trains = (32 + 40) km/hr. = 72 kmph. = $\left(72\times\frac{5}{18}\right)$ m/sec. = 20 m/sec.

Time taken by the trains in passing each other = $\left(\frac{\text{Sum of lengths}}{\text{Relative speed}}\right) = \left(\frac{240}{20}\right)$ sec. = 12 sec.

FORMULAE (BOATS & STREAMS)

1. Let the speed of a boat in still water be x km/hr and let the speed of the stream be y km/hr. Then: (*i*) Speed downstream = $(x + y)$ km/hr (*ii*) Speed upstream = $(x - y)$ km/hr.
2. Let speed downstream = u km/hr & speed upstream = v km/hr.

   Then, (*i*) *Rate in still water* = $\frac{1}{2}(u+v)$ km/hr. (*ii*) *Rate of current* = $\frac{1}{2}(u-v)$ km/hr.

SOLVED PROBLEMS

Ex. 1. ***In a stream running at 2 kmph, a motor boat goes 10 km upstream and back again to the starting point in 55 minutes. Find the speed of the motor boat in still water.***

Sol. Let the speed of motor boat in still water be x kmph.

Speed downstream = $(x + 2)$ km/hr, speed upstream = $(x - 2)$ km/hr.

$\therefore \frac{10}{x-2}+\frac{10}{x+2}=\frac{55}{60} \Leftrightarrow 11x^2-240x-44=0 \Leftrightarrow (x-22)(11x+2)=0 \Leftrightarrow x=22.$

$\therefore$ Rate in still water = 22 km/hr.

Ex. 2. ***A man can row 30 km upstream and 44 km downstream in 10 hours. Also, he can row 40 km upstream and 55 km downstream in 13 hours. Find the rate of current and the speed of the man in still water.***

Sol. Let rate upstream = x kmph & rate downstream = y kmph.

Then, $\frac{30}{x}+\frac{44}{y}=10$ & $\frac{40}{x}+\frac{55}{y}=13$

$\therefore$ $30u + 44v = 10$ & $40u + 55v = 13$, where $u = \frac{1}{x}$ & $v = \frac{1}{y}$.

On solving we get: $u = \frac{1}{5}$ and $v = \frac{1}{11}$.

$\therefore$ $x = 5$ and $y = 11$.

So, rate upstream = 5 km/hr & rate downstream = 11 km/hr.

$\therefore$ Rate of current $= \frac{1}{2}(11 - 5)$ km/hr = 3 km/hr.

$\therefore$ Rate in still water $= \frac{1}{2}(11 + 5)$ km/hr = 8 km/hr.

EXERCISE 12

1. The speeds of four cars are $2u$, $3u$, $4u$ and xu and the time taken by them to cover the same distance is xt, $4t$, $3t$ and $2t$ respectively, where x, u, t are real numbers. What is the value of x? **[CDS (I) Exam, 2024]**

(*a*) 8 (*b*) 6 (*c*) 5 (*d*) 2

2. Walking at $\frac{4}{5}$th of his usual speed, a man is 12 minutes late for his office. What is the usual time taken by him to cover that distance? **[CDS (I) Exam, 2021]**

(*a*) 48 minutes (*b*) 50 minutes (*c*) 54 minutes (*d*) 60 minutes

3. A man walks at an average speed of 3 km/hr from his home and reaches office 40 minutes early. If he walks at an average speed of 2 km/hr, he would reach office 40 minutes late. What is the distance between his home and office? **[CDS (II) Exam, 2023]**

(*a*) 6 km (*b*) 8 km (*c*) 10 km (*d*) 12 km

4. Travelling at $\frac{3}{5}$th of his usual speed, a man is late by 20 minutes. What is the usual time if he travels with his usual speed? **[CDS (I) Exam, 2025]**

(*a*) 25 minutes (*b*) 30 minutes (*c*) 32 minutes (*d*) 35 minutes

5. The time taken by a train to cross a man travelling in another train is 10 seconds, when the other train is travelling in the opposite direction. However, it takes 20 seconds, if both the trains are travelling in the same direction. The length of the first train is 200 m and that of the second train is 150 m. What is the speed of the first train ? **[CDS (I) Exam, 2023]**

(*a*) 60 km/hr (*b*) 56 km/hr
(*c*) 54 km/hr (*d*) 52 km/hr

6. The speed of a boat in still water is 15 km/hr. If it can travel 42 km downstream and 28 km upstream in the same time, then what is the speed of the stream ? **[CDS (I) Exam, 2023]**

(*a*) 2·5 km/hr (*b*) 3 km/hr (*c*) 4·5 km/hr (*d*) 6 km/hr

7. A man walks at an average speed of 3 km/hr from his home and reaches office 40 minutes early. If he walks at an average speed of 2 km/hr, he would reach office 40 minutes late. What is the distance between his home and office? **[CDS (II) Exam, 2023]**

(*a*) 6 km (*b*) 8 km (*c*) 10 km (*d*) 12 km

8. A train completely overtakes two persons, walking in the same direction with speeds 3 km/hr and 4 km/hr in 9 seconds and $\frac{75}{8}$ seconds respectively. What is the length of the train? **[CDS (II) Exam, 2023]**

(*a*) 60 m (*b*) 62·5 m (*c*) 55 m (*d*) 67·5 m

9. In covering certain distance, the average speeds of X and Y are in the ratio 4 : 5. If X takes 45 minutes more than Y to reach the destination, then what is the time taken by Y to reach the destination? **[CDS (I) Exam, 2021]**

(*a*) 135 minutes (*b*) 150 minutes (*c*) 180 minutes (*d*) 225 minutes

10. By increasing the speed of his car by 15 km/hr, a person covers a distance of 300 km by taking an hour less than before. What was the original speed of the car? **[CDS (I) Exam, 2021]**

(*a*) 45 km/hr (*b*) 50 km/hr (*c*) 60 km/hr (*d*) 75 km/hr

11. X, Y and Z travel from the same place with uniform speeds 4 km/hr, 5 km/hr and 6 km/hr respectively. Y starts 2 hours after X. How long after Y must Z start in order that they overtake X at the same instant? **[CDS (I) Exam, 2021]**

(*a*) $\frac{3}{2}$ hours (*b*) $\frac{4}{3}$ hours (*c*) $\frac{9}{8}$ hours (*d*) $\frac{11}{8}$ hours

12. A car did a journey in *t* hours. Had the average speed been *x* kmph greater, the journey would have taken *y* hours less. How long was the journey? **[CDS (II) Exam, 2020]**

(*a*) $x(t-y)ty$ (*b*) $x(t-y)ty^{-1}$ (*c*) $x(t-y)ty^{-2}$ (*d*) $x(t+y)ty$

13. Thee cares A, B and C started from a point at 5 pm 6 pm and 7 pm respectively and travelled at uniform speed of 60 km/hr, 80 km/hr and *x* km/hr respectively in the same direction. If all the three met at another point at the same instant during their journey, then what is the value of x? **[CDS (I) Exam, 2019]**

(*a*) 120 (*b*) 110 (*c*) 105 (*d*) 100

14. A race has three parts. The speed and time required to complete the individual parts for a runner is displayed on the following charts: **[CDS (I) Exam, 2019]**

	Part I	Part II	Part II
Speed (kmph)	50	80	100

What is the average speed of this runner?

(*a*) 8.17 km/ph (*b*) 8.00 km/ph (*c*) 7.80 km/ph (*d*) 7.77 km/ph

15. Radha and Hema are neighbours and study in the same school. Both of them use bicycles to go to the school. Radha's speed is 8 km/hr whereas Hema's speed is 10 km/hr. Hema takes 9 minutes less than Radha to reach the school. How for is the school from the locality of Radha and Hema? **[CDS (I) Exam, 2019]**

(*a*) 5 km (*b*) 5.5 km (*c*) 6 km (*d*) 6.5 km

16. Two towers A and B of height 23 m and 11 m respectively, stand 9 m apart. A straight rod is joined to the two tops of the towers. A monkey sitting on the top of A, climbs the rod to reach the top of B. If the monkey takes 5 minutes to reach the other end, what is the average speed of the monkey? **[CDS (II) Exam, 2024]**

(*a*) 10 m/min (*b*) 5 m/min (*c*) 10 cm/sec (*d*) 5 cm/sec

17. A car has an average speed of 60 km per hour while going from Delhi to Agra and has an average speed of *y* km per hour while returning to Delhi from Agra (by travelling the same distance). If the average speed of the car for the whole journey is 48 pm per hour, then what is the value of y? **[CDS (I) Exam, 2019]**

(*a*) 30 km per hour (*b*) 35 km per hour (*c*) 40 km per hour (*d*) 45 km per hour

18. A cyclist covers his first 20 km at an average speed of 40 km/ph, another 10 km at an average speed of 10 kmph and the last 30 km at an average speed of 40 kmpr. Then the average speed of the entire journey is: **[CDS (I) Exam, 2018]**

(*a*) 20 km/ph (*b*) 26.67 km/ph (*c*) 28.24 km/ph (*d*) 30 km/ph

19. A thief steals a car parked in a house and goes away with a speed of 40 kmph. The theft was discovered after half an hour and immediately the owner sets off in another car with a speed of 60 kmph. When will the owner meet the theft? **[CDS (I) Exam, 2018]**

(*a*) 55 km from the owner's houses and one hour after the theft

(*b*) 60 km from the owner's house and 1.5 hours after the theft

(*c*) 60 km from the owner's house and 1.5 hours after the discovery of the theft

(*d*) 55 km from the owner's house and 1.5 hours after theft

20. A car travels the first one-third of a certain distance with a speed of 10 km/hr, the next one-third distance with a speed of 20 km/hr and the last one-third distance with a speed of 60 km/hr. The average speed of the car for the whole journey is **[CDS (I) Exam, 2015]**

(*a*) 18 km/hr (*b*) 24 km/hr (*c*) 30 km/hr (*d*) 36 km/hr

21. Two persons A and B start simultaneously from two places c km apart, and walk in the same direction. If A travels at the rate of p km/hr and B travels at the rate of q km/hr. then A has travelled before he overtakes B at a distance of **[CDS (I) Exam, 2015]**

(*a*) $\frac{qc}{p+q}$ km (*b*) $\frac{pc}{p-q}$ km (*c*) $\frac{qc}{p-q}$ km (*d*) $\frac{pc}{p+q}$ km

22. A person goes to a market between 4 p.m. and 5 p.m. When he comes back, he finds that the hour hand and minute hand have interchanged their positions. For how much time (approximately) was he out of his house? **[CDS (I) Exam, 2015]**

(*a*) 55.25 minutes (*b*) 55.30 minutes (*c*) 55.34 minutes (*d*) 55.38 minutes

23. In a flight of 600 km, an aircraft was slowed down due to had weather. Its average speed for the trip was reduced by 200 km/hr and the time of flight increased by 30 minutes. The duration of the flight is **[CDS (I) Exam, 2015]**

(*a*) 1 hours (*b*) 2 hours (*c*) 3 hours (*d*) 4 hours

24. With a uniform speed, a car covers a distance in 8 hours. Had the speed been increased by 4 km/hr, the same distance could have been covered in 7 hours and 30 minutes. What is the distance covered?

(*a*) 420 km (*b*) 480 km (*c*) 520 km (*d*) 640 km

[CDS (I) Exam, 2015]

25. A runs $1\frac{2}{3}$ times as fast as B. If A gives B a start of 80 m, how far must the winning post from the starting point so that A and B might reach it at the same time?

(*a*) 200 m (*b*) 300 m (*c*) 270 m (*d*) 160 m

[CDS (I) Exam, 2015]

26. A thief is noticed by a policemen from a distance of 200 m. The thief starts running and the policeman chases him. The thief and the policeman run at the speed of 10 km/hr and 11 km/hr respectively. What is the distance between them after 6 minutes?

(*a*) 100 m (*b*) 120 m (*c*) 150 m (*d*) 160 m

[CDS (I) Exam, 2015]

27. A light was seen regularly at an interval of 13 seconds. It was seen for the first time at 1 hours 54 minutes 50 seconds (a.m.) and the last time at 3 hours 17 minutes 49 seconds (a.m.). How many times was the light seen? **[CDS (I) Exam, 2015]**
(*a*) 375 (*b*) 378 (*c*) 383 (*d*) 384

28. In a race *A, B* and *C* take part. *A* beats *B* by 30 m, *B* beats *C* by 20m and *A* beats *C* by 48 m. Which of the following is/are correct? **[CDS (II) Exam, 2015]**
1. The length of the race is 300 m. **2.** The speeds of *A, B* and *C* are in the ratio 50 : 45 : 42
Select the correct answer using the code given below:
(*a*) 1 only (*b*) 2 only (*c*) Both 1 and 2 (*d*) Neither 1 nor 2

29. By increasing the speed of his car by 15km/hour, a person covers 300 km distance by taking an hour less than before. The original speed of the car was **[CDS (II) Exam, 2015]**
(*a*) 45 km/hour (*b*) 50 km/hour (*c*) 60 km/hour (*d*) 75 km/hour

30. Three athletes run a 4 km race. Their speeds are in the ratio 16 : 15 : 11. When the winner wins the race, then the distance between the athlete in the second position to the athlete in the third position is **[CDS (II) Exam, 2015]**
(*a*) 1000 m (*b*) 800 m (*c*) 750 m (*d*) 600 m

31. In a race of 100 m, *A* beats *B* by 4 m and *A* beats *C* by 2m. By how many meters (approximately) would *C* beat *B* in another 100 m race assuming *C* and *B* run with their respective speeds as in the earlier race? **[CDS (II) Exam, 2015]**
(*a*) 2 (*b*) 2.04 (*c*) 2.08 (*d*) 3.2

32. A boy is cycling such that the wheels of the cycle are making 140 revolutions per minute. If the radius of the wheel is 30 cm, the speed of the cycle is **[CDS (II) Exam, 2015]**
(*a*) 15.5 km/hour (*b*) 15.84 km/hour (*c*) 16 km/hour (*d*) 16.36 km/hour

33. A cyclist moves non-stop from A to B, a distance of 14 km, at a certain average speed. If his average speed reduces by 1 km per hour, he takes 20 minutes more to cover the same distance. The original average speed of the cyclist is **[CDS (I) Exam, 2016]**
(*a*) 5 km/hour (*b*) 6 km/hour (*c*) 7 km/hour (*d*) None of the above

34. A clock strikes once at 1 o'clock, twice at 2 o'clock and thrice at 3 o'clock, and so on. If it takes 8 seconds to strike at 5 o'clock, the time taken by it to strike at 10 o'clock is
(*a*) 14 seconds (*b*) 16 seconds (*c*) 18 seconds (*d*) None of the above
[CDS (I) Exam, 2016]

35. A bike consumes 20 mL of petrol per kilometre, if it is driven at a speed in the range of 25-50 km/hour and consumes 40 mL of petrol per kilometre at any other speed. How much petrol is consumed by the bike in travelling a distance of 50 km, if the bike is driven at a speed of 40 km /hour for the first 10 km, at a speed of 60 km/hour for the next 30 km and at a speed of 30 km/ hour for the last 10 km? **[CDS (I) Exam, 2016]**
(*a*) 1 L (*b*) 1·2 L (*c*) 1·4 L (*d*) 1·6 L

36. A person goes to a market between 4 p.m. and 5 p.m. When he comes back, he finds that the hour hand and the minute hand of the clock have interchanged their positions. For how much time (approximately) was he out of his house? **[CDS (I) Exam, 2016]**
(*a*) 55·38 minutes (*b*) 55·48 minutes (*c*) 55·57 minutes (*d*) 55·67 minutes

37. In a race of 1000 m, A beats 15 by 100 m or 10 seconds. If they start a race of 1000 m simultaneously from the same point and if *B* gets injured after running 50 m less than half the race length and due to which his speed gets halved, then by how much time will A beat B? **[CDS (I) Exam, 2016]**
(*a*) 65 seconds (*b*) 60 seconds (*c*) 50 seconds (*d*) 45 seconds

38. Outside a meeting room, Madhukar was told by a person that each meeting takes place after 13/4 hours. The last meeting has been over just 45 minutes ago and the next meeting will take place at 2 p.m. At what time did Madhukar receive this information? **[CDS (I) Exam, 2016]**
(*a*) 10:20 a.m. (*b*) 11:30 a.m. (*c*) 11:40 a.m. (*d*) 11:50 a.m.

39. A person travels 7 km eastwards and then turns right and travels 3 km and further turns right and travels 13 km. What is the shortest distance of the present position of the person from his starting point? **[CDS (I) Exam, 2016]**
(*a*) 6 km (*b*) $3\sqrt{5}$ km (*c*) 7 km (*d*) $4\sqrt{5}$ km

40. Two men, *A* and *B* run a 4 km race on a course 0.25 km round. If their speeds are in the ratio 5 : 4, how often does the winner pass the other? **[CDS (II) Exam, 2016]**
(*a*) Once (*b*) Twice (*c*) Thrice (*d*) Four times

41. The wheels of a car are of diameter 80 cm each. The car is travelling at a speed of 66 km/hour. What is the number of complete revolution each wheel makes in 10 minutes?
(*a*) 4275 (*b*) 4350 (*c*) 4375 (*d*) 4450
[CDS (II) Exam, 2016]

42. In a 100 m race, A runs at a speed of $\frac{5}{3}$ m/s. If A gives a start of 4m to B and still beats him by 12 seconds, what is the speed of B? **[CDS (I) Exam, 2017]**
(*a*) $\frac{5}{4}$ m/s (*b*) $\frac{7}{5}$ m/s (*c*) $\frac{4}{3}$ m/s (*d*) $\frac{6}{5}$ m/s

43. A thief is spotted by a policeman from a distance of 100m. When the policeman starts the chase, the thief also starts running. If the speed of the thief is 8 km/hour and that of the policeman is 10 km/hour, then how far will the thief have to run before he is overtaken?
(*a*) 200m (*b*) 300m (*c*) 400m (*d*) 500m
[CDS (I) Exam, 2017]

44. Walking at the rate of 4 kmph a man covers a distance in 2 hours 45 min. Running at a speed of 16.5 kmph the man will cover the same distance in:
(*a*) 40 min (*b*) 45 min
(*c*) 100 min (*d*) 41 min 15 sec

Direction (Q. 45-47): Consider the following for the next three (3) items that follow:
Two trains A and B started from stations P and Q respectively towards each other. Train A started at 7 p.m. at a speed of 60 km/hr and train B started at 4 am (next day) at a speed of 90 km/hr. The distance between the two stations P and Q is 800 km. **[CDS (I) Exam, 2024]**

45. How far from station Q will the two trains meet ?
(*a*) 104 km (*b*) 144 km (*c*) 156 km (*d*) 504 km

46. At what time will the two trains meet ?
(*a*) 5 : 28 am (*b*) 5 : 44 am (*c*) 4 : 56 am (*d*) 6 : 24 am

47. If the lengths of the two trains A and B are 400 m and 500 m respectively, then what is the time taken by them to cross each other ?
(*a*) 21.6 seconds (*b*) 18.2 seconds (*c*) 17.4 seconds (*d*) 15.4 seconds

48. Two men starting from the same place walk at the rate of 4 kmph and 4.5 kmph respectively. What time will they take to be 8.5 km apart, if they walk in the same direction?
(*a*) 1 hour (*b*) 4 hrs 15 min (*c*) 17 hrs (*d*) 2 hrs 55 min.

49. Two cyclists start from the same place in opposite directions. One goes towards north at 18 kmph and the other goes towards south at 20 kmph. What time will they take to be 47.5 km apart?

(*a*) $2\frac{1}{4}$ hrs (*b*) $1\frac{1}{4}$ hrs (*c*) 2 hrs 23 min (*d*) $23\frac{1}{4}$ hrs

PROBLEMS ON TRAINS

50. A train X takes 2 hours less than a train Y to cover a distance of 192 km between two cities. Their average speeds differ by 16 km/hr. How long does the faster train take to cover the journey? **[CDS (I) Exam, 2022]**

(*a*) 3 hours (*b*) 4 hours (*c*) 5 hours (*d*) 6 hours

51. A train 200 m long passes a platform 100 m long in 10 seconds. What is the speed of the train? **[CDS (I) Exam, 2021]**

(*a*) 40 m/s (*b*) 30 m/s (*c*) 25 m/s (*d*) 20 m/s

52. A train takes two hours less for a journey of 300 km if its speed is increased by 5 km/hr from its usual speed. What is its usual speed? **[CDS (I) Exam, 2020]**

(*a*) 50 km/hr (*b*) 40 km/hr (*c*) 35 km/hr (*d*) 25 km/hr

53. A train length 110 m is moving at a uniform speed of 132 km/hr. The time required to cross a bridge of length 165 m is **[CDS (I) Exam, 2021]**

(*a*) 6.5 seconds (*b*) 7 seconds
(*c*) 7.5 seconds (*d*) 8.5 seconds

54. A train moving with a speed of 60 km per hours crosses an electric pole in 30 seconds. What is the length of the train in metres? **[CDS (I) Exam, 2018]**

(*a*) 300 (*b*) 400 (*c*) 500 (*d*) 600

55. A train 100 m long passes a platform 100 m long in 10 seconds. The speed of the train is **[CDS (II) Exam, 2018]**

(*a*) 36 kmph (*b*) 45 kmph (*c*) 54 kmph (*d*) 72 kmph

56. Two trains are moving in the same direction at 1.5 km/minute and 60 km/hour respectively. A man in the faster train observes that it takes 27 seconds to cross the slower train. The length of the slower train is **[CDS (II) Exam, 2015]**

(*a*) 225m (*b*) 230m (*c*) 240m (*d*) 250m

57. A train of certain length takes time t to pass completely through a station of length x. The same train with the same speed takes time $2t$ to pass completely through another station of length y. What is the time taken by the train to pass completely through a station of length $(x+y)$? **[CDS (I) Exam, 2025]**

(*a*) $\frac{(2yt+xt)}{(y-x)}$ (*b*) $\frac{(yt+xt)}{(y-x)}$ (*c*) $\frac{(3yt-xt)}{(2y-x)}$ (*d*) $\frac{(2yt-xt)}{(y-x)}$

58. Two trains A and B leave Delhi for Hyderabad at 7:00 a.m. and 7:50 a.m. on the same day and travel at 80 kmph and 100 kmph respectively. After how many kilometers from Delhi will the two trains be together ? **[CDS (I) Exam, 2024]**

(*a*) $\frac{200}{3}$ km (*b*) 100 km (*c*) $\frac{400}{3}$ km (*d*) $\frac{1000}{3}$ km

59. A passenger train takes 1 hour less for a journey of 120 km, if its speed is increased by 10 km/hour from its usual speed. What is its usual speed? **[CDS (I) Exam, 2016]**

(*a*) 50 km/hour (*b*) 40 km/hour (*c*) 35 km/hour (*d*) 30 km/hour

60. When the speed of a train is increased by 20%, it takes 20 minutes less to cover the same distance. What is the time taken to cover the same distance with the original speed?

(*a*) 140 minutes (*b*) 120 minutes (*c*) 100 minutes (*d*) 80 minutes

[CDS (II) Exam, 2016]

61. A train is travelling at 48 km/hour completely crosses another train having half its length and travelling in opposite direction at 42 km/hour in 12s. It also passes a railway platform in 45s. What is the length of the platform? **[CDS (II) Exam, 2016]**

(*a*) 600 m (*b*) 400 m (*c*) 300 m (*d*) 200 m

62. A passenger train departs from Delhi at 6pm from Mumbai. At 9pm, an express train, whose average speed exceeds that of the passenger train by 15km/hour leaves Mumbai for Delhi. Two trains meet each other mid–route. At what time do they meet, given that the distance between the cities is 1080 km? **[CDS (I) Exam, 2017]**

(*a*) 4 p.m. (*b*) 2 a.m. (*c*) 12 midnight (*d*) 6 a.m.

63. A 225 m long train is running at a speed of 30 km/hour. How much time does it take to cross a man running at 3 km/hour in the same direction? **[CDS (I) Exam, 2017]**

(*a*) 40 seconds (*b*) 30 seconds

(*c*) 25 seconds (*d*) 15 seconds

64. Train X crosses a man standing on the platform in 24 seconds and train Y crosses a man standing on the platform in 18 seconds. They cross each other while running in opposite directions in 20 seconds. What is the ratio of speed of X to speed of Y? **[CDS (II) Exam, 2024]**

(*a*) 1 : 2 (*b*) 2 : 3 (*c*) 1 : 3 (*d*) 3 : 4

65. A train 360 m long is running at a speed of 54 kmph. Time taken by the train to cross a tunnel 390 m long, is:

(*a*) 25 sec (*b*) 50 sec (*c*) 30 sec (*d*) 45 sec

66. The length of a train that takes 18 seconds to pass a pole when it runs at a speed of 36 km/hr, is:

(*a*) 288 m (*b*) 180 m (*c*) 200 m (*d*) 500 m

67. A train 100 m long, running at 36 kmph takes 25 seconds to pass a bridge. The length of the bride (in metres) is:

(*a*) 90 (*b*) 144 (*c*) 150 (*d*) 540

68. A train 100 m long travels at 70 kmph. A man is running at 10 kmph in the same direction in which the train is going. The train will pass the man in:

(*a*) 6 sec (*b*) $6\frac{2}{3}$ sec (*c*) 7 sec (*d*) 8 sec

69. A train 270 m long is moving at a speed of 25 kmph. It will cross a man coming from the opposite direction at a speed of 2 kmph in:

(*a*) 28 sec (*b*) 24 sec (*c*) 32 sec (*d*) 36 sec

70. A 150 m long train crosses a man walking at a speed of 6 kmph in the opposite direction in 6 seconds. The speed of the train (in kmph) is:

(*a*) 66 (*b*) 84 (*c*) 96 (*d*) 106

71. A train of length 150 m takes 10 seconds to pass over another train 100 m long coming from the opposite direction. If the speed of the first train be 30 kmph, the speed of the second train is:
(*a*) 54 km/hr (*b*) 60 km/hr (*c*) 72 km/hr (*d*) 36 km/hr

72. A train speeds past a pole in 15 seconds and a platform 100 m long in 25 seconds. Its length in metres is:
(*a*) 50 (*b*) 150 (*c*) 200 (*d*) 375

73. A train overtakes two persons who are walking in the same direction in which the train is going, at the rate of 2 kmph and 4 kmph and passes them completely in 9 and 10 seconds respectively. The length of the train (in metres) is:
(*a*) 45 (*b*) 54 (*c*) 50 (*d*) 72

74. Two trains running in the same direction at 40 kmph and 22 kmph completely pass one another in 1 minute. If the length of the first train is 125 m, the length of the second train is:
(*a*) 125 m (*b*) 150 m (*c*) 175 m (*d*) 200 m

PROBLEMS ON BOATS STREAMS

75. A motor boat has speed 30 km/hr in still water. It goes 60 km down stream and comes back in $\frac{9}{2}$ hours. What is the speed of the stream? **[CDS (I) Exam, 2021]**
(*a*) 5 km/hr (*b*) 8 km/hr (*c*) 10 km/hr (*d*) 12 km/hr

76. A man can row at a speed of x km/hr in still water. If in a stream which is flowing at a speed of y km/hr it takes him z hours to row to a place and back, then what is the distance between the two places? **[CDS (I) Exam, 2018]**
(*a*) $\frac{z(x^2 - y^2)}{2y}$ (*b*) $\frac{z(x^2 - y^2)}{2x}$ (*c*) $\frac{z(x^2 - y^2)}{2zx}$ (*d*) $\frac{z(x^2 - y^2)}{z}$

77. A man rows downstream 32 km and 14 km upstream, and he takes 6 hours to cover each distance. What is the speed of the man? **[CDS (I) Exam, 2015]**
(*a*) 0.5 km/hr (*b*) 1 km/hr (*c*) 1.5 km/hr (*d*) 2 km/hr

78. A motor boat, whose speed is 15 km/hours in still water goes 30 km downstream and comes back in a total of 4 hours and 30 minutes. The speed of the stream is
(*a*) 4 km/hour (*b*) 5 km/hour (*c*) 6 km/hour (*d*) 10 km/hour

79. A person can row downstream 20 km in 2 hours and upstream 4 km in 2 hours. What is the speed of the current? **[CDS (II) Exam, 2016]**
(*a*) 2 km/hour (*b*) 2.5 km/hour (*c*) 3 km/hour (*d*) 4 km/hour

80. A man goes upstream with an average speed of 24 kmph and comes downstream with an average speed of 36 kmph. The distance travelled in both the cases being the same, the average speed for the entire journey is:
(*a*) 30 kmph (*b*) 24 kmph (*c*) 28.8 kmph (*d*) 32.6 kmph

81. A man can row downstream at 10 kmph and upstream at 5 kmph. Man's rate in still water is:
(*a*) 2.5 km/hr (*b*) 5 km/hr (*c*) 7.5 km/hr (*d*) 15 km/hr

82. A boat moves downstream at the rate of 1 km in 6 minutes and upstream at the rate of 1 km in 10 minutes. The speed of the current (in km/hr) is:
(*a*) 1 (*b*) 1.5 (*c*) 2 (*d*) 2.5

83. A man rows upstream 11 km and downstream 26 km taking 5 hours each time. The velocity of the current is:

(*a*) 1 km/hr (*b*) 1.5 km/hr (*c*) 2.5 km/hr (*d*) 7.5 km/hr

84. If a man's rate with the current is 12 kmph and the rate of the current is 1.5 kmph, then the man's rate against the current (in km/hr) is:

(*a*) 5.25 (*b*) 7.5 (*c*) 6.75 (*d*) 9

85. A boat goes 40 km upstream in 8 hours and a distance of 36 km downstream in 6 hours. The speed of the boat in standing water (in km/hr) is:

(*a*) 5 (*b*) 5.5 (*c*) 6 (*d*) 6.5

86. A man can row 5 kmph in still water. If the river is running at 1 kmph, it takes him 75 minutes to row to a place and back. How far is the place?

(*a*) 2.5 km (*b*) 3 km (*c*) 4 km (*d*) 5 km

87. The speed of a boat in still water is 15 kmph and the rate of the current is 3 kmph. The distance travelled downstream in 12 minutes is:

(*a*) 3.6 km (*b*) 2.4 km (*c*) 1.2 km (*d*) 1.8 km

88. The speed of a boat downstream is 15 kmph and the speed of the stream is 1.5 kmph. The speed of the boat upstream (in km/hr) is:

(*a*) 8.25 (*b*) 12 (*c*) 13.5 (*d*) 16.5

89. A man can swim 3 kmph in still water. If the velocity of the stream be 2 kmph, the time taken by him to swim to a place 10 km upstream and back, is:

(*a*) $8\frac{1}{3}$ hrs (*b*) $9\frac{1}{5}$ hrs (*c*) 10 hrs (*d*) 12 hrs

90. The current of a stream runs at 1 kmph. A motor boat goes 35 km upstream and back again to the starting point in 12 hours. The speed of the motor boat in still water (in kmph) is:

(*a*) 6 (*b*) 7 (*c*) 8 (*d*) 8.5

91. A boat covers 24 km upstream and 36 km downstream in 6 hours, while it covers 36 km upstream and 24 km downstream in $6\frac{1}{2}$ hours. The velocity of the current (in km/hr) is:

(*a*) 1 (*b*) 1.5 (*c*) 2 (*d*) 2.5

92. A man can row $9\frac{1}{3}$ kmph in still water and he finds that it takes him thrice as much time to row up than as to row down the same distance in the river. The speed of the current (in km/hr) is:

(*a*) $3\frac{1}{3}$ (*b*) $3\frac{1}{9}$ (*c*) $1\frac{1}{4}$ (*d*) $4\frac{2}{3}$

93. A man can row three-quarters of a km against the stream in $11\frac{1}{4}$ minutes and return in $7\frac{1}{2}$ minutes. The speed of the man in still water is:

(*a*) 2 km/hr (*b*) 3 km/hr (*c*) 4 km/hr (*d*) 5 km/hr

ANSWERS

1. (*b*)	**2.** (*a*)	**3.** (*b*)	**4.** (*b*)	**5.** (*c*)	**6.** (*b*)	**7.** (*b*)	**8.** (*b*)	**9.** (*c*)	**10.** (*c*)
11. (*b*)	**12.** (*b*)	**13.** (*a*)	**14.** (*b*)	**15.** (*c*)	**16.** (*d*)	**17.** (*c*)	**18.** (*b*)	**19.** (*b*)	**20.** (*a*)

21. (*b*)	**22.** (*d*)	**23.** (*a*)	**24.** (*b*)	**25.** (*a*)	**26.** (*a*)	**24.** (*d*)	**28.** (*c*)	**29.** (*c*)	**30.** (*a*)
31. (*b*)	**32.** (*b*)	**33.** (*c*)	**34.** (*b*)	**35.** (*d*)	**36.** (*a*)	**37.** (*a*)	**38.** (*b*)	**39.** (*b*)	**40.** (*c*)
41. (*c*)	**42.** (*c*)	**43.** (*c*)	**44.** (*a*)	**45.** (*c*)	**46.** (*b*)	**47.** (*a*)	**48.** (*c*)	**49.** (*b*)	**50.** (*b*)
51. (*b*)	**52.** (*d*)	**53.** (*c*)	**54.** (*c*)	**55.** (*d*)	**56.** (*a*)	**57.** (*d*)	**58.** (*d*)	**59.** (*b*)	**60.** (*b*)
61. (*b*)	**62.** (*d*)	**63.** (*b*)	**64.** (*a*)	**65.** (*b*)	**66.** (*b*)	**67.** (*c*)	**68.** (*a*)	**69.** (*d*)	**70.** (*b*)
71. (*b*)	**72.** (*c*)	**73.** (*c*)	**74.** (*c*)	**75.** (*c*)	**76.** (*b*)	**77.** (*c*)	**78.** (*b*)	**79.** (*d*)	**80.** (*c*)
81. (*c*)	**82.** (*c*)	**83.** (*b*)	**84.** (*d*)	**85.** (*b*)	**86.** (*b*)	**87.** (*a*)	**88.** (*b*)	**89.** (*d*)	**90.** (*a*)
91. (*c*)	**92.** (*d*)	**93.** (*d*)							

SOLUTIONS

1. Let '*d*' be the distance

$\Rightarrow d = 2u \times xt = 2\ uxt$ ...(i)

$\Rightarrow d = 3u \times 4t = 12ut$ ...(ii)

$\Rightarrow d = 4u \times 3t = 12\ ut$ ...(iii)

$\Rightarrow d = xu \times 2t = 2xut$ ...(iv)

From eqn. (i) and (ii)

$$2uxt = 12ut \Rightarrow 2x = 12 \Rightarrow x = 6$$

2. Let the usual time be t minute

Now, according to question

$$\frac{5}{4}t = t + 12$$

$$\Rightarrow \frac{5}{4}t - t = 12 \Rightarrow \frac{t}{4} = 12 \Rightarrow t = 48 \text{ minutes}$$

3. Let required distance = *d* km

Now, according to the question,

$$\left(\frac{1}{0.5} \times 8.5\right) \Rightarrow \frac{d}{6} = \frac{4}{3} \Rightarrow d = 8 \text{ km}$$

4. Let the usual time taken by the man be T minutes.

When traveling at 3/5th of the usual speed, time taken $= T\left(\frac{5}{3}\right)$.

According to question,

$$\Rightarrow \frac{5T}{3} - T = 20 \Rightarrow \frac{5T - 3T}{3} = 20 \Rightarrow \frac{2T}{3} = 20 \Rightarrow T = \frac{20 \times 3}{2} = 30 \text{ minutes}$$

5. Let speed of train of length 200 m = *x* m/s

Let speed of train of length 150 m = *y* m/s

Now, we have

$$x - y = \frac{200}{20} = 10 \quad \text{...(i)}$$

$$x + y = \frac{200}{10} = 20 \quad \text{...(ii)}$$

Adding (i) and (ii), we get

$x = 15\text{ m/s} = 15 \times \frac{18}{5} = 54\text{ km/hr}$ $\quad [\because 1\text{ m/s} = \frac{18}{5}\text{ km/h}]$

6. Let speed of stream be s km/hr

Downstream speed = (15 + s) km/hr

Upstream speed = (15 – s) km/hr

Now, according to the question, $\frac{42}{15+s} = \frac{28}{15-s}$

$\Rightarrow \frac{6}{15+s} = \frac{4}{15-s} \quad \Rightarrow 90 - 6s = 60 + 4s \quad \Rightarrow 10s = 30 \quad \Rightarrow s = 3\text{ km/hr}$

7. Let required distance = d km

Now, according to the question,

$\left(\frac{1}{0.5} \times 8.5\right) \Rightarrow \left(\frac{1}{38} \times 47.5\right)\text{hrs} \quad \Rightarrow d = 8\text{ km.}$

8. Let length of train be d m and speed be s m/s

Now, according to the question,

$$\frac{d}{s - \left(3 \times \frac{5}{18}\right)} = 9 \Rightarrow d = 9s - 9 \times \frac{5}{6} \Rightarrow d = 9s - \frac{15}{2}$$

$\Rightarrow \quad 2d = 18s - 15 \quad$ ...(i)

$$\frac{d}{s - \left(4 \times \frac{5}{18}\right)} = \frac{75}{8} \Rightarrow 8d = 75s - 75 \times \frac{10}{9}$$

$\Rightarrow \quad 24d = 225s - 250 \quad$ ...(ii)

$\Rightarrow$ Multiplying (i) by 12 and substracting (i) and (ii), we get

$\Rightarrow \quad s = \frac{70}{9}$

By putting $s = \frac{70}{9}$ in (i), we get,

$$2d = 18 \times \frac{70}{9} - 15 = 125 \Rightarrow d = 62.5$$

9. Let the time taken by x and y to reach the destination is $5x$ and $4x$ respectively.

Now, we have,

$5x - 4x = 45$

$\Rightarrow x = 45$

$\Rightarrow$ Required time = $4x = 4 \times 45 = 180$ minutes

10. Let the original speed of the car be x km/hr

After increasing speed = $(x + 15)$ km/hr

Now, according to the question,

$$\frac{300}{x} - \frac{300}{x+15} = 1$$

$\Rightarrow 300 \times 15 = x(x + 15) \Rightarrow 4500 = x^2 + 15x \Rightarrow x^2 + 15x - 4500 = 0$

$\Rightarrow x^2 + 75x - 60x - 4500 = 0 \Rightarrow x(x + 75) - 60(x + 75) = 0 \Rightarrow (x + 75)(x - 60) = 0$

$\Rightarrow x = -75$ (not possible) $\Rightarrow x = 60$ km/hr

11. Distance covered by z in 2 hours = $4 \times 2 = 8$ km

Time required for $y = \dfrac{8}{5-4} = 8$ hr

Let z starts t hours after y

Now, we have

$8 \times 5 = (8 - t) \times 6$

$\Rightarrow 40 = 48 - 6t \Rightarrow 6t = 8 \Rightarrow t = \dfrac{4}{3}$ hours

12. Let distance be d km and the initial speed be s km/hr

$$\therefore \qquad d = st \qquad \text{...}(i)$$

Now, according to the question,

$d = (s + x)(t - y)$

$\Rightarrow st = st - sy + xt - xy \Rightarrow sy = x(t - y) \Rightarrow s = \dfrac{x(t-y)}{y}$

Put the value of s in eqn (i), we get,

$d = \dfrac{x(t-y)}{y} \times t = tx(t - y)y^{-1}$ km

13. Let car A, B, and C meet d km far from the point and time taken by car A, B and C be t_1, t_2 and t_3 respectively.

Now,

$$t = \frac{d}{60} \qquad \text{...}(i)$$

$$t - 1 = \frac{d}{80} \qquad \text{...}(ii)$$

$$t - 2 = \frac{d}{x} \qquad \text{...}(iii)$$

From eqn (i) and (ii), we get

$\dfrac{d}{60} - 1 = \dfrac{d}{80} \Rightarrow \dfrac{d}{60} - \dfrac{d}{80} = 1 \Rightarrow \dfrac{d}{240} = 1 \Rightarrow d = 240$ km

Now, from eqn (i), we have

$$t = \frac{240}{60} = 4 \text{ hr.}$$

Now, from eqn (iii), we have

$$4 - 2 = \frac{240}{x}$$

$\Rightarrow x = 120$ km/hr

14. $\therefore$ Average speed = $\dfrac{\text{Total distance}}{\text{Total time}}$

$$= \frac{\frac{5}{6}\times 9+\frac{8}{6}\times 8+\frac{10}{6}\times 7.5}{\frac{5}{6}+\frac{8}{6}+\frac{10}{6}} = \frac{45+64+75}{23} \Rightarrow \frac{184}{23} = 8 \text{ km/hr}$$

15. Let the distance be d km

Now, we have,

$$\frac{d}{8}-\frac{d}{10}=\frac{9}{60} \quad \Rightarrow \frac{d}{40}=\frac{3}{20} \quad \Rightarrow d = 6 \text{ km}$$

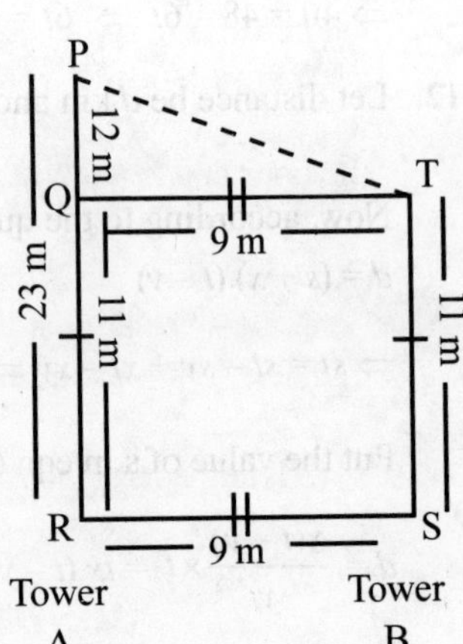

16. In ΔPQT :

$H^2 = P^2 + b^2$

$H^2 = 12^2 + 9^2$

$H = 15 \text{ m} = 1500 \text{ cm}$

If the monkey covered 1500 cm in 5 min (300 sec):

Speed = $\dfrac{1500}{300}$ = 5 cm/ sec

17. Let the distance between Delhi and Agra be d km

Time = $\dfrac{d}{60}+\dfrac{d}{y}$

Now, according to question,

$$\frac{2d}{\text{Average speed}} = \frac{d(60+y)}{60y}$$

$\Rightarrow 2 \times 60y = \text{Average speed} \times (60 + y) \Rightarrow 120y = 48 \times (60 + y)$

$\Rightarrow 10y = 240 + 4y \Rightarrow 6y = 240 \Rightarrow y = 40$ km/hr

18. $\because$ Average speed = $\dfrac{\text{Total distance}}{\text{Total time}} = \dfrac{20+10+30}{\frac{20}{40}+\frac{10}{10}+\frac{30}{40}} = \dfrac{60}{\frac{1}{2}+1+\frac{3}{4}}$

$$= \frac{60}{\frac{2+4+3}{4}} = \frac{60}{9}\times 4 = \frac{80}{3} = 26.67 \text{ km/hr}$$

19. Let the owner meet the thief d km from his house.

Now, we have

$$\frac{d}{40}=\frac{d}{60}+\frac{1}{2} \quad \Rightarrow \quad \frac{d}{40}-\frac{d}{60}=\frac{1}{2} \quad \Rightarrow \quad \frac{d}{120}=\frac{1}{2}$$

$\Rightarrow d = 60$ km

Time elapsed after theft = $\dfrac{60}{40}$ = 1.5 hrs.

20. Let the distance be 3D km

Total travel time = $\frac{D}{10} + \frac{D}{20} + \frac{D}{60}$ hr = $\frac{10D}{60}$ hr

Average speed of the car = $\frac{3D(6)}{D}$ km/hr = 18 km/hr

21. Time taken for A to overtake B = $\frac{c}{p-q}$ hours

Distance travelled by the time he overtakes B = $p\left(\frac{c}{p-q}\right)$ km = $\frac{pc}{p-q}$ km

22. Let us say the person goes to the market at 4 : *x* p.m. From the choices, it can be understand that he was out of his house for less than one hour.
∴ He must have returned at a time between 5 p.m and 6 p.m.
Let us say he returned at 5 : *y* p.m.
The hour and minute hands had interchanged their positions when he returned.
∴ It follows that
(*i*) The angular position of the hour hand at 4 : *x* p.m.
= The angular position of the minute hand at 5 : *y* p.m.
(ii) The angular position of the minute hand at 4 : *x* p.m.
= The angular position of the hour hand at 5 : *y* p.m.
(All angles are measured from 12)

$$4(30) + \frac{x}{2} = 6y \qquad ...(1)$$

$$6x = 5\,(30) + \frac{y}{2} \qquad ... (2)$$

$$\left(6x - \frac{y}{2}\right) - \left(6y - \frac{x}{2}\right) = 150 - 120$$

$$\Rightarrow \qquad \frac{13}{2}(x - y) = 30 \Rightarrow x - y = \frac{60}{13}$$

Time for which the person was out of his house = 5 : *y* p.m – 4 : *x* p.m = 1 hr + (*y* – *x*) min

$$= 1\text{hr} - \frac{60}{13}\text{ min}, = 55.38 \text{ min}$$

23. Let the duration of the flight be t hours.

Average speed of the flight = $\frac{600}{t}$ km/hr

If the average speed was 200 km/hr less, the time of flight would have been 30 minutes more.

Average speed would be = $\frac{600}{t + \frac{1}{2}}$ km/hr.

$$\frac{600}{t + \frac{1}{2}} = \frac{600}{t} - 200$$

$$200 = \frac{600}{t} - \frac{600}{t + \frac{1}{2}}$$

$$t^2 + \frac{1}{2}t - 1.5 = 0$$

$$2t^2 + t - 3 = 0$$

$(2t + 3)(t - 1) = 0$

$t = 1\ (\because t > 0)$

24. Let the speed be S km/hr.
Distance = 8 S km.
Had the speed been increased by 4 km/hr, the distance could have been covered in 7 hours and 30 minutes.

$$\text{Distance} = 8S = \frac{15}{2}(S + 4)$$

$$\Rightarrow \frac{S}{2} = 30 \Rightarrow S = 60$$

Distance = 8S km = 480 km

25. A runs $\frac{5}{3}$ times as fast as B.
A gives B a start of 80 m. Also both A and B reach the winning post at the same time. ...(1)
Let us say the winning post is d m from the starting point (of A).
(1) implies distances run by A and B when they finish simultaneously are d m and $(d - 80)$ m respectively.
Ratio of the speeds of A and B = Ratio of the distance run by A and B in the same time

$$= \frac{d}{d - 80} = \frac{5}{3} \Rightarrow d = 200$$

26. Distance run in 6 minutes by the

– thief $= 10\left(\frac{6}{60}\right)$ km = 1 km.

– policeman $= 11\left(\frac{6}{60}\right)$ km = 1.1 km

The thief is noticed by the policeman from a distance of 0.2 km.
After 6 minutes, the thief and the policeman would be 1.2 km and 1.1 km respectively.
Then, the distance between them would be 100 m

27. Time gap between the first and the last sightings of the light

	3 hr	17 min	49 s
=	–1 hr	54 min	50 s
	1 hr 22 min	59s = 4979s	

Let the number of sightings be n.
Time gap between the first and the last sightings = $13(n - 1)$ seconds.
$13(n - 1) = 4979 \Rightarrow n - 1 = 383 \Rightarrow n = 384$

28. The data is tabulated below. Let the length of the race be L m.

A	B	C
L	L – 30	L – 48
	L	L – 20

$$\frac{L - 30}{L} = \frac{L - 48}{L - 20} \Rightarrow L^2 - 50L + 600 = L^2 - 48L$$

$\Rightarrow L = 300$. (1) is true.
The speeds of the 3 people are in the ratio
300 : 270 : 252 = 50 : 45 : 42. (2) is true.

29. Let the original speed be v km / hr.

$\frac{300}{v} - 1 = \frac{300}{v+15}$ (given)

$\frac{300}{v} - \frac{300}{v+15} = 1$

$\frac{300(15)}{v(v+15)} = 1$

$\Rightarrow v(v + 15) = 4500 = 60(75) \ \Rightarrow v = 60.$

30. Let the speed of the three athletes be $16x$, $15x$ and $11x$. (all in m/s)

Time taken for the fastest runner to finish the race = $\frac{4000}{16x}$s = $\frac{250}{x}$s

In the same time, the second and the third runner would have run $\frac{250}{x}(15x)$ and $\frac{250}{x}(11x)$ metres respectively.

Distance between them = 250 (4) m = 1000m.

31. The distances covered by the 3 in equal intervals of time are shown as

A	B	C
100	96	98
	$96\left(\frac{100}{98}\right)$	100

When C covers 98 m, the gap between C and B is 2 m.

When C covers 100 m, It is $2\left(\frac{100}{98}\right) = 2 + \frac{2}{49} \approx 2 + \frac{2}{50} = 2.04$

32. The distance covered by the cycle every minute is 140 ($2\pi r$) cm

$= 140\ (2)\ \frac{22}{7}(30)$ cm = 26400 cm.

The speed is (26400) (60) $\frac{\text{cm}}{\text{hr}} = 15.84\ \frac{\text{km}}{\text{hr}}$.

33. Distance from A to B = 14 km.

Let the averages speed be x kmps.

Accordingly $\frac{14}{x-1} - \frac{14}{x} = \frac{20}{60}$

$\frac{14(x-x+1)}{x(x-1)} = \frac{1}{3}$

$x^2 - x - 42 = 0$

$(x - 7)\ (x + 6) = 0$

$x = 7\ \ (\because x \neq -6)$

x cannot be-ve

$\therefore$ The original speed of the cyclist = 7 kmph

34. When the clock took 8 seconds to strike 5 times

At 10 O'clock the clock strikes for 10 times so it takes 2 × 8 = 16 Second

35. Petrol consumed for 1st 10 km running at 40 km/hour is 10 × 20 ml = 200 *ml.*

Petrol Consumption for next 30 km with speed 60 km/hr is $30 \times 40 = 1200$ *ml*
Petrol consumption for next 10 km with a speed of 30 km/hr is $10 \times 20 = 200$ *ml.*
Total petrol consumption for 50km is $200 + 1200 + 200 = 1600$ $ml = 1.6l$

36. The position of the hour hand and the minute hand at the two times are shown below.

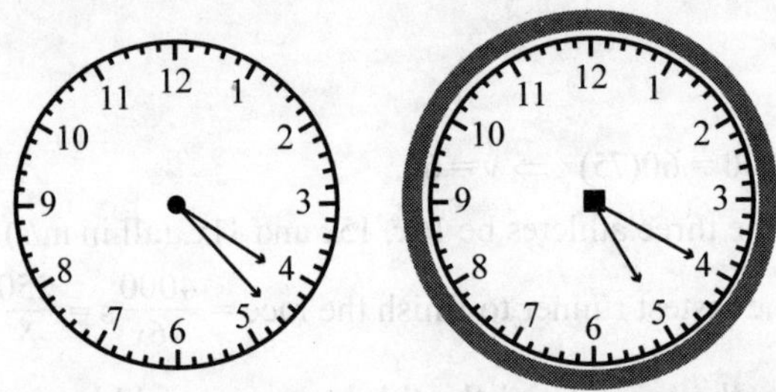

From the time the person leaves to the time he returns. The hour hand moves through a small angle *Q*, while the minute hand moves through 12θ. Together the cover one entire circle

$\therefore \quad \theta + 12\theta = 360^\circ$ or $\theta = \dfrac{360^\circ}{13}$

In one hour, the hour hand cover 30°.

$\therefore$ To cover θ, it takes $\dfrac{12}{13}$ hours, $\dfrac{12}{13}$hours $= \dfrac{720}{13}$ minutes

$= 55\dfrac{5}{13}$ minutes = 55 hr 38 minutes

37. The data is tabulated below. The distance covered by A, B are denoted by *a, b* respectively.

a	b	Time
	100	10
1000	900	90

b	Time
450	45
550	110

In the time A covers 1000 *m*, R covers 900 *m*. *B* taken 10s for the next 100*m*

$\therefore$ He must have taken 90s for 900*m* or 100s for 1000*m*.

In the given race, A takes 90s. while *B* takes 45s, for the first 450 *m* and 110s for the next 550 *m*, i.e. a total of 155s

$\therefore$ A beats *B* by 65s.

38. "Each meeting takes place after 13/4 hours." We interpret this to mean that the interval between the end of one meeting and the beginning of the next is 13/4 hours.

The next Meeting will start at 2.00 p.m. Therefore, the last meeting must have ended at 10 : 45 a.m.

Madhukar received this information 45 minutes after that, ie at 11 : 30 a.m.

39.

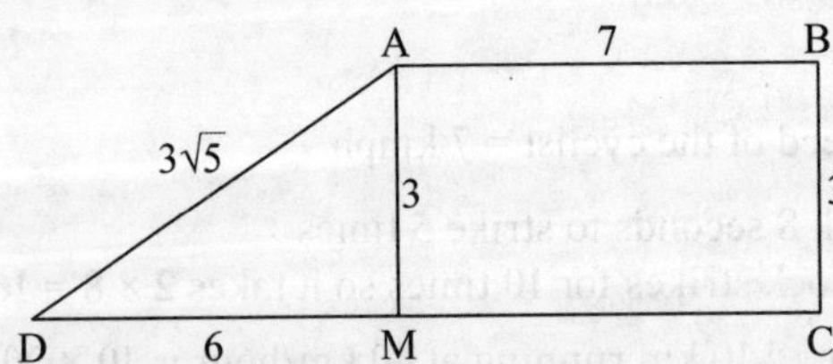

The middle position is A and the find position is D. We see that in Δ AMD, the side are 3(1), 3(2) and $3\sqrt{5}$.

40. A and B start from the same point on the circular track and run M the same direction with speeds in the ratio 5 : 4.
As 5 – 4 is 1, we divide the track into 1 'part', to get the points where A overtaken B. This means A overtaken B only at the starting point. This happen when A completes 5 rounds 10 rounds and 15 rounds. B would have completed 4, 8 and 12 rounds respectively. After that A completes 1 more round and for him the race is over. The winner (A) passes B a total of 3 times.

41. Diametre of circle = 80 cm

Radius = $\frac{80}{2} = 40$ cm

Distance covered in one revolution
$= 2\pi r = 2 \times \pi \times 40 = 80\pi$ cm

Speed = 66 km/hr, Time = 10 min. = $\frac{1}{6}$ hr

We know, speed $= \frac{\text{Distance}}{\text{Time}} \Rightarrow 66 = \frac{D}{\frac{1}{6}} \Rightarrow$ D = 11 km = 1100000 cm

$$\text{Number of revolutions} = \frac{\text{Total distance}}{\text{Distance covered in 1 revolution}} = \frac{1100000}{80\pi} = \frac{110000}{8 \times \frac{22}{7}}$$

$$= \frac{110000 \times 7}{8 \times 22} = 4375$$

The number of revolution made in covering the distance is 4375.

42.

B ← 96 m →
A ——●——
← 100 m →

$\text{Speed}_A = \frac{5}{3}$ m/s; $\text{Time}_A = \frac{100}{5} \times 3 = 60$ sec ; $\text{Time}_B = 60 + 12 = 72$ sec; $\text{Speed}_B = \frac{96}{72} = \frac{4}{3}$ m/s

43.

10 km/hr 8 km/hr
P T
← 100 m →

Relative Speed = 10 – 8 = 2 km/hr = $2 \times \frac{5}{18} = \frac{5}{9}$ m/s

$$\text{Time} = \frac{\text{Distance}}{\text{Speed}} = \frac{100}{5} \times 9 = 180 \text{ sec}$$

Speed of thief = $8 \times \frac{5}{18} = \frac{20}{9}$ m/s

Distance covered $= \frac{20}{9} \times 180 = 400\ m$

44. Distance = (Time speed) = $\left(\frac{11}{4} \times 4\right) = 11$ km.

Now, distance = 11 km and speed = 16.5 kmph.

$\therefore$ Time taken $= \frac{\text{Distance}}{\text{Speed}} = \frac{11}{16.5}$ hrs $= \left(\frac{11}{16.5} \times 60\right)$ min $= 40$ min.

45. Distance covered by train A in 9 hr = 9 × 60 = 540 km

Time when they meet $= \frac{800-540}{60+90} = \frac{260}{150} = \frac{26}{15}$ hr

$\therefore$ Required distance $= \frac{26}{15} \times 90 = 156$ km

46. From qn 45,

time when they meet = 4 am + $\frac{26}{15}$ hr

$= 4$ am $+ 1 + \frac{11}{15} \times 60 = 4$ am $+ 1 + 44 = 5:44$ am

47. Required time $= \frac{400+500}{(90+60) \times \frac{5}{18}} = \frac{900}{150 \times \frac{5}{18}} = \frac{900}{750} \times 18 = 21.6$ seconds

48. To be 0.5 km apart, they take 1 hour

To be 8.5 km apart, they would take $= \left(\frac{1}{0.5} \times 8.5\right)$ hrs = 17 hrs.

49. To be (18 + 20) km apart, they take 1 hour

To be 47.5 km apart, they would take $= \left(\frac{1}{38} \times 47.5\right)$ hrs $= 1\frac{1}{4}$ hrs.

50. Let train x takes t hours

Let train y takes $(t + 2)$ hours

Now, we have,

$\Rightarrow \frac{192}{t} - \frac{192}{t+2} = 16 \Rightarrow \frac{12}{x} - \frac{12}{t+2} = 1$

$\Rightarrow 12(t+2) - 12t = t(t+2) \Rightarrow 24 = t^2 + 2t \Rightarrow t^2 + 2t - 24 = 0$

$\Rightarrow t^2 + 6t - 4t - 24 = 0 \Rightarrow t(t+6) - 4(t+6) = 0 \Rightarrow (t+6)(t-4) = 0$

$\Rightarrow t = -6$ (not possible) and $t = 4$ hours

51. We have,

Total distance = 200 + 100 = 300 m

Time = 10 seconds

Speed $= \frac{\text{Distance}}{\text{Time}} = \frac{300}{10} = 30$ m/s

52. Let the usual speed of the train be x km/hr

Now, we have,

$$\frac{300}{x}-\frac{300}{x+5}=2 \Rightarrow 300\left(\frac{1}{x}-\frac{1}{x+5}\right)=2$$

$\Rightarrow 300\,(5) = (x^2 + 5x) \times 2 \Rightarrow\ 750 = x^2 + 5x \Rightarrow x^2 + 5x - 750 = 0$

$\Rightarrow x^2 + 30x - 25x - 750 = 0 \Rightarrow (x - 25)\,(x + 30) = 0 \Rightarrow x = 25$ km/h

53. Speed = $132 \times \frac{5}{18} = \frac{110}{3}$

Total distance = 110 + 165 = 275 m

$$\text{Time} = \frac{\text{Distance}}{\text{speed}}$$

$$= \frac{275}{110} \times 3 = \frac{15}{2} = 7.5 \text{ seconds}$$

54. We have speed = $60 \times \frac{5}{8} = \frac{50}{3}$ m/s

Length of the train = speed × time = $\frac{50}{3} \times 30$ = 500 m

55. Distance = 100 + 100 = 200 m

Time = 10 seconds

$\therefore$ Speed of the train = $\frac{\text{Distance}}{\text{Time}} = \frac{200}{10}$ = 20 m/s = $20 \times \frac{18}{5}$ = 72 km/hr

56. The speed of the first train = $1.5\,\frac{\text{km}}{\text{min}}$

The speed of the second train = $60\,\frac{\text{km}}{\text{hr}} = 1\,\frac{\text{km}}{\text{min}}$

Any point on the faster train covers the length of the slower train in 27s.

The length of slower train is $(0.5)\left(\frac{27}{60}\right)$ km $= \frac{9}{40}$km = 225 m

57. **Case 1:** Through station of length

$$\frac{L+x}{v} = t \quad \ldots\text{(i)}$$

Case 2: Through station of length y

$$\frac{L+y}{v} = 2t \quad \ldots\text{(ii)}$$

From (i): $v = \frac{L+x}{t}$

From (ii): $v = \frac{L+y}{2t}$

Equating both expressions for v :

$$\frac{L+x}{t}=\frac{L+y}{2t}$$

Multiply both sides by $2t$:

$$2(L+x)=L+y$$

$\Rightarrow$ $$2L+2x=L+y$$

$\Rightarrow$ $$L=y-2x \qquad (iii)$$

Now, time to pass station of length $x+y$:

Required time: $\frac{L+(x+y)}{v}$

Using (i): $v=\frac{L+x}{t}$

So, $=\frac{L+x+y}{\frac{L+x}{t}}=\frac{(L+x+y)\cdot t}{L+x}$

Substitute $L=y-2x$: $\frac{(y-2x+x+y)\cdot t}{y-2x+x}=\frac{(2y-x)\cdot t}{y-x}=\frac{(2yt-xt)}{(y-x)}$

58. Distance covered by train A in 50 minutes = $80\times\frac{50}{60}$

Now, time when they meet = $\frac{80\times\frac{50}{60}}{100-80}$

Required distance = $\frac{100\times\frac{200}{3}}{20}=100\times\frac{10}{3}=\frac{1000}{3}$ km

59. Let x kmph be the usual speed of the train

Accordingly $\frac{120}{x}-\frac{120}{x-10}=1$; $\frac{x-10-x}{x(x-10)}=\frac{1}{120}$

$x^2-10x-1200=0$, $(x-40)(x+30)=0$

$x=40$ (x can not be – ve)

$\therefore$ The usual speed of the train = 40 kmph

60. Let the original speed of the train be $5u$ and the time taken be 6t.

$\therefore$ If the speed is increased by $6u$, the time would be 5t.

As $6t-5t=20$ min, $t=20$ min and $6t=120$ min.

61. Let the length of the first train be 2L. The length of the second train is L. Let the length of the platform be P meters

The speed of the first train in 48 $\frac{km}{hr}$, *i.e.*, $\frac{40}{3}\frac{M}{s}$.

The speed of the second train is 42 $\frac{km}{hr}$, *i.e.*, $\frac{35}{3}\frac{M}{s}$.

The first train crosses the second in 12s.

$$\therefore \frac{2L+L}{\frac{40}{3}+\frac{35}{3}} = 12\text{ s} \Rightarrow 3L = 300s \Rightarrow L = 100$$

The first train crosses the platform in 45s.

$$\therefore \frac{2L+P}{40/3} = 45 \Rightarrow P + 200 = 40\ (15) = 600$$

Or P = 400 m

62.

Delhi		Mumbai
6 p.m.		9 p.m.
Speed = x	←1080 km→	Speed = $x + 15$

Speed of passenger train = x km/hr
Speed of express train = $(x + 15)$ km/hr
According to the question,

$$\frac{540}{x} - \frac{540}{x+15} = 3$$

$$\Rightarrow 540\left[\frac{x+15-x}{x^2+15x}\right] = 3 \Rightarrow \frac{180\times 15}{x^2+15x} = 1 \Rightarrow x^2 + 15x - 180\times 15 = 0$$

$$\Rightarrow (x+60)(x-45) = 0 \Rightarrow x = 45$$

$$\therefore \frac{540}{x+15} = \frac{540}{45+15} = \frac{540}{60} = 9\text{ hrs}$$

So, the two trains will meet at 6 a.m

63. Distance = 225 m
(Length of train)
Relative Speed = (30 – 3) = 27 km/hr = $27\times \frac{5}{18} = \frac{15}{2}$ m/s

Time $= \frac{225}{15}\times 2 = 15\times 2 = 30$ Seconds

64. Let the speed of train X be S_X and its length be a.

So, $S_X = \frac{a}{24} \Rightarrow a = 24S_X$

Let the speed of train Y be S_Y and its length be b.

So, $S_Y = \frac{b}{18} \Rightarrow b = 18S_Y$

Now, Both trains cross each other in 20 sec, so

$$S_X + S_Y = \frac{a+b}{20}$$

$$\Rightarrow 20(S_X + S_Y) = a + b \Rightarrow 20S_X + 20S_Y = 24S_X + 18S_Y$$

$$\Rightarrow 24S_X - 20S_X = 20S_Y - 18S_Y \Rightarrow 4S_X = 2S_Y$$

$$\therefore S_X : S_Y = 2:4 = 1:2$$

65. Speed $= \left(54 \times \frac{5}{18}\right)$ m/sec = 15 m/sec.

Time taken $= \frac{(360+390)}{15}$ sec = 50 sec.

66. Speed $= \left(36 \times \frac{5}{18}\right)$ m/sec = 10 m/sec.

Let the length of the train be x metres.

Then, $\frac{x}{18} = 10 \Leftrightarrow x = 180$ m.

67. Let the length of bridge be x metres.

Speed of the train $= \left(36 \times \frac{5}{18}\right)$ m/sec = 10m/sec.

$\therefore \quad \frac{100+x}{25} = 10 \Leftrightarrow x = 150$ metres.

68. Relative speed = (70 – 10) kmph $= \left(60 \times \frac{5}{18}\right)$ m/sec $= \left(\frac{50}{3}\right)$ m/sec.

Required time = Time taken to cover 100 m at $\left(\frac{50}{3}\right)$ m/sec. $= \left(100 \times \frac{3}{50}\right)$ sec = 6 sec.

69. Relative speed = (25 + 2) kmph $= \left(27 \times \frac{5}{18}\right)$ m/sec $= \frac{15}{2}$ m/sec.

$\therefore$ Required time = Time taken to cover 270 m at $\frac{15}{2}$ m/sec $= \left(270 \times \frac{2}{15}\right)$ sec = 36 sec.

70. Let the speed of the train be x km/hr.

Relative speed = $(x + 6)$ kmph $= \left[(x+6) \times \frac{5}{18}\right]$ m/sec

$$\therefore \quad \frac{150}{(x+6) \times \frac{5}{18}} = 6 \Leftrightarrow 5x + 30 = 450 \Leftrightarrow x = 84.$$

71. Relative speed $= \left(\frac{150+100}{10}\right)$ m/sec $= \left(\frac{250}{10} \times \frac{18}{5}\right)$ km/hr = 90 km/hr.

$\therefore$ Speed of second train = (90 – 30) km/hr = 60 km/hr.

72. Let the length of the train be x metres. Then,

$$\frac{x}{15} = \frac{x+100}{25} \Leftrightarrow 25x = 15(x+100) \Leftrightarrow x = 150\text{ m}.$$

73. Let the length of train be x km & its speed be y km/hr

Then, $\frac{x}{y-2} = \frac{9}{60 \times 60}$ and $\frac{x}{y-4} = \frac{10}{60 \times 60}$ $\Leftrightarrow 9y - 3600x = 18$ and $10y - 3600x = 40$

$\Leftrightarrow y - 400x = 2$ and $y - 360x = 40$

$\therefore 40x = 2$ or $x = \left(\frac{2}{40} \times 1000\right)$ m = 50m.

74. Relative speed = (40 – 22) km/hr = $\left(18 \times \frac{5}{18}\right)$ m/sec = 5 m/sec

Let the length of second train be x metres. Then, $\frac{125 + x}{5} = 60 \Leftrightarrow x = 175$ m.

75. Let speed of stream be x km/h

Now, according to the question,

$$\frac{60}{30+x} + \frac{60}{30-x} = \frac{9}{2}$$

$$\Rightarrow 3600 = \frac{9}{2}(900 - x^2) \Rightarrow 800 = 900 - x^2 \Rightarrow x^2 = 1000 \Rightarrow x = 10 \text{ km/h}$$

76. Let the distance between the two places be d km

Let speed of boat in still water = x km/s

Let speed of boat in still stream = y km/s

$\therefore$ Downstream speed = $(x + y)$ km/h

and upstream speed = $(x - y)$ km/h

Now, we have

$$\Rightarrow z = \frac{d}{x+y} + \frac{d}{x-y} \Rightarrow z = \frac{2dx}{x^2 - y^2} \Rightarrow d = \frac{z(x^2 - y^2)}{2x}$$

77. Let the speed of the boat in still water and the current be x km/hr and y km/hr respectively

$6(x + y) = 32$ and $6(x - y) = 14$

$x + y = \frac{32}{6}$ and $x - y = \frac{14}{6}$

$(x + y) - (x - y) = \frac{18}{6} \Rightarrow 2y = 3 \Rightarrow y = 1.5.$

78. Let the speed of the stream be $\underline{v}$ km/hr.

Total time taken to travel upstream and downstream = $\frac{30}{15 - v} + \frac{30}{15 + v} = 4.5$

$$\frac{30(15+v) + 30(15-v)}{15^2 - v^2} = 4.5$$

$$\frac{900}{4.5} = 15^2 - v^2 \Rightarrow v = 5.$$

79. Let the person's speed in still water be V km/hr and the Speed of the current be u, Km/hr.

$\frac{20}{V+u} = 2$ and $\frac{4}{V-u} = 2$; i.e., $V + u = 10$ and $V - u = 2$, $V = 6$, $u = 4$

80. Average speed = $\frac{2xy}{x+y} = \left(\frac{2 \times 24 \times 36}{24 + 36}\right)$ km/hr = 28.8 km/hr.

81. Rate in still water = $\frac{1}{2}(10 + 5)$ kmph = 7.5 kmph.

82. Rate downstream = $\left(\frac{1}{6}\times 60\right)$ kmph = 10 kmph.

Rate upstream = $\left(\frac{1}{10}\times 60\right)$ kmph = 6 kmph.

Speed of current = $\frac{1}{2}(10-6)$ kmph = 2 kmph.

83. Rate upstream = $\left(\frac{11}{5}\right)$ kmph = 2.2 kmph.

Rate downstream = $\left(\frac{26}{5}\right)$ kmph = 5.2 kmph.

$\therefore$ Velocity of current = $\frac{1}{2}(5.2-2.2)$ kmph = 1.5 kmph.

84. Let man's rate against the current be x kmph.

Then, $\frac{1}{2}(12-x) = 1.5 \Leftrightarrow 12-x = 3 \Leftrightarrow x = 9$ kmph.

85. Rate upstream = $\left(\frac{40}{8}\right)$ km/hr = 5 km/hr.

Rate downstream = $\left(\frac{36}{6}\right)$ km/hr = 6 km/hr.

$\therefore$ Rate in still water = $\frac{1}{2}(5+6)$ km/hr = 5.5 km/hr.

86. Rate upstream = (5 – 1) kmph = 4 kmph.
Rate downstream = (5 + 1) kmph = 6 kmph.
Let the required distance be x km.

Then, $\frac{x}{4}+\frac{x}{6} = \frac{75}{60} \Leftrightarrow 25x = 75 \Leftrightarrow x = 3$ km.

87. Rate downstream = (15 + 3) kmph = 18 kmph.

Distance travelled in 12 min = $\left(\frac{18}{60}\times 12\right)$ km. = 3.6 km.

88. Let the speed upstream be x km/hr. Then, $\frac{1}{2}(15-x) = 1.5 \Leftrightarrow x = 12$ kmph.

89. Rate upstream = 1 kmph, Rate downstream = 5 kmph.

$\therefore$ Total time taken = $\left(\frac{10}{1}+\frac{10}{5}\right)$ hrs = 12 hrs.

90. Let the speed in still water be x km/hr. Then,

$\frac{35}{x-1}+\frac{35}{x+1} = 12 \Leftrightarrow 35(x+1+x-1) = 12(x^2-1) \Leftrightarrow 12x^2 - 70x - 12 = 0 \Leftrightarrow 6x^2 - 35x - 6 = 0$

$\Leftrightarrow (x-6)(6x+1) = 0 \Leftrightarrow x = 6.$

91. Let speed upstream & speed downstream be x km/hr and y km/hr respectively. Then,

$$\frac{24}{x}+\frac{36}{y}=6 \text{ and } \frac{36}{x}+\frac{24}{y}=\frac{13}{2}$$

$\therefore\ 24u + 36v = 6$ and $36u + 24v = \frac{13}{2}$, where $u = \frac{1}{x}$ & $v = \frac{1}{y}$

Adding, we get: $60(u + v) = \frac{25}{2}$ or $u + v = \frac{5}{24}$.

Subtracting, we get: $12(u - v) = \frac{1}{2}$ or $u - v = \frac{1}{24}$.

Solving $u + v = \frac{5}{24}$ & $u - v = \frac{1}{24}$, we get $u = \frac{1}{8}$ & $v = \frac{1}{12}$

$\therefore\quad x = 8, y = 12.$

Velocity of current $= \frac{1}{2}(12-8)$km/hr $= 2$ km/hr.

92. Let speed upstream = x km/hr. Then, speed downstream = $3x$ km/hr

Speed in still water $= \frac{1}{2}(x + 3x)$ kmph $= 2x$ kmph

Speed of the current $= \frac{1}{2}(3x - x)$ kmph $= x$ kmph.

$\therefore\quad 2x = \frac{28}{3}$ or $x = 4\frac{2}{3}$ km/hr.

93. Speed upstream $= \left(\frac{3}{4}\times\frac{4}{45}\times 60\right)$ km/hr $= 4$ km/hr.

Speed downstream $= \left(\frac{3}{4}\times\frac{2}{15}\times 60\right)$ km/hr $= 6$ km/hr.

$\therefore$ Speed in still water $= \frac{1}{2}(4 + 6)$ km/hr $= 5$ km/hr.

13

CHAPTER

Problems on Ages

EXERCISE 13

1. The ratio of Meena's age and Kamla's age is 3 : 5 and the sum of their ages is 80 years. The ratio of their ages after 10 years will be:
(*a*) 2:3 (*b*) 1:2 (*c*) 3:2 (*d*) 3:5

2. The age of a father 10 years ago was thrice the age of his son. Ten years hence, the father's age will be twice that of his son. The ratio of their present ages is:
(*a*) 5:2 (*b*) 9:2 (*c*) 7:3 (*d*) 13:4

3. The sum of the ages of a son and his father is 56 years. After 4 years, the age of the father will be three times that of his son. Their ages respectively are:
(*a*) 12 years, 44 years (*b*) 16 years, 42 years (*c*) 16 years, 48 years (*d*) 18 years, 36 years

4. In 10 years, A will be twice as old as B was 10 years ago. If A is now 9 years older than B, the present age of B is:
(*a*) 19 years (*b*) 29 years (*c*) 39 years (*d*) 49 years

5. The sum of the ages of a mother and a daughter is 50 years. Also, 5 years ago, the mother's age was 7 times the age of the daughter. The present ages of mother and daughter respectively are:
(*a*) 35 years, 15 years (*b*) 38 years, 12 years (*c*) 40 years, 10 years (*d*) 42 years, 8 years

6. The age of Q exceeds the age of P by 3 years. The age of R is twice the age of P and the age of Q is twice the age of S. Further, the age difference of R and S is 30 years. What is the sum of the ages of P and Q ? **[CDS (I) Exam, 2023]**
(a) 35 years (b) 38 years (c) 39 years (d) 45 years

7. The combined age of a man and his wife is 6 times the combined age of their children. Two years ago their combined age was 10 times the combined age of their children; and six years later their combined age will be 3 times the combined age of their children. How many children do they have if each child is at least 2 years old ? **[CDS (II) Exam, 2023]**
(a) 2 (b) 3 (c) 4 (d) 5

8. Ten (10) years before, the ages of a mother and her daughter were in the ratio 3:1. Ion another 10 years from now, the ratio of their ages will be 13 : 7. What are their present ages? **[CDS (II) Exam, 2018]**
(*a*) 39 years, 21 years (*b*) 55 years, 25 years
(*c*) 75 years, 25 years (*d*) 49 years, 31 years

9. The ratio of the sum and differences of the ages of the father and the son is 11 : 3. Consider the following statements: **[CDS (II) Exam, 2018]**
1. The ratio of their ages is 8 : 5.
2. The ratio of their ages after the son attains twice the present age will be 11 : 8.
Which of the statements given above is/are correct?
(*a*) 1 only (*b*) 2 only (*c*) Both 1 and 2 (*d*) Neither 1 nor 2

10. Four years ago, the average age of A and B was 18 years. Now the average age of A, B and C is 24 years. After 8 years, the age of C will be **[CDS (II) Exam, 2015]**

(*a*) 32 years (*b*) 28 years
(*c*) 36 years (*d*) 40 years

11. There are three brothers. The sum of ages of two of them at a time are 4 years, 6 years and 8 years. The age difference between the eldest and the youngest is **[CDS (I) Exam, 2016]**

(*a*) 3 years (*b*) 4 years (*c*) 5 years (*d*) 6 years

12. Leela got married 6 years ago. Today her age is $1\frac{1}{4}$ times her age at the time of her marriage. Her son's are is $\frac{1}{10}$ times her age. What is the present age of her son? **[CDS (I) Exam, 2017]**

(*a*) 1 year (*b*) 2 years (*c*) 3 years (*d*) 4 years

13. The present age of a father is equal to sum of the ages of his 4 children. After ten years the sum of the ages of the children will be 1-6 times the age of their father. What is the present age of father? **[CDS (II) Exam, 2021]**

(*a*) 36 years (*b*) 40 years (*c*) 42 years (*d*) 45 years

14. Ena was born 4 years after her parents' marriage. Her mother is 3 years younger than her father and 24 years older than Ena, who is 13 years old. At what age did Ena's father get marriage? **[CDS (I) Exam, 2020]**

(*a*) 25 years (*b*) 24 years (*c*) 23 years (*d*) 22 years

15. Mahesh is 60 years old. Ram is 5 years younger to Mahesh and 4 years elder to Raju. Babu is a younger brother of Raju and he is 6 years younger. What is the age difference between Mahesh and Babu? **[CDS (I) Exam, 2021]**

(*a*) 18 years (*b*) 15 years (*c*) 15 years (*d*) 11 years

16. Priya's age was cube of an integral number (different from I) four years ago and square of an integral number after four years. How long should she wait so that her age becomes square of a number in the previous year and cube of a number in the next year? **[CDS (I) Exam, 2019]**

(*a*) 7 years (*b*) 12 years (*c*) 14 years (*d*) 21 years

17. Radha and Rani are sisters. Five years back, the age of Radha was three times that of Rani, but one year back the age of Radha was two times that of Rani. What is the age difference between them? **[CDS (II) Exam, 2019]**

(*a*) 8 (*b*) 9 (*c*) 10 (*d*) 11

18. Ashok's age is (1/6)th of his father's age. The father's age will be twice of Kamal's age after 10 years. If Kamal's eighth birthday was celebrated 2 years before, what is Ashok's present age?

(*a*) 24 years (*b*) 30 years (*c*) $6\frac{2}{3}$ years (*d*) None

19. The sum of the ages of a father and son is 45 years. Five years ago the product of their ages was 4 times the father's age at that time. The present ages of the father and son respectively are:

(*a*) 25 years, 10 years (*b*) 36 years, 9 years
(*c*) 39 years, 6 years (*d*) None of these

20. 10 years ago, Mona's mother was 4 times older than her daughter. After 10 years, the mother will be twice older than the daughter. Mona's present age is:

(*a*) 5 years (*b*) 10 years (*c*) 20 years (*d*) 30 years

21. The age of a woman is a two-digit integer. On reversing this integer, the new integer is the age of her husband who is elder to her. The difference between their ages is one-eleventh of their sum. What is the difference between their ages? **[CDS (I) Exam, 2018]**

(*a*) 8 years (*b*) 9 years (*c*) 10 years (*d*) 11 years

22. The sum of ages of a father, a mother, a son Sonu and daughters Savita and Sonia is 96 years. Sonu is the youngest member of the family. The year Sonu was born, the sum of the ages of all the members of the family was 66 years. If the father's age now is 6 times that of Sonu's present age, then 12 years hence, the father's age will be **[CDS (II) Exam, 2018]**

(*a*) 44 years (*b*) 45 years (*c*) 46 years (*d*) 48 years

ANSWERS

1. (*a*) **2.** (*c*) **3.** (*a*) **4.** (*c*) **5.** (*c*) **6.** (*d*) **7.** (*b*) **8.** (*b*) **9.** (*b*) **10.** (*c*)
11. (*b*) **12.** (*c*) **13.** (*b*) **14.** (*c*) **15.** (*b*) **16.** (*c*) **17.** (*a*) **18.** (*d*) **19.** (*b*) **20.** (*c*)
21. (*b*) **22.** (*d*)

SOLUTIONS

1. Let their ages be $3x$ and $5x$ years.

Then, $3x + 5x = 80 \Leftrightarrow x = 10$.

$\therefore$ Their present ages are 30 years & 50 years.

Ratio of their ages after 10 years = 40 : 60 = 2 : 3.

2. Let son's age 10 years ago be x.

Father's age 10 years ago = $3x$.

$2(x + 20) = (3x + 20) \Leftrightarrow x = 20$.

10 years ago, father's age = 60 years & son's age = 20 years.

$\therefore$ Ratio of their present ages = 70 : 30 = 7 : 3.

3. Let son's age = x. Then, father's age = $(56 - x)$.

$56 - x + 4 = 3(x + 4) \Leftrightarrow 4x = 48 \Leftrightarrow x = 12$.

So, their ages are 12 years & 44 years.

4. Let B's present age be x. Then, A's age = $(x + 9)$.

$(x + 9 + 10) = 2(x - 10) \Leftrightarrow x = 39$.

$\therefore$ B's present age = 39 years.

5. Let daughter's age = x. Then, mother's age = $(50 - x)$.

$7(x - 5) = (50 - x - 5) \Leftrightarrow 8x = 80 \Leftrightarrow x = 10$.

$\therefore$ Present age of mother = 40 years & daughter's age = 10 years

6. We have,

$Q = P + 3$ and $Q = 2S$

$\Rightarrow P + 3 = 2S \quad \Rightarrow \quad P = 2S - 3$ ...(i)

Now, $R = 2P$ and $R - S = 30$

$\therefore 2P = S + 30$

$\Rightarrow P = \dfrac{S+30}{2}$ ....(ii)

From (i) and (ii), we get

$2S - 3 = \dfrac{S+30}{2}$

$\Rightarrow 4S - 6 = S + 30 \;\Rightarrow S = 12 \;\Rightarrow P = 2S - 3 = 21$

Q = 2S = 24

$\therefore$ P + Q = 21 + 24 = 45 years

7. Let the present ages of a man and his wife together is a and their children is b.
Let number of children = c
Now, according to the question,

$a = 6b$ ...(i)

Two years ago, we have,

$a - 2 - 2 = 10(b - 2c)$

$= a - 4 = 10(b - 2c)$ ...(ii)

Now, six years later

$a + 6 + 6 = 3(b + 6c) \Rightarrow a + 12 = 3(b + 6c)$ ...(iii)

From (i) and (ii), we get,

$6b - 4 = 10b - 20c$

$\Rightarrow 20c - 4 = 4b \quad \Rightarrow y = 5c - 1$ ...(iv)

From (i) and (iii), we get,

$6b + 12 = 3(b + 6b) \Rightarrow 6b + 12 = 3b + 18c \Rightarrow 3b = 18c - 12$

$\Rightarrow b = 6c - 4$ ...(v)

From (iv) and (v), we get

$5c - 1 = 6c - 4 \Rightarrow c = 3$

8. Let 10 years before the age of the mother and daughter be $3x$ and x years respectively.

Now, we have,

$$\frac{3x+20}{x+20} = \frac{13}{7} \Rightarrow 21x + 140 = 13x + 260 \Rightarrow 8x = 120 \Rightarrow x = 15$$

$\therefore$ Present age of mother = $3x + 10$ = 55 years
Present age of daughter = $x + 10$ = 25 years

9. Let the ages of father and son be f and s years respectively

Now,

(i) $\dfrac{f+s}{f-s} = \dfrac{11}{3} \Rightarrow 3f + 3s = 11f - 11s \Rightarrow 8f = 14s \Rightarrow \dfrac{f}{s} = \dfrac{7}{4}$

Hence, statement-1 is not correct

(i) Let the father and son's ages after s year are $f + s$ and $s + s$

$$\Rightarrow \text{Ratio} = \frac{f+s}{s+s} = \frac{f}{2s} + \frac{s}{2s} = \frac{1}{2}\left[\frac{f}{s} + 1\right] = \frac{1}{2}\left[\frac{7}{4} + 1\right] \text{[from (1)]}$$

$$= \frac{1}{2}\left[\frac{11}{4}\right] = \frac{11}{8} \Rightarrow \frac{f+s}{2s} = \frac{11}{8}$$

Hence, statement – 2 is correct

10. If the average age of A and B 4 years ago was 18 years, their average age today would be 18 + 4, viz 22 years.
Sum of A's and B's ages = 2(22) viz 44 years
Sum of A's, B's and C's ages = 3(24) viz 72 years
Age of C = (72 – 44) years = 28 years
After 8 years, C will be (28 + 8) years = 36 years.

11. Let a, b, c be ages of three brothers

According

$a + b = 4$ ... (1)

$b + c = 6$ ... (2)

$a + c = 8$ ... (3)

$2(a + b + c) = 8 + 6 + 4 = 18$

$a + b + c = 9$

from (1) $c = 9 - 4 = 5$

from (2) $a = 9 - 6 = 3$

from (3) $\underline{b} = 9 - 8 = 1$

$\therefore$ The difference of the ages between the eldest and the youngest is $c - b = 5 - 1 = 4$ years

12. Let Son's age be x

$\therefore$ Leela's age be $10x$

Age during marriage $= 10x - 6$

According to the question, $10x = \frac{5}{4}(10x - 6)$

$40x = 50x - 30$ $x = 3$

$\therefore$ Son's age is 3 years

13. Let the present age of father be x years and sum of ages of his 4 children be y years

Now, we have,

$x = y$ ...(i)

$1.6(x + 10) = y + 40$

$\Rightarrow 1.6x + 16 = x + 40$ $[\because x = y]$

$\Rightarrow 0.6x = 24$ $\Rightarrow x = 40$ years

14. Present age of Ena = 13 years

Present age of Ena's mother = 13 + 24 = 37 years

Present age of Ena's father = 37 + 3 = 40 years

$\therefore$ Age of Ena's father at the time of marriage = 40 – 13 – 4 = 23 years

15. Mahesh's age = 60 years

Ram's age = 60 – 5 = 55 years

Raju's age = 55 – 4 = 51 years

Babu's age = 51 – 6 = 45 years

$\therefore$ Required difference = 60 – 45 = 15 years

16. Let Priya's present age is x years

Now, according to question,

$(x - 4)$ is cube of an integral number

$\therefore$ Possible values of $(x - 4)$ are 8, 27, 64 etc.

Possible values of x are 12, 31, 68 etc.

Possible values of $(x + 4)$ should be square of an integral number

$\therefore$ We have $x + 4 = 16$ $\Rightarrow x = 12$ years

Possible square of a number greater than x = 16, 25, 36, 49 ....

Possible cube of a number greater than x = 27, 64, 125 ....

$\therefore$ The age becomes square of a number in the previous years and cube of a number in the next year = 26 years.

$\therefore$ year, when she waits = 26 – 12 = 14 years

17. Let the present age of Radha and Rani be x and y years respectively.

Now, we have,

$x - 5 = 3 \times (y - 5)$

$\Rightarrow x - 5 = 3y - 15$

$\Rightarrow 3y - x = 10$ ...(*i*)

and

$(x - 1) = 2(y - 1)$

$= 2y - x = 1$ ...(ii)

Subtracting (i) and (ii), we get

$y = 9$ years and $x = 17$ years

$\therefore$ Required difference = 17 – 9 = 8 years

18. Kamal's present age = 10 years.

Kamal's age after 10 years = 20 years.

Father's age after 10 years = 40 years

Father's age now = 30 years.

Ashok's age now = $\left(y \times \frac{18}{5}\right)$ years = 5 years.

19. Let son's age = x. Then, father's age = $(45 - x)$

$(x - 5)(45 - x - 5) = 4(45 - x - 5) \Leftrightarrow x - 5 = 4 \Leftrightarrow x = 9.$

$\therefore$ Father's age now = 36 years & son's age = 9 years.

20. Let Mona's age 10 yrs ago = x.

Her mother's age then = $4x$.

$$2(x + 20) = 4x + 20 \Leftrightarrow 2x = 20 \Leftrightarrow x = 10.$$

$\therefore$ Mona's present age = (10 + 10) years = 20 years.

21. Let the two digits of the age of the woman be y and x.

$\therefore$ Age of woman = $10y + x$

Age of man = $10x + y$

Difference of their ages = $10x + y - 10y - x = 9x - 9y$

Sum of their ages = $11(x + y)$

Now, according to the question,

$9x - 9y = \frac{1}{11} \times 11(x + y) = 9x - 9y = x + y$

$\Rightarrow 8x = 10y \quad \Rightarrow 4x = 5y \quad \Rightarrow \frac{x}{y} = \frac{5}{4}$

$\therefore$ Required difference = $9(x - y) = 9(5 - 4) = 9 \times 1 = 9$ years

22. Let Sonu's present age = x years

Now, we have,

$66 = 96 - 5x$

$\Rightarrow \; 5x = 30$

$\Rightarrow \; x = 6$ years

$\Rightarrow$ Sonu's father's age = $6x$ = 36 years

$\therefore$ Father's age after 12 years = 36 + 12 = 48 years

14 CHAPTER Simple Interest

Important Points on Simple Interest

(*i*) The money borrowed is called **Principal** or **Sum**.

(*ii*) Extra money paid for using other's money is called **Interest**.

(*iii*) If the interest on a certain sum for a certain period is reckoned uniformly, it is called **Simple Interest**, abbreviated as S.I.

(*iv*) Amount = Principal + Interest.

(*v*) **Formulae:** Let Principal = *P*; Rate = *R*% and Time = *T* years. Then,

$$\text{S.I.} = \frac{P \times R \times T}{100};\ P = \frac{100 \times S.I.}{R \times T};\ T = \frac{100 \times S.I.}{P \times R}\ \&\ R = \frac{100 \times S.I.}{P \times T}$$

Ex. 1. ***Find the amount of ₹1400 invested at 14% during the period from 5th Feb. 1994 to 19th April 1994.***

Sol. Here $P = 1400$ & $R = 14$ %.

	Feb	March	April	
Time:	23	31	19	$= 73 \text{ days} = \frac{1}{5}$ years.

$$\therefore \text{S.I.} = ₹\left(1400 \times 14 \times \frac{1}{5} \times \frac{1}{100}\right) = ₹39.20.$$

$\therefore$ Amount = ₹(1400 + 39.20) = ₹1439.20.

Ex. 2. ***A certain sum amounts to ₹1008 in 2 years, ₹1112 in 3 years. Find the rate and the sum.***

Sol. S.I. for 1 year = ₹(1112 – 1008) = ₹104

S.I. for 2 years = ₹(2 × 104) = ₹208.

$\therefore$ Principal = ₹(1008 – 208) = ₹800.

Now, Sum = ₹800, S.I. for 1 year = ₹104.

$$\therefore \text{Rate} = \frac{100 \times \text{S.I.}}{\text{Sum} \times \text{Time}} = \left(\frac{100 \times 104}{800 \times 1}\right) = 13\%.$$

Ex. 3. ***A sum was put at simple interest at a certain rate for 2 years. Had it been put at 3% higher rate, it would have fetched ₹300 more. Find the sum.***

Sol. Let Sum = *x* and original rate = *R*%. Then,

$$\frac{x \times (R+3) \times 2}{100} - \frac{x \times R \times 2}{100} = 300 \text{ or } 2Rx + 6x - 2Rx = 300 \times 100 \text{ or } x = 5000.$$

Ex. 4. ***A shopkeeper borrowed ₹20000 from two money lenders. For one loan he paid 12% and for the other 14% per annum. After one year, he paid ₹2560 as interest. How much did he borrow at each rate?***

Sol. Suppose money at 12 % = x and that at 14 % = $(20000 - x)$.

$$\therefore \quad \frac{x \times 12 \times 1}{100} + \frac{(20000 - x) \times 14 \times 1}{100} = 2560$$

or $12x + 280000 - 14x = 256000$ or $x = 12000$.

$\therefore$ Money at 12 % = ₹12000 & Money at 14 % = ₹8000.

Ex. 5. ***What annual instalment will discharge a debt of ₹2210 due in 4 years at 7% simple interest?***

Sol. Let the annual payment be ₹x. Then,

(Amount of x for 3 years) + (Amount of x for 2 years) + (Amount of x for 1 year) + x = 2210

$$\left(x + \frac{x \times 7 \times 3}{100}\right) + \left(x + \frac{x \times 7 \times 2}{100}\right) + \left(x + \frac{x \times 7 \times 1}{100}\right) + x = 2210.$$

or $442x = 2210 \times 100$ or x = ₹500.

EXERCISE 14

1. A person borrowed ₹9,000 at 7%, ₹12000 at 8% and ₹15,000 at 9% simple interest per annum. He had to pay ₹50,700 at the end of n years. what is the value of N? **[CDS (I) Exam, 2022]**

(*a*) 3 (*b*) 4 (*c*) 5 (*d*) 6

2. A sum of money was invested a simple interest at a certain rate for 5 years. Had it been invested at a 5% higher rate, it would have fetched ₹500 more. What was the principal amount? **[CDS (I) Exam, 2021]**

(*a*) ₹2,000 (*b*) ₹1,800 (*c*) ₹1,600 (*d*) ₹1,200

3. A sum was put at simple interest at certain rate for 2 years. Had it been put at 1% higher rate of interest, it would have fetched ₹24 more. What is the sum? **[CDS (I) Exam, 2020]**

(*a*) ₹500 (*b*) ₹600 (*c*) ₹800 (*d*) ₹1200

4. The simple interest on a certain sum is one-fourth of the sum. If the number of years and the rate of annual interest are numerically equal, then the number of years is **[CDS (II) Exam, 2020]**

(*a*) 2.5 (*b*) 3 (*c*) 3.5 (*d*) 5

5. A sum of money invested at simple interest triples itself in 8 years and becomes n times in 20 years. What is the value of n ? **[CDS (I) Exam, 2024]**

(*a*) 5 (*b*) 6 (*c*) 7.5 (*d*) 9

6. Two equal amounts were borrowed at 5% and 4% simple interest. The total interest after 4 years amounted to ₹405. What was the total amount borrowed? **[CDS (II) Exam, 2019]**

(*a*) ₹1075 (*b*) ₹1100 (*c*) ₹1125 (*d*) ₹1150

7. In a business dealing, A owes B 20,000 payable after 5 years, whereas B owes A 12,000 payable after 4 years. They want to settle it now at the rate of 5% simple interest. Who gives how much money in this settlement? **[CDS (I) Exam, 2023]**

(a) Both are at par (b) B gives 6,000 to A

(c) A gives 6,000 to B (d) A gives 4,000 to

8. A lent ₹25000 to B and at the same time lent some amount to C at the same 7% simple interest. After 4 years A received ₹11200 as interest from B and C. How much did A lend to C? **[CDS (II) Exam, 2019]**

(*a*) ₹20000 (*b*) ₹25000 (*c*) ₹15000 (*d*) ₹10000

9. A person borrowed ₹5,000 at 5% rate of interest per annum and immediately lent it at 5.5%. After two years he collected the amount and settled his loan. What is the amount gained by him loan. What is the amount gained by him in this transaction? **[CDS (I) Exam, 2018]**
(*a*) ₹25 (*b*) ₹50 (*c*) ₹100 (*d*) ₹200

10. The annual income of a person decreased by ₹64 if the annual rate of interest decreases from 4% to 3.75%. What is his original annual income? **[CDS (I) Exam, 2018]**
(*a*) ₹24,000 (*b*) ₹25,000 (*c*) ₹25,600 (*d*) ₹24,600

11. In how much time would the simple interest on a principal amount be 0.125 times the principal amount at 10% per annum?
(*a*) $1\frac{1}{4}$ years (*b*) $1\frac{3}{4}$ years (*c*) $2\frac{1}{4}$ years (*d*) $2\frac{3}{4}$ years
[CDS (I) Exam, 2015]

12. If a sum of money at a certain rate of simple interest per year doubles in 5 years and at a different rate of simple interest per year becomes three times in 12 years, then the difference in the two rates of simple interest per year is **[CDS (I) Exam, 2016]**
(*a*) 2% (*b*) 3% (*c*) $3\frac{1}{3}$% (*d*) $4\frac{1}{3}$%

13. The rate of interest on two different schemes is the same and it is 20%. But in one of the schemes, the interest is compounded half yearly and in the other the interest is compounded annually. Equal amounts are invested in the schemes. If the difference of the returns after 2 years is ₹482, then what is the principal amount in each scheme? **[CDS (II) Exam, 2019]**
(*a*) ₹10,000 (*b*) ₹16,000 (*c*) ₹20,000 (*d*) ₹24,000

14. Satish borrowed ₹830 from Nitin at 12% rate of interest for 3 years. He then added some more money to the borrowed sum and lent it to Deepak for the same time at 14% simple interest. If Satish gains ₹93.90 in the whole transaction, then the sum lent by him to Deepak is
(*a*) ₹865 (*b*) ₹885 (*c*) ₹910 (*d*) ₹935

15. Kanak rent ₹600 to Monika for 2 years and ₹150 to Anil for 4 years and received altogether from both ₹90 as simple interest. The rate of interest is
(*a*) 4% (*b*) 5% (*c*) 10% (*d*) 12%

16. *A* lent ₹1200 to *B* for 3 years at certain rate of simple interest and ₹1000 to *C* for the same time at the same rate. If he gets ₹50 more from *B* than from *C*, then the rate per cent is
(*a*) $10\frac{1}{3}$% (*b*) $6\frac{2}{3}$% (*c*) $8\frac{1}{3}$% (*d*) $9\frac{2}{3}$%

17. The difference between the interests received from two different banks on ₹500 for 2 years, is ₹2.50. The differences between their rates is
(*a*) 1% (*b*) 0.5% (*c*) 2.5% (*d*) 0.25%

18. A sum was put at simple interest at a certain rate for 2 years. Had it been put at 1% higher rate, it would have fetched ₹24 more. The sum is
(*a*) ₹600 (*b*) ₹800 (*c*) ₹1200 (*d*) ₹480

19. In what time will a sum of money double itself at $6\frac{1}{4}$% p.a. simple interest?
(*a*) 5 years (*b*) 8 years (*c*) 12 years (*d*) 16 years

20. At simple interest, a sum doubles after 20 years. The rate of interest per annum is
(*a*) 5% (*b*) 10% (*c*) 12% (*d*) Data inadequate

21. A sum of money becomes $\frac{8}{5}$ of itself in 5 years at a certain rate of interest. The rate percent per annum is

(*a*) 5% (*b*) 8% (*c*) 10% (*d*) 12%

22. A sum of money trebles itself in 15 years 6 months. In how many years would it double itself?

(*a*) 6 years 3 months (*b*) 7 years 9 months (*c*) 9 years 6 months (*d*) 8 years 3 months

23. The simple interest on a sum of money at 8% per annum for 6 years is half the sum. The sum is

(*a*) ₹4800 (*b*) ₹6000 (*c*) ₹8000 (*d*) Data inadequate

24. The simple interest on a sum of money will be ₹600 after 10 years. If the principal is trebled after 5 years, what will be the total interest at the end of the tenth year?

(*a*) ₹600 (*b*) ₹900 (c) ₹1200 (*d*) Data inadequate

25. The simple interest on ₹1820 from Oct. 9, 1994 to Dec. 21, 1994 at $7\frac{1}{2}\%$ will be

(*a*) ₹22.50 (*b*) ₹27.30 (*c*) ₹28.80 (*d*) ₹29

26. In what time will the simple interest on ₹780 at 5% be equal to the simple interest on ₹600 at $6\frac{1}{2}\%$?

(*a*) 2 years (*b*) $3\frac{1}{2}$ years (*c*) 5 years (*d*) always

27. What should be the least number of years in which the simple interest on ₹2600 at $6\frac{2}{3}\%$ will be an exact number of rupees?

(*a*) 2 years (*b*) 3 years (*c*) 4 years (*d*) 5 years

28. Two equal amounts of money are deposited in two banks, each at 15% per annum, for $3\frac{1}{2}$ years and 5 years. If the difference between their interests is ₹144, each sum is

(*a*) ₹460 (*b*) ₹500 (*c*) ₹640 (*d*) ₹720

29. The rate of interest on a sum of money is 4% per annum for the first 2 years, 6% per annum for the next 4 years and 8% per annum for the period beyond 6 years. If the simple interest accrued by the sum for a total period of 9 years is ₹1120, what is the sum?

(*a*) ₹1500 (*b*) ₹2000 (*c*) ₹2500 (*d*) ₹4000

30. A sum of ₹2500 is lent out in two parts, one at 12% and another one at $12\frac{1}{2}\%$. If the total annual income is ₹306, the money lent at 12% is

(*a*) ₹1200 (*b*) ₹1300 (*c*) ₹1240 (*d*) ₹1340

31. A man invested $\frac{1}{3}$ of his capital at 7%; $\frac{1}{4}$ at 8% and the remainder at 10%. If his annual income is ₹561, the capital is

(*a*) ₹5400 (*b*) ₹6000 (*c*) ₹6600 (*d*) ₹7200

32. A man lends ₹10000 in four parts. If he gets 8% on ₹2000; $7\frac{1}{2}\%$ on ₹4000 and $8\frac{1}{2}\%$ part on ₹1400, what percent must he get for the remainder, if the average interest is 8.13%?

(*a*) 7% (*b*) 9% (*c*) $9\frac{1}{4}\%$ (*d*) $10\frac{1}{2}\%$

33. A sum was put at simple interest at a certain rate for 2 years. Had it been put at 3% higher rate, it would have fetched ₹72 more. The sum is:

(*a*) ₹1600 (*b*) ₹1500 (*c*) ₹1200 (*d*) ₹1800

34. What annual payment will discharge a debt of ₹580 due in 5 years, the rate being 8% per annum?

(*a*) ₹166.40 (*b*) ₹65.60 (*c*) ₹100 (*d*) ₹120

ANSWERS

1. (*c*)	**2.** (*a*)	**3.** (*d*)	**4.** (*d*)	**5.** (*b*)	**6.** (*c*)	**7.** (*c*)	**8.** (*c*)	**9.** (*b*)	**10.** (*c*)
11. (*a*)	**12.** (*c*)	**13.** (*c*)	**14.** (*d*)	**15.** (*b*)	**16.** (*c*)	**17.** (*d*)	**18.** (*c*)	**19.** (*d*)	**20.** (*a*)
21. (*d*)	**22.** (*b*)	**23.** (*d*)	**24.** (*c*)	**25.** (*b*)	**26.** (*d*)	**27.** (*b*)	**28.** (*c*)	**29.** (*b*)	**30.** (*b*)
31. (*c*)	**32.** (*b*)	**33.** (*c*)	**34.** (*c*)						

SOLUTIONS

1. Let time = t years

Case-I

$$\text{S.I.} = \frac{900 \times 7 \times t}{100} = 6300t$$

Amount = 9000 + 630 t ...(*i*)

Case-II

$$\text{Amount} = 12000 + \frac{12000 \times 8 \times t}{100} = 12000 + 960\,t \quad ...(ii)$$

Case-III

$$\text{Amount} = 15000 + \frac{15000 \times 9 \times t}{100} = 15000 + 1350t \quad ...(iii)$$

Adding (i), (ii) and (iii), we get

$9000 + 630t + 12000 + 960t + 15000 + 1350t \Rightarrow 36000 + 2940t$

Now, we have

$36000 + 2940t = 50{,}7000 \Rightarrow 2940t = 14{,}700 \Rightarrow t = 5$ years

2. $\therefore$ S.I. $= \dfrac{P \times R \times T}{100} \Rightarrow 500 = \dfrac{5 \times 5 \times P}{100} \Rightarrow P = 500 \times 4 =$ ₹2000

3. Let the principal be ₹x.

Time = 2 years

Now, we have

$\dfrac{x \times 2 \times 1}{100} = 24 \Rightarrow 2x = 24000 \Rightarrow x =$ ₹1200

4. We have, S.I. $= \dfrac{1}{4}p$, time = t years

Now, we have.

$\dfrac{1}{4}P = \dfrac{P \times t \times t}{100} \Rightarrow P = \dfrac{t^2 P}{25} \Rightarrow t^2 = 25 \Rightarrow t = 5$ years

5. $100 \xrightarrow{\times 3} 300$

+ 200 (S.I.)

8 yrs $\longrightarrow$ 300

$\therefore$ 1 yrs $\longrightarrow \dfrac{200}{8} = 25$

$\therefore$ 20 yrs $\longrightarrow 25 \times 20 = ₹500$

now,

$100 \xrightarrow{(+500)} 600$

(×6)

$\therefore \quad n = 6$

6. Let the equal amount borrowed be ₹x

Now, according to question,

$$\frac{x \times 5 \times 4}{100} + \frac{x \times 4 \times 4}{100} = ₹405 \Rightarrow \frac{20x}{100} + \frac{16p}{100} = ₹405 \Rightarrow 36x = ₹40500 \Rightarrow x = \frac{40500}{36} = ₹1125$$

7. Present value of A's debt $= \dfrac{20,000}{\left(1 + \dfrac{25}{100}\right)} = 20,000 \times \dfrac{100}{125} = ₹16,000$

Now, present value of B's debt $= 12,000 \times \dfrac{100}{120} = ₹10,000$

A gives to B ₹(16,000 – 10,000) = ₹6000

8. Let the amount that A lend to C be ₹x

Now, we have, $\dfrac{25,000 \times 7 \times 4}{100} + \dfrac{x \times 7 \times 4}{100} = 11200$

$\Rightarrow 7000 + \dfrac{28x}{100} = 11200 \Rightarrow 0.28\,x + 11200 - 7000$

$\Rightarrow 0.28\,x = 4200\ 1 \Rightarrow \quad x = \dfrac{42000}{28} = ₹15,000.$

9. Simple interest, when $r = 5\%$

S.I. $= \dfrac{5000 \times 5 \times 2}{100} = ₹500$

Simple interest, when $r = 5.5\%$

S.I. $= \dfrac{5000 \times 5.5 \times 2}{100} = ₹550$

$\therefore$ Required gain = 550 – 500 = ₹50

10. Let annual income be ₹x

Now, according to the question,

$$\frac{x + (4 - 3.75) \times 1}{100} = ₹64 \Rightarrow 0.25x = 6400 \Rightarrow x = 6400 \times 4 = 25600$$

11. Let the time taken be N years.

$$0.125P = \frac{PN(10)}{10^2} \Rightarrow N = 1.25 = 1\frac{1}{4}$$

12. Let the sum be ₹100 and rate interest be x%
time = 5 years amount = ₹200

$$200 = 100\left(1+\frac{5x}{100}\right) \Rightarrow x = 20\,\%$$

Let the rate be y %, Time 12 years

$$\text{Amount ₹}300 = 100\left(1+\frac{12y}{100}\right)$$

$$y = \frac{200}{12} = \frac{50}{3}\,\%$$

$$x - y = 20 - \frac{50}{3} = \frac{10}{3} = 3\frac{1}{3}\,\%$$

13. Let the principal amount be ₹x
In 1st scheme,
We have,

$$A = P\left(1+\frac{r}{2\times 100}\right)^{2t} = x\left(1+\frac{20}{2\times 100}\right)^{2\times 2} \Rightarrow x\left(1+\frac{10}{100}\right)^{4} \Rightarrow \frac{14641}{10000}x = ₹1.4641x$$

In 2nd scheme, we have

$$A = x\left(1+\frac{20}{100}\right)^2 = x\left(\frac{120}{100}\right)^2 = \frac{14400}{1000}x = 1.44x$$

Now, we have

$$1.4641x - 1.44x = 482 \Rightarrow 0.0241x = 482 \Rightarrow x = \frac{482}{0.0241} = ₹20{,}000$$

14. $\frac{x\times 14\times 3}{100} - \frac{830\times 12\times 3}{100} = 93.90 \Rightarrow .42x = 93.90 + 298.80$

$\therefore$ $x = 935.$

15. Let rate = R% per annum. Then,

$$\frac{600\times R\times 2}{100} + \frac{150\times R\times 4}{100} = 90 \Rightarrow R = 5\%$$

16. $\frac{1200\times 3\times R}{100} - \frac{1000\times 3\times R}{100} = 50 \Rightarrow R = 8\frac{1}{3}\%.$

17. $\frac{500\times 2\times R_1}{100} - \frac{500\times 2\times R_2}{100} = 2.50 \Rightarrow 10(R_1 - R_2) = 2.50.$

$\therefore R_1 - R_2 = .25\%.$

18. $\frac{x\times(R+1)\times 2}{100} - \frac{x\times R\times 2}{100} = 24 \Rightarrow \frac{2xR + 2x - 2xR}{100} = 24.$

$$\therefore x = \left(\frac{24\times 100}{2}\right) = 1200.$$

19. Suppose, sum = x. Then, S.I. = x & Rate = $6\frac{1}{4}\%$.

$$\therefore \text{ Time} = \frac{100 \times x}{x \times \frac{25}{4}} = 16 \text{ years.}$$

20. Sum = x, S.I. = x, Time = 20 years.

$$\therefore \text{ Rate} = \left(\frac{100 \times x}{x \times 20}\right)\% = 5\%.$$

21. Let sum = x. Then, S.I. = $\left(\frac{8}{5}x - x\right) = \frac{3x}{5}$, Time = 5 years.

$$\therefore \text{ Rate} = \left(100 \times \frac{3x}{5} \times \frac{1}{x \times 5}\right) = 12\%$$

22. Let sum = x. Then, S.I. = $2x$ & Time = $\frac{31}{2}$ years.

$$\therefore \text{ Rate} = \left(\frac{100 \times 2x}{x \times \frac{31}{2}}\right) = \left(\frac{400}{31}\right)\%.$$

Again, sum = x. Then, S.I. = x & Rate = $\left(\frac{400}{31}\right)\%$

$$\therefore \text{ Time} = \left(\frac{100 \times x}{x \times \frac{400}{31}}\right) = \frac{31}{4} \text{ years} = 7 \text{ years } 9 \text{ months.}$$

23. Let, sum = x. Then, S.I. = $\frac{1}{2}x$, Rate = 8% & Time = 6 years.

$$\therefore \quad \frac{x \times 8 \times 6}{100} = \frac{1}{2}x \Rightarrow \frac{12}{25} = \frac{1}{2}, \text{ which is false.}$$

So data is inadequate.

24. $\frac{P \times R \times 10}{100} = 600 \Rightarrow P \times R = 6000$

Total Interest at the end of tenth year. = $\left(300 + \frac{3P \times R \times 5}{100}\right) = \left(300 + \frac{15 \times 6000}{100}\right) = 1200$

[$\because PR = 6000$]

25. Note that the day on which money is deposited is not counted and the day on which it is withdrawn is included.

Oct. Nov. Dec.

Time : 22 + 30 + 21 = 73 days = $\frac{73}{365}$ year = $\frac{1}{5}$ year.

$$\therefore \quad \text{S.I.} = ₹\left(1820 \times \frac{15}{2} \times \frac{1}{5} \times \frac{1}{100}\right) = ₹27.30.$$

26. $\frac{780\times 5\times T}{100}=600\times\frac{13}{2}\times\frac{T}{100}=39T.$

This is true for all values of T. So, (d) is true.

27. S.I. = ₹$\left(2600\times\frac{20}{3}\times\frac{1}{100}\times T\right)$, which is an exact number of rupees when the least value of T is 3 years.

28. Let the sum be x. Then, $\frac{x\times 5\times 15}{100}-x\times 15\times\frac{7}{2}\times\frac{1}{100}=144\Rightarrow x=640.$

29. Let the sum be x. Then, $\frac{x\times 4\times 2}{100}+\frac{x\times 6\times 4}{100}+\frac{x\times 8\times 3}{100}=1120\Rightarrow x=\frac{1120\times 100}{56}=2000.$

30. Let the money lent at 12% be x.

Then, money at $12\frac{1}{2}\%=(2500-x)$.

$$\therefore\quad \frac{x\times 12\times 1}{100}+(2500-x)\times\frac{25}{2}\times\frac{1}{100}=306$$

or $24x+25(2500-x)=61200$ or $x=(62500-61200)=1300.$

31. Let the capital be ₹x. Then, $\frac{x}{3}\times\frac{7}{100}\times 1+\frac{x}{4}\times\frac{8}{100}\times 1+\frac{5x}{12}\times\frac{10}{100}\times 1=561$

or $\frac{7x}{300}+\frac{8x}{400}+\frac{50x}{1200}=561\Rightarrow 102x=561\times 1200=6600.$

32. Let the required rate be x %. Then,

$$\frac{2000\times 8\times 1}{100}+4000\times\frac{15}{2}\times\frac{1}{100}+1400\times\frac{17}{2}\times\frac{1}{100}+2600\times x\times\frac{1}{100}=\frac{10000\times 8.13\times 1}{100}$$

or $160+300+119+26x=813$ or $x=9.$

33. Let the sum be ₹x & original rate R%. Then,

$$\frac{x\times(R+3)\times 2}{100}-\frac{x\times R\times 2}{100}=72\Rightarrow\frac{2Rx+6x-2Rx}{100}=72.$$

$\therefore$ $6x=72\times 100$ or $x=1200.$

34. Let the annual payment be ₹x. Then,

(Amt. of x for 4 years) + (Amt. of x for 3 years) + (Amt. of x for 2 years) + (Amt. of x for 1 year) + x = 580

$$\left(x+\frac{x\times 4\times 8}{100}\right)+\left(x+\frac{x\times 3\times 8}{100}\right)+\left(x+\frac{x\times 2\times 8}{100}\right)+\left(x+\frac{x\times 1\times 8}{100}\right)+x=580$$

$$5x+\frac{32x+24x+16x+8x}{100}=580\Rightarrow 580x=580\times 100\Rightarrow x=100.$$

Compound Interest

Sometimes the borrower and the lender agree to fix up a certain unit of time, say *yearly* or *half-yearly* or *quarterly* to settle the previous account.

In such cases, the amount after first unit of time becomes the principal for second unit of time; the amount after second unit of time becomes the principal for third unit of time and so on.

After a specified period, the difference between the amount and the principal is called Compound Interest abbreviated as C.I.

FORMULAE:

Let, Principal = P, Time = n years & Rate = R % per annum.

I. *When Interest is compounded Annually*:

In this case, Amount = $P\left(1+\frac{R}{100}\right)^n$.

II. *When Interest is compounded Half-Yearly*:

In this case, Amount = $P\left(1+\frac{(R/2)}{100}\right)^{2n}$.

III. *When Interest is compounded Quarterly*:

In this case, Amount = $P\left(1+\frac{(R/4)}{100}\right)^{4n}$.

IV. *When Time is Fraction of a Year*:

Let the time be a fraction of a year, say $3\frac{2}{3}$ years.

Then, Amount = $P\left(1+\frac{R}{100}\right)^3 \times \left(1+\frac{\frac{2}{3}R}{100}\right)$.

V. *When Rates are different for different years, say R_1, R_2, R_3 per cent* for Ist, 2nd and 3rd year, respectively.

Then, Amount = $P\left(1+\frac{R_1}{100}\right)\left(1+\frac{R_2}{100}\right)\left(1+\frac{R_3}{100}\right)$.

SOLVED PROBLEMS

Ex. 1. *Find compound interest on ₹5000 at 16% per annum for 1 year, compounded half-yearly.*

Sol. Principal = ₹500, Time = 1 year = 2 half years, Rate = 16% per annum = 8% half-yearly.

$\therefore$ Amount = $₹\left\{5000\times\left(1+\frac{8}{100}\right)^2\right\}$ = Rs.$\left(5000\times\frac{27}{25}\times\frac{27}{25}\right)$ = ₹5832.

$\therefore$ C.I. = ₹(5832 – 5000) = ₹832.

Ex. 2. ***Find compound interest on ₹8000 at 20% per annum for 9 months, compounded quarterly.***

Sol. Principal = ₹8000, Time = 9 months = 3 quarters,

Rate = 20% per annum = 5% per quarter.

$$\therefore \text{Amount} = ₹\left\{8000 \times \left(1+\frac{5}{100}\right)^3\right\} = ₹\left(8000 \times \frac{21}{20} \times \frac{21}{20} \times \frac{21}{20}\right) = ₹9261.$$

C.I. = ₹(9261 – 8000) = ₹1261.

Ex. 3. ***The difference between the compound interest and the simple interest on a certain sum at 10% per annum for 2 years is ₹52. Find the sum.***

Sol. Let the sum be ₹x. Then,

$$\text{S.I.} = ₹\left(x \times \frac{10}{100} \times 2\right) = \text{Rs.}\frac{x}{5}.$$

$$\text{C.I.} = ₹\left[x \times \left(1+\frac{10}{100}\right)^2 - x\right] = ₹\left(\frac{21x}{100}\right).$$

$$\therefore \quad (\text{C.I.}) - (\text{S.I.}) = \left(\frac{21x}{100} - \frac{x}{5}\right) = \frac{x}{100}.$$

So, $\frac{x}{100} = 52 \Leftrightarrow x = 5200.$ Hence, Sum = ₹5200.

Ex. 4. ***If the compound interest on a certain sum for 3 years at 10% per annum be ₹331, what would be the simple interest?***

Sol. $P\left(1+\frac{10}{100}\right)^3 - P = 331 \Rightarrow \frac{1331}{1000}P - P = 331$

$$\therefore \quad \frac{331}{1000}P = 331 \text{ or } P = \left(\frac{331 \times 1000}{331}\right) = 1000.$$

$$\therefore \quad \text{S.I.} = ₹\left(1000 \times \frac{10}{100} \times 3\right) = ₹300.$$

EXERCISE 15

1. What is the least number of complete years in which a sum of money put out at 20% compound interest (compounded annually) will be more than doubled? **[CDS (I) Exam, 2025]**
   (*a*) 2 (*b*) 3 (*c*) 4 (*d*) 5
2. How much will ₹10,000 amount to in one year's time at 4% rate of interest per annum if the interest is compounded once in every three months? (take approximate value) **[CDS (I) Exam, 2025]**
   (a) ₹10,406 (b) ₹10,416 (c) ₹10,426 (d) ₹10,436
3. The difference between compound interests on ₹800 for 1 year at 20% per annum when compounded half-yearly and quarterly, is:
   (*a*) Nil (*b*) ₹2.50 (*c*) ₹4.40 (*d*) ₹6.60
4. If the rate of interest is 5%, then what would be the difference between compound interest and simple interest received on ₹10,000 (each) after 3 years from now? **[CDS (I) Exam, 2024]**

(*a*) ₹175.25 (*b*) ₹152.25 (*c*) ₹76.25 (*d*) ₹24.25

5. A sum of money on compound interest amounts to ₹9680 in 2 years and to ₹10648 in 3 years. The rate of interest per annum is:

(*a*) 5% (*b*) 10% (*c*) 15% (*d*) 20%

6. The difference between compound interest and simple interest at the same rate for ₹5000 for 2 years is ₹72. The rate of interest per annum is:

(*a*) 6% (*b*) 8% (*c*) 10% (*d*) 12%

7. A sum of money amounts to ₹6690 after 3 years and to ₹10035 after 6 years on compound interest. The sum is:

(*a*) ₹4400 (*b*) ₹4460 (*c*) ₹4520 (*d*) ₹4445

8. A sum amounts to ₹2916 in 2 years and to ₹3149.28 in 3 years at compound interest. The sum is:

(*a*) ₹1500 (*b*) ₹2000 (*c*) ₹2500 (*d*) ₹3000

9. Three amounts *x*, *y*, *z* are such that *y* is the compound interest on *x*; and *z* is the compound interest on *y*. The rate of interest per annum and the time period in years are same. Which one of the following is correct? **[CDS (I) Exam, 2024]**

(*a*) $x^2 = yz$ (*b*) $y^2 = zx$ (*c*) $z^2 = xy$ (*d*) $x = yz$

10. Let *x* be the compound interest at the end of 3 years on a sum of ₹1000 at the rate of 10% compounded annually and *y* be the simple interest at the end of 3 years on a sum of ₹1000 at the annual rate of 11%. What is the difference between *x* and *y*? **[CDS (I) Exam, 2024]**

(*a*) 16 (*b*) 15 (*c*) 5 (*d*) 1

11. A sum of money compounded annually doubles itself in 5 years. In how many years will it become four times of itself? **[CDS (II) Exam, 2021]**

(*a*) 10 years (*b*) 12 years (*c*) 15 years (*d*) 20 years

12. What is the least number of complete years in which a sum of money put out at 40% annual compound interest will be more than trebled? **[CDS (I) Exam, 2019]**

(*a*) 3 (*b*) 4 (*c*) 5 (*d*) 6

13. What is the principal amount which earns ₹210 as compound interest for the second year at 5% per annum? **[CDS (I) Exam, 2018]**

(*a*) ₹2000 (*b*) ₹3200 (*c*) ₹4000 (*d*) ₹4800

14. The difference between the compound interest (compounded annually) and the simple interest on a certain sum of money at 12% per annum for 2 years is ₹72. What is the principal amount? **[CDS (I) Exam, 2021]**

(*a*) ₹6,500 (*b*) ₹6,000 (*c*) ₹5,500 (*d*) ₹5,000

15. What is the difference between simple interest and compound interest on ₹10,000 for two years at 20% per annum compounded half-yearly ? **[CDS (I) Exam, 2023]**

(*a*) ₹842 (*b*) ₹756 (*c*) ₹641 (*d*) ₹542

16. A person borrowed ₹10,000 at 12% rate of interest per annum compounded quarterly for a period of 9 months. What is the interest paid by him to settle his account after 9 months ? **[CDS (II) Exam, 2023]**

(*a*) ₹927·27 (*b*) ₹947·47 (*c*) ₹967·67 (*d*) ₹987·87

17. A sum of money at 20% rate of compound interest per annum becomes more than 100 times in n years. What is the least value of n? (Use $\log_{10} 2 = 0{\cdot}301$, $\log_{10} 3 = 0{\cdot}477$) **[CDS (II) Exam, 2023]**

(*a*) 23 (*b*) 24 (*c*) 25 (*d*) 26

18. The difference between compound interest and simple interest at the same rate of interest *R* percent per annum on an amount to ₹15,000 for 2 years is ₹96. What is the value of *R*?

(*a*) 8 (*b*) 10

(*c*) 12 (*d*) Cannot be determined due to insufficient data

[CDS (I) Exam, 2015]

19. There is 60% increase in an amount in 6 years at simple interest. What will be the compound interest on ₹12000 after 3 years at the same rate of interest? **[CDS (I) Exam, 2015]**

(*a*) ₹2160 (*b*) ₹3120 (*c*) ₹3972 (*d*) ₹6240

20. An automobile financer claims to be lending money at simple interest, but he includes the interest every six months for calculating the principal. If he is charging an interest at the rate of 10%, the effective rate of interest becomes **[CDS (I) Exam, 2015]**

(*a*) 10.25% (*b*) 10.5% (*c*) 10.75% (*d*) 11%

21. A sum of ₹10,000 is deposited for 1 year at the rate of interest 10% compounded half yearly. What will be the interest at the end of one year? **[CDS (II) Exam, 2015]**

(*a*) ₹1000 (*b*) ₹1025 (*c*) ₹1050 (*d*) ₹1100

22. A sum of ₹8400 was taken as a loan. This is to be paid in two equal instalments. If the rate of interest is 10% per annum, compounded annually, then the value of each instalment is

[CDS (I) Exam, 2017]

(*a*) ₹4200 (*b*) ₹4480

(*c*) ₹4840 (*d*) None of the above

23. The difference between the simple and the compound interest on a certain sum of money at 4% per annum in 2 years is ₹10. What is the sum? **[CDS (I) Exam, 2017]**

(*a*) ₹5000 (*b*) ₹6000 (*c*) ₹6250 (*d*) ₹7500

24. The compound interest on a certain sum of money for 2 years at 10% per annum is ₹420. The simple interest on the same sum at the same rate and for the same time will be:

(*a*) ₹350 (*b*) ₹375 (*c*) ₹380 (*d*) ₹400

25. A sum of ₹550 was taken as a loan. This is to be paid back in two equal instalments. If the rate of interest be 20% compounded annually, then the value of each instalment is:

(*a*) ₹421 (*b*) ₹396 (*c*) ₹360 (*d*) ₹350

26. The difference between the compound interest and simple interest for 2 years on a sum of money is ₹160. If the simple interest be ₹2880, the rate per cent is:

(*a*) $5\frac{5}{9}$% (*b*) $12\frac{1}{2}$% (*c*) $11\frac{1}{9}$% (*d*) 9%

27. The difference between simple interest and the compound interest on a certain sum of money for 3 years at 10% per annum is ₹15.50. The sum is:

(*a*) ₹500 (*b*) ₹550 (*c*) ₹600 (*d*) ₹460

28. The compound interest on a sum for 2 years is ₹832 and the simple interest on the same sum for the same period is ₹800. The difference between the compound interest and the simple interest for 3 years will be:

(*a*) ₹48 (*b*) ₹66.56 (*c*) ₹98.56 (*d*) ₹96.48

29. The difference between simple interest and compound interest on a sum for 2 years at 8%, when the interest is compounded annually is ₹16. If the interest were compounded half-yearly, the difference in 2 years would be nearly:

(*a*) ₹16 (*b*) ₹16.80 (*c*) ₹21.85 (*d*) ₹24.64

30. A sum of ₹12000 deposited at compound interest becomes double after 5 years. After 20 years it will become:

(*a*) ₹120000 (*b*) ₹192000 (*c*) ₹124000 (*d*) ₹96000

ANSWERS

1. (*c*)	**2.** (*a*)	**3.** (*c*)	**4.** (*c*)	**5.** (*b*)	**6.** (*d*)	**7.** (*b*)	**8.** (*c*)	**9.** (*b*)
10. (*d*)	**11.** (*a*)	**12.** (*b*)	**13.** (*c*)	**14.** (*d*)	**15.** (*c*)	**16.** (*a*)	**17.** (*d*)	**18.** (*a*)
19. (*c*)	**20.** (*a*)	**21.** (*b*)	**22.** (*c*)	**23.** (*c*)	**24.** (*d*)	**25.** (*c*)	**26.** (*c*)	**27.** (*a*)
28. (*c*)	**29.** (*d*)	**30.** (*b*)						

SOLUTIONS

1. As per the question, A = 2P

Let's put the values in the formula:

$2P = P(1+0.2)^n \Rightarrow 2 = (1+0.2)^n \Rightarrow 2 = 1.2^n \Rightarrow n = 4$, as $1.2^n = 2.07 > 2$

∴ The least number of complete years in which an amount will be more than doubled at 20% compound interest per annum is 4 years.

2. $A = 10,000\left(1+\frac{4}{4\times100}\right)^{4\times1}$

$\Rightarrow A = 10,000\left(1+\frac{1}{100}\right)^4 \quad \Rightarrow A = 10,000(1+0.01)^4$

$\Rightarrow A = 10,000(1.01)^4 \quad \Rightarrow A = 10,000\times1.04060401$

$\Rightarrow A \approx ₹10,406$

3. **Case I**: *When compounded Half- Yearly*:

$$\text{Amount} = ₹\left[800\times\left(1+\frac{10}{100}\right)^2\right] = \text{Rs.}\left[800\times\frac{11}{10}\times\frac{11}{10}\right] = ₹968.$$

∴ C.I. = ₹(968 – 800) = ₹168

Case II: *When compounded Quarterly*:

$$\text{Amount} = ₹\left[800\times\left(1+\frac{5}{100}\right)^4\right] = ₹\left(800\times\frac{21}{20}\times\frac{21}{20}\times\frac{21}{20}\times\frac{21}{20}\right) = ₹972.40$$

∴ C.I. = ₹(972.40 – 800) = ₹172.40.

Hence, required difference = ₹(172.40 – 168),= ₹4.40.

4.

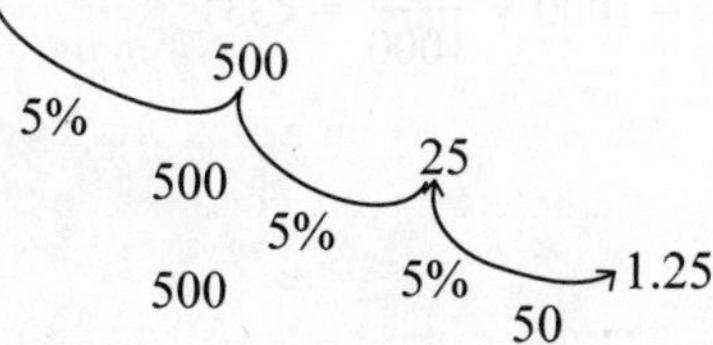

Now, difference = 50 + 25 + 1.25 = 76.25

5. Interest on ₹9680 in 1 year = ₹(10648 – 9680) = ₹968

$\therefore$ Interest on ₹100 for 1 year = $\left(\frac{968}{9680}\times 100\right) = 10\ \%$

6. $\left[5000\times\left(1+\frac{R}{100}\right)^2 - 5000\right] - \left(\frac{5000\times 2\times R}{100}\right) = 72 \Leftrightarrow 5000\times\left[\left(1+\frac{R}{100}\right)^2 - 1 - \frac{R}{50}\right] = 72$

$\Leftrightarrow 1+\frac{R^2}{10000}+\frac{2R}{100}-1-\frac{R}{50}=\frac{72}{5000} \Leftrightarrow R^2 = 144 \Leftrightarrow R = 12.$

7. $P\left(1+\frac{R}{100}\right)^3 = 6690 \ldots(i)$ and $P\left(1+\frac{R}{100}\right)^6 = 10035 \ldots(ii)$

On dividing we get: $\left(1+\frac{R}{100}\right)^3 = \frac{10035}{6690} = \frac{3}{2}.$

Putting this value in (*i*), we get:

$$P\times\frac{3}{2} = 6690 \text{ or } P = \left(6690\times\frac{2}{3}\right) = 4460.$$

8. Interest on ₹2916 for 1 year = ₹(3149.28 – 2916) = ₹233.28

Interest on ₹100 for 1 year = $\left(\frac{233.28}{2916}\times 100\right) = 8\%.$

$\therefore\ P\left(1+\frac{8}{100}\right)^2 = 2916 \Leftrightarrow P = \left(2916\times\frac{25}{27}\times\frac{25}{27}\right) = 2500.$

9. Given that

When $P = x$, then $CI = y$

$\Rightarrow y = x\left\{1+\frac{R}{100}\right\}^T - x \quad \Rightarrow \frac{x+y}{x} = \left\{1+\frac{R}{100}\right\}^T$ ...(i)

When $P = y$, then $CI = z$

$z = y\left\{1+\frac{R}{100}\right\}^T - y \quad \Rightarrow \frac{z+y}{y} = \left\{1+\frac{R}{100}\right\}^T$ ...(ii)

From equation (i) & (ii):

$$\frac{x+y}{z} = \frac{z+y}{y} \Rightarrow xy + y^2 = xz + xy \Rightarrow y^2 = xz$$

10. $\therefore$ C.I. $= P\left[\left(1+\frac{r}{100}\right)^n - 1\right]$

$\therefore\quad x = 1000\left[\left(1+\frac{10}{100}\right)^3 - 1\right] = 1000\left[\frac{1331}{1000}-1\right] = 1000\times\frac{331}{1000} = ₹331$

Now,

$\therefore$ S.I. $= \frac{P\times R\times T}{100}\quad y = \frac{1000\times 3\times 11}{100} = ₹300$

$\therefore\ x - y = ₹1$

11. Let principal amount be ₹p

$$\therefore A = p\left(1+\frac{R}{100}\right)^T \Rightarrow 2p = p\left(1+\frac{R}{100}\right)^5 = 2 = \left(1+\frac{R}{100}\right)^5 \Rightarrow (2)^{\frac{1}{5}} = \left(1+\frac{R}{100}\right)$$

Now,

$$4p = p\left(1+\frac{R}{100}\right)^T \Rightarrow 4 = \left(1+\frac{R}{100}\right)^T \Rightarrow 4 = (2)^{\frac{1}{5}T} \Rightarrow 2^2 = (2)^{\frac{T}{5}}$$

$$\Rightarrow \quad 2 = \frac{T}{5} \Rightarrow T = 10 \text{ years}$$

12. Let sum of money be ₹p

$$\because \quad A = p\left(1+\frac{R}{100}\right)^t$$

put $t = 1$

$$\Rightarrow \quad A = p\left(1+\frac{40}{100}\right)$$

If put $t = 2$

$$A = p\left(1+\frac{40}{100}\right)^2 = 1.96p$$

put $t = 3$

$$A = p\left(1+\frac{40}{100}\right)^3 = p\,(1.4)^3 = 2.74p$$

Now, put $t = 4$

$$A = p\left(1+\frac{40}{100}\right)^4 = p\,(1.4)^4 = 3.84p$$

$\therefore$ It is more than trebled

13. $\because$ C.I. $= P\left[\left(1+\dfrac{R}{100}\right)^{-1}\right]$

Let the principal amount be ₹x

Now, according to the question,

$$x\left[\left(1+\frac{5}{100}\right)^2 - 1\right] - x\left[\left(1+\frac{5}{100}\right) - 1\right] = 210$$

$$\Rightarrow x\left\{\left(\frac{21}{20}\right)^2 - 1\right\} - x\left\{\frac{21}{20} - 1\right\} = 210 \Rightarrow \left\{\left(\frac{21}{20}\right)^2 - x\right\} - \left[\frac{21}{20} - 1\right] = 210$$

$$\Rightarrow x\frac{41}{400} - \frac{x}{20} = 210 \Rightarrow \frac{41x - 20x}{400} = 210 \Rightarrow \frac{21x}{400} = 210$$

$$\Rightarrow \frac{x}{400} = 10 \Rightarrow x = 400 \times 10 = 4000$$

14. Let principal amount be ₹p

$$\because A = p\left(1+\frac{R}{100}\right)^T = p\left(1+\frac{12}{100}\right)^2 = p\,(1.12)^2 = 1.2544p$$

$\therefore$ C.I. $= A - P = 1.2544p - p = 0.2544p$

Now,

$$\because \text{S.I.} = \frac{P\times R\times T}{100} = \frac{p\times 12\times 2}{100} = 0.24p$$

Now, we have,

$0.2544p - 0.24p = 72$

$$\Rightarrow 0.0144p = 72 \Rightarrow p = \frac{720000}{144} = ₹5{,}000$$

15. $r = \frac{20}{2} = 10\%$

$T = 2 \times 2 = 4$ (half yearly)

$$\text{S.I.} = \frac{10{,}000\times 4\times 10}{100} = ₹4000$$

$$\text{C.I.} = 10{,}000\left(1+\frac{10}{100}\right)^4 = 10{,}000 = 10{,}000 \times \frac{14641}{10{,}000} - 10{,}000 = 4641$$

$\Rightarrow$ Required difference $= 4641 - 4000 =$ ₹641

16. $$\therefore \text{C.I.} = \left[P\left(1+\frac{R}{100}\right)^n - P\right] = 10000\left[\left(1+\frac{3}{100}\right)^3 - 1\right]$$

$$= 10000\left[\frac{92727}{1000000}\right] = ₹927.27.$$

17. $$\therefore \text{A} = \text{P}\,A = P\left(1+\frac{R}{100}\right)^n$$

$$100P = P\left(1+\frac{R}{100}\right)^n$$

$$\Rightarrow 100 = \left(1+\frac{20}{100}\right)^n \Rightarrow 100 = \left(\frac{12}{10}\right)^n \Rightarrow \log_m 100 = n\log\frac{12}{10} \Rightarrow 2 = n(2\log 2 + \log 3 - 1)$$

$$\Rightarrow 2 = n\,(2\times 0.301 + 0.477 - 1) \Rightarrow 2 = n\,(1.079 - 1) \Rightarrow n = \frac{2}{0.079} = \frac{2000}{79} = 25.32$$

$\therefore$ least value of $n = 26$

18. Difference between the compound interest and simple interest on a sum of P for 2 years at R% p.a. $= P\left(\frac{R}{100}\right)^2$ $\qquad \therefore 15000\left(\frac{R}{100}\right)^2 = 96$

$$\Rightarrow \left(\frac{R}{100}\right)^2 = \frac{96}{15000} = \frac{4}{625} \Rightarrow \frac{\text{R}}{100} = \frac{2}{25} \Rightarrow \text{R} = 8.$$

19. Let the amount be ₹P. The amount increases by 60% in 6 years at simple interest.

∴ Simple interest for 6 years = 0.6P.

Let the rate of interest be R % p.a.

$$0.6\text{P} = \frac{\text{P}(6)(\text{R})}{100} \Rightarrow \text{R} = 10 \text{ and } \frac{\text{R}}{100} = 0.1. \therefore 1 + \frac{\text{R}}{100} = 1.1$$

Compound interest on ₹12000 after 3 years at 10% p.a = 12000 $[(1.1)^3 - 1]$ = 12000 (0.331) = 3972.

20. Let the principal be ₹100.

Rate of interest = 10% p.a = 5 % per half year.

Interest for the first half of the year = ₹5.

The financer includes the interest every six months for calculating the principal.

∴ the principal for the second half of the first year = ₹105.

Interest for the second half of the first year = ₹$\frac{5}{100}$ (105) = ₹5.25

Interest for the first year = ₹10.25., Effective rate of interest = 10.25% p.a.

21. If the principal is p, and the half-yearly interest rate is r, the amount at the end of 6 months is $p(1 + r)$ and at the end of 1 year is $p(1 + r)^2$.

∴ Amount at the end of 1 year = 10,000 $(1.05)^2$ = 10000 (1.1025) = 11025.

The interest is ₹1025.

22. Let the installment be x, and Rate be r % ∴ $\frac{x}{\left(1+\frac{r}{100}\right)} + \frac{x}{\left(1+\frac{r}{100}\right)^2}$ = Loan (sum)

$$\Rightarrow \quad \frac{x}{\left(1+\frac{10}{100}\right)} + \frac{x}{\left(1+\frac{10}{100}\right)^2} = 8400$$

$$\Rightarrow \quad \frac{10x}{11} + \frac{100x}{121} = 8400$$

$$\Rightarrow \quad \frac{110x+100x}{121} = 8400 \Rightarrow x = \frac{8400\times121}{210} = 4840$$

23. Let Principal be ₹P, Rate = 4 %

$$\text{Difference} = P\left(\frac{R}{100}\right)^2 \Rightarrow P = \frac{10\times100\times100}{4\times4} = 6250$$

24. $P\times\left(1+\frac{10}{100}\right)^2 - P = 420 \Leftrightarrow \left(\frac{121}{100}-1\right)P = 420 \Leftrightarrow P = \left(\frac{420\times100}{21}\right) = 2000.$

$$\therefore \text{S.I.} = ₹\left(2000\times\frac{10}{100}\times2\right) = ₹400.$$

25. Let the value of each instalment be ₹x.

Then $\frac{x}{\left(1+\frac{20}{100}\right)} + \frac{x}{\left(1+\frac{20}{100}\right)^2} = 550 \Leftrightarrow \frac{5x}{6} + \frac{25x}{36} = 550 \Leftrightarrow x = 360.$

26. S.I. for 1 year = ₹1440.

Interest on ₹1440 for 1 year = ₹160.

$\therefore$ Interest on ₹100 for 1 year = $\left(\frac{160}{1440}\times 100\right) = 11\frac{1}{9}\%$.

27. Let, sum be ₹x. Then, S.I. = $\left(\frac{x\times 10\times 3}{100}\right) = \frac{3x}{10}$.

C.I. = $\left[x\left(1+\frac{10}{100}\right)^3 - x\right] = \frac{331x}{1000}$.

$\therefore \frac{331x}{1000} - \frac{3x}{10} = 15.50 \Leftrightarrow \frac{31x}{1000} = 15.50 \Leftrightarrow x = 500.$

28. S.I. for l year = ₹400.

S.I. on ₹400 for 1 year = ₹(832 – 800)

$\therefore$ Rate = $\left(\frac{32}{400}\times 100\right) = 8\%$.

Difference for 3rd year = S.I. on ₹832 = ₹$\left(832\times\frac{8}{100}\times 1\right)$ = ₹66.56.

$\therefore$ Total difference = ₹(32 + 66.56) = ₹98.56.

29. S.I. on (S.I. for 1 year) = ₹16

S.I. on $\left(\frac{P\times 1\times 8}{100}\right)$ = ₹16

S.I. on P = ₹$\left(\frac{16\times 100}{8}\right)$ = ₹200.

$\therefore$ $P = \left(\frac{100\times 200}{8\times 1}\right)$ = ₹2500.

Amount for 2 years compounded half-yearly = $\left[2500\times\left(1+\frac{4}{100}\right)^4\right]$ = ₹2924.64

$\therefore$ C.I. = ₹(2924.56 – 2500) = ₹424.64

S.I. = ₹$\left(\frac{2500\times 8\times 2}{100}\right)$ = ₹400.

$\therefore$ (C.I.) – (S.I.) = ₹24.64.

30. $P\left(1+\frac{R}{100}\right)^5 = 2P \Rightarrow \left(1+\frac{R}{100}\right)^5 = 2.$

Amount after 20 years = $P\left(1+\frac{R}{100}\right)^{20}$ = $16P$ = ₹(16 × 12000) = ₹192000.

Logarithms

Highlights: Logarithms to base 10, laws of logarithms, use of logarithmic tables.

Logarithm: *If a is a positive real number, other than 1 and $a^m = x$, then m is called the logarithm of x to the base a, written as $\log_a x$.*

Thus, $a^m = x \Leftrightarrow \log_a x = m$.

Ex. (*i*) $10^2 = 100 \Leftrightarrow \log_{10} 100 = 2$ (*ii*) $5^3 = 125 \Leftrightarrow \log_5 125 = 3$

(*iii*) $3^{-3} = \frac{1}{27} \Leftrightarrow \log_3\left(\frac{1}{27}\right) = -3$ (*iv*) $(.1)^2 = .01 \Leftrightarrow \log_{.1}(.01) = 2$.

Properties of Logarithms:

(*i*) $\log_a(xy) = \log_a x + \log_a y$ (*ii*) $\log_a\left(\frac{x}{y}\right) = \log_a x - \log_a y$

(*iii*) $\log_a(x^k) = k(\log_a x)$ (*iv*) $\log_a 1 = 0\ [\because a^0 = 1]$

(*v*) $\log_x x = 1$ (*vi*) $\log_a x = \frac{1}{\log_x a}$

(*vii*) $\log_a x = \frac{\log_b x}{\log_b a}$. So, $\log_b x = \frac{\log x}{\log b}$ (*viii*) $a^{\log_a x} = x$ e.g. $5^{\log_5 7} = 7$.

(*ix*) If $a > 1$ and $x > 1$, then $\log_a x > 0$ (*x*) If $0 < a < 1$ and $0 < x < 1$, then $\log_a x > 0$

(*xi*) If $0 < a < 1$ and $x > 1$, then $\log_a x > 0$ (*xii*) If $a > 1$ and $0 < x < 1$, then $\log_a x < 0$.

Remark: *When base is not mentioned, it will be taken as 10.*

Common Logarithm: *Logarithms to the base 10 are known as common logarithms.*

$\log_{10} x$ consists of two parts, the integral part is known as *Characteristic* and the decimal part is known as *Mantissa*.

To Find the Characteristic:

Case I: ***When the number is greater than* 1:**

In this case, *the characteristic is one less than the number of digits in the left of the decimal point in the given number.*

Case II: *When the number is less than* 1:

In this case, the characteristic is one more than the number of zeros between the decimal point and the first significant digit of the number and is negative.

Instead of – 1, – 2, – 3 etc., we write $\bar{1}$ (one-bar), $\bar{2}$ (two-bar) etc.

Ex. 1. From the table, if mantissa of 2536 is 0.4041, then

log 253.6 = 2.4041; log 25.36 = 1.4041; log 2.536 = 0.4041,

log 0.2536 = $\bar{1}$.4041; log 0.02536 = $\bar{2}$.4041 etc.

Antilogarithm: If $\log x = y$, then $x = \text{antilog}(y)$.

How to Insert Decimal Point?

Case I: *When Characteristic of logarithms is positive:*

In this case, *insert the decimal point after the* $(n + 1)$*th digit, where n is the characteristic.*

Case II: *When characteristic of logarithms is negative*:

In this case, insert the decimal paint so that the first significant digit is at nth place, where $\bar{n}$ *is the characteristic.*

Ex. 2. Find antilog (2.4567); antilog (1.4567); antilog (0.4567); antilog ($\bar{1}$.4567) and antilog ($\bar{2}$.4567).

Sol. From table of antilog, we have: antilog (0.4567) = 2863.

Also, characteristic in 2.4567 is 2.

So, insert decimal point after 3 places.

$\therefore$ antilog (2.4567) = 286.3.

Similarly, antilog (1.4567) = 28.63, antilog (0.4567) = 2.863

Further, characteristic in $\bar{1}$.4567 is $\bar{1}$.

So, insert decimal point so that first significant digit is at Ist place.

$\therefore$ antilog ($\bar{1}$.4567) = 0.2863

Similarly, antilog ($\bar{2}$.4567) = 0.02863.

SOLVED EXAMPLES

Ex. 1. ***Evaluate*: (*i*) $36^{\log_6 4}$ (*ii*) $\log_a 1$ (*iii*) $\log_{25} 25$.**

Sol. (*i*) We know that: $a^{\log_a x} = x$.

$$\therefore\ 36^{\log_6 4} = (6^2)^{\log_6 4} = 6^{2\log_6 4} = 6^{\log_6 (4^2)} = 6^{\log_6 (16)} = 16.$$

(*ii*) $\log_a = 1 = m \Leftrightarrow a^m = 1 = a^0 \Leftrightarrow m = 0$.

(*iii*) We know that $\log_a a = 1$. So, $\log_{25} 25 = 1$.

Ex. 2. ***Evaluate*:**

(*i*) $\log_9 27 - \log_{27} 9$ (*ii*) $\log_3 5 . \log_{25} 27$

Sol. (*i*) $\log_9 27 = m \Leftrightarrow 9^m = 27 \Leftrightarrow 3^{3m} = 3^3 \Leftrightarrow 3m = 3 \Leftrightarrow m = 1$.

And, $\log_{27} 9 = n \Leftrightarrow (27)^n = 9 \Leftrightarrow 3^{3n} = 3^2 \Leftrightarrow 3n = 2 \Leftrightarrow n = \frac{2}{3}$.

$$\therefore\ \log_9 27 - \log_{27} 9 = \left(1 - \frac{2}{3}\right) = \frac{1}{3}.$$

(*ii*) We know that $\log_a b = \frac{\log b}{\log a}$.

$$\therefore\ \log_3 5 . \log_{25} 27 = \frac{\log 5}{\log 3} \times \frac{\log 27}{\log 25} = \frac{\log 5}{\log 3} \times \frac{\log 3^3}{\log 5^2} = \frac{\log 5}{\log 3} \times \frac{3\log 3}{2\log 5} = \frac{3}{2}.$$

Ex. 3. ***Evaluate: $\log_2 (\log_3 81)$.***

Sol. $\log_3 81 = m \Leftrightarrow 3^m = 81 = 3^4 \Leftrightarrow m = 4$.

$\therefore\ \log_2(\log_3 81) = \log_2 4 = \log_2(2^2) = 2(\log_2 2) = 2 \times 1 = 2$.

Ex. 4. ***If $\log_{10} 2 = 0.30103$, then find the number of digits in 2^{64}.***

Sol. $\log_{10}(2^{64}) = 64 \times \log_{10} 2 = 64 \times 0.30103 = 19.26592$

$\therefore$ Its characteristic is 19.

Hence, the number of digits in 2^{64} is 20.

Ex. 5. ***If*** $\log_{10}[98+\sqrt{x^2-12x+36}] = 2$***, find x.***

Sol. $\log_{10}\left[98+\sqrt{x^2-12x+36}\right] = 2 = \log_{10}100$

$\Leftrightarrow 98+\sqrt{x^2-12x+36} = 100 \Leftrightarrow \sqrt{x^2-12x+36} = 2$

$\Leftrightarrow x^2-12x+36 = 4 \Leftrightarrow x^2-12x+32 = 0 \Leftrightarrow (x-8)(x-4) = 0 \Leftrightarrow x = 8$ or $x = 4$.

Ex. 6. ***Find the value of*** $\log_{10}20 + 1\log_{10}30 - \frac{1}{2}\log_{10}36$.

Sol. $\log_{10}20 + \log_{10}30 - \frac{1}{2}\log_{10}36 = \log_{10}20 + \log_{10}30 - \log_{10}(36^{1/2})$

$= \log_{10}20 + \log_{10}30 - \log_{10}6 = \log_{10}\left(\frac{20\times30}{6}\right) = \log_{10}100 = \log_{10}(10^2) = 2.$

Ex. 7. ***If*** $\log_{10}2 = 0.3010$ ***and*** $\log_{10}3 = 0.4717$***, find:***

(*i*) $\log_{10}25$ (*ii*) $\log_{10}4.5$

Sol. (*i*) $\log_{10}25 = \log_{10}\left(\frac{100}{4}\right) = \log_{10}100 - \log_{10}4 = 2 - 2\log_{10}2 = (2 - 2\times0.3010) = 1.398$

(*ii*) $\log_{10}4.5 = \log_{10}\left(\frac{9}{2}\right) = \log_{10}9 - \log_{10}2 = \log_{10}(3^2) - \log_{10}2 = 2\log_{10}3 - \log_{10}2$

$= (2\times0.4771 - 0.3010) = 0.6532.$

EXERCISE 16

1. What is $\log_{10}31.25$ equal to? **[CDS (I) Exam, 2021]**
   (*a*) $3-5\log_{10}2$ (*b*) $3-2\log_{10}2$ (*c*) $5-5\log_{10}2$ (*d*) $5-3\log_{10}2$
2. If $5^{x-3} = 8$, then what is x equal to? **[CDS (I) Exam, 2022]**
   (*a*) $\frac{3}{1-\log_{10}2}$ (*b*) $\frac{3}{1+\log_{10}2}$ (*c*) $1-\frac{2}{1-10\log_{10}2}$ (*d*) $\frac{5}{1-\log_{10}2}$
3. What is $\log_{10}2000+\log_{10}400+4\log_{10}25+5\log_{10}20$ equal to? **[CDS (I) Exam, 2025]**
   (*a*) 10 (*b*) 16 (*c*) 18 (*d*) 20
4. If the roots of the equation $x^2-4x-\log_{10}N = 0$ are real, then what is the minimum value of N? **[CDS (II) Exam, 2021]**
   (*a*) 0.1 (*b*) 0.01 (*c*) 0.001 (*d*) 0.0001
5. If $5^{x-1} = (2.5)^{\log_{10}5}$, then what is the value of x? **[CDS (II) Exam, 2021]**
   (*a*) 1 (*b*) $\log_{10}2$ (*c*) $\log_{10}5$ (*d*) $2\log_{10}5$
6. How many digits are there in $(54)^{10}$? (Given that $\log_{10}2 = 0.301$ and $\log_{10}3 = 0.477$) **[CDS (I) Exam, 2020]**
   (*a*) 16 (*b*) 18 (*c*) 19 (*d*) 27
7. If $\log x = 1.2500$ and $y = x^{\log x}$, then what is $\log y$ equal to? **[CDS (I) Exam, 2020]**
   (*a*) 4.2500 (*b*) 2.5625 (*c*) 1.5625 (*d*) 1.2500
8. If $6^{3-4x}\,4^{x+5} = 8$ (Given $\log_{10}2 = 0.301$ and $\log_{10}3 = 0.477$), then which one of the following is correct? **[CDS (I) Exam, 2020]**
   (*a*) $0 < x < 1$ (*b*) $1 < x < 2$ (*c*) $2 < x < 3$ (*d*) $3 < x < 4$

9. It is given that $\log_{10} 2 = 0.301$ and $\log_{10} 3 = 0.477$. How many digits are there in $(108)^{10}$?

[CDS (I) Exam, 2019]

(*a*) 19 (*b*) 20 (*c*) 21 (*d*) 22

10. What is the value of $\log_{10}(\cos\theta) + \log_{10}(\sin\theta) + \log_{10}(\tan\theta) + \log_{10}(\cot\theta) + \log_{10}(\sec\theta) + \log_{10}(\sec\theta) + \log_{10}(\text{cosec}\theta)$? **[CDS (II) Exam, 2019]**

(*a*) –1 (*b*) 0 (*c*) 0.5 (*d*) 1

11. If $\log_{10}1995 = 3.3000$, then what is the value of $(0.001995)^{\frac{1}{2}}$? **[CDS (II) Exam, 2019]**

(*a*) $\frac{1}{10^{0.3475}}$ (*b*) $\frac{1}{10^{0.3375}}$ (*c*) $\frac{1}{10^{0.3275}}$ (*d*) $\frac{1}{10^{0.3735}}$

12. Let XYZ be an equilateral triangle in which Sy = 7 cm. If a denotes the area of the triangle, then what is the value of $\log_{10}A^4$? (Given that $\log_{10}1050 - 3.0212$ and $\log_{10}35 = 1.5441$)

[CDS (I) Exam, 2022]

(*a*) 5.3070 (*b*) 5.3700 (*c*) 5.5635 (*d*) 5.6535

13. What is the number of digits in 7^{25}, 8^{23} and 9^{20} respectively?

[Given $\log_{10}2 = 0.301$, $\log_{10}3 = 0.477$, $\log_{10}7 = 0.845$] **[CDS (II) Exam, 2019]**

(*a*) 21, 20, 19 (*b*) 20, 19, 18 (*c*) 22, 21, 20 (*d*) 22, 20 21

14. For $0 < m < 1$, which one of the following is correct? **[CDS (I) Exam, 2018]**

(*a*) $\log_{10} m < m^2 < m < m^{-1}$ (*b*) $m < m^{-1} < m^2 < \log_{10} m$

(*c*) $\log_{10} m < m < m^{-1} < m^2$ (*d*) $\log_{10} m < m^{-1} < m < m^2$

15. If $\log_{10}2 = 0.3010$ and $\log_{10}3 = 0.4771$, then the value of $\log_{100}(0.72)$ is equal to

[CDS (II) Exam, 2018]

(*a*) 0.9286 (*b*) $\bar{1}.9286$ (*c*) 1.8572 (*d*) $\bar{1}.8572$

16. If $\frac{\log_{10}(100001 - 4^x)}{5 - x} = 1$, then what is x equal to? **[CDS (I) Exam, 2025]**

(*a*) 0 (*b*) 1 (*c*) 10 (*d*) 100

17. $\log_a\left(\frac{p}{q}\right)$ is equal to:

(*a*) $\log_a p - \log_a q$ (*b*) $(\log_a p) + (\log_a q)$ (*c*) $\log_a p + \log_a q$ (*d*) $\log_a q - \log_a p$

18. If $\log_a 4 = \frac{1}{4}$, then a is equal to:

(*a*) 16 (*b*) 64 (*c*) 128 (*d*) 256

19. The value of $\log_{27} 9$ is:

(*a*) $\frac{1}{3}$ (*b*) $\frac{3}{2}$ (*c*) $\frac{2}{3}$ (*d*) 3

20. The value of $\log_5\left(\frac{1}{625}\right)$ is :

(*a*) 4 (*b*) – 4 (*c*) $\frac{1}{4}$ (*d*) $-\frac{1}{4}$

21. The value of $\log_{\sqrt{2}} 16$ is:

(*a*) 4 (*b*) 8 (*c*) 16 (*d*) $\frac{1}{8}$

22. If $\log_8 x = \frac{2}{3}$, then the value of x is:

(*a*) $\frac{3}{4}$ (*b*) $\frac{4}{3}$ (*c*) 4 (*d*) 3

23. If $\log_{10000} x = -\frac{1}{4}$, then the value of x is:

(*a*) $\frac{1}{10}$ (*b*) $\frac{1}{100}$ (*c*) $\frac{1}{1000}$ (*d*) $\frac{1}{10000}$

24. The value of $\log_{.01}(1000)$ is:

(*a*) $\frac{1}{3}$ (*b*) $-\frac{1}{3}$ (*c*) $\frac{3}{2}$ (*d*) $-\frac{3}{2}$

25. The value of $\log_2 (\log_5 625)$ is:

(*a*) 2 (*b*) 5 (*c*) 10 (*d*) 15

26. The value of $\log_{10} 0.00001$ is:

(*a*) – 4 (*b*) – 5 (*c*) $-\frac{1}{4}$ (*d*) $-\frac{1}{5}$

27. If $\log_x 0.1 = -\frac{1}{3}$, then the value of x is:

(*a*) 10 (*b*) 100 (*c*) 1000 (*d*) $\frac{1}{1000}$

28. If $\log_2 [\log_3 (\log_2 x)] = 1$, then x is equal to:

(*a*) 512 (*b*) 128 (*c*) 12 (*d*) 0

29. If $\log_{10} 2x = 1$, the value of x is:

(*a*) $\frac{1}{5}$ (*b*) 100 (*c*) 5 (*d*) 20

30. If $\log_{10}x + \log_{10}y = z$, then x is equal to:

(*a*) $\frac{z}{y}$ (*b*) $\frac{10^z}{y}$ (*c*) yz (*d*) $(10^z)y$

31. The value of $\left(\frac{1}{\log_3 60} + \frac{1}{\log_4 60} + \frac{1}{\log_5 60}\right)$ is:

(*a*) 5 (*b*) 0 (*c*) 60 (*d*) 1

32. If $\log_{10} 125 + \log_{10} 8 = x$, then x is equal to:

(*a*) – 3 (*b*) 3 (*c*) $\frac{1}{3}$ (*d*) .064

33. If $\log a + \log b = \log (a + b)$, then:

(*a*) $a = b$ (*b*) $b = \frac{1}{a}$ (*c*) $b = \left(\frac{a-1}{a}\right)$ (*d*) $b = \left(\frac{a}{a-1}\right)$

34. If $\log \frac{m}{n} + \log \frac{n}{m} = \log (m + n)$, then:

(*a*) $m + n = 1$ (*b*) $\frac{m}{n} = 1$ (*c*) $m - n = 1$ (*d*) $m^2 - n^2 = 1$

35. The value of: $(\log_3 4)(\log_4 5)(\log_5 6)(\log_6 7)(\log_7 8)(\log_8 9)$ is:

(*a*) 2 (*b*) 7 (*c*) 8 (*d*) 33

36. If $\log(x+1) + \log(x-1) = \log 3$, then x is equal to:

(*a*) 1 (*b*) 2 (*c*) 3 (*d*) 4

37. If $\frac{\log 8}{\log 2} = x$, then x is equal to:

(*a*) 2 (*b*) 3 (*c*) 4 (*d*) $\frac{1}{2}$

38. The value of x satisfying $\log_{32} x = 0.8$ is:

(*a*) 25.6 (*b*) 16 (*c*) 10 (*d*) 12.8

39. If $\log_{10} x = 1.9675$, then $\log_{10}(1000x)$ is:

(*a*) 19.675 (*b*) 4.9675 (*c*) 1.9675×3 (*d*) 1967.5

40. If $\frac{\log a}{b-c} = \frac{\log b}{c-a} = \frac{\log c}{a-b}$, then the value of $a^a b^b c^c$ is:

(a) abc (*b*) $\frac{1}{abc}$ (*c*) 1 (*d*) $\log(abc)$

41. $16^{\log_4 5}$ equals:

(*a*) 5 (*b*) 16 (*c*) 25 (*d*) $\frac{5}{64}$

42. $\log_5 5 . \log_4 9 . \log_3 2$ simplifies to:

(*a*) 2 (*b*) 1 (*c*) 5 (*d*) $\frac{2}{3}$

43. $\log_2 7$ is:

(*a*) an integer (*b*) a prime number (*c*) a rational number (*d*) an irrational number

44. The value of $\left[\frac{1}{\log_{(q/p)} x} + \frac{1}{\log_{(q/r)} x} + \frac{1}{\log_{(r/p)} x}\right]$ is:

(*a*) 3 (*b*) 2 (*c*) 1 (*d*) 0

45. If $\log_4 x + \log_2 x = 6$, then the value of x is:

(*a*) 2 (*b*) 4 (*c*) 8 (*d*) 16

46. The value of $3^{-\frac{1}{2}\log_3 9}$ is:

(*a*) -1 (*b*) $-\frac{1}{3}$ (*c*) $\frac{1}{3}$ (*d*) $-\frac{3}{2}$

47. The value of $\frac{1}{5}\log_{10} 3125 - 4\log_{10} 2 + \log_{10} 32$ is **[CDS (I) Exam, 2016]**

(*a*) 0 (*b*) 1 (*c*) 2 (*d*) 3

48. What is the solution of $\log_{10}\left[1-\left\{1-\left(1-x^2\right)^{-1}\right\}^{-1}\right]^{-\frac{1}{2}} = 1$? **[CDS (II) Exam, 2016]**

(*a*) $x = 100$ (*b*) $x = 10$ (*c*) $x = 1$ (*d*) $x = 0$

49. What are the roots of the equation $2^{x+2}.27^{\frac{x}{x-1}} = 9$? **[CDS (II) Exam, 2016]**

(*a*) 2, 1 $log\left(\frac{3}{2}\right)$ (*b*) 2, $1 - log\left(\frac{2}{3}\right)$ (*c*) $-2, 1 - \left(\frac{\log 3}{\log 2}\right)$ (*d*) $-2, 1 - \left(\frac{\log 2}{\log 3}\right)$

50. The number of digits in 3^{30} is n and it is given that $\log_{10}3 = 0.4771$. What is the value of n? **[CDS (II) Exam, 2016]**

(*a*) 13 (*b*) 14 (*c*) 15 (*d*) 16

51. What is the number of digits of 2^{40}? (Given that $\log_{10} 2 = 0.301$) **[CDS (I) Exam, 2017]**

(*a*) 14 (*b*) 13 (*c*) 12 (*d*) 11

52. What is the solution of the equation $x\log_{10}\left(\frac{10}{3}\right)+\log_{10}3-\log_{10}(2+3^x)+x$?

(*a*) 10 (*b*) 3 (*c*) 1 (*d*) 0

[CDS (I) Exam, 2017]

ANSWERS

1. (*a*)	**2.** (*a*)	**3.** (*c*)	**4.** (*d*)	**5.** (*d*)	**6.** (*b*)	**7.** (*c*)	**8.** (*b*)	**9.** (*c*)
10. (*d*)	**11.** (*b*)	**12.** (*a*)	**13.** (*c*)	**14.** (*a*)	**15.** (*b*)	**16.** (*a*)	**17.** (*a*)	**18.** (*d*)
19. (*c*)	**20.** (*b*)	**21.** (*b*)	**22.** (*c*)	**23.** (*a*)	**24.** (*d*)	**25.** (*a*)	**26.** (*b*)	**27.** (*c*)
28. (*a*)	**29.** (*c*)	**30.** (*b*)	**31.** (*d*)	**32.** (*b*)	**33.** (*d*)	**34.** (*a*)	**35.** (*a*)	**36.** (*b*)
37. (*b*)	**38.** (*b*)	**39.** (*b*)	**40.** (*c*)	**41.** (*c*)	**42.** (*b*)	**43.** (*d*)	**44.** (*d*)	**45.** (*d*)
46. (*c*)	**47.** (*b*)	**48.** (*b*)	**49.** (*c*)	**50.** (*c*)	**51.** (*b*)	**52.** (*d*)		

SOLUTIONS

1. We have,

$$\log_{10} 31.25 = \log_{10}\left(\frac{3125}{100}\right) = \log_{10}(3125) - \log_{10}(100) \quad \left[\because \log_{10}\frac{p}{q} = \log_{10}(p) - \log_{10}(q)\right]$$

$$= \log_{10}(5^5) - \log_{10}(10^2) = 5\log_{10}5 - 2\log_{10}10 \quad \left[\because \log_{10}(p^q) = q\times\log_{10}(p)\right]$$

$$= 5\left[\log_{10}\left(\frac{10}{2}\right)\right]-2 = 5\left[\log_{10}(10)-\log_{10}(2)\right]-2 = 5-5\log_{10}(2)-2 \Rightarrow 3-5\log_{10}(2)$$

2. We have, $5^{x-3} = 8 \Rightarrow 5^{x-3} = 2^3$

By by taking log on both sides, we get,

$\log(5^{x-3}) = \log(2^3)$

$$\Rightarrow (x-3)\log_{10}5 = 3\log_{10}2 \Rightarrow x-3 = \frac{3\log_{10}2}{\log_{10}5} \Rightarrow x = \frac{3\log_{10}2}{\log_{10}}+3$$

$$\Rightarrow x = 3\left[\frac{\log_{10}2}{\log_{10}5}+1\right] \Rightarrow x = 3\left[\frac{\log_{10}2+\log_{10}5}{\log_{10}5}\right] \Rightarrow x = 3\left[\frac{\log_{10}(2\times 5)}{\log_{10}5}\right]$$

$$\Rightarrow x = \frac{3\times 1}{\log_{10}5} \Rightarrow x = \frac{3}{\log_{10}\left(\frac{10}{2}\right)} \Rightarrow x = \frac{3}{\log_{10}(10)-\log_{10}(2)} \Rightarrow x = \frac{3}{1-\log_{10}(2)}$$

3. $\log_{10}2000+\log_{10}400+4\log_{10}25+5\log_{10}20 = [\log_{10}(2000\times 400\times 25^4\times 20^5)]$

$\Rightarrow 2000 = 2^4\times 5^3,\ 400 = 2^4\times 5^2,\ 25^4 = (5^2)^4 = 5^8,\ 20^5 = (2^2\times 5)^5 = 2^{10}\times 5^5$

$\Rightarrow$ Product $= (2^4\times 5^3)\times(2^4\times 5^2)\times(5^8)\times(2^{10}\times 5^5) = 2^{4+4+10}\times 5^{3+2+8+5} \Rightarrow 2^{18}\times 5^{18}$

$\Rightarrow (2\times 5)^{18} = 10^{18}$

Therefore, $\log_{10}(10^{18}) = 18$

4. Given that, $x^2 - 4x - \log_{10} N = 0$

It is a quadratic equation. Here, $a = 1$, $b = -4$, $c = \log_{10}N$

Now, we have, $b^2 - 4ac \geq 0$

$\Rightarrow (-4)^2 - 4 \times 1 \times (-\log_{10} N) \geq 0 \Rightarrow 16 + 4 \log 10\ N \geq 0 \Rightarrow 4 \log 10\ N \geq -16$

$\Rightarrow \log_{10} N \geq -4 \Rightarrow N \geq 10^{-4} \Rightarrow N \geq \dfrac{1}{10000} \Rightarrow N \geq 0.0001$

5. We have, $5^{x-1} = (2.5)^{\log_{10} 5}$

$\Rightarrow \log_{10} 5 = (x-1) \log_{2.5} 5 \Rightarrow (x-1) = \dfrac{(\log_{10} 5)}{(\log_{2.5} 5)} \Rightarrow x - 1 = \log_{10}2.5$

$\Rightarrow x = \log_{10} 2.5 + 1 \Rightarrow x = \log_{10}^{\ 2.5} + \log_{10}^{\ 10} \Rightarrow x = \log_{10} (10 \times 2.5)$

$\Rightarrow x = \log_{10} 25 \Rightarrow x = \log_{10} 5^2 \Rightarrow x = 2 \log_{10} 5$

6. We have, $\log_{10} (54)^{10}$

$= 10 \log_{10} 54$ $[\because \log a^b = b \log a]$

$= 10 \log_{10} (2 \times 3^3) = 10 \log_{10} 2 + 3 \times 10 \log_{10} 3$ $[\because \log ab = \log a + \log b]$

$= 10 \times 0.301 + 30 \times 0.477 = 3.01 + 14.31 = 17.32$

$\therefore$ No. of digits = 18

7. We have, $y = x^{\log_x}$

By taking log on both sides, we get

$\log y = \log (x^{\log x}) \Rightarrow \log y = \log x \log x \Rightarrow \log y = 1.25 \times 1.25 = 1.5625$ $[\because x = 1.25]$

8. We have, $6^{3-4x}\ 4^{x+5} = 8$

$\Rightarrow 3^{3-4x} \times 2^{3-4x} \times 2^{2(x+5)} = 2^3 \Rightarrow 3^{3-4x} \times 2^{3-4x+2x+10} = 2^3 \Rightarrow 3^{3-4x} \times 2^{13-2x} = 2^3$

$\Rightarrow 3^{3-4x} = 2^{3-(13-2x)} \Rightarrow 3^{3-4x} = 2^{3-13+2x} \Rightarrow 3^{3-4x} = 2^{2x-10}$

By taking log on both sides, we get

$\log y = \log (x^{\log x})$

$\Rightarrow \log (3^{3-4x}) = \log (2^{2x-10}) \Rightarrow (3-4x) \log_3 = 2x - 10 \log 2 \Rightarrow (3-4x) \times 0.477 = (2x - 10) \times 0.301$

$\Rightarrow 1.4313 - 1.9084x = 0.602x - 0.301 \Rightarrow 0.602x + 1.9084x = 0.301 + 1.4313$

$\Rightarrow 2.5104x = 4.4412 \Rightarrow x = \dfrac{4.4413}{2.5104} = 1.7692$

9. We have,

$\log_{10} (108)^{10} = 10 \log_{10} (2^2 \times 3^3)$

$= 10 \log_{10} 2^2 + 10 \log_{10} 3^3$ $[\because \log ub = \log a + \log b]$

$= 2 \times 10 \times \log_{10} 2 + 3 \times 10 \times \log_{10} 3 = 20 \times 0.301 + 30x \times 0.477$

$\Rightarrow 6.02 + 14.31 = 20.33$ $\because$ No. of digits = 21

10. We have,

$\log_{10} (\cos\theta) + \log_{10} (\sin\theta) + \log_{10} (\tan\theta) + \log_{10} (\cot\theta) + \log_{10} (\sec\theta) + \log_{10} (\text{cosec}\theta)$

$\Rightarrow \cos\theta \times \sin\theta \times \tan\theta \times \cot\theta \times \sec\theta \times \text{cosec}\theta$ [by logarithms rule] $\Rightarrow 1$

11. Let $x = (0.001995)^{\frac{1}{8}}$

Taking log on both sides, we get

$\log x = \left((0.001995)^{\frac{1}{8}} \right) \Rightarrow \log x = \dfrac{1}{8} \log (0.001995) \Rightarrow \log x = \dfrac{1}{8} \log \left(1995 \times \dfrac{1}{10^6} \right)$

$\Rightarrow \log x = \frac{1}{8}(\log 1995 + \log 10^{-6}) \Rightarrow \log x = \frac{1}{8}(3.3 - 6)$

$\Rightarrow \log x = \frac{-2.7}{8} \Rightarrow \log x = -0.3375 \Rightarrow x = \frac{1}{10^{0.3375}}$

12. $\log_{10} A^4 = \log_{10}(A^2)^2 = 2\log_{10} A^2$

$= 2\log_{10}\left(\frac{\sqrt{3}}{4}\times 7\times 7\right)^2 = 2\log_{10}[3\times(3.5)^4] = 2\log_{10}[3\times 3.5\times(3.5)^3]$

$= 2[\log_{10} 10.5 + \log_{10} 3.5^3] = 2[\log_{10} 10.5 + 3\log_{10} 3.5]$

$= 2\left[\log_{10}\frac{1050}{100} + 3\log_{10}\frac{35}{10}\right] = 2[\log_{10} 1050 - \log_{10} 100] + 3(\log_{10} 35 - \log_{10} 10)$

$= 2[(3.0212 - 2) + 3(1.5441 - 1)] = 2[1.0212 + 3\times 0.5441] = 2[1.0212 + 1.6323]$

$= 2[2.6535] = 5.3070$

13. Let $a = 7^{25}$

By taking log on both sides, we get,

$\log a = \log 7^{25} \Rightarrow \log a = 25\log 7 \Rightarrow \log a = 25\times 0.845 \Rightarrow \log a = 21.125 \Rightarrow a = (10)^{21.125}$

$\therefore$ No. of digits = 22

Now, let $b = 8^{23} = (2^3)^{23} = 2^{69}$

By taking log on both sides,

We get, $\log b = \log 2^{69}$

$\Rightarrow \log b = 69\log 2 \Rightarrow \log b = 69\times 0.301 \Rightarrow \log b = 20.769 \Rightarrow b = (10)^{20.769}$

$\therefore$ No. of digits = 21

Now, let $c = 9^{20} = (3^2)^{20} = 3^{40}$

By taking log on both sides, we get

$\log c = \log 3^{40} \Rightarrow \log c = 40\log 3 \Rightarrow \log c = 40\times 0.477 \Rightarrow \log c = 19.08 \Rightarrow c = 10^{19.08}$

$\therefore$ Number of digits = 20

14. We have, $0 < m < 1$ [$\log m$ is –ve. So, it is the least and all other terms will be positive.]

For number less than 1, we have $-1 < 1 < 2$

$\Rightarrow m^2 < m < m^{-1}$

$\therefore \log m < m^2 < m < m^{-1}$

15. Let $x = \log_{100}(0.72)$

$$= \frac{\log_{10}(072)}{\log_{10} 100} = \frac{\log_{10}\left(2^3\times 3^2\times\frac{1}{10^2}\right)}{\log_{10} 1(0^2)} = \left[\frac{\log_{10}(2^3) + \log_{10}(3^2) + \log_{10}(10^{-2})}{\log_{10}(10^2)}\right]$$

$$= \left[\frac{3\log_{10}(2) + 2\log_{10}(3) - 2\log_{10} 10}{2\log_{10} 10}\right] = \left[\frac{3\times 0.301 + 2\times 0.4771 - 2}{2}\right]$$

$$= \frac{0.903 + 0.9542 - 2}{2} = \frac{1.8572 - 2}{2} = \frac{-0.1428}{2}$$

$= -0.0714 = -1 + 0.9286$

$= \bar{1}.9286$

16. $\log_{10}(100001-4^x)=5-x \Rightarrow 100001-4^x=10^{5-x} \Rightarrow 100001-4^x=10^{5-x}$

Check all the options: $x=0$

$100001-4^0=10^{5-0} \Rightarrow 100001-1=100000$

$x=0$ satisfies the condition.

$x=1$

$$100001-4^1=10^{5-1} \Rightarrow 100001-4=10000$$
$$99997 \neq 10000$$

$x=1$ does not satisfy the condition.

$x=10$

$100001-4^{10}=10^{5-10} \Rightarrow 100001-4^{10}\neq 10^{-5}$

$x=10$ does not satisfy the condition.

$\Rightarrow$ The correct answer is $x=0$

17. $\log_a\left(\dfrac{p}{q}\right)=\log_a p-\log_a q$

18. $\log_a 4=\dfrac{1}{4}\Rightarrow a^{1/4}=4\Rightarrow a=4^4=256.$

19. $\log_{27} 9=x \Leftrightarrow 27^x=9 \Leftrightarrow (3^3)^x=3^2 \Leftrightarrow 3^{3x}=3^2 \quad \Leftrightarrow 3x=2 \Leftrightarrow x=\dfrac{2}{3}.$

20. $\log_5\left(\dfrac{1}{625}\right)=x \Leftrightarrow 5^x=\dfrac{1}{625}=\dfrac{1}{5^4}=5^{-4} \Leftrightarrow x=-4.$

21. $\log_{\sqrt{2}}(16)=x \Leftrightarrow (\sqrt{2})^x=16=2^4 \Leftrightarrow 2^{x/2}=2^4 \Leftrightarrow \dfrac{x}{2}=4 \Leftrightarrow x=8.$

22. $\log_8 x=\dfrac{2}{3} \Leftrightarrow 8^{2/3}=x \Leftrightarrow (2^3)^{2/3}=x \Leftrightarrow x=2^2=4.$

23. $x=(10000)^{-1/4}=(10^4)^{-1/4}=10^{-1}=\dfrac{1}{10}.$

24. $\log_{0.01}(1000)=x \Leftrightarrow (0.01)^x=1000 \Leftrightarrow \left(\dfrac{1}{100}\right)^x=1000$

$\Leftrightarrow 10^{-2x}=10^3 \Leftrightarrow -2x=3 \Leftrightarrow x=-\dfrac{3}{2}.$

25. Let $\log_5 625=x$. Then $5^x=625=5^4 \Leftrightarrow x=4.$

Let $\log_2(\log_5 625)=y$. Then, $\log_2 4=y \Leftrightarrow 2^y=4=2^2 \Leftrightarrow y=2.$

26. $\log_{10}(0.00001)=x \Leftrightarrow 10^x=0.00001=\dfrac{1}{10^5}=10^{-5} \Leftrightarrow x=-5.$

27. $\log_x(0.1)=-\dfrac{1}{3} \Leftrightarrow x^{-1/3}=1 \Leftrightarrow x^{-1}=0.001 \Leftrightarrow x^{-1}=\dfrac{1}{1000} \Leftrightarrow x=1000.$

28. $\log_2[\log_3(\log_2 x)]=1=\log_2 2 \Rightarrow \log_3(\log_2 x)=2$

$\Rightarrow \log_2 x=3^2=9 \Rightarrow x=2^9=512.$

29. $\log_{10}2x=1 \Rightarrow 2x=10^1 \Rightarrow x=5.$

30. $\log_{10} x + \log_{10} y = z \Leftrightarrow \log_{10} xy = z \Leftrightarrow xy = 10^z \Leftrightarrow x = \dfrac{10^z}{y}$.

31. Given Exp. = $\log_{60} 3 + \log_{60} 4 + \log_{60} 5 = \log_{60} (3 \times 4 \times 5) = \log_{60} 60 = 1$.

32. $\log_{10} 125 + \log_{10} 8 = \log_{10}\left(\dfrac{1000}{8}\right) + \log_{10} 8 = \log_{10} (1000) - \log_{10} 8 + \log_{10} 8 = \log_{10} 10^3$

$= 3(\log_{10} 10) = 3 \times 1 = 3$.

33. $\log(a + b) = \log a + \log b = \log ab \Leftrightarrow a + b = ab \Leftrightarrow b(a - 1) = a \Leftrightarrow b = \left(\dfrac{a}{a-1}\right)$.

34. $\log(m + n) = \log\dfrac{m}{n} + \log\dfrac{n}{m} = \log\left(\dfrac{m}{n} \times \dfrac{n}{m}\right) = \log 1 \quad \therefore m + n = 1$.

35. Given Exp. = $\dfrac{\log 4}{\log 3} \cdot \dfrac{\log 5}{\log 4} \cdot \dfrac{\log 6}{\log 5} \cdot \dfrac{\log 7}{\log 6} \cdot \dfrac{\log 8}{\log 7} \cdot \dfrac{\log 9}{\log 8} = \dfrac{\log 9}{\log 3} = \dfrac{\log 3^2}{\log 3} = \dfrac{2(\log 3)}{(\log 3)} = 2$.

36. $\log(x + 1) + \log(x - 1) = \log 3 \Leftrightarrow \log [(x + 1)(x - 1)] = \log 3$

$\Leftrightarrow (x + 1)(x - 1) = 3 \Leftrightarrow x^2 - 1 = 3 \Leftrightarrow x^2 = 4 \Leftrightarrow x = 2$.

37. $x = \dfrac{\log 8}{\log 2} = \dfrac{\log 2^3}{\log 2} = \dfrac{3(\log 2)}{(\log 2)} = 3$.

38. $\log_{32} x = 0.8 \Leftrightarrow x = (32)^{0.8} = (2^5)^{4/5} = 2^4 = 16$.

39. $\log_{10} (1000x) = \log_{10} 1000 + \log_{10} x = 3 + 1.9675 = 4.9675$.

40. Let $\dfrac{\log a}{(b-c)} = \dfrac{\log b}{(c-a)} = \dfrac{\log c}{(a-b)} = x$. Then,

$\log a = (b - c)x$, $\log b = (c - a)x$ and $\log c = (a - b)x$.

$\therefore \quad \log(a^a b^b c^c) = a \log a + b \log b + c \log c = a(b - c)x + b(c - a)x + c(a - b)x = 0 = \log 1$.

$\therefore a^a b^b c^c = 1$.

41. $16^{\log_4 5} = (4^2)^{\log_4 5} = 4^{2\log_4 5} = 4^{\log_4 (5^2)} = 4^{\log_4 25} = 25. \qquad [\because a^{\log_a x} = x]$

42. Given Exp. = $\dfrac{\log 5}{\log 5} \times \dfrac{\log 9}{\log 4} \times \dfrac{\log 2}{\log 3} = \dfrac{\log 3^2}{\log 2^2} \times \dfrac{\log 2}{\log 3} = \dfrac{2\log 3}{2\log 2} \times \dfrac{\log 2}{\log 3} = 1$.

43. If possible let $\log_2 7$ be rational, which in simplest form is (p/q).Then,

$\log_2 7 = \dfrac{p}{q} \Leftrightarrow 2^{p/q} = 7 \Leftrightarrow 2^p = 7^q$, which is wrong as L.H.S. is even & R.H.S. is odd.

$\therefore \log_2 7$ is irrational.

44. Given Exp. = $\log_x \left(\dfrac{p}{q}\right) + \log_x \left(\dfrac{q}{r}\right) + \log_x \left(\dfrac{r}{p}\right) = \log_x \left(\dfrac{p}{q} \times \dfrac{q}{r} \times \dfrac{r}{p}\right) = \log_x 1 = 0$.

45. $\log_4 x + \log_2 x = 6 \Leftrightarrow \dfrac{\log x}{\log 4} + \dfrac{\log x}{\log 2} = 6 \Leftrightarrow \dfrac{\log x}{2\log 2} + \dfrac{\log x}{\log 2} = 6 \Leftrightarrow 3\log x = 12\log 2$

$\Leftrightarrow \log x^3 = \log 2^{12} \Leftrightarrow x^3 = 2^{12} \Leftrightarrow x = 2^4 = 16$.

46. $\dfrac{1}{2}\log_3 9 = \log_3 (9^{1/2}) = \log_3 3 = 1. \qquad \therefore 3^{-\frac{1}{2}\log_3 9} = 3^{-1} = \dfrac{1}{3}$.

47. $\frac{1}{5}\log_{10}3125-4\log_{10}2+\log_{10}32=\log 5-4\log 2+5\log 2=\log 5+\log 2=1$.

48. $\log_{10}\left[1-\left\{1-\left(1-x^2\right)^{-1}\right\}^{-1}\right]^{-\frac{1}{2}}=1$ or $log_{10}[1-A]^{-\frac{1}{2}}=1$

$\Rightarrow\ [1-A]^{-\frac{1}{2}}=10\Rightarrow 1-A=\frac{1}{100}\Rightarrow \text{A}=\left\{1-\left(1-x^2\right)^{-1}\right\}^{-1}=\frac{99}{100}$

$\Rightarrow\ 1-\frac{1}{1-x^2}=\frac{100}{99}\Rightarrow\frac{-1}{99}=\frac{1}{1-x^2}\Rightarrow x^2-1=99\Rightarrow x=10$

49. $2^{x+2}\ 27^{\frac{x}{x-1}}=9$ ie $2^{x+2}\ 3^{\frac{3x}{x-1}}=3^2$

$\Rightarrow\ 2^{x+2}=3^{2-\frac{3x}{x-1}}=3^{\frac{-x-2}{x-1}}\Rightarrow (x+2)\log 2=-\frac{(x+2)}{x-1}\log 3$

$\Rightarrow\ (x+2)\left[\log 2+\frac{\log 3}{x-1}\right]=0\Rightarrow x=-2$ or $x=1-\left(\frac{\log 3}{\log 2}\right)$

50. Let N = 3^{30}, $\log \text{N}=30\log 3=30\ (0.4771)=14.313$

$\therefore$ N has 15 digits (The log of a 2 digit number is 1……, of a 2 digit number is 2…… and so on)

51. $x=2^{40}$

$\log_{10}x=\log_{10}2^{40}$

$\log_{10}x=40\log_{10}2=40\times 0.301=12.040$

Anti $\log(X)=12+1=13$

So, number of digits in 2^{40} is 13

52. $x\log_{10}\left(\frac{10}{3}\right)+\log_{10}3=\log_{10}(2+3^x)+x$

$\Rightarrow\ x\left[\log_{10}10-\log_{10}3\right]+\log_{10}3=\log_{10}\left(2+3^x\right)+x$

$\Rightarrow\ x\left[1-\log_{10}3\right]+\log_{10}3=\log_{10}\left(2+3^x\right)+x$

$\Rightarrow\ x-x\log_{10}3+\log_{10}3=\log_{10}\left(2+3^x\right)+x$

$\Rightarrow\ \log_{10}3^{-x}\times 3=\log_{10}\left(2+3^x\right)\quad\left[\because \log_{10}m+\log_{10}n=\log_{10}(mn)\right]$

$\Rightarrow\ 3^{-x}\times 3=2+3^x\Rightarrow\frac{3}{3^x}=2+3^x$

Let $3^x=a$

$\Rightarrow\ \frac{3}{a}=2+a\Rightarrow 3=2a+a^2\Rightarrow a^2+2a-3=0\Rightarrow (a+3)(a-1)=0\quad a=-3$ or 1

Putting $a=1$ in $3^x=a$, we get $3^x=1$

$\Rightarrow\ 3^x=3^{\circ}\Rightarrow x=0$

UNIT 2

MENSURATION

Area : Area of squares; Rectangles; Parallelograms; Triangles and Circles. Area of figures which can be split up into these figures (Field Book).

Volume & Surface Area: Surface area and volume of cube, Cuboids, Cones, Cylinders and Spheres.

17 CHAPTER Area

Highlights: Areas of squares, rectangles, parallelograms, triangles and circles. Areas of figures which can be split up into these figures (Field Book)

FORMULAE

I. (*i*) *Area of a Rectangle* = (*Length* × *Breadth*)

$\therefore\ Length = \frac{Area}{Breadth}$ and $Breadth = \frac{Area}{Length}$.

II. Area of a square = $(\text{side})^2 = \frac{1}{2}(\text{diagonal})^2$

III. (*i*) Area of a triangle = $\frac{1}{2}$ × Base × Height.

(*ii*) Area of an equilateral triangle = $\frac{\sqrt{3}}{4} \times (\text{side})^2$.

(*iii*) If a, b, c be the sides of a triangle and $s = \frac{1}{2}(a + b + c)$, then:

Area of the triangle = $\sqrt{s(s-a)(s-b)(s-c)}$.

(*iv*) If r and R be the radii of incircle and circumcircle of an equilateral triangle of side a, then: $r = \frac{a}{2\sqrt{3}}$ and $R = \frac{a}{\sqrt{3}}$.

IV. (*i*) Area of || gm = Base × Height.

(*ii*) Area of Rhombus = $\frac{1}{2}$ × (*Product of diagonals*).

(*iii*) Area of Trapezium = $\frac{1}{2}$ × (*Sum of parallel sides*) × (*distance between them*)

V. (*i*) *Area of Room* = (*Length*) × (*Breadth*).

(*ii*) Area of 4 walls of a Room = 2 (*Length* + *breadth*) × Height.

VI. (*i*) *Area of a circle* = πR^2. (*ii*) *Circumference of a circle* = $2\pi R$.

(*iii*) *Length of an Arc* = $\frac{2\pi R\theta}{360}$. (*iv*) *Area of a Sector* = $\frac{1}{2}(\text{arc } AB \times R) = \frac{\pi R^2\theta}{360}$.

SOLVED PROBLEMS

Ex. 1. ***Find the area of an isosceles triangle, in which each one of equal sides is 5 m and the third side 6 m.***

Sol. Here $a = 5$, $b = 5$ and $c = 6$.

$\therefore\ s = \frac{1}{2}(a+b+c) = 8, (s-a) = 3, (s-b) = 3\ \&\ (s-c) = 2.$

$\therefore$ Area $= \sqrt{s(s-a)(s-b)(s-c)} = \sqrt{8\times3\times3\times2} = 12\,\text{m}^2.$

Ex. 2. ***The base of a triangular field is three times its altitude. If the cost of cultivating the field at ₹240 per hectare be ₹3240, find its base and height.***

Sol. Area of the field $= \frac{\text{Total Cost}}{\text{Rate}} = \left(\frac{3240}{240}\right)$ hectares $= (13.5 \times 10000)\ \text{m}^2 = 135000\ \text{m}^2.$

Let the height be x metres. Then, base $= 3x$ metres.

$\therefore\ \frac{1}{2}\times x\times 3x = 135000 \Rightarrow x^2 = 90000 \Rightarrow x = 300.$

$\therefore$ Height = 300 m and Base = 900 m.

Ex. 3. ***The sides of a rectangular lawn are in the ratio 3 : 2 and its area is $\frac{1}{6}$ hectares. Find the dimensions of the lawn.***

Sol. Let the sides be $3x$ and $2x$.

Then, $3x \times 2x = \frac{1}{6}\times 10000 \Rightarrow x = \frac{100}{6} = \frac{50}{3}.$

$\therefore$ Breadth $= \left(2\times\frac{50}{3}\right)\text{m} = 33\frac{1}{3}\,\text{m}$ & Length $= \left(3\times\frac{50}{3}\right)\text{m} = 50\,\text{m}.$

Ex. 4. ***Find the cost of carpeting a room 13 m long and 9 m wide with a carpet 75 cm broad at ₹8 per metre.***

Sol. Area of the carpet = (Area of room) $= (13 \times 9)\ \text{m}^2 = 117\ \text{m}^2.$

Breadth of the carpet = 75 cm $= \left(\frac{75}{100}\right)\text{m} = \frac{3}{4}\,\text{m}.$

$\therefore$ Length of the carpet $= \frac{\text{Area}}{\text{Breadth}} = \left(117\times\frac{4}{3}\right)\text{m} = 156\ \text{m}.$

Hence, the cost of the carpet = ₹(156 × 8) = ₹1248.

Ex. 5. ***A room is half as long again as it is broad. The cost of carpeting the room at ₹18 per m² is ₹972 and the cost of white-washing the 4 walls at ₹6 per m² is ₹1080. Find the dimensions of the room.***

Sol. Let breadth $= x$, length $= \frac{3}{2}x$ and height $= y$ metres.

Area of the floor $= \left(\frac{972}{18}\right)\text{m}^2 = 54\ \text{m}^2.$

$\therefore\ x \times \frac{3}{2}x = 54$ or $x^2 = 36$ or $x = 6.$

$\therefore$ Breadth = 6 m and length $= \left(\frac{3}{2}\times 6\right)\text{m} = 9\ \text{m}.$

Now area of 4 walls $= \left(\frac{1080}{6}\right)\text{m}^2 = 180\ \text{m}^2.$

$\therefore\ 2(9 + 6) \times y = 180$ or $y = 6$ m.

So, Height = 6 m.

Ex. 6. ***A rectangular grassy plot is 116 m by 68 m. It has a gravel path 2.5 m wide all round it on the inside. Find the area of the path.***

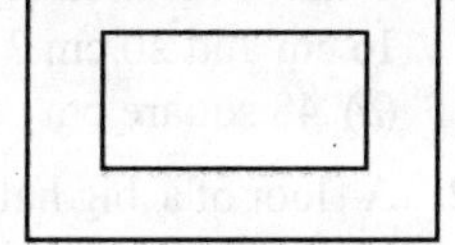

Sol. Area of the plot = $(116 \times 68)\text{m}^2 = 7888\ \text{m}^2$.

Area of the plot excluding the path

$= [(116 - 5) \times (68 - 5)]\text{m}^2$

$= [(111 \times 63)\text{m}^2 = 6993\ \text{m}^2$

$\therefore$ Area of the path $= (7888 - 6993)\text{m}^2 = 895\ \text{m}^2$.

Ex. 7. ***Find the area of a rhombus each one of whose sides measures 20 cm and one diagonal 24 cm.***

Sol. We know that the diagonals of a rhombus bisect each other at right angles. Then,

$AB = 20$ cm, $OB = \frac{1}{2}BD = 12$ cm & $\angle AOB = 90°$.

$\therefore \quad OA = \sqrt{AB^2 - OB^2}$

$= \sqrt{(20)^2 - (12)^2} = \sqrt{256} = 16\,\text{cm}.$

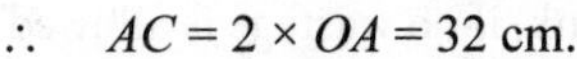

$\therefore \quad AC = 2 \times OA = 32$ cm.

Hence, area of rhombus $= \frac{1}{2} \times (AC \times BD) = \frac{1}{2} \times (32 \times 24)\text{cm}^2 = 384\,\text{cm}^2$.

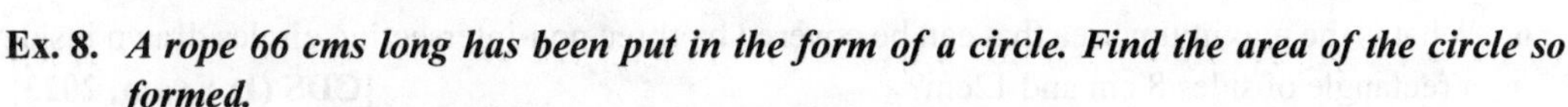

Ex. 8. ***A rope 66 cms long has been put in the form of a circle. Find the area of the circle so formed.***

Sol. Circumference of the circle = 66 cm.

Let its radius be R.

Then, $2\pi R = 66 \Rightarrow R = \left(\frac{66}{2} \times \frac{7}{22}\right)$ cm = 10.5 cm.

$\therefore$ Area of the circle $= \left(\frac{22}{7} \times 10.5 \times 10.5\right)\text{cm}^2 = 346.5\ \text{cm}^2$.

Ex. 9. ***In a circle of radius 35 cm, an arc subtends an angle of 72° at the centre. Find the length of the arc and the area of the sector.***

Sol. Length of the arc $= \frac{2\pi r\theta}{360} = \left(2 \times \frac{22}{7} \times 35 \times \frac{72}{360}\right)\text{cm} = 44\,\text{cm}$

Area of the sector $= \frac{\pi r^2\theta}{360} = \left(\frac{22}{7} \times 35 \times 35 \times \frac{72}{360}\right)\text{cm}^2 = 770\,\text{cm}^2$.

Ex. 10. ***The diameter of the wheel of a bus is 140 cm. How many revolutions per minute must the wheel make in order to keep a speed of 66 km per hour?***.

Sol. Distance covered in 1 minute $= \left(\frac{66 \times 1000 \times 100}{60}\right)$ cm = 110000 cm

Distance covered in 1 revolution = circumference of wheel $= \left(2 \times \frac{22}{7} \times \frac{140}{2}\right)$ cm = 440 cm.

$\therefore$ Number of revolutions $= \left(\frac{110000}{440}\right) = 250$.

EXERCISE 17

1. What is the area of the circle (approximately) inscribed in a triangle with side lengths 12 cm, 16 cm and 20 cm ? **[CDS (I) Exam, 2023]**

(*a*) 48 square cm (*b*) 50 square cm (*c*) 52 square cm (*d*) 54 square cm

2. A floor of a big hall has dimensions 30 m 60 cm and 23 m 40 cm. It is to be paved with square tiles of same size. What is the minimum number of tiles required ? **[CDS (I) Exam, 2023]**

(*a*) 30 (*b*) 36 (*c*) 169 (*d*) 221

3. What is the maximum area of a rectangle, in square cm, whose perimeter is 400 cm? **[CDS (I) Exam, 2024]**

(*a*) 100 (*b*) 200 (*c*) 1000 (*d*) 10,000

4. Let x be the area of a square inscribed in a circle of radius r and y be the area of an equilateral triangle inscribed in the same circle. Which one of the following is correct ? **[CDS (I) Exam, 2023]**

(*a*) $9x^2 = 16y^2$ (*b*) $27x^2 = 64y^2$ (*c*) $36x^2 = 49y^2$ (*d*) $16x^2 = 21y^2$

5. A plank of wood 4.25 m long and 3.4 m wide is to be cut into square pieces of equal size. How many square pieces of largest size can be cut from the plank, if no wastage is allowed ? **[CDS (I) Exam, 2024]**

(*a*) 20 (*b*) 90 (*c*) 400 (*d*) 500

6. What is the maximum area that can be covered by three non-intersecting circles drawn inside a rectangle of sides 8 cm and 12cm? **[CDS (I) Exam, 2023]**

(*a*) 16π square cm (*b*) 18π square cm (*c*) 20π square cm (*d*) 24π square cm

7. The perimeter and the area of a right-angled triangle are 36 cm and 54 square cm respectively. What is the length of the hypotenuse? **[CDS (I) Exam, 2023]**

(*a*) 12 cm (*b*) 14 cm (*c*) 15 cm (*d*) 16 cm

8. In a triangle ABC, AB + BC = 7.1 cm, BC + CA = 12.1 cm and CA + AB = 7.2 cm. What is the area of the triangle? **[CDS (I) Exam, 2025]**

(*a*) 3 square cm (*b*) 32 square cm

(*c*) 33 square cm (*d*) 3.3 square cm

9. A sector is cut from a circle of radius 21 cm. If the length of the arc of the sector is 55 cm, then what is the area of the sector? **[CDS (I) Exam, 2021]**

(*a*) 577.5 cm^2 (*b*) 612.5 cm^2 (*c*) 705.5 cm^2 (*d*) 725.5 cm^2

10. A square is drawn inside a square of side 14 cm in such a way that the corners of the inner square coincide with the midpoints of the sides of the outer square. What is the area lying between the two squares? **[CDS (II) Exam, 2024]**

(*a*) 98 square cm (*b*) 56 square cm

(*c*) 49 square cm (*d*) 24.5 square cm

11. If the perimeter of a right-angled triangle is 30 cm and the hypotenuse is 13 cm, then what is the area of the triangle? **[CDS (I) Exam, 2021]**

(*a*) 24 cm^2 (*b*) 27 cm^2 (*c*) 30 cm^2 (*d*) 36 cm^2

12. A person wishes to fence 375 m^2 rectangular garden. He has 65 m of barbed wire and is able to fence only three sides of the garden. What is the perimeter of the garden? **[CDS (II) Exam, 2021]**

(*a*) 80 m (*b*) 84 m (*c*) 90 m (*d*) 100 m

13. The area of a rhombus is 336 square cm. If the length of one of its diagonals is 48 cm, then what is the perimeter of the rhombus? **[CDS (II) Exam, 2021]**
(*a*) 200 cm (*b*) 120 m (*c*) 100 cm (*d*) 90 cm

Direction: Consider the following for the next two items that follow:

A chord of a circle of radius 2.1 cm subtends an angle of 120° at the centre.

(take $\pi = \frac{22}{7}$ and $\sqrt{3} = 1.732$)

14. What is the approximate area of minor segment of the circle? **[CDS (II) Exam, 2021]**
(*a*) 2.71 cm^2 (*b*) 2.42 cm^2 (*c*) 1.91 cm^2 (*d*) 1.71 cm^2

15. What is the approximate area of major segment of the circle? **[CDS (II) Exam, 2021]**
(*a*) 10.05 cm^2 (*b*) 10.15 cm^2 (*c*) 11.05 cm^2 (*d*) 11.15 cm^2

16. The length and breadth of a room are 21 m and 16 m respectively. If the length of the longest rod that can be placed in the room is 29 m, then what is the height of the room? **[CDS (II) Exam, 2021]**
(*a*) 10 m (*b*) 11 m (*c*) 12 m (*d*) 13 m

17. The base of a right-angled triangle is $\frac{4}{3}$ times the height of the triangle. If the area of the triangle is 54 square cm, then what is the perimeter of the triangle? **[CDS (II) Exam, 2024]**
(*a*) 30 cm (*b*) 32 cm (*c*) 36 cm (*d*) 40 cm

18. What is the area of a triangle having sides 4, 4, and 6 units? **[CDS (II) Exam, 2024]**
(*a*) $3\sqrt{7}$ square units (*b*) 8 square units
(*c*) 7 square units (*d*) $7\sqrt{3}$ square units

19. From a circular piece of cardboard of radius 3 cm, two sectors of 40° each have been cut off. The area of the remaining portion is **[CDS (I) Exam, 2015]**
(*a*) 11 square cm (*b*) 22 square cm
(*c*) 33 square cm (*d*) 44 square cm

20. Three equal circles each of diameter *d* are drawn on a place in such a way that each circle touches the other two circles. A big circle is drawn in such a manner that it touches each of the small circles internally. The area of the big circle is **[CDS (I) Exam, 2015]**
(*a*) πd^2 (*b*) $\pi d^2 (2-\sqrt{3})^2$ (*c*) $\frac{\pi d^2 (\sqrt{3}+1)^2}{2}$ (*d*) $\frac{\pi d^2 (\sqrt{3}+2)^2}{12}$

21. A square and an equilateral triangle have equal perimeter. If the diagonal of the square is $12\sqrt{2}$ cm, then the area of the triangle is **[CDS (II) Exam, 2015]**
(*a*) $24\sqrt{2}$ cm^2 (*b*) $24\sqrt{3}$ cm^2 (*c*) $48\sqrt{3}$ cm^2 (*d*) $64\sqrt{3}$ cm^2

22. Consider the following statements: **[CDS (II) Exam, 2015]**
1. If the height of a cylinder is doubled, the area of the curved surface is doubled.
2. If the radius of a hemispherical solid is doubled, its total surface area becomes fourfold.
(*a*) 1 only (*b*) 2 only (*c*) Both 1 and 2 (*d*) Neither 1 nor 2

23. What is the ratio of the lengths of diagonals of a rhombus.

Statement I : One diagonal of the rhombus is equal to its side.

Statement II : The longer diagonal of the rhombus is equal to $\sqrt{3}$ times its side.

[CDS (I) Exam, 2024]

(*a*) If the Question can be answered by one of the Statements alone, but not by the other.
(*b*) If the Question can be answered by either Statement alone.
(*c*) If the Question can be answered by using both the Statements together, but cannot be answered by using either Statement alone.
(*d*) If the Question cannot be answered even by using both Statements together.

24. The diameter of a wheel that makes 453 revolutions to move 2 km and 26 decameter is equal to **[CDS (II) Exam, 2015]**

(*a*) $1\frac{9}{22}$m (*b*) $1\frac{13}{22}$m (*c*) $2\frac{5}{11}$m (*d*) $2\frac{7}{11}$m

25. From a rectangular sheet of sides 18 cm and 14 cm, a semicircular portion with smaller side as diameter is taken out. Then the area of the remaining sheet will be **[CDS (II) Exam, 2015]**

(*a*) 98 cm^2 (*b*) 100 cm^2 (*c*) 108 cm^2 (*d*) 175 cm^2

26. ABCD is a square. If the sides *AB* and *CD* are increased by 30%, sides *BC* and *AD* are increased by 20%, then the area of the resulting rectangle exceeds the area of the square by

(*a*) 50% (*b*) 52% (*c*) 54% (*d*) 56%

[CDS (II) Exam, 2015]

27. The area of a trapezium is 336 cm^2. If its parallel sides are in the ratio 5 : 7 and the perpendicular distance between them is 14 cm, then the smaller of the parallel sides is

(*a*) 20 cm (*b*) 22 cm (*c*) 24 cm (*d*) 26 cm

[CDS (II) Exam, 2015]

28. The number of rounds that a wheel of diameter 7/11 metre will make in travelling 4 km will be **[CDS (I) Exam, 2016]**

(*a*) 500 (*b*) 1000 (*c*) 1700 (*d*) 2000

29. One side of a rectangular field is 4 metres and its diagonal is 4 metres. The area of the field is:

(*a*) 12 m^2 (*b*) 15 m^2 (*c*) 20 m^2 (*d*) $4\sqrt{5}$ m^2

30. The ratio between the length and breadth of a rectangular field is 5 : 4. If the breadth is 20 metres less than the length, the perimeter of the field is

(*a*) 260 m (*b*) 280 m (*c*) 360 m (*d*) None of these

31. The length of a hall is (4/3) times its breadth. If the area of the hall be 300 m^2, the difference between the length and breadth is:

(*a*) 15 m (*b*) 4 m (*c*) 3 m (*d*) None of these

32. A rectangular plot is half as long again as it is broad. The area of the plot is (2/3) hectares. The length of the plot is:

(*a*) 100 m (*b*) 66.66 m (*c*) $33\frac{1}{3}$ m (*d*) $\left(\frac{100}{\sqrt{3}}\right)$ m

33. A park is 10 metres long and 8 metres broad. The length of the longest pole that can be placed in the park is:

(*a*) 6 m (*b*) 12.8 m (*c*) 13.4 m (*d*) 18 m

34. A rectangular carpet has an area of 120 sq. m and a perimeter of 46 m. The length of its diagonal is:

(*a*) 15 m (*b*) 16 m (*c*) 17 m (*d*) 20 m

35. The cost of cultivating a square field at the rate of ₹160 per hectare is ₹1440. The cost of putting a fence around it at the rate of 75 paise per metre is:

(*a*) ₹900 (*b*) ₹1800 (*c*) ₹360 (*d*) ₹810

36. The length of a rectangle is twice its width. If the length of its diagonal is $9\sqrt{5}$ metres, the perimeter of the rectangle is:
(*a*) 27 m (*b*) 54 m (*c*) 81 m (*d*) None of these

37. A hall 40 metres long, 15 metres broad is to be paved with stones, each measuring 6 dm by 5 dm. The number of stones required is:
(*a*) 1000 (*b*) 2000 (*c*) 3000 (*d*) None of these

38. The cost of carpeting a room 15 metres long with a carpet 75 cm wide at 30 paise per metre is ₹36. The breadth of the room is:
(*a*) 6 metres (*b*) 8 metres (*c*) 9 metres (*d*) 12 metres

39. The length of a rectangle is doubled while its breadth is halved. What is the percentage change in area?
(*a*) 50 (*b*) 75 (*c*) No change (*d*) None of these

40. The length of a rectangle is increased by 60%. By what percent would the width have to be decreased to maintain the same area?
(*a*) $37\frac{1}{2}\%$ (*b*) 60% (*c*) 75% (*d*) 120%

41. A rectangular lawn 60 metres by 40 metres has two roads each 5 metres wide running in the middle of it, one parallel to length and the other parallel to breadth. The cost of gravelling the roads at 60 paise per sq. metre is:
(*a*) ₹300 (*b*) ₹280 (*c*) ₹285 (*d*) ₹250

42. A rectangle has 15 cm as its length and 150 cm^2 as its area. Its area is increased to $1\frac{1}{3}$ times the original area by increasing only its length. Its new perimeter is:
(*a*) 50 cm (*b*) 60 cm (*c*) 70 cm (*d*) 80 cm

43. Keshav rides along the boundary of a rectangular park at the rate of 12 km/hr and completes one full round in 15 minutes. If the length of the park is 4 times its breadth, the area of the park is:
(*a*) 360000 m^2 (*b*) 36000 m^2 (*c*) 3600 m^2 (*d*) None of these

44. A rectangular carpet has an area of 60 m^2. Its diagonal and longer side together equal 5 times the shorter side. The length of the carpet is:
(*a*) 5 m (*b*) 12 m (*c*) 13 m (*d*) 14.5 m

45. If the base of a rectangle is increased by 10% and the area is unchanged, then the corresponding altitude must be decreased by:
(*a*) $9\frac{1}{11}\%$ (*b*) 10% (*c*) 11% (*d*) $11\frac{1}{9}\%$

46. The dimensions of a rectangular room when increased by 4 metres, are in ratio 4 : 3 and when decreased by 4 metres, are in the ratio 2 : 1. The dimensions of the room are:
(*a*) 6 m, 4 m (*b*) 12 m, 8 m (*c*) 16 m, 12 m (*d*) 24 m, 16 m

47. What is the area of the region enclosed by three identical circles (each of radius 4 cm) touching each other? **[CDS (I) Exam, 2022]**
(*a*) $\frac{8\pi}{3}$ square cm (*b*) $\left(16\sqrt{3}-\frac{8\pi}{3}\right)$ square cm
(*c*) $\left(16\sqrt{3}-8\pi\right)$ square cm (*d*) $\frac{16\pi}{\sqrt{3}}$ square cm

48. Two sides of a triangle forming a right angle are $6x^2$ and $(2x^2 - 1)$. If the area of the triangle is 84 square units, then what is the perimeter of the triangle? **[CDS (I) Exam, 2022]**

(*a*) 51 units (*b*) 53 units (*c*) 56 units (*d*) 69 units

49. A rectangle of length 10 units and breadth 8 units is split into two squares each of area x square units and two rectangles each of area y square units. Consider the following statements:

1. y is always greater than x.
2. y can be 15 square units.

Which of the above statements is/are correct? **[CDS (I) Exam, 2022]**

(*a*) 1 only (*b*) 2 only (*c*) Both 1 and 2 (*d*) Neither 1 nor 2

50. A square sheet is formed by joining n identical square sheets of same size. If the length of the diagonal of the bigger square sheet so formed is m, then what is the side length of a smaller square sheet? **[CDS (I) Exam, 2022]**

(*a*) $\frac{m}{\sqrt{n}}$ (*b*) $\frac{m}{2\sqrt{n}}$ (*c*) $\frac{m}{\sqrt{2n}}$ (*d*) $\frac{\sqrt{2}m}{\sqrt{n}}$

51. $ABCD$ is a trapezium in which AB is parallel to DC. Let E and F be the midpoints on AD and BC respectively. If $EF = 10$ cm and $BA - DC = 4$ cm, then what is the value of $AB \times DC$? **[CDS (I) Exam, 2022]**

(*a*) 84 square cm (*b*) 96 square cm (*c*) 100 square cm (*d*) 108 square cm

52. $ABCD$ is a parallelogram with $AB = 15$ cm and $AD = 8$ cm. If θ is the acute angle between AB and AD, then what is the area of the parallelogram in square cm? **[CDS (I) Exam, 2022]**

(*a*) $60 \sin\theta$ (*b*) $120 \sin\theta$ (*c*) $60 \cos\theta$ (*d*) $120 \cos\theta$

53. A rectangle is 48 cm long and 14 cm wide. If the diagonal makes an angle θ with the longer side, then what is $(\sec\theta + \text{cosec}\theta)$ equal to? **[CDS (I) Exam, 2020]**

(*a*) $\frac{775}{168}$ (*b*) $\frac{725}{168}$ (*c*) $\frac{375}{84}$ (*d*) $\frac{325}{84}$

54. What is the area of the triangle having side lengths $\frac{y}{z}+\frac{z}{x}, \frac{z}{x}+\frac{x}{y}, \frac{x}{y}+\frac{y}{z}$? **[CDS (I) Exam, 2020]**

(*a*) $\frac{(x+y+z)^2}{xyz}$ (*b*) $\sqrt{\frac{xyz}{x+y+z}}$ (*c*) $\sqrt{\frac{x}{y}+\frac{y}{z}+\frac{z}{x}}$ (*d*) $\sqrt{\frac{xy+yz+zy}{xyz}}$

55. A square and a rectangle have equal areas. If one side of the rectangle is of length numerically equal to the square of the length of the side of the square then the other side of the rectangle is **[CDS (I) Exam, 2020]**

(*a*) square root of the side of the square
(*b*) half the side of the square
(*c*) of unit length
(*d*) double the side of the square

56. Let $PQRS$ be the diameter of circle of radius 9 cm. The length PQ, QR and RS are equal. Semi-circle is drawn with QS as diameter (as shown in the given figure). What is the ratio of the shaded region to that of the unshaded region? **[CDS (I) Exam, 2020]**

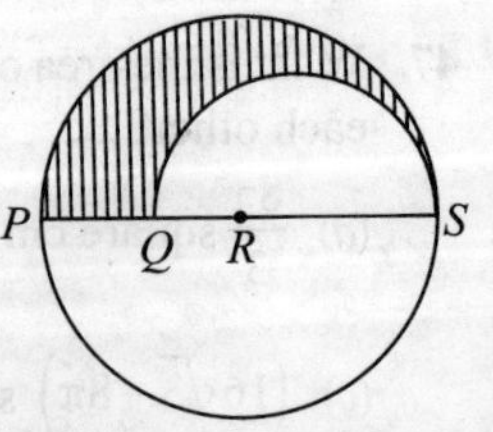

(*a*) 25 : 121 (*b*) 5 : 13
(*c*) 5 : 18 (*d*) 1 : 2

57. A road curve is to be laid out on circle. What radius should be used if the track is to change direction by 42° in distance of 44m? (Assume $\pi = \frac{22}{7}$) **[CDS (II) Exam, 2020]**

(*a*) 60 m (*b*) 66 m (*c*) 75 m (*d*) 80 m

58. The two sides of a triangle are 40 cm and 41 cm. If the perimeter of the triangle is 90 cm, what is its area? **[CDS (II) Exam, 2020]**

(*a*) 90 cm^2 (*b*) 135 cm^2 (*c*) 150 cm^2 (*d*) 180 cm^2

59. A bicycle wheel makes 5000 revolutions in moving 11 km. What is the radius of the wheel? (Assume $\pi = \frac{22}{7}$) **[CDS (II) Exam, 2020]**

(*a*) 17.5 cm (*b*) 35 cm (*c*) 70 cm (*d*) 140 cm

Direction: Consider the following for the next three items:

ABCD is a quadrilateral with AB = 9 cm, BC = 40 cm. *CD* = 28 cm, *DA* = 15 cm and angle *ABC* is a right-angle. **[CDS (I) Exam, 2019]**

60. What is the area of triangle *ADC*?

(*a*) 126 cm^2 (*b*) 124 cm^2 (*c*) 122 cm^2 (*d*) 120 cm^2

61. What is the area of quadrilateral *ABCD* ? **[CDS (I) Exam, 2019]**

(*a*) 300 cm^2 (*b*) 306 cm^2 (*c*) 312 cm^2 (*d*) 316 cm^2

62. What is the difference between perimeter of triangle *ABC* and perimeter of triangle ADC? **[CDS (I) Exam, 2019]**

(*a*) 4 cm (*b*) 5 cm (*c*) 6 cm (*d*) 7 cm

63. A 12 m long wire is cut into two pieces, one of which is bent into a circle and the other into a square enclosing the circle. What is the radius of the circle? **[CDS (I) Exam, 2019]**

(*a*) $\frac{12}{\pi+4}$ (*b*) $\frac{6}{\pi+4}$

(*c*) $\frac{3}{\pi+4}$ (*d*) $\frac{6}{\pi+2\sqrt{2}}$

64. An equilateral triangle and a square are constructed using metallic wires of equal length. what is the ratio of areas of triangle to that square? **[CDS (II) Exam, 2019]**

(*a*) 3 : 4 (*b*) 2 : 3 (*c*) $4\sqrt{3}:9$ (*d*) $2\sqrt{3}:9$

65. The length and breadth of a rectangle are increased by 20% and 10% respectively. What is the percentage increase in the area of the rectangle? **[CDS (II) Exam, 2019]**

(*a*) 32% (*b*) 30% (*c*) 25% (*d*) 15%

66. A square is drawn such that its vertices are lying on a circle of radius 201 mm. What is the ratio of area of circle to that of square? **[CDS (II) Exam, 2019]**

(*a*) 11 : 7 (*b*) 7 : 11 (*c*) 20 : 19 (*d*) 19 : 20

67. A region of area A bounded by a circle C is divided into a n regions, each of area $\frac{A}{n}$, by drawing circles of radii $r_1, r_2, r_3, \ldots r_{n-1}$ such that $r_1 < r_2 < r_3 < \ldots r_{n-1}$ concentric with the circle C. If $P_m = \frac{r_{m+1}}{r_m}$ where m = 1, 2, 3, ... (n – 2), then which one of the following is correct? **[CDS (I) Exam, 2018]**

(*a*) p_m increases as m increases

(*b*) p_m decreases as m increases

(*c*) p_m remains constant as m increases

(*d*) p_m increases for some values of m as m increases and then decreases thereafter

68. A square and an equilateral triangle have the same perimeters. If the diagonal of the square is $6\sqrt{2}$ cm, then what is the area of the triangle? **[CDS (I) Exam, 2018]**

(*a*) $12\sqrt{2}$ cm^2 (*b*) $12\sqrt{3}$ cm^2 (*c*) $16\sqrt{2}$ cm^2 (*d*) $16\sqrt{3}$ cm^2

69. What is the area of the region bounded internally by a square of side of length '*a*' and externally by a circle passing through the four corers of the square? **[CDS (I) Exam, 2018]**

(*a*) $(\pi - 1)a^2$ square units (*b*) $\frac{(\pi-1)a^2}{2}$ square units

(*c*) $(\pi - 2)a^2$ square units (*d*) $\frac{(\pi-2)a^2}{2}$ square units

70. If base and hypotenuse of a right triangle are $(u^2 - v^2)$ and $(u^2 + v^2)$ respectively and the area of the triangle is 2016 square units, then the perimeter of the triangle may be **[CDS (II) Exam, 2018]**

(*a*) 224 units (*b*) 288 units (*c*) 448 units (*d*) 576 units

71. A circle is inscribed in an equilateral triangle of side of length l. The area of any square inscribed in the circle is **[CDS (II) Exam, 2018]**

(*a*) $\frac{l^2}{2}$ (*b*) $\frac{\sqrt{3}l^2}{4}$ (*c*) $\frac{l^2}{4}$ (*d*) $\frac{l^2}{6}$

72. In the figure given below, *ABC* is a right-angled triangle where $\angle A = 90°$, $AB = p$ cm and $AC = q$ cm. On the three sides as diameters, semicircles are drawn as shown in the figure. The area of the shaded portion, in square cm. is

(*a*) pq (*b*) $\frac{\pi(p^2+q^2)}{2}$

(*c*) $\pi(p^2+q^2)$ (*d*) $\frac{pq}{2}$ **[CDS (II) Exam, 2018]**

73. Four equal – sized maximum circular plates are cut off from a square paper sheet of area 784 square cm. The circumference of each plate is **[CDS (I) Exam, 2015]**

(*a*) 11 cm (*b*) 22 cm (*c*) 33 cm (*d*) 44 cm

74. The area of a rhombus with side 13 cm and one diagonal 10 cm will be

(*a*) 140 square cm (*b*) 130 square cm

(*c*) 120 square cm (*d*) 110 square cm

[CDS (I) Exam, 2015]

75. Two circles, each of radius, r, with centres P and Q, are such that each circle passes through the centre of the other circle. Then the area common to the circles is less than one-third of the sum of the areas of the two circles by **[CDS (I) Exam, 2015]**

(*a*) $\frac{\sqrt{3}x^2}{4}$ (*b*) $\frac{\sqrt{3}x^2}{3}$ (*c*) $\frac{\sqrt{3}x^2}{2}$ (*d*) $\sqrt{3x^2}$

76. A sphere and a cube have same surface area. The ratio of square of their volumes is
(*a*) 6 : π (*b*) 5 : π (*c*) 3 : 5 (*d*) 1 : 1
[CDS (I) Exam, 2015]

77. The total outer surface area of a right circular cone of height 24 cm with a hemisphere of radius 7 cm upon its base is **[CDS (I) Exam, 2015]**
(*a*) 327π square cm (*b*) 307π square cm (*c*) 293π square cm (*d*) 273π square cm

78. A rectangular block of wood having dimensions 3 m × 2m × 1.75 has to be painted on all its faces. The layer of paint must be 0.1 mm tick. Paint comes in cubical boxes having their edges equal to 10 cm. The minimum number of boxes of paint to be purchased is
(*a*) 5 (*b*) 4 (*c*) 3 (*d*) 2
[CDS (I) Exam, 2015]

79. The base of an isosceles triangle is 300 units and each of its equal sides is 170 units. Then the area of the triangle is **[CDS (I) Exam, 2016]**
(*a*) 9600 square units
(*b*) 10000 square units
(*c*) 12000 square units
(*d*) None of the above

80. Four equal discs are placed such that each one touches two other ₹If the area of empty space enclosed by them is 150/847 square centimetre, then the radius of each disc is equal to
(*a*) 7 /6 cm (*b*) 5/6 cm (*c*) 1/2 cm (*d*) 5/11 cm
[CDS (I) Exam, 2016]

81. ABC is a triangle in which D is the midpoint of BC and E is the midpoint of AD. Which of the following statements is/are correct? **[CDS (I) Exam, 2016]**
1. The area of triangle ABC is equal to four times the area of triangle BED.
2. The area of triangle ADC is twice the area of triangle BED.
Select the correct answer using the code given below.
(*a*) 1 only (*b*) 2 only (*c*) Both 1 and 2 (*d*) Neither 1 nor 2

Direction (Q. 82-83): Consider the following for the next two (2) items that follow:

ABCD is a circle with centre and taking OC as diameter a circle is drawn as drawn in the figure given below: Let OB = 7 cm

(Take $\pi = \frac{22}{7}$) **[CDS (I) Exam, 2024]**

H
C O A
D

82. What is the area of the shaded region?
(*a*) 38.5 square cm (*b*) 48 square cm
(*c*) 62.5 square cm (*d*) 66.5 square cm

83. What is the ratio of the area of the shaded region to the area of the non-shaded region?
(*a*) $\frac{19}{25}$ (*b*) $\frac{18}{25}$ (*c*) $\frac{17}{25}$ (*d*) $\frac{16}{25}$

84. The area of four walls of a room is 120 m^2. The length of the room is twice its breadth. If the height of the room is 4m, what is area of the floor? **[CDS (II) Exam, 2016]**
(*a*) 40 m^2 (*b*) 50 m^2 (*c*) 60 m^2 (*d*) 80 m^2

85. If a square of side x and an equilateral triangle of side y are inscribed in a circle, then what is the ratio of x to y? **[CDS (II) Exam, 2016]**

(*a*) $\sqrt{\frac{2}{3}}$ (*b*) $\sqrt{\frac{3}{2}}$ (*c*) $\frac{3}{\sqrt{2}}$ (*d*) $\frac{\sqrt{2}}{3}$

86. The ratio of the curved surface area to the total surface area of a right circular cylinder is 1 : 2. If the total surface area is 616 cm^2, what is the volume of the cylinder?

(*a*) 539 cm^3 (*b*) 616 cm^3 (*c*) 1078 cm^3 (*d*) 1232 cm^3

[CDS (II) Exam, 2016]

87. What is the number of rounds that a wheel of diameter $\frac{5}{11}$ m. will make in traversing 7 km?

(*a*) 3300 (*b*) 3500 (*c*) 4400 (*d*) 4900

[CDS (II) Exam, 2016]

88. The cost of painting a spherical vessel of diameter 14 cm is ₹8008. What is the cost of painting per square centimeter? **[CDS (II) Exam, 2016]**

(*a*) ₹8 (*b*) ₹9 (*c*) ₹13 (*d*) ₹14

89. A circle and a square have the same perimeter. Which one of the following is correct?

(*a*) Their areas are equal

(*b*) The area of the circle is larger

(*c*) The area of the square is $\frac{\pi}{2}$ times area of circle.

(*d*) The area of the square is π times area of circle **[CDS (II) Exam, 2016]**

90. The diameter of the front wheel of an engine is $2x$ cm and that of rear wheel is $2y$ cm. To cover the same distance, what is the number of times the rear wheel revolves when the front wheel revolves n times? **[CDS (II) Exam, 2016]**

(*a*) $\frac{n}{xy}$ (*b*) $\frac{ny}{x}$ (*c*) $\frac{nx}{y}$ (*d*) $\frac{xy}{n}$

Direction (91-92): A tent of circus is made of canvas and is in the form of right circular cylinder and right circular cone above it. The height and diameter of the cylindrical part of the tent are 5 m and 126 m respectively. The total height of the tend is 21m.

91. What is the curved surface area of the cylinder? **[CDS (II) Exam, 2016]**

(*a*) 1980 m^2 (*b*) 2010 m^2

(c) 2100 m^2 (*d*) 2240 m^2

92. How many square metres of canvas are used? **[CDS (II) Exam, 2016]**

(*a*) 14450 (*b*) 14480 (*c*) 14580 (*d*) 14850

93. What is the area of a triangle with sides of length 12*cm*, 13 *cm* and 5 *cm*?

(*a*) 30 *cm*^{2} (*b*) 35 *cm*^{2}

(*c*) 40 *cm*^{2} (*d*) 42 *cm*^{2}

[CDS (II) Exam, 2016]

94. If the perimeter of a circle is equal to that of a square, then what is the ratio of area of circle to that of square? **[CDS (II) Exam, 2016]**

(*a*) 22 : 7 (*b*) 14 : 11 (*c*) 7 : 22 (*d*) 11 : 14

95. If the perimeter of a rectangle is 10cm and the area is 4 cm^2, then its length is

(*a*) 6 cm (*b*) 5 cm (*c*) 4.5 cm (*d*) 4 cm

[CDS (I) Exam, 2017]

96. The areas of two circular fields are in the ratio 16 : 49. If the radius of the bigger field is 14 m, then what is the radius of the smaller field? **[CDS (I) Exam, 2017]**

(*a*) 4 m (*b*) 8 m (*c*) 9 m (*d*) 10 m

97. If each of the dimensions of a rectangle is increased by 200%, the area is increased by

(*a*) 300% (*b*) 400% (*c*) 600% (*d*) 800%

[CDS (I) Exam, 2017]

98. Three circles each of radius 3.5 cm touch one another. The area subtended between them is

(*a*) $6\left(\sqrt{3}\pi - 2\right)$ square units (*b*) $6\left(2\pi - \sqrt{3}\right)$ square units

(*c*) $\frac{49}{8}\left(2\sqrt{3} - \pi\right)$ square units (*d*) $\frac{49}{8}\left(\sqrt{3} - \pi\right)$ square units

[CDS (I) Exam, 2017]

99. A copper wire when bent in the form of a square encloses an area of 121 cm^2. If the same wire is bent in the form of a circle, it encloses an area equal to **[CDS (I) Exam, 2017]**

(*a*) 121 cm^2 (*b*) 144 cm^2 (*c*) 154 cm^2 (*d*) 168 cm^2

100. The radius of a circle is increased so that its circumference increases by 15%. The area of the circle will increase by **[CDS (I) Exam, 2017]**

(*a*) 31.25% (*b*) 32.25% (*c*) 33.25% (*d*) 34.25%

101. A field is divided into four regions as shown in the given figure. What is the area of the field in square meters? **[CDS (I) Exam, 2017]**

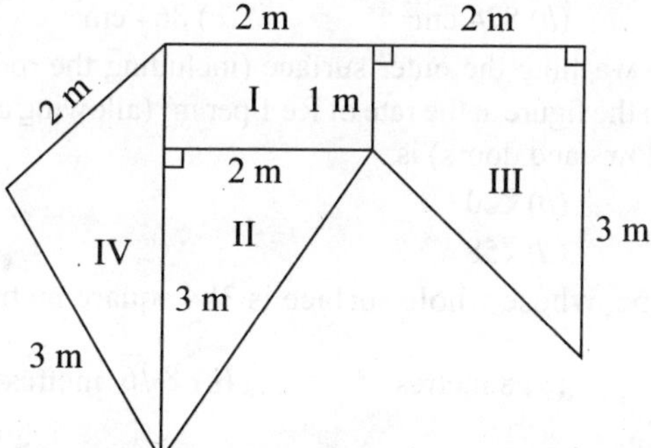

(*a*) $6+\frac{3}{4}\sqrt{5}$ (*b*) $5+\frac{3}{2}\sqrt{3}$ (*c*) $9+\frac{3}{4}\sqrt{15}$ (*d*) $7+2\sqrt{2}$

102. In the figure given below, D is diameter of each circle. What is the diameter of the shaded circle? **[CDS (I) Exam, 2017]**

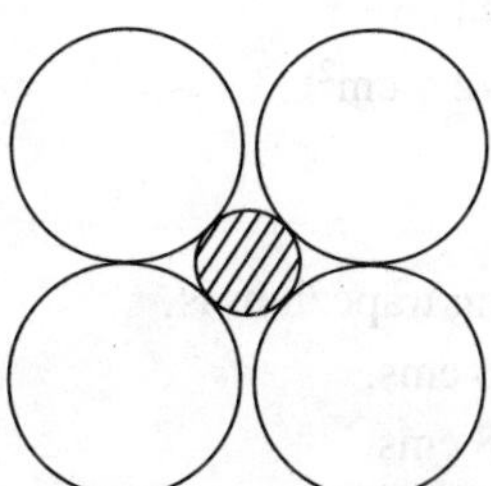

(*a*) $D\left(\sqrt{2}-1\right)$ (*b*) $D\left(\sqrt{2}+1\right)$ (*c*) $D\left(\sqrt{2}+2\right)$ (*d*) $D\left(2-\sqrt{2}\right)$

103. From the details given in the adjoining figure, it follows that the ratio of the area of the parallelogram *ABCD* and that of triangle *ABM* is:

(*a*) 4 : 1 (*b*) 6 : 1
(*c*) 8 : 1 (*d*) 7 : 2

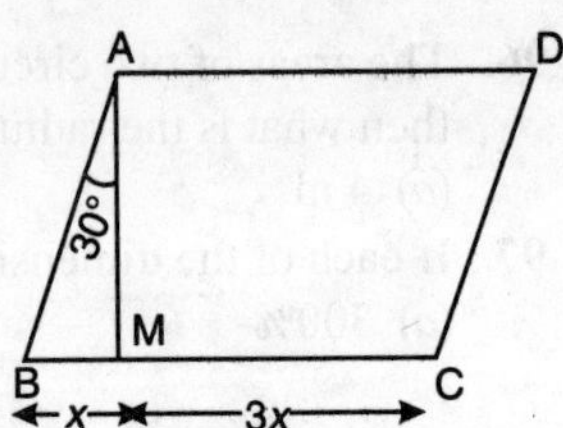

Direction (Q. 104-105): Consider the following for the next two (02) items that follow:

Let ABCD be the diameter of a circle of radius 6 cm. The lengths AB, BC and CD are equal. Semi-circles are drawn with AB and BD as diameters as shown in the figure given below.

[CDS (I) Exam, 2024]

104. What is the ratio of the area of the shaded region to that of the non-shaded region ?

(*a*) 2 : 7 (*b*) 2 : 5
(*c*) 3 : 5 (*d*) 5 : 8

105. What is the perimeter of the shaded region ?

(*a*) 24π cm (*b*) 18π cm (*c*) 15π cm (*d*) 12π cm

106. Area of the smallest square that can circumscribe a circle of area 616 cm^2 is:

(*a*) 784 cm^2 (*b*) 824 cm^2 (*c*) 864 cm^2 (*d*) None of these

107. The cost of white washing the outer surface (including the root) of a house, as shown in the figure at the rate of Re 1 per m^2 (allowing a margin of 18 m^2 for windows and doors) is

(*a*) ₹11 (*b*) ₹20
(*c*) ₹40 (*d*) ₹58

108. The edge of a cube, whose whole surface is 384 square metres is

(*a*) 6 metres (*b*) 8 metres (*c*) $8\sqrt{6}$ metres (*d*) $\frac{8\sqrt{6}}{\pi}$ metres

109. The cross section of a canal is in the shape of a trapezium. It is 10 m wide at the top and 6 m wide at the bottom. The area of its cross section is 72 square metres. The depth of the canal is

(*a*) 7.2 metres (*b*) 12 metres (*c*) 36 metres (*d*) 9 metres

110. In the adjoining figure of a trapezium *ABCD*, it is being given that *AB* = 16 cms, *CD* = 11 cms, and *BD* = 13 cms. The area of the trapezium is

(*a*) 169 cm^2 (*b*) 144 cm^2
(*c*) 175.5 cm^2 (*d*) 142.5 cm^2

111. The perimeter of the adjoining trapezium is

(*a*) 44 cms. (*b*) 46 cms.
(*c*) 52 cms. (*d*) 48 cms.

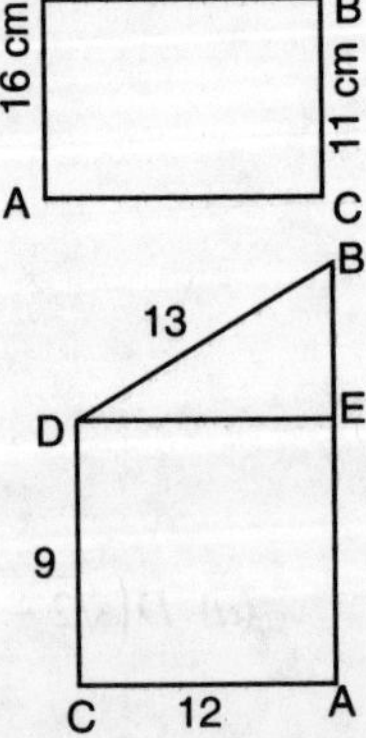

112. In the adjoining trapezium *ABCD*, it is being given that $AD = 20$ m, $AB = CD$, $AE = 12$ m and $BE = 5$ m. The perimeter of the trapezium is

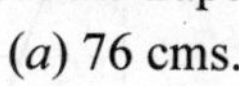
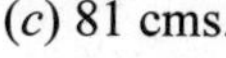

(*a*) 76 cms. (*b*) 72 cms.
(*c*) 81 cms. (*d*) 67 cms.

113. In the given framework, it being given that $OG = OF$. The area of the frame work is

(*a*) 56 m^2 (*b*) 64 m^2
(*c*) 74 m^2 (*d*) $28\sqrt{3}$ m^2

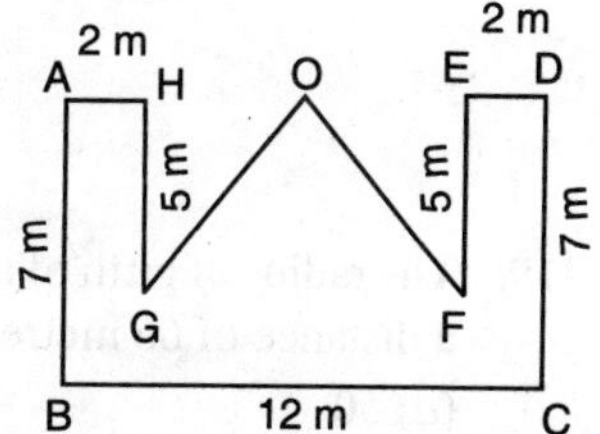

114. In the given figure the shaded region shows a lawn. On the three sides of the lawn there are flower-beds having a uniform width of 2 metres. The total area of the flower beds is

(*a*) 64 m^2 (*b*) 72 m^2
(*c*) 76 m^2 (*d*) 82 m^2

115. The area of the given figure is:

(*a*) 74 m^2 (*b*) 66 m^2
(*c*) 84 m^2 (*d*) 92 m^2

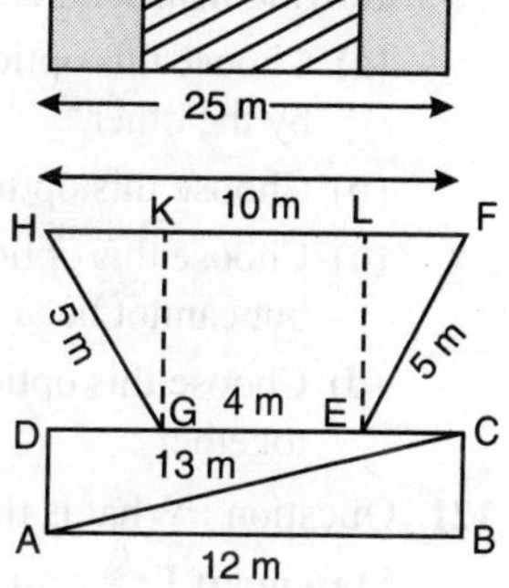

116. In the adjoining figure *ABCD* is a trapezium in which the parallel sides *AB*, *CD* are both perpendicular to *BC*. If $AB = 16$ m, $CD = 8$ m, and $AD = 17$ m the area of the trapezium is:

(*a*) 140 m^2 (*b*) 168 m^2
(*c*) 180 m^2 (*d*) 156.4 m^2

117. The dimensions of a field are given in metres in the adjoining figure. The area of the field is:

(*a*) 3000 m^2 (*b*) 3565 m^2
(*c*) 3600 m^2 (*d*) 3540 m^2

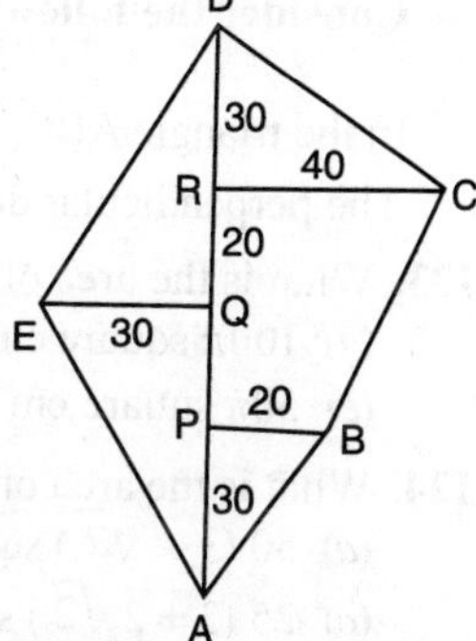

118. The adjoining figure shows the cross-section of a concrete structure with the measurements as given. The area of cross section is:

(*a*) 3.65 m^2 (*b*) 4.50 m^2 (*c*) 5.72 m^2
(*d*) 6.2 m^2

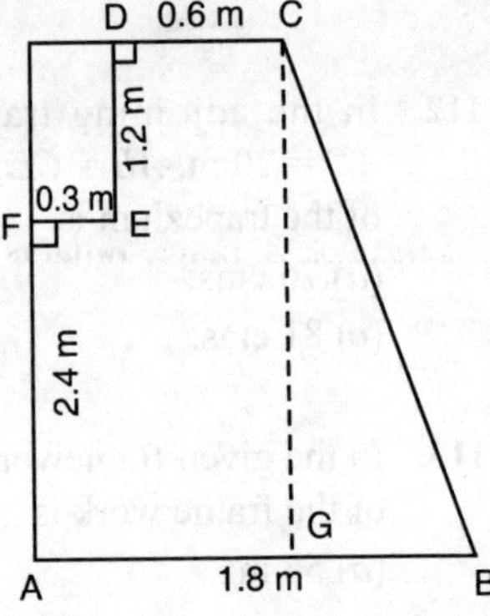

119. The radius of girth of a garden roller is 21 cms. How many revolutions does it make in moving a distance of 66 metres?

(*a*) 50 (*b*) 22 (*c*) 147 (*d*) 7

120. The area of a circle inscribed in an equilateral triangle of side 24 cms is:

(*a*) 24 π cm^2 (*b*) 36 π cm^2 (*c*) 48 π cm^2 (*d*) 18 π cm^2

Each item contains a Question followed by twoStatements. Answer each item using the following instructions :

(a) Choose this option if the Question can be answered by one of the Statements alone but not by the other.
(b) Choose this option if the Question can be answered by either Statement alone.
(c) Choose this option if the Question can be answered by using both the Statements together, but cannot be answered by using either Statement alone
(d) Choose this option if theQuestion cannot be answered even by using both Statements together

121. Question : What is the area of the circle C ? **[CDS (I) Exam, 2023]**

Statement I : An arc of length 7 cm subtends an angle 30° at the centre of C.

Statement II : A chord of length 10 cm subtends an angle 90° at the centre of C.

122. The plinth of a house has an area of 200 square metres. It is rectangular in shape and its length and breadth are in the ratio 2 : 1. The owner of the house extends the terrace by 1 m on each side. What is the percentage of area that has increased in the terrace relative to the plinth ? **[CDS (I) Exam, 2023]**

(*a*) 40% (*b*) 32% (*c*) 20% (*d*) 15·5%

Consider the following for the next three (03) items that follow :

In the triangle ABC, AB = 6 cm, BC = 8 cm and AC = 10 cm.
The perpendicular dropped from B meets the side AC a

123. What is the area of the circle ?

(*a*) 100π square cm (*b*) 96π square cm
(*c*) 50π square cm (*d*) 48π square cm

124. What is the area of the smaller square ? **[CDS (I) Exam, 2023]**

(*a*) 50 $(3 - \sqrt{2})$ square cm (*b*) 25 $(3 - 2\sqrt{2})$ square cm
(*c*) 25 $(3 + 2\sqrt{2})$ square cm (*d*) 50 $(3 - 2\sqrt{2})$ square cm

125. What is the area of the shaded region ? **[CDS (I) Exam, 2023]**
(*a*) $(96 - 25\pi)$ square cm (*b*) $(92 - 25\pi)$ square cm
(*c*) $(96 - 16\pi)$ square cm (*d*) $(92 - 16\pi)$ square cm

126. The corners of an equilateral triangular plate were cut in such a manner that it forms a regular hexagonal plate. What is the ratio of the area of the triangular plate to the area of the hexagonal plate ? **[CDS (II) Exam, 2023]**
(*a*) 2 : 1 (*b*) 3 : 2 (*c*) 4 : 3 (*d*) 5 : 3

127. The circumference of a circle exceeds the diameter by 16·8 cm. What is the diameter of the circle ? (Take $\pi = \frac{22}{7}$) **[CDS (II) Exam, 2023]**
(*a*) 6.24 cm (*b*) 6.42 cm (*c*) 7.64 cm (*d*) 7.84 cm

128. A pendulum swings through an angle of 9° and its end describes an arc of length 14.3 cm. What is the length of the pendulum? (Take $\pi = \frac{22}{7}$) **[CDS (II) Exam, 2023]**
(*a*) 88 cm (*b*) 91 cm (*c*) 95 cm (*d*) 98 cm

129. Consider the following statements : **[CDS (II) Exam, 2023]**
1. The angle in a sector greater than a semi-circle is less than a right angle.
2. If two sides of a pair of opposite sides of a cyclic quadrilateral are equal, then its diagonals are also equal.

Which of the statements given above is/are correct?
(*a*) 1 only (*b*) 2 only (*c*) Both 1 and 2 (*d*) Neither 1 nor 2

130. The area of a circle inscribed in an equilateral triangle is 462 cm^2. The perimeter of the triangle is:
(*a*) $42\sqrt{3}$ cms. (*b*) 126 cms. (*c*) 72.6 cms. (*d*) 168 cms.

131. A bicycle wheel makes 2500 revolutions in moving 5.5 km. The diameter of the wheel is:
(*a*) $40\frac{1}{7}$ cms. (*b*) 35 cms. (*c*) 64.6 cms. (*d*) 70 cms.

132. A road which is 7 metres wide runs round a circular park whose circumference is 176 metres. The area of the road is:
(*a*) 1386 m^2 (*b*) 1512 m^2 (*c*) 1472 m^2 (*d*) 1760 m^2

133. In the adjoining figure, the area of the shaded portion is
(*a*) 65.3 m^2 (*b*) 42.4 m^2
(*c*) 30.6 m^2 (*d*) 39 m^2

6 m
8 m

134. The adjoining figure shows the inner boundary of 400 m running track which has two parallel straights, each of length 100 m and two semicircular ends.
The area enclosed by the inner boundary is approximately:
(*a*) 945 m^2 (*b*) 9545 m^2
(*c*) 954.5 m^2 (*d*) 96 m^2

100 m
100 m

135. The boundary of the shaded region in the given diagram consists of three semi-circular arcs, the smaller ones being equal. If the diameter of the larger arc is 10 cm, the area of the shaded region is:

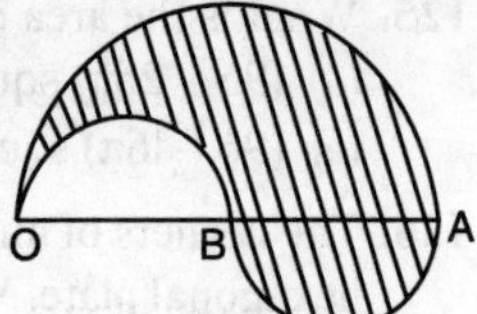

(*a*) 39.3 cm^2 (*b*) 46.4 cm^2
(*c*) 35.6 cm^2 (*d*) 37.9 cm^2

136. A bed of roses is like the adjoining diagram. In the centre is a square and on each side there is a semi-circle. The side of the square is 21 metres. If each rose plant needs 6 m^2 of space, the number of plants is

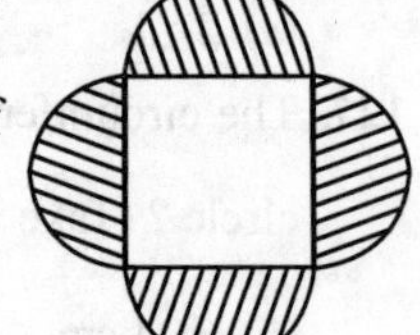

(*a*) 176 (*b*) 163
(*c*) 168 (*d*) 189

137. A child draws the figure of an aeroplane as given. Here the wings *EDCF* and *AGHB* are parallelograms, the tail *ADK* is an isosceles triangle, the cockpit *BLC* is a semi-circle and the portion *ABCD* is a square.

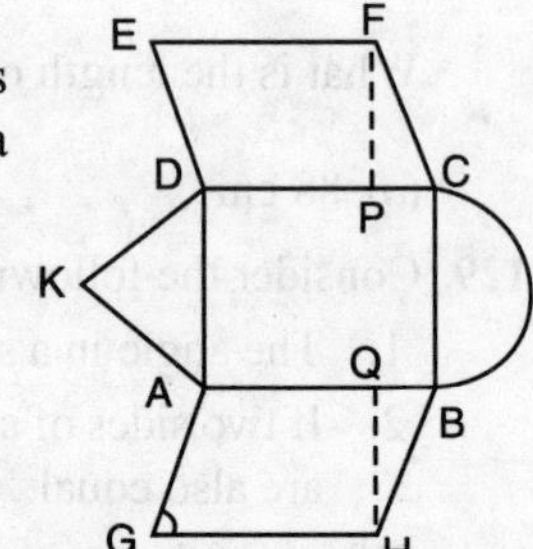

Let $FP \perp CD$ & $HQ \perp AB$,

$AB = 6$ cms, $KD = 5$ cms

$FP = HQ = 2$ cms.

The area of the figure

(*a*) 98.14 cm^2 (*b*) 87.25 cm^2
(*c*) 84.63 cm^2 (*d*) 91.56 cm^2

138. The inside perimeter of a running track shown in the adjoining figure with semi-circular ends and straight parallel sides is 312 metres. The length of straight portions of the track is 90 meter. If the track has a uniform width of 2 metres throughout, the area of the track is:

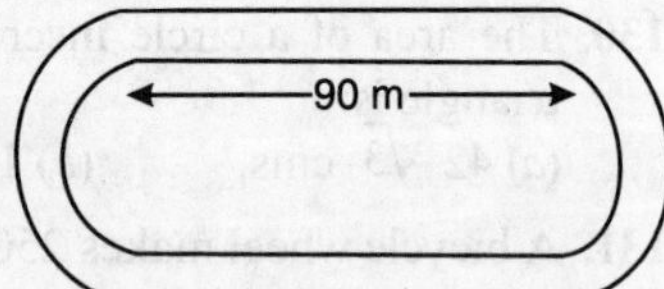

(*a*) $(72\pi + 420)$ m^2
(*b*) $(88\pi + 360)$ m^2
(*c*) $(36\pi + 640)$m^2
(*d*) $(66\pi + 240)$ m^2

139. Four horses are tethered at four corners of a square plot of side 63 metres, so that they just cannot reach one another. The area left ungrazed is:

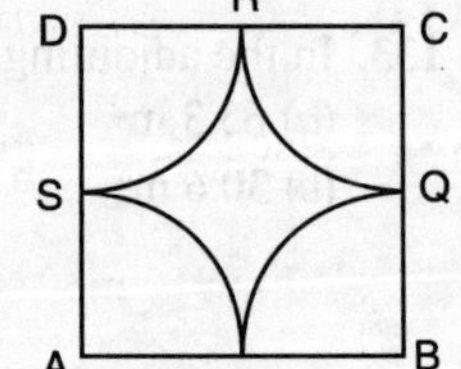

(*a*) 675.5 m^2
(*b*) 780.6 m^2
(*c*) 785.8 m^2
(*d*) 850.5 m^2

140. A rectangular piece is 20 m long and 16 m wide. From its four corners, quadrants of radii 3.5 metres have been cut. The area of the remaining part is

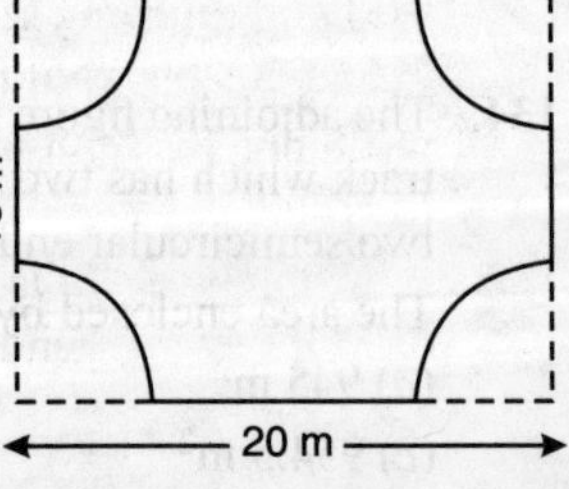

(*a*) 281.5 m^2 (*b*) 276.4 m^2
(*c*) 265.6 m^2 (*d*) 264.8 m^2

141. In the given figure, the diameter of the biggest semi-circle is 56 cms, and the radius of the smallest circle is 7 cms. The area of the shaded portion is:

(*a*) 482 cm^2 (*b*) 462 cm^2
(*c*) 654 cm^2 (*d*) 804 cm^2

142. *PQRS* is flower bed with $OP = 28$ m and $OR = 21$ m. The area of the bed is:

(*a*) 269.5 m^2 (*b*) 272.4 m^2
(*c*) 257.6 m^2 (*d*) 266.2 m^2

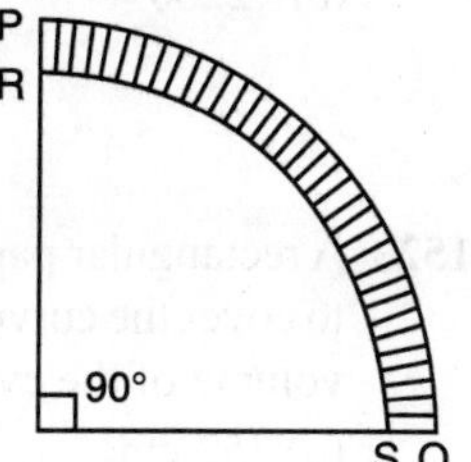

143. The perimeter of a square circumscribed about a circle of radius *r* is

(*a*) $2r$ (*b*) $4r$ (*c*) $8r$ (*d*) $2\pi r$

144. A wire in the form of a square encloses the area *A* cm^2. If the same wire is bent into the form of a circle, the area it will enclose is

(*a*) $\frac{\pi A^2}{4}$ (*b*) $\frac{4A}{\pi}$ (*c*) $4\pi A^2$ (*d*) $\frac{A}{4\pi}$

145. A circle is inscribed in an equilateral triangle and the square is inscribed in the circle. The ratio of the area of the triangle to the area of the square is

(*a*) $3 : \sqrt{2}$ (*b*) $\sqrt{3} : 1$
(*c*) $\sqrt{3} : \sqrt{2}$ (*d*) $3\sqrt{3} : 2$

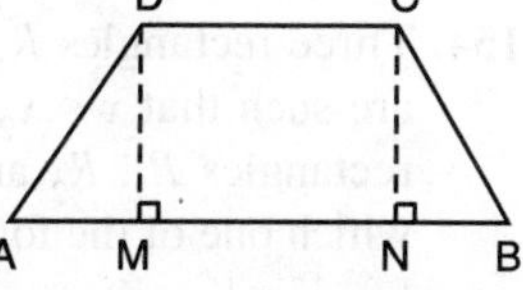

146. A cylindrical iron pillar 49 cms. high and 6 cms. in radius is surmounted by a cone 14 cms. high. The volume of pillar is

(*a*) 5981 cm^2 (*b*) 6072 cm^2 (*c*) 7012 cm^2 (d) 8154 cm^2

147. A copper wire of diameter 3 mm. is evenly wrapped on the cylinder of length 42 cm and diameter 49 cms to cover the whole surface. The length of wire is

(*a*) 215.6 m (*b*) 192.4 m
(*c*) 204.8 m (*d*) 196.5 m

148. The cross-section of a swimming pool is a trapezium *ABCD* as shown in the diagram. $AB = 25$ m, $AC = 75$ cm and $BD = 8$ m. If the length of the pool is 80 metres, the volume of water it holds, is

(*a*) 9050 m^3 (*b*) 8925 m^3
(*c*) 8750 m^3 (*d*) 9175 m^3

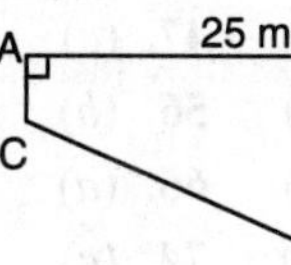

149. The cross-section of a tunnel perpendicular to its length is a trapezium *ABCD* as shown in the figure. It being given that $AM = BN$; $AB = 4.4$ m; $CD = 3$ m; the height of the tunnel is 2.4 m and the length of the tunnel is 50 m. The cost of painting the internal surface of the tunnel excluding the floor at the rate of Re 1 per m^2 is

(*a*) ₹400 (*b*) ₹432 (*c*) ₹560 (*d*) ₹380

150. A swimming bath is 300 m long and 30 m wide. It is m deep at the shallows end and 4 m deep at the other end. When it is full, the volume of water is:

(*a*) 22500 m^3 (*b*) 21650 m^3 (c) 20050 m^3 (*d*) 22725 m^3

151. The cross-section of a railway tunnel is a rectangle 6 m broad and 8 m high, surmounted by a semi-circle, as shown in the figure. The tunnel is 35 m long. The cost of plastering the internal surface of the tunnel (excluding the floor), at the rate of ₹2 per m² is

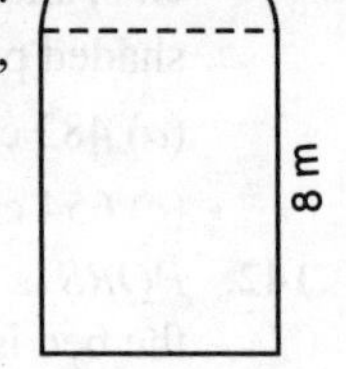

(*a*) ₹1960.75 (*b*) ₹1780
(*c*) ₹2200 (*d*) ₹1676

152. A rectangular paper 22 cm by 8 cm can be exactly wrapped to cover the curved surface of a cylinder of height 8 cm. The volume of the cylinder is:

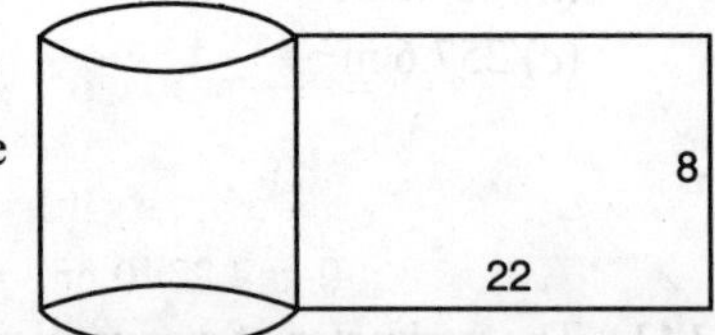

(*a*) 352 cm³ (*b*) 256 cm³
(*c*) 312 cm³ (*d*) 308 cm³

153. The figure given alongside shows the cross-section of a water channel consisting of a rectangle and a semi-circle. Assuming that the channel is always full, find the volume of water discharged through it in one minute, if water is flowing at the rate of 20 cm per second.

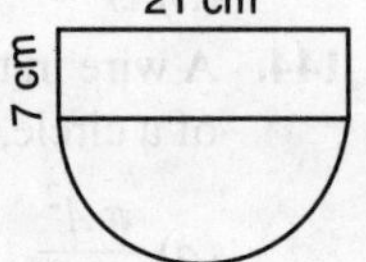

(*a*) 0.3843 m³ (*b*) 0.2963 m³
(*c*) 0.3659 m³ (*d*) 0.3685 m³

154. Three rectangles R_1 R_2 and R_3 have the same area. Their lengths x_1, x_2 and x_3 respectively are such that $x < x_2 < x_3$. If V_1 V_2 and V_3 are the volumes of the cylinders formed from the rectangles R_1, R_2 and R_3 respectively by joining the parallel sides along the breadth, then which one of the following is correct? **[CDS (II) Exam, 2015]**

(*a*) $V_3 < V_2 < V_1$ (*b*) $V_1 < V_3 < V_2$ (*c*) $V_1 < V_2 < V_3$ (*d*) $V_3 < V_1 < V_2$

ANSWERS

1. (*b*)	**2.** (*d*)	**3.** (*d*)	**4.** (*b*)	**5.** (*a*)	**6.** (*d*)	**7.** (*c*)	**8.** (*d*)	**9.** (*a*)
10. (*a*)	**11.** (*c*)	**12.** (*a*)	**13.** (*c*)	**14.** (*a*)	**15.** (*d*)	**16.** (*c*)	**17.** (*c*)	**18.** (*a*)
19. (*b*)	**20.** (*d*)	**21.** (*d*)	**22.** (*c*)	**23.** (*b*)	**24.** (*b*)	**25.** (*d*)	**26.** (*d*)	**27.** (*a*)
28. (*d*)	**29.** (*a*)	**30.** (*c*)	**31.** (*d*)	**32.** (*a*)	**33.** (*b*)	**34.** (*c*)	**35.** (*a*)	**36.** (*b*)
37. (*b*)	**38.** (*a*)	**39.** (*c*)	**40.** (*a*)	**41.** (*c*)	**42.** (*b*)	**43.** (*a*)	**44.** (*b*)	**45.** (*a*)
46. (*b*)	**47.** (*c*)	**48.** (*c*)	**49.** (*b*)	**50.** (*c*)	**51.** (*b*)	**52.** (*b*)	**53.** (*a*)	**54.** (*c*)
55. (*c*)	**56.** (*b*)	**57.** (*a*)	**58.** (*d*)	**59.** (*b*)	**60.** (*a*)	**61.** (*b*)	**62.** (*c*)	**63.** (*b*)
64. (*c*)	**65.** (*a*)	**66.** (*a*)	**67.** (*b*)	**68.** (*d*)	**69.** (*d*)	**70.** (*b*)	**71.** (*d*)	**72.** (*d*)
73. (*d*)	**74.** (*c*)	**75.** (*c*)	**76.** (*a*)	**77.** (*d*)	**78.** (*c*)	**79.** (*c*)	**80.** (*d*)	**81.** (*c*)
82. (*d*)	**83.** (*a*)	**84.** (*b*)	**85.** (*a*)	**86.** (*c*)	**87.** (*d*)	**88.** (*c*)	**89.** (*b*)	**90.** (*c*)
91. (*a*)	**92.** (*d*)	**93.** (*a*)	**94.** (*b*)	**95.** (*d*)	**96.** (*b*)	**97.** (*d*)	**98.** (*c*)	**99.** (*c*)
100. (*b*)	**101.** (*c*)	**102.** (*a*)	**103.** (*c*)	**104.** (*a*)	**105.** (*d*)	**106.** (*a*)	**107.** (*c*)	**108.** (*b*)
109. (*d*)	**110.** (*b*)	**111.** (*d*)	**112.** (*a*)	**113.** (*b*)	**114.** (*d*)	**115.** (*a*)	**116.** (*c*)	**117.** (*c*)
118. (*b*)	**119.** (*a*)	**120.** (*c*)	**121.** (*b*)	**122.** (*b*)	**123.** (*a*)	**124.** (*d*)	**125.** (*b*)	**126.** (*b*)
127. (*d*)	**128.** (*b*)	**129.** (*b*)	**130.** (*b*)	**131.** (*d*)	**132.** (*a*)	**133.** (*c*)	**134.** (*b*)	**135.** (*a*)
136. (*d*)	**137.** (*a*)	**138.** (*b*)	**139.** (*d*)	**140.** (*a*)	**141.** (*b*)	**142.** (*a*)	**143.** (*c*)	**144.** (*b*)
145. (*d*)	**146.** (*b*)	**147.** (*a*)	**148.** (*c*)	**149.** (*a*)	**150.** (*a*)	**151.** (*b*)	**152.** (*b*)	**153.** (*a*)
154. (*c*)								

SOLUTIONS

1. We know that in radius of right angled Δ $r = \dfrac{p+b-h}{2}$

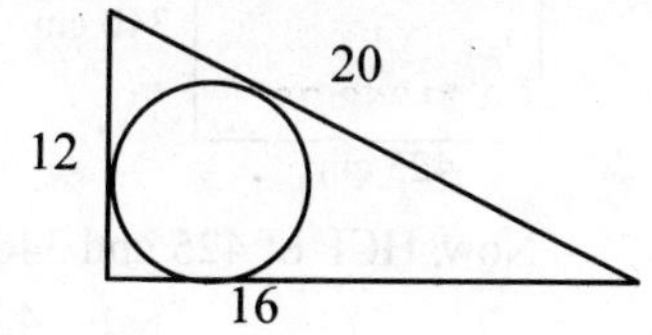

$\therefore r = \dfrac{12+16-20}{2} = 4$ cm

Area of circle = πr^2

$= \dfrac{22}{7} \times 4 \times = \dfrac{352}{7} \approx 50 \text{ cm}^2$

2. Length = 3060 cm, breadth = 2340 cm

HCF of 3060 and 2340 cm = 180

$\therefore$ required number of tiles $= \dfrac{3060 \times 2340}{180 \times 180} = 221$

3. $\Rightarrow 400 = 2(l+b)$

$\Rightarrow \qquad l + b = 200$

To maximize the area, set $l = b$

$\Rightarrow \qquad 2l = 200 \;\Rightarrow l = 100$

The area is:

$A = l \times b = 100 \times 100 = 10{,}000 \text{ cm}^2$

$\Rightarrow$ The maximum area of the rectangle is 10,000 cm^2.

4. **Case 1:** Pythagoras' theorem, we get

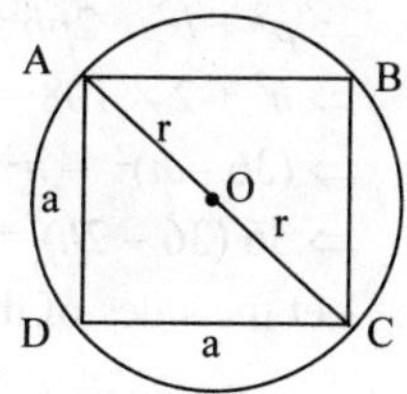

$a^2 + a^2 = 2r^2$

$\Rightarrow 2a^2 = 4r^2$

$\Rightarrow a = \sqrt{2}r$

ar (square) $= a^2 = 2r^2 = x \;\Rightarrow r = \sqrt{\dfrac{x}{2}}$

Case 2:

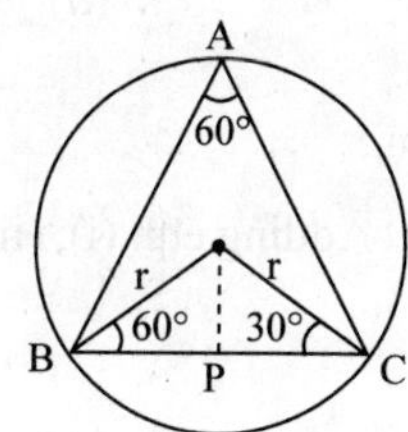

$\angle BOC = 2\angle BAC = 120°$

$\therefore \angle BOP = \angle POC = 60°$

$BP = r \sin 60° = \dfrac{r\sqrt{3}}{2}$

$=$ side of $\Delta = 2 \times \dfrac{r\sqrt{3}}{2} = \sqrt{3}\, r$

$\Rightarrow y = \dfrac{\sqrt{3}}{4}(\sqrt{3}r)^2 = \dfrac{3\sqrt{3}}{4}r^2 \qquad \left[\therefore ar \text{ (equilateral) } \Delta = \dfrac{\sqrt{3}}{4} \times \text{side}^2\right]$

$\Rightarrow y = \dfrac{3\sqrt{3}}{4} \times \left(\dfrac{\sqrt{x}}{2}\right)^2 \qquad \left[\therefore r = \sqrt{\dfrac{x}{2}}\right]$

$\Rightarrow y = \dfrac{3\sqrt{3}}{8}x \;\Rightarrow 8y = 3\sqrt{3}x \;\Rightarrow 64y^2 = 27x^2$

5. 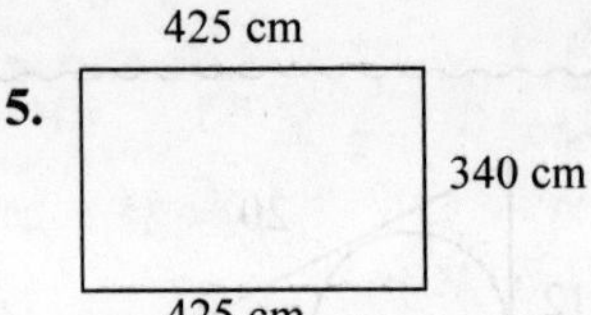

Now, HCF of 425 and 340 = 85

$\therefore$ Number of tiles = $\dfrac{425 \times 340}{85 \times 85} = 20$ tiles

6. 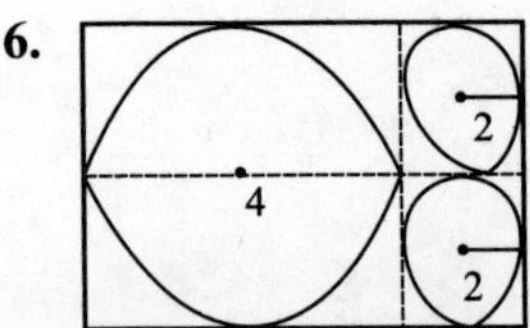

Total area = $\pi\,(4^2 + 2^2 + 2^2)$ [$\because$ area of circle = πr^2]

$= \pi\,(16 + 4 + 4) = 24\,\pi$ cm^2

7. We have,

$$\frac{1}{2} \times p \times b = 54 \;\Rightarrow pb = 108$$

and $$p + b + h = 36 - h \;\Rightarrow p + b = 36 - h$$

Squaring both side, we get

$(p + b)^2 = (36 - h)^2$

$\Rightarrow p^2 + b^2 + 2pb = (36 - h)^2$

$\Rightarrow h^2 + 2 \times 108 = (36 - h)^2$ [$\because p^2 + b^2 = h^2$]

$\Rightarrow (36 - h)^2 - h^2 = 216 \quad \Rightarrow (36 - h + h)\ (26 - h - h) = 216$

$\Rightarrow 36\,(36 - 2h) = 216 \quad \Rightarrow 36 - 2h = 6 \quad \Rightarrow h = 15$ cm

8. Let the sides of the triangle be:

$$AB = a,\ BC = b,\ CA = c$$

$$a + b = 7.1 \qquad \text{...(i)}$$

$$b + c = 12.1 \qquad \text{...(ii)}$$

$$c + a = 7.2 \qquad \text{...(iii)}$$

Adding eqn (i), (ii) and (iii)

$$(a+b)+(b+c)+(c+a) = 7.1 + 12.1 + 7.2$$

$$2a + 2b + 2c = 26.4$$

$$a + b + c = 13.2$$

Now calculate each side:

$$a = 13.2 - 12.1 = 1.1$$

$$b = 13.2 - 7.2 = 6.0$$

$$c = 13.2 - 7.1 = 6.1$$

Heron's formula for the area of a triangle is given by:

$$\text{Area} = \sqrt{s(s-a)(s-b)(s-c)}$$

$$s = \frac{a+b+c}{2} = \frac{1.1+6.0+6.1}{2} = \frac{13.2}{2} = 6.6 \text{ cm}$$

$$\text{Area} = \sqrt{6.6(6.6-1.1)(6.6-6.0)(6.6-6.1)}$$

$$= \sqrt{6.6 \times 5.5 \times 0.6 \times 0.5}$$

$$= \sqrt{6.6 \times 5.5 \times 0.3}$$

$$= \sqrt{10.89}$$

$$= 3.3 \text{ cm}^2$$

9. Length of the arc of sector $= \frac{\pi r Q}{180°}$

$\Rightarrow \quad 55 = \frac{22}{7} \times 21 \frac{Q}{180°} \Rightarrow Q = 150°$

Now,

$\therefore \quad$ Area of sector $= \frac{\pi r^2 Q}{360°} = \frac{22}{7} \times 21 \times 21 \times \frac{150°}{360°} = \frac{33 \times 7 \times 5}{2} = 33 \times 3.5 \times 5 = 577.5 \text{ cm}^2$

10. In Δ APQ:

Side AP $= \left(\frac{1}{2}\right) \times AD$

$\Rightarrow AP = AQ = 7$ cm

Area of Δ APQ $= \left(\frac{1}{2}\right) \times 7 \times 7 = \frac{49}{2}$

The area of the shaded region:

Area of (Δ APQ + ΔDPS + ΔRSC + ΔQRB) $= \left(\frac{49}{2}\right) \times 4 = 98 \text{ cm}^2$

The area lying between the two squares $= (14 \times 14) - 98 = 196 - 98 = 98 \text{ cm}^2$

11. We have,

Perimeter = 30 cm

$\Rightarrow \quad AB + BC + AC = 30$ cm

$\Rightarrow \quad AB + BC + 13 = 30$ cm

$\Rightarrow \quad AB + BC = 17$ cm ...(*i*)

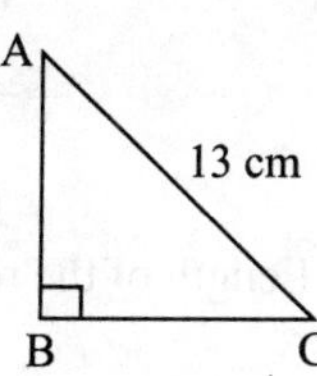

By Pythagoras Theorem,

$AC^2 = AB^2 + BC^2$

$\Rightarrow \quad 169 = AB^2 + BC^2$ ...(*ii*)

$(AB + BC)^2 = AB^2 + BC^2 + 2AB.BC$

$\Rightarrow \quad (17)^2 = 169 + 2\, AB.BC \quad \Rightarrow 289 - 169 = 2\, AB.BC$

$\Rightarrow \quad 2\, AB.BC = 120 \Rightarrow AB \times BC = 60$

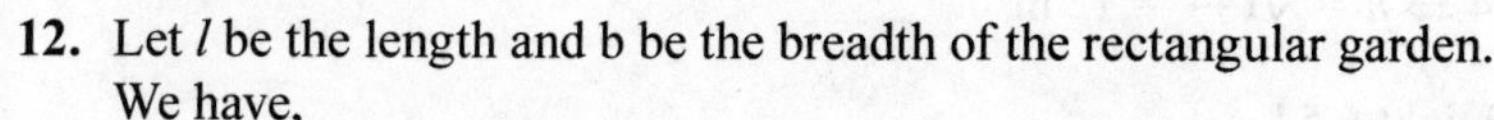

$\therefore \quad Ar\,(\Delta ABC) = \frac{1}{2} \times AB \times BC = \frac{1}{2} \times 60 = 30 \text{ cm}^2$

12. Let l be the length and b be the breadth of the rectangular garden.

We have,

$l \times b = 375$ ... (*i*)

Now,

$2b + l = 65$ ...(ii)

or $2l + b = 65$ ...(iii)

Similarly,

Solving equation (i) and (iii), we get

$l = 7.5, 25$

And $b = 50, 15$

Put $l = 25$m and $b = 15$ m $[\because l > b]$

So, perimeter of rectangle = 2(25 + 15) = 2 × 40 = 80 m

13. Let other diagonal be x cm

Now,

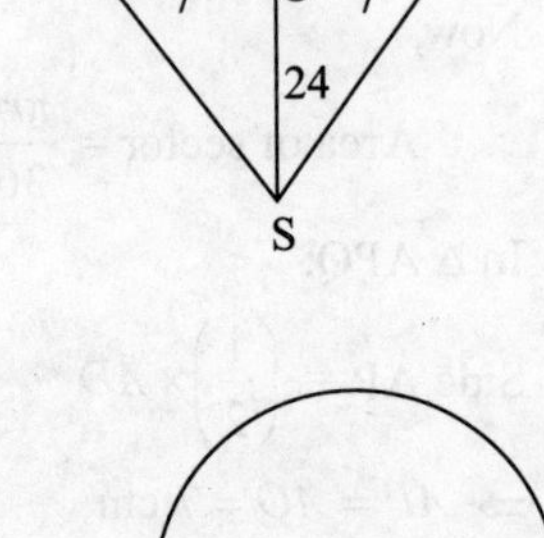

Area of rectangle = $\frac{1}{2} \times 48 \times x$

$\Rightarrow \quad 336 = \frac{1}{2} \times 48 \times x$

Now, in ΔPOQ

$PQ^2 = OP^2 + OQ^2 = 24^2 + 7^2$

$= 576 + 49 = 625$

$\Rightarrow \quad PQ = \sqrt{625} = 25$ cm

$\therefore \quad$ Perimeter of rhombus = 4 × side = 4 × 25 = 100 cm

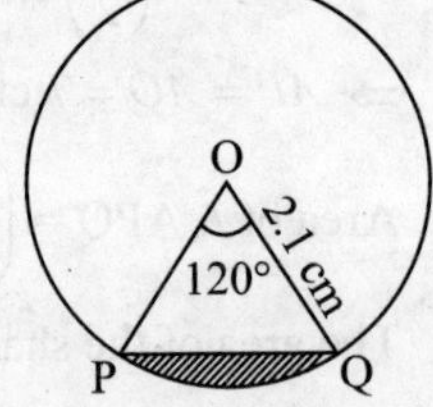

14. Area of $\Delta POQ = \frac{1}{2} r^2 \sin\theta = \frac{1}{2} \times 2.1 \times 2.1 \times \sin 120$

$= \frac{1}{2} \times 2.1 \times 2.1 \times \frac{\sqrt{3}}{2} = \frac{441 \times 1.732}{4} = 1.9 \text{ cm}^2$

Area of minor segment = Area of sector – Area of triangle corresponding to the sector

$= \pi r^2 \frac{\theta}{360} - 1.9 = \frac{22}{7} \times 2.1 \times 2.1 \times \frac{120°}{360°} - 1.9$

$= 4.62 - 1.9 = 2.71 \text{ cm}^2$

15. Area of major segment = Area of circle – Area of minor segment

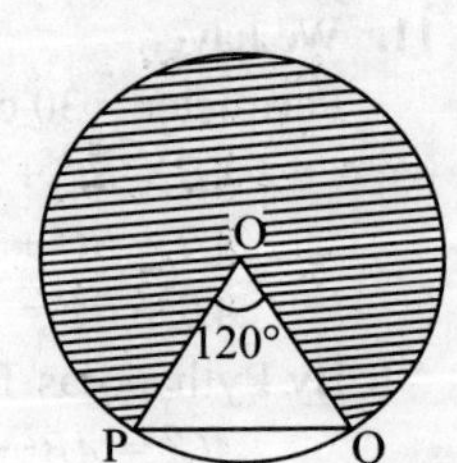

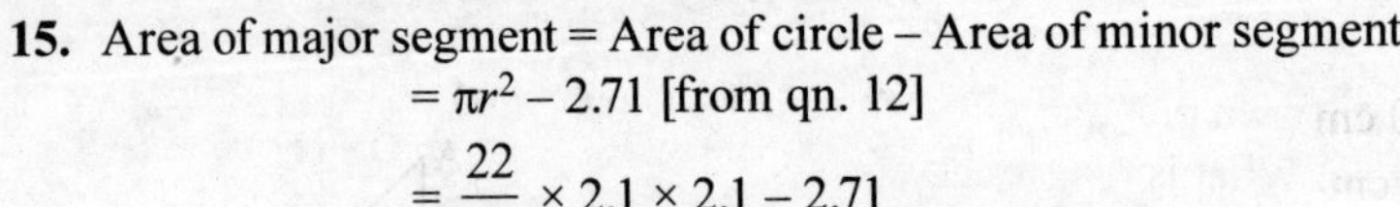

$= \pi r^2 - 2.71$ [from qn. 12]

$= \frac{22}{7} \times 2.1 \times 2.1 - 2.71$

$= 13.86 - 2.71 = 11.15 \text{ cm}^2$

16. $\therefore$ Length of the rod = diagonal of room

Here,

$l = 21$m, $b = 16$ m,

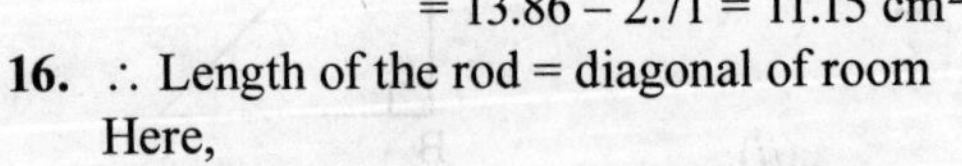

Length of the rod = $\sqrt{l^2 + b^2 + h^2}$

$\Rightarrow 29 = \sqrt{21^2 + 16^2 + h^2} \Rightarrow 29 = \sqrt{441 + 256 + h^2}$

$\Rightarrow 841 = 441 + 253 + h^2 \Rightarrow 841 = 697 + h^2$

$\Rightarrow h^2 = 841 - 697 = 144 \Rightarrow h = \sqrt{144} = 12$ m

17. $\left(\frac{1}{2}\right) \times \left(\frac{4}{3}\right) \times \text{Height} \times \text{Height} = 54$

$$\Rightarrow \quad \left(\frac{2}{3}\right)\times \text{Height}^2 = 54$$

$$\Rightarrow \quad \text{Height}^2 = 54\times\left(\frac{3}{2}\right) = 81$$

$$\Rightarrow \quad \text{Height} = \sqrt{81} = 9 \text{ cm}$$

$$\Rightarrow \quad \text{Base} = \left(\frac{4}{3}\right)\times 9 = 12 \text{ cm}$$

$$\Rightarrow \quad \text{Hypotenuse}^2 = 9^2 + 12^2 = 81 + 144 = 225$$

$$\Rightarrow \quad \text{Hypotenuse} = \sqrt{225} = 15 \text{ cm}$$

$$\Rightarrow \quad \text{Perimeter} = 9 + 12 + 15 = 36 \text{ cm}$$

18. $s = \dfrac{4+4+6}{2} = \dfrac{14}{2} = 7$ units

Using Heron's formula to find the area:

$$\text{Area} = \sqrt{7(7-4)(7-4)(7-6)}$$

$$= \sqrt{7\times3\times3\times1} = \sqrt{63} = 3\sqrt{7} \text{ square units}$$

19. Area remaining

$$= \pi r^2 - \frac{\pi r^2(80)}{360} = \pi r^2\left(\frac{7}{9}\right) \approx \frac{227}{7}(9)\left(\frac{7}{9}\right) \text{ cm}^2 = 22 \text{ cm}^2.$$

20. The circle circumscribing the triangle will have the same centre as that of the bigger circle because all the vertices are equidistant from the bigger circle.

Circumradius of the triangle $= \dfrac{2r\cdot 2r\cdot 2r}{4\dfrac{\sqrt{3}}{4}\cdot r^2} = \dfrac{2r}{\sqrt{3}}$.

$\therefore$ Radius of the bigger circle $= r + \dfrac{2r}{\sqrt{3}}$

$$= \frac{r(\sqrt{3}+2)}{\sqrt{3}} = \frac{d}{2\sqrt{3}}(\sqrt{3}+2)$$

Area of the bigger circle $= \dfrac{r}{12}d^2(\sqrt{3}+2)^2$.

21. The diagonal of the square is $12\sqrt{2}$cm.

$\therefore$ The side of the square is 12 cm. Each side of the equilateral triangle is $\dfrac{12(4)}{3}$, viz, 16 cm.

The area of the triangle is $\dfrac{\sqrt{3}}{4}(16^2)$ cm^2, viz, $64\sqrt{3}$ cm^2.

22. (1) If the height (h) of a cylinder doubles, the curved surface area also doubles. ($\because$ A = 2πrh)

(2) If the radius of a solid hemisphere doubles, the surface area becomes four-fold. ($\because$ A = $2\pi r^2 + \pi r^2$ viz. $3\pi r^2$); Both (1) (2) are true.

23. From statement I:

$$OA = a^2 - \frac{Q^2}{4} = \frac{\sqrt{3}a}{2}$$

Similarly,

$$OC = \frac{\sqrt{3}a}{2}$$

$\therefore \quad AC = 2 \times \frac{\sqrt{3}a}{2} = \sqrt{3}a$

$\therefore \quad$ ratio $= a : \sqrt{3}a = 1 : \sqrt{3}$

From statement II:

$\therefore \quad$ ratio $= a : \sqrt{3}a = 1 : \sqrt{3}$

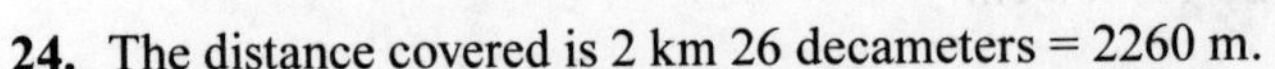

24. The distance covered is 2 km 26 decameters = 2260 m.
Let the diameter of the wheel be d m. In 452 revolutions, it covers the given distance.

$\therefore \quad 452\ (\pi d) = 2260$ m

$\Rightarrow \quad d = \frac{2260}{452} \times \frac{7}{22}$ m $= \frac{35}{22}$ m $= 1\frac{13}{22}$ m.

25. The length of the sheet (l) is 18 cm and the breadth (b) is 15 cm. The area of the semicircle whose diameter is the breadth is $\frac{\pi}{2}\left(\frac{b}{2}\right)^2 = \frac{\pi b^2}{8}$

The area of the remaining sheet is lb $- \frac{\pi b^2}{8}$

$= [18(14) - \frac{22}{7}\frac{(14)(14)}{8}]\text{cm}^2 = (252 - 77)\ \text{cm}^2 = 175\ \text{cm}^2$.

26. Let the side of the square be a. The area is a^2.
The area of the resulting rectangle is $(1.3a)(1.2a) = 1.56a^2$.
$\therefore$ The area increases from a^2 by $0.56a^2$, i.e. it increases by 56%.

27. Let the parallel sides of the trapezium be $5x$ and $7x$. The distance between the parallel sides (h) is 14 cm. The area is $\left(\frac{5x + 7x}{2}\right)h = 6xh = 336\ \text{cm}^2$.

$\therefore \quad x = \frac{336}{6(14)}$ cm = 4 cm.

The length of the smaller of the parallel sides is $5x = 20$ cm.

28. Distance covered in one complete round by the wheel = πd diameter) = $\frac{22}{7} \times \frac{7}{11} = 2$ m.

The number of rounds that wheel makes by covering 4 km.

$= 4000$ m is $\frac{4000}{2} = 2000$

29. Other side = $\sqrt{5^2 - 4^2} = \sqrt{9} = 3$ m.

$\therefore$ Area = $(4 \times 3)\ \text{m}^2 = 12\ \text{m}^2$.

30. Let length = $5x$ and breadth = $4x$.

Then, $\quad 5x - 4x = 20 \Rightarrow x = 20$.

$\therefore$ Length = 100 m & Breadth = 80 m.

$\therefore$ Perimeter = 2 (100 + 80) m = 360 m.

31. Let breadth = x & length = $\frac{4}{3}x$.

Then, $x \times \frac{4}{3}x = 300 \Rightarrow x^2 = \frac{900}{4} \Rightarrow x = \frac{30}{2} = 15$.

Diff. between length & breadth = $\left(\frac{4}{3}x - x\right) = \frac{x}{3} = \frac{15}{3} = 5$ m.

32. Let breadth = x. Then length = 1.5 x in metres.

Then $x \times \frac{3}{2}x = \frac{2}{3} \times 10000$. So, $x = \frac{2}{3} \times 100 = \frac{200}{3}$.

$\therefore$ Length of plot = $\left(\frac{3}{2} \times \frac{100}{3}\right)$ m = 100 m.

33. Length of longest pole = length of diagonal = $\sqrt{10^2 + 8^2} = \sqrt{164} = 12.8$ m.

34. Let length = x metres and breadth = y metres.

Then, $x + y = 23$ and $xy = 120$

$\therefore x^2 + y^2 = (x + y)^2 - 2xy = (23)^2 - 240 = 289$.

$\therefore$ Length of diagonal = $\sqrt{289}$ = 17 m.

35. Area = $\frac{\text{Total cost}}{\text{Rate}} = \left(\frac{1440}{160}\right)$ hectares = (9×10000) m^2.

$\therefore$ Length of each side = $\sqrt{9 \times 10000}$ = 300 m.

Perimeter = (300 × 4) m = 1200 m.

Cost of fencing = ₹$\left(1200 \times \frac{75}{100}\right)$ = ₹900.

36. Let breadth = x metres. Then, length = $2x$ metres.

$\therefore x^2 + (2x)^2 = (9\sqrt{5})^2 \Rightarrow 5x^2 = 405 \Rightarrow x^2 = 81 \Rightarrow x = 9$

$\therefore$ Breadth = 9 m & Length = 18 m.

Perimeter = 2 (9 + 18) m = 54 m.

37. Area of hall = (400 × 150) dm^2

Area of one stone = (6 × 5) dm^2

$\therefore$ Number of stones = $\left(\frac{400 \times 150}{6 \times 5}\right)$ = 2000.

38. Length of the carpet = $\frac{\text{Total Cost}}{\text{Rate}} = \frac{3600}{30} = 120$ m.

Let breadth of the room be x metres.

Then, $120 \times \frac{75}{100} = 15 \times x$ or $x = 6$ metres.

39. Let original length = l & breadth = b.

New length = $2l$ & new breadth = $\frac{1}{2}b$.

New area = $\left(2l\times\frac{1}{2}b\right)=lb$ = original area.

So, no change in area.

40. Let length = l and breadth = b.

New length = 160% of $l=\frac{8l}{5}$. Let, new breadth = z.

Then, $lb=\frac{8l}{5}\times z$ or $z=\frac{5}{8}b$

Decrease = $\left[\left(b-\frac{5}{8}b\right)\times\frac{1}{b}\times 100\right]\%=37\frac{1}{2}\%$.

41. Area of Roads = $(60\times 5+40\times 5-5\times 5)$ m^2

$=475$ m^2

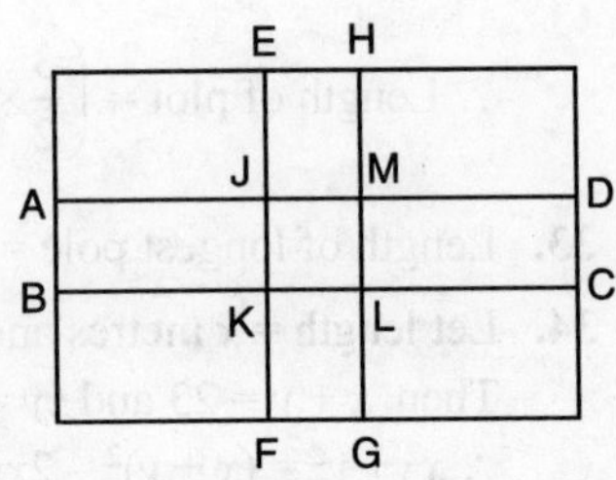

Cost of gravelling

$=₹\left(475\times\frac{60}{100}\right)=₹285$.

42. Breadth = $\frac{\text{Area}}{\text{Length}}=\left(\frac{150}{15}\right)$ cm = 10 cm.

New area = $\left(\frac{4}{3}\times 150\right)$ cm^2 = 200 cm^2, breadth = 10 cm.

New length = $\left(\frac{200}{10}\right)$ cm = 20 cm.

New perimeter = 2(20 + 10) cm = 60 cm.

43. Perimeter = Distance travelled in 15 min. = $\left(12000\times\frac{15}{60}\right)$ m = 3000 m

Let breadth = x metres. Then, length = $4x$ metres.

$\therefore$ $2(x+4x)=3000$ or $x=300$.

$\therefore$ Breadth = 300 m & length = 1200 m.

Areaa = (300×1200) m^2 = 360000 m^2.

44. Let length = x metres and breadth = y metres,

Then, $xy=60$ & $\sqrt{x^2+y^2}+x=5y$

or $xy=60$ & $x^2+y^2=(5y-x)^2=25y^2+x^2-10xy$

$\therefore$ $xy=60$ & $24y^2-10\times 60=0$ or $y^2=25$ or $y=5$.

So, $x=\left(\frac{60}{5}\right)=12$ m. Length of the carpet = 12 m.

45. Let length = l and breadth = b. Let, new breadth = x.

Then, (110% of l) $\times x=lb$ $\Rightarrow$ $\frac{110}{100}lx=lb\Rightarrow x=\frac{10}{11}b$.

$\therefore$ Decrease in breadth = $\left[\left(b-\frac{10}{11}b\right)\times\frac{1}{b}\times 100\right]\%=9\frac{1}{11}\%$.

46. Let length = l and breadth = b. Then, $\frac{l+4}{b+4}=\frac{4}{3}$ and $\frac{l-4}{b-4}=\frac{2}{1}$

$\therefore \quad 3l - 4b = 4$ and $l - 2b = -4$.

Solving these equations, we get $l = 12$ & $b = 8$.

47. From the figure, we have, ΔPQR is an equilateral triangle.

Here, $PQ = QR = RS = 8$ cm

Now,

area of shaded region = $ar(\Delta PQR)$ – area of sectors formed in three circles

$$=\frac{\sqrt{3}}{4}\times 8\times 8-3\times\frac{60}{360}\times\pi\times 4\times 4 \;=\left(16\sqrt{3}-8\pi\right)\text{cm}^2$$

48. Area of $\Delta PQR = \frac{1}{2}\times PQ \times QR$

P

$6x^2$

Q $2x^2-1$ R

$\Rightarrow 84 = \frac{1}{2}\times 6x^2 \times (2x^2-1) \Rightarrow 84 = 3x^2 \times (2x^2 - 1)$

$\Rightarrow 28 = 2x^4 \times - x^2$

Let $x^2 + t$

We have,

$x^4 - x^2 - 28 = 0$

$\Rightarrow 2t^2 - 7 - 28 \Rightarrow 2t^2 - 8t + 7t - 28 = 0$

$\Rightarrow 2t(t-4) + 7(t-4) = 0 \Rightarrow (2t+7)(t-4) = 0 \Rightarrow t = \frac{7}{2}$ (not possible) or $t = 4$

$\Rightarrow x^2 = 4 \Rightarrow x = 2$

$\therefore \quad PQ = 6x^2 = 6 \times 4 = 24$

$QR = 2x^2 - 1 = 2 \times 2^2 - 1 = 7$

$PR = \sqrt{24^2+7^2} = \sqrt{576+49} = \sqrt{625} = 25$

$\therefore \quad$ Perimeter = 7 + 24 + 25 = 56 units

49. We have,

ar(square AIHG and EFCJ) = $x = 5 \times 5 = 25$ units^2

And ar (rectangle TBFE and GHJD) = $y = 5 \times 3 = 15$ units^2

$\therefore \quad x > y$

Hence, statement (2) is correct

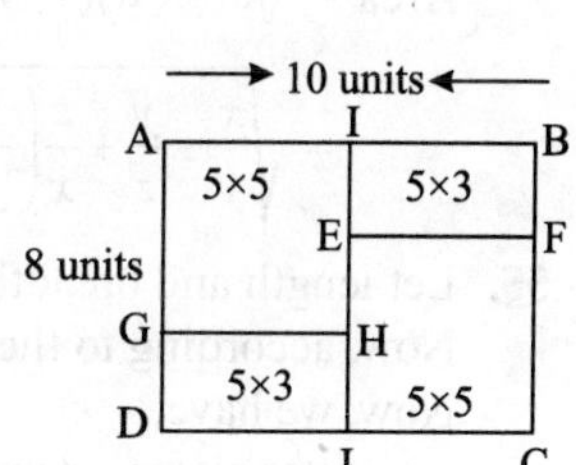

50. Let p and p be the side of n identical and large square

We have,

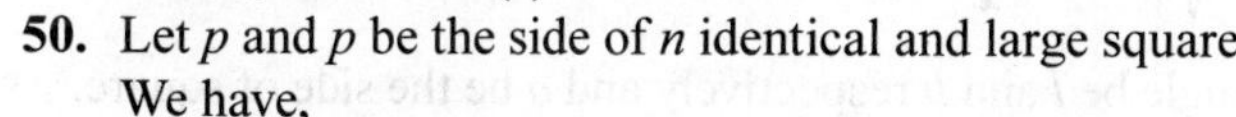

Diagonal of square = $p\sqrt{2} = m \Rightarrow p = m\sqrt{2}$

Now,

Area of large square = sum of the area of small square

$\Rightarrow p^2 = n \times p^2 \Rightarrow m^2 \times 2 = n \times p^2 \Rightarrow p = \frac{m}{\sqrt{2n}}$

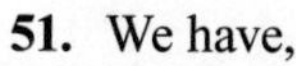

51. We have,

AB – DC = 4... (i)

We know that,

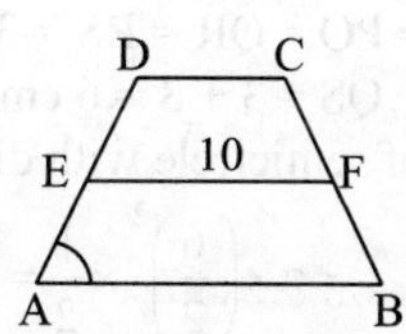

$EF = \frac{AB+CD}{2} = \frac{4+2CD}{2}$ [from (i)]

$$\Rightarrow \quad 10 = \frac{4 + 2CD}{2} \Rightarrow 20 = 4 + 2CD \Rightarrow 2CD = 16 \quad \Rightarrow CD = 8 \text{ cm}$$

And $AB = CD + 4 = 8 + 4 = 12$ cm

Now, $AB \times DC = 12 \times 8 = 96$ cm^2

52. $\therefore$ We know that

Area of parallelogram ABCD = AB × AD × $\sin\theta$

$= 15 \times 8 \times \sin\theta$

$= 120 \sin\theta$ cm^2

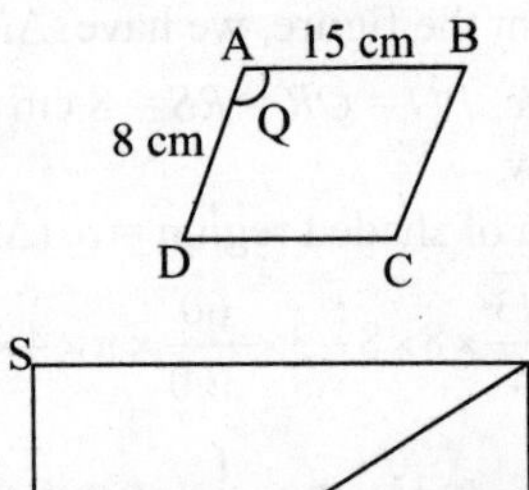

53. In ΔPQR,

$$PR = \sqrt{48^2 + 14^2} = \sqrt{2304 + 196}$$

$$= \sqrt{2500} = 50 \text{ cm}$$

$$\sec\theta = \frac{PR}{PQ} = \frac{50}{48} = \frac{25}{24}$$

$$\text{cosec}\,\theta = \frac{PR}{QR} = \frac{50}{14} = \frac{25}{7}$$

Now,

$$\sec\theta + \text{cosec}\,\theta = \frac{25}{24} + \frac{25}{7} = \frac{175 + 600}{168} = \frac{775}{168}$$

54. Here,

$$a = \frac{y}{z} + \frac{y}{x}, \; b = \frac{z}{x} + \frac{x}{y}, \; c = \frac{x}{y} + \frac{y}{z}$$

$$s = \frac{a + b + c}{2} = \frac{\frac{y}{z} + \frac{z}{x} + \frac{z}{x} + \frac{x}{y} + \frac{x}{y} + \frac{y}{z}}{2} = \frac{2\left(\frac{x}{y} + \frac{y}{z} + \frac{z}{x}\right)}{2} = \frac{x}{y} + \frac{y}{z} + \frac{z}{x}$$

Now, by Heron's formula:

$$\text{Area} = \sqrt{s(s-a)(s-b)(s-c)}$$

$$= \sqrt{\frac{x}{y} + \frac{y}{z} + \frac{z}{x}\left(\frac{x}{y} \times \frac{y}{z} \times \frac{z}{x}\right)} = \sqrt{\frac{x}{y} + \frac{y}{z} + \frac{z}{x}}$$

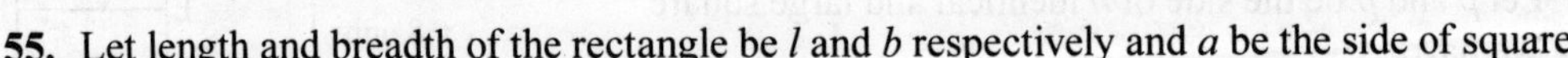

55. Let length and breadth of the rectangle be l and b respectively and a be the side of square.

Now, according to the question

Now, we have,

ar (rectangle) = ar(square)

$\Rightarrow lb = x^2 \Rightarrow x^2 \times b = x^2 \Rightarrow b = 1$

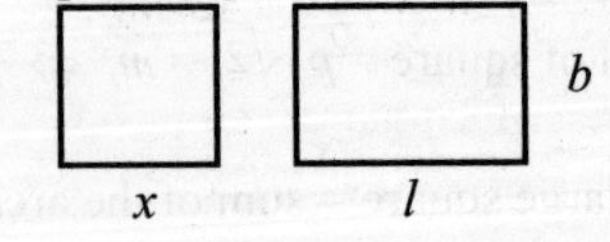

56. Area of semicircle with diameter 9 cm $= \frac{\pi r^2}{2} = \frac{\pi \times 9^2}{2 \times 2^2} = \frac{81\pi}{8}$ ... (i)

We have, PQ = QR = RS = 3 cm

QS = 3 + 3 = 6 cm

Now, area of semicircle with diameter QS = 6 cm

$$= \pi \times \left(\frac{6}{2}\right)^2 \times \frac{1}{2} = \frac{9\pi}{2} \text{ cm}^2 \qquad \text{...}(ii)$$

Now, Area of shaded region $= \frac{81\pi}{8} - \frac{9\pi}{2} - \frac{81\pi - 36\pi}{8} = \frac{45\pi}{8}$ cm^2 = ...(*iii*)

Area of unshaded region $= \frac{81\pi}{8} + \frac{9\pi}{2} = \frac{81\pi + 36\pi}{8} = \frac{117\pi}{8}$ ...(*iv*)

$\therefore$ Required ratio $= \frac{45\pi}{8} : \frac{117\pi}{8} = 5:13$

57. We have,

Length of the curve = Length of the arc of the circle

$\Rightarrow 44 = \frac{\pi r\theta}{180°} \Rightarrow 44 = \frac{22}{7} \times r \times \frac{42°}{180°} \Rightarrow r = 60$ m

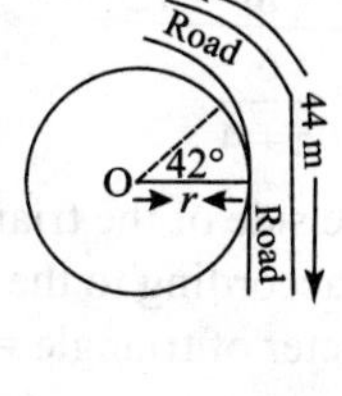

58. Perimeter $= a + b + c$

$\Rightarrow 90 = a + b + c \Rightarrow 90 = 40 + 41 + c$

$\Rightarrow 90 = 81 + c \Rightarrow c = 9$

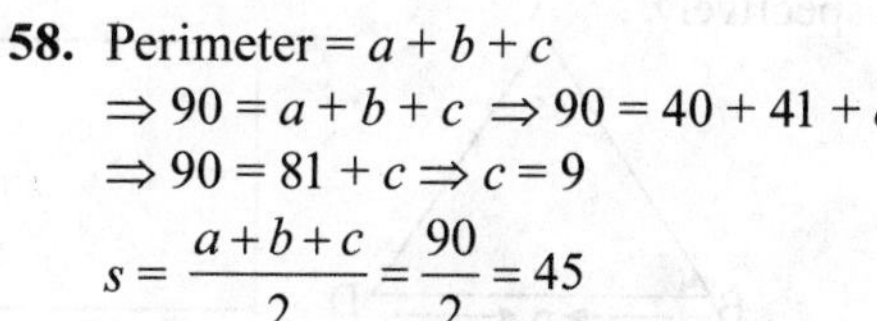

$s = \frac{a+b+c}{2} = \frac{90}{2} = 45$

Now, by Heron's formula

Area of $\Delta = \sqrt{s(s-a)(s-b)(s-c)} = \sqrt{45\,(45-41)\,(45-40)\,(45-9)}$

$= \sqrt{45 \times 4 \times 5 \times 36} = 6 \times 3 \times 5 \times 2 = 180$ cm^2

59. Let the radius of the wheel be r cm

Distance covered in one revolution $= \frac{11 \text{ km}}{5000} = \frac{1100000}{5000} = 220$

[$\because$ 1 km = 1000 m and 1m = 100 cm]

Now, $2\pi r = 220$

$\Rightarrow 2 \times \frac{22}{7} \times r = 220 \Rightarrow r = 35$ cm

By Pythagoras theorem

AC $= \sqrt{AB^2 + BC^2} = \sqrt{9^2 + 40^2} = \sqrt{81 + 1600}$

$= \sqrt{1681} = 41$ cm

60. In ΔADC

$a = 15$ cm, $b = 28$ cm, $c = 41$ cm

$s = \frac{15+28+41}{2} = \frac{84}{2} = 42$ cm

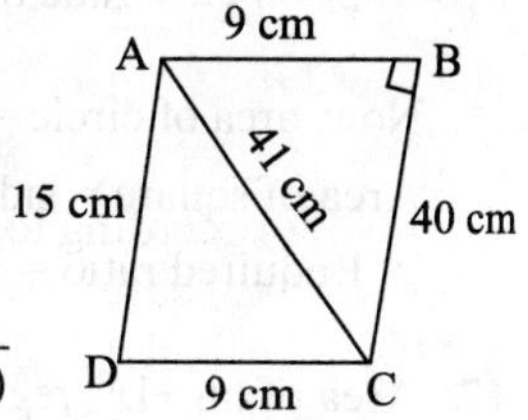

By Heron's formula

Area $= \sqrt{s(s-a)\,(s-b)\,(s-c)} = \sqrt{42 \times (42-15)\,(42-28)\,(42-41)}$

$= \sqrt{42 \times 27 \times 14 \times 1} = 3 \times 3 \times 14$

$= 126$ cm^2

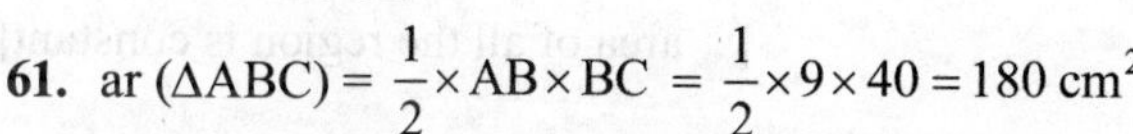

61. ar (ΔABC) $= \frac{1}{2} \times \text{AB} \times \text{BC} = \frac{1}{2} \times 9 \times 40 = 180$ cm^2

Now, ar (quadrilateral ABCD) = ar (ΔABC) + ar (ΔADC)

$= 180 + 126 = 306$ cm^2 [from qn 60]

62. Perimeter of ΔABC = 90 + 40 + 41 = 90 cm

Perimeter of $\Delta ADC = 15 + 28 + 41 = 84$ cm

$\therefore$ Required difference = 90 – 84 = 6 cm

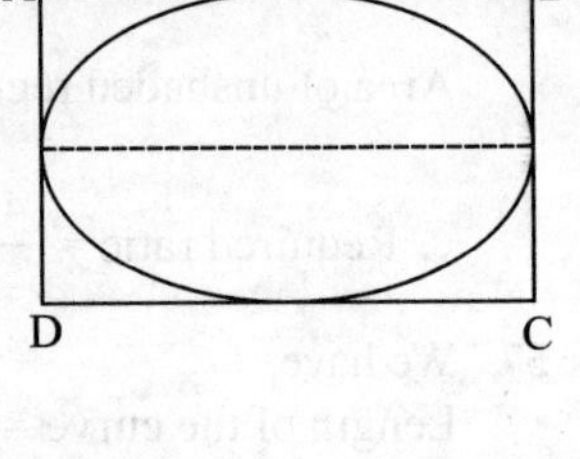

63. Let the side of the square be a and radius of circle be r

We have,

$a = 2r$

Now, according to the question,

Circumference of circle to perimeter of square = 12m

$\Rightarrow 2\pi r + 4 \times a = 12\text{m} \Rightarrow 2\pi r + 4 \times 2r = 12\text{m}$

$\Rightarrow 2\pi r + 8r = 12\text{ m} \quad \Rightarrow \pi r + 4r = 6 \Rightarrow r + (\pi + 4) = 6$

$\Rightarrow r = \dfrac{6}{\pi + 4}$

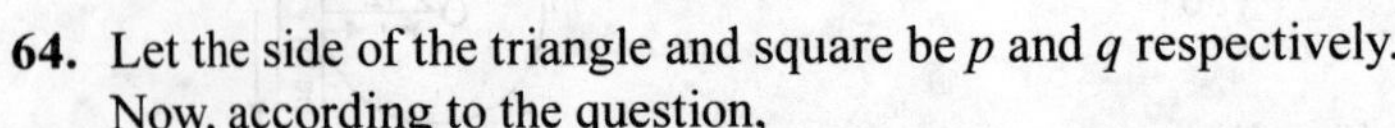

64. Let the side of the triangle and square be p and q respectively.

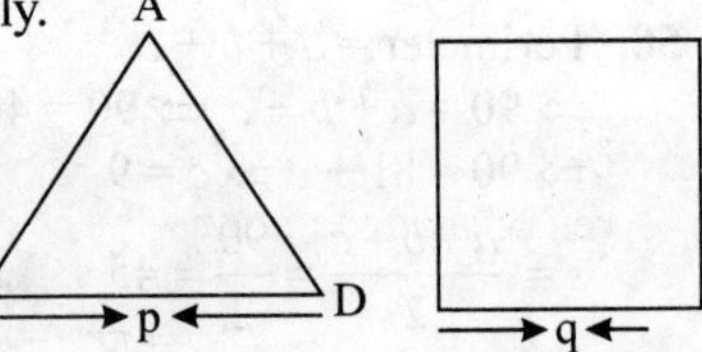

Now, according to the question,

Perimeter of triangle = Perimeter of square

$\Rightarrow 3p = 4q \Rightarrow \dfrac{p}{q} = \dfrac{4}{3}$

Let $p = 4x$ and $q = 3x$

Area of triangle = $\dfrac{\sqrt{3}}{4} \times \text{side}^2 = \dfrac{\sqrt{3}}{4} \times (4x)^2 = 4\sqrt{3}x^2$

Area of square = side$^2 = (3x)^2 = 9x^3$

$\therefore$ Required ratio = $4\sqrt{3} : 9$

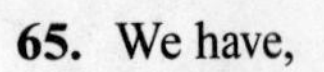

65. We have,

$a = 20, b = 10$

Percentage increase in the area of rectangle = $a + b + \dfrac{ab}{100}$

$\Rightarrow 20 + 10 + \dfrac{10 \times 20}{100} \Rightarrow 20 + 10 + 2 = 32\%$

66. From figure, we have,

AC = $2r$ = Diagonal of square

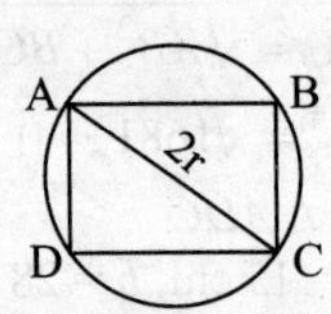

$\Rightarrow 2r = \sqrt{2} \times \text{side of square} \Rightarrow \text{Side} = \dfrac{2r}{\sqrt{2}} = \sqrt{2}r$

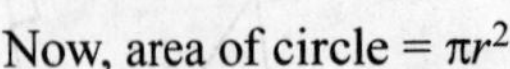

Now, area of circle = πr^2

Area of square = side × side = $\sqrt{2}r \times \sqrt{2}r = 2r^2$

$\therefore$ Required ratio = $\pi r^2 : 2r^2 = 22 : 14 = 11 : 7$

67. Area of (m +1)[th] region = $\pi\left(r_{m+1^2} - r_{m^2}\right) = \pi r_m^2 \left(\dfrac{r_{m+1}}{r_m}\right)^2 - 1 = \pi r_m^2 \times \left(p_{m^2} - 1\right)$

Now, we have

$p_{m^2} - 1 \propto \dfrac{1}{r^2}$ [$\because$ area of all the region is constant]

$\therefore$ If m increases r_m increases and p_m decreases with m increases.

68. Let p be the side of square and q be the side of equilateral triangle.

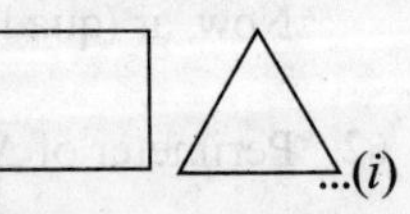

Given that,

Perimeter of square = perimeter of rectangle ...(i)

and diagonal of square = $\sqrt{2} \times p$

$\Rightarrow 6\sqrt{2} - \sqrt{2} \times p \Rightarrow p = 6$

Now, from (i)

$4 \times p = 3 \times q$

$\Rightarrow 24 = 3q \quad [\because p = 6]$

Now, area of equilateral triangle = $\dfrac{\sqrt{3}}{4} \times \text{side}^2 = \dfrac{\sqrt{3}}{4} \times q^2 \Rightarrow \dfrac{\sqrt{3}}{4} \times 8 \times 8 = 16\sqrt{3}$ cm²

69. We have,

Diameter of circle = Diagonal of square

Let r be the radius and a be the side of square

$2 \times r = \sqrt{2} \times a \Rightarrow r = \dfrac{a}{\sqrt{2}}$

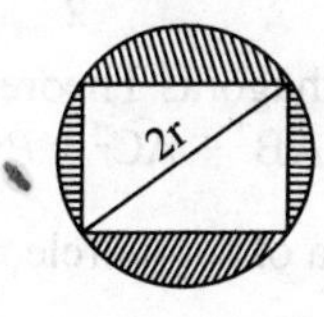

Now,

Area of shaded region = Area of circle – Area of square

$$= \pi r^2 - a^2 = \pi \times \left(\frac{a}{\sqrt{2}}\right)^2 - a^2$$

$$= \frac{\pi a^2}{2} - a^2 = \frac{(\pi - 2)a^2}{2} \text{ square units.}$$

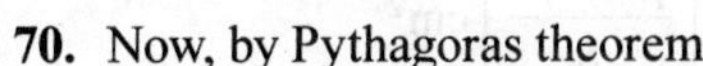

70. Now, by Pythagoras theorem,

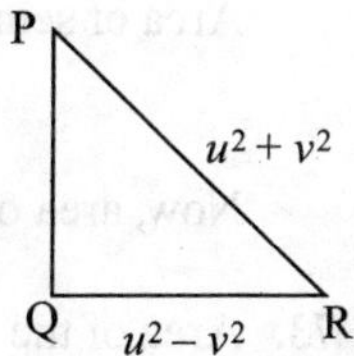

$PQ = \sqrt{PR^2 = QR^2}$

$= \sqrt{(u^2 + v^2) - (u^2 - v^2)^2} = \sqrt{4u^2v^2} = 2uv$

Now, area of triangle PQR = $\dfrac{1}{2} \times$ PQ × QR

$2016 = \dfrac{1}{2} \times (u^2 - v^2) \times 2uv$

$\Rightarrow 2016 = (u^2 - v^2) \times uv$

$\Rightarrow uv\,(u + v)\,(u - v) = 32 \times 7 \times 9$

$\Rightarrow uv\,(u + v)\,(u - v) = 16 \times 9\sqrt{2} \times 7\sqrt{2}$

Now, we have

$u = 8\sqrt{2}$

$v = \sqrt{2}$

$\therefore$ Perimeter = PQ + PR + OR

$= 2uv + u^2 - v^2 + u^2 + v^2$

$= 2 \times 8\sqrt{2} \times \sqrt{2} + 2 \times (8\sqrt{2})^2$

$= 32 + 2 \times 128 = 32 + 256 = 288$ units

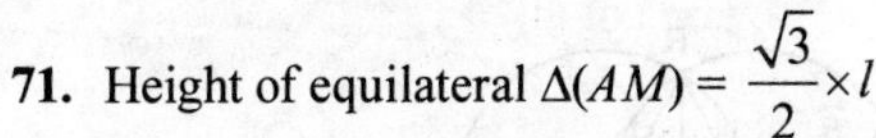

71. Height of equilateral $\Delta(AM) = \dfrac{\sqrt{3}}{2} \times l$

Now, radius of inscribed circle = $\dfrac{1}{3} \times \text{AM} = \dfrac{1}{3} \times \dfrac{\sqrt{3}}{2} \times l = \dfrac{l}{2\sqrt{3}}$

Side of wquare (*PQ*) = $\sqrt{Op^2+OQ^2}=\sqrt{r^2+r^2}=\sqrt{2}r=\sqrt{2}\times\frac{l}{2\sqrt{3}}=\frac{l}{\sqrt{6}}$

Now, area of square (PQRS) inscribed in the circle

$= (\text{side})^2 = \left(\frac{l}{\sqrt{6}}\right)^2 = \frac{l^2}{6}$

72. We have,

Area of $\Delta ABC = \frac{1}{2}\times b\times h = \frac{1}{2}\times pq$

By Pythagoras Theorem

$BC^2 = AB^2 + AC^2 = P^2 + q^2$

$\because$ Area of semicircle $= \frac{\pi}{2}(r)^2$

Now, area of semi-circle with diameter AB $= \frac{\pi}{2}\left(\frac{p}{2}\right)^2 = \frac{\pi p^2}{8}$ cm²

Area of semi-circle with diameter AC $= \frac{\pi}{2}\left(\frac{q}{2}\right)^2 = \frac{\pi q^2}{8}$ cm²

Area of semi-circle with diameter BC $= \frac{\pi}{2}\left(\frac{p^2+q^2}{4}\right)^2 = \pi\left(\frac{p^2+q^2}{8}\right)$ cm²

Now, area of shaded region $= \frac{PQ}{2}+\frac{\pi p^2}{8}+\frac{\pi q^2}{8}\left[\pi\left(\frac{p^2+q^2}{8}\right)\right] = \frac{pq}{2}$

73. Area of the square paper sheet = 784 sq cm.

Four equal – sized maximum circular plates are cut off from the square paper sheet.

$\therefore$ The sheet must have been divided into four identical parts and then the largest circle must have been cut from each part.

Each of the four parts must be a square whose area is 196 sq cm.

$\therefore$ Side of each part = 14 cm.

The diameter of each plate must be equal to the side of the square from which it was cut i.e. 14 cm.

Circumference of each plate $= \frac{22}{7}(14)$ cm = 44 cm.

74. Let x cm be the length of the other diagonal.

In any parallelogram (and not just rhombus), the sum of the squares of the diagonals is equal to twice the sum of the squares of any two adjacent sides.

$\therefore\ 2(13^2 + 13^2) = 10^2 + x^2$

$\Rightarrow\ x^2 = 576 \Rightarrow x = 24$

Area $= \frac{1}{2}(24)(10)$ cm² = 120 cm².

75. Area of segment ROSQ

= Area of sector PRQS – Area of ΔPRS

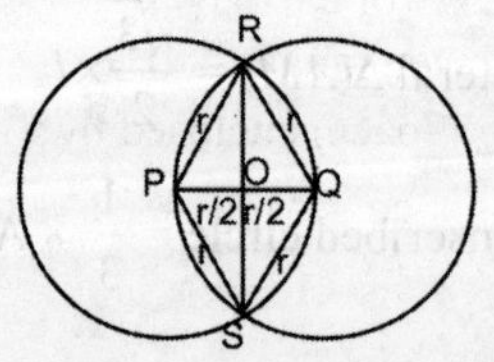

Sin (∠PRS) $= \frac{r/2}{r} = \frac{1}{2}$

$\Rightarrow \angle PRS = 30°$

$\therefore$ $\angle RPO = 60°$

$\Rightarrow$ $\angle PRS = 120°$.

Area of segment ROSQ = $\frac{\pi x^2}{3} - \frac{1}{2}(\sqrt{3}x)\left(\frac{r}{2}\right) = \frac{1}{3}\pi x^2 - \frac{\sqrt{3}r^2}{4}$

Area common to the circles = 2 (Area of segment ROSQ)

$$= \frac{2}{3}\pi x^2 - \frac{\sqrt{3x^2}}{2} = \frac{1}{3}(\text{Sum of Area}) - \frac{\sqrt{3x^2}}{2}.$$

76. Let r be the radius of the sphere and a be the length of each side of the cube.

Given that $4\pi r^2 = 6a^2$; $\frac{r}{a} = \sqrt{\frac{6}{4\pi}}$

Ratio of the square of volumes = $\frac{\left(\frac{4}{3}\pi r^3\right)^2}{(a^3)^2} = \left(\frac{4}{3}\pi (r/a)^3\right)^2$

$$= \left(\frac{4}{3}\pi \times \left(\frac{6}{4\pi}\right)^{\frac{3}{2}}\right)^2 = \frac{4^2}{3^2}\pi^2 \frac{6^3}{4^3\pi^3} = \frac{6^3}{3^2}\frac{1}{4\pi} = \frac{6}{\pi}.$$

77. Let l be the slant height of the cone. l = $\sqrt{24^2 + 7^2}$ cm = 25 cm.

T. S. A = $\pi(7)(25) + \frac{4\pi(7^2)}{2}$ cm^2 = $7\pi(39)$ cm^2 = 273π cm^2.

78. Approximate volume of paint required
= Surface area of wood (thickness of layer of paint)

$$= 2(6 + 3.5 + 5.25)\left(\frac{1}{10000}\right) \text{m}^3 = \frac{29.5}{10000} \text{m}^3.$$

Volume of paint in 1 box = 0.001 m^3.

$\therefore$ Minimum number of boxes required = $\frac{0.00295}{0.001} = 2.95$

i.e., number of boxes should be 3.

Note: The actual value of volume of paint needed will be
$(abc) - (a + x)(b + x)(c + x)$,
where a, b, c are lengths of the sides and $x = 2 \times 0.1$ mm.
Since the difference from the previous value is negligible with the given thickness (in mm), we can approximate using the previous method.

79. The base b = 300 units.
Length of equal side a = 170 units

Areas of the triangle = $\frac{b}{4}\sqrt{4a^2 - b^2} = \frac{300}{4}\sqrt{4 \times 170^2 - 300^2}$

$$= 75 \times 160 = 12000$$

80. Let r be the radius of each disc

$\Rightarrow$ Length of inner square = $2r$

$\therefore$ Area (empty space) enclosed by the discs.

= Areas of space – Area of $\angle 1$ sectors

= Area of square – 1 circle area = $4r^2 - \pi r^2$

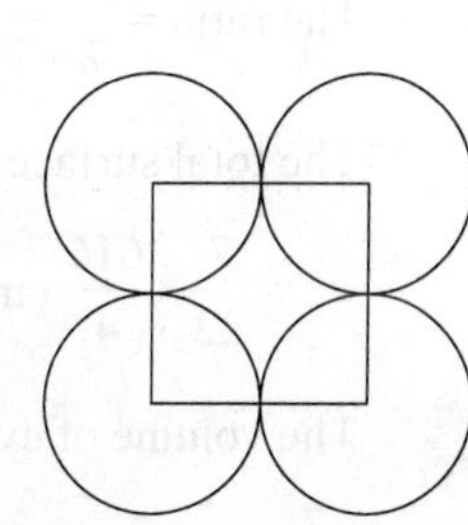

$$r^2(4-\pi)=\frac{150}{847}$$

$$r^2=\frac{150\times 7}{6\times 847}=\frac{25}{121};\ r=\frac{5}{11}$$

81.

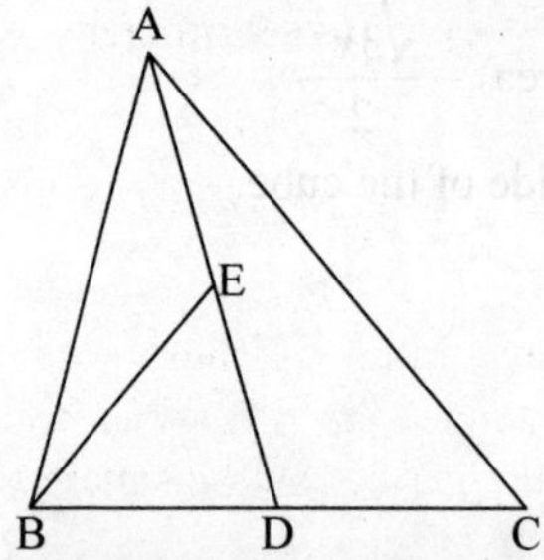

BD = DC and AE = ED.
As ΔABC = 2 As ΔABD = 4 As ΔBED (1) is true.
As ΔADC = As ΔABD = 2 As ΔBED (2) is true

82. Area of shaded region in quarter circle = $2\times\left(\frac{\pi r^2}{4}-\frac{1}{2}r^2\right)$

$=2\times\left(\frac{1}{4}\times\frac{22}{7}\times 49-\frac{1}{2}\times 49\right)=2\times\left(\frac{154}{4}-\frac{98}{4}\right)=2\times\frac{56}{4}=28\text{ cm}^2$

Area of circle of diameter 7 = $\frac{22}{7}\times\frac{49}{4}=38.5$

$\therefore$ area of shaded region = 28 + 38.5 = 66.5 cm^2

83. required ratio = $\frac{66.5}{154-66.5}=\frac{66.5}{87.5}=\frac{19}{25}$

84. Let the breadth, length and height of the room be *b, l, h* respectively
$\therefore$ $(b+l)\,h=60\ m^2,\ l=2b$ and $h=4m$
$\therefore$ $b+2b=15\ m\Rightarrow b=5m,\ l=10\ m$
The area of the floor is 50 m^2.

85. If r is the radius of the circle, the side of the inscribed square is $\sqrt{2}\,r$ and the side of the inscribed equilateral triangle is $\sqrt{3}\,r$. The required ratio is $\sqrt{2}:\sqrt{3}$

86. Let the radius and the height of the cylinder be r and h.
The curved surface area = $2\pi rh+2\pi r^2$

The ratio = $\frac{h}{h+r}=\frac{1}{2}\Rightarrow r=h.$

The total surface area = $2\,2\pi r(r+h)=4\pi r^2=616\text{ cm}^2$

$r^2=\frac{7}{22}\times\frac{616}{4}\text{ cm}^2\Rightarrow r=7\text{ cm}$

The volume of cylinder = $\pi r^3=\frac{22}{7}(7^3)\text{ cm}^3=1078\text{ cm}^3$

87. The diameter of the wheel = $\frac{5}{11}$ m. Circumference ≈ $\frac{22}{7}\times\frac{5}{11}=\frac{10}{7}$ m.

In covering a distance of 7 km, the wheel would make could make $\frac{7000}{10}(7)$, it 4900 revolutions.

88. Let the radius of the vessel be r and the cost of painting be ₹a per square centimeter.
$r = 7$cm.

$$4\pi r^2 a = 8008 \Rightarrow a = \left(\frac{7}{22}\right)\frac{(8008)}{4(49)} = \frac{7}{22}\times\frac{8(7)(11)(13)}{4(49)} = 13$$

89. Suppose we have a string of length L. We can bend it to form an equilateral triangle, a square, a regular pentagon and so on general a regular n-polygon. As n tends to infinity we get a circle Also the area keep increasing with n. The thought could have the least area, the square would have a greater area and so on. The circle would have the greatest area.

90. The radii of the front and rear wheels are x cm, y cm respectively of the front wheel makes n revolutions, the distance would be $(2\pi x)n$ to cover this distance, the rear wheel have to make r_a revolutions, where $(2\pi y)\, n = (2\pi x)\, n$ .... $m = \frac{nx}{y}$

91. The curved surface area of the cylinder = $2\pi rh \approx 2\left(\frac{22}{7}\right)(63)\,(5)\,m^2 = 1980\ m^2$

92. The total area = $2\pi rh + \pi rl = \pi r\ (2h + l) \approx \frac{22}{7}(63)\,(10 + 65)\,m^2 = 11\,(9)\,(150)\,m^2 = 14850\ m^2$

93. As 5, 12, 13 is a Pythagorean triplet, we can take the base as 12 and the height as 5. The area is 30 cm^2

94. The side of the square be a and the radius of the circle be r. As the perimeter are equal, $4a = 2\pi r\ \left(or\ a = \frac{\pi r}{2}\right) = \frac{\pi r^2(4)}{\pi^2 r^2} = \frac{4}{11} \approx \frac{28}{22} = \frac{14}{11}$

95. Perimeter of rectangle = 10
Let, length of l and breadth be b
$\therefore\ 2(l + b) = 10 \Rightarrow l + b = 5$
Also, $lb = 4$
Taking 1 = 4 and $b = 1$, $l + b = 5$, $4 + 1 = 5$
LHS = RHS
And $lb = 4$, $4 \times 1 = 4$
LHS = RHS
Thus, length of rectangle is 4cm.

96. Let r be the radius of smaller circular field and R be the radius of bigger field.
A_1 be the area of smaller field and A_2 be the area of bigger field.
According to the question,

$$\frac{A_1}{A_2} = \frac{16}{49} \Rightarrow \frac{\pi r^2}{\pi R^2} = \frac{16}{49}$$

$$\Rightarrow \frac{r^2}{14\times 14} = \frac{16}{49} \Rightarrow r = 8m$$

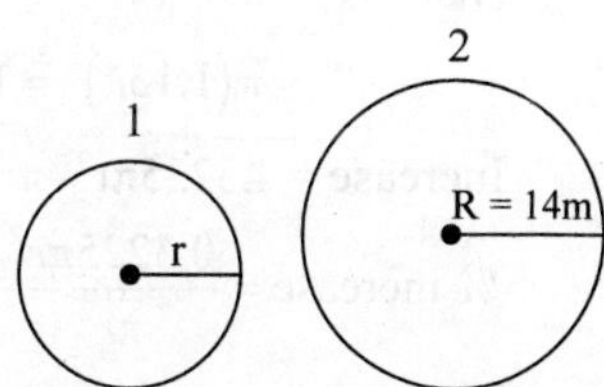

97. $l^1 = l + \frac{200}{100} l = 3l$

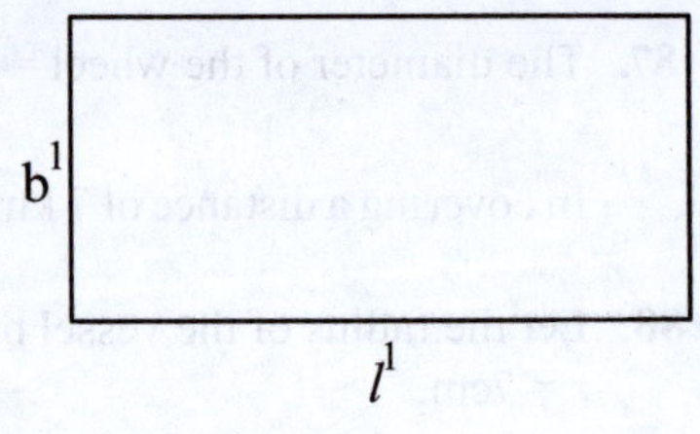

$\therefore \quad b^1 = 3b$

$A = lb$

$A^1 = 3l \times 3b = 9lb$

Increased area $= 9lb - lb = 8lb$

$\% = \frac{8lb}{lb} \times 100 = 800\%$

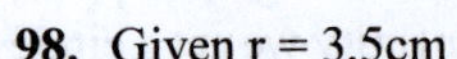

98. Given r = 3.5cm

ABC is an equilateral Δ

AB = 7cm = AC = BC

∠A = ∠B = ∠C = 60°

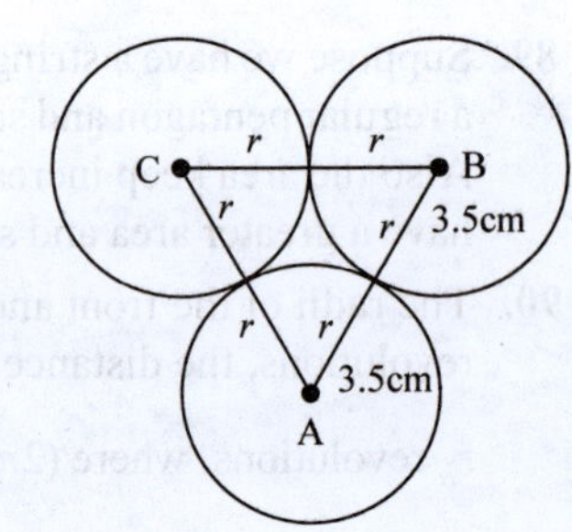

Area subtended between 3 circles $= ar(\Delta ABC) - \frac{180}{360}\pi r^2$

$$= \frac{\sqrt{3}}{4} \times 7^2 - \frac{1}{2}\pi(3.5)^2 = \frac{\sqrt{3}}{4} \times 49 - \frac{1}{2}\pi\left(\frac{7}{2}\right)^2$$

$$= \frac{\sqrt{3}}{4} \times 49 - \frac{1}{2}\pi\frac{49}{4} = \frac{49}{8}\left(2\sqrt{3} - \pi\right) cm^2$$

99. $a^2 = 121$

$a = 11$ cm

Perimeter = 44 cm

Let radius of circle be r

circumference = 44cm

$\therefore \quad 2\pi r = 44$

$r = \frac{22}{\pi}$

Area $= \pi r^2 = \pi \times \frac{22}{\pi} \times \frac{22}{\pi} = 22 \times \frac{22}{22} \times 7 = 22 \times 7 = 154 \text{ cm}^2$

100. Circumference (C) = 2πr

$C^1 = 115\% \text{ of } C = 1.15C,$

$2\pi r^1 = 1.15 \times 2\pi r$

$\Rightarrow \quad r^1 = 1.15r$

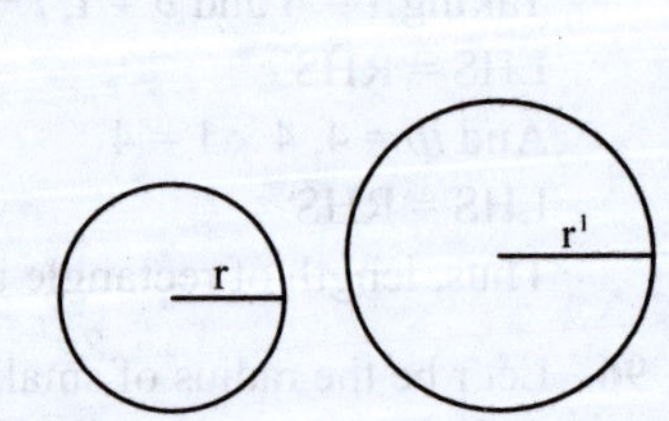

Also, $A^1 = \pi\left(r^1\right)^2$

$= \pi(1.15r)^2 = 1.3225\pi r^2$

Increase $= 1.3225\pi r^2 - \pi r^2 = 0.3225\pi r^2$

% increase $= \frac{0.3225\pi r^2}{\pi r^2} \times 100 = 32.25\%$

101. Area of I = $2 \times 1 = 2m^2$

Area of II = $\frac{1}{2} \times b \times h = \frac{1}{2} \times 2 \times 3 = 3m^2$

Area of III = $(1+3) \times \frac{1}{2} \times 2 = 4m^2$ $\left[\because \text{Area of trapezium}\right]$

$$s = \frac{a+b+c}{2}$$

$$\Rightarrow \ s = \frac{2+3+4}{2} = \frac{9}{2} = 4.5$$

Area of IV = $\sqrt{s(s-a)(s-b)(s-b)}$

$$= \sqrt{4.5(2.5)(1.5)(0.5)} = \frac{3 \times 5 \times 5\sqrt{5 \times 3}}{10 \times 10} = \frac{3}{4}\sqrt{15}$$

Area of I, II, III and IV $= 2+3+4+\frac{3}{4}\sqrt{15} = 9+\frac{3}{4}\sqrt{15}\ m^2$

102. In ΔPQR

$$PR = \sqrt{D^2 + D^2} = \sqrt{2D^2} = D\sqrt{2}$$

Diameter of the shaded circle = $PR - \left(\frac{D}{2} + \frac{D}{2}\right)$

$$= PR - D$$

$$= D\sqrt{2} - D$$

$$= D\left(\sqrt{2} - 1\right)$$

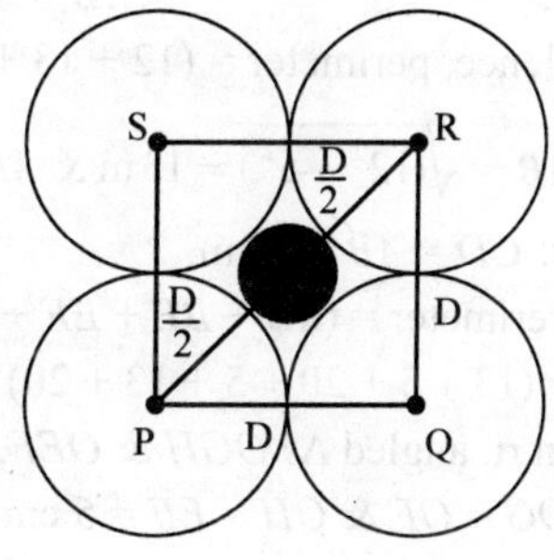

103. $\frac{\text{Area of } \| \text{ gm } ABCD}{\text{Area of } \Delta ABM} = \frac{BC \times AM}{\frac{1}{2} \times BM \times AM} \times \frac{2 \times 4x}{x} = \frac{8}{1}$, *i.e.*, $8:1$.

104. Radius = 6 cm, Diamter = 12 cm

Area of shaded region = $\frac{\pi 6^2}{2} - \left[\frac{\pi \times (2)^2}{2} + \frac{\pi (4)^2}{2}\right]$

$18\pi - (2\pi + 8\pi)$

$18\pi - 10\pi = 8\pi$

ar(unshaded region) = $18\pi + 10\pi = 28\pi$

$\therefore$ required ratio = $\frac{8\pi}{28\pi} = 2:7$

105. Perimeter of shaded region = $\pi\,(6 + 2 + 4) = 12\pi$

106. $\pi R^2 = 616 \Rightarrow R^2 = \left(616 \times \frac{7}{22}\right) = 196 \Rightarrow R = 14$ cm.

$\therefore$ Side of the square = $2R$ = 28 cm.

Area of the square = $(28 \times 28)\text{cm}^2 = 784\ \text{cm}^2$.

107. Area to be white washed = $[(2 \times 4 \times 2.5) + (2 \times 4 \times 2.5) + (2 \times 3 \times 2.5) + (2 \times \frac{1}{2} \times 3 \times 1)] - 18$

$= (20 + 20 + 15 + 3) - 18 = 40$ sq. metres.

$\therefore$ Cost of white washing = ₹40.

108. $6a^2 = 384 \Rightarrow a^2 = 64$ or $a = 8$ metres

109. $\frac{1}{2} \times$ (sum of parallel sides) $\times$ depth $= 72 \Rightarrow \frac{1}{2} \times (10 + 6) \times$ depth $= 72$. So, depth = 9 metres.

110. Clearly $AE = CD = 11$ cms.

$\therefore$ $BE = (AB - AE) = 5$ cms.

In rt. angled triangle BED, we have

$DE^2 = (BD^2 - BE^2) = [(13)^2 - 5^2] = 144$. So, $DE = 12$ cms.

$\therefore$ Area of trapezium $= \frac{1}{2} \times (16 + 8) \times 12 = 144$ cm^2.

111. $BE = 12$ cms, and $BD = 13$ cms.

$\therefore DE = \sqrt{(BD^2 - BE^2)} = \sqrt{(169 - 144)} = 5\,\text{cm}$.

Now $AE = CD = 9$ cms.

$\therefore$ $AB = AE + BE$

Hence, perimeter = (12 + 13 + 9 + 14) = 48 cms.

112. $AB = \sqrt{(12^2 + 5^2)} = 13\,\text{m}$ & $AD = 20\,\text{m}$.

$\therefore CD = AB = 13$ m.

Perimeter = $(AB + BE + EF + FC + CD + AD)$

= (13 + 5 + 20 + 5 + 13 + 20) cms = 76 cms.

113. In rt. angled Δs OGH & OEF, we have

$OG = OF$ & $GH = EF = 5$ cms.

$\therefore OH = \sqrt{(OG^2 - 5^2)} = \sqrt{(OF^2 - 5^2)} = OE$.

Now, $OH + OE = [12 - (2 + 2)] = 8$ metres.

Area of the framework = (Area of rectangle $ABCD$) – (area $\Delta\, OHG$ + area $\Delta\, OEF$)

$$= \left[(12 \times 7) - \left(\frac{1}{2} \times 4 \times 5 + \frac{1}{2} \times 4 \times 5\right)\right] = (84 - 20) = 64 \text{ sq. m.}$$

114. Area of the flower beds = [(25 × 10) – (21 × 8)] = 82 m^2.

115. $BC = \sqrt{(13^2 - 12^2)} = 5$ m.

$\therefore$ Area of rectangle $ABCD = (12 \times 5)\text{m}^2 = 60\text{ m}^2$

From G & E draw $GK \perp HF$ and $EL \perp HF$.

$KL = 4$ m. So, $HK = LF = 3$ m.

So, $GK = \sqrt{(HG^2 - HK^2)} = \sqrt{(5^2 - 3^2)} = 4\,\text{m}$.

$\therefore$ Area of trapezium $GEFH = \frac{1}{2} \times (4 + 3) \times 4 = 14$ m^2.

Hence, the area of the given figure = (60 + 14) = 74 m^2.

116. $EB = DC = 8$ m.

$\therefore AE = (AB - BE) = (16 - 8) = 8$ m.

So, $DE = \sqrt{(AD^2 - AE^2)} = \sqrt{\{(17)^2 - 8^2\}} = 15$ m.

$\therefore$ Area of trapezium $ABCD = \frac{1}{2} \times (16 + 8) \times 15 = 180$ m^2.

117. Total area of the field = (Δ DQE + Δ AQE + Δ DRC + trap. $PBCR$ + Δ APB)

$= \left[\frac{1}{2}(30+20)\times 30\right] + \left[\frac{1}{2}(30+20)\times 30\right] + \left[\frac{1}{2}\times 30\times 40\right] + \frac{1}{2}\times(40+20)(20+20) + \frac{1}{2}\times 30\times 20$

$= (750 + 750 + 600 + 1200 + 300) = 3600$ m^2.

118. Produce CD to C' such that $DC' = 0.3$ m. Also draw $CG \perp AB$.

Then, $AG = CC' = CD + EF = (0.6 + 0.3) = 0.9$

So, $BG = (1.8 - 0.9) = 0.9$.

Also, $CG = AF + ED = (2.4 + 1.2) = 3.6$ m.

$\therefore$ Required area = [area (rect. $AGCC'$) + (Δ GBC) – rect. $FEDC'$]

$= \left[0.9\times 3.6 + \frac{1}{2}\times 0.9\times 3.6 - 0.3\times 1.2\right]$ sq. m.

$= (3.24 + 1.62 - 0.36) = 4.50$ m^2.

119. Circumference $= 2 \times \frac{22}{7} \times 21 = 132$ cms. = 1.32 m.

In moving 1.32 m, it makes 1 revolution.

In moving 66 m, it makes $\left(\frac{1}{1.32}\times 66\right) = 50$ revolutions.

120. Let ABC be the given equilateral triangle in which a circle has been inscribed. Let O be the centre of this circle. Let AD & BE be the bisectors of $\angle A$ and $\angle B$ respectively.

Then, $\angle OBD = \frac{1}{2}(60°) = 30°$ & $\angle BOD = 60°$.

Let $OD = a$. Then, $OB = 2OD = 2a$.

Also, $BD = 12$ cm.

$\therefore a^2 + 12^2 = (2a)^2$ or $3a^2 = 144$ or $a^2 = 48$ or $a = 4\sqrt{3}$.

Area of the circle $= \pi a^2 = (\pi \times 48)$ cm^2.

121. Statement I:

According to the question,

$7 = \frac{\frac{\pi r}{30}}{180}$ $\left[\because \text{Length of are} = \frac{\pi r\theta}{180°}\right]$

$\Rightarrow r = \frac{42}{\pi}$

$\therefore$ Area of circle $= \pi r^2 = \pi \times \left(\frac{42}{\pi}\right)^2 = \pi \times \frac{1764}{\pi^2} = \frac{1764}{\pi}$ cm^2

Hence, statement I alone is sufficient to answer the question

Statement II:

Let c be the centre of the circle

$\therefore AC = BC = r$

$AC^2 + BC^2 = AB^2$

$\Rightarrow 2r^2 = 100$

$\Rightarrow r = 5\sqrt{2}$

$\therefore$ Area of circle = $\pi \times (5\sqrt{2})^2 = 50\pi$ cm^2

Hence, statement II alone is sufficient to answer the question.

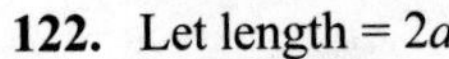

122. Let length = $2a$

Breadth = a

Area = $l \times b = 2a \times a = 200$

$\Rightarrow 2a^2 = 200$

$\Rightarrow a = 10$

$\therefore$ Length = 2a = 20 m

Breadth = 10 m

New length = 20 + 2 = 22 m

New breadth = 10 + 2 = 12 m

$\therefore$ Required % = $\dfrac{264 - 200}{200} \times 100 = 32\%$

123. $r_2 = Cr$

$OB = (r - 2)$

$AB = (r - 4)$

$OA = r$

Now, in ΔOAB

$OA^2 = OB^2 + AB^2$

$\Rightarrow r^2 = (r - 4)^2 + (r - 2)^2 \quad \Rightarrow r^2 = 2r^2 = 12r + 20$

$\Rightarrow r^2 = 12r + 20 = 0$

$\Rightarrow r = 10$ and $r = 2$ (nor possible)

$\therefore$ Area of circle = $\pi r^2 = 100\pi$ m^2

124. $AB = 10 - a = OB$

OA = 10

Now, in ΔOAB

$OA^2 = AB^2 + OB^2$

$\Rightarrow 10^2 = (10 - a)^2 + (10 - a)^2 \quad \Rightarrow 100 = 2\,(10 - a)^2$

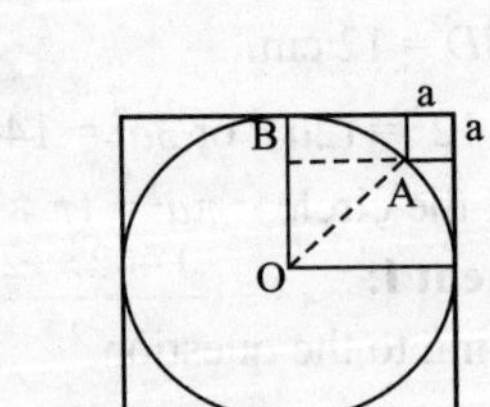

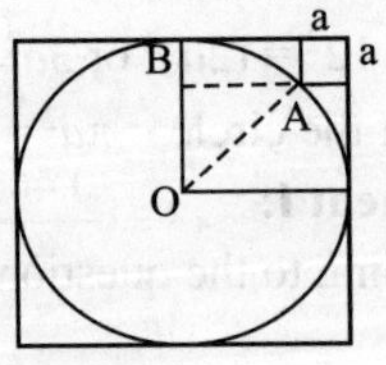

Squaring both sides, we get

$\Rightarrow \sqrt{2}\,(10 - a) = 10 \quad \Rightarrow 10 - a = 5\sqrt{2} \quad \Rightarrow a = 5(2 - \sqrt{2})$

Now, Ar (smaller square) = $[5(2-\sqrt{2})]^2 = 25\,[4 + 2 - 4\sqrt{2}] = 50\,[3 - 2\sqrt{2}]$ cm^2

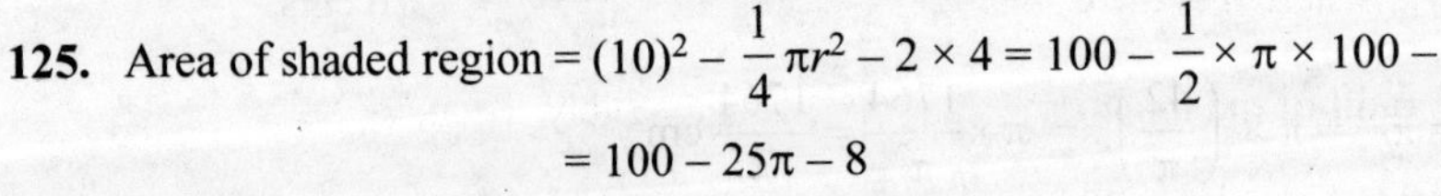

125. Area of shaded region = $(10)^2 - \dfrac{1}{4}\pi r^2 - 2 \times 4 = 100 - \dfrac{1}{2} \times \pi \times 100 - 8$

$= 100 - 25\pi - 8$

$= (92 - 25\pi)$ cm^2

126. Area of $\Delta = \frac{\sqrt{3}}{4} \times \text{side}^2$

$$= \frac{\sqrt{3}}{4} \times (3a)^2 = \frac{9\sqrt{3}}{4} a^2$$

Area of regular hexagon = $ar\,(\Delta PQR) - 3 \times ar\,(\Delta PQ'R')$

$$= \frac{9\sqrt{3}}{4} a^2 - \frac{3\sqrt{3}}{4} a^2 = \frac{\sqrt{3}}{4}(9a^2 - 3a^2) = \frac{\sqrt{3}}{4} \times 6a^2 = \frac{6\sqrt{3}}{4} a^2$$

$\therefore$ required ratio = 9 : 6 = 3 : 2

127. We have, $2\pi r - 2r = 16.8$

$$\Rightarrow 2r(\pi - 1) = 16.8 \Rightarrow 2r\left(\frac{22}{7} - 1\right) = 16.8 \Rightarrow 2r \times \frac{15}{7} = 16.8 \Rightarrow 2r = \frac{39.2}{5} = 7.84 \text{ cm}$$

128. Length of arc $= \frac{\pi r Q}{180}$

$$\Rightarrow 14.3 = \frac{22}{7} \times \frac{9}{180} \times r \quad \Rightarrow r = 91 \text{ cm}$$

129. (1) $\therefore$ (1) is not correct

(2) $\angle BAC = \angle BDC$ [$\therefore$ in same segment BC]

and $\angle ABD = \angle ACD$ [$\therefore$ in the same segment AD]

$\therefore \Delta ABO \cong \angle DCO$

$AO = DO$ ...(i)

$OC = OB$ ...(ii)

By adding (i) and (ii), we get

$AO + OC = OB + OD \Rightarrow AC = BD$

(2) is correct

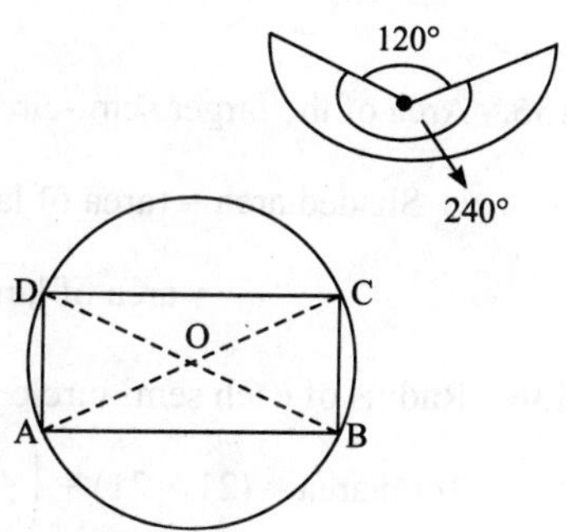

130. Let x be the side of the triangle and a be the radius of the incircle. Then,

$$a^2 + \left(\frac{x}{2}\right)^2 = 4a^2 \text{ or } 3a^2 = \frac{x^2}{4} \text{ or } a^2 = \frac{x^2}{12}.$$

Area of the circle $= \pi a^2 = \pi \frac{x^2}{12}$.

$$\frac{\pi x^2}{12} = 462 \Rightarrow x^2 = \frac{0.462 \times 12 \times 7}{22} \text{ or } x = 42.$$

$\therefore$ Perimeter of the triangle = 3 × 42 = 126 cm.

131. Number of revolutions $= \frac{\text{distance}}{\text{circumference}}$.

$\therefore 2500 = \frac{550000}{\pi \times d}$, where d is the diameter. $\therefore \quad d = \frac{550000 \times 7}{2500 \times 22} = 70$ cms.

132. Let r and R be the radii of the inner and outer circle respectively. Then,

$$2\pi r = 176 \text{ or } r = \frac{176 \times 7}{22 \times 2} = 28 \text{ m}.$$

$\therefore \quad R = (28 + 7) = 35$ m.

So, Area of road $= \pi(R^2 - r^2) = \pi(35^2 - 28^2) = \frac{22}{7}(35 + 28)(35 - 28) = 1386$ m^2.

133. Diagonal of the rectangle $= \sqrt{(8^2 + 6^2)} = 10$ m.

$\therefore$ Radius of circle = 5 m.

Area of the shaded region $= [(\pi \times 5^2) - (8 \times 6)]$ m^2 = 30.6 m^2.

134. Sum of lengths of circular parts = [400 – (100 + 100)] = 200 m.

Lengths of each circular part = 100 m.

$\therefore \pi r = 100$ or $r = \frac{100}{\pi}$.

Area of both semi-circular ends $= 2 \times \frac{1}{2}(\pi r^2) = \pi r^2 = \left(\pi \times \frac{100}{\pi^2} \times 100\right) = \frac{10000}{\pi}$.

$\therefore$ Area of the inner boundary $= \left[\frac{10000}{x} + 100 \times 2r\right] = \left[\frac{10000}{\pi} + \frac{200 \times 100}{\pi}\right]$

$= \frac{30000}{\pi} = \left(\frac{30000 \times 7}{22}\right) = 9545$ m^2.

135. Area of the larger semi-circle $= \frac{1}{2}(\pi \times 5^2)\text{ cm}^2 = \frac{25}{2}\pi\text{ cm}^2$.

$\therefore$ Shaded area = (area of larger semi-circle) – (area of smaller one) + (area of smaller one)

= area of larger semi-circle $= \left(\frac{1}{2} \times \frac{22}{7} \times 25\right) = 39.3$ cm^2.

136. Radius of each semi-circle = 10.5 metres.

Total area $= (21 \times 21) + \left(4 \times \frac{1}{2} \times \frac{22}{7} \times 10.5 \times 10.5\right) = 1134$ sq. cms.

$\therefore$ Number of rose plants = (1134 ÷ 6) = 189.

137. Total area = (|| gm *AGHB*) + (|| gm *DCFE*) + (sq. *ABCD*) + (semi-circle *BLC*) + (Δ *ADK*)

$= (6 \times 2) + (6 \times 2) + (6 \times 6) + \left(\frac{1}{2} \times \frac{22}{7} \times 3 \times 3\right) + \left(\frac{1}{2} \times 6 \times 4\right)$

= 98.14 cm^2. $\quad [\because$ Height of $\Delta ADK = \sqrt{(5^2 - 3^2)} = 4]$

138. Do yourself.

139. Area grazed = (4 × sector *BPQ*) $= \left[4 \times \frac{1}{4}\pi \times \left(\frac{63}{2}\right)^2\right] = \left(\frac{22}{7} \times \frac{63 \times 63}{2 \times 2}\right) = 3118.5$ m^2.

Area ungrazed = [(63 × 63) – 3118.5] = 850.5 m^2.

140. Required area $= \left[(20 \times 16) - \left(\frac{1}{4} \times 4 \times \frac{22}{7} \times 3.5 \times 3.5\right)\right] = 281.5$ m^2.

141. Required area $= \left(\frac{1}{2} \times \frac{22}{7} \times 28 \times 28\right) - \left[\left(2 \times \frac{1}{2} \times \frac{22}{7} \times 14 \times 14\right) + \left(\frac{22}{7} \times 7 \times 7\right)\right] = 462$ cm^2.

142. Required area $= \left[\left(\frac{1}{4} \times \frac{22}{7} \times 28 \times 28\right) - \left(\frac{1}{4} \times \frac{22}{7} \times 21 \times 21\right)\right]\text{m}^2 = 269.5$ m^2.

143. Side = $2r$. So, perimeter = $8r$.

144. Let a be the side of the square.

Then, $a^2 = A \Rightarrow a = \sqrt{A}$. So, perimeter = $4\sqrt{A}$. Now, $2\pi r = 4\sqrt{A} \Rightarrow r = \frac{2\sqrt{A}}{\pi}$.

$\therefore$ Area of the circle = $\pi r^2 = \pi \times \frac{4A}{\pi^2} = \frac{4A}{\pi}$.

145. Let each side of the triangle be $2a$ and r be the radius of the inscribed circle.

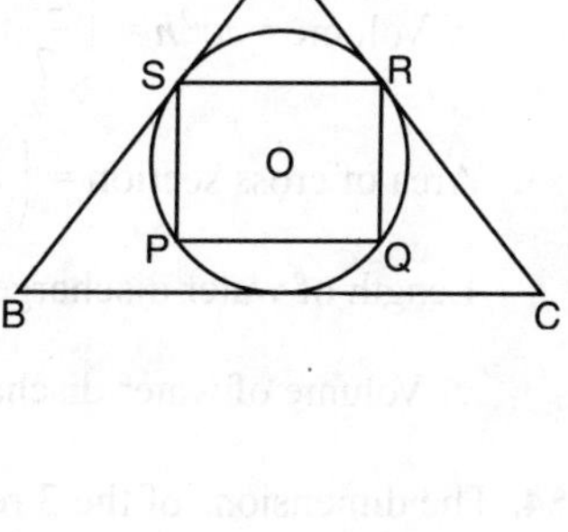

Then, $3r^2 = a^2$ or $r = \frac{a}{\sqrt{3}}$.

$\therefore \frac{1}{2} \times$ (diagonal of square) = $\frac{a}{\sqrt{3}}$, diagonal = $\frac{2a}{\sqrt{3}}$.

$\therefore$ Area of square = $\frac{1}{2} \times$ (diagoanl)^2 = $\frac{1}{2} \times \frac{4a^2}{3} = \frac{2a^2}{3}$.

Also, area of triangle = $\frac{\sqrt{3}}{4}(2a)^2 = \sqrt{3}\,a^2$.

$\therefore (\Delta ABC) : (\text{sq. } PQRS) = \sqrt{3}a^2 : \frac{2a^2}{3}$ i.e., $3\sqrt{3} : 2$.

146. Required volume = $\left[\left(\frac{22}{7} \times 6 \times 6 \times 49\right) + \left(\frac{1}{3} \times \frac{22}{7} \times 6 \times 6 \times 14\right)\right]$ = 6072 cm^3.

147. Number of turns = $\frac{\text{length of cylinder}}{\text{diameter of wire}} = \frac{420}{3} = 140.$

Length of wire in each turn = $\left(2 \times \frac{22}{7} \times \frac{49}{2}\right)$ = 154 cms.

Length of total wire = (154 × 140) cms = 215.6 m.

148. Volume = $\left[\frac{1}{2} \times (8 + 0.75) \times 25 \times 80\right]$ cu. m = $8750\,\text{m}^3$.

149. Total internal surface = (perimeter of cross-section × length)
= [(4.4 + 3 + 2.5 + 2.5) × 50] m^2 = 620 m^2.

$$\left[\begin{array}{l} \because AM = BN = \frac{1}{2}(44 - 3) = 0.7\text{m} \;\&\; DM = 2.4\text{ m} \\ \text{So, } AD = \sqrt{(AM^2 + DM^2)} = \sqrt{6.25} = 25 = BC \end{array}\right]$$

Area of floor = (AB × 50) = (4.4 × 50) m^2 = 220 m^2.

$\therefore$ Area to be painted = (620 – 220) = 400 m^2.

Hence, the cost of painting = ₹400.

150. Area of trapezium (*i.e.*, cross-section) = $\left[\frac{1}{2} \times 300 \times (1 + 4)\right]$ m^3 = $750\,\text{m}^2$.

Volume of water = (750 × 30)m^3 = 22500 m^3.

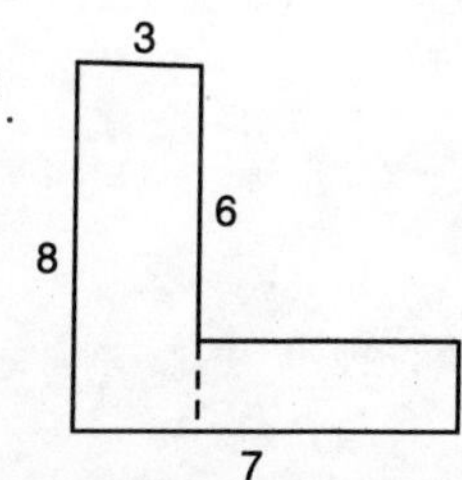

151. Perimeter of cross-section = $\left(6+8+8+\frac{22}{7}\times3\right)=\left(\frac{220}{7}\right)$ m.

Area of internal surface = $\left(\frac{220}{7}\times35\right)\text{m}^2=1100\text{m}^2$.

Area to be painted = (1100 – 6 × 35)m² = 890 m².

∴ Cost of plastering = ₹(890 × 2) = ₹1780.

152 $2\pi r\times8=22\times8\Rightarrow r=\frac{11}{\pi}=\left(11\times\frac{7}{22}\right)=\frac{7}{2}$ cm.

∴ Volume = $\pi r^2h=\left(\frac{22}{7}\times\frac{7}{2}\times\frac{7}{2}\times8\right)=308\text{ cm}^3$.

153. Area of cross section = $\left(21\times7+\frac{1}{2}\times\frac{22}{7}\times\frac{21}{2}\times\frac{21}{2}\right)\text{cm}^2.=\frac{1281}{4}\text{cm}^2=\left(\frac{1281}{4\times100\times100}\right)\text{m}^2$

Length of water discharged in 60 seconds. = (60 × 20) cm = 1200 cm = 12 m.

∴ Volume of water discharged in 1 minute = $\left(\frac{1281}{4\times100\times100}\times12\right)\text{m}^3=0.3843\text{m}^3$.

154. The dimensions of the 3 rectangles are x_1, y_1; x_2, y_2; x_3, y_3. $x_1 < x_2 < x_3$ and $x_1y_1 = x_2y_2 = x_3y_3$ (∴ $y_1 > y_2 > y_3$). By joining the parallel sides along the breadth to form a cylinder, the length becomes the circumference of the base (*i.e.*, $x = 2\pi r$) and the breadth becomes the height. The quantities are tabulated below.

	R_1	R_2	R_3
Length	x_1	x_2	x_3
Breadth	y_1	y_2	y_3
Height of cylinder	y_1	y_2	y_3
Base radius	$\frac{x_1}{2\pi}$	$\frac{x_2}{2\pi}$	$\frac{x_3}{2\pi}$
Volume	$\frac{x_1^2y_1}{4\pi}$	$\frac{x_2^2y_2}{4\pi}$	$\frac{x_3^2y_3}{4\pi}$

As $x_1y_1 = x_2y_2 = x_3y_3$, and $x_1 < x_2 < x_3$, it follows that $V_1 < V_2 < V_3$.

18 CHAPTER

Volume and Surface Areas

Highlights: Surface area and volume of cuboids, lateral surface and volume of right circular cones and cylinders. Surface area and volume of spheres. Prism and tetrahedron.

Important Concepts & Formulae:

1. (*i*) A cuboid has 8 vertices, 12 edges and 6 faces.
   (*ii*) A Prism has 6 vertices, 9 edges and 5 faces.
   (*iii*) A Pyramid has 5 vertices, 6 edges and 4 faces.
   (*iv*) A Tetrahedron has 4 vertices, 6 edges and 5 faces.

Cuboid: Let length = l, breadth = b & height = h. Then,

(*i*) Volume of cuboid = $(l \times b \times h)$ cubic units.
(*ii*) Whole surface of the cuboid = $2(lb + bh + hl)$ sq. units
(*iii*) Diagonal of the cuboid = $\sqrt{l^2 + b^2 + h^2}$ units.

Cube: Let each edge of a cube be a units. Then,

(*i*) Volume of the cube = a^3 cubic units.
(*ii*) Whole surface of the cube = $6a^2$ sq. units.
(*iii*) Diagonal of the cube = $\sqrt{3}\,a$ units.

Cylinder: Let radius of base be r & height or length be h.

(*i*) Volume of the cylinder = $(\pi r^2 h)$ cubic units.
(*ii*) Curved surface area of the cylinder = $(2\pi\, rh)$ sq. units.
(*iii*) Total surface area of the cylinder = $(2\pi\, rh + 2\pi\, r^2)$ sq. units.

Cone: Let radius of base = r, height = h & slant height = l.

(*i*) $l = \sqrt{h^2 + r^2}$.
(*ii*) Volume of the cone = $\frac{1}{3}\pi r^2 h$ cubic units.
(*iii*) Curved surface area of the cone = $(\pi\, rl)$ sq. units.
(*iv*) Total surface area of the cone = $(\pi rl + \pi r^2)$ sq. units.

Sphere: Let the radius of the sphere be r. Then:

(*i*) Volume of the sphere = $\left(\frac{4}{3}\pi r^3\right)$ cubic units.
(*ii*) Surface area of the sphere = $(4\pi\, r^2)$ sq. units.
(*iii*) Volume of a hemi-sphere = $\left(\frac{2}{3}\pi r^3\right)$ cubic units.

(*iv*) Curved surface area of the hemisphere = $2\pi r^2$ sq. units.

(*v*) Total surface area of the hemisphere = $(3\pi r^2)$ sq. units.

Prism: Volume of a Prism = (Area of the base) × Height

SOLVED PROBLEMS

Ex. 1. ***Three cubes of sides 8 cm, 6 cm and 1 cm are melted to form a new cube. Find the surface area of the cube so formed.***

Sol. Volume of new cube = Sum of volumes of given cubes = $(8^3 + 6^3 + 1^3) = 729$ cm^3.

$\therefore$ Edge of new cube = $(9 \times 9 \times 9)^{1/3} = 9$ cm.

$\therefore$ Surface area of this new cube = $6a^2 = (6 \times 9 \times 9)$ cm^2 = 486 cm^2.

Ex. 2. ***Find the volume, curved surface area and the total surface area of a cylinder of length 60 cm with diameter of the base 7 cm.***

Sol. Here $r = 3.5$ cm and $h = 60$ cm.

$\therefore$ Volume of the cylinder = $\pi r^2 h = \left(\frac{22}{7} \times 3.5 \times 3.5 \times 60\right) = 2310\text{ cm}^3$.

Curved Surface Area of the cylinder = $2\pi rh = \left(2 \times \frac{22}{7} \times 3.5 \times 60\right) = 1320\text{ cm}^2$

Total Surface Area = $(2\pi rh + 2\pi r^2) = 2\pi r(h + r)$

$$= \left[2 \times \frac{22}{7} \times 3.5 \times (60 + 3.5)\right]\text{cm}^2 = 1397\text{ cm}^2.$$

Ex. 3. ***A powder tin has a square base with side 8 cm and height 14 cm. Another tin has a circular base with diameter 8 cm and height 14 cm. Find the difference in their capacities.***

Sol. Volume of rectangular tin = $(l \times b \times h)$ cm^3

= $(8 \times 8 \times 14)$ cm^3 = 896 cm^3

Volume of cylindrical tin = $\pi r^2 h = \left(\frac{22}{7} \times 4 \times 4 \times 14\right)\text{cm}^3 = 704\text{ cm}^3$.

$\therefore$ Difference in their capacities = $(896 - 704)$ cm^3 = 192 cm^3.

Ex. 4. ***Find the volume, curved surface area and total surface area of a hemisphere of diameter 21 cm.***

Sol. Volume = $\frac{2}{3}\pi r^3 = \left(\frac{2}{3} \times \frac{22}{7} \times \frac{21}{2} \times \frac{21}{2} \times \frac{21}{2}\right) = 2425.5\text{ cm}^3$.

Curved Surface Area = $2\pi r^2 = \left(2 \times \frac{22}{7} \times 10.5 \times 10.5\right) = 693\text{ cm}^2$.

Total Surface Area = $3\pi r^2 = \left(3 \times \frac{22}{7} \times 10.5 \times 10.5\right) = 1039.5\text{ cm}^2$

Ex. 5. ***How many bullets can be made out of a lead cylinder 14 cm high and 6 cm radius, each bullet being 1.5 cm in diameter.***

Sol. Number of bullets = $\frac{\text{Volume of cylinder in cm}^3}{\text{Volume of 1 bullet in cm}^3} = \frac{\pi \times 6 \times 6 \times 14}{\frac{4}{3} \times \pi \times 0.75 \times 0.75 \times 0.75} = 896.$

Ex. 6. ***Find the number of lead balls of diameter 1 cm each that can be made from a sphere of diameter 16 cm.***

Sol. Number of balls = $\dfrac{\text{Volume of big sphere}}{\text{Volume of 1 small ball}} = \dfrac{\frac{4}{3}\pi\times8\times8\times8}{\frac{4}{3}\pi\times0.5\times0.5\times0.5} = 4096.$

Ex. 7. ***A well with 14 m inside diameter is dug 8 m deep. The earth taken out of it has been evenly spread all around it to a width of 21 m to form an embankment. Find the height of the embankment.***

Sol. Volume of earth dug out = $\left(\dfrac{22}{7}\times7\times7\times8\right) = 1232\ \text{m}^3.$

Area of embankment = $\pi(R^2 - r^2)$

$$= \frac{22}{7}\times[(28)^2-(7)^2] = \left(\frac{22}{7}\times35\times21\right)\text{m}^2 = 2310\,\text{m}^2.$$

$\therefore$ Height of embankment = $\left(\dfrac{1232}{2310}\times100\right)\text{cm} = 53.3\ \text{cm}.$

Ex. 8. ***Find the slant height, volume, curved surface area and the whole surface area of a cone of base radius 21 cm and height 28 cm.***

Sol. Here $r = 21$ cm and $h = 28$ cm.

$\therefore$ Slant height, $l = \sqrt{r^2+h^2} = \sqrt{(21)^2+(28)^2} = \sqrt{1225} = 35$ cm.

Volume = $\dfrac{1}{3}\pi r^2 h = \left(\dfrac{1}{3}\times\dfrac{22}{7}\times21\times21\times28\right) = 12936\ \text{cm}^3.$

Curved Surface Area = $\pi rl = \left(\dfrac{22}{7}\times21\times35\right)\text{cm}^2 = 2310\ \text{cm}^2$

Total Surface Area = $(\pi rl + \pi r^2) = \left(2310+\dfrac{22}{7}\times21\times21\right)\text{cm}^2 = 3696\ \text{cm}^2.$

Ex. 9. ***The radius of the base af a cone is 2.1 cm and its height is 8.4 cm. It is melted and recast into a sphere. Find the radius of the sphere.***

Sol. Clearly, Volume of the cone = Volume of the sphere.

$\therefore\quad \dfrac{1}{3}\pi\times2.1\times2.1\times8.4 = \dfrac{4}{3}\pi\times R^3$

So, $R^3 = (2.1)^3$ or $R = 2.1$ cm.

Ex. 10. ***Find the area of the iron sheet required to prepare a cone 24 cm high with base radius 7 cm.***

Sol. Here $r = 7$ cm & $h = 24$ cm

Now, $l = \sqrt{(24)^2+(7)^2} = \sqrt{625} = 25$ cm.

$\therefore$ Area of the sheet = Total surface area of the cone

$$= \pi rl + \pi r^2 = \pi r(l + r) = \left[\frac{22}{7}\times7\times(25+7)\right]\text{cm}^2 = 704\,\text{cm}^2.$$

Ex. 11. ***How many iron rods, each of length 7 m and diameter 2 cm can be made out of 0.88 cubic meter of iron?***

Sol. Number of rods $= \frac{0.88}{\text{Volume of 1 rod}} = \frac{0.88}{\pi r^2 h} = \frac{0.88}{\left(\frac{22}{7} \times \frac{1}{100} \times \frac{1}{100} \times 7\right)} = 400.$

Ex. 12. ***A closed wooden box with external dimensions 80 cm × 60 cm × 40 cm is made of 2 cm thick wood. If 1 cm³ of wood weighs 10 gms, find the weight of the empty box.***

Sol. Volume of wood $= [(80 \times 60 \times 40) - (76 \times 56 \times 36)]\ \text{cm}^3 = 38784\ \text{cm}^3$.

Weight of the empty box $= \left(\frac{38784 \times 10}{1000}\right) \text{kg} = 387.84\ \text{kg}$.

EXERCISE 18

PROBLEMS ON CUBES & CUBOIDS

1. The surface area of a cube is increased by 25%. If p is the percentage increase in its length, then which one of the following is correct? **[CDS (I) Exam, 2023]**
(*a*) $16 < p < 18$ (*b*) $14 < p < 16$ (*c*) $12 < p < 14$ (*d*) $10 < p < 12$

2. The length, breadth and height of a cuboid are increased by 10%, 20% and 50% respectively. What is the percentage increase in volume of the cuboid ? **[CDS (I) Exam, 2024]**
(*a*) 100% (*b*) 99% (*c*) 98% (*d*) 50%

3. The length of a diagonal of a cuboid is 11 cm. The surface area is 240 square cm. What is the sum of its length, breadth and height ? **[CDS (I) Exam, 2023]**
(*a*) 16 cm (*b*) 17 cm (*c*) 18 cm (*d*) 19 cm

4. The volume of a cuboid is 3600 cubic cm The areas of two adjacent faces are 225 square cm and 144 square cm. What is the area of the other adjacent face ? **[CDS (I) Exam, 2023]**
(*a*) 400 square cm (*b*) 360 square cm (*c*) 320 square cm (*d*) 300 square cm

5. The areas of the three mutually perpendicular faces of a cuboids are *x. y. z.* If *V* is the volume, then *xyz* is equal to **[CDS (II) Exam, 2015]**
(*a*) V (*b*) V^2 (*c*) $2V$ (*d*) $2V^2$

6. If the surface area of a cube is 13254 cm², then the length of its diagonal is
(*a*) $44\sqrt{2}$cm (*b*) $44\sqrt{3}$cm (*c*) $47\sqrt{2}$cm (*d*) $47\sqrt{3}$cm
[CDS (II) Exam, 2015]

7. A cube whose edge is 14 cm long has on each of its faces a circle of 7 cm radius painted yellow. What is the total area of unpainted surface ? (Take $\pi = \frac{22}{7}$) **[CDS (I) Exam, 2024]**
(*a*) 126 square cm (*b*) 189 square cm (*c*) 252 square cm (*d*) 315 square cm

8. If the length, breadth and height of a rectangular parallelopiped are in the ratio 6 : 5 : 4 and if the total surface area is 33300 sq. cm, then the length, breadth and the height of the parallelopiped (in cm) respectively are:
(*a*) 90, 85, 60 (*b*) 85, 75, 60 (*c*) 90, 75, 70 (*d*) 90, 75, 60

9. The maximum length of a pencil that can be kept in a rectangular box of dimensions 8 cm × 6 cm × 2 cm, is:
(*a*) $2\sqrt{13}$ cm (*b*) $2\sqrt{14}$ cm (*c*) $2\sqrt{26}$ cm (*d*) $10\sqrt{2}$ cm

10. How many bricks, each measuring 25 cm × 12.5 cm × 7.5 cm will be needed to construct a wall 15 m long, 1.8 m high and 37.5 cm thick?
(*a*) 4400 (*b*) 4660 (*c*) 4320 (*d*) 4575

11. The area of the base of a rectangular tank is 6500 sq. cm and the volume of water contained in it is 2.6 cubic metres. The depth of the water is:
(*a*) 3.5 m (*b*) 4 m (*c*) 5 m (*d*) 8 m

12. A tank 3 m long, 2 m wide and 1.5 m deep is dug in a field 22 m long and 14 m wide. If the earth dug out is evenly spread over the field, the rise in level of the field will be:
(*a*) 0.299 cm (*b*) 0.29 cm (*c*) 2.98 cm (*d*) 4.15 cm

13. The dimensions of an open box are 52 cm, 40 cm and 29 cm. Its thickness is 2 cm. If 1 cm^3; of metal used in the box weights 0.5 gms, the weight of the box is:
(*a*) 8.56 kg (*b*) 7.76 kg (*c*) 7.576 kg (*d*) 6.832 kg

14. The area of the cardboard needed to make a box of size 25 cm × 15 cm × 8 cm will be:
(*a*) 390 cm^2 (*b*) 1000 cm^2 (*c*) 1390 cm^2 (*d*) 2780 cm^2

15. If V be the volume and S the surface area of a cuboid of dimensions a, b, c then $\frac{1}{V}$ is equal to:
(*a*) $\frac{S}{2}(a+b+c)$ (*b*) $\frac{2}{S}\left(\frac{1}{a}+\frac{1}{b}+\frac{1}{c}\right)$ (*c*) $\frac{2S}{a+b+c}$ (*d*) $2S(a+b+c)$

16. The length of the hall is 15 m and width 12 m. The sum of the areas of the floor and the flat roof is equal to the sum of the areas of the four walls. The volume of the hall is:
(a) 1800 m^3 (*b*) 1200 m^3 (*c*) 900 m^3 (*d*) 720 m^3

17. An open box is made of wood 3 cm thick. Its external length is 1.46 m, breadth 1.16 m and height 8.3 dm. The cost of painting the inner surface of the box at 50 paise per 100 cm^2 is:
(*a*) ₹138.50 (*b*) ₹277 (*c*) ₹415.50 (*d*) ₹554

18. The surface area of a cube of length x is equal to the surface area of a sphere of radius y. Consider the following statements: **[CDS (I) Exam, 2022]**
1. $2x > 3y$
2. The volume of the cube is greater than the volume of the sphere.
Which of the above statements is/are correct?
(*a*) 1 only (*b*) 2 only (*c*) Both 1 and 2 (*d*) Neither 1 nor 2

19. A cubical block of side 14 cm is surmounted by a hemisphere of radius 7 cm. What is the total surface area of the solid thus formed? (Take $\pi = \frac{22}{7}$) **[CDS (II) Exam, 2021]**
(*a*) 1330 cm^2 (*b*) 1306 cm^2 (*c*) 1296 cm^2 (*d*) 1256 cm^2

20. How many silver coins, 3.5 in diamter and of thicknes 4 mm, must be melted to form a cuboid of dimensions 21 cm × 11 cm × 7 cm? (Take $\pi = \frac{22}{7}$) **[CDS (II) Exam, 2021]**
(*a*) 420 (*b*) 210 (*c*) 200 (*d*) 168

21. A river 3 m deep and 40 m wide is flowing at the rate of 2 km/hr and falls into the sea. What is the amount of water in litres that will fall into the sea from this river in a minute?
[CDS (I) Exam, 2020]
(*a*) 40,00,000 lites (*b*) 4,00,000 litres (*c*) 40,000 litres (*d*) 4,000 litres

22. A large water tank has the shape of a cube. If 128m^3 of water is pumped out, the water level goes down by 2m. Then the maximum capacity of the tank is **[CDS (II) Exam, 2015]**
(*a*) 512 m^3 (*b*) 480 m^3 (*c*) 324 m^3 (*d*) 256 m^3

Consider the following for the next two (02) items that follow:

The sides of an open box are 0·5 cm thick and bottom is 1 cm thick. The internal length, breadth and depth are respectively 14 cm, 10 cm and 8 cm. It is completely filled with water.

23. If the material weighs 2000 kg per cubic metre, then what is the weight of the material used in the construction of the box? **[CDS (I) Exam, 2025]**

(*a*) 360 gm (*b*) 365 gm (*c*) 720 gm (*d*) 730 gm

24. If water weighs 1000 kg per cubic metre, then what is the weight of the box with water? **[CDS (I) Exam, 2025]**

(*a*) 1.850 kg (*b*) 1.900 kg (*c*) 2.050 kg (*d*) 2.100 kg

ANSWERS

1. (*d*)	**2.** (*c*)	**3.** (*d*)	**4.** (*a*)	**5.** (*b*)	**6.** (*d*)	**7.** (*c*)	**8.** (*d*)	**9.** (*c*)
10. (*c*)	**11.** (*b*)	**12.** (*c*)	**13.** (*d*)	**14.** (*c*)	**15.** (*b*)	**16.** (*b*)	**17.** (*b*)	**18.** (*d*)
19. (*a*)	**20.** (*a*)	**21.** (*a*)	**22.** (*a*)	**23.** (*d*)	**24.** (*a*)			

SOLUTIONS

1. Surface area of cube = 600

area of each surface = 100

$\therefore$ side = $\sqrt{100} = 10$

Final surface area of cube = $600 \times \frac{125}{100} = 750$

After increase the length

Final area of each surface = 125

Final length = $\sqrt{125} = 5\sqrt{5}$

Now, according to the question

$$p = \frac{5\sqrt{5}-10}{10} \times 100 = 10(5\sqrt{5}-10) = 10(5 \times 2.236 - 10) = 11.8$$

2. Let length, breadth and height of cuboid be l, b and h respectively.

$\therefore$ Volume = lbh

New volume = $\frac{110}{100} l \times \frac{12}{10} b \times \frac{15}{10} h = \frac{198}{100}$

$\therefore$ % increase = $\frac{198-100}{100} \times 100 = 98\%$

3. Let l, b and h be the length, breadth and height respectively.

Diagonal of cuboid = $\sqrt{l^2+b^2+h^2} \Rightarrow 11 = \sqrt{l^2+b^2+h^2} \Rightarrow 121 = l^2 + b^2 + h^2$

and Surface area of cuboid = $2 \times (lb + bh + lh) \Rightarrow 240 = 2 \times (lb + bh + lh)$

Now,

$$l + b + h = \sqrt{l^2+b^2+h^2+2(lb+bh+lh)} = \sqrt{121+240} = \sqrt{361} = 19 \text{ cm}$$

4. We have,

$lbh = 3600 \text{ cm}^3$ ...(i)

$lb = a \text{ cm}^2$ ...(ii)

$bh = 225 \text{ cm}^2$ ...(iii)

$lh = 144 \text{ cm}^2$ ...(iv)

Multiplying (ii), (iii) and (iv), we get

$$l^2b^2h^2 = a \times 225 \times 144 \Rightarrow a = \frac{3600 \times 3600}{225 \times 144} = 400 \text{ cm}^2$$

5. Let the length, breadth and height of the cuboid be a, b, c respectively.

$\therefore$ $ab = x$, $ac = y$, $bc = z$ and hence $xyz = a^2b^2c^2 = V^2$

6. The total surface area of the cube is 13254 cm^2.

$\therefore$ The surface area of each face is $\frac{13254}{6}$cm^2 = 2209 cm^3. The side of the cube is 47 cm. The length of the body diagonal is $47\sqrt{3}$ cm.

7. $\therefore$ $r = 7$ cm

$\therefore$ Area of circle $= 2 \times \frac{22}{7} \times 7 = 154 \text{ cm}^2$

$\therefore$ Total area of unpainted surface = 6 (14^2 – 154) = 6(196 – 154) = 6 × 42 = 252 cm^2

8. Let length = $6x$, breadth = $5x$ & height = $4x$. Then,

$2(6x \times 5x + 5x \times 4x + 4x \times 6x) = 33300 \Rightarrow 74x^2 = 16650$

$\therefore$ $x^2 = 225$ or $x = \sqrt{225} = 15$.

So, length = 90 cm, breadth = 75 cm & height = 60 cm.

9. Maximum length = length of diagonal = $\sqrt{8^2 + 6^2 + 2^2} = \sqrt{104} = 2\sqrt{26}$ cm.

10. Number of bricks = $\frac{\text{Volume of the wall in cm}^3}{\text{Volume of 1 brick in cm}^3} = \left(\frac{1500 \times 180 \times 37.5}{25 \times 12.5 \times 7.5}\right) = 4320.$

11. Depth = $\frac{\text{Volume in cm}^3}{\text{Area in cm}^2} = \left(\frac{2.6 \times 100 \times 100 \times 100}{6500}\right)\text{cm} = 400 \text{ cm} = 4 \text{ m}.$

12. Area of the field = (22 × 14) m^2 = 308 m^2.

Area of tank = (3 × 2) m^2 = 6m^2

Area on which earth is spread = (308 – 6)m^2 = 302 m^2.

Area of earth dug = (3 × 2 × 1.5)m^3 = 9 m^3.

Rise in level = $\frac{\text{Volume}}{\text{Area}} = \left(\frac{9}{302}\right)\text{m} = \left(\frac{9 \times 100}{302}\right)\text{cm} = 2.98\,\text{cm}.$

13. Volume of metal = (52 × 40 × 29 – 48 × 36 × 27) cm^3 = 13664 cm^3.

Weight of the box = $\left(\frac{13664 \times 0.5}{1000}\right)\text{kg} = 6.832$ kg.

14. Required area = 2(25 × 15 + 15 × 8 + 25 × 8) cm^2 = 1390 cm^2.

15. $\frac{1}{V} = \frac{1}{S} \times \frac{S}{V} = \frac{2}{S}\left[\frac{(ab + bc + ca)}{abc}\right] = \frac{2}{S}\left(\frac{1}{a} + \frac{1}{b} + \frac{1}{c}\right).$

16. $2 \times 15 \times 12 = 2(15 + 12) \times h \Rightarrow h = \frac{2 \times 15 \times 12}{2 \times 27} = \frac{20}{3}\text{m}.$

$\therefore$ Volume = $\left(15 \times 12 \times \frac{20}{3}\right)\text{m}^3 = 1200\,\text{m}^3.$

17. Internal length = 140 cm, internal breadth = 110 cm and internal height = 80 cm.

Internal surface area = $(lb + 2bh + 2lh)$ cm^3.

$= (140 \times 110 + 2 \times 110 \times 80 + 2 \times 140 \times 80)$ cm^3 = 55400 cm^3.

$\therefore$ Cost of painting = $\left(\frac{50}{100} \times \frac{1}{100} \times 55400\right)$ = ₹277.

18. We have

Surface area of cube = surface area of a sphere

$\Rightarrow 6x^2 = 4\pi y^2 \Rightarrow \frac{x^2}{y^2} = \frac{4}{6} \times \frac{22}{7} \Rightarrow \frac{x^2}{y^2} = \frac{44}{21} \Rightarrow \frac{x}{y} = \sqrt{\frac{44}{21}}$

Statement -1

$2x > 3y$

$\Rightarrow 4x^2 > 9y^2 \Rightarrow 4 \times 44 > 9 \times 21 \Rightarrow 176 < 189 \neq 2x < 3y$

Statement -2

Volume of cube = $x^3 = \left(\sqrt{44}\right)^3 = 44\sqrt{44} = 88\sqrt{11}$

Volume of sphere = $\frac{4}{3}\pi y^3 = \frac{4}{3} \times \frac{22}{7} \times \left(\sqrt{21}\right)^3 = 88\sqrt{21}$

$\therefore$ Volume of sphere > volume of cube

Thus statement 2 is not correct.

19.

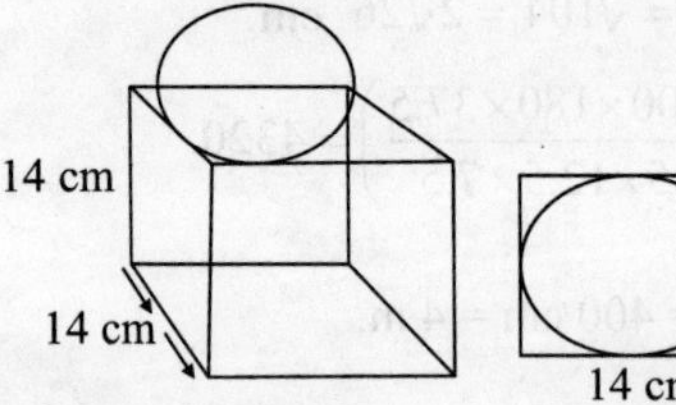

Now,

Total surface area of solid = surface area of cube + CSA of hemisphere – Area of the base

$= 6 \times \text{side}^2 + 2\pi r^2 - \pi r^2$

$= 6 \times 14^2 + \frac{22}{7} \times 7 \times 7$

$= 1176 + 154 = 1330$ cm^2

20. We have,

$r = \frac{3.5}{2} = 1.75$ cm

$h = 4$ mm $= 0.4$ cm [$\because$ 1 cm = 100 mm]

$\therefore$ Volume of silver cion = $\pi r^2 h = \frac{22}{7} \times 1.75 \times 1.75 \times 0.4 = 3.85$ cm^3

Volume of cuboid = $l \times b \times h$

$= 21 \times 11 \times 7 = 1617$

Now,

No. of silver coins = $\frac{1617}{3.85} = 420$

21. We have,

$l = \frac{2000}{60} = \frac{100}{3}$ m [∵ 1 km = 1000 m and 1 hr = 60 min]

$b = 40$ m

$h = 3$ m

Volume of cuboid = Amount of water that will fall into the sea from this river = $\frac{100}{3} \times 40 \times 3$

$= 4000 \text{ m}^3$

$= 4000 \times 1000$ [∵ 1 m³ = 1000 litres]

$= 40{,}00{,}000$

22. Let the side of the cube be a. The drop in water level (h) = 2m.

$a^2\text{h} = 128 \text{ m}^3 \Rightarrow a^2 = \frac{128}{2} \text{ m}^2 \Rightarrow a = 8$ m.

The capacity of the tank = $a^3 = 512 \text{ m}^3$.

23. Outer dimension = internal + 2 × side thickness (for L and B)

Volume of material = $\text{Volume}_{\text{outer}} - \text{Volume}_{\text{inner}}$

$1 \text{ cm}^3 = 10^{-6} \text{ m}^3$; Mass = Density × Volume

$\Rightarrow$ Outer L = 14 + 2 × 0.5 = 15cm

$\Rightarrow$ Outer B = 10 + 2 × 0.5 = 11 cm

$\Rightarrow$ Outer D = 8 + 1 = 9 cm

$\Rightarrow \text{Volume}_{\text{outer}} = 15 \times 11 \times 9 = 1485 \text{ cm}^3$

$\Rightarrow \text{Volume}_{\text{inner}} = 14 \times 10 \times 8 = 1120 \text{ cm}^3$

$\Rightarrow \text{Volume}_{\text{material}} = 1485 - 1120 = 365 \text{ cm}^3$

$\Rightarrow \text{Volume (m}^3) = 365 \times 10^{-6} = 3.65 \times 10^{-4} \text{ m}^3$

$\Rightarrow \text{Mass} = 2000 \times 3.65 \times 10^{-4} = 0.73 \text{ kg} = 730 \text{ gm}$

24. Weight of material = 730 gm.

Volume of water = $\text{Volume}_{\text{inner}} = 14 \times 10 \times 8 = 1120 \text{ cm}^3 = 1120 \times 10^{-6} \text{ m}^3$

Weight of water = $1000 \times 1120 \times 10^{-6}$ kg = 1.12 kg

Total weight of the box with water = 0.73 kg + 1.12 kg = 1.850 kg

PROBLEMS ON CYLINDER

1. The number of coins 1.5 cm in diameter and 0.2 cm thick to be melted to form a right circular cylinder of height 10 cm and diameter 4.5 cm, is:
   (*a*) 540 (*b*) 450 (*c*) 380 (*d*) 472
2. If the radius of the base of a right circular cylinder is halved, keeping the height same, what is the ratio of the volume of the reduced cylinder to that of the original one?
   (*a*) 1 : 4 (*b*) 1 : 8 (*c*) 1 : 2 (*d*) 8 : 1
3. The radius of a wire is decreased to one-third. If volume remains the same, the length will become:
   (*a*) 1 time (*b*) 3 times (*c*) 6 times (*d*) 9 times
4. The area of the curved surface area of a cylinder is 4400 cm² and the circumference of its base is 110 cm. The volume of the cylinder (in cm³) is:
   (*a*) 36000 (*b*) 38500 (*c*) 42250 (*d*) 40150

5. The radii of two cylinders are in the ratio 2 : 3 and their heights are in the ratio 5 : 3. The ratio of their curved surface areas is:
(*a*) 10:9 (*b*) 2:5 (*c*) 16:9 (*d*) 8:7

6. The altitude of a circular cylinder is increased six times and the base area is decreased to one-ninth of its value. The factor by which the lateral surface of the cylinder increases, is:
(*a*) $\frac{2}{3}$ (*b*) $\frac{1}{2}$ (*c*) $\frac{3}{2}$ (*d*) 2

7. 2.2 cubic dm. of lead is to be drawn into a cylindrical wire 0.50 cm in diameter. The length of the wire is:
(*a*) 110 m (*b*) 124 m (*c*) 112 m (*d*) 98 m

8. Two rectangular sheets of paper each 30 cm × 18 cm are made into two right circular cylinders, one by rolling the paper along its length and the other along the breadth. The ratio of the volumes of the two cylinders, thus formed, is:
(*a*) 2:1 (*b*) 3:2 (*c*) 4:3 (*d*) 5:3

9. Two cylindrical vessels with radii 15 cm & 10 cm and heights 35 cm & 15 cm respectively are filled with water. If this water is poured into a cylindrical vessel 15 cm in height, then the radius of the vessel is:
(*a*) 25 cm (*b*) 20 cm (*c*) 17.5 cm (*d*) 18 cm

10. A hollow garden roller 63 cm wide with a girth of 440 cm is made of iron 4 crn thick. The volume of the iron used is:
(*a*) 56372 cm^2 (*b*) 58752 cm^2 (*c*) 54982 cm^2 (*d*) 57636 cm^3

11. Two times the total surface area of a solid right circular cylinder is three times its curved surface area. If h is the height and r is the radius of the base of the cylinder, then which one of the following is correct ? **[CDS (I) Exam, 2023]**
(a) h = r (b) h = 2r (c) 2h = 3r (d) 3h = 4r

12. A square sheet of side length 44 cm is rolled along one of its sides to form a cylinder by making opposite edges just to touch each other. What is the volume of the cylinder ?
(Take $\pi = \frac{22}{7}$) **[CDS (I) Exam, 2023]**
(a) 6776 cubic cm (b) 6248 cubic cm (c) 5896 cubic cm (d) 5680 cubic cm

13. A cylindrical pipe has inner diameter of 14 cm. Water flows though it at a rate of 154 litres per minute. What is the speed of water in km/hr? (Take $\pi = \frac{22}{7}$) **[CDS (I) Exam, 2022]**
(*a*) 0.5 (*b*) 0.6 (*c*) 0.8 (*d*) 1

14. The ratio of the radius of base to the heights of a cylinder is 2 : 3. if the volume of the cylinder is 1617 cm^3, then what is the curved surface area of the cylinder?
(Take $\pi = \frac{22}{7}$) **[CDS (I) Exam, 2021]**
(*a*) 242 cm^2 (*b*) 385 cm^2 (*c*) 462 cm^2 (*d*) 770 cm^2

15. The difference between the outside and the inside surface area of a cylindrical pipe 14 cm long is 44 cm^2. The pipe is made of 99 cm^3 of metal. If *R* is the outer radius and r is the inner radius of the pipe, then what is $(R + r)$ equal to? (Take $\pi = \frac{22}{7}$) **[CDS (I) Exam, 2021]**
(*a*) 9 cm (*b*) 7.5 cm (*c*) 6 cm (*d*) 4.5 cm

16. A hallow right circular cylindrical vessel of volumes V whose diameter is equal to its heights, is completely filled with water. A heavy sphere of maximum possible volume is then completely immersed in the vessel. What volume of water remains in the vessel? **[CDS (II) Exam, 2019]**

(*a*) $\frac{V}{2}$ (*b*) $\frac{V}{3}$ (*c*) $\frac{2V}{3}$ (*d*) $\frac{V}{4}$

17. A solid metallic cylinder of height 10 cm and radius 6 cm is melted to make two cones in the ratio of volume 1 : 2 and of same height as 10 cm. What is the percentage increase in the flat surface area? **[CDS (II) Exam, 2019]**

(*a*) 25% (*b*) 50% (*c*) 75% (*d*) 100%

18. A solid right cylinder is of height π cm. If its lateral surface area is half its total surface area, then the radius of its base if **[CDS (II) Exam, 2018]**

(*a*) $\frac{\pi}{2}$ cm (*b*) π cm (*c*) $\frac{1}{\pi}$ cm (*d*) $\frac{2}{\pi}$ cm

19. A cylindrical vessel of radius 4 cm contains water. A solid sphere of radius 3 cm is lowered into the water until it is completely immersed. The water level in the vessel will rise by **[CDS (I) Exam, 2015]**

(*a*) 1.5 cm (*b*) 2cm (*c*) 2.25 cm (*d*) 4.5 cm

20. A pipe with square cross-section is supplying water to a cistern which was initially empty. The area of cross-section is 4 cm^2 and the nozzle velocity of water is 40 m/s. the dimensions of the cistern are 10m × 8m × 6m. then the cistern will be full in **[CDS (II) Exam, 2015]**

(*a*) 9.5 hours (*b*) 9 hours
(*c*) 8 hours 20 minutes (*d*) 8 hours

21. A hollow cylindrical drum has internal diameter of 30 cm and a height of 1 m. What is the maximum number of cylindrical boxes of diameter 10 cm and height 10 cm each that can be packed in the drum? **[CDS (II) Exam, 2015]**

(*a*) 60 (*b*) 70 (*c*) 80 (*d*) 90

22. 30 metallic cylinders of same size are melted and cast in the form of cones having the same radius and height as those of the cylinders.

Consider the following statement

Statement I: A maximum of 90 cones will be obtained.

Statement II: The curved surface of the cylinder can be flattened in the shape of a rectangle but the curved surface of the cone when flattened has the shape of triangle.

Which one of the following is correct in respect of the above?

(*a*) Both Statement I and Statement II are correct and Statement II is the correct explanation of Statement I

(*b*) Both Statement I and Statement II are correct and Statement II is not the correct explanations of Statement I

(*c*) Statement I is correct but Statement II is not correct

(*d*) Statement I is not correct but Statement II is correct. **[CDS (II) Exam, 2015]**

23. A water tank, open at the top, is hemispherical at the bottom and cylindrical above it. The radius is 12m and the capacity is $3312\pi m^3$. The ratio of the surface areas of the spherical and cylindrical portion is **[CDS (II) Exam, 2015]**

(*a*) 3 : 5 (*b*) 4 : 5 (*c*) 1 : 1 (*d*) 6 : 5

24. A building is in the form of a cylinder surmounted by a hemispherical dome on the diameter of the cylinder. The height of the building is three times the radius of the base of the cylinder. The building contains $67\frac{1}{21}$ m^3 of air. What is the height of the building?

(*a*) 6 m (*b*) 4 m (*c*) 3 m (*d*) 2 m

[CDS (II) Exam, 2016]

25. The radius of the base and the height of a solid right circular cylinder are in the ratio 2 : 3. and its volume is 1617 cm^3. What is the total surface area of the cylinder?

(*a*) 462 cm^2 (*b*) 616 cm^2 (*c*) 770 cm^2 (*d*) 786 cm^2

[CDS (II) Exam, 2016]

26. Rain water from a roof 22 *m* × 20 *m* drains into a cylindrical vessel having diameter of base 2 m and height 3.5m. If the vessel is just full, what is the rainfall? **[CDS (II) Exam, 2016]**

(*a*) 3.5 cm (*b*) 3 cm (*c*) 2.5 cm (*d*) 2 cm

27. A ball of radius 1 cm is put into a cylindrical pipe so that it fits inside the pipe. If the length of the pipe is 14 m, what is the surface area of the pipe? **[CDS (I) Exam, 2017]**

(*a*) 2200 square cm (*b*) 4400 square cm (*c*) 8800 square cm (*d*) 17600 square cm

28. Ice-cream, completely filled in a cylinder of diameter 35 cm and height 32 cm, is to be served by completely filling identical disposable cones of diameter 4 cm and height 7 cm. The maximum number of persons that can be served in this way is **[CDS (I) Exam, 2017]**

(*a*) 950 (*b*) 1000 (*c*) 1050 (*d*) 1100

29. If 1 cubic cm of cast iron weighs 21 gms, then the weight of a cast iron pipe of length 1 m with a bore of 3 cm and in which the thickness of the metal is 1 cm, is:

(*a*) 21 kg (*b*) 24.2 kg (*c*) 26.4 kg (*d*) 18.6 kg

30. A circular pipe is to be designed in such a way that water flowing through it at a velocity of 7 m per min. is collected at its open end at the rate of 11 cubic metre per min. what should be the inner radius of the pipe?

(*a*) $\sqrt{2}$ m (*b*) 2 m (*c*) $\frac{1}{\sqrt{2}}$ m (*d*) $\frac{1}{2}$ m

31. Given a solid cylinder of radius 10 cm and length 1000 cm, a cylindrical hole is made into it to obtain a cylindrical shell of uniform thickness and having volume equal to one-fourth of the original cylinder. The thickness of the cylindrical shell is:

(*a*) $5(\sqrt{5}-2)$cm (*b*) $10(2-\sqrt{3})$cm (*c*) 5 cm (*d*) $5\sqrt{2}$ cm

32. Water flows at the rate of 10 m per min. from a cylindical pipe 5 mm in diameter. How long will it take to fill up a conical vessel whose diameter at the base is 40 cm and depth 24 cm.?

(*a*) 48 min. 15 sec. (*b*) 51 min. 12 sec. (*c*) 52 min. 1 sec. (*d*) 55 min.

ANSWERS

1. (*b*)	**2.** (*a*)	**3.** (*b*)	**4.** (*b*)	**5.** (*a*)	**6.** (*d*)	**7.** (*c*)	**8.** (*d*)	**9.** (*a*)
10. (*b*)	**11.** (*b*)	**12.** (*a*)	**13.** (*b*)	**14.** (*c*)	**15.** (*d*)	**16.** (*b*)	**17.** (*b*)	**18.** (*a*)
19. (*c*)	**20.** (*c*)	**21.** (*b*)	**22.** (*c*)	**23.** (*b*)	**24.** (*a*)	**25.** (*c*)	**26.** (*c*)	**27.** (*c*)
28. (*c*)	**29.** (*c*)	**30.** (*c*)	**31.** (*a*)	**32.** (*b*)				

SOLUTIONS

1. Number of coins = $\frac{\text{Volume of cylinder}}{\text{Volume of 1 coin}} = \left(\frac{\pi \times 2.25 \times 2.25 \times 10}{\pi \times 0.75 \times 0.75 \times 0.2}\right) = 450.$

2. Let, original radius = R. Then, new radius = $(R/2)$

$$\frac{\text{Volume of reduced cylinder}}{\text{Volume of original cylinder}} = \frac{\pi\times\left(\frac{R}{2}\right)^2\times h}{\pi\times R^2\times h} = \frac{1}{4}.$$

3. Let radii be R & $(R/3)$ & heights be h & H. Then,

$$\pi R^2 h = \pi\times\left(\frac{R}{3}\right)^2\times H \Rightarrow H = 3h.$$

So, the length will become 3 times.

4. $2\pi rh = 4400$ and $2\pi r = 110$.

$\therefore$ $110\times h = 4400$ or $h = 40$.

Also, $2\pi r = 110 \Rightarrow r = \frac{110}{2\pi} = \frac{55}{\pi}$.

$$\therefore \text{Volume} = \pi\times\left(\frac{55}{\pi}\right)^2\times 40 = \left(\frac{55\times55\times40}{22}\times7\right) = 38500\text{ cm}^3.$$

5. Let the radii be $2R$, $3R$ and heights be $5h$, $3h$.

$$\therefore \text{Ratio of their Curved Surface Areas} = \frac{2\pi\times2R\times5h}{2\pi\times3R\times3h} = \frac{10}{9}.$$

6. Let original height = h & base radius = R.

Let New radius = r and New height = $6h$.

Now, $\pi r^2 = \frac{1}{9}\pi R^2 \Rightarrow \frac{r}{R} = \frac{1}{3} \Rightarrow R = 3r$.

$$\therefore \frac{\text{Original lateral surface}}{\text{New lateral surface}} = \frac{2\pi\times3r\times h}{2\pi\times r\times6h}$$

$\therefore$ New lateral surface = 2(original lateral surface).

7. Volume of wire = $(2.2\times10\times10\times10)$ cm^3 = 2200 cm^3.

$$\frac{22}{7}\times0.25\times0.25\times h = 2200 \Rightarrow h = \left(2200\times\frac{7}{22}\times\frac{1}{0.25\times0.25}\right)\text{cm}$$

$$\therefore \text{Length of wire} = \left(\frac{11200}{100}\right)\text{m} = 112\text{ m}.$$

8. $H_1 = 18$ and $2\pi R_1 = 30$ or $R_1 = \frac{15}{\pi}$.

$H_2 = 30$ and $2\pi R_2 = 18$ or $R_2 = \frac{9}{\pi}$.

$$\text{Ratio of volumes} = \frac{\pi R_1^2 H_1}{\pi R_2^2 H_2} = \frac{\frac{225}{\pi^2}\times18}{\frac{81}{\pi^2}\times30} = \frac{5}{3}.$$

9. Volume of new vessel = $[\pi\times(15)^2\times35+\pi\times(10)^2\times15] = 9375\pi$.

$\therefore$ $\pi R^2\times15 = 9375\pi \Rightarrow R^2 = 625 \Rightarrow R = 25$ cm.

10. Circumference of girth of roller = 440 cm.

$2\pi R = 440 \Rightarrow R = \left(220 \times \frac{7}{22}\right) = 70$ cm.

Outer radius = 70 cm & inner radius = (70 – 4) cm = 66 cm.

$\therefore$ Volume of iron = $\pi[(70)^2 - (66)^2] \times 63 = \left(\frac{22}{7} \times 136 \times 4 \times 63\right) \text{cm}^3 = 58752 \text{ cm}^3$.

11. We have,

$2 \times TSA = 3 \times CSA \Rightarrow 2 \times 2\pi r(r+h)\ 3 \times 2\pi rh \Rightarrow 4r + 4h = 6h \Rightarrow 4r + 2h \Rightarrow h = 2r$

12. Circumference of circle = 44 = $2\pi r$

$\Rightarrow \quad 44 = 2 \times \frac{22}{7} \times r \Rightarrow r = 7$ cm $\quad h = 44$ cm

$\therefore \quad$ Volume of cylinder = $\pi r^2 h = \frac{22}{7} \times 7 \times 7 \times 44 = 6776 \text{ cm}^2$

13. Volume = $(154 \times 60 \times 1000) \text{ cm}^3$

$r = \frac{14}{2} = 7$ cm

Now, according to the question,

$\pi r^2 h = (154 \times 60 \times 1000) \Rightarrow \frac{22}{7} \times 7^2 \times h = 154 \times 60 \times 1000$

$\Rightarrow \quad h = 60{,}000$ cm = 0.6 km $\quad$ [$\because$ 1 km = 1000 m; 1 m = 100 cm]

14. Let radius and height of a cylinder are $2x$ and $3x$ respectively

We have,

Volume of cylinder = 1617 cm^3

$\Rightarrow \quad \pi r^2 h = 1617 \Rightarrow \frac{22}{7} \times 4x^2 \times 3x = 1617 \Rightarrow r^3 = \frac{147 \times 7}{24} \Rightarrow x = \frac{7}{2}$

$\therefore \quad r = 2x = 7$ cm and $h = 3x = \frac{21}{2}$ cm

$\therefore \quad$ CS of cylinder = $2\pi rh = 2 \times \frac{22}{7} \times 7 \times \frac{21}{2} = 462 \text{ cm}^2$

15. We have, $h = 14$ cm and $2\pi rh - 2\pi rh = 44$

$\Rightarrow \quad 2\pi \times 14(R - r) = 44 \Rightarrow 2 \times \frac{22}{7} \times 14(R - r) = 44 \Rightarrow R - r = \frac{1}{2}$

Now, volume of pipe = 99 cm^3

$\Rightarrow \quad \pi R^2 h - \pi r^2 h = 99 \Rightarrow \frac{22}{7} \times 14(R^2 - r^2) = 99 \Rightarrow 44 \times \frac{1}{2}(R + r) = 99$

$\Rightarrow \quad 22(R + r) = 99 \Rightarrow R + r = \frac{9}{2} = 4.5$

16. Let height of cylinder = $2r$ = diameter of cylinder

$V = \pi r^2 h = \pi r^2 \times 2r = 2\pi r^3$

Diameter of Sphere = diameter of cylinder = $2r$

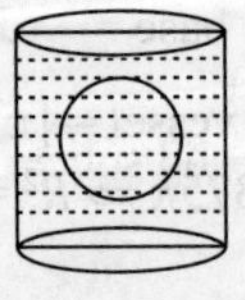

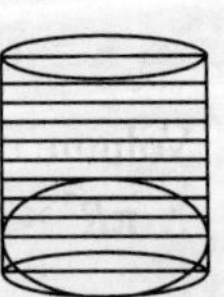

$\therefore$ Volume of Sphere $= \frac{4}{3}\pi r^3 = \frac{2}{3}\times 2\pi r^3 = \frac{2}{3}V$

$\therefore$ Volume of water remains in the vessel $= V - \frac{2}{3}V = \frac{V}{3}$

17.

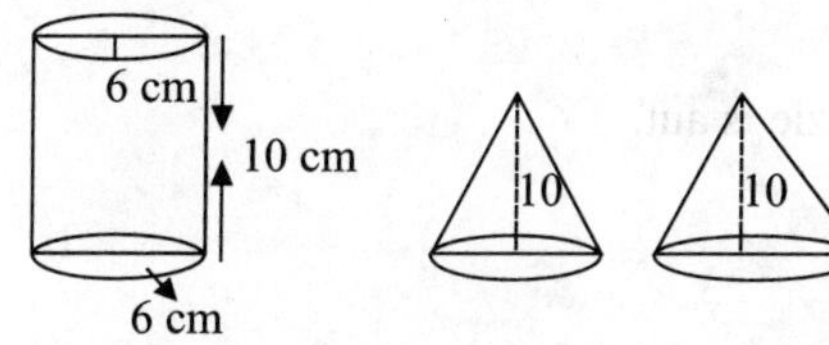

Volume of cylinder $= \pi r^2 h = 360\pi$ cm^3

Flat surface area of cylinder $= 2\pi r^2 = 2\pi \times 6 \times 6 = 72\pi$

Given that the ration of volume of cones are 1 : 2

Volume of smaller cones $= \frac{360\pi}{3}$ cm^3 $= 120\pi$ cm^3 and

Volume of larger cones $= 2 \times \frac{360\pi}{3}$ cm^3 $= 240\pi$ cm^3

Let radius of smaller cones = r cm

Let radius of larger cones = R cm

Now, $\frac{1}{3}\pi r^2 h = 120\pi$

$\Rightarrow \frac{1}{3} \times r^2 \times 10 = 120 \Rightarrow r = 6$ cm $\Rightarrow \frac{1}{3}\pi R^2 h = 240\pi$

$\Rightarrow \frac{1}{3} \times R^2 \times 10 = 240 \Rightarrow R^2 = 72 \Rightarrow R = 6\sqrt{2}$ cm

Now, flat surface area of two cones $= \pi r^2 + \pi R^2$

$= \pi \times 6^2 + \pi(6\sqrt{2})^2$

$= 36\pi + 72\pi = 108\pi$ cm^2

$\therefore$ Percentage increase $= \frac{36\pi}{72\pi}\times 100 = 50\%$

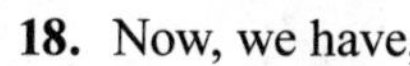

18. Now, we have,

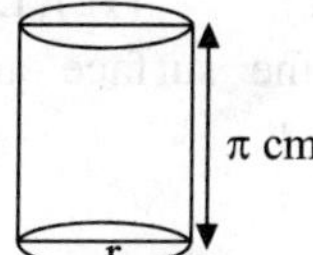

Lateral surface area of cylinder $= \frac{1}{2} \times$ T.S.A. of cylinder

$2\pi rh = \frac{1}{2} \times 2\pi rh\,(r + h)$

$\Rightarrow 2\pi^2 r = \pi r(r + \pi) \Rightarrow 2\pi r^2 = r^2\pi + r\pi^2 \Rightarrow \pi^2 r = 2r^2\pi \Rightarrow \pi = 2r \Rightarrow r = \frac{\pi}{2}$

$\Rightarrow$ radius of its base $= \frac{\pi}{2}$ cm

19. Volume of the solid sphere $= \frac{4}{3}\pi (3)^3 = 36\pi$ cm^3

The volume of the water displaced by the sphere must be equal to the volume of the solid sphere.

Volume of water displaced = $\pi (4)^2$ (Rise in the water level) = $16\pi (h)$ (say)

$$16\pi (h) = 36\pi \Rightarrow h = 2.25 \text{ cm}$$

20. The area of the cross-section of the pipe $(a) = 4 \text{ cm}^2$.
Nozzle velocity $(u) = 40$ m/s.
The volume of the cistern (V) = 480 m^3.
Let the time required to fill the cistern be *t*.
The volume of the water flowing through the nozzle is aut.
$\therefore$ aut = V

$$\Rightarrow t = \frac{V}{au} = \frac{480}{4(10^{-4})(40)} \text{ s} = 30{,}000 \text{ s}$$

$$= \frac{30000}{3600} \text{ hr} = 8\frac{1}{3} \text{ hr} = 8 \text{ hr } 20 \text{ min.}$$

21. The height of the drum is 100 cm. We have to pack cylindrical boxes, each of height 10 cm, into the drum. We can divide the drum into 10 layers. If the number of boxes that can be packed in each layer is *n*, the total number of boxes would be 10*n*.

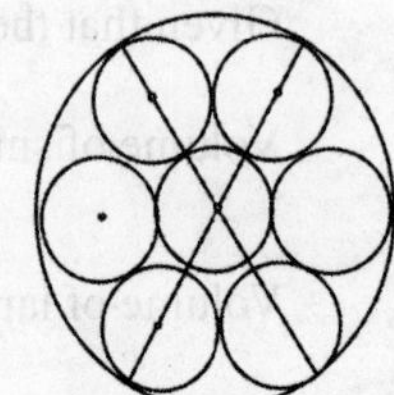

As can be seen from the figure above, $n = 7$.
$\therefore$ $10n = 70$.

Note: Carrom players are familiar with the fact that exactly 6 coins can be arranged around the central red coin.

22. One cylinder can produce 3 cones of the same radius and height.
$\therefore$ 30 cylinders can produce 90 cones. (I) is true.
(II) when a cone is flattened, we get a sector of a circle, not a triangle. (It does not matter how we interpret the word 'flatten'. We can cut the cone along its slant height and roll out the surface or we can take a hammer and flattern the cone. In either case we would get a sector.)

23. The radius of the hemispherical base of the tank (as well as the cylindrical part of height *h*) is $r = 12$ m.

The total capacity of the tank = $\frac{2}{3}\pi r^3 + \pi r^2 h = 3312\pi \text{ m}^3$

$$\therefore \frac{2}{3}r^3 + r^2h = 3312 \Rightarrow \frac{2}{3}(12)(12)(12) + (12)^2h = 3312$$

$$\Rightarrow h = \frac{3312 - 1152}{2(12)} = \frac{2160}{(12)(12)} = 15$$

The ratio of the surface areas of the hemispherical and the cylindrical portion is $\frac{2\pi r^2}{2\pi rh} = \frac{r}{h} = \frac{12}{15} = \frac{4}{5}$.

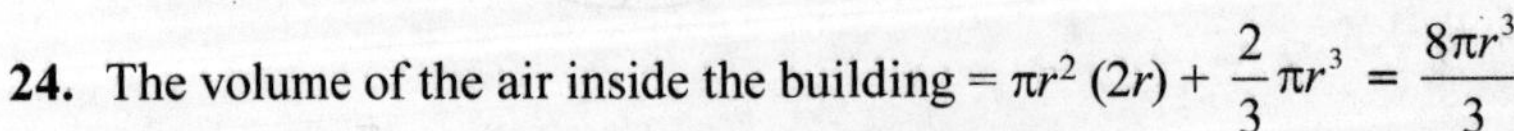

24. The volume of the air inside the building = $\pi r^2 (2r) + \frac{2}{3}\pi r^3 = \frac{8\pi r^3}{3}$

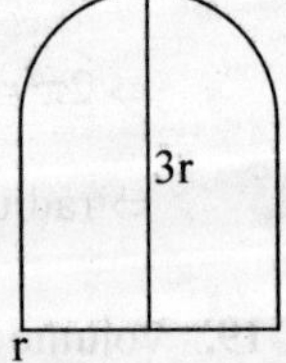

$$\therefore \quad \frac{8\pi r^3}{3} = 67\frac{1}{21} \text{ m}^3 = \frac{1408}{21} m^3$$

$$\therefore \quad r^3 = \frac{3}{1}\left(\frac{7}{22}\right)\frac{1}{8} \times \frac{1408}{21} m^3 \Rightarrow r = 2m.$$

$\therefore$ The height of the building = $3r = 6m$

25. Let the radius of the base of the solid cylinder be $2x$.

$\therefore$ The height is $3x$. The Volume $= \pi(2x)^2(3x) = 1617\text{ cm}^3$

$$\therefore\ x^3 = \frac{1}{12}\frac{7}{22}\frac{1617}{1}\text{ cm}^3 \Rightarrow x^3 = \frac{7}{12}\left(\frac{147}{2}\right)cm^3 \Rightarrow x = \frac{7}{2}\text{ cm.}$$

The total surface area $= 2\pi rh + 2\pi r^2 = 2\pi r(r+h) = 2\pi\,(2x)\,(5x)$

$$= 2\pi\,(10)x^2 = 20\left(\frac{22}{7}\right)\left(\frac{49}{4}\right)cm^2 = 770\text{ cm}^2$$

26. The dimensions of the roof are 20 m × 22 m. Let the rainfall be x.
Let the radius and height of the cylindrical vessel be r ant h.
$r = 1\ m$ and $h = 3.5$ m

$$\therefore\ 20\,(22)\,x = \pi r^2 h \text{ as } x \approx \frac{22}{7}\frac{(1^2)\,(3.5)}{20\,(22)} = \frac{1}{40}\text{ m} = 2.5\text{ cm}$$

27. Given $r = 1$cm and $h = 1400$cm
Curved surface area $= 2\pi rh$

$$= 2\times\frac{22}{7}\times1\times1400 = 8800\text{ cm}^2$$

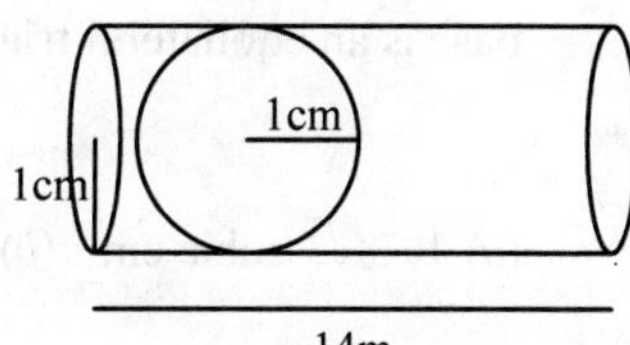

28. n be the number of persons

$$n = \frac{\pi r^2_{cylinder}\ h_{cylinder}}{\frac{1}{3}\pi r^2_{cone} h_{cone}} = \frac{\pi\times\frac{35}{2}\times\frac{35}{2}\times32}{\frac{1}{3}\times\pi\times2\times2\times7} = 1050$$

29. Inner radius = 1.5 cm & outer radius = 2.5 cm.

$\therefore$ Volume of iron $= \pi\times(2.5)^2\times100 - \pi\times(1.5)^2\times100$

$$= \frac{22}{7}\times100\times[(2.5)^2-(1.5)^2]\text{ cm}^3 = \frac{8800}{7}\text{ cm}^3.$$

$$\text{Weight of iron} = \left(\frac{8800}{7}\times\frac{21}{1000}\right)\text{kg} = 26.4\text{ kg}.$$

30. Volume flown per min. $= (\pi r^2\times7) = \left(\frac{22}{7}\times7\times r^2\right) = 22r^2$.

$$\therefore\qquad 22r^2 = 11 \Rightarrow r^2 = \frac{1}{2} \Rightarrow r = \frac{1}{\sqrt{2}}\text{m.}$$

31. Volume of original cylinder $= [\pi\times(10)^2\times1000]\text{ cm}^3$.

$$\text{Volume of shell} = \frac{1}{4}\times\pi\times100000 = (25000\pi)\text{cm}^3.$$

$\pi[R^2-(10)^2]\times1000 = 25000\pi \Rightarrow R^2 = 125 \Rightarrow R = \sqrt{125}$

$\therefore$ Thickness $= (R-10) = \sqrt{125}-10 = 5\sqrt{5}-5\times2 = 5(\sqrt{5}-2)$ cm.

32. Volume flown $= \frac{1}{3}\pi\times(20)^2\times24 = 3200\pi$

Volume flown in 1 min. = $\left(\pi \times \frac{2.5}{10} \times \frac{2.5}{10} \times 1000\right) = 62.5\pi.$

$\therefore$ Time Taken $\frac{3200\pi}{62.5\pi} = 51\ \text{min}.12\ \text{sec}.$

PROBLEMS ON CONE

1. The radius and height of a right circular cone are in the ratio 3 : 7. If the volume of the cone is 528 cm^3, then what is the height of the cone? (Take $\pi = \frac{22}{7}$) **[CDS (I) Exam, 2022]**

(*a*) 3.5 cm (*b*) 70. cm (*c*) 10.5 cm (*d*) 14.0 cm

2. A cone and a hemisphere have equal bases and equal volumes. What is the ratio of the height of the cone to the radius of the hemisphere? **[CDS (I) Exam, 2021]**

(*a*) 1 : 1 (*b*) 2 : 1 (*c*) 3 : 2 (*d*) 4 : 3

3. The section of a solid right circular cone by a plane containing vertex and perpendicular to base is an equilateral triangle of side 14 cm. What is the volume of the cone? $\left(\pi = \frac{22}{7}\right)$ **[CDS (II) Exam, 2024]**

(*a*) $1078\sqrt{3}$ cubic cm (*b*) $\frac{1078}{\sqrt{3}}$ cubic cm (*c*) $539\sqrt{3}$ cubic cm (*d*) $\frac{539}{\sqrt{3}}$ cubic cm

4. The volumes of two cones are in the ratio 1 : 4 and their diameters are in the ratio 4 : 5. What is the ratio of their heights? **[CDS (I) Exam, 2020]**

(*a*) 25 : 64 (*b*) 16 : 25 (*c*) 9 : 16 (*d*) 5 : 9

5. The lateral surface area of a cone is 462 cm^2. Its slant height is 35 cm. The radius of the base of the cone is **[CDS (II) Exam, 2018]**

(*a*) 8.4 cm (*b*) 6.5 cm (*c*) 4.2 cm (*d*) 3.2 cm

6. Two cones have their heights in the ratio 1 : 3. if the radii of their bases are in the ratio 3 : 1, then the ratio of their volumes will be **[CDS (II) Exam, 2018]**

(*a*) 1 : 1 (*b*) 2 : 1 (*c*) 3 : 1 (*d*) 9 : 1 37.

7. The height of a right circular cone is 21 cms and the diameter of the base is $2r$ cms, then its volume (in cu. cms) is:

(*a*) $21\pi r^2$ (*b*) $14\pi r^2$ (*c*) $7\pi r^2$ (*d*) $9\pi r^2$

8. The lateral surface area (in cm^2) of a cone with height 3 cm and radius 4 cm, is

(*a*) $62\frac{6}{7}$ (*b*) $52\frac{6}{7}$ (*c*) $31\frac{3}{7}$ (*d*) $15\frac{5}{7}$

9. The area of the base of a right circular cone is 154 cm^2: and its height is 14 cm. The curved surface of the cone is:

(*a*) $(154 \times \sqrt{5})$ cm^2 (*b*) 11 cm^2 (*c*) $(154 \times \sqrt{7})$ cm^2 (*d*) 5324 cm^2

10. A cone of height 8 m has a curved surface area 188.4 m^2. The radius of the base is: (Take $\pi = 3.14$)

(*a*) 4 m (*b*) 4.5 m (*c*) 5 m (*d*) 6 m

11. The radius and height of a right circular cone are in the ratio of 5 : 12 and its volume is 2512 cm^3. The slant height of the cone is: (Take $\pi = 3.14$)

(*a*) 14 cm (*b*) 16 cm (*c*) 24 cm (*d*) 26 cm

12. How many metres of cloth 2.5 m wide will be required to make a conical tent whose base radius is 7 m and height is 24 m?
(*a*) 120 m (*b*) 180 m (*c*) 220 m (*d*) 550 m

13. The length of canvas 1.1 m wide required to build a conical tent of height 14 m and the floor area 346.5 m^2, is
(*a*) 665 m (*b*) 525 m (*c*) 490 m (*d*) 860 m

14. A conical tent is to accomodate 11 persons such that each person occupies 4 m^2 space on the ground and has 220 m^3 of air to breathe. The height of the cone is:
(*a*) 14 m (*b*) 15 m (*c*) 16 m (*d*) 20 m

15. The diameters of two cones are equal. If their slant heights are in the ratio 5 : 4, the ratio of their curved surface areas is:
(*a*) 4 : 5 (*b*) 25 : 16 (*c*) 16 : 25 (*d*) 5 : 4

16. If the radius of the base of a cone is halved, keeping the height same, what is the ratio of the volume of the reduced cone to that of the original cone?
(*a*) 1 : 2 (*b*) 1 : 3 (*c*) 1 : 4 (*d*) 2 : 3

17. A cone of height 7 cm and base radius 3 cm is carved from a rectangular block of wood 10 cm × 5 cm × 2 cm. The percentage of wood wasted is:
(*a*) 34% (*b*) 46% (*c*) 54% (*d*) 66%

18. If the height of a cone is doubled, then its volume is increased by:
(*a*) 100% (*b*) 200% (*c*) 300% (*d*) 400%

19. If the volumes of two cones are in the ratio 1 : 4 and their diameters are in the ratio 4 : 5, then the ratio of their heights is:
(*a*) 1:5 (*b*) 5:4 (*c*) 5:16 (*d*) 25:64

20. The curved surface area of one cone is twice that of the other while the slant height of the latter is twice that of the former. The ratio of their radii is:
(*a*) 2:1 (*b*) 4:1 (*c*) 8:1 (*d*) 1 : 1

21. A lamp shade is in the shape of a part of a cone and its top and bottom ends are circles whose circumferences are respectively 30 cm and 40 cm. The perpendicular distance between the ends is 6 cm. If the cone were to be completed, then how far would its vertex be from the top end ? **[CDS (II) Exam, 2023]**
(*a*) 20 cm (*b*) 18 cm (*c*) 12 cm (*d*) 9 cm

Consider the following for the next three (03) items:

A cube is inscribed in a sphere. A right circular cylinder is within the cube touching all the vertical faces. A right circular cone is inside the cylinder. Their heights are same and the diameter of the cone is equal to that of the cylinder.

22. Let *V* be the volume of an inverted cone with vertex at origin and the axis of the cone is along positive *y*-axis. The one is filled with water up to half of its height. The volume of water is
(*a*) $\frac{V}{8}$ (*b*) $\frac{V}{6}$ (*c*) $\frac{V}{3}$ (*d*) $\frac{V}{2}$

23. Into a conical tent of radius 8.4 m and vertical height 3.5m, how many full bags of wheat can be emptied, if space required for the wheat in each bag is 1.96 m^3? **[CDS (II) Exam, 2016]**
(*a*) 264 (*b*) 201 (*c*) 132 (*d*) 105

24. A drinking glass of height 24 cm is in the shape of frustum of a cone and diameters of its bottom and top circular ends are 4 cm and 18 cm respectively. If we take capacity of the glass as πx cm^3, then what is the value of *x*? **[CDS (II) Exam, 2016]**
(*a*) 824 (*b*) 1236 (*c*) 1628 (*d*) 2472

Consider the following for the next two (02) items that follow:

The angle at the vertex of a conical body is 120°.

25. What is the ratio of the radius of the conical body to its slant height? **[CDS (I) Exam, 2025]**

(*a*) 1 : 2 (*b*) $\sqrt{3}:1$ (*c*) $\sqrt{3}:2$ (*d*) $\sqrt{2}:1$

26. If the sum of slant height, height and radius is $\left(9+3\sqrt{3}\right)$ cm, then what is the volume of the cone? **[CDS (I) Exam, 2025]**

(*a*) 27π cubic cm (*b*) $18\sqrt{3}\pi$ cubic cm (*c*) 24π cubic cm (*d*) $27\sqrt{3}\pi$ cubic cm

Consider the following for the next two (02) items that follow:

A frustum of a right cone has a top of diameter 2k, bottom of diameter 2·5k and height k.

27. What is the whole surface area of the frustum? **[CDS (I) Exam, 2025]**

(*a*) $39\pi k^2/8$ (*b*) $41\pi k^2/8$ (*c*) $43\pi k^2/8$ (*d*) $45\pi k^2/8$

28. What is the volume of the frustum? **[CDS (I) Exam, 2025]**

(*a*) $61\pi k^3/48$ (*b*) $59\pi k^3/48$ (*c*) $57\pi k^3/48$ (*d*) $53\pi k^3/48$

29. If the ratio of the radius of the base of a right circular cone to its slant heights is 1 : 3, what is the ratio of the total surface area of the curved surface area? **[CDS (I) Exam, 2018]**

(*a*) 5 : 3 (*b*) 3 : 1 (*c*) 4 : 1 (*d*) 4 : 3

30. If the ratio of the circumference of the base of a right circular cone of radius r to its height is 3 : 1, then what is the area of the curved surface of the cone? **[CDS (I) Exam, 2018]**

(*a*) $3\pi r^2$ (*b*) $\dfrac{2\pi r^2\sqrt{4\pi^2+9}}{3}$ (*c*) $\dfrac{\pi r^2\sqrt{\pi^2+1}}{3}$ (*d*) $\dfrac{\pi r^2\sqrt{4\pi^2+9}}{3}$

ANSWERS

1. (*d*)	**2.** (*b*)	**3.** (*b*)	**4.** (*a*)	**5.** (*c*)	**6.** (*c*)	**7.** (*c*)	**8.** (*a*)	**9.** (*a*)
10. (*d*)	**11.** (*d*)	**12.** (*c*)	**13.** (*b*)	**14.** (*b*)	**15.** (*d*)	**16.** (*c*)	**17.** (*a*)	**18.** (*a*)
19. (*d*)	**20.** (*b*)	**21.** (*b*)	**22.** (*a*)	**23.** (*c*)	**24.** (*a*)	**25.** (*c*)	**26.** (*a*)	**27.** (*a*)
28. (*a*)	**29.** (*c*)	**30.** (*d*)						

SOLUTIONS

1. Let radius and height of a right circular cone are 3*x* and 7*x* respectively.

Volume of cone = $\dfrac{1}{3}\pi r^2 h$

$\Rightarrow\ 528 = \dfrac{1}{3}\times\dfrac{22}{7}\times 9x^2\times 7x \Rightarrow 528 = 66x^3 \Rightarrow x^3 = 8 \Rightarrow x = 2$ cm

$\therefore$ height of the cone = 7*x* = 7 × 2 = 14 cm

2. We have,

Volume of cone = volume of hemisphere

$\Rightarrow\ \dfrac{1}{3}\pi r^2 h = \dfrac{2}{3}\pi r^3$ [$\because$ bases are equal] $\Rightarrow h = 2r = \dfrac{h}{r} = \dfrac{2}{1} \Rightarrow 2:1$

3. Height of cone = height of equilateral triangle

Radius of cone = half of triangle base

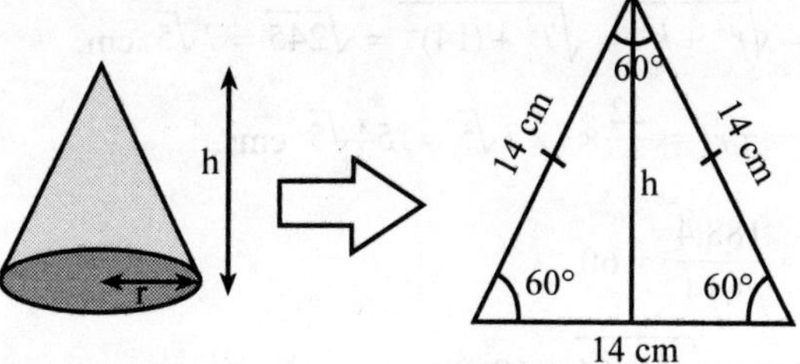

Height of equilateral triangle = $\frac{\sqrt{3}}{2} \times \text{side} \Rightarrow \frac{\sqrt{3}}{2} \times 14 = 7\sqrt{3}$ cm

Radius of cone = $\frac{\text{side of triangle}}{2} \Rightarrow r = \frac{14}{2} = 7$ cm

The volume of the cone = $\frac{1}{3}\pi r^2 h = \left(\frac{1}{3}\right) \times \left(\frac{22}{7}\right) \times 7 \times 7 \times 7\sqrt{3} \Rightarrow \frac{1078}{\sqrt{3}}$ cubic cm

4. Let diameters of two comes are $4x$ and $5x$ respectively and value be V_1 and V_2 respectively.

$\therefore r_1 = 2x$ and $r_2 = \frac{5x}{2}$

Now,

$$\frac{V_1}{V_2} = \frac{1}{4} \Rightarrow \frac{\frac{1}{3}\pi \times (2x)^2 \times h_1}{\frac{1}{3}\pi \times \left(\frac{5x}{2}\right)^2 \times h_2} = \frac{1}{4}$$ [where h_1 and h_2 be the heights of cone]

$$\Rightarrow \frac{16x^2 h_1}{25x^2 h_2} = \frac{1}{4} \Rightarrow \frac{h_1}{h_2} = \frac{25}{64}$$

5. Lateral surface area of cone = πrl

$462 = \frac{22}{7} \times r \times 35 \qquad r = \frac{21}{5} = 4.2$

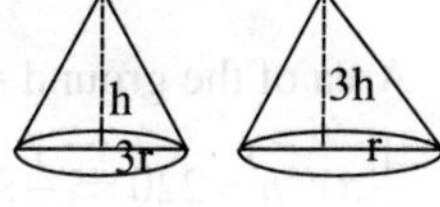

$\therefore$ radius of the base of cone = 4.2 cm

6. Let the heights of the two cones be h and $3h$ respectively and radii of their bases are $3r$ and r respectively.

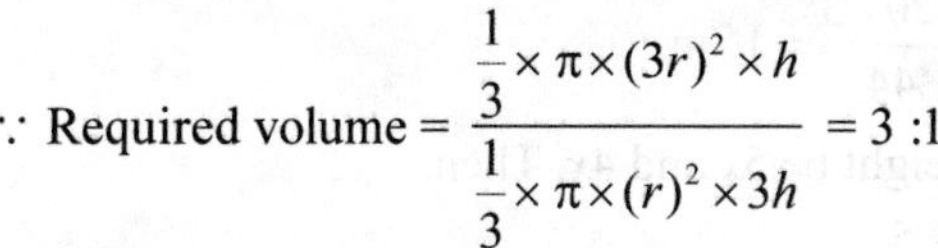

$$\because \text{Required volume} = \frac{\frac{1}{3} \times \pi \times (3r)^2 \times h}{\frac{1}{3} \times \pi \times (r)^2 \times 3h} = 3:1$$

7. Volume = $\left[\frac{1}{3}\pi \times r^2 \times 21\right] \text{cm}^3 = (7\pi r^2)\,\text{cm}^3$.

8. $l = \sqrt{r^2 + h^2} = \sqrt{4^2 + 3^2} = 5$ cm.

$\therefore$ Lateral surface area = $\pi\, rl = \left(\frac{22}{7} \times 4 \times 5\right) \text{cm}^2 = 62\frac{6}{7}\,\text{cm}^2$.

9. $\pi r^2 = 154 \Rightarrow r^2 = \left(154 \times \frac{7}{22}\right) \Rightarrow r = 7$. Also, $h = 14$ cm.

$\therefore$ Slant height, $l = \sqrt{r^2 + h^2} = \sqrt{7^2 + (14)^2} = \sqrt{245} = 7\sqrt{5}$ cm.

$\therefore$ Curved surface $= \pi rl = \frac{22}{7} \times 7 \times 7\sqrt{5} = 154\sqrt{5}$ cm^2.

10. $\pi rl = 188.4 \Rightarrow rl = \frac{188.4}{3.14} = 60.$

$\therefore\ r\sqrt{r^2 + h^2} = 60 \Rightarrow r\sqrt{r^2 + 64} = 60$

$\therefore\ r^2(r^2 + 64) = 3600$ or $r^4 + 64r^2 - 3600 = 0.$

So, $(r^2 + 100)(r^2 - 36) = 0$ or $r = 6$ m.

11. Let radius $= 5x$ and height $= 12x$.

$\frac{1}{3}\pi \times (5x)^2 \times 12x = 2512 \Rightarrow 100 \times 3.14x^3 = 2512 \Rightarrow x^3 = \frac{2512}{314} = 8.$ So, $x = 2$.

$\therefore\ l = \sqrt{r^2 + h^2} = \sqrt{25x^2 + 144x^2} = 13x = 26$ cm.

12. $l = \sqrt{7^2 + (24)^2} = \sqrt{625} = 25$ m.

Area of cloth needed $= \pi rl = \left(\frac{22}{7} \times 7 \times 25\right) = 550$ m^2.

Length of cloth $= \frac{\text{Area}}{\text{Width}} = \left(\frac{550}{2.5}\right)$ m $= 220$ m.

13. $\pi r^2 = 346.5 \Rightarrow r^2 = \left(346.5 \times \frac{7}{22}\right) = \frac{441}{4} \Rightarrow r = \frac{21}{2}.$

$\therefore\ l = \sqrt{r^2 + h^2} = \sqrt{\frac{441}{4} + (14)^2} = \sqrt{\frac{1225}{4}} = \frac{35}{2}.$

So, area of cloth needed $= \pi rl = \left(\frac{22}{7} \times \frac{21}{2} \times \frac{35}{2}\right) = \left(\frac{33 \times 35}{2}\right)$

$\therefore$ Length of cloth $= \left(\frac{33 \times 35}{2 \times 1.1}\right) = 525$ m.

14. Area of the ground $= (11 \times 4)$ m$^2 = 44$ m^2. So, $\pi r^2 = 44$.

$\frac{1}{3}\pi r^2 h = 220 \Rightarrow \frac{1}{3} \times 44 \times h = 220 \Rightarrow h = \frac{220 \times 3}{44} = 15$ m.

15. Let the radius of each be r and the slant height be $5x$ and $4x$. Then,

Ratio of their curved surfaces $= \frac{\pi r \times 5x}{\pi r \times 4x} = \frac{5}{4}.$

16. Let the original radius $= r$ & original height $= h$.

New radius $= (r/2)$ and new height $= h$.

$$\therefore \frac{\text{Volume of reduced cone}}{\text{Volume of original cone}} = \frac{\frac{1}{3}\pi \times \left(\frac{r}{2}\right)^2 \times h}{\frac{1}{3}\pi \times r^2 \times h} = \frac{1}{4}.$$

17. Original volume of wood = $(10 \times 5 \times 2)$ cm^3 = 100 cm^3.

Wood wasted = $\left(100 - \frac{1}{3} \times \frac{22}{7} \times 3 \times 3 \times 7\right)$ = 34 cm^3.

$\therefore$ Percentage of wood wasted = 34%.

18. Let base radius = r & height = h.

Then, original volume = $\frac{1}{3}\pi r^2 h = V(\text{say})$

New radius = r & New height = $2h$.

New volume = $\frac{1}{3}\pi r^2 \times 2h = 2V$.

$\therefore$ Increase % = $\left(\frac{V}{V} \times 100\right)\% = 100\%$.

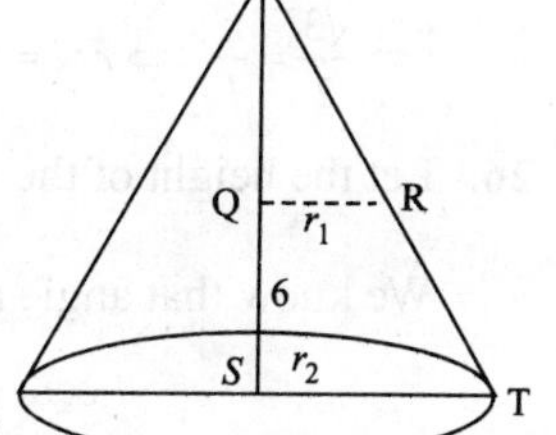

19. Let their diameters be $4x$ & $5x$ and their heights be H_1 and H_2.

Then, $\dfrac{\frac{1}{3} \times \pi(2x)^2 \times H_1}{\frac{1}{3} \times \pi \times \left(\frac{5x}{2}\right)^2 \times H_2} = \frac{1}{4} \Rightarrow \frac{H_1}{H_2} = \left(\frac{1}{4} \times \frac{25}{16}\right) = \frac{25}{64}$.

20. Let their slant heights be l and $2l$ and their radii be R_1 and R_2. Then,

$\therefore$ $\dfrac{\pi R_1 l}{\pi R_2 (2l)} = \frac{2}{1} \Rightarrow \frac{R_1}{R_2} = \frac{4}{1}$.

21. We have, $2\pi r_1 = 30$ and $2\pi r_2 = 40 \Rightarrow \frac{r_1}{r_2} = \frac{3}{4}$

$\therefore \Delta PQR \sim \Delta PST$

$\frac{PQ}{QR} = \frac{PS}{ST} \Rightarrow \frac{PQ}{r_1} = \frac{PQ+6}{r_2}$

$\Rightarrow$ $\frac{PQ}{PQ+6} = \frac{3}{4} \Rightarrow PQ = 18$ cm

22. The inverted cone is filled to half its height with water. The water itself forms a cone similar to the given cone (container). The linear dimensions are in the ratio 1 : 2 and the volumes are in the ratio 1 : 8.

Therefore, if the volume of the given cone is V, the volume of the water is $\frac{V}{8}$.

23. The volume of the conical tent is $\frac{1}{3}\pi\, r^2 h$, where r in the radius of the base and h is the height of the tent. The volume of the wheat of 1 bag is 1.96 m^3. Therefore, The number of logs which can be emptied is

$\frac{1}{3} \times \frac{22}{7} \times \frac{84}{10} \times \frac{84}{10} \times \frac{35}{10} \times \frac{100}{196} = \frac{1}{3} \times \frac{22}{1} \times \frac{84}{2} \times \frac{84}{196} = \frac{11}{3} \times \frac{12}{1} \times \frac{12}{4} = 132$

24. Let the top radius, the bottom radius and the height of the glass be R, r and h respectively. The volume = $\frac{1}{3}\pi\,(24)\left(9^2 + 18 + 2^2\right) = 8\pi(103) = 824\pi = x\pi$ (given) $\therefore x = 8$

25. In ΔAOB,

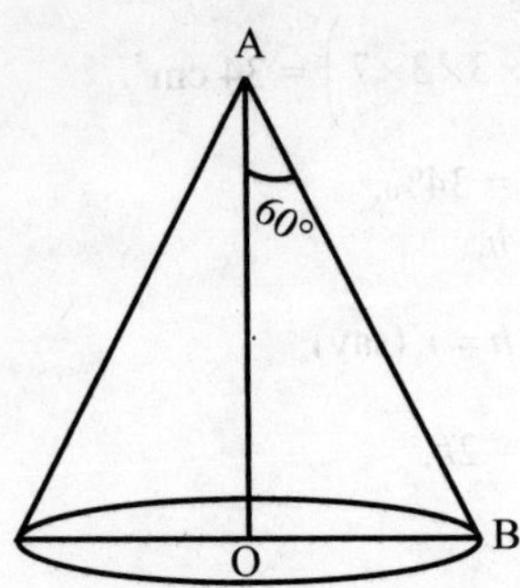

$$\sin 60^\circ = \frac{\text{perpendicular}}{\text{hypotenuse}} = \frac{OB}{AB}$$

$\Rightarrow \frac{\sqrt{3}}{2} = \frac{r}{l} \quad \Rightarrow r : l = \sqrt{3} : 2$

26. Let the height of the cone be (h), radius be (r), and slant height be (l).

We know that angle at the vertex is 120° and $\tan(60^\circ) = \frac{r}{h}$, so

$$r = \sqrt{3} \times h$$

$(\sqrt{3}h) + h + 2h = 9 + 3\sqrt{3} \Rightarrow (\sqrt{3} + 3)h = 9 + 3\sqrt{3} \Rightarrow h = \frac{(9 + 3\sqrt{3})}{(\sqrt{3} + 3)}$

Rationalize the denominator:

$$h = \frac{(9 + 3\sqrt{3})(\sqrt{3} - 3)}{(3 - 9)} \Rightarrow h = \frac{-18}{-6} = 3\ cm$$

Now, $r = \sqrt{3} \times h = \sqrt{3} \times 3 = 3\sqrt{3}\ cm$

Volume $= \frac{1}{3}\pi r^2 h = \frac{1}{3}\pi(3\sqrt{3})^2 \times 3 = 27\pi$ cubic cm

27. Top radius : $r_1 = \frac{D_1}{2} = \frac{2k}{2} = k$

Bottom radius : $r_2 = \frac{D_2}{2} = \frac{2.5k}{2} = \frac{5k}{4}$

Height: $h = k$

$$l = \sqrt{(k)^2 + \left(\frac{5k}{4} - k\right)^2}$$

$$l = \sqrt{k^2 + \left(\frac{5k - 4k}{4}\right)^2} = \sqrt{\frac{16k^2 + k^2}{16}} = \sqrt{\frac{17k^2}{16}} = \frac{k\sqrt{17}}{4}$$

$$A = \pi r_1^2 + \pi r_2^2 + \pi l(r_1 + r_2)$$

$$A = \pi k^2 + \pi\left(\frac{5k}{4}\right)^2 + \pi\left(\frac{k\sqrt{17}}{4}\right)\left(k + \frac{5k}{4}\right)$$

$$A = \pi k^2 + \frac{25\pi k^2}{16} + \frac{9\pi k^2\sqrt{17}}{16}$$

$$A = \frac{16\pi k^2 + 25\pi k^2 + 9\pi k^2\sqrt{17}}{16} = \frac{\pi k^2(41 + 9\sqrt{17})}{16}$$

$$= 9\sqrt{17} \approx 9 \times 4.123 = 37.107$$

$$A \approx \frac{\pi k^2(41 + 37.107)}{16} = \frac{78.107\,\pi k^2}{16} = \frac{39.0535\,\pi k^2}{8}$$

This value is very close to option (a) $\frac{39\pi k^2}{8}$. The small discrepancy is likely due to rounding of $\sqrt{17}$.

28. Substitute the values into the volume formula:

$$V = \frac{1}{3}\pi(k)(k^2 + (1.25k)^2 + k \times 1.25k)$$

$$\Rightarrow V = \frac{1}{3}\pi k\left(k^2 + 1.5625k^2 + 1.25k^2\right) \Rightarrow V = \frac{1}{3}\pi k\left(1k^2 + 1.5625k^2 + 1.25k^2\right)$$

$$\Rightarrow V = \frac{1}{3}\pi k\left(k^2(1 + 1.5625 + 1.25)\right) \Rightarrow V = \frac{1}{3}\pi k\left(3.8125k^2\right) \Rightarrow V = \frac{3.8125}{3}\pi k^3$$

To express 3.8125 as a fraction:

$$3.8125 = \frac{38125}{10000} = \frac{1525}{400} = \frac{305}{80} = \frac{61}{16}$$

So, $$V = \frac{1}{3}\pi k\left(\frac{61}{16}k^2\right) \Rightarrow V = \frac{61}{48}\pi k^3$$

$\therefore$ *The volume of the frustum is* $\frac{61}{48}\pi k^3$.

29. Let the radius and slant height of cone be x and $3x$ respectively

$$\therefore h = \sqrt{l^2 - r^2} = \sqrt{(3x)^2 - x^2} = 8x^2 = 2\sqrt{2}\,x$$

$$\therefore \text{Required ratio} = \frac{\pi r l(l + r)}{\pi r l} = (l + r) = 4 : 1$$

30. Let circumference of base and height of right circular cone are $3x$ and x respectively

$$\Rightarrow 2\pi r = 3x \Rightarrow 2\pi r = 3x \quad \Rightarrow r = \frac{3x}{2\pi} \Rightarrow x = \frac{2r\pi}{3}$$

Now,

$$l = \sqrt{h^2 + r^2} = \sqrt{x^2 + \left(\frac{3x}{2\pi}\right)^2} = \sqrt{x^2 + \frac{9x^2}{4\pi^2}} = \sqrt{\frac{x\sqrt{4\pi^2 + 9}}{2\pi}}$$

Now,

CSA of cone = $\pi r l$

$$= \pi \times \frac{3x}{2\pi} \times \frac{x \times \sqrt{4\pi^2 + 9}}{2\pi} = \frac{3x^2}{4\pi} \times \sqrt{4\pi^2 + 9}$$

$$= 3 \times \frac{4\pi^2 r^2}{9 \times 4\pi} \times \sqrt{4\pi^2 + 9} = \frac{\pi r^2}{3}\sqrt{4\pi^2 + 9}$$

PROBLEMS ON SPHERE

1. If three metallic spheres of radii 6 cm, 8 cm and 10 cm are melted to form a single sphere, then the diameter of the new sphere will be **[CDS (I) Exam, 2015]**
(*a*) 12 cm (*b*) 24 cm (*c*) 30 cm (*d*) 36 cm

2. If the radius of a sphere is increased by 10%, then the volume will be increased by
(*a*) 33.1% (*b*) 30% (*c*) 50% (*d*) 10%
[CDS (I) Exam, 2015]

3. The radius of a sphere is equal to the radius of the base of a right circular cone, and the volume of the sphere is double the volume of the cone. The ratio of the height of the cone to the radius of its base is **[CDS (I) Exam, 2015]**
(*a*) 2: 1 (*b*) 1: 2 (*c*) 2: 3 (*d*) 3: 2

4. How many spherical bullets each of 4 cm in diameter can be made out of a cube of lead whose edge is 44 cm? **[CDS (II) Exam, 2015]**
(*a*) 2541 (*b*) 2551 (*c*) 2561 (*d*) 2571

5. What is the volume of a sphere of radius 3 cm? **[CDS (II) Exam, 2016]**
(*a*) 36π cm^3 (*b*) 18π cm^3 (*c*) 9π cm^3 (*d*) 6π cm^3

6. If the surface area of a sphere is reduced to one-ninth of the area, its radius reduces to
(*a*) One-fourth (*b*) One-third (*c*) One-fifth (*d*) One-ninth
[CDS (I) Exam, 2017]

7. The volume of a sphere is 38808 cu.cm. The curved surface area of the sphere (in cm^2) is:
(*a*) 5544 (*b*) 1386 (*c*) 8316 (*d*) 4158

8. A cone, a hemi-sphere and a cylinder stand on equal bases of radius *R* and have equal heights *H*. Their whole surfaces are in the ratio:
(*a*) $(\sqrt{2}+1):3:4$ (*b*) $(\sqrt{3}+1):3:4$ (*c*) $(\sqrt{2}+1):7:8$ (*d*) $(\sqrt{3}+1):7:8$

9. From a solid sphere of radius 5 cm, a right circular cylindrical hole of radius 3 cm and its axis passing through the centre is removed. The total surface area of the remaining solid (in cm^2) is:
(*a*) 80π (*b*) 88π (*c*) 128π (*d*) 142π

10. The inside of a bowl is part of a sphere. When water is put into the bowl to a depth d, the water surface becomes a circle of radius 2*d*. What is the radius of the sphere?
[CDS (I) Exam, 2025]
(*a*) 2.5*d* (*b*) 2.75*d* (*c*) 3*d* (*d*) 3.25*d*

11. A spherical wooden ball of radius r is to be divided into eight identical parts by cutting by planes passing through the same diameter. What is the surface area of each final piece?
[CDS (II) Exam, 2024]
(*a*) $\frac{\pi r^2}{3}$ (*b*) $\frac{3\pi r^2}{2}$ (*c*) $\frac{2\pi r^2}{3}$ (*d*) $\frac{4\pi r^2}{3}$

12. A sphere of diameter 12.6 cm is melted and cast into a right circular cone of height 25.2 cm. The radius of the base of the cone is:
(*a*) 6.3 cm (*b*) 2.1 cm
(*c*) 2 cm (*d*) 3 cm

13. The diameter of a sphere is 6 cm. It is melted and drawn into a wire of diameter 2 mm. The length of the wire is:
(*a*) 12 m (*b*) 18 m (*c*) 36 m (*d*) 66 m

14. A hollow spherical shell is made of a metal of density 3 gm per cm^3. If its external and internal radii be 9 cm and 6 cm, the weight of the shell (in kg) is:

(*a*) 3.904π (*b*) 1.952π (*c*) 3.874π (*d*) 2.347π

15. A hemispherical bowl of internal radius 9 cm contains a liquid. This liquid is to be filled into cylindrical shaped small bottles of diameter 3 cm and height 4 cm. How many bottles will be needed to empty the bowl?

(*a*) 27 (*b*) 35 (*c*) 54 (*d*) 63

16. Two metallic right circular cones having their heights 4.1 cm and 4.3 cm and the radii of their bases 2.1 cm each. have been melted together and recast into a sphere. The diameter of the sphere is:

(*a*) 4.2 crn (*b*) 1.4 cm (*c*) 3.5 cm (*d*) 6.3 cm

17. A solid iron ball is melted and 64 smaller solid balls of equal size are made using the entire volume of iron. What is the ratio of the surface area of the larger ball to the sum of the surface areas of all the smaller balls ? **[CDS (I) Exam, 2023]**

(*a*) 0·25 (*b*) 0·5 (*c*) 0·75 (*d*) 1

18. What is the radius of the sphere passing through the corners of the cuboid with edges 8 cm, 12 cm and 24 cm ? **[CDS (II) Exam, 2023]**

(*a*) 10·5 cm (*b*) 14 cm (*c*) 21 cm (*d*) 28 cm

19. Three solid lead spheres of diameters 6 cm, 8 cm and 10 cm are melted together and recast as a solid sphere. What is the percentage diminution of the surface area as compared to the sum of the surface areas of the three spheres? **[CDS (II) Exam, 2023]**

(*a*) 25% (*b*) 26% (*c*) 27% (*d*) 28%

20. A solid sphere of radius 3 cm is melted to form a hollow cylinder of height 4 cm and external diameter 10 cm. What is the thickness of the cylinder? **[CDS (II) Exam, 2023]**

(*a*) 0·42 cm (*b*) 0·46 cm (*c*) 0·50 cm (*d*) 1·00 cm

21. A sphere of radius 5 cm is dropped in a right circular cylindrical vessel partly filled with water. The radius of the cylindrical vessel is 10 cm. If the sphere is completely submerged in water, by how much will the level of water rise in the cylindrical vessel? **[CDS (II) Exam, 2023]**

(*a*) $\frac{5}{3}$ cm (*c*) $\frac{5}{2}$ cm (*c*) 1 cm (*d*) $\frac{5}{6}$ cm

22. A spherical ball of lead 6 cm in diameter is melted and recast into three spherical balls. The diameters of the balls are 2 cm, 4 cm and *x* cm. Which one of the following is correct? **[CDS (I) Exam, 2022]**

(*a*) 5 cm < *x* cm < 5.2 cm (*b*) 5.2 cm < *x* cm < 5.4 cm
(*c*) 5.4 cm < *x* cm < 5.6 cm (*d*) 5.6 cm < *x* cm < 5.8 cm

23. Out of 4 identical balls of radius *r*, 3 balls are placed on a plane such that each ball touches the other two balls. The 4th ball is placed on them such that this ball touches all the three balls. What is the distance of centre of 4th ball from the plane? **[CDS (I) Exam, 2022]**

(*a*) $2\sqrt{\frac{2}{3}}r$ unit (*b*) $\frac{\sqrt{3}+2\sqrt{2}}{\sqrt{2}}r$ unit (*c*) $\frac{r}{3-2\sqrt{2}}$ unit (*d*) $\frac{\sqrt{3}+2\sqrt{2}}{\sqrt{3}}r$ unit

24. A hollow spherical shell is made of a metal of density 7 g /cm^3. If its internal and external radii are 3 cm and 6 cm respectively, then what is the mass of the shell?

(Take $\pi = \frac{22}{7}$) **[CDS (II) Exam, 2021]**

(*a*) 2772 g (*b*) 3322 g (*c*) 4433 g (*d*) 5544 g

Consider the following for the next three (03) items:

A cube is inscribed in a sphere. A right circular cylinder is within the cube touching all the vertical faces. A right circular cone is inside the cylinder. Their heights are same and the diameter of the cone is equal to that of the cylinder.

25. What is the ratio of the volume of the sphere to that of the cone? **[CDS (I) Exam, 2019]**

(*a*) $6\sqrt{3}:1$ (*b*) $7:2$ (*c*) $3\sqrt{3}:1$ (*d*) $5\sqrt{3}:1$

26. What is the ratio of the volume of the cube to that of the cylinder? **[CDS (I) Exam, 2019]**

(*a*) 4 : 3 (*b*) 21 : 16 (*c*) 14 : 11 (*d*) 45 : 32

27. Consider the following statements: **[CDS (I) Exam, 2019]**

1. The surface area of the sphere is $\sqrt{5}$ time the curved surface area of the cone.
2. The surface area of the cube is equal to the curved surface area of the cylinder.

Which of the above statements is/are correct?

(*a*) 1 only (*b*) 2 only (*c*) Both 1 and 2 (*d*) Neither 1 nor 2

ANSWERS

1. (*b*)	**2.** (*a*)	**3.** (*a*)	**4.** (*a*)	**5.** (*a*)	**6.** (*b*)	**7.** (*a*)	**8.** (*a*)	**9.** (*d*)
10. (*a*)	**11.** (*b*)	**12.** (*a*)	**13.** (*c*)	**14.** (*b*)	**15.** (*c*)	**16.** (*a*)	**17.** (*a*)	**18.** (*b*)
19. (*d*)	**20.** (*d*)	**21.** (*a*)	**22.** (*b*)	**23.** (*d*)	**24.** (*d*)	**25.** (*a*)	**26.** (*c*)	**27.** (*d*)

SOLUTIONS

PROBLEMS ON SPHERE

1. The three spheres are melted (together) to form a single sphere. When a solid is melted, its volume does not change. Let the radius of the new sphere be *r*.

$$\frac{4}{3}\pi r^3 = \frac{4}{3}\pi(6^3) + \frac{4}{3}\pi(8^3) + \frac{4}{3}\pi(10^3)$$

$\Rightarrow r^3 = 6^3 + 8^3 + 10^3 = 2^3(3^3 + 4^3 + 5^3) = 2^3\, 6^3 = 12^3 \Rightarrow r = 12$ and the diameter is 24.

2. Let the radius of the sphere be *r*.

If the radius of the sphere is increased by 10% the new radius will be 1.1r.

New volume will be $\frac{4}{3}\pi(1.1r)^3\%$. Increase in volume $= (1.1^3 - 1^3)100\% = 33.1\%$

3. Let r be the radius of each of the cones and spheres.

Given that, $\frac{4}{3}\pi r^3 = 2\left(\frac{1}{3}\pi r^2 h\right)$

Where h is the height of the cone $\frac{h}{r} = \frac{2}{1}$.

4. Let the radius of each bullet be *r* (*r* = 2 cm) and the number of bullets be *n*. The side of the cube of lead is *a* = 44 cm.

$$n\left(\frac{4}{3}\pi r^3\right) = a^3 \Rightarrow n = \frac{3}{4} \times \frac{7}{22} \times \frac{44}{8}(44)(44) = 3(7)(11)(11) = 121(21) = 2541.$$

5. The volume of a sphere of radius *r* in $\frac{4}{3}\pi r^3$

$\therefore\ 2 = 3$ cm, volume $= \frac{4}{3}\pi\ (27)\ \text{cm}^3 = 36\pi\ \text{cm}^3$

6. Surface area (S) = $4\pi r^2$, and $S^1 = 4\pi r^2$

According to the question, $S^1 = \frac{1}{9}S$

$$\Rightarrow \quad 4\pi r^{1^2} = \frac{1}{9}4\pi r^2 \Rightarrow r^{1^2} = \frac{r^2}{9} = \left(\frac{r}{3}\right)^2 \Rightarrow r^1 = \frac{r}{3}$$

7. $\frac{4}{3}\pi R^3 = 38808 \Rightarrow R^3 = \left(38808 \times \frac{3}{4} \times \frac{7}{22}\right) = 9261.$

$\therefore \quad R = (21 \times 21 \times 21)^{1/3}$ or $R = 21$.

$\therefore$ Curved Surface Area = $4\pi R^2 = \left(4 \times \frac{22}{7} \times 21 \times 21\right)\text{cm}^2 = 5544\,\text{cm}^2$.

8. Ratio of whole surfaces of cone, hemisphere & cylinder.

$$= (\pi R\sqrt{R^2+H^2} + \pi R^2) : 3\pi R^2 : 2\pi RH + 2\pi R^2$$

$$= (\pi H\sqrt{2H^2} + \pi H^2) : 3\pi H^2 : 4\pi H^2 \quad [\because \text{Height of hemisphere} = \text{Radius} \Rightarrow H = R]$$

$$= (\sqrt{2}+1)\pi H^2 : 3\pi H^2 : 4\pi H^2 = (\sqrt{2}+1) : 3 : 4.$$

9. Clearly, height of the cylinder = dia. of the sphere = 10 cm.

$\therefore$ Required area = $[4\pi \times (5)^2 - 2 \times (\pi \times 3^2) + 2\pi \times 3 \times 10]$ cm² = $(100\pi + 60\pi - 18\pi) = 142\pi$.

10. Using the geometry of spheres, the radius of the sphere (R) is given by:

$$R = \frac{r^2+d^2}{2d}$$

$$R = \frac{r^2+d^2}{2d} = \frac{(2d)^2+d^2}{2d} = \frac{4d^2+d^2}{2d} = \frac{5d^2}{2d} = 2.5d$$

11.

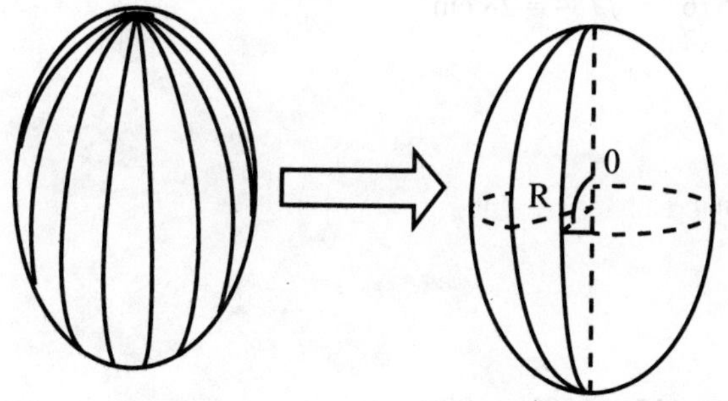

R 0

The surface area of each final piece = Area of circle + (Area of sphere/8)

$$= \pi r^2 + \left(\frac{4\pi r^2}{8}\right) = \pi r^2 + \frac{\pi r^2}{2} = \frac{3\pi r^2}{2}$$

$\therefore$ The surface area of each final piece is $\frac{3\pi r^2}{2}$.

12. $\frac{4}{3}\pi \times (6.3)^3 = \frac{1}{3}\pi R^2 \times 25.2 \Rightarrow R = 6.3$ cm.

13. $\frac{4}{3}\pi \times (3)^3 = \frac{1}{3}\pi(0.1)^2 \times H \Rightarrow H = \frac{36}{0.01}\text{cm} = 3600\,\text{cm} = 36\,\text{m}.$

14. Volume of shell = $\frac{4}{3}\pi(9^3 - 6^3) = \frac{4}{3}\pi \times 513 = 684\pi$

$\therefore$ Weight = (Volume × Density) = $\left(684\pi \times \frac{3}{1000}\right)$kg = (1.952π) kg

15. Number of bottles = $\frac{\text{Volume of the bowl}}{\text{Volume of 1 bottle}} = \frac{\frac{2}{3}\pi \times 9^3}{\pi \times \left(\frac{3}{2}\right)^2 \times 4} = 54.$

16. Sum of the volume of the cones = Volume of the sphere.

$\therefore \frac{1}{3}\pi \times (2.1)^2 \times 4.1 + \frac{1}{3}\pi \times (2.1)^2 \times 4.3 = \frac{4}{3}\pi R^3$

$\Leftrightarrow \frac{1}{3}\pi \times (2.1)^2(4.1 + 4.3) = \frac{4}{3}\pi R^3 \Leftrightarrow R^3 = (2.1)^3 \Rightarrow R = 2.1.$

$\therefore$ Diameter of the sphere = 4.2 cm.

17. Let radius of larger solid iron ball = R

Let radius of smaller solid iron ball = r

Now, according to the question

$$\frac{4}{3}\pi r^3 = 64 \times \frac{4}{3}\pi r^3 \Rightarrow \quad R^3 = 64r^3 \Rightarrow R = 4r$$

$\therefore$ Required ratio = $\frac{4\pi R^2}{64 \times 4\pi r^2} = \frac{(4r)^2}{64 \times r^2} = \frac{1}{4} = 0.25$

18. Diameter of sphere = Diagonal of cuboid = $\sqrt{l^2 + b^2 + h^2} = \sqrt{8^2 + 12^2 + 24^2}$

$= \sqrt{64 + 144 + 576} = \sqrt{784} = 28$ cm

$\therefore$ Radius of sphere = $\frac{28}{2} = 14$ cm

19. We have,

$r_1 = 6$ cm, $r_2 = 8$ cm and $r_3 = 10$ cm

Let radius of new sphere = R

$\left(346.5 \times \frac{7}{22}\right) = \frac{441}{4} \Rightarrow r = \frac{21}{2}.$

$\Rightarrow \quad R^3 = 6^3 + 8^3 + 10^3 = 216 + 512 + 1000 = 1728$

$\Rightarrow \quad R = 12$ cm

$\therefore$ Surface area = $4\pi R^2 = 4 \times \pi \times 12^2 = 576\pi$

Now,

Surface area of 3 solid lead spheres = $4\pi\,(r_1^2 + r_2^2 + r_3^2)$

$= 4\pi\,(6^2 + 8^2 + 10^2)$

$= 4\pi \times 200 = 800\pi$

$\therefore$ required percentage = $\frac{800\pi - 576\pi}{800\pi} \times 100 = \frac{224}{800} \times 100 = 28\%$

20. $r = 3$ cm
$R = 5$ cm
Now, according to the question,

$$\frac{4}{3}\pi r^2 = \pi h(R^2 - r^2)$$

$$\Rightarrow \frac{4}{3}\pi 3\times 3\times 3 = 4\times(5^2 - r^2)$$

$$\Rightarrow 9 = 25 - r^2 \Rightarrow r^2 = 16 \Rightarrow r = 4$$

$\therefore$ thickness $= 5 - 4 = 1$ cm

21. Let height of cylinder $= h$
$r = 5$ cm (Radius of sphere)
$R = 10$ cm (Radius of cylinder)
Now, according to the question,
Volume of sphere = volume of cylinder

$$\frac{4}{3}\pi r^3 = \pi R^2 h \Rightarrow \frac{4}{3}\times 5^3 = 10^2\times h \Rightarrow h = \frac{5}{3}\text{ cm}$$

22. $r = \frac{6}{2} = 3$ cm

Volume of ball $= \frac{4}{3}\pi r^3 = \frac{4}{3}\times\frac{22}{7}\times 27$

Now, according to the question,

$$\frac{4}{3}\times\frac{22}{7}\times 27 = \frac{4}{3}\times\frac{22}{7}\left(1^3 + 2^3 + \left(\frac{x}{2}\right)^3\right)$$

$$\Rightarrow 27 = 1 + 8 + \frac{x^3}{8} = 18 = \frac{x^3}{8} \Rightarrow x^3 = 144$$

$\because$ $5^3 < 144 < 63$
By checking option, we get
(a) $5.23 = 140.608 < 144$ (b) $5.4^3 = 157.464 > 144$
$\because$ 5.2 cm $< x$ cm < 5.4 cm

23. Side of triangular pyramid $= \sqrt{\frac{2}{3}}\times 2r$

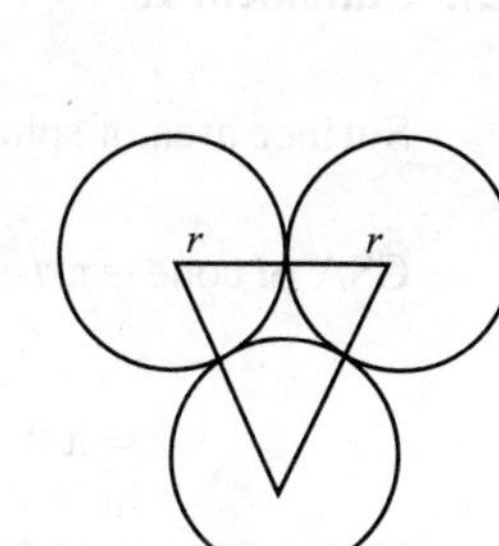

Now, distance of centre of 4th ball from the plane $= \frac{2\sqrt{2}}{\sqrt{3}}r + r$

$$= \frac{r}{\sqrt{3}}\left(2\sqrt{2} + \sqrt{3}\right)$$

24. We have, $r = 3$ and $R = 6$ cm

Mass of shell $= \frac{4}{3}\pi(R^3 - r^3)\times 7 = \frac{4}{3}\times\frac{22}{7}\left(6^3 - 3^3\right)\times 7$

$$= \frac{4}{3}\times 22\,(216 - 27) = \frac{4}{3}\times 22\times 189 = 5544\text{ g.}$$

Solution: (25 – 27)

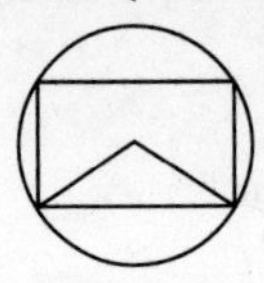 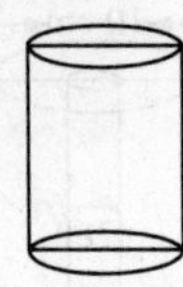 

Let length of cube = x

$\therefore$ diagonal of cube = $\sqrt{3}\,x$

$\therefore$ radius of sphere = $\dfrac{\sqrt{3}x}{2} = R$

Now, volume of cube = Side^3 = x^3

Volume of sphere = $\dfrac{4}{3}\pi R^3 = \dfrac{4}{3}\times\pi\times\dfrac{3\sqrt{3}}{8}x^3 = \dfrac{\sqrt{3}}{2}\pi x^3$

Height of cylinder = $h = x$ [$\because$ height of cylinder = radius of cube]

Radius of cylinder = $r = \dfrac{x}{2}$

$\therefore$ Volume of cylinder = $\pi r^2 h = \pi\,\dfrac{x^2}{4}\times x = \dfrac{\pi x^3}{4}$

Now, given that right circular cone is inside the cylinder with same base and height

$\therefore$ Volume of cone = $\dfrac{1}{3}\pi r^2 h = \dfrac{1}{3}\times\pi\times\dfrac{x^2}{4}\times x = \dfrac{\pi x^3}{12}$

25. Required ratio = $\dfrac{\frac{\sqrt{3}}{2}\pi x^3}{\frac{\pi}{12}x^3} = 6\sqrt{3}:1$

26. Required ration = $\dfrac{x^3}{\frac{\pi}{4}x^3} = \dfrac{4\times 7}{22} = 14:11$

27. Statement 1:

Surface area of sphere = $4\pi R^2 = 4\times\pi\times\left(\dfrac{\sqrt{3}}{2}x\right)^2 = 3\pi x^2$

CSA of cone = $\pi rl = \pi\times\dfrac{x}{2}\times\sqrt{h^2+r^2}$

$= \pi\times\dfrac{x}{2}\times\sqrt{x^2+\dfrac{x^2}{4}} = \pi\times\dfrac{x}{2}\times\dfrac{\sqrt{5}x}{2} = \dfrac{\sqrt{5}\pi x^2}{4}$

Hence, statement 1 is not correct

Statement 2 :

Surface area of cube = $6\times\text{side}^2 = 6\times x^2 = 6x^2$

CSA of cylinder = $2\pi rh = 2\pi\,\dfrac{x}{2}\times x = \pi r^2$

Hence, statement 2 is not correct

UNIT 3

ALGEBRA

Basic Operations: Basic operations, Polynomials, Factorizations, Remainder Theorem and Factorization Theorem ; H.C.F. & L.C.M. of polynomials ; Rational Expressions, Surds & indices.

Linear Equations: Simultaneous linear equations in two variables, analytical and graphical solutions. Simultaneous linear inequations. Practical problems on linear equations.

Quadratic Equations : Solution of quadratic equations, Relation between roots and coefficients (only real roots to be considered). Practical problems on quadratic equations.

Set Theory: Set language and set notation.

Union, Intersection & Complementations of sets and their properties.

UNIT 3

ALGEBRA

Basic Operations : Basic operations, Polynomials, Factorization : Remainder Theorem and Factorization Theorem, H.C.F. & L.C.M. of polynomials, Rational Expressions, Surds & indices.

Linear Equations : Simultaneous linear equations in two variables : analytical and graphical solutions, Simultaneous linear inequations, Practical problems on linear equations.

Quadratic Equations : Solution of quadratic equations, Relation between roots and coefficients (only real roots to be considered), Practical problems on quadratic equations.

Set Theory: Set language and set notation.

Union, Intersection & Complementations of sets and their properties.

19 Basic Operations and Factorization

CHAPTER

Highlights: Basic operations, Simple factors, Factor Theorem, Remainder Theorem, Division of polynomials

Formulae:

(*i*) $(x+y)^2 = x^2 + y^2 + 2xy$ and $(x-y)^2 = x^2 + y^2 - 2xy$.

(*ii*) $(x+y)^2 - (x-y)^2 = 4xy$ and $(x+y)^2 + (x-y)^2 = 2(x^2+y^2)$

(*iii*) $(x+y)^3 = x^3 + y^3 + 3xy(x+y)$

(*iv*) $(x-y)^3 = x^3 - y^3 - 3xy(x-y)$

(*v*) $(x^2 - y)^2 = (x+y)(x-y)$

(*vi*) $(x^3 + y^3) = (x+y)(x^2 + y^2 - xy)$

(*vii*) $(x^3 - y^3) = (x-y)(x^2 + y^2 + xy)$

(*viii*) $(x+y+z)^2 = [x^2 + y^2 + z^2 + 2(xy + yz + zx)]$

(*ix*) $(x^3 + y^3 + z^3 - 3xyz) = (x+y+z)(x^2 + y^2 + z^2 - xy - yz - zx)$

(*x*) $x + y + z = 0 \Rightarrow x^3 + y^3 + z^3 = 3xyz$.

Factor Theorem: If *$f(x)$ is completely divisible by $(x-a)$, then $f(a) = 0$.*

Thus, $(x-a)$ is a factor of $f(x) \Leftrightarrow f(a) = 0$.

Ex. 1. ***Show that $(x+1)$ is a factor of $f(x) = 4x^2 + 3x - 1$.***

Sol. Putting $x + 1 = 0$, i.e. $x = -1$ in the given expression we get $f(-1) = 0$.

So, $(x+1)$ is a factor of $f(x)$.

Remainder Theorem: *If an expression $f(x)$ is divided by $(x-a)$, then the remainder is $f(a)$.*

Ex. 2. ***Let $f(x) = x^3 + 3x^2 - 5x + 4$ be divided by $(x-1)$. Find the remainder.***

Sol. Remainder $= f(1) = 1^3 + 3 \times 1^2 - 5 \times 1 + 4 = 3$.

Some Important Results:

(*i*) $(x^n - a^n)$ is divisible by $(x-a)$ for all values of n.

(*ii*) $(x^n + a^n)$ is divisible by $(x+a)$ only when n is odd.

(*iii*) $(x^n - a^n)$ is divisible by $(x+a)$ only for even values of n.

(*iv*) $(x^n + a^n)$ is never divisible by $(x-a)$.

EXERCISE 19

1. If $x = a + b + \dfrac{(a-b)^2}{4a+4b}$ and $y = \dfrac{a+b}{4} + \dfrac{ab}{a+b}$, then what is the value of $(x-a)^2 - (y-b)^2$? **[CDS (I) Exam, 2023]**

(*a*) a^2 (*b*) b^2 (*c*) ab (*d*) a^2b^2

2. If $x = 97 + 56\sqrt{3}$, then what is the value of $\sqrt[4]{x} + \dfrac{1}{\sqrt[4]{x}} = ?$ **[CDS (II) Exam, 2023]**

(*a*) 7 (*b*) 6 (*c*) 5 (*d*) 4

3. If x, y, z are real numbers such that $x + y + z = 10$ and $xy + yz + zx = 18$, then what is the value of $x^3 + y^3 + z^3 - 3xyz$? **[CDS (I) Exam, 2025]**

(*a*) 400 (*b*) 440 (*c*) 460 (*d*) 500

4. Which one of the following is a factor of $3\sqrt{3}x^3+2\sqrt{2}yx^3+18xy+6\sqrt{6}$?

[CDS (II) Exam, 2023]

(*a*) $\sqrt{3}x+\sqrt{xy}-\sqrt{3}$ (*b*) $\sqrt{3x}+\sqrt{2y}-\sqrt{6}$

(*c*) $3x^2+2y^2-\sqrt{18x}-\sqrt{12y}-\sqrt{6xy}+6$ (*d*) $3x^2+2y^2+\sqrt{18x}+\sqrt{12y}-\sqrt{6xy}+6$

5. What is the minimum value of $\left(\frac{a^2+3a+1}{a}\right)\left(\frac{b^2+3b+1}{b}\right)$ for $a, b > 0$,

(*a*) 1 (*b*) 9 (*c*) 16 (*d*) 25

[CDS (I) Exam, 2024]

6. p varies directly as $\left(x^2+y^2+z^2\right)$. When $x=1, y=2, z=3$, then $p=70$. What is the value of p when $x=-1, y=1, z=5$? **[CDS (II) Exam, 2024]**

(*a*) 100 (*b*) 125 (*c*) 135 (*d*) 140

7. If $p=\frac{a^2}{(b-a)(c-a)}, q=\frac{b^2}{(c-b)(a-b)}, r=\frac{c^2}{(a-c)(b-c)}$, then what is $(p+q+r)^2$ equal to?

[CDS (II) Exam, 2023]

(*a*) 9 (*b*) 4 (*c*) 1 (*d*) 0

8. If $2s = a + b + c$, then what is $s^2 + (s - a)(s - b) + (s - b)(s - c) + (s - c)(s - a)$ equal to?

[CDS (II) Exam, 2023]

(*a*) $(a + b + c)^2$ (*b*) $ab + bc + ca$

(*c*) $2(ab + bc + ca)$ (*d*) $3(ab + bc + ca)$

9. What is $\frac{(x-y)^3+(y-z)^3+(z-x)^3}{3(x-y)(y-z)(z-x)}$ equal to? **[CDS (I) Exam, 2019]**

(*a*) 1 (*b*) 0 (*c*) $\frac{1}{3}$ (*d*) 3

10. If $a^x + b^y = c^z$ and $b^2 = ac$, then what is $\frac{1}{x}+\frac{1}{z}$ equal to? **[CDS (I) Exam, 2019]**

(*a*) $\frac{1}{y}$ (*b*) $-\frac{1}{y}$ (*c*) $\frac{2}{y}$ (*d*) $-\frac{2}{y}$

11. The expression $5^{2n} - 2^{3n}$ has a factor **[CDS (I) Exam, 2019]**

(*a*) 3 (*b*) 7 (*c*) 17 (*d*) None of the above

12. For what value of k can the expression $x^3 + kx^2 - 7x + 6$ be resolved into three linear factors? **[CDS (II) Exam, 2019]**

(*a*) 0 (*b*) 1 (*c*) 2 (*d*) 3

13. What is $(x - a)(x - b)(x - c)$ equal to? **[CDS (II) Exam, 2019]**

(*a*) $x^3 - (a + b + c)x^2 + (bc + ca + ab)x - abc$

(*b*) $x^3 + (a + b + c)x^2 + (bc + ca + ab)x + abc$

(*c*) $x^3 - (bc + ca + ab)x^2 + (a + b + c)x + abc$

(*d*) $x^3 - (bc + ca + ab)x^2 - (a + b + c)x - abc$

14. If x^2-5x+4 is a factor of x^4-px^2+q, then what are the values of p and q respectively?

[CDS (I) Exam, 2025]

(*a*) 17,16 (*b*) 16,17 (*c*) 15,16 (*d*) 16,15

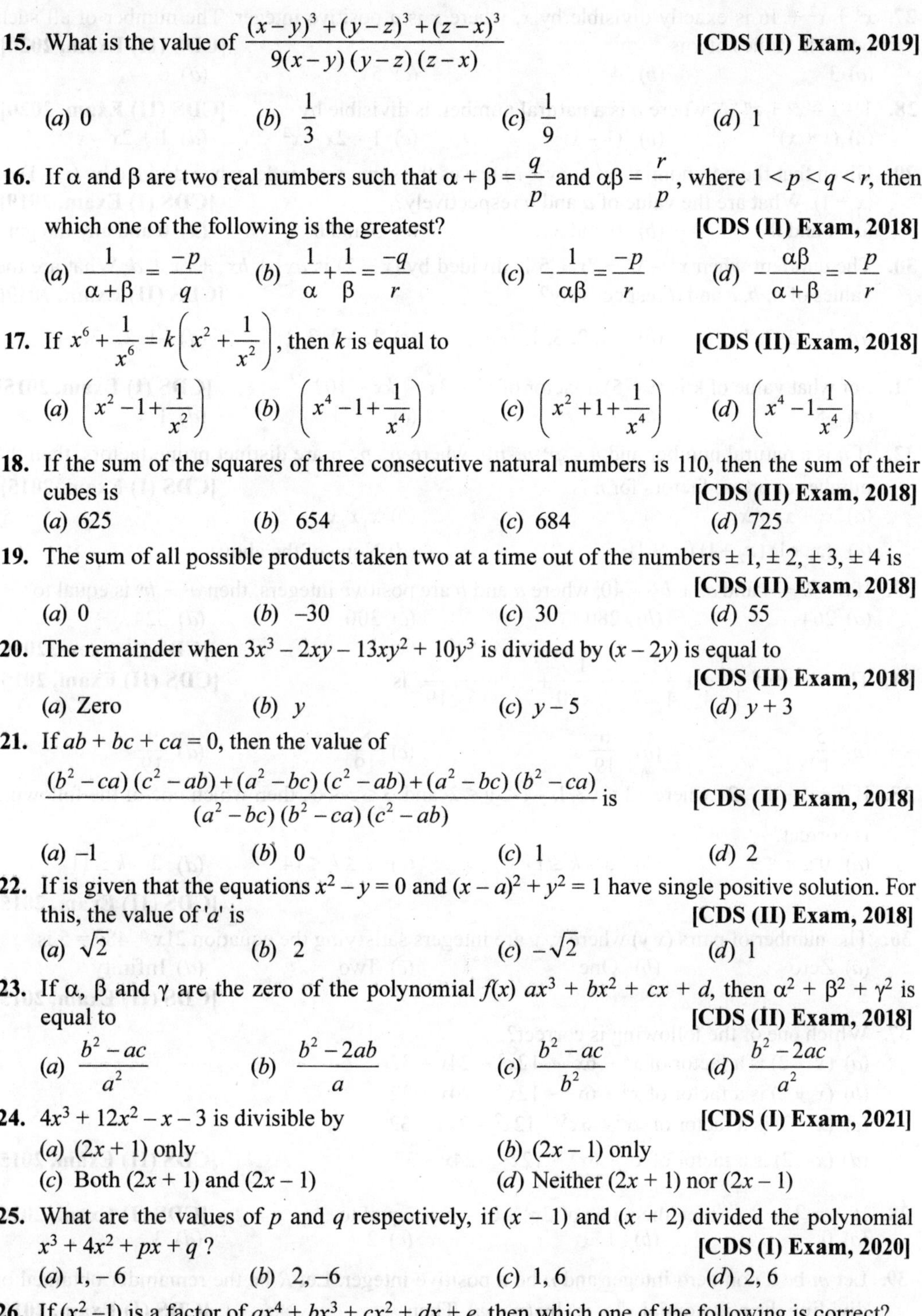

15. What is the value of $\dfrac{(x-y)^3+(y-z)^3+(z-x)^3}{9(x-y)(y-z)(z-x)}$ **[CDS (II) Exam, 2019]**

(*a*) 0 (*b*) $\frac{1}{3}$ (*c*) $\frac{1}{9}$ (*d*) 1

16. If α and β are two real numbers such that $\alpha+\beta=\frac{q}{p}$ and $\alpha\beta=\frac{r}{p}$, where $1<p<q<r$, then which one of the following is the greatest? **[CDS (II) Exam, 2018]**

(*a*) $\frac{1}{\alpha+\beta}=\frac{-p}{q}$ (*b*) $\frac{1}{\alpha}+\frac{1}{\beta}=\frac{-q}{r}$ (*c*) $-\frac{1}{\alpha\beta}=\frac{-p}{r}$ (*d*) $-\frac{\alpha\beta}{\alpha+\beta}=-\frac{p}{r}$

17. If $x^6+\frac{1}{x^6}=k\left(x^2+\frac{1}{x^2}\right)$, then k is equal to **[CDS (II) Exam, 2018]**

(*a*) $\left(x^2-1+\frac{1}{x^2}\right)$ (*b*) $\left(x^4-1+\frac{1}{x^4}\right)$ (*c*) $\left(x^2+1+\frac{1}{x^4}\right)$ (*d*) $\left(x^4-1\frac{1}{x^4}\right)$

18. If the sum of the squares of three consecutive natural numbers is 110, then the sum of their cubes is **[CDS (II) Exam, 2018]**

(*a*) 625 (*b*) 654 (*c*) 684 (*d*) 725

19. The sum of all possible products taken two at a time out of the numbers ± 1, ± 2, ± 3, ± 4 is **[CDS (II) Exam, 2018]**

(*a*) 0 (*b*) –30 (*c*) 30 (*d*) 55

20. The remainder when $3x^3-2xy-13xy^2+10y^3$ is divided by $(x-2y)$ is equal to **[CDS (II) Exam, 2018]**

(*a*) Zero (*b*) y (*c*) $y-5$ (*d*) $y+3$

21. If $ab+bc+ca=0$, then the value of

$$\frac{(b^2-ca)(c^2-ab)+(a^2-bc)(c^2-ab)+(a^2-bc)(b^2-ca)}{(a^2-bc)(b^2-ca)(c^2-ab)}$$ is **[CDS (II) Exam, 2018]**

(*a*) –1 (*b*) 0 (*c*) 1 (*d*) 2

22. If is given that the equations $x^2-y=0$ and $(x-a)^2+y^2=1$ have single positive solution. For this, the value of '*a*' is **[CDS (II) Exam, 2018]**

(*a*) $\sqrt{2}$ (*b*) 2 (*c*) $-\sqrt{2}$ (*d*) 1

23. If α, β and γ are the zero of the polynomial $f(x)$ ax^3+bx^2+cx+d, then $\alpha^2+\beta^2+\gamma^2$ is equal to **[CDS (II) Exam, 2018]**

(*a*) $\frac{b^2-ac}{a^2}$ (*b*) $\frac{b^2-2ab}{a}$ (*c*) $\frac{b^2-ac}{b^2}$ (*d*) $\frac{b^2-2ac}{a^2}$

24. $4x^3+12x^2-x-3$ is divisible by **[CDS (I) Exam, 2021]**

(*a*) $(2x+1)$ only (*b*) $(2x-1)$ only

(*c*) Both $(2x+1)$ and $(2x-1)$ (*d*) Neither $(2x+1)$ nor $(2x-1)$

25. What are the values of p and q respectively, if $(x-1)$ and $(x+2)$ divided the polynomial x^3+4x^2+px+q ? **[CDS (I) Exam, 2020]**

(*a*) 1, – 6 (*b*) 2, –6 (*c*) 1, 6 (*d*) 2, 6

26. If (x^2-1) is a factor of $ax^4+bx^3+cx^2+dx+e$, then which one of the following is correct? **[CDS (II) Exam, 2021]**

(*a*) $a+b+c=d+e$ (*b*) $a+b+e=c+d$ (*c*) $b+c+d=a+e$ (*d*) $a+c+e=b+d$

27. $x^3 + x^2 + 16$ is exactly divisible by x, where x is a positive integer. The number of all such possible values of x is **[CDS (II) Exam, 2020]**

(a) 3 (b) 4 (c) 5 (d) 6

28. $1 - x - x^n + x^{n+1}$, where n is a natural number, is divisible by **[CDS (II) Exam, 2020]**

(a) $(1+x)^2$ (b) $(1-x)^2$ (c) $1-2x-x^2$ (d) $1+2x-x^2$

29. Given that the polynomial $(x^2 + ax + b)$ leaves the same remainder when divided by $(x-1)$ or $(x+1)$. What are the value of a and b respectively? **[CDS (I) Exam, 2019]**

(a) 4 and 0 (b) 0 and 3 (c) 3 and 0 (d) 0 and any integer

30. The quotient when $x^4 - x^2 + 7x + 5$ is divided by $(x+2)$ is $ax^3 + bx^2 + cx + d$. What are the values of a, b, c and d respectively? **[CDS (II) Exam, 2019]**

(a) 1, –2, 3, 1 (b) –1, 2, 3, 1 (c) 1, – 2, 3, 1 (d) –1, 2, –3, –1

31. For what value of k is $(x-5)$ a factor of $x^3 - 3x^2 + kx - 10$? **[CDS (I) Exam, 2015]**

(a) –8 (b) 4 (c) 2 (d) 1

32. If n is a natural number and $n = p_1^{x_1} p_2^{x_2} p_3^{x_3}$, where p_1, p_2, p_3 are distinct prime factors, then the number of prime factors for n is **[CDS (I) Exam, 2015]**

(a) $x_1 + x_2 + x_3$ (b) $x_1 x_2 x_3$

(c) $(x_1+1)(x_2+1)(x_3+1)$ (d) None of the above

33. If $a - b = 4$ and $a^2 + b^2 = 40$, where a and b are positive integers, then $a^3 + b^6$ is equal to

(a) 264 (b) 280 (c) 300 (d) 324

[CDS (I) Exam, 2015]

34. The value of $\frac{1}{1\times4} + \frac{1}{4\times7} + \frac{1}{7\times10} + ... + \frac{1}{16\times19}$ is **[CDS (II) Exam, 2015]**

(a) $\frac{5}{19}$ (b) $\frac{6}{19}$ (c) $\frac{8}{19}$ (d) $\frac{9}{19}$

35. If $k = x - y + 2z$ where $-2 \le x \le 1$, $-1 \le y \le 2$ and $3 \le z \le 6$, then which one of the following is correct?

(a) $0 \le k \le 9$ (b) $5 \le k \le 11$ (c) $2 \le k \le 14$ (d) $2 \le k \le 11$

[CDS (II) Exam, 2015]

36. The number of pairs $(x\ y)$ where x, y are integers satisfying the equation $21x + 48y = 5$ is

(a) Zero (b) One (c) Two (d) Infinity

[CDS (II) Exam, 2015]

37. Which one of the following is correct?

(a) $(x+2)$ is a factor of $x^4 - 6x^3 + 12x^2 - 24x + 32$

(b) $(x+2)$ is a factor of $x^4 + 6x^3 - 12x^2 + 24x - 32$

(c) $(x-2)$ is a factor of $x^4 - 6x^3 + 12x^2 - 24x + 32$

(d) $(x-2)$ is a factor of $x^4 - 6x^3 - 12x^2 + 24x - 32$ **[CDS (II) Exam, 2015]**

38. If $x = 2^{\frac{1}{3}} + 2^{-\frac{1}{3}}$, then the value of $2x^3 - 6x - 5$ is equal to **[CDS (I) Exam, 2016]**

(a) 0 (b) 1 (c) 2 (d) 3

39. Let m be a non-zero integer and n be a positive integer. Let R be the remainder obtained on dividing the polynomial $x^n + m^n$ by $(x - m)$. Then **[CDS (I) Exam, 2016]**

(a) R is a non-zero even integer (b) R is odd, if m is odd

(c) $R = s^2$ for some integer s, if n is even (d) $R = t^3$ for some integer t, if 3 divides n

40. If $a^3 = 117 + b^3$ and $a = 3 + b$, then the value of $a + b$ is (given that $a > 0$ and $b > 0$)
(*a*) 7 (*b*) 9 (*c*) 11 (*d*) 13
[CDS (I) Exam, 2016]

41. For what value of k is $(x + 1)$ a factor of $x^3 + kx^2 - x + 2$? **[CDS (I) Exam, 2016]**
(*a*) 4 (*b*) 3 (*c*) 1 (*d*) –2

42. The solution of the inequation $1+\frac{1}{x}-\frac{1}{x^2} \geq 0$ is (given that $x \neq 0$) **[CDS (I) Exam, 2016]**
(*a*) $x > 0$ (*b*) $x < 0$
(*c*) $\frac{-1-\sqrt{5}}{2} \leq x \leq \frac{-1+\sqrt{5}}{2}$ (*d*) $x \leq \frac{-1-\sqrt{5}}{2}$ or $x \geq \frac{-1+\sqrt{5}}{2}$

43. Which of the points $P(5, -1)$, Q(3, –2) and $R(1, 1)$ lie in the solution of the system of inequations $x + y \leq 4$ and $x - y \geq 2$? **[CDS (I) Exam, 2016]**
(*a*) Q and R only (*b*) P and R only (*c*) P and Q only (*d*) P, Q and R

44. What would be the maximum value of Q in the equation 5P9 + 3R7 + 2Q8 = 1114?
(*a*) 9 (*b*) 8 (*c*) 5 (*d*) 4
[CDS (II) Exam, 2016]

45. Consider the following statements in respect of positive odd integers x and y:
1. $x^2 + y^2$ is even integer. **2.** $x^2 + y^2$ is divisible by 4
Which of the above statements is/are correct? **[CDS (II) Exam, 2016]**
(*a*) 1 only (*b*) 2 only (*c*) Both 1 and 2 (*d*) Neither 1 nor 2

46. $(x + 4)$ is a factor of which one of the following expressions? **[CDS (I) Exam, 2017]**
(*a*) $x^2 - 7x + 44$ (*b*) $x^2 + 7x - 44$ (*c*) $x^2 - 7x - 44$ (*d*) $x^2 + 7x + 44$

47. If $a^3 = 335 + b^3$ and $a = 5 + b$, then what is the value of $a + b$ (given that $a > 0$ and $b > 0$)?
(*a*) 7 (*b*) 9 (*c*) 16 (*d*) 49
[CDS (I) Exam, 2017]

48. If $a + b = 5$ and $ab = 6$, then what is the value of $a^3 + b^3$? **[CDS (I) Exam, 2017]**
(*a*) 35 (*b*) 40 (*c*) 90 (*d*) 125

49. If $x^3 + \frac{1}{x^3} = \frac{65}{8}$ and $y^3 + \frac{1}{y^3} = \frac{730}{27}$, then which one of the following is a value of xy ?
[CDS (II) Exam, 2024]
(*a*) 3 (*b*) 6 (*c*) 8 (*d*) 9

50. If $\left(x+\frac{1}{x}\right)=4$, then $\left(x^4+\frac{1}{x^4}\right)$ is equal to:
(*a*) 196 (*b*) 194 (*c*) 192 (*d*) 190

51. If $\left(x+\frac{1}{x}\right)=\sqrt{3}$, then the value of $\left(x^3+\frac{1}{x^3}\right)$ is:
(*a*) 0 (*b*) $3\sqrt{3}$ (*c*) $3(\sqrt{3}-1)$ (*d*) $3(\sqrt{3}+1)$

52. If $\left(x^4+\frac{1}{x^4}\right) = 322$, the value of $\left(x-\frac{1}{x}\right)$ is:
(*a*) 4 (*b*) 6 (*c*) 8 (*d*) $3\sqrt{2}$

53. If $\left(x^3+\frac{1}{x^3}\right)=52$, the value of $\left(x+\frac{1}{x}\right)$ is:

(*a*) 4 (*b*) 3 (*c*) 6 (*d*) 13

54. If $\left(x^3-\frac{1}{x^3}\right)=14$, the value of $\left(x-\frac{1}{x}\right)$ is:

(*a*) 5 (*b*) 4 (*c*) 3 (*d*) 2

55. If x is an integer such that $\left(x+\frac{1}{x}\right)=\left(\frac{17}{4}\right)$, then the value of $\left(x-\frac{1}{x}\right)$ is:

(*a*) 4 (*b*) $\frac{13}{4}$ (*c*) $\frac{15}{4}$ (*d*) $\frac{1}{4}$

56. If $\left(x^4+\frac{1}{x^4}\right)=727$, the value of $\left(x^3-\frac{1}{x^3}\right)$ is:

(*a*) 125 (*b*) 140 (*c*) 155 (*d*) 170

57. If $\left(2x-\frac{3}{x}\right)=5$, the value of $\left(4x^2-\frac{9}{x^2}\right)$ is:

(*a*) 25 (*b*) 30 (*c*) 35 (*d*) 49

58. If $\left(x+\frac{1}{x}\right)=3$, the value of $\left(x^6+\frac{1}{x^6}\right)$ is:

(*a*) 927 (*b*) 414 (*c*) 364 (*d*) 322

59. If $x+y=7$ and $xy=12$, the value of (x^2+y^2) is:

(*a*) 25 (*b*) 29 (*c*) 37 (*d*) 49

60. If $x+y=5$ and $xy=6$, the value of (x^3+y^3) is:

(*a*) 91 (*b*) 133 (*c*) 217 (*d*) 343

61. If $x+y=5$ and $xy=6$, the value of (x^3-y^3) is:

(*a*) –19 (*b*) 19 (*c*) – 63 (*d*) 63

62. If $a+b+c=0$, then $(a^3+b^3+c^3)$ is equal to:

(*a*) 0 (*b*) abc

(*c*) $3\,abc$ (*d*) $(ab+bc+ca)$

63. If $a+b+c=0$, then the value of $\left(\frac{a^2}{bc}+\frac{b^2}{ca}+\frac{c^2}{ab}\right)$ is:

(*a*) 1 (*b*) 0 (*c*) –1 (*d*) 3

64. If $x+y+z=9$ & $xy+yz+zx=23$, then the value of $(x^3+y^3+z^3-3xyz)$ is:

(*a*) 108 (*b*) 207 (*c*) 669 (*d*) 729

65. If $3^x-3^{x-1}=18$, then the value of x^x is:

(*a*) 3 (*b*) 8 (*c*) 27 (*d*) 216

66. The value of $\left(\frac{0.125\times0.125-0.004\times0.004}{0.125+0.004}\right)$ is:

(*a*) 0.129 (*b*) 0.121 (*c*) 0.6 (*d*) 0.03125

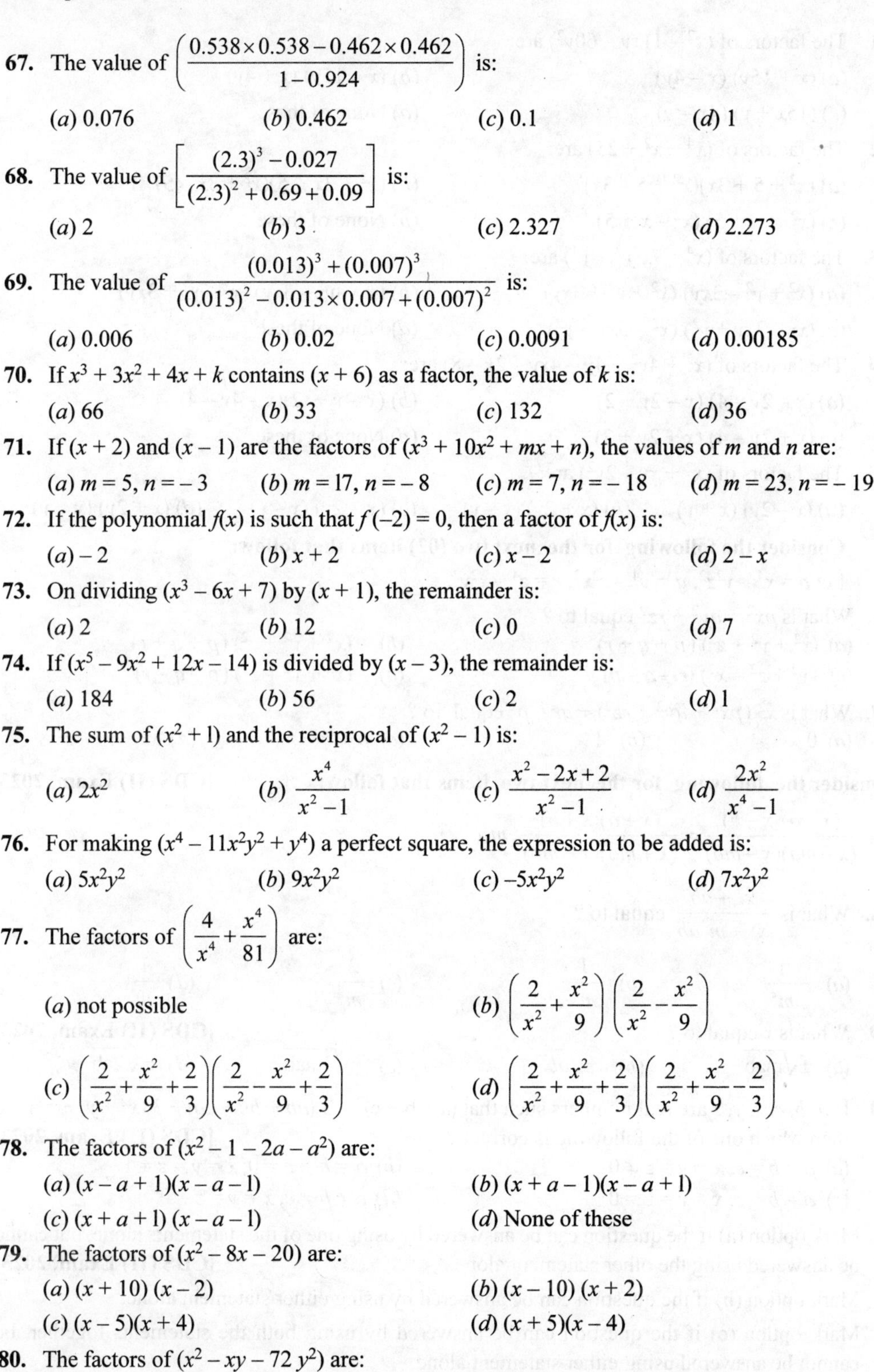

67. The value of $\left(\dfrac{0.538\times0.538-0.462\times0.462}{1-0.924}\right)$ is:

(*a*) 0.076 (*b*) 0.462 (*c*) 0.1 (*d*) 1

68. The value of $\left[\dfrac{(2.3)^3-0.027}{(2.3)^2+0.69+0.09}\right]$ is:

(*a*) 2 (*b*) 3 (*c*) 2.327 (*d*) 2.273

69. The value of $\dfrac{(0.013)^3+(0.007)^3}{(0.013)^2-0.013\times0.007+(0.007)^2}$ is:

(*a*) 0.006 (*b*) 0.02 (*c*) 0.0091 (*d*) 0.00185

70. If x^3+3x^2+4x+k contains $(x+6)$ as a factor, the value of k is:

(*a*) 66 (*b*) 33 (*c*) 132 (*d*) 36

71. If $(x+2)$ and $(x-1)$ are the factors of (x^3+10x^2+mx+n), the values of m and n are:

(*a*) $m=5, n=-3$ (*b*) $m=17, n=-8$ (*c*) $m=7, n=-18$ (*d*) $m=23, n=-19$

72. If the polynomial $f(x)$ is such that $f(-2)=0$, then a factor of $f(x)$ is:

(*a*) -2 (*b*) $x+2$ (*c*) $x-2$ (*d*) $2-x$

73. On dividing (x^3-6x+7) by $(x+1)$, the remainder is:

(*a*) 2 (*b*) 12 (*c*) 0 (*d*) 7

74. If $(x^5-9x^2+12x-14)$ is divided by $(x-3)$, the remainder is:

(*a*) 184 (*b*) 56 (*c*) 2 (*d*) 1

75. The sum of (x^2+1) and the reciprocal of (x^2-1) is:

(*a*) $2x^2$ (*b*) $\dfrac{x^4}{x^2-1}$ (*c*) $\dfrac{x^2-2x+2}{x^2-1}$ (*d*) $\dfrac{2x^2}{x^4-1}$

76. For making $(x^4-11x^2y^2+y^4)$ a perfect square, the expression to be added is:

(*a*) $5x^2y^2$ (*b*) $9x^2y^2$ (*c*) $-5x^2y^2$ (*d*) $7x^2y^2$

77. The factors of $\left(\dfrac{4}{x^4}+\dfrac{x^4}{81}\right)$ are:

(*a*) not possible

(*b*) $\left(\dfrac{2}{x^2}+\dfrac{x^2}{9}\right)\left(\dfrac{2}{x^2}-\dfrac{x^2}{9}\right)$

(*c*) $\left(\dfrac{2}{x^2}+\dfrac{x^2}{9}+\dfrac{2}{3}\right)\left(\dfrac{2}{x^2}-\dfrac{x^2}{9}+\dfrac{2}{3}\right)$

(*d*) $\left(\dfrac{2}{x^2}+\dfrac{x^2}{9}+\dfrac{2}{3}\right)\left(\dfrac{2}{x^2}+\dfrac{x^2}{9}-\dfrac{2}{3}\right)$

78. The factors of $(x^2-1-2a-a^2)$ are:

(*a*) $(x-a+1)(x-a-1)$

(*b*) $(x+a-1)(x-a+1)$

(*c*) $(x+a+1)(x-a-1)$

(*d*) None of these

79. The factors of $(x^2-8x-20)$ are:

(*a*) $(x+10)(x-2)$

(*b*) $(x-10)(x+2)$

(*c*) $(x-5)(x+4)$

(*d*) $(x+5)(x-4)$

80. The factors of $(x^2-xy-72y^2)$ are:

(*a*) $(x-8y)(x+9y)$ (*b*) $(x-9y)(x+8y)$ (*c*) $(x-y)(x+72y)$ (*d*) $(x-6y)(x+12y)$

81. The factors of $(x^2 - 11xy - 60y^2)$ are:

(*a*) $(x + 15y)(x - 4y)$ (*b*) $(x - 15y)(x + 4y)$

(*c*) $(15x + y)(4x - y)$ (*d*) None of these

82. The factors of $(x^4 + x^2 + 25)$ are:

(*a*) $(x^2 + 5 + 3x)(x^2 + 5 - 3x)$ (*b*) $(x^2 + 3x + 5)(x^2 + 3x - 5)$

(*c*) $(x^2 + x + 5)(x^2 - x + 5)$ (*b*) None of these

83. The factors of $(x^4 - 7x^2y^2 + y^4)$ are:

(*a*) $(x^2 + y^2 - 3xy)(x^2 + y^2 + 3xy)$ (*b*) $(x^2 - y^2 - 3xy)(x^2 - y^2 + 3xy)$

(*c*) $(x^2 - 3xy + y^2)(x^2 - 3xy - y^2)$ (*d*) None of these

84. The factors of $(x^2 + 4y^2 + 4y - 4xy - 2x - 8)$ are:

(*a*) $(x - 2y - 4)(x - 2y + 2)$ (*b*) $(x - y + 2)(x - 4y - 4)$

(*c*) $(x + 2y - 4)(x + 2y + 2)$ (*d*) None of these

85. The factors of $(x^2 + xy - 2y^2)$ are:

(*a*) $(x - 2y)(x + y)$ (*b*) $(x + 2y)(x - y)$ (*c*) $(x - 2y)(x - y)$ (*d*) $(x + 2y)(x + y)$

Consider the following for the next two (02) items that follow:

Let $p = x^4 - y^2z^2$, $q = y^4 - z^2x^2$, $r = z^4 - x^2y^2$.

86. What is $px^2 + qy^2 + rz^2$ equal to ?

(*a*) $(x^2 + y^2 + z^2)(p + q + r)$ (*b*) $-(x^2 + y^2 + z^2)(p + q + r)$

(*c*) $(y^2 + z^2 - x^2)(r - q - p)$ (*b*) $-(x^2 + y^2 - z^2)(p - q - r)$

87. What is $x^2(px^2 + qy^2 + rz^2) + qr - p^2$ equal to ?

(*a*) 0 (*b*) 1 (*c*) $p + q + r$ (*d*) $x^2 + y^2 + z^2$

Consider the following for the next two items that follow: **[CDS (II) Exam, 2023]**

Let $\dfrac{(x-a)(x-b)}{(x-ma)(x-mb)} = \dfrac{(x+a)(x+b)}{(x+ma)(x+mb)}$; $m, a, > 0$.

88. What is $\dfrac{x^2 + ab}{x^2 + m^2ab}$ equal to ?

(*a*) $-\dfrac{1}{m^2}$ (*b*) $\dfrac{1}{m^2}$ (*c*) $\dfrac{2}{m}$ (*d*) $\dfrac{1}{m}$

89. What is x equal to ? **[CDS (II) Exam, 2023]**

(*a*) $\pm\sqrt{mab}$ (*b*) $\pm\sqrt{ab}$ (*c*) $\pm\sqrt{3mab}$ (*d*) $\pm\sqrt{2ab}$

90. If a, b, c, x, y, z are real numbers such that $(a + b + c)^2 - 3(ab + bc + ca) + 3(x^2 + y^2 + z^2) = 0$, then which one of the following is correct ? **[CDS (II) Exam, 2023]**

(*a*) $a = b = c, x = y = z \neq 0$ (*b*) $a = b = c = 0, x = y = z = 1$

(*c*) $a = b = c, x = y = z = 0$ (*d*) $a \neq b \neq c, x = y = z = 0$

Mark option (a) if the question can be answered by using one of the statements alone, but cannot be answered using the other statement alone. **[CDS (II) Exam, 2023]**

Mark option (b) if the question can be answered by using either statement alone.

Mark option (c) if the question can be answered by using both the statements together, but cannot be answered using either statement alone.

Mark option (d) if the question cannot be answered even by using both the statements together.

91. Question : Is $m > n$ if m, n are real numbers ?

Statement-I: $m = (1 - p)(p^2 + p + 1)$ and $n = (p + 1)(p^2 - p + 1)$

Statement-II: $m = pn$

94. Question: Are x, y, z equal, where x, y, z are real numbers ?

Statement I: $x^2 + y^2 + z^2 - xy - yz - zx = 0$

Statement II: $x^3 + y^3 + z^3 - 3xyz = 0$

93. If $a + b + c = 0$, then which of the following are correct? **[CDS (I) Exam, 2021]**

1. $a^3 + b^2 + b^3 = 3\ abc$
2. $a^2 + b^2 + b^3 = -2(ab + bc + ca)$
3. $a^3 + b^2 + c^3 = -3ab\ (a + b)$

Select the correct answer using the code given below:

(*a*) 1 and 2 only (*b*) 2 and 3 only (*c*) 1 and 3 only (*d*) 1, 2, and 3

94. Which one of the following is a factor of the polynomial

$(x - 1)(x - 2)(x - 4) - 90$? **[CDS (II) Exam, 2021]**

(*a*) $x + 14$ (*b*) $x - 14$ (*c*) $x - 6$ (*d*) $x - 7$

95. What is the sum of the linear factors (in x and y) of the expression

$2x^2 + xy - 3y^2$? **[CDS (II) Exam, 2021]**

(*a*) $2x - 3y$ (*b*) $3x - 2y$ (*c*) $3x + 2y$ (*d*) $2x + 3y$

96. If $96 - 64a^3 + \frac{8}{a^6} - \frac{48}{a^3} - t^3 = 0$, then what is $a^2t + 4a^3$ equal to? **[CDS (II) Exam, 2021]**

(*a*) 0 (*b*) 1 (*c*) 2 (*d*) 3

97. If $(2ab - b^2) : (6a^2 - ab) = 1 : 6$, then what is $(a + b) : (a - b)$ equal to ? **[CDS (II) Exam, 2021]**

(*a*) 3 only (*b*) 5 only (*c*) –3 or 3 (*d*) – 5 or 5

98. Consider the following inequalities: **[CDS (II) Exam, 2021]**

1. $\frac{a^2 - b^2}{a^2 + b^2} > \frac{a - b}{a + b}$ where $a > b > 0$
2. $\frac{a^3 + b^3}{a^2 + b^2} > \frac{a^2 + b^2}{a + b}$ only when $a > b < 0$

Which of the above is/are correct?

(*a*) 1 only (*b*) 2 only (*c*) Both 1 and 2 (*d*) Neither 1 nor 2

99. If $\frac{ay - bx}{c} = \frac{cx - az}{b} = \frac{bz - cy}{a}$, then which of the following is/are correct?

1. $\frac{x}{a} = \frac{y}{b}$ 2. $\frac{x + y + z}{a + b + x} = \frac{z}{c}$ **[CDS (II) Exam, 2021]**

(*a*) 1 only (*b*) 2 only (*c*) Both 1 and 2 (*d*) Neither 1 nor 2

100. If $x^2 + y^2 + z^2 = 3$, where $x, y,$ and z are integers, then how many values can $(xy + yz + zx)$ have? **[CDS (I) Exam, 2025]**

(*a*) One (*b*) Two (*c*) Three (*d*) Four

ANSWERS

1. (*b*)	**2.** (*d*)	**3.** (*c*)	**4.** (*c*)	**5.** (*d*)	**6.** (*c*)	**7.** (*c*)	**8.** (*b*)	**9.** (*a*)
10. (*c*)	**11.** (*c*)	**12.** (*a*)	**13.** (*a*)	**14.** (*b*)	**15.** (*b*)	**16.** (*c*)	**17.** (*b*)	**18.** (*c*)
19. (*b*)	**20.** (*a*)	**21.** (*b*)	**22.** (*a*)	**23.** (*d*)	**24.** (*c*)	**25.** (*a*)	**26.** (*d*)	**27.** (*c*)
28. (*b*)	**29.** (*d*)	**30.** (*a*)	**31.** (*a*)	**32.** (*b*)	**33.** (*b*)	**34.** (*b*)	**35.** (*c*)	**36.** (*a*)
37. (*c*)	**38.** (*a*)	**39.** (*a*)	**40.** (*a*)	**41.** (*d*)	**42.** (*d*)	**43.** (*c*)	**44.** (*a*)	**45.** (*a*)
46. (*c*)	**47.** (*b*)	**48.** (*a*)	**49.** (*b*)	**50.** (*b*)	**51.** (*a*)	**52.** (*a*)	**53.** (*a*)	**54.** (*d*)
55. (*d*)	**56.** (*b*)	**57.** (*c*)	**58.** (*d*)	**59.** (*a*)	**60.** (*c*)	**61.** (*b*)	**62.** (*c*)	**63.** (*d*)
64. (*a*)	**65.** (*c*)	**66.** (*b*)	**67.** (*d*)	**68.** (*a*)	**69.** (*b*)	**70.** (*c*)	**71.** (*c*)	**72.** (*b*)
73. (*b*)	**74.** (*a*)	**75.** (*b*)	**76.** (*b*)	**77.** (*d*)	**78.** (*c*)	**79.** (*b*)	**80.** (*b*)	**81.** (*b*)
82. (*a*)	**83.** (*a*)	**84.** (*a*)	**85.** (*b*)	**86.** (*a*)	**87.** (*a*)	**88.** (*d*)	**89.** (*a*)	**90.** (*a*)
91. (*d*)	**92.** (*b*)	**93.** (*d*)	**94.** (*d*)	**95.** (*c*)	**96.** (*c*)	**97.** (*d*)	**98.** (*c*)	**99.** (*c*)
100. (*b*)								

SOLUTIONS

1. Let $a = 1$, $b = 2$ satisfied

$$\therefore x = 1 + 2 + \frac{(-1)^2}{4+8} = 3 + \frac{1}{12} = \frac{37}{12}$$

$$y = \frac{3}{4} + \frac{2}{3} = \frac{9+8}{12} = \frac{17}{12}$$

Now,

$$(x-a)^2 - (y-b)^2 = \left(\frac{37}{12} - 1\right)^2 - \left(\frac{17}{12} - 2\right)^2 = \left(\frac{25}{12}\right)^2 - \left(\frac{7}{12}\right)^2$$

$$= \frac{625-49}{144} = \frac{576}{144} = 4 = b^2$$

2. We have,

$$x = 97 + 56\sqrt{3} = 97 + 2\,(28)\sqrt{3} = 87 + 2.74\sqrt{3}$$

$$= (7)^2 + (4\sqrt{3})^2 + 2.7.4\sqrt{3} = (7 + 4\sqrt{3})^2 = (2 + \sqrt{3})^4$$

$$\text{Now, } \sqrt[4]{x} + \frac{1}{\sqrt[4]{x}} = 2 + \sqrt{3} + \frac{1}{2+\sqrt{3}} = 2 + \sqrt{3} + \frac{1}{2+\sqrt{3}} \times \frac{2-\sqrt{3}}{2-\sqrt{3}} = 2 + \sqrt{3} + 2 - \sqrt{3} = 4.$$

3. We know, $x^3 + y^3 + z^3 - 3xyz = (x+y+z)\,[(x+y+z)^2 - 3(xy+yz+zx)]$

$$x^3 + y^3 + z^3 - 3xyz = (10)\,[(10)^2 - 3(18)]$$

$$\Rightarrow x^3 + y^3 + z^3 - 3xyz = 10[100 - 54]$$

$$\Rightarrow x^3 + y^3 + z^3 - 3xyz = 10 \times 46$$

$$\Rightarrow x^3 + y^3 + z^3 - 3xyz = 460$$

4. $3\sqrt{3}x^3 + 2\sqrt{2}y^3 - 18xy + 6\sqrt{6}$

$$\left(\sqrt{3}x\right)^3 + \left(\sqrt{2}y\right)^3 + \left(\sqrt{6}\right)^3 - 3.\sqrt{3}x.\sqrt{2}y.\sqrt{6}$$

Using $a^3 + b^3 + c^3 - 3abc = (a + b + c)(a^2 + b^2 + c^2 - ab - bc - ca)$

$$= \left(\sqrt{3}.x+\sqrt{2}y+\sqrt{6}\right)\left(\sqrt{3}x\right)^2+\left(\sqrt{2}y\right)^2+\left(\sqrt{6}\right)^2-\sqrt{3}x.\sqrt{2}y-\sqrt{2}y.\sqrt{6}-\sqrt{6}.\sqrt{3}x.$$

$$= \left(\sqrt{3}x+\sqrt{2}y+\sqrt{6}\right)\left(3x^2+2y^2+6-\sqrt{6}xy-\sqrt{12}y-\sqrt{18}x\right).$$

5. We have, $\left(\dfrac{a^2+3a+1}{a}\right)\left(\dfrac{b^2+3b+1}{b}\right)$

$$\left(a+3+\frac{1}{a}\right)\left(b+3+\frac{1}{b}\right)$$

$(2 + 3)(2 + 3)$ $\left[\because a+\dfrac{1}{b}=2 \text{ and } b+\dfrac{1}{b}=2\right]$

$5 \times 5 = 25$

6. Let, $p = k\cdot\left(x^2+y^2+z^2\right)$

When $x = 1, y = 2, z = 3, p = 70 : x^2 + y^2 + z^2 = 1^2 + 2^2 + 3^2 = 1+4+9 = 14$

So, $70 = k\cdot 14$

$$k = \frac{70}{14} = 5$$

Value of p when $x = -1, y = 1, z = 5:$

$$= k((-1)^2 + (1)^2 + (5)^2) = 5 \times 27 = 135$$

7. We have,

$$p=\frac{a^2}{(b-q)(c-a)}, q=\frac{b^2}{(c-b)(a-b)}, r=\frac{c^2}{(a-c)(b-c)}$$

$$\because (p+q+r)^2=\left(\frac{a^2}{(b-a)(c-a)}+\frac{b^2}{(c-b)(a-b)}+\frac{c^2}{(a-c)(b-c)}\right)^2$$

$$=\left[\frac{(c-b)a^2+b^2(a-c)+c^2(b-a)}{(a-b)(b-c)(c-a)}\right]=\frac{(a^2c-a^2b)+ab^2-cb^2+bc^2-ac^2}{(a-b)(b-c)(c-a)}$$

$$=\left[\frac{(a^2c-b^2c)+(ab^2-a^2b)+(bc^2-ac^2)}{(a-b)(b-c)(c-a)}\right]^2=\left[\frac{(a-b)\{a+b\}c-ab=c^2}{(a-b)(b-c)(c-a)}\right]^2$$

$$=\frac{ac+bc-ab-c^2}{(b-c)(c-a)}=1$$

8. $2s = a + b + c$

$s^2 + (s - a)(s - b) + (s - b)(s - c) + (s - c)(s - a)$

$= 4s^2 - s(a + b + b + c + c + a) + ab + bc + ca$

$= 4s^2 + 2s(a + b + c) + ab + bc + ca$

$= 4s^2 + 2s(2s) + ab + bc + ca = 4s^2 + 4s^2 + ab + bc + ca = ab + bc + ca$

9. We know that if $a + b + c = 0$

Then, $a^3 + b^3 + c^3 = 3abc$

Here, $a = x - y, b = y - z, c = z - x$

$a + b + c = x - y + y - z + z - x = 0$

Now, $\dfrac{(x-y)^3+(y-z)^3+(z-x)^3}{3(x-y)(y-z)(z-x)} = \dfrac{3(x-y)(y-z)(z-x)}{3(x-y)(y-z)(z-x)} = 1$

10. Let $a^x = b^y = c^z = p$, where p is a real number

$\Rightarrow a = p^{\frac{1}{x}}, b = p^{\frac{1}{y}}, c = p^{\frac{1}{z}}$ and $b^2 = ac \Rightarrow p^{\frac{2}{y}} = p^{\frac{1}{x}} \times p^{\frac{1}{z}} \Rightarrow p^{\frac{2}{y}} = p^{\frac{1}{x}+\frac{1}{z}} \Rightarrow \dfrac{1}{x}+\dfrac{1}{z}=\dfrac{2}{y}$

11. We have, $5^{2n} - 2^{3n}$
Put $n = 1$
We get, $5^2 - 2^3 = 25 - 8 = 17$

12. We have,
$x^3 + kx^2 + 7x + 6 = 0$ ...(i)
Now, by options
(a) put $k = 0$ in eqn (i), we get
$x^3 - 7x + 6 = 0 \Rightarrow (x-2)(x-1)(x+3) = 0$
(b) put k = 1 in eqn (i), we get
$x^3 + x^2 - 7x + 6 = 0 \Rightarrow x(x^2 + x - 7) + 6 = 0$
(c) put $k = 2$, in eqn (i), we get
$\Rightarrow x^3 + 2x^2 - 7x + 6 = 0 \Rightarrow x(x^2 + 2x - 7) + 6 = 0$
(d) put $k = 3$, in eqn (i), we get
$x^3 + 3x^2 - 7x + 6 = 0 \Rightarrow x(x^2 + 3x - 7) + 6 = 0$
hence $k = 0$

13. We have,
$(x-a)(x-b)(x-c)$
$\Rightarrow (x^2 - ax - bx + ab) \times (x-c) \Rightarrow (x^2 - x(a+b) + ab) \times (x-c)$
$\Rightarrow x^3 - x^2(a+b) + abx - cx^2 + cx(a+b) - abc$
$\Rightarrow x^3 - x^2(a+b+c) + x(ab+ac+bc) - abc = 0$

14. $x^2 - 5x + 4 = 0$
$x^2 - 4x - x + 4 = 0$
$x(x-4) - 1(x-4) = 0$
$x = 4$ and 1
Substitute $x = 1 \Rightarrow x^4 - px^2 + q = 0$
$x^4 - px^2 + q = 0$
$1^4 - p(1^2) + q = 0$
$\Rightarrow 1 - p + q = 0 \Rightarrow q = p - 1$ ...(i)
Substitute $x = 4$ in $x^4 - px^2 + q = 0$
$x^4 - px^2 + q = 0 \Rightarrow 4^4 - p(4^2) + q = 0$
$\Rightarrow 256 - 16p + q = 0 \Rightarrow q = 16p - 256$ ...(ii)
From eqn (i) and (ii)
$\Rightarrow p - 1 = 16p - 256 \Rightarrow 255 = 15p \Rightarrow p = 17$
Putting the value of p in eqn (i)
$q = p - 1 \Rightarrow q = 17 - 1 \Rightarrow q = 16$
$\therefore$ The values of p and q are 17 and 16 respectively.

15. We known that, if $a + b + c = 0$, then $a^3 + b^3 + c^3 = 3abc$

Here $a = x - y$, $b = y - z$, $c = z - x$

Now, $\dfrac{(x-y)^3 + (y-z)^3 + (z-x)^3}{9(x-y)(y-z)(z-x)}$

$\therefore a + b + c = x - y + y - z + z - x = 0 \Rightarrow \dfrac{3(x-y)(y-z)(z-x)}{9(x-y)(y-z)(z-x)} \Rightarrow \dfrac{1}{3}$

16. We have

$\alpha + \beta = \dfrac{-q}{p}$ and $\alpha\beta = \dfrac{r}{p}$ and $1 < p < q > r$

Put $p = 2$, $q = 3$, $r = 4$

By options:

(a) $\dfrac{1}{\alpha+\beta} = -\dfrac{p}{q} = -\dfrac{2}{3}$ (b) $\dfrac{1}{\alpha} + \dfrac{1}{\beta} = \dfrac{\alpha+\beta}{\alpha\beta} = -\dfrac{q}{r} = -\dfrac{3}{4}$

(c) $-\dfrac{1}{\alpha\beta} = -\dfrac{p}{r} = -\dfrac{2}{4} = -\dfrac{1}{2}$ (d) $\dfrac{\alpha\beta}{\alpha+\beta} = -\dfrac{r}{q} = -\dfrac{4}{3}$

$\therefore -\dfrac{1}{2}$ is the greatest

Hence, $-\dfrac{1}{\alpha\beta}$ is the greatest

17. $x^6 + \dfrac{1}{x^6} = k\left(x^2 + \dfrac{1}{x^2}\right)$

We know that $a^3 + b^3 = (a + b)(a^2 + b^2 - ab)$

$\Rightarrow x^6 + \dfrac{1}{x^6} = \left(x^2\right)^3 + \left(\dfrac{1}{x^2}\right)^3$

$= \left(x^2 + \dfrac{1}{x^2}\right)\left(\left(x^2\right)^2 + \left(\dfrac{1}{x^2}\right)^2 - x^2 \times \dfrac{1}{x^2}\right) = \left(x^2 + \dfrac{1}{x^2}\right)\left(x^4 + \dfrac{1}{x^4} - 1\right)$

Now. $\left(x^2 + \dfrac{1}{x^2}\right)\left(x^4 + \dfrac{1}{x^4} - 1\right) = k\left(x^2 + \dfrac{1}{x^2}\right)$

$\Rightarrow k = x^4 + \dfrac{1}{x^4} - 1$

18. Let three consecutive numbers be n, $n + 1$ and $n + 2$.

Now, $(n)^2 + (n+1)^2 + (n+2)^2 = 110$

$\Rightarrow n^2 + n^2 + 1 + 2n + n^2 + 4 + 4n - 110 = 0$

$\Rightarrow 3n^2 + 6n - 105 = 0$ $\Rightarrow n^2 + 2n - 35 = 0$ $\Rightarrow n^2 + 7n - 5n - 35 = 0$

$\Rightarrow n(n + 7) - 5(n + 7) = 0$ $\Rightarrow (n - 5)(n + 7) = 0$

$n = -7$ (not possible) [$\because$ n is a natural number]

$\therefore n = 5$

So, consecutive numbers are 5, 6 and 7

Required cubes $= (5)^3 + (6)^3 + (7)^3$

$= 125 + 216 + 343 = 684$

19. Let the sum of all possible products $= a$

$\Rightarrow (1)^2 + (-1)^2 + (2)^2 + (-2)^2 + (3)^2 + (-3)^2 + (4)^2 + (-4)^2$

$\Rightarrow 1 + 1 + 4 + 4 + 9 + 9 + 16 + 16 = 60$

Square of sum of given numbers $(\pm 1 \pm 2 \pm 3 \pm 4)^2 = 0$
We have,
Square of sum of n numbers = sum of their square + 2 (sum of all possible products taken two at a time)

$$\Rightarrow\ 0 = 60 + 2a \Rightarrow\ 2a = -60\ \Rightarrow\ a\frac{-60}{2} = -30$$

20. We have, $f(x) = 3x^3 - 2x^2 - 13xy^2 + 10y^3$
$x - 2y = 0$
now, put $x = 2y$ in $f(x)$
$= 3 \times (2y)^3 - 2(2y)^2 - 13 \times 2y \times y^2 + 10y^3$
$= 3 \times 8y^3 - 2 \times 4y^2 - 26y^3 + 10y^3$
$= 24y^3 - 8y^2 - 26y^3 + 10y^3$
$= -2y^3 + 2y^3 = 0$

21. Given that, $ab + bc + ca = 0$
let $a = -1$, $b = 2$ and $c = 2$

$$\text{now, } \frac{(b^2-ca)(c^2-cb)+(a^2-bc)(c^2-ab)+(a^2-bc)(b^2-ca)}{(a^2-bc)(b^2-ca)(c^2-ab)}$$

$$\Rightarrow \frac{(4-(-2))(4-2(-2)+(1-4)(4-(-2))+(1-4)(4-(-2))}{(1-4)(4-(-2)(4-(-2))}$$

$$\Rightarrow \frac{6\times 6+(-3)(6)+(-3)(6)}{(-3)(6)(6)} \Rightarrow \frac{36-18-18}{108} = 0$$

22. We have, $x^2 - y^2 = 0$
$\Rightarrow x^2 = y^2$ ...(i)
and $(x - a)^2 + y^2 = 1$
$\Rightarrow x^2 + a^2 - 2xa + y^2 = 1 \Rightarrow x^2 + a^2 - 2xa + x^2 = 1 \Rightarrow 2x^2 + 2xa + (a^2 - 1) = 0$
here $a = 2$, $b = -2a$, $c = (a^2 - 1)$
$D = b^2 - 4ac = 0$
[$\because$ We know that quadratic eqn of from $ax^2 + bx + c = 0$ have only one positive solution then $D = b^2 - 4ac = 0$]
$\Rightarrow (-2a)^2 - 4 \times 2 \times (a^2 - 1) = 0 \Rightarrow 4a^2 - 8(a^2 - 1) = 0$
$\Rightarrow 4a^2 - 8a^2 + 8 = 0$
$\Rightarrow -4a^2 + 8 = 0 \Rightarrow 4a^2 = 8 \Rightarrow a^2 = 2 \Rightarrow a = \pm\sqrt{2}$
$= a = \sqrt{2}$ [$\because$ there is only one positive solution]

23. We have, $f(x) = ax^3 + bx^2 + cx + d$ $(a \neq 0)$
$\therefore$ α, β and γ are the zeros of the polynomial.

$$\therefore \alpha + \beta + \gamma = \frac{-b}{a}$$

$$\alpha\beta + \beta\gamma + \gamma\alpha = \frac{c}{a}$$

$$\alpha\beta\gamma = -\frac{d}{a}$$

$(\alpha + \beta + \gamma)^2 = \alpha^2 + \beta^2 + \gamma^2 + 2(\alpha\beta + \beta\gamma + \gamma\alpha)$

$$\Rightarrow \left(\frac{-b}{a}\right)^2 = \alpha^2 + \beta^2 + \gamma^2 + 2\frac{c}{a}$$

$\Rightarrow \dfrac{b^2}{a^2} = \alpha^2 + \beta^2 + \gamma^2 + \dfrac{2c}{a}$

$\Rightarrow \alpha^2 + \beta^2 + \gamma^2 = \dfrac{b^2}{a^2} - \dfrac{2c}{a} = \dfrac{b^2 - 2ac}{a^2}$

24. We have,

$f(x) = 4x^3 + 12x^2 - x - 3$

$= 4x^2 (x + 3) - 1 (x + 3)$

$= (4x^2 - 1) (x + 3)$

$= (2x + 1) (2x - 1) (x + 3)$

Hence, $(2x + 1)$ and $(2x - 1)$ are factor of given polynomial

25. Put $x = 1$ in given eqn

$x^3 + 4x^2 + px + q = 0$

$\Rightarrow 1 + 4 + p + q = 0 \quad \Rightarrow 5 + p + q = 0$

$\Rightarrow p + q = -5$ ...(i)

Now, put $x = -2$ in the given eqn

$x^3 + 4x^2 + px + q = 0$

$-8 + 16 - 2p + q = 0$

$-2p + q = -8$ ...(ii)

Subtracting (i) and (ii), we get

$3p = 3 \Rightarrow p = 1$

Now, put $p = 1$ in equ. (i), we get

$q = -5 - 1 = -6$

26. Given that $(x^2 - 1)$ is a factor of $ax^4 + bx^3 + cx^2 + dx + e$

Thus, $x = 1$ and -1, satisfies the given equ.

Put $x = -1$, we get

$ax^4 + bx^3 + cx^2 + dx + e = 0$

$\Rightarrow a - b + c - d + e = 0 \quad \Rightarrow a + c + e = b + d$

27. Let $n = x^3 + x^2 + 16$

Now, $\dfrac{n}{x} = \dfrac{x^3 + x^2 + 16}{x} = x^2 + x + \dfrac{16}{x}$

Thus, 16 must be divisible by x

$\therefore x = 1, 2, 4, 8$ and 16

28. Let $f(x) = 1 - x - x^n + x^{n+1}$

$= (1 - x) - x^n (1 - x)$

$= (1 - x) (1 - x^n)$

$= (1 - x) (1 - x) (1 + x + x^2 .... x^{n-1})$ $[\because 1 - x^n = (1 - x) (1 + x + x^2 .... x^{n-1})]$

$= (1 - x)^2 (1 + x + x^2 .... x^{n-1})$

Thus, $1 - x^n - x^{n+1}$ is divisible by $(1 - x)^2$

29. $x - 1 = 0 \Rightarrow x = 1$ and $x + 1 = 0 \Rightarrow x = -1$

Now, according to the question,

Put $x = 1$ in the given eqn. $x^2 + ax + b = 0$

$\Rightarrow 1 + a + b = 0$

$\Rightarrow a + b = -1$ ...(*i*)

Now, put $x = -1$ in the given eqn,

$x^2 + ax + b = 0$

$\Rightarrow 1 - a + b = 0$

$\Rightarrow -a + b = -1$ ...(*ii*)

Now,

$a + b = -a + b$

$\Rightarrow$ $a = 0$ and b is any integers

30. Given that when, $f(x) = x^4 - x^2 + 7x + 5$ is divided by $(x + 2)$, the quotient is $ax^3 + bx^2 + cx + d$

Now,

$$
\begin{array}{r|l}
 & x^3 - 2x^2 + 3x + 1 \\
x+2 & x^4 - x^2 + 7x + 5 \\
 & \underline{\underset{-}{x^4} + \underset{-}{2x^3}} \\
 & -2x^3 - x^2 \\
 & \underline{\underset{+}{-2x^3} \underset{+}{- 4x^2}} \\
 & 3x^2 + 7x \\
 & \underline{\underset{-}{3x^2} + \underset{-}{6x}} \\
 & x + 5 \\
 & \underline{\underset{-}{x} + \underset{-}{2}} \\
 & 3
\end{array}
$$

Thus, when $f(x)$ is divided by $(x + 2)$, the quotient is $x^3 - 2x^2 + 3x + 1$

Now, by comparing, we get

$$a = 1, b = -2, c = 3, d = 1$$

31. Factor theorem: $x - k$ is a factor of the polynomial $f(x)$, if and only if $f(k) = 0$.

Let $f(x) = x^3 - 3x^2 + kx - 10$.

$x - 5$ is a factor of $f(x)$.

$\therefore$ $f(5) = 0$

$f(5) = 5^3 - 3(5)^2 + k(5) - 10 = 0$

$40 + 5k = 0 \Rightarrow k = -8$

32. The number of distinct prime factors of n is 3. (They are P_1, P_2, P_3). Here, we interpret the phrase 'number of prime factors for n' as the number of prime factors, not necessarily distinct. Thus 4 has 2 prime factors, 8 has 3, 8(9) has 5 and so on.

$\therefore$ The number of prime factors of n is x_1, x_2, x_3.

33. $a - b = 4$ ... (1)

$(a - b)^2 = 16$

$\Rightarrow (a^2 + b^2) - 2ab = 16 \Rightarrow 40 - 2ab = 16 \Rightarrow 2ab = 24$

$\Rightarrow ab = 12 \Rightarrow b = \dfrac{12}{a}(1) \Rightarrow a - \dfrac{12}{a} = 4 \Rightarrow a^2 - 4a - 12 = 0$

$\Rightarrow (a - 6)(a + 2) = 0 \Rightarrow a = 6$ or $a = -2$.

Since a and b are positive, $a = 6$

$\Rightarrow b = 6 - 4 = 2$

$a^3 + b^6 = 216 + 64 = 280.$

34. $\dfrac{1}{1(4)} = \dfrac{1}{3}\left(\dfrac{1}{1} - \dfrac{1}{4}\right)$

$\dfrac{1}{4(7)} = \dfrac{1}{3}\left(\dfrac{1}{4} - \dfrac{1}{7}\right)$

$$\frac{1}{16(19)} = \frac{1}{3}\left(\frac{1}{16} - \frac{1}{19}\right)$$

$\therefore$ The given expression $= \frac{1}{3}(1 - \frac{1}{19}) = \frac{6}{19}$

35. $-2 \le x \le 1 \Rightarrow -2 \le x \le 1$ ... (1)

$-1 \le y \le 2 \Rightarrow -2 \le -y \le 1$ ... (2)

$3 \le z \le 6 \Rightarrow 6 \le 2z \le 12$ ... (3)

(1), (2), (3) $\Rightarrow 2 \le x - y + 2z \le 14$.

36. $21x + 48y = 5$. If x, y are integers, the LHS is a multiple of 3 while the RHS is not. There can be no integral solutions.

37. If $x - a$ is a factor of $f(x)$, then f(a) should be equal to zero (Factor Theorem). For example, in option (*a*), $x + 2$ can be seen as $(x - (-2))$ and $f(x) = x^4 - 6x^3 + 12x^2 - 24x + 32$.

$f(-2) = (-2)^4 - 6(-2)^3 + 12(-2)^2 - 24(-2) + 32 = 192$.

Since $f(-2) \neq 0$, $(x - (-2))$ is not a factor of $f(x)$.

When we check option (C), we see that $f(2)$ is zero.

$\therefore$ $x - 2$ is a factor of $x^4 - 6x^3 + 12x^2 - 24x + 32$.

38. $\frac{1}{3}$; $\frac{22}{7}$

$\frac{\pi}{\sqrt{2}}(x)$; $\frac{\pi}{\sqrt{3}}$

39. Let $f(x) = \left[\frac{92727}{1000000}\right]$

When $f(x)$ is divided by $\frac{a+b}{2\sqrt{ab}} \le 1$ then the remainder is $f(m)$

$\therefore$ $f(m) = \sqrt{ab}\ (m^n)$

Which is always even positive integer value of m, n

40. $\frac{108 \times 4}{216} = 2$ and $\frac{2a}{3} = \frac{4b}{5} = \frac{3c}{4} = \alpha$; $\frac{3\alpha}{2}, b = \frac{5\alpha}{4}, c = \frac{4\alpha}{3}$

$$\Rightarrow \frac{18}{a}\sqrt{a^2 + c^2 - b^2} \Rightarrow ab = 10; \frac{18}{\frac{3\alpha}{2}}\sqrt{\left(\frac{3\alpha}{2}\right)^2 + \left(\frac{4\alpha}{3}\right)^2 + \left(\frac{5\alpha}{4}\right)^2} = \frac{12}{\alpha}\sqrt{\frac{9\alpha^2}{4} + \frac{16\alpha^2}{9} + \frac{25\alpha^2}{16}}$$

$$; \frac{12}{\alpha} \times \sqrt{\frac{324\alpha^2 + 256\alpha^2 - 225\alpha^2}{144}}$$

41. Let $f(x) = x^3 + kx^2 - x + 2$; if $x + 1$ is a factor of $f(x)$

$\Rightarrow f(-1) = 0 \Rightarrow (-1)^3 + k(-1)^2 - (-1) + 2 = 0$

$-1 + k + 1 + 2 = 0$; $k = -2$

42. $1 + \frac{1}{x} - \frac{1}{x^2} \ge 0 \Rightarrow \frac{x^2 + x - 1}{x^2} \ge 0$ as $x^2 > 0$

$\Rightarrow x^2 + x - 1 \ge 0$; $x = \frac{-1 \pm \sqrt{1+4}}{2} = \frac{-1 \pm \sqrt{5}}{2}$

$\Rightarrow \dfrac{-1-\sqrt{5}}{2} \geq x$ or $x \leq \dfrac{-1-\sqrt{5}}{2}$ or $x \geq \dfrac{-1+\sqrt{5}}{2}$

43. $x + y \leq 4$ and $x - y \geq 2$ For P (5, – 1)
$\Rightarrow 5 - 1 \leq 4$ and $5 + 1 \geq 2$; Both are true
$\therefore$ P $\in$to the solution region
Q (3, – 2) $3 - 2 \leq 4$ and $3 - (-2) \geq 2$ both are true
$\therefore$ Q $\in$ solution region R (1, 1); $1 + 1 \leq 4$ and $1 - 1 \geq 2$.
$\therefore$ R $\notin$ solution region $\therefore$ Only P, Q $\in$ Solution region

44. We along the number in a column

5	P	9
3	R	7
2	Q	8
11	1	4

We are that P + R + Q + 2 = 11 (and net 21)
$\therefore$ P + Q + R = 9. The maximum Value of Q is 9.

45. *x, y* are positive integer.

1. x^2 is odd. y^2 is odd $\therefore$ $x^2 + y^2$ is even True

2. Let $x = 1, y = 3$ $x^2 + y^2 = 10$ which in not divisible by 4 (False).
Only (1) is true

46. $x + 4 = 0 \Rightarrow x = -4$
Putting $x = -4$ in following expressions, we get
(*i*) $x^2 - 7x + 44$

$\Rightarrow (-4)^2 - 7(-4) + 44 \Rightarrow 16 + 28 + 44 \Rightarrow 88$

(ii) $x^2 + 7x - 44$

$\Rightarrow (-4)^2 + 7(-4) - 44 \Rightarrow 16 - 28 - 44 \Rightarrow 16 - 72 \Rightarrow -56$

(iii) $x^2 - 7x - 44$

$\Rightarrow (-4)^2 - 7(-4) - 44 \Rightarrow 16 + 28 - 44 \Rightarrow 44 - 44 \Rightarrow 0$

(iv) $x^2 + 7x + 44$

$\Rightarrow (-4)^2 + 7(-4) + 44 \Rightarrow 16 - 28 + 44 \Rightarrow 60 - 28 \Rightarrow 32$

So, $(x + 4)$ is a factor of expression (iii)

47. $a^3 = 335 + b^3$

$\Rightarrow \quad a^3 - b^3 = 335$ ... (i)

and $a - b = 5$, $a = 5 + b$ ... (ii)

From equation (i), we get

$(a - b)\left(a^2 + b^2 + ab\right) = 335$

$\Rightarrow (5)\left(a^2 + b^2 + ab\right) = 335 \; [\because a - b = 5] \Rightarrow a^2 + b^2 + ab = 67$

$\Rightarrow \ (a-b)^2+3ab=67 \Rightarrow 3ab=67-25 \Rightarrow 3ab=42$

$\Rightarrow \qquad ab=14$ ... (iii)

Put (ii) in (iii)

$$(5+b)b=14 \qquad [\because a=5+b]$$

$b^2+5b-14=0$

$\Rightarrow \ b^2+7b-2b-14=0 \Rightarrow b(b+7)-2(b+7)=0$

$b=-7,2$

$\therefore$ Put $b=2$ in (ii),

$a=5+2=7$

$\therefore \ a+b=7+2=9$

48. $(a+b)^3=a^3+b^3+3ab\,(a+b)$

$a^3+b^3=(a+b)^3-3ab\,(a+b)$

$=(5)^3-3(6)\,(5)=35$ [$\because$ Given $a+b=5$ and $ab=6$]

49. $x^3+\dfrac{1}{x^3}=\dfrac{65}{8}$

Let $x = 2$ (hit and trail method) (cube of 2 = 8)

$$\text{LHS } =2^3+\frac{1}{2^3}=8+\frac{1}{8}=\frac{65}{5}$$

LHS = RHS

So, $\qquad x=2$

$$y^3+\frac{1}{y^3}=\frac{730}{27}$$

Let, $y = 3$ (cube of 3 = 27)

$$\text{LHS } =27+\frac{1}{27}=\frac{729+1}{27}=\frac{730}{27}$$

LHS = RHS

So, $\qquad y=3$

Thus $\qquad xy=2\times 3=6$

50. $x+\dfrac{1}{x}=4 \Rightarrow \left(x+\dfrac{1}{x}\right)^2=16 \Rightarrow x^2+\dfrac{1}{x^2}+2=16 \Rightarrow x^2+\dfrac{1}{x^2}=14$

Now, $x^2+\dfrac{1}{x^2}=14 \Rightarrow \left(x^2+\dfrac{1}{x^2}\right)^2=196 \Rightarrow x^4+\dfrac{1}{x^4}+2=196 \Rightarrow x^4+\dfrac{1}{4^4}=194.$

51. $\left(x+\dfrac{1}{x}\right)=\sqrt{3} \Rightarrow \left(x+\dfrac{1}{x}\right)^3=(\sqrt{3})^3 \Rightarrow x^3+\dfrac{1}{x^3}+3x\,.\,\dfrac{1}{x}\left(x+\dfrac{1}{x}\right)=3\sqrt{3}$

$$\Rightarrow x^3+\frac{1}{x^3}+3\sqrt{3}=3\sqrt{3} \Rightarrow x^3+\frac{1}{x^3}=0.$$

52. $\left(x^4+\frac{1}{x^4}\right)=322 \Rightarrow x^4+\frac{1}{x^4}+2=324 \Rightarrow \left(x^2+\frac{1}{x^2}\right)^2=(18)^2$

Now, $x^2+\frac{1}{x^2}=18 \Rightarrow x^2+\frac{1}{x^2}-2=16 \Rightarrow \left(x-\frac{1}{x}\right)^2=4^2 \Rightarrow \left(x-\frac{1}{x}\right)=4.$

53. $\left(x+\frac{1}{x}\right)^3=\left(x^3+\frac{1}{x^3}\right)+3\left(x+\frac{1}{x}\right).$

$\therefore \left(x+\frac{1}{x}\right)^3-3\left(x+\frac{1}{x}\right)=\left(x^3+\frac{1}{x^3}\right)=52 \Rightarrow y^3-3y-52=0$, where $y=\left(x+\frac{1}{x}\right)$

Clearly, $y=4$ satisfies $y^3-3y-52=0$.

$\therefore x+\frac{1}{x}=4.$

54. $\left(x-\frac{1}{x}\right)^3=\left(x^3-\frac{1}{x^3}\right)-3\left(x-\frac{1}{x}\right).$

$\therefore \left(x-\frac{1}{x}\right)^3+3\left(x-\frac{1}{x}\right)=\left(x^3-\frac{1}{x^3}\right)=14$

$\therefore y^3+3y-14=0$, where $y=\left(x-\frac{1}{x}\right).$

Clearly, $y=2$ satisfies $y^3+3y-14=0$.

$\therefore x-\frac{1}{x}=2.$

55. $\left(x+\frac{1}{x}\right)^2=\left(\frac{17}{4}\right)^2 \Rightarrow x^2+\frac{1}{x^2}+2=\frac{289}{16} \Rightarrow x^2+\frac{1}{x^2}=\frac{257}{16}.$

$\therefore x^2+\frac{1}{x^2}-2=\left(\frac{257}{16}-2\right)=\frac{225}{16} \Rightarrow \left(x-\frac{1}{x}\right)^2=\left(\frac{15}{4}\right)^2.$

$\therefore \left(x-\frac{1}{x}\right)=\frac{15}{4}.$

56. $x^4+\frac{1}{x^4}+2=729 \Rightarrow \left(x^2+\frac{1}{x^2}\right)^2=(27)^2 \Rightarrow x^2+\frac{1}{x^2}=27.$

$\therefore x^2+\frac{1}{x^2}-2=25 \Rightarrow \left(x-\frac{1}{x}\right)^2=5^2 \Rightarrow \left(x-\frac{1}{x}\right)=5.$

$\therefore \left(x-\frac{1}{x}\right)^3=5^3 \Rightarrow x^3-\frac{1}{x^3}-3\left(x-\frac{1}{x}\right)=125.$

$\therefore \left(x^3-\frac{1}{x^3}\right)=(125+3\times 5)=140.$

57. $\left(2x-\frac{3}{x}\right)^2=5^2 \Rightarrow 4x^2+\frac{9}{x^2}-2\times 2x\times\frac{3}{x}=25 \Rightarrow 4x^2+\frac{9}{x^2}=37$

$\therefore 4x^2+\frac{9}{x^2}+12=49 \Rightarrow \left(2x+\frac{3}{x}\right)^2=7^2 \Rightarrow \left(2x+\frac{3}{x}\right)=7.$

$$\therefore \left(4x^2 - \frac{9}{x^2}\right) = \left(2x - \frac{3}{x}\right)\left(2x + \frac{3}{x}\right) = (5 \times 7) = 35.$$

58. $\left(x + \frac{1}{x}\right)^2 = 3^2 \Rightarrow x^2 + \frac{1}{x^2} = 7 \Rightarrow \left(x^2 + \frac{1}{x^2}\right)^3 = 343$

$$\therefore x^6 + \frac{1}{x^6} + 3\left(x^2 + \frac{1}{x^2}\right) = 343 \Rightarrow x^6 + \frac{1}{x^6} + 3 \times 7 = 343$$

$$\therefore \left(x^6 + \frac{1}{x^6}\right) = (343 - 21) = 322.$$

59. $x^2 + y^2 = (x + y)^2 - 2xy = (49 - 24) = 25.$

60. $x^3 + y^3 = (x + y)^3 - 3xy\,(x + y) = 343 - (3 \times 6 \times 7) = 217.$

61. $(x - y)^2 = (x + y)^2 - 4xy = (25 - 24) = 1$. So, $(x - y) = 1$

$\therefore (x^3 - y^3) = (x - y)^3 + 3xy\,(x - y) = 1 + 3 \times 6 \times 1 = 19.$

62. $a + b + c = 0 \Rightarrow a^3 + b^3 + c^3 = 3abc$

63. $a + b + c = 0 \Rightarrow a^3 + b^3 + c^3 = 3abc$

$$\therefore \frac{a^3}{abc} + \frac{b^3}{abc} + \frac{c^3}{abc} = 3 \text{ or } \frac{a^2}{bc} + \frac{b^2}{ac} + \frac{c^2}{ab} = 3.$$

64. $(x^3 + y^3 + z^3 - 3xyz) = (x + y + z)\,(x^2 + y^2 + z^2 - xy - yz - zx)$

$= (x + y + z)[(x + y + z)^2 - 3(xy + yz + zx)]$

$= 9 \times (81 - 3 \times 23)$

$= (9 \times 12) = 108.$

65. $3^x - 3^{x-1} = 18 \Leftrightarrow 3^x - \frac{3^x}{3} = 18 \Leftrightarrow a - \frac{a}{3} = 18,$

where $a = 3^x \Leftrightarrow 2a = 54 \Leftrightarrow a = 27 \Leftrightarrow 3^x = 3^3 \Leftrightarrow x = 3$

$\therefore x^x = 3^3 = 27.$

66. Given Exp. $= \left(\frac{a^2 - b^2}{a + b}\right) = (a - b) = (0.125 - 0.004) = 0.121.$

67. Given Exp. $= \left(\frac{a^2 - b^2}{a - b}\right) = (a + b) = (0.538 + 0.462) = 1.$

68. Given Exp. $= \frac{(2.3)^3 - (0.3)^3}{(2.3)^2 + 2.3 \times 0.3 + (0.3)^2} = \left(\frac{a^3 - b^3}{a^2 + ab + b^2}\right) = (a - b) = (2.3 - 0.3) = 2.$

69. Given Exp. $= \frac{a^3 + b^3}{a^2 - ab + b^2} = (a + b) = 0.013 + 0.007 = 0.02.$

70. Since $x + 6$ is a factor of the given expression, it becomes zero when $x = -6$.

$\therefore (-6)^3 + 3 \times (-6)^2 + 4 \times (-6) + k = 0 \Leftrightarrow k = 132.$

71. Clearly, $x = -2$ as well as $x = 1$ will make the given expression zero.

So, $2m - n = 32$ & $m + n = -11$.

Solving these equations, we get: $m = 7, n = -18$.

72. Since $x = -2$ makes $f(x)$ zero, so $(x + 2)$ is its factor.

73. Let $f(x) = x^3 - 6x + 7$. This when divided by $(x + 1)$ gives remainder,

$f(-1) = (-1)^3 - 6 \times (-1) + 7 = 12.$

74. Let $f(x) = x^5 - 9x^2 + 12x - 14$. This when divided by $(x - 3)$ gives remainder

$f(3) = 3^5 - 9 \times 3^2 + 12 \times 3 - 14 = 184.$

75. Required sum $= (x^2 + 1) + \dfrac{1}{(x^2-1)} = \dfrac{x^4-1+1}{x^2-1} = \dfrac{x^4}{x^2-1}$.

76. $x^2 - 11x^2y^2 + y^4 = [(x^2)^2 + (y^2)^2 - 2x^2y^2] - 9x^2y^2.$

So, to make it a perfect square, we will have to add $9x^2y^2$.

77. $$\frac{4}{x^4}+\frac{x^4}{81}=\left(\frac{2}{x^2}\right)^2+\left(\frac{x^2}{9}\right)^2+2\times\frac{2}{x^2}\times\frac{x^2}{9}-\frac{4}{9}=\left(\frac{2}{x^2}+\frac{x^2}{9}\right)^2-\left(\frac{2}{3}\right)^2$$

$$=\left(\frac{2}{x^2}+\frac{x^2}{9}+\frac{2}{3}\right)\left(\frac{2}{x^2}+\frac{x^2}{9}-\frac{2}{3}\right).$$

78. $x^2 - 1 - 2a - a^2 = x^2 - (1 + 2a + a^2) = x^2 - (a + 1)^2 = (x + a + 1)(x - a - 1).$

79. $x^2 - 8x - 20 = x^2 - 10x + 2x - 20 = x(x - 10) + 2(x - 10) = (x - 10)(x + 2).$

80. $x^2 - xy - 72y^2 = x^2 - 9xy + 8xy - 72y^2 = x(x - 9y) + 8y(x - 9y) = (x - 9y)(x + 8y)$

81. $x^2 - 11xy - 60y^2 = x^2 - 15xy + 4xy - 60y^2 = x(x - 15y) + 4y(x - 15y) = (x - 15y)(x + 4y).$

82. $x^4 + x^2 + 25 = (x^2)^2 + (5)^2 + 10x^2 - 9x^2 = (x^2 + 5)^2 - (3x)^2 = (x^2 + 5 + 3x)(x^2 + 5 - 3x).$

83. $x^4 - 7x^2y^2 + y^4 = x^4 + y^4 + 2x^2y^2 - 9x^2y^2 = (x^2 + y^2)^2 - (3xy)^2 = (x^2 + y^2 - 3xy)(x^2 + y^2 + 3xy).$

84. $x^2 + 4y^2 + 4y - 4xy - 2x - 8 = (x^2 + 4y^2 - 4xy) - 2x + 4y - 8 = (x - 2y)^2 - 2(x - 2y) - 8$

$= a^2 - 2a - 8$, where $a = (x - 2y) = a^2 - 4a + 2a - 8 = a(a - 4) + 2(a - 4) = (a - 4)(a + 2)$

$= (x - 2y - 4)(x - 2y + 2).$

85. $x^2 + xy - 2y^2 = x^2 + 2xy - xy - 2y^2 = x(x + 2y) - y(x + 2y) = (x + 2y)(x - y).$

86. We have,

$p = x^4 - y^2z^2$ ...(i)

$q = y^4 - z^2x^2$ ...(ii)

and $r = z^4 - x^2y^2$ ...(iii)

Multiplying (i), (ii) and (iii) by x, y and z respectively and adding, we get

$px^2 + qy^2 + rz^2 = x^6 - x^2y^2z^2 + y^6 - x^2y^2z^2 + z^6 - x^2y^2z^2$

$= x^6 + y^6 + z^6 - 3x^2y^2z^2$

$= (x^2 + y^2 + z^2)(x^4 + y^4 + z^4 - x^2y^2 - y^2z^2 - x^2z^2) = (x^2 + y^2 + z^2)(p + q + r)$

87. $x^2(px^2 + qy^2 + rz^2) + qr - p^2$

$= px^4 + qx^2y^2 + rx^2z^2 + qr - p^2$

$= p(x^4 - p) + qx^2y^2 + r(q + x^2z^2)$

$= p(y^2z^2) + qx^2y^2 + ry^4$ [Using (i), (ii) & (iii) from qn 86]

$= y^2(pz^2 + qx^2 + ry^2)$

$= y^2(x^4z^2 - y^2z^4 + y^4x^2 - z^2x^4 + z^4y^2 + z^4y^2 - x^2y^4)$ [From qn. 86]

$= y^2 \times 0 = 0$

88. $$\frac{(x-a)(x-b)}{(x-ma)-(x-mb)}=\frac{(x+a)(x+b)}{(x+ma)(x+mb)}$$

Where $m, a, b > 0$

$$\Rightarrow \frac{(x-a)(x-b)}{(x+a)(x+b)} = \frac{(x-ma)(x-mb)}{(x+ma)(x+mb)} \Rightarrow \frac{x^2-(a+b)x+ab}{x^2+x(a+b)+ab} = \frac{x^2-x(ma+mb)+m^2ab}{x^2+(ma+mb)x+m^2ab}$$

By componendo and dividendo

$$\frac{2(x^2+ab)}{2x(a+b)} = \frac{2(x^2+m^2ab)}{2(x(ma+mb)} \Rightarrow \frac{x^2+ab}{x(a+b)} = \frac{x^2+m^2ab}{xm(a+b)} \Rightarrow \frac{x^2+ab}{x^2+m^2ab} = \frac{1}{m}$$

89. From qn 43, we have

$$\frac{x^2+ab}{x^2+m^2ab} = \frac{1}{m} \Rightarrow mx^2 + mab = x^2 + m^2ab \Rightarrow x^2(m-1) = mab(m-1)$$

$$\Rightarrow x^2 = mab \Rightarrow x = \pm\sqrt{mab}$$

90. $(a+b+c)^2 - 3(ab+bc+ca) + 3(x^2+y^2+z^2) = 0$
$\Rightarrow a^2+b^2+c^2+2ab+2bc+2ca-3ab-3bc-3ca+3(x^2+y^2+z^2) = 0$
$\Rightarrow a^2+b^2+c^2 - \text{ab} - \text{bc} - \text{ca} + 3(x^2+y^2+z^2) = 0$
when
$a = b = c$ and $x = y = z = 0$

91. (1) $m = (1-p)(p^2+p+1) = 1-p^3$
$h = (p+1)(p^2-p+1) = 1+p^3$
(2) $m = p_n$
$1-p^3 = p(1+p^3) \Rightarrow 1-p^3 = p^4 + \text{p} \Rightarrow p^4+p^3+p-1 = 0$
Both the statements can not answer the questions.

92. (1) $x^2+y^2+z^2-xy-yz-zx = 0$

$$\Rightarrow \frac{1}{2}\left[(x-y)^2+(y-z)^2+(z-x)^2\right] = 0 \Rightarrow x = y = z$$

(2) $x^3+y^3+z^3-3xyz = 0$

$$\Rightarrow \frac{1}{2}(x+y+z)\left[(x-y)^2+(y-z)^2+(z-x)^2\right] = 0$$

$\therefore\ x+y+z = 0$
$\therefore$ (b) is correct

93. Statement–1
$\because a^3+b^3+c^3-3abc = (a+b+c)(a^2+b^2+c^3-ab-bc-ca)$
Put $a+b+c = 0$
Then, $a^3+b^3+c^3-3abc = 0 \Rightarrow a^3+b^3+c^3 = 3abc$
Hence, statement–1 is correct
Statement–2
$\because$ We know that
$a^2+b^2+c^2 = (a+b+c)^2 - 2(ab+bc+ca)$
Put $a+b+c = 0$
$a^2+b^2+c^2 = -2(ab+bc+ca)$
Hence, statement-2 is correct
Statement–3
We have, $a^3+b^3+c^3 = 3abc$ [from statement 1]
$c = -(a+b)$ [$\because$ Given $a+b+c = 0$]
$\Rightarrow a^3+b^3+c^3 = -3ab(a+b)$
Hence, statement-3 is correct

94. Let $f(x) = (x - 1)(x - 2)(x - 4) - 90$

By options

Put $x = 7$, we get $(7-1)(7-2)(7-4) - 90$

$6 \times 5 \times 3 - 90$

$90 - 90 = 0$

$\therefore (x - 7)$ is a factor of $f(x)$

95. $2x^2 + xy - 3y^2$

$\Rightarrow 2x^2 + 3xy - 2xy - 3y^2 \Rightarrow x(2x + 3y) - y(2x + 3y) \Rightarrow (x - y)(2x + 3y)$

$\therefore$ Sum of linear factors $= (x - y) + (2x + 3y) = 3x + 2y$

96. We have, one equation and two variables

$\therefore$ put $a = 1$

We have, $96 - 64a^3 + \dfrac{8}{a^2} - \dfrac{48}{a^3} = t^3 = 0$

$\Rightarrow 96 - 64 + 8 - 48 - t^3 = 0 \Rightarrow 104 - 112 = t^3 \Rightarrow t^3 = -8 \Rightarrow t = -2$

Now, $a^2t + 4a^3$

$(1)^2 \times -2 + 4 \times (1)^3$

$-2 + 4 = 2$

97. We have,

$\dfrac{2ab - b^2}{6a^2 - ab} = \dfrac{1}{6}$

$\Rightarrow 12ab - 6b^2 = 6a^2 - ab \Rightarrow 6a^2 - 13ab + 6b^2 = 0$

$\Rightarrow 6a^2 - 9ab - 4ab + 6b^2 = 0 \Rightarrow 3a(2a - 3b) - 2b(2a - 3b) = 0$

$\Rightarrow (3a - 2b)(2a - 3b) = 0$

$\Rightarrow 3a - 2b = 0$ $\quad 2a - 3b = 0 \Rightarrow 3a = 2b \Rightarrow 2a = 3b$

Case 1:

$3a = 2b$

$a : b = 2 : 3 \Rightarrow \dfrac{a+b}{a-b} = \dfrac{2+3}{2-3} \Rightarrow a + b : a - b = -5$

Case 2 if $2a = 3b$

$\dfrac{a}{b} = \dfrac{3}{2} \Rightarrow \dfrac{a+b}{a-b} = \dfrac{5}{1} \Rightarrow a + b : a - b = 5$

98. (1) $\dfrac{a^2 - b^2}{a^2 + b^2} > \dfrac{a-b}{a+b}$ where $a > b > 0$

Let $a = 2$ and $b = 1$

$\Rightarrow \dfrac{2^2 - 1^2}{2^2 + 1^2} > \dfrac{2-1}{2+1} \Rightarrow \dfrac{3}{5} > \dfrac{1}{3}$

Hence, statement (1) is correct

(2) $\dfrac{a^3 + b^3}{a^2 + b^2} > \dfrac{a^2 + b^2}{a+b}$ only when $a > b > 0$

Let $a = 2$ and $b = 1$

$\Rightarrow \dfrac{2^3 + 1^3}{2^2 + 1^2} > \dfrac{2^2 + 1^2}{2+1} \Rightarrow \dfrac{9}{5} > \dfrac{5}{3}$

Hence, statement-2 is correct

99. We have, $\dfrac{ay-bx}{c}=\dfrac{cx-az}{b}=\dfrac{bz-cy}{a}$

Multiplying the above equations by c, b and a in the consecutive equations

$$\Rightarrow \frac{ay-bx}{c}\times\frac{c}{c}=\frac{cx-az}{b}\times\frac{b}{b}=\frac{bz-cy}{a}\times\frac{a}{a}$$

$$\Rightarrow \frac{acy-bcx}{c^2}=\frac{bcx-azb}{b^2}=\frac{abz-cya}{a^2}=p \text{ let} \quad ...(i)$$

By equating, we get

$\Rightarrow acy-bcx=pc^2=pc^2$ ...(ii) $\Rightarrow bcx-azb=b^2p$ ...(iii) $\Rightarrow abz-cya=a^2p$ ...(iv)

By adding (ii), (iii) & (iv), we get

$O=p\,(a^2+b^2+c^2)\Rightarrow p=O$ ...(v)

From eqn (ii) and (v), we get

$acy=bcx\Rightarrow \dfrac{x}{y}=\dfrac{a}{b}$ ...(vi)

Hence, statement-1 is true

From equ (iii) and (v), we get

$bcx=azb\Rightarrow \dfrac{x}{z}=\dfrac{a}{c}$ ...(vii)

From eqn (iv) and (v), we get

$abz=cya\Rightarrow \dfrac{y}{z}=\dfrac{b}{c}$ ...(viii)

From, (vi), (vii) and (viii), we get

$x=at$, $y=bt$ and $z=ct$ [let t be the ratio]

$$\frac{x+y+z}{a+b+c}=\frac{z}{r}$$

L.H.S.

$$\Rightarrow \frac{at+bt+ct}{a+b+c}=t$$

R.H.S

$$\frac{ct}{c}=t$$

Hence, statement-2 is true

100. We are given:

$x^2+y^2+z^2=3,$ where x, y, z are integers.

Since (x, y, z) are integers, their squares (x^2, y^2, z^2) must be non-negative integers (0, 1, 4, 9, ...).

The only way to sum three non-negative integer squares to get 3 is if each square is 1:

$\Rightarrow x^2=1,\ y^2=1,\ z^2=1$

This implies that (x, y, z) can only be either (1) or (–1).

Let $P=(xy+yz+zx)$, $S=(x+y+z)$

$$(x+y+z)^2=x^2+y^2+z^2+2(xy+yz+zx)\Rightarrow S^2=3+2P$$

$$\Rightarrow \qquad 2P = S^2 - 3 \Rightarrow P = \left(\frac{S^2 - 3}{2}\right)$$

Now, consider all possible combinations of x, y and z from (1, -1) and calculate S and P for each case

Case 1: All three integers are 1

$(x, y, z) = (1, 1, 1)$

$S = 1 + 1 + 1 = 3$

$$P = \left(\frac{3^2 - 3}{2}\right) = \left(\frac{9 - 3}{2}\right) = \left(\frac{6}{2}\right) = 3$$

Case 2: Two integers are 1, and one is –1

Possible combinations:

(1, 1, –1), (1, –1, 1), (–1, 1, 1)

For any of these combinations:

S = 1 + 1 – 1 = 1

$$P = \left(\frac{1^2 - 3}{2}\right) = \left(\frac{1 - 3}{2}\right) = \left(\frac{-2}{2}\right) = -1$$

Case 3: One integer is 1, and two are –1

Possible combinations: (1, –1, –1), (–1, 1, –1), (–1, –1, 1)

For any of these combinations:

S = 1 – 1 – 1 = –1

$$P = \left(\frac{(-1)^2 - 3}{2}\right) = \left(\frac{1 - 3}{2}\right) = \left(\frac{-2}{2}\right) = -1$$

Case 4: All three integers are –1

$(x, y, z) = (-1, -1, -1)$

$S = -1 - 1 - 1 = -3$

$$P = \left(\frac{(-3)^2 - 3}{2}\right) = \left(\frac{9 - 3}{2}\right) = \left(\frac{6}{2}\right) = 3$$

The distinct values that $(xy + yz + zx)$ can have are 3 and –1.

Therefore, there are 2 possible values for $(xy + yz + zx)$.

20 CHAPTER

H.C.F. and L.C.M. of Polynomials

Divisor: *A polynomial p (x) is called a divisor of another polynomial f (x), if f (x) = p (x) . g (x) for some polynomial g (x) .*

H.C.F. of Polynomials: *A polynomial h (x) is called the H.C.F. or G.C.D. of two or more given polynomials, if h(x) is a polynomial of highest degree dividing each one of the given polynomials.*

Remark. The coefficient of highest degree term in H.C.F. is always taken as positive.

Ex. 1. ***Find the H.C.F. of $(x^4 - y^4)$ and $(x^6 - y^6)$.***

Sol. $x^4 - y^4 = (x^2 - y^2)(x^2 + y^2) = (x - y)(x + y)(x^2 + y^2)$.

And, $x^6 - y^6 = (x^3)^2 - (y^3)^2 = (x^3 - y^3)(x^3 + y^3) = (x - y)(x^2 + xy + y^2)(x + y)(x^2 - xy + y^2)$.

$\therefore$ Required H.C.F. $= (x - y)(x + y) = (x^2 - y^2)$.

Ex. 2. ***If $(x - 6)$ is the H.C.F. of $(x^2 - 2x - 24)$ & $(x^2 - kx - 6)$ find the value of k.***

Sol. Clearly, $(x - 6)$ must divide both $(x^2 - 2x - 24)$ & $(x^2 - kx - 6)$.

So, $x = 6$ will make each polynomial zero.

$\therefore 6^2 - 2 \times 6 - 24 = 0$ and $6^2 - 6k - 6 = 0$ or $k = 5$.

L.C.M. *A polynomial p(x) is called the l.c.m of two or more given polynomials, if it is a polynomial of smallest degree which is divided by each one of the given polynomials.*

Two Important Properties:

(*i*) L.C.M. of two polynomials $= \dfrac{\text{Product of polynomials}}{\text{Their H.C.F.}}$

(*ii*) For any two polynomials $p(x)$ and $q(x)$, we have:

$p(x) \times q(x) = (\text{Their H.C.F.}) \times (\text{Their L.C.M.})$

Ex. 3. ***Find the L.C.M. of:***

(i) $(2x^2 - 3x - 2)$ & $(x^3 - 4x^2 + 4x)$ (ii) $(x^2 + x - 6)$ & $4(4 - x^2)$

Sol. (*i*) Let $p(x) = 2x^2 - 3x - 2 = (x - 2)(2x + 1)$. And, $q(x) = x(x^2 - 4x + 4) = x(x - 2)^2$.

$\therefore$ H.C.F. $= (x - 2)$.

Hence, L.C.M. $= \dfrac{p(x).q(x)}{\text{H.C.F.}} = \dfrac{x(x-2)^3(2x+1)}{(x-2)} = x(x - 2)^2(2x + 1)$.

(*ii*) Let $f(x) = x^2 + x - 6 = (x + 3)(x - 2)$, $g(x) = 4(4 - x^2) = -4(x - 2)(x + 2)$.

$\therefore$ H.C.F. of $f(x)$ and $g(x) = (x - 2)$.

So, L.C.M. $= \dfrac{f(x) . g(x)}{\text{H.C.F}} = \dfrac{-4(x+3)(x-2)^2(x+2)}{(x-2)} = -4(x + 3)(x - 2)(x + 2)$.

Ex. 4. ***The G.C.D. of two polynomials is $(x + 3)$ and their L.C.M. is $(x^3 - 7x + 6)$. If one of the polynomials is $(x^2 + 2x - 3)$, find the other.***

Sol. Other polynomial $= \dfrac{\text{H.C.F.} \times \text{L.C.M.}}{\text{Given Polynomial}} = \dfrac{(x+3)(x^3 - 7x + 6)}{(x^2 + 2x - 3)}$

$$= \frac{(x+3)(x^3-7x+6)}{(x+3)(x-1)} = \left(\frac{x^3-7x+6}{x-1}\right) = (x^2+x-6).$$

Ex. 5. ***Find the L.C.M. of $(2x^2 - 8)$, $(3x^2 - 9x + 6)$ and $(6x^2 + 18x + 12)$.***

Sol. $f(x) = (2x^2 - 8) = 2(x^2 - 4) = 2(x - 2)(x + 2)$;

$g(x) = (3x^2 - 9x + 6) = 3(x^2 - 3x + 2) = 3(x - 2)(x - 1)$;

$h(x) = (6x^2 + 18x + 12) = 6(x^2 + 3x + 2) = 6(x + 2)(x + 1)$.

$\therefore$ L.C.M. $= 6(x - 2)(x + 2)(x + 1)$.

EXERCISE 20

1. What is the HCF of the polynomials $x^8 + x^4 + 1$ and $x^4 + x^2 + 1$? **[CDS (I) Exam, 2025]**

(*a*) 1 (*b*) $x^4 - x^2 + 1$ (*c*) $x^4 + x^2 + 1$ (*d*) $x^4 - x^2 - 1$

2. What is the HCF of the polynomials $x^6 - 3x^4 + 3x^2 - 1$ and $x^3 + 3x^2 + 3x + 1$? **[CDS (I) Exam, 2020]**

(*a*) $(x + 1)$ (*b*) $(x + 1)^2$ (*c*) $x^2 + 1$ (*d*) $(x + 1)^3$

3. What is the HCF $x^4 - 13x^2y^2 - 300y^4$, $x^3 - 4x^2y - 4xy^2 - 5y^3$ and $x^3 - 125y^3$? **[CDS (I) Exam, 2024]**

(*a*) $x - 5y$ (*b*) $x + 5y$ (*c*) $x^2 + 5xy + 25y^2$ (*d*) 1

4. If HCF of 768 and x^6y^2 is $32xy$ for natural numbers $x \ge 2$, $y \ge 2$, then what is the value of $(x + y)$? **[CDS (I) Exam, 2024]**

(*a*) 5 (*b*) 7 (*c*) 9 (*d*) 11

5. What is HCF of $a^4 + 2a^3 + 3a^2 + 2a + 1$ and $a^6 - 2a^3 + 1$? **[CDS (II) Exam, 2024]**

(*a*) $a^3 + 3a^2 + 2a + 1$ (*b*) $a^3 + a^2 + a + 1$

(*c*) $(a^2 + a + 1)^2$ (*d*) $(a^2 - a + 1)^2$

6. What is the HCF of $acx^3 + bcx^2 + adx^2 + acdx + bdx + bcd$ and $adx^3 + acx^2 + bdx^2 + bcx + acdx + bcd$ if HCF $(c, d) = 1$, $c \ne d$? **[CDS (II) Exam, 2023]**

(*a*) $bx + c$ (*b*) $cx + d$ (*c*) $ax + d$ (*d*) $ax + b$

7. What is the LCM of $x^4 + x^2y^2 + y^4$, $x^3 + y^3$, $x^3 - y^3$?

(*a*) $(x^2 - y^2)(x^4 + x^2y^2 + y^4)^2$ (*b*) $(x^2 - y^2)(x^4 + 2x^2y^2 + y^4)$

(*c*) $(x^6 + y^6)$ (*d*) $(x^6 - y^6)$

8. Consider the following in respect of natural numbers a, b and c: **[CDS (I) Exam, 2016]**

1. LCM $(ab, ac) = a$ LCM (b, c) **2.** HCF $(ab, ac) = a$ HCF (b, c)

3. HCF $(a, b) <$ LCM (a, b) **4.** HCF (a, b) divides LCM (a, b).

Which of the above are correct?

(*a*) 1 and 2 only (*b*) 3 and 4 only (*c*) 1, 2 and 4 only (*d*) 1, 2, 3 and 4

9. What is LCM of $x^3 + 8$, $x^2 + 5x + 6$ and $x^3 + 4x^2 + 4x$? **[CDS (I) Exam, 2017]**

(*a*) $x(x+2)^2(x+3)(x^2-2x+4)$ (*b*) $x(x-2)^2(x-3)(x^2+2x+4)$

(*c*) $(x+2)^2(x+3)(x^2-2x+4)$ (*d*) $(x-2)^2(x-3)(x^2-2x+4)$

10. The L.C.M. of $(x^3 - x^2 - 2x)$ and $(x^3 + x^2)$ is:

(*a*) $x^3 - x^2 - 2x$ (*b*) $x^2 + x$ (*c*) $x^4 - x^3 - 2x^2$ (*d*) $x - 2$

11. The L.C.M. of $(x^2 - 1)$, $(x^2 + x)$ and $(x + 1)^2$ is:

(*a*) $(x + 1)$ (*b*) $(x^2 - 1)(x + 1)^2$ (*c*) $(x^2 - 1)(x^2 + x)$ (*d*) $(x^2 - 1)(x + 1)^2$

12. H.C.F. and L.C.M. of two polynomials are $(x + y)$ and $(3x^5 + 5x^4y + 2x^3y^2 - 3x^2y^3 - 5xy^4 - 2y^5)$ respectively. If one of the polynomials is $(x^2 - y^2)$, the other one is:

(*a*) $3x^4 - 8x^3y + 10x^2y^2 + 7xy^3 - 2y^4$ (*b*) $3x^4 - 8x^3y - 10x^2y^2 + 7xy^3 + 2y^4$

(*c*) $3x^4 + 8x^3y + 10x^2y^2 + 7xy^3 + 2y^4$ (*d*) $3x^4 + 8x^3y - 10x^2y^2 + 7xy^3 + 2y^4$

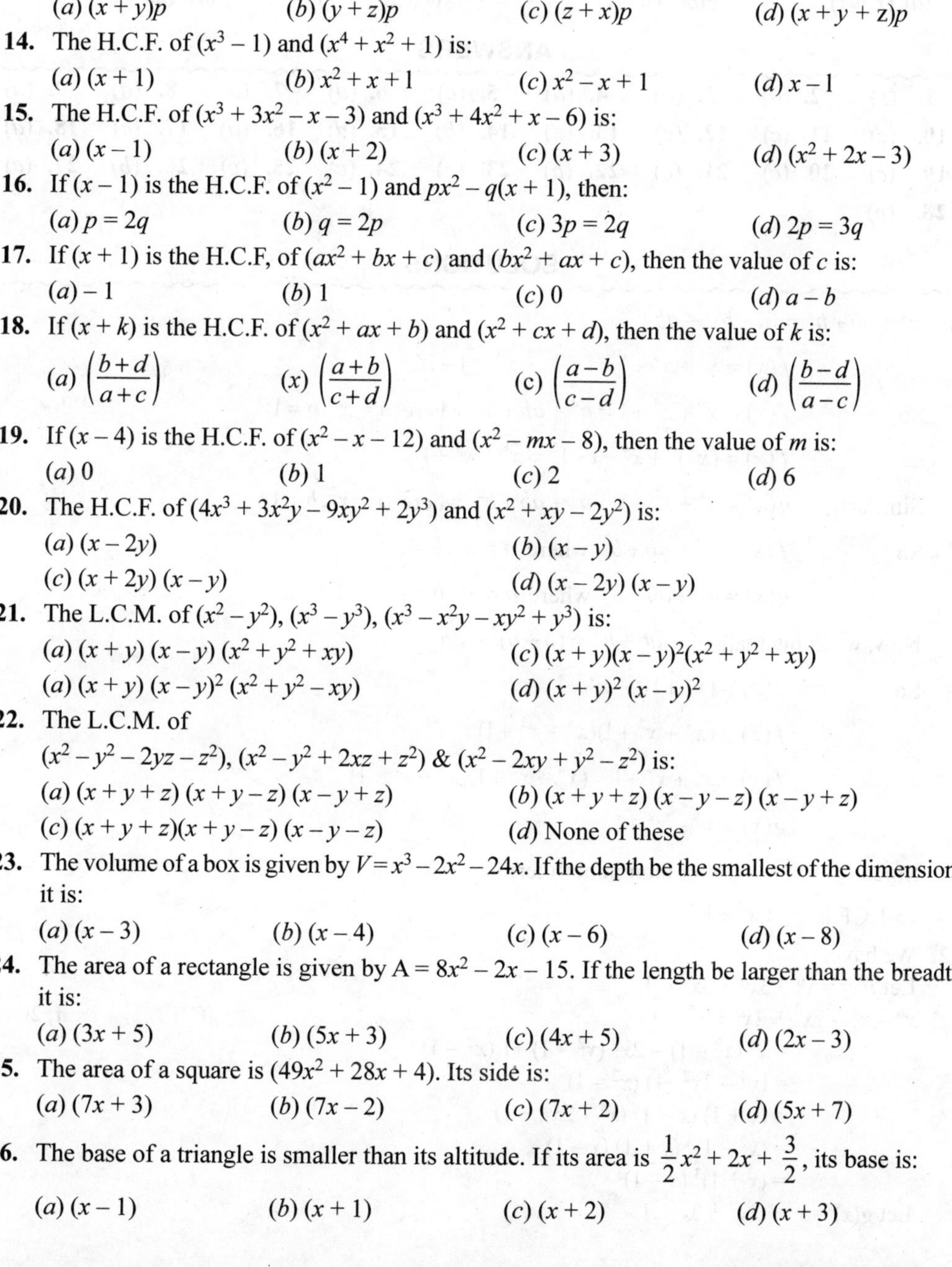

13. The product of two non-zero expressions is $(x + y + z)\,p^3$. If their H.C.F. is p^2, then their L.C.M. is:

(*a*) $(x + y)p$ (*b*) $(y + z)p$ (*c*) $(z + x)p$ (*d*) $(x + y + z)p$

14. The H.C.F. of $(x^3 - 1)$ and $(x^4 + x^2 + 1)$ is:

(*a*) $(x + 1)$ (*b*) $x^2 + x + 1$ (*c*) $x^2 - x + 1$ (*d*) $x - 1$

15. The H.C.F. of $(x^3 + 3x^2 - x - 3)$ and $(x^3 + 4x^2 + x - 6)$ is:

(*a*) $(x - 1)$ (*b*) $(x + 2)$ (*c*) $(x + 3)$ (*d*) $(x^2 + 2x - 3)$

16. If $(x - 1)$ is the H.C.F. of $(x^2 - 1)$ and $px^2 - q(x + 1)$, then:

(*a*) $p = 2q$ (*b*) $q = 2p$ (*c*) $3p = 2q$ (*d*) $2p = 3q$

17. If $(x + 1)$ is the H.C.F, of $(ax^2 + bx + c)$ and $(bx^2 + ax + c)$, then the value of c is:

(*a*) -1 (*b*) 1 (*c*) 0 (*d*) $a - b$

18. If $(x + k)$ is the H.C.F. of $(x^2 + ax + b)$ and $(x^2 + cx + d)$, then the value of k is:

(*a*) $\left(\dfrac{b+d}{a+c}\right)$ (*x*) $\left(\dfrac{a+b}{c+d}\right)$ (c) $\left(\dfrac{a-b}{c-d}\right)$ (*d*) $\left(\dfrac{b-d}{a-c}\right)$

19. If $(x - 4)$ is the H.C.F. of $(x^2 - x - 12)$ and $(x^2 - mx - 8)$, then the value of m is:

(*a*) 0 (*b*) 1 (*c*) 2 (*d*) 6

20. The H.C.F. of $(4x^3 + 3x^2y - 9xy^2 + 2y^3)$ and $(x^2 + xy - 2y^2)$ is:

(*a*) $(x - 2y)$ (*b*) $(x - y)$

(*c*) $(x + 2y)(x - y)$ (*d*) $(x - 2y)(x - y)$

21. The L.C.M. of $(x^2 - y^2)$, $(x^3 - y^3)$, $(x^3 - x^2y - xy^2 + y^3)$ is:

(*a*) $(x + y)(x - y)(x^2 + y^2 + xy)$ (*c*) $(x + y)(x - y)^2(x^2 + y^2 + xy)$

(*a*) $(x + y)(x - y)^2(x^2 + y^2 - xy)$ (*d*) $(x + y)^2(x - y)^2$

22. The L.C.M. of

$(x^2 - y^2 - 2yz - z^2)$, $(x^2 - y^2 + 2xz + z^2)$ & $(x^2 - 2xy + y^2 - z^2)$ is:

(*a*) $(x + y + z)(x + y - z)(x - y + z)$ (*b*) $(x + y + z)(x - y - z)(x - y + z)$

(*c*) $(x + y + z)(x + y - z)(x - y - z)$ (*d*) None of these

23. The volume of a box is given by $V = x^3 - 2x^2 - 24x$. If the depth be the smallest of the dimensions, it is:

(*a*) $(x - 3)$ (*b*) $(x - 4)$ (*c*) $(x - 6)$ (*d*) $(x - 8)$

24. The area of a rectangle is given by $A = 8x^2 - 2x - 15$. If the length be larger than the breadth, it is:

(*a*) $(3x + 5)$ (*b*) $(5x + 3)$ (*c*) $(4x + 5)$ (*d*) $(2x - 3)$

25. The area of a square is $(49x^2 + 28x + 4)$. Its side is:

(*a*) $(7x + 3)$ (*b*) $(7x - 2)$ (*c*) $(7x + 2)$ (*d*) $(5x + 7)$

26. The base of a triangle is smaller than its altitude. If its area is $\dfrac{1}{2}x^2 + 2x + \dfrac{3}{2}$, its base is:

(*a*) $(x - 1)$ (*b*) $(x + 1)$ (*c*) $(x + 2)$ (*d*) $(x + 3)$

27. The HCF and LCM of two polynomials are $(x+y)$ and $(3x^5+5x^4y+2x^3y^2-3x^2y^3-5xy^4-2y^5)$ respectively. If one of the polynomials is (x^2-y^2), then the other polynomial is

(*a*) $3x^4 - 8x^3y + 10x^2y^2 + 7xy^3 - 2y^4$ (*b*) $3x^4 - 8x^3y - 10x^2y^2 + 7xy^3 + 2y^4$

(*c*) $3x^4 + 8x^3y + 10x^2y^2 + 7xy^3 + 2y^4$ (*d*) $3x^4 + 8x^3y - 10x^2y^2 + 7xy^3 + 2y^4$

[CDS (I) Exam, 2015]

28. The sum and difference of two expressions are $5x^2 - x - 4$ and $x^2 + 9x - 10$ respectively. The HCF of the two expressions will be **[CDS (I) Exam, 2016]**

(*a*) $(x + 1)$ (*b*) $(x - 1)$ (*c*) $(3x + 7)$ (*d*) $(2x - 3)$

ANSWERS

1. (*c*)	**2.** (*d*)	**3.** (*a*)	**4.** (*a*)	**5.** (*c*)	**6.** (*d*)	**7.** (*d*)	**8.** (*d*)	**9.** (*a*)
10. (*c*)	**11.** (*c*)	**12.** (*c*)	**13.** (*d*)	**14.** (*b*)	**15.** (*d*)	**16.** (*a*)	**17.** (*c*)	**18.** (*d*)
19. (*c*)	**20.** (*c*)	**21.** (*c*)	**22.** (*b*)	**23.** (*c*)	**24.** (*c*)	**25.** (*c*)	**26.** (*b*)	**27.** (*c*)
28. (*b*)								

SOLUTIONS

1. $a^2+ab+b^2=(a+b)^2-ab$

$f(x)=x^8+x^4+1=(x^4)^2+x^4\times1+1^2$

So $f(x)=x^8+x^4+1=a^2+ab+b^2$ where $a=x^4, b=1$

$\Rightarrow$ $f(x)=(x^4)^2+x^4\times1+1^2=x^8+x^4+1$

Similarly, $g(x)=x^4+x^2+1=a^2+ab+b^2$ where $a=x^2, b=1$

So $f(x)=a^2+ab+b^2$ where $a=x^4, b=1$

$g(x)=a^2+ab+b^2$ where $a=x^2, b=1$

Now, we know that: $a^2+ab+b^2=(a+b)^2-ab$

So $f(x)=(x^4+1)^2-x^4$

$f(x)=(x^4+x^2+1)(x^4-x^2+1)$

$f(x)=x^8+x^4+1=(x^4+x^2+1)(x^4-x^2+1)$

$g(x)=x^4+x^2+1$

Since x^4+x^2+1 is common in both terms.

$\Rightarrow$ HCF is x^4+x^2+1

2. We have,

Let $f(x) = x^6 - 3x^4 + 3x^2 - 1$

$x^6 - x^4 - 2x^4 + 2x^2 + x^2 - 1$

$= x^4(x^2 - 1) - 2x^2(x^2 - 1) + 1(x^2 - 1)$

$= (x^4 - 2x^2 + 1)(x^2 - 1)$

$= (x + 1)(x - 1)(x^4 - 2x^2 + 1)$

$= (x^2 - 1)^2 (x + 1)(x - 1)$

$= (x + 1)^3 (x - 1)^3$

Let $g(x) = x^3 + 3x^2 + 3x + 1$

$= x^3 + x^2 + 2x^2 + 2x + x + 1$

$\Rightarrow \quad x^2(x + 1) + 2x(x + 1) + (x + 1) \Rightarrow (x^2 + 2x + 1)(x + 1) \Rightarrow (x + 1)^3$

$\therefore \quad$ H.C.F. $= (x + 1)^3$

3. We have,

$x^4 - 13x^2y^2 - 300y^4$

$= x^4 - 25x^2y^2 + 12x^2y^2 - 300y^4 = x^2(x^2 - 25y)^2 + 12y^2(x^2 - 25y^2)$

$= (x^2 + 12y^2)(x^2 - 25y^2) = (x^2 - 12y^2)(x + 5y)(x - 5y)$

$= x^3 - 4x^2y - 4xy^2 - 5y^3$

$= (x - 5y)(x^2 + xy + y^2)$ and $x^3 - 125y^3$

$= x^3 - (5y)^3$

$= (x - 5y)(x^2 + 5xy + 25y^2)$

$\Rightarrow \quad$ HCF $= x - 5y$

4. $N_1 = 768 = 2^8 \times 3$

$N_2 = x^6 y^2 = 2^6 \times 3^2$

By option (a)

Let $x = 2$ and $y = 3$

$\therefore$ HCF $= 2^6 \times 3 = 2^5 \times 2 \times 3 = 32\,xy$

5. Let's put $a = 2$,

$a^4 + 2a^3 + 3a^2 + 2a + 1$

$\Rightarrow \quad 2^4 + 2(2)^3 + 3(2)^2 + 2(2) + 1 = 49$

$a^6 - 2a^3 + 1$

$\Rightarrow \quad (2)^6 - 2(2)^3 + 1 = 49$

So, HCF of 49 and 49 is 49.

Checking by option,

$(a^2 + a + 1)^2$

$\Rightarrow \quad (2^2 + 2 + 1)^2 = (4 + 2 + 1)^2 = 7^2 = 49$ (true)

6. $acx^3 + bcx^2 + adx^2 + acdx + bdx + bcd$

$= cx^2(ax + b) + adx^2 + bdx + acdx + bcd = cx^3(ax + b) + dx(ax + b) + cd(ax + b)$

$= (cx^2 + dx + cd)(ax + b)$

and $adx^3 + acx^2 + bdx^2 + bcx + acdx + bcd$

$= adx^3 + bdx^2 + acx^2 + bcx + acdx + bcd = dx^2(ax + b) + cx(ax + b) + cd(ax + b)$

$= (dx^2 + cx + cd)(ax + b)$

$\therefore$ HCF $= ax + b$.

7. We have, $x^4 + x^2y^2 + y^4 = (x^2 + y^4 - xy)(x^2 + y^2 + xy)$

$x^3 + y^3 = (x + y)(x^2 + y^2 - xy)$

$x^3 - y^3 = (x - y)(x^2 + y^2 + xy)$

$\therefore$ required LCM $= (x + y)(x - y)(x^4 + x^2y^2 + y^4)$

$= (x^2 - y^2)(x^4 + x^2y^2 + y^4)$

$= (x^2)^3 - (y^2)^3 = x^6 - y^6$

8. All the four statement are standard results and are true

9. x^3+8 factor is $(x+2)(x^2-2x+4)$ ... (i)

(x^2+5x+6) factor is $(x+3)(x+2)$ ... (ii)

$x(x^2+4x+4)$ factor is $x(x+2)^2$ ... (iii)

$\therefore$ LCM of (*i*), (ii) and (iii) is $x(x+2)^2(x+3)(x^2-2x+4)$

10. $x^3 - x^2 - 2x = x(x^2 - x - 2) = x(x-2)(x+1)$, $x^3 + x^2 = x^2(x+1)$.

$\therefore$ L.C.M $= x^2(x+1)(x-2) = x^2(x^2 - x - 2) = (x^4 - x^3 - 2x^2)$.

11. $x^2 - 1 = (x-1)(x+1)$, $x^2 + x = x(x+1)$ & 3rd exp. is $(x+1)^2$

$\therefore$ L.C.M. $= (x-1)x(x+1)^2 = (x-1)(x+1)x(x+1) = (x^2-1)(x^2+x)$.

12. Other polynomial $= \dfrac{(x+y)(3x^5+5x^4y+2x^3y^2-3x^2y^3-5xy^4-2y^5)}{(x^2-y^2)}$

$= \dfrac{3x^5+5x^4y+2x^3y^2-3x^2y^3-5xy^4-2y^5}{(x-y)}$

$= 3x^4 + 8x^3y + 10x^2y^2 + 7xy^3 + 2y^4$

13. L.C.M. $= \dfrac{\text{Product}}{\text{H.C.F.}} = \dfrac{(x+y+z)p^3}{p^2} = (x+y+z)p$.

14. $x^3 - 1 = (x-1)(x^2+x+1)$, $x^4 + x^2 + 1 = (x^4 + 2x^2 + 1) - x^2 = (x^2+1)^2 - x^2 = (x^2+x+1)(x^2-x+1)$

$\therefore$ H.C.F. $= (x^2+x+1)$

15. $x^3 + 3x^2 - x - 3 = x^2(x+3) - 1(x+3) = (x+3)(x^2-1) = (x+3)(x-1)(x+1)$

$x^3 + 4x^2 + x - 6$ is clearly zero when $x = 1$. So, $(x-1)$ is its factor.

$x^3 + 4x^2 + x - 6 = (x-1)(x^2+5x+6) = (x-1)(x+2)(x+3)$.

$\therefore$ H.C.F. $= (x+3)(x-1) = (x^2+2x-3)$.

16. Since $(x-1)$ is the H.C.F., so it will divide each one of the given expressions. So, $x = 1$ will make each one zero.

$\therefore \quad p \times 1^2 - q(1+1) = 0$ or $p = 2q$.

17. Since $(x+1)$ is the H.C.F., it will divide each one of the given expressions. So, $x = 1$ will make each one zero.

$\therefore \ a - b + c = 0$ and $b - a + c = 0$.

Adding these equations, we get $2c = 0$ or $c = 0$.

18. Since $(x+k)$ is the H.C.F., it will divide each one of the given expressions. So, $x = -k$ will make each one zero.

$\therefore k^2 - ak + b = 0$, $k^2 - ck + d = 0$. So, $k^2 - ak + b = k^2 - ck + d$ or $k = \left(\dfrac{b-d}{a-c}\right)$.

19. Clearly, $x = 4$ will make each one of the given expressions zero. So, $16 - 4m - 8 = 0$ or $m = 2$.

20. $x^2 + xy - 2y^2 = x^2 - y^2 + xy - y^2 = (x^2 - y^2) + y(x-y) = (x-y)(x+2y)$

Putting $x = y$ in $(4x^3 + 3x^2y - 9xy^2 + 2y^3)$, it becomes zero.

So, $(x-y)$ is a factor. On dividing, we get:

$(4x^3 + 3x^2y - 9xy^2 + 2y^3) = (x-y)(4x^2 + 7xy - 2y^2) = (x-y)(x+2y)(4x-y)$

$\therefore$ H.C.F. $= (x-y)(x+2y)$.

21. $(x^2 - y^2) = (x - y)(x + y)$, $(x^3 - y^3) = (x - y)(x^2 + xy + y^2)$, $(x^3 - x^2y - xy^2 + y^3)$
$= x^2(x - y) - y^2(x - y) = (x - y)(x^2 - y^2) = (x - y)^2 (x + y)$.
$\therefore$ L.C.M. $= (x - y)^2 (x + y)(x^2 + y^2 + xy)$

22. $(x^2 - y^2 - 2yz - z^2) = x^2 - (y + z)^2 = (x + y + z)(x - y - z)$.
$(x^2 - y^2 + 2xz + z^2) = (x + z)^2 - y^2 = (x + y + z)(x - y + z)$.
$(x^2 - 2xy + y^2 - z^2) = (x - y)^2 - z^2 = (x - y + z)(x - y - z)$.
$\therefore$ L.C.M. $= (x + y + z)(x - y + z)(x - y - z)$

23. $V = x(x^2 - 2x - 24) = x(x - 6)(x + 4)$. So, depth $= (x - 6)$.

24. $A = 8x^2 - 2x - 15 = (2x - 3)(4x + 5)$. So, length $= (4x + 5)$.

25. Area $= (49x^2 + 28x + 4) = (7x + 2)^2$. So, side $= (7x + 2)$.

26. $A = \frac{1}{2}(x^2 + 4x + 3) = \frac{1}{2}(x + 3)(x + 1)$. So, base $= (x + 1)$.

27. HCF $= x + y$
LCM $= 3x^5 + 5x^4y + 2x^3y^2 - 3x^2y^3 - 5xy^4 - 2y^5$
We notice the pattern in the coefficients of the 6 terms in the LCM.
LCM $= (x^3 - y^3)(3x^2 + 5xy + 2y^2) = (x - y)(x^2 + xy + y^2)(3x + 2y)(x + y)$
Let the two polynomials be $f(x)$ and $g(x)$, where $f(x) = x^2 - y^2$.

HCF	$f(x) = (x - y)(x + y)$	LCM
$x + y$	$g(x)$	$(x - y)(3x + 2y)(x + y)(x^2 + xy + y^2)$

Of the 4 factors in the LCM, one is there only in $f(x)$ and one is there in both $f(x)$ and $g(x)$. The other two are there only in $g(x)$.

$$\therefore \quad g(x) = (x + y)(3x + 2y)(x^2 + xy + y^2)$$
$$= (x + y)(3x^3 + 5x^2y + 5xy^2 + 2y^3)$$
$$= 3x^4 + 8x^3y + 10x^2y^2 + 7xy^3 + 2y^4$$

28. Let a, b be the two given expressions
$\Rightarrow a + b = 5x^2 - x - 4$ and ... (1)
$a - b = x^2 + 9x - 10$ ... (2)
(1) + (2); $2a = 6x^2 + 8x - 14$
$a = 3x^2 + 4x - 7 = (3x + 7)(x - 1)$
and (1) – (2) $\Rightarrow 2b = 4x^2 - 10x + 6$
$b = 2x^2 - 5x + 3 = (2x - 3)(x - 1)$

H.C.F of a, b is $x - 1$
(which is the common factor)

CHAPTER

Linear Equations

Highlights: *Simultaneous linear equations in two unknowns and their solutions. Practical problems leading to two simultaneous linear equations, Linear inequations.*

Linear Equations in Two Variables: *An equation of the form $ax + by = c$, where a, b, c are real numbers is called a linear equation in two variables x and y.*

The graph of a linear equation $ax + by = c$ is a straight line.

Ex. ***Draw the graph of $3x + 2y = 11$.***

Sol. We may write the given equation as, $y = \left(\frac{11-3x}{2}\right)$.

If we put $x = 1$, then clearly $y = 4$.

If we put $x = 3$, then clearly $y = 1$.

So, we have the table:

x	1	3
y	4	1

Now, we plot the points (1, 4) and (3, l) on the graph paper, join these points and produce the line in both ways. Thus, we obtain the graph of the given equation.

System of Linear Equations:

Consistent System: *A system consisting of two simultaneous linear equations is said to be consistent, if it has at least one solution.*

Inconsistent System: *A system of two simultaneous linear equations is said to be inconsistent, if it has no solution at all.*

Ex. Consider the system of equations: $x + y = 2$ & $2x + 2y = 5$. Clearly, there are no values of x and y which may simultaneously satisfy the given equations. So, the system given above is inconsistent.

Important Rules:

I. The system $a_1x + b_1y = c_1$, $a_2x + b_2y = c_2$; has:

(i) a unique solution, if $\frac{a_1}{a_2} \uparrow \frac{b_1}{b_2}$;

(ii) an infinite number of solutions, if $\frac{a_1}{a_2} = \frac{b_1}{b_2} = \frac{c_1}{c_2}$;

(iii) no solution, if $\frac{a_1}{a_2} = \frac{b_1}{b_2} \neq \frac{c_1}{c_2}$.

II. The homogeneous system $a_1x + b_2y = c_2$ has a ***non-zero solution*** only when $\frac{a_1}{a_2} = \frac{b_1}{b_2}$ and in this case, the system has an infinite number of solutions.

III. **The graphs of $a_1x + b_1y = c_1$, $a_2x + b_2y = c_2$ will be**

(*i*) *parallel*, if the system has *no solution*;

(*ii*) *coincident*, if the system has *an infinite number of solutions*:

(*iii*) *intersecting*, if the system has a *unique solution*.

IV. The equations of the type $ax + by = c$ and $k\,ax + k\,by = kc$ are known as **dependent equations**.

SOLVED PROBLEMS

Ex. 1. ***Solve*: $37x + 29y = 95$, $29x + 37y = 103$.**

Sol. In this problem note that coefficients of x and y in the two equations are interchanged. In such a problem, once we add the equations and next time, we subtract them.

Adding the given equations, we get:

$$66x + 66y = 198 \quad \text{or} \quad x + y = 3.$$

Subtracting the given equations, we get:

$$8x - 8y = -8 \quad \text{or} \quad x - y = -1.$$

On solving $x + y = 3$, $x - y = -1$, we get $x = 1$ & $y = 2$.

Ex. 2. ***Solve:*** $\dfrac{40}{x+y} + \dfrac{2}{x-y} = 5, \ \dfrac{25}{x+y} - \dfrac{3}{x-y} = 1.$

Sol. Putting $\dfrac{1}{x+y} = u$ and $\dfrac{1}{x-y} = v$, the given equations become: $40u + 2v = 5$, $25u - 3v = 1$.

Solving these equations, we get: $u = \dfrac{1}{10}$ and $v = \dfrac{1}{2}$.

$$\therefore \ \frac{1}{x+y} = \frac{1}{10}, \ \frac{1}{x-y} = \frac{1}{2}. \quad \text{So, } x + y = 10 \ \& \ x - y = 2.$$

On solving $x + y = 10$, $x - y = 2$, we get $x = 6$ and $y = 4$.

Ex. 3. ***Find k for which the system $6x - 2y = 3$, $kx - y = 2$ has a unique solution.***

Sol. The given system will have a unique solution, if

$$\frac{a_1}{a_2} \neq \frac{b_1}{b_2} \ \ i.e. \ \frac{6}{k} \neq \frac{-2}{-1} \ \ \text{or} \ \ k \neq 3.$$

Ex. 4. ***Find k for which the system $2x + 3y - 5 = 0$, $4x + ky - 10 = 0$ has an infinite number of solutions.***

Sol. For an infinite number of solutions, we have 1:

$$\frac{a_1}{a_2} = \frac{b_1}{b_2} = \frac{c_1}{c_2} \ \ i.e. \ \frac{2}{4} = \frac{3}{k} = \frac{5}{10} \ \ \text{or} \ \ k = 6.$$

Ex. 5. ***Find whether the system $2x + 3y = 1$, $4x + 6y = 4$ has (i) a unique solution or (ii) an infinite number of solutions or (iii) no solution.***

Sol. We have: $\dfrac{a_1}{a_2} = \dfrac{2}{4} = \dfrac{1}{2}, \ \dfrac{b_1}{b_2} = \dfrac{3}{6} = \dfrac{1}{2} \ \& \ \dfrac{c_1}{c_2} = \dfrac{1}{4}.$

$$\therefore \ \frac{a_1}{a_2} = \frac{b_1}{b_2} \neq \frac{c_1}{c_2}. \quad \text{So, the given system has no solution.}$$

Ex. 6. ***Find whether the system $3x - 2y = 1$, $6x - 4y = 2$ has (i) a unique solution or (ii) an infinite number of solutions.***

Sol. We have: $\frac{a_1}{a_2} = \frac{3}{6} = \frac{1}{2}, \frac{b_1}{b_2} = \frac{-2}{-4} = \frac{1}{2} \& \frac{c_1}{c_2} = \frac{1}{2}$

$\therefore \quad \frac{a_1}{a_2} = \frac{b_1}{b_2} = \frac{c_1}{c_2}.$

So, the given system has an infinite number of solutions.

Ex. 7. ***Find k such that the system 3x + 5y = 0, kx + 10y = 0 has a non-zero solution.***

Sol. The given system has a non-zero solution, if $\frac{3}{k} = \frac{5}{10}$ or $k = 6$.

EXERCISE 21

1. A piece of cloth costs ₹10,000. If a 2 m longer piece of the same cloth is purchased for the same amount, it would cost ₹250 less per metre. What is the original length of the piece of cloth? **[CDS (I) Exam, 2022]**

(*a*) 8 m (*b*) 10 m (*c*) 12 m (*d*) 16 m

2. 5 pencils, 6 notebooks and 7 erasers cost ₹250; whereas 6 pencils, 4 notebooks and 2 erasers cost ₹180. What is the cost of 2 notebooks and 4 erasers? **[CDS (I) Exam, 2021]**

(*a*) ₹90 (*b*) ₹75 (*c*) ₹60 (*d*) ₹40

3. If the system of equations $7x + ky = 27$ and $kx + 7y = 19$ have unique solution, then which one of the following is correct? **[CDS (II) Exam, 2021]**

(*a*) $k \neq 7$ (*b*) $k \neq 13$ (*c*) $k = 7$ (*d*) $k = 13$

4. A person saves ₹1000 more than he did the previous years. If he saves ₹2000 in the first years, in how many years will he save ₹170000? **[CDS (II) Exam, 2021]**

(*a*) 16 years (*b*) 17 years (*c*) 18 years (*d*) 19 years

5. If $ab + xy - xb = 0$ and $bc + yz - cy = 0$, then what is $\frac{x}{a} + \frac{c}{z}$ **[CDS (I) Exam, 2020]**

(*a*) $\frac{y}{b}$ (*b*) $\frac{b}{y}$ (*c*) 1 (*d*) 0

6. If $(p + 2)(2q - 1) = 2pq - 10$ and $(p - 2)(2q - 1) = 2pq - 10$, then what is pq equal to ? **[CDS (I) Exam, 2020]**

(*a*) –10 (*b*) –5 (*c*) 5 (*d*) 10

7. The number of prime number which are less than 100 is **[CDS (I) Exam, 2017]**

(*a*) 24 (*b*) 25 (*c*) 26 (*d*) 27

8. If $65x - 33y = 97$ and $33x - 67y = 1$, then what is xy equal to ? **[CDS (I) Exam, 2018]**

(*a*) 2 (*b*) 3 (*c*) –2 (*d*) –3

9. If $\frac{b}{y} + \frac{z}{c} = 1$ and $\frac{c}{z} + \frac{x}{a} = 1$, then what is $\frac{ab + xy}{bx}$ equal to ? **[CDS (I) Exam, 2018]**

(*a*) 1 (*b*) 2 (*c*) 0 (*d*) –1

10. If $x + y + z = 0$, then what is $(y + z - x)^3 + (z + x - y)^3 + (x + y - z)^3$ equal to? **[CDS (I) Exam, 2018]**

(*a*) $(x + y + z)^3$ (*b*) $3(x + y)(y + z)(z + x)$

(*c*) $24xyz$ (*d*) $-24xyz$

11. Consider the following statements:

1. The equations $1990x - 173y = 11$ has no solution in integers for x and y,

2. The equal $3x - 12y = 7$ has no solution in integers for x and y,

Which of the above statements is/are correct?

(*a*) 1 only (*b*) 2 only (*c*) Both 1 and 2 (*d*) Neither 1 nor 2

[CDS (I) Exam, 2015]

12. The expression $x^3q^2 - x^3pt + 4x^2pt - 4x^2q^2 + 3xq^2 - 3xpt$ is divisible by

(*a*) $(x-1)$ only

(*b*) $(x-3)$ only

(*c*) both $(x-1)$ and $(x-3)$

(*d*) neither $(x-1)$ nor $(x-3)$

[CDS (I) Exam, 2015]

13. For which value of k does the pair of equations $x^2 - y^2 = 0$ and $(x-k)^2 + y^2 = 1$ yield a unique positive solution of x? **[CDS (I) Exam, 2015]**

(*a*) 2 (*b*) 0 (*c*) $\sqrt{2}$ (*d*) $-\sqrt{2}$

14. If $x + y + z = 0$ then $x^3 + y^3 + z^3 + 3xyz$ is equal to **[CDS (I) Exam, 2015]**

(*a*) 0 (*b*) $6\,xyz$ (*c*) $12\,xyz$ (*d*) xyz

15. Consider the following in respect of the equation $y = \dfrac{\sqrt{(x-1)^2}}{x-1}$

1. $y = 1$ if $x > 1$

2. $y = -1$ if $x < 1$

3. y exists for all values of x

Which of the above statements is/are correct? **[CDS (II) Exam, 2015]**

(*a*) 1 only (*b*) 2 only (*c*) 1 and 2 only (*d*) 1, 2 and 3

16. If a, b and c satisfy the equation $x^3 - 3x^2 + 2x + 1 = 0$ then what is the value of $\frac{1}{a} + \frac{1}{b} + \frac{1}{c}$?

(*a*) $-\frac{1}{2}$ (*b*) 2 (*c*) -2 (*d*) $\frac{1}{2}$

[CDS (II) Exam, 2015]

17. The difference of maximum values of the expressions $(6 + 5x - x^2)$ and $(y - 6 - y^2)$ for any real values of x and y is **[CDS (II) Exam, 2015]**

(*a*) 16 (*b*) 17 (*c*) 18 (*d*) 19

18. If $a^2 - by - cz = 0$, $ax - b^2 + cz = 0$ and $ax + by - c^2 = 0$, then the value of $\frac{x}{a+x} + \frac{y}{b+y} + \frac{z}{c+z}$. Will be **[CDS (I) Exam, 2016]**

(*a*) $a + b + c$ (*b*) 3 (*c*) 1 (*d*) 0

19. If the linear factors of $ax^2 - (a^2 + 1)x + a$ are p and q, then $p + q$ is equal to

(*a*) $(x - 1)(a + 1)$ (*b*) $(x + 1)(a + 1)$ (*c*) $(x - 1)(a - 1)$ (*d*) $(x + 1)(a - 1)$

[CDS (I) Exam, 2016]

20. The value of k, for which the system of equations $3x - ky - 20 = 0$ and $6x - 10y + 40 = 0$ has no solution, is **[CDS (I) Exam, 2016]**

(*a*) 10 (*b*) 6 (*c*) 5 (*d*) 3

21.

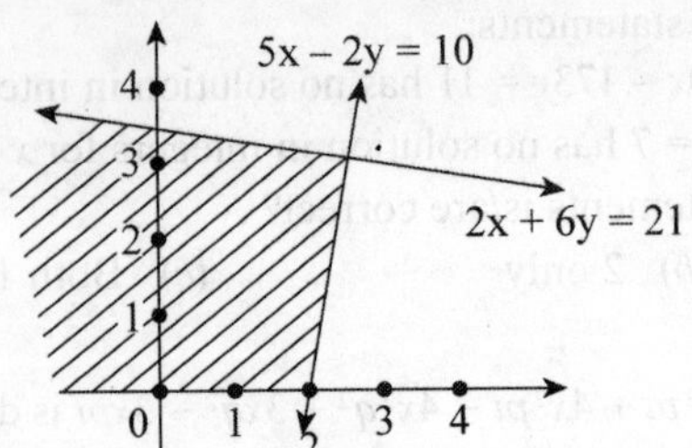

The linear inequations, for which the shaded area in the figure given above is the solution set, are **[CDS (I) Exam, 2016]**

(*a*) $2x + 6y \leq 21, 5x - 2y \leq 10$ (*b*) $2x + 6y \leq 21, 5x - 2y \geq 10$

(*c*) $2x + 6y \geq 21, 5x - 2y \leq 10$ (*d*) $2x + 6y \geq 21, 5x - 2y \geq 10$

22. If $2p + 3q = 12$ and $4p^2 + 4pq - 3q^2 = 126$, then what is the value of $p + 2q$?

(*a*) 5 (*b*) $\frac{21}{4}$ (*c*) $\frac{25}{4}$ (*d*) $\frac{99}{16}$

[CDS (II) Exam, 2016]

23. If $4x + 3a = 0$, then what is the value of $\frac{x^2 + ax + a^2}{x^3 - a^3} - \frac{x^2 - ax + a^2}{x^3 + a^3}$?

(*a*) $-\frac{4}{7a}$ (*b*) $\frac{7}{4}$ (*c*) $-\frac{32}{7a}$ (*d*) $\frac{24}{7a}$

[CDS (II) Exam, 2016]

Each item contains aQuestion followed by two Statements. Answer each item using the following instructions :

(a) Choose this option if the Question can be answered by one of the Statements alone but not by the other.

(b) Choose this option if the Question can be answered by either Statement alone.

(c) Choose this option if the Question can be answered by using both the Statements together, but cannot be answered by using either Statement alone

(d) Choose this option if the Question cannot be answered even by using both Statements together

24. x and y are consecutive odd integers. **[CDS (I) Exam, 2023]**

Question: Can the value of $(x + y)$ be determined uniquely ?

Statement I: $(x + y)^4 = 256$.

Statement II : $(x + y)^3 < 16$.

25. The system of equations $2x + 4y = 6$ and $4x + 8y = 8$ is **[CDS (I) Exam, 2017]**

(*a*) Consistent with a unique solution

(*b*) Consistent with infinitely many solutions

(*c*) Inconsistent

(*d*) None of the above

26. The solution of the system of simultaneous linear equations $4x - 3y = 7$ and $7x + 5y = 2$ is:

(*a*) $x = 1, y = 1$ (*b*) $x = -1, y = 1$ (*c*) $x = -1, y = -1$ (*d*) $x = 1, y = -1$

27. If $7x + 9y = 85$ and $4x + 5y = 48$, then:

(*a*) $x = 5, y = \frac{50}{9}$ (*b*) $x = \frac{40}{7}, y = 5$ (*c*) $x = 37, y = 8$ (*d*) $x = 7, y = 4$

28. If $12x + 17y = 53$ and $17x + 12y = 63$, then x and y are respectively equal to

(*a*) 1, 4 (*b*) 3, 1 (*c*) $4, \frac{5}{17}$ (*d*) 0, 5

29. The solution of the pair of equations $\frac{3}{x}+\frac{6}{y} = 12$ and $-\frac{4}{x}+\frac{12}{y} = 44$ is given by:

(*a*) $x = 2, y = 3$ (*b*) $x = -2, y = 3$ (*c*) $x = -\frac{1}{2}, y = 3$ (*d*) $x = -\frac{1}{2}, y = \frac{1}{3}$

30. Consider the equations P: $\frac{2}{x}+\frac{3}{y} = 6$ and Q: $\frac{1}{x}+\frac{1}{2y} = 2$. Then, $x = \frac{2}{3}, y = 1$ satisfy

(*a*) P, not Q (*b*) Q, not P (*c*) P and Q both (*d*) neither P nor Q

31. The solution to the system of equations $|x + y| = 1$ and $x - y = 0$ is given by:

(*a*) $x = y = \frac{1}{2}$ (*b*) $x = y = -\frac{1}{2}$

(*c*) $x = 1, y = 0$ (*d*) $x = y = \frac{1}{2}$ or $x = y = -\frac{1}{2}$

32. The solution of the equations $\frac{m}{x}+\frac{n}{y} = a, \frac{n}{x}+\frac{m}{y} = b$ is given by:

(*a*) $x = \frac{n^2+m^2}{am-bn}, y = \frac{m^2-n^2}{bm-an}$ (*b*) $x = \frac{m^2-n^2}{am-bn}, y = \frac{n^2-m^2}{bm-an}$

(*c*) $x = \frac{m^2-n^2}{am-bn}, y = \frac{m^2-n^2}{bm-an}$ (*d*) $x = \frac{n^2-m^2}{am-bn}, y = \frac{n^2-m^2}{bm-an}$

33. The solution set of the equations $\frac{3}{x+y}+\frac{2}{x-y} = 2$ and $\frac{9}{x+y}-\frac{4}{x-y} = 1$ is given by:

(*a*) $x = \frac{1}{2}, y = \frac{3}{2}$ (*b*) $x = \frac{5}{2}, y = \frac{1}{2}$ (*c*) $x = \frac{3}{2}, y = \frac{1}{2}$ (*d*) $x = \frac{1}{2}, y = \frac{5}{2}$

34. The solution of the equations

$$\frac{1}{2(2x+3y)}+\frac{12}{7(3x-2y)} = \frac{1}{2}; \text{ and } \frac{7}{(2x+3y)}+\frac{4}{(3x-2y)} = 2 \text{ is:}$$

(*a*) {(1, 0)} (*b*) {(1, –1)} (*c*) {(1, 2)} (*d*) {(2, 1)}

35. The solution of the equations $4x + 6y = 3xy$ and $8x + 9y = 5xy$ is given by:

(*a*) $x = 1, y = 2$ (*b*) $x = 2, y = 3$

(*c*) $x = 1, y = -1$ (*d*) $x = 3, y = 4$

36. If $\frac{2}{x}+\frac{3}{y} = \frac{9}{xy}$ and $\frac{4}{x}+\frac{9}{y} = \frac{21}{xy}$, where $x \neq 0, y \neq 0$, then the values of x and y are respectively:

(*a*) 0 and 1 (*b*) 1 and 2 (*c*) 2 and 3 (*d*) 1 and 3

37. The equations $3x - 4y = 5$ and $12x - 16y = 20$ have:

(*a*) no common solution. (*b*) exactly one common solution.

(*c*) exactly two common solutions. (*d*) more than two common solutions.

38. The equations $4x + 7y - 10 = 0$ and $20x + 35y - 50 = 0$ are:

(*a*) consistent and have a unique solution (*b*) consistent and have infinitely many solutions
(*c*) inconsistent (*d*) None of these

39. The equations $3x + y - 1 = 0$ and $6x + 2y - 5 = 0$ are

(*a*) consistent and have a unique solution (*b*) consistent and have infinitely many solutions
(*c*) inconsistent (*d*) None of these

40. Of the following,

1. $2x + 4y = 7$ 2. $3x + 6y = 10$ 3. $3x - y = 2$ 4. $6x - 2y = 4$

the two dependent equations are

(*a*) 1 and 2 (*b*) 3 and 4 (*c*) 1 and 3 (*d*) 2 and 4

41. The condition for which the system of equations $kx - y = 2$ and $6x - 2y = 3$ has a unique solution, is

(*a*) $k = 3$ (*b*) $k \neq 3$ (*c*) $k \neq 0$ (*d*) $k = 0$

42. The value of k for which the system of equations $2x + 3y = 5$ and $4x + ky = 10$ has an infinite number of solutions, is

(*a*) 1 (*b*) 3 (*c*) 6 (*d*) 0

43. The value of k for which the system of equations $x + 2y - 3 = 0$ and $5x + ky + 7 = 0$ has no solution, is

(*a*) 10 (*b*) 6 (*c*) 3 (*d*) 1

44. For what value of k will the system of equations $3x + 5y = 0$ and $kx + 10y = 0$ have a non-zero solution ?

(*a*) 0 (*b*) 2 (*c*) 6 (*d*) 8

45. A system of two simultaneous linear equations in two variables is inconsistent, if their graphs

(*a*) are parallel (*b*) are coincident (*c*) intersect in one point (*d*) None of these

46. For what value of k will the equations $2x + 32y + 3 = 0$ and $3x + 48y + k = 0$ represent coincident lines ?

(*a*) $\frac{2}{3}$ (*b*) $\frac{3}{2}$ (*c*) $\frac{9}{2}$ (*d*) 1

47. The equations $ax + b = 0$ and $cx + d = 0$ are consistent, if

(*a*) $ad = bc$ (*b*) $ad + bc = 0$ (*c*) $ab - cd = 0$ (*d*) $ab + cd = 0$

48. The sum of two numbers is 80. If the larger number exceeds four times the smaller one by 5, then the smaller number is

(*a*) 5 (*b*) 15 (*c*) 20 (*d*) 25

49. The difference between two numbers is 5 and the difference between their squares is 65. The larger number is

(*a*) 9 (*b*) 10 (*c*) 11 (*d*) 12

50. If three times the larger of two numbers is divided by the smaller, the quotient and the remainder, each is equal to 6. If five times the smaller is divided by the larger, the quotient is 2 and the remainder is 3. The smaller number is

(*a*) 6 (*b*) 7 (*c*) 8 (*d*) 9

51. Let a and b be two positive real numbers such that $a\sqrt{a} + b\sqrt{b} = 32$ and $a\sqrt{b} + b\sqrt{a} = 31$. What is the value of $\frac{5(a+b)}{7}$? **[CDS (I) Exam, 2019]**

(*a*) 5 (*b*) 7
(*c*) 9 (*d*) cannot be determined

ANSWERS

1. (*a*)	**2.** (*b*)	**3.** (*a*)	**4.** (*b*)	**5.** (*c*)	**6.** (*c*)	**7.** (*b*)	**8.** (*a*)	**9.** (*a*)
10. (*d*)	**11.** (*b*)	**12.** (*c*)	**13.** (*c*)	**14.** (*b*)	**15.** (*c*)	**16.** (*c*)	**17.** (*c*)	**18.** (*c*)
19. (*a*)	**20.** (*c*)	**21.** (*a*)	**22.** (*d*)	**23.** (*c*)	**24.** (*c*)	**25.** (*c*)	**26.** (*d*)	**27.** (*d*)
28. (*b*)	**29.** (*d*)	**30.** (*c*)	**31.** (*d*)	**32.** (*c*)	**33.** (*b*)	**34.** (*d*)	**35.** (*d*)	**36.** (*d*)
37. (*d*)	**38.** (*b*)	**39.** (*c*)	**40.** (*b*)	**41.** (*b*)	**42.** (*c*)	**43.** (*a*)	**44.** (*c*)	**45.** (*a*)
46. (*c*)	**47.** (*a*)	**48.** (*b*)	**49.** (*a*)	**50.** (*b*)	**51.** (*c*)			

SOLUTIONS

1. Let the original length of the piece of cloth is l m

Now, we have,

$$\frac{10,000}{l} - \frac{10,000}{l+2} = 250 \Rightarrow \frac{10,000 \times 2}{l(l+2)} = 250$$

$\Rightarrow l^2 + 2l = 80 \Rightarrow l^2 + 2l - 80 = 0 \Rightarrow l^2 + 10l - 8l - 80 = 0$

$\Rightarrow l(l + 10) - 8(l + 10) = 0 \Rightarrow (l - 8)(l + 10) = 0 \Rightarrow l = 8\text{m}$

2. Let the cost of 1 pencil, 1 notebook and 1 eraser be ₹x, ₹y and ₹z respectively,

Now, we have

$5x + 6y + 7z = 250$

By multiplying 6 on both sides, we get

$30x + 36y + 42z = 1500$ ...(*i*)

and $6x + 4y + 2z = 180$

By multiplying 5 on both sides, we get

$30x + 20y + 10z = 900$ ...(*ii*)

By subtracting eqn (ii) from eqn (i) we get

$16y + 32z = 600 \Rightarrow 2y + 4z = 75$

3. $\therefore$ We know that if

$a_1 x + b_1 y + c_1 = 0$

$a_2 x + b_2 y + c_2 = 0$ have unique solution than $\frac{a_2}{a_2} \neq \frac{b_1}{b_2}$

$\Rightarrow \frac{7}{k} \neq \frac{k}{7} \Rightarrow K^2 \neq 49 \Rightarrow K \neq 7$

4. We have,

$a = 2000$

$d = 1,000$

Now, $S_n = \frac{n}{2}[2a + (n-1)d]$

$170000 = \frac{n}{2}[2 \times 2000 + (n - 1) 1,000]$

$\Rightarrow 340 = n[4 + n - 1] \Rightarrow 340 = 3n + n^2 \Rightarrow n^2 + 3n - 340 = 0$

$\Rightarrow n^2 + 20n - 17n - 340 = 0 \Rightarrow n(n + 20) - 17(n + 20) = 0$

$\Rightarrow (n - 17)(n + 20) = 0 \Rightarrow n = 17$ and $n = -20$ (not possible)

5. We have,

$ab + xy - xb = 0 \Rightarrow ab = xb - xy \Rightarrow x(b - y) = 9b$

$\Rightarrow \dfrac{x}{a} = \dfrac{b}{b-y}$ and $bc + yz - cy = 0 \Rightarrow yz = cy - bc \Rightarrow yz = c(y - b)$

$\Rightarrow \dfrac{c}{z} = \dfrac{y}{y-b}$

Now, we have, $\dfrac{x}{a} + \dfrac{c}{z} \Rightarrow \dfrac{b}{b-y} + \dfrac{y}{y-b} = \dfrac{b}{b-y} - \dfrac{y}{b-y} = \dfrac{b-y}{b-y} = 1$

6. We have, $(p + 2)(2q - 1) = 2pq - 10 \Rightarrow 2pq - p + 4q - 2 = 2pq - 10$

$\Rightarrow 4q - p + 8 = 0$ ...(*i*)

and $(p - 2)(2q - 1) = 2pq - 10$

$2pq - p - 4q + 2 = 2pq - 10$

$p + 4q - 12 = 0$ ...(*ii*)

by adding eqn (i) and (ii), we get

$8q - 4 = 0 \Rightarrow 8q = 4 \Rightarrow a = \dfrac{1}{2}$

By putting $q = \dfrac{1}{2}$ in eqn (i), we get $2 - p + 8 = 0 \Rightarrow p = 10$

now, $pq = 10 \times \dfrac{1}{2} = 5$

7. The number of prime numbers which are less than 100 is 25

8. We have,

$65x - 33y = 97$ ...(i) $33x - 65y = 1$ ...(ii)

Adding eqn (i) and (ii), we get $98x - 98y = 98$

$\Rightarrow x - y = 1$ ...(iii)

Now, subtracting eqn (i) and (ii), we get $32x + 32y = 96 \Rightarrow x + y = 3$ ...(iv)

Now, adding (iii) and (iv), we get $x = 2$

Put $x = 2$ in eqn (iii), we get $y = 1$

Now, $xy = 2 \times 1 = 2$

9. We have,

$\dfrac{b}{y} + \dfrac{z}{c} = 1 \Rightarrow \dfrac{b}{y} = 1 - \dfrac{z}{c} \Rightarrow \dfrac{b}{y} = \dfrac{c-z}{c} \Rightarrow \dfrac{y}{b} = \dfrac{c}{c-z}$ ...(i)

$\dfrac{c}{z} + \dfrac{x}{a} = 1 \Rightarrow \dfrac{x}{a} = 1 - \dfrac{c}{z} \Rightarrow \dfrac{x}{a} = \dfrac{z-c}{z} \Rightarrow \dfrac{a}{x} = \dfrac{z}{z-c}$

Now, $\dfrac{ab+xy}{bx} = \dfrac{a}{x} + \dfrac{y}{b} = \dfrac{z}{z-c} + \dfrac{c}{c-z} = \dfrac{z}{z-c} - \dfrac{c}{z-c} = \dfrac{z-c}{z-c} = 1$

10. We have,

$(y + z - x)^3 + (z + x - y)^3 + (x + y - z)^3 \Rightarrow (x + y + z - 2x)^3 + (y + z + x - 2y)^3 = (x + y + z - 2z)^3$

$\Rightarrow (-2x)^3 + (-2y)^3 + (-2z)^3$ $[\because x + y + z = 0]$

$\Rightarrow -8(x^3 + y^3 + z^3)$

$\therefore$ We know that if $x + y + z = 0 \Rightarrow x^3 + y^3 + z^3 = 3xyz \Rightarrow -8 \times 3xyz = -24xyz$

11. If HCF $(a, b) = 1$, then ax ! by = N has integral solutions for all integral values of N. If HCF $(a, b) = d$, then $ax \pm$ by = N has integral solutions if N is a multiple of d and no integral solution otherwise.

(1) HCF (1990, 173) = 1. 1990 = 2(5) (199).

$\therefore$ HCF (1990, 173) = 1. Eq (1) has solutions. Statement (1) is false.

(2) HCF (3, 12) = 3 and 7 is not a multiple of 3. Eq (2) has no solutions and statement (2) is true.

12. $x^3 (q^2 - pt) + 4x^2 (pt - q^2) + 3x (q^2 - pt) = (q^2 - pt) (x^3 - 4x^2 + 3x) = (q^2 - pt)x (x - 1) (x - 3)$

The given expression is divisible by both $x - 1$ and $x - 3$.

13. $x^2 = y^2$ and $(x - k)^2 + y^2 = 1$

$\therefore \ (x - k)^2 + x^2 = 1 \Rightarrow 2x^2 - 2kx + k^2 - 1 = 0$

$$x = \frac{2k \pm 2\sqrt{(2k)^2 - 4(2)(k^2 - 1)}}{2(2)} = \frac{2k \pm 2\sqrt{(k)^2 - (2k^2 - 2)}}{4} = \frac{2k \pm 2\sqrt{2 - k^2}}{4}$$

For the given pair of equations to yield a unique solution for x, $-k^2 + 2$ must be 0 *i.e.*, $k = \pm\sqrt{2}$

$\therefore$ The unique solution for x is $\frac{k}{2}$. For this solution to be positive, k must be positive.

$\therefore \ k = \sqrt{2}$.

14. $x + y + z = 0 \ \therefore \ x^3 + y^3 + z^3 = +3xyz$ and $x^3 + y^3 + z^3 + 3xyz = 6xyz$

15. $Y = \dfrac{\sqrt{(x - 1)^2}}{x - 1}$

The expression $\sqrt{4}$ is normally used to denote 2. To refer to -2 (in relation to 4), we use the notation $-\sqrt{4}$.

If $x > 0$, $\sqrt{(x - 1)^2} = x - 1$ and $y = 1$.

If $x < 1$, $\sqrt{(x - 1)^2} = 1 - x$ and $y = -1$

Also, if $x = 1$, y is not defined $\therefore$ (1) and (2) are true.

16. $x^3 - 3x^2 + 2x + 1 = 0$ The roots are a, b, c.

$\therefore \ a + b + c = 3$, $ab + bc + ca = 2$ and $abc = -1$

$$\frac{1}{a} + \frac{1}{b} + \frac{1}{c} = \frac{ab + bc + ca}{abc} = \frac{2}{-1} = -2.$$

17. $6 + 5x - x^2 = -(x^2 - 5x - 6)$

$$= -[x^2 - 5x + 6.25 - 12.25] = -[(x - 2.5)^2 - 12.25] = 12.25 - (x - 2.5)^2$$

The maximum value is 12.25.

$y - 6 - y^2 = -[y^2 - y + 6] = -[y^2 - y + 0.25 + 5.75]$

$$= -[(y - \tfrac{1}{2})^2 + 5.75] = -5.75 - (y - \tfrac{1}{2})^2.$$

The maximum value is -5.75.

The difference of the two maximum values is 18.

18. $a^2 - by - cz = 0$ ... (i)

$ax - b^2 + cz = 0$ ... (ii)

$ax + by - c^2 = 0$ ... (iii)

$(1) + (2) + (3) \Rightarrow 2ax + a^2 - b^2 - c^2 = 0$

$$x = \frac{-a^2 + b^2 + c^2}{2a}, \ x + a = \frac{a^2 + b^2 + c^2}{2a}$$

Similarly is $y = \dfrac{a^2+c^2-b^2}{2b}$ and $z = \dfrac{a^2+b^2-c^2}{2c}$

$$y+b = \frac{a^2+b^2+c^2}{2b}\,;\; z+c = \frac{a^2+b^2+c^2}{2c}\,;\; \frac{x}{a+x}+\frac{y}{b+y}+\frac{z}{c+z}$$

$$= \frac{b^2+c^2\,a^2}{2a}+\frac{a^2+c^2\,b^2}{2b}+\frac{a^2+b^2-c^2}{2c}$$

$$\frac{\left(a^2+b^2+c^2\right)}{2a}+\frac{\left(a^2+b^2+c^2\right)}{2b}+\frac{\left(a^2+b^2+c^2\right)}{2c}$$

$$\Rightarrow \frac{b^2+c^2-a^2+a^2+c^2-b^2+a^2+b^2-c^2}{a^2+b^2+c^2} = 1$$

19. $ax^2-\left(a^2+1\right)x+a = ax^2 - a^2x - x + a = ax(x-a) - 1(ax-1)(x-a)$

$\Rightarrow$ $(ax-1)(x-a)$ are the linear factors of the given expression

Let $p = : ax{-}1$ and $q = x - a$

$$p + q = ax - 1 + x - a = x(a+1) - 1\,(a+1) = (a+1)(x-1)$$

20. $3x - ky - 20 = 0$ and $6x - 10y + 40 = 0$ has no Solution $\Rightarrow$ the two lines are parallel

$\Rightarrow \dfrac{3}{6} = \dfrac{-k}{-10} \quad \Rightarrow \quad k = 5$

21. The region is to the left of the line $5x - 2y = 10$, *i.e.*, $5x - 2y \le 10$.

In is below the line $2x + 6y = 21$. $\therefore$ $2x + 6y \le 21$

(the shaded region in actually also bound by the x – axis)

The complete description should included the condition $y \ge 0$.)

22. $2p + 3q = 12$ ... (i)

$4p^2 + 4pq - 3q^2 = 126$ ... (ii)

We can substitute $12 - 2p$ for $3q$ in (2) But first we multiply (2) by 3.

$\Rightarrow$ $12p^2+4p(12-2p)-(12-2p)^2 = 126(3)$

$\Rightarrow$ $12p^2+48p-8p^2-144+48p-4p^2 = 126(3)$ $\Rightarrow 96p = 522 \Rightarrow p = \dfrac{522}{96} = \dfrac{87}{16}$

$(1) \Rightarrow 3q = 12 - \dfrac{87}{8} = \dfrac{9}{8} \Rightarrow q = \dfrac{3}{8}$

$\therefore\; p + 2q = \dfrac{87}{16} + \dfrac{6}{8} = \dfrac{99}{16}$

23. $\dfrac{x^2+ax+a^2}{x^3-a^3} = \dfrac{x^2+ax+a^2}{(x-a)\left(x^2+ax+a^2\right)} = \dfrac{1}{x-a}$

Similarly $\dfrac{x^2-ax+a^2}{x^3+a^3} = \dfrac{1}{x+a}$

$\therefore$ The given expression, $E = \dfrac{1}{x-a} - \dfrac{1}{x+a} = \dfrac{2a}{x^2-a^2}$

$$= \frac{2a}{\left(-\dfrac{3a}{4}\right)^2 - a^2}\left(x = \frac{-3a}{4}\right) = -\frac{2a(16)}{7a^2} = \frac{-32}{7a}$$

24. Let $x = p$

$y = p + 2$

Statement I:

$(x + y)^4 = 256 \quad \Rightarrow x + y = 4$

$\therefore (x, y) = (1, 3), (-1, -3)$

Statement II: $(x + y)^3 < 16 \Rightarrow ((x + y)^3)^{1/3} < (16)^{1/3} \quad \Rightarrow x + y < 2(2)^{1/3}$

25. $\Rightarrow \dfrac{a_1}{a_2} = \dfrac{2}{4} = \dfrac{1}{2} \Rightarrow \dfrac{b_1}{b_2} = \dfrac{4}{8} = \dfrac{1}{2} \Rightarrow \dfrac{c_1}{c_2} = \dfrac{6}{4} = \dfrac{3}{4} \quad \therefore \dfrac{a_1}{a_2} = \dfrac{b_1}{b_2} \neq \dfrac{c_1}{c_2}$

The above equation is the case of no solution. So, the system of equations is inconsistent.

26. Multiplying 1st equation by 5 and 2nd by 3 and adding we get $x = 1$. Putting $x = 1$ in 1st, we get $4 \times 1 - 3y = 7$ or $y = -1$.

27. Multiplying lst eqn. by 5, 2nd by 9 and subtracting we get, $x = 7$. Putting $x = 7$ in lst, we get: $7 \times 7 + 9y = 85$ or $9y = 36$ or $y = 4$. So, $x = 7$ and $y = 4$.

28. Adding given equations we get $29\,(x + y) = 116$ or $x + y = 4$. Subtracting given equations, we get $5\,(x - y) = 10$ or $x - y = 2$.

Solving $x + y = 4$, $x - y = 2$ we get $x = 3$ and $y = 1$.

29. Putting $\dfrac{1}{x} = u$ and $\dfrac{1}{y} = v$, the given equations become: $3u + 6v = 12$ and $-4u + 12v = 44$

i.e. $u + 2v = 4$ and $u - 3v = -11$.

Solving these eqns., we get $u = -2$ and $v = 3$.

$\therefore \quad \dfrac{1}{x} = -2$ and $\dfrac{1}{y} = 3$. So, $x = -\dfrac{1}{2}$ and $y = \dfrac{1}{3}$.

30. Putting $\dfrac{1}{x} = u$ and $\dfrac{1}{y} = v$, eqns. P and Q become:

$2u + 3v = 6$ and $u + \dfrac{1}{2}v = 2$ *i.e.* $2u + v = 4$.

Solving $2u + 3v = 6$ and $2u + v = 4$,we get: $u = \dfrac{3}{2}$ & $v = 1$

$\therefore \qquad \dfrac{1}{x} = \dfrac{3}{2}$ and $\dfrac{1}{y} = 1$. So, $x = \dfrac{2}{3}$ and $y = 1$.

So. $\qquad x = \dfrac{2}{3}$ and $y = 1$ satisfy both P and Q.

31. Note that $|a| = 1$ means $a = 1$ or $a = -1$.

So, $|x + y| = 1 \Leftrightarrow x + y = 1$ or $-(x + y) = 1$ *i.e.* $x + y = -1$.

Solving $x + y = 1$, $x - y = 0$ we get $x = \dfrac{1}{2}$ & $y = \dfrac{1}{2}$.

Solving $x + y = -1$, $x - y = 0$ we get $x = -\dfrac{1}{2}$ & $y = -\dfrac{1}{2}$. $\therefore x = y = \pm\dfrac{1}{2}$.

32. Put $\dfrac{1}{x} = u$ and $= \dfrac{1}{y} = v$. Then, given eqns. are: $mu + nv = a$ and $nu + mv = b$.

Multiplying lst eqn. by m, 2nd by n & subtracting, we get:

$(m^2 - n^2)u = an - bm$. So, $u = \left(\dfrac{an - bm}{m^2 - n^2}\right)$.

Multiplying lst eqn. by n, 2nd by m & subtracting, we get:

$(m^2 - n^2)v = bm - an$. So, $v = \left(\dfrac{bm-an}{m^2-n^2}\right)$.

$\therefore\ \dfrac{1}{x} = \left(\dfrac{an-bm}{m^2-n^2}\right) \Rightarrow x = \left(\dfrac{m^2-n^2}{am-bn}\right);\ \dfrac{1}{y} = \left(\dfrac{bm-an}{m^2-n^2}\right) \Rightarrow y = \left(\dfrac{m^2-n^2}{bm-an}\right)$.

33. Putting $\dfrac{1}{x+y} = u$ and $\dfrac{1}{x-y} = v$, the given equations become: $3u + 2v = 2$ and $9u - 4v = 1$.

Solving these we get: $u = \dfrac{1}{3}$ and $v = \dfrac{1}{2}$.

$\therefore\ \dfrac{1}{x+y} = \dfrac{1}{3}$ & $\dfrac{1}{x-y} = \dfrac{1}{2}$. So, $x + y = 3$ and $x - y = 2$.

Solving these equations, we get: $x = \dfrac{5}{2}$ and $y = \dfrac{1}{2}$.

34. Putting $\dfrac{1}{2x+3y} = u$ and $\dfrac{1}{3x-2y} = v$, the given equation become:

$\dfrac{1}{2}u + \dfrac{12}{7}v = \dfrac{1}{2}$ and $7u + 4v = 2$, or $7u + 24v = 7$ and $7u + 4v = 2$.

Solving these, we get: $v = \dfrac{1}{4}$ and $u = \dfrac{1}{7}$.

$\therefore\ 2x + 3y = 7$ and $3x - 2y = 4$.

Solving these equations, we get: $x = 2$ and $y = 1$.

$\therefore$ Solution Set = {(2, 1)}.

35. Dividing each term of each one of the given equations, they become: $\dfrac{6}{x} + \dfrac{4}{y} = 3$ and $\dfrac{9}{x} + \dfrac{8}{y} = 5$

Putting $\dfrac{1}{x} = u$ and $\dfrac{1}{y} = v$, these equations become: $6u + 4v = 3$ and $9u + 8v = 5$

Solving these equations, we get: $u = \dfrac{1}{3}$ and $v = \dfrac{1}{4}$.

$\therefore\ \dfrac{1}{x} = \dfrac{1}{3}$ and $\dfrac{1}{y} = \dfrac{1}{4}$. So, $x = 3$ and $y = 4$.

36. Multiplying each equation throughout by xy, we get: $3x + 2y = 9$ and $9x + 4y = 21$.

On solving these equations, we get: $x = 1$, $y = 3$.

37. The given equations are $3x - 4y = 5$ and $3x - 4y = 5$.

Thus, there is one equation in two variables.

So, the given equations have an infinite number of solutions.

Hence, (*d*) is true.

38. Given equations are $4x + 7y = 10$ & $4x + 7y = 10$.

Clearly, the given equations, have an infinite number of solutions and hence they are consistent also.

39. Given equations are: $3x + y = 1$ and $3x + y = 5$.

Clearly, the given equations are inconsistent. $\left(\text{Note that } \dfrac{a}{a'} = \dfrac{b}{b'} \neq \dfrac{c}{c'}\right)$

40. Clearly, 3 and 4 are dependent equations.

41. For a unique solution of $a_1x + b_1y + c_1 = 0$ & $a_2x + b_2y + c_2 = 0$ we have: $\frac{a_1}{a_2} \neq \frac{b_1}{b_2}$

So, for given equations to have a unique solution, we must have: $\frac{k}{6} \neq \frac{-1}{-2} \Leftrightarrow k \neq 3.$

42. For infinite number of solutions: $\frac{a_1}{a_2} = \frac{b_1}{b_2} = \frac{c_1}{c_2}$. $\frac{2}{4} = \frac{3}{k} = \frac{5}{10}$ or $k = 6$

43. The system has no solution if $\frac{a_1}{a_2} = \frac{b_1}{b_2} \neq \frac{c_1}{c_2}$.

$\therefore$ $\frac{1}{5} = \frac{2}{k} \neq \frac{-3}{7}$. So, $k = 10$.

44. For $a_1 x + b_1 y = 0$ and $a_2x + b_2 y = 0$ to have a non-zero solution, we have $\frac{a_1}{a_2} = \frac{b_1}{b_2}$.

$\therefore$ $\frac{3}{k} = \frac{5}{10}$ or $k = 6$.

45. A system of two linear equations in two variables is inconsistent if their graphs are parallel.

46. Lines will be coincident if $\frac{a_1}{a_2} = \frac{b_1}{b_2} = \frac{c_1}{c_2}$. *i.e.* if $\frac{2}{3} = \frac{32}{48} = \frac{3}{k}$ *i.e.* if $k = \frac{9}{2}$.

47. These equations are consistent if $\frac{a}{c} = \frac{b}{d}$ *i.e.* $ad = bc$.

48. Let the numbers be x and y. Then, $x + y = 80$ and $x - 4y = 5$. Solving these we get: $y = 15$.

49. Let the numbers be x and y. Then, $x - y = 5$ and $x^2 - y^2 = 65$.

$\therefore$ $\frac{x^2 - y^2}{x - y} = \frac{65}{5} \Leftrightarrow x + y = 13.$

Solving $x - y = 5$ & $x + y = 13$, we get: $x = 9$ and $y = 4$.

50. Let the larger number be x and smaller one y.
Dividend = Divisor × Quotient + Remainder.
$\therefore$ $3x = 6y + 6$, $5y = 2x + 3$.
Solving these equations, we get: $x = 16$ and $y = 7$.
Hence, smaller number = 7.

51. Let $a = p^2$ and $b = q^2$

Now, we have $p^3 + q^3 = 32$ ...(*i*)

and $a\sqrt{b} + b\sqrt{a} = 31 \Rightarrow \sqrt{a}\sqrt{b}(\sqrt{a} + \sqrt{b}) = 31$

$\Rightarrow pq\,(p + q) = 31$ ...(*ii*)

Now, $(p + q)^3 = p^3 + q^3 + 3pq\,(p + q)$ [using identity]

$\Rightarrow (p + q)^3 = 32 + 3 \times 31 = 32 + 93 = 125 \Rightarrow (p + q)^3 = 5^3 \Rightarrow p + q = 5$

Now,

$\because p^3 + q^3 = (p = q)\,(p^2 + q^2 - pq) = 32$

$\Rightarrow 5\,(p^2 + q^2) - 31 = 32$ [from (ii)] $\Rightarrow p^2 + q^2 = \frac{63}{5} \Rightarrow a + b = \frac{63}{5}$

Now, $\frac{5(a+b)}{7} = \frac{5}{7} \times \frac{63}{5} = 9$

22 Quadratic Equations

Highlights: Solutions of quadratic equations, relation between its roots and coefficients, problems leading to quadratic equations.

Definitions & Review of Concepts

Quadratic Equation: *Let f (x) be a quadratic polynomial. Then, the equation f (x) = 0 is called a quadratic equation.*

The values of x satisfying $f(x) = 0$ are called its roots or zeros.

Facts About Quadratic Equation

I. (*i*) The general form of a quadratic equation is:
$ax^2 + bx + c = 0$, where $a, b, c \in R$ and $a \neq 0$.

(*ii*) If α, β are the roots of $ax^2 + bx + c = 0$, then

$$\alpha = \frac{-b+\sqrt{b^2-4ac}}{2a} \text{ and } \beta = \frac{-b-\sqrt{b^2-4ac}}{2a}$$

(*iii*) A quadratic equation has exactly two roots, may be real or imaginary or coincident.

(*iv*) If $p + \sqrt{q}$ is a root of a quadratic equation, then its other root is $p - \sqrt{q}$.

(*v*) In $ax^2 + bx + c = 0$, the expression $D = b^2 - 4ac$ is called its discriminant.

II. Nature of Roots of $ax^2 + bx + c = 0$:

Let $D = b^2 - 4ac$ be the discriminant of the given equation.

Then, the roots of $ax^2 + bx + c = 0$ are:

(*i*) *real and equal if $D = 0$;*

(*ii*) *real, unequal and irational, when $D > 0$ and D is a perfect square;*

(*iii*) *real, unequal and irrational, when $D > 0$ but D is not a perfect square;*

(*iv*) *imaginary, if $D < 0$.*

(*v*) *integers, when $a = 1$, b & c are integers and the roots are rational.*

III. Condition that a x^2: + $bx + c$ is factorizable into two linear factors:

$ax^2 + bx + c$ *is factorizable into two linear factors only when $b^2 - 4\,ac \geq 0$.*

IV. Relation Between Roots & Coefficients:

Let α, β be the roots of the equation, $ax^2 + bx + c = 0$.

Then, *Sum of roots*, $(\alpha + \beta) = -\frac{b}{a}$ & *Product of roots*, $\alpha\beta = \frac{c}{a}$.

V. To Form a Quadratic Equation:

A quadratic equation whose roots are α and β is given by:

$$x^2 - (\alpha + \beta)x + \alpha\beta = 0$$

i.e., $x^2 -$ (*Sum of roots*) x + *product of roots* $= 0$.

SOME SOLVED PROBLEMS

Ex. 1. ***Solve:*** $x^4 - 10x^2 + 9 = 0$

Sol. $x^4 - 10x^2 + 9 = 0 \quad \Leftrightarrow y^2 - 10y + 9 = 0$ where $x^2 = y$

$\Leftrightarrow (y-9)(y-1) = 0 \quad \Leftrightarrow y = 9$ or $y = 1$.

$\Leftrightarrow x^2 = 9$ or $x^2 = 1$

$\Leftrightarrow x = -3$ or $x = 1$.

Ex. 2. ***Form an equation whose roots are 7 and –3.***

Sol. Sum of roots $= 7 + (-3) = 4$, Product of roots $= 7 \times (-3) = -21$.

$\therefore$ Required equation is: $x^2 -$ (Sum of roots) $x +$ Product of roots $= 0$

i.e., $x^2 - 4x - 21 = 0$.

Ex. 3. ***Form a quadratic equation one of whose root is*** $3 + \sqrt{5}$.

Sol. One root $= 3 + \sqrt{5}$, other root $= 3 - \sqrt{5}$.

$\therefore$ Sum of roots $= 6$, Product of roots $= (9 - 5) = 4$.

$\therefore$ Required equation is: $x^2 - 6x + 4 = 0$.

Ex. 4. ***Find the nature of lhe roots of the equations:***

(i) $2x^2 - 5x - 4 = 0$ ***(ii)*** $2x^2 - 7x + 2 = 0$

(iii) $9x^2 + 6x + 1 = 0$ ***(iv)*** $15x^2 - 11x + 3 = 0$

Sol. (*i*) $D = (-5)^2 - 4 \times 2 \times (-4) = (25 + 32) = 57 > 0$.

Also, 57 is not a perfect square.

So, the roots are real, distinct and irrational.

(*ii*) $D = (-7)^2 - 4 \times 2 \times 2 = 49 - 16 = 33 > 0$ and 36 is a perfect square.

So, the roots are real, distinct and rational.

(*iii*) $D = 6^2 - 4 \times 9 \times 1 = 0$. So, the roots are real and equal.

(*iv*) $D = (-11)^2 - 4 \times 15 \times 3 = (121 - 180) = -59$.

So, the roots are imaginary.

Ex. 5. ***Find the values of k for which*** $9x^2 + mx + 1 = 0$ ***has both equal roots.***

Sol. For equal roots, we must have $D = 0$ *i.e.,* $m^2 - 36 = 0$.

So, $m = \pm 6$.

Ex. 6. ***If*** α, β ***are the roots of*** $2x^2 + 3x - 1 = 0$***, form an equation whose roots are*** $\frac{\alpha}{\beta}$ ***and*** $\frac{\beta}{\alpha}$.

Sol. We have: $\alpha + \beta = \frac{-3}{2}$ and $\alpha\beta = -\frac{1}{2}$.

$$\therefore \text{ Sum of new roots} = \frac{\alpha}{\beta} + \frac{\beta}{\alpha} = \frac{\alpha^2 + \beta^2}{\alpha\beta} = \frac{(\alpha+\beta)^2 - 2\alpha\beta}{\alpha\beta} = \left(\frac{\frac{9}{4}+1}{-\frac{1}{2}}\right) = \frac{13}{4} \times \left(\frac{-2}{1}\right) = \frac{-13}{2}.$$

$$\text{Product of new roots} = \left(\frac{\alpha}{\beta} \times \frac{\beta}{\alpha}\right) = 1$$

$$\therefore \text{ Required equation is } x^2 - \left(-\frac{13}{2}\right) x + 1 = 0 \text{ or } 2x^2 + 13x + 2 = 0.$$

Ex. 7. ***Which of the following quadratic polynomials can be factorized into a product of real linear factors?***

(i) $2x^2 - 5x + 8$ ***(ii)*** $3x^2 + 5x + 2$

Sol. (*i*) Since $D = (-5)^2 - 4 \times 2 \times 8 = 25 - 64 = -39 < 0$, so $2x^2 - 5x + 8$ is not factorizable into real linear factors.

(*ii*) Since $D = 5^2 - 4 \times 3 \times 2 = 1 > 0$, so $3x^2 + 5x + 2$ is factorizable into real linear factors.

Ex. 8. ***Find the set of values of p for which $2x^2 + 6x + k$ can be factorized into real linear factors.***

Sol. $2x^2 + 6x + k$ will be factorizable only when $D \geq 0$.

i.e., $6^2 - 4 \times 2 \times k \geq 0$ or $k \leq \frac{9}{2}$.

Ex. 9. ***If one root of $3x^2 + 11x + k = 0$ be reciprocal of the other, find the value of k.***

Sol. Let one root be α. Then, another root $= (1/\alpha)$.

$\therefore$ Product of roots $= \left(\alpha \times \frac{1}{\alpha}\right) = 1$.

But, product of roots $= \frac{k}{3}$. So, $\frac{k}{3} = 1$ or $k = 3$.

Ex. 10. ***A two digit number is four times the sum and three times the product of its digits. Find the number.***

Sol. Let ten's digit be x and unit's digit y.

Then, $10x + y = 4(x + y)$ and $10x + y = 3xy \Leftrightarrow y = 2x$ and $10x + y = 3xy$

$\Leftrightarrow 10x + 2x = 6x^2 \Leftrightarrow 6x^2 - 12x = 0$

$6x(x - 2) = 0 \Leftrightarrow x = 0$ or $x = 2$

So, $x = 2$ [$\because$ *Ten's digit can not be* 0]

Also, $y = 2x = 4$.

$\therefore$ Number = 24.

EXERCISE 22

1. If the roots of the equation $x^2 - bx + c = 5$ differ by 5, then which one of the following is correct? **[CDS (I) Exam, 2023]**

(*a*) $b^2 = 4c + 5$ (*b*) $c^2 = 4b - 5$

(*c*) $b^2 + c^2 = 5$ (*d*) $b^2 - c^2 = 5$

2. If *a*, *b*, *c* are non-zero real numbers such that $a + b + c = 0$, then what are the roots of the equation $ax^2 + bx + c = 0$? **[CDS (I) Exam, 2023]**

(*a*) 2, 1 + (*c*/*a*) (*b*) 1, *a*/*c* (*c*) 1, c/a (*d*) 2, (*c*/*a*) – 1

3. If $p(\neq 0)$ and $q(\neq 0)$ are the roots of the equation $x^2 + px + q = 0$, then what is $p^2 + q^2$ equal to? **[CDS (I) Exam, 2025]**

(*a*) 2 (*b*) 3 (*c*) 4 (*d*) 5

4. For what values of *m*, is $mx^2 + mx + 8x + 9$ a perfect square ? **[CDS (I) Exam, 2023]**

(*a*) 1, 4 (*b*) 4, 9 (*c*) 9, 16 (*d*) 4, 16

Consider the following for the next two (02) items that follow:

A quadratic equation is given by

$(a + b + c)\,x^2 - (2a + 2b)\,x + (a + b - c) = 0$;

where *a*, *b* and *c* are real and distinct.

5. What is the greatest value of *k* for which $2x^2 - 4x + k = 0$ has real roots? **[CDS (I) Exam, 2024]**

(*a*) 1 (*b*) 2 (*c*) 3 (*d*) 4

6. Consider the following statements: **[CDS (II) Exam, 2023]**

1. One of the roots of the equation is always less than 1 if a, b and c are all positive.
2. One of the roots of the equation is always negative if a, b and c are all negative.

Which of the statements given above is/are correct?

(*a*) 1 only (*b*) 2 only (*c*) Both 1 and 2 (*d*) Neither 1 nor 2

7. The equations: $x^2+px+q=0$ and $x^2+qx+p=0\,(p \neq q)$ have a common root.

What is the value of $(p+q)$? **[CDS (I) Exam, 2025]**

(*a*) –1 (*b*) 0 . (*c*) 1 (*d*) 2

8. If two quadratic equations $px^2+px+4=0$ and have a common root 2 , then what is $p+q$ equal to? **[CDS (I) Exam, 2025]**

(*a*) –3 (*b*) –2 (*c*) 0 (*d*) 3

9. If the sum as well as the product of the roots of the equation $px^2 - 6x + q = 0$ is 6, then what is $(p+q)$ equal to ? **[CDS (I) Exam, 2021]**

(*a*) 8 (*b*) 7 (*c*) 6 (*d*) 5

10. Which one of the following equations does not have real roots? **[CDS (II) Exam, 2021]**

(*a*) $2x^2 + 16x + 3 = 0$ (*b*) $2x^2 + 10x - 1 = 0$

(*c*) $x^2 - 8x + 1 = 0$ (*d*) $4x^2 + 9x + 6 = 0$

11. The sum and the product of the roots of a quadratic equation are 7 and 12 respectively. If the bigger root is halved and the smaller root is doubled, then what is the resulting quadratic equation ? **[CDS (II) Exam, 2021]**

(*a*) $x^2 - 6x + 12 = 0$ (*b*) $x^2 - 8x + 12 = 0$ (*c*) $x^2 + 8x + 12 = 0$ (*d*) $x^2 - 10x + 12 = 0$

12. If the roots of the equation $x^2-(k-2)x+(k+1)=0$ are equal, then what are the values of k ? **[CDS (II) Exam, 2024]**

(*a*) 0, 4 (*b*) 0, 8 (*c*) 4, 4 (*d*) 2, 6

13. If α and β are the roots of the equation $x+a+b=\dfrac{abx}{ab+ax+bx}$, then what is $(\alpha\beta+\alpha+\beta)$ equal to? **[CDS (I) Exam, 2025]**

(*a*) $ab+a+b$ (*b*) $ab-a-b$ (*c*) $a+b-ab$ (*d*) $-(ab+a+b)$

14. If one of the roots of the equation $ax^2 - 4ax + 15 = 0$ is $\dfrac{2}{3}$, then what is the sum of the squares of the roots? **[CDS (II) Exam, 2021]**

(*a*) $\dfrac{15}{2}$ (*b*) $\dfrac{17}{2}$ (*c*) $\dfrac{19}{2}$ (*d*) $\dfrac{21}{2}$

15. If α and β are the roots of the quadratic equation $x^2 + kx - 15 = 0$ such that $\alpha - \beta = 8$, then what is the positive value of k? **[CDS (I) Exam, 2020]**

(*a*) 2 (*b*) 3 (*c*) 4 (*d*) 5

16. If the area of a square is $2401x^4 + 196x^2 + 4$, then what is its side length? **[CDS (I) Exam, 2020]**

(*a*) $49x^2 + 3x + 2$ (*b*) $49x^2 - 3x + 2$ (*c*) $49x^2 + 2$ (*d*) $59x^2 + 2$

17. If the roots of the quadratic equation $x^2 - 4x - \log_{10} N = 0$ are real, then what is the minimum value of N? **[CDS (II) Exam, 2020]**

(*a*) 1 (*b*) $\frac{1}{10}$ (*c*) $\frac{1}{100}$ (*d*) $\frac{1}{1000}$

18. For $x > 0$, what is the minimum value of $x + \frac{x+2}{2x}$? **[CDS (I) Exam, 2019]**

(*a*) 1 (*b*) 2 (*c*) $2\frac{1}{2}$ (*d*) $\frac{1}{10000}$

19. If $\frac{1+px}{1-px}\sqrt{\frac{1-qx}{1+qx}} = 1$, then what are the non-zero solution of x? **[CDS (I) Exam, 2019]**

(*a*) $\pm\frac{1}{p}\sqrt{\frac{2p-q}{q}}, 2p \neq q$ (*b*) $\pm\frac{1}{pq}\sqrt{p-q}, p \neq q$

(*c*) $\pm\frac{1}{q}\sqrt{p-q}, p \neq q$ (*d*) $\pm\frac{q}{p}\sqrt{2p-q}, 2p \neq q$

20. If p and q are the roots of the equation $x^2 - 15x + r = 0$ and $p - q = 1$, then what is the value of r? **[CDS (I) Exam, 2019]**

(*a*) 55 (*b*) 56 (*c*) 60 (*d*) 64

21. For the in equation $x^2 - 7x + 12 > 0$, which one of the following is correct? **[CDS (I) Exam, 2019]**

(*a*) $3 < x < 4$ (*b*) $-\infty < x < 3$ only

(*c*) $4 < x < \infty$ (*d*) $-\infty < x < 3$ or $4 < x < \infty$

22. If $x = \frac{1+\sqrt{3}}{2}$ and $y = x^3$, then y satisfies which one of the following equations? **[CDS (II) Exam, 2019]**

(*a*) $8y^2 - 20y - 1 = 0$ (*b*) $8y^2 + 20y - 1 = 0$

(*c*) $8y^2 + 20y + 1 = 0$ (*d*) $8y^2 - 20y + 1 = 0$

23. If the equations $x^2 + 5x + 6 = 0$ and $x^2 + kx + 1 = 0$ have a common root, then what is the value of k? **[CDS (II) Exam, 2019]**

(*a*) $-\frac{5}{2}$ or $\frac{10}{3}$ (*b*) $\frac{5}{2}$ or $\frac{10}{3}$ (*c*) $\frac{5}{2}$ or $-\frac{10}{3}$ (*d*) $-\frac{5}{2}$ or $\frac{10}{3}$

24. The equation $x^2 + pr + q = 0$ has roots equal to p and q where $q \neq 0$. What are the value of p and q respectively? **[CDS (II) Exam, 2019]**

(*a*) 1, –2 (*b*) 1, 2 (*c*) –1, 2 (*d*) –1, –2

25. If $(b - 6)$ is one roots of the quadratic equation $x^2 - 6x + b = 0$, where b is an integers, then what is the maximum value of b^2? **[CDS (II) Exam, 2019]**

(*a*) 36 (*b*) 49 (*c*) 64 (*d*) 81

26. What is the maximum value of the expression $\frac{1}{x^2 + 5x + 10}$? **[CDS (II) Exam, 2019]**

(*a*) $\frac{15}{4}$ (*b*) $\frac{15}{2}$ (*c*) 1 (*d*) $\frac{4}{15}$

27. Two numbers p and q are such that the quadratic equation $px^2 + 3x + 2q = 0$ has –6 as the sum and the product of the roots. What is the value of $(p - q)$? **[CDS (II) Exam, 2019]**

(*a*) –1 (*b*) 1 (*c*) 2 (*d*) 3

28. If the roots of the equation $px^2 + x + r = 0$ are reciprocal to each other, then which one of the following is correct? **[CDS (I) Exam, 2018]**

(*a*) $p = 2r$ (*b*) $p = r$ (*c*) $2p = r$ (*d*) $p = 4r$

29. If α and β are the roots of the equation $ax^2 + bx + c = 0$, then what is the value of the expressions $(\alpha + 1)(\beta + 1)$? **[CDS (I) Exam, 2018]**

(*a*) $\frac{a+b+c}{a}$ (*b*) $\frac{b+c-a}{a}$ (*c*) $\frac{a-b+c}{a}$ (*d*) $\frac{a+b-c}{a}$

30. If $x^2 - 6x - 27 > 0$, then which one of the following is correct? **[CDS (II) Exam, 2018]**

(*a*) $-3 < x < 9$ (*b*) $x < 9$ or $x > -3$ (*c*) $x > 9$ or $x -3$ (*d*) $x < -3$ only

31. If α and β are the roots of the equation $ax^2 + bx + c = 0$, then the value of $\frac{1}{a\alpha+b} + \frac{1}{a\beta+b}$ is **[CDS (II) Exam, 2018]**

(*a*) $\frac{a}{bc}$ (*b*) $\frac{b}{ac}$ (*c*) $\frac{c}{ab}$ (*d*) $\frac{1}{abc}$

32. The minimum value of the expression $2x^2 + 5x + 5$ is **[CDS (II) Exam, 2018]**

(*a*) 5 (*b*) $\frac{15}{8}$ (*c*) $-\frac{15}{8}$ (*d*) 0

33. The sign of the quadratic polynomial $ax^2 + bx + c$ is always positive if

(*a*) *a* is positive and $b^2 - 4ac$ # 0

(*b*) *a* is positive and $b^2 - 4ac$ $ 0

(*c*) *a* can be any real number and $b^2 - 4ac$ # 0

(*d*) *a* can be any real number and $b^2 - 4ac$ $ 0 **[CDS (I) Exam, 2015]**

34. If the roots of the quadratic equation $x^2 - 4x - log_{10}N = 0$ are all real, then the minimum value of *N* is **[CDS (II) Exam, 2015]**

(*a*) $\frac{1}{100}$ (*b*) $\frac{1}{1000}$ (*c*) $\frac{1}{10000}$ (*d*) 10000

35. If the roots of the equation $lx^2 + mx + m = 0$ are in the ratio $p : q$, then $\sqrt{\frac{p}{q}} + \sqrt{\frac{q}{p}} + \sqrt{\frac{m}{l}}$ is equal to **[CDS (I) Exam, 2016]**

(*a*) 0 (*b*) 1 (*c*) 2 (*d*) 3

36. If the equations $x^2 - px + q = 0$ and $x^2 + qx - p = 0$ have a common root, then which one of the following is correct? **[CDS (I) Exam, 2016]**

(*a*) $p - q = 0$ (*b*) $p + q - 2 = 0$ (*c*) $p + q - 1 = 0$ (*d*) $p - q - 1 = 0$

37. Let *p* and *q* be non-zero integers. Consider the polynomial

$A(x) = x^2 + px + q$

It is given that $(x - m)$ and $(x - km)$ are simple factors of $A(x)$, where *m* is a non-zero integer and *k* is a positive integer, $k \geq 2$. Which one of the following is correct? **[CDS (I) Exam, 2016]**

(*a*) $(k + 1)^2 p^2 = kq$ (*b*) $(k + 1)^2 q = kp^2$

(*c*) $k^2 q = (k + 1) p^2$ (*d*) $k^2 p^2 = (k + 1)^2 q$

38. If the sum of the roots of $ax^2 + bx + c = 0$ is equal to the sum of the squares of their reciprocals, then which one of the following relations is correct? **[CDS (I) Exam, 2016]**

(a) $ab^2 + bc^2 = 2a^2c$ (b) $ac^2 + bc^2 = 2b^{2a}$
(c) $ab^2 + bc^2 = a^2c$ (d) $a^2 + b^2 + c^2 = 1$

39. Under what condition on p and q, one of the roots of the equation $x^2 + px + q = 0$ is the square of the other? **[CDS (I) Exam, 2016]**
(a) $1 + q + q^2 = 3pq$ (b) $1 + p + p^2 = 3pq$ (c) $p^3 + q + q^2 = 3pq$ (d) $q^3 + p + p^2 = 3pq$

40. If λ is an integer and α, β are the roots of $4x^2 - 16x + \frac{\lambda}{4} = 0$ such that $1 < \alpha < 2$ and $2 < \beta < 3$, then how many values can λ take? **[CDS (II) Exam, 2016]**
(a) 3 (b) 9 (c) 14 (d) 15

41. In the quadratic equation $x^2 + ax + b = 0$, *a*, and *b* can take any value from the set {1, 2, 3, 4}. How many pairs of values of *a* and *b* are possible in order that the quadratic equation has real roots? **[CDS (II) Exam, 2016]**
(a) 6 (b) 7 (c) 8 (d) 16

42. If α and β are the two zeros of the polynomial $25x^2 - 15x + 2$, then what is a quadratic polynomial whose zeros are $(2\alpha)^{-1}$ and $(2\beta)^{-1}$? **[CDS (II) Exam, 2016]**
(a) $x^2 + 30x + 2$ (b) $8x^2 - 30x + 25$ (c) $8x^2 - 30x$ (d) $x^2 + 30x$

43. If P and q are the roots of $x^2 + px + q = 0$, then which one of the following is correct?
(a) $p = 0$ or 1 (b) $p = 1$ only (c) $p = 0 -2$ or 0 (d) $p = -$ only
[CDS (II) Exam, 2016]

44. If α and β are the roots of the quadratic equation $2x^2 - 6x + k = 0$, where $k < 0$, then what is the maximum value of $\frac{\alpha}{\beta} + \frac{\beta}{\alpha}$? **[CDS (I) Exam, 2017]**
(a) 2 (b) –2 (c) 9 (d) –9

45. If one root of $(a^2 - 5a + 3)x^2 + (3a - 1)x + 2 = 0$ is twice the other, then what is the value of '*a*'? **[CDS (I) Exam, 2017]**
(a) $\frac{2}{3}$ (b) $-\frac{2}{3}$ (c) $\frac{1}{3}$ (d) $-\frac{1}{3}$

46. If α and β are the roots of the equation $x^2 + px + q = 0$, then what is $\alpha^2 + \beta^2$ equal to?
(a) $p^2 - 2q$ (b) $q^2 - 2p$ (c) $p^2 + 2q$ (d) $q^2 - q$
[CDS (I) Exam, 2017]

47. Aman and Alok attempted to solve a quadratic equation. Aman made a mistake in writing down the constant term and ended up in roots (4, 3). Alok made a mistake in writing down the coefficient of *x* to get roots (3, 2). The correct roots of the equation are
(a) –4, –3 (b) 6, 1 (c) 4, 3 (d) –6, –1
[CDS (I) Exam, 2017]

48. If $x = \frac{\sqrt{a+b} - \sqrt{a-b}}{\sqrt{a+b} + \sqrt{a-b}}$, then what is $bx^2 - 2ax + b$ equal to $(b \neq 0)$?
(a) 0 (b) 1 (c) ab (d) $2ab$
[CDS (I) Exam, 2017]

49. $x = 3$ is a solution of the equation $3x^2 + (k - 1)x + 9 = 0$ if k has value:
(*a*) 13 (*b*) –13 (*c*) 11 (*d*) –11

50. The expression $x^4 + 7x^2 + 16$ can be factorized as:
(*a*) $(x^2 + x + 1)(x^2 + x + 16)$ (*b*) $(x^2 + x + 1)(x^2 - x + 16)$
(*c*) $(x^2 + x + 4)(x^2 - x + 4)$ (*d*) $(x^2 + x - 4)(x^2 - x + 4)$

51. The solution of $2 - x = \frac{x-2}{x}$ would include:
(*a*) –2, –1 (*b*) 2, –1 (*c*) –4, 2 (*d*) 4, –2

52. The common root of the equations: $x^2 - 7x + 10 = 0$ and $x^2 - 10x + 16 = 0$ is:
(*a*) – 2 (*b*) 3 (*c*) 5 (*d*) 2

53. Let $f(x) = ax^2 + bx + c$. Then, match the following:

A. Sum of roots of $f(x) = 0$	1. $\frac{c}{a}$
B. Product of roots of $f(x) = 0$	2. $\frac{-b}{a}$
C. Roots of $f(x) = 0$ are real & distinct	3. $b^2 - 4ac = 0$
D. Roots of $f(x) = 0$ are real and identical.	4. $b^2 - 4ac > 0$
	5. $b^2 - 4ac < 0$

The correct matching is:
(*a*) $1 - B, 2 - A, 3 - D, 4 - C$ (*b*) $1 - A, 2 - B, 3 - C, 4 - D$
(*c*) $1 - C, 2 - E, 3 - B, 4 - A$ (*d*) $1 - E, 2 - D, 3 - A, 4 - B$

54. The roots of $2x^2 - 6x + 7 = 0$ are:
(*a*) real, unequal and rational (*b*) real, unequal and irrational
(*c*) real and equal (*d*) imaginary

55. The roots of $2x^2 - 6x + 3 = 0$ are:
(*a*) real, unequal and rational (*b*) real, unequal and irrational
(*c*) real and equal (*d*) imaginary

56. The equation $x^2 + 4x + k = 0$ has real roots. Then:
(*a*) $k \geq 4$ (*b*) $k \leq 4$ (*c*) $k \leq 0$ (*d*) $k \geq 0$

57. The value of k for which $x^2 - 4x + k = 0$ has coincident roots is:
(*a*) 4 (*b*) – 4 (*c*) 0 (*d*) – 2

58. If the equation $x^2 + 2(k + 2)x + 9k = 0$ has equal roots, the values of k are:
(*a*) 1, 4 (*b*) – 1, 4 (*c*) 1, – 4 (*d*) – 1, – 4

59. If the roots of $ax^2 + bx + c = 0$ be equal, then the value of c is:
(*a*) $\frac{-b}{2a}$ (*b*) $\frac{b}{2a}$ (*c*) $\frac{-b^2}{4a}$ (*d*) $\frac{b^2}{4a}$

60. If the roots of $x^2 + 4mx + 4m^2 + m + 1 = 0$ are real, then:
(*a*) $m = -1$ (*b*) $m \leq -1$ (*c*) $m \geq -1$ (*d*) $m \geq 0$

61. The positive value of k for which the equation $x^2 + kx + 64 = 0$ and $x^2 - 8x + k = 0$ will both have real roots, is:
(*a*) 4 (*b*) 8 (*c*) 12 (*d*) 16

62. If $\log_{10}(x^2 - 6x + 45) = 2$, then the values of x are:
(*a*) 6, 9 (*b*) 9, – 5 (*c*) 10, 5 (*d*) 11, – 5

63. A and B solved a quadratic equation. In solving it, A made a mistake in the constant term and obtained the roots as 5, – 3 while B made a mistake in the coefficient of x and obtained the roots as 1, – 3. The correct roots of the equation are:
(a) + 1, + 3 (b) – 1, 3 (c) – 1, – 3 (d) 1, – 1

64. The value of p so that $x^2 + 5px + 16 = 0$ has no real root:
(a) is greater than $\frac{8}{5}$ (b) is less than $-\frac{8}{5}$ (c) lies between $-\frac{8}{5}$ and $\frac{8}{5}$ (d) None of these

65. The roots of a quadratic equation are 5 and – 2. The equation is:
(a) $x^2 - 3x + 10 = 0$ (b) $x^2 - 3x - 10 = 0$ (c) $x^2 + 3x + 10 = 0$ (d) $x^2 + 3x - 10 = 0$

66. If one root of the equation $3x^2 - 10x + 3 = 0$ is $\frac{1}{3}$, the other root is:
(a) $-\frac{1}{3}$ (b) – 3 (c) 3 (d) $\frac{1}{3}$

67. The quadratic equation whose one root is $3 + 2\sqrt{3}$, is:
(a) $x^2 + 6x - 3 = 0$ (b) $x^2 - 6x - 3 = 0$ (c) $x^2 + 6x + 3 = 0$ (d) $x^2 - 6x + 3 = 0$

68. If the equation $ax^2 - 5x + c = 0$ has 10 as the sum of the roots and also as the product of the roots, which of the following is true?
(a) $a = \frac{1}{2}, c = 5$ (b) $a = 2, c = 3$ (c) $a = 5, c = \frac{1}{2}$ (d) $a = 3, c = 2$

69. If the sum of the roots of the equation $kx^2 + 2x + 3k = 0$ is equal to their product, then the value of k is:
(a) $\frac{1}{3}$ (b) $-\frac{1}{3}$ (c) $\frac{2}{3}$ (d) $-\frac{2}{3}$

70. If α, β be the roots of $ax^2 + bx + c = 0$, the value of $\alpha^2 + \beta^2$ is:
(a) $\frac{b^2 - 4ac}{2a}$ (b) $\frac{b^2 - 2ac}{2a}$ (c) $\frac{b^2 - 2ac}{a^2}$ (d) $\frac{b^2 - 4ac}{2\,ac}$

71. If α, β are the roots of the equation $x^2 - 8x + p = 0$ and $\alpha^2 + \beta^2 = 40$, then p is equal to:
(a) 8 (b) 10 (c) 12 (d) 14

72. If α, β are the roots of the equation $x^2 + x + 1 = 0$, the value of $\alpha^4 + \beta^4$ is:
(a) 0 (b) 1 (c) – 1 (d) None of these

73. If α, β are the roots of the equation $2x^2 - 4x + 3 = 0$, then the value of $\alpha^3 + \beta^3$ is:
(a) – 1 (b) 1 (c) 2 (4) 0

74. If α, β are the roots of the equation $x^2 - 5x + 6 = 0$, the value of $\alpha^2 - \beta^2$ is:
(a) ±4 (b) ±5 (c) ±6 (d) 0

75. If α, β are the roots of the equation $ax^2 + bx + c = 0$, the value of $\frac{\alpha}{\beta^2} + \frac{\beta}{\alpha^2}$ is:
(a) $\frac{b^2 - 2ac}{ac}$ (b) $\frac{-b^3 + 3abc}{a^3}$ (c) $\frac{-b^3 + 3abc}{ac^2}$ (d) $\frac{b^2 - 2ac}{a^2}$

76. If one root of $5x^2 + 13x + k = 0$ be the reciprocal of the other root, the value of k is:
(a) 0 (b) 1 (c) 2 (d) 5

77. The value of k for which the roots α, β of the equation: $x^2 - 6x + k = 0$ satisfy the relation $3\alpha + 2\beta = 20$, is
(a) 8 (b) –8 (c) 16 (d) –16

78. If α, β are the roots of the equation $2x^2 - 3x + 1 = 0$, then the equation whose roots are $\frac{\alpha}{\beta}$ and $\frac{\beta}{\alpha}$ is:

(*a*) $2x^2 + 5x + 2 = 0$ (*b*) $2x^2 - 5x - 2 = 0$ (*c*) $2x^2 + 5x - 2 = 0$ (*d*) $2x^2 - 5x + 2 = 0$

79. The quadratic equation whose roots are the reciprocals of the roots of the equation $3x^2 - 20x + 17 = 0$, is:

(*a*) $20x^2 - 17x + 3 = 0$ (*b*) $17x^2 - 20x + 3 = 0$
(*c*) $20x^2 + 17x - 3 = 0$ (*d*) $17x^2 + 20x - 3 = 0$

80. If α, β are the roots of the equation $x^2 + kx + 12 = 0$ such that $\alpha - \beta = 1$, the value of k is:

(*a*) 0 (*b*) ± 5 (*c*) ± 1 (*d*) ±7

81. Consider the following statements:

I. If the roots of the equation $ax^2 + bx + c = 0$ are negative reciprocal of each other, then $a + c = 0$.
II. A quadratic equation can have maximum two roots.
III. If α, β are the roots of a quadratic equation such that $\alpha + \beta = 22$ and $\alpha - \beta = 8$, then the equation $x^2 - 22x + 112 = 0$ has α and β as its roots.
IV. If α, β are the roots of the equation $2x^2 - 4x + 1 = 0$, the value of $\frac{1}{\alpha + 2\beta} + \frac{1}{\beta + 2\alpha}$ is $\frac{12}{17}$.

Of these statements,

(*a*) I, II and IV are correct (*b*) I, III & IV are correct
(*c*) none is correct (*d*) all are correct

82. The condition for $ax^2 + bx + c$ to be expressed as the product of linear polynomials is:

(*a*) $b^2 - 4ac = 0$ (*b*) $b^2 - 4ac < 0$ (*c*) $b^2 - 4ac \geq 0$ (*d*) $b^2 = ac$

83. The expression $x^2 - x + 1$ has:

(*a*) one proper linear factor (*b*) two proper linear factors
(*c*) no proper linear factor (*d*) None of these

84. Which of the following quadratic polynomials can be factorized into a product of real linear factors?

(*a*) $2x^2 - 5x + 9$ (*b*) $2x^2 + 4x - 5$ (*c*) $3x^2 + 4x + 6$ (*d*) $5x^2 - 3x + 2$

85. Which of the following is true for real values of k for which the polynomial $12x^2 + 6x + k$ can be factorized into a product of real linear factors?

(*a*) $k < \frac{9}{2}$ (*b*) $k \leq \frac{9}{2}$ (*c*) $k > \frac{9}{2}$ (*d*) $k \geq \frac{9}{2}$

86. Match List I with List II. List I contains quadratic polynomials and List II contains the conditions for these polynomials to be factorizable into a product of real linear factors.

List I	**List II**
A. $4x^2 + kx + 1$	$k \text{ ″ } \frac{1}{2}$
B. $kx^2 - 4x + k$	2. $k \geq 4$ or $k \leq -4$
C. $kx^2 - 2x + 2$	3. $k \geq 8$ or $k \leq 0$
D. $2x^2 - kx + k$	4. $-2 \leq k \leq 2$

	A B C D
(*a*)	3 2 1 4
(*b*)	2 4 1 3
(*c*)	4 1 3 2
(*d*)	1 3 4 2

87. The value of x in the equation $8\left(x^2+\frac{1}{x^2}\right)-42\left(x-\frac{1}{x}\right)+29=0$ is:

(*a*) 4 (*b*) -2 (*c*) $\frac{1}{2}$ (*d*) $\frac{1}{4}$

88. The value of x in the equation $\left(x+\frac{1}{x}\right)^2-\frac{3}{2}\left(x-\frac{1}{x}\right)=4$ is:

(*a*) -2 (*b*) $\frac{1}{2}$ (*c*) -1 (*d*) 0

89. The solution set of the equation $(x+1)(x+2)(x+3)(x+4)=120$ is:

(*a*) {6, 1} (*b*) {6, –1} (*c*) {–6, 1} (*d*) {–6, –1}

90. The roots of the equation $4^x - 3.2^{x+2} + 32 = 0$ would include:

(*a*) 1, 2 and 3 (*b*) 1 and 2 (*c*) 1 and 3 (*d*) 2 and 3

91. The solution set of the equation $5^{x+1} + 5^{2-x} = 126$ is:

(*a*) {1, 2} (*b*) {–1, 2} (*c*) {1, –2} (*d*) {–1, –2}

92. The two parts into which 57 should be divided so that their product is 782, are:

(*a*) 43, 14 (*b*) 33, 24 (*c*) 34, 23 (*d*) 44, 13

93. The sum of a number and its reciprocal is $2\frac{1}{20}$. The number is:

(*a*) $\frac{5}{4}$ (*b*) $\frac{3}{4}$ (*c*) $\frac{4}{3}$ (*d*) $\frac{1}{6}$

94. Let p, q be the roots of the equation

$x^2+mx-n=0$ and m, n be the roots of the equation

$x^2+px-q=0$ (m, n, p, q are non-zero numbers).

Which of the following statements is/are correct?

I. $m(m+n)=-1$

II. $p+q=1$

Select the answer using the code given below: **[CDS (II) Exam, 2024]**

(*a*) I only (*b*) II only (*c*) Both I and II (*d*) Neither I nor II

95. The perimeter of a rectangle is 82 m and its area is 400 m^2. The breadth of the rectangles is:

(*a*) 25 m (*b*) 16 m (*c*) 9 m (*d*) 20 m

96. If the sum and product of the roots of a quadratic equation are 2 and – 100 respectively, then which one of the following is correct ? **[CDS (I) Exam, 2024]**

(*a*) There are infinitely many such equations having different roots.

(*b*) There is only one such equation which is $x^2 + 2x - 100 = 0$.

(*c*) There is only one such equation which is $x^2 - 2x - 100 = 0$.

(*d*) There is no such equation.

97. Consider the following statements in respect of the polynomial $a(b-c)(x-b)(x-c)+b(c-a)(x-c)(x-a)+c(a-b)(x-a)(x-b)$: **[CDS (I) Exam, 2023]**

1. The coefficient of x^2 is 0.

2. The coefficient of x is $(a-b)(b-c)(c-a)$.

Which of the statements given above is/are correct?

(*a*) 1 only (*b*) 2 only (*c*) Both 1 and 2 (*d*) Neither 1 nor 2

98. Consider the following statements in respect of the polynomial $1 - x - x^n + x^{n+1}$, where n is a natural number : **[CDS (I) Exam, 2023]**

1. It is divisible by $1 - 2x + x^2$
2. It is divisible by $1 - x^n$.

Which of the statements given above is/are correct?

(a) 1 only (b) 2 only (c) Both 1 and 2 (d) Neither 1 nor 2

99. How many quadratic equations have the sum of their roots equal to the product of their roots? **[CDS (I) Exam, 2022]**

(*a*) Zero (*b*) One (*c*) Two (*d*) Infinitely many

100. Let a and α and β be the roots of the equation $\frac{1}{x+a+c} = \frac{1}{x} + \frac{1}{a} + \frac{1}{b}$; $a \neq 0, b \neq 0, x \neq 0$.

Which one of the following is a quadratic equation whose roots are α^2 and β^2?

(a) $x^2 + (a^2 + b^2)x + a^2b^2 = 0$ (b)$x^2 - (a^2 + b^2)x + a^2b^2 = 0$

(c) $x^2 - (a^2 + b^2)x - a^2b^2 = 0$ (b)$x^2 + (a^2 + b^2)x - a^2b^2 = 0$

101. If $x = x = \cfrac{6}{7 - \cfrac{6}{7 - \cfrac{6}{7 - \cfrac{6}{7 - x}}}}$; x > 1, then what is the value of $x^2 - 3x + 2$ equal to ?

(a) 0 (b) 1 (c) 19 (d) 20

Mark option (a) if the question can be answered by using one of the statements alone, but cannot be answered using the other statement alone.

Mark option (b) if the question can be answered by using either statement alone.

Mark option (c) if the question can be answered by using both the statements together, but cannot be answered using either statement alone.

Mark option (d) if the question cannot be answered even by using both the statements together.

102. Question : What is the other root of the quadratic equation with real coefficients if one of the roots is $\frac{-4 - \sqrt{10}}{2}$?

Statement-I:

The product of the roots is $-\frac{3}{2}(3 + \sqrt{10})$.

Statement-II:

The sum of roots of quadratic equation is – 1.

103. A question is given followed by two statements I and II. Consider the Question and the Statements and mark the correct option.

Question: Can we have a common solution which is prime?

Statement I: $x^2 - 26x + 133 = 0$

Statement II: $x^2 - 44x + 475 = 0$

Which one of the following is correct in respect of the above Question and the Statements?

[CDS (II) Exam, 2024]

(*a*) The Question can be answered by using one of the Statements alone, but cannot be answered using the other Statement alone

(*b*) The Question can be answered by using either Statement alone

(*c*) The Question can be answered by using both the Statements together, but cannot be answered using either Statement alone

(*d*) The Question cannot be answered even by using both the Statements together

104. If α and β are the roots of the quadratic equation $x^2 + ax + \beta = 0$, where $\beta \neq 0$, then what is the value of $\alpha + \beta$? **[CDS (I) Exam, 2022]**

(*a*) 4 (*b*) 3 (*c*) –1 (*d*) –3

105. If the sum of the roots of the equation $x^2 - k^2 x + 30kx - 161x - 64 = 0$ is zero, then what is the difference of the roots? **[CDS (I) Exam, 2022]**

(*a*) 15 (*b*) 16 (*c*) 17 (*d*) 18

106. What is the condition that the roots of the equation $ax^2 + ba + c = 0$ are in the ration c : 1 ? **[CDS (I) Exam, 2022]**

(*a*) $b^2 = a(c + 1)^2$ (*b*) $a^2 = b(c + 1)^2$ (*c*) $b^2 = a(c - 1)^2$ (*d*) $ab^2 = (c + 1)^2$

107. Consider a question and two statements: **[CDS (I) Exam, 2022]**

Question: Does the equation $ax^2 + bx + c = 0$ have real roots of opposite sign?

Statement-I: The discriminant D > 0

Statement-II: $c/a > 0$

Which one of the following is correct in respect of the question and the statements?

(*a*) Statement-I alone is sufficient to answer the question

(*b*) Statement-II alone is sufficient to answer the question

(*c*) Both statement-I and Statement-II are together sufficient to answer the question

(*d*) Both statement-I and Statement-II are not sufficient to answer the question

ANSWERS

1. (*a*)	**2.** (*c*)	**3.** (*d*)	**4.** (*d*)	**5.** (*b*)	**6.** (*a*)	**7.** (*a*)	**8.** (*b*)	**9.** (*b*)
10. (*d*)	**11.** (*b*)	**12.** (*b*)	**13.** (*b*)	**14.** (*b*)	**15.** (*a*)	**16.** (*c*)	**17.** (*d*)	**18.** (*c*)
19. (*a*)	**20.** (*b*)	**21.** (*d*)	**22.** (*a*)	**23.** (*b*)	**24.** (*a*)	**25.** (*d*)	**26.** (*d*)	**27.** (*c*)
28. (*b*)	**29.** (*c*)	**30.** (*c*)	**31.** (*b*)	**32.** (*b*)	**33.** (*a*)	**34.** (*c*)	**35.** (*a*)	**36.** (*d*)
37. (*b*)	**38.** (*a*)	**39.** (*c*)	**40.** (*d*)	**41.** (*b*)	**42.** (*b*)	**43.** (*a*)	**44.** (*d*)	**45.** (*a*)
46. (*a*)	**47.** (*b*)	**48.** (*a*)	**49.** (*d*)	**50.** (*c*)	**51.** (*b*)	**52.** (*d*)	**53.** (*a*)	**54.** (*d*)
55. (*b*)	**56.** (*b*)	**57.** (*a*)	**58.** (*a*)	**59.** (*d*)	**60.** (*b*)	**61.** (*d*)	**62.** (*d*)	**63.** (*b*)
64. (*c*)	**65.** (*b*)	**66.** (*c*)	**67.** (*b*)	**68.** (*a*)	**69.** (*d*)	**70.** (*c*)	**71.** (*c*)	**72.** (*c*)
73. (*a*)	**74.** (*b*)	**75.** (*c*)	**76.** (*d*)	**77.** (*a*)	**78.** (*d*)	**79.** (*b*)	**80.** (*d*)	**81.** (*a*)
82. (*c*)	**83.** (*c*)	**84.** (*b*)	**85.** (*b*)	**86.** (*b*)	**88.** (*a*)	**89.** (*c*)	**89.** (*c*)	**90.** (*d*)
91. (*c*)	**92.** (*c*)	**93.** (*a*)	**94.** (*c*)	**95.** (*b*)	**96.** (*c*)	**97.** (*a*)	**98.** (*c*)	**99.** (*d*)
100. (*b*)	**101.** (*d*)	**102.** (*b*)	**103.** (*c*)	**104.** (*c*)	**105.** (*b*)	**106.** (*a*)	**107.** (*c*)	

SOLUTIONS

1. We have,

$x^2 - bx + c = 5$

$\Rightarrow x^2 - bx + c - 5 = 0$

Let the roots of the equation be denoted be α and β.

Sum of roots $(\alpha + \beta) = b$

Product of roots $(\alpha \times \beta) = c - 5$ and $\alpha - \beta = 5$

$\therefore$ We know that

$\Rightarrow (\alpha - \beta)^2 = (\alpha + \beta)^2 - 4 * \alpha * \beta \Rightarrow 5^2 = b^2 - 4 \times (c - 5) = 25 \Rightarrow b^2 - 4c + 20 = 25$

$\therefore \quad b^2 = 4c + 5$

2. We have,

$a + b + c = 0$

$\Rightarrow b = -(a + c) \quad ...(i)$

Now, $x = \dfrac{(a+c) \pm \sqrt{[-(a+c)]^2 - 4ac}}{2a}$ $\left[\because \text{ when } ax^2 + bx^2 + c = 0 \text{ is quadratic equations, then} = x = \dfrac{-b \pm \sqrt{b^2 - 4ac}}{2a}\right]$

$\Rightarrow x = \dfrac{(a+c) \pm \sqrt{a^2 - 2ac + c^2}}{2a} \Rightarrow x = \dfrac{(a+c) \pm (a-c)}{2a} \Rightarrow x = \dfrac{2a}{2a} = 1$ and $x = \dfrac{2c}{2a} = \dfrac{c}{a}$

3. Given: $x^2 + px + q = 0$, and $ax^2 + bx + c = 0$

Here: $a = 1,\ b = p,\ c = q$

Sum of roots: $p + q = -\dfrac{b}{a} = -\dfrac{p}{1} = -p$

So $\quad p + q = -p \Rightarrow q = -p - p = -2p$

at $x = p$, $\quad p^2 + p.p + q = 0$

$p^2 + p^2 + q = 0$

$2p^2 + q = 0$

$2p\,(q - 1) = 0$

$2p = 0, p - 1 = 0$

$p = 0, p = 1$

Taking $p = 1$ and $q = -2$

$p^2 + q^2 = 1^2 + (-2)^2 = 5$

4. $mx^2 + mx + 8x + 9$

$mx^2 + x(m + 8) + 9$

$a = m, b = m + 8, c = 9$

Now, $b^2 - 4ac = 0$

$\Rightarrow (m + 8)^2 - 36\,m = 0 \Rightarrow m^2 + 16m + 64 - 36\,m = 0 \quad \Rightarrow m^2 - 20\,m + 64 = 0$

$\Rightarrow m^2 - 16\,m - 4\,m + 64 = 0 \Rightarrow (m - 16)\,(m - 4) = 0 \Rightarrow m = 4, 16$

5. According to the question,

$D \geq 0 \Rightarrow b^2 \geq 4ac$

Here, $a = 2, b = -4, c = k$

$\Rightarrow \quad 16 \geq 4 \times 2 \times k \Rightarrow 16 \geq 8k \Rightarrow 2 \geq k$

$\therefore$ greatest value of $k = 2$

6. From qn. 5, we have

$x = 1$ and $\dfrac{a+b-c}{a+b+c}$

$1, \dfrac{a+b+c}{a+b+c} - \dfrac{2c}{a+b+c}$

$\therefore 1, 1-\dfrac{2c}{a+b+c}$

$\therefore$ Statement 1 is true, but statement 2 is false

7. From the two equations:

$\alpha^2+p\alpha+q=0$ ...(i) $\alpha^2+q\alpha+p=0$ ...(ii)

$\Rightarrow (p\alpha+q)-(q\alpha+p)=0 \Rightarrow p\alpha-q\alpha+q-p=0 \Rightarrow \alpha(p-q)=p-q$

Eqn (ii) – (i)

$\Rightarrow \alpha=1$

Substitute $\alpha=1$ in the first equation:

$1^2+p(1)+q=0 \Rightarrow 1+p+q=0 \Rightarrow p+q=-1$

8. $Px^2+Px+4=0$ ...(i)

$x^2+qx+q=0$ ...(ii)

Substituting the root 2 into the eqn (i)

$$p(2)^2+p(2)+4=0 \Rightarrow 4p+2p+4=0 \Rightarrow 6p+4=0 \Rightarrow 6p=-4 \Rightarrow p=-\frac{2}{3}$$

Substituting the root 2 into eqn (ii)

$$(2)^2+q(2)+q=0 \Rightarrow 4+2q+q=0 \Rightarrow 4+3q=0 \Rightarrow 3q=-4 \Rightarrow q=-\frac{4}{3}$$

$$p+q=-\frac{2}{3}-\frac{4}{3} \Rightarrow p+q=-\frac{6}{3} \Rightarrow p+q=-2$$

9. We have, $px^2-6x+q=0$

Now, according to the question

Sum of roots = product of roots

$$\Rightarrow \quad \frac{6}{p}=\frac{q}{p} \Rightarrow q=6$$

Now, $\quad pq=6$

$\Rightarrow \quad 6p=6 \Rightarrow p=1$

$\therefore \quad (p+q)=1+6=7$

10. $\because$ We know that an equation

$ax^2+bx+c=0$ has real roots,

if $b^2-4ac>0$

By options, we get

(a) $(16)^2-4\times2\times3>0$ (real roots)

(b) $(10)^2-4\times2\times-1>0$ (real roots)

(c) $(-8)^2-4\times1\times1>0$ (real roots)

(d) $(9)^2-4\times4\times6<0$ (not real roots)

11. Let the roots of the quadratic equation

$ax^2+bx+c=0$ be α and β.

Given that

$\alpha+\beta=7$ ...(i)

$\alpha\beta = 12$ ...(ii)

$\because (\alpha - \beta)^2 = (\alpha + \beta)^2 - 4\alpha\beta = (7)^2 - 4 \times 12 = 49 - 48 = 1$

$\Rightarrow \quad \alpha - \beta = 1$ ...(iii)

From (i) and (ii), we get $\alpha = 4$ and $\beta = 3$

Now, according to the question

$$\alpha = 2 \text{ and } \beta = 6$$

Now, equation is

$x^2 - (\alpha + \beta)x = \alpha\beta = 0 \Rightarrow x^2 - (2 + 6)x + 2 \times 6 = 0 \Rightarrow x^2 - 8x + 12 = 0$

12. Discriminant $(D) = b^2 - 4ac$

$\Rightarrow \; D = [-(k-2)]^2 - 4 \times 1 \times (k+1) \Rightarrow \; D = (k-2)^2 - 4(k+1)$

$\Rightarrow \; D = k^2 - 4k + 4 - 4k - 4 \; \Rightarrow \; D = k^2 - 8k$

Set $D = 0$:

$\Rightarrow \; k^2 - 8k = 0 \; \Rightarrow \; k(k-8) = 0 \; \Rightarrow \; k = 0$ or $k = 8$

The valid values of k are 0 and 8.

13. The standard quadratic form: $Ax^2 + Bx + C = 0$

$$x + a + b = \frac{abx}{ab + ax + bx}$$

Multiply both sides by $(ab + ax + bx)$:

$\Rightarrow (x + a + b)(ab + ax + bx) = abx$

$\Rightarrow x(ab) + x(ax) + x(bx) + a(ab) + a(ax) + a(bx) + b(ab) + b(ax) + b(bx) = abx$

$\Rightarrow abx + ax^2 + bx^2 + a^2b + a^2x + abx + ab^2 + abx + b^2x = abx$

$\Rightarrow (a+b)x^2 + (a^2x + b^2x + 2abx) + (a^2b + ab^2) = 0$

$\Rightarrow (a+b)x^2 + (a^2 + b^2 + 2abx) + ab(a+b) = 0$

As we know $(a^2 + b^2 + 2ab) = (a+b)^2$

$\Rightarrow (a+b)x^2 + (a+b)^2x + ab(a+b) = 0 \Rightarrow x^2 + (a+b)x + ab = 0$

Sum of roots: $\alpha + \beta = \left(-\dfrac{B}{A}\right) = -\dfrac{(a+b)}{1} = -(a+b)$

Product of roots: $\alpha\beta = \dfrac{C}{A} = \dfrac{ab}{1} = ab$

$\Rightarrow (\alpha\beta + \alpha + \beta) = ab + (-(a+b)) \Rightarrow (\alpha\beta + \alpha + \beta) = ab - a - b$

14. Put $x = \dfrac{3}{2}$ in the given eqn, we get, $ax^2 - 4ax + 15 = 0$

$\Rightarrow \dfrac{9}{4}a - 4 \times a \times \dfrac{3}{2} + 15 = 0 \Rightarrow \dfrac{9}{4}a - 6a + 15 = 0$

$\Rightarrow = \dfrac{15}{4}a = -15 \Rightarrow a = 4$

$\therefore 4x^2 - 16x + 15 = 0$

$\Rightarrow 4x^2 - 10x - 6x + 15 = 0 \Rightarrow 2x(2x-5) - 3(2x-5) = 0$

$\Rightarrow (2x-3)(2x-5) = 0 \Rightarrow x = \frac{3}{2}, \frac{5}{2}$

$\therefore$ Sum of the squares of the roots $= \left(\frac{3}{2}\right)^2 + \left(\frac{5}{2}\right)^2 = \frac{9}{4} + \frac{25}{4} = \frac{34}{4} = \frac{17}{2}$

15. We have,

$\alpha + \beta = -k$

$\alpha\beta = -15$

$\alpha - \beta = -8$

$\therefore \quad (\alpha - \beta)^2 = (\alpha + \beta)^2 - 4\alpha\beta \Rightarrow (8)^2 = (-k)^2 - 4 \times -15$

$\Rightarrow \quad 64 = k^2 + 60 \Rightarrow k^2 = 4 \Rightarrow k = \pm 2$

$\therefore \quad k = 2$

16. We have,

Area of square $= 2401\,x^4 + 196x^2 + 4 = (49x^2)^2 + 2 \times 49x^2 \times 2 + (2)^2 = (49x^2 + 2)^2$

$[\because (a+b)^2] = a^2 + b^2 + 2ab]$

$\Rightarrow (\text{side})^2 = (49x^2 + 2)^2 \Rightarrow \text{side} = 49x^2 + 2$

17. We have, $x^2 - 4x - \log_{10} N = 0$

$b^2 - 4a\,c > 0 \quad [\because$ roots are real]

$\Rightarrow (4)^2 - 4 \times \log_{10} N \geq 0 \Rightarrow 16 + 4\log_{10} N \geq 0 \Rightarrow 4\log_{10} N \geq -16$

$\Rightarrow \log_{10} N \geq -4 \Rightarrow N \geq 10^{-4} \Rightarrow N = \frac{1}{10^4} = \frac{1}{10000}$

18. We have,

Let $y = x + \frac{x+2}{2x} \quad \Rightarrow y = \frac{2x^2 + x + 2}{2x}$

$2yx = 2x^2 + x + 2 \quad \Rightarrow 2x^2 + (1-2y)x + 2 = 0$

For roots are real

$b^2 - 4ac \geq 0$

$(1-2y)^2 - 4 \times 2 \times 2 \geq 0$

$1 + 4y^2 - 4y - 16 \geq 0$

$4y^2 - 4y - 15 \geq 0$

$4y^2 - 10y + 6y - 15 \geq 0$

$2y(2y-5) + 3(2y-5) \geq 0$

$(2y+3)(2y-5) \geq 0$

$y \geq \frac{5}{2}$ and $y \leq \frac{-3}{2}$

$\therefore$ minimum value is $= 2\frac{1}{2}$

19. We have,

$\frac{1+px}{1-px}\sqrt{\frac{1-qx}{1+qx}} = 1 \Rightarrow \sqrt{\frac{1-qx}{1+qx}} = \frac{1-px}{1+px}$

By taking square both sides, we get,

$$\frac{1-qx}{1+qx}=\left(\frac{1-px}{1+px}\right)^2 \Rightarrow \frac{1-qx}{1+qx}=\frac{1+p^2x^2-2px}{1+p^2x^2+2px}$$

By componendo and dividendo rule, we get

$$\frac{2}{-2qx}=\frac{2(1+p^2x^2)}{-4px}$$

$\Rightarrow (1 + p^2x^2) \times qx = 2px \quad \Rightarrow x \times (p^2 qx^2 + q - 2p) = 0$

$\therefore x = 0$ or $p^2qx^2 + q - 2p = 0$

$$\Rightarrow p^2qx^2 = 2p - q \Rightarrow x^2 = \frac{2p-q}{p^2q} \quad \Rightarrow x = \pm\frac{1}{p}\sqrt{\frac{2p-q}{q}}$$

$$\therefore x = 0 \text{ or } \pm\frac{1}{p}\sqrt{\frac{2p-q}{q}}$$

20. We have, $x^2 - 15x + r = 0$

and p and q are the roots of the given equation

Given equation

$\therefore p + q = 15$ ...(*i*)

$pq = r$ ...(*ii*)

Given that

$p - q = 1$ ...(*iii*)

From (i) and (ii), we have $p = 8$ and $q = 7$

$\therefore r = 0\ pq = 8 \times 7 = 56$

21. We have, $x^2 - 7x + 12 > 0$

$\Rightarrow x^2 - 4x - 3x + 12 > 0 \quad \Rightarrow (x - 4)(x - 3) > 0$

$\therefore$ we have two cases

Case I: If $(x - 4) > 0$ and $(x - 3) > 0$

$\Rightarrow x > 4$ and $x > 3$

$\therefore x \in (4, \infty)$ ... (ii)

Case II: If $(x - 4) < 0$ and $(x - 3) < 0$

$\Rightarrow x < 4$ and $x < 3$

$\therefore\ x \in (-\infty, 3)$ ...(ii)

$\therefore$ from (x) and (ii),

We have, $x \in (-\infty, 3) \cup (4, \infty)$

22. We have,

$$x = \frac{1+\sqrt{3}}{2}$$

$$\Rightarrow x^2 = \left(\frac{1+\sqrt{3}}{2}\right)^2 = \frac{4+2\sqrt{3}}{4} = \frac{2+\sqrt{3}}{2} \quad \Rightarrow y = x^3 + x^2 \times x = \frac{2+\sqrt{3}}{2} \times \frac{1+\sqrt{3}}{2} = \frac{5+3\sqrt{3}}{4}$$

$$\Rightarrow y^2 = \left(\frac{5+3\sqrt{3}}{4}\right)^2 \quad \Rightarrow y^2 = \frac{25+27+30\sqrt{3}}{16} \quad \Rightarrow y^2 = \frac{52+30\sqrt{3}}{16} \quad \Rightarrow y^2 = \frac{26+15\sqrt{3}}{8}$$

$\Rightarrow 8y^2 = 26 + 15\sqrt{3}$ and $20y = \dfrac{20(5+3\sqrt{3})}{4} = 25 + 15\sqrt{3}$

$\Rightarrow 8y^2 - 20y - 1 = (26 + 15\sqrt{3}) - (25 + 15\sqrt{3}) - 1 = 0$

23. We have, $x^2 + 5x + 6 = 0$

$\Rightarrow x^2 + 3x + 2x + 6 = 0 \Rightarrow x(x+3) + 2(x+3) = 0$

$\Rightarrow (x+2)(x+3) = 0 \Rightarrow x = -2$ and -3

Now, $x^2 + kx + 1 = 0$

$\Rightarrow (-2)^2 + kx - 2 + 1 = 0$ [put $x = -2$] $\Rightarrow 4 - 2k + 1 = 0$

$\Rightarrow k = \dfrac{5}{2}$ and $x^2 + kx + 1 = 0 \Rightarrow (-3)^2 + k \times -3 + 1 = 0$ [put $x = -3$]

$\Rightarrow 9 - 3k + 1 = 0 \Rightarrow k = \dfrac{10}{3}$

24. We have,

p and q are the roots of the equation

$x^2 + px + q = 0$ ...(*i*)

$x^2 - (p+q)x + pq = 0$ ...(*ii*)

From equations (i) and (ii), we get

$p = -(p+q)$

$\Rightarrow 2p = -q$ ...(*iii*)

and $q = pq$

$\Rightarrow p = 1$

put $p = 1$ in eqn (iii)

$\therefore q = -2$

25. put $x = b - 6$ in the given eqn

$x^2 - 6x + b = 0$

$\Rightarrow (b-6)^2 - 6x + b = 0$ [$\because (b-6)$ is a root of the given eqn]

$\Rightarrow b^2 + 36 - 12b - 6(b-6) + b = 0 \quad \Rightarrow b^2 + 36 - 12b - 6b + 36 + b = 0$

$\Rightarrow b^2 - 17b + 72 = 0 \quad \Rightarrow b^2 - 9b - 8b + 72 = 0$

$\Rightarrow b(b-9) - 8(b-9) = 0 \Rightarrow (b-9)(b-8) = 0 \Rightarrow b = 9, b = 8$

$\therefore$ Maximum value of $b^2 = 9^2 = 81$

26. Let $y = x^2 + 5x + 10$

$\Rightarrow x^2 + 5x + 10 - y = 0$

$\therefore D = b^2 - 4ac \geq 0$ v real value of x

$\Rightarrow 25 - 4(10 - y) \geq 0 \Rightarrow 25 - 40 + 4y \geq 0$

$\Rightarrow -15 + 4y \geq 0 \quad \Rightarrow y \geq \dfrac{15}{4}$

$\therefore$ Maximum value of $\dfrac{1}{y} = \dfrac{1}{\frac{15}{4}} = \dfrac{4}{15}$

27. We have,

$px^2 + 3x + 2q = 0$

Sum of roots $= -6$

$\Rightarrow \frac{-3}{p} = -6 \Rightarrow p = \frac{1}{2}$

and product of roots = -6

$\frac{2q}{p} = -6 \Rightarrow \frac{2q}{\frac{1}{2}} = 6 \quad \left[\because p = \frac{1}{2}\right]$

$\Rightarrow 4q = -6 \quad \Rightarrow q = \frac{-3}{2}$

$\therefore (p-q) = \frac{1}{2} - \left(-\frac{3}{2}\right) = \frac{4}{2} = 2$

28. We have, $px^2 + x + r = 0$

Let α and $\frac{1}{\alpha}$ are the roots of the given equation

$\alpha \times \frac{1}{\alpha} = \frac{r}{p} \quad \Rightarrow \frac{r}{p} = 1 \quad \Rightarrow p = r$

29. We have, $ax^2 + bx + c = 0$

and α and β are the roots of the given equation

$\alpha + \beta = \frac{b}{a}$

$\therefore \alpha\beta = \frac{c}{a}$

Now, $(\alpha + 1)(\beta + 1) \Rightarrow \alpha\beta + \alpha + \beta + 1 \quad \Rightarrow \frac{c}{a} - \frac{b}{a} + 1 \Rightarrow \frac{c - b + a}{a}$

30. $x^2 - 6 - 27 > 0$

$\Rightarrow x^2 - 9x + 3x - 27 > 0 \quad \Rightarrow x(x-9) + 3(x-9) > 0 \Rightarrow (x-9)(x+3) > 0$

If $(x - 9) > 0$ and $(x + 3) > 0$

$x > 9$ and $x < -3 \Rightarrow x > 9$ ...(i)

If $(x - 9) < 0$ and $(x + 3) < 0$

$\Rightarrow x > 9$ and $x > -3 \Rightarrow x < -3$ ...(ii)

From (*i*) and (ii), we get $x < -3$ or $x < 9$

31. Given that α and β are the roots of the equation $ax^2 + bx + c = 0$

$\therefore$ we have, $\alpha + \beta = -\frac{b}{a}$

$\alpha\beta = \frac{c}{a}$

Now, $\frac{1}{a\alpha + b} + \frac{1}{a\beta + b} = \frac{a(\alpha + \beta) + 2b}{a^2\alpha\beta + ab\alpha + ab\beta + b^2} = \frac{a \times -\frac{b}{a} + 2b}{a^2\alpha\beta + ab(\alpha + \beta) + b^2}$

$= \frac{b}{a^2 \times \frac{c}{a} + ab \times \frac{-b}{a} + b^2} = \frac{b}{ac - b^2 + b^2} = \frac{b}{ac}$

32. Let $y = 2x^2 + 5x + 5$

$2x^2 + 5x + 5 - y = 0$

$b^2 - 4ac \geq 0$ for all real value of x.

$\Rightarrow 25 - 4 \times 2(5 - y) = 0 \quad \Rightarrow 25 - 8(5 - y) = 0$

$\Rightarrow 25 - 40 + 8y = 0 \quad \Rightarrow -15 = -8y$

$\Rightarrow y = \dfrac{15}{8}$

$\therefore$ minimum value of $y = \dfrac{15}{8}$

33. The graph of $y = ax^2 + bx + c$ is a parabola. If $a > 0$, y has a minimum value and the parabola opens upwards.
If $a < 0$, y has a minimum value and the parabola opens downwards. (If $a = 0$, the graph is a straight line.)
As y has to be positive for all values, $a > 0$. The parabola should not intersect the x-axis, *i.e.*, the equation $ax^2 + bx + c = 0$ should not have any real roots. Therefore, $b^2 - 4ac < 0$.
Among the choices, we should expect (*e*) $a > 0$, $b^2 - 4ac < 0$.
These precise conditions are not there. Instead in choice (*a*), we have $a > 0$, $b^2 - 4ac \leq 0$.
If $b^2 - 4ac = 0$, there is exactly one value of x.
(viz $-\frac{b}{a}$), for which $y = 0$ (and therefore y is not positive.).
Choice (*a*) should not be accepted if choice (*e*) mentioned above is also available.
As all the other choices are clearly wrong, in the given conditions, the candidates best response would be (*a*).

34. $x^2 - 4x - \log_{10}N = 0$ is a quadratic equation in x. For the roots to be real, discriminant ($b^2 - 4ac$) should be greater than or equal to zero.
$(-4)^2 - 4(1)(-\log_{10}N) \geq 0$
$16 + 4\log_{10}N \geq 0$ & $\log_{10}N \geq -4$ & $N \geq 10^{-4}$.

35. Let the roots be α, β.
$\Rightarrow \alpha = pk$ and $\beta = qk$

$$\alpha + \beta = k(p + q) = -\frac{m}{l} \qquad \text{... (i)}$$

$$\alpha\beta = pqk^2 = \frac{m}{l} \qquad \text{... (ii)}$$

From (1) and (2) $(\alpha + \beta) = -\alpha\beta$

$\Rightarrow k(p + q) = -pqk^2$

$$\Rightarrow k = \left(\frac{p+q}{pq}\right) \qquad \text{... (iii)}$$

$$\therefore d = pk = \frac{-(p+q)}{p}; \quad \sqrt{\frac{p}{q}} + \sqrt{\frac{q}{p}} + \sqrt{\frac{m}{l}} = \frac{p+q}{\sqrt{pq}} + \sqrt{pqk^2}$$

$$= \frac{p+q}{\sqrt{pq}} + \sqrt{pq}\left(\frac{-(p+q)}{pq}\right) = \frac{p+q}{\sqrt{pq}} - \frac{(p+q)}{\sqrt{pq}} = 0$$

36. Let x be the common root of $x^2 - px + q = 0$ and $x^2 + qx - p = 0$

$\Rightarrow$ $x^2 - px + q = x^2 + qx - p \Rightarrow x(p+q) = p+q; x = 1$

$\therefore$ $1 - p + q = 0$ or $p - q - 1 = 0$

37. A $(x) = x^2 + px + q;$; $x - m$ is a factor of A (x)

$x = m \quad \Rightarrow \quad m^2 + pm + q = 0$ ... (*i*)

Also x – km is also a factor of A(x) $\Rightarrow$ A (km) 0

$\Rightarrow$ $k^2m^2 + pkm + q = 0$ ... (*ii*)

(2) – (1) $m^2(k^2-1) + pm(k-1) = 0$

$\Rightarrow m = \dfrac{-p}{k+1}$; From (1): $\left(-\dfrac{p}{k+1}\right)^2 + p\left(\dfrac{-p}{k+1}\right) + q = 0$

$p^2 - p^2(k+1) + q(k+1)^2 = 0$; $q(k+1)^2 = p^2 k$

38. Let α, β be the roots of $ax^2 + bx + c = 0$

$\Rightarrow$ $\alpha + \beta = \dfrac{-b}{a}, \alpha\beta = \alpha\beta = \dfrac{c}{a}$

Given $\alpha + \beta = \dfrac{1}{\alpha^2} + \dfrac{1}{\beta^2} = \dfrac{\alpha^2 + \beta^2}{\alpha^2\beta^2}$

$$\alpha + \beta = \frac{(\alpha+\beta)^2 - 2\alpha\beta}{\alpha^2\beta^2}$$

$$\frac{-b}{a} = \frac{\frac{b^2}{a^2} - 2\frac{c}{a}}{\frac{c^2}{a^2}} \Rightarrow -bc^2 = a(b^2 - 2ac)$$

$\Rightarrow$ $ab^2 + bc^2 = 2a^2c$

39. Let α, β be the roots of $x^2 + px + q = 0$

$\Rightarrow$ $\alpha + \beta = -p$ and $\alpha\beta = q$. Also $\alpha^2 = \beta$;

$\alpha^3 = \alpha\beta = q; \alpha = q^{\frac{1}{3}}$

$\Rightarrow \beta = \alpha^2 = q^{\frac{2}{3}}; \alpha + \beta = q^{\frac{1}{3}} + q^{\frac{2}{3}} = -p$

Cubing on both sides $q + q^2 + 3q^{\frac{1}{3}} \cdot q^{\frac{2}{3}} (q^{\frac{1}{3}} + q^{\frac{2}{3}}) = -p^3$

$q + q^2 + 3q(-p) = -p^3$;

$\Rightarrow$ $p^3 + q + q^2 = 3pq$

40. $f(x) = 4x^2 - 16x + \dfrac{\lambda}{4} = 0$ ... (*i*)

The roots of (*i*) are α, β where $1 < \alpha < 2 < \beta < 3$

$\therefore$ $f(1) > 0, f(2) < 0$ and $f(3) > 0$

i.e., $4-16+\frac{\lambda}{4}>0,\ 16-32+\frac{\lambda}{4}<0,\ 36-48+\frac{\lambda}{4}>0$

i.e., $\lambda>48, \lambda<64, \lambda>48$

λ could be 49,50, ...63. It can take 15 values

41. The equation $x^2+ax+b=0$ has to have real route. Therefore, $a^2-4b\geq 0$.

The values of (a, b) from $A\times A$, where $A=\{1, 2, 3, 4\}$ are listed below.

If $a=1$, Here are 0 values possible for b.

If $a=2$, b could be 1.

If $a=3$, if could be 1 or 2

If $a=4$, b could be 1, 2, 3 or 4

$\therefore$ There are 7 possible value for (a, b).

42. Let $f(x)=25x^2-15x+2$. The zeroes of $f(x)$ the roots of the equation $f(x)=0$ ... (i)

These roots are α and β.

Let $2x=y$. If we express (1) in terms of y, we get an

Equation whose roots are $(2\alpha)^{-1}$ and $(2\beta)^{-1}$

i.e., $25\left(\frac{1}{2y}\right)^2-15\left(\frac{1}{2y}\right)+2=0 \quad \Rightarrow 8y^2-30y+25=0$

The polynomial where zeroes are $(2\alpha)^{-1}$ and $(2\beta)^{-1}$ is

$g(x)=8x^2-30x+25$

43. The roots of $x^2+px+q=0$ are p and q.

$p+q=-p$ and $pq=q$

i.e., $q=-2p$ and $p(-2p)=-2p$, *i.e.*, $2p(p-1)=0$

If $p=0$, then $q=0$. If $p=1$, then $q=-2$

Among the choices, only (*a*) is correct.

44. $\alpha+\beta=\frac{-6}{2}=-3,\ \alpha\beta=\frac{k}{2}$

$$\frac{\alpha}{\beta}+\frac{\beta}{\alpha}=\frac{\alpha^2+\beta^2}{\alpha\beta}=\frac{(\alpha+\beta)^2-2\alpha\beta}{\alpha\beta}=2\frac{(9-k)}{k} \quad \text{... (i)}$$

As, $k<0$

So, put $\frac{\alpha}{\beta}+\frac{\beta}{\alpha}$ equal to -2, we get

$$\Rightarrow 2\frac{(9-k)}{k}=-2,\ 9-k=-k$$

Here, value of k is not possible Now, Put $\frac{\alpha}{\beta}+\frac{\beta}{\alpha}$ equal to -9

$$\frac{2(9-k)}{k}=-9$$

$$\Rightarrow 2(9-k)=-9k \quad \Rightarrow \quad 18-2k=-9k$$

$\Rightarrow 7k = -18 \qquad \Rightarrow k = \dfrac{-18}{7}$

So, the maximum value of $\dfrac{\alpha}{\beta}+\dfrac{\beta}{\alpha}$ is – 9, where $k < 0$

45. The two roots are, α, 2α

$$\frac{\text{Product}}{\text{sum}} = \frac{\frac{c}{a}}{\frac{-b}{a}} = \frac{c}{-b}$$

$$\therefore \quad \frac{\alpha \times 2\alpha}{\alpha + 2\alpha} = \frac{2\alpha^2}{3\alpha} = \frac{2}{-(3a-1)} \Rightarrow \alpha = \frac{3}{-(3a-1)}$$

Sum = $\alpha + 2\alpha = 3\alpha$

$$\therefore \quad 3\alpha = \frac{-(3a-1)}{a^2-5a+3}$$

$$\Rightarrow \quad \frac{3\times 3}{-(3a-1)} = \frac{-(3a-1)}{a^2-5a+3} \Rightarrow 9a^2 - 45a + 27 = 9a^2 - 6a + 1$$

$$\Rightarrow \quad 45a - 6a = 27 - 1 \qquad \Rightarrow 39a = 26 \quad \Rightarrow a = \frac{26}{39} = \frac{2}{3}$$

46. $\alpha + \beta = -p$

$\alpha\beta = q$

$\therefore \; \alpha^2 + \beta^2 = (\alpha+\beta)^2 - 2\alpha\beta = (-p)^2 - 2q = p^2 - 2q$

47. $ax^2 + bx + c = 0$

Sum $= \alpha + \beta = \dfrac{-b}{a}$

Product $= \alpha\beta = \dfrac{c}{a}$

Sum $= \alpha + \beta = 4 + 3 = 7$

Product $= \alpha\beta = 3 \times 2 = 6$

Checking through option,
The two roots are (6, 1)

48. $\dfrac{1}{x} = \dfrac{\sqrt{a+b}+\sqrt{a-b}}{\sqrt{a+b}-\sqrt{a-b}}$

Applying componendo and dividendo, we get $\dfrac{1+x}{1-x} = \dfrac{\sqrt{a+b}}{\sqrt{a-b}}$

Squaring both sides, we get $\dfrac{(1+x)^2}{(1-x)^2} = \dfrac{a+b}{a-b}$

$$\Rightarrow \frac{1+x^2+2x}{1+x^2-2x} = \frac{a+b}{a-b}$$

Applying componendo and dividendo again we get $\frac{2(1+x^2)}{2(2x)} = \frac{a}{b}$

$\Rightarrow b + bx^2 = 2ax \quad \Rightarrow bx^2 - 2ax + b = 0$

49. Putting $x = 3$, we get $27 + 3(k - 1) + 9 = 0$ or $k = -11$.

50. $x^4 + 7x^2 + 16 = (x^4 + 8x^2 + 16) - x^2 = (x^2 + 4)^2 - x^2 = (x^2 + 4 + x)(x^2 + 4 - x) = (x^2 + x + 4)(x^2 - x + 4)$.

51. Given Eqn. is $2x - x^2 = x - 2 \Leftrightarrow x^2 - x - 2 = 0 \Leftrightarrow (x + 1)(x - 2) = 0 \Leftrightarrow x = 2$ or $x = -1$.

52. $x^2 - 7x + 10 = 0 \Leftrightarrow (x - 5)(x - 2) = 0 \Leftrightarrow x = 5, 2;$
$x^2 - 10x + 16 = 0 \Leftrightarrow (x - 8)(x - 2) = 0 \Leftrightarrow x = 8, 2$
$\therefore$ Common root is 2.

53. $f(x) = 0 \Leftrightarrow ax^2 + bx + c = 0$.

Sum of roots of $f(x) = 0$ is $-\frac{b}{a}$; Product of roots $= \frac{c}{a}$

Roots of $f(x) = 0$ are real & distinct $\Leftrightarrow b^2 - 4ac > 0$.
Roots of $f(x) = 0$ are real & identical $\Leftrightarrow b^2 - 4ac = 0$.
$\therefore$ Correct matching is $1 - B$, $2 - A$, $3 - D$ & $4 - C$.

54. $D = 36 - 56 = -20 < 0$. So, roots are imaginary.

55. $D = 36 - 24 = 12 > 0$ and 12 is not a perfect square.
So, the roots are real, unequal and irrational.

56. For real roots, we have $D \geq 0$, *i.e.*, $16 - 4k \geq 0$ or $k \leq 4$.

57. For coincident roots, $D = 0$ *i.e.*, $16 - 4k = 0 \Leftrightarrow k = 4$.

58. For equal roots we must have $D = 0$.
i.e. $4(k+2)^2 - 36k = 0 \Leftrightarrow (k+2)^2 - 9k = 0 \Leftrightarrow k^2 - 5k + 4 = 0 \Leftrightarrow (k-4)(k-1) = 0 \Leftrightarrow k = 4$ or $k = 1$.

59. For equal roots, $D = 0$. So, $b^2 - 4ac = 0$.

$\therefore 4ac = b^2$ or $c = \frac{b^2}{4a}$.

60. For real roots, $D \geq 0$ *i.e.*, $16m^2 - 4(4m^2 + m + 1) \geq 0$.
$\therefore -4m - 4 \geq 0$ or $4m \leq -4$ or $m \leq -1$.

61. For real roots, we have: $k^2 - 256 \geq 0$ and $64 - 4k \geq 0$
i.e., $k^2 \geq 256$ and $k \leq 16$ *i.e.*,, $(k \geq 16$ or $k \leq -16)$ & $k \leq 16$
$\therefore$ For both the equations to have real roots, we must have $k = 16$.

62. $10g_{10}(x^2 - 6x + 45) = 2 \Leftrightarrow x^2 - 6x + 45 = 10^2 = 100$
$\Leftrightarrow x^2 - 6x - 55 = 0 \Leftrightarrow 0\ (x - 11)(x + 5) = 0 \Leftrightarrow x = 11$ or $x = -5$.

63. Sum of roots $= 5 + (-3) = 2$, Product of roots $= 1 \times (-3) = -3$.
$\therefore$ Eqn. is $x^2 - 2x - 3 = 0 \Leftrightarrow (x - 3)(x + 1) = 0 \Leftrightarrow x = 3$ or -1.

64. For no real root, we must have $D < 0$.

$\therefore \quad 25p^2 - 64 < 0 \Rightarrow p^2 < \frac{64}{25} \Rightarrow p < \frac{8}{5}$ or $p > -\frac{8}{5} \Rightarrow -\frac{8}{5} < p < \frac{8}{5}$.

65. Sum of roots $= (5 - 2) = 3$, Product of roots $= -10$.
$\therefore$ The required equation is: $x^2 - 3x - 10 = 0$.

66. Sum of roots $= \frac{10}{3}$. So other root $= \left(\frac{10}{3} - \frac{1}{3}\right) = 3$.

67. One root = $3+2\sqrt{3}$, other root = $3-2\sqrt{3}$
Sum of roots = 6 & Product of roots = $(3+2\sqrt{3})(3-2\sqrt{3}) = -3$.
$\therefore$ Required equation is: $x^2 - 6x - 3 = 0$.

68. Sum of roots = $\frac{5}{a}$ and Product of roots = $\frac{c}{a}$.

$\therefore \frac{5}{a} = 10$ and $\frac{c}{a} = 10$ *i.e.,* $a = \frac{1}{2}$ and $c = 5$

69. Sum of roots = $\frac{-2}{k}$ and Product of roots = $\frac{3k}{k} = 3$.

$\therefore \frac{-2}{k} = 3$ or $k = \frac{-2}{3}$.

70. $\alpha+\beta = -\frac{b}{a}$ and $\alpha\beta = \frac{c}{a}$.

$$\therefore \quad \alpha^2+\beta^2 = (\alpha+\beta)^2 - 2\alpha\beta = \frac{b^2}{a^2} - \frac{2c}{a} = \left(\frac{b^2-2ac}{a^2}\right).$$

71. $\alpha+\beta = 8$, $\alpha\beta = p$, $\alpha^2+\beta^2 = 40$.
Now, $\alpha^2+\beta^2 = (\alpha+\beta)^2 - 2\alpha\beta \Leftrightarrow 40 = 64 - 2p \Leftrightarrow p = 12$.

72. $\alpha+\beta = -1$ and $\alpha\beta = 1$.
$\therefore \alpha^2+\beta^2 = (\alpha+\beta)^2 - 2\alpha\beta = (-1)^2 - 2\times 1 = -1$.
$\alpha^4+\beta^4 = (\alpha^2+\beta^2)^2 - 2\alpha^2\alpha^2 = (-1)^2 - 2(1^2) = -1$

73. $\alpha+\beta = \frac{4}{2} = 2$ and $\alpha\beta = \frac{3}{2}$.

$$\therefore (\alpha^3+\beta^3) = (\alpha+\beta)^3 - 3\alpha\beta(\alpha+\beta) = \left(2^3 - 3\times\frac{3}{2}\times 2\right) = (8-9) = -1.$$

74. $\alpha+\beta = 5$ and $\alpha\beta = 6$.

$\therefore \alpha-\beta = \sqrt{(\alpha+\beta)^2 - 4\alpha\beta} = \sqrt{25-24} = \pm 1$.

$\therefore (\alpha^2-\beta^2) = (\alpha+\beta)(\alpha-\beta) = (5\times 1)$ or $[5\times(-1)] = 5$ or -5.

75. $\alpha+\beta = -\frac{b}{a}$ and $\alpha\beta = \frac{c}{a}$

$$\therefore \quad \frac{\alpha}{\beta^2}+\frac{\beta}{\alpha^2} = \frac{\alpha^3+\beta^3}{\alpha^2\beta^2} = \frac{(\alpha+\beta)^3 - 3\alpha\beta(\alpha+\beta)}{(\alpha\beta)^2} = \frac{-\frac{b^3}{a^3} - \frac{3c}{a}\left(-\frac{b}{a}\right)}{\frac{c^2}{a^2}} = \frac{-b^3+3abc}{ac^2}$$

76. Let the roots be α and $(1/\alpha)$. Then, product of roots = 1. $\therefore \frac{k}{5} = 1$. So, $k = 5$.

77. $\alpha+\beta = 6$ and $3\alpha + 2\beta = 20 \Rightarrow \alpha = 4, \beta = 2$.
Product of roots = k. So, $k = (4\times 2) = 8$.

78. $\alpha+\beta = \frac{3}{2}$ and $\alpha\beta = \frac{1}{2}$.

$$\text{Sum of new roots} = \frac{\alpha}{\beta}+\frac{\beta}{\alpha} = \frac{\alpha^2+\beta^2}{\alpha\beta} = \frac{(\alpha+\beta)^2 - 2\alpha\beta}{\alpha\beta} = \frac{\frac{9}{4}-1}{\frac{1}{2}} = \frac{5}{2}.$$

Product of new roots = $\left(\frac{\alpha}{\beta}\times\frac{\beta}{\alpha}\right)=1.$

$\therefore$ Required equation is $x^2-\frac{5}{2}x+1=0$ or $2x^2-5x+2=0.$

79. Putting $x=\frac{1}{y}$ in the given equation, we get:

$\frac{3}{y^2}-\frac{20}{y}+17=0$ or $3-20y+17y^2=0$

So, the required equation is; $17x^2-20x+3=0.$

80. $\alpha+\beta=-k$ and $\alpha\beta=12 \Rightarrow (\alpha-\beta)^2=(\alpha+\beta)^2-4\alpha\beta=k^2-48.$

$\therefore\ k^2-48=1 \Rightarrow k^2=49 \Rightarrow k=\pm 7.$

81. Let the roots be α and $(-1/\alpha)$. Then, product of roots = – 1.

$\therefore\ \frac{c}{a}=-1$ or $a+c=0.$ So, *I* is true.

II is clearly true.

$\alpha+\beta=22,\ \alpha-\beta=8 \Rightarrow \alpha=15,\ \beta=7.$

So, $\alpha+\beta=22$ & $\alpha\beta=105$

$\therefore$ An equation with α, β as roots is: $x^2-22x+105=0.$

So, *III* is incorrect.

Further in *IV* we have: $\alpha+\beta=2$ and $\alpha\beta=\frac{1}{2}$

$$\frac{1}{\alpha+2\beta}+\frac{1}{\beta+2\alpha}=\frac{(\beta+2\alpha)+(\alpha+2\beta)}{(\alpha+2\beta)(\beta+2\alpha)}=\frac{3(\alpha+\beta)}{2(\alpha^2+\beta^2)+5\alpha\beta}$$

$$=\frac{3(\alpha+\beta)}{2[(\alpha+\beta)^2-2\alpha\beta]+5\alpha\beta}=\frac{3(\alpha+\beta)}{2(\alpha+\beta)^2+\alpha\beta}=\frac{3\times 2}{2\times 4+\frac{1}{2}}=6\times\frac{2}{17}=\frac{12}{17}.$$

So, *IV* is correct. Hence, *I, II, IV* are correct.

82. The required condition is $b^2-4ac\geq 0.$

83. Since $D=1-4=-3<0$, so the given expression has no proper factor

84. In $2x^2-5x+9$ we have $D=25-72<0.$

In $2x^2+4x-5$, we have $D=16+40>0.$

In $3x^2+4x+6$, we have $D=16-72<0.$

In $5x^2-3x+2$, we have $D=9-40<0.$

So, $2x^2+4x-5$ is the only expression factorizable into linear factors

85. For factorization into linear factors, we must have: $36-8k\geq 0$ or $k\leq\frac{9}{2}.$

86. *A*: $4x^2+kx+1$ is factorizable when $k^2-16\geq 0.$ *i.e.*, when $k\geq 4$ or $k\leq -4.$

B: kx^2-4x+k is factorizable when $16-4k^2\geq 0.$ *i.e.*, when $k^2\leq 4$ or $-2\leq k\leq 2.$

C: kx^2-2x+2 is factorizable when $4-8k\geq 0$ *i.e.*, when $k\leq\frac{1}{2}.$

D: $2x^2-kx+k$ is factorizable when $k^2-8k\geq 0$

But $k(k-8)\geq 0 \Leftrightarrow (k\geq 0\ \&\ k-8\geq 0)$ or $(k\leq 0,\ k-8\leq 0) \Leftrightarrow x\geq 8$ or $k\leq 0.$

So, the correct matching is: *A B C D*

2 4 1 3

87. Put $x - \frac{1}{x} = y$ so that $x^2 + \frac{1}{x^2} - 2 = y^2$ or $x^2 + \frac{1}{x^2} = y^2 + 2$.

$\therefore$ Given Eqn. is $8(y^2 + 2) - 42y + 29 = 0$ or $8y^2 - 42y + 45 = 0$

or $8y^2 - 12y - 30y + 45 = 0 \Leftrightarrow (2y - 3)(4y - 15) = 0 \Leftrightarrow y = \frac{3}{2}$ or $\frac{15}{4}$.

Now, $x - \frac{1}{x} = \frac{3}{2} \Rightarrow 2x^2 - 3x - 2 = 0 \Leftrightarrow (2x + 1)(x - 2) = 0 \Leftrightarrow x = -\frac{1}{2}$ or 2.

And, $x - \frac{1}{x} = \frac{15}{4} \Leftrightarrow 4x^2 - 15x - 4 = 0 \Leftrightarrow (4x + 1)(x - 4) = 0 \Leftrightarrow x = -\frac{1}{4}$ or 4

So, $x = 4$ satisfies the given equation.

88. Put $x - \frac{1}{x} = y$ so that $x^2 + \frac{1}{x^2} - 2 = y^2$ *i.e.,* $x^2 + \frac{1}{x^2} + 2 = y^2 + 4$

$\therefore\ x - \frac{1}{x} = y$ and $\left(x + \frac{1}{x}\right)^2 = y^2 + 4$.

So, Given Eqn. is $y^2 + 4 - \frac{3}{2}y = 4$ or $y\left(y - \frac{3}{2}\right) = 0$.

$\therefore y = 0$ or $y = \frac{3}{2}$.

Now, $x - \frac{1}{x} = 0 \Rightarrow x^2 - 1 = 0 \Rightarrow x = \pm 1$.

$x - \frac{1}{x} = \frac{3}{2} \Rightarrow 2x^2 - 3x - 2 = 0 \Leftrightarrow (2x + 1)(x - 2) = 0 \Leftrightarrow x = -\frac{1}{2}$ or $x = 2$.

Clearly, $x = -1$ satisfies the given equation.

89. $(x + 1)(x + 4)(x + 2)(x + 3) = 120 \Leftrightarrow (x^2 + 5x + 4)(x^2 + 5x + 6) = 120$

$\Leftrightarrow (y + 4)(y + 6) = 120$, where $x^2 + 5x = y \Leftrightarrow y^2 + 10y - 96 = 0 \Leftrightarrow (y + 16)(y - 6) = 0$

$\Leftrightarrow y = -16$ or $y = 6$

Now, $x^2 + 5x = -16 \Leftrightarrow x^2 + 5x + 16 = 0 \Leftrightarrow x$ is imaginary [as $D < 0$]

$x^2 + 5x = 6 \Leftrightarrow x^2 + 5x - 6 = 0 \Leftrightarrow (x + 6)(x - 1) = 0 \Leftrightarrow x = -6$ or 1.

90. Given equation is: $2^{2x} - 3.2^x \times 2^2 + 32 = 0$

or $2^{2x} - 12 \times 2^x + 32 = 0 \Leftrightarrow y^2 - 12y + 32 = 0$, where $2^x = y$

$\Leftrightarrow (y - 8)(y - 4) = 0 \Leftrightarrow y = 8$ or $y = 4$.

$\therefore$ $[2^x = 8 = 2^3 \Rightarrow x = 3]$ And $[2^x = 4 = 2^2 \Rightarrow x = 2]$.

91. Given Eqn. is $5 \times 5^x + \frac{5^2}{5^x} = 126$ or $5y + \frac{25}{y} = 126$, where $y = 5^x$.

Thus, $5y^2 - 126y + 25 = 0 \Leftrightarrow (y - 25)(5y - 1) = 0 \Leftrightarrow y = 25$ or $y = \frac{1}{5}$.

Now, $5^x = 25 = 5^2 \Rightarrow x = 2$. And, $5^x = \frac{1}{5} = 5^{-1} \Rightarrow x = -1$.

$\therefore$ Solution set is $\{-1, 2\}$.

92. Let the parts be x and $(57 - x)$.

$\therefore$ $x(57 - x) = 782 = x^2 - 57x + 782 = 0 \Leftrightarrow (x - 34)(x - 23) = 0 \Leftrightarrow x = 34$ or $x = 23$.

So, the required parts are 34 and 23.

93. Let the number be x. Then,

$x+\frac{1}{x}=\frac{41}{20} \Leftrightarrow 20x^2-41x+20=0 \Leftrightarrow (4x-5)(5x-4)=0 \Leftrightarrow x=\frac{5}{4}$ or $x=\frac{4}{5}$.

So, $x=\frac{5}{4}$.

94. p, q be the roots of $x^2+mx-n=0$

$\Rightarrow \quad p+q=-m$

$\Rightarrow \quad p=-m-q$ ...(i)

$\Rightarrow \quad pq=-n$ ...(ii)

Similarly, m, n are the roots of $x^2+px-q=0$

$\Rightarrow \quad m+n=-p$ ...(iii)

$\Rightarrow \quad mn=-q$ ...(iv)

Now, put the value of (i) in (iii): $m+n=-p$

$\Rightarrow \quad m+n=-(-m-q)$

$\Rightarrow \quad m+n=m+q \Rightarrow n=q$

From equation (iv): $mn=-q$

$\Rightarrow \quad m(q)=-q \Rightarrow m=-1$

From equation (ii): $pq=-n$

$pq=-q$

$p=-1$

Put the value of $p=-1$ and $m=-1$ in equation (i):

$p=-m-q$

$\Rightarrow \quad -1=-(-1)-q$

$\Rightarrow \quad q=1+1=2$

Since, $n=q=2$

Now, check statements:

Statement (I): $m(m+n)=-1$

$\Rightarrow \quad -1(-1+2)=-1$

$\Rightarrow \quad -1(1)=-1$

$\Rightarrow \quad -1=-1\,(\text{true})$

Statement (II): $p+q=1$

$\Rightarrow \quad -1+2=1$

$\Rightarrow \quad 1=1\,(\text{true})$

Hence, both the options are correct.

95. Semi-perimeter = 41. Let length = x & breadth = $(41 - x)$.

$\therefore$ $x(41 - x) = 400 \Leftrightarrow x^2 - 41x + 400 = 0 \Leftrightarrow (x - 16)(x - 25) = 0 \Leftrightarrow x = 16$ or 25

$\therefore$ Breadth = 16 m.

96. We have,

$$\alpha + \beta = 2$$
$$\alpha\beta = -100$$

$\therefore$ quadratic equation $= x^2 - (\alpha + \beta)\,x + \alpha\beta = 0$

$\Rightarrow x^2 - 2x - 100 = 0$

$\therefore$ (c) is the correct option

97. $a(b - c)(x - b)(x - c) + b(c - a)(x - c)(x - a) + c(a - b)(x - a)(x - b)$

Now, by first term, we get

$\Rightarrow a(b - c)(x - b)(x - c) \Rightarrow (ab - ac)[x^2 - x(b + c) + bc]$

$\Rightarrow (ab - ac)x^2 - x(ab - ac)(b + c) + ((ab - ac)bc$

By ignoring the constant term, we get

$(ab - ac)x^2 - x(ab - ac)(b + c) \Rightarrow (ab - ac)x^2 - x(ab^2 - abc - abc - ac^2)$

$\Rightarrow (ab - ac)x^2 - x(ab^2 - ac^2)$

Now, we have,

$[(ab - ac)x^2 - x(ab^2 - ac^2)] + [(bc - ba)x^2 - x(bc^2 - ba^2)] + [(ca - cb)x^2 - x(ca^2 - cb^2)]$

(1) coefficient of x^2

$x^2(ab - ac + bc - ba + ca - cb) = 0$

(2) coefficient of x

$x(ab^2 - ac^2 + bc^2 - ba^2 + ca^2 - cb^2)$

$\therefore$ Statement (1) is correct.

98. We have

$1 - x - x^n + x^{n+1} = (1 - x) - x^n(1 - x) = (1 - x)(1 - x^n)$

$\therefore$ Statement (2) is correct. $[\because (1 - x^n)$ is a factor]

$\therefore 1 - x^n = (1 - x)(1 + x) \,.\, p$

Where p is remaining factor

$1 - x - xn + x^{n+1} = (1 - x)(1 - x)(1 + x) \,.\, p = (1 - 2x + x^2)(1 + x) \,.\, p$

$\therefore$ Statement (1) is correct.

99. Let the quadratic eqn be $px^2 + qx + r = 0$ and α and β are roots of the eqn

$$\therefore \alpha + \beta = \frac{-q}{p}$$

$$\alpha\beta = \frac{r}{p}$$

Now, according to the question,

$$-\frac{q}{p} = \frac{r}{p}$$

$\Rightarrow -q = r$ or $q = -r$

Hence, infinitely many equations are possible.

100. $\dfrac{1}{x+a+b} = \dfrac{1}{x} + \dfrac{1}{a} + \dfrac{1}{b} \Rightarrow \dfrac{1}{x+a+b} - \dfrac{1}{x} = \dfrac{1}{a} + \dfrac{1}{b} \Rightarrow \dfrac{x-x-a-a}{x(x+a+b)} = \dfrac{a+b}{ab}$

$\Rightarrow (-a-b)\,ab = [x^2 + x\,(a+b)]\,(a+b) \Rightarrow x^2 + (a+b)\,x + ab = 0$

We have,

$\Rightarrow + \Rightarrow = -(a+b)$ [$\Rightarrow \alpha$ and β are roots]

and $a = ab$

Now, roots are α^2 and β^2

$\therefore \quad \alpha^2 + \beta^2 = (\alpha+\beta)^2 - 2 \times \alpha = a^2 + b^2$

$\alpha^2 + \beta^2 = a^2b^2$

$x^2 - (\alpha^2 + \beta^2)x + \alpha^2\,\beta^2 = 0$

$x^2 - (a^2 + b^2)x + a^2\,b^2 = 0$

101. $x = \dfrac{6}{7 - \dfrac{6}{7 - \dfrac{6}{7 - \dfrac{6}{7-x}}}} \Rightarrow x = \dfrac{6}{7-x}$

$\Rightarrow 7x - x^2 = 6 \Rightarrow x^2 - 7x + 6 = 0 \Rightarrow (x-6)(x-1) = 0 \Rightarrow x = 6$ and $x = 1$

Now, $x^2 - 3x + 2 = 6^2 - 3 \times 6 + 2 = 20$

102. Let $\alpha = \dfrac{-4-\sqrt{10}}{2}$ (α and β are roots of a quadratic eqn)

(1) Product of roots $= -\dfrac{3}{2}(3+\sqrt{10}) = \dfrac{-4-\sqrt{10}}{2}\beta = \dfrac{-3}{2}(3+\sqrt{10})$

$\Rightarrow \beta = \dfrac{3(3+\sqrt{10})}{(2+\sqrt{10})}$

(2) $\alpha + \beta = -1 \Rightarrow B = -1 + \dfrac{4+\sqrt{10}}{2} = \dfrac{2+\sqrt{10}}{2}$

$\therefore$ (b) is correct

103. For Statement I: $x^2 - 26x + 133 = 0$

Here: $a = 1, b = -26, c = 133$

Discriminant: $D = b^2 - 4ac$

$\Rightarrow \quad D = (-26)^2 - 4(1)(133) = 676 - 532 = 144$

Roots: $x = \dfrac{26 \pm \sqrt{144}}{2} \Rightarrow x = \dfrac{26 \pm 12}{2} \Rightarrow x = \dfrac{38}{2} = 19$ or $x = \dfrac{14}{2} = 7$

For Statement II: $x^2 - 44x + 475 = 0$

Here: $a = 1, b = -44, c = 475$

Discriminant: $D = b^2 - 4ac = (-44)^2 - 4(1)(475) = 1936 - 1900 = 36$

Roots: $x = \dfrac{44 \pm \sqrt{36}}{2} \Rightarrow x = \dfrac{44 \pm 6}{2}$

$\Rightarrow \quad x = \dfrac{50}{2} = 25$ or $x = \dfrac{38}{2} = 19$

Both statements have a common prime solution: $x = 19$

We have used both statements to come to this conclusion.

104. We have, $x^2 + \alpha x + \beta = 0$, where, α and β are roots of the equ.

$\alpha + \beta = -\alpha$ ... (i) $\alpha\beta = \beta$ ... (ii)

$\Rightarrow \alpha = 1$

Now, put $\alpha = 1$ in eqn (i), we get

$1 + \beta = -1 \Rightarrow \beta = -2$

Now, $\alpha + \beta = 1 + (-2) = 1 - 2 = -1$

Let α and β are roots of the given

105. We have,

$x^2 - k^2x + 30kx - 161x - 64 = 0 \Rightarrow x^2 - x(k^2 + 30k - 161) - 64 = 0$

Now,

$\alpha + \beta = k^2 + 30k - 161 = 0$

$\alpha\beta = -\frac{64}{1} = -64$

Now, $(\alpha - \beta)^2 = (\alpha + \beta)^2 - 4\,\alpha\beta$

$\Rightarrow (\alpha - \beta)^2 = (0)^2 - 4 \times -64 \Rightarrow (\alpha - \beta)^2 = 4 \times 64$

$\Rightarrow \alpha - \beta = \sqrt{4 \times 64} \qquad \Rightarrow \alpha - \beta = \pm 16$

$\therefore \alpha - \beta = 16$

106. Let the roots of the given eqn be α and β.

Now, we have, $\alpha : \beta = c : 1$ and $ax^2 + bx + c$

$\therefore \alpha - \beta = -\frac{b}{a}$ and $\alpha\beta = \frac{c}{a}$

Now, $(\alpha + \beta)^2 = \left(-\frac{b}{a}\right)^2 \Rightarrow \alpha^2 + \beta^2 + 2\alpha\beta = \frac{b^2}{a^2}$

Now, $\frac{(\alpha+\beta)^2}{\alpha\beta} = \frac{b^2}{a^2} \times \frac{a}{c}$

$\Rightarrow \frac{\alpha^2 + \beta^2 + 2\,a\beta}{\alpha\beta} = \frac{b^2}{ac} \Rightarrow \frac{\alpha}{\beta} + \frac{\beta}{\alpha} + 2 = \frac{b^2}{ac} \Rightarrow c + \frac{1}{c} + 2 = \frac{b^2}{ac} \qquad \left[\because \frac{\alpha}{\beta} = c\right]$

By multiplying ac on both sides, we get $ac^2 + a + 2ac = b^2$

$\Rightarrow a\,(c^2 + 1 + 2c) = b^2 \Rightarrow a\,(c + 1)^2 = b^2$

107. We have, $ax^2 + bx + c = 0$

(I) $D > 0 \Rightarrow b^2 - 4ac > 0 \Rightarrow b^2 > 4ac$

$\therefore$ We can not determine sign of a and c.

(II) $\frac{c}{a} > 0$

Thus, roots will be the same sign as well

$\therefore ax^2 + bx + c = 0$ has real roots with same sign

Hence, both statement-I and statement-II are together sufficient to answer the question.

Rational Expressions

CHAPTER

Polynomials: An expression of the form $p(x) = a_0 + a_1x + a_2x^2 + ... + a_nx^n$, where $a_0, a_1, a_2, ..., a_n$ *are real numbers and n is a non-negative integer is called a polynomial of degree n, where* $a_n \neq 0$.

Ex. 1. ***$(7x^3 - 5x^2 + 2x - 1)$ is a polynomial of degree 3.***

Ex. 2. ***$(6x^2 + 3x + \sqrt{x} + 2)$ is not a polynomial, since in a polynomial, every power of x must be a non-negative integer.***

Rational Expression: *An expression of the form* $\frac{p(x)}{q(x)}$, *where p(x) and q(x) are polynomials and q(x) ≠ 0 is called a rational expression.*

Every polynomial is a rational expression but a rational expression need not be a polynomial.

Rational Expression in Lowest Terms: *If p(x) and q(x) are polynomials such that g.c.d.* $\{p(x), q(x)\} = 1$, *then* $\frac{p(x)}{q(x)}$ *is a rational expression in its lowest terms.*

Results on Rational Expressions:

(*i*) Sum (or difference) of two rational expressions is a rational expression.

(*ii*) Product of two rational expressions is a rational expression.

(*iii*) Addition as well as multiplication on rational expressions satisfies the commutative and associative laws.

(*iv*) 0 is the additive identity.

(*v*) Additive inverse of $\frac{p(x)}{q(x)}$ is $\frac{-p(x)}{q(x)}$.

SOLVED EXAMPLES

Ex. 1. ***What should be added to $2x^2 + 3x - 5$ to get $x^2 - x + 1$?***

Sol. Let $(2x^2 + 3x - 5) + p(x) = x^2 - x + 1$.

Then, $p(x) = (x^2 - x + 1) - (2x^2 + 3x - 5) = -x^2 - 4x + 6$.

Ex. 2. ***What must be subtracted from $x^2 - 5x + 8$ to get $2x^2 - x + 1$?***

Sol. Let $(x^2 - 5x + 8) - p(x) = 2x^2 - x + 1$.

Then, $p(x) = (x^2 - 5x + 8) - (2x^2 - x + 1) = -x^2 - 4x + 7$.

Ex. 3. ***What should be subtracted from $\frac{7x}{x^2 + x - 12}$ to get $\frac{4}{x+4}$?***

Sol. Let $\frac{7x}{x^2+x-12} - p(x) = \frac{4}{x+4}$. Then, $p(x) = \frac{7x}{x^2+x-12} - \frac{4}{x+4} = \frac{7x}{(x+4)(x-3)} - \frac{4}{(x+4)}$

$$= \frac{7x - 4(x-3)}{(x+4)(x-3)} = \frac{3x+12}{(x+4)(x-3)} = \frac{3(x+4)}{(x+4)(x-3)} = \frac{3}{x-3}.$$

Ex. 4. ***Find the product:*** $(x^2 + 3x - 4) \times \frac{1}{(x^2 + 2x - 8)}$.

Sol. $(x^2 + 3x - 4) \times \frac{1}{(x^2 + 2x - 8)} = \frac{(x+4)(x-1)}{(x+4)(x-2)} = \frac{x-1}{x-2}$.

Ex. 5. ***Simplify:*** $\left\{\frac{(6x^2 + 7x - 3)}{(2x^2 + x - 3)} \times \frac{x^2 - 5x - 14}{(3x^2 + 5x - 2)}\right\}$.

Sol. Given Exp. $= \frac{(6x^2 + 9x - 2x - 3)}{(2x^2 + 3x - 2x - 3)} \times \frac{(x^2 - 7x + 2x - 14)}{(3x^2 + 6x - x - 2)}$

$$= \frac{3x(2x+3) - (2x+3)}{x(2x+3) - (2x+3)} \times \frac{x(x-7) + 2(x-7)}{3x(x+2) - (x+2)} = \frac{(2x+3)(3x-1)}{(2x+3)(x-1)} \times \frac{(x-7)(x+2)}{(x+2)(3x-1)} = \left(\frac{x-7}{x-1}\right).$$

EXERCISE 23

1. Consider a question and two statement: **[CDS (I) Exam, 2022]**

Is $\frac{x^4 + y^4}{x^2 + y^4}$ always greater then $\frac{x^4 + y^4}{x^2 + y^4} (x \neq 0)$?

Statement-I : $x > y$

Statement-II : $x^2 + y > 2xy$

Which one of the following is correct in respect of the question and the statement?

(*a*) Statement-I alone is required to answer the question
(*b*) Statement-II alone is required to answer the question
(*c*) Both statement-I and statement-II are required to answer the question
(*d*) Neither statement-I nor statement-II is required to answer the question

2. What is the value of x, if $\frac{b + \sqrt{b^2 - 2bx}}{b - \sqrt{b^2 - 2bx}} = a$? **[CDS (I) Exam, 2021]**

(*a*) $\frac{ab}{(a+b)}$ (*b*) $\frac{2ab}{(a+1)}$ (*c*) $\frac{2ab}{(a+1)^2}$ (*d*) $\frac{ab}{(a+b)^2}$

3. The expression $\frac{(x^3 -)(x^2 - 9x + 14)}{(x^2 + x + 1)(x^2 - 8x + 7)}$ simplified to **[CDS (I) Exam, 2021]**

(*a*) $(x - 1)$ (*b*) $(x - 2)$ (*c*) $(x - 7)$ (*d*) $(x + 2)$

4. What should be added to $\frac{1}{(x-2)(x-4)}$ to get $\frac{2x-5}{(x^2 - 5x + 6)(x-4)}$? **[CDS (I) Exam, 2021]**

(*a*) $\frac{1}{(x^2 - 7x + 12)}$ (*b*) $\frac{1}{(x^2 + 7x + 12)}$ (*c*) $\frac{1}{(x^2 - 7x - 12)}$ (*d*) $\frac{1}{(x^2 + 7x - 12)}$

5. Suppose $p(x) = x^4 + a_3x^3 + a_2x^2 + a_1x + a_0$ and $q(x) = x^4 + b_3x^3 + b_2x^2 + b_1x + b_0$ are the polynomials. If $\alpha, \beta, \gamma, \delta$ are zeros of $p(x)$ and $\alpha, \beta, \gamma, \lambda$ are zeros of $q(x)$, then what is $\frac{p(x) - q(x)}{(x-\alpha)(x-\beta)(x-\gamma)}$ equal to ? **[CDS (I) Exam, 2024]**

(*a*) $-\lambda + \delta$ (*b*) $\lambda - \delta$ (*c*) $\lambda + \delta$ (*d*) $-\lambda - \delta$

6. What is $\frac{8x}{1-x^4}-\frac{4x}{x^2+1}+\frac{x+1}{x-1}-\frac{x-1}{x+1}$ **[CDS (I) Exam, 2021]**

(*a*) 0 (*b*) 1 (*c*) 2 (*d*) 4

7. For what integral value of *x* is $\cfrac{12}{7-\cfrac{6}{7-\cfrac{3}{5-x}}}=x$? **[CDS (I) Exam, 2021]**

(*a*) 4 (*b*) 3 (*c*) 2 (*d*) 1

8. If $x(x-1)(x-2)(x-3)+1=k^2$, then which one of the following is a possible expression for *k* ? **[CDS (I) Exam, 2021]**

(*a*) x^2-3x+1 (*b*) x^2-3x-1 (*c*) x^2+3x-1 (*d*) x^2-2x-1

9. What is $\frac{1}{bc(a-b)(a-c)}+\frac{1}{ca(b-c)(b-a)}+\frac{1}{ab(c-a)(c-a)}$ equal to ? **[CDS (I) Exam, 2021]**

(*a*) $a+b+c$ (*b*) 3 (*c*) $ab+bc+ca$ (*d*) 0

10. If $x+\frac{1}{x}=\frac{5}{2}$, then what is $x^4-\frac{1}{x^4}$ equal to ? **[CDS (I) Exam, 2021]**

(*a*) $\frac{195}{16}$ (*b*) $\frac{255}{16}$ (*c*) $\frac{625}{16}$ (*d*) 0

11. If $A+B=\frac{x^2-8}{x+2}$ and $A-B=\frac{-x^2+2x+4}{x+2}$, the what is *B* equal to? **[CDS (II) Exam, 2021]**

(*a*) $\frac{x^2-4}{x^2+4x+4}$ (*b*) $\frac{x^2-4}{x^2-4x+4}$ (*c*) $\frac{2x^2-7x+3}{2x^2-1}$ (*d*) $\frac{2x^2+7x-3}{2x-1}$

12. If 2 is a zero of the polynonmial $p(x)=x^3+3x^2-6x-a$, then what is the sum of the squares of the other zeros of the polynomial ? **[CDS (I) Exam, 2024]**

(*a*) 10 (*b*) 17 (*c*) 21 (*d*) 37

13. The sum of all possible products taken two at a time out of the numbers ± 1, ± 2, ± 3, ± 4, ± 5 is **[CDS (II) Exam, 2020]**

(*a*) 0 (*b*) – 30 (*c*) – 55 (*d*) 22

14. What is the value of $\frac{a^2+ac}{a^2c-c^3}-\frac{a^2-c^2}{a^2c+2ac^2+c^3}-\frac{2c}{a^2-c^2}+\frac{3}{a+c}$? **[CDS (II) Exam, 2020]**

(*a*) 0 (*b*) 1 (*c*) $\frac{ac}{a^2+c^2}$ (*d*) $\frac{6}{a+c}$

15. If $\frac{x}{b+c}=\frac{y}{c+a}=\frac{z}{b-a}$, then which one of the following is correct? **[CDS (II) Exam, 2020]**

(*a*) $x+y+z=0$ (*b*) $x-y-z=0$ (*c*) $x+y-z=0$ (*d*) $z+2y+3z=0$

16. For $x=\frac{4\sqrt{6}}{\sqrt{2}+\sqrt{3}}$, what is the value of $\frac{x+2\sqrt{2}}{x-2\sqrt{2}}+\frac{x+2\sqrt{3}}{x-2\sqrt{3}}$? **[CDS (I) Exam, 2019]**

(*a*) 1 (*b*) $\sqrt{2}$ (*c*) $\sqrt{3}$ (*d*) 2

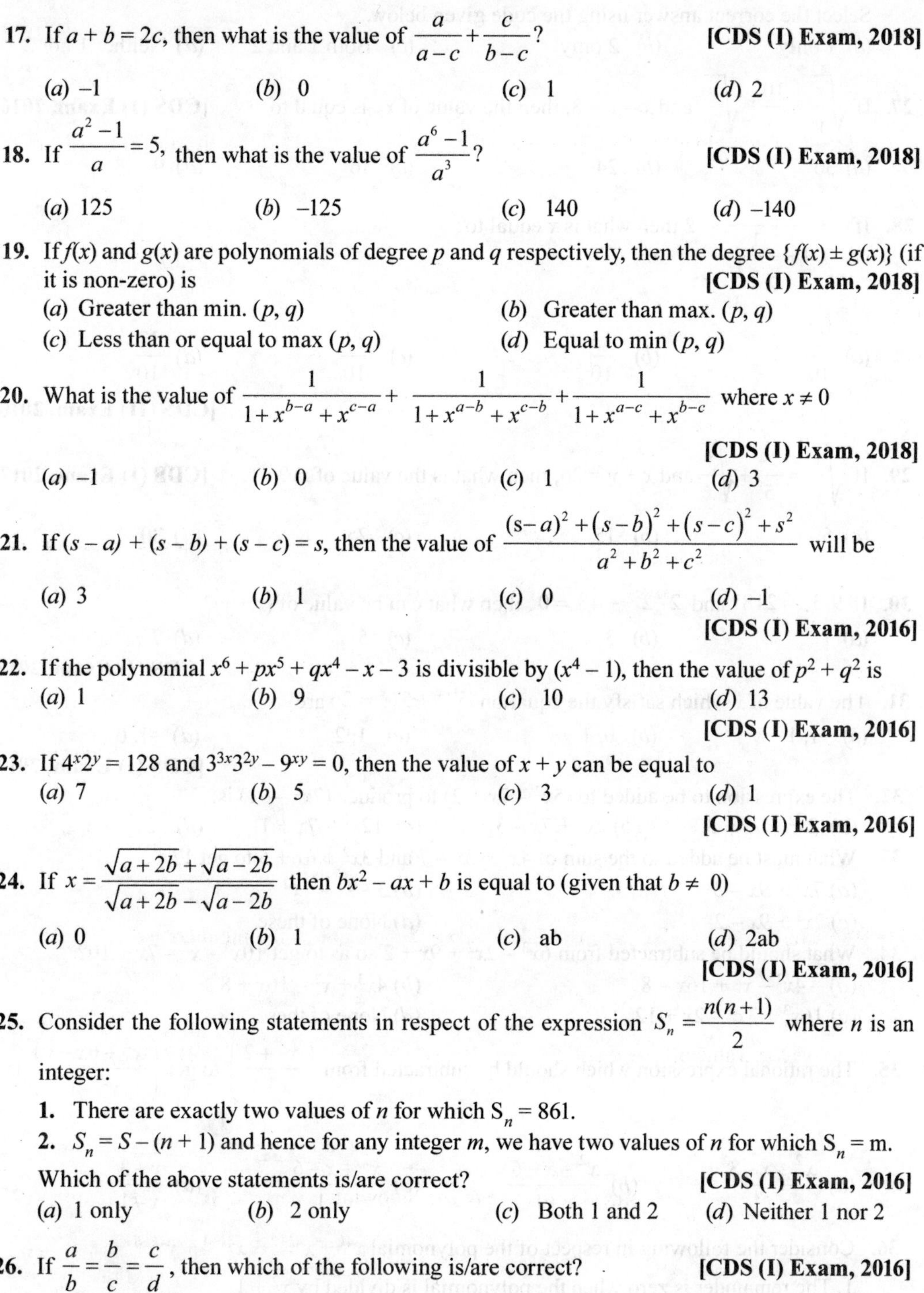

17. If $a + b = 2c$, then what is the value of $\frac{a}{a-c}+\frac{c}{b-c}$? **[CDS (I) Exam, 2018]**

(*a*) –1 (*b*) 0 (*c*) 1 (*d*) 2

18. If $\frac{a^2-1}{a}=5$, then what is the value of $\frac{a^6-1}{a^3}$? **[CDS (I) Exam, 2018]**

(*a*) 125 (*b*) –125 (*c*) 140 (*d*) –140

19. If $f(x)$ and $g(x)$ are polynomials of degree p and q respectively, then the degree $\{f(x) \pm g(x)\}$ (if it is non-zero) is **[CDS (I) Exam, 2018]**

(*a*) Greater than min. (p, q) (*b*) Greater than max. (p, q)

(*c*) Less than or equal to max (p, q) (*d*) Equal to min (p, q)

20. What is the value of $\frac{1}{1+x^{b-a}+x^{c-a}}+\frac{1}{1+x^{a-b}+x^{c-b}}+\frac{1}{1+x^{a-c}+x^{b-c}}$ where $x \neq 0$

[CDS (I) Exam, 2018]

(*a*) –1 (*b*) 0 (*c*) 1 (*d*) 3

21. If $(s-a)+(s-b)+(s-c)=s$, then the value of $\frac{(s-a)^2+(s-b)^2+(s-c)^2+s^2}{a^2+b^2+c^2}$ will be

(*a*) 3 (*b*) 1 (*c*) 0 (*d*) –1

[CDS (I) Exam, 2016]

22. If the polynomial $x^6+px^5+qx^4-x-3$ is divisible by (x^4-1), then the value of p^2+q^2 is

(*a*) 1 (*b*) 9 (*c*) 10 (*d*) 13

[CDS (I) Exam, 2016]

23. If $4^x 2^y = 128$ and $3^{3x}3^{2y} - 9^{xy} = 0$, then the value of $x + y$ can be equal to

(*a*) 7 (*b*) 5 (*c*) 3 (*d*) 1

[CDS (I) Exam, 2016]

24. If $x=\frac{\sqrt{a+2b}+\sqrt{a-2b}}{\sqrt{a+2b}-\sqrt{a-2b}}$ then bx^2-ax+b is equal to (given that $b \neq 0$)

(*a*) 0 (*b*) 1 (*c*) ab (*d*) 2ab

[CDS (I) Exam, 2016]

25. Consider the following statements in respect of the expression $S_n=\frac{n(n+1)}{2}$ where n is an integer:

1. There are exactly two values of n for which $S_n = 861$.

2. $S_n = S-(n+1)$ and hence for any integer m, we have two values of n for which $S_n = m$.

Which of the above statements is/are correct? **[CDS (I) Exam, 2016]**

(*a*) 1 only (*b*) 2 only (*c*) Both 1 and 2 (*d*) Neither 1 nor 2

26. If $\frac{a}{b}=\frac{b}{c}=\frac{c}{d}$, then which of the following is/are correct? **[CDS (I) Exam, 2016]**

1. $\frac{b^3+c^3+d^3}{a^3+b^3+c^3}=\frac{d}{a}$ **2.** $\frac{a^2+b^2+c^2}{b^2+c^2+d^2}=\frac{a}{d}$

Select the correct answer using the code given below.

(*a*) 1 only (*b*) 2 only (*c*) Both 1 and 2 (*d*) Neither 1 nor 2

27. If $\sqrt{\frac{x}{y}} = \frac{10}{3} - \sqrt{\frac{y}{x}}$ and $x - y = 8$, then the value of xy is equal to **[CDS (I) Exam, 2016]**

(*a*) 36 (*b*) 24 (*c*) 16 (*d*) 9

28. If $x + \cfrac{1}{1 + \cfrac{1}{2 + \cfrac{1}{3}}} = 2$ then what is x equal to?

(*a*) $\frac{7}{10}$ (*b*) $\frac{13}{10}$ (*c*) $\frac{11}{10}$ (*d*) $\frac{17}{10}$

[CDS (II) Exam, 2016]

29. If $\sqrt{\frac{x}{y}} = \frac{24}{5} + \sqrt{\frac{y}{x}}$ and $x + y = 26$, then what is the value of xy? **[CDS (I) Exam, 2017]**

(*a*) 5 (*b*) 15 (*c*) 25 (*d*) 30

30. If $9^x 3^y = 2187$ and $2^{3x} 2^{2y} - 4^{xy} = 0$, then what can be value of $(x + y)$?

(*a*) 1 (*b*) 3 (*c*) 5 (*d*) 7

[CDS (I) Exam, 2017]

31. The value of x which satisfy the equation $5^{1+x} + 5^{1-x} = 26$ are

(*a*) –1, 1 (*b*) 0, 1 (*c*) 1, 2 (*d*) –1, 0

[CDS (I) Exam, 2017]

32. The expression to be added to $(5x^2 - 7x + 2)$ to produce $(7x^2 - 1)$ is:

(*a*) $2x^2 + 7x + 3$ (*b*) $2x^2 + 7x - 3$ (*c*) $12x^2 - 7x + 1$ (*d*) $2x^2 - 3$

33. What must be added to the sum of $4x^2 + 3x - 7$ and $3x^2 + 6x + 5$ to get 1?

(*a*) $7x^2 + 9x - 3$ (*b*) $3 - 9x - 7x^2$

(*c*) $7x^2 + 9x - 2$ (*d*) None of these

34. What should be subtracted from $6x^3 - 2x^2 + 9x + 2$ so as to get $10x^3 - x^2 - 7x + 10$?

(*a*) $-4x^3 - x^2 + 16x - 8$ (*b*) $4x^3 + x^2 - 16x + 8$

(*c*) $16x^3 + 3x^2 - 2x + 12$ (*d*) None of these

35. The rational expression which should be subtracted from $\left(\frac{x^2+2}{x-1}\right)$ to get $\left(\frac{x^2+6x-1}{x^2-1}\right)$ is:

(*a*) $\frac{x^2+x-3}{x+1}$ (*b*) $\frac{x^2-x-6}{x+1}$ (*c*) $\frac{x^2+x-6}{x-1}$ (*d*) $\frac{x+3}{x-1}$

36. Consider the following in respect of the polynomial $x^{4k} + x^{4k+2} + x^{4k+4} + x^{4k+6}$.

1. The remainder is zero when the polynomial is divided by $x^2 + 1$.
2. The remainder is zero when the polynomial is divided by $x^4 + 1$.

Which of the statements given above is/are correct? **[CDS (I) Exam, 2024]**

(*a*) 1 only (*b*) 2 only (*c*) Both 1 and 2 (*d*) Neither 1 nor 2

37. $\left(\frac{x^2-5x+6}{x^2-9x+20}\times\frac{x^2-5x+4}{x^2-3x+2}\right)$ in simplified form is:

(*a*) $\frac{x-2}{x-3}$ (*b*) $\frac{x-3}{x-5}$ (*c*) $\frac{x-2}{x-5}$ (*d*) $\frac{x-5}{x-2}$

38. Which of the the following statements are correct?

1. Every polynomial is a rational expression. 2. Every rational expression is a polynomial.
3. Degree of $(x^5 - 7x^2 + 3)$ is 5.

Of these statements the correct ones are:

(*a*) 1 & 2 (*b*) 2 & 3 (*c*) 1 & 3 (*d*) 1, 2 & 3

39. If $x = \frac{a}{a+b}$ and $y = \frac{b}{a-b}$, then $\frac{1}{x}+\frac{1}{y}$ is equal to:

(*a*) $\frac{ab}{a^2+b^2}$ (*b*) $\frac{ab}{a^2-b^2}$ (*c*) $\frac{a^2+b^2}{ab}$ (*d*) $\frac{a^2-b^2}{ab}$

40. $\left(\frac{\sqrt{2}x-\sqrt{8}}{(x^2-4)}\times\frac{\sqrt{8}x+\sqrt{32}}{(x^2+4)}\right)$ on simplification gives:

(*a*) $\frac{4}{x^2+4}$ (*b*) $\frac{\sqrt{2}}{x-2}$ (*c*) $\frac{2\sqrt{2}}{x+2}$ (*d*) $\frac{2}{x^2-4}$

41. $\left(\frac{x^2-7x+12}{x^2-2x-35}\times\frac{x^2-13x+42}{x^2-x-6}\times\frac{x^2+7x+10}{x^2-10x+24}\right)$ in simplified form is:

(*a*) $\frac{x-1}{x-3}$ (*b*) $\frac{x-6}{x+7}$ (*c*) $\frac{x-7}{x+12}$ (*d*) 1

42. The value of $\frac{1-x^4}{1+x}+\frac{1+x^2}{x}\times\frac{1}{x(1-x)}$ is:

(*a*) $\frac{1}{x}$ (*b*) $1 + x$ (*c*) $1 - x^2$ (*d*) 1

43. The value of $\left(\frac{\frac{x}{y}-1}{1-\frac{x}{y}}\right)$ is:

(*a*) 0 (*b*) 1 (*c*) – 1 (*d*) $\frac{x}{y}$

44. $\frac{(x+y)^2-(x-y)^2}{x^2y-xy^2}$ is equal to:

(*a*) $\frac{1}{xy}$ (*b*) $\frac{1}{x-y}$ (*c*) $\frac{2}{x-y}$ (*d*) $\frac{4}{x-y}$

45. The value of $\left(\frac{x-3}{x^2-x-6}+\frac{2x-1}{2x^2+5x-3}-\frac{2x+5}{x^2+5x+6}\right)$ is:

(*a*) 1 (*b*) 0 (*c*) – 1 (*d*) $\frac{x-3}{x-2}$

46. The value of $\left[\frac{x^2-(y-z)^2}{(x+z)^2-y^2}+\frac{y^2-(x-z)^2}{(x+y)^2-z^2}+\frac{z^2-(x-y)^2}{(y+z)^2-x^2}\right]$ is:

(*a*) 0 (*b*) 1 (*c*) – 1 (*d*) None of these

47. The simplified form of $\left(\frac{x^3-27}{x^2+3x+9}\right)$ is:

(*a*) $(x-3)$ (*b*) $\frac{1}{(x-3)}$ (*c*) $\frac{x-3}{x+3}$ (*d*) $\frac{1}{x+3}$

ANSWERS

1. (*d*)	**2.** (*c*)	**3.** (*b*)	**4.** (*a*)	**5.** (*b*)	**6.** (*a*)	**7.** (*c*)	**8.** (*a*)	**9.** (*d*)
10. (*b*)	**11.** (*c*)	**12.** (*b*)	**13.** (*c*)	**14.** (*d*)	**15.** (*b*)	**16.** (*d*)	**17.** (*c*)	**18.** (*c*)
19. (*c*)	**20.** (*c*)	**21.** (*b*)	**22.** (*c*)	**23.** (*b*)	**24.** (*a*)	**25.** (*b*)	**26.** (*a*)	**27.** (*d*)
28. (*b*)	**29.** (*c*)	**30.** (*c*)	**31.** (*a*)	**32.** (*b*)	**33.** (*b*)	**34.** (*a*)	**35.** (*a*)	**36.** (*c*)
37. (*b*)	**38.** (*d*)	**39.** (*c*)	**40.** (*a*)	**41.** (*d*)	**42.** (*d*)	**43.** (*c*)	**44.** (*d*)	**45.** (*b*)
46. (*b*)	**47.** (*a*)							

SOLUTIONS

1. Let $\frac{x^6+y^6}{x^4+y^4}>\frac{x^4+y^4}{x^2+y^2}$

$\Rightarrow x^8+y^8+x^6y^2+y^6x^2>x^8+y^8+2x^4y^4$

$\Rightarrow x^2y^2(x^4+y^4)>2x^4y^4$

$\Rightarrow x^4+y^4>2x^2y^2$

$\Rightarrow x^4+y^4-2x^2y^2>0$

$\Rightarrow (x^2+y^2)^2>0$ ...(*i*)

or $(y^2-x^2)^2>0$

$\therefore$ x is not always greater than y.

now, $\{(x+y)(x-y)\}^2>0$

$\Rightarrow x+y>0$ and $x-y>0$

Thus, statement I and II are not sufficient to answer the question.

2. We have, $\frac{b+\sqrt{b^2-2bx}}{b-\sqrt{b^2-2bx}}=\frac{a}{1}$ by componendo and dividendo rule, we get $\frac{b}{\sqrt{b^2-2ab}}=\frac{a+1}{a-1}$

Squaring both sides, we get, $\frac{b^2}{b^2-2bx}=\frac{a^2+1+2a}{a^2+1-2a}$

Again, by componendo and dividendo rule, we get, $\frac{2b^2-2bx}{2bx}=\frac{2(a^2+1)}{4a}$

$\Rightarrow \frac{b^2-bx}{bx}=\frac{a^2+1}{2a}$

$\Rightarrow 2ab^2-2abx=a^2bx+bx$

$\Rightarrow 2ab^2 - x(a^2b + b + 2ab)$

$\Rightarrow x = \dfrac{2ab^2}{b(a^2+1+2a)} \quad \Rightarrow x = \dfrac{2ab}{(a+1)^2}$

3. We have, $\dfrac{(x^3-1)\,(x^2-9x+14)}{(x^2+x+1)(x^2-8x+7)}$

$\Rightarrow \dfrac{(x-1)\,(x^2+1+x)\,(x^2-9x+14)}{(x^2+x+1)\,(x^2-8x+7)}$ $[\because a^3 - b^3 = (a-b)\,(a^2+ab+b^2)]$

$\Rightarrow \dfrac{(x-1)\,(x-7)\,(x-2)}{(x-7)\,(x-1)} = x-2$

4. Let we add t

$\therefore \dfrac{1}{(x-2)(x-4)} + t = \dfrac{(2x-5)}{(x^2-5x+6)\,(x-4)}$

$\dfrac{1}{(x-2)\,(x-4)} + t = \dfrac{(2x-5)}{(x-3)(x-2)\,(x-4)} \quad \Rightarrow t = \dfrac{2x-5}{(x-3)(x-2)(x-4)} - \dfrac{1}{(x-2)(x-4)}$

$\Rightarrow t = \dfrac{(2x-5)-(x-3)}{(x-3)(x-2)(x-4)} \quad \Rightarrow t = \dfrac{(x-2)}{(x-3)(x-2)\,(x-4)} \quad \Rightarrow t = \dfrac{1}{x^2-7x+12}$

5. We have,

$$p\,(x) = (x-\alpha)\,(x-\beta)\,(x-y)\,(x-\delta)$$
$$q\,(x) = (x-\alpha)\,(x-\beta)\,(x-y)\,(x-\lambda)$$

Now, $\dfrac{p(x)q(x)}{(x-\alpha)\,(x-\beta)\,(x-y)} = (x-\delta) - (x-\lambda) = \lambda - \delta$

6. We have, $\dfrac{8x}{1-x^4} - \dfrac{4x}{x^4+1} + \dfrac{x+1}{x-1} - \dfrac{x-1}{x+1}$

Now, put $x = 0$

$1 - 1 = 0$

7. Check by options,

(a) Put $x = 4$

$$\frac{12}{7-\dfrac{6}{7-\dfrac{3}{5-4}}} = x \quad \Rightarrow \frac{12}{7-\dfrac{6}{7-3}} = x \Rightarrow x = \frac{12}{7-\dfrac{6}{4}} \Rightarrow x = \frac{12\times 4}{22} = \frac{24}{11} \neq 4$$

(b) Put $x = 3$

$$\frac{12}{7-\dfrac{6}{-7\dfrac{3}{2}}} = x \quad \Rightarrow x = \frac{12}{7-\dfrac{6\times 2}{11}} = \frac{12\times 11}{65} \neq 4$$

(c) Put $x = 2$

$$\Rightarrow x = \frac{12}{7-\dfrac{6}{7-\dfrac{3}{3}}} = \frac{12}{7-\dfrac{6}{6}} = \frac{12}{6} = 2$$

8. We have,

$x(x-1)(x-2)(x-3)+1=k^2$

Let $x=4$

$\Rightarrow 4\times 3\times 2\times 1+1=k^2$

$\Rightarrow k^2=25$

$\Rightarrow k=\pm 5$

By options, we get

(a) x^2-3x+1

$x=4$

$\Rightarrow 16-12+1$

$\Rightarrow 17-12=5$

It is possible

(b) $x^2-3x-1=16-12-1=3$ (not possible)

(c) $x^2+3x-1=16+12-1=28-1=27$ (not possible)

(d) $x^2-2x-1=16-8-1=7$ (not possible)

9. $$\frac{1}{bc(a-b)(a-c)}+\frac{1}{ca(b-c)(b-a)}+\frac{1}{ab(c-a)(b-c)}$$

$$=\frac{1}{bc(a-b)(c-a)}-\frac{1}{ca(b-c)(a-b}-\frac{1}{ab(c-a)(b-c)}$$

$$=\frac{-a(b-c)-b(c-a)-c(a-b)}{abc(a-b)(b-c)(c-a)}=\frac{-ab+ac-bc+ab-ac+bc}{abc(a-b)(b-c)(c-a)}=0$$

10. $x+\dfrac{1}{x}=\dfrac{5}{2}$

Squaring both sides, we get

$$x^2+\frac{1}{x^2}+2=\frac{25}{4} \quad\Rightarrow\quad x^2+\frac{1}{x^2}=\frac{25}{4}-2 \Rightarrow x^2+\frac{1}{x^2}=\frac{17}{4}$$

Now, $\left(x^2-\dfrac{1}{x^2}\right)^2=\left(x^2+\dfrac{1}{x^2}\right)^2-4$

$$\Rightarrow \left(x^2-\frac{1}{x^2}\right)^2=\frac{289}{16}-4 \Rightarrow \left(x^2-\frac{1}{x^2}\right)^2=\frac{225}{16} = x^2-\frac{1}{x^2}=\frac{15}{4}$$

Now,

$$\Rightarrow x^4-\frac{1}{x^4}=\left(x^2+\frac{1}{x^2}\right)\left(x^2-\frac{1}{x^2}\right)=\frac{17}{4}\times\frac{15}{4}=\frac{255}{16}$$

11. We have,

$$A+B=\frac{x^2-8}{x+2} \quad ...(i) \qquad A-B=\frac{-x^2+2x+4}{x+2} \quad ...(ii)$$

Subtracting (i) and (ii), we get

$$2B=\frac{x^2-8-(-x^2+2x+4)}{x+2}$$

$$\Rightarrow 2B=\frac{x^2-8+x^2-2x-4}{x+2}$$

$\Rightarrow 2B = \dfrac{2x^2 - 2x - 12}{x+2}$ $\Rightarrow B = \dfrac{x^2 - x - 6}{x+2}$ $\Rightarrow B = \dfrac{x^2 - 3x + 2x - 6}{x+2}$

$\Rightarrow B = \dfrac{(x+2)(x-3)}{x+2} = x - 3$

Now, by options,

(a) $\dfrac{x^2-4}{x^2+4x+4} = \dfrac{(x+2)(x-2)}{(x+2)^2} = \dfrac{x-2}{x+2}$

(b) $\dfrac{(x-2)(x+2)}{(x-2)^2} = \dfrac{x+2}{x-2}$

(c) $\dfrac{2x^2-7x+3}{2x-1} = \dfrac{2x^2-6x-x+3}{(2x-1)} = \dfrac{2x(x-3)-1(x-3)}{(2x-1)} = \dfrac{(2x-1)(x-3)}{(2x-1)} = x-3$

Hence, (c) is the correct option

12. Let 3 zeros of the equation be α, β and y

$\alpha = 2$

Now, $P(2) = (2)^3 + 3 \times (2)^2 - 6 \times 2 - a = 0$

$\Rightarrow \quad 8 + 12 - 12 - a = 0 \quad \Rightarrow a = 8$

$\alpha + \beta + y = \dfrac{-b}{a} = -3 \Rightarrow \beta + y = -5$

$\alpha\beta y = \dfrac{c}{a} = 8 \quad \Rightarrow \beta y = 4$

$\therefore \quad \beta^2 + y^2 = (\beta + y)^2 - 2\beta y = (-5)^2 - 2 \times 4 = 25 - 8 = 17$

13. Let the sum of all possible products taken two at a time $= P$

$\Rightarrow (1 + 2 + 3 + 4 + 5 - 1 - 2 - 3 - 4 - 5)^2$

$= (1)^2 + (2)^2 + (3)^2 + (4)^2 + (5)^2 + (-1)^2 + (-2)^2 + (-3)^2 + (-4^2 + (-5)^2 + 2\text{P}$

$\Rightarrow 0 = 2[1^2 + 2^2 + 3^2 + 4^2 + 5^2] + 2\text{P}$

$\Rightarrow 2P = -2\left[\dfrac{5 \times 6 \times 11}{6}\right]$ $\left[\because \text{sum of square} = \dfrac{n(n+1)(2n+1)}{6}\right]$

$\Rightarrow P = -55$

14. We have,

$\dfrac{a^2+ac}{a^2c-c^3} - \dfrac{a^2-c^2}{a^2c+2ac^2+c^3} - \dfrac{2c}{a^2-c^2} + \dfrac{3}{a+c}$

Let $a = 0$, we get

$\dfrac{-c^2}{c^3} - \dfrac{2c}{-c^2} + \dfrac{3}{c}$

$\dfrac{1}{c} + \dfrac{2}{c} + \dfrac{3}{c} = \dfrac{6}{c}$

now, in options, put $a = 0$

$\dfrac{6}{0+c} = \dfrac{6}{c}$

Hence (d) is correct otion.

15. Let $\frac{x}{b+c} = \frac{y}{c+a} = \frac{z}{b-a} = p$

$\Rightarrow x = bp + cp$

$y = cp + ap$

$z = bp - ap$

Now, from options

(a) $x + y + z = bp + cp + cp + ap + bp - ap \neq 0$

(b) $x - y - z = bp + cp - cp - ap - bp + ap = 0$

(c) $x + y - z = bp + cp + cp + ap - bp + ap \neq 0$

(d) $x + 2y - 3z = bp + cp + 2cp + 2ap + 3bp + 3ap \neq 0$

16. $x = \frac{4\sqrt{6}}{\sqrt{2}+\sqrt{3}}$

$\frac{x}{2\sqrt{2}} = \frac{2\sqrt{3}}{\sqrt{2}+\sqrt{3}}$

By componendo and dividento rule, we get

$\frac{x+2\sqrt{2}}{x-2\sqrt{2}} = \frac{2\sqrt{3}+\sqrt{2}+\sqrt{3}}{2\sqrt{3}-\sqrt{2}-\sqrt{3}} \quad \Rightarrow \quad \frac{x+2\sqrt{2}}{x-2\sqrt{2}} = \frac{3\sqrt{3}+\sqrt{2}}{\sqrt{3}-\sqrt{2}}$ ...(*i*)

Now, $\frac{x}{2\sqrt{3}} = \frac{2\sqrt{2}}{\sqrt{3}-\sqrt{2}}$

By componendo and dividento rule, we get,

$\frac{x+2\sqrt{3}}{x-2\sqrt{3}} = \frac{2\sqrt{2}+\sqrt{3}+\sqrt{2}}{2\sqrt{2}-\sqrt{2}-\sqrt{3}} \quad \Rightarrow \quad \frac{x+2\sqrt{3}}{x-2\sqrt{3}} = \frac{3\sqrt{2}+\sqrt{3}}{\sqrt{2}-\sqrt{3}}$ ...(*ii*)

Adding eqn (i) and (ii), we get

$$\frac{x+2\sqrt{2}}{x-2\sqrt{2}} + \frac{x+2\sqrt{3}}{x-2\sqrt{3}} = \frac{3\sqrt{3}+\sqrt{2}}{\sqrt{3}-\sqrt{2}} + \frac{3\sqrt{2}+\sqrt{3}}{\sqrt{2}-\sqrt{3}}$$

$$= \frac{3\sqrt{3}+\sqrt{2}}{\sqrt{3}-\sqrt{2}} - \frac{3\sqrt{2}+\sqrt{3}}{\sqrt{3}-\sqrt{2}} = \frac{2\sqrt{3}-2\sqrt{2}}{\sqrt{3}-\sqrt{2}} = \frac{2(\sqrt{3}-2)}{\sqrt{3}-\sqrt{2}} = 2$$

17. We have,

$a + b = 2c \Rightarrow a - c = c - b$

Now, $\frac{a}{a-c} + \frac{c}{b-c} = \frac{a}{c-b} + \frac{c}{b-c} = \frac{a}{c-b} - \frac{c}{c-b} = \frac{a-c}{c-b} = \frac{c-b}{c-b} = 1$

18. We have, $a - \frac{1}{a} = 5$

Cubing both sides, we get

$a^3 - \frac{1}{a^3} 3\left(a - \frac{1}{a}\right) = 125$ $\quad [\because (a-b)^3 = a^3 - b^3 - 3ab(a-b)]$

$\Rightarrow a^3 - \frac{1}{a^3} - 3 \times 5 = 125 \Rightarrow a^3 - \frac{1}{a^3} = 125 + 15 = 140$

19. $\because$ We know that degree is the highest power of the given polynomial.

So, we add $f(x)$ and $g(x)$

If $p = q$ and the coefficients are negative to each other.

Then degree will be less than maximum.

20. We have,

$$\frac{1}{1+x^{b-a}+x^{c-a}}+\frac{1}{1+x^{a-b}+x^{c-b}}+\frac{1}{1+x^{a-c}+x^{b-c}}$$

$$=\frac{1}{1+\frac{x^b}{x^b}+\frac{x^c}{x^a}}+\frac{1}{1+\frac{x^a}{x^b}+\frac{x^c}{x^b}}+\frac{1}{1+\frac{x^a}{x^c}+\frac{x^b}{x^c}}$$

$$=\frac{x^a}{x^a+x^b+x^c}+\frac{x^b}{x^a+x^b+x^c}+\frac{x^c}{x^a+x^b+x^c}=\frac{x^a+x^b+x^c}{x^a+x^b+x^c}=1$$

21. $\dfrac{(s-a)^2+(s-b)^2+(s-c)^2+s^2}{a^2+b^2+c^2}$

$$\frac{3s^2-2s(a+b+c)+a^2+b^2+c^2+s^2}{a^2+b^2+c^2}=\frac{4s^2-2s(2s)+a^2+b^2+c^2}{a^2+b^2+c^2}=1 \qquad (\because a+b+c=2s)$$

22. Let $f(x) = x^6 + px^5 + q\,x^4 - x - 3$

Given $f(x)$ is divisible by $x^4 - 1$

$\Rightarrow\ f(\pm 1) = 0 \Rightarrow f(1) \Rightarrow p + q = 4$ ...(1)

$f(-1) \Rightarrow -p + q = 2$ ...(2)

(1) + (2) $\Rightarrow q = 3$ and (1) – (2) $\Rightarrow p = 1$

$\therefore\ p^2 + q^2 = 1^2 + 3^2 = 10$

23. $4^x\, 2^y = 2^7 \Rightarrow 2^{2x}\, 2^y = 2^7$

$\Rightarrow\ 2x + y = 7 \Rightarrow y = 7 - 2x$ ...(1)

$3^{3x+2y} = 3^{2xy}$

$3x + 2y = 2xy$

$3x + 2(7 - 2x) = 2x(7 - 2x)$ from

$\Rightarrow\ x = 2$ or $\dfrac{7}{4}$. When $x = 2, y = 3$ $\quad \therefore x + y = 5$

24. $x=\dfrac{\sqrt{a+2b}+\sqrt{a-2b}}{\sqrt{a+2b}-\sqrt{a-2b}}\times\dfrac{\sqrt{a+2b}+\sqrt{a+2b}}{\sqrt{a+2b}+\sqrt{a+2b}}$

$x=\dfrac{(\sqrt{a+2b}+\sqrt{a-2b})^2}{a+2b-(a-2b)}$; $x=\dfrac{2a-2\sqrt{a^2-4b^2}}{4b}$

$2bx - a = -\sqrt{a^2-4b^2}$; $(2bx - a)^2 = (-\sqrt{a^2-4b^2})^2$

$4b^2x^2 - 4bax + a^2 = a^2 - 4b^2$;; $bx^2 - ax + b = 0$

25. (i) $S_n = \frac{n(n+1)}{2} = 861$; $n^2 + n - 1722 = 0$; $(n+42)(n-41) = 0$

$n = 41$ or -42

n can not be negative

$\therefore$ (1) is not true.

(ii) $S_n = S_{(n-1)} \Rightarrow S_{41} = S_{-(41+1)}$ (True)

$\therefore$ only second statement is true

26. Let $\frac{b}{a} = \frac{c}{b} = \frac{d}{c} = r$ i.e. $b = ar$, $c = ar^2$, $d = ar^3$

$$\therefore \quad \frac{b^2}{a^2} = \frac{c^2}{b^2} = \frac{d^2}{c^2} = \frac{b^2+c^2+d^2}{a^2+b^2+c^2} = r^2 = \frac{c}{a} \text{ and not } \frac{d}{a} \text{ (2) is false.}$$

$$\text{Also } \frac{b^3}{a^3} = \frac{c^3}{b^3} = \frac{d^3}{c^3} = \frac{b^3+c^3+d^3}{a^3+b^3+c^3} = r^3 = \frac{d}{a} \text{ (1) is true.}$$

27. $\sqrt{\frac{x}{y}} + \sqrt{\frac{y}{x}} = \frac{10}{3} = 3 + \frac{1}{3}$; $\therefore \sqrt{\frac{x}{y}} = 3$ and $\sqrt{\frac{y}{x}} = \frac{1}{3}$ or vice versa.

i.e. $x = 9y$; As $x - y = 8$, if follows that $y = 1$ and $x = 9$

$\therefore \; xy = 9$

28. $x + \cfrac{1}{1 + \cfrac{1}{2 + \cfrac{1}{3}}} = x + \cfrac{1}{1 + \cfrac{3}{7}} = x + \frac{7}{10} = 2$; $x = \frac{13}{10}$

29. $\sqrt{\frac{x}{y}} = \frac{24}{5} + \sqrt{\frac{y}{x}}$

$$\Rightarrow \sqrt{\frac{x}{y}} - \sqrt{\frac{y}{x}} = \frac{24}{5} \quad \Rightarrow \quad \frac{x-y}{\sqrt{xy}} = \frac{24}{5}$$

Squaring both sides, we get

$$\frac{x^2+y^2-2xy}{xy} = \frac{576}{25}$$

$$\Rightarrow \frac{(x+y)^2 - 4xy}{xy} = \frac{576}{25} \quad \Rightarrow \quad \frac{(x+y)^2}{xy} - \frac{4xy}{xy} = \frac{576}{25}$$

$$\Rightarrow \frac{(x+y)^2}{xy} - 4 = \frac{576}{25} \quad \Rightarrow \quad \frac{(26)^2}{xy} = \frac{576}{25} + 4 \quad [\because x + y = 26]$$

$$\Rightarrow \frac{(26)^2}{xy} = \frac{676}{25} \quad \Rightarrow xy = 25$$

30. $9^x\ 3^y = 2187$

$\Rightarrow\ 3^{2x}\ 3^y = 2187 \qquad \Rightarrow\ 3^{2x+y} = 2187$

$\Rightarrow\ 3^{2x+y} = 3^7 \qquad \Rightarrow\ 2x + y = 7$ ...(*i*)

Also, $2^{3x}\ 2^{2y} - 4^{xy} = 0$

$\Rightarrow\ 2^{3x}\ 2^{2y} - 2^{2xy} = 0 \qquad \Rightarrow\ 3x + 2y = 2xy$ ...(*ii*)

Put $x = 2$ and $y = 3$ in (*i*), we get $2 \times 2 + 3 = 7$

$\Rightarrow\ 7 = 7$

(LHS) = (RHS)

So, the value of $x + y = 2 + 3 = 5$

31. $5^{1+x} + 5^{1-x} = 26$

$\Rightarrow\ 5^1\ 5^x + 5^1\ 5^{-x} = 26 \qquad \Rightarrow\ 5.5^x + \dfrac{5}{5^x} = 26$

Let $5^x = a \qquad \Rightarrow\ 26 + \dfrac{5}{a} = 26$

$\Rightarrow\ 5a^2 - 26a + 5 = 0 \qquad \Rightarrow\ 5a^2 - 25a - a + 5 = 0$

$\Rightarrow\ 5a(a-5) - 1(a-5) = 0 \qquad \Rightarrow\ a = 5, \dfrac{1}{5}$

Putting Value of a = 5 and $\dfrac{1}{5}$,

$\therefore\ 5^x = a$	$5^x = a$
$5^x = 5$	$5^x = \dfrac{1}{5} = 5^{-1}$
$\Rightarrow x = 1$	$\Rightarrow x = -1$

So, value of x are 1, – 1

32. Let $(5x^2 - 7x + 2) + p(x) = 7x^2 - 1$.
Then, $p(x) = (7x^2 - 1) - (5x^2 - 7x + 2) = (2x^2 + 7x - 3)$.

33. Let $(4x^2 + 3x - 7) + (3x^2 + 6x + 5) + p(x) = 1$.
Then, $(7x^2 + 9x - 2) + p(x) = 1$ or $p(x) = 3 - 9x - 7x^2$.

34. Let $(6x^3 - 2x^2 + 9x + 2) - p(x) = 10x^3 - x^2 - 7x + 10$
Then, $p(x) = (6x^3 - 2x^2 + 9x + 2) - (10x^3 - x^2 - 7x + 10) = -4x^3 - x^2 + 16x - 8$.

35. Let $\dfrac{x^2+2}{x-1} - p(x) = \dfrac{x^2+6x-1}{x^2-1}$. Then,

$$p(x) = \frac{x^2+2}{x-1} - \frac{x^2+6x-1}{x^2-1} = \frac{(x+1)(x^2+2)-(x^2+6x-1)}{(x^2-1)} = \frac{x^3-4x+3}{x^2-1}$$

$$= \frac{(x-1)(x^2+x-3)}{(x^2-1)} = \left(\frac{x^2+x-3}{x+1}\right).$$

36. We have,

$x^{4k}(x^1 + x^2 + x^4 + x^6)$

$\Rightarrow \quad x^{4k}(1\,(1 + x^2) + x^4(1 + x^2)] \Rightarrow x^{4k} \times (1 + x^4)(1 + x^2)$

$\therefore$ both (1) and (2) are true

37. Given Exp. $= \dfrac{(x-3)(x-2)}{(x-5)(x-4)} \times \dfrac{(x-4)(x-1)}{(x-2)(x-1)} = \dfrac{x-3}{x-5}$.

38. Clearly, each one of the given statements is true.

39. $\dfrac{1}{x} + \dfrac{1}{y} = \dfrac{a+b}{a} + \dfrac{a-b}{b} = \dfrac{b(a+b)+a(a-b)}{ab} = \dfrac{a^2+b^2}{ab}$.

40. Given Exp. $= \dfrac{\sqrt{2}x - 2\sqrt{2}}{(x-2)(x+2)} \times \dfrac{2\sqrt{2}x + 4\sqrt{2}}{x^2+4} = \dfrac{\sqrt{2}(x-2)}{(x-2)(x+2)} \times \dfrac{2\sqrt{2}(x+2)}{(x^2+4)} = \dfrac{4}{(x^2+4)}$.

41. Given Exp. $= \dfrac{(x-3)(x-4)}{(x-7)(x+5)} \times \dfrac{(x-7)(x-6)}{(x-3)(x+2)} \times \dfrac{(x+5)(x+2)}{(x-4)(x-6)} = 1.$

42. Given Exp. $= \dfrac{(1-x)(1+x)(1+x^2)}{(1+x)} \times \dfrac{x}{(1+x^2)} \times \dfrac{1}{x(1-x)} = 1.$

43. Given Exp. $= \dfrac{x-y}{y} \times \dfrac{y}{y-x} = \dfrac{(x-y)}{-(x-y)} = -1.$

44. Given Exp. $= \dfrac{4xy}{xy(x-y)} = \dfrac{4}{x-y}$.

45. Given Exp. $= \dfrac{(x-3)}{(x-3)(x+2)} + \dfrac{(2x-1)}{(2x-1)(x+3)} - \dfrac{(2x+5)}{x^2+5x+6}$

$$= \left(\frac{1}{x+2} + \frac{1}{x+3}\right) - \frac{(2x+5)}{x^2+5x+6} = \frac{2x+5}{x^2+5x+6} - \frac{2x+5}{x^2+5x+6} = 0.$$

46. Given Exp. $= \dfrac{(x+y-z)(x-y+z)}{(x+z+y)(x+z-y)} + \dfrac{(y+x-z)(y-x+z)}{(x+y+z)(x+y-z)} + \dfrac{(z+x-y)(z-x+y)}{(y+z+x)(y+z-x)}$

$$= \frac{x+y-z}{x+y+z} + \frac{y+z-x}{x+y+z} + \frac{z+x-y}{x+y+z} = \left(\frac{x+y+z}{x+y+z}\right) = 1.$$

47. Given Exp. $= \dfrac{(x-3)(x^2+3x+9)}{(x^2+3x+9)} = (x-3).$

Surds and Indices

CHAPTER

Laws of Indices:

(i) $a^m \times a^n = a^{m+n}$ (ii) $\frac{a^m}{a^n} = a^{m-n}$ (iii) $(a^m)^n = a^{mn}$

(iv) $(ab)^n = a^n b^n$ & $\left(\frac{a}{b}\right)^n = \frac{a^n}{b^n}$ (v) $a^0 = 1$

Laws of Surds:

We write $\sqrt[n]{a} = a^{1/n}$.

(i) $(\sqrt[n]{a})^n = (a^{1/n})^n = a$ (ii) $\sqrt[n]{ab} = \sqrt[n]{a} \cdot \sqrt[n]{b}$. (iii) $\sqrt[m]{\sqrt[n]{a}} = \sqrt[mn]{a}$

(iv) $(\sqrt[n]{a})^m = \sqrt[n]{a^m}$ (v) $\sqrt[n]{\frac{a}{b}} = \frac{\sqrt[n]{a}}{\sqrt[n]{b}}$

SOLVED EXAMPLES

Ex. 1. ***Simplify:*** $\left(\frac{\mathbf{2^{1/2} \times 3^{1/3} \times 4^{1/4}}}{\mathbf{10^{-1/5} \times 5^{3/5}}} + \frac{\mathbf{3^{4/3} \times 5^{-7/5}}}{\mathbf{4^{-3/5} \times 6}}\right)$.

Sol. Given Exp. $= \left(\frac{2^{1/2} \times 3^{1/3} \times 4^{1/4}}{(2\times 5)^{-1/5} \times 5^{3/5}} \times \frac{4^{-3/5} \times 6}{3^{4/3} \times 5^{-7/5}}\right) = \frac{2^{1/2} \times 3^{1/3} \times (2^2)^{1/4} \times (2^2)^{-3/5} \times 2 \times 3}{2^{-1/5} \times 5^{-1/5} \times 5^{3/5} \times 3^{4/3} \times 5^{-7/5}}$

$$= \frac{2^{1/2} \times 3^{1/3} \times 2^{1/2} \times 2^{-6/5} \times 2 \times 3}{2^{-1/5} \times 5^{-1/5} \times 5^{3/5} \times 3^{4/3} \times 5^{-7/5}} = \frac{2\left(\frac{1}{2}+\frac{1}{2}-\frac{6}{5}+1\right) \cdot 3\left(\frac{1}{3}+1\right)}{2^{\;1/5} \cdot 3^{4/3} \cdot 5\left(-\frac{1}{5}+\frac{3}{5}-\frac{7}{5}\right)}$$

$$= \frac{2\left(\frac{4}{5}+\frac{1}{5}\right)}{5^{-1}} = (2 \times 5) = 10.$$

Ex. 2. ***If $3^x = 5^y = (75)^z$, show that $z = \frac{xy}{(2x+y)}$.***

Sol. Let $3^x = 5^y = (75)^z = k$. Then, $3 = k^{1/x}$, $5 = k^{1/y}$ & $75 = k^{1/z}$

Now, $75 = 3 \times 5^2 \Leftrightarrow k^{1/z} = k^{1/x} \,.\, k^{2/y} \Leftrightarrow k^{1/z} = k\left(\frac{1}{x}+\frac{2}{y}\right)$

$\therefore \quad \frac{1}{z} = \frac{1}{x} + \frac{2}{y} = \frac{2x+y}{xy}$. So, $z = \frac{xy}{(2x+y)}$.

Ex. 3. ***Simplify:*** $\left(\dfrac{x^b}{x^c}\right)^a \times \left(\dfrac{x^c}{x^a}\right)^b \times \left(\dfrac{x^a}{x^b}\right)^c$.

Sol. Given Exp. = $\left(x^{(b-c)}\right)^a \times \left(x^{(c-a)}\right)^b \times \left(x^{(a-b)}\right)^c = x^{a(b-c)+b(c-a)+c(a-b)} = x^0 = 1.$

Ex. 4. ***Which is larger $\sqrt[3]{3}$ or $\sqrt[4]{4}$?***

Sol. Given surds are of order 3 & 4, whose l.c.m. is 12.

We change each one into a surd of order 12.

$$\therefore \quad \sqrt[3]{3} = 3^{1/3} = 3^{\left(\frac{1}{3}\times\frac{4}{4}\right)} = 3^{4/12} = (3^4)^{1/12} = (81)^{1/12}$$

And, $$\sqrt[4]{4} = 4^{1/4} = 4^{\left(\frac{1}{4}\times\frac{3}{3}\right)} = 4^{3/12} = (4^3)^{1/12} = (64)^{1/12}$$

Clearly, $(81)^{1/2} > (64)^{1/12}$. So, $\sqrt[3]{3} > \sqrt[4]{4}$.

EXERCISE 24

1. If $x = 7 + 4\sqrt{3}$, then what is the value of $\sqrt{x} + \dfrac{1}{\sqrt{x}}$? **[CDS (I) Exam, 2022]**

   (*a*) 1 (*b*) 2 (*c*) 3 (*d*) 4

2. If $x = 9999$, then what is the value of the following? **[CDS (I) Exam, 2022]**

   $\dfrac{4x^3 - x}{(2x+1)(6x-3)}$

   (*a*) 1111 (*b*) 2222 (*c*) 3333 (*d*) 6666

3. If $9x + \sqrt{1+x^2}\,(y + \sqrt{1+y^2}) = 1$, x and y are real numbers, then what is the value of $(x + y)^2$? **[CDS (I) Exam, 2022]**

   (*a*) 0 (*b*) 1 (*c*) 4 (*d*) 9

4. If $\dfrac{x-y}{x\sqrt{y}+y\sqrt{x}} = \dfrac{1}{\sqrt{x}}$; $(x > 0, y > 0)$ then what is the value of $\dfrac{x}{y}$? **[CDS (I) Exam, 2022]**

   (*a*) 1 (*b*) 2 (*c*) 4 (*d*) 8

5. If $a + b = 2$, $\dfrac{1}{a} + \dfrac{1}{b} = 2$ then what is the value of $x^3 + b^3$? **[CDS (I) Exam, 2022]**

   (*a*) 2 (*b*) 4 (*c*) 6 (*d*) 8

6. If $p = \sqrt[3]{\left(a+\sqrt{a^2+b^3}\right)} + \sqrt[3]{\left(a-\sqrt{a^2+b^3}\right)}$, then what is $p^3 + 3bp$ equal to ? **[CDS (I) Exam, 2024]**

   (*a*) $-2a$ (*b*) a (*c*) $2a$ (*d*) $3a$

7. Which one of the following is the largest divisor of $x^3 + 3^{x+1} + 3^{x+2}$, if x is any natural number? **[CDS (I) Exam, 2020]**

   (*a*) 3 (*b*) 13 (*c*) 39 (*d*) 117

8. If $5^{x+1} - 5^{x-1} = 600$, then what is the value of 10^{2x}? **[CDS (I) Exam, 2020]**

   (*a*) 1 (*b*) 1000 (*c*) 100000 (*d*) 1000000

9. If m and n are positive integers such that $m^n = 1331$, then what is the value of $(m-1)^{n-1}$ **[CDS (I) Exam, 2020]**

(*a*) 1 (*b*) 100 (*c*) 121 (*d*) 125

10. What is $\frac{1}{a^{m-n}-1}+\frac{1}{a^{n-m}-1}$ equal to ? **[CDS (I) Exam, 2020]**

(*a*) 1 (*b*) –1 (*c*) 0 (*d*) $2a^{m-n}$

11. If $x=\sqrt{2}, y=\sqrt[3]{3}$ and $z=\sqrt[6]{6}$, then which one of the following is correct? **[CDS (I) Exam, 2020]**

(*a*) $y<x<z$ (*b*) $z<x<y$ (*c*) $z<y<x$ (*d*) $x<y<z$

12. If $\left(x^8+\frac{1}{x^8}\right)=47$, what is the value of $\left(x^6+\frac{1}{x^6}\right)$? **[CDS (I) Exam, 2020]**

(*a*) 36 (*b*) 27 (*c*) 18 (*d*) 9

13. If $x^m = \sqrt[14]{x\sqrt{x\sqrt{x}}}$, then what is the value of m ? **[CDS (II) Exam, 2020]**

(*a*) $\frac{1}{8}$ (*b*) $\frac{1}{4}$ (*c*) $\frac{3}{4}$ (*d*) $\frac{7}{4}$

14. What is the square root of $4x^4 + 8x^3 - 4x + 1$? **[CDS (II) Exam, 2020]**

(*a*) $2x^2 - 2x - 1$ (*b*) $2x^2 - x - 1$ (*c*) $2x^2 + 2x + 1$ (*d*) $2x^2 + 2x - 1$

15. If $3^x = 4^y = 12^z$, then z is equal to **[CDS (I) Exam, 2019]**

(*a*) xy (*b*) $x+y$ (*c*) $\frac{xy}{x+y}$ (*d*) $4x + 3y$

16. If $x=y^{1/a}, y=z^{1/b}$ and $z = x^{1/c}$ where $x \neq 1, y \neq 1, z \neq 1$, then what is the value of abc? **[CDS (I) Exam, 2018]**

(*a*) –1 (*b*) 1 (*c*) 0 (*d*) 3

17. If $2b = a + c$ and $y^2 = xz$, then what is $x^{b-c}\, y^{c-a}\, z^{a-b}$ equal to? **[CDS (I) Exam, 2018]**

(*a*) 3 (*b*) 2 (*c*) 1 (*d*) –1

18. If $x = 2 + 2^{2/3} + 2^{1/2}$, then the value of the expression $x^3 - 6x^2 + 6x$ will be **[CDS (II) Exam, 2018]**

(*a*) 2 (*b*) 1 (*c*) 0 (*d*) –2

19. If $a^x + b^y = c^z$ and $abc = 1$, then the value of $\frac{1}{x}+\frac{1}{y}+\frac{1}{z}$ will be equal to **[CDS (II) Exam, 2018]**

(*a*) –1 (*b*) 0 (*c*) 1 (*d*) 3

20. If $a = xy^{p-1}, b = yz^{q-1}$, $c = cx^{r-1}$ then $a^{q-r}\, b^{r-p}\, c^{p-q}$ is equal to **[CDS (I) Exam, 2018]**

(*a*) abc (*b*) xyz (*c*) 0 (*d*) None of the above

21. What is $\frac{5+\sqrt{10}}{5\sqrt{5}-2\sqrt{20}-\sqrt{32}+\sqrt{50}}$ equal to? **[CDS (I) Exam, 2015]**

(*a*) 5 (*b*) $5\sqrt{2}$ (*c*) $5\sqrt{5}$ (*d*) $\sqrt{5}$

22. What are the possible solutions for x of the equation $x^{\sqrt{x}}, \sqrt[1]{x^x}, \sqrt[2]{x^x}$ where x and n are positive integers? **[CDS (I) Exam, 2015]**

(*a*) 0, n (*b*) 1, n (*c*) n, n^2 (*d*) 1, n^2

23. What is $\sqrt{4+\sqrt{4-\sqrt{4+\sqrt{4-\ldots\ldots\ldots}}}}$ equal to? **[CDS (II) Exam, 2015]**

(*a*) 3 (*b*) $\frac{\sqrt{13}-1}{2}$ (*c*) $\frac{\sqrt{13}+1}{2}$ (*d*) 0

24. If $x=\sqrt{3}+\sqrt{2}$, then the value of $x^3+x+\frac{1}{x}+\frac{1}{x^3}$ is **[CDS (II) Exam, 2015]**

(*a*) $10\sqrt{3}$ (*b*) $20\sqrt{3}$ (*c*) $10\sqrt{2}$ (*d*) $20\sqrt{2}$

25. If $x=\frac{91}{216}$, then the value of $3-\frac{1}{(1-x)^{1/3}}$ is **[CDS (II) Exam, 2015]**

(*a*) $\frac{9}{5}$ (*b*) $\frac{5}{9}$ (*c*) $\frac{4}{9}$ (*d*) $\frac{4}{5}$

26. $7^{10}-5^{10}$ is divisible by **[CDS (I) Exam, 2016]**

(*a*) 5 (*b*) 7 (*c*) 10 (*d*) 11

27. The value of the expression $\frac{(243+647)^2+(243-647)^2}{243\times 243\times 647\times 647}$ is equal to **[CDS (I) Exam, 2016]**

(*a*) 0 (*b*) 1 (*c*) 2 (*d*) 3

28. What is $\frac{6^2+7^2+8^2+9^2+10^2}{\sqrt{7+4\sqrt{3}}-\sqrt{4+2\sqrt{3}}}$ equal to? **[CDS (II) Exam, 2016]**

(*a*) 330 (*b*) 340 (*c*) 355 (*d*) 366

29. What is the remainder when 2^{100} is divided by 101? **[CDS (II) Exam, 2016]**

(*a*) 1 (*b*) 11 (*c*) 99 (*d*) 100

30. What is $\sqrt{1+\frac{1}{1^2}+\frac{1}{2^2}}+\sqrt{1+\frac{1}{2^2}+\frac{1}{3^2}}+\ldots.+\sqrt{1+\frac{1}{2007^2}+\frac{1}{2008^2}}$ equal to?

[CDS (II) Exam, 2016]

(*a*) $2008\frac{1}{2008}$ (*b*) $2007\frac{1}{2007}$ (*c*) $2007-\frac{1}{2008}$ (*d*) $2008-\frac{1}{2009}$

31. $2^{122}+4^{62}+8^{42}+4^{64}+2^{130}$ Divisible by which one of the following integers?

[CDS (II) Exam, 2016]

(*a*) 3 (*b*) 5 (*c*) 7 (*d*) 11

32. If $x=t^{\frac{1}{t-1}}$ and $y=t^{\frac{1}{t-1}}$, $t>0$, $t\neq 1$, then what is the relation between x and y?

[CDS (I) Exam, 2017]

(*a*) $y^x=x^{1/y}$ (*b*) $x^{1/y}=y^{1/x}$ (*c*) $x^y=y^x$ (*d*) $x^y=y^{1/x}$

33. If $x=2+2^{2/3}+2^{1/3}$, then what is the value of x^3-6x^2+6x? **[CDS (I) Exam, 2017]**

(*a*) 3 (*b*) 2 (*c*) 1 (*d*) 0

34. The value of $\left(\frac{243}{32}\right)^{-4/5}$ is:

(*a*) $\frac{81}{16}$ (*b*) $\frac{16}{81}$ (*c*) $\frac{4}{9}$ (*d*) $\frac{9}{4}$

35. If $\sqrt[3]{32} = 2^x$, then x is equal to:

(*a*) 5 (*b*) 3 (*c*) $\frac{3}{5}$ (*d*) $\frac{5}{3}$

36. If $5^{x+3} = (25)^{3x-4}$, then the value of x is:

(*a*) $\frac{5}{11}$ (*b*) $\frac{11}{5}$ (*c*) $\frac{11}{3}$ (*d*) $\frac{13}{5}$

37. If $3^{4x-2} = 729$, then the value of x is:

(*a*) 1 (*b*) 1.5 (*c*) 2 (*d*) 2.5

38. If $2^{2x-1} = \frac{1}{8^{x-3}}$, then *n* the value of x is:

(*a*) 3 (*b*) 2 (*c*) 0 (*d*) – 2

39. If $2^x - 2^{x-1} = 4$, then the value of x^x is:

(*a*) 27 (*b*) 4 (*c*) 1 (*d*) 256

40. The value of x for which $2^{x+4} - 2^{x+2} = 3$, holds, is:

(*a*) 0 (*b*) –2 (*c*) 2 (*d*) –1

41. The value of $\frac{3^{(12+n)} \times 9^{(2n-7)}}{3^{5n}}$ is:

(*a*) $\frac{1}{3}$ (*b*) $\frac{9}{13}$ (*c*) $\frac{1}{9}$ (*d*) $\frac{2}{3}$

42. If $a = (l - m)$, $b = (m - n)$ and $c = (n - l)$, then the expression $(x^a)^n \cdot (x^b)^l \cdot (x^c)^m$ in the simplified form is:

(*a*) 0 (*b*) x (*c*) x^{-1} (*d*) 1

43. What is the minimum value of $\frac{(a^8 + a^4 + 1)(b^8 + b^4 + 1)}{a^4 b^4}$, where $a > 0$, $b > 0$?

(*a*) 1 (*b*) 4 (*c*) 9 (*d*) 16

[CDS (II) Exam, 2024]

44. The value of $\left(\frac{x^a}{x^b}\right)^{(a+b)} \times \left(\frac{x^b}{x^c}\right)^{(b+c)} \times \left(\frac{x^c}{x^a}\right)^{(c+a)}$ is:

(*a*) 0 (*b*) x^{abc} (*c*) x^{a+b+c} (*d*) 1

45. The expression:

$$\frac{1}{1+x^{(b-a)}+x^{(c-a)}}+\frac{1}{1+x^{(a-b)}+x^{(c-b)}}+\frac{1}{1+x^{(b-c)}+x^{(a-c)}}$$ is equal to:

(*a*) x^{a-b-c} (*b*) 1 (*c*) 0 (*d*) None of these

46. If $2^x = 3^y = 6^{-z}$, then $\left(\frac{1}{x}+\frac{1}{y}+\frac{1}{z}\right)$ is equal to:

(*a*) 0 (*b*) 1 (*c*) $\frac{3}{2}$ (*d*) $-\frac{1}{2}$

47. If $\left(x+\frac{1}{yz}\right)-\left(y+\frac{1}{zx}\right)=\left(y+\frac{1}{zx}\right)-\left(z+\frac{1}{xy}\right)$ and $x+z \neq 2y$, then what is xyz equal to?

(*a*) –3 (*b*) –1 (*c*) 1 (*d*) 3

48. $\left(\dfrac{1}{1+a^{n-m}}+\dfrac{1}{1+a^{m-n}}\right)$ is equal to:

(*a*) 0 (*b*) 1 (*c*) $\dfrac{1}{2}$ (*d*) a^{m+n}

49. If $m = a^x$, $n = a^y$ and $m^y n^x = a^{2/z}$, then the value of xyz is:

(*a*) 1 (*b*) 2 (*c*) 3 (*d*) 4

50. The simplified form of $\dfrac{x^{7/2}\cdot\sqrt{y^3}}{x^{5/2}\cdot\sqrt{y}}$ is:

(*a*) $\dfrac{x^2}{y}$ (*b*) $\dfrac{x^3}{y^2}$ (*c*) $\dfrac{x^6}{y^3}$ (*d*) xy

51. $\dfrac{2^{n+4}-2(2^n)}{2(2^{n+3})}+2^{-3}$ is equal to:

(*a*) 2^{n+1} (*b*) $-2^{n+1}+\dfrac{1}{8}$ (*c*) $\dfrac{9}{8}-2^n$ (*d*) 1

52. If $pqr = 1$, then $\left(\dfrac{1}{1+p+q^{-1}}+\dfrac{1}{1+q+r^{-1}}+\dfrac{1}{1+r+p^{-1}}\right)$ is equal to:

(*a*) 0 (*b*) $\dfrac{1}{pq}$ (*c*) pq (*d*) 1

53. The value of $\left(\dfrac{x^b}{x^c}\right)^{(b+c-a)}\cdot\left(\dfrac{x^c}{x^a}\right)^{(c+a-b)}\cdot\left(\dfrac{x^a}{x^b}\right)^{(a+b-c)}$ is:

(*a*) x^{abc} (*b*) x^{a+b+c} (*c*) $x^{ab+bc+ca}$ (*d*) 1

54. The value of $\left(\dfrac{x^a}{x^b}\right)^{1/ab}\cdot\left(\dfrac{x^b}{x^c}\right)^{1/bc}\cdot\left(\dfrac{x^c}{x^a}\right)^{1/ca}$ is:

(*a*) 1 (*b*) 0 (*c*) $x^{1/abc}$ (*d*) $x^{1/(ab+bc+ca)}$

55. If *x, y, z* are positive real numbers, the value of $\left(\sqrt{x^{-1}y}\cdot\sqrt{y^{-1}\cdot z}\cdot\sqrt{z^{-1}x}\right)$ is:

(*a*) xyz (*b*) $\dfrac{1}{xyz}$ (*c*) 1 (*d*) $\sqrt{xyz}$

56. $\sqrt{2}, \sqrt[3]{4}$ and $\sqrt[4]{6}$ in ascending order are:

(*a*) $\sqrt{2},\sqrt[3]{4},\sqrt[4]{6}$ (*b*) $\sqrt[4]{6},\sqrt[3]{4},\sqrt{2}$ (*c*) $\sqrt{2},\sqrt[4]{6},\sqrt[3]{4}$ (*d*) $\sqrt[4]{6},\sqrt{2},\sqrt[3]{4}$

ANSWERS

1. (*d*)	**2.** (*c*)	**3.** (*a*)	**4.** (*c*)	**5.** (*a*)	**6.** (*a*)	**7.** (*d*)	**8.** (*d*)	**9.** (*b*)
10. (*b*)	**11.** (*b*)	**12.** (*c*)	**13.** (*a*)	**14.** (*d*)	**15.** (*c*)	**16.** (*b*)	**17.** (*c*)	**18.** (*a*)
19. (*b*)	**20.** (*d*)	**21.** (*c*)	**22.** (*d*)	**23.** (*c*)	**24.** (*b*)	**25.** (*a*)	**26.** (*d*)	**27.** (*c*)
28. (*a*)	**29.** (*a*)	**30.** (*a*)	**31.** (*d*)	**32.** (*c*)	**33.** (*b*)	**34.** (*b*)	**35.** (*d*)	**36.** (*b*)

37. (*c*) **38.** (*b*) **39.** (*a*) **40.** (*b*) **41.** (*c*) **42.** (*d*) **43.** (*c*) **44.** (*d*) **45.** (*b*)
46. (*a*) **47.** (*c*) **48.** (*b*) **49.** (*a*) **50.** (*d*) **51.** (*d*) **52.** (*d*) **53.** (*d*) **54.** (*a*)
55. (*c*) **56.** (*c*)

SOLUTIONS

1. $x = 7 + 4\sqrt{3}$

$$\sqrt{x} = \sqrt{(2+\sqrt{3})^2} = 2+\sqrt{3}$$

Now, $\sqrt{x} + \dfrac{1}{\sqrt{x}}$

$$2+\sqrt{3}+\frac{1}{(2+\sqrt{3})} = \sqrt{3}+2+2-\sqrt{3} = 4$$

2. $$\frac{4x^8 - x}{(2x+1)\,(6x-3)} = \frac{x\left[4x^2-1\right]}{(2x+1)\,(6x-3)}$$

$$= \frac{x\left[(2x)^2-1\right]}{(2x+1)\,(6x-3)} = \frac{x[(2x+1)(2x-1)]}{(2x+1)\,3\,(2x-1)} = \frac{x}{3} = \frac{9999}{3} = 3333$$

3. We have,

$$\left(x+\sqrt{1+x^3}\right)\left(y+\sqrt{1+y^2}\right) = 1$$

Let $x + \sqrt{1+x^2} = p$ ...(i) and $y + \sqrt{1+y^2} = \dfrac{1}{p}$ ...(ii)

$\sqrt{1+x^2} = p - x$

Squaring both sides, we get

$$1 + x^2 = p^2 + x^2 - 2px$$
$$x = \frac{p^2-1}{2p}$$

Now, from eqn (ii),

$$\sqrt{1+y^2} = \frac{1}{p} - y$$

Squaring both sides, we get

$$1 + y^2 = y^2 + \frac{1}{p^2} - \frac{2y}{p} \Rightarrow \frac{2y}{p} = \frac{1}{p^2} - 1 \Rightarrow \frac{2y}{p} = \frac{1-p^2}{p^2} \Rightarrow y = \frac{1-p^2}{2p}$$

Now, $(x + y)^2 = \left(\dfrac{p^2-1}{2p} + \dfrac{1-p^2}{2p}\right)^2 = \dfrac{p^2-1+1-p^2}{2p} = 0$

4. We have,

$$\frac{x-y}{x\sqrt{y}+y\sqrt{x}} = \frac{1}{\sqrt{x}} \Rightarrow \frac{x-y}{\sqrt{x}\sqrt{y}\left(\sqrt{x}+\sqrt{y}\right)} = \frac{1}{\sqrt{x}} \Rightarrow \frac{x-y}{\sqrt{y}(\sqrt{x}+\sqrt{y})} = 1$$

$$\Rightarrow \frac{(\sqrt{x}+\sqrt{y})(\sqrt{x}-\sqrt{y})}{\sqrt{y}(\sqrt{x}+\sqrt{y})} = 1 \qquad [\because a^2 - b^2 = (a+b)(a-b)]$$

$$\Rightarrow \sqrt{x}-\sqrt{y} = \sqrt{y} \quad \Rightarrow \sqrt{x} = 2\sqrt{y} \;\Rightarrow \frac{\sqrt{x}}{\sqrt{y}} = 2 \Rightarrow \frac{x}{y} = 4$$

5. We have, $a + b = 2$ ...(i)

$$\frac{1}{a}+\frac{1}{b} = 2 \quad \Rightarrow \frac{a+b}{ab} = 2$$

$\Rightarrow ab = 1$ $[\because a + b = 2]$

Now, $\because (a + b)^3 = a^3 + b^3 + 3ab\,(a + b)$

$$(2)^3 = a^3 + b^3 \times 2 \quad \Rightarrow \quad a^3 + b^3 = 8 - 6 \Rightarrow a^3 + b^3 = 2$$

6. Let $b = 0$

$$p = \sqrt[3]{a+\sqrt{a^2}} + \sqrt[3]{a-\sqrt{a^2}}$$

$$p = \sqrt[3]{29} \quad \Rightarrow p^3 + 3bp = (\sqrt[3]{29})^3 + 0 = 2a$$

7. We have, $3x + 3^{x+1} + 3^{x+2} = 3^x\,(1 + 3 + 9) = 3^x\,(13)$

Put $x = 1$, we get

$$x^3\,(13) = 3 \times 13 = 39 \quad [\because x \in n]$$

Now, put $x = 2$, we get $x^3\,(13) = 3^2 \times 13 = 9 \times 13 = 117$

Thus, 117 is the largest divisor.

8. We have, $5^{x+1} - 5^{x-1} = 600$

$$\Rightarrow 5^x.5 - \frac{5^x}{5} = 600 \quad \Rightarrow 5^x\left(5-\frac{1}{5}\right) = 600$$

$$\Rightarrow 5^x = \left(\frac{24}{5}\right) = 600 \Rightarrow 5^x = 125 \Rightarrow 5^x = 5^3 \Rightarrow x = 3$$

Now, $10^{2x} = 10^{2\times 3} = 10^6 = 1000000$

9. We have,

$m^n = 1331$

$11^3 = 1331$

Here $m = 11$ and $n = 3$

Now, $(m - 1)^{n-1} = (11 - 1)^{3-1} = 10^2 = 100$

10. $\dfrac{1}{a^{m-n}-1} + \dfrac{1}{a^{n-m}-1}$

$$= \frac{1}{\frac{a^m}{a^n}-1} + \frac{1}{\frac{a^n}{a^m}-1} = \frac{a^n}{a^m-a^n} + \frac{a^m}{a^n-a^m} = \frac{a^n}{a^m-a^n} - \frac{a^m}{a^m-a^n} = \frac{a^n-a^m}{a^m-a^n} = \frac{-(a^m-a^n)}{a^m-a^n} = -1$$

11. We have, $x = \sqrt{2},\ y = \sqrt[3]{3}$ and $z = \sqrt[6]{6}$

LCM of 2, 3, and 6 = 6

$$x = \sqrt{2} = (2^3)^{\frac{1}{6}} = (8)^{\frac{1}{6}}$$

$$y = \sqrt[3]{3} = \left(3^2\right)^{\frac{1}{6}} = (9)^{\frac{1}{6}}$$

$$z = (6)^{\frac{1}{6}}$$

Hence, $(9)^{\frac{1}{6}} > (8)^{\frac{1}{6}} > (6)^{\frac{1}{6}} \Rightarrow y > x > z$

12. We have,

$$x^8 + \frac{1}{x^8} = 47$$

Adding 2 on both sides, we get

$$\left(x^4 + \frac{1}{x^4}\right)^2 = 49 \quad \Rightarrow x^4 + \frac{1}{x^4} = 7$$

Again adding 2 on both sides, we get

$$\Rightarrow \qquad \left(x^2 + \frac{1}{x^2}\right)^2 = 9 \Rightarrow x^2 + \frac{1}{x^2} = 3$$

Now, $$\left(x + \frac{1}{x}\right)^2 = x^2 + \frac{1}{x^2} + 2 = 5$$

$$x + \frac{1}{x} = \sqrt{5}$$

Cubing both sides, we get

$$x^3 + \frac{1}{x^3} + 3\left(x + \frac{1}{x}\right) = 5\sqrt{5} \Rightarrow x^3 + \frac{1}{x^3} + 3\sqrt{5} = 5\sqrt{5} \Rightarrow x^3 + \frac{1}{x^3} = 2\sqrt{5}$$

Now, $$x^6 + \frac{1}{x^6} = \left(x^3 + \frac{1}{x^3}\right)\left(x^3 + \frac{1}{x^3}\right) - 2 = 2\sqrt{5} \times 2\sqrt{5} - 2$$

$$= 4 \times 5 - 2 = 20 - 2 = 18$$

13. $x^m = \sqrt[14]{x\sqrt{x\sqrt{x}}} \Rightarrow x^m = \sqrt[14]{x\sqrt{x \times x^{\frac{1}{2}}}} \Rightarrow x^m = \sqrt[14]{x\sqrt{x^{\frac{3}{2}}}} \Rightarrow x^m = \sqrt[14]{x \times x^{\frac{3}{4}}}$

$\Rightarrow \quad x^m = \sqrt[14]{x^{1+\frac{3}{4}}} \Rightarrow x^m = \sqrt[14]{x^{\frac{7}{4}}} \Rightarrow x^m = x^{\frac{7}{4} \times \frac{1}{14}} \quad \Rightarrow \quad x^m = x^{\frac{1}{8}} \Rightarrow x^m = \frac{1}{8}$

14. Let $y = 4x^4 + 8x^3 - 4x + 1$

$\Rightarrow y = 4x^4 + 8x^3 - 4x + 1 + 4x^2 - 4x^2 \Rightarrow y = 4x^4 + 4x^2 + 1 + 8x^3 - 4x - 4x^2$

$\Rightarrow y = (2x^2)^2 + (2x)^2 + (1)^2 + 2 \times 2x^2 \times 2x - 2 \times 2x \times 1 - 2 \times 2x^2 \times 1$

$\Rightarrow y = (2x^2 + 2x - 1)^2$ $\qquad [\because (x + y + z)^2 = x^2 + y^2 + z^2 + 2xy + 2yz + 2zx]$

$\because \sqrt{y} = \sqrt{(2x^2 + 2x - 1)^2} = 2x^2 + 2x - 1$

15. Let $3^x = 4^y = 12^z = k \Rightarrow 3 = k^{\frac{1}{x}}, 4 = k^{\frac{1}{y}}$ and $12 = k^{\frac{1}{z}}$

Now, 3 × 4 = 12

$k^{\frac{1}{x}} \times k^{\frac{1}{y}} = k^{\frac{1}{z}} \Rightarrow k^{\frac{1}{x}+\frac{1}{y}} = k^{\frac{1}{z}} \Rightarrow \frac{1}{x}+\frac{1}{y}=\frac{1}{z} \Rightarrow \frac{x+y}{xy} = \frac{1}{z} \Rightarrow z = \frac{xy}{x+y}$

16. $x^a = \left(y^{\frac{1}{a}}\right)^a = y \Rightarrow (x^a)^b = y^b = \left(z^{\frac{1}{b}}\right)^b = z \Rightarrow (x^{ab})^c = z^c = \left(z^{\frac{1}{c}}\right)^c = x \Rightarrow abc = 1$

17. We have, $2b = a + c \Rightarrow b - c = a - b$

Now, $x^{b-c}\, y^{c-a}\, z^{a-b} = x^{b-c}\, y^{c-a}\, z^{b-c} = (xz)^{b-c}\, y^{c-a}$

$= (y^2)^{b-c}\, y^{c-a} = y^{2b-2c+c-a} = y^{2c-2c} = y^0 = 1$

18. We have,

$x = 2 + 2^{2/3} + 2^{1/3} \Rightarrow x - 2 = 2^{2/3} + 2^{1/3}$

Cubing both sides, we get

$x^3 - 8 - 6x\,(x - 2) = (2^{2/3})^3 + (2^{1/3})^3 + 3.2^{2/3}.2^{1/3}\,(2^{2/3} + 2^{1/3})$

$\Rightarrow\ x^3 - 8 - 6x^2 + 12x = 4 + 2 + 6\,(x - 2)$

$\Rightarrow\ x^3 - 8 - 6x^2 + 12x = 6 + 6x - 12$

$\Rightarrow x^3 - 6x^2 + 6x = 2$

19. Let $a^x = b^y = c^z = p \Rightarrow a = p^{1/x}, b = p^{1/y}, c = p^{1/z}$

$abc = 1$

$(k)^{\frac{1}{x}+\frac{1}{y}+\frac{1}{z}} = p \Rightarrow \frac{1}{x}+\frac{1}{y}+\frac{1}{z} = 0$

20. Let $p : q : r = 1$

Now, $a^{q-r}\, b^{r-p}\, c^{p-q} = a^0\, b^0\, c^0 = 1$

$a = xy^{p-1} = x$

$b = xz^{q-1} = y$

$c = xz^{r-1} = z$

Hence, $a^{q-r}\, b^{r-p}\, c^{p-q} = 1$

21. $\frac{5+\sqrt{10}}{5\sqrt{5} - 2\sqrt{20} - \sqrt{32} + \sqrt{50}} = \frac{5+\sqrt{10}}{5\sqrt{5} - 2(2\sqrt{5}) - 4\sqrt{2} + 5\sqrt{2}}$

$= \frac{5+\sqrt{10}}{\sqrt{5}+\sqrt{2}} = \frac{\sqrt{5}\,(\sqrt{5}+\sqrt{2})}{\sqrt{5}+\sqrt{2}} = \sqrt{5} = \sqrt{5}.$

22. For such questions, we should go by the choices. The values of x, $x^{\sqrt{x}}$ $n\sqrt{x^x}$ for $x = 0, 1, n, n^2$ are tabulated below.

x,	$x^{\sqrt{x}}$	$\sqrt[n]{x^x} = x^{\frac{x}{n}}$
0	0	undefined
1	1	1
n	$n^{\sqrt{n}}$	n
n^2	$(n^2)^n$	$(n^2)^n$

We see that x = 1 and n^2 satisfy the equation.

23. Let the given expression be x.

$\therefore\ x = \sqrt{4+\sqrt{4-x}} \Rightarrow x^2 - 4 = \sqrt{4-x} \Rightarrow (x^2-4)^2 = 4 - x \Rightarrow x + (x^2-4)^2 = 4.$

All the choices are positive. By trial, $x = \dfrac{\sqrt{13}+1}{2}$

24. $x = \sqrt{3}+\sqrt{2} \therefore \dfrac{1}{x} = \dfrac{1}{\sqrt{3}+\sqrt{2}} = \sqrt{3}-\sqrt{2} \quad \therefore x + \dfrac{1}{x} = 2\sqrt{3}$

$x^2 = 5 + 2\sqrt{6}$ and $\dfrac{1}{x^2} = 5 - 2\sqrt{6} \quad \therefore x^2 + \dfrac{1}{x^2} = 10$

$x^3 + \dfrac{1}{x^3} = \left(x + \dfrac{1}{x}\right)\left(x^2 - 1 + \dfrac{1}{x^2}\right) = 2\sqrt{3}(9) = 18\sqrt{3}$

$\therefore$ The given expression $\left(x + \dfrac{1}{x}\right) + \left(x^3 + \dfrac{1}{x^3}\right) = 20\sqrt{3}$

25. $3 - \dfrac{1}{\left(1-\dfrac{91}{216}\right)^{1/3}} \dfrac{1}{\left(1-\dfrac{91}{216}\right)^{1/3}} = 3 = 3 - \dfrac{1}{\left(\dfrac{125}{216}\right)^{1/3}} = 3 - \dfrac{1}{\dfrac{5}{6}} = 3 - \dfrac{6}{5} = \dfrac{9}{5}.$

26. We have to check weather $7^{10} - 5^{10}$ in divisible by 5, 7, 10 and 11. It in not divisible by 5 $(\because 5^{10}$ is but 7^{10} is not$)$

It is not divisible by 7. It is not divisible by 10 ($\because$ it is not divisible by 5)

By formulas theorem $7^{10} = 11p+1$ and $5^{10} = 11q+1$, where pq are integers

$\therefore\ 7^{10} - 5^{10}$ is divisible by 11.

27. $\dfrac{(a+b)^2 + (a-b)^2}{a^2+b^2} = \dfrac{2(a^2+b^2)}{a^2+b^2} = 2$

28. $\sqrt{7+4\sqrt{3}} = \sqrt{7+2\sqrt{12}} = 2+\sqrt{3}$ and $\sqrt{4+2\sqrt{3}} = \sqrt{3}+1$

$\therefore$ The denominator is 1. The given expression is

$6^2 + 7^2 + 8^2 + 9^2 + 10^2 = 36 + 49 + 64 + 81 + 100 = 330$

29. By Formant's theorem, if P is a prime number and a is relatively prime to P, there a^{p-1} leaves a remainder of 1, when divided by P.

$\therefore$ from $\dfrac{2^{100}}{101} = 1$

30. $1^2 + 2^2 + 2^2 = 3^2$

$2^2 + 3^2 + 6^2 = 7^2$

$3^2 + 4^2 + 12^2 = 13^2$ and so on.

$$1 + \frac{1}{n^2} + \frac{1}{(nH)^2} = \frac{n^2(nH)^2 + (nH)^2 + n^2}{n^2(nH)^2}$$

and $\sqrt{1 + \dfrac{1}{n^2} + \dfrac{1}{(nH)^2}} = \dfrac{n^2+nH}{n(nH)} = 1 + \dfrac{1}{n(nH)} = 1 + \dfrac{1}{n} - \dfrac{1}{nH}$

$\therefore\ \sqrt{1 + \dfrac{1}{1^2} + \dfrac{1}{2^2}} = 1 + \dfrac{1}{1} - \dfrac{1}{2} \Rightarrow \sqrt{1 + \dfrac{1}{2^2} + \dfrac{1}{3^2}} = 1 + \dfrac{1}{2} - \dfrac{1}{3}$

$$\sqrt{1+\frac{1}{2007^2}+\frac{1}{2008^2}}=1+\frac{1}{2007}-\frac{1}{2008}$$

The given expression = 2007 + 1

$$-\frac{1}{2008}=2008-\frac{1}{2008}$$

31. All the number are powers of 2.

$2^{122}+4^{62}+8^{42}+4^{64}+2^{130} = 2^{122}+2^{124}+2^{126}+2^{128}+2^{130}$

$= 2^{122}\,[1+4+16+64+256] = 2^{122}\,(341) = 2^{122}\,(11)\,(31)$

Among the number in the option, the given number is divisible by 11.

32. $y=\left(t^{\frac{1}{t-1}}\right)^t$

$$y=x^t \qquad \left[\because x=t^{\frac{1}{t-1}}\right] \qquad \text{... }(i)$$

$$\Rightarrow \frac{y}{x}=\frac{t^{\frac{t}{t-1}}}{t^{\frac{1}{t-1}}}=t^{\frac{t}{t-1}-\frac{1}{t-1}}=t^{\frac{t-1}{t-1}}\Rightarrow \frac{y}{x}=t \qquad \text{... }(ii)$$

Put (*ii*) in (*i*), we get, $y=x^{\frac{y}{x}}=(x^y)^{\frac{1}{x}}\Rightarrow y^x=x^y$ $\left[\begin{array}{l}\because a=b^{\frac{1}{2}}\\ \Rightarrow a^2=b\end{array}\right]$

33. $x=2+2^{\frac{2}{3}}+2^{\frac{1}{3}} \;\Rightarrow\; x-2=2^{\frac{2}{3}}+2^{\frac{1}{3}}$

Cubing both sides

$$x^3-8-3\,(x)\,(2)\,(x-2)=4+2+3\left(2^{\frac{2}{3}}\right)(2)^{\frac{1}{3}}\left(2^{\frac{2}{3}}+2^{\frac{1}{3}}\right)$$

$$\Rightarrow x^3-8-6x\,(x-2)=6+6\,(x-2) \qquad \left[\because 2^{\frac{2}{3}}+2^{\frac{1}{3}}=x-2\right]$$

$$\Rightarrow x^3-8-6x^2+12x=6+6(x-2) \quad \Rightarrow x^3-8-6x^2+12x-6-6x+12=0$$

$$\Rightarrow x^3-6x^2+6x=2$$

34. $\left(\frac{243}{32}\right)^{\frac{-4}{5}}=\left(\frac{32}{243}\right)^{\frac{4}{5}}=\left(\frac{2^5}{3^5}\right)^{\frac{4}{5}}=\left[\left(\frac{2}{3}\right)^5\right]^{\frac{4}{5}}=\left(\frac{2}{3}\right)^{\left(5\times\frac{4}{5}\right)}=\left(\frac{2}{3}\right)^4=\frac{16}{81}$

35. $3\sqrt{32}=2^x \Leftrightarrow (2^5)^{1/3}=2^x \Rightarrow 2^{5/3}=2^x \Leftrightarrow x=\frac{5}{3}.$

36. $5^{x+3}=(25)^{3x-4}=5^{2(3x-4)}=5^{6x-8}\Leftrightarrow x+3=6x-8\Leftrightarrow x=\ x=\frac{11}{5}.$

37. $3^{4x-2}=729=3^6\Leftrightarrow 4x-2=6\Leftrightarrow x=2.$

38. $2^{2x-1}=\frac{1}{8^{x-3}}=8^{3-x}=2^{3(3-x)}=2^{9-3x}\Leftrightarrow 2x-1=9-3x\Leftrightarrow x=2$

39. $2^x-2^{x-1}=4\Leftrightarrow 2^{x-1}(2-1)=2^2\Leftrightarrow 2^{x-1}=2^2\Leftrightarrow x-1=2\Leftrightarrow x=3.$

$\therefore x^x=3^3=27.$

40. $2^{x+4} - 2^{x+2} = 3 \Leftrightarrow 2^x \,.\, 2^4 - 2^x \,.\, 2^2 = 3 \Leftrightarrow 2^x \times (16-4) = 3 \Leftrightarrow 2^x = \dfrac{3}{12} = \dfrac{1}{4} = 2^{-2} \Leftrightarrow x = -2.$

41. $\dfrac{3^{(12+n)} \times 9^{(2n-7)}}{3^{5n}} = \dfrac{3^{(12+n)} \times 3^{2(2n-7)}}{3^{5n}} = 3^{(12n+4n-14-5n)} = 3^{-2} = \dfrac{1}{3^2} = \dfrac{1}{9}.$

42. Given Exp. $= x^{(l-m)\,n} \cdot x^{(m-n)l} \cdot x^{(n-l)m} = x^{\ln - mn + nl + nm - lm} = x^0 = 1.$

43. Let $x = a^4$ and $y = b$, so the expression becomes:

$$\Rightarrow \frac{(x^2+z+1)\,(y^2+y+1)}{xy}$$

Apply AM-GM on $(x^2, x, 1)$: $(x^2 + x + 1 \geq 3\sqrt[3]{x^2 \cdot x \cdot 1} = 3\sqrt[3]{x^3} = 3x.$

Equality holds when $x = 1$.

Similarly, $(y^2 + y + 1 \geq 3y)$.

$$\Rightarrow \frac{(x^2+x+1)\,(y^2+y+1)}{xy} \geq \frac{(3x)\,(3y)}{xy}. \quad \Rightarrow \frac{(x^2+x+1)\,(y^2+y+1)}{xy} \geq 9.$$

The minimum value is achieved when $x = y = 1$, i.e., $a = b = 1 \Rightarrow$ Minimum value = 9.

44. Given Exp. $= (x^{a-b})^{(a+b)} \times (x^{b-c})^{(b+c)} \times (x^{c-a})^{(c+a)} = x^{(a-b)(a+b)} \times x^{(b-c)(b+c)} \times x^{(c-a)(c+a)}$

$= x^{(a2 - b2 + b2 - c2 + c2 - a2)} = x^0 = 1.$

45. Given Exp. $= \dfrac{1}{1+\dfrac{x^b}{x^a}+\dfrac{x^c}{x^a}} + \dfrac{1}{1+\dfrac{x^a}{x^b}+\dfrac{x^c}{x^b}} + \dfrac{1}{1+\dfrac{x^b}{x^c}+\dfrac{x^a}{x^c}} = \left(\dfrac{x^a+x^b+x^c}{x^a+x^b+x^c}\right) = 1.$

46. $2^x = 3^y = 6^{-z} = k \Rightarrow 2 = k^{1/x}, 3 = k^{1/y}$ & $6 = k^{-1/z}$

$2 \times 3 = 6 \Leftrightarrow k^{1/x}.\, k^{1/y} = k^{-1/z} \Rightarrow \dfrac{1}{x} + \dfrac{1}{y} = -\dfrac{1}{z} \Rightarrow \dfrac{1}{x} + \dfrac{1}{y} + \dfrac{1}{z} = 0$

47. $\left(\dfrac{x}{yz}\right) = \left(\dfrac{1}{xz}\right) + \left(\dfrac{1}{xy}\right)$

$$\frac{x}{yz} = \frac{(y+z)}{xyz}$$

$$x^2 = y + z$$

$$x^2 + x = x + y + z \text{ (by adding } x \text{ in both side)}$$

$$(x+1) = \frac{(x+y+z)}{x} \qquad \ldots(1)$$

Similarly,

$$(y+1) = \frac{(x+y+z)}{y} \qquad \ldots(2)$$

$$(z+1) = \frac{(x+y+y)}{z} \qquad \ldots(3)$$

Now, $\left[\dfrac{(x-1)}{(x^2-1)}\right] + \left[\dfrac{(y-1)}{(y^2-1)}\right] + \left[\dfrac{(z-1)}{(z^2-1)}\right]$

$$\left[\frac{(x-1)}{(x+1)(x-1)}\right]+\left[\frac{(y-1)}{(y+1)(y-1)}\right]+\left[\frac{(z-1)}{(z+1)(z-1)}\right] \Rightarrow \left[\frac{1}{(x+1)}\right]+\left[\frac{1}{(y+1)}\right]+\left[\frac{1}{(z+1)}\right]$$

From (1), (2) and (3)

$$\left[\frac{x}{(x+y+z)}\right]+\left[\frac{y}{(x+y+z)}\right]+\left[\frac{z}{(x+y+z)}\right] \Rightarrow \frac{(x+y+z)}{(x+y+z)}=1$$

48. Given Exp. $= \dfrac{1}{1+\dfrac{a^n}{a^m}}+\dfrac{1}{1+\dfrac{a^m}{a^n}}=\left(\dfrac{a^m+a^n}{a^m+a^n}\right)=1.$

49. $m^y \,.\, n^x = a^{2/z} \Rightarrow (a^x)^y \,.\, (a^y)^x = a^{2/z} \Rightarrow a^{2xy} = a^{2/z} \Rightarrow a^{xy} = a^{1/z}$

$\therefore \qquad (a^{xy})^z = (a^{1/z})^z = a \Rightarrow xyz = 1.$

50. Given Exp. $= x\left(\dfrac{7}{2}-\dfrac{5}{2}\right)\cdot y\left(\dfrac{3}{2}-\dfrac{1}{2}\right) = xy.$

51. Given Exp. $= \dfrac{2^{n+4}-2^{n+1}}{2^{n+4}}+\dfrac{1}{8}=\dfrac{2^{n+1}(2^3-1)}{2^{n+4}}+\dfrac{1}{8}=\dfrac{7}{3^2}+\dfrac{1}{8}=\dfrac{8}{8}=1.$

52. Given Exp. $= \dfrac{q}{q+pq+1}+\dfrac{1}{1+q+pq}+\dfrac{pq}{pq+1+q} \qquad \left[\because\ r=\dfrac{1}{pq}\right]$

$= \left(\dfrac{q+1+pq}{q+1+pq}\right)=1.$

53. Given Exp. $= x^{(b-c)(b+c-a)+(c-a)(c+a-b)+(a-b)(a+b-c)} = x^0 = 1$

54. Given Exp. $= x\left(\dfrac{a-b}{ab}+\dfrac{b-c}{bc}+\dfrac{pq}{pq+1+q}\right) = x^{\left(\frac{1}{b}-\frac{1}{a}+\frac{1}{c}-\frac{1}{b}+\frac{1}{a}-\frac{1}{c}\right)} = x^0 = 1.$

55. Given Exp. $= \sqrt{\dfrac{x}{y}},\sqrt{\dfrac{z}{y}},\sqrt{\dfrac{x}{z}}=\dfrac{\sqrt{y}}{\sqrt{x}}\times\dfrac{\sqrt{z}}{\sqrt{y}}\times\dfrac{\sqrt{x}}{\sqrt{z}}=1.$

56. Given surds are of order 2, 3 and 4 whose l.c.m. is 12. So, we change each into a surd of order 12.

$\sqrt{2} = 2^{1/2} = 2^{\left(\frac{1}{2}\times\frac{6}{6}\right)} = 2^{6/12} = (2^6)^{1/12} = (64)^{1/2} \Rightarrow \sqrt[3]{4} = 4^{1/3} = 4^{\left(\frac{1}{3}\times\frac{4}{4}\right)} = 4^{4/12} = (256)^{1/12}$

$\sqrt[4]{6} = 6^{1/4} = 6^{\left(\frac{1}{4}\times\frac{3}{3}\right)} = 6^{3/12} = (6^3)^{1/12} = (216)^{1/12}$

Clearly, $(64)^{1/12} < (216)^{1/12} < (256)^{1/12}$, *i.e.*, $\sqrt{2} < \sqrt[4]{6} < \sqrt[3]{4}$.

25 CHAPTER Set Theory

Set: *A well defined collection of objects is called a set.*

The objects in a set are called its *members,* or *elements.*

If A is a set and x is an element of A, we write, $x \in A$. If x does not belong to A, we write, $x \notin A$.

Ex. ***(i) If A is the set of vowels in English alphabet, we have A = {a, e, i, o, u}.***

Clearly, $a \in A, b \notin A, e \in A$ etc.

(*ii*) Let B be the set of all integers lying between – 3 and 3, then $B = \{-2, -1, 0, 1, 2\}$.

(*iii*) The collection of all honest persons of India is not a set, since the term '*honest persons*' is not well defined.

Similarly, '*Beautifill girls,*' '*Good players,*' '*Intelligent students,*' etc. do not form sets.

How to describe or specify a set ?

I. Roster Method: *Under this method, we just make a list of the objects of the set and put them within braces { }.*

Ex. (*i*) If A is set of first eight prime numbers, then

$A = \{2, 3, 5, 7, 11, 13, 17, 19\}$.

(ii) If B is the set of all counting numbers between 50 and 500, then

$B = \{5l, 52, 53, ..., 498, 499\}$.

II. Set Builder Form: *This method consists in listing the properties common to all the elements of the set.*

We write, $\{x : x$ has properties $P\}$.

Ex. (*i*) If A is the set of all natural numbers less than 6, then we write, $A = \{x : x \in N, x < 6\}$.

(*ii*) If $B = \{3, 5, 7,9, 11\}$, then we can write:

$B = (x : x = 2n + 1,$where $n \in N, n < 6\}$.

Singleton Set: *A set consisting of a single element is called a singleton set.*

Ex. $\{x : x + 3 = 3\} = \{0\}$, which is a singleton set.

Empty Set: *A set containing no element at all is called an empty set or a null set or a void set and it is denoted by* (ϕ). *In set builder form* $\phi = \{ \}$.

Ex. (*i*) $\{x : x \in N, 3 < x < 4\} = \phi$, since there is no natural number lying between 3 and 4.

(*ii*) $\{x : x \in R, x^2 = -1\} = \phi$. since there is no real number whose square is – 1.

(*iii*) $\{x : x$ is a number, $x \neq x\} = \phi$.

Finite & Infinite Sets: *A set in which the process of counting of elements surely comes to an end, is called a finite set.*

A set which is not finite is known as an infinite set.

Ex. 1. ***Each one of the sets given below is a finite set.-***

(*i*) Set of all persons on earth. (*ii*) $\{x \in N, x < 5$ croresl (*iii*) $\{x : x \in I, x$ is a factor of $5000\}$.

Ex. 2. ***Each one of the sets given below is an infinite set:***

(*i*) Set of all points on the arc of a circle. (*ii*) Set of all concentric circles with a given centre.

(*iii*) $\{x : x \in I, x < 5\}$ (*iv*) $\{x : x \in Q, 0 < x < 1\}$

(v) $\{x : x \in N, x > 10000\}$

Equal Sets: *Two sets A and B are said to be equal, if every element of A is in B and every element of B is in A and we write, A = B.*

Remarks: (*i*) *The elements of a set may be listed in any order.*

Thus, {1, 2. 3} = {1, 3, 2} = {2, 3, 1} etc.

(*ii*) *The repetition of elements in a set is meaningless.* Thus,

{1, 2, 3, 2, 3, 1, 1, 2, 3. 3, 2, 1} = {1, 2, 3}.

(*iii*) $\phi \neq \{0\}$ since ϕ is a set containing no element at all, {0} is a singleton set and 0 is not a set.

Cardinal Number of a Set: *The number of distinct elements contained in a set A is called the cardinal number of A, to be denoted by n(A).*

Equivalent Sets: *Two sets A and B are said to be equivalent if n(A) = n(B).*

Remark: Equivalent sets are not always equal but equal sets are always equivalent. Thus, {1, 2, 3} and {*a*, *b*, *c*} are equivalent but not equal.

Subsets: Let *A* and *B* be two sets given in such a way that every element of *A* is in *B*, then we say that *A* is a subset of *B* and we write, $A \subseteq B$.

Super Set: *If $A \subseteq B$, we say that B is a superset of A and we write, $B \supseteq A$.*

Proper Subset: If $A \subseteq B$ and $A \neq B$, then *A* is called a proper subset of *B* and we write. $A \subset B$.

Ex. (*i*) $\{1\} \subset \{1, 2, 3\}$ (*ii*) $N \subset W \subset I \subset Q \subset R$

Comparable Sets: Two sets *A* and *B* are said to be comparable if either $A \subseteq B$ or $B \subseteq A$.

Results on Subsets:

(*i*) ϕ is a subset of every set.

(*ii*) Every set is a subset of itself.

(*iii*) The number of all possible subsets of a set containing *n* elements is 2^n.

(iv) The number of all proper subsets of a set containing *n* elements is $(2^n - 1)$.

(*v*) The set of all subsets of a given set *A* is called the power set of *A*, denoted by *P*(*A*).

If *A* has *n* elements, *P*(*a*) has 2^n elements.

Ex. 1. ***Write down the power set of A = {l, {2, 3}}.***

Sol. We have: $A = \{1, B\}$, where $B = \{2, 3\}$.

$\therefore$ $P(A) = \{\phi, \{1\}, \{B\}, \{1, B\}\}$

$= \{\phi, \{1\}, \{\{2, 3\}\}, \{1\{2, 3\}\}\}$.

OPERATIONS ON SETS

I. Union of Sets: *The union of two sets A and B, denoted by $A \cup B$ is the set of all those elements which are either in A or in B or in both A and B.*

Thus, $A \cup B = \{x : x \in A \text{ or } x \in B\}$.

Clearly, $x \in A \cup B \Rightarrow x \in A$ or $x \in B$.

And, $x \notin A \cup B \Rightarrow x \notin A$ and $x \in B$.

Ex. 1. **If $A = \{a, b, c, d\}$ and $B = \{b, d, e, f, g\}$, then**
$A \cup B = \{a, b, c, d, e, f, g\}$.

Ex. 2. If A is the set of all positive integers and B is the set of all negative integers, then clearly $A \cup B$ is the set of all integers except zero.

II. Intersection of Sets: *The intersection of two sets A and B, denoted by $A \cap B$ is the set of all those elements which are common to both A and B.*

Thus, $A \cap B = \{x : x \in A \text{ and } x \in B\}$.

Clearly, $x \in A \cap B \Rightarrow x \in A$ and $x \in B$.

and, $x \in A \cap B \Rightarrow x \in A$ or $x \notin B$.

Ex. 1. If $A = \{b, d, e, f\}$ and $B = \{b, e, g, h\}$, then $A \cap B = \{b, e\}$.

Ex. 2. If $A = \{x : x \in N, x \text{ is a multiple of } 4\}$ and $B = \{x : x \in N, x \text{ is a multiple of } 6\}$, then $A \cap B = \{x : x \in N, x \text{ is a multiple of } 12\}$.

Note that l.c.m. of 4 and 6 is 12.

III. Difference of Sets: *For any two sets A and B, we define:*

$A - B = \{x : x \in A \text{ and } x \notin B\}$.

Ex. 1. If $A = \{2, 3, 5, 1, 11\}$ and $B = \{5, 7, 9, 11, 13\}$, then $A - B = \{2, 3\}$ and $B - A = \{9, 13\}$.

In general, $A - B \neq B - A$.

Ex. 2. If $A = \{1.2,3\}$ and $B = \{l, 2, 3. 4\}$, then $A - B = \phi$.

Note that: $A \subseteq B \Rightarrow A - B = \phi$.

Also, $A - B = B - A \Leftrightarrow A = B$.

IV. Symmetric Difference of Sets: The symmetric difference of two sets A and B is defined by:

$A \Delta B = (A - B) \cup (B - A)$.

Ex. 3. If $A = \{a, b, c, d\}$ and $B = \{b, d, e, f\}$, then

$A = B = \{a, c\}$ and $B - A = \{e, f\}$.

$\therefore A \Delta B = (A - B) \cup (B - A) = \{a, c, e, f\}$.

Universal Set: In a given problem on sets there happens to be a set which is a superset of each one of the given sets. This set is known as the universal set, to be denoted by U.

V. Complement of a Set: *Let A be a subset of the universal set U, then the complement of A, denoted by A^c is defined by:*

$A^c = \{x : x \in U, x \notin A\}$.

Clearly, $x \in A^c \Leftrightarrow x \notin A$.

Ex. 1. If $U = \{a, b, c, d, e, f\}$, and $A = \{b, d, e\}$, then $A^c = \{a, c, f\}$.

Ex. 2. If U is the set of all letters in English alphabet and A is the set of all vowels, then A^c is the set of all consonants. .

Some Results on Complementatiun:

(*a*) $\phi^c = U$ (*b*) $U_c = \phi$ (*c*) $A \cup A^c = U$ (*d*) $A \cap A^c = \phi$

Laws of operations:

I. (*i*) $A \cup A = A$ (*ii*) $A \cup \phi = A$ (*iii*) $A \cap A = A$ (*iv*) $A \cap \phi = \phi$

II. *Commutative Laws:*

(*i*) $(A \cup B) = B \cup A$ (*ii*) $A \cap B = B \cap A$

III. *Associative Laws:*

(*i*) $(A \cup B) \cup C = A \cup (B \cup C)$ (*ii*) $(A \cap B) \cap C = A \cap (B \cap C)$

IV. *Distributive Laws:*

$(i)\ A \cup (B \cap C) = (A \cup B) \cap (A \cup C)$

$(ii)\ A \cap (B \cup C) = (A \cap B) \cup (A \cap C)$

V. *De-Morgan 's Laws:*

$(i)\ (A \cup B)^c = A^c \cap B^c$ $(ii)\ (A \cap B)^c = A^c \cup B^c$.

Some More Important Results:

(i) $A \cup B = \phi \Leftrightarrow A = \phi$ and $B = \phi$. (ii) $A - B = \phi \Leftrightarrow A \subseteq B$

(iii) $A - B = A \cap B^C$, (iv) $A \subseteq B \Leftrightarrow B^C \subseteq A^C$.

(v) $A - (B \cup C) = (A - B) \cap (A - C)$ (vi) $A - (B \cap C) = (A - B) \cup (A - C)$

(vii) $A \cap (B - C) = (A \cap B) - (A \cap C)$

Venn Diagrams : *We represent sets by closed bounded figures, known as Venn diagrams.*

The universal set is denoted by a rectangular region and each subset of it is shown by a closed bounded figure within this rectangle.

Venn Diagrams in Different Situations:

Case I: *When* $A \subseteq U, B \subseteq U$ and $A \cap B \neq \phi$. In this case Venn diagram oonsists of two intersecting bounded figures within a rectangular region, representing A and B respecitively.

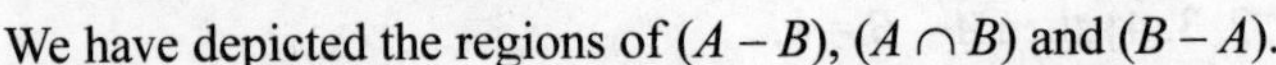

We have depicted the regions of $(A - B)$, $(A \cap B)$ and $(B - A)$.

Case II: *When* $A \subseteq U, B \subseteq U$ and $A \cap B = \phi$. In this case, the Venn diagram consists of two closed bounded non-intersecting figures lying within a rectangular region.

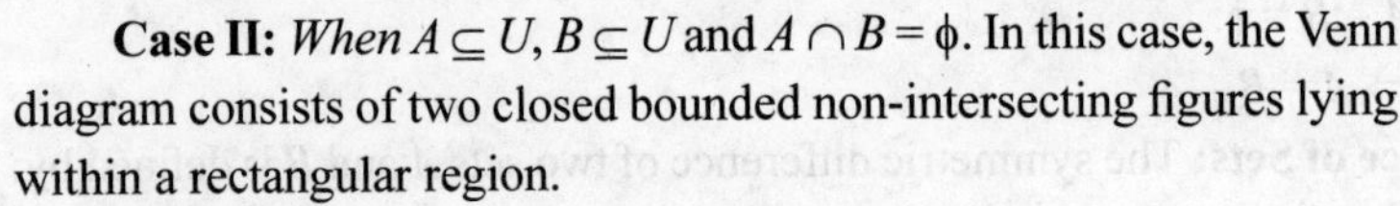

In this case, $A \cap B = \phi$, $A - B = A$ & $B - A = B$

Case III: When $A \subseteq B \subseteq U$.

In this case, the Venn diagram has been shown in the adjoining figure.

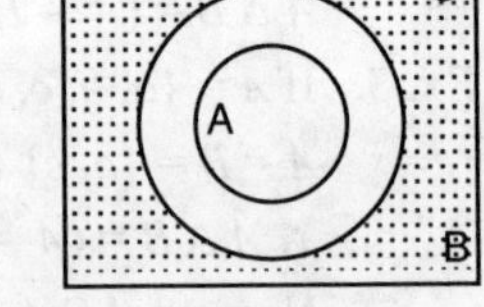

Clearly, $A \cup B = B, A \cap B = A$ & $A - B = \phi$.

Important Results from Venn Diagrams:

Let A and B be two intersecting sets. While counting the elements of $A \cup B$, we count the elements of $A \cap B$ twice, once in counting the elements of A and again in counting the elements of B. So:

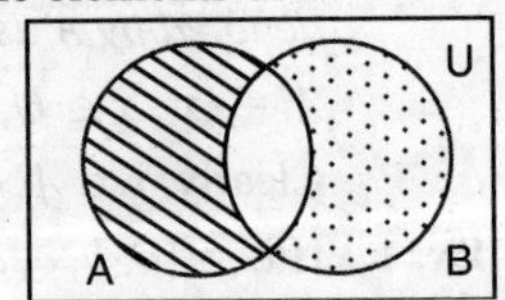

(i) $n(A \cup B) = n(A) + n(B) - n(A \cap B)$

(ii) $n(A \cup B) = n(A) + n(B)$, when $A \cap B = \phi$

(iii) $n(A - B) + n(A \cap B) = n(A)$

(iv) $n(B - A) + n(A \cap B) = n(B)$

(v) $n(A - B) + n(A \cap B) + n(B - A) = n(A \cup B)$

Ex. 1. ***A survey shows that 74% of the Indians like apples whereas 68% like oranges. What percentage of Indians like both apples and oranges ?***

Sol. Let A and B denote the sets of Indians who like apples and oranges respectively. Then,,clearly:

$n(A \cup B) = 100$, $n(A) = 74$ and $n(B) = 68$.

Using: $n(A \cup B) = n(A) + n(B) - n(A \cup B)$, we get:

$n(A \cap B) = n(A) + n(B) - n(A \cup B) = (74 + 68 - 100) = 42$.

Hence 42% of the Indians like both apples and oranges.

Ex. 2. ***In a group of 1000 persons , 760 can speak Hindi and 430 can speak Bengali. Find:***

(i) how many can speak both Hindi and Bengali;

(ii) how many can speak Hindi only;

(iii) how many can speak Bengali only.

Sol. Let A and B denote the sets of persons who can speak Hindi and Bengali respectively. Then,

$n(A \cup B) = 1000, n(A) = 760$ and $n(B) = 430$.

$\therefore n(A \cap B) = n(A) + n(B) - n(A \cap B) = (760 + 430 - 1000) = 190$.

Thus, 190 persons can speak both Hindi and Bengali.

Also, $n(A - B) + n(A \cap B) = n(A)$.

$\therefore n(A - B) = n(A) - n(A \cap B) = (760 - 190) = 570$.

Thus, 570 persons can speak Hindi only.

Again, $n(B - A) + n(A \cap B) = n(B)$

$\therefore n(B - A) = n(B) - n(A \cap B) = (430 - 190) = 240$.

Thus, 240 persons can speak Bengali only.

Ordered Pair: *A pair of two elements a and b listed in a specific order, is called an ordered pair (a, b).*

Note that ; $(2, 3) \neq (3. 2)$, while $\{2, 3\} = \{3, 2\}$

We have: $(a_1, b_1) = (a_2, b_2) \Leftrightarrow a_1 = a_2$ and $b_1 = b_2$.

Cartesian Product of Two Sets: *The cartesian product $A \times B$ of two non-empty sets A and B is defined as:*

$\boldsymbol{A \times B} = \{(a, b): a \in A \text{ and } b \in B\}$.

(*i*) If $A = \phi$ or $B = \phi$, we define $A \times B = \phi$.

(*ii*) In general $A \times B \neq B \times A$.

(*iii*) $n(A \times B) = n(A) . n(B)$.

(*iv*) If either A or B is an infinite set, then $A \times B$ is an infinite set.

Ex. 1. If $A = \{1, 2, 3\}$, $B = \{3, 4\}$ & $C = \{4, 5, 6\}$, *find*:

(*i*) $A \times (B \cap C)$ (*ii*) $(A \cap B) \times C$

Sol. (*i*) $A = \{1, 2, 3\}$ & $B \cap C = \{4\}$.

$\therefore A \times (B \cap C) = \{(1, 4), (2, 4), (3, 4)\}$

(*ii*) $A \cap B = \{3\}$ and $C = \{4, 5, 6\}$.

$\therefore (A \cap B) \times C = \{(3, 4), (3, 5), (3, 6)\}$.

Functions: *Let A and B be twa non-empty sets. Then a rule f which associates ta each x e A, a unique element $f(x) \in B$, is called a fimction or a mapping from A to B. And, we write, $f : A \rightarrow B$. $f(x)$ is called the image of x. And, x is called the pre-image ti f (x). The set A is called the domain of f while B is called the cu-domain of f.*

Also, $f(A) = \{f(x) : x \in A\}$ is called the range of f.

Ex. ***Let A* = {- 1, 1, 2, 3} and *B* = {1, 4, 9, 16}. Let *f*(*x*) = x^2. Find the image of *A*.**

Sol. $f(-1) = (-1)^2 = 1, f(1) = 1^2 = 1, f(2) = 2^2 = 4$ & $f(3) = 3^2 = 9$.

$\therefore f(a) = \{1, 4, 9\}$.

One-One & Many-One Functions: Let $f : A \rightarrow B$. We say that f is one-one, if distinct elements in A have distinct images in B.

Thus, f is one-one, if $f(x_1) = f(x_2) \Rightarrow x_1 = x_2$.

On the other hand, if *two or more than two elements in A have the same image in B, we say that f is many-one.*

Ex. 1. ***Let f: R → R : f(x) = x². Show that f is many-one.***

Sol. We have $= f(-1) = (-1)^2 = 1$ & $f(1) = 1^2 = 1$.

Thus, two distinct elements in R have the same image in R.

So, f is many-one.

Ex. 2. ***Let f : N → N : f(x) = 3x. Show that f is one-one.***

Sol. $f(x_1) = f(x_2) = 3x_1 = 3x_2 \Rightarrow x_1 = x_2$.

Hence, f is one-one.

Onto & Into Functions: *Let $f : A \rightarrow B$. We say that f is onto, if every element in B has at least one pre-image in A.*

Clearly, f is *onto* $\Leftrightarrow$ *range* $(f) = B$.

On the other hand, *if there exists even a single element in B having no pre-image in A, we say that f is into.*

Ex. 3. ***Show that: f : N → N : f(x) = 2x is not onto.***

Sol. Let $y = 2x$. Then, $x = \frac{1}{2}y$.

If we put $y = 3$, then $x = \frac{3}{2} \notin N$.

So, $3 \in N$ has no pre-image in N.

Hence, f is not onto.

Ex. 4. ***Let X be the set of all non-zero real numbers. Show that: $f : X \rightarrow X : f(x) = \frac{1}{x}$ is an onto function.***

Sol. Let $y = \frac{1}{x}$. Then, $x = \frac{1}{y}$.

Now, if any non-zero real number may be the value of y, the corresponding value of x will always be a non-zero real number. So, f is onto.

EXERCISE 25

1. If A = {1, 2, {3, 4}}, which of the following is a correct statement ?
   (*a*) $3 \in A$ (*b*) $\{1\} \subseteq A$ (*c*) $\{2\} \in A$ (*d*) $1 \subseteq A$
2. If A = {l, (2, 3), 5}, which of the following statements is incorrect ?
   (*a*) $(2, 3) \in A$ (*b*) $\{\{2, 3\}\} \subseteq A$ (*c*) $\{3\} \subseteq A$ {*d*} $\{5\} \subseteq A$
3. If A = {2, {3, 4}, 6), which of the following is correct ?
   (*a*) $\{2\} \in A$ (*b*) $\{3\} \in A$ (*c*) $\{4\} \in A$ (*d*) $\{3, 4) \in A$
4. If A = {1, (2, 3}, 4}, which of the following is a subset of A ?
   (*a*) {1, 2} (*b*) {2.3} (*c*) {3, 4} (*d*) {1, 4}
5. The number of all possible subsets of a set containing *n* elements is:
   (*a*) n (*b*) $2n$ (*c*) 2^n (*d*) $n\,!$
6. The number of all possible proper subsets of (2, 3, 5} is:
   (*a*) 3 (*b*) 3! (*c*) 7 (*d*) 8
7. The number of all possible subsets of the set (1, (2, 3}} is:
   (*a*) 2 (*b*) 4 (*c*) 6 (*d*) 8

8. If $A = \{a, b\}$, then the power set of A is:

(*a*) $\{a^b, b^a\}$ (*b*) $\{a^2, b^2\}$

(*c*) $\{\phi. \{a\}, \{b\}\}$ (d) $\{\phi, \{a\}, \{b\}, \{a, b\}\}$

9. The set which is comparable with the set (a, b, c, d} out of the following is:

(*a*) $\{x, y, z, t\}$ (*b*) $\{b, c, d, e\}$ (*c*) $\{a, b, d, e\}$ (*d*) $\{b, c, d\}$

10. Which one of the following is a correct statement ?

(*a*) $\phi = 0$ (*b*) $\phi = \{0\}$ (*c*) $\phi = \{\phi\}$ (*d*) $\phi = \{\}$

11. Which one of the following is a correct statement ?

(*a*) $\{a\} \in \{a, b, c\}$ (*b*) $a \subseteq \{a, b, c\}$ (*c*) $\phi \in \{a, b, c\}$ (*d*) $a \in \{\{a\}, b\}$

(*e*) None of these

12. Which one of the following is a correct statement ?

(*a*) $\{a\} \subseteq \{\{a\}, b, c\}$ (*b*) $\{a\} \in \{\{a\}, b, c\}$

(*c*) $\{a, b\} \subseteq \{\{a\}, b, c\}$ (*d*) $a \in \{\{a\}, b, c\}$

13. Which one of the following is a correct statement ?

(*a*) Every subset of an infinite set is infinite

(*b*) Every set has a proper subset

(*c*) $\{a, b, c, 1, 2, 3, a, b, c, 1, 2, 3,$ is an infinite set

(*d*) Every subset of a finite set is finite

14. Which of the following is a correct statement ?

(*a*) $\{\phi\} \subset \{a, b, c\}$ (*b*) $\phi \in \{a, b, c\}$ (*c*) $\phi \subset \{a, b, c\}$ (*d*) None of these

15. Which of the following is a correct statement ?

(*a*) $\{x : x + 5 = 5\} = \phi$ (*b*) $\{x : x + 5 = 5\}$is not a set

(*c*) $\{x : x + 5 = 5\} = 0$ (*d*) $\{x : x + 5 = 5\} = \{0\}$

16. Which of the following is a singleton set ?

(*a*) $\{x \in R : x^2 = x\}$ (*b*) $\{x \in \text{N} : 3x = 4\}$

(*c*) $x \in R : x^2 = -1\}$ (*d*) $(x : x$ is an integer which is neither + ve nor negative}

17. Which of the following is a true statement 7

(*a*) $\{a, b, c, 1, 2. 3\}$ is not a set (*b*) $\{a, b\} \subset \{\{a\}, b, c\}$

(*c*)$\{x \in R: |x| = -1\} = \phi$ (d)$\{x \in R : 1 < x < 2\}$is finite

18. Which one of the following is an infinite set ?

(*a*) $\{x : x \in N, x < 50\}$ (*b*)$\{x : x \in I, x < 50\}$

(*c*) $\{x : x \in I, x$ is a factor of 500} (d) $\{x : x$ is a whole number, $x < 1000\}$

19. Which one of the following is a finite set ?

(*a*) $\{x : x \in I, x < 1\}$ (*b*)$\{x : x \in R, 0 < x < 1\}$

(*c*) $\{x : x \in N, x > 5\}$ (*d*) $(x : x$ is an even prime)

20. Which one of the following is an empty set ?

(*a*) $\{0\}$ (*b*) $\{x \in N : 5x = 9\}$

(*c*) $\{x \in 0 < x < 4$ & x is even (*d*) $\{x \in I : (x - 1)(x - 2) = 0$ & x is odd}

21. Which of the following pairs of sets is disjoint ?

(*a*) $\{0, 1, 2\}$ & $\{0, -1, -2\}$ (*b*) $\{1, 3, 1, 4, 5\}$ & $\{3, 5, 7\}$

(*c*) $\{1, 3, 5, 7\}$ & $\{2, 4, 7, 9\}$ (d) $\{1, 2, 3\}$ & $\{-1, -2, -3\}$

22. Let $A = \{1, 2, 3, 4, 5, 6\}$. Then, the incorrect statement is:

(*a*) $\phi \subseteq A$ (*b*) $A \subseteq A$ (*c*) $\{1, 5, 7\} \subseteq A$ (*d*) $\{1\} \subseteq A$

23. Let $A = \{x \in R : 0 < x < 1\}$. Which of the following is not an element of A?

(*a*) 1 (*b*) $\frac{1}{2}$ (*c*) $\frac{1}{3}$ (*d*) $\frac{1}{4}$

24. Let A be the set of squares of natural numbers and let $x \in A, y \in A$, Then:

(*a*) $x + y \in A$ (*b*) $x - y \in A$ (*c*) $\left(\frac{1}{2}\right)^{\circ}$ (d) $xy \in A$

25. If A is the set of all positive integers and B is the set of all negative integers, then $A \cup B$ is:

(*a*) the set of all integers (*b*) $\{0\}$

(*c*) the set of all integers except zero (*d*) None of these

26. Let $A = \{0, 1, 3, 4\}$, $B = \{5, 6, 1, 3, 9\}$ and $C = \{0, 1, 2, 3, 9, 13\}$. Then $(A \cap B) \cup C$ is:

(*a*) $\{0, 1, 2, 3, 9, 13\}$ (*b*) $(0, 1, 2, 3, 4, 9, 13\}$

(*c*) $\{0, 1, 3\}$ (*d*) $\{1, 3\}$

27. Let $A = \{1,2,3,4,5\}$, $B = \{2,4,6,8\}$ and $C = \{3, 4, 5, 6\}$. Then, $(A \cup B) \cap C$ is:

(*a*) $\{3, 4)$ (*b*) $(3, 4, 5, 6\}$ (*c*) $(1, 2, 3, 4, 6, 8)$ (d) $\{4, 6\}$

28. If $A \cap B = A$, then:

(*a*) $A \subset B$ (*b*) $B \subset A$ (*c*) $A \not\subset B$ & $B \not\subset A$ (d) None

29. If $A \cup B = \phi$, then:

(*a*) $A = \phi$ or $B = \phi$ (*b*) $A = \phi$ and $B = \phi$ (*c*) $A = \phi$ & $B \neq \phi$ (d) $A \neq \phi$ and $B = \phi$

30. Which of the following is a true statement ?

(*a*) $(A \cup B)^C = (A^C \cup B^C)$ (*b*) $(A \cup B)^C = A \cap B^C$

(*c*) $(A \cup B)^C = A^C \cap B$ (d) $(A \cup B)^C = A^C \cap B^C$

31. Which one of the following is a true statement ?

(*a*) $A^C = \{x : x \in U \wedge x \in A\}$ (*b*) $A^C = \{x : x \in U \wedge x \notin A\}$

(*c*) $A^C = \{x : x \in UV\ x \in A\}$ (*d*) $A^C = \{x \in UV\ x \notin A\}$

where '$\wedge$' stands for 'and' while 'V' stands for 'or' and U is the universel set.

32. If A and B are two subsets of X, then $A^C \subseteq B$ if and only if:

(*a*) $A \cup B = X$ (*b*) $A \cap B = \phi$ (*c*) $A \cap B = A$ (*d*) None of these

33. The shaded region in the adjoining diagram is:

(*a*) A^C (*b*) B^C

(*c*) $A - B$ (*d*) $B - A$

U

A B

34. In a class containing 200 students, n students prefer both tea and coffee; $2n$ students prefer coffee; $3n$ students prefer tea; $4n$ students prefer neither tea nor coffee. What is the value of n?

[CDS (II) Exam, 2024]

(*a*) 20 (*b*) 25 (*c*) 30 (*d*) 35

35. In a class of 160 students, each of them opt at least one language from among English, Hindi and Sanskrit. It is found that 130 students opt English, 120 students Hindi and 110 Sanskrit. If the students opt either only one language or all three languages, then what is the number of students who study all three languages? **[CDS (II) Exam, 2024]**

(*a*) 40 (*b*) 60 (*c*) 80 (*d*) 100

36. In a competitive examination, 250 students have registered. Out of these, 50 students have registered for Physics, 75 students for Mathematics and 35 students for both Mathematics and Physics. What is the number of students who have registered neither for Physics nor for Mathematics? **[CDS (I) Exam, 2020]**

(*a*) 90 (*b*) 100 (*c*) 150 (*d*) 160

37. The maximum marks in a Test are converted from 250 to 50 for the purpose of an Internal Assessment. The highest marks scored were 170 and lowest marks were 70. What is the difference between the maximum and minimum marks scored in the Internal Assessment?

(*a*) 15 (*b*) 17 (*c*) 20 (*d*) 24

[CDS (I) Exam, 2020]

38. In an examination, 53% candidates failed in English and 42% failed in Mathematics. If 17% failed in both the subjects, then what percent passed in both the subjects?

(*a*) 77 (*b*) 85 (*c*) 48 (*d*) 23

[CDS (I) Exam, 2019]

Consider the following for the next three (03) items:

In a certain town of population size 1,00,000 three types of newspapers (I, II and III) are available. The percentages of the people in the town who read these papers are as follows:

Newspaper	Proportion of readers
I	10%
II	30%
III	5%
Both I and II	8%
Both II and III	4%
Both I and III	2%
All the three (I, II and III)	

39. What is the number of people who read only one newspaper? **[CDS (I) Exam, 2019]**

(*a*) 20,000 (*b*) 25,000 (*c*) 30,000 (*d*) 35,000

40. What is the number of people who read at least two newspaper? **[CDS (I) Exam, 2019]**

(*a*) 12,000 (*b*) 13,000 (*c*) 14,000 (*d*) 15,000

41. What is the number of people who do not red any of these three newspaper?

(*a*) 62,000 (*b*) 64,000 (*c*) 66,000 (*d*) 68,000

[CDS (I) Exam, 2019]

42. If $X = \{a, \{b\}, c\}$, $Y = \{\{a\}, b, c\}$ and $Z = \{a, b, \{c\}\}$ then $(X \cap Y) \cap Z$ equal to

(*a*) $\{a, b, c\}$ (*b*) $\{\{a\}, \{b\}, \{c\}\}$ (*c*) $\{\Phi\}$ (*d*) Φ

[CDS (I) Exam, 2019]

43. Out of 85 children playing badminton of table tennis or both the total number of girls in the group in 79% of the total number of boys in the group. The number of boys playing only badminton is 50% of the number of boys and the total number of boys playing badminton is 60% of the total number of boys. The number of children playing only table tennis is 40% of the total number of children and a total tennis both. The number of girls playing only badminton is **[CDS (I) Exam, 2018]**

(*a*) 14 (*b*) 16 (*c*) 17 (*d*) 25

44. In a class of 60 boys, there are 45 boys who play chess and 30 days who play carrom. If every boy of the class plays at least one of the two games, then how many boys play carrom only?

(*a*) 30 (*b*) 20 (*c*) 15 (*d*) 10

Consider the following for the next 04 (four) items that follow: [CDS (I) Exam, 2018]

In an examination of class XII, 55% students, passed in Biology, 62% passed in Physics, 60% passed in Chemistry, 25% passed in Physics and Biology, 30% passed in Physics and Chemistry, 28% passed in biology and Chemistry. Only 2% failed in all the subjects.

45. What percentage of students passed in all the three subjects.

(*a*) 6 (*b*) 5 (*c*) 4 (*d*) 3

46. What percentage of students passed in exactly one subject?

(*a*) 21 (*b*) 23 (*c*) 25 (*d*) 27

47. If the number of students in 360, then how many passed in at least two subjects?

(*a*) 270 (*b*) 263 (*c*) 265 (*d*) 260

48. What is the ratio of number of students who passed in both Physics and Chemistry to number of students who passed in both Biology and Physics but not Chemistry?

(*a*) 7 : 10 (*b*) 10 : 7 (*c*) 9 : 7 (*d*) 7 : 9

49. Let A and B be finite non-empty sets with the number of elements in A = m and number of elements in B = n. Let $m > n$. If for some integer $k \geq 1$, the number of non-empty subsets of A = 2^k + the number of non-empty subsets of B, then which one of the following is correct?

(*a*) $m = n + 2$

(*b*) $m = n + 1$

(*c*) $m = n + p$ for some odd prime number p

(*d*) $m = n + t$ for some composite number t **[CDS (I) Exam, 2015]**

50. In a gathering of 100 people, 70 of them can speak Hindi, 60 can speak English and 30 can speak French. Further, 30 of them can speak both Hindi and English, 20 can speak both Hindi and French. If x is the number of people who can speak both English and French, then which one of the following is correct? (Assume that everyone can speak at least one of the three languages.) **[CDS (I) Exam, 2016]**

(*a*) $9 < x \leq 30$ (*b*) $0 \leq x < 8$ (*c*) $x = 9$ (*d*) $x = 8$

51. Let A = {7, 8, 9, 11, 12} and B = {7, 10, 14, 15}. What is the number of elements in (A – B) and (B – A) respectively? **[CDS (II) Exam, 2016]**

(*a*) 2 and 4 (*b*) 4 and 2 (*c*) 2 and 2 (*d*) 4 and 4

52. If A = {x : x is a multiple of 7}, B = {x : x is a multiple of 5} and C = {x : x is a multiple of 35}, then which one of the following is a null set? **[CDS (I) Exam, 2017]**

(*a*) $(A-B)\cup C$ (*b*) $(A-B)-C$ (*c*) $(A\cap B)\cap C$ (*d*) $(A\cap B)-C$

53. In a group of persons, each one knows either Hindi or Tamil. If 100 know Hindi, 50 know Tamil and 30 know both, how many persons are there in the group ?

(*a*) 120 (*b*) 130 (*c*) 150 (*d*) 180

54. If 63% of Indians like milk and 76% like tea, how many like both ?

(*a*) 13% (*b*) 26% (*c*) 39% (*d*) $6\frac{1}{2}\%$

55. In a group of 52 persons, 16 drink tea but not coffee and 33 drink tea. How many drink coffee but not tea ?

(*a*) 3 (*b*) 7 (*c*) 17 (*d*) 19

ANSWERS

1. (*b*)	**2.** (*c*)	**3.** (*d*)	**4.** (*d*)	**5.** (*c*)	**6.** (*c*)	**7.** (*b*)	**8.** (*d*)	**9.** (*d*)
10. (*d*)	**11.** (*e*)	**12.** (*b*)	**13.** (*d*)	**14.** (*c*)	**15.** (*d*)	**16.** (*d*)	**17.** (*c*)	**18.** (*b*)
19. (*d*)	**20.** (*b*)	**21.** (*d*)	**22.** (*c*)	**23.** (*a*)	**24.** (*d*)	**25.** (*c*)	**26.** (*a*)	**27.** (*b*)
28. (*d*)	**29.** (*b*)	**30.** (*d*)	**31.** (*b*)	**32.** (*a*)	**33.** (*c*)	**34.** (*b*)	**35.** (*d*)	**36.** (*d*)
37. (*c*)	**38.** (*d*)	**39.** (*a*)	**40.** (*a*)	**41.** (*d*)	**42.** (*d*)	**43.** (*a*)	**44.** (*c*)	**45.** (*c*)
46. (*b*)	**47.** (*a*)	**48.** (*b*)	**49.** (*b*)	**50.** (*a*)	**51.** (*b*)	**52.** (*d*)	**53.** (*a*)	**54.** (*c*)
55. (*d*)								

SOLUTIONS

1. $\{1\} \subseteq A$ is true.

2. Since $3 \notin A$, so, $\{3\} \subseteq A$ is incorrect.

3. Clearly, $\{3, 4\} \in A$.

4. Since $1 \in A$ and $4 \in A$, so $\{1, 4\} \subseteq A$.

5. Number of all possible subsets of a set containing n elements is 2^n.

6. Number of all possible proper subsets $= (2^3 - 1) = 7$.

7. The given set contains two elements, namely, 1 and (2, 3).
So, the number of subsets of the given set is 2^2 *i.e.,* 4.

8. The proper set of A is the set of all possible subsets of A, given by $\{\phi. \{a\}, \{b\}, \{a, b\}\}$.

9. Note that two sets A and B are comparable if $A \subset B$ or $B \subset A$.
So, $(b, c, d\}$ is comparable with the given set.

10. Clearly, $\phi = \{ \}$ is a correct statement.

11. None of the first four statements is true.

12. Clearly, $\{a\} \in \{\{a\}, b, c\}$.

13. (*a*) is false, since the set N of all natural numbers is an infinite set having a finite subset $\{1)$.
(*b*) is false, since ϕ has no proper subset.
(*c*) is false, since $\{a, b, c, 1, 2, 3, a, b, c, 1, 2, 3, ...\} = \{a, b, c, 1, 2, 3, ...\}$. which is finite.
(*d*) Clearly, every subset of a finite set is finite.

14. Since ϕ is a subset of every set, so (*c*) is true.

15. Clearly, 0 is the only number satisfying $x + 5 = 5$.
So, its solution set is $\{0\}$.

16. Since 0 is the only integer which is neither positive nor negative, so (*d*) is true.

17. Since there is no real number whose modulus is negative, so (*c*) is true.

18. The set of all integers less than 50 is clearly an infinite set. So, (*b*) is true.

19. $\{x : x \text{ is an even prime}) = \{2\}$, which is finite.

20. Clearly, $\{x \in N: 5x = 9\} = \phi$.

21. Clearly, no element is common to the sets given in (*d*).

22. Since $7 \notin A$, so (*c*) is false.

23. Clearly, $1 \notin A$.

24. Let $x = m^2$ and $y = n^2$ for some $m, n \in N$.
Then, $xy = m^2n^2 = (mn)^2$, where $mn \in N$.
Note that $4 \in A$, $16 \in A$ but $\frac{4}{16} \notin A$.

25. Since 0 is neither positive nor negative, so $A \cup B$ is the set of all integers except zero.

26. $A \cap B = \{1, 3\}$ So, $(A \cap B) \cup C = \{0, 1, 2, 3, 9, 13\}$.

27. $A \cup B = \{1, 2, 3, 4, 5, 6, 8\}$. So, $(A \cup B) \cap C = \{3, 4, 5, 6\}$.

28. $A \cap B = A \Rightarrow A \subseteq B$.

29. $A \cup B = \phi \Rightarrow A = \phi$ and $B = \phi$.

30. By De-Morgan's law, $(A \cup B)^C = A^C \cap B^C$.

31. By definition: $A^C = \{x \in U \wedge x \notin A\}$.

32. $A^C \subseteq B = A \cup A^C \subseteq A \cup B \Rightarrow X \subseteq A \cup B \Rightarrow A \cup B = X$.

33. Shaded region is $A - B$.

34. Students prefer only Tea = students prefer Tea − students prefer both tea and coffee
$\Rightarrow 3n - n = 2n$
Students prefer only coffee = students prefer coffee − students prefer both tea and coffee
$\Rightarrow 2n - n = n$
Students prefer only Tea + only coffee + both + neither tea nor coffee = 200
$\Rightarrow 2n + n + n + 4n = 200 \Rightarrow 8n = 200 \Rightarrow n = \frac{200}{8} \Rightarrow n = 25$

35. Total students, $|U| = 160$
$|E| = 130$
$|H| = 120$
$|S| = 110$
$n(E \cup H \cup S) = n(E) + n(H) + n(S) - (3 \times n(E \cap H \cap S)) - n(E \cap H \cap S)$
$\Rightarrow 160 = 130 + 120 + 110 - (2 \times n(E \cap H \cap S))$
$\Rightarrow 160 = 360 - (2 \times n(E \cap H \cap S)) \Rightarrow 2 \times n(E \cap H \cap S) = 200 \Rightarrow n(E \cap H \cap S) = 100$

36. We have,
$n(P) = 50$, $n(M) = 75$, $n(P \cap M) = 35$
$\because n(P \cup M) = n(P) + n(M) - n(P \cap M) = 50 + 75 - 35 = 125 - 35 = 90$
Total students = 250
$\because$ Number of students who have registered neither for physics nor for mathematics
$= 250 - 90 = 160$

37. We have,
Maximum marks scored in the internal assessment $= 170 \times \frac{50}{250} = 34$
and lowest marks scored in the internal assessment $= 70 \times \frac{50}{250} = 14$
Now, required difference $= 34 - 14 = 20$

38. We have,
$n(E) = 52\%$, $n(M) = 42\%$ and $n(M \cap E) = 17\%$
$\because n(M \cup E) = n(M) + n(E) - n(M \cap E) \Rightarrow n(M \cup E) = 42 + 52 - 17 = 94 - 17 = 77\%$
$\because$ Percentage of candidates who passed in both the subject $= 100 - 77 = 23\%$

For Sol. 39-41

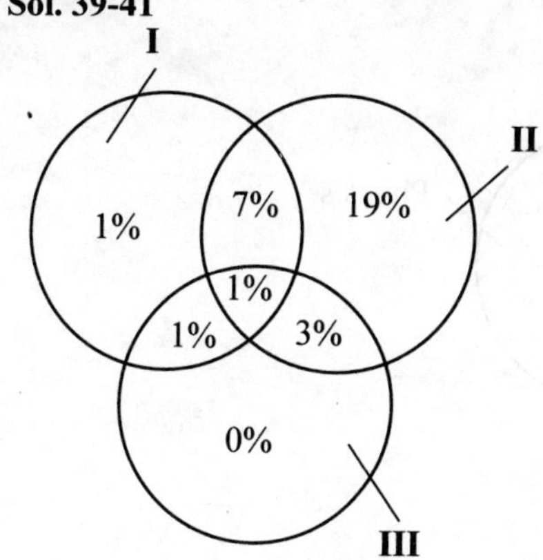

39. No. of people who read only one newspaper = (1 + 19)% of 1,00,000

$$= 1,00,000 \times \frac{20}{100} = 20,000$$

40. From the above venn diagram = No. of people who read at least two newspaper

$$= (1 + 7 + 3 + 1)\% \times 1,00,000$$

$$= 1,00,000 \times \frac{12}{100} = 12,000$$

41. From the above venn diagram = No. of people who do not read any of these three newspapers

$$= 1,00,000 \times (100 - (20 + 11 + 1\%)$$

$$= 1,00,000 \times (100 - 32)\% = 1,00,000 \times \frac{68}{100} = 68000$$

42. We have,

X = $\{a, \{b\}, c\}$

Y = $\{\{a\}, b, c\}$

Z = $\{a, b, \{c\}\}$

Now, $X \cap Y \{c\}$

$\therefore (X \cap Y) \cap Z = \phi$

43. We have,

Total number of children = 85

Let total number of boys = x

Total number of girls = y

Now, according to the question

$$y = 70\frac{x}{100} \quad \Rightarrow 85 - x = \frac{7x}{10} \quad \Rightarrow 85 = \frac{17x}{10} \quad \Rightarrow x = 50$$

$y = 85 - 50 = 35$

Now, we have number of children playing only table tennis = $85 \times \frac{40}{100} = 34$

Number of children playing badminton and table tennis both = 12

Number of children playing badminton only = 85 –34 – 12 = 85 – 46 = 39

Number of girls playing only badminton = 39 – 25 = 14

44. Number of boys who play carrom only = 60 –45 = 15

For Sol. 45-48

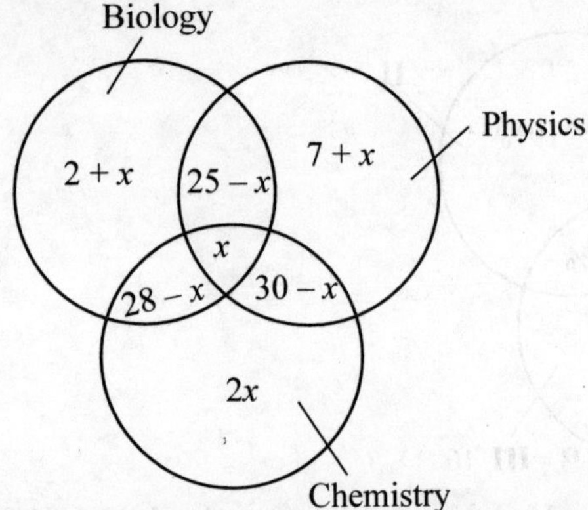

Let the percentage of students who passed in all three subjects = x%

45. We have, from the above venn diagram

Total no. of passed students = 98%

$\Rightarrow (7 + x + 2 + x + 2 + x + 28 - x + 30 - x + 25 - x + x\%) = 98\%$

$\Rightarrow (7 + x + 2 + x + 2 + x + 28 - x + 30 - x + 25 - x + x)\% = 98\%$

$\Rightarrow x + 94 = 98 \Rightarrow x = 98 - 94 = 4$

46. We have $x = 4\%$ [from qn 45]

Hence, percentage of students passed in exactly one subject $= (2 + x + 2 + x + 7 + x)\%$

$= 11 + 3x$

$= 11 + 3 \times 4 = 23\%$

47. No. of students passed in atleast two subject $= \dfrac{25 - x + 28 - x + 30 - x + x}{100} \times 360°$

$= \dfrac{83 - 2x}{10} \times 36$

$= \dfrac{83 - 2 \times 4}{5} \times 18$ [$\because x = 4$ from qn 45]

$= \dfrac{75}{5} \times 18 = 270$

48. Required ratio $= \dfrac{30}{25 - x} = \dfrac{30}{25 - 4} = \dfrac{30}{21} = 10 : 7$ [$\because x = 4$ from qn 45]

49. Number of non-empty subset of A = $2^m - 1$.

Number of non–empty subset of B = $2^n - 1$

$2^m - 1 = 2^k + 2^n - 1, 2^m - 2^n = 2^k, 2^n (2^{m-n} - 1) = 2^k$ ... (1)

Since k is an integer (≥ 1), RHS of (1) is a power of 2.

$\therefore$ LHS of (1) is also a power of 2.

$2^{m-n} - 1$ is odd. As LHS is the power of two,

$2^{m-n} - 1$ must be 1.

$\Rightarrow 2^{m-n} = 2^1 \Rightarrow m - n = 1.$

50. n (HUEUF) = 100

$\therefore$ $n(H \cap E) = 30$, $n(H \cap F) = 20$, $n(E \cap F) = x$

$\therefore$ $100 = 70 + 60 + 30 - 30 - 20 - x + 20$

$x = 30$

For x to be minimum $n(H \cap E \cap F) = 0$

then $100 = 70 + 60 + 30 - 30 - 20 - x$

$x = 10$

$\therefore$ $10 \le x \le 30$

or $9 < x \le 30$

For x to be maximum $n(H \cap E \cap F)$

Must be 20.

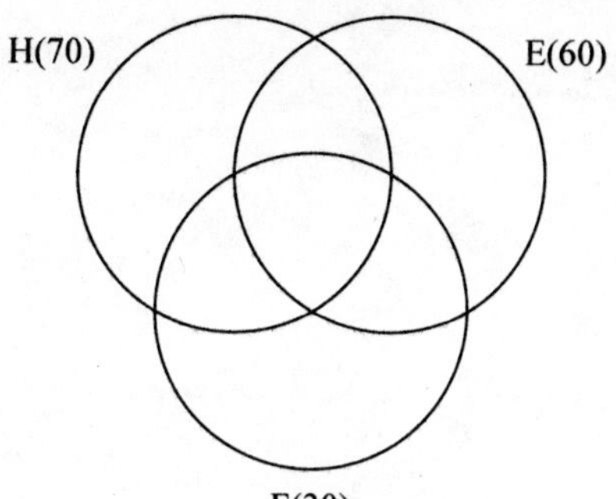

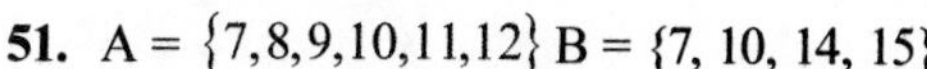

51. A = $\{7, 8, 9, 10, 11, 12\}$ B = $\{7, 10, 14, 15\}$

$\therefore$ A – B = { 8, 9, 11, 12} and B – A = {14, 15}

A – B and B – A have 4 and 2 elements respectively

52. $A = \{7, 14, 21, 28, 35.......\}$

$B = \{5, 10, 15, 20, 25, 30, ...\}$

$C = \{35, 70, ...\}$

$\therefore$ $(A \cap B) - C = \phi$,which is a null set

So, the answer is option (*d*)

53. Given: $n(H) = 100$, $n(T) = 50$ & $n(H \cap T) = 30$.

$\therefore n(H \cup T) = n(H) + n(T) - n(H \cap T) = (100 + 50 - 30) = 120.$

54. $n(A) = 63$, $n(B) = 76$ and $n(A \cup B) = 100$.

$\therefore n(A \cap B) = n(A) + n(B) - n(A \cup B) = (63 + 76 - 100) = 39.$

55. $n(A \cup B) = 52$, $n(A - B) = 16$ and $n(A) = 33$.

$n(A) = n(A - B) + n(A \cap B) = n(A \cap B) = 33 - 16 = 17.$

$n(A \cup B) = n(A - B) + n(A \cap B) + n(B - A)$

$\Rightarrow n(B - A) = n(A \cup B) - n(A - B) - n(A \cap B) = (52 - 16 - 17) = 19.$

UNIT 4

TRIGONOMETRY

Trigonometrical Ratios: Six Trigonometric Ratios, Values of these ratios for 0°, 30°, 45°, 60° and 90°. Simple Trigonometrical Identities.

Heights & Distances: Simple Problems on Heights and Distances.

26 Identities and T-Ratios

Definitions & Review of Concepts

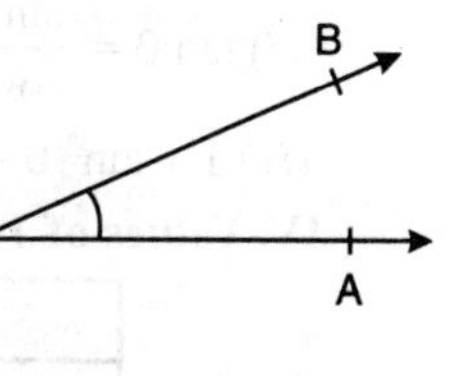

Angle: *When a ray OA rotates about the point O from its initial position OA to final position OB, we say that ∠AOB has been formed.*

OA is called the *initial side* and *OB*, the *terminal side* of *∠AOB*.

Positive & Negative Angles: *When the ray OA rotates in anticlockwise direction, a positive angle is formed, and when it moves in clockwise direction, a negative angle is formed.*

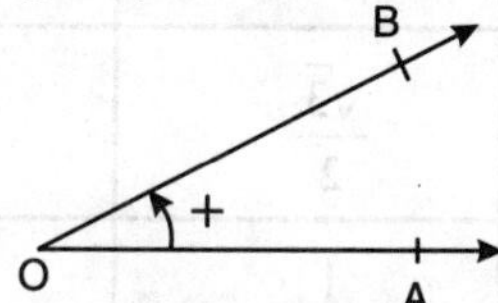

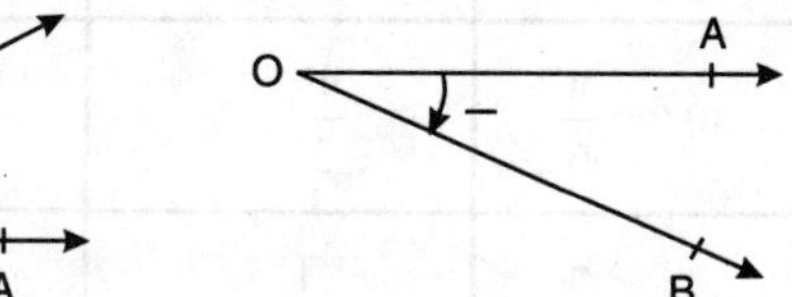

Quadrants: Let *X′ OX* and *YOY′* be two lines at right angles to each other. We call *X′OX* and *YOY′* as *x*-axis and *y*-axis respectively.

These lines divide the plane into 4 parts. The parts *XOY, YOX′, X′OY′* and *Y′OX are known as 1st, 2nd, 3rd and 4th quadrant respectively.*

An angle is said to be in a particular quadrant, if the terminal side of the angle lies in that quadrant.

II I
III IV

Measurement of Angles:

Mainly there are three systems for measuring angles.

Sexagesimal system: In this system, angle is measured in degrees.

1 right angle = 90°, 1° = 60° and 1′ = 60″.

Centesimal system: In this system angle is measured in grades.

1 right angle = 100^g, 1^g = 100′ and 1′ = 100″.

Circular Measure: In this system, angle is measured in radians.

The angle subtended at the centre of a circle by an arc of length equal to its radius is 1 radian, written as 1^C.

$$\pi^C = 180° = 200^g = 2 \text{ right angles}$$

FORMULAE

I. Relation Between Angle, Radius & Arc Length:

If θ^C is the angle made by an arc of length *s* at the centre of a circle of radius *r*, then $\theta = \dfrac{s}{r}$.

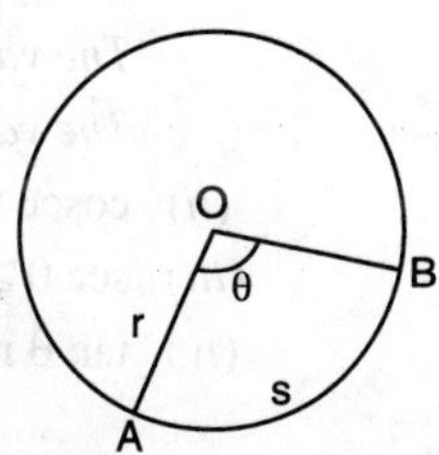

II. Trigonometric Rations:

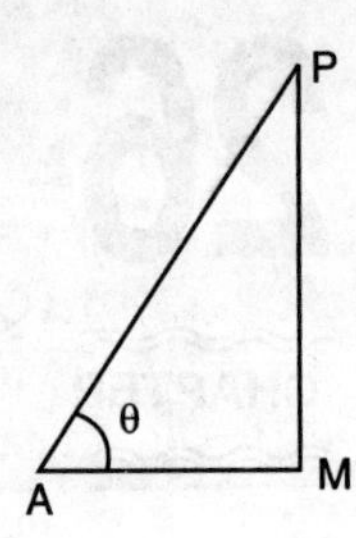

$$\sin\theta = \frac{\text{Perp.}}{\text{Hyp.}} = \frac{PM}{AP}; \cos\theta = \frac{\text{Base}}{\text{Hyp.}} = \frac{AM}{AP};$$

$$\tan\theta = \frac{\text{Perp.}}{\text{Base}} = \frac{PM}{AM};$$

$$\text{cosec}\,\theta = \frac{1}{\sin\theta}; \sec\theta = \frac{1}{\cos\theta}; \cot\theta = \frac{\cos\theta}{\sin\theta}.$$

III. Identifies:

(*i*) $\tan\theta = \dfrac{\sin\theta}{\cos\theta}$ (*ii*) $\cot\theta = \dfrac{\cos\theta}{\sin\theta}$ (*iii*) $\sin^2\theta + \cos^2\theta = 1$

(*iv*) $1 + \tan^2\theta = \sec^2\theta$ (*v*) $1 + \cot^2\theta = \text{cosec}^2\theta$

IV. Values of T-Ratios:

Angle	sin θ	cos θ	tan θ
0°	0	1	0
$30° = \frac{\pi}{6}$	$\frac{1}{2}$	$\frac{\sqrt{3}}{2}$	$\frac{1}{\sqrt{3}}$
$45° = \frac{\pi}{4}$	$\frac{1}{\sqrt{2}}$	$\frac{1}{\sqrt{2}}$	1
$60° = \frac{\pi}{3}$	$\frac{\sqrt{3}}{2}$	$\frac{1}{2}$	$\sqrt{3}$
$90° = \frac{\pi}{2}$	1	0	∞
180° = π	0	– 1	0

V. Sign of T-Ratios:

1st Quadrant: All positive.

2nd Quadrant: sin θ & cosec θ positive.

3rd Quadrant: tan θ & cot θ positive.

4th Quadrant: cos θ & sec θ positive.

Remember:	I	II	III	IV
	All	sin	tan	cos

VI. Range of T-Ratios:

(*i*) $-1 \le \sin\theta \le 1$ and $-1 \le \cos\theta \le 1$

Thus, $|\sin\theta| \le |\cos\theta| \le 1$.

The value of sin θ is never greater than 1 and never less than –1.

The value of cos θ is never greater than 1 and never less than –1.

(*ii*) $\text{cosec}\,\theta \ge 1$ and $\text{cosec}\,\theta \le -1$.

(*iii*) $\sec\theta \ge 1$ and $\sec\theta \le -1$.

(*iv*) tan θ may assume any value.

VII. Increasing & Decreasing Functions:

(*i*) The value of $\sin\theta$ increases from $-\frac{\pi}{2}$ to $\frac{\pi}{2}$. As θ increases in this interval, then $\sin\theta$ also increases.

(*ii*) 1st Quadrant: $\sin\theta$ increases from 0 to 1; $\cos\theta$ decreases from 1 to 0 and $\tan\theta$ increases from 0 to ∞.

(*iii*) 2nd Quadrant: $\sin\theta$ decreases from 1 to 0; $\cos\theta$ decreases from 0 to –1; $\tan\theta$ decreases from ∞ to 0.

(*iv*) 3rd Quadrant: $\sin\theta$ decreases from 0 to –1; $\cos\theta$ increases from – 1 to 0; $\tan\theta$ increases from 0 to ∞.

(*v*) 4th Quadrant: $\sin\theta$ increases from – 1 to 0; $\cos\theta$ increases from 0 to 1; $\tan\theta$ decreases from ∞ to 0.

VIII. T-Ratios of Negative, Complementary, Supplementary Angles etc.

(*i*) $\sin(-\theta) = -\sin\theta$, $\cos(-\theta) = \cos\theta$, $\tan(-\theta) = -\tan\theta$.

(*ii*) $\sin(90° - \theta) = \cos\theta$; $\cos(90° - \theta) = \sin\theta$; $\tan(90° - \theta) = \cot\theta$.

(*iii*) $\sin(90° - \theta) = \cos\theta$; $\cos(90° + \theta) = -\sin\theta$; $\tan(90° + \theta) = -\cot\theta$.

(*iv*) $\sin(180° - \theta) = \sin\theta$; $\cos(180° - \theta) = -\cos\theta$; $\tan(180° - \theta) = -\tan\theta$.

(*v*) $\sin(180° + \theta) = -\sin\theta$; $\cos(180° + \theta) = -\cos\theta$; $\tan(180° + \theta) = \tan\theta$.

(*vi*) $\sin(360° - \theta) = -\sin\theta$; $\cos(360° - \theta) = \cos\theta$; $\tan(360° - \theta) = -\tan\theta$.

(*vii*) $\sin(360° + \theta) = \sin\theta$; $\cos(360° + \theta) = \cos\theta$; $\tan(360° + \theta) = \tan\theta$.

IX. Sum & Difference Formula:

(*i*) $\sin(x + y) = \sin x\cos y + \cos x\sin y$. (*ii*) $\sin(x - y) = \sin x\cos y + \cos x\sin y$.

(*iii*) $\cos(x + y) = \cos x\cos y - \sin x\sin y$. (*iv*) $\cos(x - y) = \cos x\cos y + \sin x\sin y$.

(*v*) $2\sin x\cos y = \sin(x + y) + \sin(x - y)$. (*vi*) $2\cos x\sin y = \sin(x + y) - \sin(x - y)$.

(*vii*) $2\cos x\cos y = \cos(x + y) + \cos(x - y)$.

(*viii*) $2\sin x\sin y = \cos(x - y) - \cos(x + y)$. (*ix*) $\sin^2 x - \sin^2 y = \sin(x + y)\,.\,\sin(x - y)$,

(*x*) $\cos^2 x - \sin^2 y = \cos(x + y)\,.\,\cos(x - y)$,

X. Some More Formulae:

(*i*) $\sin C + \sin D = 2\sin\left(\frac{C+D}{2}\right)\cos\left(\frac{C-D}{2}\right)$.

(*ii*) $\sin C - \sin D = 2\cos\left(\frac{C+D}{2}\right)\sin\left(\frac{C-D}{2}\right)$.

(*iii*) $\cos C + \cos D = 2\cos\left(\frac{C+D}{2}\right)\cos\left(\frac{C-D}{2}\right)$.

(*iv*) $\cos C - \cos D = -2\sin\left(\frac{C+D}{2}\right)\sin\left(\frac{C-D}{2}\right)$.

XI. Tangent Formulae:

(*i*) $\tan(x + y) = \dfrac{\tan x + \tan y}{1 - \tan x\tan y}$ (*ii*) $\tan(x - y) = \dfrac{\tan x - \tan y}{1 + \tan x\tan y}$

XII. T-Ratios of Multiple Angles:

(*i*) $\sin 2x = 2\sin x\cos x$

(*ii*) $\cos 2x = \cos^2 x - \sin^2 x = 2\cos^2 x - 1 = 1 - 2\sin^2 x$.

(*iii*) $1 + \cos 2x = 2\cos^2 x$ and $1 - \cos 2x = 2\sin^2 x$.

(*iv*) $\sin 3x = 3\sin x - 4\sin^3 x$ (*v*) $\cos 3x = 4\cos^3 x - 3\cos x$

(*vi*) $\tan 2x = \dfrac{2\tan x}{1-\tan^2 x}$ (*vii*) $\tan 3x = \left(\dfrac{3\tan x - \tan^3 x}{1-3\tan^2 x}\right)$.

(*viii*) $\cos 2x = \dfrac{1-\tan^2 x}{1+\tan^2 x}$ (*ix*) $\sin 2x = \dfrac{2\tan x}{1+\tan^2 x}$

XIII. T-Ratios of Sub-multiple Angles:

(*i*) $\sin x = 2\sin(x/2)\cos(x/2)$

(*ii*) $\cos x = \cos^2\dfrac{x}{2} - \sin^2\dfrac{x}{2} = 2\cos^2\dfrac{x}{2} - 1 = 1 - 2\sin^2\dfrac{x}{2}$

(*iii*) $\tan x = \dfrac{2\tan(x/2)}{1-\tan^2(x/2)}$ (*iv*) $1-\cos x = 2\sin^2\dfrac{x}{2}$ and $1+\cos x = 2\cos^2\dfrac{x}{2}$

(*v*) $\sin x = \dfrac{2\tan(x/2)}{1+\tan^2(x/2)}$ (*vi*) $\cos x = \dfrac{1-\tan^2(x/2)}{1+\tan^2(x/2)}$

XIV. T-Ratios of Some Special Angles:

(*i*) $\sin 15° = \dfrac{\sqrt{3}-1}{2\sqrt{2}} = \cos 75°$ (*ii*) $\cos 15° = \dfrac{\sqrt{3}+1}{2\sqrt{2}} = \sin 75°$

(*iii*) $\sin 18° = \dfrac{\sqrt{5}-1}{4} = \cos 72°$ (*iv*) $\cos 18° = \dfrac{\sqrt{10+2\sqrt{5}}}{4} = \sin 72°$

(*v*) $\sin 36° = \dfrac{\sqrt{10-2\sqrt{5}}}{4} = \cos 54°$ (*vi*) $\cos 36° = \dfrac{\sqrt{5}+1}{4} = \sin 54°$

(*vii*) $\sin 22\dfrac{1°}{2} = \dfrac{\sqrt{2-\sqrt{2}}}{2}$ (*viii*) $\cos 22\dfrac{1°}{2} = \dfrac{\sqrt{2+\sqrt{2}}}{2}$

XV. Trigonometric Equations:

(*i*) $\sin\theta = 0 \Rightarrow \theta = n\pi$ (*ii*) $\cos\theta = 0 \Rightarrow \theta = (2n+1)\dfrac{\pi}{2}$

(*iii*) $\tan\theta = 0 \Rightarrow \theta = n\pi$

XVI. Periodicity: *A function f(x) has periodicity p, if f (x + p) = f(x),*

$\sin x$ has periodicity 2π, since $\sin(x + 2\pi) = \sin x$.

$\cos x$ has periodicity 2π, since $\cos(x + 2\pi) = \cos x$.

Things To Remember:

1. $\dfrac{\pi}{6} = 30°, \dfrac{\pi}{4} = 45°, \dfrac{\pi}{3} = 60°, \dfrac{\pi}{2} = 90°, \pi = 180° \;\&\; 2\pi = 360°$.

2. (*i*) All T-ratios are positive in 1st quadrant.

(*ii*) $\sin\theta$ and $\operatorname{cosec}\theta$ are +ve in 1st & 2nd quadrants.

(*iii*) $\tan\theta$ and $\cot\theta$ are + ve in 1st & 3rd quadrants.

(*iv*) $\cos\theta$ and $\sec\theta$ are +ve in 1st & 4th quadrants.

3. (*i*) $\sin(180° - \theta) = \sin\theta$; $\tan(180° + \theta) = \tan\theta$; $\cos(360° - \theta) = \cos\theta$.

(*ii*) $\sin(\pi - \theta) = \sin\theta$; $\tan(\pi + \theta) = \tan\theta$; $\cos(2\pi - \theta) = \cos\theta$.

4. (*i*) 2 sin *A* cos *B* = sin (*A* + *B*) + sin (*A* – *B*)
(*ii*) 2 cos *A* sin *B* = sin (*A* + *B*) – sin (*A* – *B*)
(*iii*) 2 cos *A* cos *B* = cos (*A* + *B*) + cos (*A* – *B*)
(*iv*) 2 sin *A* sin *B* = cos (*A* – *B*) – cos (*A* + *B*)

SOLVED PROBLEMS

Ex. 1. ***If*** $\sin\theta = \frac{12}{13}$ ***and*** $\frac{\pi}{2} < \theta < \pi$, ***find the values of cos θ, tan θ, cosec θ, sec θ and cot θ.***

Sol. Since $\frac{\pi}{2} < \theta < \pi$, so θ lies in 2nd quadrant.
In 2nd quadrant, cos θ is negative.

$$\therefore \quad \cos\theta = -\sqrt{1-\sin^2\theta} = -\sqrt{1-\frac{144}{169}} = -\sqrt{\frac{25}{169}} = \frac{-5}{13}.$$

$$\tan\theta = \frac{\sin\theta}{\cos\theta} = \frac{-12}{5};\ \text{cosec}\,\theta = \frac{1}{\sin\theta} = \frac{13}{12};$$

$$\sec\theta = \frac{1}{\cos\theta} = \frac{-13}{5}\ \text{and}\ \cot\theta = \frac{1}{\tan\theta} = -\frac{5}{12}.$$

Ex. 2. ***Evaluate:***
cosec² 90° + sec² 30° + tan² 30° + tan² 60° + sin² 0° + cosec² 60°.

Sol. We know that $\text{cosec}\,90° = \frac{1}{\sin 90°} = 1, \sec 30° = \frac{1}{\cos 30°} = \frac{2}{\sqrt{3}};$

$$\tan 30° = \frac{1}{\sqrt{3}}, \tan 60° = \sqrt{3}, \sin 0° = 0,\ \text{cosec}\,60° = \frac{1}{\sin 60°} = \frac{2}{\sqrt{3}}.$$

$\therefore$ cosec² 90° + sec² 30° + tan² 30° + tan² 60° + sin² 0° + cosec² 60°

$$= \left[1^2 + \left(\frac{2}{\sqrt{3}}\right)^2 + \left(\frac{1}{\sqrt{3}}\right)^2 + (\sqrt{3})^2 + 0^2 + \left(\frac{2}{\sqrt{3}}\right)^2\right] = \left(1 + \frac{4}{3} + \frac{1}{3} + 3 + 0 + \frac{4}{3}\right) = 7.$$

Ex. 3. ***Evaluate:*** $\left(\sin\frac{\pi}{6} + \cos\frac{\pi}{3} - \tan^3\frac{\pi}{4} + \text{cosec}^2\frac{\pi}{2}\right).$

Sol. We know that: $\sin\frac{\pi}{6} = \sin 30° = \frac{1}{2}, \cos\frac{\pi}{3} = \cos 60° = \frac{1}{2};$

$$\tan\frac{\pi}{4} = \tan 45° = 1\ \text{and}\ \text{cosec}\frac{\pi}{2} = \frac{1}{\sin\frac{\pi}{2}} = \frac{1}{\sin 90°} = 1.$$

$$\therefore\ \sin\frac{\pi}{6} + \cos\frac{\pi}{3} - \tan^3\frac{\pi}{4} + \text{cosec}^2\frac{\pi}{2} = \left(\frac{1}{2} + \frac{1}{2} - 1^3 + 1^2\right) = 1.$$

Ex. 4. ***Find the values of:***
(i) sin 150° (ii) cos 135° (iii) tan (– 210°) (iv) cot 240° (v) sec 330° (vi) cosec 390°

Sol. (*i*) $\sin 150° = \sin(180° - 30°) = \sin 30° = \frac{1}{2}.$

(*ii*) $\cos 135° = \cos(180° - 45°) = -\cos 45° = -\frac{1}{\sqrt{2}}.$

(*iii*) $\tan(-210°) = -\tan 210° = -\tan(180° + 30°) = -\tan 30° = -\frac{1}{\sqrt{3}}$.

(*iv*) $\cot 240° = \cot(180° + 60°) = \cot 60° = \frac{1}{\sqrt{3}}$.

(*v*) $\sec 330° = \sec(360° - 30°) = \sec 30° = \frac{1}{\cos 30°} = \frac{2}{\sqrt{3}}$.

(*vi*) $\operatorname{cosec} 390° = \operatorname{cosec}(360° + 30°) = \operatorname{cosec} 30° = \frac{1}{\sin 30°} = 2$.

Ex. 5. ***If $\sec\theta = 2$ and $0 < \theta < 2\pi$, find the values of θ.***

Sol. $\sec\theta = 2 \Rightarrow \cos\theta = \frac{1}{2} = \cos\frac{\pi}{3} \Rightarrow \theta = \frac{\pi}{3}$.

But, $\cos\theta = \cos(2\pi - \theta)$.

$\therefore \quad \cos\frac{\pi}{3} = \cos\left(2\pi - \frac{\pi}{3}\right) = \cos\frac{5\pi}{3}$.

Hence, the required values of θ are $\frac{\pi}{3}$ and $\frac{5\pi}{3}$.

Ex. 6. ***If $\tan\theta = \sqrt{3}$ and $0 < \theta < 2\pi$, find the values of θ.***

Sol. $\tan\theta = \sqrt{3} = \tan\frac{\pi}{3} \Rightarrow \theta = \frac{\pi}{3}$.

But, $\tan\theta = \tan(\pi + \theta)$.

So, $\tan\frac{\pi}{3} = \tan\left(\pi + \frac{\pi}{3}\right) = \tan\frac{4\pi}{3}$.

Hence, the required values of θ are $\frac{\pi}{3}$ and $\frac{4\pi}{3}$.

Ex. 7. ***Find the values of:***

(i) $\sin 56° - \cos 34°$ ***(ii) $\sin^2 41° + \sin^2 59°$*** ***(iii) $\frac{\tan 57°}{\cot 33°}$***

(iv) $\dfrac{\cos(90° + \theta) \times \sec(-\theta) \times \tan(180° - \theta)}{\sec(360° - \theta) \times \sin(180° + \theta) \times \cot(90° - \theta)}$ ***(v) $\cos 780° + \sin 480°$***

Sol. (*i*) $\sin 56° - \cos 34° = \sin 56° - \cos(90° - 56°) = \sin 56° - \sin 56° = 0$.

(*ii*) $\sin^2 41° + \sin^2 59° = \sin^2 41° + \sin^2(90° - 41°) = \sin^2 41° + \cos^2 41° = 1$.

(*iii*) $\frac{\tan 57°}{\cot 33°} = \frac{\tan(90° - 33°)}{\cot 33°} = \frac{\cot 33°}{\cot 33°} = 1$.

(*iv*) $\dfrac{\cos(90° + \theta) \cdot \sec(-\theta) \cdot \tan(180° - \theta)}{\sec(360° - \theta) \cdot \sin(180° + \theta) \cdot \cot(90° - \theta)}$

$$= \frac{(-\sin\theta) \cdot \sec\theta \cdot (-\tan\theta)}{\sec(-\theta) \cdot (-\sin\theta) \cdot \tan\theta} = \frac{-\sin\theta \cdot \sec\theta \cdot \tan\theta}{\sec\theta \cdot \sin\theta \cdot \tan\theta} = -1.$$

(*v*) $\cos 780° + \sin 480° = \cos(360° \times 2 + 60°) + \sin(360° + 120°)$

$= \cos 60° + \sin 120° = \cos 60° + \sin(180° - 60°)$

$$= \cos 60° + \sin 60° = \left(\frac{1}{2} + \frac{\sqrt{3}}{2}\right) = \left(\frac{\sqrt{3}+1}{2}\right).$$

Ex. 8. ***If $x^2 + y^2 + z^2 = r^2$ and $x = r \cos A \sin B$, $y = r \sin A \sin B$, find the value of z.***

Sol. $x^2 + y^2 = r^2 \cos^2 A \sin^2 B + r^2 \sin^2 A \sin^2 B = r^2 \sin^2 B(\cos^2 A + \sin^2 A) = r^2 \sin^2 B.$

Now, $x^2 + y^2 + z^2 = r^2 \Rightarrow z^2 = r^2 - (x^2 + y^2)$

$\therefore z^2 = r^2 - r^2 \sin^2 B = r^2 (1 - \sin^2 B) = r^2 \cos^2 B$. Hence, $z = r \cos B$.

Ex. 9. ***Evaluate:***

(i) sin 79° cos 19° – cos 79° sin 19° ***(ii) cos 48° cos 12° – sin 48° sin 12°***

(iii) cos 70° cos 40° + sin 70° sin 40° ***(iv) sin 36° cos 9° + cos 36° sin 9°***

Sol. (*i*) $\sin 79° \cos 19° - \cos 79° \sin 19° = \sin (79° - 19°) = \sin 60° = \frac{\sqrt{3}}{2}.$

$[\because \sin x \cos y - \cos x \sin y = \sin (x - y)]$

(*ii*) $\cos 48° \cos 12° - \sin 48° \sin 12° = \cos (48° + 12°) = \cos 60° = \frac{1}{2}.$

$[\because \cos x \cos y - \sin x \sin y = \cos (x + y)]$

(*iii*) $\cos 70° \cos 40° + \sin 70° \sin 40° = \cos (70° - 40°) = \cos 30° = \frac{\sqrt{3}}{2}.$

$[\because \cos x \cos y + \sin x \sin y = \cos (x - y)]$

(*iv*) $\sin 36° \cos 9° + \cos 36° \sin 9° = \sin(36° + 9°) = \sin 45° = \frac{1}{\sqrt{2}}.$

Ex. 10. ***If 0 < θ < 90°, solve the following equations:***

(i) $2 \cos^2 \theta + \sin \theta - 2 = 0$ ***(ii) $\tan^2 \theta + 3 = 3 \sec \theta$***

Sol. (*i*) $2 \cos^2 \theta + \sin \theta - 2 = 0 \Leftrightarrow 2(1 - \sin^2\theta) + \sin \theta - 2 = 0$

$\Leftrightarrow 2 \sin^2 \theta - \sin \theta = 0 \Leftrightarrow \sin \theta (2 \sin \theta - 1) = 0$

$\Leftrightarrow \sin \theta = 0$ or $\sin \theta = \frac{1}{2} \Leftrightarrow \theta = \frac{\pi}{6}.$ $[\because \theta \neq 0]$

(*ii*) $\tan^2 \theta + 3 = 3 \sec \theta \Rightarrow (\sec^2 \theta - 1) + 3 = 3 \sec \theta$

$\Leftrightarrow \sec^2 \theta - 3 \sec \theta + 2 = 0 \Leftrightarrow (\sec \theta - 2) (\sec \theta - 1) = 0$

$\Leftrightarrow \sec \theta = 2$ or $\sec \theta = 1 \Leftrightarrow \theta = \frac{\pi}{3}.$ $[\because \theta \neq 0]$

Ex. 11. ***Find the values of tan 15° and cos 105°.***

Sol. $$\tan 15° = \tan (45° - 30°) = \frac{\tan 45° - \tan 30°}{1 + \tan 45° \tan 30°} = \frac{1 - \frac{1}{\sqrt{3}}}{1 + \frac{1}{\sqrt{3}}} = \left(\frac{\sqrt{3}-1}{\sqrt{3}+1}\right).$$

$$\cos 105° = \cos (60° + 45°) = \cos 60° \cos 45° - \sin 60° \sin 45° = \left(\frac{1}{2} \times \frac{1}{\sqrt{2}} - \frac{\sqrt{3}}{2} \times \frac{1}{\sqrt{2}}\right) = \left(\frac{1-\sqrt{3}}{2\sqrt{2}}\right).$$

Ex. 12. ***If x = a sin θ and y = b tan θ, prove that: $\frac{a^2}{x^2} - \frac{b^2}{y^2} = 1$.***

Sol. Clearly, $\frac{a}{x}$ = cosec θ and $\frac{b}{y}$ = cot θ.

Squaring and subtracting, we get: $\left(\frac{a^2}{x^2}-\frac{b^2}{y^2}\right) = (\text{cosec}^2\,\theta - \cos^2\theta) = 1.$

Ex. 13. ***Which of the following relations is true?***

(i) $\sec^2\theta + \tan^2\theta = 1$ ***(ii)*** $\tan^2\theta + \cot^2\theta = 1$

(iii) $\text{cosec}^2\,\theta - \cot^2\theta = 1$ ***(iv)*** $\cot^2\theta - \text{cosec}^2\,\theta = 1$

Sol. Since $1 + \cot^2\theta = \text{cosec}^2\,\theta$, so $\text{cosec}^2\,\theta - \cot^2\theta = 1$ is the true statement.

Ex. 14. ***If*** $\sin\theta = \frac{4}{5}$, ***find the values of*** $\sin 2\theta, \cos 2\theta, \tan 2\theta, \sin 3\theta$ ***and*** $\cos 3\theta$.

Sol. $\cos\theta = \sqrt{1-\sin^2\theta} = \sqrt{1-\frac{16}{25}} = \sqrt{\frac{9}{25}} = \frac{3}{5}.$

$\therefore \sin 2\theta = 2\sin\theta\cos\theta = \left(2\times\frac{4}{5}\times\frac{3}{5}\right) = \frac{24}{25};$

$\cos 2\theta = (1-2\sin^2\theta) = \left(1-2\times\frac{16}{25}\right) = \frac{-7}{25};$

$\tan 2\theta = \frac{\sin 2\theta}{\cos 2\theta} = \frac{24}{25}\times\left(\frac{25}{-7}\right) = \frac{-24}{7};$

$\sin 3\theta = 3\sin\theta - 4\sin^3\theta = \left[3\times\frac{4}{5}-4\times\left(\frac{4}{5}\right)^3\right] = \frac{44}{125};$

$\cos 3\theta = (4\cos^3\theta - 3\cos\theta) = \left[4\times\left(\frac{3}{5}\right)^3 - 3\times\frac{3}{5}\right] = -\frac{117}{125}.$

Ex. 15. ***If*** $\tan\phi = \frac{3}{4}$, ***find the values of*** $\cos 2\phi$ ***and*** $\sin 2\phi$.

Sol. $\cos 2\phi = \frac{1-\tan^2\phi}{1+\tan^2\phi} = \left(\frac{1-\frac{9}{16}}{1+\frac{9}{16}}\right) = \left(\frac{7}{16}\times\frac{16}{25}\right) = \frac{7}{25}.$

$\sin 2\phi = \frac{2\tan\phi}{1+\tan^2\phi} = \left(\frac{2\times\frac{3}{4}}{1+\frac{9}{16}}\right) = \left(\frac{3}{2}\times\frac{16}{25}\right) = \frac{24}{25}.$

Ex. 16. ***Evaluate:*** $\cos^2 45° - \sin^2 15°$.

Sol. We know that $\cos^2\theta - \sin^2\phi = \cos(\theta+\phi)\,.\,\cos(\theta-\phi)$

$\therefore \cos^2 45° - \sin^2 15° = \cos(45° + 15)\,.\,\cos(45° - 15°) = \cos 60°\,.\,\cos 30° = \frac{1}{2}\times\frac{\sqrt{3}}{2} = \frac{\sqrt{3}}{4}.$

Ex. 17. ***Prove that:***

(i) $\frac{\sin 5\theta - \sin 3\theta}{\cos 3\theta + \cos 5\theta} = \tan\theta$ ***(ii)*** $\frac{\sin\phi - \sin 3\phi}{\cos\phi + \cos 3\phi} = \tan 2\phi$

(iii) $\dfrac{\sin 3x - \sin x}{\cos x\,\theta - \cos 3x} = \cot 2x$

Sol. (*i*) $$\frac{\sin 5\theta - \sin 3\theta}{\cos 3\theta + \cos 5\theta} = \frac{2\cos\left(\frac{5\theta+3\theta}{2}\right)\sin\left(\frac{5\theta-3\theta}{2}\right)}{2\cos\left(\frac{5\theta+3\theta}{2}\right)\cos\left(\frac{5\theta-3\theta}{2}\right)} = \tan\theta.$$

$$\left[\begin{array}{l}\text{Using}\sin C - \sin D = 2\cos\left(\frac{C+D}{2}\right)\sin\left(\frac{C-D}{2}\right)\\ \quad\cos C + \cos D = 2\cos\left(\frac{C+D}{2}\right)\cos\left(\frac{C-D}{2}\right)\end{array}\right]$$

(*ii*) $$\frac{\sin\phi + \sin 3\phi}{\cos\phi + \cos 3\phi} = \frac{2\sin\left(\frac{3\phi+\phi}{2}\right)\cos\left(\frac{3\phi-\phi}{2}\right)}{2\cos\left(\frac{3\phi+\phi}{2}\right)\cos\left(\frac{3\phi-\phi}{2}\right)} = \tan 2\phi.$$

$$\left[\begin{array}{l}\text{Using}\sin C + \sin D = 2\sin\left(\frac{C+D}{2}\right)\cos\left(\frac{C-D}{2}\right)\\ \text{and}\cos C + \cos D = 2\cos\left(\frac{C+D}{2}\right)\cos\left(\frac{C-D}{2}\right)\end{array}\right]$$

(*iii*) $$\frac{\sin 3x - \sin x}{\cos x - \cos 3x} = \frac{2\cos\left(\frac{3x+x}{2}\right)\sin\left(\frac{3x-x}{2}\right)}{2\sin\left(\frac{3x+x}{2}\right)\sin\left(\frac{3x-x}{2}\right)} = \cot 2x$$

Ex. 18. ***If cos*** $(\alpha+\beta) = \frac{4}{5}$, **sin** $(\alpha-\beta) = \frac{5}{13}$ ***and*** α, β ***lie between*** **0** ***and*** $\frac{\pi}{4}$, ***then find the value of tan*** 2α.

Sol. $\cos(\alpha+\beta) = \frac{4}{5} \Rightarrow \sec(\alpha+\beta) = \frac{5}{4}.$

$\Rightarrow \quad \tan^2(\alpha+\beta) = [\sec^2(\alpha+\beta) - 1] = \left(\frac{25}{16} - 1\right) = \frac{9}{16}. \quad \Rightarrow \tan(\alpha+\beta) = \frac{3}{4}.$

Again, $\sin(\alpha-\beta) = \frac{5}{13} \Rightarrow \cos(\alpha-\beta) = \sqrt{\left(1 - \frac{25}{169}\right)} = \frac{12}{13}.$

$\therefore \tan(\alpha-\beta) = \left(\frac{5}{13}\times\frac{13}{12}\right) = \frac{5}{12}.$

So, $\tan 2\alpha = \tan[(\alpha+\beta)] + (\alpha-\beta)] = \dfrac{\tan(\alpha+\beta) + \tan(\alpha-\beta)}{1 - \tan(\alpha+\beta)\tan(\alpha-\beta)} = \left(\dfrac{\frac{3}{4}+\frac{5}{12}}{1-\frac{3}{4}\times\frac{5}{12}}\right) = \dfrac{56}{33}.$

EXERCISE 26

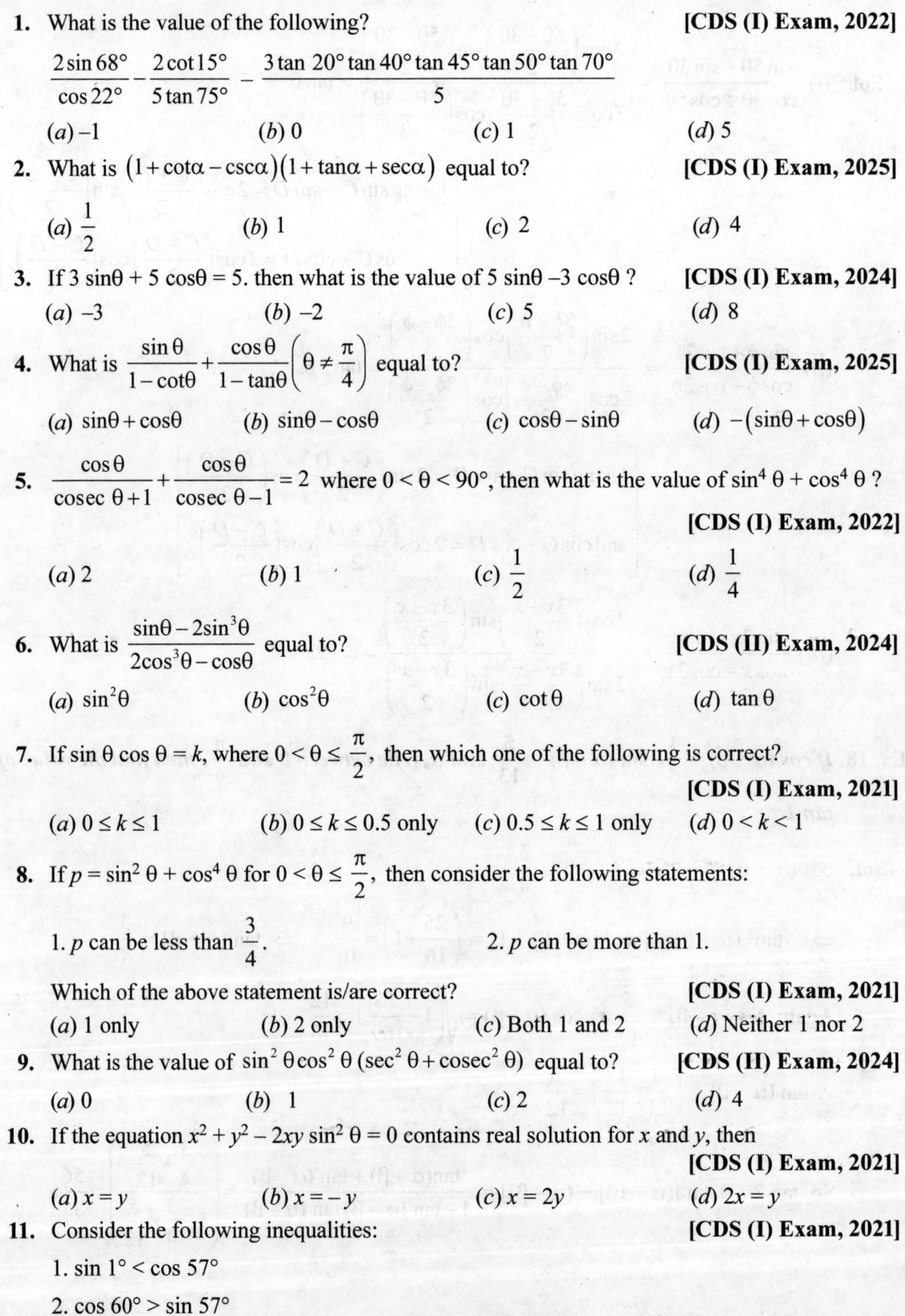

1. What is the value of the following? **[CDS (I) Exam, 2022]**

$$\frac{2\sin 68^\circ}{\cos 22^\circ}-\frac{2\cot 15^\circ}{5\tan 75^\circ}-\frac{3\tan 20^\circ\tan 40^\circ\tan 45^\circ\tan 50^\circ\tan 70^\circ}{5}$$

(*a*) –1 (*b*) 0 (*c*) 1 (*d*) 5

2. What is $(1+\cot\alpha-\csc\alpha)(1+\tan\alpha+\sec\alpha)$ equal to? **[CDS (I) Exam, 2025]**

(*a*) $\frac{1}{2}$ (*b*) 1 (*c*) 2 (*d*) 4

3. If 3 sinθ + 5 cosθ = 5. then what is the value of 5 sinθ –3 cosθ ? **[CDS (I) Exam, 2024]**

(*a*) –3 (*b*) –2 (*c*) 5 (*d*) 8

4. What is $\frac{\sin\theta}{1-\cot\theta}+\frac{\cos\theta}{1-\tan\theta}\left(\theta\neq\frac{\pi}{4}\right)$ equal to? **[CDS (I) Exam, 2025]**

(*a*) $\sin\theta+\cos\theta$ (*b*) $\sin\theta-\cos\theta$ (*c*) $\cos\theta-\sin\theta$ (*d*) $-(\sin\theta+\cos\theta)$

5. $\frac{\cos\theta}{\text{cosec }\theta+1}+\frac{\cos\theta}{\text{cosec }\theta-1}=2$ where $0<\theta<90^\circ$, then what is the value of $\sin^4\theta+\cos^4\theta$? **[CDS (I) Exam, 2022]**

(*a*) 2 (*b*) 1 (*c*) $\frac{1}{2}$ (*d*) $\frac{1}{4}$

6. What is $\frac{\sin\theta-2\sin^3\theta}{2\cos^3\theta-\cos\theta}$ equal to? **[CDS (II) Exam, 2024]**

(*a*) $\sin^2\theta$ (*b*) $\cos^2\theta$ (*c*) $\cot\theta$ (*d*) $\tan\theta$

7. If $\sin\theta\cos\theta=k$, where $0<\theta\leq\frac{\pi}{2}$, then which one of the following is correct? **[CDS (I) Exam, 2021]**

(*a*) $0\leq k\leq 1$ (*b*) $0\leq k\leq 0.5$ only (*c*) $0.5\leq k\leq 1$ only (*d*) $0<k<1$

8. If $p=\sin^2\theta+\cos^4\theta$ for $0<\theta\leq\frac{\pi}{2}$, then consider the following statements:

1. p can be less than $\frac{3}{4}$. 2. p can be more than 1.

Which of the above statement is/are correct? **[CDS (I) Exam, 2021]**

(*a*) 1 only (*b*) 2 only (*c*) Both 1 and 2 (*d*) Neither 1 nor 2

9. What is the value of $\sin^2\theta\cos^2\theta\,(\sec^2\theta+\text{cosec}^2\theta)$ equal to? **[CDS (II) Exam, 2024]**

(*a*) 0 (*b*) 1 (*c*) 2 (*d*) 4

10. If the equation $x^2+y^2-2xy\sin^2\theta=0$ contains real solution for x and y, then **[CDS (I) Exam, 2021]**

(*a*) $x=y$ (*b*) $x=-y$ (*c*) $x=2y$ (*d*) $2x=y$

11. Consider the following inequalities: **[CDS (I) Exam, 2021]**

1. sin 1° < cos 57°

2. cos 60° > sin 57°

Which of the above is/ are correct?

(*a*) 1 only (*b*) 2 only (*c*) Both 1 and 2 (*d*) Neither 1 nor 2

12. What is $\frac{\sin^3\theta+\cos^3\theta}{\sin\theta+\cos\theta}+\frac{\sin^3\theta-\cos^3\theta}{\sin\theta-\cos\theta}$ equal to? **[CDS (II) Exam, 2021]**

(*a*) 0 (*b*) 1 (*c*) 2 (*d*) 4

13. If $2\sin^4\alpha+2\cos^4\alpha-1=0$, where $0\le\alpha<\frac{\pi}{2}$, then what is $\sin 2\alpha+\cos 2\alpha$ equal to?

[CDS (I) Exam, 2025]

(*a*) 0 (*b*) 1 (*c*) $\frac{\sqrt{3}+1}{2}$ (*d*) $\frac{\sqrt{3}-1}{2}$

14. If $\frac{\cos^2\theta-3\cos\theta+2}{\sin^2\theta}=1$, where $0°<\theta<90°$ then what is $\sin^2\theta+\cos\theta$ equal to ?

[CDS (II) Exam, 2021]

(*a*) $\frac{5}{4}$ (*b*) $\frac{3}{2}$ (*c*) $\frac{7}{4}$ (*d*) 2

15. Consider the following: **[CDS (II) Exam, 2021]**

1. $\sin^4\theta-\sin^2\theta=\cos^4\theta-\cos^2\theta$
2. $\sin^4\theta-\sin^4\theta=1+2\sin^2\theta\cos^2\theta$
3. $\tan^4\theta-\tan^2\theta=\cos^4\theta-\sec^2\theta$

Which of the above are identities?

(*a*) 1 and 2 only (*b*) 2 and 3 only (*c*) 1 and 3 only (*d*) 1, 2, and 3

16. For how many values of α does the expression $(\sin\alpha+2)(\sin\alpha+4)(\sin\alpha-2)(\sin\alpha-4)$ become zero? **[CDS (I) Exam, 2025]**

(*a*) No value (*b*) One (*c*) Two (*d*) Four

17. What is the value of x, where $0\le x<30^\circ$, satisfying $\tan 3x\cdot\tan 6x=1$?

[CDS (I) Exam, 2025]

(*a*) 0° (*b*) 10° (*c*) 12° (*d*) 15°

18. If $\theta+\sec\theta=k$, then what is the value of $\sin^2\theta-\tan^2\theta$? **[CDS (I) Exam, 2020]**

(*a*) $4-k$ (*b*) $4-k^2$ (*c*) k^2-4 (*d*) k^2+2

19. *ABC* is a triangle inscribed in a semicircle of diameter *AB*. What is $\cos(A+B)+\sin(A+B)$ equal to ? **[CDS (I) Exam, 2020]**

(*a*) 0 (*b*) $\frac{1}{4}$ (*c*) $\frac{1}{2}$ (*d*) 1

20. Consider the following statements: **[CDS (I) Exam, 2020]**

1. $\sin\theta=x+\frac{1}{x}$ is possible for some real value of x.

2. $\cos\theta=x+\frac{1}{x}$ is possible for some real value of x.

Which of the above statements is/are correct?

(*a*) 1 only (*b*) 2 only (*c*) Both 1 and 2 (*d*) Neither 1 nor 2

21. If $\operatorname{cosec}\theta-\sin\theta=p^3$ and $\sec\theta-\cos\theta=q^3$, then what is the value of $\tan\theta$?

[CDS (I) Exam, 2020]

(*a*) $\frac{p}{q}$ (*b*) $\frac{q}{p}$ (*c*) pq (*d*) p^2q^2

22. What is the least value of $(25 \operatorname{cosec}^2 x + \sec^2 x)$? **[CDS (I) Exam, 2020]**

(*a*) 40 (*b*) 36 (*c*) 26 (*d*) 24

23. If the equation $x \cos \theta = x^2 + p$ has a real solution for every θ where $0 < \theta \le \frac{\pi}{4}$, then which one of the following is correct ? **[CDS (I) Exam, 2024]**

(*a*) $p = \frac{1}{8}$ (*b*) $p \le \frac{1}{8}$ (*c*) $p \ge \frac{1}{8}$ (*d*) $p \le \frac{1}{4}$

24. If $\tan 6\theta = \cot 2\theta$, where $0 < 6\theta < \frac{\pi}{2}$, then what is the value of $\sec 4\theta$? **[CDS (I) Exam, 2020]**

(*a*) $\sqrt{2}$ (*b*) 2 (*c*) $\frac{2}{\sqrt{3}}$ (*d*) $\frac{4}{3}$

25. What is the value of $\sin^2 6° + \sin^2 12° + \sin^2 18° + ... + \sin^2 84° + \sin^2 90°$? **[CDS (I) Exam, 2020]**

(*a*) 1 (*b*) 2 (*c*) 4 (*d*) 8

26. If $\cos^2 x + \cos x = 1$, then what is the value of $\sin^{12} x + 3 \sin^{10} x + 3 \sin^8 x + \sin^6 x$? **[CDS (I) Exam, 2019]**

(*a*) 1 (*b*) 2 (*c*) 4 (*d*) 8

27. If $0 < \theta < 90°$, $\sin \theta = \frac{3}{5}$ and $x = \cot \theta$, then what is the value of $1 + 3x + 9x^2 + 27x^3 + 81x^4 + 243x^5$? **[CDS (I) Exam, 2019]**

(*a*) 941 (*b*) 1000 (*c*) 1220 (*d*) 1365

28. If $\frac{\sin\theta+\cos\theta+1}{\sin\theta+\cos\theta-1} = p \sec \theta + q \tan \theta$, where $0 < \theta < \frac{\pi}{2}$, then what is the $p + q$ equal to ? **[CDS (I) Exam, 2024]**

(*a*) 0 (*b*) 1 (*c*) 2 (*d*) 4

29. What is $\sin^4 \theta - \cos^4 \theta$ equal to for any real number θ ? **[CDS (I) Exam, 2018]**

(*a*) 1 (*b*) $1 - 2 \sin^2 \theta$ (*c*) $2 \cos^2 \theta + 1$ (*d*) $1 - 2 \cos^2 \theta$

30. What is $\cot 1° \cot 23° \cot 45° \cot 67° \cot 89°$ equal to ? **[CDS (I) Exam, 2018]**

(*a*) 0 (*b*) 1 (*c*) $\frac{1}{2}$ (*d*) $\frac{1}{3}$

31. If $\tan (3A) = \cot (A - 22°)$, where 3A is an acute angle, then what is the value of A ? **[CDS (I) Exam, 2024]**

(*a*) 25° (*b*) 27° (*c*) 28° (*d*) 30°

32. Consider the following statements: **[CDS (I) Exam, 2018]**

1. $(\sec^2 \theta - 1)(1 - \text{coces}^2 \theta) = 1$

2. $\sin \theta (1+ \cos \theta)^{-1} + (1 + \cos \theta)(\sin \theta^{-1}) = 2 \operatorname{cosec} \theta$

Which of the above is/are correct?

(*a*) 1 only (*b*) 2 only (*c*) Both 1 and 2 (*d*) Neither 1 nor 2

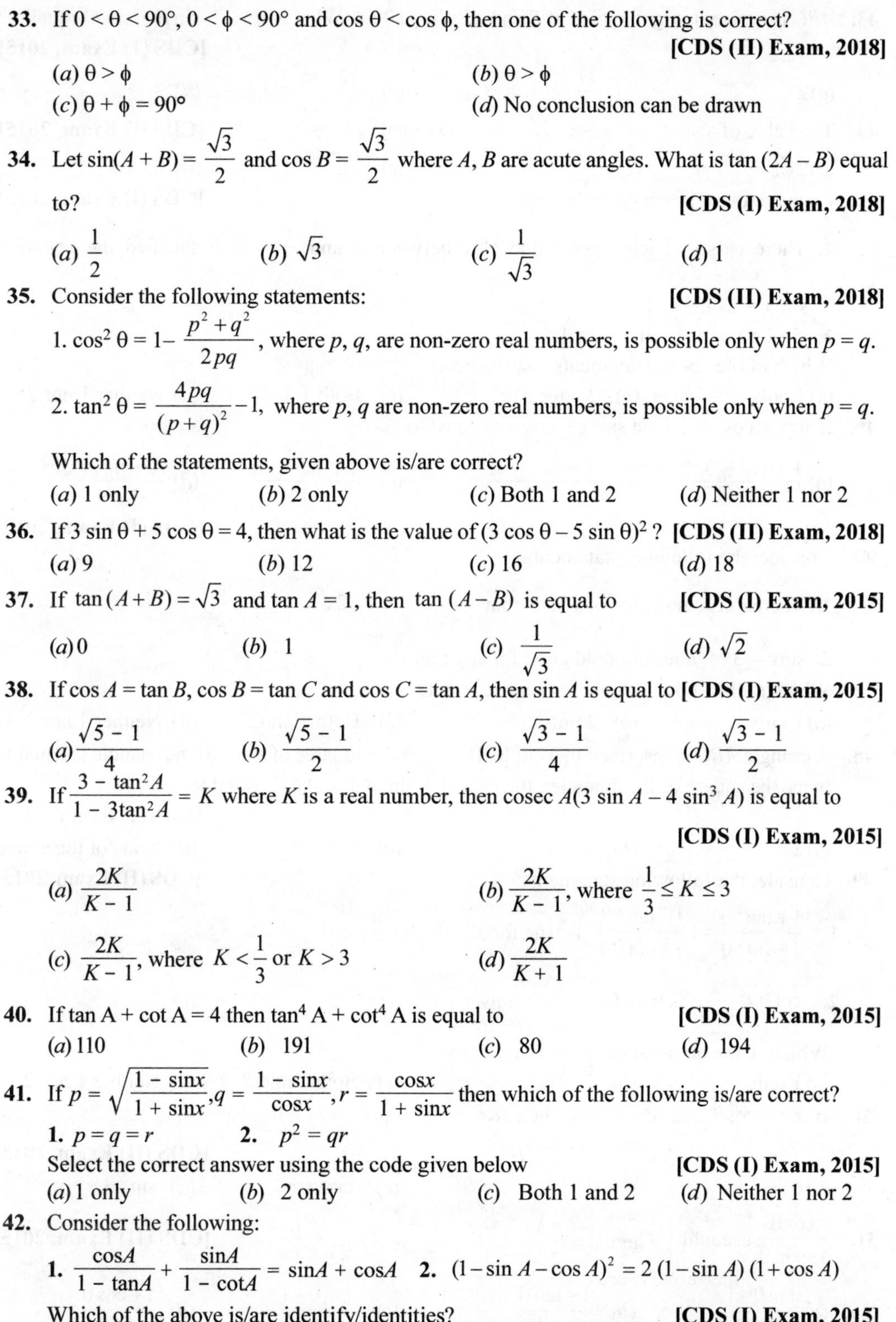

33. If $0 < \theta < 90°$, $0 < \phi < 90°$ and $\cos\theta < \cos\phi$, then one of the following is correct? **[CDS (II) Exam, 2018]**

(*a*) $\theta > \phi$ (*b*) $\theta > \phi$

(*c*) $\theta + \phi = 90°$ (*d*) No conclusion can be drawn

34. Let $\sin(A+B) = \dfrac{\sqrt{3}}{2}$ and $\cos B = \dfrac{\sqrt{3}}{2}$ where A, B are acute angles. What is $\tan(2A - B)$ equal to? **[CDS (I) Exam, 2018]**

(*a*) $\dfrac{1}{2}$ (*b*) $\sqrt{3}$ (*c*) $\dfrac{1}{\sqrt{3}}$ (*d*) 1

35. Consider the following statements: **[CDS (II) Exam, 2018]**

1. $\cos^2\theta = 1 - \dfrac{p^2+q^2}{2pq}$, where p, q, are non-zero real numbers, is possible only when $p = q$.

2. $\tan^2\theta = \dfrac{4pq}{(p+q)^2} - 1$, where p, q are non-zero real numbers, is possible only when $p = q$.

Which of the statements, given above is/are correct?

(*a*) 1 only (*b*) 2 only (*c*) Both 1 and 2 (*d*) Neither 1 nor 2

36. If $3\sin\theta + 5\cos\theta = 4$, then what is the value of $(3\cos\theta - 5\sin\theta)^2$? **[CDS (II) Exam, 2018]**

(*a*) 9 (*b*) 12 (*c*) 16 (*d*) 18

37. If $\tan(A+B) = \sqrt{3}$ and $\tan A = 1$, then $\tan(A-B)$ is equal to **[CDS (I) Exam, 2015]**

(*a*) 0 (*b*) 1 (*c*) $\dfrac{1}{\sqrt{3}}$ (*d*) $\sqrt{2}$

38. If $\cos A = \tan B$, $\cos B = \tan C$ and $\cos C = \tan A$, then $\sin A$ is equal to **[CDS (I) Exam, 2015]**

(*a*) $\dfrac{\sqrt{5}-1}{4}$ (*b*) $\dfrac{\sqrt{5}-1}{2}$ (*c*) $\dfrac{\sqrt{3}-1}{4}$ (*d*) $\dfrac{\sqrt{3}-1}{2}$

39. If $\dfrac{3-\tan^2 A}{1-3\tan^2 A} = K$ where K is a real number, then $\operatorname{cosec} A(3\sin A - 4\sin^3 A)$ is equal to **[CDS (I) Exam, 2015]**

(*a*) $\dfrac{2K}{K-1}$ (*b*) $\dfrac{2K}{K-1}$, where $\dfrac{1}{3} \le K \le 3$

(*c*) $\dfrac{2K}{K-1}$, where $K < \dfrac{1}{3}$ or $K > 3$ (*d*) $\dfrac{2K}{K+1}$

40. If $\tan A + \cot A = 4$ then $\tan^4 A + \cot^4 A$ is equal to **[CDS (I) Exam, 2015]**

(*a*) 110 (*b*) 191 (*c*) 80 (*d*) 194

41. If $p = \sqrt{\dfrac{1-\sin x}{1+\sin x}}, q = \dfrac{1-\sin x}{\cos x}, r = \dfrac{\cos x}{1+\sin x}$ then which of the following is/are correct?

1. $p = q = r$ **2.** $p^2 = qr$

Select the correct answer using the code given below **[CDS (I) Exam, 2015]**

(*a*) 1 only (*b*) 2 only (*c*) Both 1 and 2 (*d*) Neither 1 nor 2

42. Consider the following:

1. $\dfrac{\cos A}{1-\tan A} + \dfrac{\sin A}{1-\cot A} = \sin A + \cos A$ **2.** $(1-\sin A-\cos A)^2 = 2(1-\sin A)(1+\cos A)$

Which of the above is/are identify/identities? **[CDS (I) Exam, 2015]**

(*a*) 1 only (*b*) 2 only (*c*) Both 1 and 2 (*d*) Neither 1 nor 2

43. ABC is a triangle right angled at B and $AB : BC = 3 : 4$. What is $\sin A + \sin B + \sin C$ equal to? **[CDS (I) Exam, 2015]**

(*a*) 2 (*b*) $\frac{11}{5}$ (*c*) $\frac{12}{5}$ (*d*) 3

44. The value of $\text{cosec}^2 67° + \sec^2 57° - \cot^2 33° - \tan^2 23°$ is **[CDS (I) Exam, 2015]**

(*a*) $2\sqrt{2}$ (*b*) 2 (*c*) $\sqrt{2}$ (*d*) 0

45. Consider the following statements: **[CDS (I) Exam, 2015]**

1. There exists at least one value of x between 0 and $\frac{\pi}{2}$ which satisfied the equation $\sin^4 x - 2\sin^2 x - 1 = 0$.

2. sin 1.5 is greater than cos 1.5.

Which of the above statements is/are correct?

(*a*) 1 only (*b*) 2 only (*c*) Both 1 and 2 (*d*) Neither 1 nor 2

46. If $\sin x + \cos x = c$ then $\sin^6 x + \cos^6 x$ is equal to

(*a*) $\frac{1 + 6c^2 - 3c^4}{16}$ (*b*) $\frac{1 + 6c^2 - 3c^4}{4}$ (*c*) $\frac{1 + 6c^2 + 3c^4}{16}$ (*d*) $\frac{1 + 6c^2 + 3c^4}{4}$

[CDS (I) Exam, 2015]

47. Consider the following statements:

1. There exists no value of x such that $\frac{1}{1-\sin x} = 4 + 2\sqrt{3}, 0 < x < \frac{\pi}{2}$

2. $\sin x = 3^{\sin^2 x}$ does not hold good for any real x

Which of the above statements is/are correct?

(*a*) 1 only (*b*) 2 only (*c*) Both 1 and 2 (*d*) Neither 1 nor 2

48. A triangle ABC is inscribed in a circle. If sum of the squares of sides of the triangle is equal to twice the square of the diameter, then $\sin^2 A + \sin^2 B + \sin^2 C$ is equal to

[CDS (I) Exam, 2015]

(*a*) 2 (*b*) 3 (*c*) 4 (*d*) None of the above

49. Consider the following statements: **[CDS (II) Exam, 2015]**

1. $\frac{1+\tan^2\theta}{1+\cot^2\theta} = \left(\frac{1-\tan\theta}{1-\cot\theta}\right)^2$ is true for all $0 < \theta < \frac{\pi}{2}, \theta \neq \frac{\pi}{4}$

2. $\cot\theta = \frac{1}{\tan\theta}$ is true for $\theta = 45°$ only

Which of the above statements is/are correct?

(*a*) 1 only (*b*) 2 only (*c*) Both 1 and 2 (*d*) Neither 1 nor 2

50. If $x = a\cos\theta$ and $y = b\cot\theta$, then $(ax^{-1} - by^{-1})(ax^{-1} + by^{-1})$ is equal to

[CDS (II) Exam, 2015]

(*a*) 0 (*b*) 1 (*c*) $\tan^2\theta$t (*d*) $\sin^2\theta$

51. $\frac{\cos\theta}{1-\sin\theta}$ is equal $\left(\text{where } \theta \neq \frac{\pi}{2}\right)$ **[CDS (II) Exam, 2015]**

(*a*) $\frac{\tan\theta - 1}{\tan\theta + 1}$ (*b*) $\frac{1+\tan\theta}{\cos\theta}$ (*c*) $\frac{\tan\theta+1}{\tan\theta-1}$ (*d*) $\frac{1+\cos\theta}{\sin\theta}$

52. If $\tan(x+40)^\circ \tan(x+20)^\circ \tan(3x)^\circ \tan(70-x)^\circ \tan(50-x)^\circ = 1$ then the value of x is equal to **[CDS (II) Exam, 2015]**

(*a*) 30 (*b*) 20 (*c*) 15 (*d*) 10

53. If θ is an acute angle and sini cos θ = 2 cos³ θ – 1.5 cos θ, then what is sini equal to? **[CDS (II) Exam, 2015]**

(*a*) $\frac{\sqrt{5}-1}{4}$ (*b*) $\frac{1-\sqrt{5}}{4}$ (*c*) $\frac{\sqrt{5}+1}{4}$ (*d*) $-\frac{\sqrt{5}+1}{4}$

54. Consider the following statements: **[CDS (II) Exam, 2015]**

1. sin 66° is less than cos 66°.

2. sin 26° is less than cos 26°.

Which of the above statements is/are correct?

(*a*) 1 only (*b*) 2 only (*c*) Both 1 and 2 (*d*) Neither 1 nor 2

55. If *a* and *b* are positive, then the relation $\sin\theta = \frac{2a+3b}{3b}$ is **[CDS(II)Exam,20105]**

(*a*) not possible (*b*) possible only if $a = b$

(*c*) possible if $a > b$ (*d*) possible if $a < b$

56. If tanθ + sec θ = 2, then tanθ is equal to **[CDS (II) Exam, 2015]**

(*a*) $\frac{3}{4}$ (*b*) $\frac{5}{4}$ (*c*) $\frac{3}{2}$ (*d*) $\frac{5}{2}$

57. An equilateral triangle *BOC* is drawn inside a square *ABCD*. If angle *AOD* = 2θ, what is tanθ equal to? **[CDS (II) Exam, 2015]**

(*a*) $2-\sqrt{3}$ (*b*) $1+\sqrt{2}$ (*c*) $4-\sqrt{3}$ (*d*) $2+\sqrt{3}$

58. The minimum value of $\cos^2 x + \cos^2 y - \cos^2 z$ is **[CDS (II) Exam, 2015]**

(*a*) –1 (*b*) 0 (*c*) 1 (*d*) 2

59. The value of $32\cot^2\left(\frac{\pi}{4}\right) - 8\sec^2\left(\frac{\pi}{3}\right) + 8\cos^3\left(\frac{\pi}{6}\right)$ is equal to **[CDS (II) Exam, 2015]**

(*a*) $\sqrt{3}$ (*b*) $2\sqrt{3}$ (*c*) 3 (*d*) $3\sqrt{3}$

60. $\left(\frac{\sin 35^\circ}{\cos 55^\circ}\right)^2 - \left(\frac{\cos 55^\circ}{\sin 35^\circ}\right)^2 + 2\sin 30^\circ$ is equal to **[CDS (I) Exam, 2016]**

(a) –1 (*b*) 0 (*c*) 1 (*d*) 2

61. Consider the following statements:

1. There exists a positive real number m such that $\cos x = 2^{m+1}$.

2. $mn \geq m + n$ for all *m*, *n* belonging to set of natural numbers.

Which of the above statements is/are correct?

(*a*) 1 only (*b*) 2 only (*c*) Both 1 and 2 (*d*) Neither 1 nor 2

[CDS (I) Exam, 2016]

62. If $\frac{x}{a} - \frac{y}{b}\tan\theta = 1$ and $\frac{x}{a}\tan\theta + \frac{y}{b} = 1$, then the value of $\frac{x^2}{a^2} + \frac{y^2}{b^2}$ is

(*a*) 2 sec² θ (*b*) sec² θ (*c*) cos² θ (*d*) 2cos² θ

[CDS (I) Exam, 2016]

63. Consider the following:

1. $\sqrt{\frac{1-\cos\theta}{1+\cos\theta}} = \operatorname{cosec}\theta - \cot\theta$ **2.** $\sqrt{\frac{1+\cos\theta}{1-\cos\theta}} = \operatorname{cosec}\theta + \cot\theta$

Which of the above is/are identity/identities? **[CDS (I) Exam, 2016]**

(*a*) 1 only (*b*) 2 only (*c*) Both 1 and 2 (*d*) Neither 1 nor 2

64. If $p = \cot\theta + \tan\theta$ and $q = \sec\theta - \cos\theta$, then $(p^2q)^{\frac{2}{3}} - (q^2p)^{\frac{2}{3}}$ is equal to

(*a*) 0 (*b*) 1 (*c*) 2 (*d*) 3

[CDS (I) Exam, 2016]

65. If $\dfrac{\cos^2\theta - 3\cos\theta + 2}{\sin^2\theta} = 1$ where $0 < \theta < \dfrac{\pi}{2}$, then which of the following statements is/are correct? **[CDS (I) Exam, 2016]**

1. There are two values of θ satisfying the above equation.

2. θ = 60° is satisfied by the above equation.

Select the correct answer using the code given below.

(*a*) 1 only (*b*) 2 only (*c*) Both 1 and 2 (*d*) Neither 1 nor 2

66. Which of the following is correct in respect of the equation $3 - \tan^2\theta = \alpha(1 - 3\tan^2\theta)$? (Given that is a real number.) **[CDS (I) Exam, 2016]**

(*a*) $\alpha \in \left[\dfrac{1}{3}, 3\right]$ (*b*) $\alpha \in \left(-\infty, \dfrac{1}{3}\right] \cup [3, \infty)$

(*c*) $\alpha \in \left(-\infty, \dfrac{1}{3}\right) \cup [3, \infty)$ (*d*) None of the above

67. If $\tan\theta + \cot\theta = \dfrac{4}{\sqrt{3}}$, where $0 < \theta < \dfrac{\pi}{2}$, then $\sin\theta + \cos\theta$ is equal to **[CDS (I) Exam, 2016]**

(*a*) 1 (*b*) $\dfrac{\sqrt{3}-1}{2}$ (*c*) $\dfrac{\sqrt{3}+1}{2}$ (*d*) $\sqrt{2}$

68. If $\sin\theta + \cos\theta = \dfrac{\sqrt{7}}{2}$, then what is $\sin\theta - \cos\theta$ equal to? **[CDS (II) Exam, 2016]**

(*a*) 0 (*b*) $\dfrac{1}{2}$ (*c*) 1 (*d*) $\sqrt{2}$

69. If $\sin x + \sin^2 x = 1$, then what is the value of $\cos^8 x + 2\cos^6 x + \cos^4 x$?

(*a*) 0 (*b*) 1 (*c*) 2 (*d*) 4

[CDS (II) Exam, 2016]

70. What is the value of $\text{cosec}^2 68° + \sec^2 56° - \cot^2 34° - \tan^2 22°$? **[CDS (II) Exam, 2016]**

(*a*) 0 (*b*) $\dfrac{1}{2}$ (*c*) 1 (*d*) 2

71. If $2y\cos\theta = x\sin\theta$ and $2x\sec\theta - y\,\text{cosec}\,\theta = 3$, then what is $x^2 + 4y^2$ equal to?

(*a*) 1 (*b*) 2 (*c*) 4 (*d*) 8

[CDS (II) Exam, 2016]

72. If $\sin\theta + \cos\theta = \dfrac{1+\sqrt{3}}{2}$ where $0 < \theta < \dfrac{\pi}{2}$, then what is $\tan\theta + \cot\theta$ equal to?

(*a*) $\dfrac{\sqrt{3}}{4}$ (*b*) $\dfrac{1}{\sqrt{3}}$ (*c*) $\sqrt{3}$ (*d*) $\dfrac{4}{\sqrt{3}}$

[CDS (II) Exam, 2016]

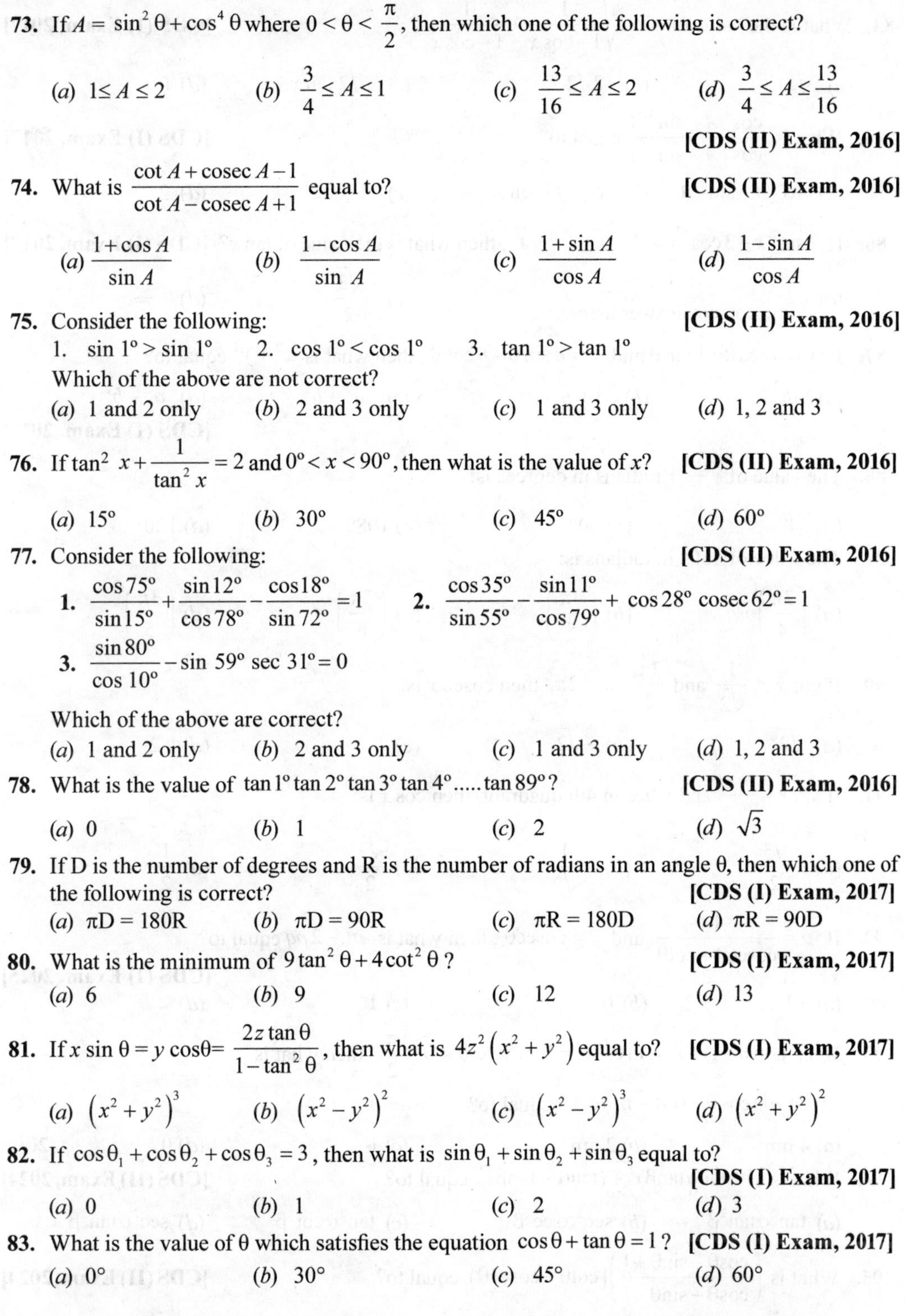

73. If $A = \sin^2\theta + \cos^4\theta$ where $0 < \theta < \dfrac{\pi}{2}$, then which one of the following is correct?

(*a*) $1 \le A \le 2$ (*b*) $\dfrac{3}{4} \le A \le 1$ (*c*) $\dfrac{13}{16} \le A \le 2$ (*d*) $\dfrac{3}{4} \le A \le \dfrac{13}{16}$

[CDS (II) Exam, 2016]

74. What is $\dfrac{\cot A + \operatorname{cosec} A - 1}{\cot A - \operatorname{cosec} A + 1}$ equal to? **[CDS (II) Exam, 2016]**

(*a*) $\dfrac{1+\cos A}{\sin A}$ (*b*) $\dfrac{1-\cos A}{\sin A}$ (*c*) $\dfrac{1+\sin A}{\cos A}$ (*d*) $\dfrac{1-\sin A}{\cos A}$

75. Consider the following: **[CDS (II) Exam, 2016]**

1. $\sin 1^\circ > \sin 1^\circ$ 2. $\cos 1^\circ < \cos 1^\circ$ 3. $\tan 1^\circ > \tan 1^\circ$

Which of the above are not correct?

(*a*) 1 and 2 only (*b*) 2 and 3 only (*c*) 1 and 3 only (*d*) 1, 2 and 3

76. If $\tan^2 x + \dfrac{1}{\tan^2 x} = 2$ and $0^\circ < x < 90^\circ$, then what is the value of x? **[CDS (II) Exam, 2016]**

(*a*) 15° (*b*) 30° (*c*) 45° (*d*) 60°

77. Consider the following: **[CDS (II) Exam, 2016]**

1. $\dfrac{\cos 75^\circ}{\sin 15^\circ} + \dfrac{\sin 12^\circ}{\cos 78^\circ} - \dfrac{\cos 18^\circ}{\sin 72^\circ} = 1$ **2.** $\dfrac{\cos 35^\circ}{\sin 55^\circ} - \dfrac{\sin 11^\circ}{\cos 79^\circ} + \cos 28^\circ \operatorname{cosec} 62^\circ = 1$

3. $\dfrac{\sin 80^\circ}{\cos 10^\circ} - \sin 59^\circ \sec 31^\circ = 0$

Which of the above are correct?

(*a*) 1 and 2 only (*b*) 2 and 3 only (*c*) 1 and 3 only (*d*) 1, 2 and 3

78. What is the value of $\tan 1^\circ \tan 2^\circ \tan 3^\circ \tan 4^\circ \ldots.. \tan 89^\circ$? **[CDS (II) Exam, 2016]**

(*a*) 0 (*b*) 1 (*c*) 2 (*d*) $\sqrt{3}$

79. If D is the number of degrees and R is the number of radians in an angle θ, then which one of the following is correct? **[CDS (I) Exam, 2017]**

(*a*) $\pi D = 180R$ (*b*) $\pi D = 90R$ (*c*) $\pi R = 180D$ (*d*) $\pi R = 90D$

80. What is the minimum of $9\tan^2\theta + 4\cot^2\theta$? **[CDS (I) Exam, 2017]**

(*a*) 6 (*b*) 9 (*c*) 12 (*d*) 13

81. If $x \sin\theta = y \cos\theta = \dfrac{2z\tan\theta}{1-\tan^2\theta}$, then what is $4z^2\left(x^2+y^2\right)$ equal to? **[CDS (I) Exam, 2017]**

(*a*) $\left(x^2+y^2\right)^3$ (*b*) $\left(x^2-y^2\right)^2$ (*c*) $\left(x^2-y^2\right)^3$ (*d*) $\left(x^2+y^2\right)^2$

82. If $\cos\theta_1 + \cos\theta_2 + \cos\theta_3 = 3$, then what is $\sin\theta_1 + \sin\theta_2 + \sin\theta_3$ equal to?

[CDS (I) Exam, 2017]

(*a*) 0 (*b*) 1 (*c*) 2 (*d*) 3

83. What is the value of θ which satisfies the equation $\cos\theta + \tan\theta = 1$? **[CDS (I) Exam, 2017]**

(*a*) 0° (*b*) 30° (*c*) 45° (*d*) 60°

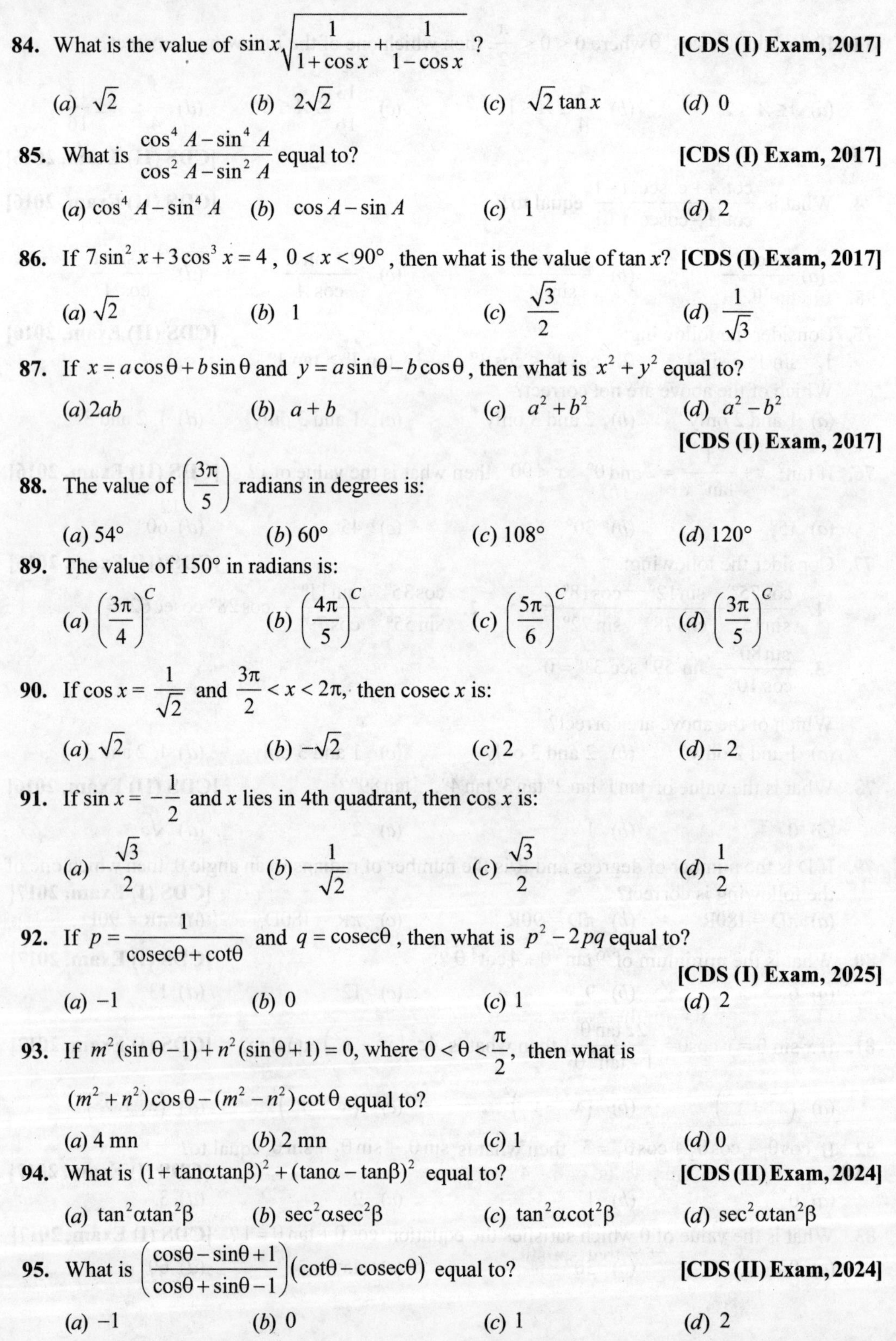

84. What is the value of $\sin x\sqrt{\frac{1}{1+\cos x}+\frac{1}{1-\cos x}}$? **[CDS (I) Exam, 2017]**

(*a*) $\sqrt{2}$ (*b*) $2\sqrt{2}$ (*c*) $\sqrt{2}\tan x$ (*d*) 0

85. What is $\frac{\cos^4 A-\sin^4 A}{\cos^2 A-\sin^2 A}$ equal to? **[CDS (I) Exam, 2017]**

(*a*) $\cos^4 A-\sin^4 A$ (*b*) $\cos A-\sin A$ (*c*) 1 (*d*) 2

86. If $7\sin^2 x+3\cos^3 x=4$, $0<x<90°$, then what is the value of $\tan x$? **[CDS (I) Exam, 2017]**

(*a*) $\sqrt{2}$ (*b*) 1 (*c*) $\frac{\sqrt{3}}{2}$ (*d*) $\frac{1}{\sqrt{3}}$

87. If $x=a\cos\theta+b\sin\theta$ and $y=a\sin\theta-b\cos\theta$, then what is x^2+y^2 equal to?

(*a*) $2ab$ (*b*) $a+b$ (*c*) a^2+b^2 (*d*) a^2-b^2

[CDS (I) Exam, 2017]

88. The value of $\left(\frac{3\pi}{5}\right)$ radians in degrees is:

(*a*) 54° (*b*) 60° (*c*) 108° (*d*) 120°

89. The value of 150° in radians is:

(*a*) $\left(\frac{3\pi}{4}\right)^C$ (*b*) $\left(\frac{4\pi}{5}\right)^C$ (*c*) $\left(\frac{5\pi}{6}\right)^C$ (*d*) $\left(\frac{3\pi}{5}\right)^C$

90. If $\cos x=\frac{1}{\sqrt{2}}$ and $\frac{3\pi}{2}<x<2\pi$, then cosec x is:

(*a*) $\sqrt{2}$ (*b*) $-\sqrt{2}$ (*c*) 2 (*d*) -2

91. If $\sin x=-\frac{1}{2}$ and x lies in 4th quadrant, then $\cos x$ is:

(*a*) $-\frac{\sqrt{3}}{2}$ (*b*) $-\frac{1}{\sqrt{2}}$ (*c*) $\frac{\sqrt{3}}{2}$ (*d*) $\frac{1}{2}$

92. If $p=\frac{1}{\operatorname{cosec}\theta+\cot\theta}$ and $q=\operatorname{cosec}\theta$, then what is p^2-2pq equal to?

[CDS (I) Exam, 2025]

(*a*) -1 (*b*) 0 (*c*) 1 (*d*) 2

93. If $m^2(\sin\theta-1)+n^2(\sin\theta+1)=0$, where $0<\theta<\frac{\pi}{2}$, then what is

$(m^2+n^2)\cos\theta-(m^2-n^2)\cot\theta$ equal to?

(*a*) 4 mn (*b*) 2 mn (*c*) 1 (*d*) 0

94. What is $(1+\tan\alpha\tan\beta)^2+(\tan\alpha-\tan\beta)^2$ equal to? **[CDS (II) Exam, 2024]**

(*a*) $\tan^2\alpha\tan^2\beta$ (*b*) $\sec^2\alpha\sec^2\beta$ (*c*) $\tan^2\alpha\cot^2\beta$ (*d*) $\sec^2\alpha\tan^2\beta$

95. What is $\left(\frac{\cos\theta-\sin\theta+1}{\cos\theta+\sin\theta-1}\right)(\cot\theta-\operatorname{cosec}\theta)$ equal to? **[CDS (II) Exam, 2024]**

(*a*) -1 (*b*) 0 (*c*) 1 (*d*) 2

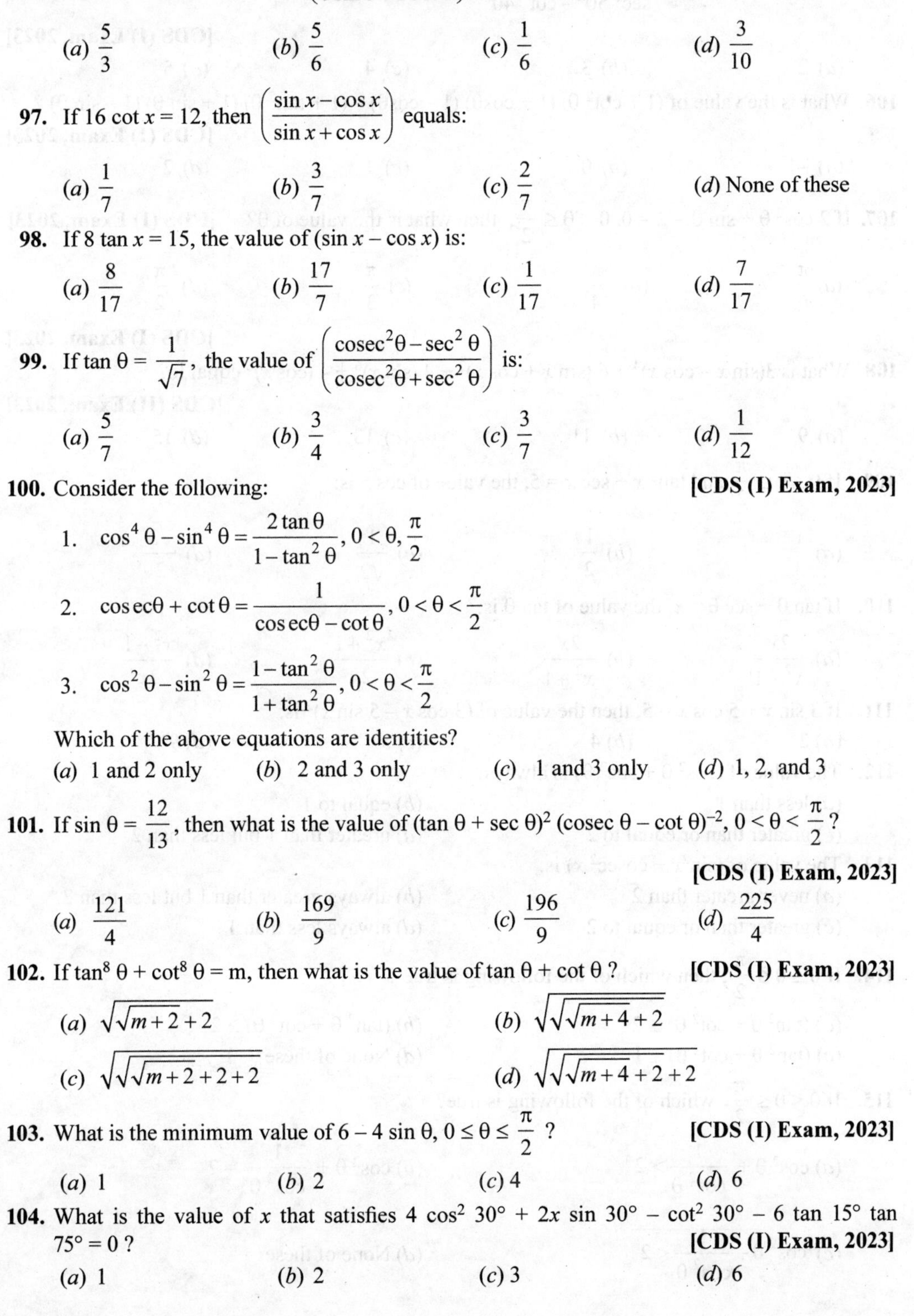

96. If 5 tan θ = 4, the value of $\left(\frac{5\sin\theta - 3\cos\theta}{5\sin\theta + 2\cos\theta}\right)$ is:

(*a*) $\frac{5}{3}$ (*b*) $\frac{5}{6}$ (*c*) $\frac{1}{6}$ (*d*) $\frac{3}{10}$

97. If 16 cot *x* = 12, then $\left(\frac{\sin x - \cos x}{\sin x + \cos x}\right)$ equals:

(*a*) $\frac{1}{7}$ (*b*) $\frac{3}{7}$ (*c*) $\frac{2}{7}$ (*d*) None of these

98. If 8 tan *x* = 15, the value of (sin *x* – cos *x*) is:

(*a*) $\frac{8}{17}$ (*b*) $\frac{17}{7}$ (*c*) $\frac{1}{17}$ (*d*) $\frac{7}{17}$

99. If $\tan\theta = \frac{1}{\sqrt{7}}$, the value of $\left(\frac{\text{cosec}^2\theta - \sec^2\theta}{\text{cosec}^2\theta + \sec^2\theta}\right)$ is:

(*a*) $\frac{5}{7}$ (*b*) $\frac{3}{4}$ (*c*) $\frac{3}{7}$ (*d*) $\frac{1}{12}$

100. Consider the following: **[CDS (I) Exam, 2023]**

1. $\cos^4\theta - \sin^4\theta = \frac{2\tan\theta}{1-\tan^2\theta}, 0 < \theta, \frac{\pi}{2}$
2. $\text{cosec}\,\theta + \cot\theta = \frac{1}{\text{cosec}\,\theta - \cot\theta}, 0 < \theta < \frac{\pi}{2}$
3. $\cos^2\theta - \sin^2\theta = \frac{1-\tan^2\theta}{1+\tan^2\theta}, 0 < \theta < \frac{\pi}{2}$

Which of the above equations are identities?

(*a*) 1 and 2 only (*b*) 2 and 3 only (*c*) 1 and 3 only (*d*) 1, 2, and 3

101. If $\sin\theta = \frac{12}{13}$, then what is the value of $(\tan\theta + \sec\theta)^2 (\text{cosec}\,\theta - \cot\theta)^{-2}$, $0 < \theta < \frac{\pi}{2}$?

[CDS (I) Exam, 2023]

(*a*) $\frac{121}{4}$ (*b*) $\frac{169}{9}$ (*c*) $\frac{196}{9}$ (*d*) $\frac{225}{4}$

102. If $\tan^8\theta + \cot^8\theta = m$, then what is the value of tan θ + cot θ ? **[CDS (I) Exam, 2023]**

(*a*) $\sqrt{\sqrt{m+2}+2}$ (*b*) $\sqrt{\sqrt{\sqrt{m+4}+2}}$

(*c*) $\sqrt{\sqrt{\sqrt{m+2}+2}+2}$ (*d*) $\sqrt{\sqrt{\sqrt{m+4}+2}+2}$

103. What is the minimum value of 6 – 4 sin θ, $0 \le \theta \le \frac{\pi}{2}$? **[CDS (I) Exam, 2023]**

(*a*) 1 (*b*) 2 (*c*) 4 (*d*) 6

104. What is the value of *x* that satisfies 4 cos² 30° + 2*x* sin 30° – cot² 30° – 6 tan 15° tan 75° = 0 ? **[CDS (I) Exam, 2023]**

(*a*) 1 (*b*) 2 (*c*) 3 (*d*) 6

105. What is the value of $\dfrac{\cos^2 32° + \cos^2 58°}{\sec^2 50° - \cot^2 40°} + 4 \tan 13° \tan 37° \tan 53° \tan 77°$?

[CDS (I) Exam, 2023]

(*a*) 2 (*b*) 3 (*c*) 4 (*d*) 5

106. What is the value of $(1 + \cot^2 \theta)(1 + \cos\theta)(1 - \cos\theta) - (1 + \tan^2 \theta)(1 + \sin \theta)(1 - \sin \theta)$?

[CDS (I) Exam, 2023]

(*a*) –1 (*b*) 0 (*c*) 1 (*d*) 2

107. If $2 \cos^2 \theta + \sin \theta - 2 = 0$, $0 < \theta \leq \dfrac{\pi}{2}$, then what is the value of θ? **[CDS (I) Exam, 2023]**

(*a*) $\dfrac{\pi}{6}$ (*b*) $\dfrac{\pi}{4}$ (*c*) $\dfrac{\pi}{3}$ (*d*) $\dfrac{\pi}{2}$

[CDS (I) Exam, 2023]

108. What is $3(\sin x - \cos x)^4 + 6(\sin x + \cos x)^2 + 4(\sin x)^6 + 4(\cos x)^6$ equal to?

[CDS (II) Exam, 2023]

(*a*) 9 (*b*) 11 (*c*) 13 (*d*) 15

109. If $0 \leq x \leq \dfrac{\pi}{2}$ and $\tan^2 x + \sec x = 5$, the value of $\cos x$ is:

(*a*) $\dfrac{1}{3}$ (*b*) $\dfrac{1}{2}$ (*c*) $\dfrac{1}{\sqrt{2}}$ (*d*) $\dfrac{\sqrt{3}}{2}$

110. If $\tan \theta + \sec \theta = x$, the value of $\tan \theta$ is:

(*a*) $\dfrac{2x}{x^2 - 1}$ (*b*) $\dfrac{2x}{x^2 + 1}$ (*c*) $\dfrac{x^2 + 1}{2x}$ (*d*) $\dfrac{x^2 - 1}{2x}$

111. If $3 \sin x + 5 \cos x = 5$, then the value of $(3 \cos x - 5 \sin x)^2$ is:

(*a*) 2 (*b*) 4 (*c*) 5 (*d*) 9

112. The value of $(\cos^2 \theta + \sec^2 \theta)$ is always

(*a*) less than 1 (*b*) equal to 1

(*c*) greater than or equal to 2 (*d*) greater than 1 but less than 2

113. The value of $(\sin^2 x + \text{cosec}^2 x)$ is:

(*a*) never greater than 2 (*b*) always greater than 1 but less than 2

(*c*) greater than or equal to 2 (*d*) always less than 1

114. If $0 \leq \theta \leq \dfrac{\pi}{2}$, then which of the following is true ?

(*a*) $(\tan^2 \theta + \cot^2 \theta) \leq 2$ (*b*) $(\tan^2 \theta + \cot^2 \theta) \geq 2$

(*c*) $(\tan^2 \theta + \cot^2 \theta) \leq 1$ (*d*) None of these

115. If $0 < \theta \leq \dfrac{\pi}{2}$, which of the following is true?

(*a*) $\cos^2 \theta + \dfrac{1}{\cos^2 \theta} > 2$ (*b*) $\cos^2 \theta + \dfrac{1}{\cos^2 \theta} = 2$

(*c*) $\cos^2 \theta + \dfrac{1}{\cos^2 \theta} < 2$ (*d*) None of these

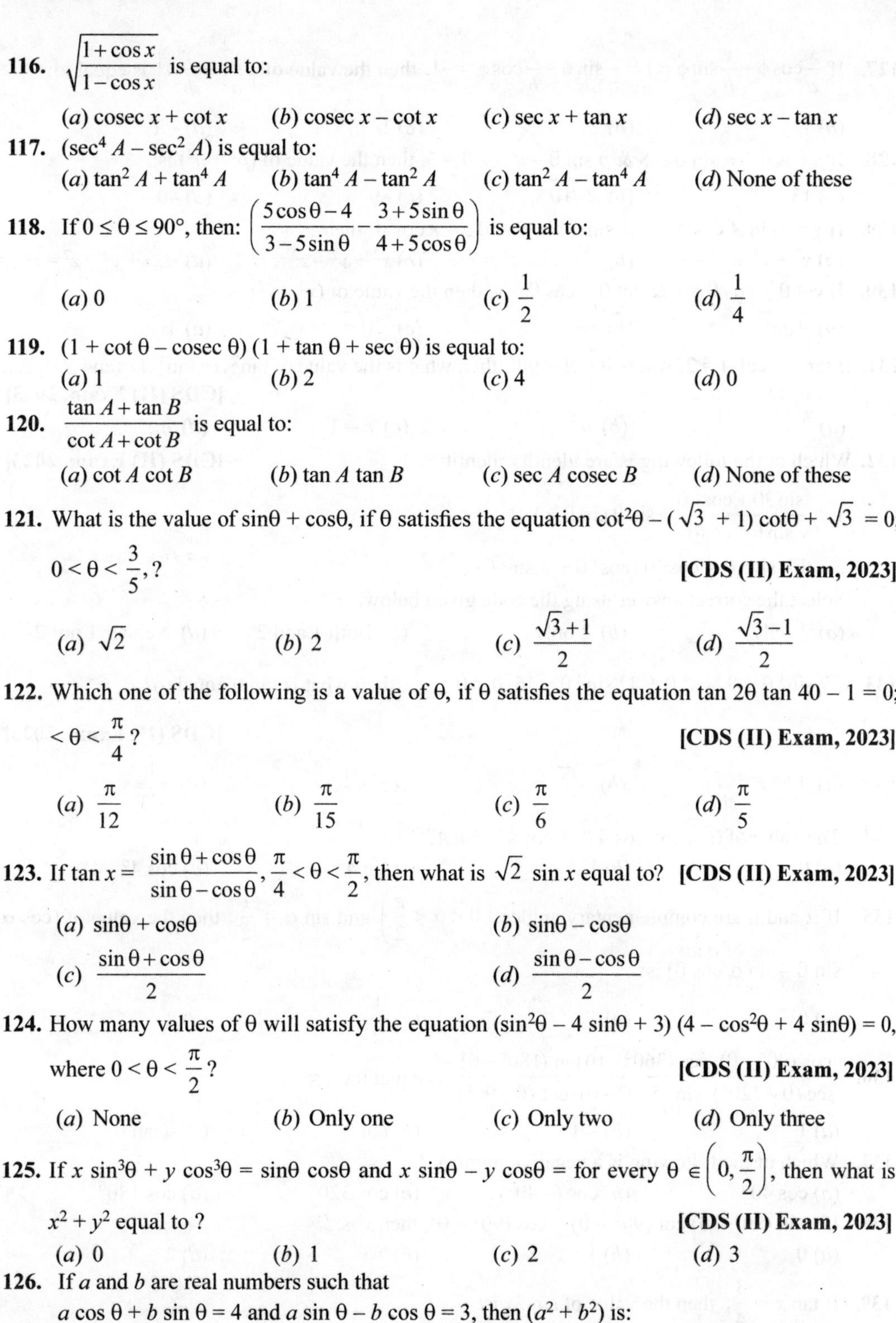

116. $\sqrt{\dfrac{1+\cos x}{1-\cos x}}$ is equal to:

(*a*) cosec x + cot x (*b*) cosec x – cot x (*c*) sec x + tan x (*d*) sec x – tan x

117. ($\sec^4 A - \sec^2 A$) is equal to:

(*a*) $\tan^2 A + \tan^4 A$ (*b*) $\tan^4 A - \tan^2 A$ (*c*) $\tan^2 A - \tan^4 A$ (*d*) None of these

118. If $0 \le \theta \le 90°$, then: $\left(\dfrac{5\cos\theta - 4}{3 - 5\sin\theta} - \dfrac{3 + 5\sin\theta}{4 + 5\cos\theta}\right)$ is equal to:

(*a*) 0 (*b*) 1 (*c*) $\dfrac{1}{2}$ (*d*) $\dfrac{1}{4}$

119. $(1 + \cot\theta - \text{cosec}\,\theta)(1 + \tan\theta + \sec\theta)$ is equal to:

(*a*) 1 (*b*) 2 (*c*) 4 (*d*) 0

120. $\dfrac{\tan A + \tan B}{\cot A + \cot B}$ is equal to:

(*a*) cot A cot B (*b*) tan A tan B (*c*) sec A cosec B (*d*) None of these

121. What is the value of sinθ + cosθ, if θ satisfies the equation $\cot^2\theta - (\sqrt{3} + 1)\cot\theta + \sqrt{3} = 0$; $0 < \theta < \dfrac{3}{5}$,? **[CDS (II) Exam, 2023]**

(*a*) $\sqrt{2}$ (*b*) 2 (*c*) $\dfrac{\sqrt{3}+1}{2}$ (*d*) $\dfrac{\sqrt{3}-1}{2}$

122. Which one of the following is a value of θ, if θ satisfies the equation tan 2θ tan 40 – 1 = 0; $< \theta < \dfrac{\pi}{4}$? **[CDS (II) Exam, 2023]**

(*a*) $\dfrac{\pi}{12}$ (*b*) $\dfrac{\pi}{15}$ (*c*) $\dfrac{\pi}{6}$ (*d*) $\dfrac{\pi}{5}$

123. If $\tan x = \dfrac{\sin\theta + \cos\theta}{\sin\theta - \cos\theta}, \dfrac{\pi}{4} < \theta < \dfrac{\pi}{2}$, then what is $\sqrt{2}\,\sin x$ equal to? **[CDS (II) Exam, 2023]**

(*a*) sinθ + cosθ (*b*) sinθ – cosθ

(*c*) $\dfrac{\sin\theta + \cos\theta}{2}$ (*d*) $\dfrac{\sin\theta - \cos\theta}{2}$

124. How many values of θ will satisfy the equation $(\sin^2\theta - 4\sin\theta + 3)(4 - \cos^2\theta + 4\sin\theta) = 0$, where $0 < \theta < \dfrac{\pi}{2}$? **[CDS (II) Exam, 2023]**

(*a*) None (*b*) Only one (*c*) Only two (*d*) Only three

125. If $x\sin^3\theta + y\cos^3\theta = \sin\theta\cos\theta$ and $x\sin\theta - y\cos\theta =$ for every $\theta \in \left(0, \dfrac{\pi}{2}\right)$, then what is $x^2 + y^2$ equal to ? **[CDS (II) Exam, 2023]**

(*a*) 0 (*b*) 1 (*c*) 2 (*d*) 3

126. If a and b are real numbers such that

$a\cos\theta + b\sin\theta = 4$ and $a\sin\theta - b\cos\theta = 3$, then $(a^2 + b^2)$ is:

(*a*) 7 (*b*) 12 (*c*) 25 (*d*) $\sqrt{12}$

127. If $\frac{x}{a}\cos\phi+\frac{y}{b}\sin\phi=1$ & $\frac{x}{a}\sin\phi-\frac{y}{b}\cos\phi=-1$, then the value of $\left(\frac{x^2}{a^2}+\frac{y^2}{b^2}\right)$ is equal to:

(*a*) 1 (*b*) 2 (*c*) 1 (*d*) – 1

128. If $p\cos\theta+q\sin\theta=8$ & $p\sin\theta-q\cos\theta=5$, then the value of (p^2+q^2) is:

(*a*) 13 (*b*) $2\sqrt{10}$ (*c*) 89 (*d*) 40

129. If $x=r\sin A\cos B$, $y=r\sin A\sin B$ and $z=r\cos A$, then

(*a*) $x^2+y^2+z^2=r^2$ (*b*) $x^2-y^2+z^2=r^2$ (*c*) $x^2+y^2-z^2=r^2$ (*d*) $-x^2+y^2-z^2=r^2$

130. If $\cot\theta+\cos\theta=x$ & $\cot\theta-\cos\theta=y$, then the value of (x^2-y^2) is:

(*a*) $4\sqrt{xy}$ (*b*) $2xy$ (*c*) $2\sqrt{xy}$ (*d*) $4xy$

131. If $\tan A+\cot A=2$, where $0<A<90°$, then what is the value of $\tan^2 A+\tan^3 A+\tan4 A+$ .... $+\tan^3 A$? **[CDS (II) Exam, 2023]**

(*a*) 1 (*b*) $n-2$ (*c*) $n-1$ (*d*) n

132. Which of the following is/are identity/identities ? **[CDS (II) Exam, 2023]**

1. $\frac{\sin 3\theta+\cos^2\theta}{\sin\theta+\cos\theta}+\sin\theta\cos\theta=1;\ 0<\theta<\frac{\pi}{2}$

2. $1-\sin^6\theta=\cos^2\theta\,(\cos^4\theta+3\sin^2\theta)$

Select the correct answer using the code given below:

(*a*) 1 only (*b*) 2 only (*c*) Both 1 and 2 (*d*) Neither 1 nor 2

133. If $7\sin^4\theta+9\cos^4\theta+42\sin^2\theta=16,\ 0<\theta<\frac{\pi}{2}$, then what is $\tan\theta$ equal to?

[CDS (II) Exam, 2023]

(*a*) 1 (*b*) $\sqrt{2}$ (*c*) $\sqrt{3}$ (*d*) $\frac{1}{\sqrt{3}}$

134. The value of (sin 43° cos 47° + cos 43° sin 47°) is:

(*a*) 0 (*b*) 1 (*c*) sin 4° (*d*) cos 4°

135. If α and β are complementary angles $\left(0<\alpha<\frac{\pi}{2}\right)$ and $\sin\alpha=\frac{1}{2}$, then the value of $(\cos\alpha\sin\beta-\sin\alpha\cos\beta)$ is:

(*a*) 0 (*b*) 1 (*c*) $\frac{1}{2}$ (*d*) 2

136. $\frac{\cos(90°+\theta)\cdot\sec(360°-\theta)\tan(180°-\theta)}{\sec(\theta-720°)\cdot\sin(540°+\theta)\cdot\cot(\theta-90°)}$ is equal to:

(*a*) 1 (*b*) – 1 (*c*) $\tan\theta$ (*d*) $-\tan\theta$

137. Which of the following is a negative number ?

(*a*) cos 40° (*b*) cos (– 40°) (*c*) cos 320° (*d*) cos 140°

138. If $x\sin(90°-\theta)\cot(90°-\theta)=\cos(90°-\theta)$, then x is:

(*a*) 0 (*b*) 1 (*c*) – 1 (*d*) 2

139. If $\tan x=\frac{3}{4}$, then the value of $\sin 2x$ is:

(*a*) $\frac{24}{25}$ (*b*) $\frac{1}{2}$ (*c*) $\frac{7}{16}$ (*d*) $\frac{5}{12}$

140. If $\tan x = \frac{p}{q}$, then the value of $(p\cos 2x + q\sin 2x)$ is:

(*a*) $\frac{q(3q^2 - p^2)}{(p^2 + q^2)}$ (*b*) $\frac{p(3q^2 - p^2)}{(p^2 + q^2)}$ (*c*) p (*d*) q

141. If $\cos x = \frac{1}{2}$, the value of $\cos 3x$ is:

(*a*) $\frac{3}{2}$ (*b*) $\frac{-1}{2}$ (*c*) 1 (*d*) – 1

142. The value of sin 75° is:

(*a*) $\frac{\sqrt{2}+1}{2\sqrt{2}}$ (*b*) $\frac{\sqrt{3}+1}{2\sqrt{2}}$ (*c*) $\frac{\sqrt{3}-1}{2\sqrt{2}}$ (*d*) $\frac{\sqrt{3}+\sqrt{2}}{2}$

143. The value of cos 75° is:

(*a*) $\frac{\sqrt{3}-1}{2\sqrt{2}}$ (*b*) $\frac{\sqrt{3}+1}{2\sqrt{2}}$ (*c*) $\frac{\sqrt{3}+\sqrt{2}}{2}$ (*d*) $\frac{\sqrt{2}-1}{2\sqrt{2}}$

144. If $\cos\alpha\cos\beta = 1 + \sin\alpha\sin\beta$, the value of $\tan\alpha\cot\beta$ is:

(*a*) 1 (*b*) – 1 (*c*) 0 (*d*) 2

145. If $\cot^2\theta - 3\sqrt{3}\cot\theta + 6 = 0$, where $\frac{\pi}{6} \le \theta < \frac{\pi}{2}$, then what is the value of $\sin\theta + \cos 2\theta$?

[CDS (I) Exam, 2025]

(*a*) 0 (*b*) 1 (*c*) $\sqrt{3}$ (*d*) $1+\sqrt{2}$

146. If $\cos\theta = \frac{1}{7}$ and $\cos\phi = \frac{13}{14}$, the value of $(\theta - \phi)$ is:

(*a*) 30° (*b*) 45° (*c*) 60° (*d*) 120°

147. (tan 20° + 2 tan 50°) is equal to:

(*a*) tan 30° (*b*) tan 10° (*c*) tan 15° (*d*) tan 70°

148. $\sin A = \frac{4}{5}$, $\cos B = \frac{5}{13}$ and $0 < A \le \frac{\pi}{2}, 0 < B \le \frac{\pi}{2}$, then the value of $\sin(A + B)$ is:

(*a*) $\frac{55}{56}$ (*b*) $\frac{20}{65}$ (*c*) $\frac{56}{65}$ (*d*) $\frac{12}{65}$

149. If $\tan A + \tan B = a$ and $\cot A + \cot B = b$, then the value of $\cot(A + B)$ is:

(*a*) $\frac{1}{a}+\frac{1}{b}$ (*b*) $\frac{1}{a}-\frac{1}{b}$ (*c*) $a + b$ (*d*) $a - b$

150. The value of $\cot\left(\frac{\pi}{4}+\theta\right)\cdot\cot\left(\frac{\pi}{4}-\theta\right)$ is:

(*a*) 0 (*b*) 1 (*c*) – 1 (*d*) 2

151. If $(1 + \tan\alpha)(1 + \tan\beta) = 2$, then $(\alpha + \beta)$ equals:

(*a*) $\frac{\pi}{6}$ (*b*) $\frac{\pi}{3}$ (*c*) $\frac{\pi}{4}$ (*d*) $\frac{\pi}{2}$

152. $\frac{\sin(\alpha+\beta)}{\cos(\alpha-\beta)}$ is equal to:

(*a*) $\frac{\tan\alpha + \tan\beta}{1 + \tan\alpha\tan\beta}$ (*b*) $\frac{\tan\alpha\tan\beta}{1 + \tan\alpha\tan\beta}$ (*c*) $\frac{\tan\alpha - \tan\beta}{1 - \tan\alpha\tan\beta}$ (*d*) None of these

153. tan 45° can be expressed as:

(*a*) $\frac{3\tan 15° + \tan^3 15°}{1 - 3\tan^2 15°}$ (*b*) $\frac{3\tan 15° + \tan^3 15°}{1 + 3\tan^2 15°}$

(*c*) $\frac{3\tan 15° - \tan^3 15°}{1 - 3\tan^2 15°}$ (*d*) $\frac{\tan^3 15° - 3\tan 15°}{1 - 3\tan^2 15°}$

154. The value of $\left[\frac{1}{\sin 10°} - \frac{\sqrt{3}}{\cos 10°}\right]$ is:

(*a*) 1 (*b*) 2 (*c*) 3 (*d*) 4

155. $\left(\frac{\sin 7x - \sin 5x}{\cos 7x + \cos 5x}\right)$ is equal to:

(*a*) tan 2*x* (*b*) cot 2*x* (*c*) tan *x* (*d*) cot *x*

156. $\left(\frac{\sin x + \sin 3x + \sin 5x + \sin 7x}{\cos x + \cos 3x + \cos 5x + \cos 7x}\right)$ is equal to:

(*a*) tan 16*x* (*b*) tan 8*x* (*c*) tan 4*x* (*d*) tan 2*x*

157. The value of (sin 50° – sin 70° + sin 10°) is:

(*a*) 0 (*b*) 1 (*c*) 2 (*d*) 3

158. The value of (tan 10° + tan 70° – tan 50°) is:

(*a*) $\frac{1}{3}$ (*b*) $\frac{1}{\sqrt{3}}$ (*c*) 3 (*d*) $\sqrt{3}$

159. If sin *x* + sin *y* = *a* and cos *x* + cos *y* = *b*, then the value of sin (*x* + *y*) is:

(*a*) $\frac{2ab}{a^2 - b^2}$ (*b*) $\frac{2ab}{a^2 + b^2}$ (*c*) $2(a^2 - b^2)$ (*d*) 4 *ab*

160. If cot *x* + cosec *x* = 5, the value of cos *x* is:

(*a*) $\frac{5}{12}$ (*b*) $\frac{12}{13}$ (*c*) $\frac{5}{13}$ (*d*) $\frac{1}{3}$

161. If $A + B = \frac{\pi}{4}$, then the value of (1 + tan *A*) (1 + tan *B*) is:

(*a*) 2 (*b*) 1 (*c*) $\frac{1}{2}$ (*d*) $\frac{3}{2}$

162. The value of $\left(\sin^2\frac{\pi}{8} + \sin^2\frac{3\pi}{8} + \sin^2\frac{5\pi}{5} + \sin^2\frac{7\pi}{8}\right)$ is:

(*a*) 1 (*b*) 2 (*c*) $1\frac{1}{8}$ (*d*) $2\frac{1}{8}$

163. The correct relation is:

(*a*) $\cos 2\theta = \frac{2\tan\theta}{1 - \tan^2\theta}$ (*b*) $\cos 2\theta = \frac{2\tan\theta}{1 + \tan^2\theta}$

(*c*) $\cos 2\theta = \frac{1 + \tan^2\theta}{1 - \tan^2\theta}$ (*d*) $\cos 2\theta = \frac{1 - \tan^2\theta}{1 + \tan^2\theta}$

164. The correct relation is:

(*a*) $\sin 2\theta = \dfrac{2\tan\theta}{1-\tan^2\theta}$ (*b*) $\sin 2\theta = \dfrac{2\tan\theta}{1+\tan^2\theta}$

(*c*) $\sin 2\theta = \dfrac{1-\tan^2\theta}{1+\tan^2\theta}$ (*d*) $\sin 2\theta = \dfrac{1+\tan^2\theta}{1-\tan^2\theta}$

165. The value of $\sin\dfrac{\pi}{10}\cdot\cos\dfrac{\pi}{5}$ is:

(*a*) $\dfrac{1}{2}$ (*b*) $-\dfrac{1}{2}$ (*c*) $-\dfrac{1}{4}$ (*d*) $\dfrac{1}{4}$

166. If $\tan\theta = \dfrac{1}{7}$ and $\tan\phi = \dfrac{1}{3}$, then the correct relation is:

(*a*) $\cos 2\theta = \sin\phi$ (*b*) $\cos 2\theta = \sin 2\phi$ (*c*) $\cos 2\theta = \sin 3\phi$ (*d*) $\cos 2\theta = \sin 4\phi$

167. The value of θ lying between 0° and 360° for which $\sin\theta = -\dfrac{1}{2}$, is:

(*a*) 210° (*b*) 300° (*c*) 30° (*d*) 120°

168. If $\tan\theta = \dfrac{x\sin\phi}{1-x\cos\phi}$, $\tan\phi = \dfrac{y\sin\theta}{1-y\cos\theta}$, then $\dfrac{\sin\theta}{\sin\phi}$ is equal to:

(*a*) $\dfrac{y}{x}$ (*b*) $\dfrac{x}{y}$ (*c*) $\dfrac{x^2}{y^2}$ (*d*) $\dfrac{y^2}{x^2}$

169. The difference between two consecutive values of θ for which $\sin\dfrac{\theta}{3}$ vanishes, is:

(*a*) 3π (*b*) π (*c*) $\dfrac{2\pi}{3}$ (*d*) $\dfrac{\pi}{3}$

170. $\sqrt{2+\sqrt{2+2\cos 4\theta}} = ?$

(*a*) $2\cos\theta$ (*b*) $2\sin\theta$ (*c*) 0 (*d*) 1

171. If sin 120° – sin 60° = sin x°, then x is:

(*a*) 30° (*b*) 60° (*c*) 90° (*d*) 180°

172. Arrange the following in ascending order and choose the correct answer from those given below:

1. sin 30° 2. sin 45° 3. cos 30° 4. tan 30°

(*a*) 1, 2, 3, 4 (*b*) 3, 1, 2, 4 (*c*) 1, 4, 2, 3 (*d*) 3, 4, 2, 1

173. If $\sin A + \dfrac{1}{\sin A} = \dfrac{5}{2}$ and angle A is acute, then the measure of angle A is:

(*a*) $\dfrac{\pi}{6}$ (*b*) $\dfrac{\pi}{4}$ (*c*) $\dfrac{\pi}{3}$ (*d*) None of these

174. If $\sec\theta - \tan\theta = x, x \neq 0$, then $\sin\theta$ will be

(*a*) $\dfrac{1}{2}\left(x+\dfrac{1}{x}\right)$ (*b*) $\dfrac{1-x^2}{1+x^2}$ (*c*) $\dfrac{1}{2}\left(x-\dfrac{1}{x}\right)$ (*d*) $\dfrac{1-x}{1+x}$

175. In ΔABC, angle C is a right angle. Then, $\tan A + \tan B$ equals:

(*a*) $\tan C$ (*b*) c (*c*) $\dfrac{c}{ab}$ (*d*) $\dfrac{c^2}{ab}$

176. If $A + B + C = \pi$, then the value of $\tan(A + B) + \tan C$ is:

(*a*) $\sqrt{2}$ (*b*) -1 (*c*) 1 (*d*) 0

177. Which one of the following is incorrect in case of a ΔABC?

(*a*) $R = \frac{abc}{4\Delta}$ (*b*) $\sin A + \sin B + \sin C = \frac{s}{R}$

(*c*) $\cos A + \cos B + \cos C = \left(1 + \frac{r}{R}\right)$ (*d*) None of these

178. Consider the following statements: **[CDS (I) Exam, 2025]**

I. $(\operatorname{cosec}\alpha - \sec\alpha)$ is always positive in the first quadrant.

II. $(\tan\alpha - \cot\alpha)$ is always negative in the first quadrant.

Which of the statements given above is/are correct?

(*a*) I only (*b*) II only (*c*) Both I and II (*d*) Neither I nor II

179. Consider the following: **[CDS (I) Exam, 2025]**

I. $1 - \sin^6\alpha = \cos^2\alpha(\cos^4\alpha - 3\cos^2\alpha + 3)$

II. $\cos^8\alpha - \sin^8\alpha = 2\sin^2\alpha(1 - \cos^4\alpha + \sin^2\alpha\cos^2\alpha)$

Which of the above is/are identities?

(*a*) I only (*b*) II only (*c*) Both I and II (*d*) Neither I nor II

180. If $A + B + C = 180°$, then $\sin A + \sin B + \sin C$ is:

(*a*) $4\sin\frac{A}{2}\sin\frac{B}{2}\sin\frac{C}{2}$ (*b*) $4\sin\frac{A}{2}\sin\frac{B}{2}\cos\frac{C}{2}$

(*c*) $4\tan\frac{A}{2}\tan\frac{B}{2}\tan\frac{C}{2}$ (*d*) None of these

181. Which of the following equations is/are possible? **[CDS (I) Exam, 2025]**

I. $\sin^2\theta = \frac{(x+y)^2}{4xy}$, where x, y are positive unequal real quantities.

II. $\sin\theta + \cos\theta = x + \frac{1}{x}$, where x is a positive real quantity.

Select the correct answer using the code given below:

(*a*) I only (*b*) II only (*c*) Both I and II (*d*) Neither I nor II

182. Consider the following statements:

I. $\tan 50° - \cot 50°$ is positive

II. $\cot 25° - \tan 25°$ is negative

Which of the statements is/are correct? **[CDS (II) Exam, 2024]**

(*a*) I only (*b*) II only (*c*) Both I and II (*d*) Neither I nor II

183. Which of the following is an identity?

(*a*) $\sin^4\theta\cos^4\theta = \sin^2\theta - \cos^2\theta$

(*b*) $\tan^2\theta - \sin^2\theta = \tan 2\theta \,.\, \sin^2\theta$

(*c*) $\frac{1 - \tan^2\theta}{1 + \tan^2\theta} = 1 - 2\sin^2\theta$

(*d*) $\sin^6\theta + \cos^6\theta - 1 - 3\sin^2\theta\cos^2\theta = 0$

ANSWERS

1. (c)	**2.** (*c*)	**3.** (*a*)	**4.** (*a*)	**5.** (*b*)	**6.** (*d*)	**7.** (*b*)	**8.** (*d*)	**9.** (*b*)
10. (*a*)	**11.** (*a*)	**12.** (*c*)	**13.** (*b*)	**14.** (*a*)	**15.** (*c*)	**16.** (*a*)	**17.** (*b*)	**18.** (*b*)
19. (*d*)	**20.** (*d*)	**21.** (*b*)	**22.** (*b*)	**23.** (*a*)	**24.** (*a*)	**25.** (*d*)	**26.** (*a*)	**27.** (*d*)
28. (*c*)	**29.** (*d*)	**30.** (*b*)	**31.** (*b*)	**32.** (*b*)	**33.** (*b*)	**34.** (*c*)	**35.** (*c*)	**36.** (*d*)
37. (*c*)	**38.** (*b*)	**39.** (*a*)	**40.** (*d*)	**41.** (*b*)	**42.** (*a*)	**43.** (*c*)	**44.** (*b*)	**45.** (*b*)
46. (*b*)	**47.** (*b*)	**48.** (*a*)	**49.** (*a*)	**50.** (*b*)	**51.** (*b*)	**52.** (*c*)	**53.** (*a*)	**54.** (*b*)
55. (*a*)	**56.** (*a*)	**57.** (*d*)	**58.** (*a*)	**59.** (*d*)	**60.** (*c*)	**61.** (*d*)	**62.** (*d*)	**63.** (*c*)
64. (*b*)	**65.** (*c*)	**66.** (*c*)	**67.** (*b*)	**68.** (*b*)	**69.** (*b*)	**70.** (*d*)	**71.** (*c*)	**72.** (*d*)
73. (*b*)	**74.** (*a*)	**75.** (*d*)	**76.** (*c*)	**77.** (*d*)	**78.** (*b*)	**79.** (*a*)	**80.** (*c*)	**81.** (*b*)
82. (*a*)	**83.** (*a*)	**84.** (*a*)	**85.** (*c*)	**86.** (*d*)	**87.** (*c*)	**88.** (*c*)	**89.** (*c*)	**90.** (*b*)
91. (*c*)	**92.** (*a*)	**93.** (*d*)	**94.** (*b*)	**95.** (*c*)	**96.** (*c*)	**97.** (*a*)	**98.** (*d*)	**99.** (*b*)
100. (*b*)	**101.** (*d*)	**102.** (*c*)	**103.** (*b*)	**104.** (*d*)	**105.** (*d*)	**106.** (*b*)	**107.** (*a*)	**108.** (*c*)
109. (*b*)	**110.** (*d*)	**111.** (*d*)	**112.** (*c*)	**113.** (*c*)	**114.** (*d*)	**115.** (*a*)	**116.** (*a*)	**117.** (*a*)
118. (*a*)	**119.** (*b*)	**120.** (*b*)	**121.** (*c*)	**122.** (*a*)	**123.** (*a*)	**124.** (*a*)	**125.** (*b*)	**126.** (*c*)
127. (*b*)	**128.** (*c*)	**129.** (*a*)	**130.** (*a*)	**131.** (*c*)	**132.** (*c*)	**133.** (*d*)	**134.** (*b*)	**135.** (*c*)
136. (*a*)	**137.** (*d*)	**138.** (*b*)	**139.** (*a*)	**140.** (*b*)	**141.** (*d*)	**142.** (*b*)	**143.** (*a*)	**144.** (*b*)
145. (*b*)	**146.** (*c*)	**147.** (*d*)	**148.** (*c*)	**149.** (*b*)	**150.** (*b*)	**151.** (*c*)	**152.** (*a*)	**153.** (*c*)
154. (*d*)	**155.** (*c*)	**156.** (*c*)	**157.** (*a*)	**158.** (*d*)	**159.** (*b*)	**160.** (*b*)	**161.** (*a*)	**162.** (*b*)
163. (*d*)	**164.** (*b*)	**165.** (*d*)	**166.** (*d*)	**167.** (*a*)	**168.** (*b*)	**169.** (*a*)	**170.** (*a*)	**171.** (*d*)
172. (*c*)	**173.** (*a*)	**174.** (*b*)	**175.** (*d*)	**176.** (*d*)	**177.** (*d*)	**178.** (*d*)	**179.** (*a*)	**180.** (*d*)
181. (*d*)	**182.** (*a*)	**183.** (*c*)						

SOLUTIONS

1. $$\frac{2\sin 68^\circ}{\cos 22^\circ}-\frac{2\cot 15^\circ}{5\tan 75^\circ}-\frac{3\tan 20^\circ\tan 40^\circ\tan 45^\circ\tan 50^\circ\tan 70^\circ}{5}$$

$$\Rightarrow \frac{2\sin 68^\circ}{\cos(90^\circ-68^\circ)}-\frac{2\cot 15^\circ}{5\tan(90^\circ-15^\circ)}-\frac{3\tan 20^\circ\tan 40^\circ\tan 45^\circ\tan(90^\circ-40^\circ)\tan(90-20^\circ)}{5}$$

$$\Rightarrow \frac{2\sin 68^\circ}{\sin 68^\circ}-\frac{2\cot 15^\circ}{5\cot 15^\circ}-\frac{3\tan 20^\circ\tan 40^\circ\times 1\times\cot 40^\circ\cot 20^\circ}{5}$$

$$\Rightarrow 2-\frac{2}{5}-\frac{3}{5}=\frac{10-2-3}{5}=\frac{5}{5}=1$$

2. $$\left(1+\frac{\cos A}{\sin A}-\frac{1}{\sin A}\right)\left(1+\frac{\sin A}{\cos A}+\frac{1}{\cos A}\right)$$

$$=\left(\frac{(\sin A+\cos A)-1}{\sin A}\right)\times\left(\frac{(\cos A+\sin A)+1}{\cos A}\right)$$

$$=\frac{(\sin A+\cos A)^2-(1)^2}{\sin A\times\cos A}$$

$$=\frac{\sin^2 A+\cos^2 A+2\sin A\cos A-1}{\sin A\cos A}$$

$$=\frac{1+2\sin A\cos A-1}{\sin A\cos A}$$

$$=\frac{2\sin A\cos A}{\sin A\cos A}=2$$

3. Let $5\sin\theta - 3\cos\theta = x$ and $3\sin\theta + 5\cos\theta = 5$

here, $a = 3, b = 5$

$\therefore \quad a^2 + b^2 = x^2 + 5^2 \Rightarrow 3^2 + 5^2 = x^2 + 5^2 \Rightarrow x = \pm 3$

4. $\dfrac{\sin\theta}{1-\dfrac{\cos\theta}{\sin\theta}} + \dfrac{\cos\theta}{1-\dfrac{\sin\theta}{\cos\theta}} = \dfrac{\sin\theta}{\dfrac{\sin\theta-\cos\theta}{\sin\theta}} + \dfrac{\cos\theta}{\dfrac{\cos\theta-\sin\theta}{\cos\theta}} = \dfrac{\sin^2\theta}{\sin\theta-\cos\theta} + \dfrac{\cos^2\theta}{\cos\theta-\sin\theta}$

$= \dfrac{\sin^2\theta}{\sin\theta-\cos\theta} - \dfrac{\cos^2\theta}{\sin\theta-\cos\theta} = \dfrac{\sin^2\theta-\cos^2\theta}{\sin\theta-\cos\theta} = \dfrac{(\sin\theta-\cos\theta)(\sin\theta+\cos\theta)}{\sin\theta-\cos\theta} = \sin\theta+\cos\theta$

5. We have,

$\dfrac{\cos\theta}{\operatorname{cosec}\theta+1} + \dfrac{\cos\theta}{\operatorname{cosec}\theta-1} = 2$

$\Rightarrow \dfrac{2\cos\theta\operatorname{cosec}\theta}{\operatorname{cosec}^2\theta-1} = 2 \Rightarrow \dfrac{2\cos\theta\operatorname{cosec}\theta}{\cos^2\theta} = 2 \qquad [\because \operatorname{cosec}^2\theta - \cot^2\theta = 1]$

$\Rightarrow 2\tan\theta = 2 \Rightarrow \tan\theta = 1 \Rightarrow \theta = 45°$

Now, $\sin^4\theta + \cos^4\theta = \sin^4 45° + \cos^4\theta\, 45° = \left(\dfrac{1}{\sqrt{2}}\right)^4 + \left(\dfrac{1}{\sqrt{2}}\right)^4 = \dfrac{1}{2}+\dfrac{1}{2} = \dfrac{2}{2} = 1$

6. $\dfrac{\sin\theta-2\sin^3\theta}{2\cos^3\theta-\cos\theta} = \dfrac{\sin\theta(1-2\sin^2\theta)}{\cos\theta(2\cos^2\theta-1)}$

$= \tan\theta\left(\dfrac{1-2\sin^2\theta}{2(1-\sin^2\theta)-1}\right) \quad \{\because \sin^2\theta+\cos^2\theta=1\}$

$= \tan\theta\left(\dfrac{1-2\sin^2\theta}{2-2\sin^2\theta-1}\right) = \tan\theta\left(\dfrac{1-2\sin^2\theta}{1-2\sin^2\theta}\right) = \tan\theta$

7. Given that

$\sin\theta\cos\theta = k \Rightarrow \dfrac{1}{2}(2\sin\theta\cos\theta) = k \Rightarrow \sin 2\theta = 2k \qquad [\because \sin 2\theta = 2\sin\theta\,\text{cost}\,\theta]$

$\because -1 \le \sin 2\theta \le 1 \Rightarrow -1 \le 2k \le 1 \Rightarrow -\dfrac{1}{2} \le k \le \dfrac{1}{2} \Rightarrow -0.5 \le k \le k\, 0.5 \Rightarrow 0 < k \le 0.5$

8. Given that

$p = \sin^2\theta + \cos^4\theta$

If $\theta - 0° \Rightarrow p - 0 + 1 = 1$

If $\theta = 30° \Rightarrow p = \sin^2 30° + \cos^4 30° = \left(\dfrac{1}{2}\right)^2 + \left(\dfrac{\sqrt{3}}{2}\right)^4 = \dfrac{1}{4}+\dfrac{9}{16} = \dfrac{13}{16} = 0.8125$

If $\theta = 45° \Rightarrow p = \sin^2 45° + \cos^4 45° = \left(\dfrac{1}{\sqrt{2}}\right)^2 + \left(\dfrac{1}{\sqrt{2}}\right)^4 = \dfrac{1}{2}+\dfrac{1}{4} = \dfrac{3}{4} = 0.75$

If $\theta = 60° \Rightarrow p = \sin^2 60° + \cos^4 60° = \left(\dfrac{\sqrt{3}}{2}\right)^2 + \left(\dfrac{1}{2}\right)^4 = \dfrac{3}{4}+\dfrac{1}{16} = \dfrac{13}{16} = 0.8125$

$\theta = 90°, p = \sin^2 90° + \cos^4 90° = 1 + 0 = 1$

Hence, $0.75 \le p \le 1 \Rightarrow \frac{3}{4} \le p \le 1$

Hence, neither statement (1) nor statement (2) is correct.

9. $\sin^2\theta\cos^2(\sec^2\theta + \text{cosec}^2\theta)$

Using the identities; $\sec\theta = \frac{1}{\cos\theta}$ and $\text{cosec}\,\theta = \frac{1}{\sin\theta}$;

$$\sin^2\theta\cos^2\theta\left(\frac{1}{\cos^2\theta} + \frac{1}{\sin^2\theta}\right) = \sin^2\theta\cos^2\theta\left(\frac{\sin^2\theta + \cos^2\theta}{\sin^2\theta\cos^2\theta}\right) = \sin^2\theta + \cos^2\theta = 1$$

10. Given that

$x^2 + y^2 - 2xy\sin^2\theta = 0 \Rightarrow x^2 + y^2 = 2xy\sin^2\theta$

If $\theta = 90°$

$\Rightarrow x^2 + y^2 = 2xy$ [$\because$ we get real solution for x and y only if $\theta = 90°$]

for $x = y$

We have, $x^2 + x^2 = 2x^2$

11. Statement-1

We have, $\sin 1° < \cos 57° \Rightarrow \sin 1° = \sin(90° - 89°) = \cos 89° < \cos 57°$

Hence statement-1 is correct

Statement-2

We have,

$\cos 60° > \sin 57°$

$\cos 60° > \sin(90° - 33°) = \cos 33°$

Hence, statement-2 is not correct

12. We have, $\frac{\sin^3\theta + \cos^3\theta}{\sin\theta + \cos\theta} + \frac{\sin^3\theta + \cos^3\theta}{\sin\theta + \cos\theta}$

Using

$a^3 + b^3 = (a + b)(a^2 - ab + b^2)$

$a^3 - b^3 = (a - b)(a^2 + ab + b^2)$

We get,

$$= \frac{(\sin\theta + \cos\theta)(\sin^2\theta + \cos^2\theta - \sin\theta\cos\theta)}{(\sin\theta + \cos\theta)} + \frac{(\sin\theta - \cos\theta)(\sin^2\theta - \cos^2\theta\sin\theta\cos\theta)}{\sin\theta - \cos\theta}$$

$= (1 - \sin\theta\cos\theta) + 1 + (\sin\theta\cos\theta) = 2$

13. $2\sin^4(\alpha) + 2\cos^4(\alpha) - 1 = 0$

$$\Rightarrow 2\left(\sin^4(\alpha) + \cos^4(\alpha)\right) = 1 \Rightarrow \sin^4(\alpha) + \cos^4(\alpha) = \frac{1}{2}$$

$$(\sin^2(\alpha) + \cos^2(\alpha))^2 - 2\sin^2(\alpha)\cos^2(\alpha) = \frac{1}{2} \Rightarrow (1)^2 - 2\sin^2(\alpha)\cos^2(\alpha) = \frac{1}{2}$$

$$\Rightarrow 1 - 2\sin^2(\alpha)\cos^2(\alpha) = \frac{1}{2} \Rightarrow 2\sin^2(\alpha)\cos^2(\alpha) = \frac{1}{2} \Rightarrow \sin^2(\alpha)\cos^2(\alpha) = \frac{1}{4}$$

$$\Rightarrow \sin(\alpha)\cos(\alpha) = \frac{1}{2}$$

Using the identity $\sin(2\alpha) = 2\sin(\alpha)\cos(\alpha) \Rightarrow \sin(2\alpha) = 2 \times \frac{1}{2} = 1$

$\sin(2\alpha)=1 \Rightarrow 2\alpha=90^\circ$

$\alpha=45^\circ$

$\sin(2\alpha)=\sin(90^\circ)=1$

$\cos(2\alpha)=\cos(90^\circ)=0$

$\sin 2\alpha+\cos 2\alpha=1+0=1$

14. Given that $\dfrac{\cos^2\theta-3\cos\theta+2}{\sin^2\theta}=1$

$\Rightarrow \cos^2\theta - 3\cos\theta + 2 = \sin^2\theta \quad \Rightarrow \cos^2\theta - 3\cos\theta + 2 = 1 - \cos^2\theta$

$\Rightarrow 2\cos^2\theta - 3\cos\theta + 1 = 0 \quad \Rightarrow 2\cos^2\theta - 2\cos\theta - \cos\theta + 1 = 0$

$\Rightarrow 2\cos\theta(\cos\theta - 1) - 1(\cos\theta - 1) = 0 \Rightarrow (2\cos\theta - 1)(\cos\theta - 1) = 0$

$\Rightarrow \cos = \dfrac{1}{2} \Rightarrow \theta = 60^\circ$ and $\cos\theta = 1$ (not possible) $[\because 0^\circ < \theta < 90^\circ]$

Now, $\sin^2\theta + \cos\theta = \sin^2 60^\circ + \cos 60^\circ = \left(\dfrac{\sqrt{3}}{2}\right)^2 + \dfrac{1}{2} = \dfrac{3}{4} + \dfrac{1}{2} = \dfrac{3+2}{4} = \dfrac{5}{4}$

15. Statement-1

We have, $\sin^4\theta - \sin^2\theta = \cos^4\theta - \cos^2\theta$

$\Rightarrow \sin^2\theta(\sin^2\theta - 1) = \cos^2\theta(\cos^2\theta - 1) \Rightarrow \sin^2\theta(-\cos^2\theta) = \cos^2\theta(-\sin^2\theta)$

$\Rightarrow -\sin^2\theta\cos^2\theta = -\sin^2\theta\cos^2\theta$

Hence, statement-1 is identities

Statement-2

We have, $\sin^4\theta + \cos^4\theta = 1 + 2\sin^2\theta\cos^2\theta$

$\Rightarrow \sin^4\theta + \cos^4\theta - 2\sin^2\theta\cos^2\theta = 1 \Rightarrow (\sin^2\theta - \cos^2\theta)^2 = 1$

Hence, statement-2 is not identities

Statement-3

$\tan^4\theta + \tan^2\theta = \sec^4\theta - \sec^2\theta \Rightarrow \tan^2\theta(\tan^2\theta + 1) = \sec^2\theta(\sec^2\theta - 1)$

$\Rightarrow \tan^2\theta\sec^2\theta = \sec^2\theta\tan^2\theta$ $[\because \sec^2\theta - \tan^2\theta = 1]$

Hence statement-3 is identities

16. $\sin\alpha + 2 = 0 \Rightarrow \sin\alpha = -2$ (not possible)

$\Rightarrow \sin\alpha + 4 = 0 \Rightarrow \sin\alpha = -4$ (not possible) $\Rightarrow \sin\alpha - 2 = 0 \Rightarrow \sin\alpha = 2$ (not possible)

$\Rightarrow \sin\alpha - 4 = 0 \Rightarrow \sin\alpha = 4$ (not possible)

Number of real values of $\alpha = 0$.

17. $\tan A \times \tan B = 1 \Rightarrow \tan A = \cot B = \tan(90^\circ - B + k \times 180^\circ)$

$\Rightarrow \tan 3x = \tan(90^\circ - 6x + n \times 180^\circ)$

$\Rightarrow 3x = 90^\circ - 6x + n \times 180^\circ \Rightarrow 9x = 90^\circ + n \times 180^\circ \Rightarrow x = 10^\circ + n \times 20^\circ$

Check n values for $0 \le x \le 30^\circ$: $n = 0 \rightarrow x = 10^\circ$ (valid)

$n = 1 \rightarrow x = 30^\circ$ (but $3x = 90^\circ \rightarrow$ tan3x undefined $\rightarrow$ reject)

$n = -1 \rightarrow x = -10^\circ$ (out of range)

$\Rightarrow x = 10^\circ$.

18. Given that

$$\cos\theta + \sec\theta = k \Rightarrow \cos\theta + \frac{1}{\cos\theta} = k \Rightarrow \frac{\cos^2\theta + 1}{\cos\theta} = k \Rightarrow \cos\theta + \frac{1}{\cos\theta} = k$$

Squaring both sides, we get

$$\cos^2\theta + \frac{1}{\cos^2\theta} + 2 = k^2 \Rightarrow \cos^2\theta + \frac{1}{\cos^2\theta} = k^2 - 2$$

Now, $\sin^2\theta - \tan^2\theta = (1 - \cos^2\theta) - (\sec^2\theta - 1) = 2 - (\cos^2\theta + \sec^2\theta)$

$$= 2 - \left(\cos^2\theta + \frac{1}{\cos^2\theta}\right) = 2 - (k^2 - 2) = 4 - k^2$$

19. ΔABC is inscribed in a semicircle, where AB is a diameter

$\therefore \angle ABC = 90°$

Now, In ΔABC

$\angle A + \angle B = 180° - 90° = 90°$

Now, $\cos(A + B) + \sin(A + B) = \cos 90° = \sin 90° = 1$

20. Statement-1

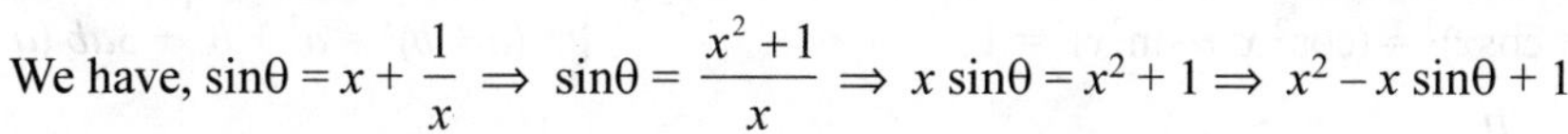

We have, $\sin\theta = x + \frac{1}{x} \Rightarrow \sin\theta = \frac{x^2 + 1}{x} \Rightarrow x\sin\theta = x^2 + 1 \Rightarrow x^2 - x\sin\theta + 1$

It is a quadratic equation

$\therefore$ for real value of x $b^2 > 4ac$

$\Rightarrow (-\sin\theta)^2 > 4 \times 1 \times 1 \Rightarrow \sin^2\theta > 4 \Rightarrow \sin\theta > 2$

$\therefore$ real value of x does not exist $[\therefore -1 \le \sin\theta \le 1]$

Statement-2

$$\cos\theta = x + \frac{1}{x} \Rightarrow \cos\theta = \frac{x^2 + 1}{x} \Rightarrow x^2 - x\cos\theta + 1 = 0$$

$\therefore$ It is a quadratic equation

$\therefore$ for any real value of x

$b^2 > 4ac$ $(-\cos\theta)^2 > 4 \times 1 \times 1 \Rightarrow \cos^2\theta > 4 \Rightarrow \cos\theta > 2$

$\therefore$ real value of x does not exist $[\because -1 \le \cos\theta < 1]$

21. We have,

$\text{cosec}\theta - \sin\theta = p^3$

$$\Rightarrow \frac{1}{\sin\theta} - \sin\theta = p^3 \Rightarrow \frac{1 - \sin^2}{\sin\theta} = p^3 \Rightarrow \frac{\cos^2\theta}{\sin\theta} = p^3 \quad ...(i)$$

Now, $\sec\theta - \cos\theta = q^3$

$$\Rightarrow \frac{1 - \cos^2\theta}{\cos\theta} = q^3 \Rightarrow \frac{\sin^2\theta}{\cos\theta} = q^3 \quad ...(ii)$$

Dividing (i) from (ii), we get

$$\frac{\sin^3\theta}{\cos^3\theta} = \frac{q^3}{p^3} \Rightarrow \tan^3\theta = \frac{q^3}{p^3} \Rightarrow \tan\theta = \frac{q}{p}$$

22. We have,

$25\,\text{cosec}^2 x = \sec^2 x$

$\Rightarrow 25\,(1 + \cot^2 x) + (1 + \tan^2 x)$ $[\because \operatorname{cosec}^2 x = 1 + \cot^2 x = 1 + \tan^2 x = \sec^2 x]$

$\Rightarrow 26 + 25\cot^2 x + \tan^2 x$

$\because$ least value $= 26 + 2\sqrt{25\times 1} = 26 + 2\times 5 = 36$

23. We have,

$x\cos\theta = x^2 + p \Rightarrow x^2 - x\cos\theta + p = 0$

$b^2 \geq 4ac$

$\cos^2\theta \geq 4\times 1\times p \Rightarrow 4p \leq \cos^2\theta \Rightarrow 4p \leq 1 \Rightarrow p \leq \dfrac{1}{4}$

24. $\tan 6\theta = \cot 2\theta$

$\tan 6\theta = \tan(90° - 2\theta) \Rightarrow 6\theta = 90° - 2\theta \Rightarrow 8\theta = 90° \Rightarrow 4\theta = 45° \Rightarrow \sec 4\theta = \sec 45° = \sqrt{2}$

25. $\sin^2 6° + \sin^2 12° + \sin^2 18° + \ldots\ldots + \sin^2 84° + \sin^2 90°$

$= \sin^2 6° + \sin^2 12° + \sin^2 18° + \ldots. + \sin^2(90 - 12)° + \sin^2(90 - 6)° + \sin^2 90°$

$= \sin^2 6° + \sin^2 12° + \sin^2 18° + \ldots + \cos^2 12° + \sin^2 6° + \sin^2 90° = 8$

26. We have, $\cos^2 x + \cos x = 1 \Rightarrow \cos x = 1 - \cos^2 x \Rightarrow \cos x = \cos^2 x$

Now, $\sin^2 x + 3\sin^{10}x + 3\sin^8 x + \sin^6 x$

$\cos^6 x + 3\cos^5 x + 3\cos^4 x + \cos^3 x$

$= (\cos^2 x + \cos x)^3 = (\cos^2 x + \sin^2 x)^3 = 1$ $[\because (a+b)^3 = a^3 + b^3 + 3ab\,(a+b)]$

27. $\sin\theta = \dfrac{3}{5} = \dfrac{p}{h}$

$b = \sqrt{h^2 - p^2} = \sqrt{5^2 - 3^2} = \sqrt{16} = 4$

$x = \cot\theta = \dfrac{p}{b} = \dfrac{4}{3}$

Now, $1 + 3x + 9x^2 = 27x^3 + 81x^4 + 243x^5$

$= 1 + 3\times\dfrac{4}{3} + 9\times\dfrac{16}{9} + 27\times\dfrac{64}{27} + 81\times\dfrac{256}{81} + 243\times\dfrac{1024}{243}$

$= 1 + 4 + 16 + 64 + 256 + 1024 = 1365$

28. $\dfrac{\sin\theta + \cos\theta + 1}{\sin\theta + \cos\theta - 1} = p\sec\theta + q\tan\theta$

where $\theta = 45°$

$$\frac{\frac{1}{\sqrt{2}} - \frac{1}{\sqrt{2}} + 1}{\frac{1}{\sqrt{2}} + \frac{1}{\sqrt{2}} - 1} = p\sqrt{2} + q \Rightarrow \frac{1}{\sqrt{2}-1} = p\sqrt{2} + q \Rightarrow \sqrt{2} + 1 = p\sqrt{2} + q$$

by comparing, we get

$p = 1, q = 1$

$\therefore\ p + q = 1 + 1 = 2$

29. $\sin^4\theta - \cos^4\theta = (\sin^2\theta)^2 - (1 - \sin^2\theta)^2 = \sin^4\theta - (1 + \sin^4\theta - 2\sin^2\theta) = 2\sin^2\theta - 1$

$= 2 - 2\cos^2\theta - 1 = 1 - 2\cos^2\theta$

30. $\cot 1° \cot 23° \cot 45° \cot 67° \cot 89° = \cot 1° \cot 23° \times 1 \times \cot(90° - 23°)\cot(90° - 1°)$

$\cot 1° \cot 23° \times \tan 23° \times \tan 1° = 1$

31. tan (3A) = cot (A – 22°)

$\Rightarrow$ cot (90° – 3A) = cot (A – 22°) $\Rightarrow$ 90° – 3A = A – 22° $\Rightarrow$ 4A = 112° $\Rightarrow$ A = 28°

32. Statement I

$(\sec^2\theta - 1)(1 - \text{cosec}^2\theta)$ [$\because \sec^2\theta - 1 = \tan\theta$ and $\text{cosec}^2\theta - 1 = \cot\theta$]

$\tan^2\theta \times -\cot^2\theta = -1$ [$\because \tan\theta \times \cot\theta = 1$]

Statement II

We have, $\sin\theta(1+\cos\theta)^{-1} + (1+\cos\theta)(\sin\theta^{-1})$

$$= \frac{\sin\theta}{1+\cos\theta} + \frac{1+\cos\theta}{\sin\theta} = \frac{\sin^2\theta + 1 + \cos^2\theta + 2\cos\theta}{\sin\theta\,(1+\cos\theta)} = \frac{2(1+\cos\theta)}{\sin(1+\cos\theta)} = 2\,\text{cosec}\,\theta$$

33. $\because 0 < \theta < 90° \Rightarrow \cos 0 < \cos\theta < \cos 90° \quad \Rightarrow 1 < \cos\theta < 0$ and $\because 0 < \phi < 90°$

$\cos 0° < \cos\phi < \cos 90°$

$1 < \cos\phi < 0$ and $\cos\phi > \cos\theta$ [given] $\Rightarrow \theta > \phi$

34. $\cos B = \dfrac{\sqrt{3}}{2} \quad \cos B = \cos 30°$

$B = 30°$ and $\sin(A+B) = \dfrac{\sqrt{3}}{2} \Rightarrow \sin(A+B) = \sin 60° \quad \Rightarrow A + B = 60° \quad \Rightarrow A = 30°$

Now, $\tan(2A - B) = \tan(60° - 30°) = \tan 30° = \dfrac{1}{\sqrt{3}}$

35. Statement I

$\because 0 \le \cos^2\theta \le 1$

$\Rightarrow 0 \le 1 - \cos^2\theta \le 1 \Rightarrow 0 \le \dfrac{p^2+q^2}{2pq} \le 1 \qquad \left[\because \cos^2\theta = 1 - \dfrac{p^2+q^2}{2pq}\right]$

$\Rightarrow 0 \le p^2 + q^2$

and

$\Rightarrow \dfrac{p^2+q^2}{2pq} \le 1 \quad \Rightarrow p^2 + q^2 \le 2pq \Rightarrow (p^2+q^2) - 2pq \le 0 \Rightarrow (p-q)^2 \le 0 \quad \Rightarrow p = q$

Statement II

We have,

$\tan^2\theta = \dfrac{4pq}{(p+q)^2} - 1 \Rightarrow \sec^2\theta - 1 = \dfrac{4pq}{(p+q)^2} - 1 \Rightarrow \sec^2\theta = \dfrac{4pq}{(p+q)^2} \Rightarrow \cos^2\theta = \dfrac{(p+q)^2}{4pq}$

$\therefore 0 \le \cos^2\theta \le 1 \quad \Rightarrow 0 \le \dfrac{(p+q)^2}{4pq} \le 1 \Rightarrow 0 \le p+q)^2$ and $\dfrac{(p+q)^2}{4pq} \le 1$

$\Rightarrow (p+q)^2 \le 4pq \Rightarrow p^2 + q^2 + 2pq - 4pq \le 0 \Rightarrow (p-q)^2 \le 0 \Rightarrow p = q$

36. We have, $3\sin\theta + 5\cos\theta = 4$ ...(i)

Let $3\cos\theta - 5\sin\theta = x$ ...(ii)

Squaring (i), we get

$$9\sin^2\theta + 25\cos^2\theta + 30\sin\theta\cos\theta = 16 \qquad \text{...(ii)}$$

Squaring (iii), we get

$$9\cos^2\theta + 25\sin^2\theta - 30\sin\theta\cos\theta = x^2 \qquad \text{...(iv)}$$

adding (iii) and (iv), we get

$9(\sin^2\theta + \cos^2\theta) + 25(\sin^2\theta + \cos^2\theta) = 16 + x^2$

$\Rightarrow 9 + 25 = 16 + x^2 \Rightarrow 34 = 16 + x^2 \Rightarrow x^2 = 34 - 16 = 18$

$\therefore (3\cos\theta - 5\sin\theta)^2 = 18$

37. $\tan(A + B) = \sqrt{3}$ and $\tan A = 1$

$\therefore A + B = 60°$ and $A = 45°$. $\quad \therefore B = 15°$

$$\tan(A - B) = \tan 30° = \frac{1}{\sqrt{3}}$$

38. $\cos A = \tan B$ ...(1)

$\cos B = \tan C$

$$\tan B = \frac{\sin B}{\cos B} = \sqrt{\frac{1-\tan^2 C}{\tan C}} \quad ...(2)$$

From (1) & (2), $\cos A = \sqrt{\dfrac{1-\tan^2 C}{\tan C}}$ ...(3)

Given that $\cos C = \tan A$

$$\tan C = \frac{\sin C}{\cos C} = \sqrt{1 - \frac{\tan^2 A}{\tan A}} \quad ...(4)$$

From (3) & (4), $\cos A = \dfrac{\sqrt{1-\left(\dfrac{\tan^2 A}{\tan^2}\right)^2}}{\sqrt{\dfrac{1-\tan^2 A}{\tan A}}}$; $\cos A = \sqrt{\dfrac{2\tan^2 A - 1}{1-\tan^2 A}}$

$\cos^2 A = \dfrac{2\sin^2 A - \cos^2 A}{\cos^2 - \sin^2 A}$; Let $\sin^2 A = x$

$\Rightarrow \cos^2 A = 1 - x$; *i.e.*, $1 - x = \dfrac{2x - (1-x)}{1-x-x}$

$1 - 3x + 2x^2 = 3x - 1, \; x^2 - 3x + 1 = 0, \qquad x = \dfrac{3 \pm \sqrt{5}}{2}.$

$\therefore \sin^2 A = \dfrac{3-\sqrt{5}}{2}$ $\left(\because \dfrac{3+\sqrt{5}}{2}\right.$ is greater than 1)

$$\Rightarrow \sin A = \sqrt{\frac{3-\sqrt{5}}{2}} = \sqrt{\frac{6-2\sqrt{5}}{2}} = \sqrt{\frac{(\sqrt{5})^2 + (1)^2 - 2\sqrt{5}\times 1}{4}} = \frac{\sqrt{5}-1}{2}$$

39. $\operatorname{cosec} A(3\sin A - 4\sin^3 A) = 3 - 4\sin^2 A$.

We need $\sin^2 A$ in terms of K

$$K = \frac{3-\tan^2 A}{1-3\tan^2 A} = \frac{3\cos^2 A - \sin^2 A}{\cos^2 A - 3\sin^2 A}.$$

Let $\sin^2 A = x$

$\Rightarrow \cos^2 A = 1 - x \qquad \Rightarrow K = \dfrac{3-4x}{1-4x}$

$\Rightarrow K - 4Kx = 3 - 4x \qquad \Rightarrow x(4 - 4K) = 3 - K$

$$x = \frac{3-K}{4(1-K)} \qquad ...(1)$$

$$\therefore\ 3 - 4\sin^2 A = 3 - 4x = 3 - \frac{4(3-K)}{4(1-K)} = \frac{2K}{K-1}.$$

We know that $0 \le \sin^2 A \le 1 \Rightarrow 0 \le x \le 1.$

From (1) $0 \le \dfrac{3-k}{4-4k} \le 1.$

When $4 - 4K > 0$, or $K < 1$,

$3 - K \ge 0$ $\quad$ $3 - K \le 4 - 4K$

$K \le 3$ $\quad$ $3K \le 1$

$K \le 1/3$

$K = 1, K \le 1/3.$

When $4 - 4K > 0$, or $K > 1$,

$3 - K \ge 0; K \ge 3$; $\quad 3 - K \ge 4 - 4K$; $\quad K \ge 1/3$; $\quad$ *i.e.*, $K \ge 3.$

$\therefore$ The range of K is $(-3, 1/3] \cup [3, 3).$

40. $\tan A + \cot A = 4.$

Squaring both sides, $\tan^2 A + \cot^2 A + 2\tan A \text{ Cot } A = 16$

$\tan^2 A + \cot^2 A = 16 - 2 \Rightarrow \tan^2 A + \cot^2 A = 14$

Squaring on both sides, $\tan^4 A + \cot^4 A + 2 = 196 \Rightarrow \tan^4 A + \cot^4 A = 194.$

41. $$p = \sqrt{\frac{1-\sin x}{1+\sin x}} = \sqrt{\frac{(1-\sin x)(1+\sin x)}{(1+\sin x)(1+\sin x)}} = \sqrt{\frac{1-\sin^2 x}{(1+\sin x)}} = \pm\frac{\cos x}{1+\sin x} \qquad ...(1)$$

$$r = \frac{(\cos x)(1-\sin x)}{(1+\sin x)(1-\sin x)} = \frac{\cos(1-\sin x)}{\cos^2 x} = \frac{1-\sin x}{\cos x} = q \qquad ...(2)$$

From (1) & (2), $q = r$, but $p = q$ or $p = -q$

(1) is false

Since $p = 1 = r$, but (2) is true.

42. $$\frac{\cos A}{1-\tan A} + \frac{\sin A}{1-\cot A} = \frac{\cos A}{1-\dfrac{\sin A}{\cos A}} + \frac{\sin A}{1-\dfrac{\cos A}{\sin A}}$$

$$= \frac{\cos^2 A}{\cos A - \sin A} - \frac{\sin^2 A}{\cos A - \sin A} = \frac{(\cos A + \sin A)(\cos A - \sin A)}{(\cos A - \sin A)}$$

$= \cos A + \sin A \therefore$ (1) is true $\qquad [1 - (\sin A + \cos A)]^2$

$= 1 + \sin^2 A + \cos^2 A + 2\sin A\cos A - 2(\sin A + \cos A)$

$= 2(1 - \sin A - \cos A + \sin A\cos A) = 2(1 - \sin A)(1 - \cos A)$

(2) is false.

43.

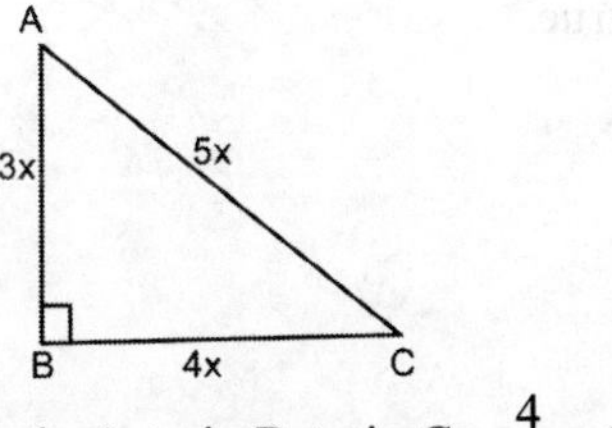

$$\sin A + \sin B + \sin C = \frac{4}{5} + 1 + \frac{3}{5} = \frac{12}{5}.$$

44. $\tan^2 23° = \cot^2 67°$ and $\cot^2 33° = \tan^2 57°$.

$\therefore$ The given expression is

$(\text{cosec}^2 67° - \cot^2 67°) + (\sec^2 57° - \tan^2 57°) = 1 + 1 = 2.$

45. Let $\sin^2 x = a \Rightarrow a^2 - 2a - 1 = 0$.

The solution for the above equation is, $a = \dfrac{2 \pm \sqrt{4+4}}{2} \Rightarrow a = 1 \pm \sqrt{2}$.

$\sin^2 x = 1 \pm \sqrt{2}$ is not possible.

since $0 \le \sin^2 x \le 1$

(1) is false.

Here, the angles are given in radians,

$1 \text{ radian} = \dfrac{180}{\pi} \text{ degrees} \approx \dfrac{630}{11} \text{ degrees},$

$\therefore 1.5 \text{ radian} \approx \dfrac{945}{11} \text{ degrees} = 85\dfrac{10}{11} \text{ degrees}$

$\sin\theta > \cos\theta$ when $45° < \theta \le 90°$

(2) is true.

46. $\sin^2 x + \cos^2 x = 1 \Rightarrow (\sin^2 x + \cos^2 x)^3 = 1$

$\sin^6 x + \cos^6 x + 3\sin^2 x \cos^2 x (\sin^2 x + \cos^2 x) = 1$

$\sin^6 x + \cos^6 x = 1 - 3\sin^2 x \cos^2 x$ ...(1)

Given that, $\sin x + \cos x = c$.

Squaring on both sides,

$\sin^2 x + \cos^2 x + 2 \sin x \cos x = c^2$

$\sin x \cos x = \dfrac{c^2 - 1}{2}$ ...(2)

From (1) & (2), $\sin^6 x + \text{sos}^6 x = 1 - 3\left(\dfrac{c^2-1}{2}\right)^2 = \dfrac{1 + 6c^2 - 3c^4}{4}$.

47. $\dfrac{1}{1 - \sin x} = 4 + 2\sqrt{3} \Rightarrow 1 = 4 + 2\sqrt{3} - \sin x(4 + 2\sqrt{3})$

$\sin x = \dfrac{4 + 2\sqrt{3} - 1}{4 + 2\sqrt{3}}; \ \sin x = 1 - \dfrac{1}{4 + 2\sqrt{3}}$

i.e., $0 < x < 1$. Therefore there exists a value of x satisfying the given condition such that $0 < x < \pi/2$.

(1) is not true.

$$\sin x = 3^{\sin^2 x}$$

If $x = 0$, RHS = 1 while LHS = 0.

If $x > 0$, RHS > 1 while LHS ≤ 1.

There is no value of x which satisfies the equation. (2) is true.

48.

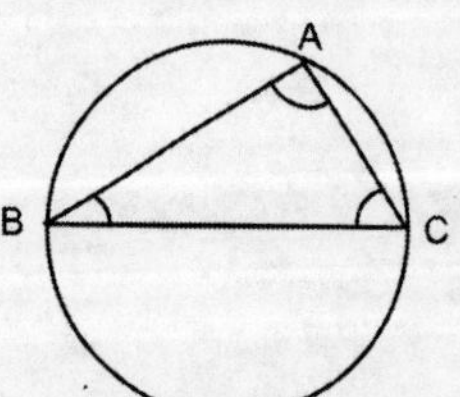

Let R be the circumradius.

AB = 2R sin C ...(1)

BC = 2 R sin A ...(2)

AC = 2R sin B ...(3)

Given that, $AB^2 + BC^2 + AC^2 = 2(2R)^2$

i.e., $AB^2 + BC^2 + AC^2 = 8R^2$ from (1), (2), & (3)

$4R^2(\sin^2A + \sin^2B + \sin^2C) = 8R^2$.

$\sin^2A + \sin^2B + \sin^2C = 2$.

49. **Statement 1:** LHS = $\dfrac{1+\tan^2\theta}{1+\cot^2\theta} = \dfrac{\sec^2\theta}{\operatorname{cosec}^2\theta} = \dfrac{\sin^2\theta}{\cos^2\theta} = \tan^2\theta$

RHS = $\left(\dfrac{1-\tan\theta}{1-\cot\theta}\right)^2 = \left(\dfrac{1-\tan\theta}{1-\dfrac{1}{\tan\theta}}\right)^2 = \left[\tan\theta\left(\dfrac{1-\tan\theta}{\tan\theta-1}\right)\right]^2$

$= (-\tan\theta)^2 = \tan^2\theta$ (R.H.S)

$\therefore$ Statement 1 is true when $\theta \neq \dfrac{\pi}{4}$.

Statement 2: $(\cot\theta)(\tan\theta) = 1$ for all values of θ except when $\theta = 0, 90°, 180°, \ldots\ldots\ldots$

$\therefore$ Statement 2 is not true.

50. $(ax^{-1} - by^{-1})(ax^{-1} + by^{-1}) = \left(\dfrac{a}{x}-\dfrac{b}{y}\right)\left(\dfrac{a}{x}+\dfrac{b}{y}\right) = \left(\dfrac{1}{\cos\theta}-\dfrac{1}{\cos\theta}\right)\left(\dfrac{1}{\cos\theta}+\dfrac{1}{\cot\theta}\right)$

$= (\sec\theta - \tan\theta)(\sec\theta + \tan\theta) = \sec^2\theta - \tan^2\theta = 1.$

51. $\dfrac{\cos\theta}{1-\sin\theta} \times \dfrac{1+\sin\theta}{1+\sin\theta} = \dfrac{\cos\theta(1+\sin\theta)}{\cos^2\theta} = \dfrac{1+\sin\theta}{\cos\theta}$.

52. $\tan(40+x)° \tan(20+x)° \tan 3x° \tan[90-(20+x)]° \tan[90-(40+x)]° = 1$

$\tan(40+x)° \tan[90-(40+x)]° = 1$ $[\because \tan\theta = \cot(90-\theta)]$

$\tan(20+x)° \tan[90-(20+x)]° = 1$

$\therefore \tan 3x° = 1 \Rightarrow 3x = 45, x = 15.$

53. $\sin\theta \cos\theta = 2\cos^3\theta - 1.5\cos\theta$

$\sin\theta = 2\cos^2\theta - 1.5$

$\sin\theta = 2 - 2\sin^2\theta - 1.5$

$2\sin^2\theta + \sin\theta - 0.5 = 0$

$\sin\theta = \dfrac{-1\pm\sqrt{5}}{4}$

since θ is acute, $\sin\theta = \dfrac{\sqrt{5}-1}{4}$.

54. From 0 to 45°, $\sin\theta < \cos\theta$. At 45° $\sin\theta = \cos\theta$ and from 45° to 90°, $\sin\theta > \cos\theta$.

$\therefore \sin 66° > \cos 66°$ and $\sin 26° < \cos 26°$.

55. $\sin\theta = \dfrac{2a+3b}{3b} = \dfrac{2a}{3b} + 1$

which is greater than 1, since $\dfrac{2a}{3b}$ is positive.

$\therefore$ The relation is not possible.

56. $\sec\theta + \tan\theta = 2$

$\sec\theta - \tan\theta = \frac{1}{2}$ $(\because \sec^2\theta - \tan^2\theta = 1)$

$2\sec\theta = \frac{5}{2} \Rightarrow \sec\theta = \frac{5}{4} \Rightarrow \tan\theta = \frac{3}{4}$

57. In the figure, M and N are the midpoints of AD and BC respectively $\angle AOM = \angle DOM = \theta$.

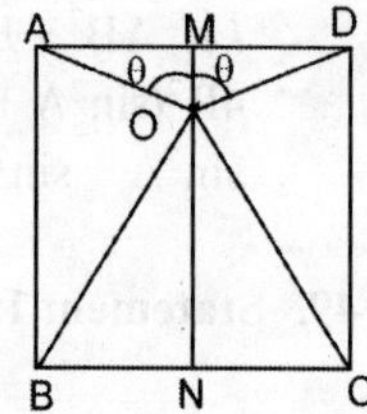

Let AM = 1 $\therefore$ AB = 2, ON = $\sqrt{3}$

($\because$ BOC is equilateral and ON is the altitude)

$\tan\theta = \frac{AM}{MO} = \frac{1}{2-\sqrt{3}} = 2 + \sqrt{3}$

58. $\cos^2 x + \cos^2 y - \cos^2 z$ will take the minimum value when $\cos^2 x$ and $\cos^2 y$ take the least possible values and cos z takes the highest possible value.

$\Rightarrow 0 + 0 - 1 = -1.$

59. $\cot\frac{\pi}{4} = \frac{1}{\tan\frac{\pi}{4}} = 1 \Rightarrow 32\cot^2\frac{\pi}{4} = 32$

$\sec\frac{\pi}{3} = \frac{1}{\cot\frac{\pi}{3}} = 2 \Rightarrow 8\sec^2\frac{\pi}{3} = 32$

$\cos\frac{\pi}{6} = \frac{\sqrt{3}}{2} \Rightarrow 8\cos^3\frac{\pi}{6} = 3\sqrt{3}$

$32\cot^2\frac{\pi}{4} - 8\sec^2\frac{\pi}{3} + 8\cos^3\frac{\pi}{6} = 3\sqrt{3}.$

60. $\left(\frac{\sin 35}{\cos(90-35)}\right)^2 - \left(\frac{\cos 55}{\sin(90-55)}\right)^2 + 2\frac{1}{2} = \left(\frac{\sin 35^\circ}{\sin 35^\circ}\right)^2 - \left(\frac{\cos 55^\circ}{\cos 55^\circ}\right)^2 + 1$ $\quad (\because \sin 30 = \frac{1}{2})$

$1 - 1 + 1 = 1$

61. **Statement (1):** For any positive real value *m*.

$m + 1 > 1 \Rightarrow 2^{m+1} > 2$

cos $x > 2$ is not possible

$\therefore$ Statement (1) is false.

Statement (2):

if $m = 1$ and $n = 2$, $mn = 2$ $m + n = 3$ $\quad\therefore$ Statement (2) is false.

62. $\frac{x}{a} - \frac{y}{b}\tan\theta = 1$ and $\frac{x}{a}\tan\theta + \frac{y}{b} = 1$

$\left(\frac{x}{a} - \frac{y}{b}\tan\theta\right)^2 + \left(\frac{x}{a}\tan\theta + \frac{y}{b}\right)^2 = 2$

$\frac{x^2}{a^2} + \frac{y^2}{b^2}\tan^2\theta + \frac{x^2}{a^2}\tan^2\theta + \frac{y^2}{b^2} = 2$

$\left(\frac{x^2}{a^2} + \frac{y^2}{b^2}\right) + \tan\theta\left(\frac{x^2}{a^2} + \frac{y^2}{b^2}\right) = 2;\ \left(\frac{x^2}{a^2} + \frac{y^2}{b^2}\right)(1 + \tan^2\theta) = 2$

$$\frac{x^2}{a^2}+\frac{y^2}{b^2}=\frac{2}{\sec^2\theta} \qquad (\because\ 1+\tan^2\theta=\sec^2\theta)$$

$$\frac{x^2}{a^2}+\frac{y^2}{b^2}=2\cos^2\theta$$

63. $\sqrt{\frac{1-\cos\theta}{1+\cos\theta}}\times\sqrt{\frac{1-\cos\theta}{1-\cos\theta}}=\sqrt{\frac{(1-\cos\theta)^2}{1-\cos^2\theta}}$

$$=\frac{1-\cos\theta}{\sin\theta}=\frac{1}{\sin\theta}-\frac{\cos\theta}{\sin\theta}\Rightarrow \operatorname{cosec}\theta-\cot\theta \qquad (\because\ 1-\cos^2\theta=\sin^2\theta)$$

Statement (1) is true

$$\sqrt{\frac{1+\cos\theta}{1-\cos\theta}}\times\sqrt{\frac{1+\cos\theta}{1+\cos\theta}}=\frac{1+\cos\theta}{\sin\theta}=\operatorname{cosec}\theta+\cot\theta$$

$\therefore$ Statement (2) is also true

64. $p=\cot\theta+\tan\theta$ and $q=\sin\theta-\cos\theta$

$$p=\frac{\cos\theta}{\sin\theta}+\frac{\sin\theta}{\cos\theta} \text{ and } q=\frac{1}{\cos\theta}-\frac{\cos\theta}{1}$$

$$p=\frac{\cos^2\theta+\sin^2\theta}{\sin\theta\cos\theta} \quad q=\frac{1-\cos^2\theta}{\cos\theta}$$

$$p=\frac{1}{\sin\theta\cos\theta},\ q=\frac{\sin^2\theta}{\cos\theta}$$

$$p^2q=\frac{1}{\sin^2\theta\cos^2\theta}\cdot\frac{\sin^2\theta}{\cos\theta}=\sec^3\theta$$

$$q^2p=\frac{\sin^4\theta}{\cos^2\theta}\cdot\frac{1}{\sin\theta\cos\theta}=\tan^3\theta$$

$$(p^2q)^{\frac{2}{3}}-(q^2p)^{\frac{2}{3}}=(\text{Sec}^3\theta)^{\frac{2}{3}}-(\tan^3\theta)^{\frac{2}{3}}=\sec^2\theta-\tan^2\theta=1$$

65. $\cos^2\theta-3\cos\theta+2=\sin^2\theta$

$$2\cos^2\theta-3\cos\theta+1=0 \qquad (\because \sin^2\theta=1-\cos^2\theta)$$

$$(2\cos\theta-1)(\cos\theta-1)=0$$

$$\cos\theta=\frac{1}{2} \text{ or } \cos\theta=1\Rightarrow\theta=60^\circ \text{ or } \theta=0$$

Statement (1) is true and statement (2) is true.

66. We need the range of values of α. We rearrange the terms to express $\tan^2\theta$ in term of α.

$$3-\tan^2\theta=\alpha-3\alpha\tan^2\theta\Rightarrow\tan^2\theta(3\alpha-1)=\alpha-3$$

$$\tan^2\theta=\frac{\alpha-3}{3\left(\alpha-\frac{1}{3}\right)}$$

As the LHS ≥ 0, the RHS ≥ 0, i.e. Both numerator and denominator are positive or both are negative.

$\alpha \in \left(-\infty, \frac{1}{3}\right) \cup [3, \infty)$

Note: $\frac{1}{3}$ has to be omitted

67. $\tan\theta + \cos\theta = \frac{4}{\sqrt{3}}$ where $0 < \theta < \frac{\pi}{2}$; $\tan\theta + \cot\theta = \frac{3+1}{\sqrt{3}} = \sqrt{3} + \frac{1}{\sqrt{3}}$

$\therefore \tan\theta = \frac{1}{\sqrt{3}}$ and $\cot\theta = \sqrt{3}$ or vice versa, *i.e.* $\theta = 30^\circ$ on 60°

In either case, $\sin\theta + \cos\theta = \frac{1}{2} + \frac{\sqrt{3}}{2} = \frac{1+\sqrt{3}}{2}$

68. $\sin\theta + \cos\theta = \frac{\sqrt{7}}{2} \Rightarrow \sin^2\theta + 2\sin\theta\cos\theta + \cos^2\theta = \frac{7}{4}$

$\therefore\ 2\sin\theta\cos\theta = \frac{3}{4} \Rightarrow \sin^2\theta - 2\sin\theta\cos\theta + \cos^2\theta = \frac{1}{4}$

or $\sin\theta - \cos\theta = \frac{1}{2}\ \left(or -\frac{1}{2}\right)$ Among the option, only $\frac{1}{2}$ is there.

69. $\sin x + \sin^2 x = 1$ ... (1)

$\therefore\ \sin x = \cos^2 x$ ... (2)

$\cos^8 x + 2\cos^6 x + \cos^4 x = \cos^4 x\left(\cos^2 x + 1\right)^2 = \sin^2 x(1+\sin x)^2 = \left[\sin x\,(1+\sin x)\right]^2 = 1$

70. $\text{Cosec}^2 68^\circ - \tan^2 22^\circ = \text{cosec}^2 68^\circ - \cot^2 68^\circ = 1$

$\text{Sec}^2 56^\circ - \cot^2 34^\circ = \sec^2 56^\circ - \tan^2 56^\circ = 1$

$\therefore$ The given expression = 2.

71. We think of the two given equation as 2 near equation in the unknowns x and y.

$(\sin\theta)x - (2\cos\theta)y = 0 \Rightarrow (2\sec\theta)x - (\text{cosec}\,\theta)y = 3.$

By Grammar's rule (as otherwise)

$$x = \frac{6\cos\theta}{-1+4} = 2\cos\theta \text{ and } y = \sin\theta$$

$$x^2 + 4y^2 = 4\cos^2\theta + 4\sin^2\theta = 4$$

72. $\sin\theta + \cos\theta = \frac{1+\sqrt{3}}{2}$ where $0 < \theta < \frac{\pi}{2}$

if $\theta = 30^\circ$, $\sin\theta = \frac{1}{2}$, $\cos\theta = \frac{\sqrt{3}}{2}$

If $\theta = 60^\circ$, $\sin\theta = \frac{\sqrt{3}}{2}$, $\cos\theta = \frac{1}{2}$

In either case, $\tan\theta + \cot\theta = \frac{1}{\sqrt{3}} + \sqrt{3} = \frac{4}{\sqrt{3}}$

73. $A = \sin^2\theta + \cos^4\theta$, where $0 \le \theta < \frac{\pi}{2}$

If $\theta = 0$, it follows that $A = 1$

Also $\cos^4\theta \le \cos^2\theta$

$\therefore$ $\sin^2\theta + \cos^4\theta \le \sin^2\theta + \cos^2\theta = 1$

Among the option, only *b* can be considered

74. $E = \dfrac{\cot A + \text{cosec } A - 1}{\cot A - \text{cosec } A + 1} = \dfrac{\cos A + 1 - \sin A}{\cos A - 1 + \sin A}$

Comparing the numerator and denominator, both 1 and sin A are changing their sign. We consider the identity

$\dfrac{\sin A}{1 - \cos A} = \dfrac{1 + \cos A}{\sin A}$, *i.e.* the identity in which sin A appear alone (contract the 1)

Both expressions are equal to $\dfrac{1 + \cos A - \sin A}{\sin A - 1 + \cos A}$

75. $\pi^\circ = 180^\circ$, $1^\circ = \dfrac{180^\circ}{\pi} \approx \dfrac{180(7)^\circ}{22} = \dfrac{630^\circ}{11} = 57\dfrac{3^\circ}{11}$

1. $\sin 1^\circ > \sin 57\dfrac{3^\circ}{11}$ False **2.** $\cos 1^\circ < \cos 57\dfrac{3^\circ}{11}$ False

3. $\tan 1^\circ \tan 57\dfrac{3^\circ}{11}$ False

All three statements are incorrect.

76. If the sum of a number and its reciprocal is 2, the number has to be 1.

$\therefore$ $\tan^2 x = 1$ or $\tan x = \pm 1$

But as $0^\circ < x < 90^\circ$, the only value of x in 45°

77. 1. $\dfrac{\cos 75^\circ}{\sin 15^\circ} + \dfrac{\sin 12^\circ}{\cos 78^\circ} - \dfrac{\cos 18^\circ}{\sin 72^\circ} = 1 + 1 - 1$ True

2. $\dfrac{\cos 35^\circ}{\sin 55^\circ} - \dfrac{\sin 11^\circ}{\cos 79^\circ} + \cos 28^\circ \ \text{cosec}\ 62^\circ = 1 - 1 + 1 = 1$ True

3. $\dfrac{\sin 80^\circ}{\cos 10^\circ} - \sin 59^\circ \sec 31^\circ = 1 - 1 = 0$ True

All three statements are true.

78. $\tan 1^\circ \tan 2^\circ \tan 3^\circ \ldots\ldots \tan 89^\circ = (\tan 1^\circ \tan 89^\circ)(\tan 2^\circ \tan 85^\circ)$

$(\tan 44^\circ \tan 46^\circ \tan 45^\circ) = 1\,(1)\ldots\ldots(1)\,1 = 1.$

79. We know, π radian = 180 degree

$\therefore \dfrac{\pi \textit{ Radian}}{180 \text{ degree}} = \dfrac{R \textit{ radian}}{D \text{ degree}} \Rightarrow \pi D = 180R$

80. $9\tan^2\theta + 4\cot^2\theta$

$(3\tan\theta)^2 + (2\cot\theta)^2 \Rightarrow (3\tan\theta)^2 + (2\cot\theta)^2 - 2(3\tan\theta)(2\cot\theta) + 2(3\tan\theta)(2\cot\theta)$

$\Rightarrow (3\tan\theta - 2\cot\theta)^2 + 12$

Assume, $3\tan\theta - 2\tan\theta = 0$

$\therefore (3\tan\theta - 2\cot\theta)^2 + 12 = 0 + 12 = 12$

81. Let $x\sin\theta = y\cos\theta = \dfrac{2z\tan\theta}{1 - \tan^2\theta} = k$

$$\therefore \quad x=\frac{k}{\sin\theta}, y=\frac{k}{\cos\theta} \Rightarrow \frac{z2\tan\theta}{1-\tan^2\theta}=k$$

$$\Rightarrow z\tan 2\theta = k \qquad \left[\because \tan 2\theta = \frac{2\tan\theta}{1-\tan^2\theta}\right]$$

$$\Rightarrow z=\frac{k}{\tan^2\theta}$$

Put value of x, y, z in $4z^2\left(x^2+y^2\right)$

$$\therefore \quad \frac{4k^2}{\tan^2 2\theta}\left(\frac{k^2}{\sin^2\theta}+\frac{k^2}{\cos^2\theta}\right)=\frac{4k^4}{\tan^2 2\theta}\left(\frac{1}{\sin^2\theta}+\frac{1}{\cos^2\theta}\right)=\frac{4k^4\cos^2 2\theta}{\sin^2 2\theta}\left(\frac{\cos^2\theta+\sin^2\theta}{\sin^2\theta\cos^2\theta}\right)$$

$$=\frac{4k^4\left(\cos^2\theta-\sin^2\theta\right)^2}{\left(2\sin\cos\theta\right)^2}\times\frac{1}{\left(\sin\theta\cos\theta\right)^2}=\frac{4k^4}{4}\left(\frac{\cos^2\theta-\sin^2\theta}{\sin^2\theta\cos^2\theta}\right)^2$$

$$=k^4\left(\frac{\cos^2\theta}{\sin^2\theta\cos^2\theta}-\frac{\sin^2\theta}{\sin^2\theta\cos^2\theta}\right)^2=k^4\left(\frac{1}{\sin^2\theta}-\frac{1}{\cos^2\theta}\right)^2$$

$$=\left(\frac{k^2}{\sin^2\theta}-\frac{k^2}{\cos^2\theta}\right)^2=\left(x^2-y^2\right)^2 \qquad \left[\because x=\frac{k}{\sin\theta}\ and\ y=\frac{k}{\cos\theta}\right]$$

82. Given,

$\cos\theta_1+\cos\theta_2+\cos\theta_3=3$

$\Rightarrow \cos\theta_1+\cos\theta_2+\cos\theta_3=1+1+1$

$\therefore \cos\theta_1=1$

$\theta_1=0°$

$\cos\theta_2=1$

$\Rightarrow \theta_2=0°$

$\cos\theta_3=1$

$\Rightarrow \theta_3=0°$

Put value of $\theta_1,\theta_2,\theta_3$ in $\sin\theta_1,\sin\theta_2$ and $\sin\theta_3$

$\therefore \sin\theta_1+\sin\theta_2+\sin\theta_3=\sin 0°+\sin 0°+\sin 0°=0+0+0=0$

83. Put $\theta=0°$ in $\cos\theta+\tan\theta=1$

$\therefore \cos\theta+\tan\theta=1 \Rightarrow \cos 0°+\tan 0=1 \Rightarrow 1+0=1$

LHS = RHS

So, value of $\theta=0°$

84. $\sin x\sqrt{\frac{1}{1+\cos x}+\frac{1}{1-\cos x}} \Rightarrow \sin x\sqrt{\frac{1-\cos x+1+\cos x}{1-\cos^2 x}}=\sin x\sqrt{\frac{2}{1-\cos^2 x}}$

$$=\sin x\frac{\sqrt{2}}{\sqrt{\sin^2 x}}=\frac{\sin x\sqrt{2}}{\sin x}=\sqrt{2}$$

85. $\dfrac{\cos^4 A-\sin^4 A}{\cos^2 A-\sin^2 A}=\dfrac{(\cos^2 A)^2-(\sin^2 A)^2}{\cos^2 A-\sin^2 A}$

$=\dfrac{(\cos^2 A-\sin^2 A)(\cos^2 A+\sin^2 A)}{\cos^2 A-\sin^2 A}\left[\because a^2-b^2=(a-b)(a+b)\right]$

$=\cos^2 A+\sin^2 A=1$

86. $7\sin^2 x+3\cos^2 x=4$

$\Rightarrow\ 4\sin^2 x+3\sin^2 x+3\cos^2 x=4\ \Rightarrow 4\sin^2 x+3(\sin^2 x+\cos^2 x)=4$

$\Rightarrow\ 4\sin^2 x=1\quad\Rightarrow \sin x=\pm\dfrac{1}{2}$

$\sin x=\dfrac{1}{2}=\sin 30°$

$\therefore\ \tan 30°=\dfrac{1}{\sqrt{3}}$

87. $x=a\cos\theta+b\sin\theta$

$y=a\cos\theta-b\sin\theta$

$x^2+y^2=(a\cos\theta+b\sin\theta)^2+(a\sin\theta+b\cos\theta)^2$

$=a^2\cos^2\theta+b^2\sin^2\theta+2ab\cos\theta\sin\theta+a^2\sin^2\theta+b^2\cos^2\theta-2ab\cos\theta\sin\theta$

$=a^2(\cos^2\theta+\sin^2\theta)+b^2(\sin^2\theta+\cos^2\theta)+0\ =a^2(1)+b^2(1)=a^2+b^2$

88. $\pi^c=180°\Rightarrow\left(\dfrac{3\pi}{5}\right)^C=\left(\dfrac{180}{\pi}\times\dfrac{3\pi}{5}\right)^\circ=108°.$

89. $180°=\pi^C\Rightarrow 150°=\left(\dfrac{\pi}{180}\times 150\right)^C=\left(\dfrac{5\pi}{6}\right)^C.$

90. Clearly, x lies in 4th quadrant, where cosec x is negative and therefore $\sin x$ is negative.

$\sin x=-\sqrt{1-\cos^2 x}=-\sqrt{1-\dfrac{1}{2}}=-\dfrac{1}{\sqrt{2}}.$

$\therefore\ \text{cosec}\,x=-\sqrt{2}.$

91. In 4th quadrant, $\cos x$ is positive.

$\therefore\ \cos x=\sqrt{1-\sin^2 x}=\sqrt{1-\dfrac{1}{4}}=\dfrac{\sqrt{3}}{2}.$

92. We have, $p=\dfrac{1}{\text{cosec}\,\theta+\cot\theta}=\dfrac{1}{\left(\dfrac{1}{\sin\theta}+\dfrac{\cos\theta}{\sin\theta}\right)}=\dfrac{1}{\dfrac{1+\cos\theta}{\sin\theta}}=\dfrac{\sin\theta}{1+\cos\theta}$

$q=\text{cosec}\,\theta=\dfrac{1}{\sin\theta}$

So, p^2-2pq

$=\left(\dfrac{\sin\theta}{1+\cos\theta}\right)^2-2\times\dfrac{\sin\theta}{1+\cos\theta}\times\dfrac{1}{\sin\theta}$

$$=\frac{\sin^2\theta}{(1+\cos\theta)^2}-\frac{2}{1+\cos\theta}=\frac{\sin^2\theta-2(1+\cos\theta)}{(1+\cos\theta)^2}=\frac{(1-\cos^2\theta)-2(1+\cos\theta)}{(1+\cos\theta)^2}$$

$$=\frac{1-\cos^2\theta-2-2\cos\theta}{(1+\cos\theta)^2}=\frac{-1-\cos^2\theta-2\cos\theta}{(1+\cos\theta)^2}=\frac{-(1+\cos\theta)^2}{(1+\cos\theta)^2}=-1$$

93. $m^2(\sin\theta-1)+n^2(\sin\theta+1)=0$

$m^2\sin\theta-m^2+n^2\sin\theta+n^2=0\Rightarrow(m^2+n^2)\sin\theta+(-m^2+n^2)=0$

$(m^2+n^2)\sin\theta=(m^2-n^2)\Rightarrow\sin\theta=\dfrac{m^2-n^2}{m^2+n^2}$

Now, substitute $\sin\theta$ into the given expression:

$$(m^2+n^2)\cos\theta-(m^2-n^2)\cot\theta=(m^2+n^2)\cos\theta-(m^2-n^2)\times\frac{\cos\theta}{\sin\theta}$$

$$=(m^2+n^2)\cos\theta-(m^2-n^2)\times\left(\frac{\cos\theta}{\frac{m^2-n^2}{m^2+n^2}}\right)=(m^2+n^2)\cos\theta-(m^2-n^2)\times\left(\cos\theta\times\frac{m^2+n^2}{m^2-n^2}\right)$$

$$=(m^2+n^2)\cos\theta-(m^2+n^2)\cos\theta=0$$

94. $\Rightarrow(1+\tan\alpha\tan\beta)^2=1+\tan^2\alpha\tan^2\beta+2\tan\alpha\tan\beta$ ...(i)

$\Rightarrow(\tan\alpha-\tan\beta)^2=\tan^2\alpha+\tan^2\beta-2\tan\alpha\tan\beta$ ...(ii)

Eqn (i) + (ii)

$=(1+\tan\alpha\tan\beta)^2+(\tan\alpha-\tan\beta)^2$

$=1+\tan^2\alpha\tan^2\beta+2\tan\alpha\tan\beta+\tan^2\alpha+\tan^2\beta-2\tan\alpha\tan\beta$

$=1+\tan^2\alpha\tan^2\beta+\tan^2\alpha+\tan^2\beta=(1+\tan^2\alpha)+\tan^2\beta(1+\tan^2\alpha)$

$=(1+\tan^2\alpha)(1+\tan^2\beta)$

$=\sec^2\alpha\sec^2\beta$

95. Multiply the numerator and denominator of the first term by $(\cos\theta-\sin\theta-1)$.

$$\frac{\cos\theta-\sin\theta+1}{\cos\theta+\sin\theta-1}=\frac{(\cos\theta-(\sin\theta-1))}{(\cos\theta+(\sin\theta-1))}\times\frac{(\cos\theta-(\sin\theta-1))}{(\cos\theta-(\sin\theta-1))}$$

$$=\frac{(\cos\theta-(\sin\theta-1))^2}{\cos^2\theta-(\sin\theta-1)^2}=\frac{\cos^2\theta+(\sin\theta-1)^2-2\cos\theta(\sin\theta-1)}{\cos^2\theta-(\sin^2\theta-2\sin\theta+1)}$$

Using the identity $\cos^2\theta+\sin^2\theta=1$:

$$=\frac{\cos^2\theta+\sin^2\theta-2\sin\theta+1-2\cos\theta\sin\theta+2\cos\theta}{\cos^2\theta-\sin^2\theta+2\sin\theta-1}$$

$$=\frac{1-2\sin\theta+1-2\cos\theta\sin\theta+2\cos\theta}{(1-\sin^2\theta)-\sin^2\theta+2\sin\theta-1}=\frac{2-2\sin\theta-2\cos\theta\sin\theta+2\cos\theta}{-2\sin^2\theta+2\sin\theta}$$

$$=\frac{2(1-\sin\theta-\cos\theta\sin\theta+\cos\theta)}{2\sin\theta(1-\sin\theta)}=\frac{(1+\cos\theta)-\sin\theta(1+\cos\theta)}{\sin\theta(1-\sin\theta)}$$

$$= \frac{(1+\cos\theta)(1-\sin\theta)}{\sin\theta(1-\sin\theta)} = \frac{1+\cos\theta}{\sin\theta} = \frac{1}{\sin\theta} + \frac{\cos\theta}{\sin\theta} = \operatorname{cosec}\theta + \cot\theta$$

Thus, $(\operatorname{cosec}\theta + \cot\theta)(\cot\theta - \operatorname{cosec}\theta)$

$(\cot\theta + \operatorname{cosec}\theta)(\cot\theta - \operatorname{cosec}\theta) = \cot^2\theta - \operatorname{cosec}^2\theta = 1$

96. Given $\tan\theta = (4/5)$.

$$\therefore \quad \frac{5\sin\theta - 3\cos\theta}{5\sin\theta + 2\cos\theta} = \frac{5\tan\theta - 3}{5\tan\theta + 2} \qquad \textit{[Dividing Nr \& Dr by } \cos\theta]$$

$$= \frac{\left(5\times\frac{4}{5}-3\right)}{\left(5\times\frac{4}{5}+2\right)} = \frac{1}{6}.$$

97. Given $\cot x = \frac{12}{16} = \frac{3}{4}$.

$$\therefore \quad \frac{\sin x - \cos x}{\sin x + \cos x} = \frac{1-\cot x}{1+\cot x} \qquad \textit{[Dividing Nr \& Dr by } \sin x]$$

$$= \frac{\left(1-\frac{3}{4}\right)}{\left(1+\frac{3}{4}\right)} = \left(\frac{1}{4}\times\frac{4}{7}\right) = \frac{1}{7}.$$

98. $\tan x = (15/8)$

$$\therefore \sec x = \sqrt{1+\tan^2 x} = \sqrt{1+\frac{225}{64}} = \sqrt{\frac{289}{64}} = \frac{17}{8}.$$

$$\operatorname{cosec} x = \sqrt{1+\cot^2 x} = \sqrt{1+\frac{64}{225}} = \sqrt{\frac{289}{225}} = \frac{17}{15}.$$

$$\therefore \cos x = \frac{8}{17} \text{ and } \sin x = \frac{15}{17}.$$

So, $(\sin x - \cos x) = \left(\frac{15}{17} - \frac{8}{17}\right) = \frac{7}{17}$.

99. $\sec^2\theta = (1 + \tan^2\theta) = \left(1+\frac{1}{7}\right) = \frac{8}{7}$, $\operatorname{cosec}^2\theta = (1 + \cot^2\theta) = (1 + 7) = 8$.

$$\therefore \quad \frac{\operatorname{cosec}^2\theta - \sec^2\theta}{\operatorname{cosec}^2\theta + \sec^2\theta} = \frac{\left(8-\frac{8}{7}\right)}{\left(8+\frac{8}{7}\right)} = \left(\frac{48}{7}\times\frac{7}{64}\right) = \frac{3}{4}.$$

100. (1) L.H.S. $\cos^4\theta - \sin^4\theta = (\cos^2\theta + \sin^2\theta)(\cos^2\theta - \sin^2\theta) = \cos 2\theta$ (by using identities)

R.H.S.

$$\frac{2\tan\theta}{1-\tan^2\theta} = \frac{\frac{2\sin\theta}{\cos\theta}}{\frac{\cos^2\theta - \sin^2\theta}{\cos^2\theta}} = \frac{2\tan\theta}{\cos 2\theta}$$

(2) L.H.S. = $\operatorname{cosec}\theta + \cot\theta = \frac{1}{\sin\theta} + \frac{\cos\theta}{\sin\theta} = \frac{1+\cos\theta}{\sin\theta}$

RHS

$$\frac{1}{\operatorname{cosec}\theta - \cot\theta} = \frac{1}{\frac{1}{\sin\theta} - \frac{\cos\theta}{\sin\theta}} = \frac{\sin\theta}{1-\cos\theta} \times \frac{1+\cos\theta}{1+\cos\theta} = \frac{1+\cos\theta}{\sin\theta}$$

(3) L.H.S. $\cos^2\theta - \sin^2\theta = \cos 2\theta$

$$\frac{1-\tan^2\theta}{1+\tan^2\theta} = \frac{\cos^2\theta - \sin^2\theta}{\cos^2\theta + \sin^2\theta} = \cos 2\theta$$

Hence, (2) and (3) are correct

101. $\sin\theta = \frac{12}{13} = \frac{p}{h} \Rightarrow b = \sqrt{13^2 - 12^2} = 5$

Now, $(\tan\theta + \sec\theta)^2 (\operatorname{cosec}\theta - \cot\theta)^2$

$$\left(\frac{12}{5}+\frac{13}{5}\right)^2 \left(\frac{13}{12}-\frac{5}{12}\right)^{-2} \Rightarrow (5)^2 \times \left(\frac{8}{12}\right)^{-2} \Rightarrow 25 \times \left(\frac{3}{2}\right)^2 = 25 \times \frac{9}{4} = \frac{225}{4}$$

102. We have,

$\tan^8\theta + \cot^8\theta = m$

$\Rightarrow (\tan^4\theta + \cot^4\theta)^2 = m + 2 \quad \Rightarrow (\tan^4\theta + \cot^4\theta) = \sqrt{m+2}$

$\Rightarrow (\tan^2\theta + \cot^2\theta)^2 = \sqrt{m+2} + 2 \Rightarrow (\tan^2\theta + \cot^2\theta)^2 = \sqrt{\sqrt{m+2}+2}$

$\Rightarrow (\tan\theta + \cot\theta) = \sqrt{\sqrt{\sqrt{m+2}+2}+2}$

103. We have, $x = 6 - 4\sin\theta$

For making it minimum, $(4\sin\theta)$ should be maximum

$\because \theta \Sigma \left[0, \frac{\pi}{2}\right]$

$\because$ Maximum value of $\sin\theta = 1$

Minimum value of $x = 6 - 4 \times 1 = 2$

104. $4\cos^2 30° + 2x \sin 30° - \cot^2 30° - 6 \tan 15° \tan 75° = 0$

$\Rightarrow 4\cos^2 30° + 2x \sin 30° - \cot^2 30° - 6 \tan (90° - 75°) \tan 75°$ $[\because \tan(90° - \theta) = \cot\theta]$

$\Rightarrow 4\cos^2 30° + 2x \sin 30° - \cot^2 30° - 6$ $[\because \tan\theta \cot\theta = 1]$

$$\Rightarrow 4\left(\frac{\sqrt{3}}{2}\right)^2 + 2x\frac{1}{2} - (\sqrt{3})^2 - 6 = 0 \Rightarrow 3 + x - 3 - 6 = 0 \Rightarrow x = 6$$

105. $\frac{\cos^2 32° + \cos^2 58°}{\sec^2 50° - \cot^2 40°} + 4 \tan 13° \tan 37° \tan 53° \tan 77°$

$$= \frac{\cos^2(90° - 58°) + \cos^2 58°}{\sec^2(90° - 40°) - \cot^2 40°} + 4\tan 13° \tan 37° \tan(90° - 37°)\tan(90° - 13°)$$

$$\frac{\sin^2 58° + \cos^2 58°}{\csc^2 40° - \cot^2 40°} + 4\tan 13° \tan 37° \cot 37° \cot 13°$$

$\left[\begin{array}{l}\therefore \cos(90° - \theta) = \sin\theta \\ \sec(90° - \theta) = \operatorname{cosec}\theta \\ \tan(90° - \theta) = \cot\theta\end{array}\right]$

Using identities = 1 + 4 = 5

106. $(1 + \cot^2\theta)(1 + \cos\theta)(1 - \cos\theta) - (1 + \tan^2\theta)(1 + \sin\theta)(1 - \sin\theta)$
$= \text{cosec}^2\theta(1 - \cos^2\theta) - (1 + \tan^2\theta)(1 - \sin^2\theta)$

$= \text{cosec}^2\theta \sin^2\theta - \sec^2\theta \cos^2\theta = 0$ $\left[\begin{array}{l}\therefore 1+\cot^2\theta = \text{cosec}^2\theta \\ 1-\cos^2\theta = \sin^2\theta \\ 1+\tan^2\theta = \sec^2\theta \\ 1-\sin^2\theta = \cos^2\theta\end{array}\right]$

107. $2\cos^2\theta = \sin\theta - 2 = 0$

$\Rightarrow 2 - 2\sin^2\theta + \sin\theta - 2 = 0 \Rightarrow 2\sin^2\theta - \sin\theta = 0 \Rightarrow \sin\theta(2\sin\theta - 1) = 0 \Rightarrow \sin\theta = 0, \frac{1}{2}$

$\Rightarrow \theta = 0$ (not possible) $\left[0 < \theta \le \frac{\pi}{2}\right]$

and $\theta = \frac{\pi}{6}$

108. We have,
$3(\sin x - \cos x)^2 + 6(\sin x + \cos x)^2 + 4(\sin x)^6 + 4(\cos x)^6$
$= 3(\sin^2 x - \cos^2 x - 2\sin x)^2 + 6(\sin^2 x + \cos^2 x + 2\sin x\cos x + 4(\sin^2 x)^3 + 4(\cos^2 x)^3$
$= 3(1 - 2\sin x\cos x)^2 + 6(1 + \sin 2x) + 4((\sin^2 x)^3 + (\cos^2 x)^3)$
$[\because 2\sin x\cos x = \sin 2x \text{ and } \sin^2 x + \cos^2 x = 1]$
$= 3(1 + \sin^2 2x - 2\sin 2x) + 6 + 6\sin 2x + 4(\sin^2 x + \cos^2 x)(\sin^4 x + \cos^4 x - \sin^2 x\cos^2 x)$
$= 3 + \sin^2 2x - 6\sin 2x + 6 + 6\sin 2x + 4(1 - 2\sin^2 x\cos^2 x - \sin^2 x\cos^2 x)$
$= 9 + 3\sin^2 2x + 4 - 12\sin^2 x\cos^2 x$
$= 13 + 3.4\sin^2 x\cos^2 x - 12\sin^2 x\cos^2 x) = 13$

109. $\tan^2 x + \sec x = 5 \Rightarrow \sec^2 x - 1 + \sec x = 5$
$\Rightarrow \sec^2 x + \sec x - 6 = 0 \quad \Rightarrow (\sec x + 3)(\sec x - 2) = 0$
$\Rightarrow \sec x = 2 \Rightarrow \cos x = \frac{1}{2}.$ $\left[0 \le x \le \frac{\pi}{2} \Rightarrow \sec x \text{ is} + ve\right]$

110. $\sec^2\theta = (x - \tan\theta)^2 \Rightarrow 2x\tan\theta = x^2 - 1$
$\therefore \tan\theta = \left(\frac{x^2-1}{2x}\right).$

111. $3\sin x + 5\cos x = 5$ ... (*i*) (Given)
$3\cos x - 5\sin x = a$ ... (*ii*) (say)
Squaring (*i*) and (*ii*) and adding, we get
$9(\sin^2 x + \cos^2 x) + 25(\cos^2 x + \sin^2 x) = 25 + a^2$ or $a^2 = 9$.

112. $\cos^2\theta + \sec^2\theta = \cos^2\theta + \sec^2\theta - 2\cos\theta \,.\, \sec\theta + 2\cos\theta\sec\theta$
$= (\cos\theta - \sec\theta)^2 + 2 \ge 2.$

113. $\sin^2 x + \text{cosec}^2 x = \sin^2 x + \text{cosec}^2 x - 2\sin x\,\text{cosec}\, x + 2\sin x\,\text{cosec}\, x$
$= (\sin x - \text{cosec}\, x)^2 + 2 \ge 2.$

114. $\tan^2\theta + \cot^2\theta = \tan^2\theta + \cot^2\theta - 2\tan\theta\cot\theta + 2\tan\theta\cot\theta = (\tan\theta - \cot\theta)^2 + 2 \ge 2.$

115. $\cos^2\theta + \frac{1}{\cos^2\theta} = \cos^2\theta + \frac{1}{\cos^2\theta} - 2 + 2 = \left(\cos\theta - \frac{1}{\cos\theta}\right)^2 + 2 > 2 \quad \left\{\because 0 < \theta \le \frac{\pi}{2} \Rightarrow \cos\theta \ne 1\right\}$

116. $\sqrt{\dfrac{1+\cos x}{1-\cos x}} = \sqrt{\dfrac{(1+\cos x)}{(1-\cos x)} \times \dfrac{(1+\cos x)}{(1+\cos x)}} = \dfrac{(1+\cos x)}{\sqrt{1-\cos^2 x}}$

$$= \frac{1+\cos x}{\sin x} = \left(\frac{1}{\sin x} + \frac{\cos x}{\sin x}\right) = (\operatorname{cosec} x + \cot x).$$

117. $\sec^4 A - \sec^2 A = \sec^2 A(\sec^2 A - 1) = (1 + \tan^2 A)\tan^2 A = \tan^2 A + \tan^4 A.$

118. Given Exp. $= \dfrac{(5\cos\theta - 4)(4 + 5\cos\theta) - (3 + 5\sin\theta)(3 - 5\sin\theta)}{(3 - 5\sin\theta)(4 + 5\cos\theta)}$

$$= \frac{(25\cos^2\theta - 16) - (9 - 25\sin^2\theta)}{(3 - 5\sin\theta)(4 + 5\cos\theta)} = \frac{25(\cos^2\theta + \sin^2\theta) - 25}{(3 - 5\sin\theta)(4 + 5\cos\theta)} = 0.$$

119. Given Exp. $= \left(1 + \dfrac{\cos\theta}{\sin\theta} - \dfrac{1}{\sin\theta}\right)\left(1 + \dfrac{\sin\theta}{\cos\theta} + \dfrac{1}{\cos\theta}\right) = \left[\dfrac{(\sin\theta + \cos\theta) - 1}{\sin\theta}\right]\left[\dfrac{(\sin\theta + \cos\theta) + 1}{\cos\theta}\right]$

$$= \frac{(\sin\theta + \cos\theta)^2 - 1}{\sin\theta\cos\theta} = \frac{(\sin^2\theta + \cos^2\theta) + 2\sin\theta\cos\theta - 1}{\sin\theta\cos\theta} = \frac{2\sin\theta\cos\theta}{\sin\theta\cos\theta} = 2.$$

120. Given Exp. $= \dfrac{\dfrac{\sin A}{\cos A} + \dfrac{\sin B}{\cos B}}{\dfrac{\cos A}{\sin A} + \dfrac{\cos B}{\sin B}} = \dfrac{\sin A\cos B + \cos A\sin B}{\cos A\cos B} \times \dfrac{\sin A\sin B}{\cos A\sin B + \cos B\sin A}$

$$= \frac{\sin A\sin B}{\cos A\cos B} = \tan A\tan B.$$

121. $\cot^2\theta - (\sqrt{3} + 1)\cot\theta + \sqrt{3} = 0$

$\Rightarrow \cot^2\theta - \sqrt{3}\cot\theta - \cot\theta + \sqrt{3} = 0 \Rightarrow \cot\theta(\cot\theta - \sqrt{3}) - 1(\cot\theta - \sqrt{3}) = 0$

$\Rightarrow (\cot\theta - 1)(\cot\theta - \sqrt{3}) = 0 \Rightarrow \cot\theta = \sqrt{3} \Rightarrow \theta = \dfrac{\pi}{6}$

or $\cot\theta = 1$ (not possible) $\left[\because 0 < \theta < \dfrac{\pi}{4}\right]$

Now, $\sin\theta + \cos\theta = \dfrac{1}{2} + \dfrac{\sqrt{3}}{2} = \dfrac{\sqrt{3}+1}{2}$

122. $\tan 2\theta \tan 4\theta - 1 = 0$

Let $\tan 2\theta = p$

$\because \quad \tan 4\theta = \tan 4\theta = \dfrac{2\tan 2\theta}{1 - \tan^2 2\theta}$

$\Rightarrow \quad \dfrac{2p}{1-p^2} - 1 = 0 \Rightarrow \dfrac{2p^2}{1-p^2} = 1 \Rightarrow 2p^2 - 1 + p^2 = 0 \Rightarrow 3p^2 - 1 = 0 \Rightarrow 3p^2 = 1$

$\Rightarrow \quad p = \pm\dfrac{1}{\sqrt{3}} \Rightarrow \tan 2\theta = \pm\dfrac{1}{\sqrt{3}}$

We have,

$0 < 2\theta < \pi$ $\left[\because 0 < \theta < \dfrac{\pi}{2}\right]$

$\Rightarrow 2\theta = \dfrac{\pi}{6} \Rightarrow \theta = \dfrac{\pi}{12}$

123. $\tan x = \tan x = \dfrac{\sin\theta + \cos\theta}{\sin\theta - \cos\theta} \left[\because \dfrac{\pi}{4} < \theta < \dfrac{\pi}{2}\right]$

$\Rightarrow \tan x = \dfrac{1+\tan\theta}{1-\tan\theta}$ [Dividing numerator and denominator by cosθ.]

$\Rightarrow \tan x = \dfrac{1+\tan\theta}{1-\tan\theta}$

$\Rightarrow \tan x = \tan\dfrac{\left(\dfrac{\pi}{4}+\theta\right)}{1} = \dfrac{p}{b} \quad \therefore \dfrac{\pi}{4} < \theta < \dfrac{\pi}{2}$

$\dfrac{\pi}{4}+\dfrac{\pi}{4} < \theta + \dfrac{\pi}{4} < \dfrac{\pi}{2} + \dfrac{\pi}{4} \Rightarrow \dfrac{\pi}{2} < \theta + \dfrac{\pi}{4} < \dfrac{3\pi}{4}$

$\therefore \sin x = \dfrac{\tan\left(\dfrac{\pi}{4}+\theta\right)}{\sqrt{1+\tan^2\left(\dfrac{\pi}{4}+\theta\right)}} = \dfrac{\tan\left(\dfrac{\pi}{4}+\theta\right)}{\sec\left(\dfrac{\pi}{4}+\theta\right)} \Rightarrow \sin x = \sin\left(\dfrac{\pi}{4}+\theta\right)$

$\Rightarrow \sin x = \dfrac{\sin\theta + \cos\theta}{\sqrt{2}} \Rightarrow \sqrt{2}\sin x = \sin\theta + \cos\theta$

124. We have,

$(\sin^2\theta - 4\sin\theta + 3)(4 - \cos^2\theta + 4\sin\theta) = 0 \Rightarrow \sin^2\theta + 4\sin\theta + 3 = 0$

$\Rightarrow \sin^2\theta - 3\sin\theta - \sin^2\theta + 3 = 0 \Rightarrow \sin\theta(\sin\theta - 3) - 1(\sin\theta - 3) = 0$

$\Rightarrow \sin\theta = -1, 3$ (not possible) $[\because 0 < \theta < \dfrac{\pi}{2}]$

So, there is no value of θ.

125. We have,

$x\sin^3\theta + y\cos^3\theta = \sin\theta\cos\theta$ ... (i)

and $x\sin\theta = y\cos\theta = 0$ ... (ii)

$\Rightarrow x\sin\theta = y\cos\theta$ ... (iii)

By putting,

$(x\sin\theta)\sin^2\theta + y\sin^3\theta = \sin\theta\cos\theta$

$y\cos\theta(\sin^2\theta + \cos^2\theta) = \sin\theta\cos\theta$

$\Rightarrow y = \sin\theta$ $[\because \sin^2\theta = \cos^2\theta = 1]$

Now,

$x\sin\theta = \sin\theta\cos$ [By putting $y = \sin\theta$ in (ii)]

$\Rightarrow x = \cos\theta$

Now, $x^2 + y^2 = \cos^2\theta + \sin^2\theta = 1$

126. Squaring and adding, we get: $a^2(\cos^2\theta + \sin^2\theta) + b^2(\sin^2\theta + \cos^2\theta) = (4^2 + 3^2)$

$\Rightarrow a^2 + b^2 = 25.$

127. Squaring and adding the given equations, we get:

$$\frac{x^2}{a^2}(\cos^2\phi + \sin^2\phi) + \frac{y^2}{b^2}(\sin^2\phi + \cos^2\phi) = 1^2 + (-1)^2.$$

$$\therefore \quad \frac{x^2}{a^2} + \frac{y^2}{b^2} = 2.$$

128. Squaring and adding, we get:

$p^2(\cos^2\theta + \sin^2\theta) + q^2(\sin^2\theta + \cos^2\theta) = 8^2 + 5^2 \Rightarrow p^2 + q^2 = 89.$

129. $x^2 + y^2 + z^2 = r^2\sin^2 A\cos^2 B + r^2\sin^2 A\sin^2 B + r^2\cos^2 A$

$= r^2\sin^2 A(\cos^2 B + \sin^2 B) + r^2\cos^2 A$

$= r^2\sin^2 A + r^2\cos^2 A = r^2(\sin^2 A + \cos^2 A) = r^2.$

130. $x^2 - y^2 = (\cot\theta + \cos\theta)^2 - (\cot\theta - \cos\theta)^2 = 4\cot\theta\cos\theta = \dfrac{4\cos^2\theta}{\sin\theta}$

Also, $xy = (\cot\theta + \cos\theta)(\cot\theta - \cos\theta)$

$$= \cot^2\theta - \cos^2\theta = \cos^2\theta \cdot \left(\frac{1}{\sin^2\theta} - 1\right) = \cos^2\theta \cdot \frac{(1-\sin^2\theta)}{\sin^2\theta} = \frac{\cos^4\theta}{\sin^2\theta}$$

$\therefore \quad \dfrac{\cos^2\theta}{\sin\theta} = \sqrt{xy}.$

Consequently, $(x^2 - y^2) = 4\sqrt{xy}$.

131. $\tan A + \cot A = 2$

$\therefore A = 45°$

$\therefore \tan^2 A + \tan^3 A + \ldots\ldots \tan^n A = 1 + 1 + \ldots\ldots + 1 = (n-1)$

132. (1) $\dfrac{\sin^3\theta + \cos^3\theta}{\sin\theta + \cos\theta} + \sin\theta\cos\theta$

$$= \frac{(\sin\theta + \cos\theta)(\sin^2\theta + \cos^2\theta - \sin\theta\cos\theta)}{(\sin\theta + \cos\theta)} + \sin\theta\cos\theta$$

$= 1 - \sin\theta\cos\theta + \sin\theta\cos\theta = 1$

$\therefore$ (1) is correct

(2) $1 - \sin^6\theta = 1 - (\sin^2\theta)^3 = (1 - \sin^2\theta)(1 + \sin^4\theta + \sin^2\theta)$

$= \cos^2\theta(1 + (1 - \cos^2\theta)^2 + \sin^2\theta) = \cos^2\theta(1 + 1 + \cos^4\theta - 2\cos^2\theta + \sin^2\theta)$

$= \cos^2\theta(\cos^4\theta + 2(1 - \cos^2\theta) + \sin^2\theta) = \cos^2\theta(\cos^4\theta + 2\sin^2\theta + \sin^2\theta)$

$= \cos^2\theta(\cos^4\theta + 2\sin^2\theta)$

$\because$ (2) is corect

133. $7\sin^4\theta + 9\cos^2\theta + 42\sin^2\theta = 16$

$\Rightarrow 7\sin^4\theta + 9(1 - \sin^4\theta)^2 + 42\sin^2\theta = 16$

$\Rightarrow 7\sin^4\theta + 9(1 + \sin^4\theta - 2\sin^2\theta) + 42\sin^4\theta = 16 \Rightarrow 16\sin^4\theta + 24\sin^2\theta - 7 = 0$

Let $\sin^2\theta = t$

$\Rightarrow 16t^2 + 24t - 7 = 0 \Rightarrow 16t^2 + 28t - 4t - 7 = 0 \Rightarrow 4t(4t + 7) - 1(4t + 7) = 0$

$\Rightarrow (4t - 7)(4t + 7) = 0 \Rightarrow t = \dfrac{-7}{4}, \dfrac{1}{4}$

$\sin^2\theta = \dfrac{-7}{4}$ (not possible)

$\therefore \sin^2\theta = \dfrac{1}{4} \Rightarrow \sin\theta = \dfrac{1}{2} \Rightarrow \theta = \dfrac{\pi}{6}$

now, $\tan\theta = \dfrac{1}{\sqrt{3}}$

134. $\sin 43°\cos 47° + \cos 43°\sin 47° = \sin^2 43° + \cos^2 43° = 1.$

$[\because \cos 47° = \cos(90° - 43°) = \sin 43° \ \&\ \sin 47° = \cos 43°]$

135. $\sin\alpha = \frac{1}{2} \Rightarrow \cos\alpha = \sqrt{1-\sin^2\alpha} = \sqrt{1-\frac{1}{4}} = \frac{\sqrt{3}}{2}.$

$\alpha + \beta = 90° \Rightarrow \beta = (90° - \alpha).$

$\therefore \sin\beta = \sin(90° - \alpha) = \cos\alpha = \frac{\sqrt{3}}{2}$ & $\cos\beta = \cos(90° - \alpha) = \sin\alpha = \frac{1}{2}$

$\therefore \cos\alpha\sin\beta - \sin\alpha\cos\beta = \left(\frac{\sqrt{3}}{2} \times \frac{\sqrt{3}}{2} - \frac{1}{2} \times \frac{1}{2}\right) = \frac{1}{2}.$

136. We have:

$\cos(90° + \theta) = -\sin\theta$, $\sec(360° - \theta) = \sec\theta$; $\tan(180° - \theta) = -\tan\theta$;

$\sec(-\theta) = \sec\theta$, $\cot(-\theta) = -\cot\theta$ etc.

$\therefore$ Given Exp. $= \frac{(-\sin\theta)\cdot\sec\theta\cdot(-\tan\theta)}{\sec(720° - \theta)\sin(540° + \theta)[-\cot(90° - \theta)]} = \frac{\tan^2\theta}{-\sec\theta\cdot\sin(180° + \theta)\tan\theta}$

$= \frac{\tan^2\theta}{\sec\theta\cdot\sin\theta\cdot\tan\theta} = \frac{\tan^2\theta}{\tan^2\theta} = 1.$

137. $\cos 40°$ is clearly + ve, $\cos(-40°) = \cos 40° > 0$;

$\cos 320° = \cos(360° - 40°) = \cos 40° > 0$. And, $\cos 140° = \cos(180° - 40°) = -\cos 40° < 0$.

138. $x = \frac{\cos(90° - \theta)}{\sin(90° - \theta)\cot(90° - \theta)} = \frac{\sin\theta}{\cos\theta\cdot\tan\theta} = 1.$

139. $\sin 2x = \frac{2\tan x}{1 + \tan^2 x} = \frac{2 \times \frac{3}{4}}{1 + \frac{9}{16}} = \left(\frac{3}{2} \times \frac{16}{25}\right) = \frac{24}{25}.$

140. $p\cos 2x + q\sin 2x = p.\left(\frac{1 - \tan^2 x}{1 + \tan^2 x}\right) + q\cdot\left(\frac{2\tan x}{1 + \tan^2 x}\right) = p\cdot\left(\frac{1 - \frac{p^2}{q^2}}{1 + \frac{p^2}{q^2}}\right) + q\cdot\left(\frac{\frac{2p}{q}}{1 + \frac{p^2}{q^2}}\right)$

$= p\cdot\left(\frac{q^2 - p^2}{q^2 + p^2}\right) + \left(\frac{2pq^2}{q^2 + p^2}\right) = \frac{p(3q^2 - p^2)}{(p^2 + q^2)}.$

141. $\cos 3x = 4\cos^3 x - 3\cos x = 4 \times \left(\frac{1}{2}\right)^3 - 3 \times \frac{1}{2} = \left(\frac{1}{2} - \frac{3}{2}\right) = -1.$

142. $\sin 75° = \sin(45° + 30°) = \sin 45° \cos 30° + \cos 45° \sin 30° = \frac{1}{\sqrt{2}}\cdot\frac{\sqrt{3}}{2} + \frac{1}{\sqrt{2}}\cdot\frac{1}{2} = \frac{\sqrt{3}+1}{2\sqrt{2}}.$

143. $\cos 75° = \cos(45° + 30°) = \cos 45° \cos 30° - \sin 45° \sin 30° = \left(\frac{\sqrt{3}-1}{2\sqrt{2}}\right).$

144. $\cos\alpha\cos\beta - \sin\alpha\sin\beta = 1 \Rightarrow \cos(\alpha + \beta) = 1.$

$\therefore \sin(\alpha + \beta) = \sqrt{1 - \cos^2(\alpha + \beta)} = 0.$ or $\sin\alpha\cos\beta + \cos\alpha\sin\beta = 0$

or $\sin\alpha\cos\beta = -\cos\alpha\sin\beta$ or $\frac{\sin\alpha\cos\beta}{\cos\alpha\sin\beta} = -1.$

$\therefore \tan\alpha\cot\beta = -1.$

145. Given, $\cot^2\theta - 3\sqrt{3}\cot\theta + 6 = 0$

Put $\cot\theta = x$

So, $x^2 - 3\sqrt{3}x + 6 = 0$

$x^2 - 2\sqrt{3}x - \sqrt{3}x + 6 = 0$

$x\left(x - 2\sqrt{3}\right) - \sqrt{3}\left(x - 2\sqrt{3}\right) = 0$

$\left(x - 2\sqrt{3}\right)\left(x - \sqrt{3}\right) = 0$

So, $x = \sqrt{3}$ and $2\sqrt{3}$

i.e., $\cot\theta = \sqrt{3} = \cot 30^\circ$

So, $\theta = 30^\circ$

Now, $\sin\theta + \cos 2\theta = \sin 30^\circ + \cos 60^\circ = \frac{1}{2} + \frac{1}{2} = 1$

146. $\sin\theta = \sqrt{1 - \cos^2\theta} = \frac{4\sqrt{3}}{7}$, $\sin\phi = \sqrt{1 - \cos^2\phi} = \frac{3\sqrt{3}}{14}$

$\therefore \cos(\theta - \phi) = \cos\theta\cos\phi + \sin\theta + \sin\phi = \left(\frac{1}{7} \times \frac{13}{14} + \frac{4\sqrt{3}}{7} \times \frac{3\sqrt{3}}{14}\right) = \frac{49}{98} = \frac{1}{2} = \cos 60^\circ.$

$\therefore \theta - \phi = 60^\circ$

147. $\tan 70^\circ = \tan(20^\circ + 50^\circ) = \frac{\tan 20^\circ + \tan 50^\circ}{1 - \tan 20^\circ \tan 50^\circ}$

$\therefore \tan 70^\circ (1 - \tan 20^\circ \tan 50^\circ) = \tan 20^\circ + \tan 50^\circ$

$\Rightarrow \tan 70^\circ - \tan 20^\circ \tan 50^\circ \tan 70^\circ = \tan 20^\circ + \tan 50^\circ$

$\Rightarrow \tan 70^\circ - \cot 70^\circ \tan 70^\circ \tan 50^\circ = \tan 20^\circ + \tan 50^\circ$

$\Rightarrow \tan 70^\circ - \tan 20^\circ - 2\tan 50^\circ = 0 \Rightarrow \tan 20^\circ + 2\tan 50^\circ = \tan 70^\circ$

148. $\cos A = \sqrt{1 - \sin^2 A} = \frac{3}{5}$, $\sin B = \sqrt{1 - \cos^2 B} = \frac{12}{13}$.

$\therefore \sin(A + B) = \sin A\cos B + \cos A\sin B = \left(\frac{4}{5} \times \frac{5}{13} + \frac{3}{5} \times \frac{12}{13}\right) = \frac{56}{65}.$

149. $\cot(A + B) = \frac{1}{\tan(A+B)} = \frac{1 - \tan A\tan B}{\tan A + \tan B} = \frac{1}{\tan A + \tan B} - \frac{1}{\frac{1}{\tan B} + \frac{1}{\tan A}}$

$= \frac{1}{\tan A + \tan B} - \frac{1}{\cot B + \cot A} = \left(\frac{1}{a} - \frac{1}{b}\right).$

150. $\cot\left(\frac{\pi}{4} + \theta\right) \cdot \cot\left(\frac{\pi}{4} - \theta\right) = \frac{1}{\tan\left(\frac{\pi}{4} + \theta\right)} \cdot \frac{1}{\tan\left(\frac{\pi}{4} - \theta\right)} = \frac{1 - \tan\frac{\pi}{4}\tan\theta}{\tan\frac{\pi}{4} + \tan\theta} \times \frac{1 + \tan\frac{\pi}{4} \cdot \tan\theta}{\tan\frac{\pi}{4} - \tan\theta}$

$= \left(\frac{1 - \tan\theta}{1 + \tan\theta} \times \frac{1 + \tan\theta}{1 - \tan\theta}\right) = 1.$

151. $(1 + \tan\alpha)(1 + \tan\beta) = 2 \quad \Rightarrow \quad \tan\alpha + \tan\beta = 1 - \tan\alpha\tan\beta$

$$\Rightarrow \quad \frac{\tan\alpha + \tan\beta}{1 - \tan\alpha\tan\beta} = 1 \Rightarrow \tan(\alpha+\beta) = \tan\frac{\pi}{4} \quad \Rightarrow \quad \alpha + \beta = \frac{\pi}{4}.$$

152. $\dfrac{\sin(\alpha+\beta)}{\cos(\alpha-\beta)} = \dfrac{\sin\alpha\cos\beta + \cos\alpha\sin\beta}{\cos\alpha\cos\beta + \sin\alpha\sin\beta} = \dfrac{\tan\alpha + \tan\beta}{1 + \tan\alpha\tan\beta}$. [*Dividing Nr & Dr by cos* α cos β]

153. $\tan 3\theta = \dfrac{3\tan\theta - \tan^3\theta}{1 - 3\tan^2\theta}$. So, (*c*) is true.

154. $\dfrac{1}{\sin 10°} - \dfrac{\sqrt{3}}{\cos 10°} = \dfrac{\cos 10° - \sqrt{3}\sin 10°}{\sin 10°\cos 10°} = \dfrac{\frac{1}{2}\cdot\cos 10° - \frac{\sqrt{3}}{2}\sin 10°}{\frac{1}{2}(\sin 10°\cos 10°)}$

$$= \frac{2[\sin 30°\cos 10° - \cos 30°\sin 10°]}{(\sin 10°\cos 10°)} = \frac{2\sin 20°}{\frac{1}{2}(\sin 20°)} = 4.$$

155. $\dfrac{\sin 7x - \sin 5x}{\cos 7x + \cos 5x} = \dfrac{2\cos\left(\frac{7x+5x}{2}\right)\sin\left(\frac{7x-5x}{2}\right)}{2\cos\left(\frac{7x+5x}{2}\right)\cos\left(\frac{7x-5x}{2}\right)} = \tan x.$

156. Given Exp. $\dfrac{(\sin x + \sin 7x) + (\sin 3x + \sin 5x)}{(\cos x + \cos 7x) + (\cos 3x + \cos 5x)}$

$$= \frac{2\sin\left(\frac{7x+x}{2}\right)\cos\left(\frac{7x-x}{2}\right) + 2\sin\left(\frac{5x+3x}{2}\right)\cos\left(\frac{5x-3x}{2}\right)}{2\cos\left(\frac{7x+x}{2}\right)\cos\left(\frac{7x-x}{2}\right) + 2\cos\left(\frac{5x+3x}{2}\right)\cos\left(\frac{5x-3x}{2}\right)}$$

$$= \frac{\sin 4x(\cos 3x + \cos x)}{\cos 4x(\cos 3x + \cos x)} = \tan 4x.$$

157. $\sin 50° + \sin 10° - \sin 70° = 2\sin\left(\dfrac{50° + 10°}{2}\right)\cos\left(\dfrac{50° - 10°}{2}\right) - \sin 70°$

$$= 2\sin 30°\cos 20° - \sin(90° - 20°) = 2 \times \frac{1}{2}\cos 20° - \cos 20° = 0.$$

158. We shall show that: $\tan 10° + \tan 70° = \tan 60° + \tan 50°$.

We know that: $\tan A + \tan B = \dfrac{\sin(A+B)}{\sin A\cos B}$.

$\tan 10° + \tan 70° = \tan 60° + \tan 50°$

$$\Leftrightarrow \frac{\sin 80°}{\cos 10°\cos 70°} = \frac{\sin 110°}{\cos 60°\cos 50°} \Leftrightarrow \frac{\cos 10°}{\cos 10°\cos 70°} = \frac{\cos 20°}{\cos 60°\cos 50°} = \frac{2\cos 20°}{\cos 50°}$$

$$\Leftrightarrow \frac{1}{\cos 70°} = \frac{2\cos 20°}{\cos 50°}$$

$\Leftrightarrow \cos 50° = 2\cos 20°\cos 70° = \cos 90° + \cos 50° = \cos 50°$, which is true.

$\therefore \tan 10° + \tan 70° - \tan 50° = \tan 60° = \sqrt{3}.$

159. $\dfrac{\sin x+\sin y}{\cos x+\cos y}=\dfrac{a}{b}\Rightarrow \dfrac{2\sin\left(\dfrac{x+y}{2}\right)\cos\left(\dfrac{x-y}{2}\right)}{2\cos\left(\dfrac{x+y}{2}\right)\cos\left(\dfrac{x-y}{2}\right)}=\dfrac{a}{b}\Rightarrow \tan\left(\dfrac{x+y}{2}\right)=\dfrac{a}{b}.$

$$\therefore \sin(x+y)=\frac{2\tan\left(\dfrac{x+y}{2}\right)}{1+\tan^2\left(\dfrac{x+y}{2}\right)}=\frac{\dfrac{2a}{b}}{1+\dfrac{a^2}{b^2}}=\frac{2ab}{a^2+b^2}.$$

160. $\dfrac{\cos x}{\sin x}+\dfrac{1}{\sin x}=5\Rightarrow \cos x+1=5\sin x$

$\therefore$ 25 $\sin^2 x=(1+\cos x)^2$ or 25 $(1-\cos^2 x)=(1+\cos x)^2$

or 25 $(1-\cos x)=(1+\cos x)\Rightarrow 26\cos x=24.$

$\therefore\quad \cos x=\dfrac{24}{26}=\dfrac{12}{13}.$

161. $A+B=\dfrac{\pi}{4}\Rightarrow \tan(A+B)=\tan\dfrac{\pi}{4}=1$

$\Rightarrow \dfrac{\tan A+\tan B}{1-\tan A\tan B}=1\Rightarrow \tan A+\tan B=1-\tan A\tan B$

$\Rightarrow \tan A+\tan B+\tan A\tan B+1=2\ \Rightarrow (1+\tan A)(1+\tan B)=2$

162. $\sin^2\dfrac{\pi}{8}+\sin^2\dfrac{3\pi}{8}+\sin^2\dfrac{5\pi}{8}+\sin^2\dfrac{7\pi}{8}=2\sin^2\dfrac{\pi}{8}+2\sin^2\dfrac{3\pi}{8}$

$\left\{\because \sin\dfrac{7\pi}{8}=\sin\left(\pi-\dfrac{\pi}{8}\right)=\sin\dfrac{\pi}{8}\ \&\ \sin\dfrac{5\pi}{8}=\sin\left(\pi-\dfrac{3\pi}{8}\right)=\sin\dfrac{3\pi}{8}\right\}$

$=1-\cos\left(2\times\dfrac{\pi}{8}\right)+1-\cos\left(2\times\dfrac{3\pi}{8}\right)=1-\cos\dfrac{\pi}{4}+1-\cos\dfrac{3\pi}{4}=2-\cos\dfrac{\pi}{4}+\cos\dfrac{\pi}{4}=2.$

163. Clearly, (*d*) is true.

164. Clearly, (*b*) is true.

165. $\sin\dfrac{\pi}{10}\cdot\cos\dfrac{\pi}{5}=\sin 18^\circ\cos 36^\circ=\left(\dfrac{\sqrt{5}-1}{4}\right)\times\left(\dfrac{\sqrt{5}+1}{4}\right)=\dfrac{4}{16}=\dfrac{1}{4}.$

166. $\cos 2\theta=\dfrac{1-\tan^2\theta}{1+\tan^2\theta}=\dfrac{1-\dfrac{1}{49}}{1+\dfrac{1}{49}}=\dfrac{48}{50}=\dfrac{24}{25}.$

$$\tan 2\phi=\frac{2\tan\phi}{1-\tan^2\phi}=\frac{\dfrac{2}{3}}{1-\dfrac{1}{9}}=\left(\frac{2}{3}\times\frac{9}{8}\right)=\frac{3}{4}.$$

$$\therefore \sin 4\phi=\frac{2\tan 2\phi}{1+\tan^2 2\phi}=\frac{2\times\dfrac{3}{4}}{1\dfrac{9}{16}}=\left(\frac{6}{4}\times\frac{16}{25}\right)=\frac{24}{25}.$$

Hence, $\cos 2\theta=\sin 4\phi$.

167. $\sin\theta = -\frac{1}{2} = -\sin 30° = \sin(180° + 30°) = \sin 210°.$

168. $\frac{\sin\theta}{\cos\theta} = \frac{x\sin\theta}{1 - x\cos\phi}, \frac{\sin\phi}{\cos\phi} = \frac{y\sin\theta}{1 - y\cos\theta}$

$\Rightarrow \sin\theta - x\sin\theta\cos\phi = x\sin\phi\cos\theta$ & $\sin\phi - y\cos\theta\sin\phi = y\sin\theta\cos\phi$

$\Rightarrow x(\sin\phi\cos\theta + \sin\theta\cos\phi) = \sin\theta,$

$y(\sin\theta\cos\phi + \cos\theta\sin\phi) = \sin\phi \Rightarrow \frac{\sin\theta}{\sin\phi} = \frac{x}{y}.$

169. $\sin\frac{\theta}{3} = 0 \Rightarrow \frac{\theta}{3} = 0$ or $\pi \Rightarrow \theta = 0$ or 3π.

$\therefore$ Required diff. $= (3\pi - 0) = 3\pi$.

170. Given Exp. $= \sqrt{2 + \sqrt{2(1 + \cos 4\theta)}} = \sqrt{2 + \sqrt{2 \times 2\cos^2 2\theta}} = \sqrt{2 + 2\cos 2\theta} = \sqrt{2(1 + \cos 2\theta)}$

$= \sqrt{2 \times 2\cos^2\theta} = 2\cos\theta.$

171. $\sin x° = \sin 120° - \sin 60° = \sin(180° - 60°) - \sin 60° = \sin 60° - \sin 60° = 0$

$\therefore \quad x = 180°.$

172. $\sin 30° = \frac{1}{2}, \sin 45° = \frac{1}{\sqrt{2}}, \cos 30° = \frac{\sqrt{3}}{2}, \tan 30° = \frac{1}{\sqrt{3}}.$

$\frac{1}{2} < \frac{1}{\sqrt{3}} < \frac{1}{\sqrt{2}} < \frac{\sqrt{3}}{2} \Rightarrow \sin 30° < \tan 30° < \sin 45° < \cos 30°.$

173. $\frac{\sin^2 A + 1}{\sin A} = \frac{5}{2} \Rightarrow 2\sin^2 A - 5\sin A + 2 = 0. \Rightarrow (2\sin A - 1)(2\sin A - 2) = 0 \Rightarrow \sin A = \frac{1}{2}.$

$\therefore A = \frac{\pi}{6}.$

174. $\sec\theta = x + \tan\theta \Rightarrow \sec^2\theta = x^2 + \tan^2\theta + 2x\tan\theta$

$\Rightarrow 1 + \tan^2\theta = x^2 + \tan^2\theta + 2x\tan\theta \Rightarrow \tan\theta = \left(\frac{1 - x^2}{2x}\right) \Rightarrow \cot\theta = \frac{2x}{(1 - x^2)}$

$\text{cosec}\,\theta = \sqrt{1 + \cot^2\theta} = \sqrt{1 + \frac{4x^2}{(1 - x^2)^2}} = \frac{1 + x^2}{1 - x^2}. \quad \therefore \sin\theta = \frac{1 - x^2}{1 + x^2}.$

175. $A + B = \frac{\pi}{2} \Rightarrow \sin(A + B) = \sin\frac{\pi}{2} = 1.$

$\tan A + \tan B = \frac{\sin A}{\cos A} + \frac{\sin B}{\cos B} = \frac{\sin A\cos B + \cos B\sin B}{\cos A\cos B} = \frac{\sin(A + B)}{\cos A\cos B} = \frac{1}{\cos A\cos B}$

$= \frac{2bc}{b^2 + c^2 - a^2} \times \frac{2ac}{a^2 + c^2 - b^2} = \frac{2bc}{2b^2} \times \frac{2ac}{2a^2} = \frac{c^2}{ab}.$ $[\because c^2 - a^2 = b^2$ & $c^2 - b^2 = a^2]$

176. $A + B + C = \pi \Rightarrow A + B = \pi - C \Rightarrow \tan(A + B) = \tan(\pi - C)$

$\Rightarrow \tan(A + B) = -\tan C \Rightarrow \tan(A + B) + \tan C = 0.$

177. Clearly, none of the given statements is incorrect.

178. For Statement I:

$$(\operatorname{cosec}\alpha - \sec\alpha) = \frac{1}{\sin\alpha} - \frac{1}{\cos\alpha}$$

$$(\operatorname{cosec}\alpha - \sec\alpha) = \frac{\cos\alpha - \sin\alpha}{\cos\alpha + \sin\alpha}$$

Since $\cos\alpha$ and $\sin\alpha$ are both positive in the first quadrant, the numerator $(\cos\alpha - \sin\alpha)$ can be positive, zero, or negative depending on the specific value of α.

Thus, Statement I is not always true.

For Statement II:

$$(\tan\alpha - \cot\alpha) = \frac{\sin\alpha}{\cos\alpha} - \frac{\cos\alpha}{\sin\alpha}$$

$$(\tan\alpha - \cot\alpha) = \frac{\sin^2\alpha - \cos^2\alpha}{\sin\alpha \times \cos\alpha}$$

In the first quadrant, $\sin\alpha$ and $\cos\alpha$ are positive.

However, $\sin^2\alpha - \cos^2\alpha$ is negative because $\cos\alpha > \sin\alpha$ for α in the first quadrant.

Hence, $(\tan\alpha - \cot\alpha)$ is always negative in the first quadrant.

179. From I

$$\text{LHS} = 1 - \sin^6\alpha = (1 - \sin^2\alpha)(1 + \sin^2\alpha + \sin^4\alpha)$$

$$= \cos^2\alpha(1 + (1 - \cos^2\alpha) + (1 - \cos^2\alpha)^2) = \cos^2\alpha(3 - 3\cos^2\alpha + \cos^4\alpha) = \text{RHS}$$

So, (I) is true.

From II

$$\text{LHS} = \cos^8\alpha - \sin^8\alpha = (\cos^4\alpha - \sin^4\alpha)(\cos^4\alpha + \sin^4\alpha)$$

$$= (\cos^2\alpha - \sin^2\alpha)(\cos^2\alpha + \sin^2\alpha)(\cos^4\alpha + \sin^4\alpha) = (\cos 2\alpha)(1)(\cos^4\alpha + \sin^4\alpha)$$

$$\text{RHS} = 2\sin^2\alpha(1 - \cos^4\alpha + \sin^2\alpha\cos^2\alpha)$$

LHS = RHS

Thus, only (I) is an identity.

180. If $A + B + C = \pi$, then: $\sin A + \sin B + \sin C = 4\cos\frac{A}{2}\cos\frac{B}{2}\cos\frac{C}{2}$.

181. I: $\sin^2\theta = \frac{(x+y)^2}{4xy}$

Since $x, y > 0$ and $x \neq y$, $(x+y)^2 > 4xy \Rightarrow \frac{(x+y)^2}{4xy} > 1$

But $\sin^2\theta \le 1$ always.

Hence, equation I is not possible.

II: $\sin\theta + \cos\theta = x + \frac{1}{x}$

By AM–GM inequality, $x + \frac{1}{x} \ge 2$ for $x > 0$

Also, $|\sin\theta + \cos\theta| \le \sqrt{2} \approx 1.414$

So, $x+\dfrac{1}{x}\geq 2>\sqrt{2}\Rightarrow$ equation II is also not possible.

$\Rightarrow$ Neither I nor II is possible.

182. Statement I : $\tan 50^\circ - \cot 50^\circ$

$$=\frac{\sin 50^\circ}{\cos 50^\circ}-\frac{\cos 50^\circ}{\sin 50^\circ}=\frac{\sin^2 50^\circ-\cos^2 50^\circ}{\sin 50^\circ\times\cos 50^\circ}=-\frac{\cos^2 50^\circ-\sin^2 50^\circ}{\sin 50^\circ\times\cos 50^\circ}=-\frac{\cos 100^\circ}{\sin 50^\circ\times\cos 50^\circ}$$

Since $-\cos 100^\circ$ is positive and $\sin 50^\circ\times\cos 50^\circ$ is positive,

Overall value is Positive $\Rightarrow$ Statement I is True.

Statement II : $\cot 25^\circ-\tan 25^\circ=\dfrac{\cos 25^\circ}{\sin 25^\circ}-\dfrac{\sin 25^\circ}{\cos 25^\circ}=\dfrac{\cos^2 25^\circ-\sin^2 25^\circ}{\sin 25^\circ\times\cos 25^\circ}=\dfrac{\cos 50^\circ}{\sin 25^\circ\times\cos 25^\circ}$

Since $\cos 50^\circ$ is positive and $\sin 25^\circ\times\cos 25^\circ$ is positive,

Overall value is Positive $\Rightarrow$ Statement II is False.

183. $$\frac{1-\tan^2\theta}{1+\tan^2\theta}=\frac{1-\dfrac{\sin^2\theta}{\cos^2\theta}}{1+\dfrac{\sin^2\theta}{\cos^2\theta}}=(\cos^2\theta-\sin^2\theta)=(1-2\sin^2\theta).$$

$\therefore$ (*c*) is an identity.

27 CHAPTER Height and Distance

Problems On Heights and Distances

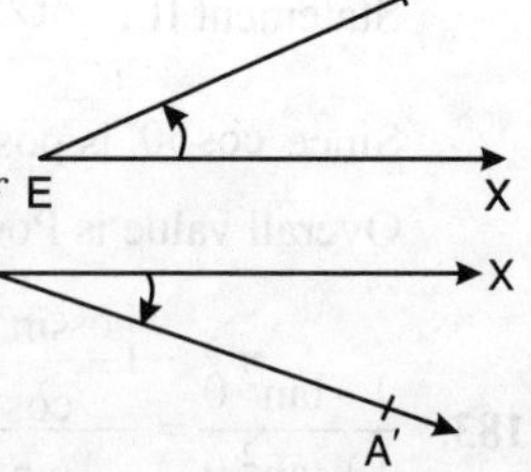

Angle of Elevation: *Let A be an object placed above the level of the eye E of the observer and let EX be the horizontal line. Then,*

∠XEA is known as the angle of elevation.

Angle of Depression: *Let A' be an object placed below the level of the eye of the observer and let EX be the horizontal line. Then, ∠XEA' is called the angle of depression.*

SOLVED PROBLEMS

Ex. 1. ***A ladder leaning against a vertical wall makes an angle of 45° with the ground. The foot of the ladder is 3 m from the wall. Find the length of the ladder.***

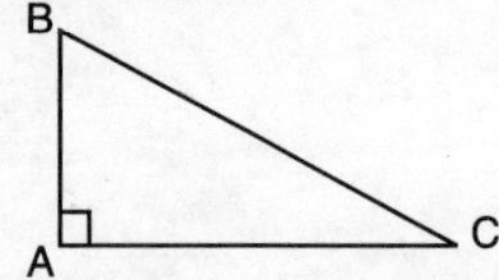

Sol. Let AB be the wall and CB, the ladder.

Then, $AC = 3$ m and $\angle ACB = 45°$

Now, $\frac{CB}{AC} = \sec 45° = \sqrt{2} \Rightarrow \frac{CB}{3} = \sqrt{2}.$

$\therefore$ Length of the ladder = $CB = 3\sqrt{2} = (3 \times 1.41)$ m

$= 4\ 23$ m

Ex. 2. ***A balloon is connected to a meteorological station by a cable of length 200 m, inclined at 60° to the horizontal. Find the height of the balloon from the ground. Assume that there is no slack in the cable.***

Sol. Let B be the balloon and AB be the vertical height. Let C be the meteorological station and CB be the cable.

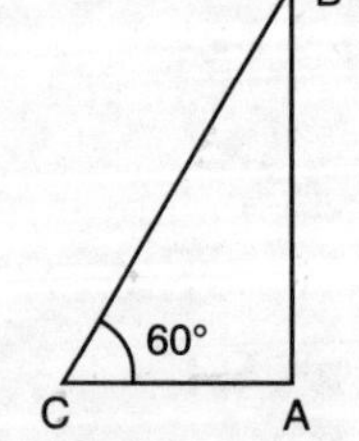

Then, $BC = 200$ m and $\angle ACB = 60°$.

Then, $\frac{AB}{BC} = \sin 60° = \frac{\sqrt{3}}{2} \Rightarrow \frac{AB}{200} = \frac{\sqrt{3}}{2}$

$\Rightarrow \quad AB = \left(\frac{200 \times \sqrt{3}}{2}\right) \text{m} = 173.2\text{m}.$

Ex. 3. ***The pilot of a helicopter at an altitude of 1200 rn finds that two ships are sailing towards it in the same direction. The angles of depression of the ships as observed from the helicopter are 60' and 45 respectively. Find the distance between the two ships.***

Sol. Let B be the position of the helicopter and let C, D be the ships. Let AB be the vertical height.

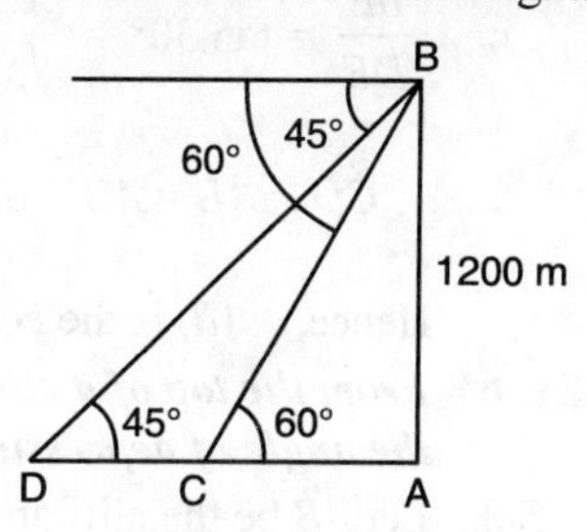

Then, $AB = 1200$ m.

$\angle ACB = 60° \ \& \ \angle ADB = 45°$.

Then, $\dfrac{AD}{AB} = \cot 45° = 1$

$\Rightarrow \quad \dfrac{AD}{1200} = 1 \Rightarrow AD = 1200$ m

And, $\dfrac{AC}{AB} = \cot 60° = \dfrac{1}{\sqrt{3}}$

$\Rightarrow \quad \dfrac{AC}{1200} = \dfrac{1}{\sqrt{3}} \Rightarrow AC = \dfrac{1200}{\sqrt{3}} = 400\sqrt{3}$m.

Distance between the ships $= CD = AD - AC$

$= (1200 - 400\sqrt{3})$ m

$= (1200 - 400 \times 1.732)$m $= 507.2$ m

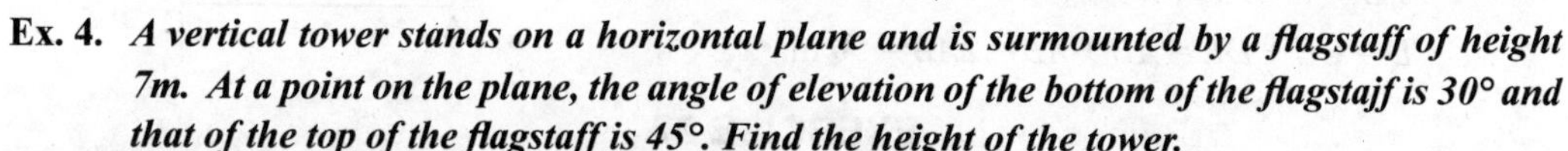

Ex. 4. ***A vertical tower stands on a horizontal plane and is surmounted by a flagstaff of height 7m. At a point on the plane, the angle of elevation of the bottom of the flagstajf is 30° and that of the top of the flagstaff is 45°. Find the height of the tower.***

Sol. Let AB be the tower and BC be the flagstaff. Then, $BC = 7$ m. Let $AB = h$.

Let O be the point of observation.

Then, $\angle AOB = 30°$ and $\angle AOC = 45°$.

Now, $\dfrac{OA}{AC} = \cot 45° = 1 \quad \Rightarrow \quad OA = AC + h + 7.$

And, $\dfrac{OA}{AB} = \cot 30° = \sqrt{3}$

$\Rightarrow \quad \dfrac{OA}{h} = \sqrt{3} \Rightarrow OA = h\sqrt{3}.$

$\therefore \quad h + 7 = h\sqrt{3}$

$\Rightarrow \quad h = \dfrac{7}{\sqrt{3}-1} \times \dfrac{\sqrt{3}+1}{\sqrt{3}+1} = \dfrac{7(\sqrt{3}+1)}{2} = 9.562$ m.

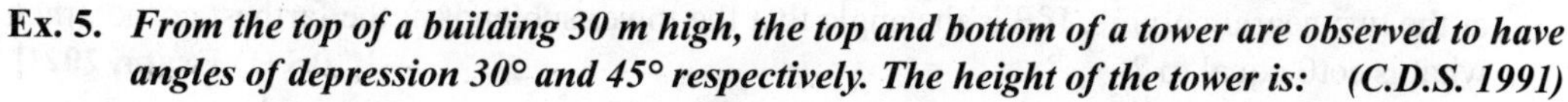

Ex. 5. ***From the top of a building 30 m high, the top and bottom of a tower are observed to have angles of depression 30° and 45° respectively. The height of the tower is:*** **(C.D.S. 1991)**

(*a*) $\mathbf{15(1+\sqrt{3})}$***m*** (*b*) $\mathbf{30(\sqrt{3}-1)}$***m*** (*c*) $\mathbf{30\left(1+\dfrac{1}{\sqrt{3}}\right)}$***m*** (*d*) $\mathbf{30\left(1-\dfrac{1}{\sqrt{3}}\right)}$***m***

Sol. Let AB be the building and CD be the tower. Then, $AB = 30$ m. Let $DC = x$

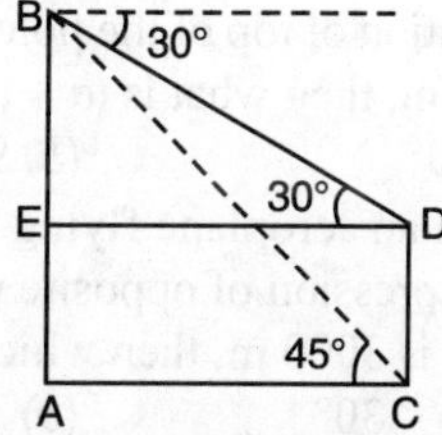

Draw $DE \perp AB$. Then, $AE = CD = x$.

$\therefore BE = (30 - x)$m

Now, $\dfrac{AC}{AB} = \cot 45° = 1 \Rightarrow \dfrac{AC}{30} = 1 \Rightarrow AC = 30$ m.

$\therefore \ DE = AC = 30$ m.

$$\frac{BE}{DE} = \tan 30° = \frac{1}{\sqrt{3}} \Rightarrow \frac{BE}{30} = \frac{1}{\sqrt{3}} \Rightarrow BE = \frac{30}{\sqrt{3}}.$$

$$\therefore CD = AE = AB - BE = \left(30 - \frac{30}{\sqrt{3}}\right) = 30\left(1 - \frac{1}{\sqrt{3}}\right) \text{m.}$$

Hence, (*d*) is the correct answer.

Ex. 6. ***From the top of a cliff 25 m high the angle of elevation of a tower is found to be equal to the angle of depression of the foot of the tower. Find the height of the tower.***

Sol. Let AB be the cliff and CD be the tower. Then, $AB = 25$ m. From B draw $BE \perp CD$.

Let $\angle EBD = \angle ACB = \alpha$.

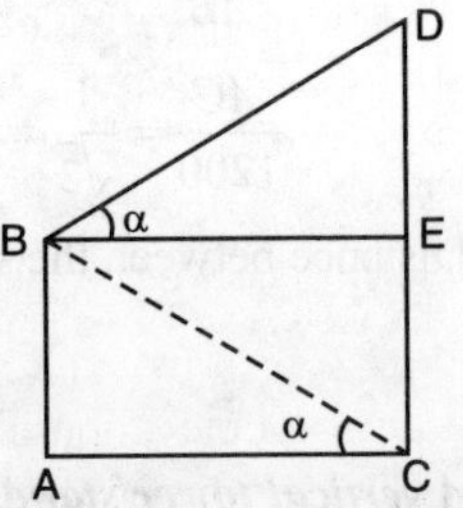

Now, $\frac{DE}{BE} = \tan\alpha$ and $\frac{AB}{AC} = \tan\alpha$

$\therefore \frac{DE}{BE} = \frac{AB}{AC}$. So, $DE = AB$ $[\because BE = AC]$

$\therefore CD = CE + DE = AB + AB = 2AB = 50$ m.

EXERCISE 27

1. A person on the top of a vertical tower observes a car moving at a uniform speed coming directly towards it. If it takes 6 minutes for the angle of depression to change from 30° to 45°, and further t minutes to reach the tower, which one of the following is correct ? **[CDS (I) Exam, 2023]**

(*a*) $7{\cdot}7 < t < 8$ (*b*) $8 < t < 8{\cdot}3$ (*c*) $8{\cdot}3 < t < 8{\cdot}6$ (*d*) $8{\cdot}6 < t < 8{\cdot}9$

2. A tower subtends an angle of 60° at a point A on the same level as the foot of the tower. B is a point vertically above A and AB = h. The angle of depression of the foot of the tower, measured from B, is 30°. What is the height of the tower? **[CDS (I) Exam, 2025]**

(*a*) 2 h (*b*) 2.5 h (*c*) 3 h (*d*) 3.5 h

3. ABCD is a square field with AB = x. A vertical pole OP of height $2x$ stands at the centre O of the square field. If L APO = 0, then what is cot 0 equal to ? **[CDS (I) Exam, 2023]**

(*a*) $\sqrt{2}$ (*b*) 2 (*c*) $2\sqrt{2}$ (*d*) $3\sqrt{2}$

4. A vertical tower standing at the corner of a rectangular field subtends angle of 60° and 45° at the two nearer corners. If θ is the angle that the tower subtends at the farthest corner, then what is cotθ equal to ? **[CDS (I) Exam, 2021]**

(*a*) $\frac{1}{2}$ (*b*) 2 (*c*) $\frac{2}{\sqrt{3}}$ (*d*) $\frac{4}{\sqrt{3}}$

5. A vertical pole of length 80 m is situated on the horizontal plane. The base of the pole is at P. There are two points A and B such that P, A, B are on the same straight line. Let the angles of elevation of top of the pole from A and B be α and β (α > β) respectively. If PA = 64 m and AB = 36 m, then what is (α + β) equal to? **[CDS (II) Exam, 2024]**

(*a*) 60° (*b*) 90° (*c*) 120° (*d*) 135°

6. From an aeroplane flying above a river at an altitude of 1200 m, it is observed that the angles of depression of opposite points on the two banks of a river are 30° and θ. If the width of the river is 3000 m, then which one of the following is correct? **[CDS (II) Exam, 2021]**

(*a*) θ < 30° (*b*) 30° < θ < 45° (*c*) 45° < θ < 60° (*d*) 60° < θ < 90°

7. A tree of height 15 cm is broken by wind in such a way that its top touches the ground and makes an angle 30° with the ground. What is the height from the ground to the point where tree is broken? **[CDS (I) Exam, 2020]**

(*a*) 10 m (*b*) 7 m (*c*) 5 m (*d*) 3 m

8. On a plane area there are two vertical towers separated by 100 feet apart. The shorter town is 40 feet tall. A pole of length 6 feet stands on the line joining the base of two towers so that the tip of the towers sand tip and the pole are also on the same line. If the distance of the pole from the shorter tower is 75 feet, then what is the height of the taller tower (approximately) **[CDS (I) Exam, 2020]**

(*a*) 85 feet (*b*) 110 feet (*c*) 125 feet (*d*) 140 feet

9. The angle of elevation of the top of a tower from two points at distances p and q from the base and on the same straight line are 27° and 63° respectively. What is the height of the tower? **[CDS (I) Exam, 2020]**

(*a*) PQ (*b*) $\sqrt{pq}$ (*c*) $\frac{pq}{2}$ (*d*) $\frac{\sqrt{pq}}{2}$

10. A plane is going in circles around an airport. The plane takes 3 minutes to complete one round The angle of elevation of the plane from a point P on the ground at time t second is equal to that at time $(t + 30)$ seconds. At time $(t + x)$ seconds, the plane filed vertically above the point P. What is x equal to? **[CDS (I) Exam, 2019]**

(*a*) 75 seconds (*b*) 90 seconds (*c*) 105 seconds (*d*) 135 seconds

11. Consider the regular hexagon ABCDEF. Two tower are situated at B and C. The angle of elevation from A to the top of the tower at B is 30°, and the angle of elevation to the top of the tower at C is 45°. What is the ratio of the height of towers at B and C? **[CDS (I) Exam, 2019]**

(*a*) $1 : \sqrt{3}$ (*b*) 1 : 3 (*c*) 1 : 2 (*d*) $1 : 2\sqrt{3}$

12. There are two parallel streets each directed North to South. A person in the first street travelling from South to North wishes to take the second street which is on his right side. At some place, he makes a 150° turn to the right and he travels for 15 minutes at the speed of 20 km/hr. After that he takes a left turn of 60° and travels for 20 minutes at the speed of 30 km/hr in order to meet the second street. What is the distance between the two streets? **[CDS (I) Exam, 2019]**

(*a*) 7.5 km (*b*) 10.5 km (*c*) 12.5 km (*d*) 15 km

13. A ladder is resting against a vertical wall and its bottom is 2.5 m away from the wall. If it slips 0.8 m down the well, then its bottom will move away from the wall by 1.4 m. What is the length of the ladder? **[CDS (I) Exam, 2019]**

(*a*) 6.2 m (*b*) 6.5 m (*c*) 6.8 m (*d*) 7.5 m

Consider the following for the next two (02) items that follow: [CDS (I) Exam, 2023]

A flagstaff stands on the top of a vertical tower. The angle of elevation of the top of the flagstaff from a certain place on the same horizontal level with the base of the tower is found to be α. Advancing a distance d towards the tower in the same horizontal plane, the angle of elevation of the top of the flagstaff is observed to be β and that of the top of the tower is observed to be γ. Let H be the height of the top of the flagstaff from the base of the tower and h be the height of the tower.

14. Which one of the following is correct

(*a*) $H \tan \gamma - h \tan \beta = 0$ (*b*) $h \tan \gamma - H \tan \alpha = 0$

(*c*) $H \tan \gamma - h \tan \beta = 0$ (*d*) $h \tan \gamma - H \tan \alpha = 0$

15. Which one of the following is correct ?

(*a*) $d = \frac{H (\cot \theta - \cot \beta)}{2}$ (*b*) $d = \frac{H (\tan \alpha - \tan \beta)}{2}$

(*c*) $d = H (\cot \alpha - \cot \beta)$ (*d*) $d = H (\tan \alpha - \tan \beta)$

Consider the following for the next two (02) items that follow: **[CDS (I) Exam, 2023]**

AB is a straight road leading to the foot P of a tower of height h. Q is at distance x from P and R is at a distance y from Q (R is farther from P than Q; R, Qare on the same side). The angle of elevation of the top of the tower at Q is twice of that at R.

(Use the formula $\tan 2\theta = \dfrac{2\tan\theta}{1-\tan^2\theta}$

16. Which one of the following is correct ?

(*a*) $x = y$ (*b*) $x < y$

(*c*) $x > y$ (*d*) Cannot be concluded due to insufficient data

17. Which one of the following is correct ?

(*a*) $h^2 = x^2 - y^2$ (*b*) $h^2 = x^2 + y^2$ (*c*) $h^2 = 2(y^2 - x^2)$ (*d*) $h^2 = y^2 - x^2$

18. The angles of elevation of the top of a tower from two points *P* and *Q* at distances m^2 and n^2 respectively, from the base and in the same straight line with it are complementary. The height of the tower is **[CDS (I) Exam, 2015]**

(*a*) $(mn)^{1/2}$ (*b*) $mn^{1/2}$ (*c*) $m^{1/2}n$ (*d*) mn

19. The angle of elevation of a cloud from a point 200 m above a lake is 30° and the angle of depression of its reflection in the lake is 60°. The height of the cloud is **[CDS (I) Exam, 2015]**

(*a*) 200 m (*b*) 300 m (*c*) 400 m (*d*) 600 m

20. From a point *X* on a bridge across a river, the angles of depression of two points *P* and *Q* on the banks on opposite side of the river are α and β respectively. If the point *X* is at a height *h* above the surface of the river, what is the width of the river if α and β are complementary?

[CDS (II) Exam, 2024]

(*a*) $2h(\tan\alpha + \cot\alpha)$ (*b*) $h\tan\alpha \cdot \tan\beta$

(*c*) $h\cot\alpha \cdot \cot\beta$ (*d*) $h\sec\alpha \cdot \operatorname{cosec}\alpha$

21. An aeroplane flying at a height of 3000 m passes vertically above another aeroplane at an instant when the angles of elevation of the two planes from some point o the ground are 60° and 45° respectively. Then the vertical distance between the two planes is

(*a*) $1000(\sqrt{3}-1)\,m$ (*b*) $1000\sqrt{3}\,m$ (*c*) $1000(3-\sqrt{3})\,m$ (*d*) $3000\sqrt{3}\,m$

[CDS (II) Exam, 2015]

22. A pole is standing erect on the ground which is horizontal. The tip of the pole is tied tight with a rope of length $\sqrt{12}$m to a point on the ground. If the rope is making 30° with the horizontal, then the height of the pole is **[CDS (II) Exam, 2015]**

(*a*) $2\sqrt{3}\,m$ (*b*) $3\sqrt{2}\,m$ (*c*) 3m (*d*) $\sqrt{3}\,m$

23. The angles of elevation of the top of a tower from two points A and B at a distance of *x* m and (*x* + 5) m from the base of the tower of height 6 m and in the same straight line with it are complementary. What is the value of *x* ? **[CDS (I) Exam, 2024]**

(*a*) 4 m (*b*) 5 m (*c*) 6 m (*d*) 9 m

24. Two observers are stationed due north of a tower (of height *x* metre) at a distance y metre from each other. The angles of elevation of the tower observed by them are 30° and 45° respectively. Then *x*/*y* is equal to **[CDS (I) Exam, 2016]**

(*a*) $\dfrac{\sqrt{2}-1}{2}$ (*b*) $\dfrac{\sqrt{3}-1}{2}$ (*c*) $\dfrac{\sqrt{3}+1}{2}$ (*d*) 1

25. A man from the top of a 100m high tower sees a car moving towards the tower at an angle of depression 30º. After some time, the angle of depression becomes 60º. What is the distance travelled by the car during this time? **[CDS (II) Exam, 2016]**

(*a*) $100\sqrt{3}\ m$ (*b*) $\frac{200\sqrt{3}}{3}\ m$ (*c*) $\frac{100\sqrt{3}}{3}\ m$ (*d*) $200\sqrt{3}\ m$

26. Two men on either side of a tower 75 m high observe the angle of elevation of the top of the tower to be 30º and 60º. What is the distance between the two men?

(*a*) $100\sqrt{3}$ m (*b*) $75\sqrt{3}$ m (*c*) $\frac{100\sqrt{3}}{3}$ m (*d*) $60\sqrt{3}$ m

[CDS (II) Exam, 2016]

27. If the length of the shadow of a tower is equal to its height, then what is the Sun's altitude at that time? **[CDS (II) Exam, 2016]**

(*a*) 15º (*b*) 30º (*c*) 45º (*d*) 60º

28. An aeroplane flying at a height of 300m above the ground passes vertically above another plane at an instant when the angles of elevation of the two planes from the same point on the ground are 60° and 45° respectively. What is the height of the lower plane from the ground?

(*a*) $100\sqrt{3}m$ (*b*) $\frac{100}{\sqrt{3}}m$ (*c*) $50\sqrt{3}m$ (*d*) $50\left(\sqrt{3}+1\right)m$

[CDS (I) Exam, 2017]

29. From the top of a building 90 m high, the angles of depression of the top and the bottom of a tree are 30° and 45° respectively. What is the height of the tree?

(*a*) $30\sqrt{3}m$ (*b*) $90-30\sqrt{3}m$ (*c*) $90+30\sqrt{3}m$ (*d*) $60+30\sqrt{3}m$

[CDS (I) Exam, 2017]

30. From an aeroplane vertically over a straight horizontal road, the angles of depression of two consecutive kilometer-stones on the opposite sides of the aeroplane are observed to be α and β. The height of the aeroplane above is road is **[CDS (I) Exam, 2017]**

(*a*) $\frac{\tan\alpha+\tan\beta}{\tan\alpha\tan\beta}$ (*b*) $\frac{\tan\alpha\tan\beta}{\tan\alpha+\tan\beta}$ (*c*) $\frac{\cot\alpha\cot\beta}{\cot\alpha+\cot\beta}$ (*d*) $\frac{\cot\alpha+\cot\beta}{\cot\alpha\cot\beta}$

31. The ratio of the length of a rod and its shadow is $1 : \sqrt{3}$. The angle of elevation of the sun is:

(*a*) 30° (*b*) 45° (*c*) 60° (*d*) 90°

32. The altitude of the sun at any instant is 60°. The height of the vertical pole that will cast a shadow of 30 m is:

(*a*) $30\sqrt{3}$ m (*b*) 15 m (*c*) $\frac{30}{\sqrt{3}}$ m (*d*) $15\sqrt{2}$ m

33. When the sun is 30° above the horizontal, the length of shadow cast by a building 50 m high is:

(*a*) $\frac{50}{\sqrt{3}}$ m (*b*) $50\sqrt{3}$ m (*c*) 25 m (*d*) $25\sqrt{3}$ m

34. If the elevation of the sun changed from 30° to 60°, then the difference between the lengths of shadows of a pole is 15 m high, made at these two positions, is:

(*a*) 7.5 m (*b*) 15 m (*c*) $10\sqrt{3}$ m (*d*) $\frac{15}{\sqrt{3}}$ m

35. In a rectangle, if the angle between a diagonal and a side is 30° and the length of diagonal is 6 cm, the area of the rectangle is:

(*a*) 9 cm² (*b*) $9\sqrt{3}$ cm² (*c*) 27 cm² (*d*) 36 cm²

36. On the level ground, the angle of elevation of a tower is 30°. On moving 20 m nearer, the angle of elevation is 60°. The height of the tower is:

(*a*) 10 m (*b*) $10\sqrt{3}$ m (*c*) 15 m (*d*) 20 m

37. The length of a string between a kite and a point on the ground is 90 m. The string makes an angle of 60° with the level ground. If there is no slack in the string, the height of the kite is:

(*a*) $90\sqrt{3}$ m (*b*) $45\sqrt{3}$ m (*c*) 180 m (*d*) 45 m

38. The length of a string between a kite and a point on the ground is 85 m. If the string makes an angle θ with the level ground such that tan θ = (15/8), how high is the kite, when there is no slack in the string?

(*a*) 78.05 m (*b*) 75 m (*c*) 316 m (*d*) None of these

39. Two persons standing on the same side of a tower measure the angles of elevation of the top of the tower as 30° and 45°. If the height of the tower is 30 m, the distance between the two persons is approximately:

(*a*) 52 m (*b*) 26 m (*c*) 82 m (*d*) 22 m

40. The angles of elevation of the top of a tower 40 m high from two points on the level ground on its opposite sides are 45° and 60°. The distance between the two points in nearest metres is:

(*a*) 60 m (*b*) 61 m (*c*) 62 m (*d*) 63 m

41. The distance between two multistoreyed buildings is 60 m. The angle of depression of the top of the first building as seen from the top of the second building, which is 150 m high, is 30°. The height of the first building is:

(*a*) 115.36 m (*b*) 117.85 m (*c*) 125.36 m (*d*) 128.34 rn

42. If from the top of a cliff 100 m high, the angles of depression of two ships out at sea are 60° and 30°, then the distance between the ships is approximately

(*a*) 173 m (*b*) 346 m (*c*) 57.6 m (*d*) 115.3 m

43. From the foot of a tower the angle of elevation of the top of a column is 60° and from the top of the tower which is 25 m high, the angle of elevation is 30°. The height of the column is:

(*a*) 37.5 m (*b*) 42.5 m (*c*) 43.3 m (*d*) 14.4 m

44. An observer standing 72 m away from a building notices that the angles of elevation of the top and the bottom of a flagstaff on the building are respectively 60° and 45°. The height of the flagstaff is:

(*a*) 124.7 m (*b*) 52.7 m (*c*) 98.3 m (*d*) 73.2 m

45. A vertical tower stands on a horizontal plane and is surmounted by a vertical flagstaff of height *h*. At a point on the plane, the angle of elevation of the bottom of the flagstaff is α and that of the top of the flagstaff is β. The height of the tower is:

(*a*) $\frac{h\tan\beta}{\tan b - \tan\alpha}$ (*b*) $\frac{h\sin\beta}{\sin\beta - \sin\alpha}$ (*c*) $\frac{h\cot\alpha}{\cot\beta - \cot\alpha}$ (*d*) $\frac{h\tan\alpha}{\tan\beta - \tan\alpha}$

46. On the same side of a tower 300 m high the angles of depression of two objects are 45' and 60° respectively. The distance between the objects is:

(*a*) 126.8 m (*b*) 216.4 m (*c*) 116.5 m (*d*) 473.2 m

47. Two poles of equal heights are standing opposite to each other on either side of a road, which is 30 m wide. From a point between them on the road. the angles of elevation of the tops are 30° and 60". The height of each pole is:

(*a*) 4.33 m (*b*) 6.5 m (*c*) 13 m (*d*) 15 m

48. The shadow of a tower standing on a level plane is found to be 60 m longer when the sun's altitude is 30° than when it is 45°. The height of the tower is:

(*a*) 81.96 m (*b*) 51.96 m (*c*) 21.96 m (*d*) None of these

49. The angles of elevation of the top of a tower from two points distant 30 m and 40 m on either side from the base and in the same straight line with it are complementary. The height of the tower is:

(*a*) 34.64 m (*b*) 69.28 m (*c*) 23.09 m (*d*) 11.54 m

50. From the top of a lighthouse, the angles of depression of two ships on the opposite sides of it are observed to be α and β. If the height of the lighthouse be *h* metres and the line joining the ships passes through the foot of the lighthouse, the distance between the ships is:

(*a*) $\frac{h(\cot\alpha+\cot\beta)}{\cot\alpha\cot\beta}$ (*b*) $\frac{h(\tan\alpha+\tan\beta)}{\tan\alpha\tan\beta}$ (*c*) $h(\tan\alpha+\tan\beta)$ (*d*) $\frac{h\tan\alpha\tan\beta}{\tan\alpha+\tan\beta}$

51. Two boats approach a lighthouse in mid-sea from opposite directions. The angles of elevation of the top of the lighthouse from the two boats are 30° and 45" respectively. If the distance between the two boats is 100 m. the height of the lighthouse is:

(*a*) 36.6 m (*b*) 73.2 m

(*c*) 136.6 m (*d*) 68.3 m

52. The angle of elevation of an aeroplane from a point on the ground is 45°. After 15 seconds flight, the elevation changes to 30°. If the aeroplane is flying at a height of 3000 m, the speed of the plane in krn per hour is:

(*a*) 304.32 (*b*) 152.16 (*c*) 527 (*d*) 263.5

53. Two observers are stationed due north of a tower at a distance of 10 rn from each other. The elevations of the tower observed by them are 30° and 45° respectively. The height of the tower is:

(*a*) 5 m (*b*) 8.66 m (*c*) 13.66 m (*d*) 10 m

54. A flagstaff of height (1/5) of the height of a tower is mounted on the top of the tower. If the angle of elevation of the top of the flagstaff as seen from the gound is 45° and the angle of elevation of the top of the tower as seen from the same place is θ, then the value of tan θ is:

(*a*) $\frac{6}{5}$ (*b*) $\frac{5\sqrt{3}}{6}$ (*c*) $\frac{5}{6}$ (*d*) $\frac{4}{5}$

55. A boat is being rowed away from a cliff 150 m high. At the top of the cliff the angle of depression of the boat changes from 60° to 45° in 2 minutes. The speed of the boat is:

(*a*) 2 km/hr (*b*) 1.9 km/hr (*c*) 2.4 km/hr (*d*) 3 km/hr

56. An aeroplane is observed to the approaching the airport. It is at a distance of 10 km from the point of observation on the ground and makes an angle of elevation θ. If the aeroplane is at a height of 8 km above the ground, then which one of the following is correct?

[CDS (I) Exam, 2022]

(*a*) $0 < \theta < 30°$ (*b*) $30° < \theta < 45°$ (*c*) $45° < \theta < 60°$ (*d*) $60° < \theta < 90°$

57. The angle of elevation of the top of a tower of height *x* metre from a point on the ground is found to be 60°. By ground *y* metre away from that point, it becomes 30°. Which one of the following relations is correct? **[CDS (I) Exam, 2022]**

(*a*) $x = y$ (*b*) $2x = 3y$ (*c*) $2x = \sqrt{3}\, y$ (*d*) $2x = \sqrt{3}\, x$

58. A pole on the ground leans at 60° with the vertical. At a point x metre away from the base of the pole on the ground, two halves of the pole subtend the same angle. If the pole and the point are in the same vertical plane, then what is the length of the pole? **[CDS (I) Exam, 2021]**

(*a*) $\sqrt{2}\,x$ metre (*b*) $\sqrt{3}\,x$ metre (*c*) $2x$ metre (*d*) $2\sqrt{2}\,x$ metre

59. The angles of elevation of the tops of two pillars of heights h and 2h from a point P on the line joining the feet of the two pillars are complementary. if the distance of the foot of the pillars from the point P and x and y respectively, then which one of the following is correct? **[CDS (I) Exam, 2015]**

(*a*) $2h^2 = x^2y$ (*b*) $2h^2 = xy^2$ (*c*) $2h^2 = xy$ (*d*) $2h^2 = x^2y^2$

60. Each side of a square subtends an angle of 60° at the top of tower of height h metres standing at the centre of the square. If I is the length of each side of the square, then what is h^2 equal to? **[CDS (I) Exam, 2018]**

(*a*) $2I^2$ (*b*) $\frac{I^2}{2}$ (*c*) $\frac{3I^2}{2}$ (*d*) $\frac{2I^2}{3}$

61. From a height of h units, a man observes the angle of elevation as a and angle of depression as α of the top and the bottom respectively of a tower of height H (> 4h). To what further height should he climb so that the values of angle of elevation and angle of depression get interchanged for the top and bottom of the tower? **[CDS (I) Exam, 2018]**

(*a*) H – h units (*b*) H – 2h units (*c*) H – 3h units (*d*) H – 4h units

62. On the top of a hemispherical dome of radius r, there stands a flag of height h. From a point on the ground, the elevation of the top of the flag is 30°. After moving a distance d towards the dome, when the flag is just visible, the elevation is 45°. The ratio of h to r is equal to **[CDS (II) Exam, 2018]**

(*a*) $\sqrt{2}-1$ (*b*) $\frac{\sqrt{3}+1}{2\sqrt{2}}$ (*c*) $\frac{\sqrt{3}+1}{2\sqrt{2}}d$ (*d*) $\frac{(\sqrt{3}+1)(\sqrt{2}-1)}{2\sqrt{2}}d$

ANSWERS

1. (b)	**2.** (c)	**3.** (c)	**4.** (c)	**5.** (b)	**6.** (c)	**7.** (c)	**8.** (a)	**9.** (b)
10. (c)	**11.** (b)	**12.** (c)	**13.** (b)	**14.** (a)	**15.** (c)	**16.** (b)	**17.** (d)	**18.** (d)
19. (c)	**20.** (d)	**21.** (c)	**22.** (d)	**23.** (a)	**24.** (c)	**25.** (b)	**26.** (a)	**27.** (c)
28. (a)	**29.** (b)	**30.** (b)	**31.** (a)	**32.** (a)	**33.** (b)	**34.** (c)	**35.** (b)	**36.** (b)
37. (b)	**38.** (b)	**39.** (d)	**40.** (d)	**41.** (a)	**42.** (d)	**43.** (a)	**44.** (b)	**45.** (d)
46. (a)	**47.** (c)	**48.** (a)	**49.** (a)	**50.** (b)	**51.** (a)	**52.** (c)	**53.** (c)	**54.** (c)
55. (b)	**56.** (c)	**57.** (d)	**58.** (b)	**59.** (c)	**60.** (b)	**61.** (b)	**62.** (a)	

SOLUTIONS

1. Now, in ΔABD

$$\tan 45° = \frac{AB}{BD}$$

$$1 = \frac{AB}{BD} \Rightarrow AB = BD$$

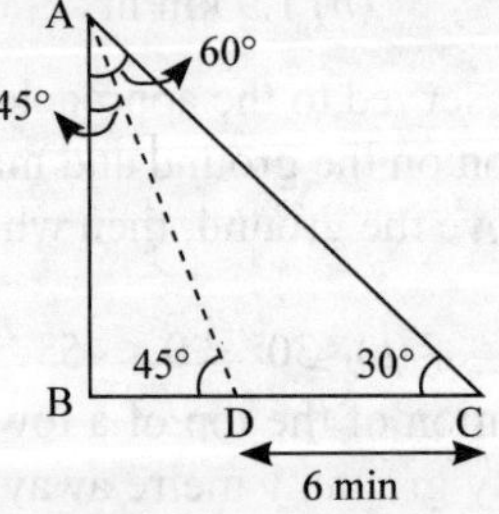

In ΔABC

$$\tan 30° = \frac{AB}{BC}$$

$$\frac{1}{\sqrt{3}} = \frac{BD}{BD+6}$$

$$\Rightarrow \text{BD} + 6 = \sqrt{3}\,\text{BD} \quad \Rightarrow 6 = \text{BD}\left(\sqrt{3}-1\right) \quad \Rightarrow \text{BD} = \frac{6}{\sqrt{3}-1} \times \frac{\sqrt{3}+1}{\sqrt{3}+1} = 3\,(\sqrt{3}+1)$$

$$= 3 \times (1.732 + 1) = 8.19$$

2. Let OP be the height of the tower,

In ΔABP,

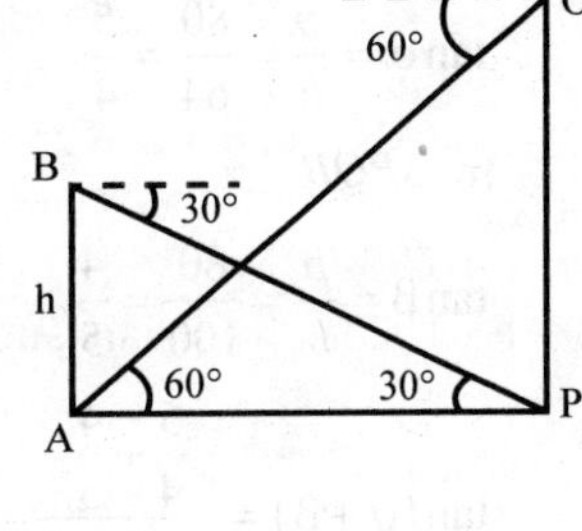

$$\tan 30^\circ = \frac{p}{b} = \frac{AB}{AOP} \Rightarrow \frac{1}{\sqrt{3}} = \frac{h}{AOP} \Rightarrow AP = h\sqrt{3}$$

Now, In ΔAOP,

$$\tan 60^\circ = \frac{p}{b} = \frac{OP}{AP} \Rightarrow \sqrt{3} = \frac{OP}{h\sqrt{3}} \Rightarrow OP = h\sqrt{3} \times \sqrt{3}$$

$$\Rightarrow OP = 3h$$

Thus height of tower = $3h$

3. $\therefore$ Diagonal of square $ABCD = \sqrt{2}a$

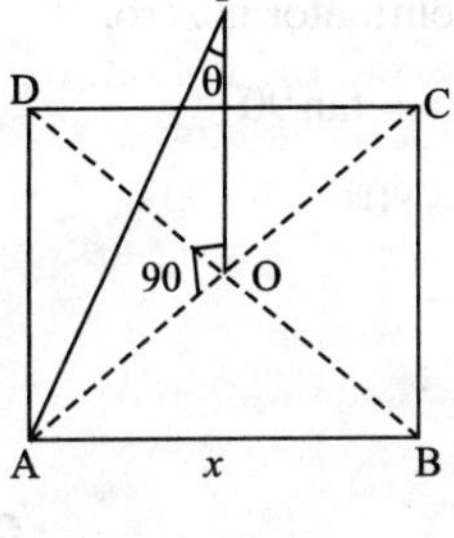

$$\Rightarrow AC = \sqrt{2}x$$

$$OA = \frac{x}{\sqrt{2}}$$

Now, in ΔOAP

$$\cot\theta = \frac{OP}{OA} = \frac{2x}{\frac{x}{\sqrt{2}}} = 2\sqrt{2}$$

4. Let the height of rectangular plot $PQRS$ be TS

$$\Rightarrow \cot 60^\circ = \frac{PS}{TS}$$

$$\Rightarrow \frac{1}{\sqrt{3}} = \frac{PS}{TS} \qquad \text{...(i)}$$

$$\cot 45^\circ = \frac{RS}{TS} \quad \Rightarrow 1 = \frac{RS}{TS}$$

$$\Rightarrow RS = TS \qquad \text{...(ii)}$$

We have,

$PS = QR$ and $PQ = SR$ [$\therefore$ $PQRS$ is a rectangle and QS is diagonal]

Now,

$QS^2 = PQ^2 + PS^2$ [By Pythagoras theorem]

$\Rightarrow QS^2 = PQ^2 = SR^2$ [$\therefore$ $PQ = SR$]

Now, $\cot\theta = \frac{QS}{TS} \Rightarrow \cot\theta = \frac{\sqrt{PQ^2 + SR^2}}{TS}$

$$\Rightarrow \cot\theta = \frac{\sqrt{\frac{PS^2}{3} + PS^3}}{PS} \qquad \text{[from eqn (i) and (ii)]}$$

$$\Rightarrow \cot\theta = \frac{\sqrt{4PS^2}}{\sqrt{3}PS} \quad \Rightarrow \cot\theta = \frac{2}{\sqrt{3}}$$

5. In ΔPQA

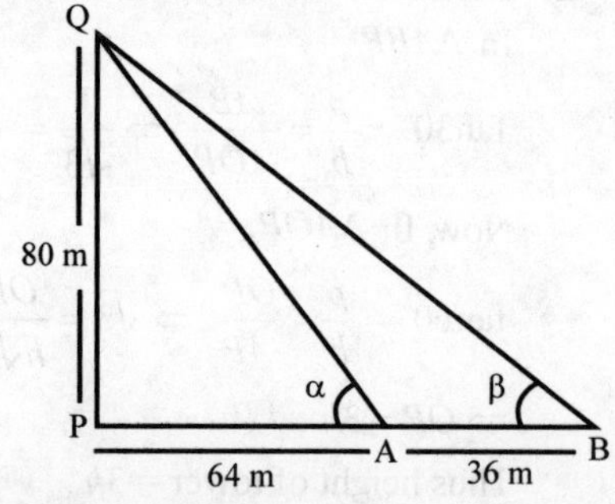

$$\tan\alpha = \frac{p}{b} = \frac{80}{64} = \frac{5}{4}$$

In ΔPQB

$$\tan\beta = \frac{p}{b} = \frac{80}{100} = \frac{4}{5}$$

$$\tan(\alpha+\beta) = \frac{\frac{5}{4}+\frac{4}{5}}{1-\frac{5}{4}\times\frac{4}{5}} = \frac{\frac{41}{20}}{(1-1)} = \infty$$

Since the denominator is zero,

$$\tan(\alpha+\beta) = \infty = \tan 90^\circ$$

Thus $(\alpha+\beta) = 90^\circ$

6. In ΔPQS

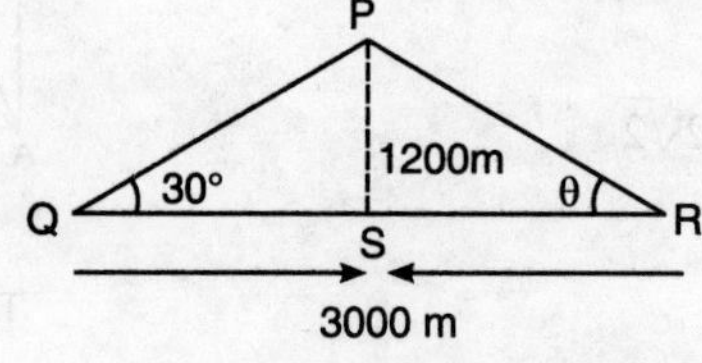

$$\tan 30^\circ = \frac{PS}{QS}$$

$$\frac{1}{\sqrt{3}} = \frac{1200}{QS}$$

$$\Rightarrow QS = 1200\sqrt{3}$$

Now, $SR = 3000 - 1200\sqrt{3} = 600(5 - 2\sqrt{3})$

$= 600(5 - 2 \times 1.732)$

$= 922$ m

Now, In ΔPSR

$$\tan\theta = \frac{PS}{SR} = \frac{1200}{922} = 1.301$$

$\therefore \tan 45^\circ = 1$ and $\tan 60^\circ = \sqrt{3} = 1.732$

$\therefore 45^\circ < \theta < 60^\circ$

7. We have,

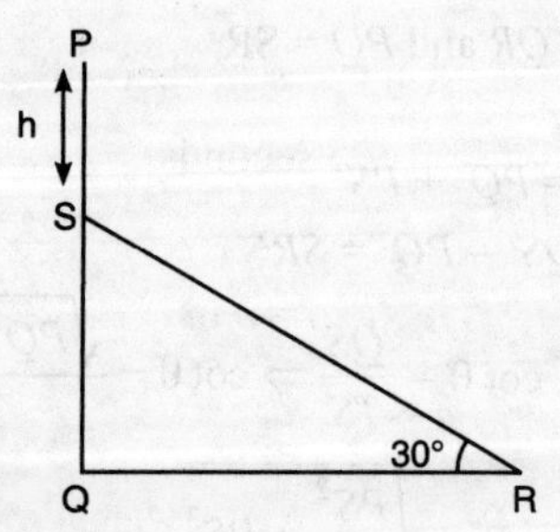

$$\sin 30^\circ = \frac{SQ}{SR}$$

$$\Rightarrow \frac{1}{2} = \frac{15-h}{h}$$

$\Rightarrow h = 30 - 2h$

$\Rightarrow 3h = 30$

$\Rightarrow h = 10$ m

$\therefore SQ = 15 - 10 = 5$m

8. In ΔRUV

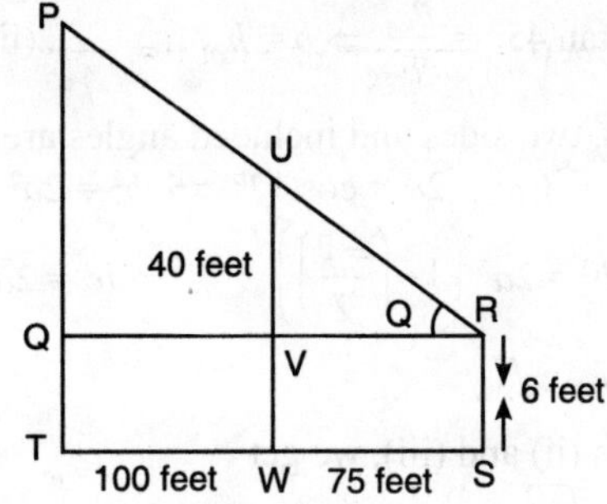

$UV = 40 - 6 = 34$

$RV = SW = 75$ feet

$\therefore \quad \tan\theta = \dfrac{UV}{RV} = \dfrac{34}{75}$ ...(i)

In ΔPQR

$QR = TS = 100 + 75 = 175$ feet

$\therefore \quad \tan\theta = \dfrac{PQ}{QR} = \dfrac{PQ}{175}$ ...(ii)

Now,

from eqn (i) and (ii) $\dfrac{34}{75} = \dfrac{PQ}{175} \Rightarrow PQ = \dfrac{175 \times 34}{75} = \dfrac{238}{3} = 79.33$ feet

$\therefore \quad PT = 79.33 + 6 = 85.33$ feet

9. In ΔABC

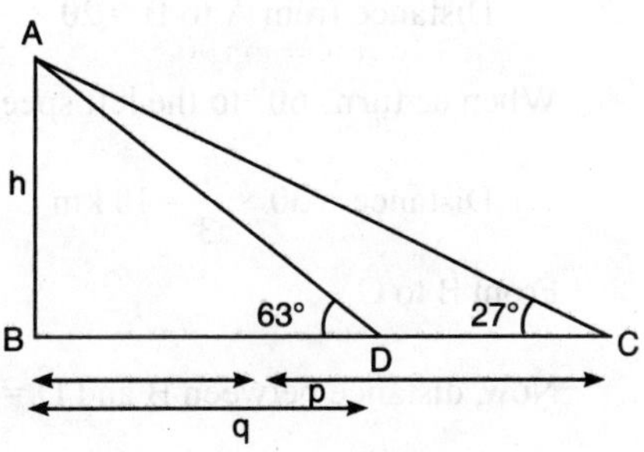

$\tan 27° = \dfrac{AB}{BC}$

$\tan 27° = \dfrac{h}{p}$ ...(i)

In ΔABD

$\tan 63° = \dfrac{AB}{BD}$

$\tan 63° = \dfrac{AB}{Q}$

$\cot 27° = \dfrac{h}{q}$ ...(ii)

Multiplying (i) and (ii), we get

$1 = \dfrac{h}{p} \times \dfrac{h}{q} \Rightarrow h^2 = pq \quad \Rightarrow h = \sqrt{pq}$

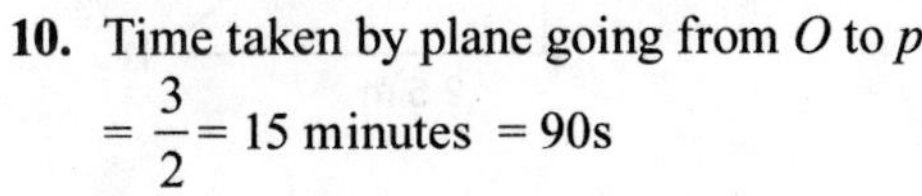

10. Time taken by plane going from O to p

$= \dfrac{3}{2} = 15$ minutes $= 90$s

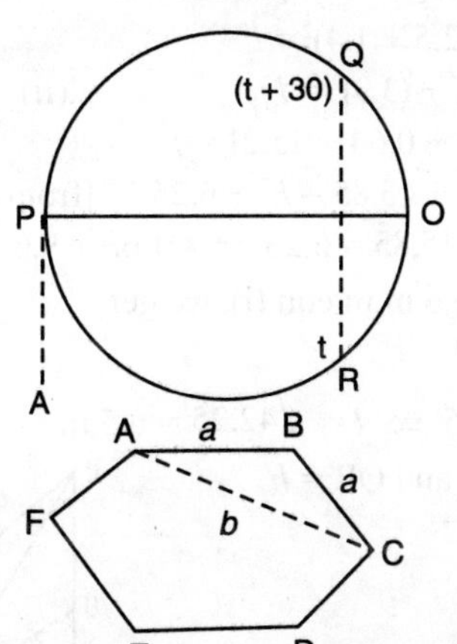

From O to R it takes 15 seconds

So, 90 + 15 = 105 seconds

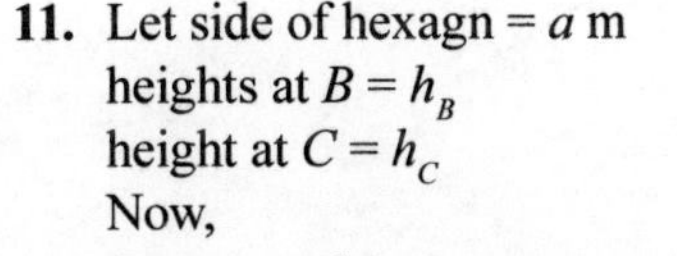

11. Let side of hexagn = a m

heights at $B = h_B$

height at $C = h_C$

Now,

$\tan 30° = \dfrac{h_B}{a} \Rightarrow a = \sqrt{3}h_B$ ... (i) and $\tan 45° = \frac{h_C}{b} \Rightarrow b = h_C$ ...(ii)

Now, two sides and included angles are given

$b^2 = a^2 + a^2 - 2a \times \cos 12° \Rightarrow b^2 = 2a^2 - 2a^2 \cos 120° \Rightarrow b^2 = 2a^2 (1 - \cos 120°)$

$\Rightarrow b^2 = 2a^2 \left(1 - \left(\frac{-1}{2}\right)\right) \Rightarrow b^2 = 2a^2 \times \frac{3}{2}$

$\Rightarrow b = \sqrt{3}a$ ...(iii)

From (ii) and (iii), we get

$h_c = \sqrt{3}a$

Now, $\frac{h_B}{h_C} = \frac{a}{\sqrt{3} \times 3a} = 1 : 3$ [from (ii) and (iii)]

12. When the person turns 150°; Speed = 20 km/h, time = 15 mins

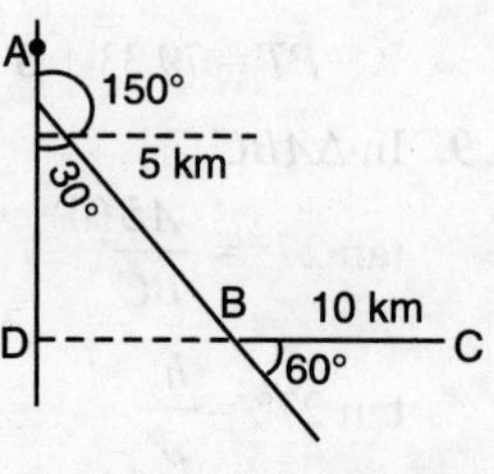

$\therefore$ Distance from A to B = $20 \times \frac{1}{4} = 5$ km

When he turns 60° to the left speed = 30 km/h, time = 20 mins

$\therefore$ Distance = $30 \times \frac{1}{3} = 10$ km

From B to C

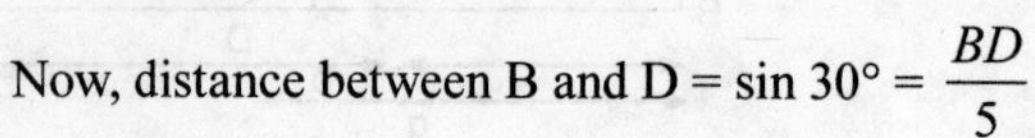

Now, distance between B and D = $\sin 30° = \frac{BD}{5}$

$\Rightarrow \frac{1}{2} = \frac{BD}{5} \Rightarrow BD = \frac{5}{2} = 2.5$ km

$\therefore$ Distance between the two streets = DB + BC = 2.5 + 10 = 12.5 km

13. We have,

$l^2 = h^2 + (2.5)^2$ [by Pythagoras theorem]

$\Rightarrow l^2 = h^2 + 6.25$ ...(i)

Now, according to the question,

$(h - 0.8)^2 + (2.5 + 1.4)^2 = l^2$

$\Rightarrow (h - 0.8)^2 + (3.9)^2 = l^2$ ...(ii)

$\Rightarrow h^2 - 1.6h + 0.64 + 15.21 = l^2$

$\Rightarrow h^2 - 1.6h + 15.85 = h^2 + 6.25$ [from (i)]

$\Rightarrow -1.6h + 15.85 = 6.25 \Rightarrow -1.6h + 9.6 \Rightarrow h = 6$m

Now, put $h = 6$ m in eqn (i), we get

$36 + 6.25 = l^2$

$\Rightarrow l^2 = 42.25 \Rightarrow l = \sqrt{42.25} = 6.5$ m

P, l, h, Q, R, 2.5 m

14. Let PR = H and QR = h

In ΔPRT

$\tan \alpha = \frac{PR}{RT}$

$\Rightarrow \frac{H}{D + d} = \tan \alpha$ ...(i)

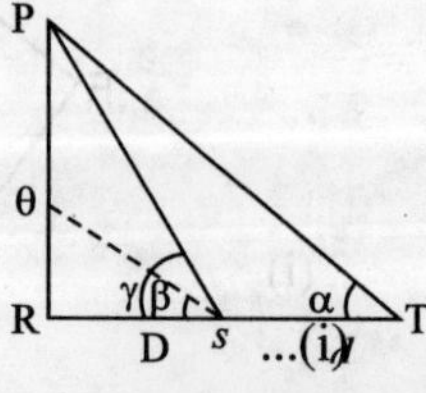

In ΔPRS

$\tan\beta = \dfrac{PR}{RS}$ ... (ii)

$\Rightarrow \tan B = \dfrac{H}{D}$

In ΔQRS $\tan\lambda = \dfrac{QR}{RS}$

$\Rightarrow \tan\lambda = \dfrac{h}{d}$ ...(iii)

From (ii) and (iii), we get,

Now,

$\dfrac{H}{h} = \dfrac{\tan\beta}{\tan y} \Rightarrow H\tan y - h\tan\beta = 0$

15. From qn. 31, we get

From (i) and (ii), we get,

$\dfrac{\tan\alpha}{\tan\beta} = \dfrac{D}{d+D} \Rightarrow \dfrac{\tan\beta}{\tan\alpha} = \dfrac{d+D}{D} \Rightarrow 1+\dfrac{d}{D} = \dfrac{\tan\beta}{\tan\alpha} \Rightarrow \dfrac{d}{D} = \dfrac{\tan\beta}{\tan\alpha} - 1$

From (ii)

$\dfrac{d}{H} = \dfrac{\tan\beta - \tan\alpha}{\tan\alpha - \tan\beta}$

$\therefore \dfrac{d}{H} = \cot\alpha - \cot\beta \Rightarrow d = H(\cot\alpha - \cot\beta)$

16. In ΔARP

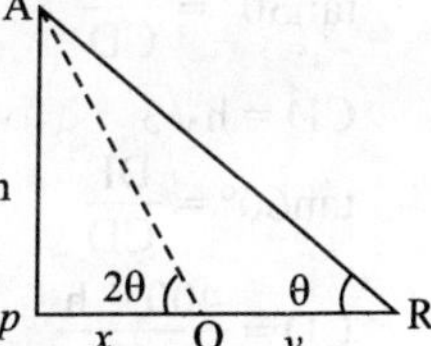

$\tan\theta = \dfrac{h}{x+y}$ ...(i)

In ΔAPQ

$\tan\theta = \dfrac{h}{x}$ ...(ii)

Dividing (i) and (ii), we get

$\dfrac{\tan\theta}{\tan 2\theta} = \dfrac{x}{x+y}$

$\Rightarrow \dfrac{\tan\theta\times(1-\tan^2\theta)}{2\tan\theta} = \dfrac{x}{x+y} \Rightarrow \dfrac{1-\tan^2\theta}{2} = \dfrac{x}{x+y} \Rightarrow \dfrac{2}{1-\tan^2\theta} = \dfrac{x+y}{x}$

$\Rightarrow \dfrac{2}{1-\tan^2\theta} = 1+\dfrac{y}{x} \Rightarrow \dfrac{2}{1-\tan^2\theta} - 1 = \dfrac{y}{x}$

$= \dfrac{2-1+\tan^2\theta}{1-\tan^2\theta} = \dfrac{y}{x} \Rightarrow \dfrac{1+\tan^2\theta}{1-\tan^2\theta} = \dfrac{y}{x} \Rightarrow \dfrac{x}{y} = \dfrac{1-\tan^2\theta}{1+\tan^2\theta}$

$\dfrac{x}{y} < 1$ $[\because \cos 2\theta < 1]$

$\Rightarrow x < y$

17. From qn 33, we get

$\frac{x}{y} = \cos 2\theta \Rightarrow x = y\cos 2\theta$

$\because \frac{h}{x} = \tan 2\theta$

$\Rightarrow h = x\tan 2\theta \Rightarrow y = y\cos 2\theta \times \frac{\sin 2\theta}{\cos 2\theta} \Rightarrow h = y\sin 2\theta$

Now,

$h^2 + x^2 = (y^2\sin^2\theta + y^2\cos^2\theta) = y^2(\sin^2\theta + \cos^2\theta)$

$\Rightarrow h^2 = x^2 = y^2 \quad \Rightarrow h^2 = y^2 - x^2$

18.

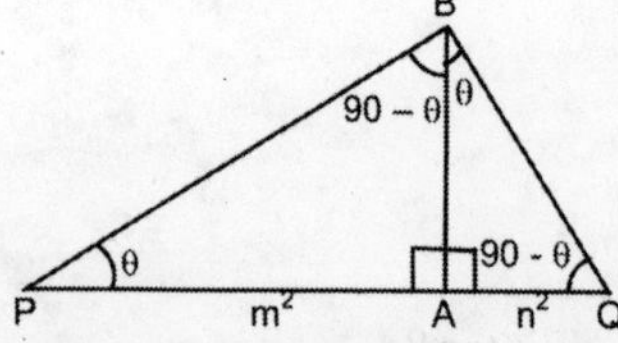

Let AB be the tower.

$\angle PBQ = 180° - (\angle BPQ + \angle BQP) = 180° - 90° = 90°$

ΔPAB & ΔBAQ are similar $\frac{\text{AP}}{\text{AB}} = \frac{\text{AB}}{\text{AQ}}$

$AB^2 = AP\,(AQ) = m^2 n^2$ or $AB = mn$.

19. Let AB denote the surface of the lake.

$\tan 30° = \frac{\text{h}}{\text{CD}}$

$\text{CD} = \text{h}\sqrt{3}$ ——— (1)

$\tan 60° = \frac{\text{DF}}{\text{CD}}$

$\text{CD} = \frac{400 + \text{h}}{\sqrt{3}}$ ——— (2)

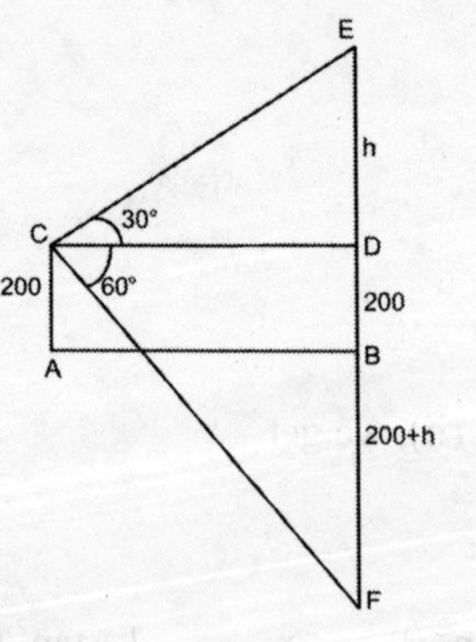

From (1) & (2),

$\text{h}\sqrt{3} = \frac{400 + \text{h}}{\sqrt{3}} \Rightarrow h = 200$ m

$\therefore$ Height of the cloud

$= (h + 200)$ m $= 400$ m.

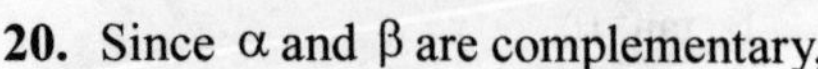

20. Since α and β are complementary,

$\beta = 90° - \alpha$

Let the width of the river be V.

From point X,

$\tan\alpha = \frac{h}{d_1}$ and $\tan\beta = \frac{h}{d_2}$

Here, d_1 and d_2 are the distances from X to points P and Q respectively.

Since $\beta = 90^\circ - \alpha$, $\tan\beta = \cot\alpha$

So, $\tan\alpha = \dfrac{h}{d_1}$ and $\cot\alpha = \dfrac{h}{d_2}$

Width of the river:

$V = d_1 + d_2 \quad \Rightarrow d_1 = \dfrac{h}{\tan\alpha}$ and $d_2 = \dfrac{h}{\cot\alpha} \quad \Rightarrow V = \dfrac{h}{\tan\alpha} + \dfrac{h}{\cot\alpha}$

$\Rightarrow V = \dfrac{h(\tan\alpha + \cot\alpha)}{\tan\alpha\ \cot\alpha} \quad \Rightarrow V = h(\tan\alpha + \cot\alpha)$

$= h\left(\dfrac{\sin\alpha}{\cos\alpha} + \dfrac{\cos\alpha}{\sin\alpha}\right) = h\left(\dfrac{\sin^2\alpha + \cos^2\alpha}{\sin\alpha\ \cos\alpha}\right) = h\left(\dfrac{1}{\sin\alpha\ \cos\alpha}\right) = h\sec\alpha \cdot \operatorname{cosec}\alpha$

21. $CP_1 = 3000\text{m}$

$OC = $ ÄABD $= \dfrac{3000}{\sqrt{3}}\text{m}$

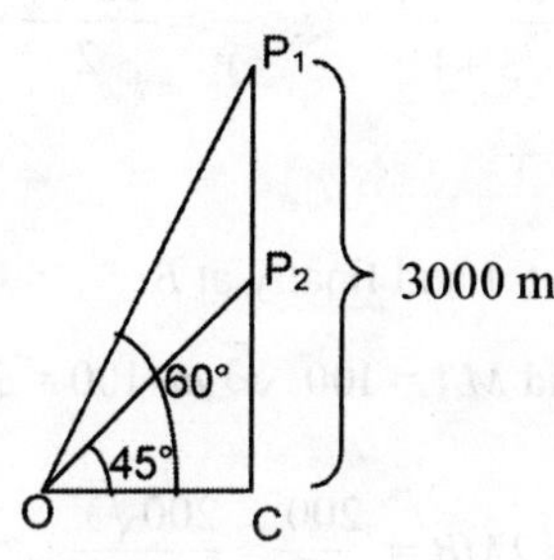

$CP_2 = OC\ (\because \tan45^\circ) = \dfrac{3000}{\sqrt{3}}\text{m}$

$P_1P_2 = CP_1 - CP_2 = \tan30^\circ = \dfrac{p}{b} = \dfrac{AB}{AD}$

$= (3000 - 1000\sqrt{3})\ \text{m}$

$= 1000(3 - \sqrt{3})\ \text{m}$

22. Height of pole = OP

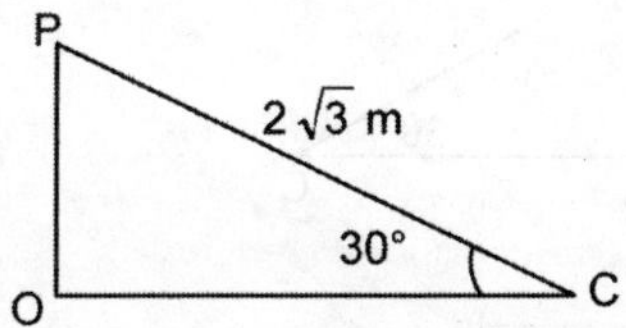

$\sin30^\circ = \dfrac{OP}{2\sqrt{3}} = \dfrac{1}{2} \Rightarrow \text{OP} = \sqrt{3}.$

23. $\tan(90 - \theta) = \dfrac{6}{x}$

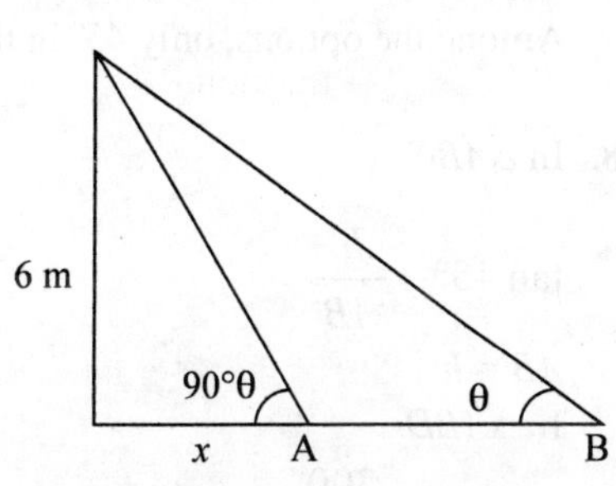

$\cot = \dfrac{6}{x}$ ...(i)

and $\tan\theta = \dfrac{6}{x+5}$ ... (ii)

Multiplying (i) and (ii), we get

$1 = \dfrac{6}{x} \times \dfrac{6}{x+5}$

$x^2 + 5x - 36 = 0 \Rightarrow x^2 + 9x - 4x - 36 = 0 \Rightarrow x(x + 9) - 4(x + 9) = 0 \Rightarrow x = 4$

$x \neq -9$ (not possible)

24. Let AD be the tower

From ΔACD, $\tan 45° = \dfrac{x}{CD}$

$\Rightarrow \quad CD = x \quad$ .... (i)

From ΔABD, $\tan 30° = \dfrac{AD}{BD}$.

$$\frac{1}{\sqrt{3}} = \frac{x}{x+y}$$

$$x\left(\sqrt{3}-1\right) = y$$

$$\frac{x}{y} = \frac{1}{\sqrt{3}-1} \times \frac{\sqrt{3}+1}{\sqrt{3}+1} \qquad \therefore \frac{x}{y} = \frac{\sqrt{3}+1}{2}$$

25. The car is initially at A and finally at B.

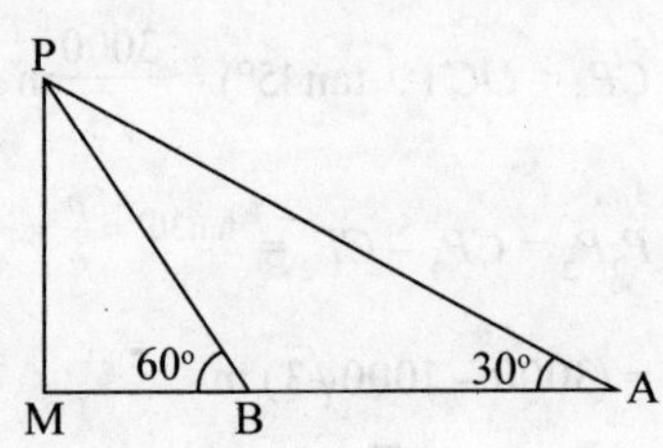

As $MB = \dfrac{100}{\sqrt{3}}\ m$ and $MA = 100\ \sqrt{3}\ m$, $100 = 3MB$

It follows that $BA = 2MB = \dfrac{200}{\sqrt{3}} = \dfrac{200\sqrt{3}}{3}$

26. In ΔAMB, A = 30°, M = 90°, B = 60°

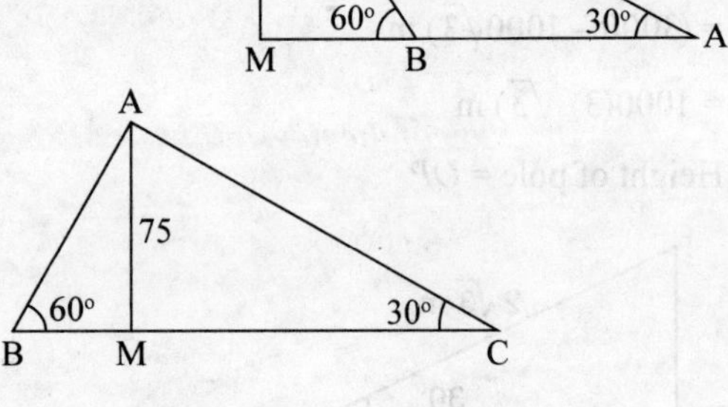

$AM = 75$, it follows that $BM = 25\sqrt{3}$

In ΔABC, A = 60°, M = 90°, C = 30°

As $AM = 75$, if follows that MC = $75\sqrt{3}$.

$\therefore$ BC = $100\sqrt{3}$

27. If the length of the shadow of a tower is equal to the height, the altitude of the sun has to be 45° (or 135°)

Among the options, only 45° in these.

28. In ΔABC

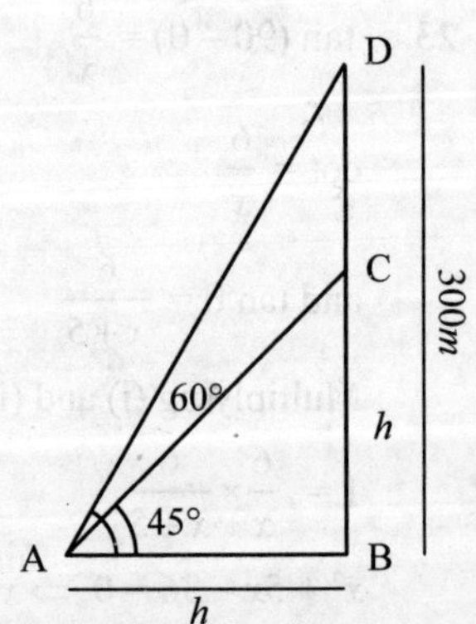

$$\tan 45° = \frac{h}{AB}$$

$AB = h$

In ΔABD

$$\tan 60° = \frac{300}{h}$$

$$\sqrt{3} = \frac{300}{h} \Rightarrow h = \frac{300}{\sqrt{3}} = 100\sqrt{3}m$$

29. Let height of tree be h

In ΔADC

$$\tan 45° = \frac{90}{x}$$

$$x = 90$$

In ΔBED

$$\tan 30° = \frac{90-h}{90}$$

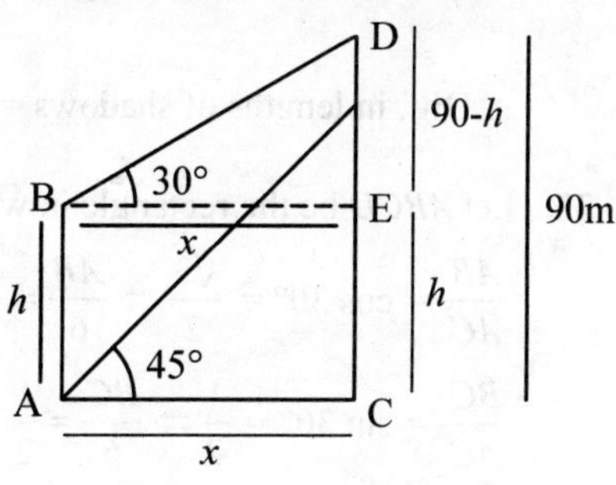

$$\Rightarrow \frac{1}{\sqrt{3}} = \frac{90-h}{90} \Rightarrow 90 = 90\sqrt{3} - \sqrt{3}h$$

$$\Rightarrow \sqrt{3}h = 90\left(\sqrt{3}-1\right)$$

$$h = \frac{90\left(\sqrt{3}-1\right)}{\sqrt{3}} \times \frac{\sqrt{3}}{\sqrt{3}} = 30\left(3-\sqrt{3}\right) = 90 - 30\sqrt{3}m$$

30. In ΔACB, $\tan\alpha = \frac{h}{a}$

$$a = \frac{h}{\tan\alpha}$$

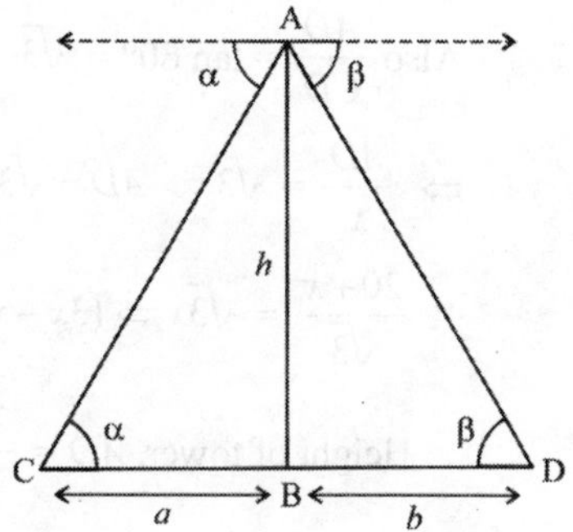

In ΔABD, $\tan\beta = \frac{h}{b}$, $b = \frac{h}{\tan\beta}$

$a + b = 1$ [Given]

$$\frac{h}{\tan\alpha} + \frac{h}{\tan\beta} = 1 \Rightarrow h\left[\frac{1}{\tan\alpha} + \frac{1}{\tan\beta}\right] = 1$$

$$h\left[\frac{\tan\alpha + \tan\beta}{\tan\alpha\tan\beta}\right] = 1 \Rightarrow h = \frac{\tan\alpha\tan\beta}{\tan\alpha + \tan\beta}$$

31. Let AB be the rod and AC be its shadow.

Let $\angle ACB = \theta$. Let $AB = x$. Then, $AC = \sqrt{3}x$.

$$\therefore \tan\theta = \frac{AB}{AC} = \frac{x}{\sqrt{3}x} = \frac{1}{\sqrt{3}} \Rightarrow \theta = 30°.$$

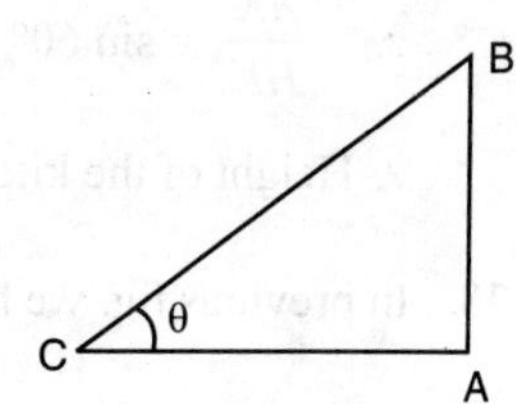

32. Let AB be the pole and AC be its shadow.

Then, $\theta = 60°$ and $AC = 30$ m.

$$\therefore \frac{AB}{AC} = \tan 60° \Rightarrow \frac{AB}{30} = \sqrt{3} = AB = 30\sqrt{3}\text{m}.$$

33. Let AB be the building and AC be its shadow.

Then, $AB = 50$ m and $\theta = 30°$.

$$\therefore \frac{AC}{AB} = \cot 30° = \sqrt{3} \Rightarrow \frac{AC}{50} = \sqrt{3} \Rightarrow AC = 50\sqrt{3}\text{ m}.$$

34. When AB =15 m, $\theta = 30°$, then $\frac{AC}{AB} = \tan 30° \Rightarrow AC = \frac{15}{\sqrt{3}}$ m.

When $AB = 15$ m, $\theta = 60°$, then $\frac{AC}{AB} = \tan 60° \Rightarrow AC = 15\sqrt{3}$ m.

$\therefore$ Diff. in lengths of shadows $= \left(15\sqrt{3} - \frac{15}{\sqrt{3}}\right) = \frac{30}{\sqrt{3}} = 10\sqrt{3}$ m.

35. Let $ABCD$ be the rectangle in which $\angle BAC = 30°$ and $AC = 6$ cm.

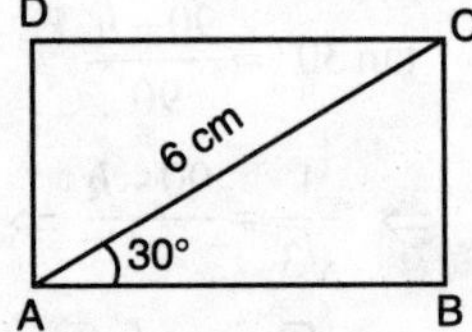

$\frac{AB}{AC} = \cos 30° = \frac{\sqrt{3}}{2} \Rightarrow \frac{AB}{6} = \frac{\sqrt{3}}{2} \Rightarrow AB = 3\sqrt{3}$ cm.

$\frac{BC}{AC} = \sin 30° = \frac{1}{2} \Rightarrow \frac{BC}{6} = \frac{1}{2} \Rightarrow BC = 3$ cm.

$\therefore$ Area of the rectangle $= AB \times BC = 9\sqrt{3}$ cm^2.

36. Let AD be the tower and B and C be the points of observation such that $BC = 20$ m, $\angle DBA = 30°$ and $\angle DCA = 60°$. Let $CD = x$.

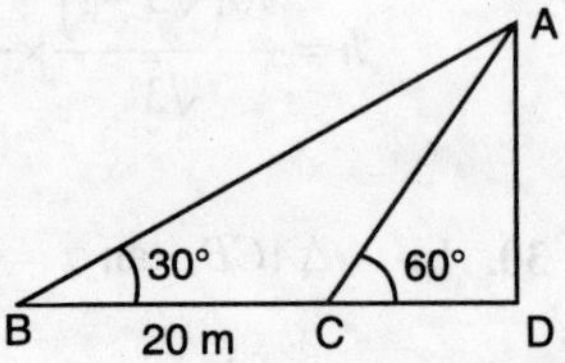

Now, $\frac{AD}{BD} = \tan 30° = \frac{1}{\sqrt{3}} \Rightarrow \frac{AD}{20+x} = \frac{1}{\sqrt{3}} \Rightarrow AD = \frac{(20+x)}{\sqrt{3}}$.

Also $\frac{AD}{CD} = \tan 60° = \sqrt{3}$

$\Rightarrow \frac{AD}{x} = \sqrt{3} \Rightarrow AD = \sqrt{3}x$.

$\therefore \frac{20+x}{\sqrt{3}} = \sqrt{3}x \Rightarrow (3x - x) = 20 \Rightarrow x = 10$.

$\therefore$ Height of tower, $AD = \frac{20+x}{\sqrt{3}} = \left(\frac{20+10}{\sqrt{3}}\right) = \frac{10\times 3}{\sqrt{3}} = 10\sqrt{3}$ m.

37. Let K be the position of the kite and HK be the string. HK= 90 m & $\angle AHK = 60°$.

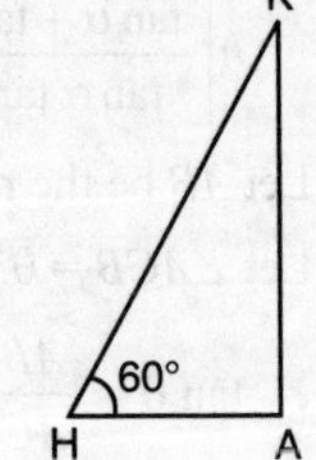

$\therefore \frac{AK}{HK} = \sin 60° = \frac{\sqrt{3}}{2} \Rightarrow \frac{AK}{90} = \frac{\sqrt{3}}{2} \Rightarrow AK = 45\sqrt{3}$

$\therefore$ Height of the kite $= 45\sqrt{3}$ m.

38. In previous fig. we have $\angle AHK$= 60°, $HK = 85$ m & $\tan\theta = \frac{15}{8}$.

Now $\tan\theta = \frac{15}{8} \Rightarrow \cos\theta = \frac{8}{15}$.

$\therefore \operatorname{cosec}\theta = \sqrt{1+\cot^2\theta} = \sqrt{1+\frac{64}{225}} = \sqrt{\frac{289}{225}} = \frac{17}{15}$.

So, $\sin\theta = \frac{15}{17}$.

Now, $\frac{AK}{HK} = \sin\theta = \frac{15}{17} \Rightarrow \frac{AK}{85} = \frac{15}{17} \Rightarrow AK = \left(\frac{85\times 15}{17}\right) = 75$ m.

$\therefore$ Height of the kite from the ground = 75 m.

39. Let AB be the tower and C, D be the persons so that
$AB = 30$ m, $\angle ACB = 30°$ and $\angle ADB = 45°$.

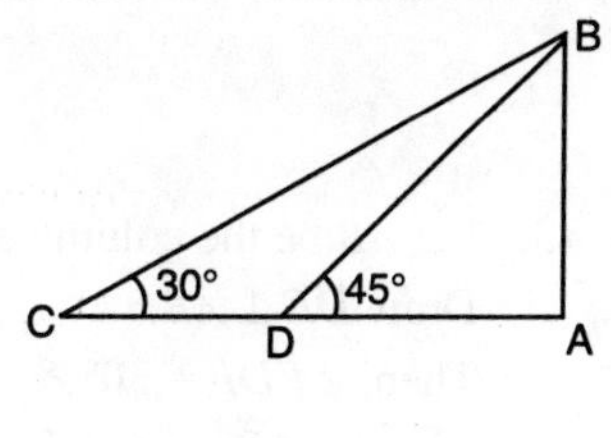

Then, $\dfrac{AD}{AB} = \cot 45° = 1 \Rightarrow \dfrac{AD}{30} = 1 \Rightarrow AD = 30\,\text{m}.$

$\dfrac{AC}{AB} = \cot 30° = \sqrt{3} \Rightarrow \dfrac{AC}{30} = \sqrt{3}$

$\Rightarrow AC = 30\sqrt{3}.$

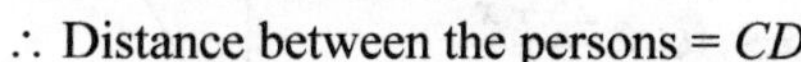

$\therefore$ Distance between the persons $= CD$

$= AC - AD = (30\sqrt{3} - 30)\text{ m} = 30(\sqrt{3} - 1)\text{ m}$

$= (30 \times 0.732)\text{ m} = 21.96\text{ m} = 22\text{ m (app.)}$

40. Let AB be the tower and C and D be points such that
$AB = 40$ m, $\angle ACB = 45°$ and $\angle ADB = 60°$.

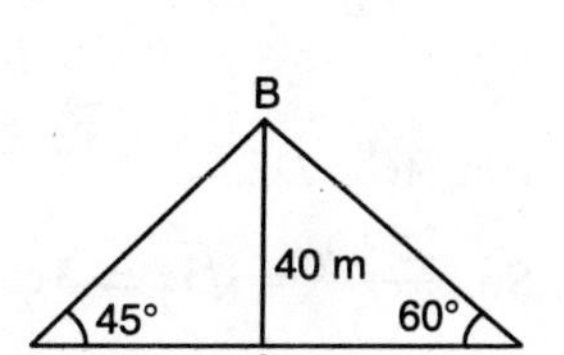

Then, $\dfrac{AC}{AB} = \cot 45° = 1$

$\Rightarrow \quad AC = AB = 40$ m.

And, $\dfrac{AD}{AB} = \cot 60° = \dfrac{1}{\sqrt{3}}$

$\Rightarrow \dfrac{AD}{40} = \dfrac{1}{\sqrt{3}} \quad \Rightarrow AD = \dfrac{40}{\sqrt{3}}\text{ m.}$

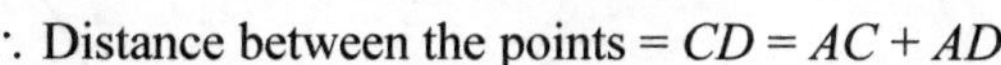

$\therefore$ Distance between the points $= CD = AC + AD$

$$= \left(40 + \frac{40 \times \sqrt{3}}{3}\right)\text{m} = \left(40 + \frac{40 \times 1.732}{3}\right)\text{m} = 63\,\text{m (app.)}$$

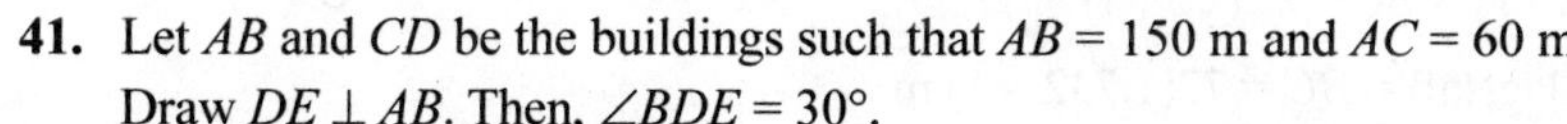

41. Let AB and CD be the buildings such that $AB = 150$ m and $AC = 60$ m
Draw $DE \perp AB$. Then, $\angle BDE = 30°$.

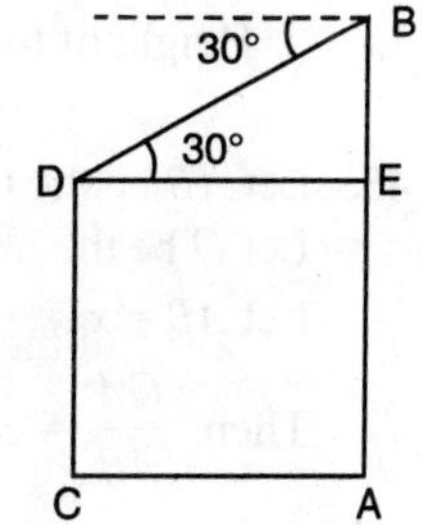

Now, $\dfrac{BE}{DE} = \tan 30° = \dfrac{1}{\sqrt{3}} \Rightarrow BE = \dfrac{DE}{\sqrt{3}} = \dfrac{AC}{\sqrt{3}} = \dfrac{60}{\sqrt{3}}.$

Height of the first building $= CD = AE = AB - BE$

$= \left(150 - \dfrac{60}{\sqrt{3}}\right)\text{m}$

$= (150 - 20\sqrt{3})\text{ m}$

$= (150 - 20 \times 1.732)\text{ m} = 115.36\text{ m.}$

42. Let AB be the cliff and C, D be the ships such that $AB = 100$ m, $\angle ACB = 30°$ & $\angle ADB = 60°$.

$\dfrac{AD}{AB} = \cot 60° = \dfrac{1}{\sqrt{3}} \Rightarrow \dfrac{AD}{100} = \dfrac{1}{\sqrt{3}}$

$\Rightarrow \quad AD = \dfrac{100}{\sqrt{3}}\text{ m.}$

$\dfrac{AC}{AB} = \cot 30° = \sqrt{3} \Rightarrow \dfrac{AC}{100} = \sqrt{3}$

$\Rightarrow \quad AC = 100\sqrt{3}\text{ m.}$

$\therefore$ Distance between the ships $= CD = AC - AD$

$$= \left(100\sqrt{3} - \frac{100}{\sqrt{3}}\right) = \frac{200\sqrt{3}}{3} = \left(\frac{200 \times 1.73}{3}\right) \text{m} = 115.3\text{m}.$$

43. Let AB be the column and CD be the tower, where $CD = 25$ m.

Draw $DE \perp AB$.

Then, $\angle EDB = 30°$ & $\angle ACB = 60°$

Clearly, $AE = CD = 25$ m. Let $BE = x$.

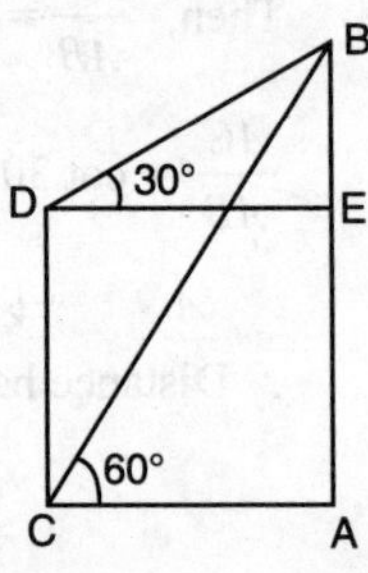

Now, $\dfrac{AC}{AB} = \cot 60° = \dfrac{1}{\sqrt{3}} \Rightarrow \dfrac{AC}{25+x} = \dfrac{1}{\sqrt{3}} \Rightarrow AC = \dfrac{25+x}{\sqrt{3}}.$

$\dfrac{DE}{EB} = \cot 30° = \sqrt{3} \Rightarrow \dfrac{DE}{x} = \sqrt{3}$

$\Rightarrow DE = \sqrt{3x}$

But, $AC = DE$.

So, $\dfrac{25+x}{\sqrt{3}} = \sqrt{3}x \Rightarrow 3x - x = 25 \Rightarrow x = 12.5\,\text{m}.$

$\therefore$ Height of the column $= (25 + 12.5)$ m $= 37.5$ m.

44. Let AB be the building and BC the flagstaff and let O be the observer.

Then, $OA = 72$ m. $\angle AOB = 45°$ & $\angle AOC = 60°$.

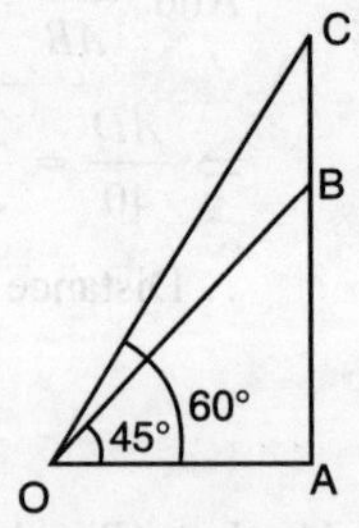

Now, $\dfrac{OB}{OA} = \tan 45° = 1 \Rightarrow \dfrac{AB}{72} = 1 \Rightarrow AB = 72$ m

$\dfrac{AC}{OA} = \tan 60° = \sqrt{3} \Rightarrow \dfrac{AB+BC}{72} = \sqrt{3}$

$\Rightarrow BC + 72 = 72\sqrt{3} \Rightarrow BC = 72(\sqrt{3}-1)$

$\therefore$ Height of the flagstaff $= BC = 72(1.732 - 1)$ m

$= (72 \times 0.732)$ m $= 52.7$ m.

45. Let AB be the tower and BC be the flagstaff.

Let O be the observer. Then, $BC = h$, $\angle AOB = \alpha$ and $\angle AOC = \beta$.

Let $AB = x$.

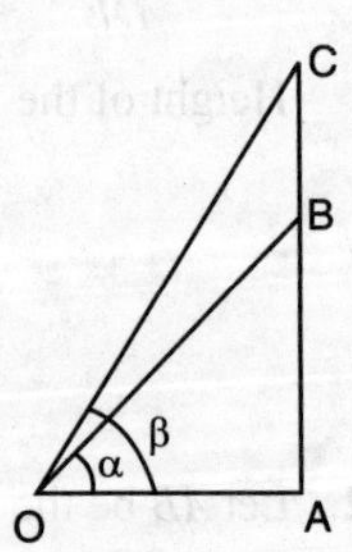

Then, $\dfrac{OA}{AB} = \cot\alpha = OA = x\cot\alpha$

$\dfrac{OA}{AC} = \cot\beta \Rightarrow \dfrac{OA}{x+h} = \cot\beta \Rightarrow OA = (x+h)\cot\beta.$

$\therefore \quad x\cot\alpha = (x+h)\cot\beta.$

$\therefore \quad x\cot\alpha = (x+h)\cot\beta$

$\Rightarrow \quad x(\cot\alpha - \cot\beta) = h\cot\beta.$

or $\quad x = \dfrac{h\cot\beta}{\cot\alpha - \cot\beta} = \dfrac{h\cot\beta}{\dfrac{1}{\tan\alpha} - \dfrac{1}{\tan\beta}} = \dfrac{h\cot\beta \times \tan\alpha\tan\beta}{\tan\beta - \tan\alpha}$

$\therefore x = \dfrac{h\tan\alpha}{\tan\beta - \tan\alpha}.$

46. Let AB be the tower and C, D be the objects such that $AB = 300$ m, $\angle ACB = 45°$ & $\angle ADB = 60°$.

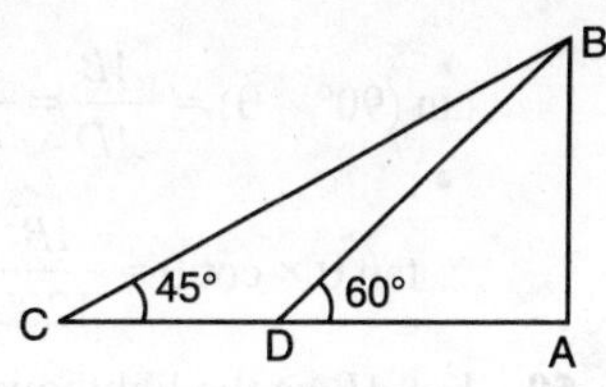

Now, $\frac{AC}{AB} = \cot 45° = 1 \Rightarrow AC = AB = 300$ m.

And, $\frac{AD}{AB} = \cot 60° = \frac{1}{\sqrt{3}}$

$\Rightarrow \frac{AD}{300} = \frac{1}{\sqrt{3}} \Rightarrow AD = \frac{300}{\sqrt{3}} = 100\sqrt{3}$ m.

$\therefore CD = AC - AD = (300 - 100\sqrt{3})$ m $= (300 - 100 \times 1.732)$m $= 126.8$ m.

47. Let AB and CD be the poles and let AC be the road.
Let E be the position of the observer.
Then, $\angle AEB = 30°$ and $\angle CED = 60°$.
Let $AE = x$ and $EC = (30 - x)$.
Let $AB = CD = h$. Then,

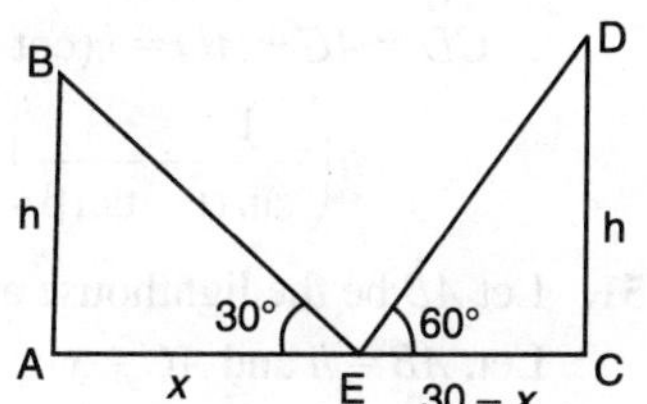

$\frac{h}{x} = \tan 30° = \frac{1}{\sqrt{3}} \Rightarrow x = \sqrt{3}h.$

And, $\frac{h}{30 - x} = \tan 60° = \sqrt{3} \Rightarrow x = \left(30 - \frac{h}{\sqrt{3}}\right).$

$\therefore \sqrt{3}h = 30 - \frac{h}{\sqrt{3}} \Rightarrow \left(\sqrt{3} + \frac{1}{\sqrt{3}}\right)h = 30 \Rightarrow h = \frac{30}{4}\sqrt{3}$ m.

$\therefore h = (7.5 \times 1.732)$ m $= 12.975$ m $= 13$ m (approx.)

48. Let OP be the tower and OA and OB be its shadows when the altitude of the sun is 45° and 30° respectively.

Then, $AB = 60$ m ; $\angle OAP = 30°$ and $\angle OBP = 45°$.
Let $OP = h$ and $OB = x$.

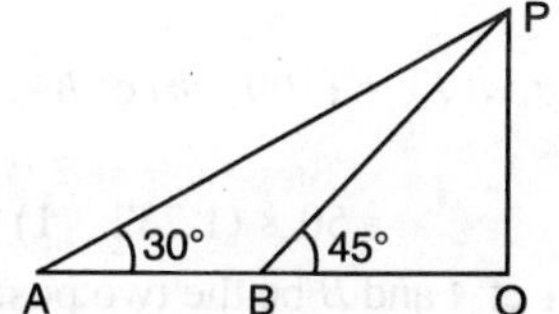

Now, $\frac{OP}{OB} = \tan 45° = 1 \Rightarrow OP = OB \Rightarrow h = x.$

$\frac{OP}{OA} = \tan 30° = \frac{1}{\sqrt{3}} \Rightarrow \frac{OP}{60 + x} = \frac{1}{\sqrt{3}}$

$\Rightarrow x = \sqrt{3}h - 60.$

$\therefore \quad h = \sqrt{3}h - 60$

or $h = \frac{60}{\sqrt{3} - 1} \times \frac{\sqrt{3} + 1}{\sqrt{3} + 1} = 30(\sqrt{3} + 1) = 30 \times 2.732 = 81.96$ m.

49. Let AB be the tower and C, D be the points of observation.
Then, $AC = 30$ m & $AD = 40$ m.
Let $\angle ACB = \theta$. Then,
$\angle ADB = 90° - \theta$.

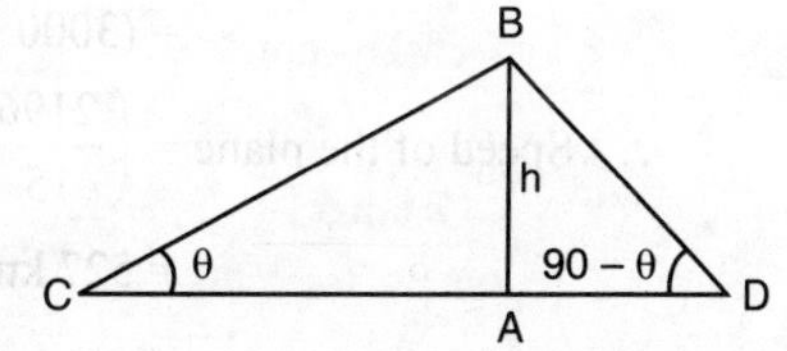

Now, $\tan \theta = \frac{AB}{AC} = \frac{AB}{30}$

$\tan(90° - \theta) = \frac{AB}{AD} = \frac{AB}{40}$ or $\cot\theta = \frac{AB}{40}$.

$\therefore \tan\theta \times \cot\theta = \frac{AB^2}{1200}$ or $AB = \sqrt{1200} = 20\sqrt{3} = (20 \times 1.732)\text{ m} = 34.64$ m.

50. Let AB be the lighthouse and C and D be the positions of the ships.
Then, $AB = h$; $\angle ACB = \alpha$ and $\angle ADB = \beta$.

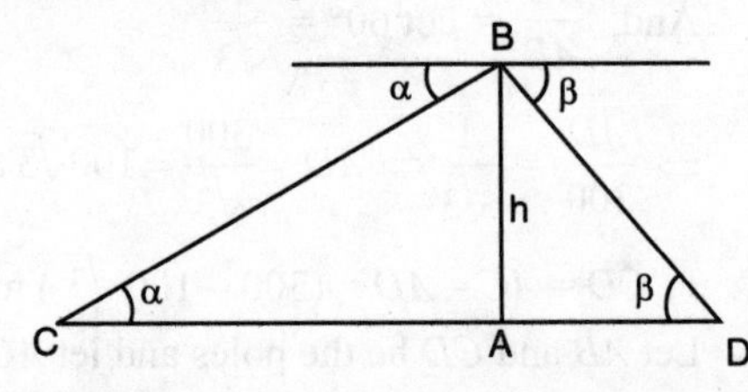

Now, $\frac{AC}{AB} = \cot\alpha \Rightarrow AC = h\cot\alpha$.

And, $\frac{AD}{AB} = \cot\beta \Rightarrow AD = h\cot\beta$.

$\therefore CD = AC + AD = h(\cot\alpha + \cot\beta)$

$= h\left(\frac{1}{\tan\alpha} + \frac{1}{\tan\beta}\right) = \frac{h(\tan\alpha + \tan\beta)}{\tan\alpha\tan\beta}$.

51. Let AB be the lighthouse and C and D be the boats.
Let, $AB = h$ and $AC = x$
Then, $AD = (100 - x)$.

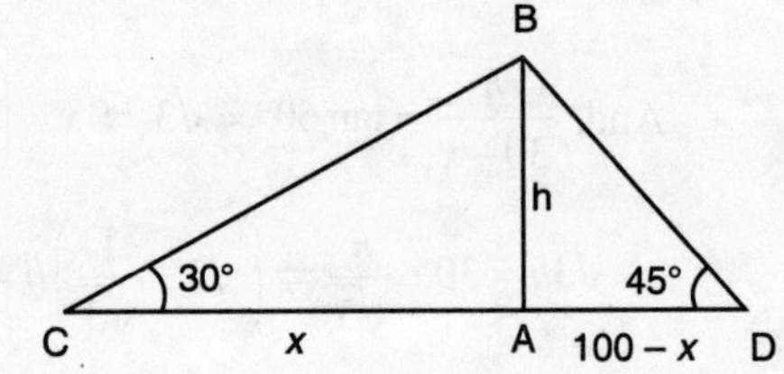

Also, $\angle ACB = 30°$ and $\angle ADB = 45°$.

Now, $\frac{AC}{AB} = \cot 30° = \sqrt{3}$

$\Rightarrow \frac{x}{h} = \sqrt{3} \Rightarrow x = h\sqrt{3}$

Also, $\frac{AD}{AB} = \cot 45° = 1 \Rightarrow \frac{100 - x}{h} = 1 \Rightarrow x = (100 - h)$.

$\therefore h\sqrt{3} = (100 - h)$ or $h = \frac{100}{\sqrt{3}+1} \times \frac{(\sqrt{3}-1)}{(\sqrt{3}-1)} = 50(\sqrt{3}-1)$

$= 50 \times (1.732 - 1) = 36.6$ m.

52. Let A and B be the two positions of the plane and let O be the point of observation and OD be the horizontal. Draw $AC \perp OD$ & $BD \perp OD$.
Then, $\angle DOB = 30°$, $\angle DOA = 45°$ & $AC = BD = 3000$ m.

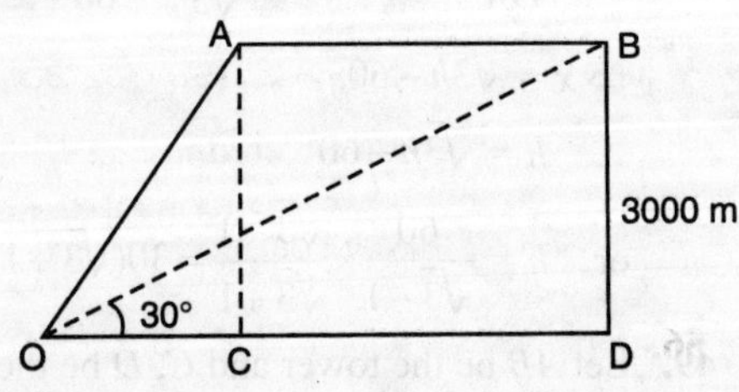

$\therefore \frac{OD}{DB} = \cot 30° = \sqrt{3} \Rightarrow OD = (3000 \times \sqrt{3})$m

$\frac{OC}{AC} = \cot 45° = 1 \Rightarrow OC = 3000$ m

Distance covered in 15 sec.

$= AB = CD = OD - OC$

$= (3000\sqrt{3} - 3000)\text{m} = 2196$ m

$\therefore$ Speed of the plane $= \left(\frac{2196}{15} \times \frac{1}{1000} \times 60 \times 60\right)$ km/hr

$= 527$ km/hr.

53. Let AB be the tower and C and D be the observers so that

$$CD = 10 \text{ m}, \angle ACB = 30° \text{ and } \angle ADB = 45°.$$

Let $AD = x$ metres. Let $AB = h$.

$$\frac{AD}{AB} = \cot 45° = 1 \Rightarrow \frac{x}{h} = 1 \Rightarrow x = h.$$

And, $\dfrac{AC}{AB} = \cot 30° = \sqrt{3}$

$$\Rightarrow \frac{x+10}{h} = \sqrt{3} \Rightarrow x = \sqrt{3}h - 10. \quad \therefore h = \sqrt{3}h - 10$$

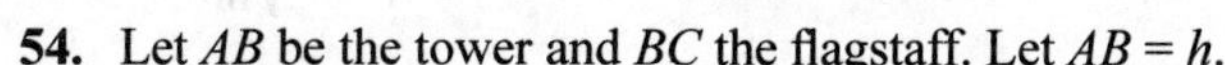

$$\Rightarrow h(\sqrt{3} - 1 = 10) \Rightarrow h = \frac{10}{\sqrt{3}-1} \times \frac{\sqrt{3}+1}{\sqrt{3}+1} \Rightarrow h = 5(\sqrt{3}+1) = (5 \times 2.732) = 13.66 \text{ m}.$$

54. Let AB be the tower and BC the flagstaff. Let $AB = h$.

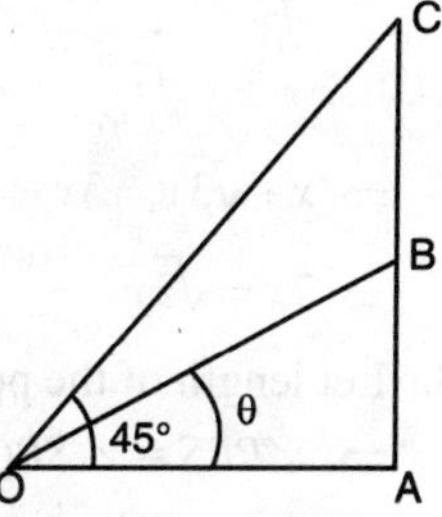

Then, $BC = \dfrac{1}{5}h$. Let O be the observer.

Then $\angle AOC = 45°$ and $\angle AOB = \theta$.

Now, $\dfrac{OA}{AC} = \cot 45° = 1 \Rightarrow \dfrac{OA}{h + \frac{1}{5}h} = 1$

$$\Rightarrow OA = \frac{6}{5}h \quad \therefore \tan\theta = \frac{AB}{OA} = \frac{h}{\frac{6}{5}h} = \frac{5}{6}.$$

55. Let AB be the cliff and C and D be the two positions of the ship.

Then, $AB = 150$ m,

$\angle ACB = 60°$ & $\angle ADB = 45°$.

Now, $\dfrac{AD}{AB} = \cot 45° = 1 \Rightarrow \dfrac{AD}{150} = 1 \Rightarrow AD = 150$ m.

$$\frac{AC}{AB} = \cot 60° = \frac{1}{\sqrt{3}} \Rightarrow \frac{AC}{150} = \frac{1}{\sqrt{3}}$$

$$\Rightarrow AC = \frac{150}{\sqrt{3}} = 50\sqrt{3} = 86.6 \text{ m}.$$

$\therefore CD = AD - AC = (150 - 86.6)\text{m} = 63.4$ m

Thus, distance covered in 2 min. = 63.4 m.

$$\therefore \text{Speed of the boat} = \left(\frac{63.4}{2} \times \frac{60}{1000}\right) \text{km/hr} = 1.9 \text{ km/hr}.$$

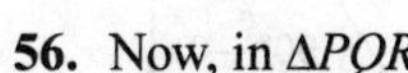

56. Now, in ΔPQR

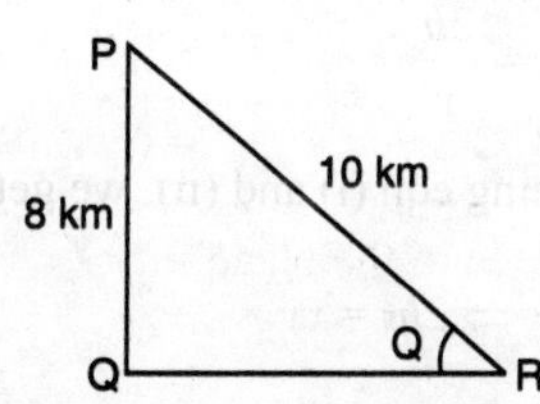

$$\sin\theta = \frac{PQ}{PR} = \frac{8}{10} = 0.8 \text{ km}$$

$$\therefore \sin 45° = \frac{1}{\sqrt{2}} = 0.71$$

$\sin 60° = \frac{\sqrt{3}}{2} = \frac{1.732}{2} = 0.87$

$0.71 \le 0.8 \le 0.87$

$45° \le \theta \le 60°$

57. In ΔPQR $\tan 60° = \frac{PQ}{QR}$

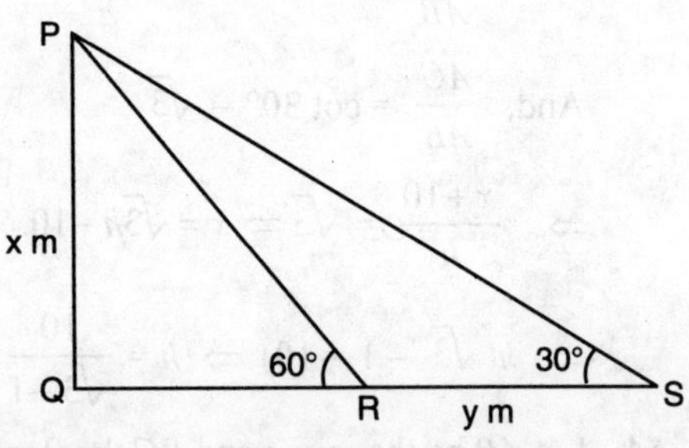

$\sqrt{3} = \frac{x}{QR} \quad \Rightarrow QR = \frac{x}{\sqrt{3}}$

In ΔPQS $\tan 30° = \frac{PQ}{QS}$

$\Rightarrow \frac{1}{\sqrt{3}} = \frac{x}{\frac{x}{\sqrt{3}} + y}$

$\Rightarrow x + \sqrt{3}y = 3x$

$\Rightarrow 2x = \sqrt{3}y$

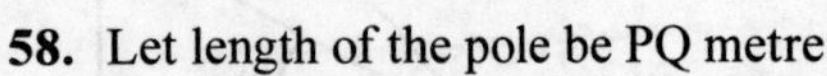

58. Let length of the pole be PQ metre

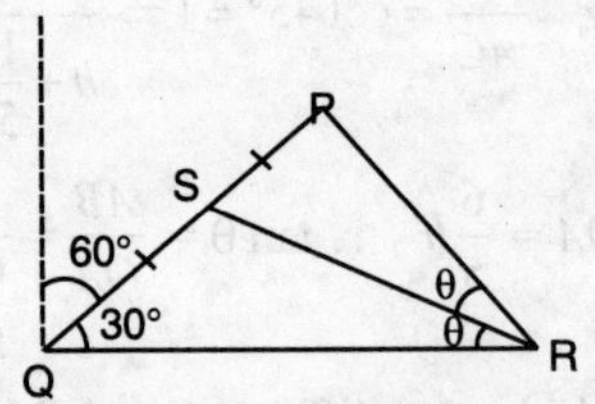

$\Rightarrow \angle PRS = \angle SRQ = Q$

$PS = SQ$

Now, in ΔPSR and ΔSRQ are congruents

$\therefore \angle PSR = \angle QSR$

$\angle PSR + \angle QSR = 180°$ [liner pair]

$\Rightarrow 2\angle PSR = 180° \Rightarrow \angle PSR = 90° = \angle QRS$

In ΔQSR

$\angle QRS = 180° - 90° - 30° = 60° = Q$

Now, $\sin 60° = \frac{QS}{QR} \Rightarrow \frac{\sqrt{3}}{2} = \frac{QS}{x} \Rightarrow QS = \frac{\sqrt{3}}{2}x$

$\therefore PQ = 2 \times QS = 2 \times \frac{\sqrt{3}}{2}x = \sqrt{3}x$

59. $\tan\theta = \frac{h}{x}$ ...(i)

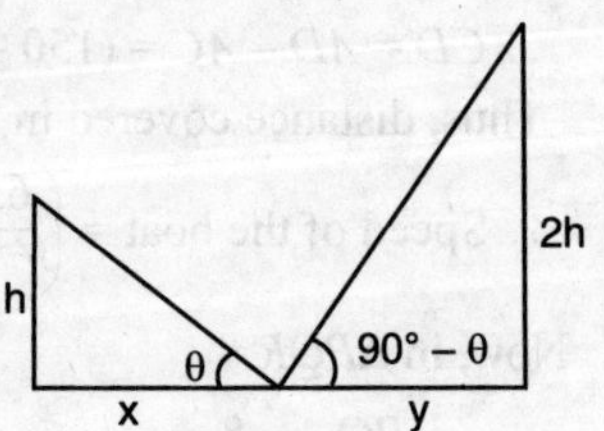

and $\tan(90° - \theta) = \frac{2h}{y}$

$\Rightarrow \cot\theta = \frac{2h}{y}$ ...(ii)

Multiplying eqn (i) and (ii), we get

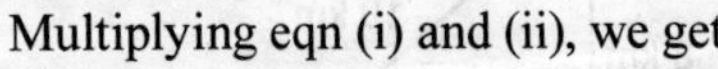

$1 = \frac{2h^2}{xy} \Rightarrow 2h^2 = xy$

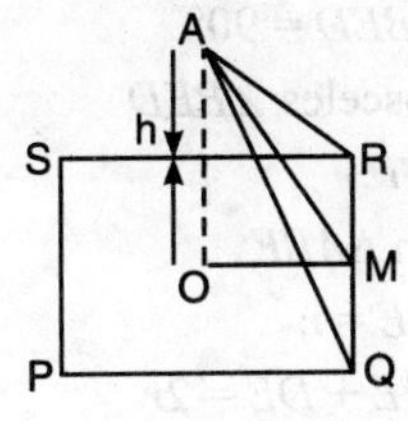

60. $OM = \dfrac{1}{2}$

$\angle QAR = 60°$

$AR = AQ$ [$\therefore$ symmetry]

AQR is an equilateral triangle of side l

AM = attitude of an equilateral triangle = $\dfrac{\sqrt{3}}{2}l$

In ΔMAO

$AO^2 = AM^2 - OM^2$ [by Pythagoras theorem]

$\Rightarrow h^2 = \left(\dfrac{\sqrt{3}\,l}{2}\right)^2 - \left(\dfrac{l}{2}\right)^2 \quad \Rightarrow h^2 = \dfrac{2l^2}{4} = \dfrac{l^2}{2}$

61. In ΔQRS

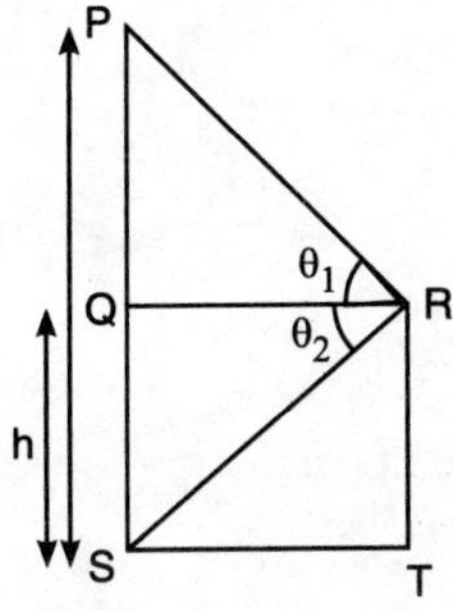

$\tan Q_2 = \dfrac{QS}{QR}$

$\Rightarrow QR = \cot Q_2\, QS = h \cot Q_2$

In ΔPQR

$\tan Q_1 = \dfrac{PQ}{QR}$

$PQ = OR \tan Q_1$

$(H - h) = h \cot Q_2 \times \tan Q_1$ ...(i)

$H = h(\cot Q_2 \tan Q_1 + 1)$ ...(ii)

Now, height changes to h

$\therefore$ angle interchanged

$H = h^1 (\cot Q_1 \tan Q_2 + 1)$ ...(iii)

now, from(ii) and (iii)

$h(\cot Q_2 \tan Q_1 + 1) = h^1 (\cot Q_1 \tan Q_2 + 1)$

$\Rightarrow h\, \dfrac{(\tan Q_1 + \tan Q_2)}{\tan Q_2} = h^1 \left(\dfrac{\tan Q_1 + \tan Q_1}{\tan Q_1}\right) \Rightarrow h^1 = \dfrac{h}{\tan Q_2} \times \tan Q_1$

$\Rightarrow h \cot Q_2 \tan Q_1 = H = h$ [from eqn (i)] $\Rightarrow h^1 = H - h$

Now, further heights to climb $h^1 - h$

$= H - h - h$

$= (H - 2h)$ units

62. In ΔABD

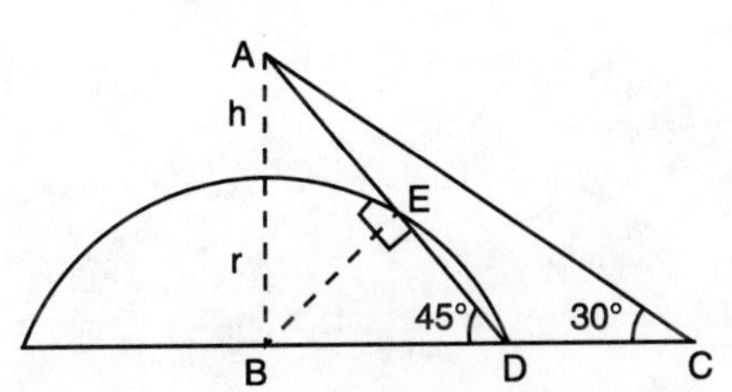

$\tan 45° = \dfrac{AB}{BD}$

$\Rightarrow 1 = \dfrac{h + r}{BD}$

$\Rightarrow BD = h + r$

Let $BE = r$

$\angle BEA = \angle BED = 90°$

Now, in isosceles $\angle BED$

$BE = ED = r$

Similarly in ΔABE

$\Rightarrow\ AE = BE = r$

$\therefore\ AD = AE + DE = 2r$

Now, in ΔABD

$AD^2 = AB^2 + BD^2$ [using Pythagoras theorem]

$\Rightarrow\ (2r)^2 = (h + r)^2 + (h + r)^2 \ \Rightarrow\ 4r^2 = 2(h + r)^2$

$\Rightarrow 2r^2 = (h + r)^2 \quad \Rightarrow r\sqrt{2} = (h + r)$

$\Rightarrow h = r(\sqrt{2} - 1) \ \Rightarrow \dfrac{h}{r} = \sqrt{2} - 1$

UNIT 5

GEOMETRY

Lines and Angles: Theorems on properties of angles at a point and those on parallel lines.

Triangles: Sides and Angles of a triangle and their properties, Congruency of triangles, Similar Triangles, Concurrence of medians, altitudes, right bisectors of the sides and angle bisectors of a triangle.

Quadrilaterals & Parallelograms: Properties of angles, sides and diagonals of a parallelogram, a rectangle, a square, a rhombus and a trapezium.

Circles: Circle and its properties including tangents and normals.

Loci: Simple problems on Loci.

28 CHAPTER Lines and Angles

Highlights: Lines and Angles, Planes, Properties ol angles at a point, parallel lines.

Definitions and Review of Concepts

1. (*i*) A line contains infinitely many points.
   (*ii*) Through a given point, there pass an infinite number of lines.
   (*iii*) Given two distinct points *A* and *B*, there is one and only one line that contains both the points.
   (*iv*) Three or more than three points are said to be collinear if there is a line which contains them all.
   (*v*) Two lines having a common point are known as intersecting lines. Two lines can intersect at the most at one point.
   (*vi*) Two lines which are both parallel to the same line are parallel to each other.
   (*vii*) The part of a line with two end points *A* and *B*, is called a line segment $\overrightarrow{AB}$
   A B
   (*viii*) A part of a line which has only one end point is called a ray.
2. (*i*) *Inclination between two rays having a common endpoint is called an angle.*
   (*ii*) An angle θ is said to be:
   I. Acute, if $\theta < 90°$ *II. Obtuse*, if $\theta > 90°$
   III. Straight, if $\theta = 180°$ *IV. Right Angle*, if $\theta = 90°$
   V. Reflex, if $180° < \theta < 360°$.
   (*iii*) Complement of θ is $(90° - \theta)$. Supplement of θ is $(180° - \theta)$
3. (*i*) **Adjacent Angles:** *Two angles are said to be adjacent, if they have a common vertex and a common arm, and the other arm of one angle is on one side of the common arm and that of the other is on the opposite side.*

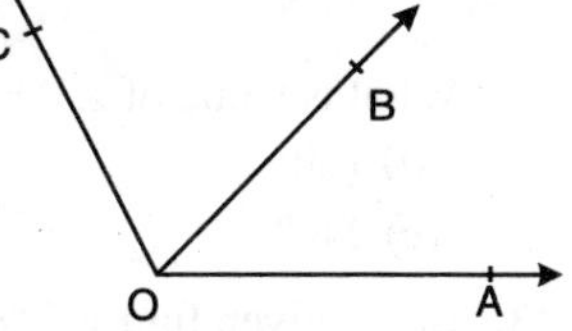

   (*ii*) **Linear Pair:** *Two adjacent angles are said to form a linear pair of angles, if their non-common arms: are two opposite rays.*
   (*iii*) The sum of a linear pair of angles is 180°.
   (*iv*) The sum of all the angles around a point is 360°.
4. **Vertically opposite Angles:** If two lines intersect, then vertically opposite angles are equal.
5. Suppose $AB \parallel CD$ and a transversal *EF* cuts them.
   Then:
   I. Pairs of corresponding $\angle s$ are:
   $(\angle 1 \ \& \ \angle 5)$, $(\angle 2 \ \& \ \angle 6)$, $(\angle 4 \ \& \ \angle 8)$, $(\angle 3 \ \& \ \angle 7)$,

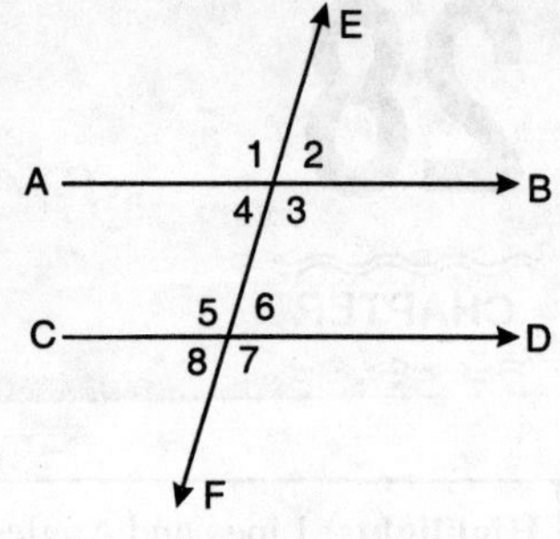

II. Pairs of Alternate angles are:

($\angle 3$ & $\angle 5$), ($\angle 4$ & $\angle 6$).

III. Pairs of interior angles on the same side of the transversal are: ($\angle 3$ & $\angle 6$), ($\angle 4$ & $\angle 5$).

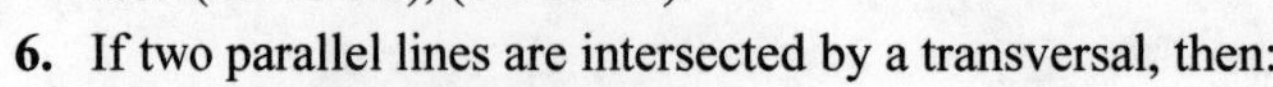

6. If two parallel lines are intersected by a transversal, then:
   (*i*) each pair of corresponding angles are equal;
   (*ii*) each pair of alternate angles are equal;
   (*iii*) interior angles on the same side of the transversal are supplementary.

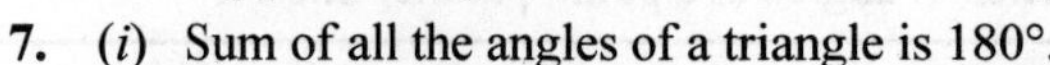

7. (*i*) Sum of all the angles of a triangle is 180°.
   (*ii*) If one side of a triangle is produced, then the exterior angle so formed is equal to sum of the interior opposite angles.

EXERCISE 28

1. In the given figure, if $\dfrac{y}{x} = 6$ and $\dfrac{z}{x} = 5$, then what is the value of x?

   (*a*) 45° (*b*) 30°
   (*c*) 15° (*d*) 10°

   [CDS (I) Exam, 2020]

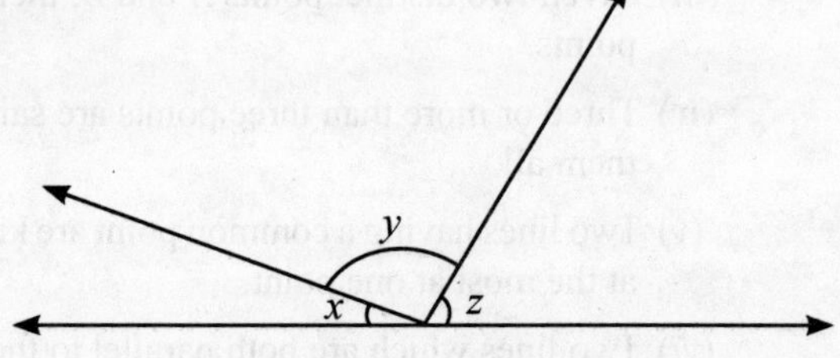

2. Angles are shown in the given figure. **[CDS (I) Exam, 2020]**

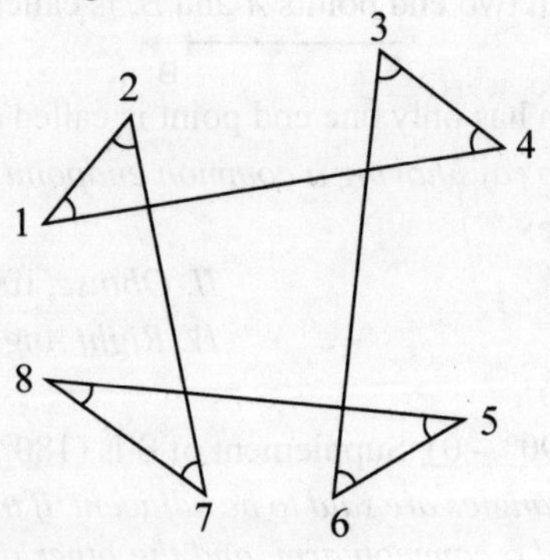

What is value of $\angle 1 + \angle 2 + \angle 3 + \angle 4 + \angle 5 + \angle 6 + \angle 7 + \angle 8$?

(*a*) 120° (*b*) 360°
(*c*) 240° (*d*) 720°

3. In the given figure PQ is parallel to RS, $\angle AEF = 90°$, $\angle BHS = 110°$, and $\angle ABC = x°$. Then what is the value of x? **[CDS (I) Exam, 2020]**

   (*a*) 15 (*b*) 25
   (*c*) 30 (*d*) 35

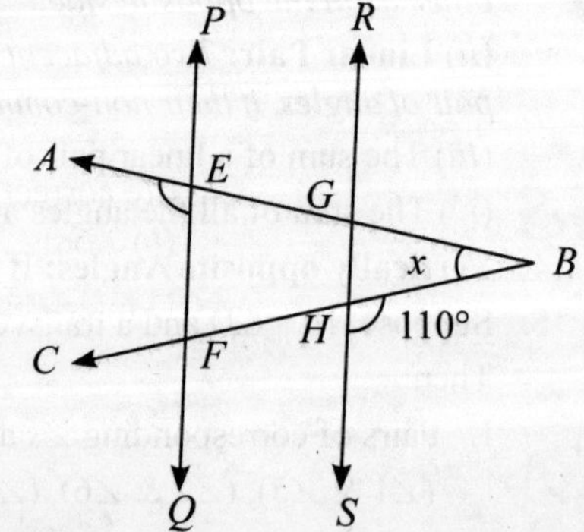

4. In the given figure AB is parallel to CD and AC is parallel to BD. If $\angle EAC = 40°$, $\angle FDG = 55°$, $\angle HAB = x°$, then what is the value of x?

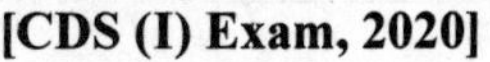

[CDS (I) Exam, 2020]

(*a*) 85 (*b*) 80
(*c*) 75 (*d*) 65

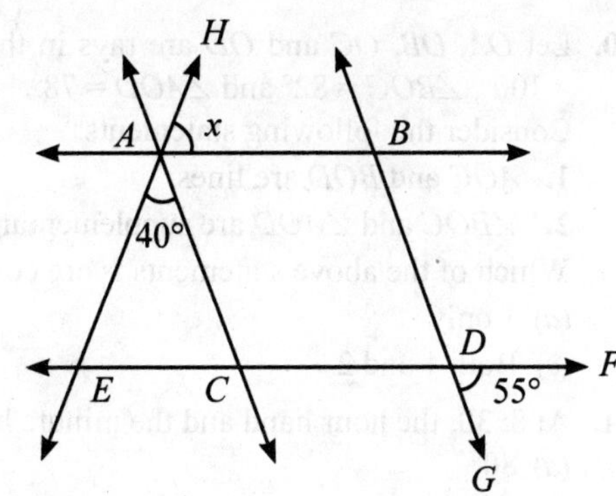

5. The difference between two angles is 15° and the sum of the angles in radian is $\frac{5\pi}{12}$. The bigger angle is k times the smaller angles. What is k equal to ? **[CDS (II) Exam, 2020]**

(*a*) $\frac{4}{3}$ (*b*) $\frac{3}{2}$ (*c*) $\frac{6}{5}$ (*d*) $\frac{7}{6}$

6. Three parallel lines x, y and z are cut by two transversals m and n. Transversal m cuts the lines x, y, z at P, Q, R respectively; and transversal n cuts the line x, y z and L, M, N respectively. If $PQ = 3$ cm, $QR = 9$ cm and $MN = 10.5$ cm, then what is the length of LM? **[CDS (I) Exam, 2019]**

(*a*) 3 cm (*b*) 3.5 cm (*c*) 4 cm (*d*) 4.5 cm

7. The length of a line segment AB is 2 cm. It is divided into two parts at a point C such that $AC^2 = AB \times BC$. What is the length of CB? **[CDS (I) Exam, 2018]**

(*a*) 0 $3\sqrt{5}$ cm (*b*) $3-\sqrt{5}$ cm (*c*) $3\sqrt{5}$ cm (*d*) $\sqrt{5}-1$ cm

8. In the figure given below, p, q, r are parallel lines; I and II m are two transversals.

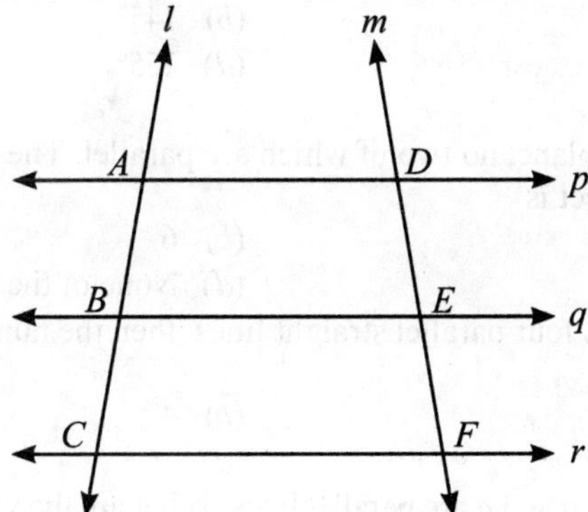

1. AB : AC = DE ; EF **[CDS (I) Exam, 2018]**
2. AB × EF = BC × DE

Which of the above is/are correct?

(*a*) 1 only (*b*) 2 only (*c*) Both 1 and 2 (*d*) Neither 1 nor 2

9. If two lines AB and CD intersect at O such that ∠AOC = 5 ∠AOD, then the four angles at O are **[CDS (I) Exam, 2018]**

(*a*) 40°, 40° 140° 140° (*b*) 30°, 30° 150° 150°
(*c*) 30°, 45° 75° 210° (*d*) 60°, 60° 120° 120°

10. Let OA, OB, OC and OD are rays in the anticlockwise direction such that $\angle AOB = \angle COD = 100°$, $\angle BOC = 82°$ and $\angle AOD = 78°$. **[CDS (I) Exam, 2015]**

Consider the following statements:

1. AOC and BOD are lines.

2. $\angle BOC$ and $\angle AOD$ are supplementary.

Which of the above statements is/are correct?

(*a*) 1 only (*b*) 2 only
(*c*) Both 1 and 2 (*d*) Neither 1 nor 2

11. At 8: 30, the hour hand and the minute hand of a clock form an angle of

(*a*) 80° (*b*) 75°
(*c*) 70° (*d*) 60° **[CDS (I) Exam, 2015]**

12. The complement angle of 80° is **[CDS (I) Exam, 2015]**

(*a*) $\frac{18}{\pi}$ radian (*b*) $\frac{5\pi}{9}$ radian
(*c*) $\frac{\pi}{18}$ radian (*d*) $\frac{9}{5\pi}$ radian

13. If X is any point within a square $ABCD$ and on AX a square $AXYZ$ is present, which of the following is/are correct? **[CDS (II) Exam, 2015]**

1. $BX = DZ$ or $BZ = DX$

2. $\angle ABX = \angle ADZ$ or $\angle ADX = \angle ABZ$

Select the correct answer using the code given below:

(*a*) 1 only (*b*) 2 only
(*c*) Both 1 and 2 (*d*) Neither 1 nor 2

14. A clock is started at noon. By 10 minutes past 5, through what angles, the hour hand moves?

(*a*) 160° (*b*) 145°
(*c*) 150° (*d*) 155°

[CDS (II) Exam, 2015]

15. There are five lines in a plane, no two of which are parallel. The maximum number of points in which they can intersect is **[CDS (I) Exam, 2016]**

(*a*) 4 (*b*) 6
(*c*) 10 (*d*) None of the above

16. If a transversal intersects four parallel straight lines, then the number of distinct values of the angles formed will be **[CDS (I) Exam, 2016]**

(*a*) 2 (*b*) 4
(*c*) 8 (*d*) 16

17. In the figure given above, p and q are parallel lines. What are the values of the angles x, y and z?

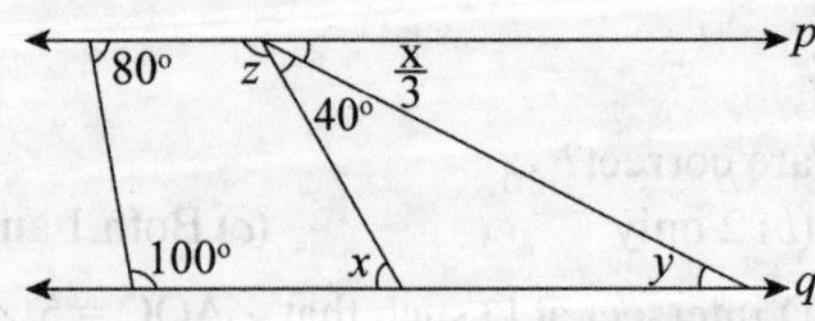

(*a*) $x = 80°, y = 40°, z = 100°$ (*b*) $x = 80°, y = 50°, z = 105°$
(*c*) $x = 70°, y = 40°, z = 110°$ (*d*) $x = 60°, y = 20°, z = 120°$

[CDS (I) Exam, 2016]

18. In the figure given below, PQ is parallel to RS and PR is parallel to QS. If ∠LPR = 35° and ∠UST = 70°, then what is ∠MPQ equal to?

(*a*) 55° (*b*) 70°
(*c*) 75° (*d*) 80°

[CDS (I) Exam, 2017]

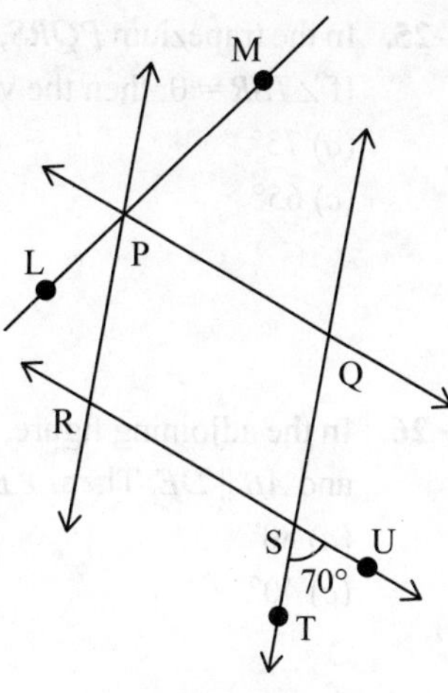

19. Two straight lines *AB* and *CD* cut each other at *O*. If ∠*BOD* = 63°, then ∠*BOC* is:

(*a*) 63° (*b*) 117° (*c*) 17° (*d*) I53“

20. In the given figure, *AOB* is a straight line. If ∠*AOC* + ∠*BOD* = 85°, then ∠*COD* is:

(*a*) 85° (*b*) 90°
(*c*) 95° (*d*) 100°

21. In the given figure, the value of *x*, that would make *POQ* a straight line, is:

(*a*) 50° (*b*) 44°
(*c*) 34° (*d*) 33°

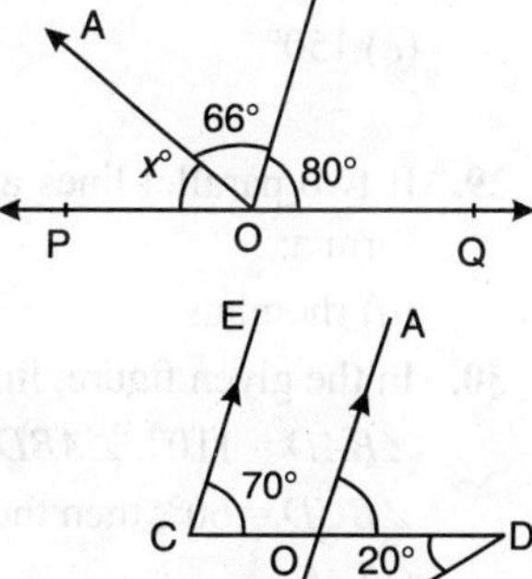

22. In the given figure, if *EC* || *AB*, ∠*ECD* = 70° and ∠*BDO* = 20°, then ∠*OBD* is:

(*a*) 20° (*b*) 50°
(*c*) 60° (*d*) 70°

23. In the given figure, if *AB* || *CD*, then ∠*FXE* is equal to:

(*a*) 30° (*b*) 50°
(*c*) 70° (*d*) 80°

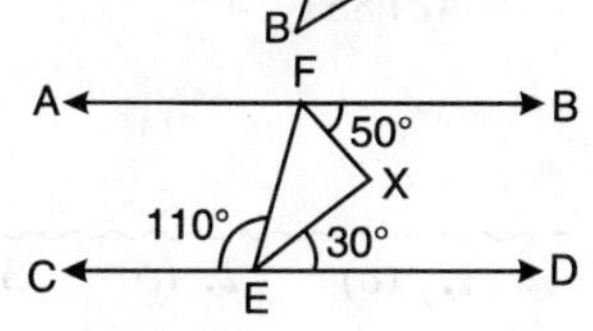

24. In the given figure *AB* || *CD*, ∠*A* = 128°, ∠*E* = 144°. Then, ∠*FCD* is equal to:

(*a*) 72° (*b*) 64°
(*c*) 136° (*d*) 92°

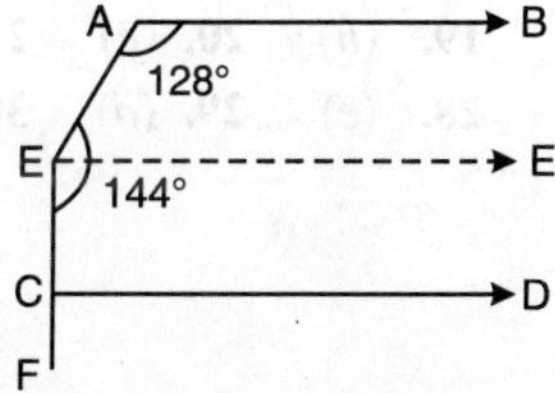

25. In the trapezium $PQRS$, $QR \parallel PS$, $\angle Q = 90°$, $PQ = QR$ and $\angle PRS = 20°$. If $\angle TSR = \theta$, then the value of θ is:

(*a*) 75° (*b*) 55°
(*c*) 65° (*d*) 45°

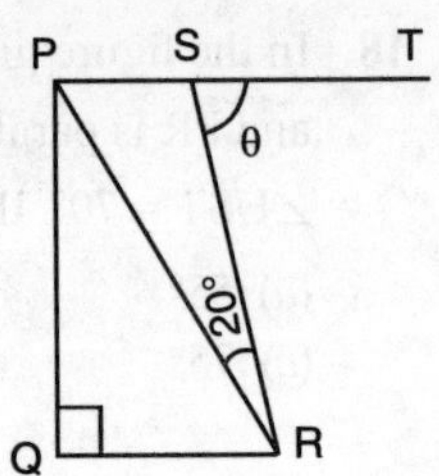

26. In the adjoining figure, $\angle ABC = 100°$, $\angle EDC = 120°$ and $AB \parallel DE$. Then, $\angle BCD$ is equal to:

(*a*) 80' (*b*) 60°
(*c*) 40° (*d*) 20°

27. In the given figure, $AB \parallel CD$, $\angle ABO = 40°$ and $\angle CDO = 30°$. If $\angle DOB = x°$, then the value of x is:

(*a*) 35 (*b*) 110
(*c*) 70 (*d*) 140

28. In the given figure, $AB \parallel CD$, $\angle ALC = 60°$, EC is the bisector of $\angle LCD$ and $\angle EF \parallel AB$. Then, $\angle CEF$ is equal to:

(*a*) 120° (*b*) 140°
(*c*) 150° (*d*) None of these

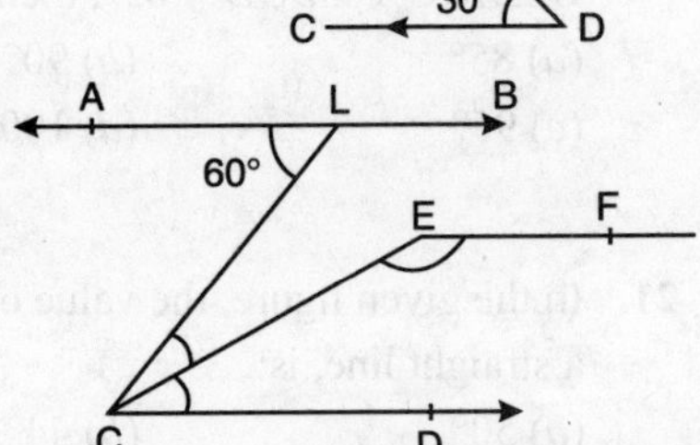

29. If two parallel lines are intersected by a transversal then the bisectors of the interior angles form a:

(*a*) rhombus (*b*) ll gm (*c*) square (*d*) rectangle

30. In the given figure, line CE is drawn parallel to DB. If $\angle BAD = 110°$, $\angle ABD = 30°$. $\angle ADC = 75°$ and $\angle BCD = 60°$, then the value of x is:

(*a*) 45 (*b*) 75
(*c*) 85 (*d*) 120

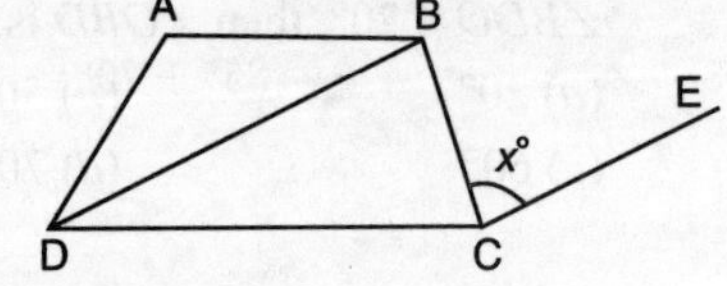

ANSWERS

1. (*c*) **2.** (*a*) **3.** (*b*) **4.** (*a*) **5.** (*b*) **6.** (*b*) **7.** (*b*) **8.** (*c*) **9.** (*b*)
10. (*d*) **11.** (b) **12.** (c) **13.** (c) **14.** (d) **15.** (c) **16.** (a) **17.** (d) **18.** (c)
19. (*b*) **20.** (*c*) **21.** (*c*) **22.** (*b*) **23.** (*d*) **24.** (*d*) **25.** (*c*) **26.** (*c*) **27.** (*c*)
28. (*c*) **29.** (*d*) **30.** (*c*)

SOLUTIONS

1. We have, $\frac{y}{x} = 6$

$\Rightarrow y = 6x$ and $\frac{z}{x} = 5$

$\Rightarrow z = 5x$

Now, $x + y + z = 180°$ [$\because$ linear pair]

$\Rightarrow x + 6x + 5x = 180°$

$\Rightarrow 12x = 180°$

$\Rightarrow x = 15°$

2. We have,

$\angle A + \angle 1 + \angle 2 = 180$

$\Rightarrow \angle A = 180° - (\angle 1 + \angle 2)$ ...(i)

Similarly,

$\angle B = 180° - (\angle 3 + 4)$ ...(ii)

$\angle C = 180° - (\angle 5 + \angle 6)$ ...(iii)

$\angle D = 180° - (\angle 7 + \angle 8)$ ...(iv)

adding (i), (ii), (iii) and (iv), we get

$\angle A + \angle B + \angle C + \angle D = 540° - (\angle 1 + \angle 2 + \angle 3 + \angle 4 + \angle 5 + \angle 6 + \angle 7 + \angle 8)$

$\Rightarrow 360° = 540° - (\angle 1 + \angle 2 + \angle 3 + \angle 4 + \angle 5 + \angle 6 + \angle 7 + \angle 8)$

[$\because$ sum of all angles of quadrilateral = 360°]

$\Rightarrow \angle 1 + \angle 2 + \angle 3 + \angle 4 + \angle 5 + \angle 6 + \angle 7 + \angle 8 = 540° - 360° = 120°$

3. We have, $PQ \parallel RS$

$\angle AEF = \angle EGH = 95°$ [corresponding angle]

$\therefore \angle GEF = 180° - 95° = 85°$ [$\because$sum of interior angles on the same side is 180°]

$\angle BHS = \angle HFQ = 110°$ [corresponding angles]

$\Rightarrow \angle BFE = 180° - 110° = 70°$

Now, $\angle ABC = 180° - (\angle BEF + \angle BFE)$

$= 180° - (85° + 70°)$

$= 180° - 155° = 25°$

4. $\angle FDG = \angle$BDC = 55° [vertically opposite angles]

$\angle CAB = 55°$ [$\because$ interior opposite angles]

$x + 40° + \angle CAB = 180° \Rightarrow x + 40° + 55° = 180° \Rightarrow x = 180° - 95° = 85°$

5. Let two angles are $\angle A$ and $\angle B$

$\angle A - \angle B = 15°$ ...(i)

$\angle A + \angle B = \frac{5 \times 180°}{12} = 75°$ ...(ii)

Adding (i) and (ii), we get

$2\angle A = 90 \Rightarrow \angle A = 45°$

$\angle B = 45° - 15° = 30°$

Now, according to the question

$45° = k\, 30° \Rightarrow k = \frac{3}{2}$

6.

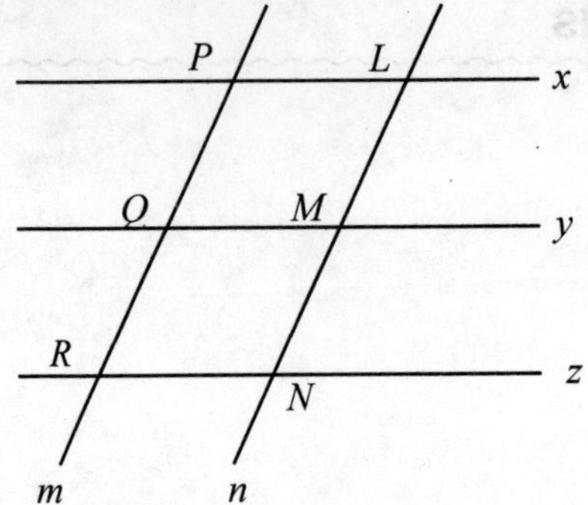

Now, according to the question

$$\frac{PQ}{QR} = \frac{LM}{MN} \Rightarrow \frac{3}{9} = \frac{LM}{10.5} \Rightarrow LM = 3.5$$

7 Let $BC = x$

$AC = AB - CB = 2 - x$

Now, $AC^2 = AB \times CB$

$(2-x)^2 = 2x \Rightarrow 4 + x^2 - 4x = 2x \Rightarrow x^2 - 6x + 4 = 0$

Using quadratic formula

$a = 1, b = -6, c = 4$

$$x = \frac{6 \pm \sqrt{36-16}}{2 \times 1} = \frac{6 \pm \sqrt{20}}{2} = \frac{6 \pm 2\sqrt{5}}{2} = 3 \pm \sqrt{5}$$

$\Rightarrow CB = 3 - \sqrt{5}$

8. From basic proportionality theorem

$$\frac{AB}{BC} = \frac{DF}{EF}$$

$\Rightarrow AB \times EF = BC \times DE$

Thus, statement – 2 is true

Again, $\frac{AB}{BC} = \frac{DE}{EF} \Rightarrow \frac{BC}{AB} = \frac{EF}{DE}$

Now, adding (1) on both sides, we get

$$\frac{AC}{AB} = \frac{DF}{DE} \Rightarrow \frac{AB}{AC} = \frac{DE}{DF}$$

Thus, statement-1 is true.

9. Let $\angle AOD = x$

$\angle AOC = 5x$

$\therefore \angle AOD + \angle AOC = 180°$ [$\because$ CD is straight line]

$\Rightarrow x + 5x = 180°$

$\Rightarrow 6x = 180°$

$\Rightarrow x = 30°$

$\therefore \angle AOD = 30° = \angle BOC$ [$\because$ vertically opposite angle]

$\angle AOC = 5 \times 30° = 150 = \angle BOD$

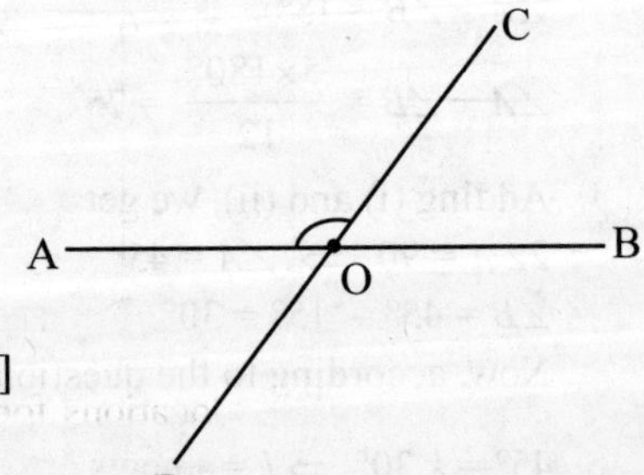

10.

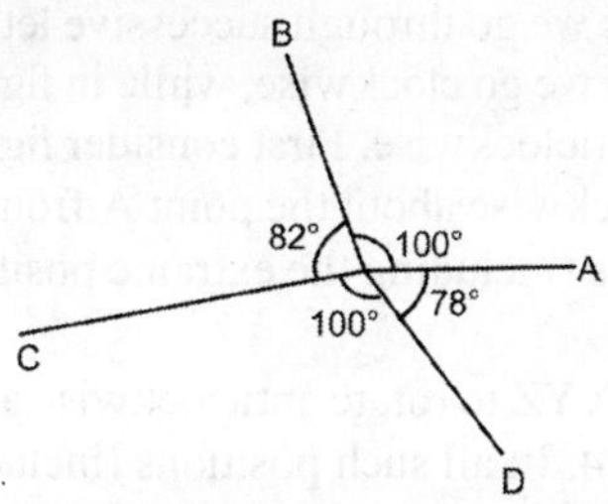

AOC and BOD are not straight lines since
$\angle AOD \neq \angle BOC$ (1) is false.
$\angle BOC + \angle AOD = 78 + 82° = 160°$.
Sum of angles is not 180°. Hence the angles are not supplementary. (2) is false.

11. Angle swept by hour hand in 60 mins = 30°.

Angle swept by hour hand in 1 min = $\frac{1}{2}°$.

$\therefore$ Angle made by hour hand (measured from 12:00) = Number of hours (30°)

$+\frac{1}{2}°$ number of minutes $= = 8(30°) + \frac{1}{2}(30°) = 255°$

Angle made by minute hand (measured from 12 : 00) = $360°\left(\frac{30°}{60}\right) = 180°$.

$\therefore$ Angle between minute hand and hour hand = 255° – 180° = 75°.

12. Complement angle of 80° = 90° – 80° = 10°
180° = π radians

$1° = \frac{\pi}{180°}$ radians

$10° = \frac{\pi}{180} \times 10$ radians $= \frac{\pi}{18}$ radians.

13.

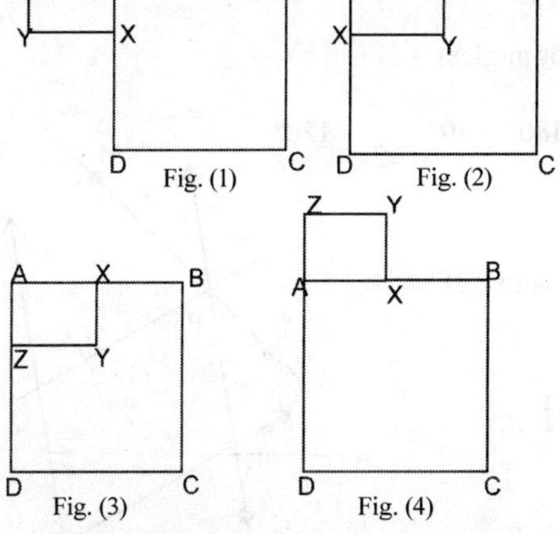

Fig. (1) Fig. (2) Fig. (3) Fig. (4)

Various possible locations for X with respect to the bigger square are shown in the figures. Which pairs of segments are equal (or which pair of angles are equal) would depend on how we label the vertices.

In all cases, for the bigger square, we are going clockwise as we go through successive letters. In figures 1 and 2, the points A, X, Y, Z occur in that order if we go clockwise, while in figures 3 and 4, the points A, X, Y, Z occur in that order if we go anticlockwise. First consider figures 1 and 2. We can imagine the square AXYZ to rotate anticlockwise about the point A from the position in Fig 1. to the position in Fig 2. In all such positions (including the extreme positions shown), we see that BX = DZ and ∠ABX = ∠ADZ.

Now consider figures 3 and 4. We can imagine the square AXYZ to rotate anticlockwise about the point A from the position in Fig 3 to the position in Fig 4. In all such positions (including the extreme positions shown), BZ = DX and ∠ADX = ∠ABZ.

∴ Both statements are correct.

Note: In the examination, it might not be vital to understand clearly in which cases BX = DZ and in which cases BZ = DX. We can consider any one figure and satisfy ourselves that the disjunction BX = DZ or BZ = DX is true.

14. The hour hand moves $\frac{1}{12}(360°)$ or 30° every 1 hour.

Speed of hour hand $= \frac{30}{60}$ deg / min $= \frac{1}{2}$ deg / minute.

From 12 : 00 noon to 5 : 10, the hour hand moves $(310)\left(\frac{1}{2}\right)^{\circ} = 155°$.

15. Five line can be considered as below

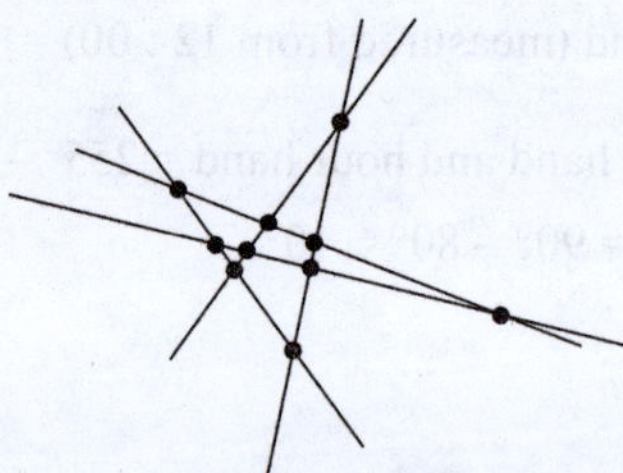

∴ Maximum five lines can intersect in 10 points

16. If a transversal intersects 4 parallel straight lines, then the number of distinct values of angles formed is 2. (If the transversal is perpendicular two the lines, all the angles would be right angles.)

17. From the figure $\frac{x}{3} + 40° = x$ (Alternate interior angles)

∴ $x = 60°$ Also $y = x - 40° = 20°$ and $z = 180° - 40° - \frac{x}{3} = 120°$

18. $\angle UST = 70°$

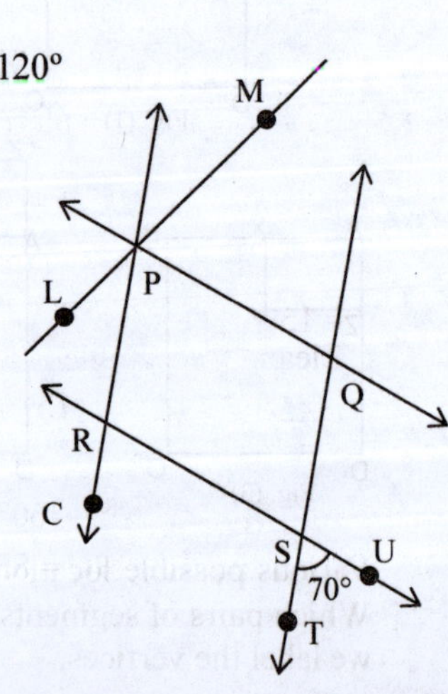

∴ $\angle QSR = \angle UST = 70°$ (vertically opposite ∠s)

$\angle QSR = \angle SRC = 70°$ (Alternate ∠s)

$\angle SRC = \angle RPQ = 70°$ (corresponding ∠s)

LPM is a straight line

∴ $\angle LPR + \angle RPQ + \angle MPQ = 180°$

$\Rightarrow 35° + 70° + \angle MPQ = 180°$

$\Rightarrow \angle MPQ = 180° - 105° = 75°$

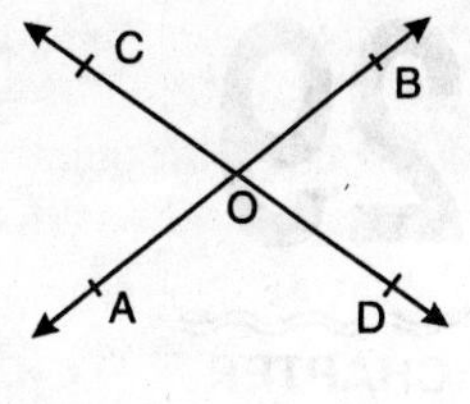

19. Since COD is a straight line, we have:
$\angle BOC + \angle BOD = 180°$. So, $\angle BOC = (180° - 63°) = 117°$.

20. Clearly,
$\angle AOC + \angle BOC = 180°$ i.e., $\angle AOC + \angle COD + \angle BOD = 180°$
$\therefore\ 85° + \angle COD = 180°$. So, $\angle COD = (180° - 85°) = 95°$.

21. POQ will be a straight line,
if $80 + 66 + x = 180$, i.e.$x = 34$.

22. $\angle AOD = \angle ECO = 70°$ (Corr. $\angle s$)
In $\Delta\ OBD$, Ext. $\angle AOD = \angle OBD + \angle BDO$.
$\therefore\ \angle OBD + 20° = 70°$. S0, $\angle OBD = 50°$.

23. $\angle BFE = \angle CEF = 110°$ (alt $\angle s$).
So, $\angle XFE = \angle BFE - \angle BFX = (110° - 50°) = 60°$.
$110° + \angle FEX + 30° = 180° \Rightarrow \angle FEX = 40°$.
Now, $\angle XFE + \angle FEX + \angle FXE = 180° \Rightarrow 60° + 40° + \angle FXE = 180°$.
$\therefore\ \angle FXE = 80°$.

24. Through E draw $EE'' \parallel AB \parallel CD$.
Then, $\angle AEE' = 180° - \angle BAE = (180° - 128°) = 52°$.
$\angle E'EC = (144° - 52°) = 92°$.
$\angle FCD = \angle E'EC = 92°$ (Corr. $\angle s$).

25. $PQ = QR$ and $\angle PQR = 90° \Rightarrow \angle QPR = \angle QRP = 45°$.
$\therefore\ \angle QRS = (45° + 20°) = 65°$.
$\therefore\ \theta = \angle QRS = 65°$ (alt. $\angle s$)

26. Produce AB to meet CD at F.
$\angle BFD = \angle EDF = 120°$ (alt. $\angle s$)
$\angle BFC = (180° - 120°) = 60°$.
$\angle CBF = (180° - 100°) = 80°$.
$\therefore\ \angle BCF = 180° - (60° + 80°) = 40°$.

27. Through O draw EOE' parallel to AB & so to CD.
$\therefore\ \angle BOE' = \angle ABO = 40°$ (alt. $\angle s$)
$\angle E'OD = \angle COD = 30°$ (alt. $\angle$s)
$\therefore\ \angle BOD = (40° + 30°) = 70°$. So, $x = 70$.

28. $\angle LCD = \angle ALC = 60°$ (alt. $\angle s$)
$\angle DCE = \frac{1}{2}\ \angle LCD = 30°$.
$\therefore\ \angle FEC = (180° - 30°)$
$= 150°$. (int. opp. $\angle s$)

29. Clearly, a rectangle is formed.

30. $\angle ADB = 180° - (110° + 30°) = 40°$.
So, $\angle BDC = (75° - 40°) = 35°$.
$\therefore\ \angle DBC = 180° - (60° + 35°) = 85°$.
$\therefore\ \angle BCE = \angle DBC = 85°$ (alt. $\angle s$).
So, $x = 85$.

29

CHAPTER

Polygons

Definitions & Useful Concepts

(*i*) **Polygon:** *It is a closed plane figure bounded by some straight lines.*

A polygon is called a *Triangle, Quadrilateral, Pentagon, Hexagon, Heptagon, Octagon, Nonagon* and *Decagon* according as it contains 3, 4, 5, 6, 7, 8, 9, 10 sides respectively.

(*ii*) **Convex & Concave Polygon:** *A polygon in which none of its interior angles is more than 180°, is known as a convex polygon.*

On the other hand, *if at least one angle ofa polygon is more than I80°, then it is said to be concave.*

(*iii*) **Regular Polygon:** *A polygon having all sides equal and all angles equal is called a regular polygon.*

I. *Each exterior angle of a regular polygon* $= \left(\dfrac{360}{\text{Number of sides}}\right)^0$.

II. *Each interior angle* = 180° – (*exterior angle*).

(*iv*) **In a convex polygon of *n* sides, we have:**

I. *Sum of all interior angles* = (2*n* – 4) *right angles*;

II. *Sum of all exterior angles* = 4 *right angles*.

(*v*) Number of diagonals of a polygon of *n* sides $= \dfrac{n(n-1)}{2} - n.$.

EXERCISE 29

1. What is the ratio of interior angle to exterior angle of a regular polygon of *n* sides?

(*a*) n (*b*) $\dfrac{n-1}{2}$ (*c*) $\dfrac{n-2}{2}$ (*d*) $\dfrac{2(n-2)}{3}$

[CDS (I) Exam, 2022]

2. What is the magnitude (in radian) or the interior angle of a regular pentagon?

(*a*) $\dfrac{\pi}{5}$ (*b*) $\dfrac{2\pi}{5}$ (*c*) $\dfrac{3\pi}{5}$ (*d*) $\dfrac{4\pi}{5}$

[CDS (I) Exam, 2020]

3. If the sum of all interior angles of a regular polygon is twice the sum of all its exterior angles, then the polygon is **[CDS (I) Exam, 2020]**

(*a*) Hexagon (*b*) Octagon (*c*) Nonagon (*d*) Decagon

4. The corners of a square of side 'a' are cut away so as to form a regular octagon. What is the side of the octagon? **[CDS (I) Exam, 2019]**

(*a*) $a(\sqrt{2}-1)$ (*b*) $a(\sqrt{3}-1)$ (*c*) $\frac{a}{\sqrt{2}+2}$ (*d*) $\frac{1}{3}$

5. Given that the angles of a polygon are all equal and each angle is a right angle.

Statement-1 : The polygon has exactly four sides. **[CDS (I) Exam, 2018]**

Statement-2: The sum of the angles of a polygon having n sides is (3n – 8) right angles.

Which one of the following is correct in respect of the above statements?

(*a*) Both statement-1 and statement-2 are true the statement-2 is the correct explanation of statement-1

(*b*) Both statement-1 and statement-2 are true but statement-2 is not the correct explanation of statement-1

(*c*) statement-1 is true but statement-2 is false

(*d*) statement-1 is false but statement-2 is true

6. The number of sides of two regular polygons are in the ratio 5 : 4. The difference between their interior angles is 9°. Consider the following statement: **[CDS (II) Exam, 2018]**

1. One of them is a pentagon and the other is a rectangle.
2. One of them is a pentagon and the other is a rectangle.
3. The sum of the their exterior angles is 720°.

Which of the above statement is/are correct?

(*a*) 1 only (*b*) 2 only (*c*) 1 and 2 (*d*) 2 and 3

7. If the difference between the interior and exterior angles of a regular polygon is 144°, then what is the number of sides of the polygon ? **[CDS (I) Exam, 2024]**

(*a*) 12 (*b*) 16 (*c*) 18 (*d*) 20

8. If a star figure is formed by elongating the sides of a regular pentagon, the the measure of each angle at the angular points of the start figure is **[CDS (I) Exam, 2015]**

(*a*) 36° (*b*) 35° (*c*) 32° (*d*) 30°

9. A circle of radius *r* is inscribed in a regular polygon with *n* sides (the circle touches all sides of the polygon). If the perimeter of the polygon is *p*, then the area of the polygon is

(*a*) $(p+n)r$ (*b*) $(2-n)r$ (*c*) $\frac{pr}{2}$ (*d*) None of the above

[CDS (II) Exam, 2015]

10. How many right angled triangles can be formed by joining the vertices of a cuboid?

(*a*) 24 (*b*) 28 (*c*) 32 (*d*) None of the above

[CDS (II) Exam, 2015]

11. If each interior angle of a regular polygon is 140°, then the number of vertices of the polygon is equal to **[CDS (I) Exam, 2016]**

(*a*) 10 (*b*) 9 (*c*) 8 (*d*) 7

12. Consider the following statements:

1. If n ≥ 3 and m ≥ 3 are distinct positive integers, then the sum of the exterior angles of a regular polygon of m sides is different from the sum of the exterior angles of a regular polygon of n sides.

2. Let m, n be integers such that m > n ≥ 3. Then the sum of the interior angles of a regular polygon of m sides is greater than the sum of the interior angles of a regular polygon of n sides, and their sum is $(m+n)\frac{\pi}{2}$

Which of the above statements is/are correct? **[CDS (I) Exam, 2016]**

(*a*) 1 only (*b*) 2 only (*c*) Both 1 and 2 (*d*) Neither 1 nor 2

13. Consider the following statements:

1. There exists a regular polygon whose exterior angle is 70°. **[CDS (I) Exam, 2016]**

2. Let $n \geq 5$. Then the exterior angle of any regular polygon of n sides is acute.

Which of the above statements is/arc correct?

(*a*) 1 only (*b*) 2 only (*c*) Both 1 and 2 (*d*) Neither 1 nor 2

14. The area of a regular hexagon of side 'a' is equal to **[CDS (I) Exam, 2017]**

(*a*) $\frac{\sqrt{2}}{3}a^2$ square units (*b*) $\frac{3\sqrt{3}}{2}a^2$ square units

(*c*) $\frac{1}{3}a^2$ square units (*d*) $\frac{\sqrt{3}}{2}a^2$ square units

15. ABCDEF is a regular polygon. Two poles at C and D are standing vertically and subtend angles of elevation 30° and 60° at A respectively. What is the ratio of the height of the pole at C to that of the pole at D? **[CDS (I) Exam, 2017]**

(*a*) 1 : 1 (*b*) $1:2\sqrt{3}$ (*c*) $2\sqrt{3}:1$ (*d*) $2:\sqrt{3}$

16. The ratio of sum of interior angles to sum of exterior angles of a regular polygon of n sides is $\frac{7}{2}$. What is the measure of an interior angle of the polygon?

(*a*) 110° (*b*) 120° (*c*) 130° (*d*) 140°

17. Each interior angle of a regular polygon is 135°. The number of sides of the polygon is:

(*a*) 6 (*b*) 8 (*c*) 5 (*d*) 9

18. The ratio of the measure of an angle of a regular octagon to the measure of its exterior angle is:

(*a*) 1:2 (*b*) 1:3 (*c*) 2:3 (*d*) 3:l

19. The sum of the interior angles of a polygon is 1080°. The number of sides of the polygon is:

(*a*) 8 (*b*) 6 (*c*) 10 (*d*) 9

20. The angles of a pentagon are in the ratio l : 2 : 3 : 5 : 9. The smallest angle is:

(*a*) 72° (*b*) 45° (*c*) 54° (*d*) 27°

21. If each interior angle of a regular polygon is 11 times its exterior angle, the number of sides of the polygon is:

(*a*) 22 (*b*) 24 (*c*) 18 (*d*) 11

22. One angle of a hexagon is 100° and all the other five angles are equal. What is the measure of each one of the equal angles?

(*a*) 36° (*b*) 144° (*c*) 124° (*d*) 72°

23. The ratio between the number of sides of two regular polygons is 1 : 2 and the ratio between their interior angles is 2 : 3. The number of sides of these polygons are respectively:

(*a*) 4, 8 (*b*) 5, 10 (*c*) 6, 12 (*d*) 8, 16

24. How many diagonals are there in a hexagon?

(*a*) 6 (*b*) 4 (*c*) ll (*d*) 9

25. How many diagonals are there in a decagon?

(*a*) 19 (*b*) 29 (*c*) 35 (*d*) 45

26. A polygon has 27 diagonals. The number of sides of the polygon is:

(*a*) 9 (*b*) 10 (*c*) 11 (*d*) 12

27. The ratio of an interior angle to the exterior angle of a regular polygon is (7 : 2). The number of sides of the polygon is:

(*a*) 6 (*b*) 9 (*c*) 7 (*d*) 8

28. The sum of the interior angles of a regular polygon is twice the sum of its exterior angles. The polygon is:

(*a*) an octagon (*b*) a nonagon (*c*) a decagon (*d*) a hexagon

29. Each exterior angle of a regular polygon is 40°. The number of sides of the polygon is:

(*a*) 8 (*b*) 9 (*c*) 6 (*d*) 10

ANSWERS

1. (*c*)	**2.** (*c*)	**3.** (*a*)	**4.** (*a*)	**5.** (*c*)	**6.** (*d*)	**7.** (*d*)	**8.** (*a*)	**9.** (*c*)
10. (*d*)	**11.** (*b*)	**12.** (*d*)	**13.** (*b*)	**14.** (*b*)	**15.** (*b*)	**16.** (*d*)	**17.** (*b*)	**18.** (*d*)
19. (*a*)	**20.** (*d*)	**21.** (*b*)	**22.** (*c*)	**23.** (*a*)	**24.** (*c*)	**25.** (*c*)	**26.** (*a*)	**27.** (*b*)
28. (*d*)	**29.** (*b*)							

SOLUTIONS

1. Sum of Interior angle of a regular polygon of n sides = $(n-2) \times 180°$

Sum of exterior angle = 360°

$\therefore$ Required ratio = $\dfrac{(n-2)\times 180°}{360°} = \dfrac{n-2}{2}$

2. Number of sides of pentagon = 5

$\therefore$ Magnitude = $\dfrac{(n-2)\times\pi}{n} = \dfrac{(5-2)\pi}{n} = \dfrac{3\pi}{5}$

3. Now, according to the question,

$(n-2) \times 180° = 2 \times 360° \Rightarrow n = 2 = 4 \Rightarrow n = 6$

Thus, the polygon is Hexagon.

4. Let side of square = a

and length of regular octagon = y

In right angle ΔABC

$AC^2 = AB^2 + BC^2$

$\because AB = BC\ [\angle A = \angle C]$

$\therefore AC^2 = AB^2 + AB^2$

$\Rightarrow AC^2 = 2AB^2 \quad \Rightarrow 2AB^2 = y^2$

$\Rightarrow AB = \dfrac{y}{\sqrt{2}} \quad \Rightarrow 2n - 4 - n = 0 \Rightarrow n = 4$

Thus, statement –1 is true

a A 45° B C p

5. Statement-2

Sum of interior angles of a polygon = $(n-2) \times 180° \Rightarrow (2n-4) \times 90$

Thus, statement-2 is false

Statement–1 : $(2n-4) \times 90° = n \times 904$ [$\because$ all angles are 90°]

6. Let the number of sides of two regular polygons are $5x$ and $4x$ respectively

Now, according to the question

$$\frac{360°}{4x}-\frac{360°}{5x}=9° \Rightarrow \frac{1}{4x}-\frac{1}{5x}=\frac{9°}{360°} \Rightarrow \frac{1}{20x}=\frac{1}{40} \quad \Rightarrow x=2$$

$\therefore$ Number of side of 1st polygon = 5 × 2 = 10°

Number of sides of 2nd polygon = 4 × 2 = 8°

Thus statement-2 and 3 are correct.

Now, side of square $= y+\frac{y}{\sqrt{2}}+\frac{y}{\sqrt{2}}$

$$\Rightarrow a=\frac{\sqrt{2}y+2y}{\sqrt{2}} \Rightarrow a=\frac{\sqrt{2}y(1+\sqrt{2})}{\sqrt{2}} \Rightarrow a=y(1+\sqrt{2}) \Rightarrow y=\frac{a}{1+\sqrt{2}} \Rightarrow y=a(\sqrt{2}-1)$$

7. Let interior angle = x and exterior angle = y

$$\begin{aligned} x+y &= 180° \\ x-y &= 144° \\ - \quad + & \quad - \\ \hline \therefore \quad y &= 180° \end{aligned}$$

$$\therefore \quad n=\frac{360}{\text{Exterior angle}}=\frac{360°}{18}=20$$

8. Each interior angle of the pentagon $=\frac{180°\,(5-2)}{5}=108°$

The 'angular points' of the star figure are A, B, C, D, E. Each of these 5 angles is an angle of the triangle whose other two angles are exterior angles of the regular pentagon.

$\angle A=\angle B=\angle C=\angle D=\angle E=180°-2\,(180°-108°)=36°.$

9. The n-sided polygon can be divided into n triangles with O, the centre of the circle as one vertex for each triangle. The altitude of each triangle is r. Let the sides of the polygon be a_1, a_2,……..a_n (Given $a_1 = a_2 =$ …….. $= a_n$).

$\therefore$ The area of the polygon is $\frac{nar}{2}=\frac{pr}{2}$.

Note: This result would follow even if the polygon is not regular.

Area of the polygon $=\frac{a_1 r}{2}+\frac{a_2 r}{2}+\ldots\ldots+\frac{a_n r}{2}=\frac{pr}{2}$.

10. A cuboid has 8 vertices, 12 edges and 6 faces. By selectively choosing any 3 vertices of the cuboid, we can form 2 types of right triangles:

(*a*) Right triangles with all three vertices on the same face.

(*b*) Right triangles with 2 vertices along an edge and the third one along the diagonally opposite edge.

Number of triangles of type (*a*) $= 6\,{}^4C_3 = 24$

Number of triangles of type (*b*) $= {}^{12}C_1\,{}^2C_1 = 24$

Total number of right triangles = 48.

11. Each exterior angle of the polygon = 180 – 140 = 40°

$$\therefore \text{ The number of sides of he polygon } = \frac{360°}{\text{exterior angle}}=\frac{360}{40}=9$$

12. The sum of the exterior angles of any converse polygon is 360°. All regular polygon are convex and for all of them, the sum of the exterior angles is 360°.
Statement I is false.
Let the sum of the interior angles of any converse polygon with sides be S_i, then S_m (for any convex polygon and not merely a regular polygon) is greater than S_n. But the second part of statement 2 is false in gendered.
$S_m = (M-2)\,\pi$ and $S_n = (n-2)\,\pi$.
$\therefore\ S_m + S_n = (M+n-4)\pi$
This would be equal to $(m+n)\dfrac{\pi}{2}$ if and only if m + n = 8.
$\therefore$ Neither 1 nor 2 is true

13. For any regular polygon, if each of the exterior angles in 0°, then 360/θ hen to be an integer. As 360/70 is an integer. We cannot have such regular polygon.
Statement: 1 is false.
If $n \geq 5$, then the exterior angle of any regular polygon of n sides is acute. So, statement 2 is true.

14. Area of regular hexagon

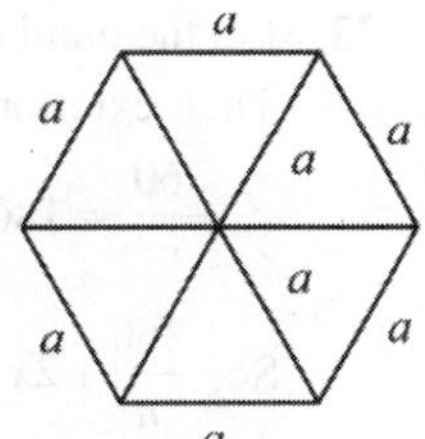

$= 6\times\dfrac{\sqrt{3}}{4}a^2$

$= \dfrac{3\sqrt{3}}{2}a^2$

15. Applying sine Rule in ΔABC

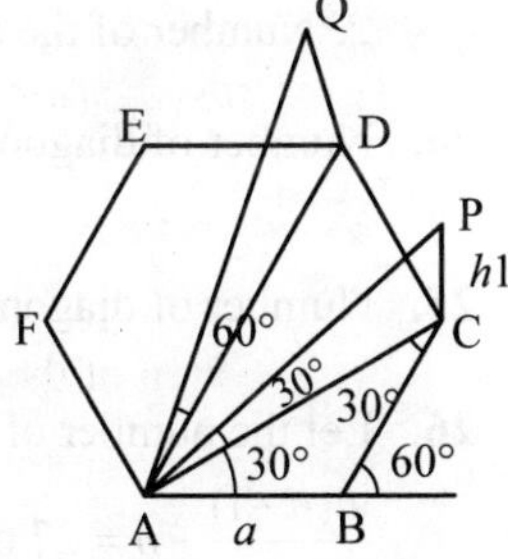

$\dfrac{\sin 120°}{AC} = \dfrac{\sin 30°}{a} \Rightarrow \dfrac{\sqrt{3}}{2AC} = \dfrac{1}{2a} \Rightarrow AC = \sqrt{3}a$

In ΔACP

$\Rightarrow\ \tan 30° = \dfrac{h_1}{AC} \Rightarrow \dfrac{1}{\sqrt{3}} = \dfrac{h_1}{\sqrt{3}a}$

$h_1 = a$

In ΔADC

$\Rightarrow\ AD = \sqrt{3a^2 + a^2} = \sqrt{4a^2} = 2a$

In ΔADQ

$\Rightarrow\ \tan 60° = \dfrac{h_2}{AD} = \dfrac{h_2}{2a} \Rightarrow h_2 = 2\sqrt{3}a \Rightarrow \dfrac{h_1}{h_2} = \dfrac{a}{2\sqrt{3}a} = \dfrac{1}{2\sqrt{3}} = 1:2\sqrt{3}$

16. $\dfrac{\text{Sum of interior angles}}{\text{Sum of exterior angles}} = 7:2 \Rightarrow \dfrac{180^{\circ}(n-2)}{360^{\circ}} = \dfrac{7}{2} \Rightarrow n = 9$

Interior angle $= \dfrac{180^{\circ}(9-2)}{9} = 20°\times 7 = 140°$

17. Each exterior angle = (180° – 135°) = 45°.

$\therefore\ \dfrac{360}{n} = 45 \Leftrightarrow n = \dfrac{360}{45} = 8.$

18. Each exterior angle of a regular octagon = $\frac{360}{8} = 45°$.

Each interior angle of a regular octagon = (180° – 45°) = 135°.

$\therefore$ Required ratio = $\frac{135}{45} = 3:1$.

19. Let the number of sides be n. Then, $(2n - 4) \times 90 = 1080 \Leftrightarrow n = 8$.

20. Sum of all angles of a pentagon = $[(2 \times 5 - 4) \times 90]° = 540°$.

Let these angles be x, $2x$, $3x$, $5x$ & $9x$. Then,

$(x + 2x + 3x + 5x + 9x) = 540 \Leftrightarrow x = 27°$.

21. Each exterior angle = $\frac{360}{n}$, each interior angle = $\left(180 - \frac{360}{n}\right)$.

$\therefore\ 180 - \frac{360}{n} = 11 \times \frac{360}{n} \Leftrightarrow 12 \times \frac{360}{n} = 180 \Leftrightarrow n = 24$.

22. Sum of all interior angles of a hexagon

$= [(2 \times 6 - 4) \times 90]° = 720°$. $\therefore\ 100 + 5x = 720 \Leftrightarrow x = 124°$.

23. Let the number of sides be n, $2n$ and their interior angles be $2x^0$ and $3x^0$ respectively.

Their exterior $\angle s$ are $(180 - 2x)^0$ and $(180 - 3x)^0$.

$\therefore\ \frac{360}{n} = 180 - 2x$ and $\frac{360}{2n} = 180 - 3x$

So, $\frac{360}{n} + 2x = 180$ and $\frac{180}{n} + 3x = 180$

Solving these equations, we get $n = 4$.

$\therefore$ Number of the sides of the polygons are 4, 8.

24. Number of diagonals = $\left[\frac{n(n-1)}{2} - n\right] = \left(\frac{6 \times 5}{2} - 6\right) = 9$.

25. Number of diagonals = $\left(\frac{10 \times 9}{2} - 10\right) = 35$.

26. Let the number of sides be n. Then,

$\frac{n(n-1)}{2} - n = 27$ or $n^2 - 3n - 54 = 0 \Leftrightarrow (n-9)(n+6) = 0 \Leftrightarrow n = 9\ [n \neq -6]$

27. Let the number of sides be n.

Then, each interior angle = $\frac{(2n-4) \times 90}{n}$.

Each exterior angle = $\frac{360}{n}$.

$\therefore\ \frac{(2n-4) \times 90}{n} \times \frac{n}{360} = \frac{7}{2} \Leftrightarrow 2n - 4 = 14 \Leftrightarrow n = 9$.

28. $(2n - 4) \times 90 = 2 \times 360 \Leftrightarrow n = 6$.

29. $\frac{360}{n} = 40 \Leftrightarrow n = 9$.

30

CHAPTER

Triangles

Useful Facts for Solving Problems on Triangles

1. (*i*) The sum of all the angles of a triangle is 180°.

(*ii*) If one side of a triangle is produced, then the exterior angle so formed is equal to the sum of two interior opposite angles.

(*iii*) If the three sides of a triangle be produced in order, then the sum of all the exterior angles so formed is 360°.

(*iv*) In a ΔABC, if the bisectors of $\angle B$ and $\angle C$ meet at O, then $\angle BOC = 90° + \frac{1}{2}\angle A$.

(*v*) In ΔABC, if sides AB and AC are produced to P and Q respectively and the bisectors of $\angle PBC$ and $\angle QCB$ intersect at Q, then $\angle BOC = 90° - \frac{1}{2}\angle A$.

(*vi*) In ΔABC, if AM is the bisector of $\angle BAC$ and $AN \perp BC$, then $\angle MAN = \frac{1}{2}(\angle B - \angle C)$.

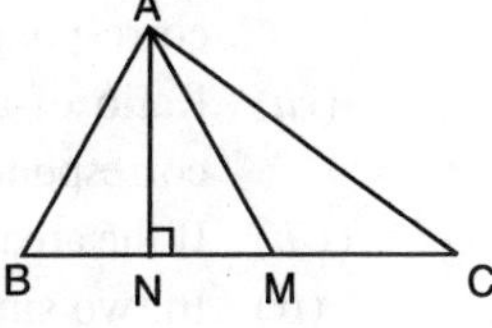

(*vii*) In ΔABC, if BC is produced to D and AL is the bisector of $\angle A$, then $\angle ABC + \angle ACD = 2\ \angle ALC$.

(*viii*) In a ΔABC, if side BC is produced to D and bisectors of $\angle ABC$ and $\angle ACD$ meet at E, then $\angle BEC = \frac{1}{2}\angle A$.

2. (*i*) **In a triangle:**

(*a*) **Orthocentre** is the point of intersection of the altitudes.

(*b*) **Circumcentre** is the point of intersection of the perpendicular bisectors of the sides.

(*c*) **Incentre** is the point of intersection of the angular bisectors

(*d*) **Centroid** is the point of intersection of the medians.

(*ii*) The circumcentre of a triangle is equidistant from its vertices.

(*iii*) The incentre of a triangle is equidistant from its sides.

(*iv*) The centroid divides a median in the ratio 2 : 1.

(*v*) The orthocentre of a right angled mangle lies at the vertex containing the right angle.

3. (*i*) If two angles of a triangle are unequal, the greater angle has the greater side opposite to it.

(*ii*) In a ΔABC, $\angle B$ is a right angle, an obtuse angle or an acute angle according as: $AC^2 = AB^2 + BC^2$, $AC^2 > AB^2 + BC^2$ or $AC^2 < AB^2 + BC^2$ respectively.

4. (*i*) In ΔABC, if $\angle B$ is obtuse, then $AC^2 = AB^2 + BC^2 + 2BC . AD$.

(*ii*) In ΔABC, if $\angle C$ is acute, then $AB^2 = BC^2 + AC^2 - 2\ BC . CD$, where $AD \perp BC$.

5. Two Δs are congruent if:

(*i*) two sides and the included angle of the one is equal to the corresponding two sides and the included angle of the other (*SAS*).

(*ii*) two angles and a side of the one is equal to the corresponding two angles and a side of the other (*SAA*).

(*iii*) three sides of the one are respectively equal to the three sides of the other (SSS).

(*iv*) Two right angled triangles are congruent, if their heights are equal and one side of the one is equal to the corresponding side of the other.

6. Similar Triangles: *Two Δs are said to be similar, if their corresponding angles are equal and their corresponding sides are proportional.*

(*i*) The internal bisector of an angle of a triangle divides the opposite side intemally in the ratio of sides containing the angle.

(*ii*) The line joining the mid-points of any two sides of a triangle is parallel to the third side and equal to half of it.

(*iii*) The diagonals of a trapezium divide each other proportionally.

(*iv*) Ratio of areas of two similar triangles is equal to the ratio of the squares of any two corresponding sides.

(*v*) Ratio of areas of two similar triangles is equal to the ratio of the squares of the corresponding altitudes.

(*vi*) Ratio of areas of two similar triangles is equal to the ratio of the squares of the corresponding medians.

(*vii*) Ratio of areas of two similar triangles is equal to the ratio of the squares of the corresponding angle bisector segments.

(*viii*) If the areas of two similar triangles are equal, then the triangles are congruent.

(*ix*) In two similar triangles, the ratio of two corresponding sides is equal to the ratio of corresponding medians.

(*x*) In two similar triangles, the ratio of two corresponding sides is equal to the ratio of their corresponding heights.

EXERCISE 30

1. The perpendicular dropped from a vertex of a right-angled triangle upon the hypotenuse divides it into two segments of lengths 9 units and 16 units respectively. What is the length of the perpendicular? **[CDS (I) Exam, 2022]**

(*a*) 6 units (*b*) 8 units (*c*) 10 units (*d*) 12 units

2. In a triangle ABC, the bisector of angle A cuts BC at D. If AB + AC = 10 cm and BD : DC = 3 : 1, then what is the length of AC? **[CDS (I) Exam, 2025]**

(*a*) 2.5 cm (*b*) 6 cm (*c*) 7.5 cm (*d*) 8 cm

3. In a triangle ABC, AB = 2 cm, BC = 4 cm, AC = 3 cm. The bisector of $\angle$A meets BC at D and the bisector of $\angle$B meets AD at E. What is AE : ED equal to? **[CDS (I) Exam, 2025]**

(*a*) 5 : 4 (*b*) 5 : 3 (*c*) 4 : 3 (*d*) 3 : 2

4. In a triangle *ABC*, *AB* = *AC* and *BC* is produced to *D* such that $\angle ACD = x$, then what is $\angle BAC$ equal to ? **[CDS (I) Exam, 2022]**

(*a*) $2x - 90°$ (*b*) $2x - 180°$ (*c*) $180° - 2x$ (*d*) $\frac{x}{2}$

5. In a triangle ABC, $AB = 16$ cm, $AC = 12$ cm and AD is the bisector of $\angle A$. If BD = 4 cm, then what is CD equal to? **[CDS (I) Exam, 2022]**

(*a*) 2 cm (*b*) 2.5 cm (*c*) 3 cm (*d*) 3.5 cm

6. An equilateral triangle os side x is inscribed in a circle of radius y. Which one of the following is correct? **[CDS (I) Exam, 2022]**

(*a*) $2y = x$ (*b*) $2y = \sqrt{3}x$ (*c*) $\sqrt{3}y = 2x$ (*d*) $\sqrt{3}y = x$

7. The measure of an angle formed by the bisectors of the angles A and C of the triangle ABC is 130°. What is the measure of the angle B? **[CDS (I) Exam, 2025]**

(*a*) 65° (*b*) 75° (*c*) 80° (*d*) 85°

8. Two isosceles triangles have equal vertical angles and their areas are in the ratio 4.84 : 5.29. What is the ratio of their corresponding heights? **[CDS (I) Exam, 2021]**

(*a*) 11 : 23 (*b*) 23 : 25 (*c*) 22 : 23 (*d*) 484 : 529

9. ABC is triangle right angles at A and AD is perpendicular to BC. If $BD = 8$ cm and $DC = 12.5$ cm, then what is AD equal to? **[CDS (I) Exam, 2021]**

(*a*) 7.5 cm (*b*) 8. 5 cm (*c*) 9 cm (*d*) 10 cm

10. In a triangle ABC, D is a point on BC. If $AB.DC = AC.BD$, $\angle BAD = \alpha$, and $\angle CAD = \beta$ then which one of the following is correct? **[CDS (II) Exam, 2024]**

(*a*) $\alpha = \beta$ (*b*) $\alpha = 2\beta$ (*c*) $2\alpha = \beta$ (*d*) $2\alpha = 3\beta$

11. The radius of circum-cercle of a right-angled triangle is 10 cm and the altitude drawn to the hypotenuse is altitude drawn to the hypotenuse is 8 cm. What is the area of the triangle? **[CDS (II) Exam, 2021]**

(*a*) 60 cm^2 (*b*) 80 cm^2 (*c*) 100 cm^2 (*d*) 120 cm^2

Consider the following for the next three (3) items that follow:

ABC is a triangle with $AB = 1.6$ cm, $BC = 6.3$ cm and $CA = 6.5$ cm. Let P and Q be the mid-point of AB and BC respectively.

12. What is $AB^2 + 4BQ^2$ equal to? **[CDS (II) Exam, 2021]**

(*a*) 41.25 cm^2 (*b*) 42.25 cm^2 (*c*) 43.75 cm^2 (*d*) 44.25 cm^2

13. What is $AQ^2 + CP^2$ equal to ? **[CDS (II) Exam, 2021]**

(*a*) AC^2 (*b*) $1.2AC^2$ (*c*) $1.25\,AC^2$ (*d*) $1.5\,AC^2$

14. What is $4(CP^2 - AQ^2)$ equal to ? **[CDS (II) Exam, 2021]**

(*a*) 101.39 cm^2 (*b*) 111.39 cm^2 (*c*) 121.29 cm^2 (*d*) 131.39 cm^2

15. ABC is an equilateral triangle. The side BC is trisected at D such that BC = 3 BD. What is the ratio of AD^2 to AB^2? **[CDS (I) Exam, 2020]**

(*a*) 7 : 9 (*b*) 1 : 3 (*c*) 5 : 7 (*d*) 1 : 2

16. In a triangle, values of all the angles are integers (in degree measure). Which one of the following cannot be the proportion of their measures? **[CDS (I) Exam, 2020]**

(*a*) 1 : 2 : 3 (*b*) 3 : 4 : 5 (*c*) 5 : 6 : 7 (*d*) 6 : 7 : 8

17. In triangle ABC, if $2\angle A = 3\angle B = 6\angle C$, then what is $\angle A + \angle C$ equal to? **[CDS (I) Exam, 2020]**

(*a*) 90° (*b*) 120° (*c*) 130° (*d*) 150°

18. If a, b and c are the sides of a triangle ABC then $\sqrt{a}+\sqrt{b}-\sqrt{c}$ is always:
(*a*) Negative (*b*) Positive (*c*) Non-negative (*d*) Non-positive
[CDS (I) Exam, 2023]

19. Let ABC be a right-angled triangle with sides 5 cm, 12 cm and 13 cm. If p is the length of the perpendicular drawn from vertex A on the hypotenuse BC, then what is the value of 13p?
(*a*) 24 (*b*) 48 (*c*) 60 (*d*) 90
[CDS (I) Exam, 2023]

20. A triangle ABC has been divided into four smaller triangles P, Q, R, S whose perimeters are 16 cm, 12 cm, 4 cm and 12 cm respectively. P, R and S contain the vertices A, B and C respectively. What is the perimeter of the triangle ABC ? **[CDS (I) Exam, 2023]**
(*a*) 18 cm (*b*) 20 cm (*c*) 22 cm (*d*) 24 cm

Consider the following for the next two (02) items that follow:

ABC is a triangle right-angled at A. Further, AB = 8 cm, BC = 10 cm. D is the point on BC such that AD is perpendicular to BC. **[CDS (I) Exam, 2025]**

21. What is AD equal to?
(*a*) 4·8 cm (*b*) 5·0 cm (*c*) 5·2 cm (*d*) 5·4 cm

22. What is ratio of area of triangle ADC to area of triangle ADB?
(*a*) 7 : 15 (*b*) 9 : 16 (*c*) 2 : 3 (*d*) 3 : 4

23. The hypotenuse AC of a right-angled ABC is 3 5 cm. If AB is doubled and BC is tripled such that ABC remains a right-angled triangle, the hypotenuse becomes 15 cm. What is AB+ BC equal to? **[CDS (II) Exam, 2023]**
(*a*) 10 cm (*b*) 9 cm (*c*) 25 cm (*d*) 8 cm

24. If a, b and c are the sides of a right-angled triangle, where $a > b > c$, then what is the value of the expression $(a+b+c)(a+b-c)(a-b+c)(a-b-c)$? **[CDS (II) Exam, 2023]**
(*a*) $4b^2c^2$ (*b*) $-4b^2c^2$ (*c*) $-2a^2b^2$ (*d*) $-4a^2b^2$

25. An isosceles triangle has its base length $2a$ and its height is h. On each side of the triangle a square is drawn external to the triangle. What is the area of the figure thus formed ?
[CDS (II) Exam, 2023]
(*a*) $6a^2+2h^2+2ah$ (*b*) $6a^2+2h^2+ah$ (*c*) $4a^2+2h^2+ah$ (*d*) $6a^2+h^2+ah$

26. In a triangle ABC, angle B = 90° and pis the length of the perpendicular from B to AC. If BC = 10 cm and AC = 12 cm, then what is the value of p? **[CDS (II) Exam, 2023]**
(*a*) $\frac{5\sqrt{11}}{3}$ (*b*) $\frac{10\sqrt{11}}{3}$ (*c*) $\frac{40}{\sqrt{61}}$ (*d*) $\frac{12}{25}$

Each item contains a Question followed by two Statements. Answer each item using the following instructions **[CDS (I) Exam, 2023]**

(a) Choose this option if the Question can be answered by one of the Statements alone but not by the other.

(b) Choose this option if the Question can be answered by either Statement alone.

(c) Choose this option if the Question can be answered by using both the Statements together, but cannot be answered by using either Statement alone

(d) Choose this option if theQuestion cannot be answered even by using both Statements together

27. Question : Is triangle Δ right angled ?

Statement I : The length of the line segment joining the mid-points of two sides of Δ is half of the third side of Δ.

Statement II : The angles of Δ are in the ratio 1 : 2 : 3.

28. The lengths of two longer sides of the triangle Δ are 25 cm and 24 cm.

Question : What is the length of the shortest side ?

Statement I : The angles of Δ are in the ratio 1 : 2 : 3.

Statement II :The length of the perpendicular drawn on the longest side of Δ from its opposite vertex is 6·72 cm.

29. In the figure given below, *ABC* is a triangle with *AB* perpendicular to *BC*. Further *BD* is perpendicular to *AC*. If *AD* = 9 cm, and *DC* = 4 cm, Then that is the length of *BD*?

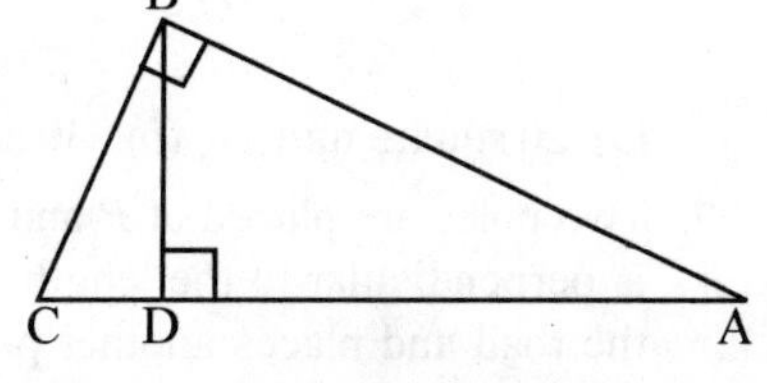

(*a*) 13.36 cm (*b*) $\frac{36}{13}$ cm

(*c*) $\frac{13}{2}$ cm (*d*) 6 cm **[CDS (II) Exam, 2018]**

30. The sides of a triangle are k, 1.5k and 2.25k. What is the sum of the squares of its medians?

[CDS (I) Exam, 2015]

(*a*) $\frac{359k^2}{64}$ (*b*) $\frac{379k^2}{64}$ (*c*) $\frac{389k^2}{64}$ (*d*) $\frac{399k^2}{64}$

31. Let *a, b, c* be the sides of a right triangle, where *c* is the hypotenuse. The radius of the circle which touches the sides of the triangle is **[CDS (I) Exam, 2015]**

(*a*) $\frac{(a+b-c)}{2}$ (*b*) $\frac{(a+b+c)}{2}$ (*c*) $\frac{(a+2b+2c)}{2}$ (*d*) $\frac{(2a+2b-c)}{2}$

32. Consider the following statements: **[CDS (I) Exam, 2015]**

1. Let *D* be a point on the side *BC* of a triangle *ABC*. If area of triangle *ABD* = are of triangle *ACD*, then for all points *O* on *AD*, area of triangle *ABO* = area of triangle *ACO*.

2. If *G* is the point of concurrence of the medians of a triangle *ABC*, then area of triangle *ABG* = area of triangle *BCG* = area of triangle *ACG*.

Which of the above statements is/are correct?

(*a*) 1 only (*b*) 2 only (*c*) Both 1 and 2 (*d*) Neither 1 nor 2

33. The sides of a triangle are 25 cm, 39 cm and 56 cm. The perpendicular from the opposite vertex on the side of 56 cm is **[CDS (I) Exam, 2015]**

(*a*) 10 cm (*b*) 12 cm (*c*) 15 cm (*d*) 16 cm

34. The point *O* is equidistant from the three sides of a triangle *ABC*. Consider the following statements: **[CDS (II) Exam, 2015]**

1. $\angle OAC + \angle OCB + \angle OBA = 90°$

2. $\angle BOC = 2\angle BAC$

3. The perpendiculars drawn from any point on *OA* to *AB* and *AC* are always equal

Which of the above statements are correct?

(*a*) 1 and 2 only (*b*) 2 and 3 only (*c*) 1 and 3 only (*d*) 1, 2 and 3

35. In a triangle ABC, ∠ABC = 60° and AD is the altitude. If AB = 6 cm and BC = 8 cm, then what is the area of the triangle? **[CDS (II) Exam, 2024]**

(*a*) 12 square cm (*b*) $12\sqrt{3}$ square cm (*c*) 24 square cm (*d*) $24\sqrt{3}$ square cm

36. In the above figure, *ABCD* is a parallelogram. *P* is a point on *BC* such that *PB: PC* = 1 : 2. *DP* and *AB* when both produced meet at *Q*. If area of triangle *BPQ* is 20 square unit, the area of triangle *DCP* is **[CDS (II) Exam, 2015]**

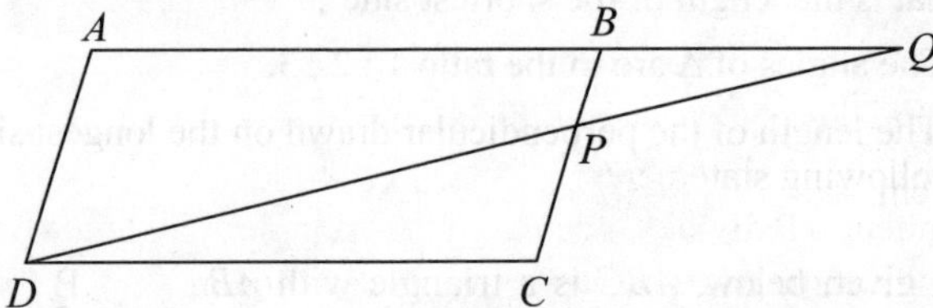

(*a*) 20 square unit (*b*) 30 square unit (*c*) 40 square unit (*d*) None of the above

37. Two poles are placed at *P* and *Q* on either side of a road such that the line joining *P* and *Q* is perpendicular to the length of the road. A person moves x metre away from *P* parallel to the road and places another pole at R. Then the person moves further x metre in the same direction and turns and moves a distance y metre away from the road perpendicularly, where he finds himself, *Q* and R on the same line. The distance between *P* and *Q* (*i.e.*, the width of the road) in metre is **[CDS (I) Exam, 2016]**

(*a*) *x* (*b*) $\frac{x}{2}$ (*c*) *y* (*d*) 2y

38. Suppose chords AB and CD of a circle intersect at a point *P* inside the circle. Two right-angled triangles A′P′B′ and C′Q′D′ are formed as shown in the figures below such that *A′P′* = AP, B′P′ = BP, C′Q′ = CP, D′Q′ = DP and ∠A′P′B′ = 90° = ∠C′Q′D′: **[CDS (I) Exam, 2016]**

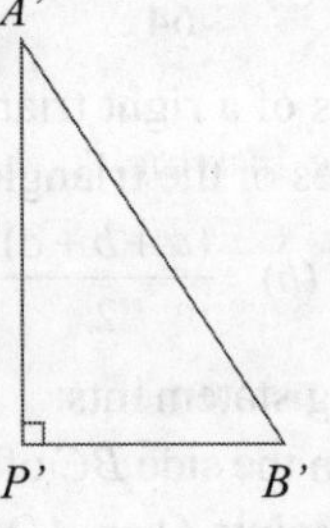

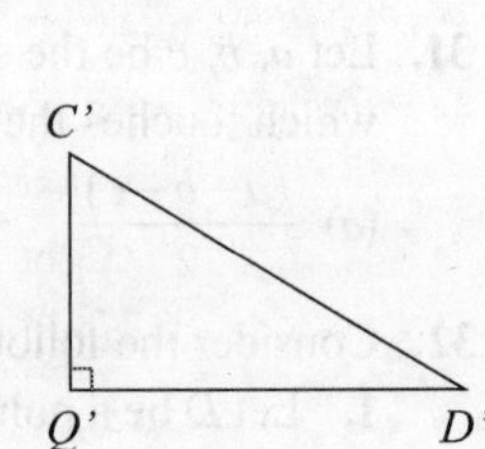

Which of the following statements are ***not*** correct?

1. *A′P′B′* and *C′Q′D′* are similar triangles, but need not be congruent.
2. *A′B′* and *C′Q′D′* are congruent triangles,
3. *A′P′B′* and *C′Q′D′* are triangles of same area.
4. *A′P′B′* and *C′Q′D′* are triangles of same perimeter.

Select the correct answer using the code given below.

(*a*) 2 and 3 only (*b*) 3 only (*c*) 1,2 and 4 only (*d*) 1, 2, 3 and 4

39. Suppose *ABC* is a triangle with *AB* of unit length. *D* and *E* are the points lying on *AB* and *AC* respectively such that *BC* and *DE* are parallel. If the area of triangle *ABC* is twice the area of triangle *ADE*, then the length of *AD* is **[CDS (I) Exam, 2016]**

(*a*) $\frac{1}{2}$ unit (*b*) $\frac{1}{3}$ unit (*c*) $\frac{1}{\sqrt{2}}$ unit (*d*) $\frac{1}{\sqrt{3}}$ unit

40. Let the triangles *ABC* and *DEF* be such that ∠*ABC* = ∠*DEF*, ∠*ACB* = ∠*DFE* and ∠*BAC* = ∠*EDF*. Let L be the midpoint of *BC* and *M* be the midpoint of *EF*. Consider the following statements: **[CDS (II) Exam, 2016]**

Statement I: Triangles *ABL* and *DEM* are similar.

Statement II: Triangle *ALC* is congruent to triangle *DMF* even if $AC \neq DF$.

Which one of the following is correct in respect of the above statements?

(*a*) Both Statement I and Statement II are true and Statement II is the correct explanation of Statement I

(*b*) Both Statement I and Statement II are true but Statement II is ***not*** the correct explanation of Statement I

(*c*) Statement I is true but Statement II is false

(*d*) Statement I is false but Statement II is true

41. Consider the following statements:

I. The line segment AP divides the area of the triangle ABC into two equal parts

II. The perimeter of the triangle APC is more than 46 cm

III. The area of the triangle APC is 50 square cm

Which of the statements given above are correct? **[CDS (II) Exam, 2024]**

(*a*) I and II only (*b*) II and III only (*c*) I and III only (*d*) I, II and III

Consider the following for the next two (02) items that follow:

The perimeter of a triangle ABC is 105 cm.

The altitudes AD, BE and CF are in the ratio 3 : 5 : 6. **[CDS (I) Exam, 2025]**

42. What is AB : BC : CA equal to?

(*a*) 10 : 6 : 5 (*b*) 5 : 10 : 6 (*c*) 6 : 5 : 3 (*d*) 3 : 5 : 6

43. What is the approximate area of the triangle ABC?

(*a*) 175 square cm (*b*) 190 square cm

(*c*) 205 square cm (*d*) 285 square cm

44. Let *ABC* be a triangle in which *AB* = *AC*. Let *L* be the locus of points *X* inside or on the triangle such that *BX* = *CX*. Which of the following statements are correct?

1. *L* is a straight line passing through A and in-centre of triangle *ABC* is on *L*.
2. *L* is a straight line passing through A and orthocentre of triangle *ABC* is a point on *L*.
3. *L* is a straight line passing through A and centroid of triangle *ABC* is a point on *L*.

Select the correct answer using the code given below.

(*a*) 1 and 2 only (*b*) 2 and 3 only (*c*) 1 and 3 only (*d*) 1, 2 and 3

[CDS (II) Exam, 2016]

45. In a triangle PQR, point X is on PQ and point Y is on PR such that *XP* = 1·5 units, XQ = 6 units, PY = 2 units and YR = 8 units. Which of the following are correct?

1. QR=5XY **[CDS (II) Exam, 2016]**
2. QR is parallel to XY.
3. Triangle PYX is similar to triangle PRQ.

Select the correct answer using the code given below.

(*a*) 1 and 2 only (*b*) 2 and 3 only (*c*) 1 and 3 only (*d*) 1, 2 and 3

46. In a right-angled triangle ABC, AB = 15 cm, BC = 20 cm and AC = 25 cm. Further, BP is the perpendicular on AC. What is the difference in the area of triangles PAB and PCB ?

[CDS (I) Exam, 2024]

(*a*) 40 square cm (*b*) 42 square cm (*c*) 45 square cm (*d*) 48 square cm

47. In the figure given above, $AD = CD = BC$.
What are the values of $\angle CDB$? **[CDS (II) Exam, 2016]**
(*a*) 32°
(*b*) 64°
(*c*) 78°
(*d*) Cannot be determined due to insufficient data

48. *ABC* is an equilateral triangle and *X*, *Y* and *Z* are the points on *BC*, *CA* and *AB* respectively such that $BX = CY = AZ$. Which of the following is/are correct? **[CDS (II) Exam, 2016]**
1. *XYZ* is an equilateral triangle.
2. Triangle *XYZ* is similar to triangle *ABC*.
Select the correct answer using the code given below.
(*a*) 1 only (*b*) 2 only (*c*) Both 1 and 2 (*d*) Neither 1 or 2

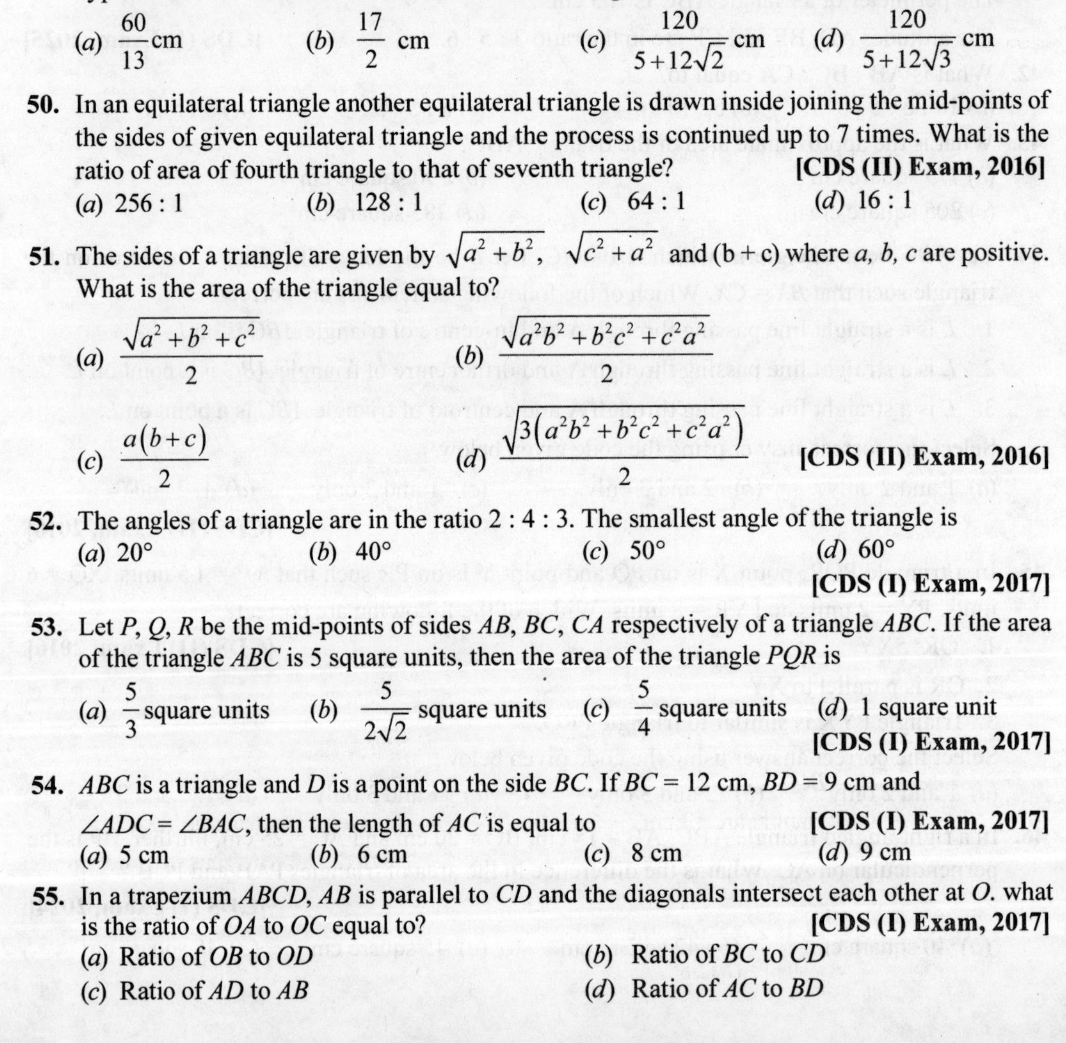

49. Let ABC be a right angled triangle with $BC = 5$ cm and $AC = 12$ cm. Let *D* be a point on the hypotenuse. *AB* such that $\angle BCD = 30°$. What is length of *CD*? **[CDS (II) Exam, 2016]**
(*a*) $\frac{60}{13}$ cm (*b*) $\frac{17}{2}$ cm (*c*) $\frac{120}{5+12\sqrt{2}}$ cm (*d*) $\frac{120}{5+12\sqrt{3}}$ cm

50. In an equilateral triangle another equilateral triangle is drawn inside joining the mid-points of the sides of given equilateral triangle and the process is continued up to 7 times. What is the ratio of area of fourth triangle to that of seventh triangle? **[CDS (II) Exam, 2016]**
(*a*) 256 : 1 (*b*) 128 : 1 (*c*) 64 : 1 (*d*) 16 : 1

51. The sides of a triangle are given by $\sqrt{a^2+b^2}$, $\sqrt{c^2+a^2}$ and (b + c) where *a*, *b*, *c* are positive. What is the area of the triangle equal to?
(*a*) $\frac{\sqrt{a^2+b^2+c^2}}{2}$ (*b*) $\frac{\sqrt{a^2b^2+b^2c^2+c^2a^2}}{2}$
(*c*) $\frac{a(b+c)}{2}$ (*d*) $\frac{\sqrt{3(a^2b^2+b^2c^2+c^2a^2)}}{2}$ **[CDS (II) Exam, 2016]**

52. The angles of a triangle are in the ratio 2 : 4 : 3. The smallest angle of the triangle is
(*a*) 20° (*b*) 40° (*c*) 50° (*d*) 60°
[CDS (I) Exam, 2017]

53. Let *P*, *Q*, *R* be the mid-points of sides *AB*, *BC*, *CA* respectively of a triangle *ABC*. If the area of the triangle *ABC* is 5 square units, then the area of the triangle *PQR* is
(*a*) $\frac{5}{3}$ square units (*b*) $\frac{5}{2\sqrt{2}}$ square units (*c*) $\frac{5}{4}$ square units (*d*) 1 square unit
[CDS (I) Exam, 2017]

54. *ABC* is a triangle and *D* is a point on the side *BC*. If $BC = 12$ cm, $BD = 9$ cm and $\angle ADC = \angle BAC$, then the length of *AC* is equal to **[CDS (I) Exam, 2017]**
(*a*) 5 cm (*b*) 6 cm (*c*) 8 cm (*d*) 9 cm

55. In a trapezium *ABCD*, *AB* is parallel to *CD* and the diagonals intersect each other at *O*. what is the ratio of *OA* to *OC* equal to? **[CDS (I) Exam, 2017]**
(*a*) Ratio of *OB* to *OD* (*b*) Ratio of *BC* to *CD*
(*c*) Ratio of *AD* to *AB* (*d*) Ratio of *AC* to *BD*

56. In the figure given below, $\angle A = 80°$ and $\angle ABC = 60°$. BD and CD bisect angle B and C respectively. What are the values of x and y respectively? **[CDS (I) Exam, 2017]**

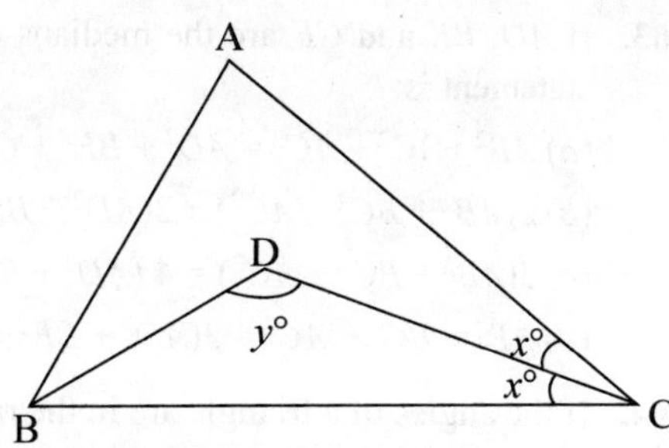

(*a*) 10 and 130 (*b*) 10 and 125
(*c*) 20 and 130 (*d*) 20 and 125

57. In the figure given below, PQR is a non-isosceles right-angled triangle, right angled at Q. If LM and QT are parallel and $QT = PT$, then what is $\angle RLM$ equal to? **[CDS (I) Exam, 2017]**

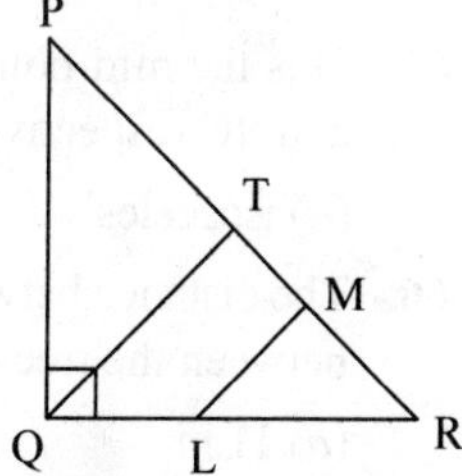

(*a*) ∠PQT (*b*) ∠LRM
(*c*) ∠RML (*d*) ∠QPT

58. In the figure given below, ABC is a triangle with $AB = BC$ and D is an interior point of the triangle ABC such that $\angle DAC = \angle DCA$.
Consider the following statements:
1. Triangle ADC is an isosceles triangle.
2. D is the centroid of the triangle ABC.
3. Triangle ABD is congruent to the triangle CBD.
Which of the above statements are correct?
(*a*) 1 and 2 only (*b*) 2 and 3 only
(*c*) 1 and 3 only (*d*) 1, 2 and 3 **[CDS (I) Exam, 2017]**

59. In the figure given below, M is the mid-point of AB and $\angle DAB = \angle CBA$ and $\angle AMC = \angle BMD$. Then the triangle ADM is congruent to the triangle BCM by **[CDS (I) Exam, 2017]**

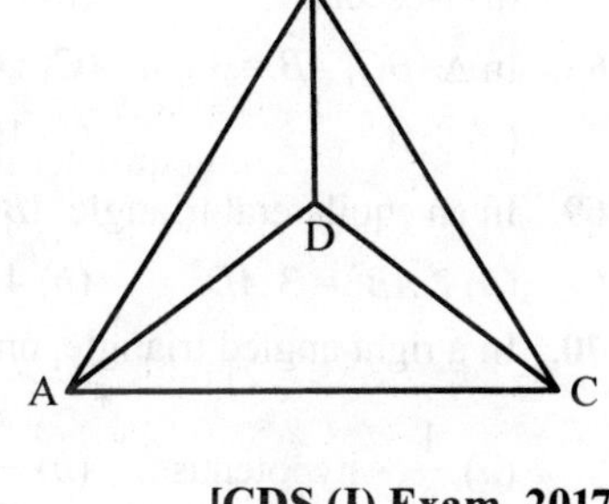

(*a*) SAS rule
(*b*) SSS rule
(*c*) ASA rule
(*d*) AAA rule

60. For a triangle ABC, the true statement is:
(*a*) $AC^2 = AB^2 + BC^2$ (*b*) $AC = AB + BC$
(*c*) $AC > AB + BC$ (*d*) $AC < AB + BC$

61. The sides of a triangle are 13 cm, 16 cm and 10 cm respectively. The triangle is:
(*a*) acute (*b*) obtuse (*c*) right angled (*d*) can't be determined

62. In a $\Delta\, ABC$, the sides AB and AC have been produced to D and E. Bisectors of $\angle CBD$ and $\angle BCE$ meet at O. If $\angle A = 64°$, then $\angle BOC$ is:
(*a*) 52° (*b*) 58° (*c*) 26° (*d*) 112°

63. If AD, BE and CF are the medians of a triangle ABC, then the true statement is:

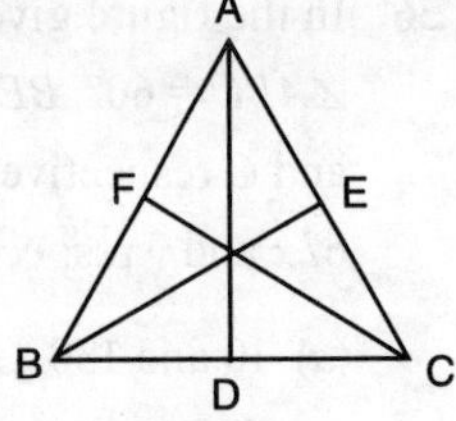

(*a*) $AB^2 + BC^2 + AC^2 = AD^2 + BE^2 + CF^2$
(*b*) $2(AB^2 + BC^2 + AC^2) = 3(AD^2 + BE^2 + CF^2)$
(*c*) $3(AB^2 + BC^2 + AC^2) = 4(AD^2 + BE^2 + CF^2)$
(*d*) $AB^2 + BC^2 + AC^2 = 3(AD^2 + BE^2 + CF^2)$

64. If the angles of a triangle are in the ratio of 2 : 3 : 4, then the greatest angle of the triangle is:
(*a*) 75° (*b*) 80° (*c*) 90° (*d*) 120°

65. D is the mid-point of the base BC of a triangle ABC. DM and DN are perpendiculars on AB and AC respectively. If $DM = DN$, the triangle is:
(*a*) isosceles (*b*) equilateral (*c*) right angled (*d*) scalene

66. The distance between the tops of two trees 20 m and 28 m high is 17 m. The horizontal distance between the trees is:
(*a*) 11 m (*b*) 31 m (*c*) 15 m (*d*) 9 m

67. In a triangle, the perpendicular from vertex to the base bisects the base. The triangle is:
(*a*) isosceles (*b*) right angled (*c*) equilateral (*d*) scalene

68. In ΔABC, $AB = 3$ cm, $AC = 4$ cm and AD is the bisector of $\angle A$. Then, $BD : DC$ is:
(*a*) 9:16 (*b*) 16:9 (*c*) 3:4 (*d*) 4:3

69. In an equilateral triangle ABC, if $AD \perp BC$, then:
(*a*) $2\,AB^2 = 3\,AD^2$ (*b*) $4AB^2 = 3\,AD^2$ (*c*) $3AB^2 = 4AD^2$ (*d*) $3AB^2 = 2AD^2$

70. In a right angled triangle, one of the angles is 60°. The side opposite to this angle is:
(*a*) $\frac{1}{2} \times$ hypotenuse (*b*) $\frac{1}{\sqrt{2}} \times$ hypotenuse (*c*) $\frac{2}{3} \times$ hypotenuse (*d*) $\frac{\sqrt{3}}{2} \times$ hypotenuse

71. Which of the following statements is true?
(*a*) The incentre of a triangle is equidistant from the sides of the triangle.
(*b*) The circumcentre of a triangle is equidistant from the sides of the triangle.
(*c*) The orthocentre of a triangle is equidistant from the sides of the triangle.
(*d*) None of these

72. In the adjoining figure, $AB = AC$ and $AP \perp BC$. Then,

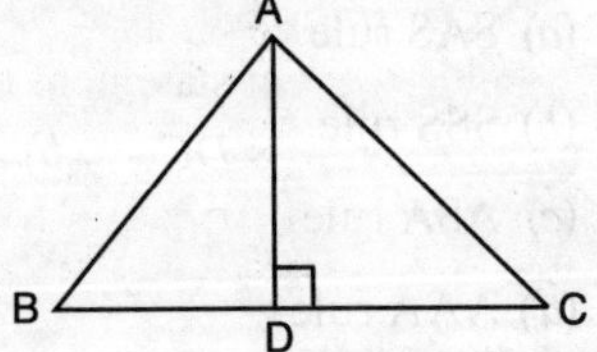

(*a*) $AB = AP$ (*b*) $AB < AP$
(*c*) $AB > AP$ (*d*) $AB \leq AP$

73. The circumcentre of a right angled triangle lies:
(*a*) in the interior of the triangle. (*b*) in the exterior of the triangle.
(*c*) at the mid-point of the hypotenuse. (*d*) at the mid-point of the base.

74. The mid-points of the sides of a triangle along with any of the vertices as the fourth point make a:
(*a*) parallelogram (*b*) rhombus (*c*) rectangle (*d*) square

75. For a triangle ABC, which of the following is true?
(*a*) $BC^2 - AB^2 = AC^2$ (*b*) $AB - AC = BC$ (*c*) $(AB - AC) > BC$ (*d*) $(AB - AC) < BC$

76. If $ABCD$ is a square and DCE is an equilateral triangle in the given figure, then $\angle DAE$ is equal to:

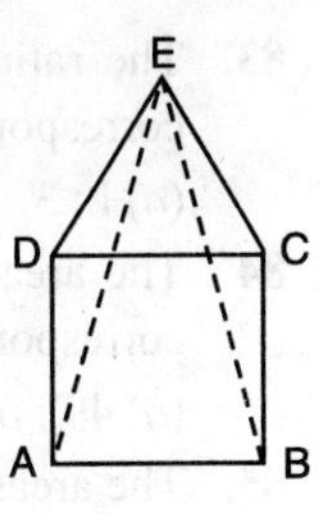

(*a*) 45° (*b*) 30°
(*c*) 15° (*d*) 22°

77. In the adjoining figure, $AD = BD = AC$; $\angle CAE = 75°$ and $\angle ACD = x^0$. Then, the value of x is:

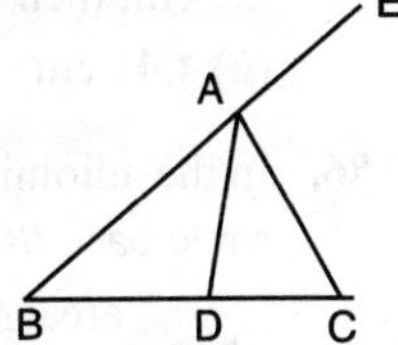

(*a*) 45° (*b*) 50°
(*c*) 60° (*d*) $37\frac{1}{2}°$

78. In the adjoining figure, if $BC = a$, $AC = b$, $AB = c$ and $\angle CAB = 120°$, then the correct relation is:

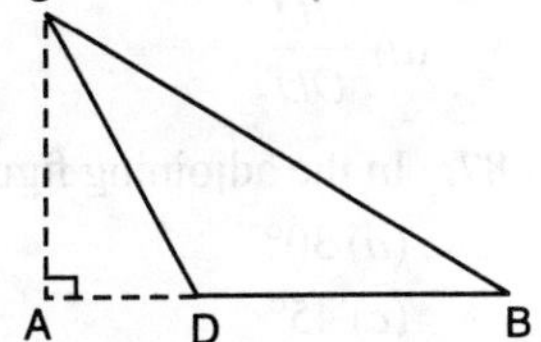

(*a*) $a^2 = b^2 + c^2 + 2bc$ (*b*) $a^2 = b^2 + c^2 - 2bc$
(*c*) $a^2 = b^2 + c^2 + bc$ (*d*) $a^2 = b^2 + c^2 - bc$

79. In the adjoining figure, if $BC = a$, $AC = b$, $AB = c$ and $\angle CAB = 60”$, then the correct relation is:

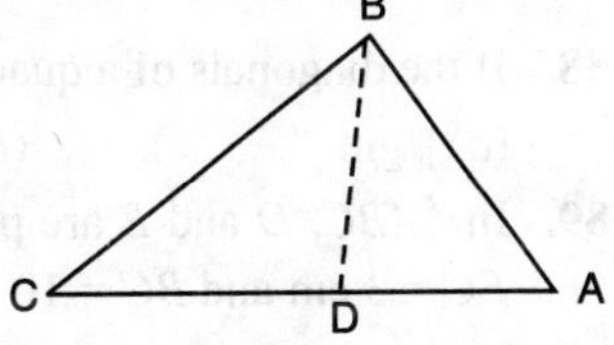

(*a*) $a^2 = b^2 + c^2 + bc$ (*b*) $a^2 = b^2 + c^2 - bc$
(*c*) $a^2 = b^2 + c^2 + 2bc$ (*d*) $a^2 = b^2 + c^2 - 2bc$

80. In a right angled ΔABC, right angled at A, if $AD \perp BC$ such that $AD = p$, If $BC = a$, $CA = b$ and $AB = c$, then:

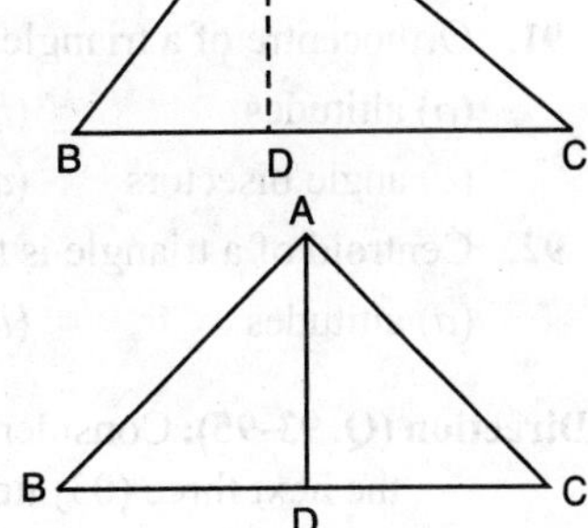

(*a*) $p^2 = b^2 + c^2$ (*b*) $\frac{1}{p^2} = \frac{1}{b^2} + \frac{1}{c^2}$
(*c*) $\frac{p}{a} = \frac{p}{b}$ (*d*) $p^2 = b^2c^2$

81. In a right angled triangle ABC in which $\angle A = 90°$, if $AD \perp BC$, then the correct statement is:

(*a*) $AB^2 = BD \times DC$ (*b*) $AB = BD \times AD$
(*c*) $AB^2 = BC \times DC$ (*d*) $AB^2 = BC \times BD$

82. In the adjoining figure, XY is parallel to AC. If XY divides the triangle into equal parts, then the value of $\frac{AX}{AB}$ is equal to:

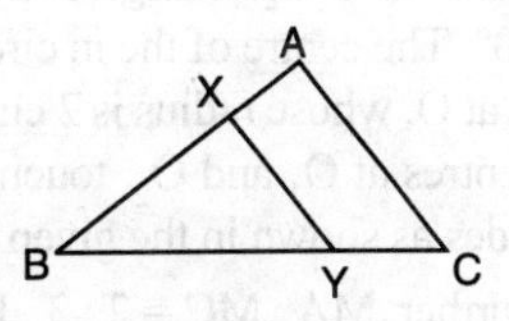

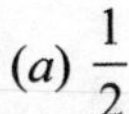

(*a*) $\frac{1}{2}$ (*b*) $\frac{1}{\sqrt{2}}$
(*c*) $\frac{\sqrt{2}+1}{\sqrt{2}}$ (*d*) $\frac{\sqrt{2}-1}{\sqrt{2}}$

83. The ratio of the corresponding sides of two similar triangles is 1 : 3. The ratio of their corresponding heights is:

(*a*) 1 : 3 (*b*) 3 : 1 (*c*) 1 : 9 (*d*) 9 : 1

84. The areas of two similar triangles are 49 cm^2 and 64 cm^2 respectively. The ratio of their corresponding sides is:

(*a*) 49 : 64 (*b*) 7 : 8 (*c*) 64 : 49 (d) None

85. The areas of two similar triangles are 12 cm^2 and 48 cm^2. If the height of the smaller one is 2.1 cm, then the corresponding height of the bigger one is:

(*a*) 4.41 cm (*b*) 8.4 cm (*c*) 4.2 cm (*d*) 0.525 cm

86. In the adjoining figure, ABC and DBC are two triangles on the same base BC. $AL \perp BC$ and $DM \perp BC$.

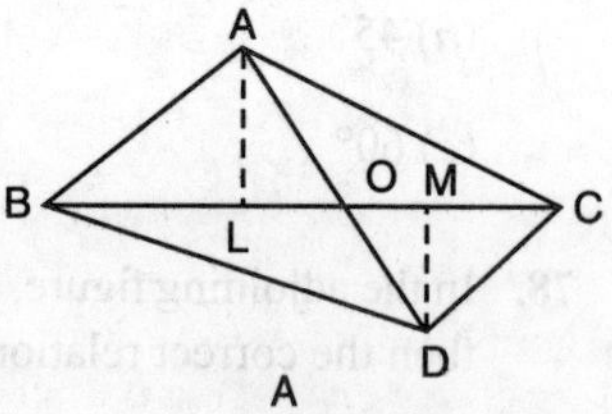

Then, $\dfrac{\text{area}(\Delta ABC)}{\text{area}(\Delta DBC)}$ is equal to:

(*a*) $\dfrac{AO}{OD}$ (*b*) $\dfrac{AO^2}{OD^2}$ (*c*) $\dfrac{AO}{AD}$ (*d*) $\dfrac{OD^2}{AO^2}$

87. In the adjoining figure, $AD : DC = 2 : 3$, then $\angle ABC$ is equal to:

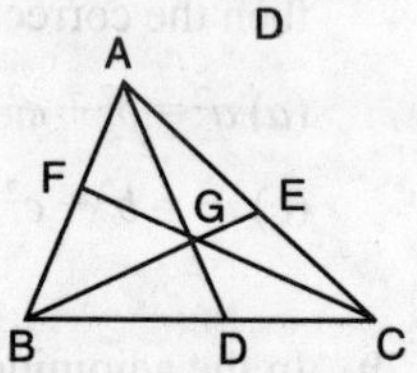

(*a*) 30° (*b*) 40°
(*c*) 45° (*d*) 110°

88. If the diagonals of a quadrilateral intersect each other proportionally, the quadrilateral is a:

(*a*) || gm (*b*) rhombus (*c*) trapezium (*d*) rectangle

89. In ΔABC, D and E are points on AB and AC respectively such that $DE \parallel BC$. If $AE = 2$ cm, $EC = 3$ cm and $BC = 10$ cm, Then DE is equal to:

(*a*) 5 cm (*b*) 4 cm (*c*) 15 cm (*d*) $\dfrac{20}{3}$ cm

90. If the medians of a triangle are equal, then the triangle is:

(*a*) isosceles (*b*) equilateral (*c*) scalene (*d*) right angled

91. Orthocentre of a triangle is determined by the:

(*a*) altitudes (*b*) perpendicular bisectors of the sides
(*c*) angle bisectors (*d*) medians

92. Centroid of a triangle is the point of intersection of its:

(*a*) altitudes (*b*) medians (*c*) angle bisectors (*d*) none of these

Direction (Q. 93-95): Consider the following for the next three (03) items that follow:

ABC is a right-angled triangle with ∠ABC = 90°. The centre of the in circle of the given triangle is at O, whose radius is 2 cm. Two more circle with centres at O_1 and O_2, touch this circle and the two sides as shown in the given figure.

Further, MA : MC = 2 : 3. **[CDS (I) Exam, 2024]**

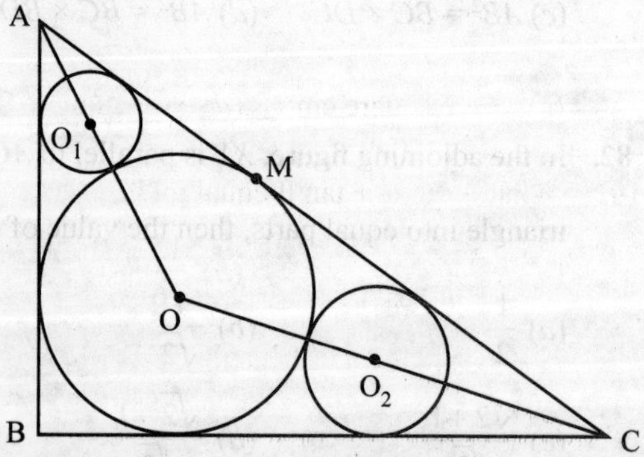

93. What is $AB + BC$ equal to?

(*a*) 10 cm (*b*) 12 cm
(*c*) 13 cm (*d*) 14 cm

94. What is the radius of the circle with centre at O_1 ?

(*a*) $4 - \sqrt{5}$ (*b*) $1 + \sqrt{5}$ (*c*) $2 + \sqrt{5}$ (*d*) $3 - \sqrt{5}$

95. What is the radius of the circle with centre at O_2 ?

(*a*) $5 - \sqrt{10}$ (*b*) $1 + 2\sqrt{5}$ (*c*) $\frac{22 - 4\sqrt{10}}{9}$ (*d*) $\frac{22 - 2\sqrt{10}}{9}$

96. *ABC* is a triangle and *O* is the point of intersection of its medians. Then :

(*a*) $3(AB^2 + BC^2 + CA^2) = 4(OA^2 + OB^2 + OC^2)$

(*b*) $4(AB^2 + BC^2 + CA^2) = 3(OA^2 + OB^2 + OC^2)$

(*c*) $\sqrt{AB^2 + BC^2 + CA^2} = 3\sqrt{OA^2 + OB^2 + OC^2}$

(*d*) $AB^2 + BC^2 + CA^2 = 3(OA^2 + OB^2 + OC^2)$

97. If *AB*, $m\angle A$ and $m\angle C$ are given, the number of triangles that can be constructed with this data is:

(*a*) one (*b*) two (*c*) three (*d*) nil

98. In a right angled triangle, if the square of the hypotenuse is twice the product of the other two sides, then one of the angles of the triangle is:

(*a*) 15° (*b*) 30° (*c*) 45° (*d*) 60°

99. In the given figure, if $\angle ABD = \angle ACE$, then ΔABC is:

(*a*) isosceles (*b*) equilateral

(*c*) right angled (*d*) obtuse angled

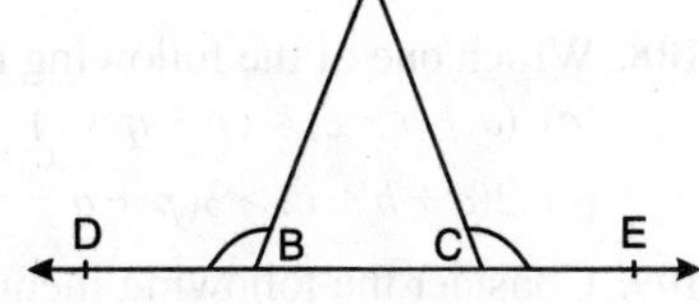

100. A 25 m long ladder is placed against a vertical wall inside a room such that the foot of the ladder is 7 m from the foot of the wall. If the top of the ladder slides 4 m downwards, then the foot of the ladder will slide by:

(*a*) 16 m (*b*) 8 m (*c*) 4 m (*d*) 2 m

Direction (Q. 101-103): Consider the following for the next three (3) items that follow:

A triangle CEF is drawn inside a square ABCD as shown in the figure given below. Given : CF = 8 cm, EF = 6 cm and CE = 10 cm.

[CDS (I) Exam, 2024]

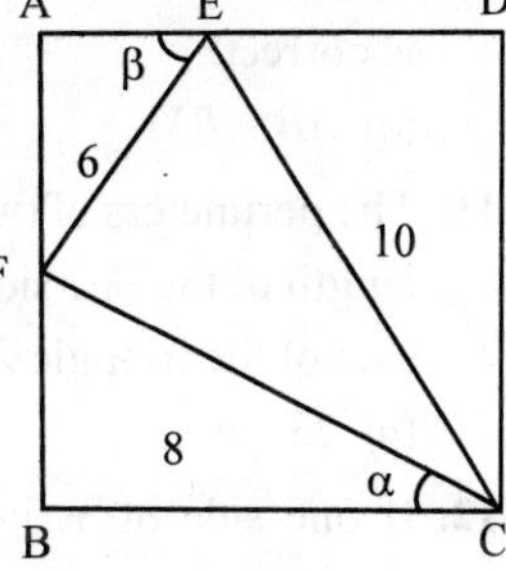

101. What is the area of the square ?

(*a*) $\frac{512}{17}$ square cm (*b*) $\frac{625}{13}$ square cm

(*c*) $\frac{1024}{17}$ square cm (*d*) $\frac{1296}{13}$ square cm

102. What is $\tan \alpha + \tan \beta$ equal to ?

(*a*) $\frac{13}{16}$ (*b*) $\frac{15}{16}$ (*c*) $\frac{17}{16}$ (*d*) $\frac{17}{4}$

103. What is the area of triangle CDE ?

(*a*) $\frac{416}{17}$ square cm (*b*) $\frac{312}{13}$ square cm (*c*) $\frac{208}{17}$ square cm (*d*) $\frac{156}{13}$ square cm

104. Δ *EFG* and Δ *LMN* are congruent to each other if:

(*a*) $\angle EFG = \angle LMN$, $\angle FEG = \angle MLN$ and $\angle FGE = \angle MNL$

(*b*) $EF = LM$, $EG = LN$ and $\angle EFG = \angle LMN$

(*c*) $EF = LM$, $\angle EFG = \angle LMN$ and $\angle FEG = \angle LNM$

(*d*) $FG = MN$, $EG = LN$ and $\angle EGF = \angle LNM$

105. *ABC* is a triangle in which $AB = AC$. The base *BC* is produced to *D* and $\angle ACD = 130°$. Then, angle *A* is equal to:

(*a*) 80° (*b*) 60° (*c*) 50° (*d*) 40°

106. ABC is a triangles right-angled at *C*. Let *P* be any point on *AC* and *Q* be any point on *BC*. Which of the following statements is/are correct? **[CDS (I) Exam, 2020]**

1. $AQ^2 + BP^2 = AB^2 + PQ^2$

2. $AB = 2\,PQ$

Select the correct answer using the code given below:

(*a*) 1 only (*b*) 2 only (*c*) Both 1 and 2 (*d*) Neither 1 nor 2

107. Three consecutive integers from the lengths of a right-angled triangle. How many sets of such three consecutive integers is/are possible? **[CDS (I) Exam, 2019]**

(*a*) Only one (*b*) Only two (*c*) Only three (*d*) Infinitely many

108. Which one of the following is correct? **[CDS (I) Exam, 2018]**

(*a*) $(a + b + c) < (p + q + r)$ (*b*) $3(a + b + c) < 4(p + q + r)$

(*c*) $2(a + b + c) > 3(p + q + r)$ (*d*) $3(a + b + c) > 4(p + q + r)$

109. Consider the following inequalities in respect of any triangle ABC:

1. AC – AB < BC

2. BC – AC < AB

3. AB – BC < AC

Which of the above are correct?

(*a*) 1 and 2 only (*b*) 2 and 3 only (*c*) 1 and 3 only (*d*) 1, 2, and 3

110. Let the bisector of the angle *BAC* of a triangle *ABC* meet *BC* in *X*. Which one of the following is correct? **[CDS (II) Exam, 2019]**

(*a*) $AB < BX$ (*b*) $AB > BX$ (*c*) $AX = CX$ (*d*) None of the above

111. The perimeters of two similar triangles *ABC* and *PQR* are 75 cm and 50 cm respectively. If the length of the one side of the triangle *PQR* is 20 cm, then what is the length of corresponding side of the triangle *ABC*? **[CDS (II) Exam, 2019]**

(*a*) 25 cm (*b*) 30 cm (*c*) 40 cm (*d*) 45 cm

112. If one side of a right-angled triangle (with all sides integers) is 15 cm, then what is the maximum perimeter of the triangle? **[CDS (II) Exam, 2019]**

(*a*) 240 cm (*b*) 225 cm (*c*) 113 cm (*d*) 112 cm

113. *ABC* is a triangle right angles at *C* with $BC = a$ and $AC = b$. If *p* is the length of the perpendicular from *C* on *AB*, then which one of the following is correct? **[CDS (I) Exam, 2018]**

(*a*) $a^2b^2 = p^2(a^2 + b^2)$ (*b*) $a^2b^2 = p^2(b^2 + a^2)$

(*c*) $2a^2b^2 = p^2(a^2 + b^2)$ (*d*) $a^2b^2 = 2p^2(a^2 + b^2)$

ANSWERS

1. (*d*)	**2.** (*a*)	**3.** (*a*)	**4.** (*b*)	**5.** (*a*)	**6.** (*d*)	**7.** (*c*)	**8.** (*c*)	**9.** (*d*)
10. (*a*)	**11.** (*b*)	**12.** (*b*)	**13.** (*c*)	**14.** (*b*)	**15.** (*a*)	**16.** (*d*)	**17.** (*b*)	**18.** (*b*)
19. (*c*)	**20.** (*b*)	**21.** (*a*)	**22.** (*b*)	**23.** (*b*)	**24.** (*b*)	**25.** (*b*)	**26.** (*a*)	**27.** (*b*)
28. (*b*)	**29.** (*d*)	**30.** (*d*)	**31.** (*a*)	**32.** (*c*)	**33.** (*c*)	**34.** (*c*)	**35.** (*b*)	**36.** (*d*)
37. (*c*)	**38.** (*c*)	**39.** (*c*)	**40.** (*c*)	**41.** (*d*)	**42.** (*a*)	**43.** (*d*)	**44.** (*d*)	**45.** (*d*)
46. (*b*)	**47.** (*b*)	**48.** (*c*)	**49.** (*d*)	**50.** (*c*)	**51.** (*c*)	**52.** (*b*)	**53.** (*c*)	**54.** (*b*)
55. (*a*)	**56.** (*c*)	**57.** (*b*)	**58.** (*c*)	**59.** (*c*)	**60.** (*d*)	**61.** (*a*)	**62.** (*b*)	**63.** (*c*)
64. (*b*)	**65.** (*a*)	**66.** (*c*)	**67.** (*a*)	**68.** (*c*)	**69.** (*c*)	**70.** (*d*)	**71.** (*a*)	**72.** (*c*)
73. (*c*)	**74.** (*a*)	**75.** (*d*)	**76.** (*c*)	**77.** (*b*)	**78.** (*c*)	**79.** (*b*)	**80.** (*b*)	**81.** (*d*)
82. (*b*)	**83.** (*a*)	**84.** (*b*)	**85.** (*c*)	**86.** (*a*)	**87.** (*b*)	**88.** (*c*)	**89.** (*b*)	**90.** (*b*)
91. (*a*)	**92.** (*b*)	**93.** (*d*)	**94.** (*d*)	**95.** (*c*)	**96.** (*d*)	**97.** (*b*)	**98.** (*c*)	**99.** (*a*)
100. (*b*)	**101.** (*c*)	**102.** (*d*)	**103.** (*a*)	**104.** (*d*)	**105.** (*a*)	**106.** (*a*)	**107.** (*a*)	**108.** (*b*)
109. (*d*)	**110.** (*a*)	**111.** (*b*)	**112.** (*a*)	**113.** (*a*)				

SOLUTIONS

1. In ΔABC

$BC^2 = AB^2 + AC^2$ [By Pythagoras theorem]

$AC^2 = (25)^2 - AB^2 \Rightarrow AC^2 = 625 - AB^2$ ...(i)

Now, in ΔADC

$AC^2 = AD^2 + CD^2 \Rightarrow AC^2 = AD^2 + 9^2$

$\Rightarrow AC^2 = AD^2 + 81$ ...(ii)

From (i) and (ii), We get

$625 - AB^2 = AD^2 + 81$

$\Rightarrow AB^2 = 544 - AD^2$ ...(iii)

In, ΔADB

$AB^2 = AD^2 + BD^2$

$\Rightarrow AB^2 = AD^2 + 16^2$

$\Rightarrow AB^2 = AD^2 + 256$ ...(iv)

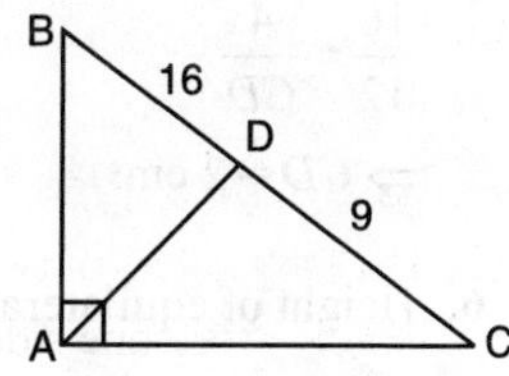

From (iii) and (iv), we get

$2AD^2 = 288$

$\Rightarrow AD^2 = 144 \Rightarrow AD = \sqrt{144} = 12$ cm

2. Since AD is the angle bisector of ∠A:

So, $\dfrac{AB}{AC} = \dfrac{BD}{DC}$

Given: $\dfrac{AB}{AC} = \dfrac{3}{1}$

Also, $AB + AC = 10$ cm

So, $AC = \dfrac{1}{(1+3)} \times 10 = \dfrac{1}{4} \times 10 = 2.5$ cm

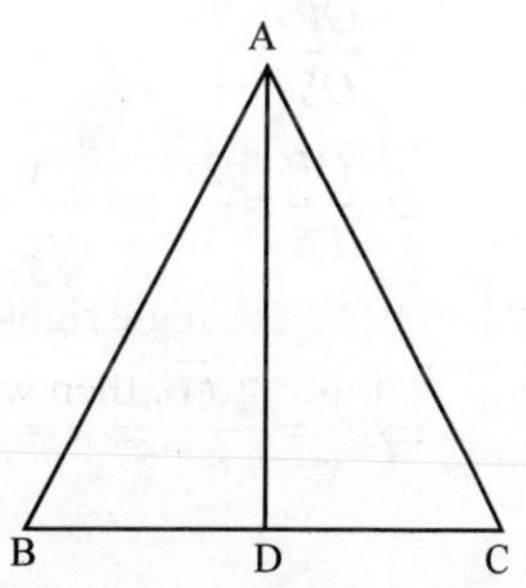

3. In ΔABC:

AD is the bisector of $\angle A$.

So, $\dfrac{BD}{DC} = \dfrac{AB}{AC} = \dfrac{2}{3}$

Since BC = 4 cm

So, $BD = \dfrac{2}{5} \times 4 = \dfrac{8}{5}$ cm

Now, in ΔABD:

BE is the bisector of $\angle B$.

So, $\dfrac{BD}{AB} = \dfrac{ED}{AE} \Rightarrow \dfrac{\frac{8}{5}}{2} = \dfrac{ED}{AE} \Rightarrow \dfrac{4}{5} = \dfrac{ED}{AE} \Rightarrow \dfrac{5}{4} = \dfrac{AE}{ED}$

So, $AE : ED = 5 : 4$

4. We have,

$\angle ABC = \angle ACB$ $[\because AB = AC]$

Now,

$\angle ACB + \angle ACD = 180°$

$\Rightarrow \angle ACB + x = 180° \Rightarrow \angle ACB = 180° - x = \angle ABC$

Now, in ΔABC

$\angle ABC + \angle ACB + \angle BAC = 180°$

$\Rightarrow \angle C = 2 \times 30° = 2 \times x = 2x$

$180° - x + 180° - x + \angle BAC = 180°$

$\Rightarrow \angle BAC = 2x - 180°$

5. We have,

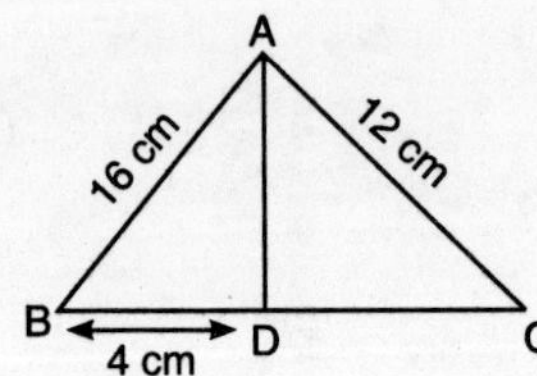

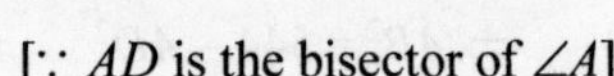
$\dfrac{AB}{AC} = \dfrac{BD}{CD}$ [$\because$ AD is the bisector of $\angle A$]

$\dfrac{16}{12} = \dfrac{4}{CD}$

$\Rightarrow CD = 2$ cm.

6. Height of equilateral $\Delta(PS) = \dfrac{\sqrt{3}}{2}x$

O is the centroid of triangle PQR

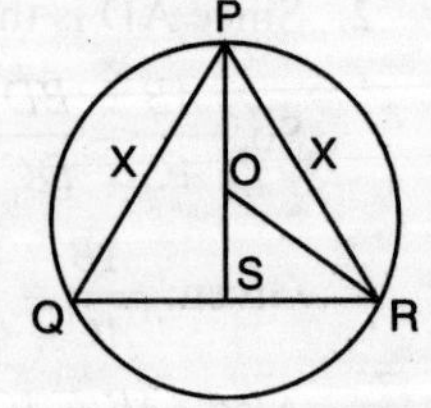

$\therefore \dfrac{OP}{OS} = \dfrac{2}{1}$

$\Rightarrow \dfrac{OP}{PS} = \dfrac{2}{3} \Rightarrow \dfrac{y}{\sqrt{3}\frac{x}{2}} = \dfrac{2}{3} \Rightarrow \dfrac{2y}{\sqrt{3}x} = \dfrac{2}{3}$

$\Rightarrow \dfrac{y}{x} = \dfrac{2\sqrt{3}}{6} \Rightarrow x = \sqrt{3}\,y$

7. Given angle between bisectors = 130°

Angle between bisectors of A and C inside triangle

$$= 90^\circ + \frac{\angle B}{2} = 130^\circ$$

$\angle B = 40° \times 2 = 80°$

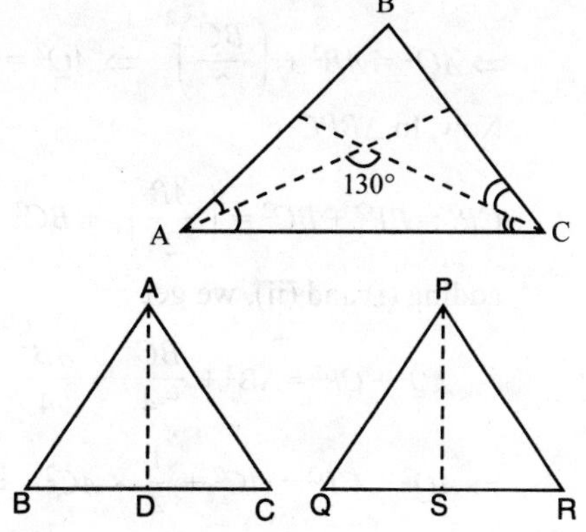

8. We have, $\dfrac{ar(\Delta ABC)}{ar(\Delta PQR)} = \dfrac{AD^2}{PS^2}$

$$\Rightarrow \frac{4.84}{5.29} = \frac{AD^2}{PS^2} \Rightarrow \left(\frac{22}{23}\right)^2 = \left(\frac{AD}{PS}\right)^2$$

$$\Rightarrow \frac{AD}{PS} = \frac{22}{23}$$

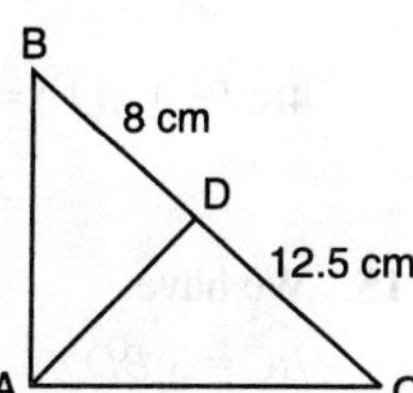

9. We have,

$AD^2 = BD \times CD$ [$\because$ ABC is a right angles Δ at A and $AD \perp BC$]

$= 8 \times 12.5 = 100$

$\Rightarrow AD = 10$ cm

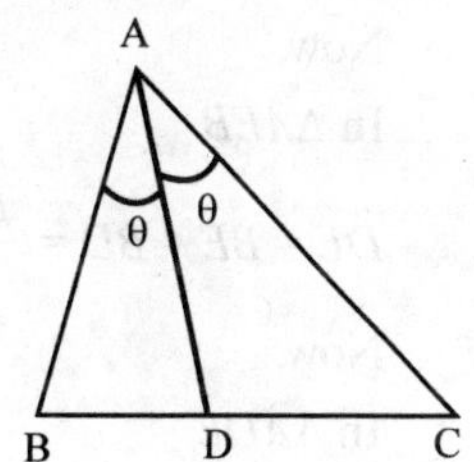

10. By Angle Bisector Theorem:

A line that bisects an angle divides the opposite side into two segments that are proportional to the other two sides of the triangle.

$$AB \cdot DC = AC \cdot BD \Rightarrow \frac{AB}{BD} = \frac{AC}{DC}$$

If $\dfrac{AB}{BD} = \dfrac{AC}{DC}$ then we can say that D bisects $\angle BAC$

$\Rightarrow$ Then, $\angle BAD = \angle CAD$ which is $\alpha = \beta$

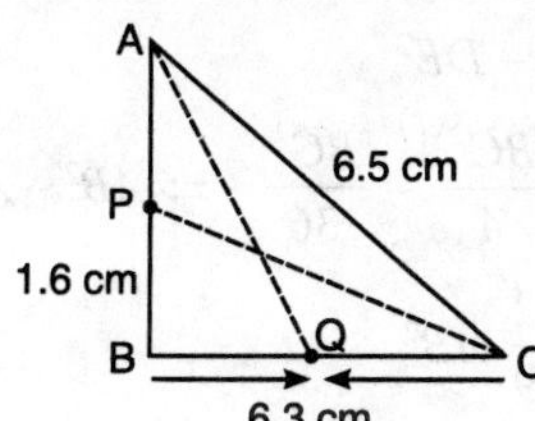

11. $BM = 8$ cm

$OC = OA = OB = 10$ cm

$AC = 20$ cm

$\therefore ar(\Delta ABC) = \dfrac{1}{2} \times 20 \times 8 = 80 \text{ cm}^2$

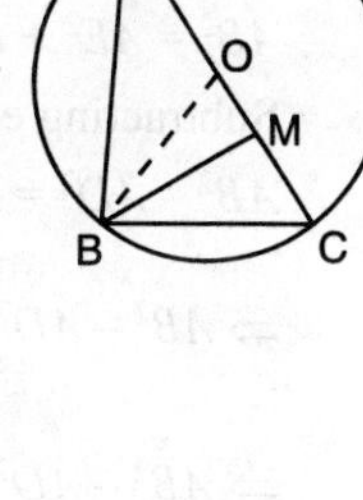

12. We have,

$AB^2 + BC^2$

$= (1.6)^2 + (6.5)^2 = 2.56 + 39.69$

$= 42.25 = (6.5)^2 = (BC)^2$

Thus, ΔABC is a right angle Δ.

Now, given that, $AP = PB$ and $BQ = QC$

Now, $AB^2 + 4BQ^2 = AB^2 + 4\left(\dfrac{BC}{2}\right)^2 = AB^2 + BC^2 = AC^2 = (6.5)^2 = 42.25 \text{ cm}^2$

13. From figure,

In ΔABQ

We have, $AQ^2 = AB^2 + BQ^2$

$$\Rightarrow AQ^2 = AB^2 + \left(\frac{BC}{2}\right)^2 \quad \Rightarrow AQ^2 = AB^2 + \frac{BC^2}{4} \quad \Rightarrow AQ^2 = AB^2 + \frac{BC^2}{4} \qquad \text{...(i)}$$

Now, In ΔBPC

$$CP^2 = BP^2 + BC^2 = \left(\frac{AB}{2}\right)^2 + BC^2 = \frac{AB^2}{4} + BC^2 \qquad \text{...(ii)}$$

adding (i) and (ii), we get

$$AQ^2 + CP^2 = \text{AB}^2 + \frac{BC^2}{4} + \frac{AB^2}{4} + BC^2$$

$$\Rightarrow AQ^2 + CP^2 = AC^2 + \frac{1}{4} \times AC^2 \quad \Rightarrow AQ^2 = CP^2 = \frac{5}{4}AC^2 \quad \Rightarrow AQ^2 + CP^2 = 1.25\ AC^2$$

14. From qn 13

$$4(CP^2 - AQ^2) = 4\left(\frac{AB^2}{4} + BC^2 - AB^2 - \frac{BC^2}{4}\right) = 3\ (BC^2 - AB^2) = 3\ (6.3)^2 - (1.6)^2)$$

$$= 3(39.69 - 2.56) = 3 \times 37.13 = 111.39 \text{ cm}^2$$

15. We have,

$BC = 3\ BD$

$$\Rightarrow BD = \frac{BC}{3} \text{ and } BE = \frac{BC}{2} \qquad [\because AE \text{ is median of } \Delta ABC]$$

Now,

In ΔAEB

$$DE = BE - BD = \frac{BC}{2} - \frac{BC}{3} = \frac{BC}{6}$$

Now,

In ΔADE

$AD^2 = AE^2 + DE^2$ ...(i) [By Pythagoras theorem]

and in ΔABE

$AB^2 = AE^2 + BE^2$ ...(ii)

Subtracting equation (i) and (i), we get

$AB^2 - AD^2 = BE^2 - DE^2$

$$\Rightarrow AB^2 - AD^2 = \frac{BC^2}{4} - \frac{BC^2}{36} \quad \Rightarrow AB^2 - AD^2 = \frac{8BC^2}{36}$$

$$\Rightarrow AB^2 - AD^2 = \frac{2}{9}AB^2 \qquad [\because \text{AB} = \text{BC} = \text{AC}]$$

$$\Rightarrow AD^2 = AB^2 = \frac{2}{9}AB^2 \Rightarrow AD^2 = \frac{7}{9}AB^2 \Rightarrow \frac{AD^2}{AB^2} = \frac{7}{9}$$

16. By checking options, we get

(a) 1 : 2 : 3

$\therefore x + 2x + 3x = 180° \Rightarrow 6x = 180° \Rightarrow x = 30°$

Thus, angles are 30°, 60° and 90°.

Hence, It is the proportion of their measures.

(b) Let angles are $3x$, $4x$ and $5x$

$\therefore$ We have, $3x + 4x + 5x = 180° \Rightarrow 12x = 180° \Rightarrow x = 15°$

Thus, angles are 45°, 60°, 75°

Hence, it is the proportion of their measures

(c) Let angles are $5x$, $6x$ and $7x$ respectively.

$\therefore$ We have, $5x + 6x + 7x = 180° \Rightarrow x = 10°$

$\therefore$ angles are 50°, 60° and 70°.

Thus, it is the proportion of their measures.

(d) Let angles are $6x$, $7x$ and $8x$ respectively.

$$6x + 7x + 8x = 180° \Rightarrow 21x = 108° \Rightarrow x = \frac{180°}{21} = \frac{60°}{7}$$

Thus, it is not the proportion of their measures.

17. We have,

$$2\angle A = 3\angle B = 6\angle C$$

$\Rightarrow$ Let $\angle C = x$

$$\angle A = 3x \text{ and } \angle B = 2x$$

Now, we have,

$$\angle A + \angle B + \angle C = 180$$

$\therefore$ $3x + 2x + x = 180°$

$\Rightarrow$ $6x = 180° \Rightarrow x = 30°$

$\therefore$ $\angle A = 3x = 90°$

$$\angle A = x = 30°$$

and $\therefore$ $\angle C = x = 30°$

$$\angle A + \angle C = 90° + 30 = 120°$$

18. $\therefore$ a, b, c are positive numbers.

$a^{\frac{1}{p}}, b^{\frac{1}{p}}, c^{\frac{1}{p}}$ are positive numbers $(\because p > 1)$

$a^{\frac{1}{p}} + b^{\frac{1}{p}} > c^{\frac{1}{p}}$ ($\because$ in a triangle sum of any two sides must be greater than 3rd side)

$\Rightarrow a^{\frac{1}{p}} + b^{\frac{1}{p}} - c^{\frac{1}{p}} > 0$

19. $ar(\Delta ABC) = ar(\Delta ABC)$

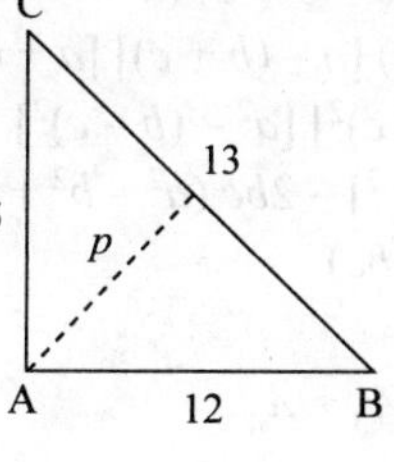

$\Rightarrow \frac{1}{2} \times AC \times AB = \frac{1}{2} \times p \times BC$

$\Rightarrow AC \times AB = p \times BC$

$\Rightarrow 5 \times 12 = p \times 13$

$\Rightarrow p = \frac{60}{13}$

$\Rightarrow 13p = 60$

20. $\therefore$ Required perimeter = Perimeter (P + R + S) – Perimeter of Q

$= (16 + 4 + 12) - 12$

$= 20$ cm

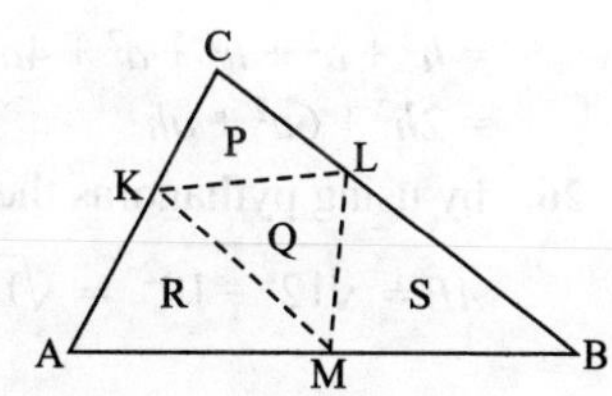

21. $AC^2 = BC^2 - AB^2$

$AC = \sqrt{(100 - 64)} = \sqrt{36} = 6$ cm

$AD = \frac{(AB \times AC)}{BC} = \frac{(6 \times 8)}{10} = 4.8$ cm

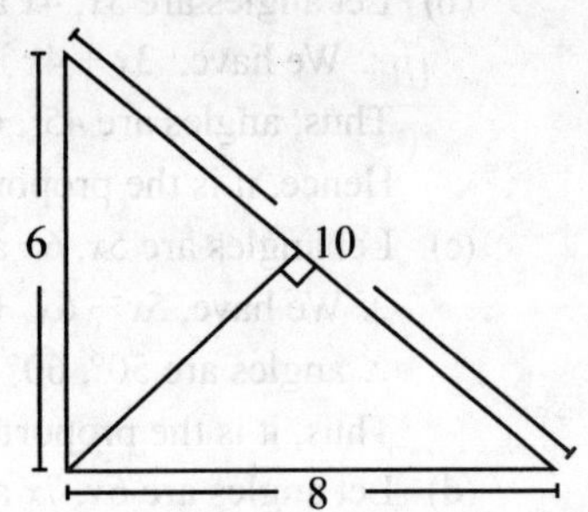

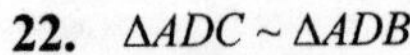

22. $\Delta ADC \sim \Delta ADB$

So, $\frac{\text{Ar } \Delta ADC}{\text{Ar } \Delta ADB} = \frac{6^2}{8^2} = \frac{36}{64} = \frac{9}{16}$

23.

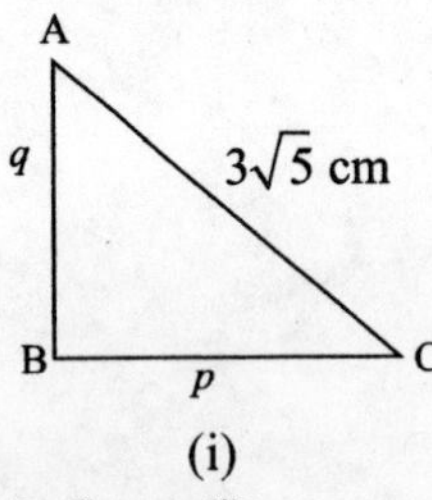

(i)

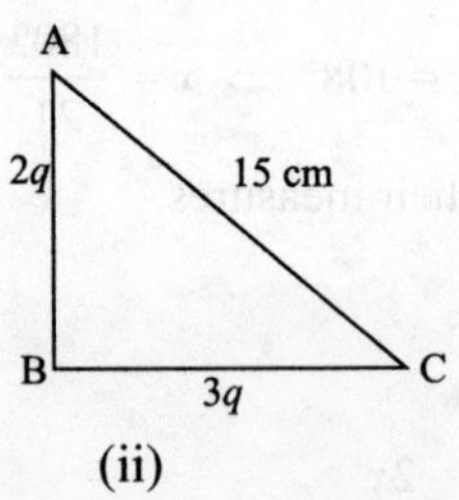

(ii)

In figure (i)

$\frac{3}{6} = \frac{5.2}{AR} \Rightarrow p^2 + q^2 = 45$ ...(i)

In figure (ii)

$(3p)^2 + (2q)^2 = (15)^2 \Rightarrow 9p^2 + 4q^2 = 225$ ...(ii)

Multiplying eqn (i) by 4 and substracting (i) and (ii), we get

$$p^2 = 9 \Rightarrow p = 3$$

by putting $p = 3$ in (i), we get

$q = 6$

$\therefore AB + BC = 9$ cm

24. a, b and c are sides of a triangle

$\because a > b > c$

$\therefore a^2 = b^2 + c^2 \Rightarrow a^2 - b^2 - c^2 = 0$

$(a + b + c)(a + b - c)(a - b + c)(a - b - c)$

$= (a + b + c)[a - (b + c)][a + (b - c)][(a - (b - c)]$

$= [a^2 - (b + c)^2][a^2 - (b - c)^2]$

$= [a^2 - b - c^2] - 2bc(a^2 - b^2 + 2bc]$

$= (-2bc)(2bc)$

$= -4b^2c^2$

25. Total area $= A_1 + A_2 + A_3 + A_4$

$= \sqrt{(h^2 + a^2)^2} + \sqrt{(h^2 + a^2)^2} + (2a)^2 + \frac{1}{2} \times 2a \times h$

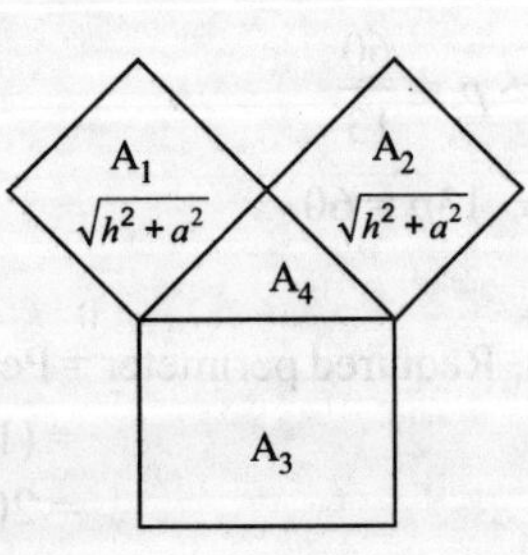

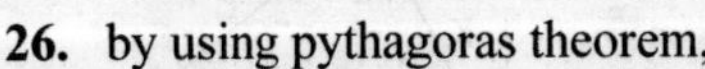

$= h^2 + a^2 + h^2 + a^2 + 4a^2 + ah$

$= 2h^2 + 6a^2 + ah$

26. by using pythagoras theorem,

$AB = \sqrt{12^2 - 10^2} = \sqrt{144 - 100} = \sqrt{44} = 2\sqrt{11}$

In ΔABC and ΔABO

$$\frac{AB}{AO} = \frac{BC}{BO} = \frac{AC}{AB}$$

$$\Rightarrow \frac{2\sqrt{11}}{AO} = \frac{10}{p} = \frac{12}{2\sqrt{11}} \Rightarrow \frac{10}{p} = \frac{12}{2\sqrt{11}} \Rightarrow p = \frac{5\sqrt{11}}{3}$$

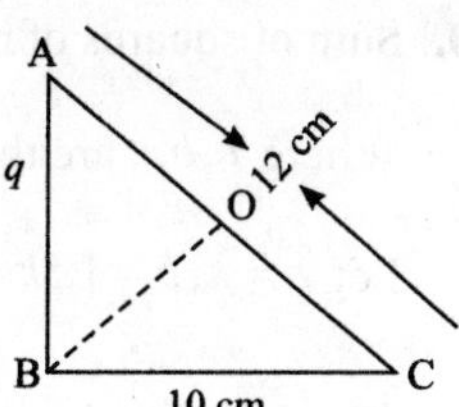

27. From Statement I, Triangle is not right angles.

Hence, statement I is not sufficient to answer the question.

Statement II:

We have, $x + 2x + 3x = 180°$ [$\because$ sum of all angles of a triangle is 180°]

$\Rightarrow 6x = 180° \Rightarrow x = 30°$

$\therefore$ Angles are 30°, 60° and 90°

$\therefore$ The triangle is right angled triangle.

28. Statement I:

$x + 2x + 3x = 180°$

$\Rightarrow 6x = 180°$ [$\because$ sum of three angles of a triangle is 18°]

$\Rightarrow x = 30°$

$\therefore$ Angles are 30°, 60° and 90°.

$\therefore$ Sides opposite to this angle = $1 : \sqrt{3} : 2$

By Pythagoras' theorem,

$$a^2 + 24^2 = 25^2 \Rightarrow a^2 = 625 - 576 = 49 \Rightarrow a = 7$$

But, 25, 24 and 7 are not in the ratio $1 : \sqrt{3} : 2$

$\therefore$ Statement I is not sufficient to answer the question

Statement II

$\therefore a = 1$

$\therefore QR \times PM = PQ \times PR$

$\Rightarrow 7 \times 24 = 25 \times PQ$

$\Rightarrow PQ = \frac{7 \times 24}{25} = 6.72$

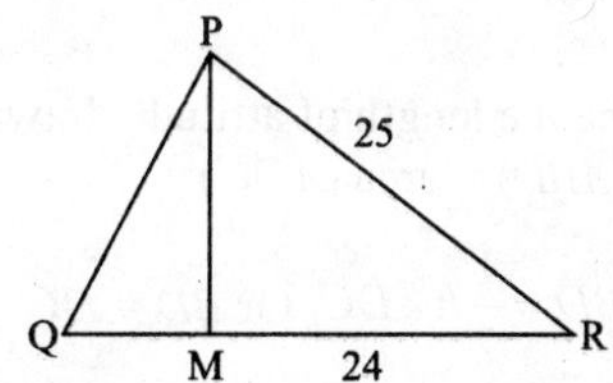

Statement II is sufficient to answer the question.

29. We have, $AC = AD + CD = 9 + 4 = 13$ cm

In ΔABC & ΔADB

$\angle ABC = \angle ADB = 90°$

$\angle A = \angle A$ [common]

$AB = AB$ [common]

$\therefore \quad \Delta ABC \cong \Delta ADB$

$\Rightarrow \quad \frac{AB}{AD} = \frac{AC}{AB}$ [$\therefore$ Corresponding sides of similar Δs are in equal ratio]

$\Rightarrow AB^2 = AD \times CA = 9 \times 13 = 117$ cm

Now, in ΔADB

$AB^2 = AD^2 + BD^2$ [By pythagoras theorem]

$\Rightarrow \; 117 = 9^2 + BD^2 \Rightarrow \; BD^2 = 117 - 81 = 36$

$= BD = 6$ cm

30. Sum of squares of medians $= \frac{3}{4}\left(a^2 + b^2 + c^2\right)$

Where a, b, c are the sides of the triangle.

Let $a = k, b = 1.5k = \frac{3k}{2}, c = 2.25k = \frac{9k}{4}$

$a^2 + b^2 + c^2 = k^2 + \frac{9}{4}k^2 + \frac{81}{16}k^2$

$a^2 + b^2 + c^2 = \frac{16}{16}k^2 + \frac{36}{16}k^2 + \frac{81}{16}k^2 = \frac{133}{16}k^2$

Sum of squares of medians $= \frac{3}{4} \times \frac{133}{16}k^2 = \frac{3 \times 133}{4 \times 16}k^2 = \frac{399}{64}k^2$

Therefore, the sum of the squares of the medians is: $\frac{399}{64}k^2$

31. The area of any triangle is equal to the product of its in radius and its semi-perimeter.

Area of the given right-angled triangle $= \frac{1}{2}ab = \frac{(a+b+c)}{2}r$

$\frac{1}{2}ab = \left(\frac{a+b+c}{2}\right)r$

$r = \frac{ab}{a+b+c} = \frac{ab(a+b-c)}{(a+b+c)(a+b-c)}$

$\frac{ab(a+b-c)}{(a+b)^2 - (c)^2} = \frac{ab(a+b-c)}{2ab} = \frac{a+b-c}{2}$

32. (1)→

Let h be the length of attitude drawn from A to BC.

Area (ABD) = area (ADC) ________ (1)

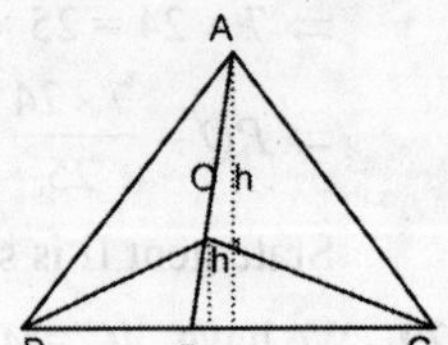

$\frac{1}{2}h \times BD = \frac{1}{2}h \times DC$ Or $BD = DC$

i.e., AD is a median.

Let h^1 be the length of the attitude O drawn to BC.

Area (OBD) = $\frac{x}{a} = 1 - \frac{c}{z}$ = $1/2h^1 \times CD$ = Area (ODC) ________ (2)

Area (ABO) = Area (ABD) – Area (OBD)

From (1) & (2)

Area (ABO) = Area (ADC) – Area (ODC) = area (AOC).

(1) is true.

(2)→

Let D, E and F be the points where the median meets BC, AC and AB respectively.

Area (ABD) = Area (ADC)

= 1/2 Area (ABC)

(∴ BD = DC = $\frac{BC}{2}$ and heights are equal).

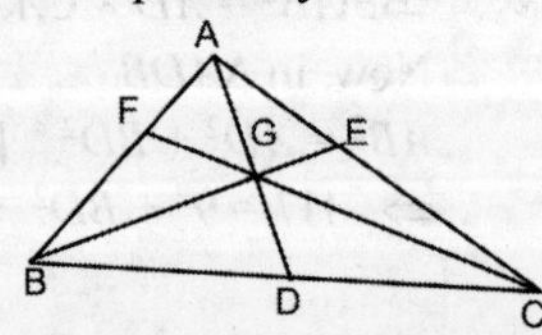

Since G is the centroid,

$AG = 2GD = \frac{2}{3}AD$

$\Rightarrow$ Area (ΔABG) = $\frac{2}{3}$ Area (ABD) = $\frac{2}{3}\times\frac{1}{2}\times$ Area (ABC) = $\frac{1}{3}$ Area (ABC).

Similarly, we can find that

Area (AGC) = Area (BGC) = 1/3 Area (ABC)

(2) is true.

33.

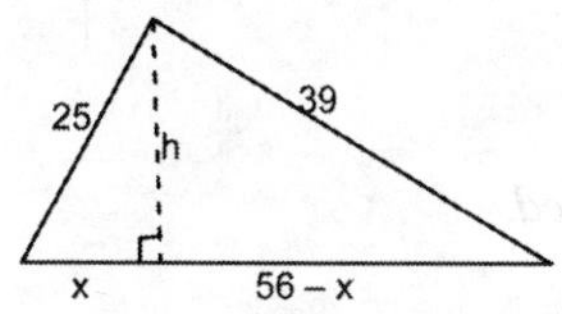

$25^2 = x^2 + h^2$ ... (1)

$39^2 = (56 - x)^2 + h^2$ ... (2)

(2) – (1) $\Rightarrow$

$\Rightarrow (56 - x)^2 - x^2 = (39 + 25)(39 - 25) \Rightarrow -112x + 56^2 = 64(14) \Rightarrow 112x = 2240 \Rightarrow x = 20.$

$\therefore\ h^2 = 25^2 - 20^2 = 225.$

$\therefore\ h = 15.$

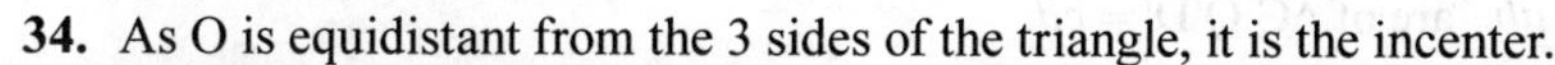

34. As O is equidistant from the 3 sides of the triangle, it is the incenter.

(**Note:** O is generally used for the orthocenter. We should be careful not to think of the orthocenter)

$\therefore$ OA, OB, OC bisect $\angle A$, $\angle B$, $\angle C$ respectively.

(1) $\angle OAC + \angle OCB + \angle OBA = x + y + z = 90°$.

(2) $\angle BOC = 2x + y + z = x + (x + y + z) = 90° + \frac{\angle A}{2}$.

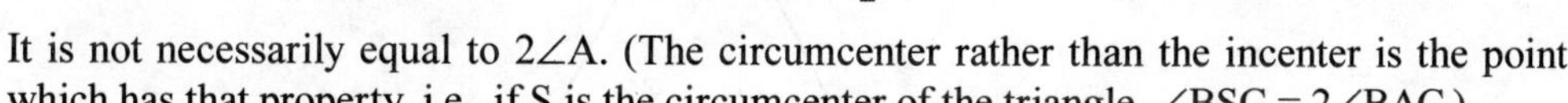

It is not necessarily equal to 2$\angle$A. (The circumcenter rather than the incenter is the point which has that property, i.e., if S is the circumcenter of the triangle, $\angle BSC = 2\angle BAC$.)

(3) This is a property of the angle bisector.

$\therefore$ (1), (3) are true.

35. Since AB = 6 cm and BC = 8 cm

$\angle ABC = 60^\circ$

$\sin\angle ABC = \sin 60^\circ = \frac{\sqrt{3}}{2}$

Area $= \frac{1}{2}\times AB\times BC\times\sin 60^\circ$

$= \frac{1}{2}\times 6\times 8\times\frac{\sqrt{3}}{2} = 12\sqrt{3}$

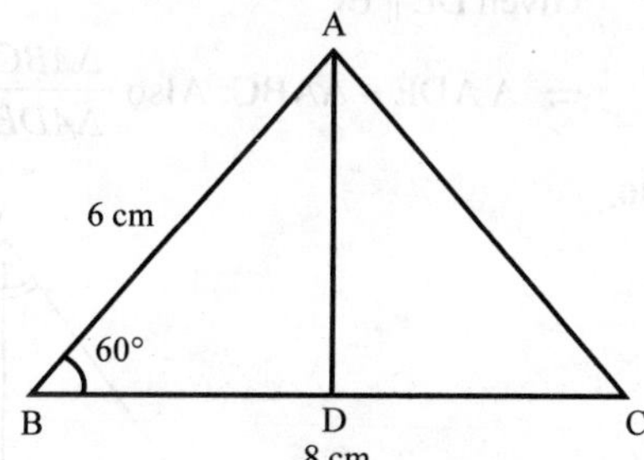

36. ΔBPQ + Δ CPD

The corresponding sides are in the ratio 1 : 2.

$\therefore$ The areas are in the ratio 1 : 4.

As the area of ΔBPQ is 20, that of ΔCPD is 80.

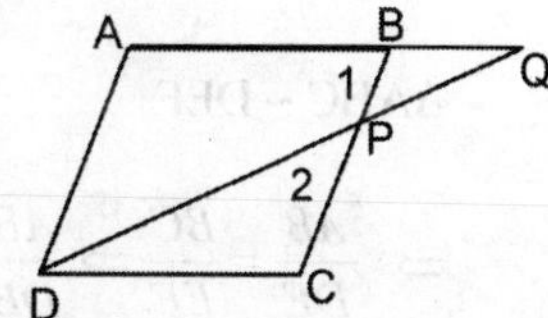

37. P be the starting and T be the final position of the persons.
From Δ RPQ and ΔRST
RP = RS = x
∠P = LS = 90º
∠PRQ = ∠SRT (Vertically opposite)
∴ Δ RPQ ~ ΔRST angles)

$$\Rightarrow \frac{PQ}{TS} = \frac{RP}{RS} \Rightarrow \frac{PQ}{y} = \frac{x}{x} \Rightarrow PQ = y$$

38. Let AP = a, PB = b, CP = c, PD = d
As chords AB, CD interect at P, AP(PB) = CP(PD), i.e. $ab = cd$.

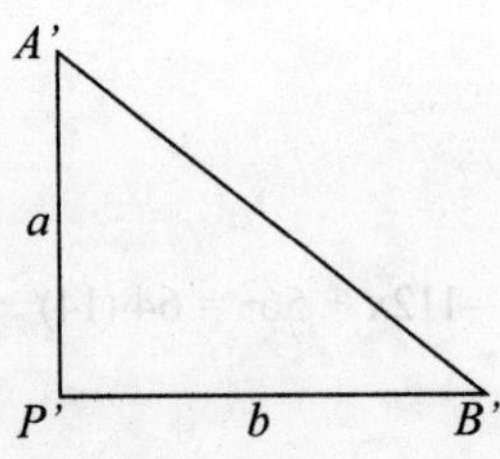

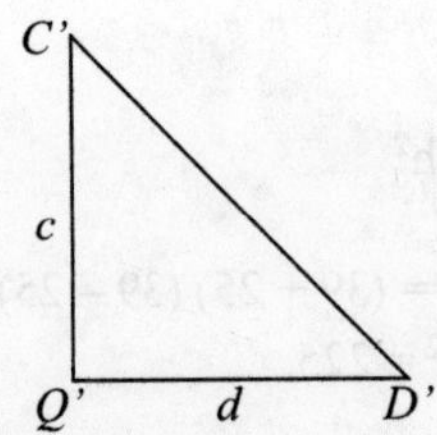

Area of $\Delta A'P'B' = ab$, are of ΔC'Q'D' = cd.
Only (3) is true
∴ 1, 2, 4 are not correct.

39.

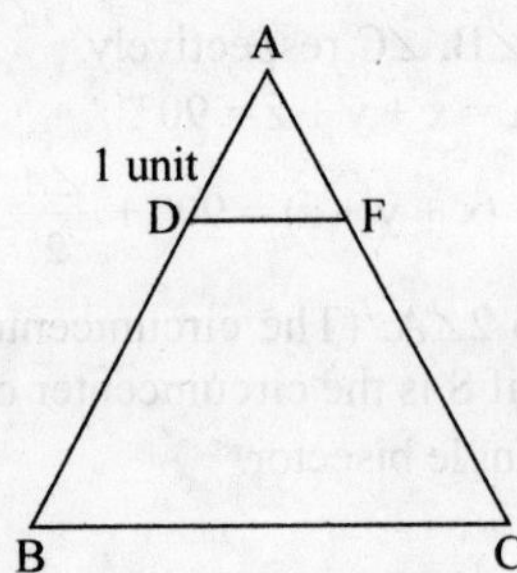

Given DE || BC

$$\Rightarrow \Delta ADE \sim \Delta ABC; \text{Also } \frac{\Delta ABC}{\Delta ADE} = \frac{AB^2}{AD^2} = 2 \Rightarrow \frac{1}{AD^2} = 2 \Rightarrow AD = \frac{1}{\sqrt{2}}$$

40.

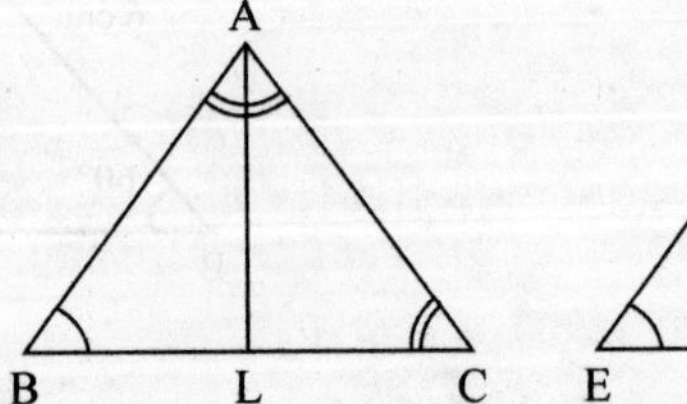

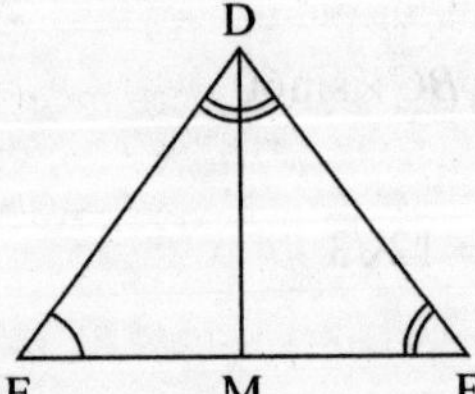

ΔABC ~ DEF

$$\Rightarrow \frac{AB}{DE} = \frac{BC}{EF} \Rightarrow \frac{AB}{DE} = \frac{\frac{1}{2}BC}{\frac{1}{2}MF} = \frac{BC}{EM} \text{ and } \angle B = \angle E$$

∴ ΔABL ~ ΔDFM (SAS)

∴ Statement (1) is true.

If two triangles are congruent then their corresponding sides are always equal.

∴ ΔALC is not congruent to Δ DMF as AC ≠ DF.

∴ Statement (ii), is false.

41. Statement I:

∵ P divides BC into two equal parts, So, AP is a median

⇒ AP divides ΔABC into two equal parts.

Hence, statement I follows.

Statement II: AB = 10 cm, BC = 20 cm, BP = 10 cm

By Pythagoras theorem:

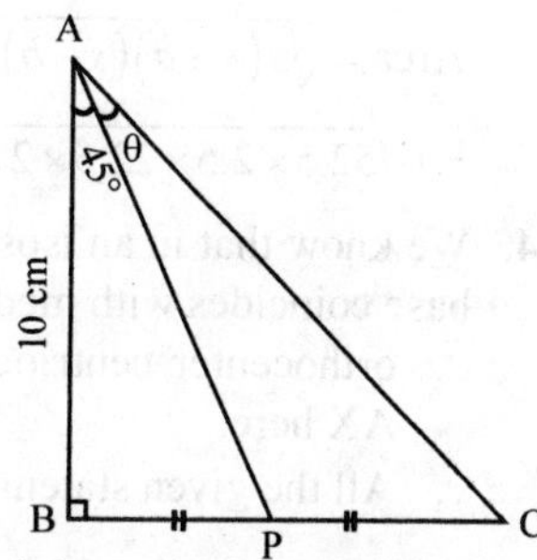

$$AC^2 = AB^2 + BC^2$$

$$\Rightarrow \quad AC^2 = 100 + 400 = 500$$

$$\Rightarrow \quad AC = \sqrt{500} = 22.36$$

$$AP^2 = AB^2 + BP^2$$

$$\Rightarrow \quad AP^2 = 100 + 100 = 200$$

$$\Rightarrow \quad AP = \sqrt{200} = 14.14$$

∴ The perimeter of ΔAPC = $10 + 14.14 + 22.36 = 46.50$ cm which is more than 46 cm.

Hence, statement II also follows.

Statement III: Area of ΔABC = $\frac{1}{2} \times BC \times AB = \frac{1}{2} \times 20 \times 10 = 100 \text{ cm}^2$

Area of $\Delta APC = \frac{100}{2} = 50 \text{ cm}^2$

Hence, statement III also follows.

All statements follow.

42. Let the sides of the triangle be *a*, *b*, *c*, where:

$a = BC$ (side opposite to vertex *A*) $b = CA$ (side opposite to vertex *B*)

$c = AB$ (side opposite to vertex *C*)

Let the altitudes be (h_a, h_b, h_c), where: $h_a = AD, h_b = BE, h_c = CF$

Given the ratio of altitudes: $h_a : h_b : h_c = 3 : 5 : 6$

Using the property that sides are inversely proportional to altitudes:

$$a : b : c = \left(\frac{1}{3}\right) : \left(\frac{1}{5}\right) : \left(\frac{1}{6}\right) \Rightarrow a : b : c = \left(\frac{1}{3}\right) \times 30 : \left(\frac{1}{5}\right) \times 30 : \left(\frac{1}{6}\right) \times 30 \Rightarrow a : b : c = 10 : 6 : 5$$

This means the ratio of the sides BC : CA : AB is 10 : 6 : 5.

Therefore, AB : BC : CA = c : a : b = 5 : 10 : 6.

43. The ratio of the sides AB : BC : CA = 5 : 10 : 6.

Let the sides be AB = 5x, BC = 10x, and CA = 6x.

The perimeter is the sum of the sides:

Perimeter = $AB + BC + CA = 5x + 10x + 6x = 21x$

Given Perimeter = 105 cm.

$21x = 105 \Rightarrow x = \left(\frac{105}{21}\right) = 5$

$a(BC) = 10x = 10 \times 5 = 50$ cm

$b(CA) = 6x = 6 \times 5 = 30$ cm

$c(AB) = 5x = 5 \times 5 = 25$ cm

The semi-perimeter (s): $s = \left(\frac{\text{Perimeter}}{2}\right) = \left(\frac{105}{2}\right) = 52.5$ cm

Area using Heron's formula:

Area $= \sqrt{s(s-a)(s-b)(s-c)} = (\sqrt{(52.5)\times(52.5-50)\times(52.5-30)\times(52.5-25)})$

$= (\sqrt{52.5\times 2.5\times 22.5\times 27.5}) = (\sqrt{99738.28125}) \approx 285 \text{ cm}^2$

44. We know that in an isosceles triangle the perpendicular drawn on the base coincides with median and perpendicular bisector of base (BC)

$\therefore$ orthocenter, centriod and in centre of the triangle lies on median ie, AX here.

$\therefore$ All the given statements are true

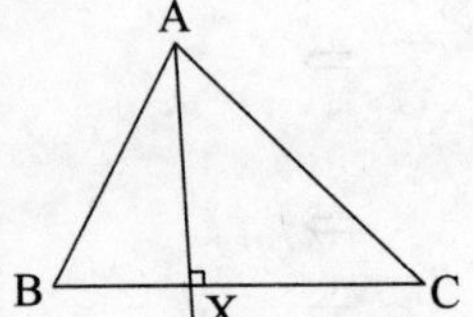

45.

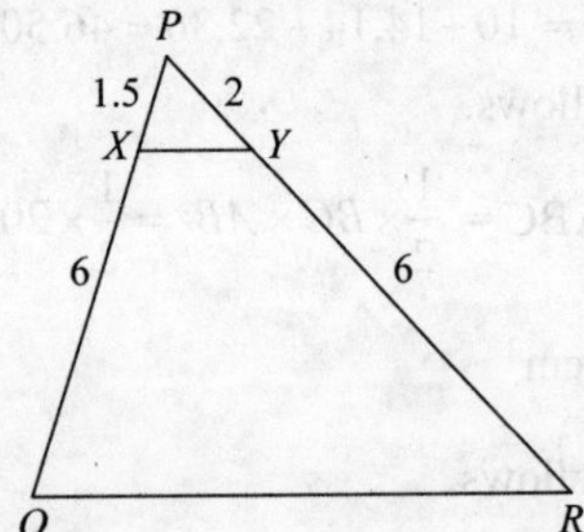

As $\frac{PX}{PQ} = \frac{PY}{PR} = \frac{1}{5}$ if follows that XY is parallel to QR and $\frac{XY}{QR} = \frac{1}{5}$

All 3 statements are true.

46. $AP = \frac{AB^2}{AC} = \frac{15^2}{25} = \frac{225}{25} = 9$ cmn

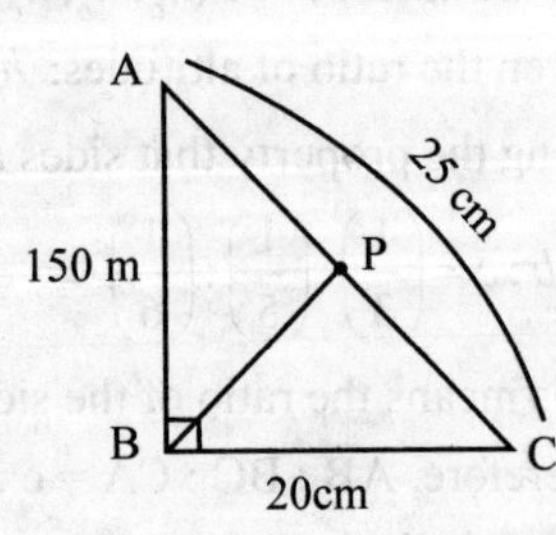

$PC = \frac{BC^2}{AC} = \frac{20^2}{25} = \frac{400}{25} = 16$ cm

$BP = \frac{AB \times BC}{AC} = \frac{15 \times 20}{25} = 12$ cm

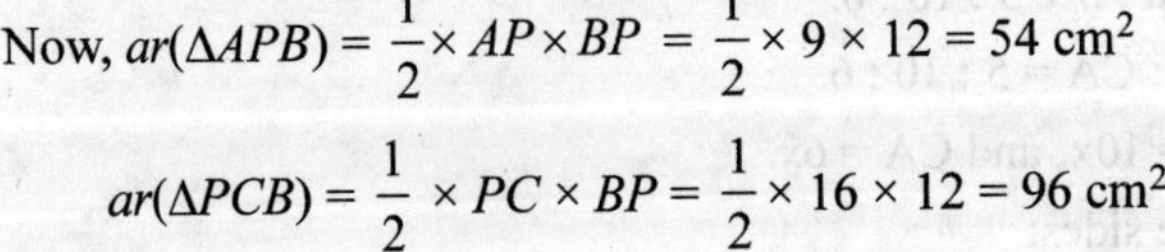

Now, $ar(\Delta APB) = \frac{1}{2} \times AP \times BP = \frac{1}{2} \times 9 \times 12 = 54 \text{ cm}^2$

$ar(\Delta PCB) = \frac{1}{2} \times PC \times BP = \frac{1}{2} \times 16 \times 12 = 96 \text{ cm}^2$

$\therefore$ required difference $= 96 - 54 = 42 \text{ cm}^2$

47. DA = DC $\therefore$ $\angle A = \angle DCA = Q$ (say)
$\angle CDB = \angle A + \angle DCA = 2Q$. Also $\angle B = 2Q$ ($\because$ CD = CB)
Also $\angle A + \angle B = 96^\circ$ $\therefore$ $Q + 2Q = 96^\circ$ or $Q = 32^\circ$
$\angle CDB = 2Q = 64^\circ$

48. BX = CY = AZ

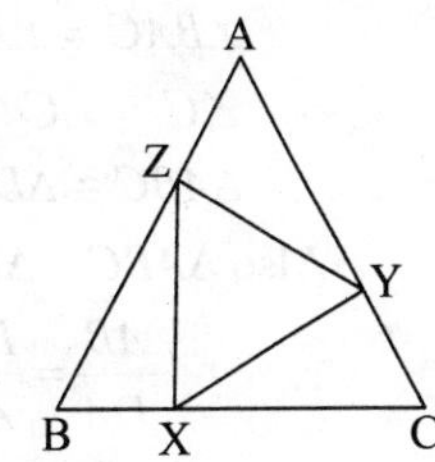

$\therefore$ BZ = CX = AY
Also $\angle B = \angle C = \angle A = 60^\circ$
$\therefore$ $\Delta BXZ \cong \Delta XYC \cong AZY$ and
Hence XZ = YX = ZY
Both statements 1 and 2 are true.

49.

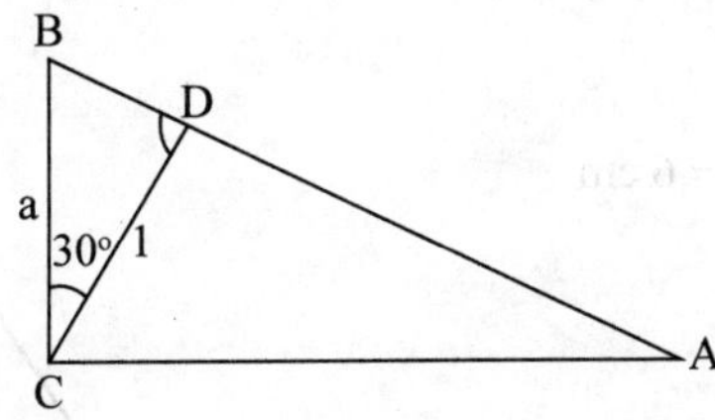

From the sine rule $\frac{a}{\sin D} = \frac{l}{\sin B}$ (Given a = 5 cm)

As D and B + 30° are supplements, sin α = sin (β + 30°)

$$\therefore\ 1 = \frac{a\sin B}{\sin(B+30^\circ)} = \frac{a\sin B}{\sin B\cos 30^\circ + \cos B\sin 30^\circ} = a\left(\frac{12}{13}\right)\frac{1}{\frac{12}{13}\left(\frac{\sqrt{3}}{2}\right)+\frac{5}{13}\frac{1}{2}} = \frac{120}{5+12\sqrt{3}}$$

50. If the triangles are denoted as T_1, T_2, T_3, ..., the ratio of the sides is $1 : \frac{1}{2} : \frac{1}{4} :$ ......and the ratio of the areas in $1 : \frac{1}{4} : \frac{1}{16}$ .... The ratio of the area of T_1, to that of T_4, is or of T_4 to T_7 is 64 : 1.

51. As clear from the figure, If the sides of the triangle are $\sqrt{a^2+b^2}$, $\sqrt{a^2+c^2}$, b + c, the area is $\frac{a(b+c)}{2}$

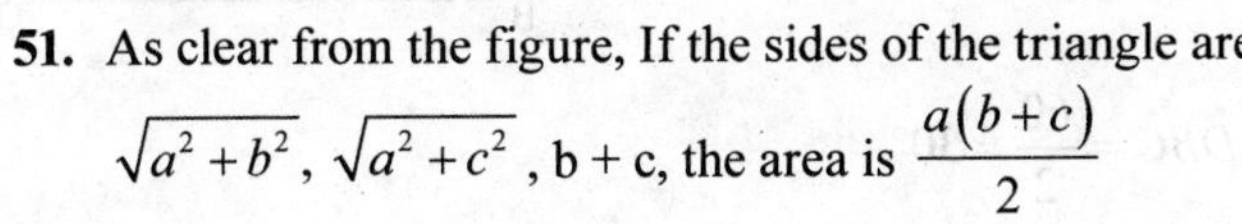

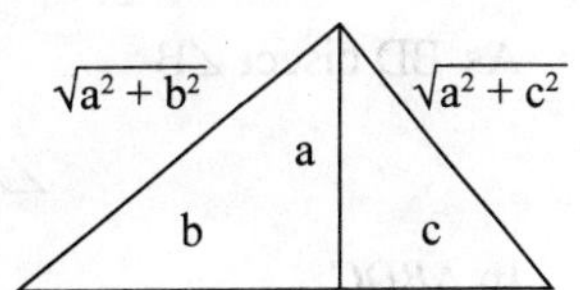

52. Smallest angle = $\frac{2}{9} \times 180 = 40^\circ$

53.

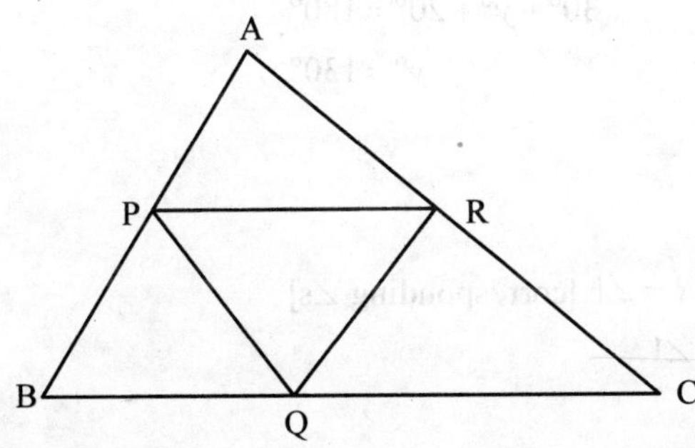

$$ar(PQR) = \frac{1}{4}ar(ABC) = \frac{1}{4} \times 5 = \frac{5}{4} \text{square units}$$

54. In ΔABC and ΔDAC

$\angle BAC = \angle ADC$

$\angle C = \angle C$

$\therefore \quad \Delta ABC \cong \Delta DAC$

Also $\Delta ABC \sim \Delta DAC$

$\therefore \quad \frac{AB}{DA} = \frac{BC}{AC} = \frac{AC}{Dc}$

$\frac{12}{AC} = \frac{AC}{3}$

$AC^2 = 36$, $AC = 6$ cm

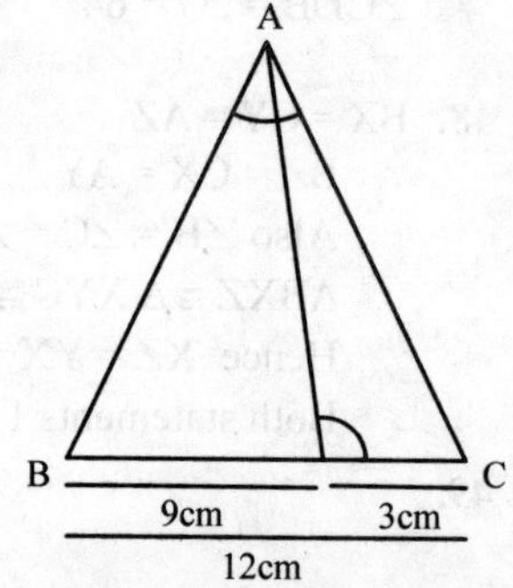

55. In ΔOAB and ΔOCD

$\angle 1 = \angle 2$ (Alt $\angle$s)

$\angle 3 = \angle 4$ (Alt $\angle$s)

$\angle 5 = \angle 6$ (Vertically opposite $\angle$s)

$\therefore$ ΔOAB ~ ΔOCD (AAA)

$\therefore \quad \frac{OA}{OC} = \frac{OB}{OD}$

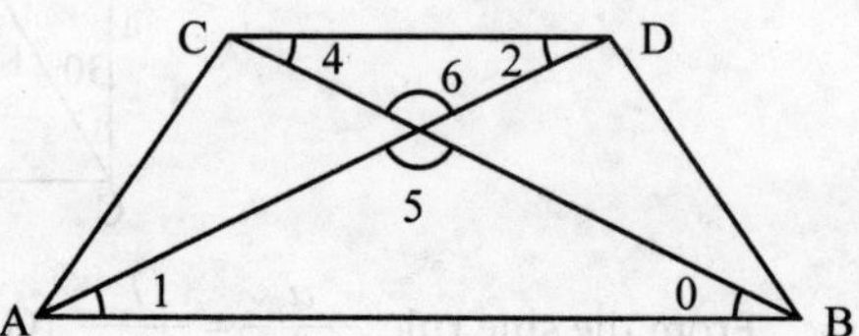

56. In ΔABC

$\angle A + \angle B + \angle C = 180°$

$80° + 60° + \angle C = 180°$

$\therefore \quad \angle C = 180° - 140° = 40°$

As CD bisect $\angle$C,

$\angle ACD = x = \frac{40}{2} = 20°$

As BD bisect $\angle$B

$$\angle ABD = \angle DBC = \frac{60}{2} = 30°$$

In ΔBDC

$$\angle DBC + y° + x° = 180°$$
$$30° + y° + 20° = 180°$$
$$y° = 130°$$

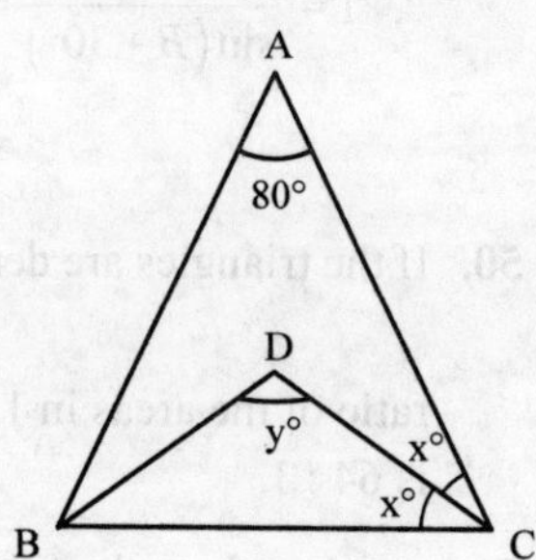

57. Let $\angle MLR = \angle 1$

$\therefore$ *As* $ML \parallel TQ$

$\angle MLR = \angle TQR = \angle 1$ [corresponding $\angle$s]

$\therefore \quad \angle PQR = 90° - \angle 1$

As $QT = PT$

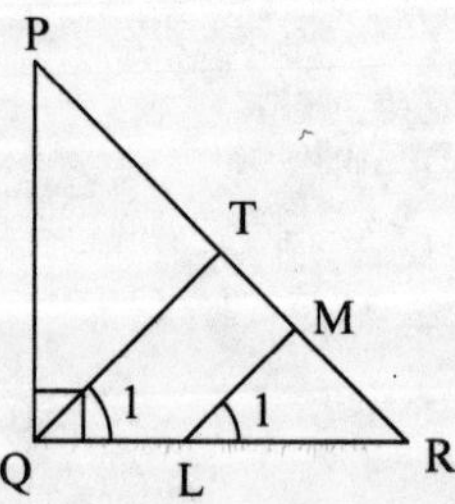

$\therefore \quad \angle PQT = \angle QPR = 90° - \angle 1$

In ΔPQR

$$\angle P + \angle R + 90° = 180° \Rightarrow \angle P + \angle R = 90° \Rightarrow \angle QPR + \angle R = 90°$$

$\Rightarrow \; 90° - \angle 1 + \angle R = 90°$

$\Rightarrow \; \angle 1 = \angle R$

$\angle RLM = \angle LRM$

58. As $\angle 1 = \angle 2$

$\therefore$ ΔADC is an isosceles Δ

Let BE ⊥ AC

$\therefore$ D can lie anywhere on BE

So, D is not the centroid of ΔABC

In ΔABD and ΔCBD

AB = BC (Given)

AD = DC ($\because \angle 1 = \angle 2$)

BD = BD (common)

$\therefore \quad \Delta ABD \cong \Delta CBD$ [*SSS*]

So, statement 1 and 3 are correct.

59. In ΔADM and ΔBCM

$\angle 1 = \angle 2$ (given)

$\angle 3 + \angle 5 = \angle 4 + \angle 5$

$AM = MB$ [M is midpoint of AB]

$\therefore \quad \Delta ADM \cong \Delta BCM$ [*ASA*]

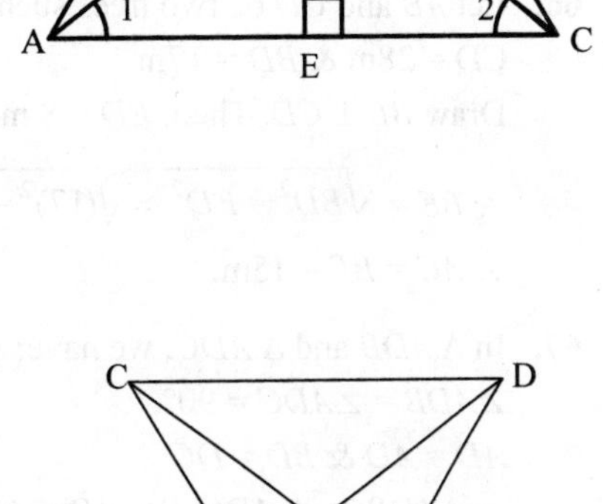

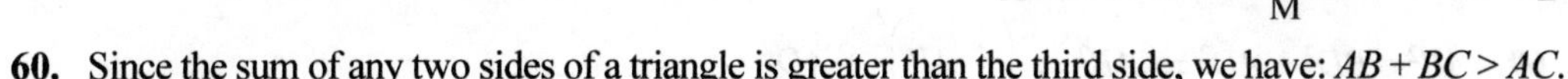

60. Since the sum of any two sides of a triangle is greater than the third side, we have: $AB + BC > AC$.

61. Clearly, $(13)^2 + 10)^2 > (16)^2$ *i.e.* $(16)^2 < (13)^2 + (10)^2$.

So, the triangle is acute: [$\because AC^2 < AB^2 + BC^2$]

62. $\angle CBD = \angle A + \angle C$, $\angle BCE = \angle B + \angle A$.

$\therefore \angle CBD + \angle BCE = (\angle A + \angle B + \angle C) + \angle A = 180° + A$

$$\frac{1}{2}\angle CBD + \frac{1}{2}\angle BCE = 90° + \frac{A}{2}.$$

But $\frac{1}{2}\angle CBD + \frac{1}{2}\angle BCE + \angle BOC = 180°$.

$$\therefore \angle BOC = 180° - \left(90° + \frac{A}{2}\right) = \left(90° - \frac{A}{2}\right) = \left(90° - \frac{64}{2}\right) = 58°.$$

63. $AB^2 + AC^2 = 2AD^2 + 2\left(\frac{1}{2}BC\right)^2 \Rightarrow AB^2 + AC^2 = 2AD^2 + \frac{1}{2}BC^2$.

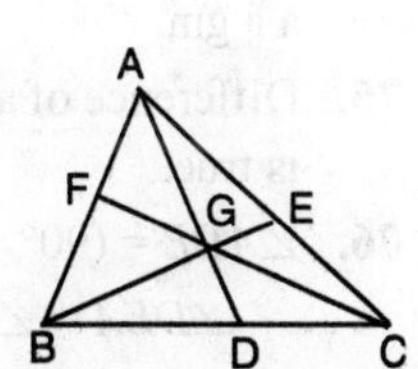

$$BC^2 + AB^2 = 2BE^2 + \frac{1}{2}AC^2$$

and $BC^2 + AC^2 = 2CF^2 + \frac{1}{2}AB^2$

Adding we get $3(AB^2 + BC^2 + AC^2) = 4(AD^2 + BE^2 + CF^2)$.

64. Let the angles be $(2x)^0$, $(3x)^0$ & $(4x)^0$. Then,

$2x + 3x + 4x = 180 \Rightarrow x = 20$.

$\therefore$ Greatest angle $= 4x = 80°$.

65. Since $DM = DN$, so D lies on the bisector of $\angle A$.

Clearly, $\Delta ABD \cong \Delta ADC$. So, $AB = AC$.

So, the triangle is isosceles.

66. Let AB and CD be two trees such that $AB = 20$m,

CD = 28m & $BD = 17$m

Draw $BE \perp CD$. Then, $ED = 8$ m.

$\therefore BE = \sqrt{BD^2 - ED^2} = \sqrt{(17)^2 - 8^2} = \sqrt{225} = 15$ m.

$\therefore AC = BE = 15$m.

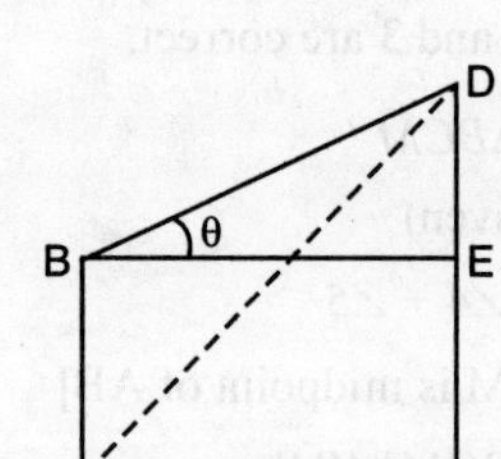

67. In ΔADB and ΔADC, we have:

$\angle ADB = \angle ADC = 90°$,

$AD = AD$ & $BD = DC$.

$\therefore \Delta ADB \cong \Delta ADC$. So, $AB = AC$.

Hence, the triangle is isosceles.

68. $\frac{BD}{DC} = \frac{AB}{AC} = \frac{3}{4}$. So, $BD : DC = 3 : 4$.

69. $AB^2 = BD^2 + AD^2 \Rightarrow AB^2 \Rightarrow \left(\frac{1}{2}AB\right)^2 + AD^2$

$\therefore 4AB^2 = AB^2 + 4AD^2 \Rightarrow 3AB^2 = 4AD^2$

70. $\frac{AB}{AC} = \sin 60° = \frac{\sqrt{3}}{2}$

$\therefore AB = \left(\frac{\sqrt{3}}{2} \times \text{hypotenuse}\right)$.

71. Clearly, (*a*) is true.

72. Since perpendicular distance is smallest, we have $AB > AP$.

73. The circumcentre of a right angled triangle lies at the mid-point of the hypotenuse.

74. The mid-points of the sides of a triangle along with any of the vertices as the fourth point make a || gm.

75. Difference of any two sides of a triangle is always less than the third side. So, $(AB - AC) < BC$ is true.

76. $\angle ADE = (90° + 60°) = 150°$. Also, $DE = AD\ (= DC)$

$\therefore \angle DEA = \angle DAE = x^0$(say) $\therefore x + x + 150 = 180 \Rightarrow x = 15°$.

77. Let $\angle BAD = y°$. Then, $\angle ABD = y°$ and $\angle ADC = x^0$.

$\therefore \angle DAC = (180° - 2x)$.

Ext. $\angle ADC = \angle DBA + \angle BAD \Rightarrow x = 2y$

$$\Rightarrow \quad y = \frac{1}{2}x.$$

$$\therefore \quad \frac{1}{2}x + 180 - 2x + 75 = 180 \Rightarrow x = 50°.$$

78. Since $\angle A$ is an obtuse angle in ΔABC, so

$$BC^2 = AB^2 + AC^2 + 2AB.AD = AB^2 + AC^2 + 2AB \cdot \frac{1}{2}AC \qquad \left[\because AD = AC\cos 60° = \frac{1}{2}AC\right]$$

$$= AB^2 + AC^2 + AB.AC.$$

$\therefore \; a^2 = b^2 + c^2 + bc.$

79. Since $\angle A$ is acute, we have:

$$BC^2 = AC^2 + AB^2 - 2AC \,.\, AD = AC^2 + AB^2 - 2AC \cdot \frac{1}{2}AB \qquad \left[\because AD = AB\cos 60° = \frac{1}{2}AB\right]$$

$$= AC^2 + AB^2 - AC \,.\, AB$$

$\therefore a^2 = b^2 + c^2 - bc.$

80. Area $(\Delta ABC) = \frac{1}{2}BC \times AD = \frac{1}{2}ap.$

Area $(\Delta ABC) = \frac{1}{2}AB \times AC = \frac{1}{2}bc.$

$$\therefore \; \frac{1}{2}ap = \frac{1}{2}bc \Rightarrow a^2p^2 = b^2c^2 \Rightarrow p^2(b^2 + c^2) = b^2c^2$$

$$\therefore \; \frac{b^2 + c^2}{b^2c^2} = \frac{1}{p^2} \text{ or } \frac{1}{p^2} = \frac{1}{b^2} + \frac{1}{c^2}.$$

81. ΔABC is similar to ΔDBA.

$\therefore \; \frac{BD}{AB} = \frac{AB}{BC}$. Hence. $AB^2 = BC \times BD.$

82. Clearly, Δs *AXY* and *ABC* are similar.

Area $(\Delta ABC) = 2 \times$ Area (ΔAXY) (*Given*)

$$\therefore \; \frac{AX^2}{AB^2} = \frac{\text{Area}(\Delta AXY)}{\text{Area}(\Delta ABC)} = \frac{1}{2}. \text{ So, } \frac{AX}{AB} = \frac{1}{\sqrt{2}}.$$

83. Ratio of heights = Ratio of sides = 1 : 3.

84. $\frac{(\text{Side of first } \Delta)^2}{(\textit{corr.} \text{ side of 2nd } \Delta)^2} = \frac{49}{64} \Rightarrow$ Ratio of corr. sides = 7 : 8.

85. $\frac{(\text{Height of smaller } \Delta)^2}{(\text{Height of bigger } \Delta)^2} = \frac{12}{48} = \frac{1}{4}.$

$$\therefore \; \frac{(2.1)^2}{x^2} = \frac{1}{4} \Leftrightarrow x = \sqrt{4 \times (2.1)^2} = (2 \times 2.1) = 4.2 \text{ cm}.$$

86. $\dfrac{\Delta ABC}{\Delta DBC} = \dfrac{\frac{1}{2}\times BC\times AL}{\frac{1}{2}\times BC\times DM} = \dfrac{AL}{DM} = \dfrac{AO}{OD}$ [$\because$ ΔALO is similar to ΔDMO]

87. Clearly, $\dfrac{AD}{DC} = \dfrac{2}{3}$ and $\dfrac{AB}{BC} = \dfrac{10}{15} = \dfrac{2}{3}$. So, $\dfrac{AD}{DC} = \dfrac{AB}{BC}$.

Thus, BD divides AC in the ratio of the other two sides.

$\therefore$ BD is the bisector of $\angle B$.

Now, $\angle CBD = 180° - (130° + 30°) = 20°$

$\therefore$ $\angle B = 2(\angle CBD) = 40°$.

88. If the diagonals of a quadrilateral intersect proportionally, then the quadrilateral is a trapezium.

89. $\dfrac{DE}{BC} = \dfrac{AB}{AC} \Rightarrow \dfrac{DE}{10} = \dfrac{2}{5}$ or $DE = 4$ cm.

90. If the medians of a triangle are equal, then the triangle is equilateral.

91. Orthocentre of a triangle is the point of intersection of its altitudes.

92. Centroid of a triangle is the point of intersection of its medians.

93. By pythagorous theorem

$AC^2 = AB^2 + BC^2$

$(5x)^2 = (2x+2)^2 + (3x+2)^2$

$25x^2 = 4x^2 + 4 + 8x + 9x^2 + 4 + 12x \Rightarrow 12x^2 = 20x + 8 \Rightarrow 12x^2 - 20x - 8 = 0 \Rightarrow 3x^2 - 5x - 2 = 0$

$\Rightarrow 3x^2 - 6x + x - 2 = 0 \Rightarrow 3x(x-2) + 1(x-2) = 0 \Rightarrow (3x+1)(x-2) = 0 \Rightarrow x = 2$

$x \neq -\dfrac{1}{3}$ (not possible)

$\therefore$ $AB + BC = 2x + 2 + 3x + 2 = 5x + 4 = 5 \times 2 + 4 = 14$ cm

94. $OA = \sqrt{4^2 + 2^2} = \sqrt{16+4} = 2\sqrt{5}$ cm

$\therefore$ $\dfrac{r_1}{r_2} = \dfrac{1-\sin\theta}{1+\sin\theta} \Rightarrow \dfrac{r_1}{2} = \dfrac{1-\frac{2}{2\sqrt{5}}}{1+\frac{2}{2\sqrt{5}}} \Rightarrow \dfrac{r_1}{2} = \dfrac{\sqrt{5}-1}{\sqrt{5}+1}$

$\Rightarrow \dfrac{r_1}{2} = \dfrac{(\sqrt{5}-1)\times(\sqrt{5}-1)}{4} \Rightarrow r_1 = \dfrac{6-2\sqrt{5}}{2} \Rightarrow r_1 = 3-\sqrt{5}$

95. $OC = \sqrt{6^2 + 2^2} = 2\sqrt{10}$

$\dfrac{r_2}{2} = \dfrac{1-\frac{1}{\sqrt{10}}}{1+\frac{1}{\sqrt{0}}}$ $\left[\because \dfrac{r_1}{r_2} = \dfrac{1-\sin\theta}{1+\sin\theta}\right]$

$\dfrac{r_2}{2} = \dfrac{\sqrt{10}-1}{\sqrt{10}+1}$

$\dfrac{r_2}{2} = \dfrac{(\sqrt{10}-1)\times(\sqrt{10}-1)}{9} \Rightarrow \dfrac{r_2}{2} = \dfrac{11-2\sqrt{10}}{9} \Rightarrow r_2 = \dfrac{22-4\sqrt{10}}{9}$

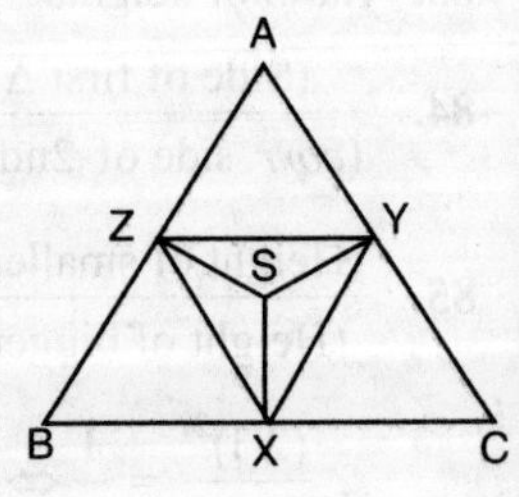

96. Since 3 times the sum of the squares on the sides is equal to 4 times the sum of the squares of the medians.

$\therefore 3\,(AB^2 + BC^2 + AC^2) = 4(AD^2 + BE^2 + CF^2)$

$\therefore 3(AB^2 + BC^2 + AC^2) = 4\left[\left(\frac{3}{2}OA\right)^2 + \left(\frac{3}{2}OB\right)^2 + \left(\frac{3}{2}OC\right)^2\right]$

$\therefore AB^2 + BC^2 + AC^2 = 3(OA^2 + OB^2 + OC^2)$ $\left[\because AD = \frac{3}{2}OA \text{ etc.}\right]$

97. $\angle B = 180' - (\angle A + \angle C)$

Now, two Δs can be formed, 1 one above AB and one below AB.

98. Given: $p^2 + q^2 = 2pq \Rightarrow p^2 + q^2 - 2pq = 0 \Rightarrow (p - q)^2 = 0 \Rightarrow p = q$.

So, one of the angles is 45°.

99. $\angle ABC = 180° - \angle ABD$ & $\angle ACB = 180° - \angle ACE = 180° - \angle ABD$.

$\angle ABC = \angle ACB$. So, $\Delta\, ABC$ is isosceles.

100. Initial height $= \sqrt{25^2 - 7^2} = \sqrt{625 - 49} = \sqrt{576} = 24$ m.

New height = (24 – 4) m = 20 m.

New distance between foot of the wall & that of ladder $= \sqrt{(25)^2 - (20)^2} = \sqrt{225} = 15$ m.

So, it will slide by (15 – 7) m = 8 m.

101. Now,

Area of square $= (4x)^2 = 16x^2 = 16 \times \frac{64}{17} = \frac{1024}{17}$ cm²

102. $\tan \alpha + \tan \beta$

$\tan \alpha + \tan (90° - \alpha) = \tan \alpha + \cot \alpha$

$\frac{1}{4} + 4 = \frac{17}{4}$ $\left[\because \tan \alpha = \frac{p}{b} = \frac{1}{4}\right]$

103. We have,

$CD = 4x = \frac{4 \times 8}{\sqrt{17}} = \frac{32}{\sqrt{17}}$

$AE = \sqrt{EF^2 - AF^2} = \sqrt{6^2 - \left(\frac{24}{\sqrt{17}}\right)^2} = \sqrt{36 - \frac{576}{17}} = \sqrt{\frac{612 - 576}{17}} = \sqrt{\frac{36}{17}}$

$= \frac{6}{\sqrt{17}}$

Now, $ED = \frac{32}{\sqrt{17}} - \frac{6}{\sqrt{17}} = \frac{26}{\sqrt{17}}$

$\therefore \quad ar(\Delta CDE) = \frac{1}{2} \times \frac{26}{\sqrt{17}} \times \frac{32}{\sqrt{17}} = \frac{416}{17}$ cm²

104. (*d*) is true by SAS.

105. $AB = AC \Rightarrow \angle ABC = \angle ACB = (180° - 130°) = 50°$.

$\therefore 50° + \angle BAC = 130° \Rightarrow \angle BAC = 80°$.

106. Statement-1

Now, in right angled ΔAQC

$AQ^2 = AC^2 + QC^2$ ...(i)

In right angled ΔBPC

$BP^2 = BC^2 + PC^2$ ...(ii)

Adding eqn (i) and (i), we get

$AQ^2 + BP^2 = QC^2 + PC^2 + AC^2 + BC^2$

$\Rightarrow AQ^2 + BP^2 = PQ^2 + AB^2$ $[\because PQ^2 = OC^2 + PC^2 \text{ and } AB^2 = AC^2 + BC^2]$

Statement-2

Given that P and Q are the general points of AC and BC respectively

$AB = 2\ PQ$ is possible only when P and Q are mid points of AC and BC respectively.

Hence, statement-2 is not correct

107. Let the side of right angle triangle be a, $a + 1$ and $a + 2$

Now, by Pythagoras theorem

$$(a)^2 + (a+1)^2 = (a+2)^2$$

$$\Rightarrow \quad a^2 + a^2 + 1 + 2a = a^2 + 4a + 4 \Rightarrow a^2 - 2a - 3 = 0$$

$$\Rightarrow \quad a^2 - 3a + a - 3 = 0$$

$$\Rightarrow \quad a(a-3) + 1\,(a-2) = 0 \Rightarrow (a+1)(a-3) = 0$$

$$\Rightarrow \quad a = -1 \text{ (not possible)} \qquad (\because \text{ the side of triangle cannot be } -ve)$$

and $\quad 1 = 3$

$\therefore$ three consecutive integers are (3, 4, 5)

108. Let the centroid be 0.

In ΔAOB

$AO + BO > AB$

$\dfrac{2p}{3} + \dfrac{2q}{3} > c$ (i) [$\therefore$ Centroid divides median in the ratio 2 : 1]

Similarly, In ΔBOC

$\dfrac{2q}{3} + \dfrac{2r}{3} > b$ ...(ii)

Similarly, In ΔAOC

$\dfrac{2p}{3} + \dfrac{2r}{3} < b$ ...(iii)

Adding (i), (ii) and (iii), we get,

$$\frac{4}{3}(p+q+r) > a+b+c \Rightarrow 4(p+q+r) > 3\,(a+b+c)$$

109. We know that the sum of two sides is greater than the 3rd side

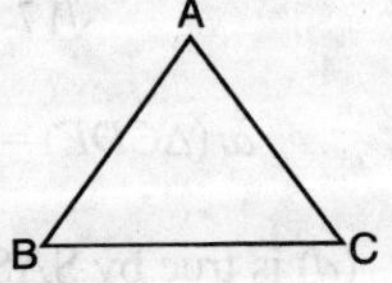

(1) $AB + AC > BC$

$\Rightarrow BC - AC < AB$ correct

(2) $AB + BC > AC$

$AC - B < BC$ correct

(3) $AC + BC < AB$

$AB - BC < AC$ correct

110. $BX = BC \times \dfrac{AB}{AB + AC}$ [$\because$ AX divides BC is in the ratio of sides AB and AC respectively]

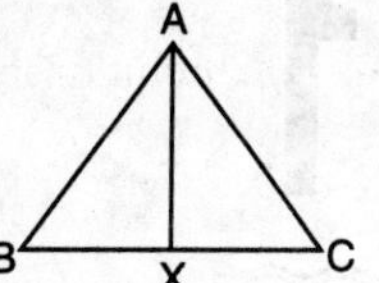

$AB + AC > BC$ [$\because$ Sum of any two sides of Δ is greater than that of third side]

Hence, $AB < BX$

111. We have,

$\Delta ABC \sim \Delta PQR$

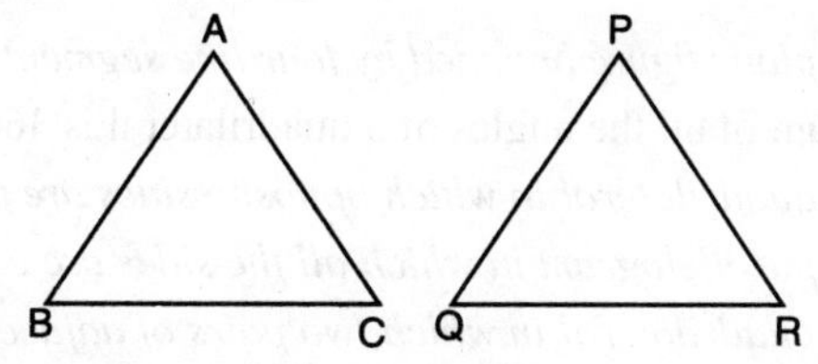

$\therefore \dfrac{\text{Perimeter } (\Delta ABC)}{\text{Perimeter } (\Delta PQR)} = \dfrac{AB}{PQ}$

$\Rightarrow \dfrac{75}{50} = \dfrac{AB}{20}$

$\Rightarrow AB = 30$ cm

112. $\because$ (112, 15, 113) is a Pythagoras triplet with one number 15.

$\therefore$ We have, $(113)^2 = (112)^2 + (15)^2$

$\therefore$ Maximum perimeter = 113 + 112 + 15 = 240 cm

113. $ar(\Delta ABC)$ with base $AB = \dfrac{1}{2} \times AB \times CD = \dfrac{1}{2} p \times AB$ ...(i)

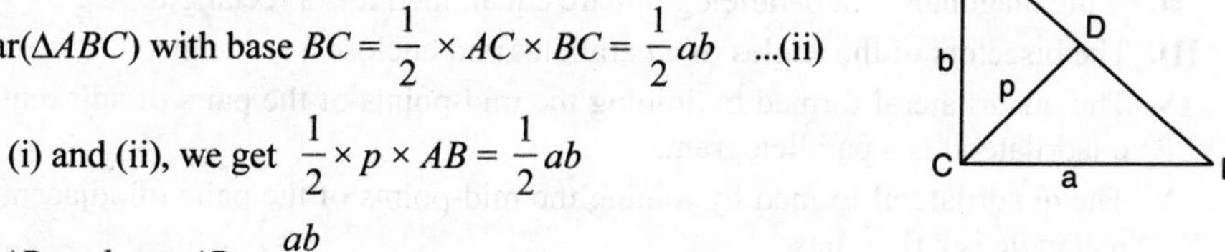

and $ar(\Delta ABC)$ with base $BC = \dfrac{1}{2} \times AC \times BC = \dfrac{1}{2} ab$ ...(ii)

from (i) and (ii), we get $\dfrac{1}{2} \times p \times AB = \dfrac{1}{2} ab$

$\Rightarrow PAB = ab \Rightarrow AB = \dfrac{ab}{p}$

Now, In right angles ΔABC

$AB^2 = AC^2 + BC^2 \Rightarrow AB^2 = b^2 + a^2 \Rightarrow \left(\dfrac{ab}{p}\right)^2 = b^2 + a^2 \Rightarrow a^2b^2 = p^2(a^2 + b^2)$

31

CHAPTER

Quadrilaterals and Parallelograms

1. (*i*) *A plane figure bounded by four line segments is called a* ***quadrilateral***.
   (*ii*) Sum of all the angles of a quadrilateral is 360°.
   (*iii*) *A quadrilateral in which opposite sides are parallel is called a* ***parallelogram***.
   (*iv*) *A parallelogram in which all the sides are equal is called a* ***rhombus***.
   (*v*) *A quadrilateral in which two pairs of adjacent sides are equal is called a* ***kite***.
   (*vi*) *A quadrilateral in which two opposite sides are parallel and two opposite sides are not parallel, is known as a* ***trapezium***.

2. **Results on Parallelogram:**

   I. *In a parallelogram, we have:*
   (*i*) Opposite sides are equal; (*ii*) Opposite angles are equal;
   (*iii*) Diagonals bisect each other; (*iv*) Diagonals need not be equal.

   II. If the diagonals of a parallelogram are equal, then it is a rectangle.

   III. The bisectors of the angles of a parallelogram enclose a rectangle.

   IV. The quadrilateral formed by joining the mid-points of the pairs of adjacent sides of a quadrilateral is a parallelogram.

   V. The quadrilateral formed by joining the mid-points of the pairs of adjacent sides of a rectangle is a rhombus.

   VI. The quadrilateral formed by joining the mid-points of the pairs of adjacent sides of a rhombus is a rectangle.

   VII. Two parallelograms on the same base and between the same parallels are equal in area.

3. **Results on Rhombus:**

   I. *In a rhombus*:
   (*i*) All sides are equal;
   (*ii*) The diagonals bisect each other at right angles.
   (*iii*) Sum of the squares of sides is equal to the sum of the squares of its diagonals

   II. If all sides of a quadrilateral are equal and its diagonals are equal, then it is a square.

EXERCISE 31

1. In a quadrilateral ABCD, AB = BC and CD = DA; AC and BD are diagonals such that AC = 6 cm and BD = 12 cm. What is the area of the quadrilateral ? **[CDS (I) Exam, 2024]**

(*a*) 24 square cm (*b*) 30 square cm (*c*) 36 square cm (*d*) 40 square cm

2. The adjacent sides of a parallelogram are 10 cm and 8 cm and the angle between them is 150°. What is the area of the parallelogram? **[CDS (I) Exam, 2025]**

(*a*) $40\sqrt{3}$ square cm (*b*) 40 square cm (*c*) $20\sqrt{3}$ square cm (*d*) 20 square cm

3. *ABCD* is a quadrilateral such that *AD* = *DC* = *CA* = 20 units, *BC* = 12 units and ΔABC = 90°. What is the approximate area of the quadrilateral *ABCD*? **[CDS (I) Exam, 2022]**

(*a*) 269 sq units (*b*) 300 sq units (*c*) 325 sq units (*d*) 329 sq units

4. *ABCD* is a parallelogram and *E* is the mid-point of *BC*. *DE* and *AB* when produced meet at *F*. Then,

(*a*) $AF = \frac{3}{2} AB$ (*b*) $AF = 2\,AB$

(*c*) $AF = 3AB$ (*d*) $AF^2 = 2AB^2$

5. In a trapezium *ABCD*, *AB* is parallel to *DC*. The diagonals *AC* and *BD* intersect at *P*. If $AP : CP = 4 : (4x - 4)$ and $BP : PD = (2x - 1) : (2x + 4)$, then what is the value of *x*.

(*a*) 4 (*b*) 3 (*c*) $\frac{3}{2}$ (*d*) 2

[CDS (I) Exam, 2021]

6. In a quadrilateral ABCD, AB = 6 cm. BC = 18 cm, CD = 6 cm and DA = 10 cm. If the diagonal BD = *x*, then which one of the following is correct ? **[CDS (I) Exam, 2024]**

(*a*) $8 < x < 12$ (*b*) $12 < x < 16$ (*c*) $16 < x < 18$ (*d*) $18 < x < 20$

7. In a quadrilateral *ABCD*, $\angle B = 90°$ and $AB^2 + BC + CD^2 – 0$, then what is ΔACD equal to ?

(*a*) 30° (*b*) 60° (*c*) 90° (*d*) 120°

[CDS (I) Exam, 2020]

8. ABCD is a cyclic quadrilateral. The bisectors of the angles A, B, C and D cut the circle of P, Q, R and S respectively. What is $\angle PQR + \angle RSP$ equal to? **[CDS (I) Exam, 2020]**

(*a*) 90° (*b*) 135° (*c*) 180° (*d*) 270°

9. Consider the following statements: **[CDS (I) Exam, 2020]**

1. The diagonals of a trapezium divide each other proportionally.

2. Any line drawn parallel to the parallel sides of a trapezium divided the non-parallel sides proportionally.

Which of the above statement is/are correct?

(*a*) 1 only (*b*) 2 only (*c*) Both 1 and 2 (*d*) Neither 1 nor 2

10. A triangle and a parallelogram have equal areas and equal bases. If the altitude of the triangle is *k* times the altitude of the parallelogram, then what is the value of *k*? **[CDS (I) Exam, 2020]**

(*a*) 4 (*b*) 2 (*c*) 1 (*d*) $\frac{1}{2}$

11. Consider the following statements:

1. The isosceles trapezium is always cyclic.

2. Any cyclic parallelogram is a rectangle.

Which of the above statements is/are correct?

(*a*) 1 only (*b*) 2 only (*c*) Both 1 and 2 (*d*) Neither 1 nor 2

[CDS (I) Exam, 2019]

12. The sides *AD*, *BC* of a trapezium *ABCD* and parallel and the diagonals *AC* and *BD* meet at *O*. If the area of triangle *AOB* is 3 cm² K and the area of triangle *BDC* is 8 cm², then what is the area of triangle *AOD*? **[CDS (II) Exam, 2019]**

(*a*) 8 cm² (*b*) 5 cm² (*c*) 3.6 cm² (*d*) 1.8 cm²

13. Let *PQRS* be a parallelogram whose diagonals *PR* and *QS* intersect at *O*. if triangle *ORS* is an equilateral triangle having a side of length 10 cm, then what is the length of the diagonal *PR*? **[CDS (II) Exam, 2019]**

(*a*) $5\sqrt{3}$ cm (*b*) $10\ \sqrt{3}$ cm (*c*) $15\ \sqrt{3}$ cm (*d*) $20\ \sqrt{3}$ cm

14. All the four sides of a parallelogram area of equal length. The diagonals are in the ratio 1 : 2. If the sum of the lengths of the diagonals is 12 cm, then what is the area of the parallelogram? **[CDS (II) Exam, 2019]**

(*a*) 9 cm^2 (*b*) 12 cm^2 (*c*) 16 cm^2 (*d*) 25 cm^2

15. Consider a trapezium *ABCD,* in which *AB* is parallel to *CD* and *AD* is perpendicular to *AB.* If the trapezium has an in circle which touches *AB* at *E* and *CE* at *F*, where *EB* = 25 cm and FC = 16 cm, then what is the diameter of the circle? **[CDS (II) Exam, 2019]**

(*a*) 16 cm (*b*) 25 cm (*c*) 36 cm (*d*) 40

16. Let S be the parallelogram obtained by joining the mid-point of the parallelogram. T. consider the following statements: **[CDS (II) Exam, 2018]**

1. The ratio of area of T so that of S is 2 : 1.

2. The perimeter of S is half of the sum of diagonals of T.

Which of the above statements is/are correct?

(*a*) 1 only (*b*) 2 only (*c*) Both 1 and 2 (*d*) Neither 1 nor 2

17. ABCD is a parallelogram, where AB : AD = 2 : 1. One of the angles of the parallelogram is 60º. The two diagonals are in the ratio **[CDS (I) Exam, 2015]**

(*a*) 7: 3 (*b*) $\sqrt{7}:\sqrt{3}$ (*c*) 7: 5 (*d*) None of the above

18. The diagonals of a trapezium are at right angles, and the slant sides, if produced, from an equilateral triangle with the greater of the two parallel sides. If the area of the trapezium is 16 square cm, then the distance between the parallel sides is **[CDS (I) Exam, 2015]**

(*a*) 2 cm

(*b*) 4 cm

(*c*) 8 cm

(*d*) Cannot be determined due to insufficient data

19. Consider the following statements: **[CDS (I) Exam, 2015]**

1. If non-parallel sides of a trapezium are equal, then it is cyclic.
2. If the chord of a circle is equal to its radius, then the angle subtended by this chord at a point in major segment is 30º.

Which of the above statements is/are correct?

(*a*) 1 only (*b*) 2 only (*c*) Both 1 and 2 (*d*) Neither 1 nor 2

20. *ABCD* is a parallelogram with *AB* and *AD* as adjacent sides. If $\angle A = 60°$ and $AB = 2AD$, then the diagonal *BD* will be equal to **[CDS (II) Exam, 2015]**

(*a*) $\sqrt{2}\,AD$ (*b*) $\sqrt{3}\,AD$ (*c*) $2\,AD$ (*d*) $3\,AD$

21. The two adjacent sides of a cyclic quadrilateral are 2 cm and 5 cm and the angle between them is 60º. If the third side is 3cm, then the fourth sides is of length **[CDS (II) Exam, 2015]**

(*a*) 2 cm (*b*) 3 cm (*c*) 4 cm (*d*) 5 cm

22. Let *A* denote the set of quadrilaterals having two diagonals equal and bisecting each other. Let *B* denote the set of quadrilaterals having diagonals bisecting each other at 90º. Then $A \cap B$ denotes **[CDS (II) Exam, 2015]**

(*a*) the set of parallelograms (*b*) the set of rhombuses

(*c*) the set of square (*d*) the set of rectangles

23. A rhombus is formed by joining midpoints of the sides of a rectangle in the suitable order. If the area of the rhombus is 2 square units, then the area of the rectangle is

(*a*) $2\sqrt{2}$ square units (*b*) 4 square units

(c) $4\sqrt{2}$ square units (*d*) 8 square units **[CDS (II) Exam, 2016]**

24. If a quadrilateral has an inscribed circle, then the sum of a pair of opposite sides equals

(*a*) Half the sum of the diagonals (*b*) Sum of the other pair of opposite sides

(*c*) Sum of the two adjacent sides (*d*) None of the above **[CDS (II) Exam, 2016]**

25. Let ABCD be a rectangle. Let P, Q, R, S be the mid-points of sides AB, BC, CD, DA respectively. Then the quadrilateral PQRS is a **[CDS (I) Exam, 2017]**

(*a*) square (*b*) rectangle, but need not be a square

(*c*) rhombus, but need not be a square (*d*) parallelogram, but need not be a rhombus

26. ABCD is a rectangle. The diagonals AC and BD intersect at O. If AB = 32 cm and AD = 24 cm, then what is OD equal to? **[CDS (I) Exam, 2017]**

(*a*) 22 cm (*b*) 20 cm (*c*) 18 cm (*d*) 16 cm

27. In the figure given below, AC is parallel to ED and AB = DE = 5cm and BC = 7cm. What is the area ABDE : area BDE : area BCD equal to? **[CDS (I) Exam, 2017]**

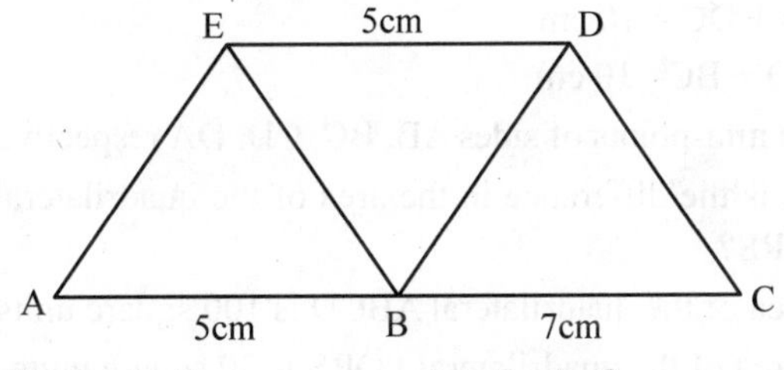

(*a*) 10 : 5 : 7 (*b*) 8 : 4 : 7 (*c*) 2 : 1 : 2 (*d*) 8 : 4 : 5

28. In the figure given below, PQRS is a parallelogram. PA bisects angle P and SA bisects angle S. What is angle PAS equal to? **[CDS (I) Exam, 2017]**

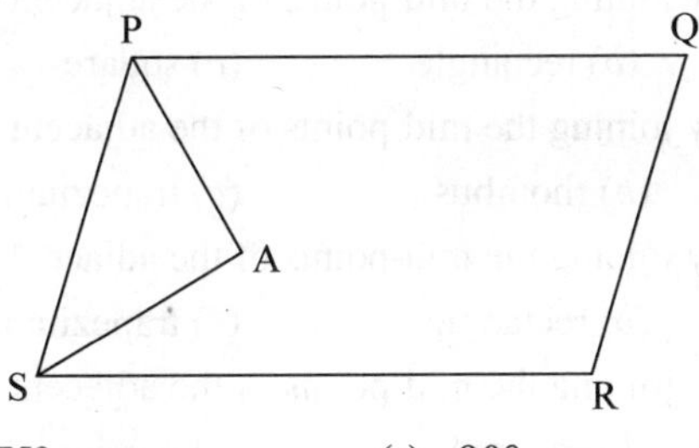

(*a*) 60° (*b*) 75° (*c*) 90° (*d*) 100°

29. ABCD is a square. X is the mid-point of AB and Y is the mid-point of BC.
Consider the following statements: **[CDS (I) Exam, 2017]**

1. Triangle ADX and BAY are congruent.
2. $\angle DXA = \angle AYB$
3. DX is inclined at an angle 60° with AY.
4. DX is not perpendicular to AY.

Which of the above statements are correct?

(*a*) 2, 3 and 4 only (*b*) 1, 2 and 4 only (*c*) 1, 3 and 4 only (*d*) 1 and 2 only

Direction (Q. 30-32): Consider the following for the next ten (10) items that follow:

Each item contains a question followed by two Statements. Answer each item using the following instructions:

Choose option

(a) If the Question can be answered by one of the Statements alone, but not by the other.

(b) If the Question can be answered by either Statement alone.

(c) If the Question can be answered by using both the Statements together, but cannot be answered by using either Statement alone.

(d) If the Question cannot be answered even by using both Statements together.

[CDS (I) Exam, 2024]

30. Area of a rectangle with length x and breadth y is P and area of a parallelogram (which is strictly not a rectangle) with adjacent sides of length x and y is Q.

Question : Is $P > Q$?

Statement I : $x : y = 2 : 1$

Statement II : The angle between the two adjacent sides of the parallelogram is 60°.

31. A circle touches all the four sides AB, BC, CD, DA of a quadrilateral ABCD.

Question : What is the perimeter of the quadrilateral ?

Statement I : AB + DC = 10 cm

Statement II : AD + BC= 10 cm

32. P, Q, R, S are the mid-points of sides AB, BC, CD, DA respectively of a quadrilateral ABCD.

Question : What is the difference in the area of the quadrilateral ABCD and the area of the quadrilateral PQRS?

Statement I : Area of the quadrilateral ABCD is 100 square units.

Statement II : Area of the quadrilateral PQRS is 50 square units.

33. The bisectors of the angles of a || gm enclose a:

(*a*) parallelogram (*b*) rhombus (*c*) rectangle (*d*) square

34. The figure formed by joining the mid-points of the adjacent sides of a quardilateral is a:

(*a*) parallelogram (*b*) rectangle (*c*) square (*d*) rhombus

35. The figure formed by joining the mid-points of the adjacent sides of a rectangle is a:

(*a*) square (*b*) rhombus (*c*) trapezium (*d*) none

36. The figure formed by joining the mid-points of the adjacent sides of a rhombus is a:

(*a*) square (*b*) rectangle (*c*) trapezium (*d*) none

37. The figure formed by joining the mid-points of the adjecent sides of a square is a:

(*a*) rhombus (*b*) square (*c*) rectangle (*d*) parallelogram

38. The figure formed by joining the mid-points of the adjacent sides of a parallelogram is a:

(*a*) rectangle (*b*) parallelogram (*c*) rhombus (*d*) square

39. If one angle of a parallelogram is 24 less than twice the smallest angle, then the largest angle of the parallelogram is:

(*a*) 176° (*b*) 68° (*c*) !12° (*d*) 102°

40. If an angle of a parallelogram is two-third of its adjacent angle, the smallest angle of the parallelogram is:

(*a*) 108° (*b*) 54° (*c*) 72° (*d*) 81°

41. In the given figure, $ABCD$ is a parallelogram in which $\angle DAB = 75°$ and $\angle DBC = 60°$. Then, $\angle BDC$ is equal to:

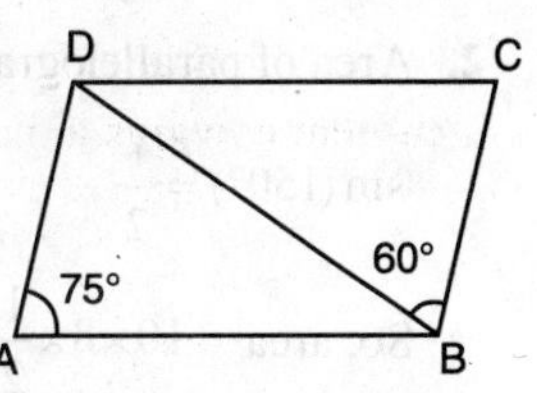

(*a*) 75° (*b*) 65°
(*c*) 45° (*d*) 55°

42. If $ABCD$ is a rectangle, E, F are the mid-points of BC and AD respectively and G is any point on EF, then $\Delta\, GAB$ equals:

(*a*) $\frac{1}{2}$ (□ $ABCD$) (*b*) $\frac{1}{3}$ (□ $ABCD$) (*c*) $\frac{1}{4}$ (□ $ABCD$) (*d*) $\frac{1}{6}$ (□ $ABCD$)

43. $ABCD$ is a parallelogram, E, F are the mid-points of BC and AD respectively and G is any point on EF. Then, area of $\Delta\, GAB$ equals:

(*a*) $\frac{1}{3}$ (|| gm $ABCD$) (*b*) $\frac{1}{4}$ (|| gm $ABCD$) (*c*) $\frac{1}{2}$ (|| gm $ABCD$) (*d*) $\frac{1}{6}$ (|| gm $ABCD$)

44. If P is a point within a rectangle $ABCD$, then:

(*a*) $AP + PC = BP + PD$ (*b*) $AP^2 + PC^2 = BP^2 + PD^2$
(*c*) $AP \times PC = BP \times PD$ (*d*) $PC^2 + PD^2 = PA^2 + PB^2$

45. If $ABCD$ is a quadrilateral such that its diagonals AC and BD intersect O to four Δs equal in area. Then $ABCD$ must be a:

(*a*) || gm (*b*) rhombus (*c*) square (*d*) rectangle

46. $ABCD$ is a parallelogram in which $\angle A = 80°$. How many points P are there in the plane of parallelogram such that it is equidistant from its vertices ?

(*a*) 1 (*b*) 2 (*c*) 3 (*d*) None of these

47. If area of *a* || *gm* with sides *a* and *b* is *A* and that of a rectangle with sides *a* and *b* is *B*, then:

(*a*) $A > B$ (*b*) $A = B$ (*c*) $A < B$ (*d*) None of these

ANSWERS

1. (*c*)	**2.** (*b*)	**3.** (*a*)	**4.** (*b*)	**5.** (*b*)	**6.** (*d*)	**7.** (*c*)	**8.** (*c*)	**9.** (*c*)
10. (*b*)	**11.** (*c*)	**12.** (*d*)	**13.** (*b*)	**14.** (*c*)	**15.** (*d*)	**16.** (*c*)	**17.** (*b*)	**18.** (*b*)
19. (*c*)	**20.** (*b*)	**21.** (*c*)	**22.** (*c*)	**23.** (*b*)	**24.** (*b*)	**25.** (*c*)	**26.** (*b*)	**27.** (*a*)
28. (*c*)	**29.** (*d*)	**30.** (*b*)	**31.** (*b*)	**32.** (*b*)	**33.** (*c*)	**34.** (*a*)	**35.** (*b*)	**36.** (*b*)
37. (*b*)	**38.** (*b*)	**39.** (*c*)	**40.** (*c*)	**41.** (*c*)	**42.** (*c*)	**43.** (*b*)	**44.** (*b*)	**45.** (*a*)
46. (*a*)	**47.** (*c*)							

SOLUTIONS

1. $ar(\Delta BCD) = \frac{1}{2} \times BD \times OC$

$= \frac{1}{2} \times 3 \times 12 = 18$

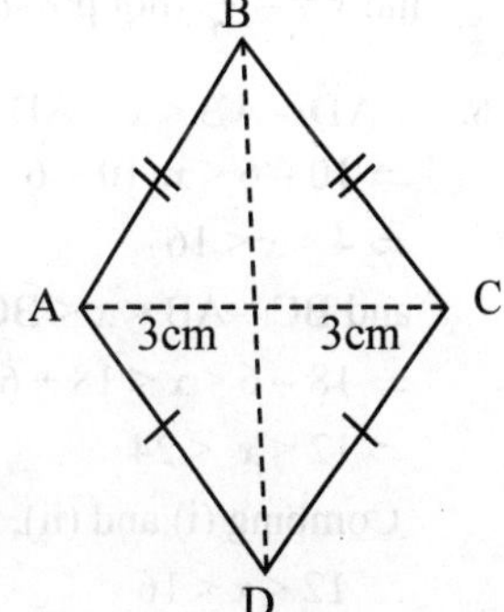

$\therefore$ ar(quadrilateral ABCD) = 2 × 18 = 36 cm³

2. Area of parallelogram $= a \times b \times (\sin \theta)$

$\sin(150°) = \frac{1}{2}$

So, area $= 10 \times 8 \times \frac{1}{2} = 40 \text{ cm}^2$

3. In ΔABC

$AC = 20$ units, $AB = 12$ units

By Pythagoras theorem,

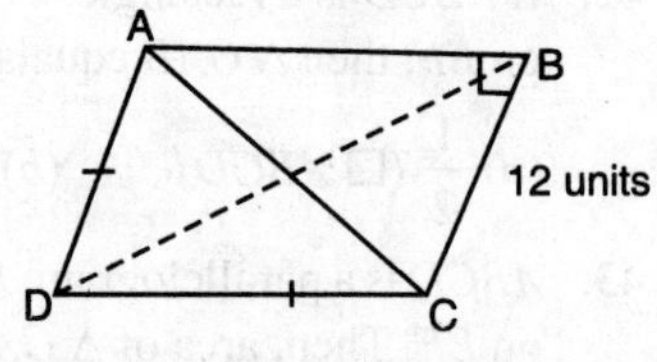

$$BC = \sqrt{AC^2 - AB^2} = \sqrt{20^2 = 12^2}$$

$$= \sqrt{400 - 144} = \sqrt{256} = 16 \text{ units}$$

$\because ar(\Delta ABC) = \frac{1}{2} \times b \times h = \frac{1}{2} \times 12 \times 16 = 96$ square units

Now, we have

ΔADC is an equilateral triangle

$ar(\Delta ADC) = \frac{\sqrt{3}}{4} \times \text{side}^2 = \frac{\sqrt{3}}{4} \times 20 \times 20 = 100\sqrt{3} = 100 \times 1.732 = 173.2$ sq

$\therefore ar(\text{quad. } ABCD) = 96 + 173.2 = 269.2 \text{ units} \approx 260$ sq units

4. In Δs EDC & EFB, we have:

$\angle DCE = \angle EBF$ (alt $\angle s$), $\angle DEC = FEB$ (vert. opp $\angle s$) and $EC = EB$.

$\therefore \Delta EDC \cong \Delta EFB$. So, $BF = DC$.

$\therefore AF = AB + BF = AB + DC = 2AB$.

5. We have,

$\frac{AP}{PC} = \frac{BP}{PD}$ [$\because$ Diagonals of Δ divides each other proportionally]

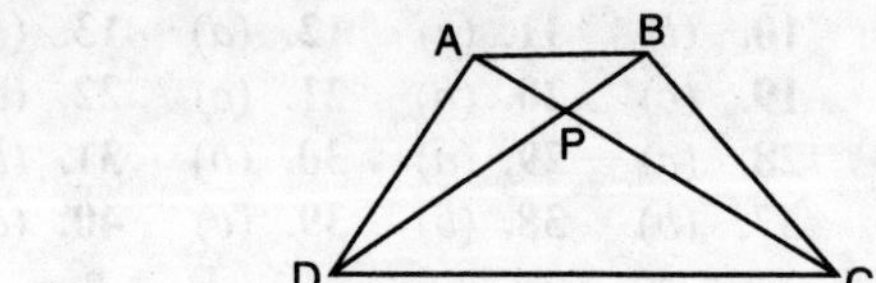

$\Rightarrow \frac{4}{4x-4} = \frac{2x-1}{2x+4}$

$\Rightarrow \frac{1}{x-1} = \frac{2x-1}{2x+4}$

$\Rightarrow 2x + 4 = 2x^2 - 3x + 1 \Rightarrow 2x^2 - 5x - 3 = 0 \Rightarrow 2x^2 - 6x + x - 3 = 0 \Rightarrow 2x(x-3) + 1(x-3) = 0$

$\Rightarrow (2x + 1)(x - 3) = 0 \Rightarrow x = 3$

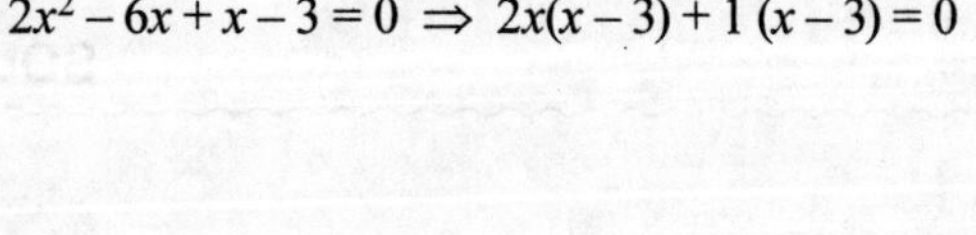

and $x = -\frac{1}{2}$ (not possible)

6. $\therefore$ AD – AB < x < AD + AB

$\Rightarrow 10 - 6 < x \;\; 10 + 6$

$\Rightarrow 4 < x < 16$ ...(i)

and BC – AB < x < BC + AB

$\Rightarrow 18 - 6 < x < 18 + 6$

$\Rightarrow 12 < x < 24$ ...(ii)

Combing (i) and (ii), we get

$\therefore 12 < x < 16$

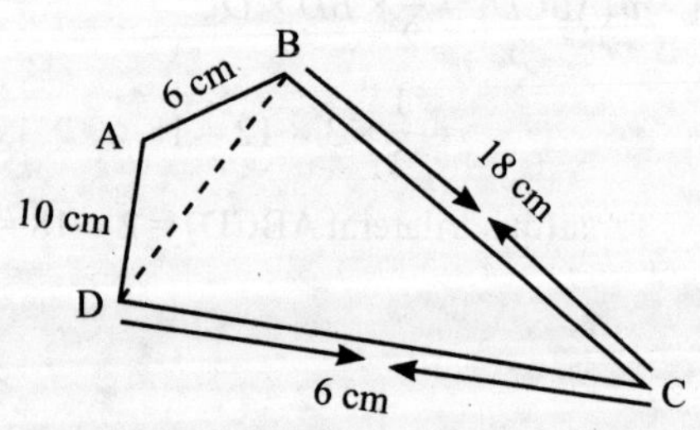

7. We have,

$AB^2 + BC^2 + CD^2 - AD^2 = O$

$\Rightarrow (AB^2 + BC^2) + CD^2 - AD^2 = 0$

$\Rightarrow AC^2 + CD^2 - AD^2 = 0 \Rightarrow AC^2 + CD^2 = AD^2$

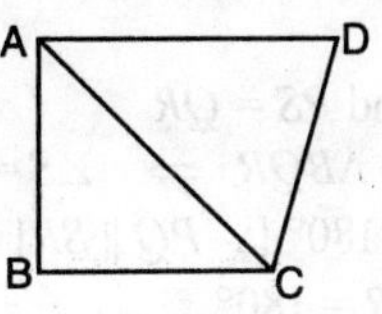

This is true and form right angled ΔACD.

When AD will be hypotenuse of ΔACD

$\therefore \angle ACD = 90°$ [$\because AD$ is hypotenuse]

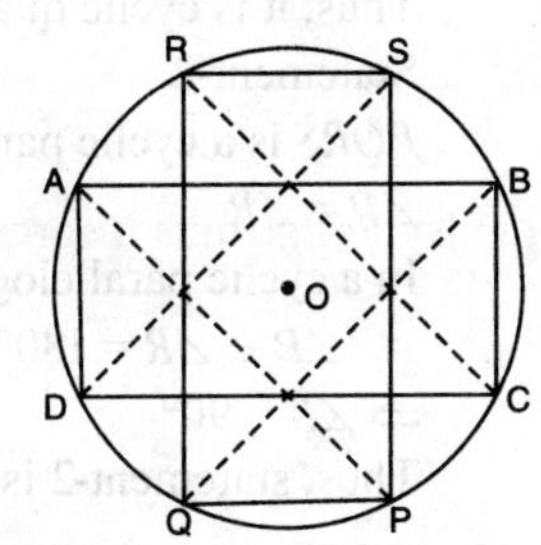

8. $\angle PQR$ is opposite to $\angle RSP$

Given that $PQRS$ is a cyclic quadrilateral,

$\Rightarrow$ Sum of opposite angles is 180°

$\therefore PQR + \angle RSP = 180°$

9. Statement-1

AC and BD are diagonals of trapezium $ABCD$ and $OP \parallel AB \parallel CD$

In ΔADC,

$OP \parallel DC$

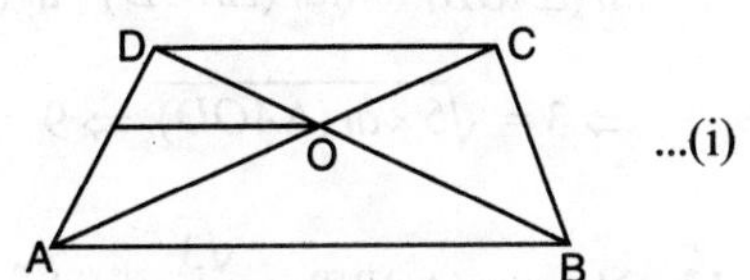

$$\Rightarrow \frac{AP}{PD} = \frac{OA}{OC} \qquad \text{...(i)}$$

In ΔADB, $OP \parallel AB$

$$\frac{AP}{PD} = \frac{OB}{OD} \qquad \text{...(ii)}$$

From (i) and (ii), we get

$$\frac{OA}{OC} = \frac{OB}{OD}$$

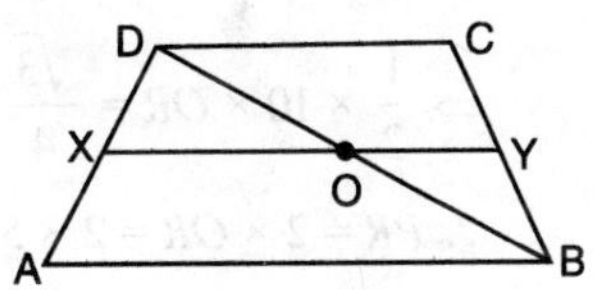

Hence, diagonals of trapezium divide each other proportionally

Statement-2

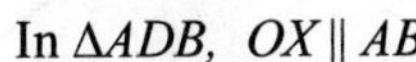

In ΔADB, $OX \parallel AB$

$$\frac{DX}{XA} = \frac{OD}{OB} \qquad \text{...(i)}$$

In ΔBDC, $OY \parallel CD$

$$\frac{BY}{CY} = \frac{OB}{OD} \text{ or } \frac{CY}{BY} = \frac{OD}{OB} \qquad \text{...(ii)}$$

from (i) and (ii), we get

$$\frac{DX}{XA} = \frac{CY}{BY}$$

Hence statement-2 is correct

10. Let the length of the base of a triangle and a parallelogram be p.

Now,

altitude of triangle = h

altitude of parallelogram = kh

Now, according to the question

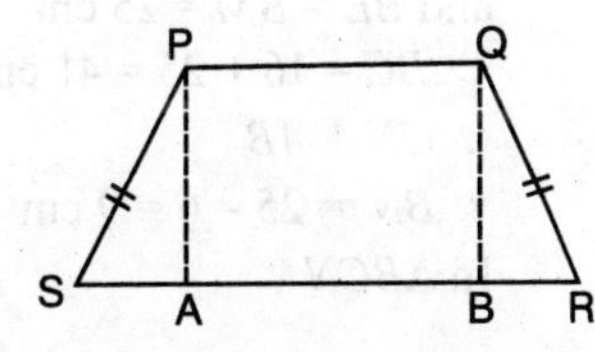

$$ph = \frac{1}{2} \times kh \times p \Rightarrow k = 2$$

11. Statement-1

$AP = BQ$ and $PS = QR$

$\therefore\ \Delta APS \cong \Delta BQR \Rightarrow \angle S = \angle R$ ...(i)

$\angle P + \angle S = 180°$ [$\because PQ \parallel SR$]

$\Rightarrow \angle P = \angle R = 180°$

Thus, it is cyclic quadrilateral

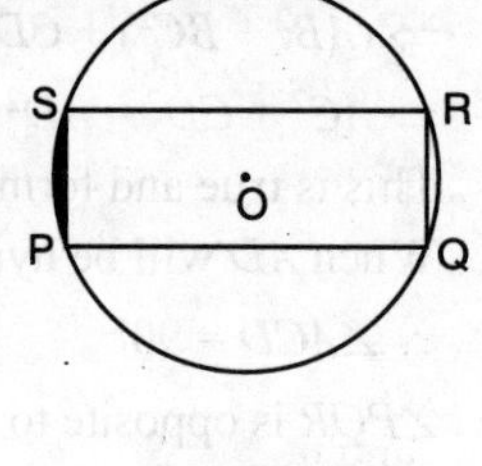

Statement-2

$PQRS$ is a cyclic parallelogram

$\angle P = \angle R$ [$\because$ $PQRS$ is a parallelogram]

In a cyclic parallelogram $PQRS$

$\therefore\ \angle P + \angle R = 180° \Rightarrow 2\angle R = 180°$ [$\therefore \angle P = \angle R$]

$\Rightarrow \angle R = 90°$

Thus, statement-2 is true

12. $ABCD$ is a trapezium $ar(\Delta AOB) = ar(\Delta COD) = 3 \text{ cm}^2$

and $ar(\Delta BOC) + ar(\Delta COD) = 8 \text{ cm}^2 \Rightarrow ar(\Delta BOC) = 8 - 3 = 5 \text{ cm}^2$

$ar(\Delta AOB) = \sqrt{ar(\Delta AOD) \times ar(\Delta BOC)}$

$\Rightarrow 3 = \sqrt{5 \times ar(\Delta AOD)} \Rightarrow 9 = 5 \times ar(\Delta AOD) \Rightarrow ar(\Delta AOD) = \frac{9}{5} = 1{:}8 \text{ cm}^2$

13. Now, $ar(\Delta QRS) = \frac{\sqrt{3}}{4} \times \text{side}^2$

$\Rightarrow \frac{1}{2} \times QS \times OR = \frac{\sqrt{3}}{4} \times 10 \times 10$

$\Rightarrow \frac{1}{2} \times 10 \times OR = \frac{\sqrt{3}}{4} \times 10 \times 10 \Rightarrow OR = 5\sqrt{3} \text{ cm}$

$\therefore PR = 2 \times OR = 2 \times 5\sqrt{3} = 10\sqrt{3} \text{ cm}$

14. The given parallelogram is rhombus.

Since, all the four sides of a parallelogram are of equal length

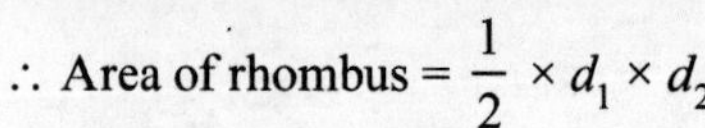

$\therefore$ Area of rhombus $= \frac{1}{2} \times d_1 \times d_2$

$\therefore d_1 = 12 \times \frac{1}{3} = 4 \text{ cm}$

$d_2 = 12 \times \frac{2}{3} = 8 \text{ cm}$ [$\because$ ratio of diagonals are 1 : 2]

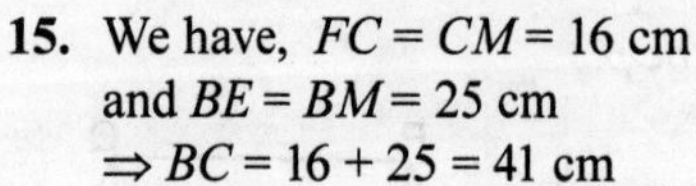

$\therefore$ Area of parallelogram $= \frac{1}{2} \times 4 \times 8 = 16 \text{ cm}^2$

15. We have, $FC = CM = 16$ cm

and $BE = BM = 25$ cm [$\because$ CF and CM are tangents to the circle from the same point C.]

$\Rightarrow BC = 16 + 25 = 41$ cm

$\because CN \perp AB$

$\therefore BN = 25 - 6 = 9$ cm

In ΔBCN

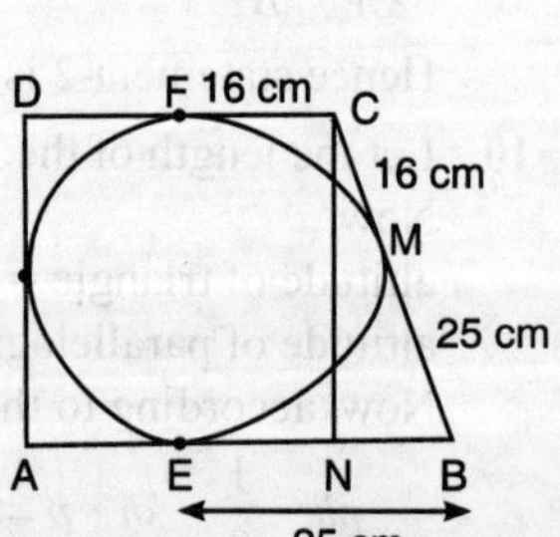

$$CN = \sqrt{BC^2 - BN^2} = \sqrt{41^2 - 9^2} = \sqrt{1681 - 81} = \sqrt{1600} = 40 \text{ cm}$$

$\therefore$ Diameter of the circle = 40 cm

16. Statement-1

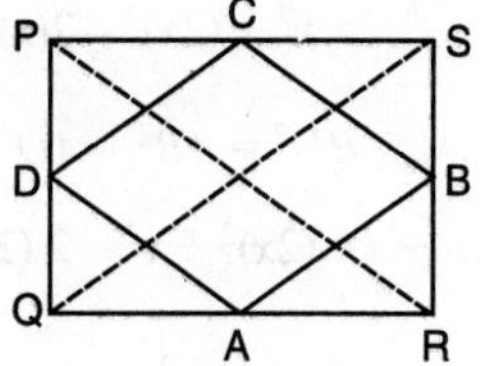

In ΔPQS

$PD = DQ$ and $PC = CS$ [by mid-point theorem]

$\Rightarrow ar = (\Delta PDC) = \frac{1}{4} \times ar\Delta PQS$ ...(i)

and in ΔQRS

$QA = AR$ and $BR = BS$

$\Rightarrow ar = (\Delta ARB) = \frac{1}{4} \times ar(\Delta QRS)$ ... (ii)

by adding (i) and (ii), we get

$ar(\Delta PDC) + ar(\Delta ARB) = \frac{1}{4} \times [ar(\Delta PQS) + ar(\Delta QRS)$

$ar(\Delta PDC) + ar(\Delta ARB) = \frac{1}{4} \times ar(PQRS)$ ... (iii)

Similarly,

$ar(\Delta CSB) + ar(\Delta AQD) = \frac{1}{4} \times ar(PQRS)$ ...(iv)

Now, by adding (iii) and (iv), we get

$\Rightarrow ar(\Delta PDC) + ar(\Delta ARB) + ar(\Delta CSB) + ar(\Delta AQD) = \frac{1}{2} \times ar(PQRS)$

$\Rightarrow ar(PQRS) - ar(ABCD) = \frac{1}{2} \times ar(PQRS) \Rightarrow \frac{1}{2} \times ar(PQRS) = ar(ABCD)$

$\because \frac{ar(\text{parallelogram } T)}{ar(\text{parallelogram } S)} = \frac{2}{1}$

Statement-2

Consider the given diagram

In ΔPQS, $CD = \frac{1}{2} \times QS$ ...(i) [By mid point theorem]

Similarly,

In ΔPQR, $AD = \frac{1}{2} \times PR$ ...(ii)

In ΔQRS, $AB = \frac{1}{2} \times QS$ ...(iii)

In ΔPSR, $BC = \frac{1}{2} \times PR$ ...(iv)

By adding (i), (ii), (iii) and (iv), we get

$CD + AD + AB + BC = PR + QS$

Perimeter $(ABCD)$ = sum of diag. $(ABCD)$

Thus, statement-2 is correct

17. AC and BD are diagonals of the parallelogram.

Let $\angle D = 60°$ $\angle A = 120°$ Let $AB = 2x$ and $AD = x$.

$CD = AB = 2x$.

$AC^2 = AD^2 + CD^2 - 2(AD)(CD)\cos\angle D$

$= x^2 + (2x)^2 - 2(x)(2x)\cos 60° = 3x^2 \qquad \left(\because \cos 60° = \dfrac{1}{2}\right)$

$BD^2 = AB^2 + AD^2 - 2(AB)(AD)\cos\angle A$

$= (2x)^2 + x^2 - 2(2x)(x)\cos 120° = 7x^2 \qquad \left(\because \cos 120° = -\dfrac{1}{2}\right)$

$\therefore \left(\dfrac{AC}{BD}\right)^2 = \dfrac{3}{7};\ \dfrac{AC}{BD} = \dfrac{\sqrt{3}}{\sqrt{7}}$ or $\dfrac{BD}{AC} = \dfrac{\sqrt{7}}{\sqrt{3}}$

18. Since the trapezium is isosceles, $AO = BO$ and $DO = OC$

i.e., $\angle OAB = \angle OBA = \angle ODC = \angle OCD = 45°$.

Let $EO = x$ and $OF = y$.

$\therefore$ $AE = EB = x$ $(\because \angle OAB = \angle OBA = 45°)$.

$DF = FC = y$ $(\because \angle OCD = \angle ODC = 45°)$.

Area of trapezium

$= \dfrac{1}{2}(2x + 2y)(x + y)$.

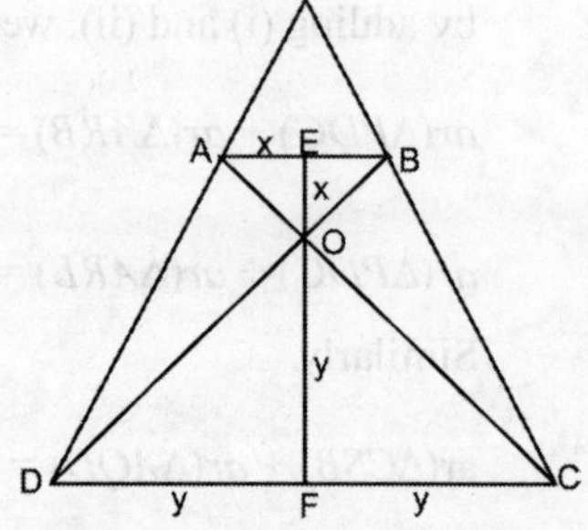

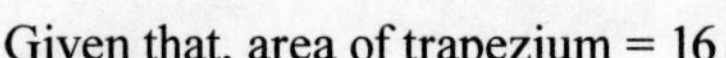

Given that, area of trapezium = 16

$(x + y)(x + y) = 16 \Rightarrow x + y = 4.$

$\therefore$ Height = 4 cm.

19.

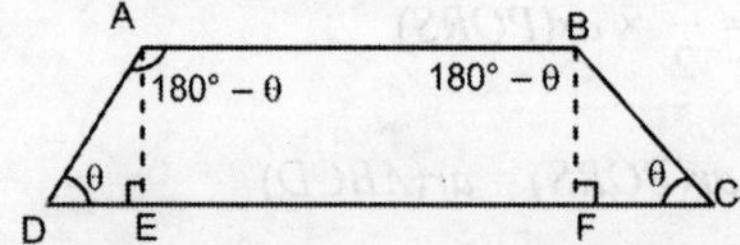

Let E, F be the feet of the perpendiculars from A and B to DC respectively. $AE = BF$ $(\because AB \parallel DC)$ $\angle E = \angle F = 90°$, $AD = BC$.

$\therefore$ ΔADE, ΔBCF and $\angle ADE = \angle BCF$.

$\angle B + \angle C = 180°$, it follows that $\angle B + \angle D = 180°$.

$\therefore$ The trapezium is cyclic. (1) is true.

ΔADB is equilateral $\Rightarrow \angle AOB = 60°$

$\angle ACB = \dfrac{1}{2} < AOB = 30°$

(2) is true.

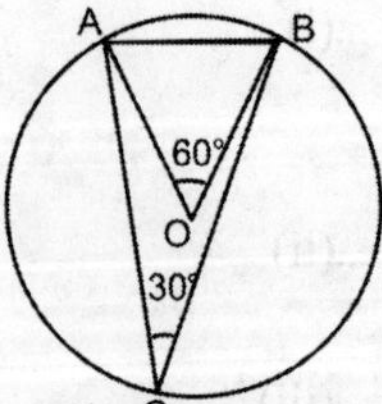

20. Let $AD = a$. $\therefore$ $AB = 2a$.

Let $BD = x$.

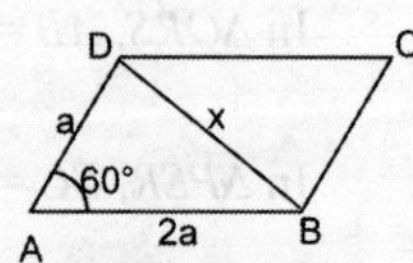

$\therefore$ $x^2 = a^2 + (2a)^2 - 2a(2a)\cos 60° = a^2 + 4a^2 - 2a^2 = 3a^2$.

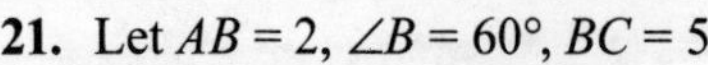

$\therefore$ $x = \sqrt{3}a$, *i.e.* $BD = \sqrt{3}AD$.

21. Let $AB = 2$, $\angle B = 60°$, $BC = 5$.

The third side is given to be 3. We can take this to be CD. But we can also take this to refer to any of the two remaining sides, (after two sides have been specified) *i.e.* we can also take this to refer to AD (It turns out that no matter how we interpret this, the fourth side can be uniquely determined.)

First, let us take $CD = 3$.

Let $DA = x$.

$AB^2 + BC^2 - 2(AB)(BC)\cos B = CD^2 + x^2 - 2(CD)(x)\cos D$

As $\angle B = 60°$, it follows that $\angle D = 120°$,

i.e. $\cos B = \frac{1}{2}$ and $\cos D = -\frac{1}{2}$.

$\therefore\ AB^2 + BC^2 - AB(BC) = CD^2 + x^2 + (CD)(x)$

$\Rightarrow\ 2^2 + 5^2 - 2(5) = 3^2 + x^2 + 2x \Rightarrow x^2 + 2x - 10 = 0 \Rightarrow (x+5)(x-2) = 0 \Rightarrow x = 2$

Now take $AD = 3$, and let $CD = x$.

$$AB^2 + BC^2 - 2(AB)(BC)\cos B = 3^2 + x^2 - 2(CD)\,x\cos D.$$

We get the same equation and the same value of x. But now x refers to CD.

22. If the diagonals of a quadrilateral bisect each other, the quadrilateral is a parallelogram. If the diagonals of a parallelogram are equal, the parallelogram has to be a rectangle. Set A is the set of rectangles.

Set B is the set of rhombuses.

$A \cap B$ is the set of squares.

23. $ABCD$ is the given rectangle and $EFGH$ is the given Rhombus

HF are mid points of AD, BC

$\Rightarrow$ Area of $\Delta\, HGF = \frac{1}{2}$ area of $HFCD$

$\Rightarrow\ \frac{1}{2}$ Area of Rhombus $= \frac{1}{2}\cdot\frac{1}{2}\ ABCD.\ \ \frac{2}{2}\times 4 = ABCD$

$\therefore$ Area of $ABCD = 2 \times 2$ units

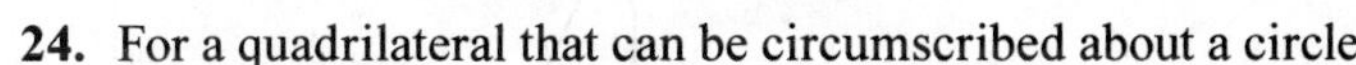

24. For a quadrilateral that can be circumscribed about a circle,

The sum of the length of one pair of opposite sides is equal to the sum of the length of the other pair of opposite sides.

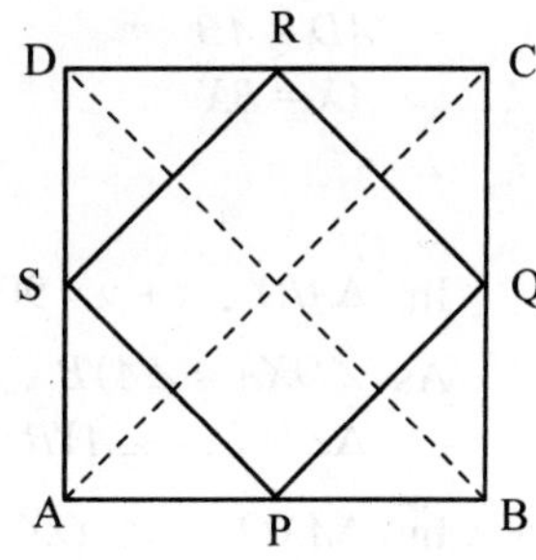

25. $PQ \parallel AC$ [$\because$ Through mid point theoram]

$PQ = \frac{1}{2}AC$

$PS \parallel BD$ [$\because$ Through mid point theoram]

$PS = \frac{1}{2}BD$

and $AC = BD$ [$\because$ Diagonals of rectangle are equal point theoram]

$\therefore\ PQ = PS$ *(i)*

Also $PQ \parallel AC$ and $SR \parallel AC$

$\therefore\ SR \parallel PQ$

So, $PQRS$ is a parallelogram (ii)

Through (*i*) and (ii), we can conclude that $PQRS$ is a Rhombus and $PQRS$ is not a square because its angles are not given to be 90°.

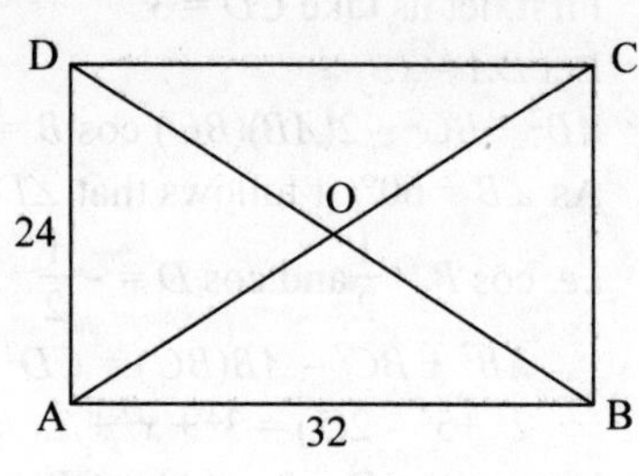

26. $BD = \sqrt{24^2 + 32^2}$

$$\left[\begin{array}{l} \because (24)^2 = (8\times3)^2 = 8^2\times3^2 \\ (32)^2 = (8\times4)^2 = 8^2\times4^2 \end{array}\right]$$

$\therefore \quad BD = \sqrt{8^2\left(3^2+4^2\right)} = \sqrt{8^2\times5^2}$

$= 40cm \left[\because 3^2 + 4^2 = 5^2 \text{ pythagoram theorems}\right]$

$OD = \dfrac{40}{2} = 20cm$

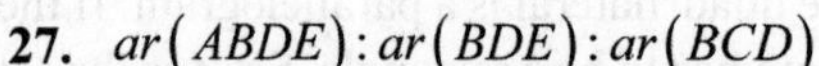

27. $ar(ABDE) : ar(BDE) : ar(BCD)$

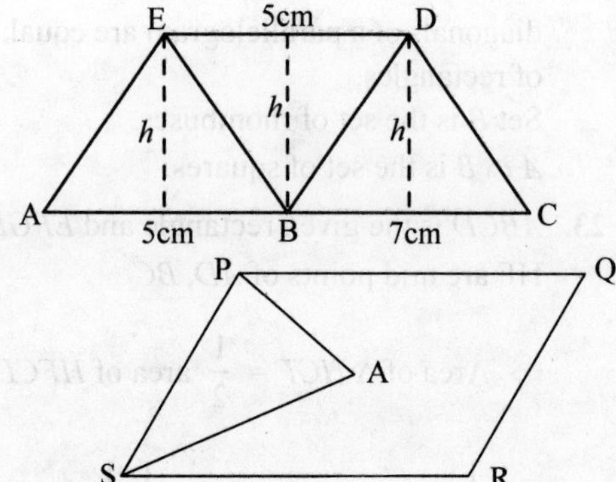

$5\times h : \dfrac{1}{2}\times5\times h : \dfrac{1}{2}\times7\times h$

$10 : 5 : 7$

28. As $PQRS$ is a parallelogram

$\therefore \quad \angle P + \angle S = 180°$

$\Rightarrow \quad \dfrac{1}{2}\angle P + \dfrac{1}{2}\angle S = \dfrac{180}{2}$

In ΔPAS, $\angle PAS = 180° - 90° = 90°$

29. As $\quad AB = BC$

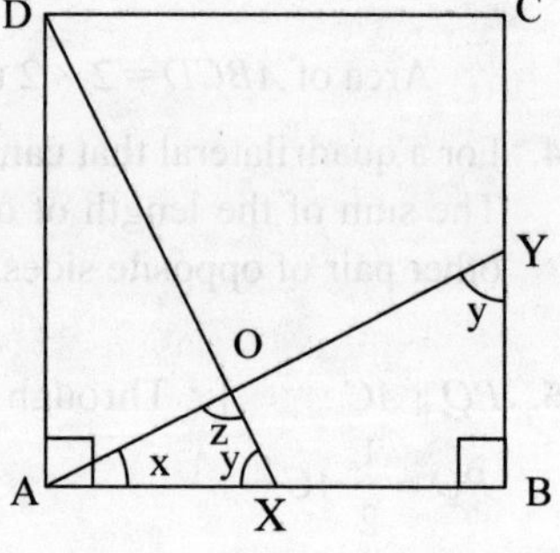

$\dfrac{1}{2}AB = \dfrac{1}{2}BC$

$\Rightarrow \quad AX = BY$

In ΔADX and ΔBAY

$\angle A = \angle B = 90°$

$AD = AB$ [sides of square are equal]

$AX = BX$

$\therefore \quad \Delta ADX \cong \Delta BAY$ [*SAS*]

$\therefore \quad \angle DXA = \angle AYB$ [*cpct*]

In ΔABY, $x + y = 90°$

As $\angle DXA = \angle AYB$

$\Delta\angle DXA = \angle AYB = y$

In ΔAXO, $\quad \angle AOX = 180 - (x + y) = 180° - 90° = 90°$

$\therefore \quad \angle AOD = 90°$

$\therefore \; DX \perp AY$ So, statement 1 and 2 are correct

30. y Area of rectangle $= xy = p \; 2a^2 = p$

x

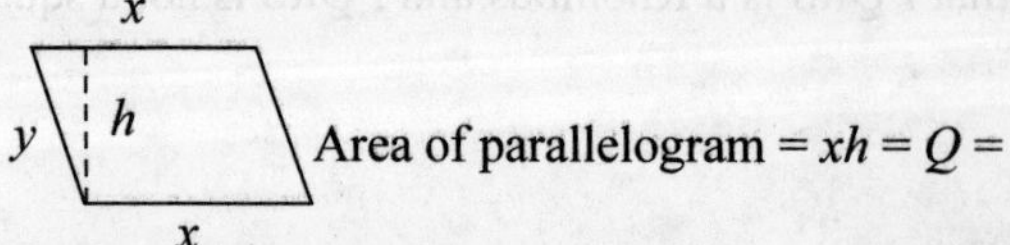

Area of parallelogram $= xh = Q = 2ah$

$\therefore \quad a > h$

$\therefore \quad P > Q$

31. From statement I

AB + DC = 10 cm

$\Rightarrow$ AD + BC = 10 cm

$\therefore$ Perimeter = 20 cm

From statement II

AD + BC = 10 cm

AB + DC = 10 cm

$\therefore$ perimeter = 20 cm

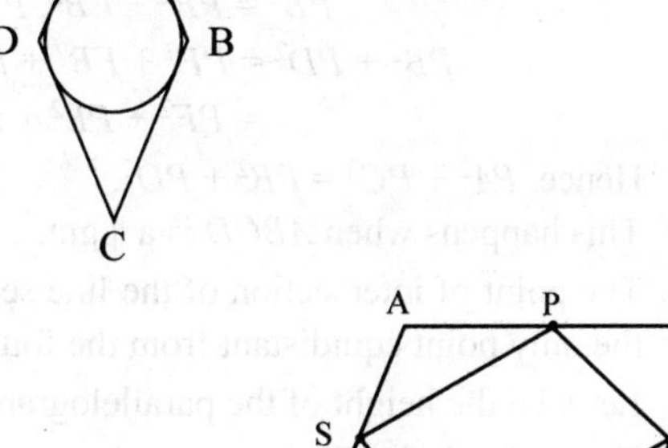

32. From statement I

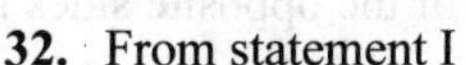

$ar(\text{quad } PQRS) = \dfrac{100}{2} = 50 \text{ cm}^2$

From statement II

$ar(ABCD) = 50 \times 2 = 100 \text{ cm}^2$

33. The bisectors of the angles of a parallelogram enclose a rectangle.

34. The figure formed by joining the mid-points of the adjacent sides of a quadrilateral is a parallelogram.

35. The figure formed by joining the mid-points of the adjacent sides of a rectangle is a rhombus.

36. The figure formed by joining the mid-points of the adjacent sides of a rhombus is a rectangle.

37. The figure formed by joining the mid-points of the adjacent sides of a square is a square.

38. The figure formed by joining the mid-points of the adjacent sides of a parallelogram is a parallelogram.

39. Let smallest angle be $x°$.

Then, $(2x - 24) + x = 180 \Rightarrow 3x = 204 \Rightarrow = x = 68$.

$\therefore$ Largest angle = (180° – 68°) = 112°.

40. Let the smallest angle be $x°$.

Then, $x + \dfrac{2}{3}x = 180 \Rightarrow x = 108$.

$\therefore$ Smallest angle $= \dfrac{2}{3}x = \left(\dfrac{2}{3} \times 108\right) = 72°$.

41. $\angle C = \angle A = 75°$ (opposite $\angle s$ of a || gm are equal)

$\therefore \angle BDC = 180° - (60° + 75°) = 45°$.

42. $AB \parallel EF \parallel CD$. So, $ABEF$ is a rectangle.

$\therefore \Delta AGB = \dfrac{1}{2}$ (area of rect. $ABEF$)

$= \dfrac{1}{2} \times \left(\dfrac{1}{2} \text{area of rect. } ABCD\right)$

$= \dfrac{1}{4}$(area of rect. $ABCD$)

43. Proof is same as given in Q. 42.

44. Draw $EPF \parallel AB$.

In right ΔPEA & ΔPFC:

$$PA^2 = PE^2 + EA^2, PC^2 = PF^2 + FC^2$$

$$\therefore \quad PA^2 + PC^2 = PE^2 + EA^2 + PF^2 + FC^2$$

In right Δs PFB and PED :

$$PB^2 = PF^2 + FB^2, PD^2 = PE^2 + ED^2$$

$$\therefore \quad PB^2 + PD^2 = PF^2 + FB^2 + PE^2 + ED^2$$

$$= PF^2 + PE^2 + EA^2 + FC^2 \; [\because FB = EA \;\&\; ED = FC]$$

Hence, $PA^2 + PC^2 = PB^2 + PD^2$.

45. This happens when $ABCD$ is a $\parallel$ gm.

46. The point of intersection of the line segments joining the mid-points of the opposite sides is the only point equidistant from the four vertices of the parallelogram.

47. Let h be the height of the parallelogram.

Then, clearly $h < b$.

$\therefore A = a \times h < a \times b = B$. So $A < B$.

32 CHAPTER

Circles

Facts Useful for Solving Questions on Circles

1. The perpendicular from the centre of a circle to a chord bisects the chord.
2. The line joining the centre of a circle to the mid-point of a chord is perpendicular to the chord.
3. Perpendicular bisector of a chord passes through the centre.
4. Equal chords of a circle (or of congruent circles) are equidistant from the centre.
5. Chords which are equidistant from the centre are equal.
6. There is one and only one circle passing through three non-collinear points.
7. The angle subtended by an arc of a circle at the centre is double the angle subtended by it at any point on the remaining part of the circle.
8. Angles in the same segment of a circle are equal.
9. The angle in a semi circle is a right angle.
10. The sum of a pair of opposite angles of a cyclic quadrilateral is 180°
11. If a side of a cyclic quadrilateral is produced, then the exterior angle is equal to the interior opposite angle.
12. Equal chords are made by equal arcs.
13. Equal chords (or equal arcs) subtend equal angles at the centre.
14. The quadrilateral formed by angle bisectors of a cyclic quadrilateral is also cyclic.
15. If two chords AB and CD of a circle intersect inside or outside the circle when produced at a point E, then $AE \times EB = DE \times EC$.

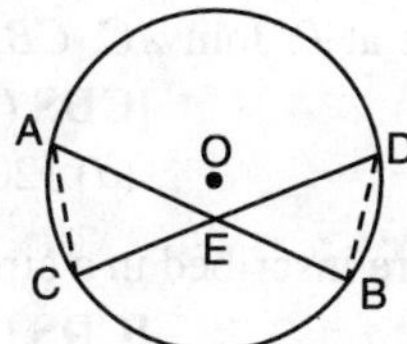

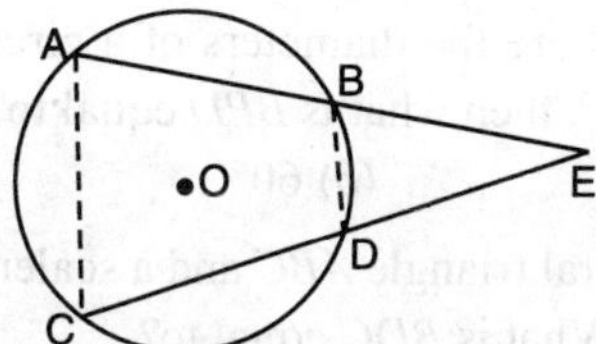

16. Suppose a circle of radius r is inscribed in a triangle of area A. If the semiperimeter of the triangle be s, then $r = \dfrac{A}{s}$.
17. A tangent at any point of a circle is perpendicular to the radius through the point of contact.
18. The lengths of two tangents drawn from an external point to a circle are equal.
19. From an extemal point, if two tangents are drawn to a circle,
    (*a*) they subtend equal angles at the centre
    (*b*) they are equally inclined to the line segment joining the centre to that point.
20. If A is an external point from which the tangents to the circle with centre O touch it at B and C, then OA is the perpendicular bisector of BC.
21. If PAB is a secant intersecting the circle at A and B, and PT is a tangent, then $PA \times PB = PT^2$.

22. If all the sides of a parallelogram touch a circle, the parallelogram is a rhombus,

23. If from the point of contact of tangent a chord is drawn, then the angles which the chord makes with the tangent line are equal respectively to the angles formed in the corresponding alternate segments.

∴ In the adjoining fig.

$\angle BAT = \angle ACB$ and $\angle BAP = \angle ADB$

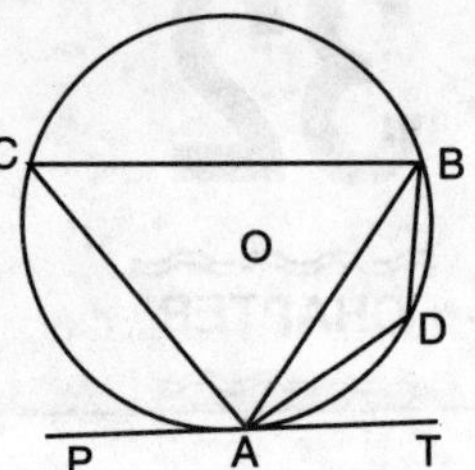

EXERCISE 32

1. What is the length of the chord of a unit circle which subtends an angle 20 at the centre, where $\theta < 45°$? **[CDS (I) Exam, 2022]**

(*a*) sin 2θ (*b*) cos 2θ (*c*) 2 sinθ (*d*) 2 cosθ

2. *CD* is a direct common tangent to two circles intersecting each other at *A* and *B*. Then, $\angle CAD + \angle CBD = ?$

(*a*) 90° (*b*) 180°

(*c*) 360° (*d*) 120°

3. C is the centre of a circle of radius 20 cm. AB is a chord of length 32 cm. E is a point on AB such that CE = 13 cm. What is AE × EB equal to? **[CDS (I) Exam, 2025]**

(*a*) 231 square cm (*b*) 256 square cm (*c*) 272 square cm (*d*) 297 square cm

4. What is the radius of the circle inscribed in a triangle whose sides are 4 cm, 7.5 cm and 8.5 cm? **[CDS (I) Exam, 2022]**

(*a*) 1.5 cm (*b*) 2 cm (*c*) 2.5 cm (*d*) 3 cm

5. AB and *CD* are the diameters of a circle which intersect at *P*. Join *AC*, *CB*, *BD* and *DA*. If $\angle PAD = 60°$, then what is *BPD* equal to? **[CDS (I) Exam, 2021]**

(*a*) 30° (*b*) 60° (*c*) 90° (*d*) 120°

6. An equilateral triangle *ABC* and a scalene triangle *DBC* are inscribed in a circle on same side of the arc. What is *BDC* equal to? **[CDS (I) Exam, 2021]**

(*a*) 30° (*b*) 45° (*c*) 60° (*d*) 90°

7. The length of an arc of a circle of radius 4 cm is π cm. What is the magnitude of the angle subtended by the arc at the centre? **[CDS (I) Exam, 2025]**

(*a*) π (*b*) π/2 (*c*) π/3 (*d*) π/4

8. Two circles touch externally. The sum of their areas is 417 square cm. If the distance between their centres is 9 cm, then what is difference between their diameters? **[CDS (II) Exam, 2021]**

(*a*) 1 cm (*b*) 1.5 cm (*c*) 2 cm (*d*) 4 cm

Consider the following for the next three (3) items that follows:

ABC is a triangle with sides *AB* = 6 cm, *BC* = 10 cm and *CA* = 8 cm. With vertices *A*, *B* and *C* as centres, three circles are drawn each touching the other two externally.

9. What is the sum of the radii of the circles? **[CDS (II) Exam, 2021]**
(*a*) 10.4 cm (*b*) 11.2 cm (*c*) 12 cm (*d*) 13 cm (*a*)

10. What is the length of the altitude of the triangle drawn from vertex A on BC?
(*a*) 2.4 cm (*b*) 3 cm (*c*) 4 cm (*d*) 4.8 cm
[CDS (II) Exam, 2021]

11. If P, Q and R are the areas of sectors at B and C within the triangle respectively, then which of the following is/are correct? **[CDS (II) Exam, 2021]**
1. $P = \pi$ cm²
2. $9Q + 4R = 36\pi$ cm²

Select the correct answer using the code given below :
(*a*) 1 only (*b*) 2 only (*c*) Both 1 and 2 (*d*) Neither 1 nor 2

12. Two circles of radii 20 cm and 16 cm intersect and the length of common chord is 24 cm. If d is the distance between their centres, then which one of the following is correct?
(*a*) $d < 26$ cm (*b*) 26 cm $< d <$ 27 cm
(*c*) 27 cm $< d <$ 28 cm (*d*) $d >$ 28 cm **[CDS (I) Exam, 2020]**

13. A circle is inscribed in an equilateral triangle. The radius of the circle is 2 cm. What is the area of the triangle? **[CDS (I) Exam, 2025]**
(*a*) $12\sqrt{3}$ square cm (*b*) 12 square cm (*c*) $9\sqrt{3}$ square cm (*d*) 9 square cm

14. A circle of diameter 8 cm is placed in such a manner that it touches two perpendicular lines. Then another smaller circle is placed in the gap such that it touches the lines and the circle. What is the diameter of the smaller circle? **[CDS (I) Exam, 2020]**
(*a*) $4(3-\sqrt{2})$ cm (*b*) $4(3-2\sqrt{2})$ cm (*c*) $8(3-\sqrt{2})$ cm (*d*) $8(3-2\sqrt{2})$ cm

15. A circle is inscribed in a triangle ABC right-angled at B. If $AB = 5$ cm and $BC = 12$ cm, then what is the radius of the circle? **[CDS (II) Exam, 2024]**
(*a*) 1 cm (*b*) 1.5 cm (*c*) 2 cm (*d*) 2.5 cm

16. If the lengths of two parallel chords in a circle of radius 10 cm are 12 cm and 16 cm, then what is the distance between these two chords? **[CDS (I) Exam, 2019]**
(*a*) 1 cm or 7 cm (*b*) 2 cm or 14 cm (*c*) 3 cm or 21 cm (*d*) 4 cm or 28 cm

17. In a circle of radius 8 cm. AB and AC are two chords such that $AB = AC = 12$ cm. What is the length of chord BC? **[CDS (I) Exam, 2019]**
(*a*) $2\sqrt{6}$ cm (*b*) $3\sqrt{6}$ cm (*c*) $3\sqrt{7}$ cm (*d*) $6\sqrt{7}$ cm

18. Two equal circles intersect such that each passes through the centre of the other. If the length of the common chord of the circles is $10\sqrt{3}$ cm, then what is the diameter of the circle?
(*a*) 10 cm (*b*) 15 cm (*c*) 20 cm (*d*) 30 cm
[CDS (I) Exam, 2019]

19. A line segment AB is the diameter of a circle with centre at O having radius 6.5 cm. Point P is in the plane of the circle such that $AP = x$ and $BP = y$ In which one of the following cases the point P does not lie on the circle? **[CDS (II) Exam, 2019]**
(*a*) $x = 6.5$ cm and $y = 6.5$ cm (*b*) $x = 12$ cm and $y = 5$ cm
(*c*) $x = 5$ cm and $y = 12$ cm (*d*) $x = 0$ cm and $y = 13$ cm

20. The radii of two circles are 4.5 cm and 3.5 cm respectively. The distance between the centres of the circles is 10 cm. What is the length of the transverse common tangent?
[CDS (I) Exam, 2018]
(*a*) 4 cm (*b*) 5 cm (*c*) 6 cm (*d*) 7 cm

21. Consider the following statements: **[CDS (I) Exam, 2018]**

1. The orthocentre of a triangle always lies inside the triangle.
2. The centroid of a triangle always lies inside the triangle.
3. The orthocentre of a right angled triangle lies on the triangle.
4. The centroid of a right angled triangle lies on the triangle.

Which of the above statements are correct?

(*a*) 1 and 2 (*b*) 1 and 4 (*c*) 2 and 3 (*d*) 2 and 4

22. In the figure given below, *XA* and *XB* are two tangents to a circle, If angle $AXB = 50°$ and AC is parallel to *XB*, then what is $\angle ACB$ equal to? **[CDS (I) Exam, 2018]**

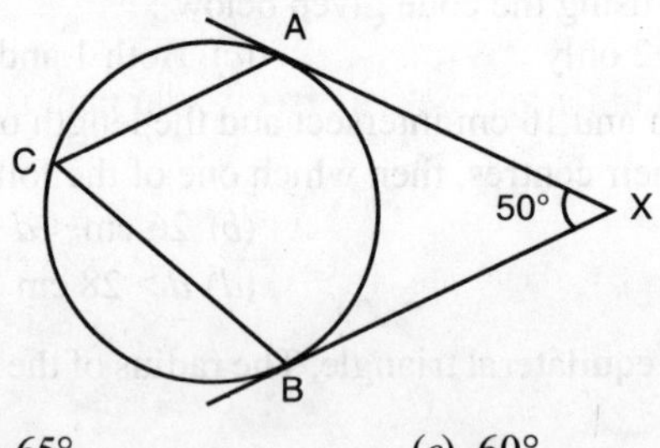

(*a*) 70° (*b*) 65° (*c*) 60° (*d*) 55°

23. The chord of a circle is $\sqrt{3}$ times its radius. The angle subtended by this chord at the minor arc is k times the angle subtended at the major arc. What is the value of k?

[CDS (II) Exam, 2018]

(*a*) 5 (*b*) 2 (*c*) $\frac{1}{2}$ (*d*) $\frac{1}{5}$

24. In the figure given below, the radius of the circle is 6 cm and AT 4 cm. The = length of tangent PT is

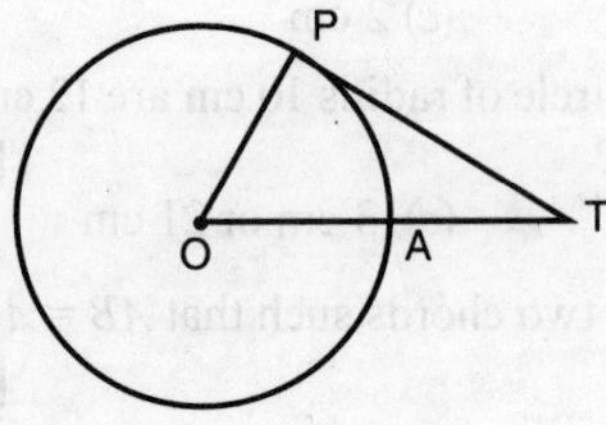

[CDS (II) Exam, 2018]

(*a*) 6 cm (*b*) 8 cm (*c*) 9cm (*d*) 10 cm

25. In the figure given below, what is $\angle BCD$ equal to? **[CDS (II) Exam, 2018]**

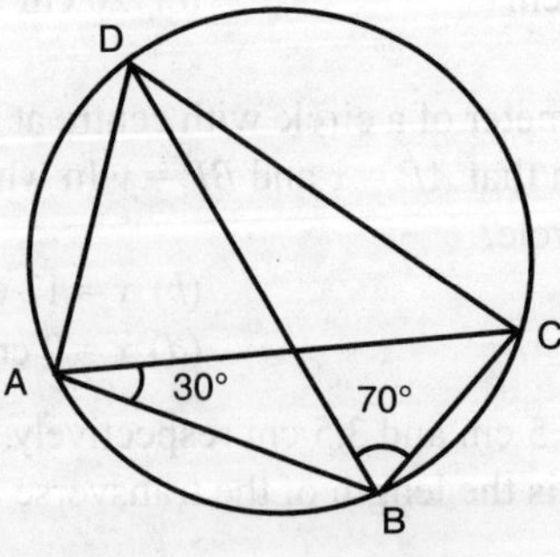

(*a*) 70° (*b*) 75° (*c*) 80° (*d*) 90°

26. In the figure given below, AB is the diameter of the circle whose centre is at O. Given that angle ECD= angle EDC = 32 deg then CEF and ∠COF respectively are **[CDS (II) Exam, 2018]**

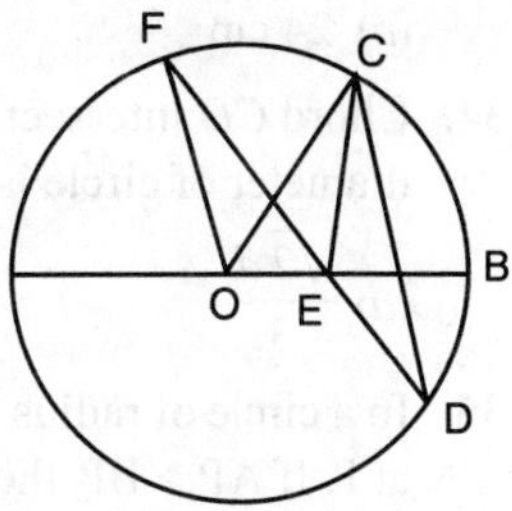

(*a*) 32°, 64° (*b*) 64°, 64°

(*c*) 32°, 32° (*d*) 64°, 32°

27. The ratio of the outer and inner perimeters of a circular path is 23: 22. If the path is 5 m wide, the diameter of the inner circle is **[CDS (I) Exam, 2015]**

(*a*) 55m (*b*) 110m (*c*) 220m (*d*) 230m

28. The area of the largest triangle that can be inscribed in a semicircle of radius *r* is

(*a*) r^2 (*b*) $2r^2$ (*c*) $3r^2$ (*d*) $4r^2$

[CDS (I) Exam, 2015]

29. AD is the diameter of a circle and AB is a chord. If AD = 34 cm, AB = 30 cm, the distance of AB from the centre of the circle is **[CDS (I) Exam, 2015]**

(*a*) 17 cm (*b*) 15 cm (*c*) 13 cm (*d*) 8 cm

Consider the following for the next two (02) items that follow:

A circle M of radius 8 cm touches externally with another circle N of radius 16 cm. Let P, Q be the points where the common tangent touches the circles M and N respectively.

30. What is the length of the common tangent PQ? **[CDS (I) Exam, 2025]**

(*a*) 16 cm (*b*) $16\sqrt{2}$ cm (*c*) 24 cm (*d*) $24\sqrt{2}$ cm

31. If U, V are the centres of the circles M and N respectively, then what is the area of the quadrilateral formed by the points P, Q, V and U?

(*a*) $192\sqrt{2}$ square cm (*b*) 192 square cm

(*c*) $96\sqrt{2}$ square cm (*d*) 96 square cm

32. *AB* is a line segment of length 2*a*, with *M* as mid-point. Semicircles are drawn on one side with *AM, MB* and *AB* as diameter as shown in the above figure. A circle with centre *O* and radius *r* is drawn such that this circle touches all the three semicircles. The value of *r* is

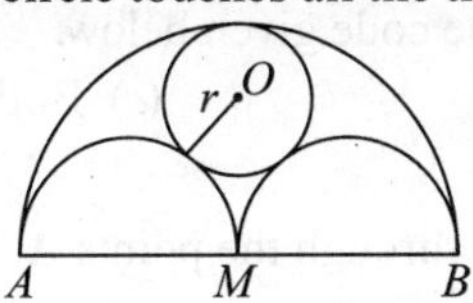

(*a*) $\frac{2a}{3}$ (*b*) $\frac{a}{2}$ (*c*) $\frac{a}{3}$ (*d*) $\frac{a}{4}$

[CDS (II) Exam, 2015]

33. From an external point *P* tangents *PA* and *PB* are drawn to the circle as shown in the above figure. *CD* is the tangent to the circle at *E*. If *AP* = 16 cm, then the perimeter of the triangle *PCD* is equal to

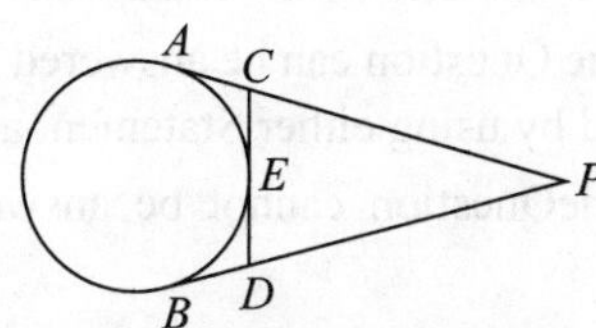

(*a*) 24 cm (*b*) 28 cm (*c*) 30 cm (*d*) 32 cm

34. Chord *CD* intersects the diameter *AB* of a circle at right angle at a point *P* in the ratio 1 : 2. If diameter of circle is *d*, then *CD* is equal to **[CDS (II) Exam, 2015]**

(*a*) $\frac{\sqrt{2}d}{3}$ (*b*) $\frac{2d}{3}$ (*c*) $\frac{2\sqrt{2}d}{3}$ (*d*) $\frac{2\sqrt{3}d}{3}$

35. In a circle of radius 2 units, a diameter AB intersects a chord of length 2 units perpendicularly at P. If AP > BP, then AP is equal to **[CDS (I) Exam, 2016]**

(*a*) $(2+\sqrt{5})$ units (*b*) $(2+\sqrt{3})$ units (*c*) $(2+\sqrt{2})$ units (*d*) 3 units

36. A truck moves along a circular path and describes 100 m when it has traced out 36° at the centre. The radius of the circle is equal to **[CDS (I) Exam, 2016]**

(*a*) $\frac{100}{\pi}$ m (*b*) $\frac{250}{\pi}$ m (*c*) $\frac{500}{\pi}$ m (*d*) $\frac{600}{\pi}$ m

37. Consider a circle with centre at O and radius 7 cm. Let QR be a chord of length 2 cm and let P be the midpoint of QR. Let CD be another chord of this circle passing through P such that ∠CPQ is acute. If *M* is the midpoint of CD and MP = $\sqrt{24}$ cm, then which of the following statements are correct? **[CDS (I) Exam, 2016]**

1. ∠QPD = 135°

2. If CP = m cm and PD = n cm, then *m* and n are the roots of the quadratic equation $x^{2-10}x + 1 = 0$.

3. The ratio of the area of triangle OPR to the area of triangle OMP is 1 : $2\sqrt{2}$,

Select the correct answer using the code given below.

(*a*) 1 and 2 only (*b*) 2 and 3 only (*c*) 1 and 3 only (*d*) 1, 2 and 3

38. Consider a circle with centre at C. Let OP, OQ denote respectively the tangents to the circle drawn from a point *O* outside the circle. Let R be a point on OP and *S* be a point on OQ such that OR × SQ = OS × RP. Which of the following statements is/are correct?

1. If X is the circle with centre at O and radius OR, and Y is the circle with centre at *O* and radius *OS*, then X = Y.

2. ∠POC+ ∠QCO= 90°

Select the correct answer using the code given below.

(*a*) 1 only (*b*) 2 only (*c*) Both 1 and 2 (*d*) Neither 1 nor 2

[CDS (I) Exam, 2016]

39. Question: Can a circle be drawn through the points A, B and C ? **[CDS (I) Exam, 2023]**

Statement I :

AB = 5 cm, BC = 5 cm, CA= 6 cm.

Statement II: AB = 3 cm, BC = 4 cm, CA= 7 cm.

(*a*) Choose this option if the Question can be answered by one of the Statements alone but not by the other.

(*b*) Choose this option if the Question can be answered by either Statement alone.

(*c*) Choose this option if the Question can be answered by using both the Statements together, but cannot be answered by using either Statement alone

(*d*) Choose this option if theQuestion cannot be answered even by using both Statements together

Consider the following for the next three (03) items that follow:

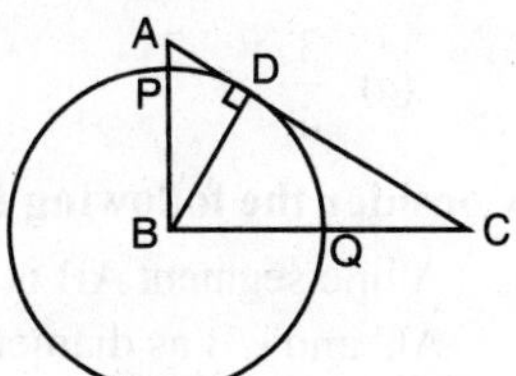

In the triangle ABC, AB = 6 cm, BC = 8 cm and AC = 10 cm. The perpendicular dropped from B meets the side AC at D. A circle of radius BD (with centre B) cuts AB and BC at P and Q respectively as shown in the figure.

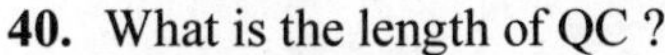

40. What is the length of QC ? **[CDS (I) Exam, 2023]**

(*a*) 4·4 cm (*b*) 4·2 cm (*c*) 3·6 cm (*d*) 3·2 cm

41. If L ABD = 0, then what is sin 0 equal to ? **[CDS (I) Exam, 2023]**

(*a*) 0·4 (*b*) 0·5 (*c*) 0·6 (*d*) 0·8

42. What is the radius of the circle ? **[CDS (I) Exam, 2023]**

(*a*) 5 cm (*b*) 4.8 cm (*c*) 4.4 cm (*d*) 4 cm

Consider the following for the next two (02) items that follow:

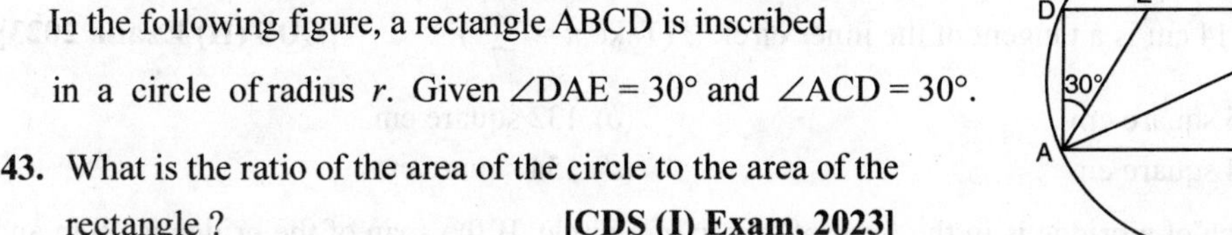

In the following figure, a rectangle ABCD is inscribed in a circle of radius *r*. Given ∠DAE = 30° and ∠ACD = 30°.

43. What is the ratio of the area of the circle to the area of the rectangle ? **[CDS (I) Exam, 2023]**

(*a*) $\frac{\pi}{\sqrt{2}}$ (*b*) $\frac{\pi}{\sqrt{3}}$ (*c*) $\frac{2\pi}{\sqrt{3}}$ (*d*) $\frac{3\pi}{\sqrt{3}}$

44. What is the area of ΔAEC ? **[CDS (I) Exam, 2023]**

(*a*) $\frac{r^2}{\sqrt{3}}$ (*b*) $\frac{r^2}{2\sqrt{3}}$ (*c*) $\frac{r^2}{3\sqrt{3}}$ (*d*) $\frac{2r^2}{\sqrt{3}}$

Consider the following for the next two (02) items that follow : [CDS (I) Exam, 2023]

In the following figure, a triangle *ABC* is inscribed in a circle with centre at 0. Let ∠POA = *x*° and ∠*OQB* = *y*° Further, *OB* = *BQ*.

45. What is the relation between *x* and *y* ?

(*a*) *x* = *y* (*b*) 2*x* = 3*y*

(*c*) *x* = 3*y* (*d*) 3*x* = 4*y*

46. If *y* = 15, then what is ∠ABC equal to ?

(*a*) 30° (*b*) 40°

(*c*) 45° (*d*) 60°

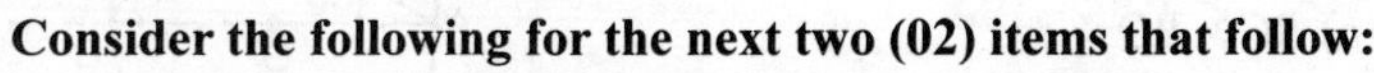

Consider the following for the next two (02) items that follow:

Two parallel chords AB and CD of a circle are of lengths 60 cm and 80 cm respectively. They are on the same side of the centre O and 10 cm apart

47. What is the diameter of the circle ? **[CDS (II) Exam, 2023]**

(*a*) 120cm (*b*) 110cm (*c*) 100cm (*d*) 90cm

48. If the chord AB subtends an angle a and chord CD subtends an angle at the centre 0, then what is the value of $\tan\left(\frac{\beta}{2}\right) - \tan\left(\frac{\alpha}{2}\right)$? **[CDS (II) Exam, 2023]**

(*a*) $\frac{3}{4}$ (*b*) $\frac{5}{12}$ (*c*) $\frac{1}{2}$ (*d*) $\frac{7}{12}$

Consider the following for the next two (02) items that follow:

A line segment AB is bisected at C and semi-circles S_1, S_2 and S_3 are drawn respectively on AB, AC and CB as diameters such that they all lie on same side of AB. A circle S is drawn touching internally S_1 and externally S_2 and S_3.

49. If r is the radius of S and R is the radius of S_2, then which one of the following is correct ?

(*a*) $R = 3r$ (*b*) $R = 2r$ (*c*) $3R = 4r$ (*d*) $2R = 3r$

[CDS (II) Exam, 2023]

50. If m is the area of the circle S and n is the area of semi-circle S_1, then which one of the following is correct ? **[CDS (II) Exam, 2023]**

(*a*) 9m = 2n (*b*) 9m = 4n (*c*) 3m = 2n (*d*) 7m = 3n

51. What is the area of the region between two concentric circles if the chord of the outer circle of length 14 cm is a tangent of the inner circle ? (Take $\pi = \frac{22}{7}$) **[CDS (II) Exam, 2023]**

(*a*) 125 square cm (*b*) 132 square cm
(*c*) 144 square cm (*d*) 154 square cm

52. The arch of a bridge is in the form of an arc of a circle. If the span of the bridge is 40 m and height in the middle is 8 m, then what is the radius of curvature of the bridge?

[CDS (II) Exam, 2023]

(*a*) 25 m (*b*) 27 m (*c*) 29 m (*d*) 31 m

53. In a quarter circle of radius *R*, a circle of radius r is inscribed. What is the ratio of *R* to *r* ?

[CDS (I) Exam, 2024]

(*a*) $(\sqrt{2}+1):1$ (*b*) $(\sqrt{3}+1):1$ (*c*) 3 : 2 (*d*) 5 : 4

54. In the adjoining figure, *A*, *B*, *C* are three points on a circle with centre *O*. If $\angle AOB = 90°$ and $\angle BOC = 120°$, then $\angle ABC$ is:

(*a*) 60° (*b*) 75°
(*c*) 90° (*d*) None of these

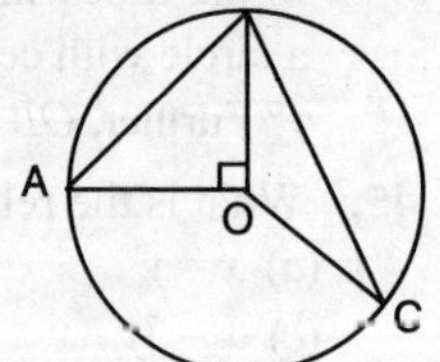

55. *AB* is a diameter and *AC* is a chord of a circle such that $\angle BAC = 30°$. The tangent at *C* intersects *AB* produced in *D*. Then,

(*a*) $BC < BD$ (*b*) $BC > BD$
(*c*) $BC = BD$ (*d*) Cannot say

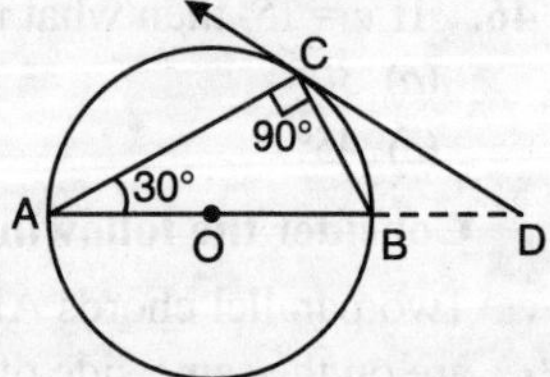

56. The area of the circle circumscribing three identical circles touching each other is $\frac{\pi(2+\sqrt{3})^2}{3}$ square cm. What is the radius of one of the smaller circles? **[CDS (I) Exam, 2025]**

(*a*) 0.5 cm (*b*) 1 cm (*c*) 1.5 cm (*d*) $\sqrt{3}$ cm

Direction (Q. 57-59): Consider the following for the next three (03) items that follow:

Consider two identical semicircles and one circle inscribed in a rectangle of length 10 cm as shown in the figure given below.

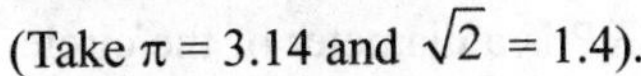

(Take $\pi = 3.14$ and $\sqrt{2} = 1.4$).

57. What is the area of triangle EOF ?

(*a*) $12.5\sqrt{3}$ square cm

(*b*) $6.25\sqrt{3}$ square cm

(*c*) 12.5 square cm

(*d*) 6.25 square cm

58. What is the area of trapezium AEFB ?

(*a*) 30 square cm (*b*) 25 square cm (*c*) 20 square cm (*d*) 18.75 square cm

59. What is the area of the shaded region ?

(*a*) 14.75 square cm (*b*) 14.25 square cm (*c*) 7.225 square cm (*d*) 7.625 square cm

60. *O* is the centre of a circle. If tangent $PQ = 12$ cm and $BQ = 8$ cm, then chord *AB* is:

(*a*) 10 cm (*b*) $4\sqrt{5}$ cm

(*c*) 4 cm (*d*) 18 cm

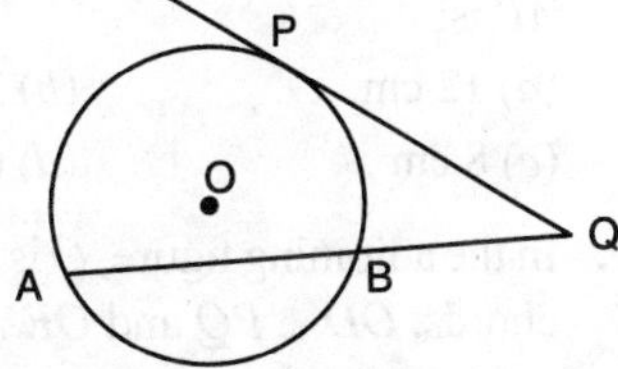

61. *AB* and *CD* are two parallel chords of a circle with centre *O* such that $AB = 6$ cm and $CD = 12$ cm. The chords are on the same side of the centre and the distance between them is 3 cm. Then, the radius of the circle is:

(*a*) 6 cm (*b*) $5\sqrt{2}$ cm (*c*) 7 cm (*d*) $3\sqrt{5}$ cm

62. In the adjoining figure, $\angle ADC = 140°$ and *AOB* is the diameter of the circle. Then, $\angle BAC$ is equal to:

(*a*) 40° (*b*) 50°

(*c*) 70° (*d*) 75°

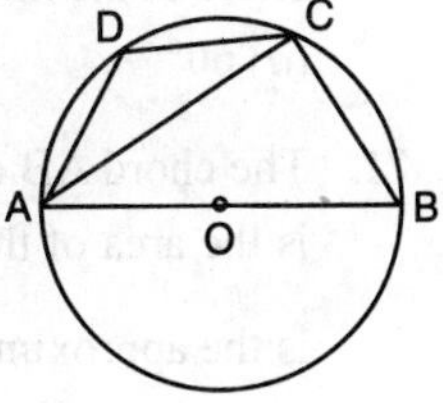

Consider the following for the next three (03) items that follow:

Let *MN* be a chord of length 16 cm of a circle with centre at *O* and radius 10 cm. The tangents at *M* and *N* intersect at a point *P*. Further, *OP* intersects *MN* perpendicularly at *Q*.

63. What is *OQ* equal to? **[CDS (II) Exam, 2024]**

(*a*) 5 cm (*b*) 6 cm (*c*) 7 cm (*d*) 8 cm

64. What is *PM* equal to?

(*a*) 10 cm (*b*) 12 cm (*c*) $\frac{40}{3}$ cm (*d*) $\frac{50}{3}$ cm

65. What is the area of triangle *OMN*?

(*a*) 36 square cm (*b*) 40 square cm (*c*) 45 square cm (*d*) 48 square cm

66. Two circles touch internally. The sum of their areas is $116\,\pi$ sq cm and the distance between their centres is 6 cm. The radius of the larger circle is:

(*a*) 7 cm (*b*) 8 cm (*c*) 10 cm (*d*) 12 cm

67. In the adjoining figure, if $\angle QPR = 67°$ and $\angle SPR = 72°$ and RP is a diameter of the circle, then $\angle QRS$ is equal to:

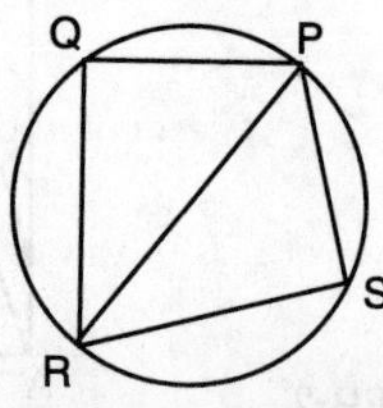

(*a*) 41° (*b*) 23° (*c*) 67° (*d*) 18°

68. In the adjoining figure, O is the centre of a circle. Chords AB and CD intersect at P. If $AB = 16$ cm, $CP = 6$ cm, $PD = 8$ cm and $AP > PB$, then AP is:

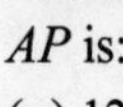

(*a*) 12 cm (*b*) 24 cm

(*c*) 8 cm (*d*) 6 cm

69. In the adjoining figure, O is the centre of the circle and PQ, RS, are its equal chords, $OD \perp PQ$ and $OE \perp RS$. If $\angle DOE = 130°$, then $\angle PDE$ is:

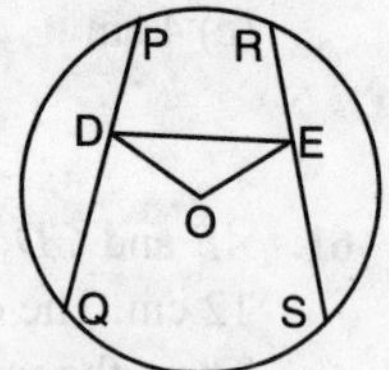

(*a*) 50° (*b*) 65°

(*c*) 40° (*d*) 70°

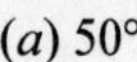

70. The chord of a circle is equal to its radius. The angle subtended by this chord at the minor arc of the circle is:

(*a*) 60° (*b*) 75° (*c*) 120° (*d*) 150°

71. The chord AB of a circle with centre at O is $2\sqrt{3}$ time the height of the monor segment. If P is the area of the sector OAB and Q is the area of the minor segment of the circle, then what is the approximate value of $\frac{P}{Q}$? **[CDS (I) Exam, 2024]**

(Take $\sqrt{3} = 1{\cdot}7$ and $\pi = 3.14$)

(*a*) 1.4 (*b*) 1.7 (*c*) 2.2 (*d*) 2.6

72. What is the area of the region between two concentric circles, if the length of a chord of the outer circle touching the inner circle at a particular point of its circumference is 14 cm ?

(Take $\pi = \frac{22}{7}$) **[CDS (I) Exam, 2024]**

(*a*) 154 square cm

(*b*) 144 square cm

(*c*) 132 square cm

(*d*) Cannot be determined due to insufficient data

Direction (Q. 73-75): Consider the following for the next three (03) items that follow:

Two circles with centres at O_1 and O_2 touching each ether are placed inside a rectangle of sides 9 cm and 8 cm as shown in the given figure: **[CDS (I) Exam, 2024]**

73. What is the sum of the areas of the two circles?

(*a*) 17π square unit (*b*) 16.75π square unit

(*c*) 16.5π square unit (*d*) 16.25π square unit

74. Which one of the following is correct in respect of angle 0?

(*a*) $0 < \theta < 30°$ (*b*) $30° < \theta < 45°$

(*c*) $45° < \theta < 60°$ (*d*) $60° < \theta < 90°$

75. What is the area of the shaded region ?

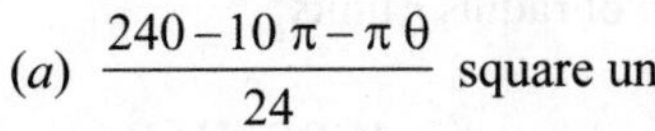

(*a*) $\frac{240 - 10\pi - \pi\theta}{24}$ square unit (*b*) $\frac{240 - 6\pi - \pi\theta}{24}$ square unit

(*c*) $\frac{120 - 12\pi - \pi\theta}{24}$ square unit (*d*) $\frac{240 - 12\pi - \pi\theta}{24}$ square unit

76. *AB* is a diameter of the circle with centre *O*. *CD* cuts *AB* at *E* such that $OE = EB$. If $CE = 6$ cm and $ED = 2$ cm, then the radius of the circle is:

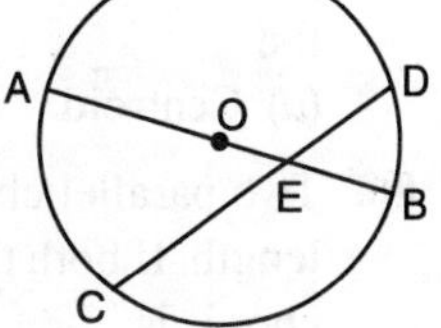

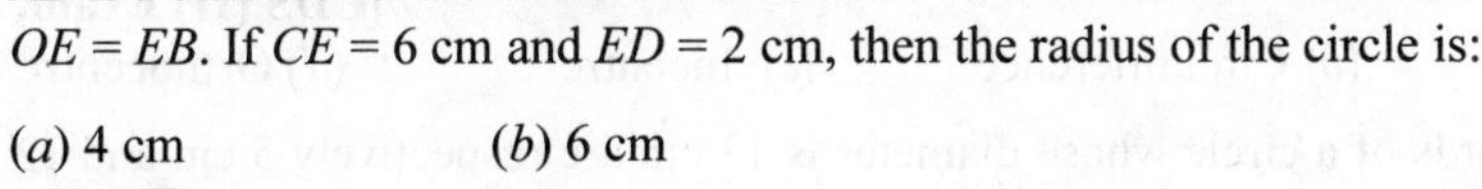

(*a*) 4 cm (*b*) 6 cm

(*c*) $4\sqrt{3}$ cm (*d*) 8 cm

77. *A* is a point outside the circle with centre *O*, *AT* and *AT′* are the tangents to the circle and P is a point on the circle as shown in the figure. Then, $\angle TPT'$ equals

(*a*) 55° (*b*) 70°

(*c*) 125° (*d*) 140°

78. Two circles of radii *R* and *r* touch each other externally and *PQ* is the direct common tangent. Then PQ^2 is equal to:

(*a*) $R - r$ (*b*) $R + r$ (*c*) $2rR$ (*d*) $4rR$

79. *ABC* is a right angled triangle with $BC = 6$ cm and $AB = 8$ cm. A circle with centre *O* is inscribed in ΔABC. The radius of the circle is:

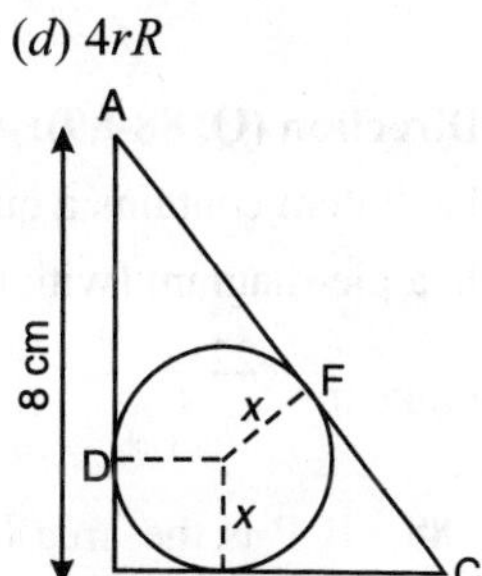

(*a*) 1 cm (*b*) 2 cm

(*c*) 3 cm (*d*) 4 cm

80. In the adjoining figure, the larger circle with radius 4 cm is touched internally by two smaller circles which also touch each other externally at the centre *O* of the larger circle. The area of the shaded portion is:

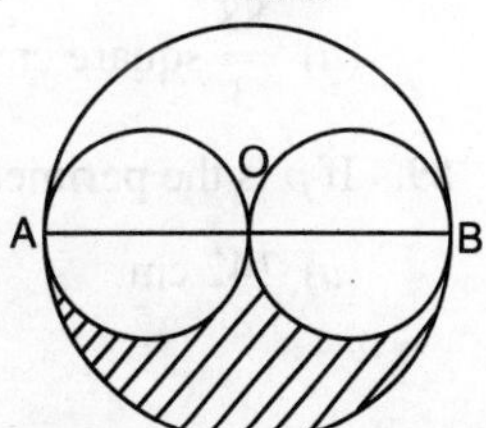

(*a*) 4π (*b*) 7π

(*c*) 12π (*d*) 16π

81. Two circles touch externally and sum of their areas is 130π cm^2 and the distance between their centres is 14 cm. What is the difference in the radii of the circles?

(*a*) 5 cm (*b*) 6 cm (*c*) 7 cm (*d*) 8 cm

[CDS (II) Exam, 2016]

82. In a circle of radius 3 units, a diameter *AB* intersects a chord of length 2 units perpendicularly at *P*. If *AP* > *BP*, then what is the ratio of *AP* to *BP*?

(*a*) $3+\sqrt{10} : 3-\sqrt{10}$ (*b*) $3+\sqrt{8} : 3-\sqrt{8}$

(c) $3+\sqrt{3} : 3-\sqrt{3}$ (*d*) $3 : \sqrt{3}$

[CDs (II) Exam, 2016]

83. If two tangents inclined at an angle 60° are drawn to a circle of radius 3 cm, then what is the length of each tangent?

[CDS (II) Exam, 2016]

(*a*) $3\sqrt{3}$ cm (*b*) $\sqrt{3}$ cm (*c*) 6 cm (*d*) $2\sqrt{2}$ cm

84. What is area of largest triangle inscribed in a semi-circle of radius *r* units?

(*a*) r^2 square units (*b*) $2r^2$ square units

(c) $3r^2$ square units (*d*) $4r^2$ square units

[CDS (II) Exam, 2016]

85. A pole stands vertically inside a triangular park *ABC*. If the angle of elevation of the top of the pole from each corner of the park is same, then in the triangle *ABC*, the foot of the pole is at the

[CDS (II) Exam, 2016]

(*a*) Centroid (*b*) Circumference (*c*) Incentre (*d*) Orthocentre

86. Two parallel chords of a circle whose diameter is 13 cm are respectively 5 cm and 12 cm in length. If both the chords are on the same side of the diameter, then the distance between these chords is

[CDS (I) Exam, 2017]

(*a*) 5.5cm (*b*) 5 cm (*c*) 3.5 cm (*d*) 3 cm

87. An equilateral triangle *ABC* is inscribed in a circle with centre *O*. Then, $\angle BOC$ is equal to:

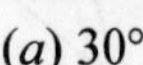

(*a*) 30° (*b*) 60°

(*c*) 90° (*d*) 120°

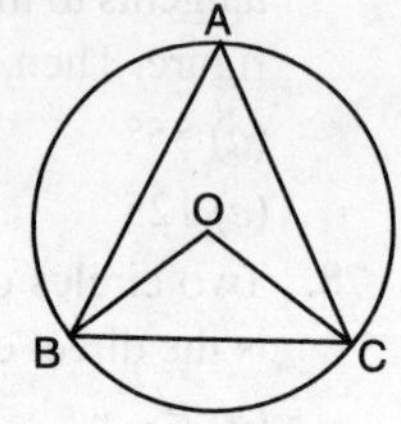

Direction (Q. 88-89): Consider the following for the next two (2) items that follow:

Each item contains a question followed by two Statements. Answer each item using the follow:

In a pie-diagram (with radius 7 cm), the central angles of the sectors are in ratio 2 : 3 : 7 : 5 : 1, (Take $\pi = \frac{22}{7}$)

[CDS (I) Exam, 2024]

88. If P is the area of the smallest sector and Q is the area of the largest sector, then what is P + Q equal to ?

(*a*) $\frac{88}{3}$ square cm (*b*) $\frac{77}{3}$ square cm (*c*) $\frac{149}{6}$ square cm (*d*) $\frac{616}{9}$ square cm

89. If *p* is the perimeter of the smallest sector, then what is the value of 9*p* ?

(*a*) 142 cm (*b*) 148 cm (*c*) 156 cm (*d*) 221 cm

ANSWERS

1. (*c*)	**2.** (*b*)	**3.** (*a*)	**4.** (*a*)	**5.** (*d*)	**6.** (*c*)	**7.** (*d*)	**8.** (*c*)	**9.** (*c*)
10. (*d*)	**11.** (*c*)	**12.** (*b*)	**13.** (*a*)	**14.** (*b*)	**15.** (*c*)	**16.** (*b*)	**17.** (*d*)	**18.** (*c*)
19. (*a*)	**20.** (*a*)	**21.** (*c*)	**22.** (*b*)	**23.** (*b*)	**24.** (*b*)	**25.** (*c*)	**26.** (*b*)	**27.** (*c*)
28. (*a*)	**29.** (*d*)	**30.** (*b*)	**31.** (*a*)	**32.** (*c*)	**33.** (*d*)	**34.** (*c*)	**35.** (*b*)	**36.** (*c*)
37. (*a*)	**38.** (*c*)	**39.** (*a*)	**40.** (*d*)	**41.** (*c*)	**42.** (*b*)	**43.** (*b*)	**44.** (*a*)	**45.** (*c*)
46. (*d*)	**47.** (*c*)	**48.** (*d*)	**49.** (*d*)	**50.** (*a*)	**51.** (*d*)	**52.** (*c*)	**53.** (*a*)	**54.** (*b*)
55. (*c*)	**56.** (*b*)	**57.** (*c*)	**58.** (*a*)	**59.** (*b*)	**60.** (*a*)	**61.** (*d*)	**62.** (*b*)	**63.** (*b*)
64. (*c*)	**65.** (*d*)	**66.** (*c*)	**67.** (*a*)	**68.** (*a*)	**69.** (*b*)	**70.** (*d*)	**71.** (*b*)	**72.** (*a*)
73. (*a*)	**74.** (*c*)	**75.** (*d*)	**76.** (*a*)	**77.** (*c*)	**78.** (*d*)	**79.** (*b*)	**80.** (*b*)	**81.** (*d*)
82. (*b*)	**83.** (*a*)	**84.** (*a*)	**85.** (*a*)	**86.** (*c*)	**87.** (*d*)	**88.** (*d*)	**89.** (*b*)	

SOLUTIONS

1. In figure,

$\angle COP$ and $\angle DOP$

$$OP = OP \text{ (common)}$$

$$\angle CPO = \angle DPO \; (90°)$$

$$OC = OD \text{ radius}$$

$$\therefore \quad \Delta COP \cong \Delta POD \quad \text{[By RHS]}$$

$$\therefore \quad \angle COP = \angle POD = \frac{2\theta}{2} = \theta$$

$$\text{and} \quad CP = PD = \frac{CD}{2} \qquad \text{...(i)}$$

Now, In ΔDOP $\sin\theta = \dfrac{DP}{OD} = \dfrac{DP}{1} \Rightarrow DP = \sin\theta$

now, from (i) $\dfrac{CD}{2} = \sin\theta \Rightarrow CD = 2\sin\theta$

2. Joint *AB*. Then, $\angle CAB = \angle BCD$ (*∠s in alternate segments*)

And, $\angle DAB = \angle CDB$ (*∠s in alternate segments*)

$\angle CAD = \angle CAB + \angle DAB = \angle BCD + \angle CDB$

$\Rightarrow \angle CAD + \angle CBD = \angle BCD + \angle CDB + \angle CBD = 180°$ (*∠s* of a Δ)

3. Radius = 20 cm, so Radius² = $20^2 = 400$

Distance from center to chord (CE) = 13 cm So, CE² = $13^2 = 169$

$$AE \times EB = 400 - 169 = 231 \text{ cm}^2$$

4. Since, $PQ^2 + QR^2$

$4^2 + (7.5)^2 = 16 + 56.25 = 72.25 = PR^2$

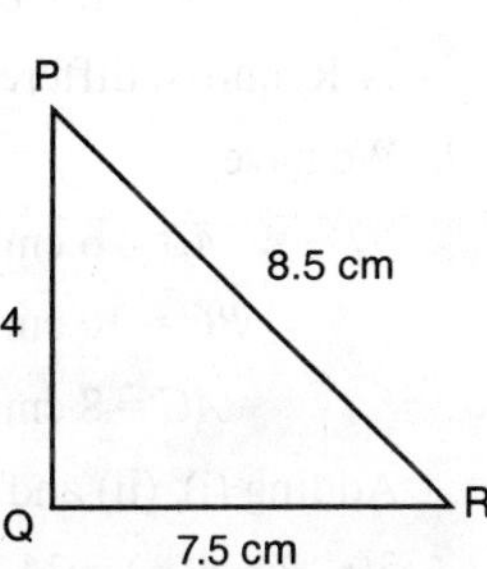

Hence, the sides of the triangle are pythagorean triplets.

According to the question,

$$\Rightarrow \quad r = \frac{PQ + QR - PR}{2}$$

$$= \frac{4 + 7.5 - 8..5}{2} = \frac{11.5 - 8.5}{2} = \frac{3}{2} = 1.5$$

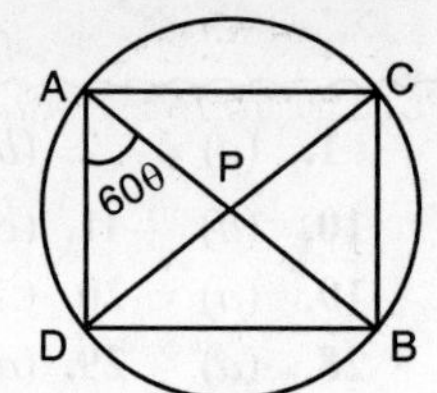

5. $\angle DAC = \angle BCD = \angle CBD$

$\angle BAD = 90°$ [All are angles in semicircle]

$\Rightarrow \quad AP = BP = CP = DP$ [radius of circle]

Now In ΔAPD and $\angle DPB$

$\angle PAD = \angle PDA = 60°$ [isosceles Δ]

$\Rightarrow \quad \angle APO = 180° - 120° = 60°$

$\Rightarrow \quad \angle APD + \angle BPD = 180°$ [Linear pair]

$\therefore \quad \angle BPD = 180° - 60° = 120°$

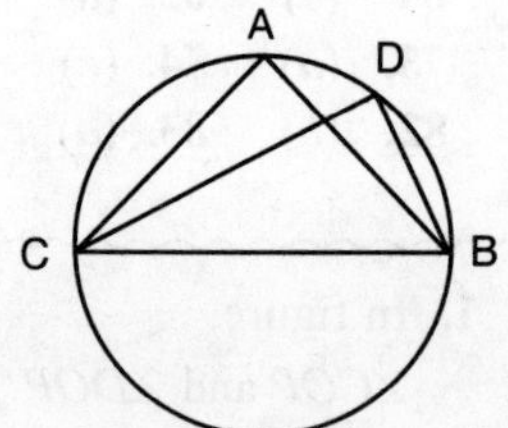

6. Given that ABC is an equilateral Δ

$\angle CAB = \angle CDB = 60°$

[Angle subtended by are BC]

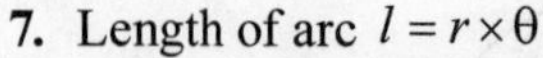

7. Length of arc $l = r \times \theta$

$$\pi = 4 \times \theta \Rightarrow \theta = \frac{\pi}{4}$$

Hence the magnitude of the angle subtended by the arc at the center is $\pi/4$

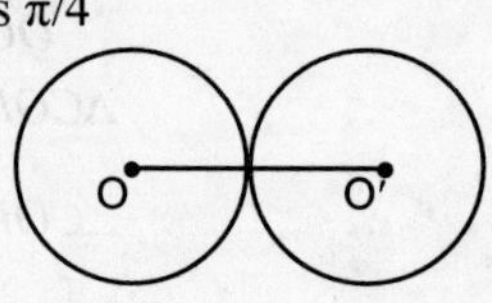

8. Let radius or two circles are R and r'

Respectively with centre O and O'

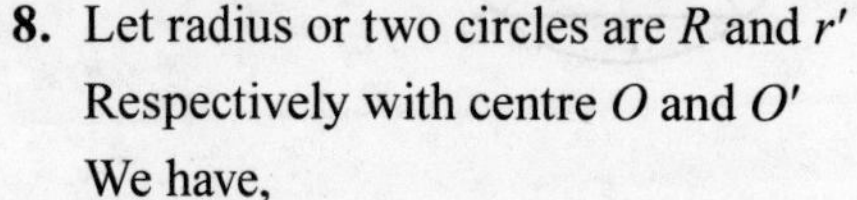

We have,

$R + r = 9\text{cm}$ ...(i)

Now, according to the question

$\pi r^2 + \pi r^2 = 41\pi$

$\Rightarrow \quad \pi(r^2 + r^2) = 41\pi$

$\Rightarrow \quad R^2 + r^2 = 41\pi$ ...(ii)

Now,

$(R + r)^2 = R^2 + r^2 + 2Rr$

$\Rightarrow \quad (9)^2 = 41 + 2Rr$ [Using (i) and (ii)]

$\Rightarrow \quad 2Rr = 40$ ...(ii)

Now,

$(R - r)^2 = R^2 + r^2 - 2Rr$

$= 41 - 40$ [using (ii) and (iii)]

$= 1$

$R - r = 1$

$\therefore$ Required difference $= 2(R - r) = 2 \times 1 = 2$ cm

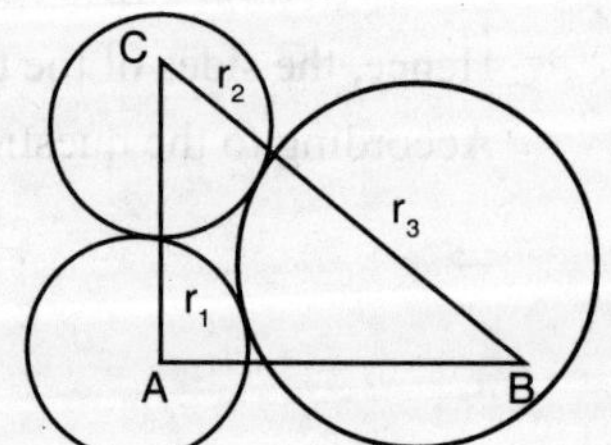

9. We have,

$AB = 6 \text{ cm} = r_1, + r_3$ ...(i)

$BC = 10 \text{ cm} = r_2 + r_3$ ...(ii)

$AC = 8 \text{ cm} = r_1 + r_2$ ...(iii)

Adding (i), (ii) and (iii), we get

$2(r_1 + r_2 + r_3) = 24 \Rightarrow r_1 + r_2 + r_3 = 12$ cm

10. We have,

ar (ΔABC) with base AB = *Ar* (ΔABC) with Base × AB

$$\Rightarrow \frac{1}{2}\times AB\times AC = \frac{1}{2}\times BC\times AD \Rightarrow \frac{1}{2}\times 6\times 8 = \frac{1}{2}\times 10\times AD \Rightarrow AD = \frac{24}{5} = 4.8 \text{ cm}$$

11. We have,

$AB^2 + AC^2 = 6^2 + 8^2 = 36 + 64 = 100 = BC^2$

$\Rightarrow$ ABC is a right angles Δ and $\angle A = 90°$

Now,

$$AB = 6 \text{ cm} = r_1 + r_3 \quad \text{...(i)}$$
$$BC = 10 \text{ cm} = r_2 + r_3 \quad \text{...(ii)}$$
$$AC = 8 \text{ cm} = r_1 + r_3 \quad \text{...(iii)}$$

Adding (i), (ii) and (iii), we get

$r_1 + r_2 + r_3 = 12$ cm ...(iv)

From (i) and (iv), we get

$$r_2 = 6 \text{ cm}$$

From (ii) and (iv), we get

$$r_1 = 2 \text{ cm}$$

From (iii) and (iv), we get

$$r_3 = 4 \text{ cm}$$

now, area of sector widh $r_1 = 2$ cm

and centre $A(P) = \pi r^2 \dfrac{\theta}{360°} \Rightarrow P = \pi \times 2\times 2\times \dfrac{90°}{360} \Rightarrow P = \pi \text{ cm}^2$

Area of sector at B and radius 4 = $\dfrac{16\pi(90° - \theta)}{360}$

Area of sector at C and radius = $\dfrac{36\pi(90° - \theta)}{360°}$

Now, $9Q + 4R = 9\times \dfrac{16\pi\,\theta}{360°} + 4\times\dfrac{36\pi(90°-\theta)}{360°} = \dfrac{144\pi(\theta + 90° - \theta)}{360°} = 36\pi \text{ cm}^2$

12. We have,

$AB = 24$ cm
$OA = 20$ cm
$O'A = 16$ cm
$OO' = d$

In ΔOAM

$$OM = \sqrt{OA^2 - AM^2} = \sqrt{20^2 - 12^2} = \sqrt{400 - 144} = \sqrt{256} = 16 \text{ cm}$$

In $\Delta AMO'$

$$O'M = \sqrt{O^1A^2 - AM^2} = \sqrt{16^2 - 12^2} = \sqrt{256 - 144} = \sqrt{112} = 4\sqrt{7}$$

$$O'M = 4 \times 2.65 = 10.6 \text{ cm}$$

$\therefore \quad d = 16 + 10.6 = 26.6$ cm

$\therefore \quad 26 \text{ cm} < d < 27 \text{ cm}$

13. Radius of circle $= \dfrac{a}{2\sqrt{3}}$

Given that the radius $r = 2$ cm, we can solve for the side length a:

$$2 = \frac{a}{2\sqrt{3}} \Rightarrow a = 2 \times 2\sqrt{3} = 4\sqrt{3} \text{ cm}$$

$$A = \frac{\sqrt{3}}{4}(16 \times 3) = 12\sqrt{3} \text{ cm}^2$$

14. In figure, O and O' is the centre of the circle

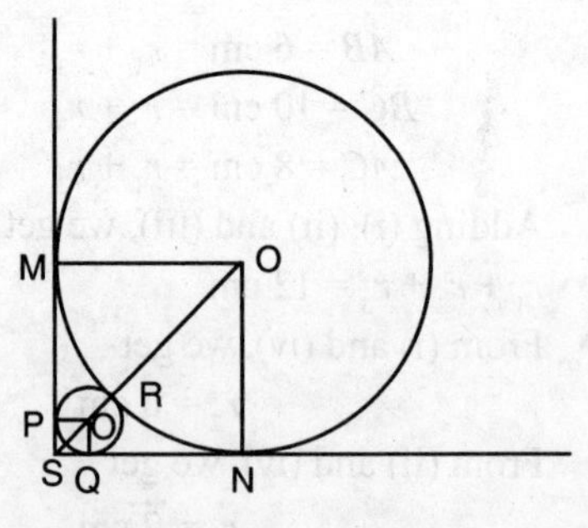

$$OM = ON = \frac{8}{2} = 4 \text{ cm}$$

Let $\quad O'R = r$ cm

In square $OMSN$

Diagonal OS = side $\times \sqrt{2} = 4\sqrt{2}$

Ins square $O'PSQ$

Diagonal $O^1s = r\sqrt{2}$

$OR = 4$ cm and $OR = r$ cm

$$\Rightarrow \quad 4\sqrt{2} = 4 + r + r\sqrt{2} \Rightarrow r(1 + \sqrt{2}) = 4(\sqrt{2} - 1)$$

$$\Rightarrow \quad r = \frac{4(\sqrt{2}-1)}{\sqrt{2}+1} = 4(\sqrt{2}-1)^2 = 4(3 - 2\sqrt{2}) \text{ cm}$$

15. By Pythagoras theorem:

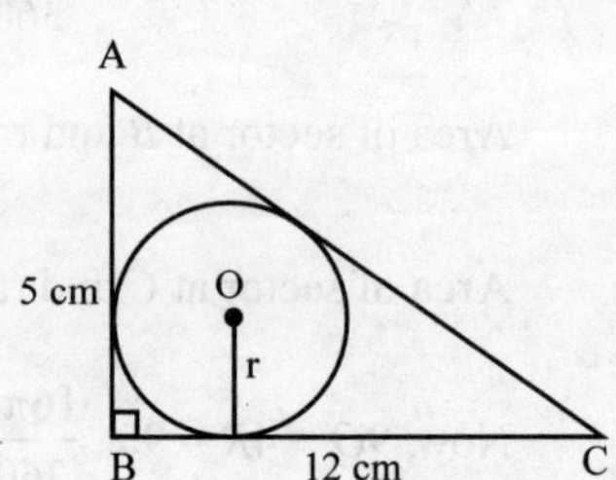

$AC^2 = AB^2 + BC^2$

$AC^2 = 5^2 + 12^2 = 25 + 144 = 169$

$AC = \sqrt{169} = 13$ cm

Area of $\Delta ABC = \frac{1}{2} \times b \times h = \frac{1}{2} \times 5 \times 12 = 30 \text{ cm}^2$

Semi-perimeter:

$$s = \frac{AB + BC + AC}{2} = \frac{5 + 12 + 13}{2} = 15 \text{ cm}$$

Then, $r = \dfrac{30}{15} = 2$ cm

$\therefore$ The radius of the in-circle is 2 cm.

16. In fig.

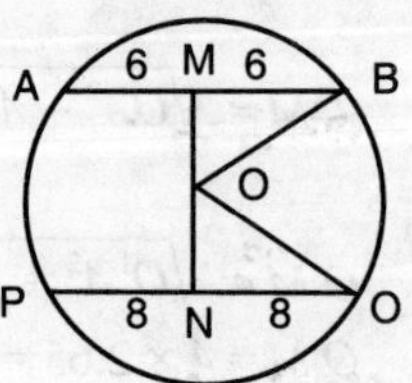

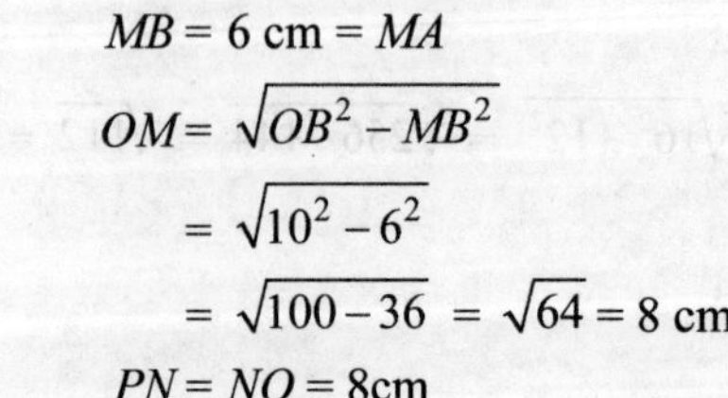

$OB = 10$ cm $= OQ$

$MB = 6$ cm $= MA$

$$OM = \sqrt{OB^2 - MB^2}$$

$$= \sqrt{10^2 - 6^2}$$

$$= \sqrt{100 - 36} = \sqrt{64} = 8 \text{ cm}$$

Now, $\quad PN = NQ = 8$cm

$$ON = \sqrt{OQ^2 - NQ^2} = \sqrt{10^2 - 8^2} = 6 \text{ cm}$$
$$MN = OM + ON = 8 + 6 = 14 \text{ cm}$$
$$MN = 8 - 6 = 2 \text{ cm}$$

17. We have,

$$AB = AC = 12 \text{ cm}$$
$$OA = OB = OC = 8 \text{ cm}$$

Let $BM = MC = x$ cm

and let $OM = p$ cm

now, In ΔABM and ΔACM

$$x^2 = 12^2 - (8 + p)^2 \qquad \text{...(i)}$$

and In ΔOMB and ΔOCM

$$x^2 = 8^2 - p^2 \qquad \text{...(ii)}$$

from eqn (i) and (ii), we get

$12^2 - (8 + p)^2 = 8^2 - p^2$

$\Rightarrow 144 - (64 + p^2 + 16p) = 64 - p^2 \Rightarrow 80 - p^2 - 16p = 64 - p^2 \Rightarrow 16p = 16 \Rightarrow p = 1$

now, put $p = 4$ in eqn (ii), we get

$$x^2 = 8^2 - 1^2 = 64 - 1 = 63 \Rightarrow x = BM = 3\sqrt{7} \text{ cm}$$
$$BC = 2x = 2 \times 3\sqrt{7} = 6\sqrt{7} \text{ cm}$$

18. We have,

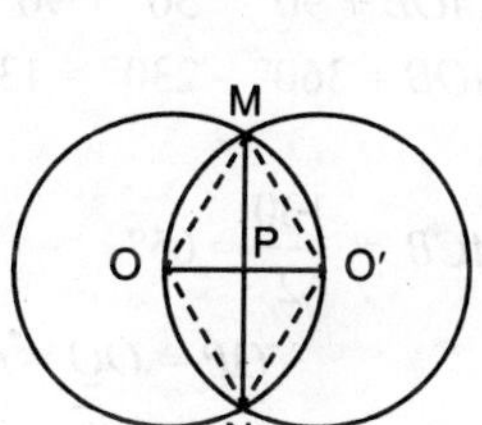

$$MN = 10\sqrt{3}$$
$$OM = O'M = ON = O'N = r = OO'$$

$\therefore$ $OMO'N$ is a rhombus

$\therefore$ $\Delta OMO'$ and ONO' are equilateral Δ

$\therefore$ $ar(\Delta OMO') + ar(\Delta ONO') = 2 \times \frac{\sqrt{3}}{4} \times r^2 = \frac{\sqrt{3}}{2} r^2$

ar (rhombus $OMO'N$) $= \frac{1}{2} \times MN \times OO' = \frac{1}{2} \times r \times 10\sqrt{3} = 5\sqrt{3}r$

Now, we have, $\frac{\sqrt{3}}{2} r^2 = 5\sqrt{3}\, r \Rightarrow r = 10$ cm

$\therefore$ Diameter = 20 cm

19. In figure

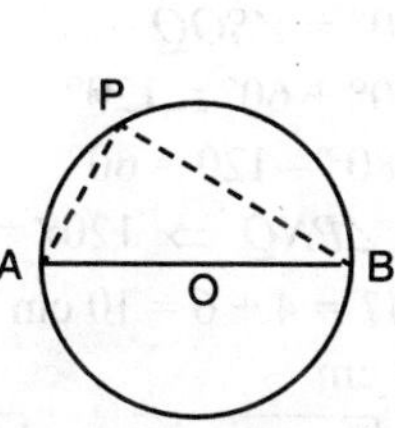

Let $AP = x$

$BP = y$

and $AB = 2 \times 6.5 = 13$ cm

Now, In ΔAPB

$\angle APB = 90°$ [$\because$ AB is diameter]

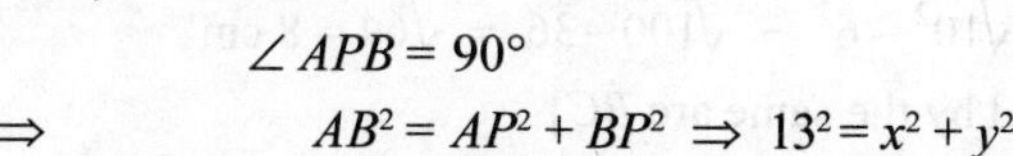

$\Rightarrow \quad AB^2 = AP^2 + BP^2 \Rightarrow 13^2 = x^2 + y^2$

Now, option (b), (c) and (d) satisfy the above equation

$\therefore$ The point P lies on the circle in such cases.

20. We have,

$$r_1 = 3.5 \text{ cm}$$
$$d = 10 \text{ cm and } r_2 = 4.5 \text{ cm}$$
$$AB = \sqrt{d^2 - (r_1 + r_2)^2}$$

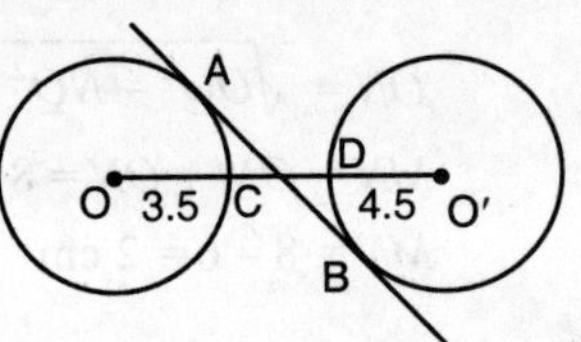

[∵ Transverse common tangent = $\sqrt{d^2 - (r_1 + r_2)^2}$]

$$= \sqrt{10^2 - (4.5 + 3.5)^2} = \sqrt{10^2 - 8^2} = \sqrt{100 - 64} = \sqrt{16} = 4 \text{ cm}$$

21. Statement-1: The orthocenter lies outside the triangle, thus it is obtuse-angled triangle.

Statement-2: The intersection point of all medians is called centroid. Thus, centroid lies inside the triangle.

Statement-3: The intersection of all altitudes of a triangle is called orthocenter. In right angled triangle, except hypotenuse, both sides are altitudes. Thus orthocenter lies on the triangle.

Statement-4: The centroid of a triangle is always inside the triangle.

22. We have,

$$\angle OAX = \angle OBX = 90°$$

[∵ AX and BX are tangents]

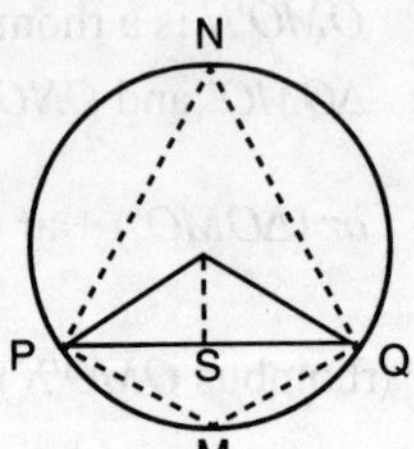

∴ In quadrilateral $OAXB$

$$\angle AOB + \angle OAX + \angle AXB + \angle OBX = 360°$$
$$\Rightarrow \angle AOB + 90° + 50° + 90° = 360°$$
$$\Rightarrow \angle AOB = 360° - 230° = 130°$$

[∵ angle subtended by chord on the circle at any point is half the angle subtended by the center.]

$$\Rightarrow \angle ACB = \frac{130}{2} = 65°$$

23. Let $OP = OQ = r$

$$PQ = \sqrt{3}\, r$$

$$PS = SQ = \frac{PQ}{2} = \frac{\sqrt{3}r}{2} \quad [\because OS \perp PQ]$$

$$\angle POS = \angle SOQ \quad [\because \Delta POS \cong OQS]$$

N
P S Q
M

In ΔPOS

$$\sin \angle POS = \frac{PS}{OP} = \frac{\sqrt{3}r}{2r} = \frac{\sqrt{3}}{2} \Rightarrow \sin \angle POS = \sin 60°$$

$$\Rightarrow \quad \angle POS = 60° = \angle SOQ$$
$$\therefore \quad \angle POQ = 60° + 60° = 120°$$
$$\angle PMQ = 180° - 120 = 60°$$
$$\angle PMQ = k\,\angle PNQ \Rightarrow 120° = k(60°) \Rightarrow k = 2$$

24. We have, $OT = OA + AT = 4 + 6 = 10$ cm

$$OP = 6 \text{ cm}$$
$$PT = \sqrt{OT^2 - OP^2} = \sqrt{10^2 - 6^2} = \sqrt{100 - 36} = \sqrt{64} = 8 \text{ cm}$$

25. $\angle BAC = \angle BDC = 30°$ [Angle subtended by the same arc BC]

∴ In ΔBDC

$$\angle BDC + \angle BCD + \angle DBC = 180°$$
$$\Rightarrow 30° + \angle BCD + 70° = 180° \Rightarrow \angle BCD = 180° - 100° = 80°$$

26. We have, $\angle CDF = 32°$

$\angle COF = 2\ \angle CDF = 2 \times 32° = 64°$

[Angle at the centre is double the angle subtended by the same chord at the circumference of the circle]

$\angle CEF = \angle ECD + \angle EDC = 32° + 32° = 64°$

[$\because$ Exterior angle of a triangle is equal to the sum of opposite interior angle]

27. Let the diameter of the inner and outer circle be d_1 m and d_2 m respectively.
Ratio of diameters = Ratio of perimeters

$\therefore\ \frac{d_2}{d_1} = \frac{23}{22}$. As the width of the path in 5 m, $d_2 - d_1 = 10$

$\therefore\ \frac{d_2}{d_1} = \frac{23}{22} = \frac{230}{220}$. As $d_2 - d_1 = 10$, $d_2 = 230$, $d_1 = 220$.

28. Any triangle inscribed in a semicircle is right angled.

Its area $= \frac{1}{2}$ (Diameter of the semicircle) (Length of the altitude from the right angled vertex to the diameter) $= \frac{1}{2}(2r)(h)$ (say) $= rh$

rh has its maximum value when h has the maximum value. As the maximum value of h is r, the maximum value rh is r^2.

Area of the largest triangle that can be inscribed in a semicircle of radius r is r^2.

29. Let the midpoint of AB be M.

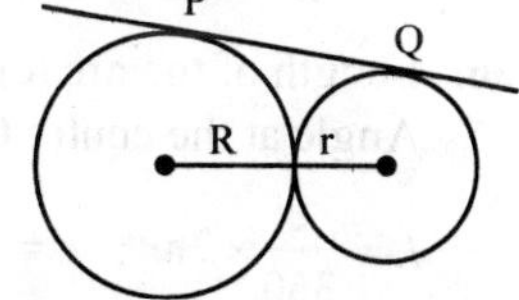

M is the foot of the perpendicular from the centre O of the circle to AB.

Let the distance of AB from the centre O of the circle be x.

$OA^2 = AM^2 + MO^2$

$17^2 = 15^2 + x^2 \qquad \therefore\ x = 8.$

30. $d = 24, r_1 = 16, r_2 = 8$

$PQ = \sqrt{(24)^2 - (16-8)^2} = \sqrt{(576-64)} = \sqrt{512} = 16\sqrt{2}$ cm

$\therefore$ The length of common tangent $PQ = 16\sqrt{2}$ cm.

31. UP = radius of M = 8 cm (UP $\perp$ PQ).

$\Rightarrow$ VQ = radius of N = 16 cm (VQ $\perp$ PQ).

$\Rightarrow$ UP || VQ and the distance between these parallel segments = length PQ $= 16\sqrt{2}$ cm.

$\Rightarrow$ Area of quadrilateral PQVU $= \frac{(UP + VQ) \times PQ}{2} = \frac{(8+16)}{2} \times 16\sqrt{2}$

$$= \frac{24}{2} \times 16\sqrt{2} = 12 \times 16\sqrt{2} = 192\sqrt{2}\ \text{cm}^2$$

$\therefore$ Area of quadrilateral PQVU $= 192\sqrt{2}$ cm^2.

32. AB = 2a, *i.e.* AL = LM = MN = NB $= \frac{a}{2}$.

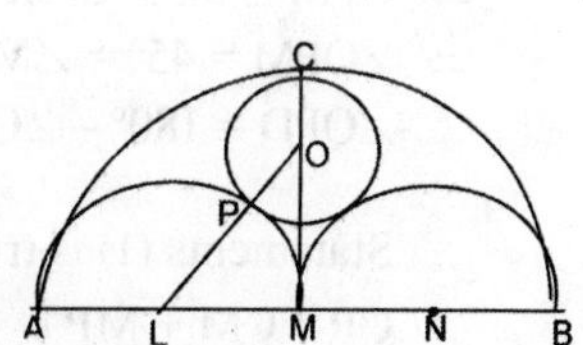

Let OC = r.

In ΔOML, OM = a – r and

OL = OP + PL $= r + \frac{a}{2}$.

$\therefore \quad \left(\frac{a}{2}\right)^2 + (a-r)^2 = (r + \frac{a}{2})^2$

$\Rightarrow \quad \frac{a^2}{4} + a^2 - 2ar + r^2 = r^2 + ar + \frac{a^2}{4} \Rightarrow a^2 = 3ar \Rightarrow r = \frac{a}{3}.$

33. CE = CA and DB = DE.

$\therefore$ CD = CE + ED = CA + DB.

Perimeter of ΔPCD = PC + PD + CD

= PC + PD + CA + DB

= PA + PB = 2PA = 2(16) cm = 32 cm.

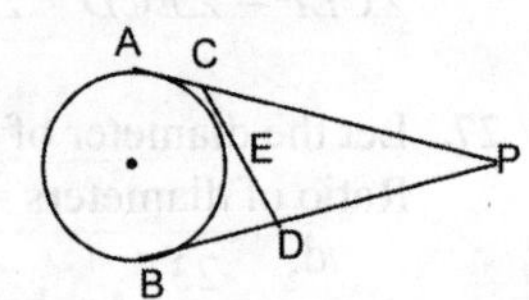

34. Let OA = r and OP = x.

$\therefore$ BP = r + x and PA = r – x.

$\frac{r+x}{r-x} = \frac{2}{1} \Rightarrow 3x = r$

$OP^2 + PC^2 = OC^2$ ($\because$ CP $\neq$ AB)

$\frac{r^2}{9} + PC^2 = r^2; \qquad PC^2 = \frac{8r^2}{9}$

$CP = \frac{2\sqrt{2}}{3} r$ and $CD = \frac{4\sqrt{2}}{3} r = \frac{2\sqrt{2}}{3} d$, where d = 2r.

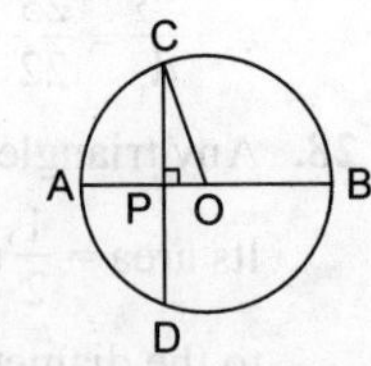

35. Let XY be the chord AB be the diameter of circle with centre O since

A P $\perp$ X Y

XP = PY = 1

In right Δ OXP ; $OP^2 = OX^2 - XP^2 = 2^2 - 1^2$

$OP = \sqrt{3}$

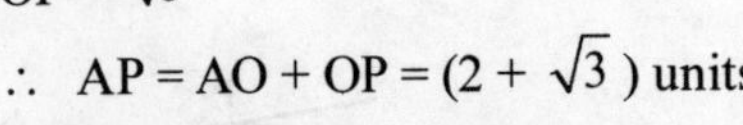

$\therefore$ AP = AO + OP = $(2 + \sqrt{3})$ units

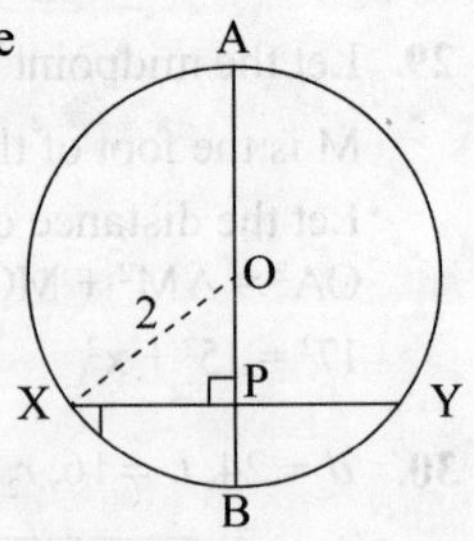

36. Length of the arc (e) = 100m.

Angle at the centre Q = 36°

$l = \frac{\theta}{360} \times 2\pi r; \; r = \frac{100 \times 7}{2 \times 22} \times \frac{360}{36}; \; r = \frac{500}{\pi} m.$

37. OQ = 7, MP = $\sqrt{24}$

QR = 2, QP = PR = 1

In right Δ OPQ. $OP = \sqrt{OQ^2 - Qp^2} = \sqrt{7^2 - 1^2}$

$= \sqrt{48} = 4\sqrt{3}$ units.

In right Δ OMP, $OP^2 = OM^2 + MP^2$

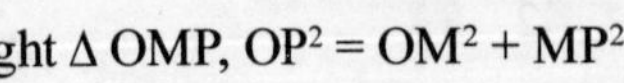

$\left(\sqrt{48}\right)^2 = OM^2 + \left(\sqrt{24}\right)^2$

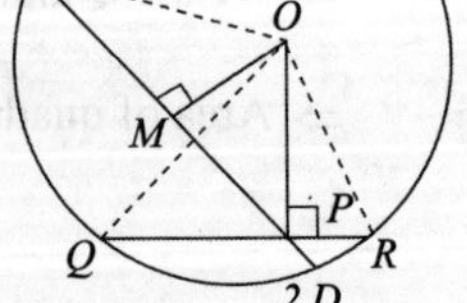

$\therefore$ OM = $\sqrt{24}$ $\qquad \therefore$ OM = MP ... (1)

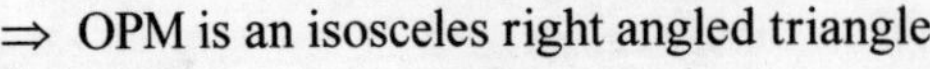

$\Rightarrow$ OPM is an isosceles right angled triangle

$\Rightarrow \angle OPM = 45° = \angle MPQ$

$\angle QPD = 180° - \angle QPM = 180° - 45° = 135°$ ... (2)

$\therefore$ Statements (1) is true $CM = \sqrt{OC^2 - OM^2} = \sqrt{7^2 - \left(\sqrt{24}\right)^2} = 5$

CP = CM + MP

$m = 5 + \sqrt{24}$ ... (3)

CM = MD = 5

PD = MD – MP

$n = 5 - \sqrt{24}$ ... (4)

$x^2 - 10x + 1 = 0$

$$x = \frac{10 \pm \sqrt{100-4}}{2} = 5 \pm 2\sqrt{6} = 5 \pm \sqrt{24}$$

∴ m, n are the roots of $x^2 - 10x + 1 = 0$

∴ Statement (2) is also true

Area of Δ OPR = $\frac{1}{2} \times OP \times PR = \frac{1}{2} \times \sqrt{48} \times 1 = 2\sqrt{3}$ units²

Area of Δ OMP = $\frac{1}{2} \times OM \times MP = \frac{1}{2} \times \sqrt{24} \times \sqrt{24} = 12$ units²

Area of Δ OPR : Area of ΔOMP.

$2\sqrt{3} : 12$

$\sqrt{3} : 6$

$1 : \sqrt{2 \times 3 \times 2 \times 3.}$

$1 : 2\sqrt{3.}$

∴ Statement (3) is not true

38. Since OP, OQ are lengths from O to the circle

⇒ OP = OQ

Also OR × SQ = OS × RP

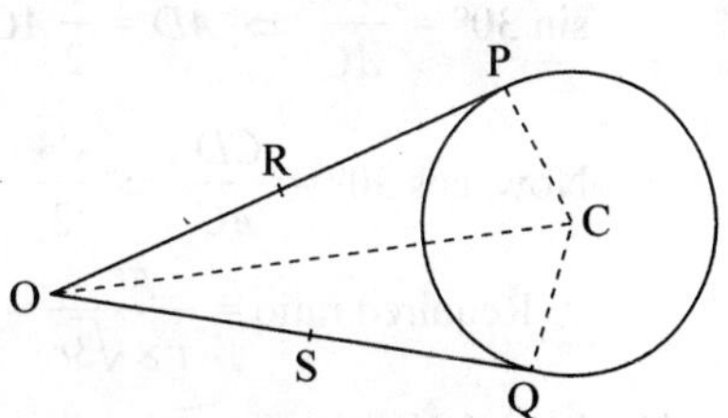

$$\frac{OR}{RP} = \frac{OS}{SQ}$$

$$\frac{OR}{RP} + 1 = \frac{OS}{SQ} + 1$$

$$\frac{OR + RP}{RP} = \frac{OS + SQ}{SQ}$$

$$\frac{OP}{RP} = \frac{OQ}{SQ} \Rightarrow RP = SQ\ (OP = OQ) \Rightarrow OR = OS$$

∴ If a circle is drawn with centre O and radius OR passes through.

O as OR = OS

∴ X = Y is true

Statement (1) is true Δ OPC ≅ Δ OQC

⇒ ∠POC = ∠COQ but ∠COQ + ∠OCQ = 90°

(OQ ⊥ CQ)

⇒ Statement (2) is also true

39. Statement I : 5 + 5 > 6 and 5 + 6 > 5

∴ A, B, and C form a triangle

Hence, circle can be drawn from these points

Statement II : $3 + 4 = 7$

$\therefore$ A, B & C are collinear and a circle cannot be drawn

40. $\therefore$ We have, $10^2 = 6^2 + 8^2$

$\therefore \angle ABC = 90°$

$\therefore \Delta ABC \sim \Delta BDC$

$\therefore \dfrac{AB}{BD} = \dfrac{AC}{BC} \Rightarrow \dfrac{6}{BD} = \dfrac{10}{8} \Rightarrow BD = 4.8 \text{ cm} = BQ$

$\therefore CQ = BC - BQ = 8 - 4.8 = 3.2 \text{ cm}$

41. We have,

$\dfrac{AB}{AD} = \dfrac{AC}{AB}$ [$\because \Delta ABC \sim \Delta ABD$]

$\Rightarrow \dfrac{6}{AD} = \dfrac{10}{6} \Rightarrow AD = 3.6 \text{ cm}$

$\therefore \sin\theta = \dfrac{AD}{AB} = \dfrac{3.6}{6} = 0.6$

42. We have,

$\dfrac{AB}{BD} = \dfrac{AC}{BC}$ [$\because \Delta ABC \sim \Delta BDC$]

$\Rightarrow \dfrac{6}{BD} = \dfrac{10}{8} \Rightarrow BD = \dfrac{6 \times 8}{10} = 4.8 \text{ cm} = \text{radius}$

43. $AC = 2r$

In ΔACD

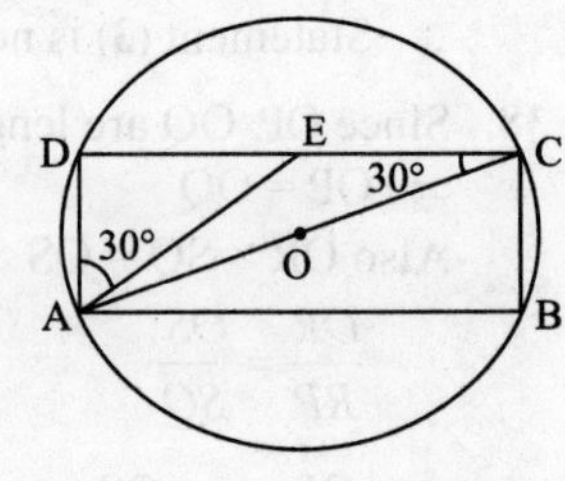

$\sin 30° = \dfrac{AD}{AC} \Rightarrow AD = \dfrac{1}{2}AC = \dfrac{1}{2} \times 2r = r$

Now, $\cos 30° = \dfrac{CD}{AC} \Rightarrow \dfrac{\sqrt{3}}{2} = \dfrac{CD}{AC} \Rightarrow CD = \dfrac{\sqrt{3}}{2} \times 2r = \sqrt{3}\, r$

$\therefore$ Required ratio $= \dfrac{\pi r^2}{r \times \sqrt{3}r} = \dfrac{\pi}{\sqrt{3}}$

44. In ΔADE

$\tan 30° = \dfrac{DE}{AD} = \dfrac{1}{\sqrt{3}} \Rightarrow \dfrac{DE}{r} = \dfrac{1}{\sqrt{3}} \Rightarrow DE = \dfrac{r}{\sqrt{3}}$

$ar\,(\Delta ACE) = \dfrac{1}{2} \times r \times \sqrt{3}r - \dfrac{1}{2} \times r \times \dfrac{r}{\sqrt{3}} = \dfrac{r^2}{r}\left(\sqrt{3} - \dfrac{1}{\sqrt{3}}\right) = \dfrac{r^2}{2} \times \dfrac{2}{\sqrt{3}} = \dfrac{r^2}{\sqrt{3}}$

45. We have

$\angle BOQ = \angle BQO = y$ [$\because$ OB = BQ]

$\therefore \angle ABO = \angle OAB = y + y = 2y$ [$\because$ OA = OB]

Now, $\angle$POA + $\angle AOB$ + $\angle BOQ$ = 180° [$\because$ PQ is a straight line]

$\Rightarrow x + 180° - 4y + y = 180° \Rightarrow x = 3y$

46. Let $\angle AOB = 180°\ 180° - 4 \times 15° = 120$

$\angle ABC = \dfrac{1}{2}\ \angle AOB = 60°$

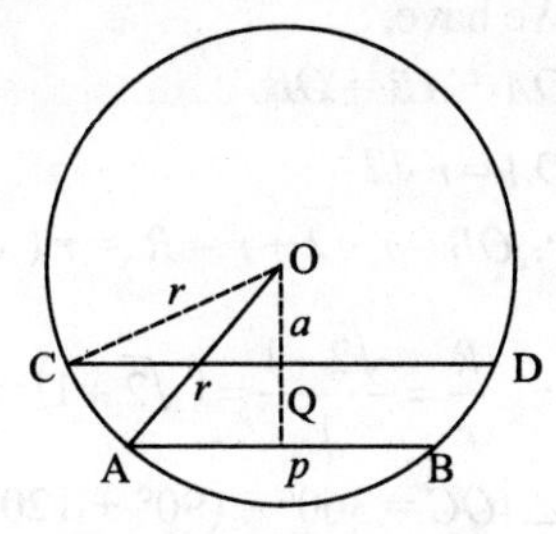

47. In ΔOQC

$Q^2 + 40^2 = r^2$ ...(i)

In ΔOAP

$(1 + 10)^2 + (30)^2 = r^2$

From (i) and (ii), we get

$\Rightarrow 2a = 600 \Rightarrow a = 30$

$(30)^2 + (40)2 = r^2 \Rightarrow r^2 = 2500 \Rightarrow r = 50$ cm

$\therefore$ diamter $= 2r = 100$ cm

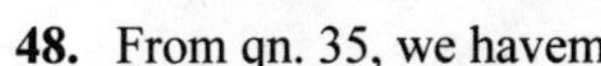

48. From qn. 35, we havem

$$\therefore \quad \angle AOB = \alpha \Rightarrow \angle AOP = \frac{\alpha}{2}$$

$$\angle COD = \beta \Rightarrow \angle COQ = \frac{\beta}{2}$$

In ΔAOP $\tan\frac{\alpha}{2} = \frac{30}{40} = \frac{3}{4}$

In ΔCOQ $\tan\frac{\beta}{2} = \frac{40}{30} = \frac{4}{3}$

Now, $\tan\left(\frac{\beta}{2}\right) - \tan\left(\frac{\alpha}{2}\right) = \frac{4}{3} - \frac{3}{4} = \frac{7}{12}$

49. We have, $OP = 2R - r$

Now, in ΔOPU

$OC^2 + CP^2 = OP^2$

$\Rightarrow (2R - r)^2 + R^2 = (R + r)^2 \Rightarrow 4R^2 - r^2 - 4Rr + R^2 = R^2 + r^2 + 2Rr$

$\Rightarrow 4R^2 = 6Rr \Rightarrow 2R = 3r$

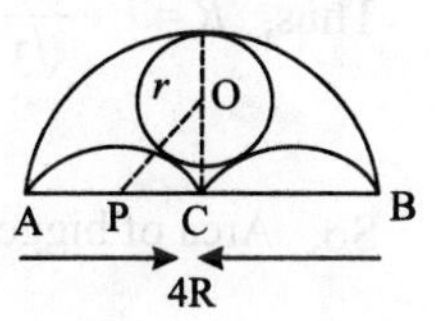

50. We have,

$m = \pi r^2$ and $n = \sqrt{10^2 - (4.5 + 3.5)^2}$ $\pi (2R^2)$ [From qn. 49]

$\Rightarrow \sqrt{10^2 - 8^2} \Rightarrow 9m = 2n$

51. In ΔOPQ

$R^2 = r^2 = 49$

Area of ring $= \pi (R^2 - r^2) = \frac{22}{7} \times 49 = 154$ cm^2

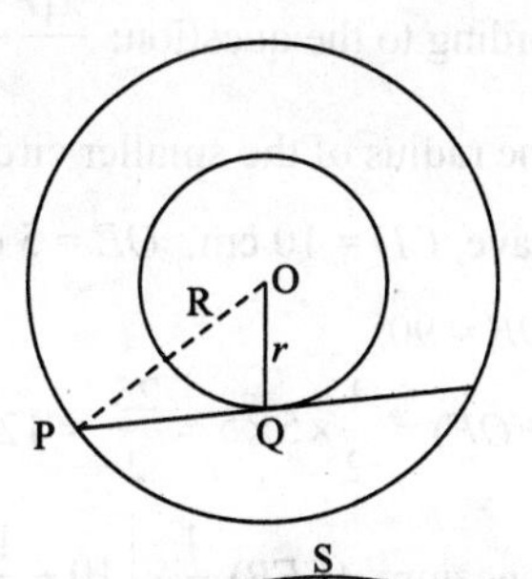

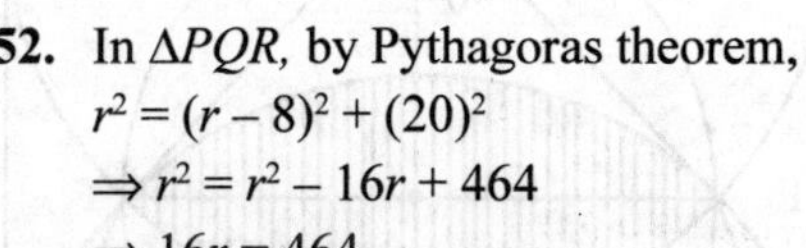

52. In ΔPQR, by Pythagoras theorem,

$r^2 = (r - 8)^2 + (20)^2$

$\Rightarrow r^2 = r^2 - 16r + 464$

$\Rightarrow 16r = 464$

$\Rightarrow r = 29$

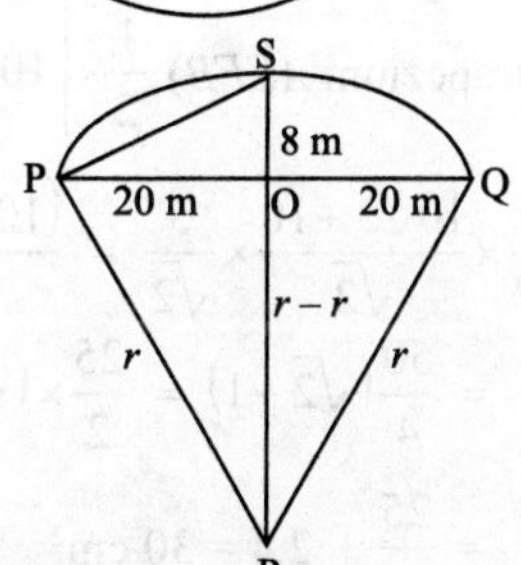

53. We have,

$OA + AB = OB$

$OA = r\sqrt{2}$

$\therefore\ OB = r\sqrt{2} + r = R = r(\sqrt{2} + 1) = R$

$\therefore\ \dfrac{R}{r} = \dfrac{\sqrt{2}+1}{1} = (\sqrt{2}+1):1$

54. $\angle AOC = 360° - (90° + 120°) = 150°.$

$\therefore\ \angle ABC = \dfrac{1}{2}\angle AOC = 75°.$

55. $\angle ACB = 90°$ ($\angle$ in a semi-circle).

So, $\angle ABC = 60°$ & $\angle CBD = 120°$.

Now, $\angle BCD = \angle BAC = 30°$ ($\angle s$ in alt. segments)

$\therefore\quad \angle BDC = 180° - (30° + 120°) = 30°$

Thus, $\angle BCD = \angle BDC$. So, $BC = BD$.

56. Radius of bigger circle (R) = r + circumradius of the equilateral triangle

Circumradius of the equilateral triangle = $\dfrac{\text{side}}{\sqrt{3}} = \dfrac{2r}{\sqrt{3}}$

Thus, $R = r + \dfrac{2r}{\sqrt{3}} = r\left(1 + \dfrac{2}{\sqrt{3}}\right) = \dfrac{r(\sqrt{3}+2)}{\sqrt{3}}$

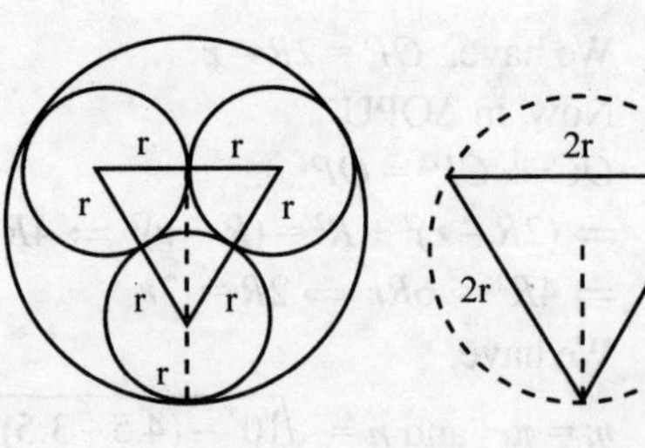

So, Area of bigger circle $= \pi\left[\dfrac{r(2+\sqrt{3})}{\sqrt{3}}\right]^2$

$= \pi\left[\dfrac{r^2(2+\sqrt{3})^2}{3}\right]$

According to the question: $\dfrac{\pi[r^2(2+\sqrt{3})^2]}{3} = \dfrac{\pi(2+\sqrt{3})^2}{3} \Rightarrow r^2 = 1 \Rightarrow r = 1$ cm

$\therefore$ The radius of the smaller circles is 1 cm each.

57. We have, $CD = 10$ cm, $OE = 5$ cm $= OD = DC$

$\angle EOF = 90°$

$ar(\Delta EOF) = \dfrac{1}{2} \times 5 \times 5 = \dfrac{25}{2} = 12.5\ \text{cm}^2$

58. ar(trapezium $AEFB$) $\dfrac{1}{2} \times \left[10 + \dfrac{10}{\sqrt{2}}\right] \times \dfrac{5}{\sqrt{2}}$

$\dfrac{1}{2} \times \dfrac{10\sqrt{2}+10}{\sqrt{2}} \times \dfrac{5}{\sqrt{2}} = \dfrac{(10\sqrt{2}+10)\times 5}{4}$

$= \dfrac{50}{4}(\sqrt{2}+1) = \dfrac{25}{2} \times (\sqrt{2}+1)$

$= \dfrac{25}{2} \times 2.4 = 30\ \text{cm}^2$

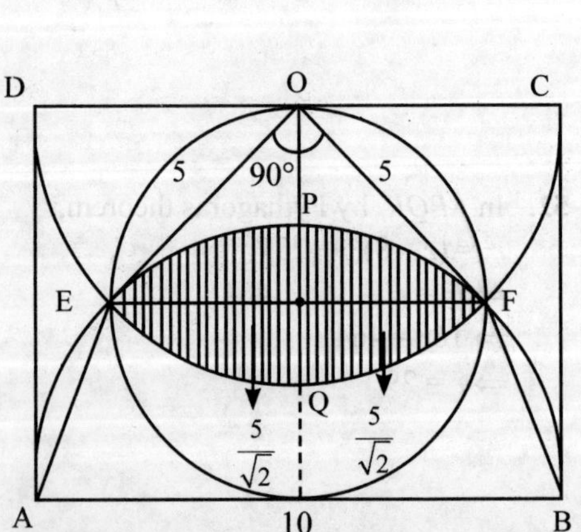

59. Area of shaded region $= 2 \times \left[\dfrac{\pi r^2\theta}{360} - \dfrac{1}{2}r^2\right] = 2\times\left[\dfrac{\pi\times 25\times 90}{360} - \dfrac{1}{2}\times 5\times 5\right]$

$= 2\times\left[\dfrac{3.14\times 25}{4} - \dfrac{25}{2}\right] = 2\times\dfrac{25}{2}\,[1.57-1] = 25\times 0.57 = 14.25 \text{ cm}^2$

For (qn 60 – 61)

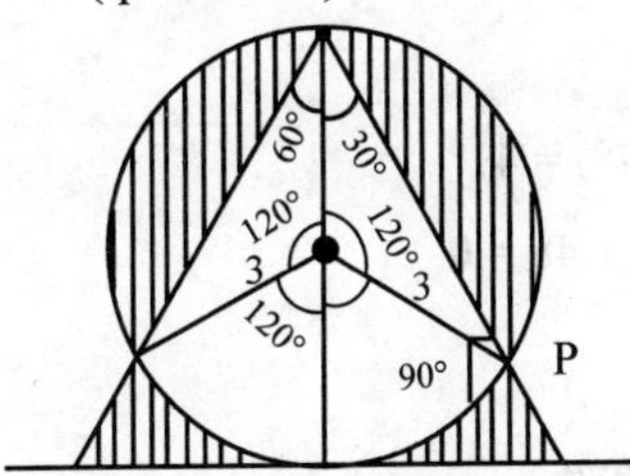

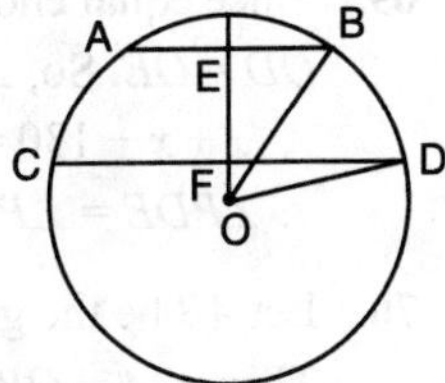

60. $BQ \times AQ = PQ^2 \Rightarrow AQ \times 8 = 144 \Rightarrow AQ = 18$ cm.

$\therefore\ AB = (AQ - BQ) = (18 - 8)$ cm $= 10$ cm.

61. Draw $OE \perp AB$, meeting CD at F. Then,

$BE = \dfrac{1}{2}AB = 3$ cm, $EF = 3$ cm, $FD = \dfrac{1}{2}CD = 6$ cm.

Let $OF = x$. Then, $OE = (3 + x)$.

$OB^2 = OE^2 + BE^2 \Rightarrow r^2 = (3 + x)^2 + 3^2 = 18 + 6x + x^2$

$OD^2 = OF^2 + FD^2 = r^2 = x^2 + 6^2 = x^2 + 36.$

$\therefore\ 18 + 6x + x^2 = x^2 + 36 = x = 3.$

so, $r^2 = (3^2 + 6^2) = 45 \Rightarrow r = \sqrt{45} = 3\sqrt{5}$ cm.

62. $\angle ABC = (180° - \angle ADC) = (180° - 140°) = 40°$ and $\angle ACB = 90°$ ($\angle$ in a semi-circle).

$\therefore\ \angle BAC = 180° - (90° + 40°) = 50°$

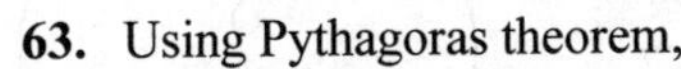

63. Using Pythagoras theorem,

$OM^2 = OQ^2 + MQ^2$

$OQ^2 = OM^2 - MQ^2$

$OQ^2 = 10^2 - 8^2 = 100 - 64 = 36$

$OQ = 6$ cm

$\therefore$ The length of OQ is 6 cm.

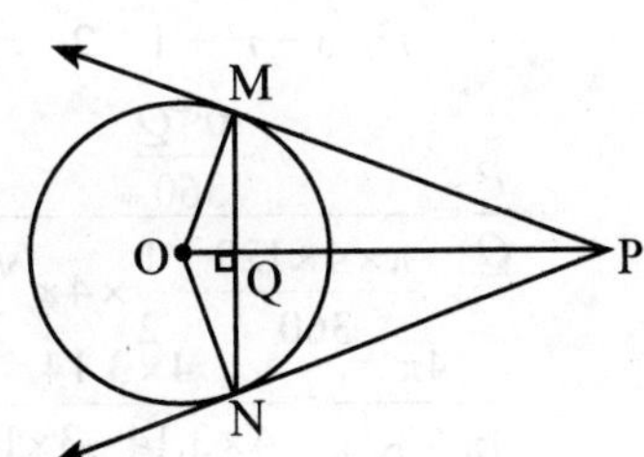

64. Let $\angle MOQ = \angle MOP = \theta$

In ΔOMQ: $\tan\theta = \dfrac{MQ}{OQ} = \dfrac{8}{6} = \dfrac{4}{3}$

In ΔPMO: $\tan\theta = \dfrac{PM}{OM} \Rightarrow \dfrac{4}{3} = \dfrac{PM}{10} \Rightarrow PM = \dfrac{40}{3}$

$\therefore$ The length of PM is $\dfrac{40}{3}$ cm.

65. Area of $\Delta OMN = \dfrac{1}{2}\times MN\times OQ = \dfrac{1}{2}\times 16\times 6 = 48 \text{ cm}^2$

66. Let $OA = r$. Then,

$OO' + O'A = r \Rightarrow O'A = (r - 6)$

$\therefore \pi r^2 + \pi(r-6)^2 = 116\pi$

or $r^2 - 6r - 40 = 0$

So, $r = 10$ cm.

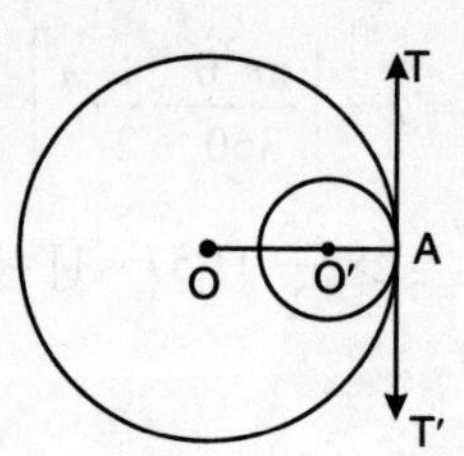

67. Clearly, it is a cyclic quadrilateral.

So, $\angle QRS = 180° - (\angle QPR + \angle SPR) = 180° - (67° + 72°) = 41°$.

68. $AP \times PB = CP \times PD \Rightarrow x \times (16 - x) = 6 \times 8 \Rightarrow x^2 - 16x + 48 = 0$.

$\therefore (x - 12)(x - 4) = 0$. So, $x = 12$ or $x = 4$.

But $AP > PB$. So, $AP = 12$ cm.

69. Since equal chords are equidistant from the centre, we have

$OD = OE$. So, $\angle ODE = \angle OED = x$.

$\therefore\ x + x + 130 = 180°$ or $x = 25$. So, $\angle ODE = 25°$.

$\therefore \angle PDE = \angle PDO - \angle ODE = (90° - 25°) = 65°$.

70. Let AB be the given chord.

Then, $OA = OB = AB$. So, $\angle AOB = 60°$.

Joint AO and produce it to meet the circle at C. Join BC.

$\angle ACB = \frac{1}{2}\angle AOB = 30°$.

But $\angle PBC$ is cyclic. So, $\angle APB = (180° - 30°) = 150°$.

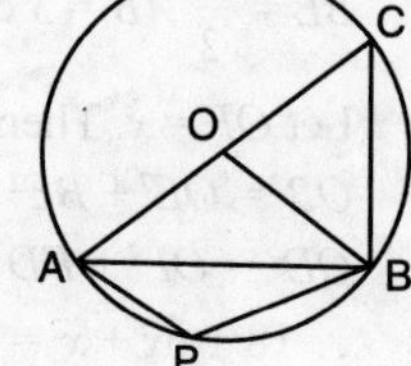

71. OD = r = OA = OB

OC = r – 1

$(OA)^2 = (AC)^2 + (OC)^2$

$r^2 = 3 + r^2 + 1 - 2r \Rightarrow\ 2r = 4 \Rightarrow r = 2$

$$\frac{P}{Q} = \frac{\frac{\pi r^2 Q}{360}}{\frac{\pi \times 4 \times 120}{360} - \frac{1}{2} \times 4 \times \frac{\sqrt{3}}{2}} = \frac{\frac{\pi \times 4 \times 120}{360}}{\frac{4\pi}{3} - \sqrt{3}} = \frac{\frac{4\pi}{3}}{\frac{4\pi - 3\sqrt{3}}{3}}$$

$$\frac{4\pi}{4\pi - 3\sqrt{3}} = \frac{4 \times 3.14}{4 \times 3.14 - 3 \times 1.7} = \frac{12.56}{12.56 - 5.1}\ \frac{12.56}{7.46} = 1.67 \approx 1.7$$

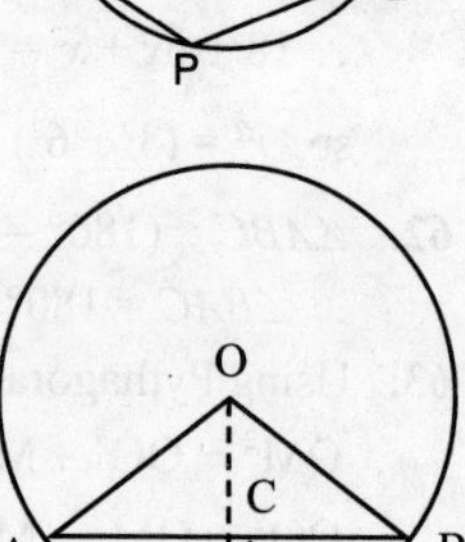

72. $OQ^2 = OP^2 + PQ^2$

$\Rightarrow R^2 = r^2 + 49 \Rightarrow\ R^2 = r^2 + 49$

Required area $= \pi(R^2 - r^2) = \frac{22}{7} \times 49 = 154$ cm^2

73. Now, sum of areas of two circles

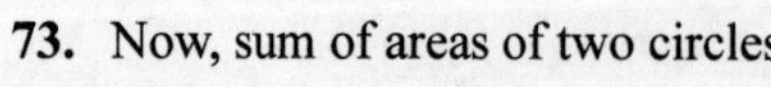

$\pi r^2 + \pi r^2 = \pi(16 + r^2) \Rightarrow\ \pi = (16 + 1) = 17\pi$

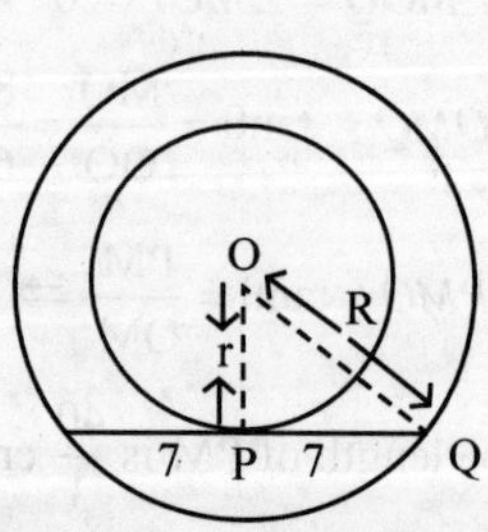

74. $\tan\theta = \frac{4}{3} = 1.33$

$\therefore\ (45° < \theta < 60°)$

75. Area of shaded region = Area of trapezium – area of sectors

$$\left[\frac{1}{2}\times 5\times 4\right]-\left[\pi\times 16\times\frac{\pi}{360^\circ}+\frac{\pi\times 1^2(180^\circ-\theta)}{360^\circ}\right]=10-\frac{16\pi\theta}{360}+\frac{\pi}{2}-\frac{\pi\theta}{360^\circ}=10-\left[\frac{15\pi\theta}{360}+\frac{\pi}{2}\right]$$

$$10-\frac{\pi}{24}-\frac{\pi}{2}=\frac{240-\pi\theta-12\pi}{24}\text{ square unit}$$

76. Let r be the radius of the circle.
Then, $OE = OB$ and $OE + OB = r \Rightarrow OE = EB = (r/2)$.

Now, $AE \times EB = CE \times ED \Rightarrow \left(r+\frac{r}{2}\right)\times\frac{r}{2}=(6\times 2)\Rightarrow r = 4\text{cm}.$

77. Join OT and OT'. Clearly, $ATOT'$ is cyclic.
$\therefore\ \angle TOT' = (180^\circ - 70^\circ) = 110^\circ.$
So, reflex $\angle TOT' = (360^\circ - 110^\circ) = 250^\circ.$
$\therefore\ \angle TPT' = \frac{1}{2}\times 250^\circ = 125^\circ.$

78. $PQ^2 = (R + r)^2 - (R - r)^2 = 4Rr.$

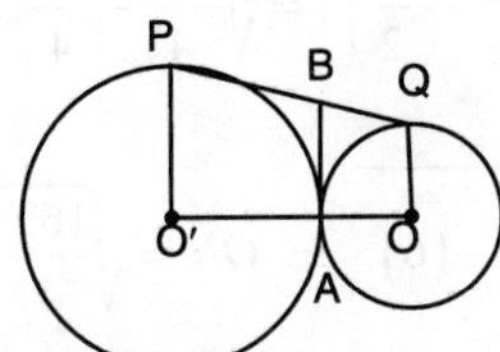

79. Draw $OD \perp AB$, $OE \perp BC$ and $OF \perp AC$. Let $OD = OE = OF = x$. Since the tangents to a circle from a point outside it are equal, we have:

$$AF = AD = (8 - x)\ ;\ CF = CE = (6 - x)$$

$\therefore$ $$AC = \sqrt{AB^2 + BC^2} = \sqrt{8^2 + 6^2} = 10\text{ cm.}$$

So, $$AC = AF + CF \Rightarrow (8 - x) + (6 - x) = 10 \text{ or } x = 2.$$

Since the tangents drawn to a circle from a point outside it are equal, we have:

80. Since the two smaller circles touch internally at the centre of larger circle, they have equal radius ; the diameter of each being 2 cm and radius of each is 1 cm.
Required area = (Area of semi-circle with radius 4 cm) – (area of semi-circle with radius 1 cm)

$$=\frac{1}{2}\times\pi\times(4)^2-2\times\frac{1}{2}\times\pi\times 1^2=7\pi.$$

81. Let the radii of the two circles be x cm and y cm.
$\therefore\ x + y = 14$ and $x^2 + y^2 = 130$. By trial (as otherwise),
$[x, y] = \{3, 11\}$ The difference in the radii is 8 cm.

82. In the figure, diameter AB = 6 and chord CD = 2.
Also $\angle$CPA = 90°
$OP^2 = OC^2 - CP^2 = 3^2 - 1^2$

$\therefore$ $$\frac{AP}{PB}=\frac{AO+OP}{BO-OP}=\frac{3+\sqrt{8}}{3-\sqrt{8}}$$

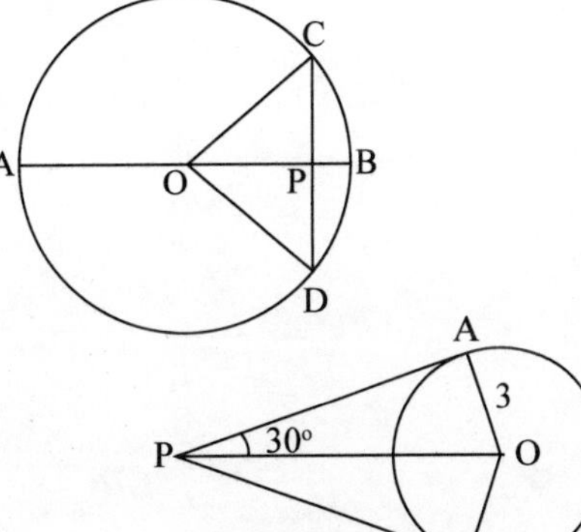

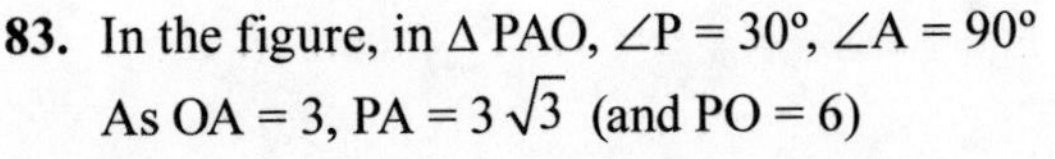

83. In the figure, in Δ PAO, $\angle$P = 30°, $\angle$A = 90°
As OA = 3, PA = $3\sqrt{3}$ (and PO = 6)

84. In the semicircle of radius r, the largest triangle that can be inscribed would have a height r (and base $2r$) its area in $\frac{1}{2}(2r)\,r = r^2$

85. Let the foot of the pole be Q and the top be P.
As $\angle PAQ = \angle PBQ = \angle PCQ$, if follows that $QA = QB = QC$
$\therefore$ Foot of the pole Q is at the centroid.

86. As the perpendicular ON and OM cuts CD and AB respectively, in equal parts

$$\therefore \quad AM = \frac{5}{2}\text{ cm} \Rightarrow CN = \frac{12}{2} = 6\text{ cm}$$

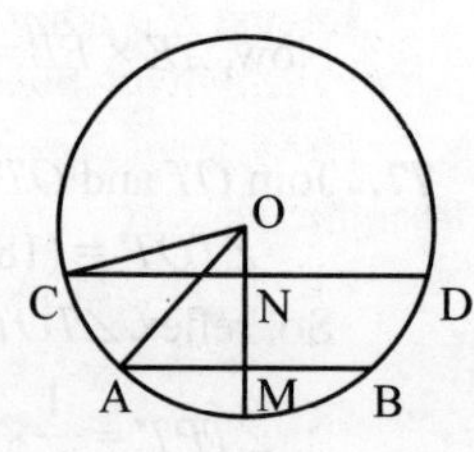

Given, diameter = 13cm

$$\therefore \quad OC = \frac{13}{2}\text{ cm}$$

In ΔOAM

$$OM = \sqrt{\left(\frac{13}{2}\right)^2 - \left(\frac{5}{2}\right)^2} = \sqrt{\frac{169}{4} - \frac{25}{4}} = \sqrt{\frac{144}{4}} = \frac{12}{2} = 6$$

In ΔOCN

$$ON = \sqrt{\left(\frac{13}{2}\right)^2 - (6)^2} \Rightarrow ON = \sqrt{\frac{169}{4} - \frac{36 \times 4}{4}} = \sqrt{\frac{25}{4}} = \frac{5}{2} = 2.5$$

$$MN = OM - ON = 6 - 2.5 = 3.5\text{ cm}$$

87. $\angle BAC = 60°$. So, $\angle BOC = 2\,\angle BAC = 120°$.

88. $P + Q$

$$\text{Area} = \pi r^2 = \frac{22}{7} \times 7 \times 7 = 154\text{ cm}^2$$

$$\text{Now,} \quad P + Q = 154 \times \frac{8}{18} = \frac{616}{9}\text{ cm}^2$$

89. Circumference $= 2\pi r = 2 \times \frac{22}{7} \times 7 = 44$ cm

$$P = \left(44 \times \frac{1}{18}\right) + 14 = \frac{22}{9} + 14 \Rightarrow 9p = 22 + 126 = 148\text{ cm}$$

33

CHAPTER

Loci

Locus: *If a point moves according to some given geometrical conditions, then the path traced out by the moving point is called its locus.*

Ex. 1. If a point moves in such a way that its distance from a given fixed point is always constant, then the locus of the path is a circle.

Ex. 2. The locus of a point equidistant from two given points A and B is the perpendicular bisector of AB.

Ex. 3. The locus of a point equidistant from two intersecting lines is the angle bisector of the angle made by these lines.

Ex. 4. If l and m be two lines such that $l \parallel m$. If a point moves in such a way that it is equidistant from land m , then locus is a line in just mid-way between l and m and parallel to each one of them.

Ex. 5. Locus of the centres of all circles passing through two given points A and B is the perpendicular bisector of AB.

Ex. 6. If the bisectors of $\angle B$ and $\angle C$ of a quadrilateral $ABCD$ intersect in P, then P is equidistant from AB and CD.

Ex. 7. If AB is a fixed line segment, then the locus of a point P such that $\angle APB = 90°$, is a circle with AB as diameter.

Ex. 8. The locus of the vertices of isosceles triangles having a common base is the perpendicular bisector of the base.

Ex. 9. AB is a fixed line. The locus of a point P so that $AB^2 = AP^2 + BP^2$ is a circle with AB as diameter.

Ex. 10. The locus of mid-points of equal chords of a circle is the circle concentric with the given circle and radius equal to the distance of equal chords from the centre.

Ex. 11. The locus of mid-points of parallel chords of a circle is the diameter of the circle perpendicular to the parallel chords.

EXERCISE 33

1. The locus of the mid-points of the radii of length 16 cm of a circle is **[CDS (I) Exam, 2018]**
   (*a*) A concentric circle of radius 8 cm
   (*b*) A concentric circle of radius 16 cm
   (*c*) The diameter of the circle
   (*d*) A straight line passing through the centre of the circle
2. The locus of a point equidistant from two intersecting lines is **[CDS (I) Exam, 2018]**
   (*a*) A straight line
   (*b*) A circle
   (*c*) A pair of straight lines
   (*d*) None of the above

3. If a point P moves such that the sum of the squares of its distances from two fixed points A and B is a constant, then the locus of the point P is **[CDS (I) Exam, 2018]**

(*a*) A straight line

(*b*) A circle

(*c*) Perpendicular bisector of AB

(*d*) An arbitrary curve

4. **_ABCDA_ is a non-cyclic quadrilateral of a circle _ABCD_ with radius r and centre at _O_. If _AB_ is the diameter and _CD_ is parallel and half of _AB_ and if the circle completes one rotation about the centre _O_, then the locus of the middle point of _CD_ is a circle of radius**

(*a*) $\frac{3r}{2}$ (*b*) $\frac{2r}{3}$ (*c*) $\frac{2\sqrt{3}r}{3}$ (*d*) $\frac{\sqrt{3}}{2}r$

[CDS (II) Exam, 2016]

5. The locus of a point equidistant from two intersecting lines is:

(*a*) a single line (*b*) a pair of lines (*c*) a circle (*d*) a parabola

6. Which of the following statements is incorrect?

(*a*) A circle is the locus of a point.

(*b*) The locus of the points which are equidistant from three non-collinear points is the centre of the circle passing through the given points.

(*c*) The locus of the points which are equidistant from three distinct points on a line is a line parallel to the given line.

(*d*) None of these

7. The locus of a point inside a circle and equidistant from two given points on the circle:

(*a*) is the chord joining the two points.

(*b*) is a circle with the line joining the two points, as its diameter.

(*c*) is the perpendicular bisector of the chord joining the two points.

(*d*) does not exist.

8. If *AB* is a line segment, then the locus of a point *P* such that $\angle APB = 90°$, is:

(*a*) a circle with *AB* as diameter. (*b*) the triangle *APB*.

(*c*) a semi-circle with *AB* as diameter. (*d*) the perpendicular bisector of *AB*.

9. The locus of the mid-points of parallel chords of a circle is:

(*a*) a circle.

(*b*) the line joining the mid-points

(*c*) the diameter perpendicular to the given chords.

(*d*) None of these

10. In a ΔABC, the point on ΔABC which is equidistant from sides *AB* and *AC* is the one where:

(*a*) the median fiom *A* meets *BC*

(*b*) the bisector of $\angle BAC$ meets *BC*

(*c*) the altitude on *BC* meets *BC*

(*d*) the per perpendicular bisector of *BC* intersects *BC*.

11. Two fixed circles are in the same plane. If *P* is a point in that plane and the tangents from *P* to the two circles are equal, then the locus of *P*:

(*a*) is a straight line perpendicular to the line joining the centres.
(*b*) is a circle
(*c*) is a parabola
(*d*) cannot be determined

12. Let C be a circle. A point P moves such that the tangents from P to C include an angle of 60°. The locus of P is:
(*a*) a straight line (*b*) a circle concentric with C
(*c*) a circle touching C (*d*) a circle intersecting C at two points

13. A number of triangles on the same base and on the same side of it are of equal areas. Then, the locus of their vertices is:
(*a*) a straight line parallel to the base (*b*) a circle with the base as the diameter
(*c*) a straight line parallel to the altitude of a triangle on the given base
(*d*) None of these

14. The locus of the vertices of the isosceles triangles formed on a given base BC is:
(*a*) a circle (*b*) a line per perpendicular to BC
(*c*) bisector of BC (*d*) right bisector of BC

15. A wheel of diameter 80 cm rolls on a plane straight road. The locus of the centre of the wheel is:
(*a*) a circle of radius 40 cm (*b*) a circle of radius 20 cm
(*c*) a straight line parallel to the path of the wheel and 40 cm distant from it
(*d*) a straight line parallel to the path of the wheel and 80 cm distant from it.

ANSWERS

1. (*a*) **2.** (*c*) **3.** (*b*) **4.** (*d*) **5.** (*b*) **6.** (*c*) **7.** (*c*) **8.** (*a*) **9.** (*c*)
10. (*b*) **11.** (*a*) **12.** (*b*) **13.** (*a*) **14.** (*d*) **15.** (*c*)

SOLUTIONS

1. The locus of all points at the same distance from the centre is known as a circle. All of the radii's midpoints will be eight centimetres away from the centre. Thus, it is circle with radius 8 cm.

2. In ΔPOQ and QOR
$PQ = QR$ [∵ it is equidistant]
$OQ = OQ$ [common]
$\angle QPO = \angle QRO = 90°$
∴ $\Delta POQ \cong \Delta QOR$
The angular bisector of the lines is the locus of all the points at the equidistant from the lines. Hence, locus is a pair of straight lines.

3. Let the co-ordinates of the points P be (p, q) A be (a, o) and B be (o, b)
We have,
$PA^2 + PB^2 = k$ (constant)
$(p - a)^2 + q^2 + p^2 + (q - b)^2 = k$
$\Rightarrow p^2 - 2ap + a^2 + q^2 + p^2 + q^2 + b^2 - 2bq = k$
$\Rightarrow 2p^2 - 2ap + a^2 + 2q^2 - 2bq + b^2 = k$

$$\Rightarrow p^2 - ap + \frac{a^2}{2} + q^2 - bq + \frac{b^2}{2} = \frac{k}{2}$$

$$\Rightarrow p^2 - ap + \frac{a^2}{4} + \frac{a^2}{4} + q^2 - bq + \frac{b^2}{4} + \frac{b^2}{4} + \frac{b^2}{4} = \frac{k}{2}$$

$$\Rightarrow \left(p - \frac{a}{2}\right)^2 + \frac{a^2}{4} + \left(q - \frac{b}{2}\right)^2 + \frac{b^2}{4} = \frac{k}{2}$$

$$\Rightarrow \left(p - \frac{a}{2}\right)^2 + \left(q - \frac{b}{2}\right)^2 = \frac{k}{2} - \frac{a^2}{4} - \frac{b^2}{4}$$

$$\Rightarrow \left(p - \frac{a}{2}\right)^2 + \left(q - \frac{b}{2}\right)^2 = t \text{ (constant)}$$

It is the eqn of a circle.

Thus, the locus of the points P is a circle.

4. As C completes one revolution on the circle, D would also complete one revolution and M, the midpoint of CD Would move on G circle of radius $\sqrt{3}\,r/2$ and centre O.

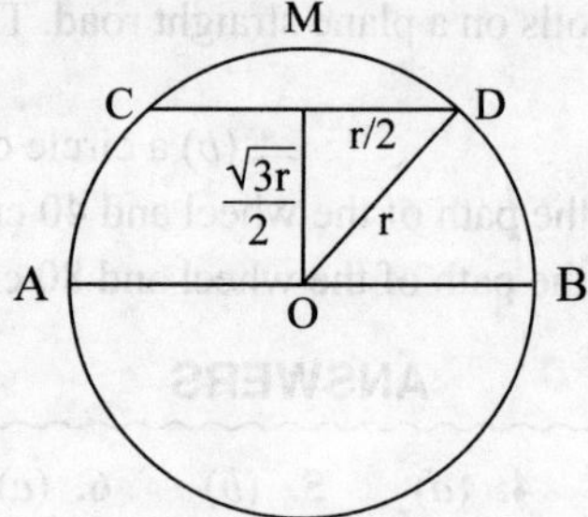

5. Locus of a point equidistant from two intersecting lines is a pair of lines, which are the bisectors of the angles between the given straight lines.

6. Clearly, (*c*) is incorrect. Actually, there is no point which is equidistant from three collinear points.

7. Clearly, the locus is the perpendicular bisector of the chord joining the points.

8. Clearly, the locus of *P* is a circle with *AB* as diameter.

9. The locus of the mid-points of the parallel chords of a circle is a diameter of the circle perpendicular to the parallel chords.

10. Clearly, the point on *BC* equidistant from *AB* and *AC,* lies on the bisector of $\angle BAC$.

11. Clearly, the locus of *P* is a line perpendicular to the line joining the centres.

12. (*b*).

13. Clearly, the locus is a straight line parallel to the base, since Δs on the same base and between the same parallels will be equal in area.

14. Let *P* be a vertex. Then, $PB = PC$ holds only when *P* lies on the right bisector of *BC*. So, the locus of *P* is the right bisector of *BC*.

15. Clearly, the centre of the wheel moves along the path and remains at a distance of 40 cm from the path of the wheel. So, the required locus is a line parallel to the path of the wheel and at a distance of 40 cm from it.

UNIT 6

COORDINATE GEOMETRY

Coordinate Geometry: Equations of lines and circles. Points of intersection of a Curve and a line & Points of intersection of two curves.

CHAPTER

34 Coordinate Geometry

Review of Concepts

1. General Results

(*i*) Distance between two points $P(x_1, y_1)$ and $Q(x_2, y_2)$ is given by:

$PQ = \sqrt{(x_2 - x_1)^2 + (y_2 - y_1)^2}$.

(*ii*) The co-ordinates of the point dividing the join of $P(x_1, y_1)$ and $Q(x_2, y_2)$ internally in the ratio $m : n$ are $\left(\frac{mx_2 + nx_1}{m+n}, \frac{my_2 + ny_1}{m+n}\right)$.

(*iii*) The co-ordinates of the point dividing the join of $P(x_1, y_1)$ and $Q(x_2, y_2)$ externally in the rartio $m : n$ are $\left(\frac{mx_2 - nx_1}{m-n}, \frac{my_2 - ny_1}{m-n}\right)$.

(*iv*) The co-ordinates of the mid-point of the line joining $P(x_1, y_1)$ and $Q(x_2, y_2)$ are $\left(\frac{x_1 + x_2}{2}, \frac{y_1 + y_2}{2}\right)$.

(*v*) The area of a triangle whose vertices are $A(x_1, y_1)$, $B(x_2, y_1)$ and $C(x_3, y_3)$ is given by:

$\Delta = \frac{1}{2}[x_1(y_2 - y_3) + x_2(y_3 - y_1) + x_3(y_1 - y_2)]$.

Remark: *Three points* $A(x_1 + y_1)$, $B(x_2, y_2)$ & $C(x_3, y_3)$ *are collinear if* $x_1(y_2 - y_3) + x_2(y_3 - y_1) + x_3(y_1 - y_2) = 0$.

(*vi*) The co-ordinates of the centroid of a triangle with vertices $A(x_1, y_1)$, $B(x_2, y_2)$ and $C(x_3, y_3)$ are $\left(\frac{x_1 + x_2 + x_3}{3}, \frac{y_1 + y_2 + y_3}{3}\right)$.

(*vii*) The co-ordinates of incentre of a triangle with sides a, b, c and vertices $A(x_1, y_1)$, $B(x_2, y_2)$ and $C(x_3, y_3)$ are:

$\left(\frac{ax_1 + bx_2 + cx_3}{a+b+c}, \frac{ay_1 + by_2 + cy_3}{a+b+c}\right)$.

(*viii*) A quadrilateral is a:

I. *Rectangle*, if its opposite sides are equal and the diagonals are equal ;

II. *Parallelogram but not a rectangle*, if its opposite sides are equal but diagonals are not equal;

III. *Square*, if all sides are equal & diagonals are equal;

IV. *Rhombus but not a square*, if all sides are equal but diagonals are not equal.

2. Equations of Lines:

(*i*) The equation of x-axis is $y = 0$.

(*ii*) The equation of y-axis is $x = 0$.

(*iii*) The equation of a line parallel to y-axis at a distance a from it, is $x = a$.

(*iv*) The equation of a line parallel to x-axis and at a distance b from it is $y = b$.

(*v*) Equation of a line passing through the points $A(x_1, y_1)$ & $B(x_2, y_2)$ is $\frac{y - y_1}{x - x_1} = \frac{y_2 - y_1}{x_2 - x_1}$.

Slope of such a line is $\left(\frac{y_2 - y_1}{x_2 - x_1}\right)$.

(*vi*) Slope-intercept form of an equation of a line is $y = mx + c$, where slope $= m$ and y-intercept $= c$.

(*vii*) General equation of a line is $ax + by + c = 0$.

(*viii*) The equation of line which makes intercepts a and b with x-axis and y-axis respectively is $\frac{x}{a} + \frac{y}{b} = 1$.

(*ix*) Equation of a line in normal form is $x \cos \alpha + y \sin \alpha = p$, where p is the length of perpendicular from the origin to the line and α is the angle made by this perpendicular with the x-axis.

(*x*) Length of perpendicular from a point $P(\alpha, \beta)$ to the line $ax + by + c = 0$ is $\frac{a\alpha + b\beta + c}{\sqrt{a^2 + b^2}}$.

(*xi*) Two lines with slopes m_1 and m_2 are:

(*a*) *parallel if* $m_1 = m_2$ (*b*) *perpendicular if* $m_1 m_2 = -1$.

(*xii*) Let m_1, m_2 be the slopes of two lines and θ be the angle between them, then $\tan \theta = \pm\left(\frac{m_1 - m_2}{1 + m_1 m_2}\right)$.

(*xiii*) The equations of the bisectors of the angles between the lines

$a_1x + b_1y + c_1 = 0$ & $a_2x + b_2y + c_2 = 0$ are:

$$\frac{a_1x + b_1y + c_1}{\sqrt{a_1^2 + b_1^2}} = \pm \frac{a_2x + b_2y + c_1}{\sqrt{a_1^2 + b_1^2}}$$

(*xiv*) Two lines $a_1x + b_1y + c_1 = 0$ & $a_2x + b_2y + c_2 = 0$ are:

I. *Coincident*, if $\frac{a_1}{a_2} = \frac{b_1}{b_2} = \frac{c_1}{c_2}$. II. *Parallel*, if $\frac{a_1}{a_2} = \frac{b_1}{b_2} \neq \frac{c_1}{c_2}$.

III. *Perpendicular*, if $a_1a_2 + b_1b_2 = 0$.

IV. *Intersecting*, if they are neither coincident nor parallel.

3. Pairs of Lines:

(*i*) The equation $ax^2 + 2hxy + by^2 = 0$ represents a pair of lines passing through the origin if $h^2 \geq ab$.

(*ii*) The angle between two lines given by $ax^2 + 2hxy + by^2 = 0$ is:

$$\theta = \tan^{-1}\left(\frac{2\sqrt{h^2 - ab}}{a + b}\right).$$

(*iii*) Two lines represented by $ax^2 + 2hxy + by^2 = 0$ will:

I. Coincide, if $h^2 = ab$;

II. be perpendicular, if $a + b = 0$.

(*iv*) The equation of the bisectors of the angles between the lines

$ax^2 + 2hxy + by^2 = 0$ are: $\frac{x^2 - y^2}{a-b} = \pm\left(\frac{xy}{h}\right)$.

(*v*) The equation $ax^2 + 2hxy + by^2 + 2gx + 2fy + c = 0$ will represent a pair of lines, if $abc + 2fgh - af^2 - bg^2 - ch^2 = 0$.

4. Equation of a Circle:

(*i*) The equtaion of a circle with centre (0, 0) and radius a is given by $x_2 + y^2 = a^2$.

(*ii*) The equation of a circle with centre (h, k) and radius a is given by

$$(x - h)^2 + (y - k)^2 = a^2.$$

(*iii*) General equation of a circle is: $x^2 + y^2 + 2gx + 2fy + c = 0$.

Its *centre is* $(-g, -f)$ and *radius is* $\sqrt{g^2 + f^2 - c}$.

(*iv*) Equation of a circle with (x_1, y_1) and (x_2, y_2) as diameter is $(x - x_1)(x - x_2) + (y - y_1)(y - y_2) = 0$.

(*v*) Equation of tangent to a circle $x^2 + y^2 = a^2$ at the point (x_1, y_1) is $xx_1 + yy_1 = a^2$.

(*vi*) Equation of normal to the circle $x^2 + y^2 = a^2$ at (x_1, y_1) is $xy_1 - x_1y = 0$

(*vii*) Length of tangent to a circle $x^2 + y^2 = a^2$ from a point $P(x_1, y_1)$ outside it, is $\sqrt{x_1^2 + y_1^2 - a^2}$.

(*viii*) Equation of tangent to a circle $x^2 + y^2 + 2\,gx + 2fy + c = 0$ at a point

$$(x_1, y_1) \text{ is } xx_1 + yy_1 + g(x + x_1) + f(y + y_1) + c = 0.$$

(*ix*) Equation of normal to a circle $x^2 + y^2 + 2gx + 2fy + c = 0$ at a point (x_1, y_1) is $\frac{x - x_1}{x_1 + g} = \frac{y - y_1}{y_1 + f}$.

(*x*) Length of tangent to a circle $x^2 + y^2 + 2gx + 2fy + c = 0$ from a point $P(x_1, y_1)$ outside it is $\sqrt{x_1^2 + y_1^2 + 2gx_1 + 2fy_1 + c}$.

(*xi*) Two circles

$x^2 + y^2 + 2g_1x + 2f_1y + c_1 = 0$ and $x^2 + y^2 + 2g_2x + 2f_2y + c_2 = 0$ cut orthogonally if $2g_1g_2 + 2f_1f_2 = c_1 + c_2$.

(*xii*) Let $S_1 = x^2 + y^2 + 2g_1x + 2f_1y + c_1 = 0$ and $S_2 = x^2 + y^2 + 2g_2x + 2f_2\,y + c_2 = 0$ be two intersecting circles. Then, the equation of their common chord is $S_1 - S_2 = 0$.

i.e., $2(g_1 - g_2)x = 2(f_1 - f_2)y + (c_1 - c_2) = 0$.

Points of Intersection of Two Curves:

If a line and a curve intersect or two curves intersect, then their points of intersection may be obtained by solving their equations simultaneously.

INEQUATIONS

Graph: Let us consider an inequation $ax + by \leq c$.

Step I: Consider the equation $ax + by = c$. Draw the graph of this equation, which is a line.

In case of strict inequalities < or > draw the line dotted, otherwise mark it thick.

Step 2: Choose a point [if possible (0, 0)], not lying on this line. Substitute its co-ordinates in the given inequation. If this point satisfies the given inequation, then shade the portion of the plane which contains the chosen point, otherwise shade the portion which does not contain this point.

The shaded portion represents the solution set. The dotted line is not a part of the solution set, while thick line is a part of it.

Ex. *Graph the solution set af 2x – y ≥ 1.*

Sol. Consider the equation, $2x - y = 1$.

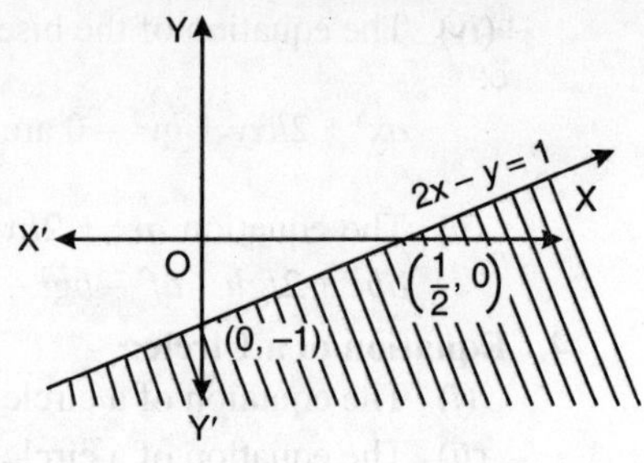

We may write this equation as $\frac{x}{(I/2)} + \frac{y}{-1} = 1$.

This line makes intercepts of $\frac{1}{2}$ and – 1 with x-axis and y-axis respectively. So, it means x-axis at $A\left(\frac{1}{2}, 0\right)$ and y-axis at $B(0, -1)$, plot these points and join them with a thick line AB. Consider (0, 0). Clearly, (0, 0) does not satisfy $2x - y \geq 1$. So, shade the portion of the plane not containing (0, 0). Shaded portion constitutes the solution set.

EXERCISE 34

1. What is the point on the xy-plane satisfying $5x + 2y = 7xy$ and $10x + 3y = 8xy$?

   (*a*) $\left(-1, \frac{1}{6}\right)$ (*b*) $\left(\frac{1}{6}, -1\right)$ (*c*) $\left(1, \frac{1}{6}\right)$ (*d*) $\left(-\frac{1}{6}, -1\right)$

   [CDS (I) Exam, 2020]

2. How many integral values of x and y satisfy the equation $5x + 9y = 7$, where – 500 x 500 an $-500 < y < 500$? **[CDS (I) Exam, 2020]**

   (*a*) 110 (*b*) 111 (*c*) 112 (*d*) None of the above

3. The tangent to the circle $x^2 + yz = 16$ and parallel to the line $y = x$ are:

   (*a*) $y = x \pm 2\sqrt{2}$ (*b*) $y = -x \pm 4\sqrt{2}$ (*c*) $y = x \pm 4\sqrt{2}$ (*d*) $y = 4\sqrt{2} \pm x$

4. The co-ordinates of a point situated on x–axis at a distance of 3 units from y–axis is:

   (*a*) (0, 3) (*b*) (3, 0) (*c*) (3, 3) (*d*) (–3, 3)

5. The co-ordinates of a point situated on y-axis at a distance of 3 units from x–axis is:

   (*a*) (0,6) (*b*) (6,0) (*c*) (6, 6) (*d*) (– 6,6)

6. The co-ordinates of a point below x–axis at a distance of 4 units from x–axis but lying on y–axis is:

   (*a*) (0, 4) (*b*) (–4, 0) (*c*) (0, – 4) (*d*) (4, –4)

7. The figure formed by the points $A(3, -2)$, $B(-2, -2)$, $C(-2, 1)$ and $D(3, 1)$ is a:

   (*a*) Parallelogram (*b*) Square (*c*) Rectangle (*d*) Rhombus

8. The figure formed by the points $P(0, 5)$, $Q(-2, -2)$, $R(5, 0)$ and $S(7, 7)$ is a:

   (*a*) Parallelograrn (*b*) Rectangle (*c*) Square (*d*) Rhombus

9. The points $A(0, 6)$, $B(-5, 3)$ and $C(3, 1,)$ are the vertices of a triangle which is:

   (*a*) Isosceles (*b*) Right angled (*c*) Equilateral (*d*) None of these

10. In which quadrant does (– 4, 3) lie ?

    (*a*) I (*b*) II (*c*) III (*d*) IV

11. What point on x-axis is equidistant from the points $A(7, 6)$ and $B(-3, 4)$?

    (*a*) (0, 4) (*b*) (–4, 0) (*c*) (3, 0) (*d*) (0, 3)

12. The area of the triangle whose vertices are $A(3, 8)$, $B(-4, 2)$ and $C(5, -1)$ (in square units) is:

    (*a*) $28\frac{1}{2}$ (*b*) $37\frac{1}{2}$ (*c*) 75 (*d*) 57

13. The co-ordinates of the point P which divides the join of $A(5, -2)$ and $B(9, 6)$ in the ratio 3 : 1 are:

(*a*) (4, – 7) (*b*) $\left(\frac{7}{2}, 4\right)$ (*c*) (8, 4) (*d*) (12, 8)

14. ABC is a triangle with vertices $A(7, -3)$, $B(3, -1)$ and $C(5, 3)$. If AD is one of its medians, then the length of this median is:

(*a*) 7 units (*b*) 5 units (*c*) 8 units (*d*) 6 units

15. The ratio in which the point $P(1, 2)$ divides the join of the points $A(-2, 1)$ and $B(7,4)$ is:

(*a*) 1 : 2 (*b*) 2 : 1 (*c*) 3 : 2 (*d*) 2 : 3

16. The ratio in which the line segment joining $A(2, -3)$ and $B(5, 6)$ is divided by x-axis is:

(*a*) 1 : 2 (*b*) 2 : 1, (*c*) 3 : 5 (*d*) 2 : 3

17. In what ratio is the line segment joining the points $P(-4, 2)$ and $Q(8, 3)$ divided by y-axis ?

(*a*) 3 : 1 (*b*) 1 : 3 (*c*) 2 : 1 (*d*) 1 : 2

18. $A(-3, b)$ and $B(1, b + 4)$ are two points and the co-ordinates of the middle point of AB are $(-1, 1)$. The value of b is:

(*a*) 1 (*b*) -1 (*c*) 2 (*d*) 0

19. If the points $A(2, 3)$, $B(5, k)$ and $C(6, 7)$ are collinear, the value of k is:

(*a*) 4 (*b*) $\frac{11}{4}$ (*c*) 6 (*d*) $-\frac{3}{2}$

20. The co-ordinates of the centroid of ΔABC with vertices $A(-1, 0)$, $B(5, -2)$ and $C(8, 2)$ is:

(*a*) (12, 0) (*b*) (6, 0) (*c*) (0, 6) (*d*) (4, 0)

21. Two vertices of a triangle ABC are $A(-1, 4)$ and $B(5, 2)$ and its centroid is $(0, -3)$. The co-ordinates of C are:

(*a*) (4, 3) (*b*) (4, 15) (*c*) (–4, –15) (*d*) (–15, – 4)

22. The area of quad. $ABCD$ with $A(1, 1)$, $B(7, -3)$, $C(12, 2)$ and $D(7, 21)$ as its vertices is:

(*a*) 35 sq. units (*b*) 65 sq. units (*c*) 85 sq. units (*d*) 132 sq. units

23. The co-ordiantes of the point of intersection of the medians of a triangle with vertices $A(-2, 2)$, $B(-1, -3)$ and $C(5. 7)$ are:

(*a*) $\left(\frac{2}{3}, 2\right)$ (*b*) (2, 6) (*c*) (1, 3) (*d*) $\left(\frac{2}{3}, 3\right)$

24. The distance of the point (4, – 3) from the origin is:

(*a*) 1 unit (*b*) 7 units (*c*) 5 units (*d*) 3 units

25. The points of intersection of the circle $x^2 + y^2 = 25$ and the line $y = 4$ are:

(*a*) (3.4). (4. 3) (*b*) (3. 4).(3.–4) (*c*) (3. 4). (–3. 4) (*d*) (–4, 3). (3. –4)

26. The points of intersection of the lines $3x + 5y = 4$ and $2x + 3y = 2$, is:

(*a*) (0, 0) (*b*) (2, –2) (*c*) (– 2, 2) (*d*) (5, – 3)

27. The line $2x + 3y = 6$ meets x-axis at the point:

(*a*) (3, 0) (*b*) (0, 3) (*c*) (3, 2) (*d*) (2, 3)

28. The line $3x - 7y = 10$ meets y-axis at the point:

(*a*) $\left(\frac{10}{3}, 0\right)$ (*b*) $\left(0, \frac{-10}{7}\right)$ (*c*) $\left(0, \frac{10}{3}\right)$ (*d*) $\left(\frac{-10}{7}, 0\right)$

29. The equation of a line parallel to y-axis and at a distance of 4 units to the right of y-axis, is:

(*a*) $x = 4$ (*b*) $y = 4$ (*c*) $x = 4y$ (*d*) $y = 4x$

30. The equation of a line parallel to y-axis at a distance of 3 units to the left of y-axis, is:
(*a*) $y = 3$ (*b*) $x = -3$ (*c*) $x + 3y = 0$ (*d*) $y + 3x = 0$

31. The equation of a line parallel to x-axis at a distance of 9 units and above x-axis, is:
(*a*) $x = 9$ (*b*) $y = 9$ (*c*) $x = 9y$ (*d*) $y = 9x$

32. The equation of a line parallel to x-axis at a distance of 5 units below x-axis is:
(*a*) $x = 5$ (*b*) $x = -5$ (*c*) $y = -5$ (*d*) $y = -5x$

33. The slope of the line $2x + 3y + 5 = 0$ is:
(*a*) 2 (*b*) 3 (*c*) $\frac{3}{2}$ (*d*) $-\frac{2}{3}$

34. The slope of the line joining $A(-3, 5)$ and $B(4, 2)$ is:
(*a*) $\frac{3}{7}$ (*b*) $\frac{-3}{7}$ (*c*) $\frac{7}{3}$ (*d*) $\frac{-7}{3}$

35. The length of perpendicular from the origin to the line $3x + 4y + 5 = 0$ is:
(*a*) 5 units (*b*) $\frac{4}{3}$ units (*c*) $\frac{5}{3}$ units (*d*) 1 unit

36. The length of perpendicular from the point (4, 1) to the line $3x - 4y + 12 = 0$, is:
(a) 4 units (*b*) $\frac{4}{3}$ units (*c*) 3 units (*d*) 1 unit

37. The lines whose equations are $3x - 7y + 9 = 0$, $9x - 21y + 27 = 0$ are:
(*a*) parallel (*b*) perpendicular (*c*) coincident (*d*) intersecting

38. The lines whose equations are $x - 2y = 7$ and $4y - 2x = 13$ are:
(*a*) parallel (*b*) coincident (*c*) intersecting (*d*) perpendicular

39. Let A (3, 4), B (4, 4) and C (4, 5) be any three points. The angle between AB and AC is:
(*a*) 30° (*b*) 60° (*c*) 45° (*d*) 90°

40. The angle between the lines represented by the equations $y - \sqrt{3}x - 5 = 0$ and $\sqrt{3}\,y - x + 6 = 0$, is:
(*a*) 30° (*b*) 45° (*c*) 60° (*d*) $22\frac{1}{2}°$

41. The intercepts made by the line $3x - 2y - 6 = 0$ with x-axis and y-axis respectively, are:
(*a*) 3, 2 (*b*) 2, 3 (*c*) 2, –3 (*d*) –2, 3

42. The angle which the line joining the points $(1, \sqrt{3})$ and $(\sqrt{2}, \sqrt{6})$ makes with x-axis is:
(*a*) 30° (*b*) 45° (*c*) 60° (*d*) 90°

43. The angle between the lines represented by the equation $6x^2 + xy - y^2 = 0$, is:
(*a*) 30° (*b*) 45° (*c*) 60° (*d*) 90°

44. The centre and radius of the circle $x^2 + y^2 - 6x + 4y - 36 = 0$ are respectively:
(*a*) $(-6, 4)$, 6 (*b*) $(-3, 2)$ $\sqrt{13}$ (*c*) $(3, -2)$, 7 (*d*) None

45. The equation of a circle with centre $(-3, 2)$ and radius 7 is:
(*a*) $x^2 + y^2 - 3x + 2y + 7 = 0$ (*b*) $x^2 + y^2 + 6x - 4y - 36 = 0$
(*c*) $x^2 + y^2 - 6x + 4y = 36 = 0$ (d) None of these

46. The equation of a circle, the co-ordinates of the end points of whose diameter are (3, 4) & (– 3, –4), is:
(*a*) $x^2 + y^2 + 25 = 0$ (*b*) $x^2 + y^2 = 16$ (*c*) $x^2 + y^2 = 25$ (*d*) None

47. The equation of tangent to a circle $x^2 + y^2 = a^2$ at the point (x_1, y_1) is:

(*a*) $x_1^2 + y_1^2 = a^2$ (*b*) $xx_1 + yy_1 = 0$ (*c*) $xx_1 + yy_1 + a^2 = 0$ (*d*) $xx_1 + yy_1 = a^2$

48. The equation of tangent to the circle $x^2 + y^2 = 25$ at the point $(-3, -4)$ is:

(*a*) $3x + 4y + 25=0$ (*b*) $3x + 4y - 25 = 0$ (*c*) $4x + 3y + 25 = 0$ (*d*) $4x + 3y - 25 = 0$

49. The length of tangent to the circle $x^2 + y^2 = 4$ from a point $(-2, 3)$ is:

(*a*) 9 units (*b*) 3 units (*c*) 5 units (*d*) 1 unit

50. The length of tangent to the circle $3x^2 + 3y^2 - 7x - 6y = 12$ from the point $(6, -7)$ is:

(*a*) 9 units (*b*) 13 units (*c*) 5 units (*d*) 1 unit

51. The region specified by $x \geq 0, x + y \geq 0$ includes:

(*a*) 1st quadrant (*b*) 2nd quadrant (*c*) 3rd quadrant (*d*) 4th quadrant

ANSWERS

1. (*b*)	**2.** (*b*)	**3.** (*c*)	**4.** (*b*)	**5.** (*a*)	**6.** (*c*)	**7.** (*c*)	**8.** (*d*)	**9.** (*a*)
10. (*b*)	**11.** (*c*)	**12.** (*b*)	**13.** (*c*)	**14.** (*b*)	**15.** (*a*)	**16.** (*a*)	**17.** (*d*)	**18.** (*b*)
19. (*c*)	**20.** (*d*)	**21.** (*c*)	**22.** (*c*)	**23.** (*a*)	**24.** (*c*)	**25.** (*c*)	**26.** (*c*)	**27.** (*a*)
28. (*b*)	**29.** (*a*)	**30.** (*b*)	**31.** (*b*)	**32.** (*c*)	**33.** (*d*)	**34.** (*b*)	**35.** (*d*)	**36.** (*a*)
37. (*c*)	**38.** (*a*)	**39.** (*c*)	**40.** (*a*)	**41.** (*c*)	**42.** (*c*)	**43.** (*b*)	**44.** (*c*)	**45.** (*b*)
46. (*c*)	**47.** (*d*)	**48.** (*a*)	**49.** (*b*)	**50.** (*a*)	**51.** (*a*)			

SOLUTIONS

1. We have,

$5x + 2y = 7xy$ ...(i)

$10x + 3y = 8xy$ ...(ii)

Multiplying equation (i) by 2 and subtracting eqn (i) and (ii), we get

$$\Rightarrow y = 6xy \Rightarrow x = \frac{1}{6}$$

Put $x = \frac{1}{6}$ in eqn (i) we get

$$5 \times \frac{1}{6} + 2y = 7 \times \frac{1}{6}y \Rightarrow \frac{5}{6} = \frac{7}{6}y - 2y \Rightarrow \frac{-5}{6}y = \frac{5}{6} \Rightarrow y = -1$$

$$\therefore \text{Required point} = \left(\frac{1}{6}, -1\right)$$

2. We have

$5x + 9y = 7$ ... (i)

Put $x = -1$ in eqn (i), we get

$$5x - 1 + 9y = 7 \Rightarrow 9y = 12 \Rightarrow y = \frac{4}{3}$$

$\therefore$ it is not an integer.

When $x = -2$ in eqn (i), we get

$$5x - 2 + 9y = 7 \qquad \Rightarrow 9y = 17 \Rightarrow y = \frac{17}{9}$$

It is not an integer.

When $x = -4$ in eqn (i), we get

$-20 + 9y = 7 \Rightarrow y = 3$

It is an integer

When $x = -5$

$-25 + 9y = 7 \Rightarrow 9y = 32 \Rightarrow y = \frac{32}{9}$

It is not an integer

When $x = -13$, in eqn (i), we get

$-65 + 9y = 7 \Rightarrow y = 8$

It is an integer and when $x = -22$ is eqn (i), we get

$-110 + 9y = 7 \Rightarrow y = 113$

It is an integer

Thus, we want the multiples of 9 between -500 and 500

AP. = $-495, -486 ...., 495$

We have,

$a = -495, l = 495, d = 9$

We have,

$a_n = a + (n-1)d \Rightarrow 495 = -495 + (n-1)9 \Rightarrow 990 = -9n - 9 \Rightarrow 9n - 999 \Rightarrow n - 111$

$\therefore$ Required no. of integral values of x and $y = 111$

3. Centre of given circle is (0, 0) and its radius is 4. Any line parallel to $y = x$ is of the form $x - y + k = 0$.

Let it be tangent to the given circle.

Then, length of perp. from centre to this line must be equal to the radius of the circle.

$$\therefore \frac{0-0+k}{\sqrt{1^2+(-1)^2}} = \pm 4 \text{ i.e., } k = \pm 4\sqrt{2}$$

Hence, the required equation is $x - y \pm 4\sqrt{2} = 0$.

4. Clearly, the point on x-axis has ordinate 0 and abscissa 3. So, the point is (3, 0).

5. The point lies on y-axis, so its abscissa is 0. So, the point is (0, 6).

6. Clearly, the point is (0, – 4).

7. $AB = \sqrt{(3+2)^2+(-2+2)^2} = 5, BC = \sqrt{(-2+2)^2+(-2-1)^2} = 3;$

$CD = \sqrt{(3+2)^2+(1-1)^2} = 5, DA = \sqrt{(3-3)^2+(1+2)^2} = 3$

$AC = \sqrt{(3+2)^2+(-2-1)^2} = \sqrt{34}$ & $BD = \sqrt{(3+2)^2+(1+2)^2} = \sqrt{34}.$

$\therefore$ Opposite sides are equal and diagonals are equal. So, it is a rectangle.

8. $PQ = \sqrt{(-2-0)^2+(-2-5)^2} = \sqrt{53}, QR = \sqrt{(5+2)^2+(0+2)^2} = \sqrt{53},$

$RS = \sqrt{(7-5)^2+(7-0)^2} = \sqrt{53}, PS = \sqrt{(7-0)^2+(7-5)^2} = \sqrt{53},$

$PR = \sqrt{(5-0)^2+(5-0)^2} = \sqrt{50}, \& QS = \sqrt{(7+2)^2+(7+2)^2} = \sqrt{162}$

Thus, the sides are equal and diagonals are not equal. So, $PQRS$ is a rhombus.

9. $AB^2 = (0+5)^2 + (6-3)^2 = 34$, $BC^2 = (-5-3)^2 + (3-1)^2 = 72$

And $AC^2 = (0-3)^2 + (6-1)^2 = 34$.

$\therefore$ $AB = AC \neq BC$ & $AB^2 + AC^2 \neq BC^2$.

So, it is an isosceles triangle.

10. Clearly, $(-4, 3)$ lies in 2nd quadrant.

11. Let $P(x, 0)$ be the required point. Then,

$PA = PB \Leftrightarrow PA^2 = PB^2 \Leftrightarrow (x-7)^2 + (0-6)^2 = (x+3)^2 + (0-4)^2$

$\Leftrightarrow x^2 - 14x + 85 = x^2 + 6x + 25 \Leftrightarrow 20x = 60 \Leftrightarrow x = 3$ $\therefore$ The perpendicular point is (3, 0).

12. $\Delta = \frac{1}{2}[x_1(y_2 - y_3) + x_2(y_3 - y_1) + x_3(y_1 - y_2)]$

$= \frac{1}{2}[3(2+1) - 4(-1-8) + 5(8-2)] = 37\frac{1}{2}$ sq. units.

13. Required point is $\left(\frac{3\times 9 + 1\times 5}{3+1}, \frac{3\times 6 + 1\times(-2)}{3+1}\right) = (8, 4)$.

14. Clearly, D is the mid-point of BC.

The co-ordinates of D are $\left(\frac{3+5}{2}, \frac{-1+3}{2}\right)$ i.e., (4, 1).

$\therefore AD = \sqrt{(7-4)^2 + (-3-1)^2} = \sqrt{25} = 5$ units.

15. Let this ratio be $k : 1$. Then,

$\frac{7k - 2\times 1}{k+1} = 1 \Leftrightarrow 7k - 2 = k + 1 \Leftrightarrow k = \frac{1}{2}$.

$\therefore$ Required ratio is $\frac{1}{2} : 1$, i.e., 1 : 2.

16. Let this ratio be $k : 1$. Clearly, the ordinate of a point lying on x-axis must be zero.

$\therefore$ $\frac{6k + 1\times(-3)}{k+1} = 0 \Leftrightarrow 6k = 3 \Leftrightarrow k = \frac{1}{2}$.

$\therefore$ Required ratio is $\frac{1}{2} : 1 = 1 : 2$.

17. Let the ratio be $k : 1$.

The abscissa of a point lying on y-axis is zero.

$\therefore$ $\frac{8k - 4\times 1}{k+1} = 0 \Leftrightarrow 8k - 4 = 0 \Leftrightarrow k = \frac{1}{2}$.

So, the required ratio is $\frac{1}{2} : 1$ *i.e.,* 1 : 2.

18. Clearly, $\frac{b + b + 4}{2} = 1 \Leftrightarrow 2b + 4 = 2 \Leftrightarrow b = -1$.

19. Clearly, area of $\Delta\, ABC$ must be zero

$\therefore$ $\frac{1}{2}[2(k-7) + 5(7-3) + 6(3-k)] = 0$

$\Leftrightarrow -4k + 24 = 0 \Leftrightarrow k = 6$.

20. The co-ordinates of the centroid of $\Delta\, ABC$ are: $\left(\dfrac{-1+5+8}{3}, \dfrac{0-2+2}{3}\right) = (4, 0)$.

21. Let the co-ordinates of C be (x, y). Then,

$\dfrac{-1+5+x}{3} = 0$ and $\dfrac{4+2+y}{3} = -3$ *i.e.*, $x = -4$ & $y = -15$.

Hence, the co-ordinates of C are $(-4, -15)$.

22. Area of $\Delta\, ABC = \frac{1}{2}[1.(-3-2)+7\cdot(2-1)+12\cdot(1+3)] = 25$ sq. units.

Area of $\Delta\, BCD = \frac{1}{2}[7\cdot(2-21)+12\cdot(21+3) \quad +7\cdot(-3-2)] = 60$ sq. units

$\therefore$ Area of Quad. $ABCD$ = ar ($\Delta\, ABC$) + ar ($\Delta\, BCD$) = 85 sq. units.

23. The required point is clearly the centroid.

Its co-ordinates are $\left(\dfrac{-2-1+5}{3}, \dfrac{2-3+7}{3}\right) = \left(\dfrac{2}{3}, 2\right)$.

24. Required distance $= \sqrt{(4-0)^2+(-3-0)^2} = 5$ units.

25. Putting $y = 4$ in $x^2 + y^2 = 25$, we get $x = \pm 3$.

$\therefore$ The required points are (3, 4) and (– 3, 4).

26. On solving $3x + 5y = 4$, $2x + 3y = 2$ we get $x = -2$, $y = 2$. So, the required point is (– 2, 2).

27. The equation of x–axis is $y = 0$;

On solving $2x + 3y = 6$ & $y = 0$, we get $x = 3$ & $y = 0$. So, the required point is (3, 0).

28. The equation of the y-axis is $x = 0$.

Solving $3x - 7y = 10$ & $x = 0$, we get the point $\left(0, \dfrac{-10}{7}\right)$.

$\therefore$ The required point is $\left(0, \dfrac{-10}{7}\right)$.

29. Clearly, the equation of the line is $x = 4$.

30. Clearly, the equation of the line is $x = -3$.

31. Clearly, the equation of the line is $y = 9$.

32. Clearly, the equation of the line is $y = -5$.

33. $2x + 3y + 5 = 0 \Rightarrow 3y = -2x - 5 \Rightarrow y = -\dfrac{2}{3}x - \dfrac{5}{3}$.

$\therefore$ Slope of the line is $-\dfrac{2}{3}$.

34. Slope $= \dfrac{y_2 - y_1}{x_2 - x_1} = \dfrac{2-5}{4+3} = \dfrac{-3}{7}$.

35. Length of perpendicular $= \dfrac{3\times 0+4\times 0+5}{\sqrt{3^2+4^2}} = \dfrac{5}{5} = 1$ unit.

36. Length of perpendicular $= \dfrac{3\times 4-4\times 1+12}{\sqrt{3^2+(-4)^2}} = 4$ units

37. Here $\frac{a_1}{a_2} = \frac{3}{9} = \frac{1}{3}, \frac{b_1}{b_2} = \frac{-7}{-21} = \frac{1}{3} \& \frac{c_1}{c_2} = \frac{9}{27} = \frac{1}{3}$.

$\therefore \frac{a_1}{a_2} = \frac{b_1}{b_2} = \frac{c_1}{c_2}$. So, the given lines are coincident.

38. Given equations are: $x - 2y - 7 = 0$ & $2x - 4y + 13 = 0$.

$\frac{a_1}{a_2} = \frac{1}{2}, \frac{b_1}{b_2} = \frac{-2}{-4} = \frac{1}{2} \& \frac{c_1}{c_2} = \frac{-7}{13}$. So, $\frac{a_1}{a_2} = \frac{b_1}{b_2} \neq \frac{c_1}{c_2}$.

Hence, the given lines are parallel.

39. Slope of AB is $m_i = \frac{4-4}{4-3} = 0$, slope of AC is $m_2 = \frac{5-4}{4-3} = 1$.

$\therefore \tan\theta = \left|\frac{m_1 - m_2}{1 + m_1 m_2}\right| = \left|\frac{0-1}{1+0}\right| = 1$. So, $\theta = 45°$.

40. $y - \sqrt{3}x - 5 = 0 \Rightarrow y = \sqrt{3}x + 5 \Rightarrow m_1 = \sqrt{3}$.

$\sqrt{3}y - x + 6 = 0 \Rightarrow y = \frac{1}{\sqrt{3}}x - 2\sqrt{3} \Rightarrow m_2 = \frac{1}{\sqrt{3}}$.

$$\therefore \tan\theta = \frac{m_1 - m_2}{1 + m_1 m_2} = \frac{\sqrt{3} - \frac{1}{\sqrt{3}}}{1 + \sqrt{3} \times \frac{1}{\sqrt{3}}} = \frac{1}{\sqrt{3}}. \text{ So, } \theta = 30°.$$

41. $3x - 2y - 6 = 0 \Rightarrow 3x - 2y = 6 \Rightarrow \frac{x}{2} - \frac{y}{3} = 1 \Rightarrow \frac{x}{2} + \frac{y}{-3} = 1$.

So, the required intercepts are 2, –3.

42. The slope of the line is $\frac{\sqrt{6} - \sqrt{3}}{\sqrt{2} - 1} = \sqrt{3}. \therefore \tan\theta = \sqrt{3}$ or $\theta = 60°$.

43. The equation is of the form $ax^2 + 2hxy + by^2 = 0$, where $a = 6$, $b = -1$ and $h = \frac{1}{2}$.

$$\therefore \quad \tan\theta = \frac{2\sqrt{h^2 - ab}}{a+b} = \frac{2\sqrt{\frac{1}{4} + 6}}{6-1} = 1. \text{ So, } \theta = 45°.$$

44. The given equation is of the form $x^2 + y^2 + 2gx + 2fy + c = 0$, where $g = 3, f = 2$ and $c = -36$.

$\therefore$ Centre is $(-g, -f) = (3, -2)$.

Radius is $\sqrt{g^2 + f^2 - c} = \sqrt{9 + 4 + 36} = 7$.

45. The required equation is $(x + 3)^2 + (y - 2)^2 = 7^2$. *i.e.*, $x^2 + y^2 + 6x - 4y - 36 = 0$.

46. The required equation is $(x - 3)(x + 3) + (y - 4)(y + 4) = 0$ *i.e.*, $x^2 + y^2 = 25$.

47. Clearly, the equation of the tangent is $xx_1 + yy_1 = a^2$.

48. Required equation is $-3x + (-4)y = 25$ or $3x + 4y + 25 = 0$.

49. Length of tangent = $\sqrt{x_1^2 + y_1^2 - r^2} = \sqrt{(-2)^2 + 3^2 - 4} = 3$ units.

50. Given equation is $x^2 + y^2 - \frac{7}{3}x - 2y - 4 = 0$.

Here $2g = -\frac{7}{3}, 2f = -2, c = -4$

Length of tangent $= \sqrt{x_1^2 + y_1^2 + 2gx_1 + 2fy_1 + c} = \sqrt{6^2 + (-7)^2 - \frac{7}{3} \times 6 - 2 \times (-7) - 4} = \sqrt{81} = 9$.

51. Consider the line $x + y = 0$.
Clearly, $(x = 0, y = 0)$ & $(x = 1, y = -1)$ satisfy it.

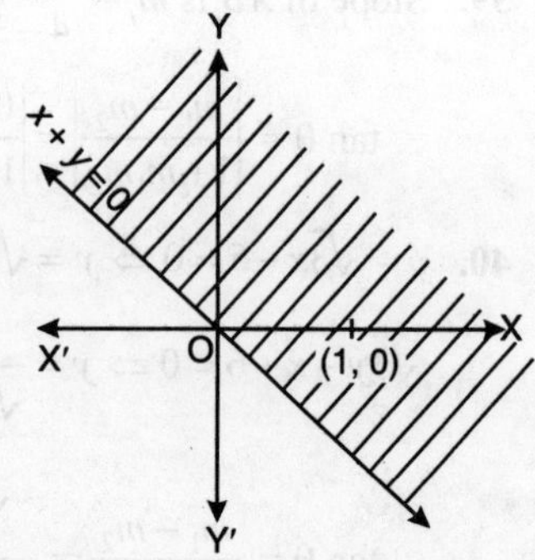

Plot the points (0, 0) & (1, – 1) and join them to get a line, which is the graph of $x + y = 0$.

Consider any point, say (1, 0). Clearly, (1, 0) satisfies the inequality, $x + y \geq 0$. Shade the part of the plane containing (1,0).

When $x \geq 0$, clearly, the first quadrant will be included as a whole.

UNIT 7

STATISTICS

Collection and Tabulation of Statistical Data; Graphical Representation–Frequency Polygons, Histograms, Bar charts, Pie-charts etc. Measure of Central Tendency: Mean, Mode, Median and Standard Deviation.

Data interpretation.

35 CHAPTER

Statistical Data

Highlights: Frequency Distribution, Mean, Mode, Median and Standard Deviation.

Definitions and Formulae:

1. Tabulation: *The arrangement of the raw data under various heads in the form of a table is called tabulation.*

Range of a Data: *The difference between the largest and the smallest observations is known as the range.*

Frequency: *Number of observations falling in a particular class is called the frequency of that class.*

Cumulative Frequency: *The cumulative frequency of a particular class is the sum of the frequencies of this class and of those prior to it.*

2. Frequency Distribution: *The distribution of frequency in various classes is known as frequency distribution.*

Types of Frequency Distribution:

(*i*) **Continuous Form:** When classes are of the form *1-11, 11-21, 21-31, etc., the distribution is said to be in continuous form.*

Note that class 1-11 means 1 or more but less than 11. Note that 11 does not belong to the class 1-11.

In class 1-1 1, we have *lower limit* = 1 & *upper limit* = 11.

Class size = (*Upper limit*) – (*Lower limit*).

Class mark = $\frac{1}{2}$ (*Upper limit* + *Lower limit*).

Ex. 1. ***Consider the table given below:***

Age (in years)	Number of Employees (Frequency)	Cumulative Frequency
30-35	5	5
35-40	7	12
40-45	6	18
45-50	9	27
50-55	4	31

Find: (*i*) Class-size (*ii*) Mid-value of 35 – 40

(*iii*) *Lower limit and upper limit of 40-45*

(*iv*) *How many employees are less than 50 years of age ?*

(*v*) *The frequency of 50 – 55.*

Sol. (*i*) Class size = (35 – 30) = 5.

(*ii*) Mid-value of 35 – 40 is $\left(\frac{35+40}{2}\right) = 37.5$.

(*iii*) Lower limit = 40, Upper limit = 45.

(*iv*) Clearly, 27 employees are of age less than 50 years.

(*v*) Frequency of 50 – 55 is 4.

3. Arithmetic Mean, Geometric Mean and Harmonic Mean:

(*i*) For any two numbers *a* and *b*, we have:

$$\text{A.M.} = \frac{1}{2}(a+b),\ \text{G.M.} = \sqrt{ab} \text{ and H.M.} = \frac{2ab}{a+b}.$$

(*ii*) For *n* observations $x_1, x_2, ..., x_n$, we have:

$$\text{A.M.} = \left(\frac{x_1 + x_2 + ... + x_n}{n}\right);\ \text{G.M.} = (x_1 x_2 \ldots xn)^{1/n}$$

$$\text{H.M.} = \frac{n}{\left(\frac{1}{x_1}+\frac{1}{x_2}+...+\frac{1}{x_n}\right)}.$$

4. Results on A.M.:

(*i*) Let there be *n* observations $x_1, x_2, ..., x_n$, then:

$$\text{A.M.} = \left(\frac{x_1 + x_2 + ... + x_n}{n}\right).$$

(*ii*) When some terms have frequency more than 1, then:

$$\text{A.M.} = \frac{\Sigma(f \times x)}{\Sigma f} = \frac{\Sigma(f \times x)}{n}.$$

(*iii*) When data is classified, then:

$\text{A.M.} = \left[A + \frac{\Sigma(f \times d)}{\Sigma f}\right]$, where *A* is the assumed mean, *d* is deviation of the term from *A* whose frequency is *f*.

(*iv*) If observations $x_1, x_2, ..., x_n$ have weights $w_1, w_2, ..., w_n$ respectively, then:

$$\text{Weighted A.M.} = \left(\frac{w_1 x_1 + w_2 x_2 + ... + w_n x_n}{w_1 + w_2 + ... + w_n}\right).$$

(*v*) Let A.M. of $x_1, x_2, ..., x_n$ be *M*, then:

(*a*) A.M. of $(x_1 + C), (x_2 + C), ..., (x_n + C)$ is $(M + C)$.

(*b*) A.M. of $Cx_1, Cx_2, ..., Cx_n$ is $C \times M$.

(c) A.M. of $\frac{x_1}{C}, \frac{x_2}{C}, ..., \frac{x_n}{C}$ is $\frac{M}{C}$, where $C \neq 0$.

5. Results on Mode:

(*i*) *In an ungrouped data, the most occurring term form gives the mode.*

(*ii*) *If some terms have frequencies more than 1, then the term having largest frequency is the mode.*

(*iii*) In a grouped data, we have:

Mode $= L_1 + \dfrac{(f - f_1)}{(2f - f_1 - f_2)} \times (L_2 - L_1)$, where L is the lower limit of modal class. L_2 is the upper limit of the modal class, f is the frequency of modal class and f_1, f_2 are the respective frequencies of the preceding and succeeding classes of the modal class.

6. Results on Median:

(*i*) In an ungrouped data, the data is arranged in ascending or descending order of magnitude.

If n is odd, then median $= \left(\dfrac{n+1}{2}\right)$ th term.

If n is even, then median $= \dfrac{1}{2}\left[\left(\dfrac{n}{2}\right)\text{th term} + \left(\dfrac{n}{2}+1\right)\text{th term}\right]$.

(*ii*) Median $= L_1 + \dfrac{(L_2 - L_1)}{f} \times \left(\dfrac{n}{2} - C\right)$, where L_1 and L_2 are respectively the lower limit and the upper limit of the median class ; f is the frequency of this class ; n is the sum of frequencies and C is the cumulative frequency of the class preceding the median class.

7. Relation Between Mean, Mode and Median:

We have: 3 (Median) – 2 (A.M.) = Mode.

8. Mean Deviation:

(*i*) Mean deviation $= \dfrac{\Sigma|\delta|}{n}$, where $\delta = |x - M|$, were M is the A.M.

(*ii*) Mean deviation $= \dfrac{\Sigma f \times |\delta|}{\Sigma f}$

9. Results on Standard Deviation:

(*i*) S.D. is given, by $\sigma = \sqrt{\dfrac{\Sigma|\delta|^2}{n}}$. (*ii*) $\sigma = \sqrt{\dfrac{\Sigma f \times \delta^2}{n} - \left(\dfrac{\Sigma f\delta}{n}\right)^2}$, where $\delta = |x - M|$.

(*iii*) S.D. $= \sqrt{Variance}$.

10. Results on Diagrams:

Histogram: It is a graphical representation of a frequency distribution in the form of rectangles with class intervals as bases and the corresponding frequencies as heights, there being no gap between any two successive rectangles.

Frequency Polygon: When a grouped frequency distribution with equal class intervals is being given, we represent the class-mark on x-axis and the frequencies on y-axis.

We complete the diagram in the form of a polygon by taking two more classes one at the beginning and the other at the end, each with frequency 0.

Frequency Polygon from Histogram: By joining the mid-points of tops of rectangles of histogram, we obtain the frequency polygon.

Ogive (Cumulative Frequency Curve): We mark the upper class limits along x-axis and cumulative frequencies along y-axis. Plot these points and join them to get a curve, called an ogive.

SOLVED PROBLEMS

Ex. 1. ***Find (i) A.M. of 52, 48, 56, 42 & 35. (ii) G.M. of 16, 24, 27, & 32. (iii) H.M. of 4, 8, 9.***

Sol. (*i*) $\text{A.M.} = \left(\dfrac{52+48+56+42+35}{5}\right) = \dfrac{233}{5} = 46.6.$

(*ii*) $\text{G.M.} = (16 \times 24 \times 27 \times 32)^{1/4} = (2^{12} \times 3^4)^{1/4} = (2^3 \times 3) = 24.$

(*iii*) $\text{H.M.} = \dfrac{3}{\left(\dfrac{1}{4}+\dfrac{1}{8}+\dfrac{1}{9}\right)} = \left(\dfrac{3\times 72}{35}\right) = 6.17.$

Ex. 2. ***The weights of 40 students are given below:***

Weight	35	36	40	45
Number of Students	12	15	7	6

Find the mean weight.

Sol.

Weight (in kg) x x	Frequency f f	$f \times x$
35	12	420
36	15	540
40	7	280
45	6	270
	$\Sigma f = 40$	$\Sigma(f \times x) = 1510$

$\therefore \text{Mean} = \dfrac{\Sigma(f \times x)}{\Sigma f} = \dfrac{1510}{40} = 37.75 \text{ kg}.$

Ex. 3. ***Find A.M. from the data given below:***

Class interval	5 – 15	15 – 25	25 – 35	35 – 45	45 – 55
Frequency	3	7	12	20	8

Sol. We have:

Class	*Mid-value* x	*Frequency* f	*Deviation* $d = x - A$	$f \times d$
5 – 15	10	3	– 20	–60
15 – 25	20	7	– 10	– 70
25 – 35	30 = A	12	0	0
35 – 45	40	20	10	200
45 – 55	50	20	20	160
		$\Sigma f = 50$		$\Sigma(f \times d) = 230$

$\therefore \text{A.M.} = A + \dfrac{\Sigma(f \times d)}{\Sigma f} = \left(30 + \dfrac{230}{50}\right) = 34.6.$

Ex. 4. ***Marks obtained by a student in 5 subjects are given as History 23, Civics 35, Economics 46, Hindi 30 and English 32. If the weightage given to these subjects be 1, 1, 2, 3, 5 respectively, then find the weighted mean.***

Sol. Weighted Mean = $\left(\dfrac{23\times1+35\times1+46\times2+30\times3+32\times5}{1+1+2+3+5}\right)=\dfrac{400}{12}=33.3$

Ex. 5. ***Find the mode of the following frequency distribution:***

Term	25	35	45	55	65
Frequency	4	9	16	15	6

Sol. Since 45 occurs most often, so mode = 45.

Ex. 6. ***Find the mode from the data given below:***

Marks obtained	0 – 5	5 – 10	10 – 15	15 – 20	20 – 25	25 – 30
Number of students	18	20	25	30	16	14

Sol. The maximum frequency is 30. So, Modal class is 15 – 20.

$\therefore L_1 = 15, L_2 = 20, f = 30, f_1 = 25$ and $f_2 = 16$.

Mode = $L_1+\dfrac{f-f_1}{2f-f_1-f_2}\times(L_2-L_1)=15+\dfrac{(30-25)}{(60-25-16)}\times(20-15)=16.3.$

Ex. 7. ***Find the median of 30, 17, 43, 39, 18, 40.***

Sol. The observations in ascending order are: 17, 18, 30, 39, 40, 43.

Here $n = 6$. So, median = $\dfrac{1}{2}$ (3rd + 4th terms) = $\dfrac{1}{2}$ (30 + 39) = 34.5.

Ex. 8. ***Find the median of the frequency distribution:***

Marks obtained	Number of Students	Cumulative Frequency
15 – 20	6	6
20 – 25	7	13
25 – 30	9	22
30 – 35	8	30
35 – 40	10	40

Sol. Here $n = 40$, So, $\dfrac{n}{2} = 20$.

Now, 20 lies in the class 25 – 30.

So, Median Class is 25 – 30.

$\therefore L_1 = 25, L_2 = 30, n = 40, C = 13$ & $f = 9$.

$\therefore$ Median = $L_1+\dfrac{(L_2-L_1)}{f}\times\left(\dfrac{n}{2}-C\right)=25+\dfrac{(30-25)}{9}\times(20-13)=\left(25+\dfrac{35}{9}\right)=28.88.$

Ex. 9. ***Find the mean deviation of 10, 9, 21, 16, 24.***

Sol. The A.M. of given data is, $M=\left(\dfrac{80}{5}\right)=16$.

$\therefore \Sigma|\delta| = |10-16|+|9-16|+|21-16|+|16-16|+|24-16| = (6+7+5+0+8) = 26.$

$\therefore$ Mean deviation = $\dfrac{\Sigma|\delta|}{n}=\dfrac{26}{5}=5.2.$

Ex. 10. ***Find the mean deviation of the frequency distribution:***

Class	0 – 10	10 – 20	20 – 30	30 – 40	40 – 50
Frequency	6	10	9	4	11

Sol. We have:

Classes	Mid-value x	Frequency f	$f \times x$	$\lvert\delta\rvert = \lvert x - M\rvert$	$f \times \lvert\delta\rvert$
0 –10	5	6	30	21	126
10 – 20	15	10	150	9	90
20 – 30	25	9	225	1	9
30 –40	35	4	140	9	36
40 –50	45	11	495	19	209
		$\Sigma f = 40$	$\Sigma fx = 1040$		$\Sigma\lvert\delta\rvert = 470$
			$M = \frac{1040}{40} = 26$		

$\therefore$ Mean Deviation $= \frac{470}{26} = 18.07.$

Ex. 11. ***Find the standard deviation of 35, 40, 42, 36, 27.***

Sol. $M = \frac{35+40+42+36+27}{5} = \frac{180}{5} = 36.$

$\Sigma\delta^2 = |35-36|^2 + |40-36|^2 + |42-36|^2 + |36-36|^2 + |27-36|^2, = (1+16+36+0+81) = 134.$

$\therefore \sigma = \frac{\Sigma\delta^2}{n} = \frac{134}{5} = 26.8.$

EXERCISE 35

1. A pie chart is drawn to show the areas in millions of square kms. of several continents. The area 11.7 sq. km. of Africa is shown by a sector subtending an angle of 82°. If the subtended angle corresponding to North America be 66°, its area is:

(*a*) 10.8 km² (*b*) 9.4 km² (*c*) 8.8 km² (*d*) 5.6 km²

2. In a histogram the area of each rectangle is proportional to:

(*a*) the class mark of the corresponding class interval

(*b*) the class size of the corresponding class interval

(*c*) frequency of the corresponding class interval

(*d*) cumulative frequency of the corresponding class interval:

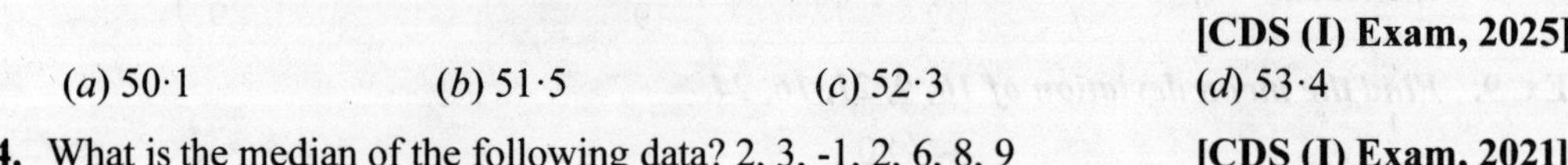

3. The arithmetic mean of 100 observations was 50. Later on, it was found that two observations were misread as 82 and 6 instead of 182 and 56. What is the value of the corrected mean?

[CDS (I) Exam, 2025]

(*a*) 50·1 (*b*) 51·5 (*c*) 52·3 (*d*) 53·4

4. What is the median of the following data? 2, 3, -1, 2, 6, 8, 9 **[CDS (I) Exam, 2021]**

(*a*) 2 (*b*) 3 (*c*) 4 (*d*) 5

5. What is the arithmetic mean of the first ten composite numbers? **[CDS (I) Exam, 2021]**

(*a*) 8.5 (*b*) 9.5 (*c*) 10.2 (*d*) 11.2

Directions (Q. 6-7): Consider the following for the next two question that follow:
Collect all the sequences of five consecutive integers such that their product is equal to one of these integers. Let X be the collection of all possible such sequences. Let P be the smallest integer and Q be the largest integer occurring in these sequences. **[CDS (II) Exam, 2021]**

6. How many such sequences of five consecutive integers are possible ?
(*a*) One (*b*) Two (*c*) Three (*d*) Five

7. What is the arithmetic mean of P and Q?
(*a*) 0 (*b*) 1 (*c*) 2
(*d*) Cannot be determined due to in- sufficient data

8. What is the median of the data 3, 5, 9, 4, 6, 11, 18? **[CDS (I) Exam, 2020]**
(*a*) 6 (*b*) 6.5 (*c*) 7 (*d*) 7.5

9. For two observations, the sum is S and product is P. What is the harmonic mean of these two observations? **[CDS (I) Exam, 2020]**
(*a*) $\frac{2S}{P}$ (*b*) $\frac{S}{(2P)}$ (*c*) $\frac{2P}{S}$ (*d*) $\frac{P}{(2S)}$

10. If the sizes in a frequency distribution are given in an ascending order of magnitude, then the median *M* is given by:

(*a*) $M = l + \frac{\left(\frac{N}{2} + C\right) \times i}{f}$ (*b*) $M = l + \frac{\left(\frac{N}{2} - C\right) \times i}{f}$

(*c*) $M = l - \frac{\left(\frac{N}{2} - C\right) \times i}{f}$ (*d*) $M = l - \frac{\left(\frac{N}{2} + C\right) \times i}{f}$

11. If the sizes in a frequency distribution are given in descending order of magnitude, then the median is given by:

(*a*) $M = l' - \frac{\left(\frac{N}{2} - C\right) \times i}{f}$ (*b*) $M = l' - \frac{\left(\frac{N}{2} + C\right) \times i}{f}$

(*c*) $M = l' - \frac{\left(\frac{N}{2} - C'\right) \times i}{f}$ (*d*) $M = l' - \frac{\left(\frac{N}{2} + C\right) \times i}{f}$

12. Consider the following data : 110, 41, 43, 95, 127, 99, 61, 92, 71, 93, 110, 36. If 93 is replaced by 94, then consider the following statements :
1. The difference between new median and old median is 1.
2. The difference between new mean and old mean is less than 0:1.
3. The difference between new mode and old mode is zero.

Which of the statements given above are correct ? **[CDS (I) Exam, 2024]**
(*a*) 1 and 2 only (*b*) 2 and 3 only (*c*) 1 and 3 only (*d*) 1, 2 and 3

13. The ages of 7 family members are 2, 5, 12, 18, 38, 40 and 60 years respectively. After 5 years a new member aged x years is added. If the mean age of the family now goes up by 1-5 years, then what is the value of *x* ? **[CDS (II) Exam, 2020]**
(*a*)1 (*b*) 2 (*c*) 3 (*d*) 4

14. Fifteen candidates appeared in an examination. The marks of the candidates who passed in the examination are 9, 6, 7, 8, 8, 9, 6, 5, 4 and 7. What is the median of marks of all the fifteen candidates? **[CDS (II) Exam, 2020]**

(*a*) 6 (*b*) 6.5 (*c*) 7 (*d*) 7.5

15. If the yield (in gm) of barley from 7 plots of size one square yard each, were found to be 180, 191, 175, 111, 154, 141 and 176, then what is the median yield? **[CDS (II) Exam, 2020]**

(*a*) 111 gm (*b*) 154 gm (*c*) 175 gm (*d*) 176 gm

16. Which one of the following measures of central tendency will be used to determine the average size of the shoe sold in the shop? **[CDS (II) Exam, 2020]**

(*a*) Arithmetic mean (*b*) Geometric mean (*c*) Median (*d*) Mode

17. Consider the following grouped frequency distribution : **[CDS (I) Exam, 2020]**

x	f
0 –10	8
10 – 20	12
20 – 30	10
30 – 40	p
40 – 50	9

If the mean of the above data is 25-2, then what is the value of p?

(*a*) 9 (*b*) 10 (*c*) 11 (*d*) 12

18. If the mean of m observations out of n observations is n and the mean of remaining observations is m, then what is the mean of all n observations? **[CDS (I) Exam, 2020]**

(*a*) $2m-\frac{m^2}{n}$ (*b*) $2m+\frac{m^2}{n}$ (*c*) $m-\frac{m^2}{n}$ (*d*) $m+\frac{m^2}{n}$

Directions (Q. 19-20)

Read the following frequency distribution for two series of observations and answer the two question that follow:

Class interval	***Frequency***	
	Series-II	Series-I
10-20	20	4
20-30	15	8
30-40	10	4
40-50	x	$2x$
50-60	y	y
Total	100	100

19. What is the mean of frequency distribution of Series-I ? **[CDS (II) Exam, 2019]**

(*a*) 33.6 (*b*) 35.6 (*c*) 37.6 (*d*) 39.6

20. What is the mode of the frequency distribution of Series-II ? **[CDS (II) Exam, 2019]**

(*a*) 26 (*b*) 36 (*c*) 46 (*d*) 56

21. The arithmetic mean of two numbers is 10 and their geometric mean is 8. What are the two numbers? **[CDS (I) Exam, 2018]**

(*a*) 15, 5 (*b*) 12, 8 (*c*) 16, 4 (*d*) 18, 2

22. The arithmetic mean of 11 observations is 11. The arithmetic mean of the first 6 observation is 10-5 and the arithmetic mean of the last 6 observations is 11.5. What is the sixth observation?
(*a*) 10.0 (*b*) 10.5 (*c*) 11.0 (*d*) 11.5

23. Data on ratings of hotels in a city is measured on **[CDS (II) Exam, 2018]**
(*a*) Nominal scale (*b*) Interval scale (*c*) Ordinal scale (*d*) Ratio scale

Directions (Q. 24-27) For the next four (4) questions that follow:

Consider the following frequency distribution:

Class	Frequency	Class	Frequency
0 – 10	4	40 – 50	12
10 – 20	5	50 – 60	8
20 – 30	7	60 – 70	4
30 – 40	10		

24. What is the mean of the distribution? **[CDS (I) Exam, 2015]**
(*a*) 37.2 (*b*) 38.1 (*c*) 39.2 (*d*) 40.1

25. What is the median class? **[CDS (I) Exam, 2015]**
(*a*) 20 – 30 (*b*) 30 – 40 (*c*) 40 – 50 (*d*) 50 – 60

26. What is the median of the distribution? **[CDS (I) Exam, 2015]**
(*a*) 37 (*b*) 38 (*c*) 39 (*d*) 40

27. What is the mode of the distribution? **[CDS (I) Exam, 2015]**
(*a*) 38.33 (*b*) 40.66 (*c*) 42.66 (*d*) 43.33

28. If a variable takes discrete value $a + 4$, $a - 3.5$, $a - 2.5$, $a - 3$, $a - 3$, $a - 2$, $a + 0.5$, $a + 5$ and $a - 0.5$ where $a > 0$, then the median of the data set is **[CDS (II) Exam, 2015]**
(*a*) $a - 2.5$ (*b*) $a - 1.25$ (*c*) $a - 1.5$ (*d*) $a - 0.75$

29. If each of n numbers $x_i = i\ (i = 1, 2, 3, ...n)$ is replaced by $(i+1)\, x_i$ then the new mean is
(*a*) $\frac{n+3}{2}$ (*b*) $\frac{n(n+1)}{2}$ (*c*) $\frac{(n+1)(n+2)}{3n}$ (*d*) $\frac{(n+1)(n+2)}{3}$
[CDS (II) Exam, 2015]

30. The arithmetic mean of n numbers is M. If the sum of first $(n-1)$ terms is k, then what is the nth number?
(*a*) $M - k$ (*b*) $nM - k$ (*c*) $n(M - k)$ (*d*) $M - nk$

31. The mean and median of 5 observations are 9 and 8 respectively. If 1 is subtracted from each observation, then the new mean and the new median will respectively be
(*a*) 8 and 7
(*b*) 9 and 7
(*c*) 8 and 9
(*d*) Cannot be determined due to insufficient data **[CDS (I) Exam, 2016]**

32. The age distribution of 40 children is as follows:

Age (in years)	5–6	6–7	7–8	8–9	9–10	10–11
No. of children	4	7	9	12	6	12

Consider the following statements in respect of the above frequency distribution:

1. The median of the age distribution is 7 years.

2. 70% of the children are in the age group 6–9 years.

3. The modal age of the children is 8 years.

Which of the above statements are correct? **[CDS (I) Exam, 2016]**

(*a*) 1 and 2 only (*b*) 2 and 3 only (*c*) 1 and 3 only (*d*) 1, 2 and 3

33. Suppose $x_{i=\lambda}{}^{i}$ for $0 \leq i \leq 10$, where $\lambda > 1$. Which one of the following is correct?

(*a*) AM < Median (*b*) GM < Median

(c) GM = Median (*d*) AM = Median **[CDS (I) Exam, 2016]**

34. Suppose $x_i, = \frac{1}{i}$ for $i = 1, 2, 3, ..., 11$. Which one of the following is not correct?

(*a*) AM >1/6 (*b*) GM > 1/6

(*c*) HM > 1/6 (*d*) Median = HM **[CDS (I) Exam, 2016]**

35. For $x > 0$ if a variable takes discrete values **[CDS (II) Exam, 2016]**

$x + 4, x - 3.5, x - 2.5, x - 3, x - 2, x + 0.5, x - 0.5, x + 5,$ then what is the median?

(*a*) $x - 1.25$ (*b*) $x - 0.5$ (*c*) $x + 0.5$ (*d*) $x + 1.25$

36. The median of set of 9 distinct observations is 20.5. If each of the largest 4 observations of the set is increased by 2, then the median of the new set **[CDS (II) Exam, 2016]**

(*a*) is increased by 2 (*b*) is decreased by 2

(*c*) is two times the original median (*d*) remains the same as that of original set

37. The mean of 20 observations is 17. On checking it was found that the two observations were wrongly copied as 3 and 6. If wrong observations are replaced by correct values 8 and 9, then what is the correct mean? **[CDS (II) Exam, 2016]**

(*a*) 17.4 (*b*) 16.6 (*c*) 15.8 (*d*) 14.2

38. The sum of two positive numbers is 40. If the GM of these two numbers is lower than their AM by 20%, then what is the difference between the two numbers ? **[CDS (I) Exam, 2024]**

(*a*) 12 (*b*) 18 (*c*) 24 (*d*) 28

39. Consider the following statements: **[CDS (II) Exam, 2016]**

1. The classes of type 15 – 19, 20 – 24, 25 – 29 etc. are exclusive classes.

2. The classes of type 15 – 20, 20 – 25, 25 – 30 etc. are inclusive classes.

Which of the above statements is/are correct?

(*a*) 1 only (*b*) 2 only (*c*) Both 1 and 2 (*d*) Neither 1 nor 2

40. Suppose the class interval 10 – 15 has frequency 30, then what is the frequency density of this class interval? **[CDS (II) Exam, 2016]**

(*a*) 2 (*b*) 3 (*c*) 5 (*d*) 6

41. Consider the following distribution: **[CDS (I) Exam, 2017]**

Class	Frequency
0–20	17
20–40	28
40–60	32
60–80	f
80–100	19

If the mean of the above distribution is 50, what is the value of f?

(*a*) 24 (*b*) 34 (*c*) 56 (*d*) 96

42. In an asymmetrical distribution, if the mean and median of the distribution are 270 and 220 respectively, then the mode of the data is **[CDS (I) Exam, 2017]**

(*a*) 120 (*b*) 220 (*c*) 280 (*d*) 370

Consider the following for the next two (02) items that follow:

The frequency distribution of marks obtained by students in an English examination is given below:

Marks obtained	Number of Students
Below 40	50
Below 50	125
Below 60	210
Below 70	315
Below 80	350

[CDS (I) Exam, 2025]

43. What is the number of students who scored between 60 and 70 marks?

(*a*) 105 (*b*) 110 (*c*) 205 (*d*) 210

44. What is the number of students who scored more than 50 marks?

(*a*) 100 (*b*) 125 (*c*) 200 (*d*) 225

45. Let the positive numbers $a_1, a_2, a_3, ..., a_{3n}$ be in GP. If P is the GM of $a_1, a_2, a_3, .. a_n$, and Q is the GM of $a_{n+1}, a_{n+2}, a_{n+3} ..., a_{3n}$, then what is the GM of $3n$ numbers ? **[CDS (I) Exam, 2024]**

(*a*) P^2Q (*b*) PQ^2 (*c*) $\sqrt{PQ}$ (*d*) $P^{1/3}\ Q^{2/3}$

46. If the arithmetic mean of 7, 5, 13, x and 9 be 10, then the value of x is:

(*a*) 10 (*b*) 12 (*c*) 14 (*d*) 16

47. What is the geometric mean of 3, 9, 27, 81, 243, 729, 2187? **[CDS (II) Exam, 2024]**

(*a*) 81 (*b*) 105 (*c*) 144 (*d*) 243

Consider the following for the next five (05) items that follow:

The frequency distribution of 205 observations on X is given below: **[CDS (I) Exam, 2025]**

X	3	5	6	7
Frequency	f	$f+2$	$f-3$	$f+6$

48. What is the value of f?

(*a*) 50 (*b*) 60 (*c*) 70 (*d*) 80

49. What is the median of the frequency distribution?

(*a*) 3 (*b*) 5 (*c*) 6

(*d*) It cannot be determined from the given data

50. What is the mode of the frequency distribution?

(*a*) 5 (*b*) 6 (*c*) 7

(*d*) It cannot be determined from the given data

51. What is the most appropriate graphical representation of the given frequency distribution of X?

(*a*) Bar diagram (*b*) Histogram

(*c*) Frequency polygon (*d*) Pie Chart

52. What is the mean of the frequency distribution?

(*a*) 3.29 (*b*) 4.29 (*c*) 5.29 (*d*) 6.29

53. The following is the data of wages per day:

5,4, 7,5, 8, 8, 8,5,7, 9,5, 7, 9, 10, 8

The mode of the data is:

(*a*) 7 (*b*) 5 (*c*) 8 (*d*) 10

54. If the geometric mean of x, 16, 50, be 20, then the value of x is:

(*a*) 40 (*b*) 20 (*c*) 10 (*d*) 4

55. The median of a given frequency distribution is found graphically with the help of:

(*a*) Histogram (*b*) Pie Chart (*c*) Frequency Curve (*d*) Ogive

56. A frequency distribution can be presented graphically by a:

(*a*) Pie diagram (*b*) Histogram (*c*) Pictogram (*d*) Cartogram

57. The correct formula is:

(*a*) Median + A.M. = 2 × Mode (*b*) Median – A.M. = Mode

(*c*) 3 (A.M.) – 2 (Median) = Mode (*d*) 3 (Median) – 2 (A.M.) = Mode

Directions (Q. 58-59) For the next two questions that follow:

Consider the following data for the next two items that follow : **[CDS (I) Exam, 2023]**

Class	0 – 30	30 – 60	60 – 90	90 – 120
Frequency	4	5	7	4

58. What is the mode of the distribution ?

(*a*) 60 (*b*) 72 (*c*) 75 (*d*) 80

59. If the median (P) and mode (Q) satisfy the relation 7(Q – P) = 9R, then what is the value of *R*?

(*a*) 6 (*b*) 5 (*c*) 3 (*d*) 1

Consider the following data for the next two items that follow :

Class	40 – 50	50 – 60	60 – 70	70 – 80
Frequency	4	3	1	2

60. What is the mean of the distribution? **[CDS (I) Exam, 2023]**

(*a*) 51 (*b*) 52 (*c*) 54 (*d*) 56

61. If *M* is the median, then what is the value of 3M? **[CDS (I) Exam, 2023]**

(*a*) $53\frac{1}{3}$ (*b*) 60 (*c*) 160 (*d*) 180

62. The sum of deviations of *n* numbers from 10 and 20 are *a*, *b* respectively. If $\frac{b}{a} = -4$, *a* then what is the mean of these *n* numbers ? **[CDS (II) Exam, 2023]**

(*a*) 12 (*b*) 14 (*c*) 16 (*d*) 18

Consider the following for the next two (02) items that follow:

A frequency distribution is as follows: **[CDS (II) Exam, 2024]**

Marks	**18–26**	**27–35**	**36–44**	**45–53**	**54–62**	**63–71**	**72–80**
Number of students	5	7	10	15	8	3	2

63. What is the median of the distribution?

(*a*) 44.9 (*b*) 45.5 (*c*) 45.9 (*d*) 46.3

64. What is the mode of the distribution?

(*a*) 47.25 (*b*) 47.75 (*c*) 48.25 (*d*) 48.75

ANSWERS

1. (*b*)	**2.** (*c*)	**3.** (*b*)	**4.** (*b*)	**5.** (*d*)	**6.** (*d*)	**7.** (*a*)	**8.** (*a*)	**9.** (*c*)
10. (*b*)	**11.** (*c*)	**12.** (*d*)	**13.** (*b*)	**14.** (*a*)	**15.** (*d*)	**16.** (*d*)	**17.** (*c*)	**18.** (*a*)
19. (*c*)	**20.** (*c*)	**21.** (*c*)	**22.** (*c*)	**23.** (*c*)	**24.** (*a*)	**25.** (*b*)	**26.** (*c*)	**27.** (*d*)
28. (*b*)	**29.** (*d*)	**30.** (*b*)	**31.** (*a*)	**32.** (*b*)	**33.** (*b*)	**34.** (*c*)	**35.** (*a*)	**36.** (*d*)
37. (*a*)	**38.** (*c*)	**39.** (*d*)	**40.** (*d*)	**41.** (*a*)	**42.** (*a*)	**43.** (*a*)	**44.** (*d*)	**45.** (*d*)
46. (*d*)	**47.** (*a*)	**48.** (*a*)	**49.** (*c*)	**50.** (*c*)	**51.** (*a*)	**52.** (*c*)	**53.** (*c*)	**54.** (*c*)
55. (*d*)	**56.** (*b*)	**57.** (*d*)	**58.** (*b*)	**59.** (*a*)	**60.** (*d*)	**61.** (*c*)	**62.** (*a*)	**63.** (*d*)
64. (*c*)								

SOLUTIONS

1. 82° represents 11.7 km².

$\therefore$ 66° will represent $\left(\frac{11.7}{82}\times 66\right)$ km² = 9.4 km².

2. In a histogram the area of each rectangle is proportional to the frequency of the corresponding class interval.

3. Initial sum of observations = Mean × Number of observations = 50 × 100 = 5000

Difference in sum due to misread observations = (182 + 56) − (82 + 6) = 238 − 88 = 150

Corrected sum of observations = 5000 + 150 = 5150

Corrected mean = Corrected sum ÷ Number of observations = 5150 ÷ 100 = 51.5

4. We have, $n = 7$ (odd)

$$\therefore \quad \text{Median} = \left(\frac{n+1}{2}\right)^{\text{th}} \text{term} = \left(\frac{7+1}{2}\right)^{\text{th}} \text{term} = 4^{\text{th}} \text{ term}$$

arrange in ascending order – 1, 2, 2, 3, 6, 8, 9

$\therefore$ Median = 3

5. Required arithmetic mean $= \frac{4+6+8+9+10+12+14+15+16+18}{10} = \frac{112}{10} = 11.2$

6. We have following cases:

Case 1 : –4, –3, –2, –1, 0

Case 2 : –3, –2, –1, 0, 1

Case 3 : –2, –1, 0, 1, 2

Case 4 : –1, 0, 1, 2, 3

Case 5 : 0, 1, 2, 3, 4

7. We have

$x = \{-4, -3, -2, -1, 0, 1, 2, 3, 4\} \Rightarrow P = -4$ and $Q = 4$

$\therefore$ Required arithmetic mean $= \frac{-4+4}{2} = 0$

8. We have, $n = 7$

Arranging the given data in ascending order:

3, 4, 5, 6, 9, 11, 18

$$\therefore \quad \text{Median} = \left(\frac{n+1}{2}\right)^{\text{th}} \text{term} = \left(\frac{7+1}{2}\right)^{\text{th}} \text{term} = 4^{\text{th}} \text{ term} = 6$$

9. Required harmonic mean = $\frac{2P}{S}$

10. Clearly, (*b*) is true. **11.** Clearly, (*c*) is true.

12. Arranging in ascending order
36, 41, 43, 61, 71, 92, 93, 95, 99, 100, 110, 127

(1) Old median = $\left(\frac{n}{2}+1\right)^{th}$ term [$\because n = 12$ is even] = (6 + 1)th term = 93

New median = 94

(2) $\frac{+1}{12} = 0.08 < 0.1$ (3) New mode = 110 = old mode

13. Let the age of new member be '*x*' years now, we have

Mean age of 7 members of the family = $\frac{2+5+12+18+38+40+60}{7} = \frac{175}{7} = 25$

After 5 years, sum of their age = 175 + 7 × 5 = 210
Now, New mean = 25 + 1.5 = 26.5
∴ Total age of 8 family members = 26.5 × 8 = 212
So, x = 212 – 210 = 2 years

14. Let the marks of the candidates who failed in the examination be x_1, x_2, x_3, x_4 and x_5
Now, arranging ascending order x_1, x_2, x_3, x_4, x_5, 4, 5, 6, 6, 6 7, 7, 8, 8, 9, 9
$n = 15$ (odd)

Now, Median = $\left(\frac{n+1}{2}\right)^{th}$ term = $\left(\frac{15+1}{2}\right)^{th}$ term = 8th terms = 6

15. Arrange in ascending order:
111 gm, 141 gm, 154 gm, 175 gm, 176 gm 180 gm and 191 gm
$n = 7$

Median = $\left(\frac{n+1}{2}\right)^{th}$ term = $\left(\frac{7+1}{2}\right)^{th}$ term = 4th terms = 176 gm

16. Mode will be used to determine the average size of the shoes sold in the shop.

17.

x	f	a	fa
0 – 10	8	5	40
10 – 20	12	15	180
20 – 30	10	25	250
30 – 40	p	35	$35p$
40 – 50	9	45	405
	$39 + p$		$875 + 35p$

Now, Mean = $\frac{\Sigma fa}{\Sigma f} \Rightarrow 25.2 = \frac{875+35p}{39+p}$

⇒ $982.8 + 25.2p = 875 + 35P$

⇒ $35P - 25.2p = 982.8 - 875$

⇒ $9.8p = 107.8 \Rightarrow p = 11$

18. Let n observations are $x_1, x_2, x_3, \ldots\ldots\ x_m, \ldots\ldots\ x_n$

Mean of 'm' observations = n

$$\Rightarrow \qquad \frac{x_1 + x_2 + x_3 \ldots\ldots 4x_m}{m} = n \qquad \text{...(i)}$$

Mean of remaining on servation = m

$$\Rightarrow \qquad \text{Mean} = \frac{x_{m+1} + \ldots\ldots + x_n}{n-m} = m \qquad \text{...(ii)}$$

Adding (i) and (ii), we get

$x_1 + x_2 + \ldots\ldots\ x_{m+1}\ \ldots\ldots + x_n = nm + m\,(n - m) \Rightarrow 2mn - m^2$

Now, total observation $= \dfrac{2mn - m^2}{n} = 2m - \dfrac{m^2}{n}$

19.

Class interval	x_i	f_i	fx_i
10 –20	15	20	300
20 – 30	25	15	375
30 – 40	35	10	350
40 – 50	45	29	1305
50 – 50	55	26	1430

We have,

$x + y = 100 - (20 + 15 + 10)$
$= 100 - 45 = 55$ ....(i)

$2x + y = 100 - (4 + 8 + 4)$
$= 100 - 16 = 84$ .....(ii)

From eqn (i) and (ii), we get

$x = 29$ and $y = 26$

$$\Rightarrow \text{Required mean} = \frac{\Sigma_{fi}\, x_1}{\Sigma_{fi}} = \frac{37.60}{100} = 37.6$$

20.

Class interval	Series II (f)
10 –20	4
20 – 30	8
30 – 40	$4 = f_0$
40 – 50	$2x = 58$ = Model class $= f_1$
50 – 60	$y = 26 = f_2$

We have, $h = 10$

$l = 40$

Now,

$$\text{Mode} = l + \frac{f_1 - f_0}{(f_1 - f_0) + (f_1 - f_2)} \times h$$

$$\text{Mode} = 40 + \frac{58 - 4}{(58 - 4) + (58 - 26)} \times 10$$

$$= 40 + \frac{54}{54 + 32} \times 10 = 40 + \frac{54}{86} \times 100$$

$$= 40 + 6.28 = 46.29 \approx 46$$

21. Let the numbers be x and y respectively

Now, we have

$$\frac{x+y}{2} = 10 \Rightarrow x+y = 20 \quad ...(i)$$

$$\sqrt{xy} = 8 \Rightarrow xy = 64$$

Now, $x - y = \sqrt{(x+y)^2 - 4xy} = \sqrt{(20)^2 - 4\times 64} = \sqrt{400-256} = \sqrt{144} = 12$

$\Rightarrow \quad x - y = 12 \quad ...(ii)$

adding (i) and (ii), we get

$2x = 32 \Rightarrow x = 16$

$y = 4$

22. Sum of 11 observations = 11 = $\frac{10.5\times 6 + 11.5\times 6 - \text{six}^{\text{th}} \text{ number}}{11}$

$\Rightarrow$ 121 = 63 + 69 – sixth number $\Rightarrow$ Sixth number = 132 – 121 = 11

23. Since, using ranks in terms of stars, which serve as a measure of hotel's quality, ratings are categorised. Hence, ratings of hotel in a city is measured on ordinal scale.

24. Mean = = $\frac{\Sigma a_i f_i}{\Sigma f_i}$, where a_i is the average of the lower and upper limits of the ith class interval and f is the frequency of the ith class interval

Mean = $\frac{1}{50}[5(4)+15(5) + 25(7) + 35(10) + 45(12) + 55(8)+65(4)] = \frac{1860}{50} = 37.2$

25. The median class is the class which contains the middle values or the middle values.

Total frequency = 50.

The middle values are the 25th and the 26th values of the 50 values. Both these are part of 30 – 40.

26. Median = $L + \frac{\left(\frac{N}{2} - cmf\right)c}{f}$ (where L is the. lower limit of the median class, N is the total frequency, cmf is the cumulative frequency up to the class preceding the median class, f is the frequency of the median class and c is the width of each class interval.)

$$= 30 + \frac{\frac{50}{2} - 16}{10}(10) = 39$$

27. Mode = $l_i \frac{\Delta_1}{\Delta_1 + \Delta_2}(C)$ (where l_1 is the lower limit of the modal class, c is the width of each class interval.

Δ_1 = frequency of the modal class – frequency of the class which immediately precedes the modal class,

Δ_2 = frequency of the modal class – frequency of the class which immediately succeeds the modal class = $40 + \frac{2}{2+4}(10) = 43.\overline{3}$

28. When the values are arranged in an ascending order, they are:

a – 3.5, a – 3, a – 2.5, a – 2, a – 0.5, a + 0.5, a +4, a + 5

Since there are 8 values, median of the set would be the average of the 4th and the 5th values.

Median = $\dfrac{(a-2)+(a-0.5)}{2} = a - 1.25.$

29. The original set of numbers = {1, 2, 3, 4,..........n}

The new set of numbers = {1(2), 2(3), 3(4), ...n (n +1)}

The kth term is of the form $k(k + 1)$.

$$T_1 + T_2 + T_3 + \ldots\ldots\ldots\ldots T_n = \sum_{k=1}^{n} k\,(k+1) = \sum_{k=1}^{n} k^2 + \sum_{k=1}^{n} k = \frac{n(n+1)\,(2n+1)}{6} + \frac{n(n+1)}{2}$$

$$= \frac{n(n+1)\,(2)\,(n+2)}{6} = \frac{n(n+1)\,(n+2)}{3}$$

$$\text{Average of n terms} = \frac{\frac{n(n+1)\,(n+2)}{3}}{n} = \frac{(n+1)(n+2)}{3}.$$

30. The arithmetic mean of n numbers = M

Sum of n numbers = $n \times M = nM$

Sum of $(n-1)$ numbers = k

Then, n^{th} number = Sum of n numbers − Sum of $(n-1)$ numbers

n^{th} number = $nM - k$

31. The mean of x_1, x_2, x_3, x_4 x_5 is 9.

$\therefore$ The mean of $x_i - 1$ (i = 1 to 5) is 8.

The median of x_1 to x_5 is 8.

$\therefore$ The median of $x_i - 1$ (I = 1 to 5) is 7.

32. From the table, the number of children in the age group 5 – 8 is 20 and in the group 8 – 11 is 20. The median age is 8 years. (1) is false.

The total number of children in the age group 6 – 9 is 7 + 9 + 12, *i.e.* 28 which is 70% of the total. (2) is true.

The modal group is 8 – 9, is the modal age is 8, (3) is true.

33. Let $\lambda = 2$.

$\therefore$ x_1 to x_{10} are 2^1. 2^2..... 2^{10}

The sum = $2\left(2^{10} - 1\right) = 2046$ AM = 204 – 6

$\therefore$ $2^7 < AM < 2^8$

GM = $\sqrt[10]{2^1\ 2^2....2^{10}} = 2^{5.5} = 32\sqrt{2} \approx 45.3$

$\therefore$ $2^5 < GM < 2^6$

The median = $\dfrac{2^5 + 2^6}{2} = \dfrac{32+64}{2} = 48$

Consider the choices.

(*a*) AM < Median False (*b*) GM < Median True

(*c*) GM > Median False (*d*) AM = Median False.

34. $x = 1,\ x_2 = \dfrac{1}{2}, x_{11} = \dfrac{1}{11}.$

The median is the 6th term, ie $\frac{1}{6}$. The 5 numbers greater than $\frac{1}{6}$ carry more 'weight' than the 5 number.

$\therefore$ AM > $\frac{1}{6}$. (*a*) is true

If *n* number are in AP. The AM is equal to the median.

Similarly, if *n* numbers are n HP. The HM is equal to the Median, ie HM = $\frac{1}{6}$

$\therefore$ (*d*) is correct and (*c*) is false
As GM > HM, it follows GM > 1/6 (*b*) true.

35. We can subtract *x* from all the 8 values, in ascending order are –3.5, –3, –2.5, –2, –0.5, 0.5, 4, 5

The median of these 8 values is $\frac{-2-0.5}{2} = -1.25$

The median of the 8 given values is $x - 1.25$

36. The median of 9 distinct observations is 20.5.
If each of 4 largest observations are increased by 2, these is no change in the median.

37. The wrong values of 3, 6 are replaced by the correct values 8, 9.
The total increases by 8 and the average by 8/20
The correct average = 17 + 0.4 = 17.4

38. Let two numbers be *a* and *b*

$a + b = 40$

$$\frac{AM}{GM} = \frac{100}{80} = \frac{5}{4} \Rightarrow \frac{a+b}{2\times\sqrt{ab}} = \frac{5}{4} \Rightarrow \frac{20}{\sqrt{ab}} = \frac{5}{4} \Rightarrow \sqrt{ab} = 16 \Rightarrow ab = 256$$

$$a - b = \sqrt{(a+b)^2 - 4ab} = \sqrt{1600-1024} = \sqrt{576} = 24$$

39. The classes 15 – 19, 20 -24, ……are inclusive. The end pointer are included.
The classes 15 – 20, 20 – 25, ….. are exclusive. The right end points are excluded.
Neither statement is correct.

40. The frequency density = $\frac{Frequency}{Class\ width} = \frac{30}{5} = 6$

41.

Midpoint of class (x_i)	$x_i f_i$	Frequency (f_i)
10	170	17
30	840	28
50	1600	32
70	70f	f
90	1710	19

$\sum x_i f_i = 4320 + 70f$, $\sum f_i = 96 + f$

$$\therefore \quad Mean = \frac{\sum x_i f_i}{\sum f_i} \Rightarrow 50 = \frac{4320 + 70f}{96 + f} \Rightarrow 5f + 480 = 432 + 7f \Rightarrow 2f = 48 \Rightarrow f = 24$$

42. Given, mean = 270, Median = 220

3 median = mode + 2 mean ⇒ 3 × 220 = mode + 2 × 270 ⇒ Mode = 660 − 540 = 120

43. Number of students in a class = Cumulative frequency of upper limit − Cumulative frequency of lower limit

Students between 60 and 70 = 315 − 210 = 105

44. Number of students scoring more than a mark = Total students − Cumulative frequency of that mark

Total students = 350 (Below 80)

Students scoring more than 50 = 350 − 125 = 225

45. $P = \sqrt[n]{a_1 \times a_2 \times a_3 \times ......\times a_n} \Rightarrow P^n = a_1 \times a_2 \times a_3 \times ..... \times a_n$ ...(i)

$Q = \sqrt[2n]{a_{n+1} \times a_{n+2} \times a_{n+3} \times ......\times a_{3n}} \Rightarrow a^{2n} = a_{n+1} \times + a_{n+2} \times a_{n+3} \times ..... \times a_{3n}$ ...(ii)

G.M. = $\sqrt[3n]{a_1 \times a_2 \times a_3 \times ......\times a_{3n}} = (a_1 \times a_2 \times a_3 \times ..... a_{3n})^{1/3n}$

By multiplying (i) and (ii) get $\left(p^n \times q^{2n}\right)^{\frac{1}{3}n} = P^{\frac{1}{3}} \times Q^{\frac{2}{3}}$

46. $\dfrac{7+5+13+x+9}{5} = 10 \Leftrightarrow x = 16.$

47. $\text{GM} = (3\times 9\times 27\times 81\times 243\times 729\times 2187)^{\frac{1}{7}} = (3^1\times 3^2\times 3^3\times 3^4\times 3^5\times 3^6\times 3^7)^{\frac{1}{7}} = (3^{28})^{\frac{1}{7}} = 3^4 = 81$

48. $f+(f+2)+(f-3)+(f+6) = 205 \Rightarrow 4f+5 = 205$

$\Rightarrow 4f = 200 \Rightarrow f = 50$

49. Cumulative frequencies:

3 → 50

5 → 50 + 52 = 102

6 → 102 + 47 = 149

7 → 149 + 56 = 205

$\dfrac{N}{2} = \dfrac{205}{2} = 102.5 \rightarrow$ lies in class $x = 6$ (since CF before – 102)

Assuming class width $h = 1$, $L = 6$:

$$\text{Median} = 6 + \left[\frac{(102.5-102)}{47}\right]\times 1$$

$$\text{Median} = 6 + \frac{0.5}{47} \approx 6 + 0.106 \approx 6$$

50. Highest frequency = 56 → modal class $x = 7$

$f_1 = 56, f_0 = 47$ (previous), $f_2 = 0$ (no class after 7), $h = 1, L = 7$

$$\text{Mode} = 7 + \left[\frac{(56-47)}{(2\times 56-47-0)}\right]\times 1 \Rightarrow \text{Mode} = 7 + \left[\frac{9}{(112-47)}\right]$$

$$\Rightarrow \text{Mode} = 7 + \frac{9}{65} \approx 7 + 0.138 \approx 7$$

51. Plot x-values on x-axis (3, 5, 6, 7). Plot frequencies on y-axis (50, 52, 47, 56)

Draw bars corresponding to each x-value with height equal to its frequency.

Graphical representation is a bar graph.

52. $f \times x$:

$3 \times 50 = 150$

$5 \times 52 = 260$

$6 \times 47 = 282$

$7 \times 56 = 392$

$$\Sigma(f \times x) = 150 + 260 + 282 + 392 = 1084$$

$$\text{Mean} = \frac{1084}{205} \approx 5.29$$

53. Since 8 occurs most often, mode = 8.

54. $(x \times 16 \times 50)^{1/3} = 20 \Rightarrow x \times 16 \times 50 = (20)^3 \quad \Rightarrow \quad x = \left(\frac{20 \times 20 \times 20}{16 \times 50}\right) = 10.$

55. The median of a given frequency distribution can be found with the help of an ogive.

56. A frequency distribution can be presented graphically by a Histogram.

57. Clearly, 3 (Median) — 2 (A.M.) = Mode.

For (qn 58-59)

Class	*f* (frequency)	C.F.
0 – 30	4	4
30 – 60	5	9
60 – 90	7	16
90 – 120	4	20

58. $\because$ Mode $= l + \dfrac{f_1 - f_0}{f_1 - f_0 - f_2} \times h$

$l = 60, f_0 = 5, f_1 = 7, \ f_2 = 4 = 60 + \dfrac{7-5}{14-5-4} \times 30 = 60 + 12 = 72$

59. Median $= \dfrac{l + \frac{N}{2} - CF}{f} \times h = 60 + \dfrac{10-9}{7} \times 30 = 60 + \dfrac{30}{7} = \dfrac{450}{7}$

Now, 7 (Q – P) = 9R

$\Rightarrow 7\left(72 - \dfrac{450}{7}\right) = 9R$ $\qquad$ [$\because$ Mode = Q = 72 and P = Median = $\dfrac{450}{7}$]

$\Rightarrow 54 = 9R \Rightarrow R = 6$

For qn (60- 61)

Class	*f* (frequency)	*x*	*f*(*x*)	C.F.
40 – 50	4	45	180	4
50 – 60	3	55	165	7
60 – 70	1	65	65	8
70 – 80	2	75	150	40
	10		560	

60. Mean $= \dfrac{\Sigma fx}{\Sigma f} = \dfrac{560}{10} = 56.$

61. Median = $\dfrac{l+\dfrac{N}{2}-CF}{f}\times h = 50+\dfrac{5-4}{3}\times 10=\dfrac{160}{3} \Rightarrow 3M = 3\times\dfrac{160}{3} = 160$

62. We have,

$\Sigma x_i - 10 = 1$ and_ $\Sigma x_i - 2 = b \Rightarrow a - b = 10$ ...(i)

and $\dfrac{b}{a} = -4 \Rightarrow b = -4a$

From (i), we get,

$5a = 10 \Rightarrow a = 2.$ Then, $b = 8$

$\therefore Ex_i = a + 10 = 12.$

63. Median = $l+\left(\dfrac{\dfrac{n}{2}-cf}{f}\right)\times h$

Mark (irregular)	Mark (regular)	F	CF
18 – 26	17.5 – 26.5	5	5
27 – 35	26.5 – 35.5	7	12
36 – 44	35.5 – 44.5	10	22
45 – 53	44.5 – 53.5	15	37
54 – 62	53.5 – 62.5	8	45
63 – 71	62.5 – 71.5	3	48
72 – 80	71.5 – 80.5	2	50

Median lies on $\left(\dfrac{n}{2}\right)^{th}$ observation, where, n = 50, $\dfrac{n}{2} = 25$

Comparing 25 with the cf value,

Then, the class lies in the interval (44.5 – 53.5), as 25 is under 37.

Also, l = 44.5, f = 15, cf = 22

$h = (53.5 - 44.5) = 9$

Median $= 44.5+\left(\dfrac{25-22}{15}\right)\times 9 = 44.5+\left(\dfrac{3}{15}\right)\times 9 = 44.5+1.8 = 46.3$

64. Class (44.5 – 53.5) is the class with the maximum number of students.

So, $l = 44.5, f_1 = 15, f_0 = 10, f_2 = 8$ and $h = (53.5 - 44.5) = 9$

$\Rightarrow$ Mode $= 44.5 + \dfrac{15-10}{(2\times 15-10-8)}\times 9 = 44.5 + \dfrac{5}{12}\times 9 = 44.5 + 3.75 = 48.25$

36

CHAPTER

Data Interpretation

(Pie Charts and Bar Diagrams)

I. TABULATION

This segment comprises inquiries derived from data concerning typical fields like Production trends over several years in a factory, Imports, Exports, Employee salaries within a factory, and Enrollment figures across different departments in a college, among others. This data is presented in tabular format. Following the table, specific questions are posed based on the provided information, and candidates are expected to respond to these queries. Rows represent the horizontal divisions in the table, columns signify the vertical divisions, and the unique headings are referred to as captions. Units of measurement are provided alongside the captions.

Direction (1-2): Consider the following data for the two (02) items that follow:

The expenditure (in lakhs of rupees) of a company for the years 2011 to 2017 is as under:

Year	Expenditure
2011	13.8
2012	15.4
2013	10.4
2014	13.1
2015	15.8
2016	17.2
2017	19.4

1. How many times the increase in expenditure in a year exceeded by more than 15% as compared to previous year? **[CDS (I) Exam, 2022]**

(*a*) 2 (*b*) 3 (*c*) 4 (*d*) 5

2. In which year, the percentage increase in expenditure is maximum compared to its previous year? **[CDS (I) Exam, 2022]**

(*a*) 2012 (*b*) 2014 (*c*) 2015 (*d*) 2017

3. In Consider the following data with regard to production of cars (in lakhs): **[CDS (I) Exam, 2021]**

	Year 2015	Year 2016
Country A	35	38
Country B	45	47
Country C	88	93
Country D	75	79
Country E	58	60.9

In which of the countries, the production of cars has increased by more than or equal to 5% in 2016 over 2015?

(*a*) B and E (*b*) A, C and D only (*c*) A, C, D and E (*d*) A, D and E only

Directions (4–7): For the next four (4) items, consider the following data with regard to different types (I, II, III, IV, V) of multivitamin tablets produced in a company (in lakhs):

Year	I	II	III	IV	V
2000	160	80	70	90	75
2001	200	150	85	160	100
2002	135	35	44	95	85
2003	240	95	120	80	120
2004	180	110	85	95	115
2005	210	150	100	92	110

4. Which product is produced least over the years 2000-2005?

(*a*) Type II (*b*) Type III (*c*) Type IV (*d*) Type V

5. In which one of the following pairs of years, the difference in total number of tablets produced between them is minimum?

(*a*) (2003, 2005) (*b*) (2001, 2005) (*c*) (2003, 2004) (*d*) (2000, 2002)

6. The ratio of percentage drop in total production in 2004 compared to 2001 to that in 2000 compared to 2001, is

(*a*) $\frac{1}{3}$ (*b*) $\frac{1}{4}$ (*c*) $\frac{1}{2}$ (*d*) $\frac{1}{5}$

7. In which year, the production of Type I is more than the sum of the production of Type III and Type IV?

(*a*) 2001 (*b*) 2002 (*c*) 2003 (*d*) 2004

Directions (Q. 8-9): Consider the following data for the two questions follows:

The table below gives the age-wise population percentage of a city: **[CDS (I) Exam, 2022]**

Age group	*Percentage*
Below 30 years	14.00
30-34.99	29.75
35-39.99	26.25
40-44.99	0
45-49.99	18.50
50 years and above	11.50

The number of persons below the age of 40 years is 10.5 lakhs.

8. What is the total population of the city (in lakhs)?

(*a*) 21 (*b*) 18 (*c*) 15 (*d*) 12

9. If the ratio of taxpayers to other persons in the same age group below 30 years is 1:2, then what is the number of taxpayers (in lakhs) in that age group?

(*a*) 0-4 (*b*) 0.7 (*c*) 0-85 (*d*) 1.05

10. The following table shows the marks of 90 students in a test of 80 marks

[CDS (I) Exam, 2021]

Marks	***Number of students***
1-10	5
11-20	8
21-30	10
31-40	13
41-50	18
51-60	17
61-70	12
71-80	7

The percentage of students who have obtained less than or equal to 50% marks is

(*a*) 30% (*b*) 40% (*c*) 45% (*d*) 60%

Directions (Q 11-13): Consider the following for the next three (03) items that follow: Circulation figures (in thousands) of different newspapers (A, B, C, D, E) for five years are given below:

[CDS (I) Exam, 2025]

Year	**A**	**B**	**C**	**D**	**E**
2019	20	10	15	8	20
2020	12	12	18	12	12
2021	24	14	17	14	15
2022	26	10	16	15	9
2023	22	16	14	16	11

11. In which of the given years was the circulation of the newspaper D close to its average circulation over all the years?

(*a*) 2020 and 2021 (*b*) 2022 and 2023

(*c*) 2022 only (*d*) 2020 only

12. In which of the years from 2019 to 2022 was the circulation of the newspaper D close to the average circulation of all the newspapers in that year?

(*a*) 2019 (*b*) 2020 (*c*) 2021 (*d*) 2022

13. How many cases are there in which the average of the circulation for an individual newspaper was more than the average of the circulation of all the newspapers?

(*a*) One (*b*) Two (*c*) Three (*d*) Four

Answers & Solutions:

1. (*a*) Percentage increase in the following years:

2011 = 13.8

$$2012 = \frac{15.4-13.8}{13.8} \times 100 = \frac{1.6}{13.8} \times 100 = 11.59\%$$

$$2013 = \frac{10.4-15.4}{15.4} \times 100 = \frac{-5}{15.4} \times 100 = -32.46\%$$

$$2014 = \frac{13.1-10.4}{10.4} \times 100 = \frac{2.7}{10.4} \times 100 = 25.96\%$$

$$2015 = \frac{15.8-13.1}{13.1} \times 100 = \frac{2.7}{13.1} \times 100 = 20.6\%$$

$$2016 = \frac{17.2-15.8}{15.8} \times 100 = \frac{1.4}{15.8} \times 100 = 8.86\%$$

$$2017 = \frac{19.4-17.2}{17.2} \times 100 = 12.79\%$$

$\therefore$ Required times = 2

2. (*b*) From qn (1)
During 2014
% increase = 25.9 – (– 32.4) = 25.9 + 32.4 = 58. 3

3. (*c*) Increased production of cars from 2015 to 2016 in following country:

$$\text{Country A} = \frac{38-35}{35} \times 100 = \frac{300}{35} = 8.5\%$$

$$\text{Country B} = \frac{47-45}{45} \times 100 = \frac{200}{45} = 4.4\%$$

$$\text{Country C} = \frac{93-88}{88} \times 100 = \frac{500}{88} = 5.68\%$$

$$\text{Country D} = \frac{79-75}{75} \times 100 = \frac{400}{75} = 5.33\%$$

$$\text{Country E} = \frac{60.9-58}{58} \times 100 = \frac{290}{58} = 5\%$$

Hence, in countries A, C, D and E, the production of cars has increased by more then or equal to 5% in 2016 over 2015.

4. (*b*) Production of different product over the years 2000 – 2005 are following:
Type I = 160 + 200 + 135 + 240 + 180 + 210 = 1125
Type II = 80 + 150 + 35 + 95 + 110 + 150 = 620
Type III = 70 + 85 + 44 + 120 + 85 + 100 = 504
Type IV = 90 + 160 + 95 + 80 + 95 + 92 = 612
Type V = 75 + 100 + 58 + 120 + 115 + 110 = 605
Hence, Type III is produced least over the years 2000-2005

5. (*a*) Total tablets produced in different years are followings:
2000 = 160 + 80 + 70 + 90 + 75 = 475
2001 = 200 + 150 + 85 + 160 + 100 = 695
2002 = 135 + 35 + 44 + 95 + 85 = 394
2003 = 240 + 95 + 120 + 80 + 120 = 655
2004 = 180 + 110 + 85 + 95 + 115 = 585
2005 = 210 + 150 + 100 + 92 + 100 = 662

Now, by checking options
(a) (2003, 2005) difference = 662 – 655 = 7
(b) (2001, 2005) difference = 695 – 662 = 33
(c) (2003, 2004) difference = 665 – 585 = 70
(d) (2000, 2002) difference = 475 – 394 = 81
Hence, option (a) is minimum.

6. (*c*) Percentage drop in total production in 2004 compared to 2001 = $\frac{695-585}{695} \times 100$

$$[\text{From qn 5}] = \frac{110}{695} \times 100$$

Percentage drop in total production in 2000 compared to 2001 = $\frac{695-475}{695} \times 100$

$$= \frac{220}{695} \times 100$$

Now, required ration = 1 : 2

7. (*c*) By comparing the production of type I to the sum of the production of type III and type IV, we get

For 2000,
160 = 90 + 70
For 2001
85 + 160 = 245 > 200
For 2002
44 + 95 = 139 > 135
For 2003,
120 + 80 = 200 < 240
For 2004,
85 + 95 = 180
In 2003, the production of Type I is more than the sum of the production of type III and type IV.

8. Percentage of people below 40 years = 14 + 29.75 + 26.25 = 70.00%
Now, according to the question,
70% of total population = 10.5

$$\Rightarrow \text{ Total population} = \frac{10.5 \times 100}{70} = 15 \text{ lakhs}$$

9. Taxpayers of age group below 30 years = 14%
Total population = 15 lakhs [from qn 8]

$$\therefore \text{ Taxpayers of age group below 30 years} = \frac{14}{100} \times 15{,}00{,}000 = 2{,}10{,}000$$

$$\therefore \text{ Required number of taxpayers} = 210000 \times \frac{1}{3} = 70000 = 0.7 \text{ lakhs}$$

10. We have,

50% of total marks = $80 \times \frac{50}{100} = 40$

⇒ Number of students who have obtained less than or equal to 50% marks

$= 5 + 8 + 10 + 13 = 36$

⇒ Required percentage = $\frac{36}{90} \times 100 = 40\%$

11. (*a*) Sum of D = $8 + 12 + 14 + 15 + 16 = 65$ (thousands)

Average of D = $\frac{65}{5} = 13$ (thousands)

Differences from average:

2019: $|8 - 13| = 5$

2020: $|12 - 13| = 1$

2021: $|14 - 13| = 1$

2022: $|15 - 13| = 2$

2023: $|16 - 13| = 3$

The circulation of D was closest to its average (13 thousands) in the years 2020 and 2021 (difference = 1 thousand).

12. (*d*) 2019: Total = $20 + 10 + 15 + 8 + 20 = 73$

Average = $\frac{73}{5} = 14.6$; D = 8 → Difference = $14.6 - 8 = 6.8$

2020: Total = $12 + 12 + 18 + 12 + 12 = 66$

Average = $\frac{66}{5} = 13.2$; D = 12 → Difference = $13.2 - 12 = 1.2$

2021: Total = $24 + 14 + 17 + 14 + 15 = 84$

Average = $\frac{84}{5} = 16.8$; D = 14 → Difference = $16.8 - 14 = 2.8$

2022: Total = $26 + 10 + 16 + 15 + 9 = 76$

Average = $\frac{76}{5} = 15.2$; D = 15 → Difference = $15.2 - 15 = 0.2$

The circulation of newspaper D was closest to the average in 2022.

13. (*b*) Total circulation for each newspaper:

A = $20 + 12 + 24 + 26 + 22 = 104$ Average = $\frac{104}{5} = 20.8$

B = $10 + 12 + 14 + 10 + 16 = 62$ Average = $\frac{62}{5} = 12.4$

C = $15 + 18 + 17 + 16 + 14 = 80$ Average = $\frac{80}{5} = 16$

D = 8 + 12 + 14 + 15 + 16 = 65 Average = $\frac{65}{5} = 13$

E = 20 + 12 + 15 + 9 + 11 = 67 Average = $\frac{67}{5} = 13.4$

Total circulation of all newspapers= 104 + 62 + 80 + 65 + 67 = 378

Average circulation of all newspapers = $\frac{378}{25} = 15.12$

There are 2 newspapers (A and C) whose average circulation is more than the average of all newspapers.

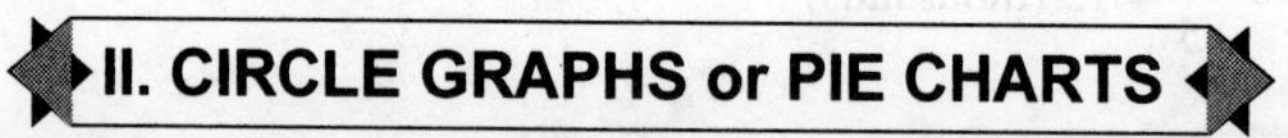

II. CIRCLE GRAPHS or PIE CHARTS

This section consists of questions based on the data regarding common disciplines such as Production over a period of few years in a factory; Imports; Exports; Salary of employees in a factory; Number of students in a college in various faculties etc. These data are in the form of a table. This table is followed by certain questions based on the information given in the table and the candidate is required to answer those questions.

The horizontal lines in a table are called its rows, the vertical lines are called columns and the distinctive heads are known as captions. The units of measurements are given with the captions.

Pie Charts: In such a diagram, the given data is distributed over a circle. Each part of the data makes a certain central angle.

For example, if out of total income of a family 24% is spent on food, then central angle made by this item = $\left(\frac{24}{100} \times 360\right) = (86.4)°$.

Ex. 1. ***The following pie chart shows the annual agricultural yield of a certain place.***

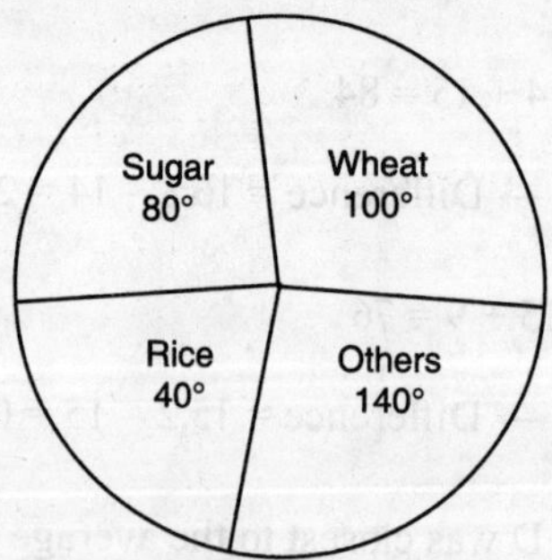

Study the above chart and answer the questions given below:

1. If the total production is 8100 tons, then the yield of rice (in tons) is:
   (*a*) 3240 (*b*) 900 (*c*) 4860 (*d*) None
2. If the yield of sugar is 2400 tons, then the yield of wheat (in tons) is:
   (*a*) 3000 (*b*) 1920 (*c*) $666\frac{2}{3}$ (*d*) None
3. The production of wheat exceeds that of rice by:
   (*a*) 50% (*b*) 100% (*c*) 150% (*d*) 250%

Answers & Solutions:

1. (*b*): Yield of rice = $\left(\frac{40}{360}\times 8100\right) = 900$ tons.

2. (*a*): Ratio of their yields = Ratio of their central angles

Let 80 : 100 :: 2400 : *x*. Then, $x = \frac{100\times 2400}{80} = 3000$.

3. (*c*): Excess on 40 = 60

Excess on 100 = $\left(\frac{60}{40}\times 100\right)\% = 150\%$.

EXERCISE

Directions : *Consider the following for the next three (03) items that follow :*

Study the following Pie Charts :

Chart-I : Number of houses : 1,20,000

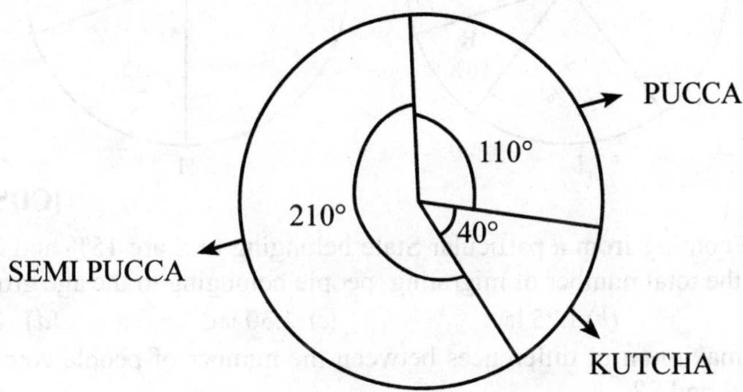

Chart-II : Population : 21,00,000

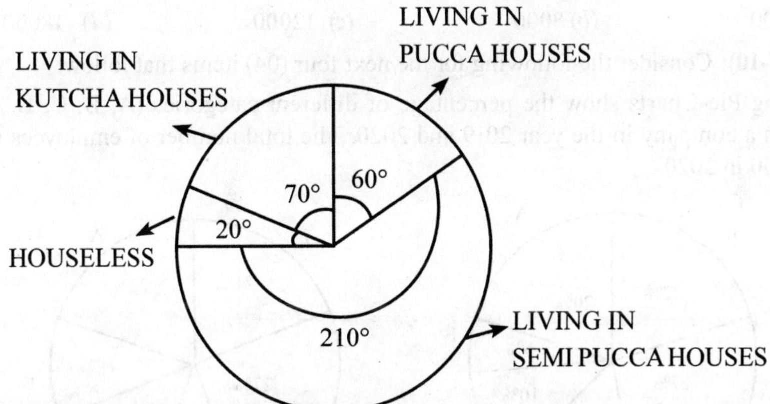

Study the graph and answer the questions given below:

1. On an average, how many persons live in every pucca house? **[CDS (I) Exam, 2025]**

(*a*) 8 (*b*) 8.54 (*c*) 9.54 (*d*) 7.54

2. If 5000 more Kutcha houses are built, then what will be approximate change in angle for Kutcha houses in Pie Chart-I? **[CDS (I) Exam, 2025]**
(*a*) 11° (*b*) 13° (*c*) 15° (*d*) 23°

3. If 300 families from the "Houseless" category shift into Kutcha houses, what will be the average number of families in every Kutcha house? **[CDS (I) Exam, 2025]**
(*a*) 15 (*b*) 13·9 (*c*) 12·9 (*d*) 10·5

Direction (Q. 4-6): Consider the following for the next three (3) items that follow:

The following Pie-Chart-I shows the people migrating to Delhi from different Indian States (P Q and R are three different States and S is the combined group of other States) and Pie-Chart-II indicates the different age groups A, B, C and D of these migrating people for each State.

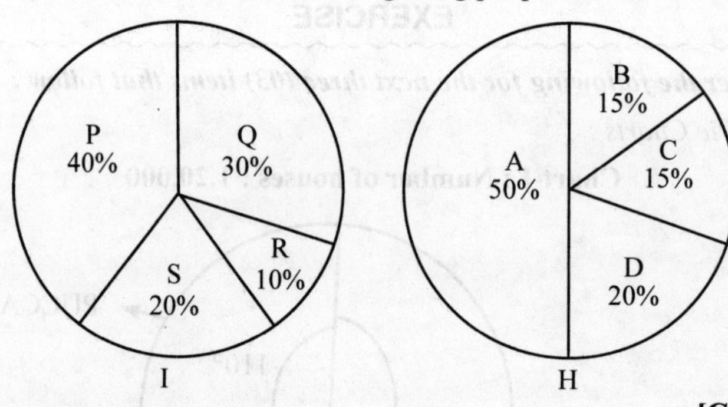

[CDS (I) Exam, 2024]

4. If the people coming from a particular State belonging to S are 15% and 24,000 in number, then what is the total number of migrating people belonging to the age group B ?
(*a*) 1.2 lac (*b*) 1.25 lac (*c*) 1.30 lac (*d*) 1.50 lac

5. What is the maximum of differences between the number of people coming from different groups P, Q, R and S?
(*a*) 1.6 lac (*b*) 1.8 lac (*c*) 2.4 lac (*d*) 2.6 lac

6. What is the difference between number of people coming from R having age group A and those coming from Q having age group D ?
(*a*) 6000 (*b*) 8000 (*c*) 12000 (*d*) 18000

Direction (7-10): Consider the following for the next four (04) items that follow:

The following Pie-Charts show the percentage of different categories (A, B, C, D, E and F) of employees in a company in the year 2019 and 2020. The total number of employees was 4000 in 2019 and 5000 in 2020.

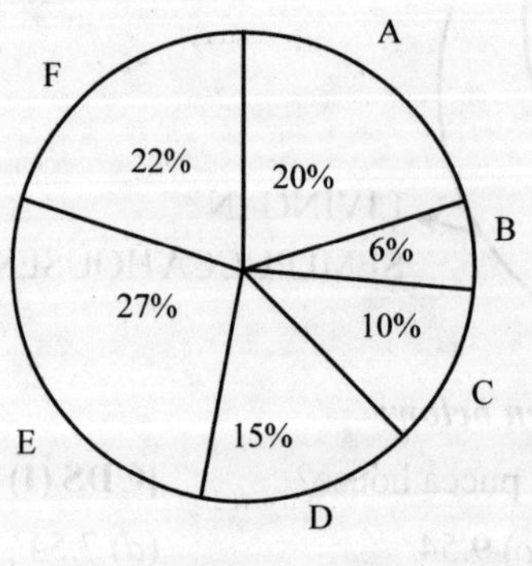

2019

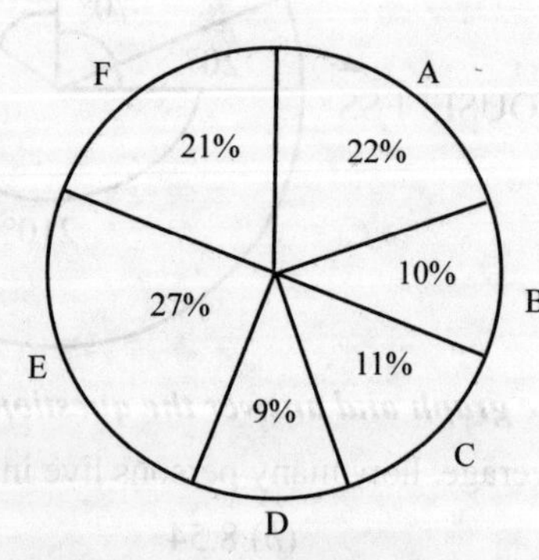

2020

7. There are two categories of employees in 2019 whose total strength remained the same in 2020. What are these two categories ?

(*a*) C and B (*b*) E and A (*c*) E and F (*d*) C and D

8. In which one of the following categories of employees, the percentage change in number of employees in 2020 was maximum as compared to maximum 2019?

(*a*) A (*b*) B (*c*) C (*d*) D

9. What was the percentage increase of category-E employees in 2020 as compared to that in 2019?

(*a*) 20% (*b*) 25% (*c*) 30% (*d*) 40%

10. There are two categories of employees whose strength increased by same percentage in 2020. What are the two categories?

(*a*) D and F (*b*) C and E (*c*) A and E (*d*) D and E

Answers & Solutions:

1. (*c*): Number of Pucca Houses $= \left(\frac{110}{360}\right) \times 1,20,000 = \frac{110000}{3}$

Population Living in Pucca Houses $= \left(\frac{60}{360}\right) \times 21,00,000 = 350000$

Average Persons per Pucca House $= \frac{\text{Population Living in Pucca Houses}}{\text{Number of Pucca Houses}}$

$$= \frac{3,50,000}{\left(\frac{1,10,000}{3}\right)} = \frac{105}{11} \approx 9.54$$

2. (*b*): Initial number of Kutcha houses $= \left(\frac{40}{360}\right) \times 1,20,000 = \frac{40000}{3}$

New total number of houses = Initial total number of houses + 5000
= 1,20,000 + 5000 = 1,25,000

New number of Kutcha houses = Initial number of Kutcha houses + 5000

$$= \left(\frac{1,20,000}{9}\right) + 5000$$

$$= \left(\frac{1,20,000 + 45,000}{9}\right)$$

$$= \frac{1,65,000}{9}$$

New angle for Kutcha houses $= \left(\frac{\text{New number of Kutcha houses}}{\text{New total number of houses}}\right) \times 360^\circ$

$$\left(\frac{\frac{1,65,000}{9}}{1,25,000}\right) \times 360^\circ = \left(\frac{165,000}{9 \times 125,000}\right) \times 360^\circ = 52.8^\circ$$

Change in angle = New angle for Kutcha houses – Initial angle for Kutcha houses

$$= 52.8^\circ - 40^\circ = 12.8^\circ \approx 13^\circ$$

3. (*b*): Initial Families in Kutcha Houses $\left(\frac{185}{360}\right) \times 3,60,000 = 185000$

Initial Houseless Families $= \left(\frac{20}{360}\right) \times 3,60,000 = 20000$

New Families in Kutcha Houses = Initial Families in Kutcha Houses + 300

$$= 1,85,000 + 300 = 1,85,300$$

Number of Kutcha Houses $= \left(\frac{40}{360}\right) \times 1,20,000 = \frac{1,20,000}{9}$

$$\text{Average families per Kutcha house} = \frac{\text{New Families in Kutcha Houses}}{\text{Number of Kutcha Houses}}$$

$$= \frac{1,85,300}{\left(\frac{1,20,000}{9}\right)} = \frac{16677}{1200} \approx 13.9$$

4. (*a*): We have, Total number of people who migrate Delhi $\times \frac{20}{100} \times \frac{15}{100} = 24000$

$\Rightarrow$ Total number of people who migrate Delhi = 8 lakh

Total number of migrating people belonging to the age group B = 8 lakh $\times \frac{15}{100} = 1.2$ lac

5. (*c*): Maximum differences $= \frac{8 \text{ lakh} \times (40-10)}{100} = 8 \text{ lac} \times \frac{30}{100} = 2.4$ lac

6. (*b*): Number of people coming from R having age group $800000 \times \frac{10}{100} \times \frac{50}{100} = 40,000$

Number of people coming from Q having age group D $= 800000 \times \frac{30}{100} \times \frac{20}{100} = 48,000$

Required difference = 48000 – 40,000 = 8000

7. (*d*) : Total number of employees for category A in the following year:

$$2019 = \frac{20 \times 4000}{100} = 800$$

$$2020 = \frac{22 \times 5000}{100} = 1100$$

For category B total number of employees in:

$$2019 = \frac{4000 \times 6}{100} = 240$$

$$2020 = \frac{5000 \times 10}{100} = 500$$

For category C total number of employees in:

$$2019 = \frac{10 \times 4000}{100} = 400$$

$$2020 = \frac{5000 \times 11}{100} = 550$$

For category D total number of employees in:

$$2019 = \frac{4000 \times 15}{100} = 600$$

$$2020 = \frac{5000 \times 9}{100} = 450$$

For category E total number of employees in:

$$2019 = \frac{4000 \times 27}{100} = 1080$$

$$2020 = \frac{5000 \times 27}{100} = 1350$$

For category F total number of employees in:

$$2019 = \frac{4000 \times 22}{100} = 880$$

$$2020 = \frac{5000 \times 21}{100} = 1050$$

Number of employees in C and D in 2019 = 400 + 600 = 1000

Number of employees in C and D in 2020 = 550 + 450 = 1000

$\Rightarrow$ Required categories = C and D

8. (*b*) : Percentage change in 2020 as compared to 2019 for different categories are following:

$$A = \frac{300}{800} \times 100 = 37.5\%;$$

$$B = \frac{260}{240} \times 100 = 108.33\%$$

$$C = \frac{150}{400} \times 100 = 37.5\%$$

$$D = \frac{-150}{600} \times 100 = -25\%$$

$$E = \frac{270}{1080} \times 100 = 25\%$$

$$F = \frac{170}{880} \times 100 = 19.3\%$$

Hence, percentage change for B is maximum.

9. (*b*) : Required percentage $= \frac{1350 - 1080}{1080} \times 100 = \frac{270}{1080} \times 100 = 25\%$

10. (*d*) : Hence, from solution 8, percentage change in categories D and E are same.

III. BAR DIAGRAMS

Multiple Bar Diagrams. In such bar diagrams two or more adjacent vertical bars are drawn to represent two or more phenomenon for the same place or period.

Ex. 1. ***The following bar diagram shows the registration of cars and total vehicles during first six months of the year 1994 in Delhi.***

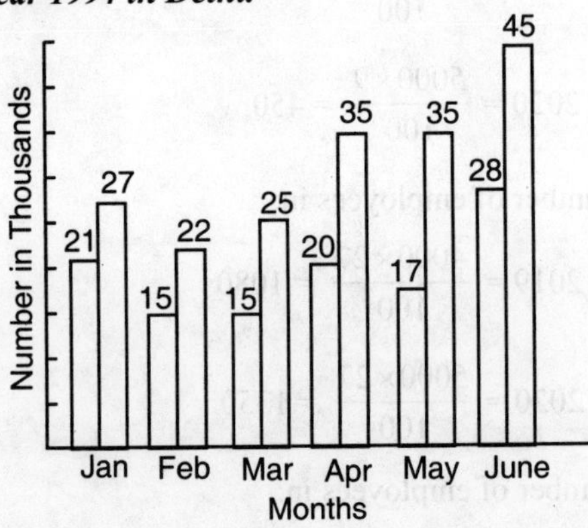

Study the above diagram and answer the questions given below:

1. What was the increase in registration of vehicles other than cars from January to April I994?

(*a*) 5000 (*b*) 10000 (*c*) 15000 (*d*) 9000

2. What was the percentage increase in registration of ears from May to June I994?

(*a*) 11% (*b*) 39.28% (*c*) 64.7% (*d*) None

3. What was the number of vehicles other than cars registered in March I994?

(*a*) 5000 (*b*) 10000 (*c*) 40000 (*d*) 37000

Answers & Solutions:

1. (*d*): Registration of other vehicles in January = 6000.

Registration of other vehicles in April = 15000

Increase = (15000 – 6000) = 9000.

2. (*c*): Registration of cars in May = 17000

Registration of cars in June = 28000

$$\text{Increase} = \left(\frac{11000}{17000}\times 100\right)\% = 64.7\%$$

3. (*c*): Number of vehicles other than cars registered in March 1994 = (25000— 15000) = 10000.

EXERCISE

Direction (1-4): Consider the following for the next four (04) items that follow:

The following Bar-Chart gives the production of cars by three different companies X, Y and Z in different years:

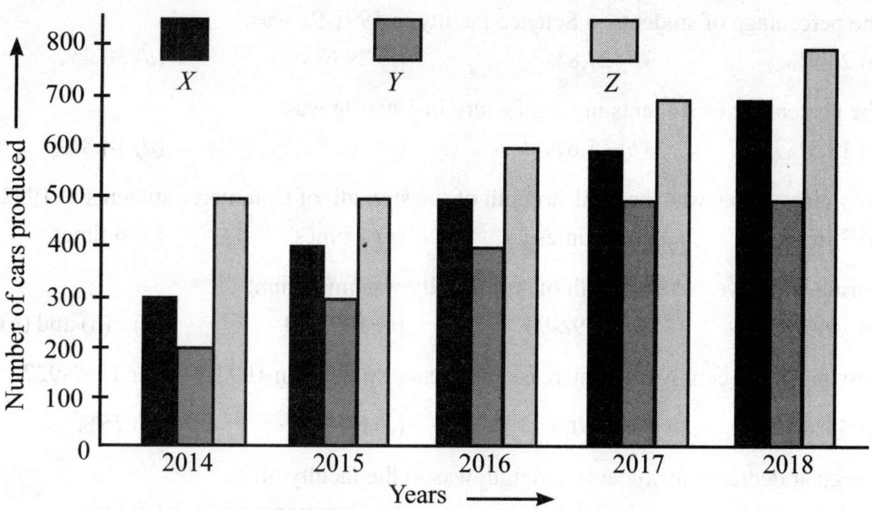

1. The percentage increase in the total production of cars from 2014 to 2018 was

(*a*) 50% (*b*) 100% (*c*) 150% (*d*) 200%

2. Consider the production of each company separately. In how many instances was the percentage increase in production of cars over the previous year's production greater than 20%?

(*a*) 5 (*b*) 6 (*c*) 7 (*d*) 8

3. In which year was the percentage in- crease in production of cars minimum as compared to its previous year?

(*a*) 2015 (*b*) 2016 (*c*) 2017 (*d*) 2018

4. In one of the years the percentage increase in production of cars was minimum as compared to its previous year. What was the minimum percentage?

(*a*) $11\frac{1}{9}\%$ (*b*) $11\frac{2}{9}\%$ (*c*) $13\frac{1}{3}\%$ (*d*) 20%

Direction (5.10): ***Shown below is the multiple bar diagram depicting the changes in the students' strength of a college in four faculties from 1991-92 to 1993-94. (scale 1 cm = 100)***

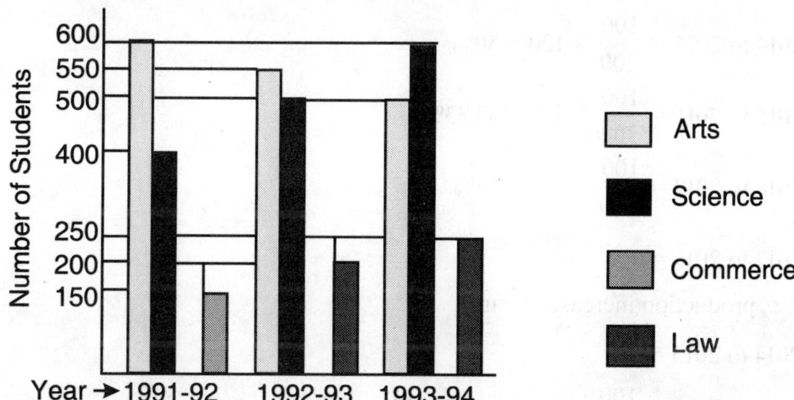

Study the above multiple bar chart and mark a tick (✓) against the correct answer in each of the following questions.

5. The percentage of students in Science faculty in 1991-92 was:

(*a*) 26.9% (*b*) 27.8% (*c*) 29.6% (*d*) 30.2%

6. The percentage of students in Law faculty in 1993-94 was:

(*a*) 18.5% (*b*) 15.6% (*c*) 16.7% (*d*) 14.8%

7. How many times was the total strength of the strength of Commerce students in 1992-93?

(*a*) 3 times (*b*) 4 times (*c*) 5 times (*d*) 6 times

8. During which year the strength of Arts faculty was minimum?

(*a*) 1991-92 (*b*) 1992-93 (*c*) 1993-94 (*d*) (*a*) and (*b*)

9. How much per cent was the increase in Science students in 1993-94 over 1991-92?

(*a*) 50% (*b*) 150% (*c*) $66\frac{2}{3}\%$ (*d*) 75%

10. A regular decrease in students' strength was in the faculty of:

(*a*) Arts (*b*) Science . (*c*) Commerce (*d*) Law

Answers & Solutions:

1. (*b*): Required percentage $= \frac{(700+500+800)-(300+200+500)}{300+200+500} = \frac{2000-1000}{1000} \times 100$

$= \frac{1000}{1000} \times 100 = 100\%$

2. (*a*): For car X, production increase of car from 2014 to 2015 $= \frac{100}{300} \times 100 = 33.33\%$

From 2015 to 2016 $= \frac{100}{400} \times 100 = 25\%$

From 2016 to 2017 $= \frac{100}{500} \times 100 = 20\%$

From 2017 to 2018 $= \frac{100}{600} \times 100 = 16.66\%$

For car Y, production increase of car:

From 2014 to 2015 $= \frac{100}{200} \times 100 = 50\%$

From 2015 to 2016 $= \frac{100}{300} \times 100 = 33.33\%$

From 2016 to 2017 $= \frac{100}{400} \times 100 = 25\%$

From 2017 to 2018 = 0

For car Z, production increase of car:

From 2014 to 2015 = 0

From 2015 to 2016 $= \frac{100}{500} \times 100 = 20\%$

From 2016 to 2017 = $\frac{100}{600} \times 100 = 16.66\%$

From 2017 to 2018 = $\frac{100}{700} \times 100 = 14.28\%$

Hence, required number of instances = 5 times

3. (*b*): Percentage of production increase from 2014 to 2015 = $\frac{(400+300+500)\ (300+200+500)}{300+200+500}$

$= \frac{1200-1000}{1000} \times 100 = \frac{200}{1000} \times 100 = 20\%$

Percentage of production increase from 2015 to 2016 = $\frac{(500+400+600)-(400+300+500)}{(400+300+500)}$

$= \frac{300}{1200} \times 100 = 25\%$

Percentage of production increase from 2016 to 2017 = $\frac{(600+500+700)-1500}{1500} \times 100$

$= \frac{300}{1500} \times 100 = 20\%$

Percentage of production increase from 2017 to 2018 = $\frac{(700+500+800)-1800}{1800} \times 100$

$= \frac{200}{1800} \times 100 \frac{100}{9} = 11.11\%$

Hence, in 2018, the percentage increase in production of cars was minimum as compared to its previous years.

4. (*a*): The minimum percentage = $\frac{100}{9}\% = 11\frac{1}{9}\%$

5. (*c*): Total number of students in 1991-92 = (600+400+200+ 150) = 1350.
Number of Science students in 1991-92 was 400.

Percentage of Science students in 1991-92 = $\left(\frac{400}{1350} \times 100\right)\% = 29.6\%$

6. (*b*): Total number of students in 1993-94 = (500 + 600 + 250 + 250) = 1600.
Number of Law students in 1993-94 is 250.

Percentage of Law students in 1993-94 = $\left(\frac{250}{1600} \times 100\right)\% = 15.6\%$

7. (*d*): Total strength in 1992–93 = (550 + 500 + 250 + 200)1 = 1500.

$\therefore \frac{\text{Total strength}}{\text{(Strength of Commerce students)}} = \frac{1500}{250} = 6.$

8. (*c*): A slight look indicates that the strength in Arts faculty in 1991-92, 1992-93 & 1993-94 was 550, 600 and 500 respectively. So, it was minimum in 1993-94.

9. (*a*): Number of Science students in 1991-92 was 400.

Number of Science students in 1993-94 was 600.

Percentage increase = $\left(\frac{200}{400}\times100\right)\% = 50\%$.

10. (*a*): As the diagram shows, the decrease every year is in Arts faculty.

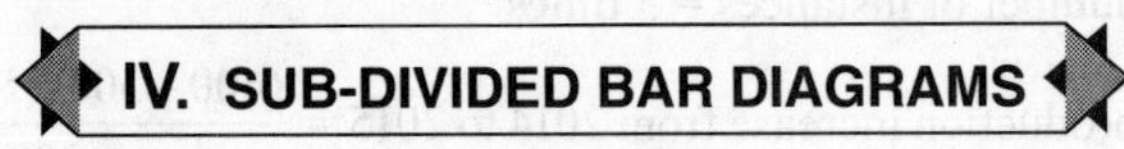

IV. SUB-DIVIDED BAR DIAGRAMS

EXERCISE

Sub-divided bar diagrams: In such diagrams every column is divided into certain parts to represent different phenomenon for the same period or place.

1. The monthly expenditure of a person is 6,000. The distribution of expenditure on various items is as follows: **[CDS (I) Exam, 2019]**

Item of expenditure	Amount (in ₹)
1. Food	2,000
2. Clothing	660
3. Fuel and rent	1,200
4. Education	480
5. Miscellaneous.	1,660

If the above data is represented by a percentage bar diagram of height 15 cm, then what are the lengths of the two segments of the bar diagram corresponding to education and miscellaneous respectively?

(*a*) 1.25 cm and 5 cm (*b*) 1.2 cm and 4.15 cm

(*c*) 1.2 cm and 3.5 cm (*d*) 4.15 cm and 6 cm

Direction (2-7): ***The sub-divided bar diagram given alongside depicts the result of B.Sc. students of a college for three years.***

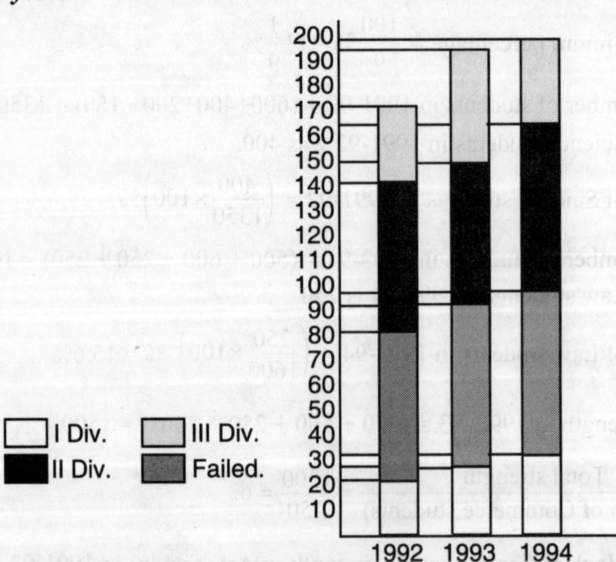

Study the bar diagram and mark a tick (✓) against the correct answer in each question.

2. How many per cent passed in 1st division in 1992?

(*a*) 20% (*b*) 34% (*c*) $14\frac{2}{7}$ % (*d*) $11\frac{13}{17}$ %

3. What was the pass percentage in 1992?

(*a*) 65% (*b*) 70% (*c*) 74.6% (*d*) 88.8%

4. In which year the college had the best result for B.Sc.?

(*a*) 1992 (*b*) 1993 (*c*) 1994 (*d*) Cannot be determined

5. What is the number of third divisioners in 1994?

(*a*) 165 (*b*) 75 (*c*) 70 (*d*) 65

6. What is the percentage of students in 1994 over 1992?

(*a*) 30% (*b*) $17\frac{11}{17}$ % (*c*) $117\frac{11}{17}$ % (*d*) 85%

7. What is the aggregate pass percentage during three years 'T

(*a*) $51\frac{2}{3}$ % (*b*) 82.7% (*c*) 80.4% (*d*) 71.6%

Direction (8-11): ***Adjoining bar diagram shows the monthly expenditure of two families on food, clothing, education, fuel, house rent and miscellaneous (in percentage).***

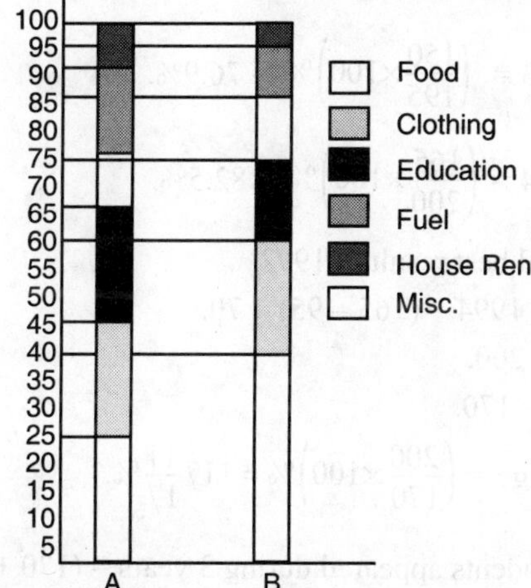

Study the above diagram and mark a tick (✓) against the correct answer in each question.

8. What fraction of the total expenditure is spent on Education in family *A*?

(*a*) $\frac{13}{20}$ (*b*) $\frac{2}{3}$ (*c*) $\frac{9}{13}$ (*d*) None of these

9. If the total annual expenditure of family *B* is ₹ 10,000 then money spent on clothes during the year is:

(*a*) ₹ 200 (*b*) ₹ 2000 (*c*) ₹ 600 (*d*) ₹ 6000

10. If the total annual expenditure of family *A* is ₹ 30,000 then money spent on food, clothes and house rent is:

(*a*) ₹ 18,500 (*b*) ₹ 18,000 (*c*) ₹ 21,000 (*d*) ₹ 15,000

11. If both the families have the same expenditure, which one spends more on education and miscellaneous together?

(*a*) Family *A* (*b*) Family *B*

(*c*) None (*d*) Cannot be determined

ANSWERS & SOLUTIONS

1. (*b*) We have maximum heights = 15 cm

Total expenditure = ₹6000

$\Rightarrow$ Expenditure on education = $\frac{15}{6000} \times 4.80 = 1.2$ cm

Expenditure on miscellaneous = $\frac{15}{6000} \times 1660 = \frac{166}{40} = 4.15$ cm

2. (*d*): Percentage of Ist divisioners = $\left(\frac{20}{170} \times 100\right) = 11\frac{13}{17}\%$.

3. (*d*): Total students passed = 140.

Total students appeared = 170.

Pass percentage = $\left(\frac{140}{170} \times 100\right)\% = 88.8\%$.

4. (*a*): Pass percentage in 1992 = $\left(\frac{140}{172} \times 100\right)\% = 88.8\%$.

Pass percentage in 1993 = $\left(\frac{150}{195} \times 100\right)\% = 76.9\%$.

Pass percentage in 1994 = $\left(\frac{165}{200} \times 100\right)\% = 82.5\%$.

So, the college recorded best result in 1992.

5. (*c*): Third diversions in 1994 = (165 – 95) = 70.

6. (*c*): Students in 1994 = 200.

Students in 1992 = 170.

Required percentage $= \left(\frac{200}{170} \times 100\right)\% = 117\frac{11}{17}\%$.

7. (*c*): Total number of students appeared during 3 years = (170 +195 + 200) = 565.

Total number of students passed during 3 years = (140 +150 + 165) = 455.

Aggregate pass percentage = $\left(\frac{455}{565} \times 100\right)\% = 80.4\%$.

8. (*d*): In family *A*, money spent on education = 20% = $\frac{20}{100} = \frac{1}{5}$

9. (*b*): In family *B*, the money spent on clothes = (20% of total expenditure)

$= ₹\left(\frac{20}{100} \times 10{,}000\right) = ₹2000.$

10. (*b*): Money spent on food, clothes and house rent in family *A*

= (30 +15 + 15) = 60% of total expenditure = $₹\left(\frac{60}{100} \times 30{,}000\right) = ₹18{,}000.$

11. (*a*): Family *A* spends on education & miscellaneous = (20 + 10) = 30%.

Family *B* spends on education & miscellaneous = (15 + 5) = 20%.

So, family *A* spends more on these heads.

CDS Examination (II) 2025

1. Let $p(x)$ be a polynomial. When $p(x)$ is divided by $(x-1)$, it leaves 2 as the remainder. When $p(x)$ is divided by $(x-2)$, it leaves 1 as the remainder. What is the remainder when $p(x)$ is divided by $(x-1)(x-2)$?

(*a*) 3 (*b*) -3 (*c*) $3-x$ (*d*) $3-2x$

2. Consider the following in respect of a positive real number x:

I. $x+\frac{1}{x}>1$ II. $\left(x+\frac{1}{x}\right)^2>2$ III. $\left(x+\frac{1}{x}\right)^4>9$

Which of the above are correct?

(*a*) I and II only (*b*) II and III only (*c*) I and III only (*d*) I, II and III

3. Let p and q be natural numbers such that $q > p$. What is the largest value of p such that q^2-5p-4 is negative?

(*a*) 3 (*b*) 4 (*c*) 5 (*d*) 6

4. Let x and y be natural numbers, each less than 20, such that x, y, $x+y$ and $x-y$ are prime numbers. How many such combinations of $(x, y, x+y, x-y)$ are possible?

(*a*) One (*b*) Two (*c*) Three (*d*) None of these

5. If $(x+1)(x+p)(x^2+p^2)=x^4-1$, then what is the value of p?

(*a*) -1 (*b*) 0
(*c*) 1 (*d*) Cannot be determined

6. If $(2+\sqrt{3})^x+(2-\sqrt{3})^x=2$, then what is $(2+\sqrt{3})^x-(2-\sqrt{3})^x$ equal to?

(*a*) 0 (*b*) 0.5 (*c*) 1 (*d*) 1.5

7. If $\frac{1}{a}+\frac{1}{b}=\frac{1}{6}$ and $\frac{1}{a^2}+\frac{1}{b^2}=\frac{13}{36}$, then what is $\frac{1}{a^3}+\frac{1}{b^3}$ equal to?

(*a*) $\frac{31}{216}$ (*b*) $\frac{35}{216}$ (*c*) $\frac{37}{216}$ (*d*) $\frac{41}{216}$

8. What is the remainder when x^6 is divided by x^2+1

(*a*) -1 (*b*) 0 (*c*) 1 (*d*) $x+1$

9. $(x+2)$ is a factor of which one of the following?

(*a*) $x^5-4x^4-3x^3+8x^2-14x+12$ (*b*) $x^5+4x^4-3x^3+8x^2-14x+12$
(*c*) $x^5-4x^4+3x^3+8x^2-14x+12$ (*d*) $x^5-4x^4-3x^3+8x^2+14x+12$

10. If $\log_{10}2=0.301$ and $\log_{10}3=0.477$, then what is the number of digits in the expansion of 60^{60}?

(*a*) 105 (*b*) 106 (*c*) 107 (*d*) 108

11. What is the remainder when $17^{25}+19^{25}$ is divided by 18?

(*a*) 0 (*b*) 1 (*c*) 3 (*d*) 9

12. The HCF of x and y is H. Consider the following statements in respect of the HCF of

$p=\dfrac{x^3+y^3}{x^2-xy+y^2}$ and $q=\dfrac{x^3-y^3}{x^2+xy+y^2}$:

I. The HCF of p and q can be H.

II. The HCF of p and q can be 2H.

Which of the statements given above is/are correct?

(*a*) I only (*b*) II only (*c*) Both I and II (*d*) Neither I nor II

13. If $x^4=x^2+1$, where $x>0$, then what is $2x^4$ equal to?

(*a*) $2+\sqrt{3}$ (*b*) $3+\sqrt{5}$ (*c*) $5-2\sqrt{3}$ (*d*) $3-\sqrt{5}$

14. If $\dfrac{p+q}{q+r}=\dfrac{r+s}{s+p}$; $(q+r)\neq 0, (s+p)\neq 0$,, then which one of the following is correct?

(*a*) $p+q+r+s=0$ (*b*) $p=r$

(c) Either $p+q+r+s=0$ or $p=r$ (*d*) None of the above

15. If n is a natural number less than 7, then what is the number of values of n for which $(12n+2)$ and $(8n+1)$ are relatively prime?

(*a*) 6 (*b*) 5 (*c*) 4 (*d*) 3

16. What is the HCF of $x^3+y^3+3xy-1$ and $(x+y)^4-1$?

(*a*) $x+y$ (*b*) $x+y+1$ (*c*) $x+y-1$ (*d*) 1

17. Let $x=n(n+1)(n+2)$, where n is an even natural number. Which of the following statements is/are correct?

I. x is always divisible by 48.

II. x^2 is always divisible by 144.

Select the answer using the code given below.

(*a*) I only (*b*) II only (*c*) Both I and II (*d*) Neither I nor II

18. What is the LCM of $x^4+x^2y^2+y^4, x^3y+y^4$ and $x^4y^2-x^3y^3$?

(*a*) $x^3y^3(x^6-y^6)$ (*b*) $x^3y^2(x^6-y^6)$ (*c*) $x^3y(x^6-y^6)$ (*d*) $xy(x^6-y^6)$

19. Let XYZ be a 3-digit number. Let *D* be the difference between *XYZ* and *ZYX*. What is the remainder when *D* is divided by 99?

(*a*) 0 (*b*) 1 (*c*) 7 (*d*) 9

20. Let p and q be two natural numbers such that $(p+q)^{p+q}$ is divisible by 512. What is the least value of $p+q$?

(*a*) 4 (*b*) 6 (*c*) 8 (*d*) 12

21. What is $\dfrac{(a+b)^2}{(c-a)(c+a+b)}+\dfrac{(a+b)c}{c^2+bc-a^2-ab}-\dfrac{(a+2b+c)}{2(c-a)}$, $a\neq b, b\neq c, c\neq a$ equal to?

(*a*) $-\dfrac{1}{2}$ (*b*) 0 (*c*) $\dfrac{1}{2}$ (*d*) 1

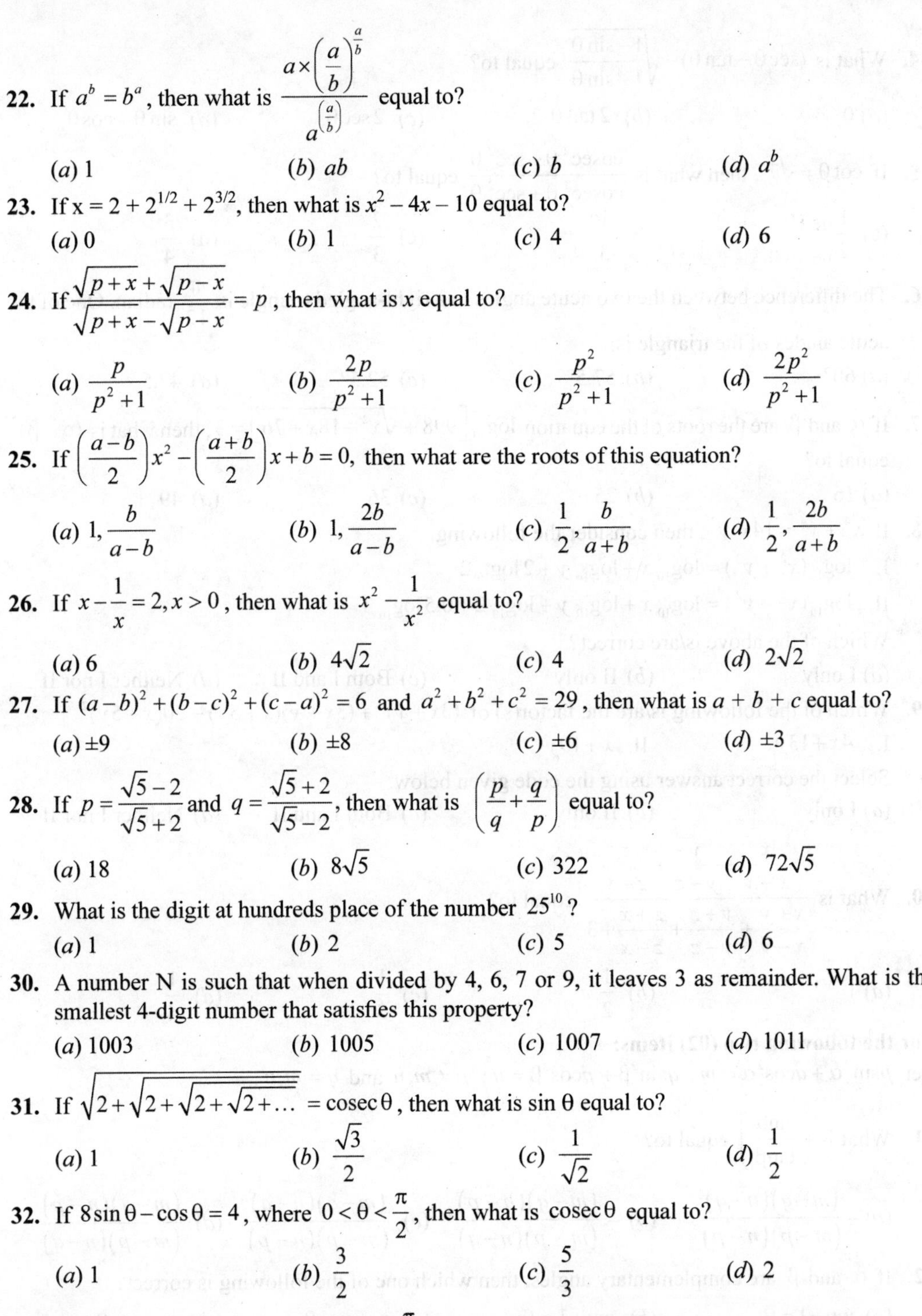

22. If $a^b = b^a$, then what is $\dfrac{a \times \left(\dfrac{a}{b}\right)^{\frac{a}{b}}}{a^{\left(\frac{a}{b}\right)}}$ equal to?

(*a*) 1 (*b*) *ab* (*c*) *b* (*d*) a^b

23. If $x = 2 + 2^{1/2} + 2^{3/2}$, then what is $x^2 - 4x - 10$ equal to?

(*a*) 0 (*b*) 1 (*c*) 4 (*d*) 6

24. If $\dfrac{\sqrt{p+x}+\sqrt{p-x}}{\sqrt{p+x}-\sqrt{p-x}} = p$, then what is *x* equal to?

(*a*) $\dfrac{p}{p^2+1}$ (*b*) $\dfrac{2p}{p^2+1}$ (*c*) $\dfrac{p^2}{p^2+1}$ (*d*) $\dfrac{2p^2}{p^2+1}$

25. If $\left(\dfrac{a-b}{2}\right)x^2 - \left(\dfrac{a+b}{2}\right)x + b = 0,$ then what are the roots of this equation?

(*a*) $1, \dfrac{b}{a-b}$ (*b*) $1, \dfrac{2b}{a-b}$ (*c*) $\dfrac{1}{2}, \dfrac{b}{a+b}$ (*d*) $\dfrac{1}{2}, \dfrac{2b}{a+b}$

26. If $x - \dfrac{1}{x} = 2, x > 0$, then what is $x^2 - \dfrac{1}{x^2}$ equal to?

(*a*) 6 (*b*) $4\sqrt{2}$ (*c*) 4 (*d*) $2\sqrt{2}$

27. If $(a-b)^2 + (b-c)^2 + (c-a)^2 = 6$ and $a^2 + b^2 + c^2 = 29$, then what is $a + b + c$ equal to?

(*a*) ±9 (*b*) ±8 (*c*) ±6 (*d*) ±3

28. If $p = \dfrac{\sqrt{5}-2}{\sqrt{5}+2}$ and $q = \dfrac{\sqrt{5}+2}{\sqrt{5}-2}$, then what is $\left(\dfrac{p}{q} + \dfrac{q}{p}\right)$ equal to?

(*a*) 18 (*b*) $8\sqrt{5}$ (*c*) 322 (*d*) $72\sqrt{5}$

29. What is the digit at hundreds place of the number 25^{10}?

(*a*) 1 (*b*) 2 (*c*) 5 (*d*) 6

30. A number N is such that when divided by 4, 6, 7 or 9, it leaves 3 as remainder. What is the smallest 4-digit number that satisfies this property?

(*a*) 1003 (*b*) 1005 (*c*) 1007 (*d*) 1011

31. If $\sqrt{2+\sqrt{2+\sqrt{2+\sqrt{2+\ldots}}}} = \operatorname{cosec}\theta$, then what is $\sin\theta$ equal to?

(*a*) 1 (*b*) $\dfrac{\sqrt{3}}{2}$ (*c*) $\dfrac{1}{\sqrt{2}}$ (*d*) $\dfrac{1}{2}$

32. If $8\sin\theta - \cos\theta = 4$, where $0 < \theta < \dfrac{\pi}{2}$, then what is $\operatorname{cosec}\theta$ equal to?

(*a*) 1 (*b*) $\dfrac{3}{2}$ (*c*) $\dfrac{5}{3}$ (*d*) 2

33. If $2\tan\theta = \sec^2\theta - 2$, where $0 < \theta < \dfrac{\pi}{2}$, then what is $\cot\theta$ equal to?

(*a*) $\sqrt{2}-1$ (*b*) $\sqrt{2}+1$ (*c*) $\sqrt{3}-1$ (*d*) $\sqrt{3}+2$

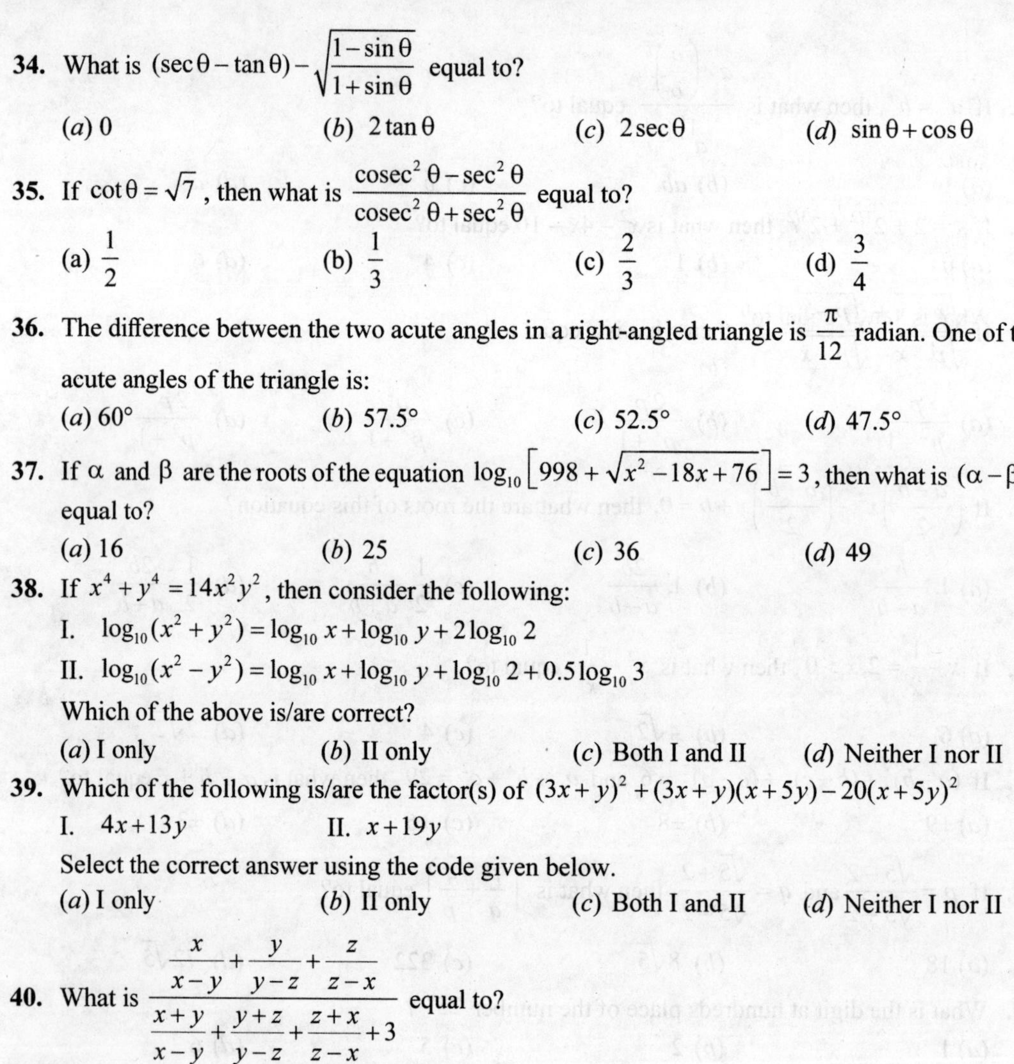

34. What is $(\sec\theta - \tan\theta) - \sqrt{\dfrac{1-\sin\theta}{1+\sin\theta}}$ equal to?

(*a*) 0 (*b*) $2\tan\theta$ (*c*) $2\sec\theta$ (*d*) $\sin\theta + \cos\theta$

35. If $\cot\theta = \sqrt{7}$, then what is $\dfrac{\operatorname{cosec}^2\theta - \sec^2\theta}{\operatorname{cosec}^2\theta + \sec^2\theta}$ equal to?

(a) $\dfrac{1}{2}$ (b) $\dfrac{1}{3}$ (c) $\dfrac{2}{3}$ (d) $\dfrac{3}{4}$

36. The difference between the two acute angles in a right-angled triangle is $\dfrac{\pi}{12}$ radian. One of the acute angles of the triangle is:

(*a*) 60° (*b*) 57.5° (*c*) 52.5° (*d*) 47.5°

37. If α and β are the roots of the equation $\log_{10}\left[998 + \sqrt{x^2 - 18x + 76}\right] = 3$, then what is $(\alpha - \beta)^2$ equal to?

(*a*) 16 (*b*) 25 (*c*) 36 (*d*) 49

38. If $x^4 + y^4 = 14x^2y^2$, then consider the following:

I. $\log_{10}(x^2 + y^2) = \log_{10} x + \log_{10} y + 2\log_{10} 2$

II. $\log_{10}(x^2 - y^2) = \log_{10} x + \log_{10} y + \log_{10} 2 + 0.5\log_{10} 3$

Which of the above is/are correct?

(*a*) I only (*b*) II only (*c*) Both I and II (*d*) Neither I nor II

39. Which of the following is/are the factor(s) of $(3x + y)^2 + (3x + y)(x + 5y) - 20(x + 5y)^2$

I. $4x + 13y$ II. $x + 19y$

Select the correct answer using the code given below.

(*a*) I only (*b*) II only (*c*) Both I and II (*d*) Neither I nor II

40. What is $\dfrac{\dfrac{x}{x-y} + \dfrac{y}{y-z} + \dfrac{z}{z-x}}{\dfrac{x+y}{x-y} + \dfrac{y+z}{y-z} + \dfrac{z+x}{z-x} + 3}$ equal to?

(*a*) 1 (*b*) $\dfrac{1}{2}$ (*c*) $\dfrac{1}{3}$ (*d*) $\dfrac{1}{4}$

For the following two (02) items:

Let $p\sin^2\alpha + q\cos^2\alpha = m$, $q\sin^2\beta + p\cos^2\beta = n$; $p \neq m, n$ and $q \neq m, n$.

41. What is $\left(\dfrac{\tan\alpha}{\tan\beta}\right)^2$ equal to?

(*a*) $-\dfrac{(m-q)(n-q)}{(m-p)(n-p)}$ (*b*) $-\dfrac{(m-q)(n-p)}{(m-p)(n-q)}$ (*c*) $\dfrac{(m-q)(n-q)}{(m-p)(n-p)}$ (*d*) $\dfrac{(m-q)(n-p)}{(m-p)(n-q)}$

42. If α and β are complementary angles, then which one of the following is correct?

(*a*) $mn - 1 = 0$ (*b*) $mn + 1 = 0$ (*c*) $m + n = 0$ (*d*) $m - n = 0$

For the following two (02) items: Let $\operatorname{cosec}\theta - \sin\theta = p$ and $\sec\theta - \cos\theta = q$.

43. What is $p\sin\theta + q\cos\theta$ equal to?

(*a*) –1 (*b*) 0 (*c*) 1 (*d*) 2

44. What is $p^2q^2\left(p^2+q^2+3\right)$ equal to?

(*a*) 0 (*b*) 1 (*c*) 2 (*d*) 4

For the following two (02) items:

Let $\dfrac{\sin\alpha}{\sin\beta}=\dfrac{4\sqrt{2}}{3}$ and $\dfrac{\cos\alpha}{\cos\beta}=\dfrac{2\sqrt{3}}{9}$

45. What is $\tan^2\alpha$ equal to?

(*a*) 8 (*b*) 6 (*c*) 4 (*d*) 3

46. What is $\tan^2\beta$ equal to?

(*a*) $\dfrac{1}{\sqrt{2}}$ (*b*) $\dfrac{3}{\sqrt{2}}$ (*c*) $\dfrac{1}{3}$ (*d*) $\dfrac{2}{3}$

For the following two (02) items:

Let $\dfrac{1+\sin\theta}{\cos\theta}=p+\sqrt{p^2+1}.$

47. What is sec θ equal to?

(*a*) p (*b*) $\sqrt{p^2+1}$ (*c*) $\dfrac{1}{\sqrt{p^2+1}}$ (*d*) $\dfrac{p}{\sqrt{p^2+1}}$

48. What is $\tan\theta$ equal to?

(*a*) p (*b*) $\sqrt{p^2+1}$ (*c*) $\dfrac{\sqrt{p^2+1}}{1}$ (*d*) $\dfrac{p}{\sqrt{p^2+1}}$

For the following two (02) items:

Let $\sin\theta+\cos\theta=p$ and $\sec\theta+\operatorname{cosec}\theta=q$. Where $p\neq 1$.

49. What is the relation between p and q?

(*a*) $p=q\left(p^2-1\right)$ (*b*) $2p=q\left(p^2-1\right)$ (*c*) $q=p^2-1$ (*d*) $2q=p\left(p^2-1\right)$

50. What is $\tan\theta+\cot\theta$ equal to?

(*a*) $\dfrac{p}{q}$ (*b*) $\dfrac{q}{p}$ (*c*) $\dfrac{2p}{q}$ (*d*) $\dfrac{2q}{p}$

51. Two trains X and Y are travelling in the same direction at 100 km/hr and 60 km/hr respectively. Train X crosses a man in train Y in 9 seconds. What is the length of train X?

(*a*) 80 m (*b*) 100 m (*c*) 120 m (*d*) 150 m

52. Two persons X and Y leave place P for place Q at 7:00 a.m. and 7:10 a.m. respectively along the same path. X walks at a speed of 4.8 km/hr and Y walks at a speed of 6 km/hr. How many kilometres from place P will X meet Y?

(*a*) 3 km (*b*) 3.5 km (*c*) 4 km (*d*) 4.5 km

53. There are two employees X and Y. X's salary is first increased by 12% and then decreased by 10%, and Y's salary is first increased by 10% and then decreased by 12%. If their salaries at present are equal, then what was the ratio of initial salary of X to initial salary of Y?

(*a*) 50 : 53 (*b*) 51 : 53 (*c*) 121 : 126 (*d*) 121 : 125

54. In a village consisting of p persons, $x\%$ can read and write. Of the males, only $y\%$ can read and write. Of the females, only $z\%$ can read and write. If $x, y > z$, then what is the number of males in the village?

(*a*) $\frac{p(x-z)}{y-z}$ (*b*) $\frac{p(y-z)}{x-z}$ (*c*) $\frac{px}{y}$ (*d*) $\frac{py}{x}$

55. X and Y are two alloys of copper (Cu) and zinc (Zn). Alloy X is prepared by mixing Cu and Zn in the ratio 5:4, and alloy Y is prepared by mixing Cu and Zn in the ratio 5:13 respectively. If equal quantities of alloys X and Y are melted to form a third alloy Z, then what is the ratio of Cu to Zn in Z?

(*a*) 5 : 8 (*b*) 5 : 7 (*c*) 6 : 7 (*d*) 7 : 8

56. Two students X and Y appeared in a test. The score of X is 20 more than that of Y. If the score of *X* is 75% of the sum of the scores of X and Y, then what is the ratio of score of X to score of Y?

(*a*) 5 : 1 (*b*) 4 : 1 (*c*) 3 : 1 (*d*) 3 : 2

57. If one root of the equation $2x^2 - 5px + 2p^2 = 0$ exceeds the other by 4, then what is the value of p?

(*a*) $\frac{8}{3}$ (*b*) $\frac{4}{3}$ (*c*) $\frac{2}{3}$ (*d*) $\frac{1}{3}$

58. An amount of ₹10,000 is borrowed at 10% per annum on compound interest for 3 years, compounded annually, and paid back in 3 equal annual installments during these years. What is the amount of each installment (approximately)?

(*a*) ₹4,437 (*b*) ₹4,237 (c ₹4,021 (*d*) ₹3,811

59. What is the solution of the inequalities $5x+3<8x-9$ and $2x+20>5x+2$?

(*a*) $4<x<6$ (*b*) $3<x<5$ (*c*) $x<3$ or $x>5$ (*d*) $x<4$ or $x>6$

60. A shopkeeper gives three consecutive discounts 10%, 20% and 25% after which he sells the article at a profit of 8% on the cost price. Had he sold the article after the first discount, how much profit would he have got?

(*a*) 20% (*b*) 40% (*c*) 50% (*d*) None of the above

61. Let AD be the altitude of triangle ABC. If $(AB+AC)=p$, $(AB-AC)=q$ and $(BD-CD)=r$, then what is BC equal to?

(*a*) $\frac{qr}{p}$ (*b*) $\frac{pr}{q}$ (*c*) $\frac{pq}{r}$ (*d*) $p+q-r$

62. The sum of the height and the radius of a right circular cylinder is 21 cm, and the radius is less than the height. If the curved surface area of the cylinder is 616 cm^2, then what is the volume of the cylinder? (Take $\pi = \frac{22}{7}$)

(*a*) 1078 cm^3 (*b*) 1617 cm^3 (*c*) 1927 cm^3 (*d*) 2156 cm^3

63. *ABC* is a triangle right angled at *B*. *P* is the midpoint of *AB* and *Q* is the midpoint of *BC*. Consider the following:

I. $AQ = \sqrt{73}$ units

II. $CP = \sqrt{52}$ units

Which of the above is/are required to determine the area of the triangle?

(*a*) I only (*b*) II only

(*c*) Both I and II (*d*) More information is needed

64. The length, breadth and height of a cuboid are consecutive integers. If the volume of the cuboid is 336 cubic units, then what is the total surface area of the cuboid?

(*a*) 288 sq. units

(*b*) 292 sq. units

(*c*) 296 sq. units

(*d*) Cannot be determined due to insufficient data

65. In a circle of radius 14 cm, APB is a shorter arc and P is the midpoint of the arc. Let C be the midpoint of the chord AB and $PC = 7$ cm. What is the length of the chord AP?

(*a*) 3.5 cm (*b*) 7 cm (*c*) 10.5 cm (*d*) 14 cm

66. Two poles of heights 10 m and 15 m are 25 m apart. What is the height of the point of intersection of the lines joining the tip of each pole to the foot of the other pole?

(*a*) 4.8 m (*b*) 5 m (*c*) 6 m (*d*) 6.4 m

67. ABC is a triangle right angled at B. Further, $(AB + BC)$ exceeds AC by 10 units. If the perimeter of the triangle is 60 units, then what is the area of the triangle?

(*a*) 75 sq. units (*b*) 100 sq. units (*c*) 125 sq. units (*d*) 150 sq. units

68. Two poles are situated 24 m apart and their heights differ by 10 m. What is the distance between their tips?

(*a*) 25 m

(*b*) 26 m

(*c*) 30 m

(*d*) Cannot be determined due to insufficient data

69. Let *X*, *Y* and *Z* be the midpoints of the sides *BC*, *CA* and *AB* of triangle *ABC* respectively. Consider the following statements:

I. The quadrilateral *AZXY* is a parallelogram.

II. The area of the quadrilateral *AZXY* is half of the area of triangle *ABC*.

Which of the statements given above is/are correct?

(*a*) I only (*b*) II only (*c*) Both I and II (*d*) Neither I nor II

70. Consider the following angles:

I. 4° II. 5° III. 6° IV. 8°

How many of the above can be the exterior angle of a regular polygon?

(*a*) One (*b*) Two (*c*) Three (*d*) All four

71. In a triangle ABC, $\angle A = 30^\circ$, $AB = 7$ cm and $AC = 12$ cm. What is the area of triangle *ABC*?

(*a*) 21 cm² (*b*) $21\sqrt{3}$ cm² (*c*) 42 cm² (*d*) $42\sqrt{3}$ cm²

72. ABC is a triangle right angled at B. D is a point on AC such that BD is perpendicular to AC. If $AB = p$ and $BC = \sqrt{3}p$, then what is BD equal to?

(*a*) $\frac{p}{3}$ (*b*) $\frac{p}{2}$ (*c*) $\frac{\sqrt{3}p}{2}$ (*d*) $\frac{\sqrt{3}p}{4}$

73. The difference between an interior angle and an exterior angle of a regular polygon is 120°. What is the number of sides of the polygon?

(*a*) 9 (*b*) 10 (*c*) 11 (*d*) 12

74. An angle θ is exactly one-fourth of its complementary angle. What is the value of angle θ ?

(*a*) 12° (*b*) 15° (*c*) 18° (*d*) 20°

75. The sides of a triangle are 11 cm, 60 cm and 61 cm. What is the area of the triangle formed by joining the midpoints of the sides of the triangle?

(*a*) 165 cm² (*b*) 110 cm² (*c*) 82.5 cm² (*d*) 72.5 cm²

76. A pendulum swings through an angle of 30° and its end describes an arc of length 55 cm. What is the length of the pendulum? (Take $\pi = \frac{22}{7}$)

(*a*) 90 cm (*b*) 100 cm (*c*) 105 cm (*d*) 110 cm

77. A conical tent has an angle of 60° at the vertex. If the curved surface area is 100, then what is the volume of the tent?

(*a*) $\frac{250\sqrt{2}}{\sqrt{3\pi}}\,m^3$ (*b*) $\frac{500\sqrt{3}}{\sqrt{\pi}}\,m^3$ (*c*) $\frac{1000\sqrt{3}}{\sqrt{2\pi}}\,m^3$ (*d*) $\frac{1000\sqrt{3}}{\sqrt{\pi}}\,m^3$

78. A right circular cone and a hemisphere have equal base and equal volume. What is the ratio of the height of the cone to the radius of the hemisphere?

(*a*) 1 : 1 (*b*) 1 : 2 (*c*) 2 : 1 (*d*) 3 : 2

79. A wire is in the form of an equilateral triangle with an area of $36\sqrt{3}$. If it is changed into a semicircle, then what is its radius?

(*a*) $\frac{9}{\pi}$ cm (*b*) $\frac{18}{\pi+2}$ cm (*c*) $\frac{18}{\pi}$ cm (*d*) None of these

80. Let the area of the largest possible square inscribed in a circle of unit radius be *x*. Let the area of the largest possible circle inscribed in a square of unit side length be *y*. What is the relation between *x* and *y*?

(*a*) $\pi x = 2y$ (*b*) $2\pi x = y$ (*c*) $\pi x = 4y$ (*d*) $\pi x = 8y$

81. A question is given followed by two statements I and II. Consider the question and the statements.

Question : What is the remainder when $x^{2n} - y^{2n} + 1$ is divided by $x^n + y^n$, where *n* is a natural number?

Statement I: *n* is odd.

Statement II: *n* is even.

Which one of the following is correct in respect of the above question and the statements.

(*a*) The question can be answered by using one of the statements alone, but cannot be answered using the other statement alone

(*b*) The Question can be answered by using either Statement alone

(*c*) The Question can be answered by using both the Statements together, but cannot be answered using either Statement alone

(*d*) The Question can be answered even without using any of the Statements

82. A Question is given followed by two Statements I and II. Consider the Question and the Statements.

Question : The product of a natural number *N* and the number *M* written by the same digits of N in the reverse order is 252. What is the number N.

Statement I: $N + M = 33$

Statement II: $N > M$

Which one of the following is correct in respect of the above Question and the Statements?

(*a*) The Question can be answered by using one of the Statements alone, but cannot be answered using the other Statement alone

(*b*) The Question can be answered by using either Statement alone

(*c*) The Question can be answered by using both the Statements together, but cannot be answered using either Statement alone

(*d*) The Question can be answered even without using any of the Statements

83. A Question is given followed by two Statements I and II. Consider the Question and the Statements.

Question : The last digit in the expansion of the number $(54D)^{100}$ is 1. What is the value of digit D?

Statement I: $D > 5$

Statement II: D is a multiple of 3

Which one of the following is correct in respect of the above Question and the Statements?

(*a*) The Question can be answered by using one of the Statements alone, but cannot be answered using the other Statement alone

(*b*) The Question can be answered by using either Statement alone

(*c*) The Question can be answered· by using both the Statements together, but cannot be answered using either Statement alone

(*d*) The Question can be answered even without using any of the Statements

84. A Question is given followed by two Statements I and II. Consider the Question and the Statements.

Question : In a triangle ABC, $\angle A = \angle B - \angle C$. Is angle A acute?

Statement I: Triangle ABC is not an obtuse-angled triangle.

Statement II: Angle C is acute

Which one of the following is correct in respect of the above Question and the Statements?

(*a*) The Question can be answered by using one of the Statements alone, but cannot be answered using the other Statement alone

(*b*) The Question can be answered by using either Statement alone

(*c*) The Question can be answered by using both the Statements together, but cannot be answered using either Statement alone

(*d*) The Question can be answered even without using any of the Statements

85. A Question is given followed by two Statements I and II. Consider the question and the Statements.

Question : In triangle ABC right-angled at B, $AC = 20$ cm. What is the circumradius of the triangle?

Statement I: $AB = 12$ cm

Statement II: $BC = 16$ cm

Which one of the following is correct in respect of the above Question and the Statements?

(*a*) The Question can be answered by using one of the Statements alone, but cannot be answered using the other Statement alone

(*b*) The Question can be answered by using either Statement alone

(*c*) The Question can be answered by using both the Statements together, but cannot be answered using either Statement alone

(*d*) The Question can be answered even without using any of the Statements

86. A Question is given followed by two Statements I and II. Consider the Question and the Statements.

Question : *ABCD* is a parallelogram with $\angle ABC = 60^\circ$. If the area is $7\sqrt{3}$ sq. units, what is the perimeter of the parallelogram?

Statement I: The lengths of the sides AB and DA are prime numbers.

Statement II: The lengths of the sides are natural numbers each greater than 1 unit.

Which one of the following is correct in respect of the above Question and the Statements?

(*a*) The Question can be answered by using one of the Statements alone, but cannot be answered using the other Statement alone

(*b*) The Question can be answered by using either Statement alone

(*c*) The Question can be answered by using both the Statements together, but cannot be answered using either Statement alone

(*d*) The Question can be answered even without using any of the Statements

87. A Question is given followed by two Statements I and II. Consider the Question and the Statements.

Question: *AB* and CD are chords of a circle intersecting at *P. If* $AP \times PB = 48$, what is $CP \times PD$ equal to?

Statement I: $AP = 8$ units

Statement II: $CP = 10$ units

Which one of the following is correct in respect of the above Question and the Statements?

(*a*) The Question can be answered by using one of the Statements alone, but cannot be answered using the other Statement alone

(*b*) The Question can be answered by using either Statement alone

(*c*) The Question can be answered by using both the Statements together, but cannot be answered using either Statement alone

(*d*) The Question can be answered even without using any of the Statements

88. A Question is given followed by two Statements I and II. Consider the Question and the Statements.

Question : In a quadrilateral *ABCD*, $AB = 6$, $BC = 18$, $CD = 6$, $DA = 9$. What is the length of diagonal *BD*?

Statement I: The length of *BD* is an integer greater than 13.

Statement II: The length of *BD* is an even integer.

Which one of the following is correct in respect of the above Question and the Statements?

(*a*) The Question can be answered by using one of the Statements alone, but cannot be answered using the other Statement alone

(*b*) The Question can be answered by using either Statement alone

(*c*) The Question can be answered by using both the Statements together, but cannot be answered using either Statement alone

(*d*) The Question can be answered even without using any of the Statements

89. A Questions is given followed by two Statements I and II. Consider the Question and the Statements.

Question : *ABC* is an isosceles triangle with $AB = AC = 10$ units. If the area is 48 square units, what is the length of base *BC*?

Statement I: The length of *BC* is an even integer.

Statement II: The height of the triangle is greater than the length of half of the base.

Which one of the following is correct in respect of the above Question and the Statements?

(*a*) The Question can be answered by using one of the Statements alone, but cannot be answered using the other Statement alone

(*b*) The Question can be answered by using either Statement alone

(*c*) The Question can be answered by using both the Statements together, but cannot be answered using either Statement alone

(*d*) The Question can be answered even without using any of the Statements

90. A Question is given followed by two Statements I and II. Consider the Question and the Statements.

Question : The diagonals of rhombus *ABCD* are in the ratio 5:12. Is one of the diagonals equal to side of rhombus?

Statement I: The sum of the diagonals = 34 cm.

Statement II: The length of a side = 13 cm.

Which one of the following is correct in respect of the above Question and the Statements?

(*a*) The Question can be answered by using one of the Statements alone, but cannot be answered using the other Statement alone

(*b*) The Question can be answered by using either Statement alone

(*c*) The Question can be answered by using both the Statements together, but cannot be answered using either Statement alone

(*d*) The Question can be answered even without using any of the Statements

91. Which measure of central tendency is least affected by the presence of extreme observations in the data?

(*a*) Arithmetic mean (*b*) Harmonic mean

(*c*) Geometric mean (*d*) Median

92. To find the average ratio like price/unit, work done/hour, kilometre/hour under certain conditions, the suitable measure of central tendency applicable is

(*a*) arithmetic mean (*b*) geometric mean

(*c*) harmonic mean (*d*) mode

93. The frequency distribution of marks of 100 candidates in a particular examination is as follows :

Marks	Number of Candidates
More than 10	100
More than 20	75
More than 30	60
More than 40	40

What are the average marks of the candidates?

(*a*) 20.5 (*b*) 22.5 (*c*) 30.5 (*d*) 32.5

94. The arithmetic mean of 200 observations is 60. If 5 is multiplied to each observation, then what will be the new arithmetic mean?

(*a*) 500 (*b*) 300 (*c*) 60 (*d*) 40

95. A distribution consists of 3 components with frequencies 45, 40 and 55 having their means 2, 2.5 and 2 respectively. What is the mean of the combined distribution?

(*a*) 2.14 (*b*) 2.25 (*c*) 2.37 (*d*) 2.50

96. Which one of the following is a positional average?

(*a*) Arithmetic mean (*b*) Median (*c*) Mode (*d*) Geometric mean

For the following **two (02)** items

The following data represent the distance covered (in metres) by two groups of athletic children. It is known that the median distance in the first group is 20.8 metres while the mean distance in the second group is 17.3 metres. Some frequencies in both the groups are missing :

Distance Class	First Group	Second Group
0-5	u	$3u$
5-10	v	$2v$
10-15	11	40
15-20	52	50
20-25	75	30
25-30	22	28

97. What is the value of u?

(*a*) 1 (*b*) 2 (*c*) 3 (*d*) 4

98. What is value of v?

(*a*) 5 (*b*) 6 (*c*) 7 (*d*) 8

For the following **two (02)** items:

Consider the following distribution having median value 24 :

Marks	Number of Students
Less than 10	5
Less than 20	30
Less than 30	$30 + k$
Less than 40	$48 + k$
Less than 50	$55 + k$

99. What is the value of k?

(*a*) 20 (*b*) 22 (*c*) 25 (*d*) 30

100. What is mean of above distribution?

(*a*) 21.625 (*b*) 22.225 (*c*) 23.225 (*d*) 24.625

ANSWERS

1. (c)	**2.** (d)	**3.** (a)	**4.** (a)	**5.** (a)	**6.** (a)	**7.** (b)	**8.** (a)	**9.** (a)
10. (c)	**11.** (a)	**12.** (c)	**13.** (b)	**14.** (c)	**15.** (a)	**16.** (c)	**17.** (b)	**18.** (b)
19. (a)	**20.** (c)	**21.** (a)	**22.** (a)	**23.** (c)	**24.** (d)	**25.** (b)	**26.** (b)	**27.** (a)
28. (c)	**29.** (d)	**30.** (d)	**31.** (d)	**32.** (c)	**33.** (a)	**34.** (a)	**35.** (d)	**36.** (c)
37. (c)	**38.** (c)	**39.** (c)	**40.** (b)	**41.** (c)	**42.** (d)	**43.** (c)	**44.** (b)	**45.** (a)
46. (c)	**47.** (b)	**48.** (a)	**49.** (b)	**50.** (b)	**51.** (b)	**52.** (c)	**53.** (c)	**54.** (a)
55. (b)	**56.** (c)	**57.** (a)	**58.** (c)	**59.** (a)	**60.** (d)	**61.** (c)	**62.** (d)	**63.** (c)

64. (b)	**65.** (d)	**66.** (c)	**67.** (d)	**68.** (b)	**69.** (c)	**70.** (d)	**71.** (a)	**72.** (c)
73. (d)	**74.** (c)	**75.** (c)	**76.** (c)	**77.** (a)	**78.** (c)	**79.** (d)	**80.** (d)	**81.** (d)
82. (b)	**83.** (c)	**84.** (d)	**85.** (d)	**86.** (b)	**87.** (d)	**88.** (b)	**89.** (a)	**90.** (b)
91. (d)	**92.** (c)	**93.** (d)	**94.** (b)	**95.** (a)	**96.** (b)	**97.** (c)	**98.** (d)	**99.** (c)
100. (d)								

SOLUTIONS

1. Given that,

Now, According to the remainders theorem,

If a polynomial $p(x)$ is divided by $(x-a)$, the remainders is $p(a)$.

i.e., Here, $p(1) = 2$

$p(2) = 1$

Now, $$p(x) = q(x)(x-1)(x-2) + (ax+b) \quad ...(1)$$

where $q(x)$ is quotient.

Put $x = 1$ in Eqn (1), we get

$$p(1) = q(1).0.(-1) + (a+b) \Rightarrow 2 = a+b \quad ...(2)$$

Again, put $x = 2$ in Eqn (1), we get

$$p(2) = q(2)\cdot 1\cdot 0 + 1(2a+b) \Rightarrow 1 = 2a+b \quad ...(3)$$

From Eqn (2) and (3), we get

$$a = -1 \text{ and } b = 3$$

Hence, remainder is $3 - x$.

We have to check these conditions hold for a positive real number x or not.

2. I. $x + \frac{1}{x} > 1$

We know that, For any positive real number x :

A.M. $\geq$ G.M.

$$\Rightarrow \quad \frac{x+\frac{1}{x}}{2} \geq \sqrt{x\cdot\frac{1}{x}} \Rightarrow x + \frac{1}{x} \geq 2$$

Equality holds iff $x = 1$.

$\Rightarrow$ For any positive real number x $x + \frac{1}{x} \geq 2$

$\Rightarrow$ For any positive real number x, $x + \frac{1}{x} > 1$ $\quad (\because 2 > 1)$

II. $\left(x + \frac{1}{x}\right)^2 \geq 2$

$\because \quad x + \frac{1}{x} \geq 2$

Squaring both sides, we get

$$\left(x + \frac{1}{x}\right)^2 \geq 2^2 \Rightarrow \left(x + \frac{1}{x}\right)^2 \geq 4$$

$\Rightarrow$ For any positive real number x, $\left(x+\frac{1}{x}\right)^2 \geq 4$

For any positive real number x, $\left(x+\frac{1}{x}\right)^2 > 2$ $(\because 4>2)$

III. $\left(x+\frac{1}{x}\right)^4 \geq 9 \because \left(x+\frac{1}{x}\right)^2 \geq 4$

Squaring both sides, we get

$$\left(x+\frac{1}{x}\right)^4 \geq 16$$

For any positive real number x, $\left(x+\frac{1}{x}\right)^4 \geq 16$

For any positive real number x, $\left(x+\frac{1}{x}\right)^4 > 9$ $(\because 16 > 9)$

3. We have, $q^2 - 5p - 4 < 0 \Rightarrow q^2 < 5p + 4$ ...(1) (Since, $q > p$)

$\Rightarrow$ $q \geq p+1 \Rightarrow q^2 \geq (p+1)^2$ ...(2)

From Eqn (1) and eq (2), we get

$$(p+1)^2 < 5p+4 \Rightarrow p^2 + 2p + 1 < 5p + 4 \Rightarrow p^2 - 3p - 3 < 0$$

By using the quadratic formula, we get

$$p = \frac{-(-3) \pm \sqrt{(-3)^2 - 4\cdot 1\cdot(-3)}}{2\cdot 1} = \frac{3 \pm \sqrt{9+12}}{2}$$

$\Rightarrow$ $\frac{3-\sqrt{21}}{2} < p < \frac{3+\sqrt{21}}{2} \Rightarrow -0.79 < p < 3.79$

$\Rightarrow p$ can be 1, 2 and 3 ($\because p$ be a natural no.)

So, largest value of p is 3.

4. Let, $x > y$

$\because x - y$ be a prime no.

$\Rightarrow x - y \geq 2$

Possible combinations are :

(i) Both x and y are even.

$\Rightarrow$ $x = 2$ and $y = 2 \Rightarrow x + y = 2 + 2 = 4$ (Not prime)

So, this is not possible.

(ii) Both x and y are odd.

$\Rightarrow x + y$ and $x - y$ are even.

$\Rightarrow$ $x + y = 2$ and $x - y = 2$

This is also not possible.

(iii) x is odd and y is even.

$\Rightarrow$ $y = 2$ and $x \geq 3$.

When $x = 3$ and $y = 2$ then

$$x + y = 3 + 2 = 5 \text{ (prime)}$$
$$x - y = 3 - 2 = 1 \text{ (not prime)}$$

So, this is possible.
When $x = 5$ and $y = 2$ then

$$x + y = 5 + 2 = 7 \text{ (prime)}$$
$$x - y = 5 - 2 = 3 \text{ (prime)}$$

So, this is possible.
When $x = 7$ and $y = 2$ then

$$x + y = 9 \text{ (Not prime)}$$

So, this is not possible.
when $x = 11$ and $y = 2$ then

$$x + y = 11 + 2 = 13 \text{ (prime)}$$
$$x - y = 11 - 2 = 9 \text{ (Not prime)}$$

When $x = 13$ and $y = 2$ then

$$x + y = 13 + 2 = 15 \text{ (Not prime)}$$

So, this is not possible.
When $x = 17$ and $y = 2$ then

$$x + y = 17 + 2 = 19 \text{ (prime)}$$
$$x - y = 17 - 2 = 15 \text{ (Not prime)}$$

So, this is not possible.
When $x = 19$ and $y = 2$ then

$$x + y = 19 + 2 = 21 \text{ (Not prime)}$$

So, this is not possible.
Hence, only one combination (5, 2, 7, 3) is possible.

5. We have, $$(x+1)(x+p)(x^2+p^2) = x^4 - 1$$

$$\Rightarrow \quad (x+1)(x+p)(x^2+p^2) = (x^2-1)(x^2+1)$$

$$\Rightarrow \quad (x+1)(x+p)(x^2+p^2) = (x+1)(x-1)(x^2+(-1)^2)$$

Comparing both sides, we get $p = -1$

6. We have, $$(2+\sqrt{3})^x + (2-\sqrt{3})^x = 2$$

Now, Let, $$a = (2+\sqrt{3})^x \text{ and } b = (2-\sqrt{3})^x$$

Now, $$ab = (2+\sqrt{3})^x(2-\sqrt{3})^x = [4-3]^x = 1^x = 1$$

$$(a-b)^2 = (a+b)^2 - 4ab = 2^2 - 4\cdot 1 = 0$$

$$\Rightarrow \quad (2+\sqrt{3})^x - (2-\sqrt{3})^x = 0$$

7. We have, $$\frac{1}{a} + \frac{1}{b} = \frac{5}{6}$$

Squaring both sides, we get

$$\left(\frac{1}{a} + \frac{1}{b}\right)^2 = \left(\frac{5}{6}\right)^2$$

$$\Rightarrow \quad \frac{1}{a^2}+\frac{1}{b^2}+2\cdot\frac{1}{a}\cdot\frac{1}{b}=\frac{25}{36} \Rightarrow \frac{13}{36}+\frac{2}{ab}=\frac{25}{36} \Rightarrow \frac{2}{ab}=\frac{12}{36}=\frac{1}{3} \quad \Rightarrow ab=6$$

Now, again, $\frac{1}{a}+\frac{1}{b}=\frac{5}{6}$

Cube both sides, we get

$$\left(\frac{1}{a}+\frac{1}{b}\right)^3=\left(\frac{5}{6}\right)^3 \quad \Rightarrow \frac{1}{a^3}+\frac{1}{b^3}+3\cdot\frac{1}{a}\cdot\frac{1}{b}\left(\frac{1}{a}+\frac{1}{b}\right)=\frac{125}{216}$$

$$\Rightarrow \quad \frac{1}{a^3}+\frac{1}{b^3}+3\cdot\frac{1}{6}\cdot\left(\frac{5}{6}\right)=\frac{125}{216} \Rightarrow \frac{1}{a^3}+\frac{1}{b^3}=\frac{125}{216}-\frac{15}{36}=\frac{35}{216}$$

8. Now, we find the remainder by the substitution method (Remainder Theorem).
Let, $x^2+1=0 \Rightarrow x^2=-1$
Put $x^2=-1$ into x^6, we get $x^6=(x^2)^3=(-1)^3=-1$

9. If $(x+2)$ is a factor of a given polynomial $P(x)$, then $x=-2$ must satisfy the polynomial, i.e., $P(-2)=0$. By checking options, we get
Now,
(a) $x^5-4x^4-3x^3+8x^2-14x+12$
Put $x=-2$:
$(-2)^5-4(-2)^4-3(-2)^3+8(-2)^2-14(-2)+12=-32-64+24+32+28+12=0$ (satisfy)
(b) $x^5+4x^4-3x^3+8x^2-14x+12$
Put $x=-2$:
$(-2)^5+4(-2)^4-3(-2)^3+8(-2)^2-14(-2)+12$
$=-32+64+24+32+28+12=128\neq 0$ (Not satisfy)
(c) $x^5-4x^4+3x^3+8x^2-14x+12$
Put $x=-2$
$(-2)^5-4(-2)^4+3(-2)^3+8(-2)^2-14(-2)+12=-32-64-24+32+28+12=-48\neq 0$
(d) $x^5-4x^4-3x^3+8x^2+14x+12$
Substitute $x=-2$:
$(-2)^5-4(-2)^4-3(-2)^3+8(-2)^2+14(-2)+12$
$=-32-64+24+32-28+12=-56\neq 0$ (Not satisfy)

Hence, (x + 2) is a factor of **Option (a)**.

10. $\log_{10}2=0.301$ and $\log_{10}3=0.477$

Let $N=60^{60}$

Taking log both sides, we get

$$\log_{10}N=\log_{10}(60^{60})$$

$$\Rightarrow \quad \log_{10}N=60\log_{10}60$$

$$\Rightarrow \quad \log_{10}N=60\cdot\log_{10}(2^2\cdot 3\cdot 5)$$

$$\Rightarrow \quad \log_{10}N=60\cdot[2\log_{10}2+\log_{10}3+\log_{10}5]$$

$$\Rightarrow \quad \log_{10}N=60\left(2\log_{10}2+\log_{10}3+\log_{10}\left(\frac{10}{2}\right)\right)$$

$\Rightarrow \log_{10} N = 60\left(2\log_{10} 2 + \log_{10} 3 + \log_{10} 10 - \log_{10} 2\right)$

$\Rightarrow \log_{10} N = 60\left(0.602 + 0.477 + 1 - 0.301\right)$

$\Rightarrow \log_{10} N = 60\left(1.778\right) \Rightarrow \log_{10} N = 106.68$

So, the number of digits in the expansion of $(60^{60}) = [106.68] + 1 = 106 + 1 = 107$

11. Now, $17 \equiv -1 \pmod{18}$ and $19 \equiv 1 \pmod{18}$

$\Rightarrow 17^{25} \equiv (-1)^{25} \pmod{18} \Rightarrow 19^{25} \equiv 1^{25} \pmod{18}$

$\Rightarrow 17^{25} \equiv -1 \pmod{18}$ ...(1) $\Rightarrow 19^{25} \equiv 1 \pmod{18}$ ...(2)

From (1) + (2), we get:

$$17^{25} + 19^{25} \equiv (-1+1) \pmod{18} \Rightarrow 17^{25} + 19^{25} \equiv 0 \pmod{18} \Rightarrow \text{Remainder is 0.}$$

12. Now,
$$P = \frac{x^3 + y^3}{x^2 - xy + y^2} = \frac{(x+y)(x^2 - xy + y^2)}{x^2 - xy + y^2} = x + y$$

$$Q = \frac{x^3 - y^3}{x^2 + xy + y^2} = \frac{(x-y)(x^2 + xy + y^2)}{x^2 + xy + y^2} = x - y$$

Since H is the HCF of x and y,

$\Rightarrow$ $x = Ha$ and $y = Hb$

where a and b are coprime, i.e., $\text{HCF}(a,b) = 1$

Put the value of $x = Ha$ and $y = Hb$ into P and Q:

$$P = Ha + Hb = H(a+b) \text{ and } Q = Ha - Hb = H(a-b)$$

Now, HCF of P and Q = HCF of $H(a+b)$ and $H(a-b) = H \times \text{HCF}(a+b, a-b)$

Let HCF of $(a + b)$ and $(a - b)$ is h.

$\Rightarrow$ h divides $a + b$ and h divides $a - b$

$\Rightarrow$ h divides $(a + b) + (a - b)$ and h divides $(a + b) - (a - b)$ also

$\Rightarrow$ h divides $2a$ and h divides $2b$.

$\because$ HCF $(a, b) = 1$

$\Rightarrow \text{HCF}(2a, 2b) = 2 \cdot \text{HCF}(a,b) = 2$

$\Rightarrow h$ is a divisor of 2 i.e., h can be 1 or 2.

Hence, HCF of P and $Q = H \cdot h = H$ or $2H$

13. Let $x^2 = y$, where, $y > 0$

Put the value of $x^2 = y$ in Eqn. $x^4 = x^2 + 1$, we get

$$y^2 = y + 1 \Rightarrow y^2 - y - 1 = 0$$

By quadratic formula, we get

$$y = \frac{-(-1) \pm \sqrt{(-1)^2 - 4(1)(-1)}}{2(1)} = \frac{1 \pm \sqrt{5}}{2}$$

$$\Rightarrow \quad x^2 = \frac{1+\sqrt{5}}{2} \Rightarrow (x^2)^2 = \left(\frac{1+\sqrt{5}}{2}\right)^2 \Rightarrow x^4 = \frac{6+2\sqrt{5}}{4}$$

$$\Rightarrow \quad 2x^4 = 2\cdot\frac{2(3+\sqrt{5})}{4} \Rightarrow 2x^4 = 3+\sqrt{5}$$

14. Given that, $\dfrac{q+r}{p+q} = \dfrac{s+p}{r+s}, \quad (q+r) \neq 0, \quad (s+p) \neq 0$

$$\Rightarrow \quad (p+q)(s+p) = (r+s)(q+r) \Rightarrow ps + p^2 + qs + qp = rq + r^2 + sq + sr$$

$$\Rightarrow \quad ps + p^2 + qs + qp = rq + r^2 + sq + sr \Rightarrow ps - sr + p^2 - r^2 + qp - rq = 0$$

$$\Rightarrow \quad s(p-r) + (p-r)(p+r) + q(p-r) = 0 \Rightarrow (p-r)(s+p+r+q) = 0$$

$\Rightarrow$ Either $p = r$ or $p+q+r+s = 0$

15. Now, $\gcd(12n+2, 8n+1) = 1$

Put $n = 1$ then

$\gcd(14, 9) = 1$ (Satisfy)

Put $n = 2$ then

$\gcd(26, 17) = 1$ (Satisfy)

Put $n = 3$ then

$\gcd(38, 25) = 1$ (Satisfy)

Put $n = 4$ then

$\gcd(50, 33) = 1$ (Satisfy)

Put $n = 5$ then

$\gcd(62, 41) = 1$ (Satisfy)

Put $n = 6$ then

$\gcd(74, 49) = 1$ (Satisfy)

$\Rightarrow$ No. of values of n are 6.

16. We have, $x^3 + y^3 + 3xy - 1 = x^3 + y^3 + (-1)^3 - 3xy(-1) = (x+y-1)(x^2+y^2+1-xy+x+y)$

$$(x+y)^4 - 1 = \left((x+y)^2\right)^2 - 1^2 = \left((x+y)^2 - 1\right)\left((x+y)^2 + 1\right) = (x+y-1)(x+y+1)$$
$$= (x^2 + y^2 + 2xy + 1)$$

Now, $\text{HCF}\left((x^3+y^3+3xy-1), (x+y)^4 - 1\right) = x+y-1$

17. I. x is always divisible by 48.

Put $n = 2$ then $x = 2\cdot 3\cdot 4 = 24$ (Not divisible by 48)

II. x^2 is always divisible by 144.

$x > 24$ and $x = 24\text{m}$, $m \geq 1 \Rightarrow x^2 = 576\,\text{m}^2 \mid 144$

So, Only statement II is correct.

18. We have to find the LCM of $x^4 + x^2y^2 + y^4, x^3y + y^4$, and $x^4y^2 - x^2y^4$.

Now, $\text{HCF} = x+y-1$

$$x^4 + x^2y^2 + y^4 = (x^2+y^2)^2 - (xy)^2 = (x^2+y^2+xy)(x^2+y^2-xy)$$

$$x^3y + y^4 = y(x^3+y^3) = y(x+y)(x^2+y^2-xy)$$

$$x^4y^2 - x^3y^3 = x^3y^2(x-y)$$

So, LCM $(x^4 + x^2y^2 + y^4, x^3y + y^4, x^4y^2 - x^3y^3)$

$$= x^3y^2(x-y)(x+y)(x^2+y^2-xy)(x^2+y^2+xy)$$
$$= x^3y^2(x^2-y^2)\{(x^2+y^2)^2-x^2y^2\}$$
$$= x^3y^2(x^2-y^2)(x^4+y^4+x^2y^2) = x^3y^2(x^6-y^6)$$

19. Now, $\therefore$ XYZ is a 3-digit number $\Rightarrow XYZ = 100\,X + 10\,Y + Z$

Similarly, $= 100\,Z + 10\,Y + X$

Now, $D = (100X+10Y+Z)-(100Z+10Y+X)$

$$= 100X+10Y+Z-100Z-10Y-X = 99X-99Z = 99(X-Z)$$

$\Rightarrow$ D is divisible by 90 $\Rightarrow$ Remainder = 0

20. By checking options, we get

(a) Let $p+q=4$ then

$$(p+q)^{p+q} = 4^4 = (2^2)^4 = 2^8 \text{ (Not divisible by } 2^9\text{)}$$

(b) Let $p+q=6$ then

$$(p+q)^{p+q} = 6^6 = (2\cdot 3)^6 = 2^6\cdot 3^6 \text{ (Not divisible by } 2^9\text{)}$$

(c) Let $p+q=8$ then

$$(p+q)^{p+q} = 8^8 = (2^3)^8 = 2^{24} \text{ (Divisible by } 2^9\text{)}$$

(d) Let $p+q=12$ then

$$(p+q)^{p+q} = 12^{12} = (2^2\cdot 3)^{12} = 2^{24}\cdot 3^{12} \text{ (Divisible by } 2^9\text{)}$$

Here, least value of $p+q$ is 8.

21. We have,

$$\frac{(a+b)^2}{(c-a)(c+a+b)}+\frac{(a+b)c}{c^2+bc-a^2-ab}-\frac{a+2b+c}{2(c-a)},\quad a\neq b, b\neq c, c\neq a$$

$$= \frac{(a+b)^2}{(c-a)(c+b)}+\frac{(a+b)c}{(c-a)(c+a+b)}-\frac{a+2b+c}{2(c-a)}$$

$$= \frac{2(a+b)^2+2c(a+b)-(c+a+b)(a+2b+c)}{2(c-a)(c+a+b)}$$

$$= \frac{2(a+b)(a+b+c)-(c+a+b)(a+2b+c)}{2(c-a)(c+a+b)}$$

$$= \frac{(a+b+c)[2(a+b)-(a+2b+c)]}{2(c-a)(a+b+c)} = \frac{a-c}{2(c-a)} = -\frac{1}{2}$$

22. We have, $\dfrac{a\times\left(\dfrac{a}{b}\right)^{\frac{a}{b}}}{a^{\left(\frac{a}{b}\right)}} = \dfrac{a^{1+\frac{a}{b}-\frac{a}{b}}}{b^{\frac{a}{b}}} = \dfrac{a}{(b^a)^{\frac{1}{b}}} = \dfrac{a}{a^{b\cdot\frac{1}{b}}}$ $(\because a^b = b^a) = 1$

23. Now, $x^2 = (2+2^{1/2}+2^{3/2})^2 = 4+2+8+2\cdot 2\cdot 2^{1/2}+2\cdot 2^{1/2}\cdot 2^{3/2}+2\cdot 2^{3/2}\cdot 2$

$$= 14+4\sqrt{2}+8+8\sqrt{2} = 22+12\sqrt{2}$$

Now, $x^2-4x-10 = 22+12\sqrt{2}-4(2+\sqrt{2}+2\sqrt{2})-10$

$$= 22+12\sqrt{2}-8-4\sqrt{2}-8\sqrt{2}-10 = 4$$

24. We have, $$\frac{\sqrt{p+x}+\sqrt{p-x}}{\sqrt{p+x}-\sqrt{p-x}}=p$$

Now LHS,

Multiplying by $\sqrt{p+x}+\sqrt{p-x}$ on numerator and denominator, we get

$$\frac{\sqrt{p+x}+\sqrt{p-x}}{\sqrt{p+x}-\sqrt{p-x}}\times\frac{\sqrt{p+x}+\sqrt{p-x}}{\sqrt{p+x}+\sqrt{p-x}}=p$$

$$\Rightarrow \frac{p+x+p-x+2\sqrt{p+x}\sqrt{p-x}}{(p+x)-(p-x)}=p \Rightarrow \frac{2p+2\sqrt{(p+x)(p-x)}}{2x}=p$$

$$\Rightarrow \frac{p+\sqrt{(p+x)(p-x)}}{x}=p \Rightarrow p+\sqrt{(p+x)(p-x)}=px \Rightarrow \sqrt{(p+x)(p-x)}=px-p$$

Squaring both sides, we get

$$p^2-x^2=p^2x^2+p^2-2p^2x \Rightarrow p^2x^2+x^2-2p^2x=0$$

$$\Rightarrow (p^2+1)x^2-2p^2x=0 \Rightarrow x((p^2+1)x-2p^2)=0$$

Either x = 0 or

$$(p^2+1)x-2p^2=0 \Rightarrow (p^2+1)x=2p^2 \Rightarrow x=\frac{2p^2}{p^2+1}$$

25. We have to find the roots of the given equation.

Now, Let x_1 and x_2 be two roots of the given equation.

We know that,

Sum of the roots $=\frac{-\text{Coeff. of } x}{\text{Coeff. of } x^2}$

$$\Rightarrow x_1+x_2=\frac{-\left(-\left(\frac{a+b}{2}\right)\right)}{\left(\frac{a-b}{2}\right)}=\frac{a+b}{a-b} \quad \text{...(1)}$$

and Product of the roots $=\frac{\text{Constant term}}{\text{Coeff. of } x^2} \Rightarrow x_1x_2=\frac{b}{\frac{a-b}{2}}=\frac{2b}{a-b}$

Now, $$(x_1-x_2)^2=(x_1+x_2)^2-4x_1x_2$$

$$\Rightarrow (x_1-x_2)^2=\left(\frac{a+b}{a-b}\right)^2-4\cdot\frac{2b}{a-b}$$

$$=\frac{(a+b)^2}{(a-b)^2}-\frac{8b}{a-b}=\frac{(a+b)^2-8b(a-b)}{(a-b)^2}$$

$$=\frac{a^2+b^2+2ab-8ab+8b^2}{(a-b)^2}=\frac{a^2+9b^2-6ab}{(a-b)^2}$$

$$\Rightarrow (x_1-x_2)^2=\frac{(a-3b)^2}{(a-b)^2}$$

$$\Rightarrow \quad x_1 - x_2 = \frac{a-3b}{a-b} \quad \text{...(2)}$$

From eq^n (1) + (2), we get

$$2x_1 = \frac{a+b}{a-b} + \frac{a-3b}{a-b} = \frac{2a-2b}{a-b} = 2 \Rightarrow x_1 = 1$$

From eq^n (1) − (2), we get

$$2x_2 = \frac{a+b}{a-b} - \frac{a-3b}{a-b} = \frac{4b}{a-b} \Rightarrow x_2 = \frac{2b}{a-b}$$

Hence, Roots are 1 and $\frac{2b}{a-b}$.

26. Given that,

$$x - \frac{1}{x} = 2, \; x > 0 \quad \text{...(1)}$$

Now, $$\left(x+\frac{1}{x}\right)^2 = \left(x-\frac{1}{x}\right)^2 + 4 \cdot x \cdot \frac{1}{x} \Rightarrow \left(x+\frac{1}{x}\right)^2 = 4+4 = 8$$

$$\Rightarrow \quad x + \frac{1}{x} = 2\sqrt{2} \quad \text{...(2)}$$

From eq^n (1) + (2), we get

$$2x = 2 + 2\sqrt{2} \Rightarrow x = 1 + \sqrt{2} \Rightarrow \frac{1}{x} = \frac{1}{1+\sqrt{2}} = \frac{1-\sqrt{2}}{1-2} = \sqrt{2} - 1$$

Now, $$x^2 - \frac{1}{x^2} = (1+\sqrt{2})^2 - (\sqrt{2}-1)^2 = 3 + 2\sqrt{2} - 3 + 2\sqrt{2} = 4\sqrt{2}$$

27. Now, $\therefore (a-b)^2 + (b-c)^2 + (c-a)^2 = 6$

$$\Rightarrow \quad a^2 + b^2 - 2ab + b^2 + c^2 - 2bc + c^2 + a^2 - 2ca = 6$$

$$\Rightarrow \quad 2(a^2+b^2+c^2) - 2(ab+bc+ca) = 6$$

$$\Rightarrow \quad 2 \cdot 29 - ((a+b+c)^2 - (a^2+b^2+c^2)) = 6$$

$$\Rightarrow \quad 58 - ((a+b+c)^2 - 29) = 6$$

$$\Rightarrow \quad 58 + 29 - 6 = (a+b+c)^2 \Rightarrow 81 = (a+b+c)^2$$

$$\Rightarrow \quad \pm 9 = a+b+c$$

28. We have, $$p = \frac{\sqrt{5}-2}{\sqrt{5}+2} \cdot \frac{\sqrt{5}-2}{\sqrt{5}-2} = \frac{5+4-4\sqrt{5}}{5-4} = 9 - 4\sqrt{5}$$

$$q = \frac{\sqrt{5}+2}{\sqrt{5}-2} \cdot \frac{\sqrt{5}+2}{\sqrt{5}+2} = \frac{5+4+4\sqrt{5}}{5-4} = 9 + 4\sqrt{5}$$

Now, $$\frac{p}{q} + \frac{q}{p} = \frac{9-4\sqrt{5}}{9+4\sqrt{5}} + \frac{9+4\sqrt{5}}{9-4\sqrt{5}} = \frac{(9-4\sqrt{5})^2 + (9+4\sqrt{5})^2}{(9+4\sqrt{5})(9-4\sqrt{5})}$$

$$= \frac{81+80-72\sqrt{5}+81+80+72\sqrt{5}}{81-80} = 322$$

29. We have,

	Hundreds place
$(25)^1 = 25$	0
$(25)^2 = 625$	6
$(25)^3 = 15625$	6
$(25)^4 = 390625$	6
....	
....	
$(25)^{10} = \ldots 625$	6

30. Now, $$\text{LCM }(4, 6, 7, 9) = 252$$

So, $$N = 252k + 3, \quad k \in \mathbf{N} \text{ and } N \geq 1000$$

$$1000 \leq 252k + 3 \Rightarrow 997 \leq 252k$$

$$\Rightarrow \quad \frac{997}{252} \leq k$$

$$\Rightarrow \quad k \geq 3.956 \Rightarrow k = 4$$

Hence, $$N = 252 \times 4 + 3 = 1008 + 3 = 1011$$

31. We have,

$$\sqrt{2 + \sqrt{2 + \sqrt{2 + \ldots}}} = \operatorname{cosec}\theta$$

Now, Squaring both sides of the given expression, we get

$$2 + \operatorname{cosec}\theta = \operatorname{cosec}^2\theta \Rightarrow \operatorname{cosec}^2\theta - \operatorname{cosec}\theta - 2 = 0$$

Using quadratic formula, we get

$$\operatorname{cosec}\theta = \frac{-(-1) \pm \sqrt{(-1)^2 - 4 \cdot 1 \cdot (-2)}}{2 \cdot 1} \Rightarrow \operatorname{cosec}\theta = \frac{1 \pm 3}{2}$$

$$\Rightarrow \quad \operatorname{cosec}\theta = \frac{4}{2}, \ \frac{-2}{2} \Rightarrow \frac{1}{\sin\theta} = 2, \ -1 \Rightarrow \sin\theta = \frac{1}{2}, \ -1$$

32. Given that, $$8\sin\theta - \cos\theta = 4, \quad 0 < \theta < \frac{\pi}{2}$$

$$8\sin\theta - \cos\theta = 4$$

Dividing both sides by $\cos\theta$, we get

$$8\tan\theta - 1 = 4\sec\theta$$

Squaring both sides, we get

$$(8\tan\theta - 1)^2 = (4\sec\theta)^2$$

$$\Rightarrow \quad 64\tan^2\theta + 1 - 2 \cdot 8\tan\theta \cdot 1 = 16\sec^2\theta$$

$$\Rightarrow \quad 64\tan^2\theta + 1 - 16\tan\theta = 16(1 + \tan^2\theta)$$

$$\Rightarrow \quad 64\tan^2\theta - 16\tan\theta + 1 - 16 - 16\tan^2\theta = 0$$

$$\Rightarrow \quad 48\tan^2\theta - 16\tan\theta - 15 = 0$$

Using quadratic formula, we get

$$\tan\theta = \frac{-(-16) \pm \sqrt{(-16)^2 - 4 \cdot 48 \cdot (-15)}}{2 \cdot 48}$$

$$\Rightarrow \quad \tan\theta = \frac{16 \pm \sqrt{256+2880}}{96} \Rightarrow \tan\theta = \frac{16 \pm \sqrt{3136}}{96} \Rightarrow \tan\theta = \frac{16 \pm 56}{96}$$

$$\Rightarrow \quad \tan\theta = \frac{72}{96} \text{ and } \frac{-40}{96}$$

$$\Rightarrow \quad \tan\theta = \frac{3}{4} \text{ and } \frac{-5}{12} \Rightarrow \text{Not possible } \left(\because 0 < \theta < \frac{\pi}{2}\right)$$

$$\Rightarrow \quad \tan\theta = \frac{3}{4}$$

By Pythagorean theorem, $x^2 = 3^2 + 4^2 = 25 \Rightarrow x = 5$

Now, $\quad \text{cosec}\theta = \frac{x}{3} = \frac{5}{3}$

33. Given that, $\quad 2\tan\theta = \sec^2\theta - 2, \quad 0 < \theta < \frac{\pi}{2}$

$$\Rightarrow \quad 2\tan\theta = 1 + \tan^2\theta - 2 \quad (\because \sec^2\theta = 1 + \tan^2\theta)$$

$\Rightarrow \tan^2\theta - 2\tan\theta - 1 = 0$

Using quadratic formula, we get

$$\tan\theta = \frac{-(-2) \pm \sqrt{(-2)^2 - 4 \cdot 1 \cdot (-1)}}{2 \cdot 1}$$

$$\Rightarrow \tan\theta = \frac{2 \pm \sqrt{4+4}}{2} \Rightarrow \tan\theta = 1 \pm \sqrt{2} \Rightarrow \cot\theta = \frac{1}{1 \pm \sqrt{2}}$$

$$\Rightarrow \cot\theta = \frac{1}{1+\sqrt{2}} \text{ and } \frac{1}{1-\sqrt{2}} \Rightarrow \cot\theta = \frac{1}{1+\sqrt{2}} \cdot \frac{1-\sqrt{2}}{1-\sqrt{2}} \text{ and also, } \cot\theta = \frac{1}{1-\sqrt{2}} \cdot \frac{1+\sqrt{2}}{1+\sqrt{2}}$$

$$\Rightarrow \cot\theta = \sqrt{2} - 1 \quad \Rightarrow \cot\theta = -(1+\sqrt{2})$$

34. $$(\sec\theta - \tan\theta)\sqrt{\frac{1-\sin\theta}{1+\sin\theta}}$$

Now, $$(\sec\theta - \tan\theta)\sqrt{\frac{1-\sin\theta}{1+\sin\theta} \cdot \frac{1-\sin\theta}{1-\sin\theta}} = (\sec\theta - \tan\theta)\sqrt{\frac{(1-\sin\theta)^2}{1-\sin^2\theta}}$$

$$= (\sec\theta - \tan\theta) \cdot \frac{1-\sin\theta}{\cos\theta} = \sec\theta - \tan\theta - (\sec\theta - \tan\theta) = 0$$

35. Now, $\frac{\text{cosec}^2\theta - \sec^2\theta}{\text{cosec}^2\theta + \sec^2\theta}$

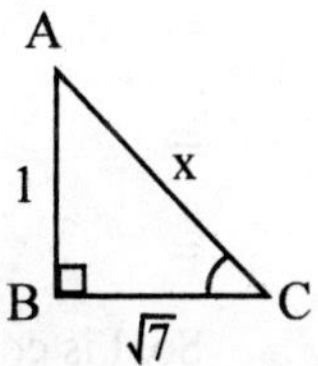

Now, $\quad \cot\theta = \sqrt{7} = \frac{BC}{AB} \quad \Rightarrow BC = \sqrt{7}, \quad AB = 1$

Now, $\quad x^2 = 1^2 + (\sqrt{7})^2 \quad$ (By Pythagorean theorem)

$$x^2 = 1 + 7 \Rightarrow x = \sqrt{8} = 2\sqrt{2}$$

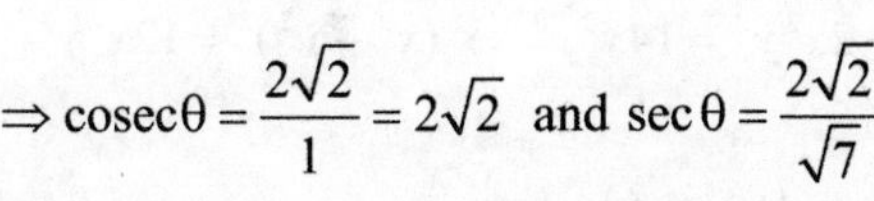

$$\Rightarrow \text{cosec}\theta = \frac{2\sqrt{2}}{1} = 2\sqrt{2} \text{ and } \sec\theta = \frac{2\sqrt{2}}{\sqrt{7}}$$

Now, $\dfrac{\text{cosec}^2\theta - \sec^2\theta}{\text{cosec}^2\theta + \sec^2\theta} = \dfrac{(2\sqrt{2})^2 - \left(\dfrac{2\sqrt{2}}{\sqrt{7}}\right)^2}{(2\sqrt{2})^2 + \left(\dfrac{2\sqrt{2}}{\sqrt{7}}\right)^2} = \dfrac{8 - \dfrac{8}{7}}{8 + \dfrac{8}{7}} = \dfrac{\dfrac{48}{7}}{\dfrac{64}{7}} = \dfrac{3}{4}$

36. Let α and β are two acute angles in a right-angled triangle

$$\alpha + \beta = \frac{\pi}{2} \text{ radian} \quad ...(1)$$

$$\alpha - \beta = \frac{\pi}{12} \text{ radian (given)} \quad ...(2)$$

From Eqn (1) + (2), we get

$$2\alpha = \frac{\pi}{2} + \frac{\pi}{12} \text{ radian} \Rightarrow 2\alpha = \frac{6\pi + \pi}{12} \text{ radian}$$

$$\Rightarrow \quad \alpha = \frac{7\pi}{24} \text{ radian} \Rightarrow \alpha = \frac{7\pi}{24} \times \frac{180°}{\pi} \Rightarrow \alpha = \frac{105°}{2} = 52.5° \text{ and } \beta = 90 - 52.5° = 37.5°$$

37. We have, $\log_{10}[998 + \sqrt{x^2 - 18x + 76}] = 3 \Rightarrow 998 + \sqrt{x^2 - 18x + 76} = 10^3$

$$\Rightarrow \quad \sqrt{x^2 - 18x + 76} = 1000 - 998 \Rightarrow \sqrt{x^2 - 18x + 76} = 2$$

Square both sides, we get

$$x^2 - 18x + 76 = 4 \Rightarrow x^2 - 18x + 72 = 0$$

Using quadratic formula, we get

$$x = \frac{-(-18) \pm \sqrt{(-18)^2 - 4 \cdot 1 \cdot 72}}{2 \cdot 1} \Rightarrow x = \frac{18 \pm \sqrt{324 - 288}}{2} \Rightarrow x = \frac{18 \pm 6}{2} \Rightarrow x = 12, 6$$

Let $\alpha = 12$ and $\beta = 6 \Rightarrow (\alpha - \beta)^2 = (12 - 6)^2 = 6^2 = 36$

38. We have check the following expressions :

I. Now, $\because x^4 + y^4 = 14x^2y^2$

$$(x^2 + y^2)^2 = 16x^2y^2$$

Taking log both sides, we get

$$\log_{10}(x^2 + y^2)^2 = \log_{10}\left(16x^2y^2\right)$$

$$\Rightarrow \quad 2\log_{10}\left(x^2 + y^2\right) = \log_{10}16 + \log_{10}x^2 + \log_{10}y^2$$

$$\Rightarrow \quad 2\log_{10}\left(x^2 + y^2\right) = \log_{10}2^4 + 2\log_{10}x + 2\log_{10}y$$

$$\Rightarrow \quad 2\log_{10}\left(x^2 + y^2\right) = 4\log_{10}2 + 2\log_{10}x + 2\log_{10}y$$

$$\Rightarrow \quad \log_{10}\left(x^2 + y^2\right) = 2\log_{10}2 + \log_{10}x + \log_{10}y$$

So, I is correct.

II.

Now, $\quad x^4 + y^4 = 14x^2y^2 \Rightarrow (x^2 - y^2)^2 + 2x^2y^2 = 14x^2y^2 \Rightarrow (x^2 - y^2)^2 = 12x^2y^2$

Taking log both sides, we get

$$\log_{10}(x^2 - y^2)^2 = \log_{10}\left(12x^2y^2\right)$$

$$\Rightarrow \quad 2\log_{10}\left(x^2 - y^2\right) = \log_{10}12 + \log_{10}x^2 + \log_{10}y^2$$

$$\Rightarrow \quad 2\log_{10}\left(x^2 - y^2\right) = \log_{10}\left(2^2 \cdot 3\right) + 2\log_{10}x + 2\log_{10}y$$

$$\Rightarrow \quad 2\log_{10}\left(x^2 - y^2\right) = 2\log_{10}2 + \log_{10}3 + 2\log_{10}x + 2\log_{10}y$$

$$\Rightarrow \quad \log_{10}\left(x^2 - y^2\right) = \log_{10}2 + 0.5\log_{10}3 + \log_{10}x + \log_{10}y$$

So, II is correct. Hence, both I and II are correct.

39. We have, $(3x+y)^2 + (3x+y)(x+5y) - 20(x+5y)^2$

Now, $(3x+y)^2 + 5\,(3x+y)\,(x+5y) - 4\,(3x+y)\,(x+5y) - 20\,(x+5y)^2$

$$= (3x+y)\,(3x+y+5(x+5y)) - 4(x+5y)\,(3x+y+5(x+5y))$$

$$= (3x+y-4(x+5y))\,(3x+y+5(x+5y))$$

$$= (-x-19y)\,(8x+26y) \;= -2\,(x+19y)\,(4x+13y)$$

So, I and II both are factors of given plynomia.

40. We have,

$$\frac{\dfrac{x}{x-y}+\dfrac{y}{y-z}+\dfrac{z}{z-x}}{\dfrac{x+y}{x-y}+\dfrac{y+z}{y-z}+\dfrac{z+x}{z-x}+3}$$

$$\frac{\dfrac{x}{x-y}+\dfrac{y}{y-z}+\dfrac{z}{z-x}}{\dfrac{x+y}{x-y}+1+\dfrac{y+z}{y-z}+1+\dfrac{z+x}{z-x}+1} = \frac{\dfrac{x}{x-y}+\dfrac{y}{y-z}+\dfrac{z}{z-x}}{\dfrac{x+y+x-y}{x-y}+\dfrac{y+z+y-z}{y-z}+\dfrac{z+x+z-x}{z-x}}$$

$$= \frac{\dfrac{x}{x-y}+\dfrac{y}{y-z}+\dfrac{z}{z-x}}{\dfrac{2x}{x-y}+\dfrac{2y}{y-z}+\dfrac{2z}{z-x}} = \frac{1}{2}$$

41. $p\sin^2\alpha + q\cos^2\alpha = m$

Dividing by $\cos^2\alpha$ both sides, we get

$$p\tan^2\alpha + q = m\sec^2\alpha$$

$$\Rightarrow \quad p\tan^2\alpha + q = m\,(1+\tan^2\alpha) \qquad (\because \sec^2\alpha = 1+\tan^2\alpha)$$

$$\Rightarrow \quad p\tan^2\alpha + q = m + m\tan^2\alpha$$

$$\Rightarrow \quad p\tan^2\alpha - m\tan^2\alpha = m - q$$

$$\Rightarrow \quad (p-m)\tan^2\alpha = m - q$$

$$\Rightarrow \quad \tan^2\alpha = \frac{m-q}{p-m}$$

$q\sin^2\beta + p\cos^2\beta = n$

Dividing by $\cos^2\beta$ both sides, we get

$$q\tan^2\beta + p = n\sec^2\beta$$

$$\Rightarrow \quad q\tan^2\beta + p = n(1+\tan^2\beta) \qquad (\because \sec^2\beta = 1+\tan^2\beta)$$

$$\Rightarrow \quad q\tan^2\beta + p = n + n\tan^2\beta$$

$$\Rightarrow \quad q\tan^2\beta - n\tan^2\beta = n - p$$

$$\Rightarrow \quad (q-n)\tan^2\beta = n-p$$

$$\Rightarrow \quad \tan^2\beta = \frac{n-p}{q-n}$$

Now, $$\left(\frac{\tan\alpha}{\tan\beta}\right)^2 = \frac{\tan^2\alpha}{\tan^2\beta} = \frac{\frac{m-q}{p-m}}{\frac{n-p}{q-n}} = \frac{(m-q)(n-q)}{(m-p)(n-p)}$$

42. Now, $\tan\alpha = \sqrt{\dfrac{m-q}{p-m}}$ and $\tan\beta = \sqrt{\dfrac{n-p}{q-n}}$ (from previous)

$\because$ α and β are complementary angles.

$$\Rightarrow \quad \alpha+\beta = \frac{\pi}{2} \Rightarrow \tan(\alpha+\beta) = \tan\frac{\pi}{2}$$

$$\Rightarrow \quad \frac{\tan\alpha+\tan\beta}{1-\tan\alpha\cdot\tan\beta} = \infty \Rightarrow 1-\tan\alpha\cdot\tan\beta = 0$$

$$\Rightarrow \quad 1-\sqrt{\frac{m-q}{p-m}}\cdot\sqrt{\frac{n-p}{q-n}} = 0$$

$$\Rightarrow \quad \frac{\sqrt{p-m}\,\sqrt{q-n}-\sqrt{m-q}\,\sqrt{n-p}}{\sqrt{p-m}\,\sqrt{q-n}} = 0$$

$$\Rightarrow \quad \sqrt{(p-m)(q-n)} - \sqrt{(m-q)(n-p)} = 0$$

$$\Rightarrow \quad \sqrt{pq-mq-np+mn} = \sqrt{mn-qn-mp+qp}$$

$$\Rightarrow \quad pq-mq-np+mn = mn-qn-mp+qp$$

$$\Rightarrow \quad -mq-np+qn+mp = 0$$

$$\Rightarrow \quad mp-mq+nq-np = 0$$

$$\Rightarrow \quad m(p-q)-n(-q+p) = 0$$

$$\Rightarrow \quad (m-n)(p-q) = 0$$

Either $m-n=0$ or $p-q=0$

43. Given that, $\operatorname{cosec}\theta - \sin\theta = p$ and $\sec\theta - \cos\theta = q$

$$\Rightarrow \quad \frac{1}{\sin\theta} - \sin\theta = p \quad \text{and} \quad \frac{1}{\cos\theta} - \cos\theta = q$$

$$\Rightarrow \quad \frac{1-\sin^2\theta}{\sin\theta} = p \quad \text{and} \quad \frac{1-\cos^2\theta}{\cos\theta} = q$$

$$\Rightarrow \quad p\sin\theta = \cos^2\theta \quad \text{and} \quad q\cos\theta = \sin^2\theta$$

Now, $p\sin\theta + q\cos\theta = \cos^2\theta + \sin^2\theta = 1$

44. From previous $p = \dfrac{\cos^2\theta}{\sin\theta}$ and $q = \dfrac{\sin^2\theta}{\cos\theta}$

Now, $p^2q^2\,(p^2+q^2+3)$

$$= \left(\frac{\cos^2\theta}{\sin\theta}\right)^2 \left(\frac{\sin^2\theta}{\cos\theta}\right)^2 \left[\left(\frac{\cos^2\theta}{\sin\theta}\right)^2 + \left(\frac{\sin^2\theta}{\cos\theta}\right)^2 + 3\right]$$

$$= \cos^2\theta\sin^2\theta\left(\frac{\cos^4\theta}{\sin^2\theta} + \frac{\sin^4\theta}{\cos^2\theta} + 3\right)$$

$$= \cos^2\theta\sin^2\theta\left(\frac{\cos^6\theta + \sin^6\theta + 3\sin^2\theta\cos^2\theta}{\sin^2\theta\cos^2\theta}\right)$$

$$= (\sin^2\theta + \cos^2\theta)^3 = 1$$

45. $\dfrac{\sin\alpha}{\sin\beta} = \dfrac{4\sqrt{2}}{3}$ and $\dfrac{\cos\theta}{\cos\beta} = \dfrac{2\sqrt{3}}{9} \Rightarrow \sin\beta = \dfrac{3}{4\sqrt{2}}\sin\alpha$ and $\cos\beta = \dfrac{9}{2\sqrt{3}}\cos\alpha$

Now, We know that,

$$\sin^2\beta + \cos^2\beta = 1$$

$$\Rightarrow \left(\frac{3}{4\sqrt{2}}\right)^2 \sin^2\alpha + \left(\frac{9}{2\sqrt{3}}\right)^2 \cos^2\alpha = 1$$

$$\Rightarrow \frac{9}{32}\sin^2\alpha + \frac{81}{12}\cos^2\alpha = 1$$

Dividing by $\cos^2\alpha$ both sides, we get

$$\frac{9}{32}\tan^2\alpha + \frac{27}{4} = \sec^2\alpha \Rightarrow \frac{9}{32}\tan^2\alpha + \frac{27}{4} = 1 + \tan^2\alpha$$

$$\Rightarrow \left(\frac{9}{32} - 1\right)\tan^2\alpha = 1 - \frac{27}{4} \Rightarrow \frac{9-32}{32}\tan^2\alpha = \frac{4-27}{4}$$

$$\Rightarrow \frac{-23}{32}\tan^2\alpha = \frac{-23}{4} \Rightarrow \tan^2\alpha = 8$$

46. $\because \dfrac{\sin\alpha}{\sin\beta} = \dfrac{4\sqrt{2}}{3}$ and $\dfrac{\cos\alpha}{\cos\beta} = \dfrac{2\sqrt{3}}{9} \Rightarrow \sin\alpha = \dfrac{4\sqrt{2}}{3}\sin\beta$ and $\cos\alpha = \dfrac{2\sqrt{3}}{9}\cos\beta$

Now, we know that,

$$\sin^2\alpha + \cos^2\beta = 1$$

$$\Rightarrow \left(\frac{4\sqrt{2}}{3}\right)^2 \sin^2\beta + \left(\frac{2\sqrt{3}}{9}\right)^2 \cos^2\beta = 1$$

$$\Rightarrow \frac{32}{9}\sin^2\beta + \frac{12}{81}\cos^2\beta = 1$$

Dividing by $\cos^2\beta$ both sides, we get

$$\frac{32}{9}\tan^2\beta + \frac{4}{27} = \sec^2\beta$$

$$\Rightarrow \quad \frac{32}{9}\tan^2\beta + \frac{4}{27} = 1 + \tan^2\beta$$

$$\Rightarrow \quad \left(\frac{32}{9} - 1\right)\tan^2\beta = 1 - \frac{4}{27}$$

$$\Rightarrow \quad \frac{32-9}{9}\tan^2\beta = \frac{27-4}{27}$$

$$\Rightarrow \quad \frac{23}{9}\tan^2\beta = \frac{23}{27} \Rightarrow \tan^2\beta = \frac{1}{3}$$

47. $\dfrac{1+\sin\theta}{\cos\theta} = p + \sqrt{p^2+1}$

Now, $\quad \because \dfrac{1+\sin\theta}{\cos\theta} = p + \sqrt{p^2+1}$

$$\Rightarrow \quad \sec\theta + \tan\theta = p + \sqrt{p^2+1} \quad \text{...(1)}$$

We know that, $\quad \sec^2\theta - \tan^2\theta = 1$

$$\Rightarrow \quad (\sec\theta + \tan\theta)(\sec\theta - \tan\theta) = 1$$

$$\Rightarrow \quad (p + \sqrt{p^2}+1)(\sec\theta - \tan\theta) = 1$$

$$\Rightarrow \quad \sec\theta - \tan\theta = \frac{1}{p + \sqrt{p^2+1}}$$

$$\Rightarrow \quad \sec\theta - \tan\theta = \frac{1}{p + \sqrt{p^2+1}} \times \frac{p - \sqrt{p^2+1}}{p - \sqrt{p^2+1}}$$

$$\Rightarrow \quad \sec\theta - \tan\theta = \frac{p - \sqrt{p^2+1}}{p^2 - (p^2+1)}$$

$$\Rightarrow \quad \sec\theta - \tan\theta = \sqrt{p^2+1} - p \quad \text{...(2)}$$

From Eqn (1) + (2), we get

$$2\sec\theta = \sqrt{p^2+1} - p + p + \sqrt{p^2+1}$$

$$\Rightarrow \quad \sec\theta = \sqrt{p^2+1}$$

48. From previous question,
From Eqn (1) – (2), we get

$$2\tan\theta = p + \sqrt{p^2+1} - \sqrt{p^2+1} + p \Rightarrow \tan\theta = p$$

49. Now, $\quad \because \sec\theta + \operatorname{cosec}\theta = q$

$$\Rightarrow \quad \frac{1}{\cos\theta} + \frac{1}{\sin\theta} = q \Rightarrow \frac{\sin\theta + \cos\theta}{\sin\theta\cos\theta} = q$$

$$\Rightarrow \quad \sin\theta\cos\theta = \frac{p}{q} \ (\because \sin\theta + \cos\theta = p).$$

Now, $(\sin\theta + \cos\theta)^2 = \sin^2\theta + \cos^2\theta + 2\sin\theta\cos\theta$

$\Rightarrow \qquad p^2 = 1 + 2 \cdot \frac{p}{q} \Rightarrow p^2 - 1 = \frac{2p}{q} \Rightarrow q\,(p^2 - 1) = 2p$

50. We have to find the value of $\tan\theta + \cot\theta$.

Now, $\tan\theta + \cot\theta = \frac{\sin\theta}{\cos\theta} + \frac{\cos\theta}{\sin\theta}$

$= \frac{\sin^2\theta + \cos^2\theta}{\sin\theta\cos\theta} = \frac{1}{\frac{p}{q}}$ (From previous question)

$= \frac{q}{p}$

51. Relative speed = Speed of train X – Speed of train

$= (100 - 60)$ km/hr = 40 km/hr $= 40 \times \frac{5}{18}$ m/s $= \frac{100}{9}$ m/s

Now, Length of train X $= \frac{100}{9}$ m/s × 90 s = 100 m

52. Distance covered by X in 10 minutes = 4.8 km/hr × $\frac{10}{60}$ hr = 0.8 km

Now, let t be the time after 7 : 10
Now, Let t be the time after 7 : 10 a.m. when Y meets.
Total distance covered by Y = Total distance covered by X

$\Rightarrow \qquad 6t = 0.8 + 4.8\,t \Rightarrow 1.2\,t = 0.8 \Rightarrow \quad t = \frac{0.8}{1.2} = \frac{2}{3}$ hrs.

Now, Distance from P when X meet $Y = 6 \times \frac{2}{3}$ km = 4 km.

53. Now, let, initial salary of X and Y are x and y respectively.

According, $x \times \frac{112}{100} = \frac{28x}{25}$

and $\frac{28x}{25} - \frac{28x}{25} \times \frac{10}{100} = \frac{28x}{25} \times \frac{9}{10}$

$y + y \times \frac{10}{100} = \frac{11y}{10}$ and $\frac{11y}{10} - \frac{11y}{10} \times \frac{12}{100} = \frac{11y}{10} \times \frac{22}{25}$

∵ Their present salaries are equal.

$\Rightarrow \qquad \frac{28x}{25} \times \frac{9}{10} = \frac{11y}{10} \times \frac{22}{25} \Rightarrow \frac{x}{y} = \frac{121}{126}$

54. Let, m be the of males in the village
and f be the no. of females in the village.
So, $P = m + f$ where p is the no. of persons in village.

Now, A/q, No. of persons who can read and write is x % 0, i.e., $\frac{xp}{100}$

No. of males who can read and write is y% of m i.e., $\frac{ym}{100}$

and No. of females who can read and write in z% i.e., $\frac{zf}{100}$.

⇒ Total no. of persons who can read and write is

$$\frac{xp}{100}=\frac{ym}{100}+\frac{zf}{100}$$

$$\frac{xp}{100}=\frac{ym}{100}+\frac{z(p-m)}{100}\ (\because p=m+f)$$

⇒ $$xp=ym+zp-zm \Rightarrow xp-zp=(y-z)m$$

⇒ $$m=\frac{xp-zp}{y-z}=\frac{(x-z)}{(y-z)}p$$

55. Let's assume we take 1 unit of alloy X and 1 unit of alloy Y.

So, total Cu in Z = Cu in X + Cu in Y $=\frac{5}{9}+\frac{5}{18}=\frac{10+5}{18}=\frac{15}{18}$

and Total Zn in Z = Zn in X + Zn in Y $=\frac{4}{9}+\frac{13}{18}=\frac{8+13}{18}=\frac{21}{18}$

Now, The ratio of Cu to Zn in Z is $\frac{15}{18}:\frac{21}{18}=5:7$

56. Let, score of X be is x and score of Y is y.

According to question $\quad x=y+20 \quad$ ...(1)

$$x=\frac{75}{100}(x+y) \quad ...(2)$$

From (1) and (2), we get

$$y+20=\frac{75}{100}(x+y) \Rightarrow y+20=\frac{3}{4}x+\frac{3}{4}y$$

⇒ $$y-\frac{3}{4}y=\frac{3}{4}x-20 \Rightarrow \frac{y}{4}=\frac{3x-80}{4} \Rightarrow y=3x-80$$

Put the value of y in eq. (1), we get

$$x=3x-80+20 \Rightarrow 60=2x \Rightarrow x=30$$

Now, put the value of x in Eqn (1), we get

$$30=y+20 \Rightarrow 10=y$$

Hence, Ratio of score of X to Y is 30 : 10 = 3 : 1.

57. Let x_1 and x_2 are two roots of the equation

$$2x^2-5px+2p^2=0$$

A/q, $\quad x_1=x_2+4 \quad$ ...(1)

We know that, $\quad x_1+x_2=\frac{5p}{2}$ ... (2) and $x_1x_2=\frac{2p^2}{2}=p^2$ ...(3)

From Eqn (1) and (2), we get

$$x_2+4+x_2=\frac{5p}{2} \Rightarrow 2x_2=\frac{5p}{2}-4 \Rightarrow x_2=\frac{5p}{4}-2$$

Put the value of x_2 in Eqn. (3), we get

$$x_1\left(\frac{5p}{4}-2\right)=p^2 \Rightarrow x_1=\frac{p^2}{\frac{5p}{4}-2}$$

Now, put the value of x_1 and x_2 in Eqn. (1), we get

$$\frac{p^2}{\frac{5p}{4}-2}=\frac{5p}{4}-2+4 \Rightarrow p^2=\left(\frac{5p}{4}-2\right)^2+4\left(\frac{5p}{4}-2\right)$$

$$\Rightarrow \quad p^2=\frac{25p^2}{16}+4-2\cdot\frac{5p}{4}\cdot 2+5p-8$$

$$\Rightarrow \quad \frac{25p^2}{16}-p^2-4=0$$

$$\Rightarrow \quad \frac{25p^2-16p^2}{16}=4 \Rightarrow 9p^2=64$$

$$\Rightarrow \quad p^2=\frac{64}{9} \Rightarrow p=\frac{8}{3}$$

58. Let, p be the principal, r is rate and X be the installment then using equal annual install-ment formula for 3 years, we get

$$\Rightarrow \quad P=X\left[\frac{1}{1+r}+\frac{1}{(1+r)^2}+\frac{1}{(1+r)^3}\right]$$

$$\Rightarrow \quad 10000=x\left[\frac{1}{1+0.1}+\frac{1}{(1+0.1)^2}+\frac{1}{(1+0.1)^3}\right]$$

$$\Rightarrow \quad 10000=x\left[\frac{1}{1\cdot 1}+\frac{1}{1\cdot 21}+\frac{1}{1\cdot 331}\right]$$

$$\Rightarrow \quad 10,000=x\left[\frac{10}{11}+\frac{100}{121}+\frac{1000}{1331}\right]$$

$$\Rightarrow \quad 10,000=x\left[\frac{1210+1100+1000}{1331}\right]$$

$$\Rightarrow \quad 10000=x\cdot\frac{3310}{1331}$$

$$x=\frac{1331}{331}\times 1000$$

$$x=4.021\cdot 13 \Rightarrow x\approx 4021$$

59. We have,

$$5x+3<8x-9 \text{ and } 2x+20>5x+2$$

$$\Rightarrow \quad 3+9<8x-5x \text{ and } 20-2>5x-2x$$

$$\Rightarrow \quad 12<3x \text{ ...(1) and } 18>3x \text{ ...(2)}$$

From Eqn (1) and (2), we get

$$12<3x<18 \Rightarrow 4<x<6$$

60. Let, the cost price of the article be x and the marked price of the article by y.

So, Final selling price $= y\times\left(1-\frac{10}{100}\right)\left(1-\frac{20}{100}\right)\left(1-\frac{25}{100}\right) = y\times\frac{9}{10}\times\frac{4}{5}\times\frac{3}{4} = \frac{27}{50}y$ ...(1)

Also, the shopkeeper makes a profit of 8% on the cost price.

So, Final selling price is $x\times\left(1+\frac{8}{100}\right) = \frac{27}{25}x$ ...(2)

From (1) and (2), we get

$$\frac{27}{50}y = \frac{27}{25}x \Rightarrow x = \frac{y}{2} \Rightarrow y = 2x$$

Now, If the article was then sold after the first discount new selling price is

$$y\left(1-\frac{10}{100}\right) = \frac{9}{10}y = \frac{9}{10}\times 2x = \frac{9}{5}x \ (\because\ y = 2x)$$

Now, Profit $= \frac{9}{5}x - x = \frac{4}{5}x$ and profit % $= \frac{\frac{4}{5}x\times 100}{x}\% = 80\%$

61. Given that,

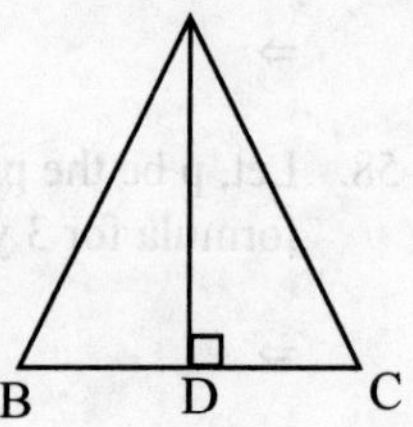

$$AD \perp BC,$$
$$AB + AC = p \quad ...(1)$$
$$AB - AC = q \quad ...(2)$$

and
$$BD - CD = r \quad ...(3)$$

From fig.,
$$BD + CD = BC \quad ...(4)$$

Now, From Eqn (1) + (2), we get

$$2AB = p+q \Rightarrow AB = \frac{p+q}{2}$$

So,
$$AC = \frac{p-q}{2}$$

Now, ΔABD,

$$AB^2 = AD^2 + BD^2 \text{ (By Pythagorean theorem)}$$

$\Rightarrow$ $$AD^2 = AB^2 - BD^2$$

$\Rightarrow$ $$AD^2 = \left(\frac{p+q}{2}\right)^2 - BD^2 \quad ...(5)$$

And In ΔADC,

$$AC^2 = AD^2 + DC^2 \text{ (By Pythagorean theorem)}$$

$\Rightarrow$ $$\left(\frac{p-q}{2}\right)^2 = AD^2 + DC^2$$

$\Rightarrow$ $$AD^2 = \left(\frac{p-q}{2}\right)^2 - DC^2 \quad ...(6)$$

From Eqn (5) and (6), we get

$\Rightarrow$ $$\left(\frac{p+q}{2}\right)^2 - BD^2 = \left(\frac{p-q}{2}\right)^2 - DC^2$$

$$\Rightarrow \quad \frac{p^2+q^2+2pq-p^2-q^2+2pq}{4}=(BD+CD)\,(BD-CD)$$

$$\Rightarrow \quad \frac{4pq}{4}=BC\cdot r \qquad \text{(from Eqn (3))}$$

$$\Rightarrow \quad BC=\frac{pq}{r}$$

62. Let, h be the height and r be the radius of the right circular cylinder.
According to the following

$$h+r=21 \qquad \text{... (1)}$$

Let A be the curved surface area and V be the volume of cylinder.

Now, A/q, $\quad A=616 \Rightarrow 2\pi rh=616 \Rightarrow rh=\frac{308}{22}\times 7$

$$\Rightarrow \quad rh=14\times 7=98 \qquad \text{...(2)}$$

Now, $\quad (h-r)^2=(h+r)^2-4rh$

$$\Rightarrow \quad (h-r)^2=21^2-4\cdot 98$$

$$\Rightarrow \quad (h-r)^2=441-392$$

$$\Rightarrow \quad (h-r)^2=49 \Rightarrow h-r=7 \qquad \text{...(3)}$$

From Eqn (1) + (3), we get

$$2h=28 \Rightarrow h=14 \text{ cm}$$

So, $r=7$ cm

Now, $\quad V=\pi r^2h=\frac{22}{7}\times 7\times 7\times 14=2156 \text{ cm}^3$

63. I. $AQ=\sqrt{73}$ units.

In ΔABQ, $AQ^2=AB^2+BQ^2$ (By Pythagorean theorem)

$$\Rightarrow \quad (\sqrt{73})^2=(AP+PB)^2+BQ^2 \Rightarrow 73=(PB+PB)^2+BQ^2$$

$$[\because AP=PB]$$

$$\Rightarrow \quad 73=4PB^2+BQ^2 \quad \text{...(1)}$$

II. $CP=\sqrt{52}$ units

In ΔPBC, $\quad PC^2=PB^2+BC^2$ (By Pythagorean theorem)

$$\Rightarrow \quad \left(\sqrt{52}\right)^2=PB^2+\left(BQ+QC\right)^2$$

$$\Rightarrow \quad 5^2=PB^2+\left(BQ+BQ\right)^2 \Rightarrow 52=PB^2+4BQ^2 \qquad [\because BQ=CQ]\text{...(2)}$$

From Eqn (1) + (2), we get

$$73+52=5PB^2+5BQ^2 \Rightarrow 125=5\,(PB^2+BQ^2)$$

$$PB^2+BQ^2=25 \qquad \text{...(3)}$$

From Eqn. (1) – (2), we get

$$73-52=3PB^2-3BQ^2 \Rightarrow 21=3\left(PB^2-BQ^2\right)$$

$$\Rightarrow \quad 7=PB^2-BQ^2 \qquad \text{...(4)}$$

From Eqn (3) + (4), we get

$$2PB^2 = 32 \Rightarrow PB^2 = 16 \Rightarrow PB = 4$$

From Eqn. (3)- (4), we get

$$2BQ^2 = 18 \Rightarrow BQ = 3$$

Now, $\text{Area}(\Delta ABC) = \frac{1}{2} \times BC \times AB = \frac{1}{2} \times (BQ + QC)(AP + PB)$

$$= \frac{1}{2} \times (BQ + BQ)(PB + PB) = \frac{1}{2} \times 2BQ \times 2PB$$

$$= 3 \times 2 \times 4 = 24 \text{ square units}$$

Hence, Both I and II are required.

64. Let $x, x+1, x+2$ are the length, breadth and height of a cuboid and V be the volume the

$$V = x(x+1)(x+2) \Rightarrow 336 = x(x+1)(x+2)$$

$\Rightarrow$ $23 \times 2 \times 3 \times 7 = x(x+1)(x+2) \Rightarrow 6 \times 7 \times 8 = x(x+1)(x+2)$

$\Rightarrow$ $x = 6$

Now, Surface area $= 2\,(x\,(x + 1) + (x + 1)\,(x + 2) + (x + 2)\,x) = 2(6 \times 7 + 7 \times 8 + 8 \times 6)$

$$= 2 \times 146 = 292 \text{ square units}$$

65. Now, P is the midpoint arc APB and C is the midpoint of chord AB, the Line segment OCP is $\perp r$ to chord AB. i.e.,

OC $\perp$ AB

where O be the centre of the circle.

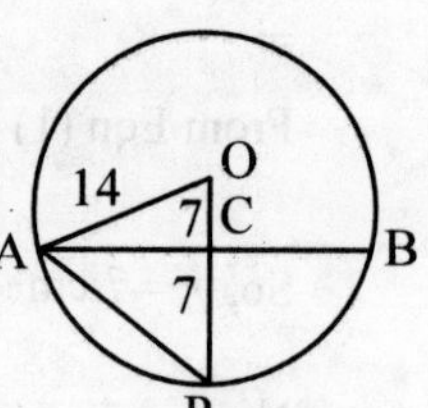

Now, $OP = OC + CP \Rightarrow 14 = OC + 7 \Rightarrow OC = 7$ cm

In ΔOAC,

$$OA^2 = OC^2 + AC^2 \text{ (By Pythagorean theorem)}$$

$\Rightarrow$ $196 - 49 = AC^2 \Rightarrow AC^2 = 147$

Now, In ΔACP,

$$AP^2 = AC^2 + CP^2 \Rightarrow AP^2 = AC^2 + CP^2$$

$\Rightarrow$ $AP^2 = 147 + 7^2 \Rightarrow AP^2 = 196$

$\Rightarrow$ $AP = 14$ cm

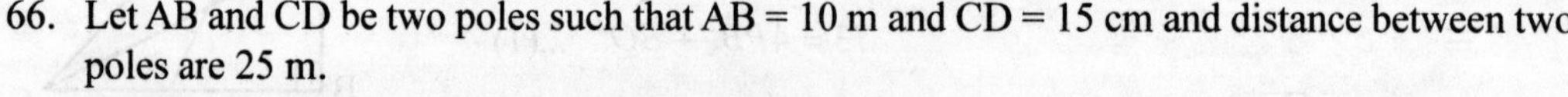

66. Let AB and CD be two poles such that AB = 10 m and CD = 15 cm and distance between two poles are 25 m.

Let P be the intersection point of AD and BC.

Now, let A be (0, 10), B be (0, 0), C be (25, 15) and D be (25, 0)

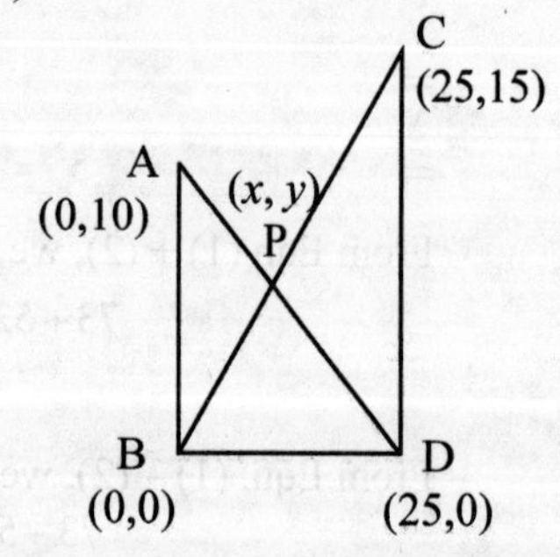

Then $AD = \text{is } y - 10 = \frac{0-10}{25-0}(x - 0)$

$$\Rightarrow y - 10 = -\frac{10}{25}x$$

$$\Rightarrow y - 10 = -\frac{2}{5}x \Rightarrow y = 10 - \frac{2}{5}x$$

and BC is $y - 0 = \frac{15-0}{25-0}(x-0) \Rightarrow y = \frac{15}{25}x \Rightarrow y = \frac{3}{5}x$

$\because P$ is intersection point of AD and BC.

$\therefore \frac{3}{5}x = 10 - \frac{2}{5}x \Rightarrow \frac{3}{5}x + \frac{2}{5}x = 10 \Rightarrow x = 10$

Now, $y = \frac{3}{5} \times 10 = 6$ m

Height of the point of intersection is 6 m .

67. We have, AB + BC + AC = 60

$\Rightarrow$ AB + BC + AB + BC – 10 = 60 ($\because$ AB + BC = AC + 10)

$\Rightarrow$ 2AB + 2BC = 60 + 10

$\Rightarrow$ 2 (AB + BC) = 70

$\Rightarrow$ AB + BC = 35

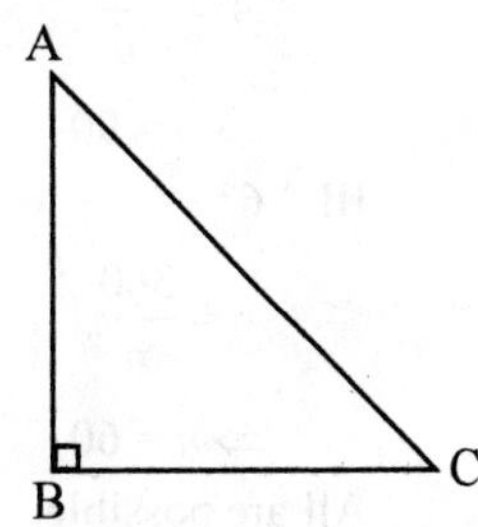

By pythagorean theorem,

$\Rightarrow$ $AB^2 + BC^2 = AC^2$

$\Rightarrow$ $(AB + BC)^2 - 2\,AB \cdot BC = (AB + BC - 10)^2$

$\Rightarrow$ $35^2 - 2AB \,.\, BC = (35 - 10)^2$

$\Rightarrow$ $1225 - 625 = 2AB \,.\, BC$

$\Rightarrow$ $2AB \cdot BC = 600 \Rightarrow AB \cdot BC = 300$

Now, Area $= \frac{1}{2} \times AB \times BC = \frac{1}{2} \times 300 = 150$ square units

68. Let AB and CD be two poles such that BD = 24 cm = AE and CE = 10m

Now, In Δ ACE

$AC^2 = CE^2 + AE^2 \Rightarrow AC^2 = 10^2 + 24^2$ (By Pythagorean theorem)

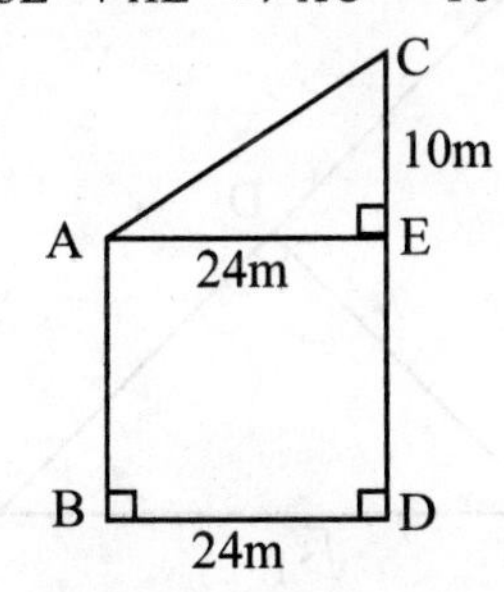

$\Rightarrow$ $AC^2 = 100 + 576$

$\Rightarrow$ $AC^2 = 676$

$\Rightarrow$ $AC = 26$ m

69. I. By midpoint theorem,

$ZY \parallel BC$ and $ZY = \frac{1}{2}BC$

$XY \parallel AB$ and $XY = \frac{1}{2}AB$

& $XZ \parallel AC$ and $XZ = \frac{1}{2}AC$

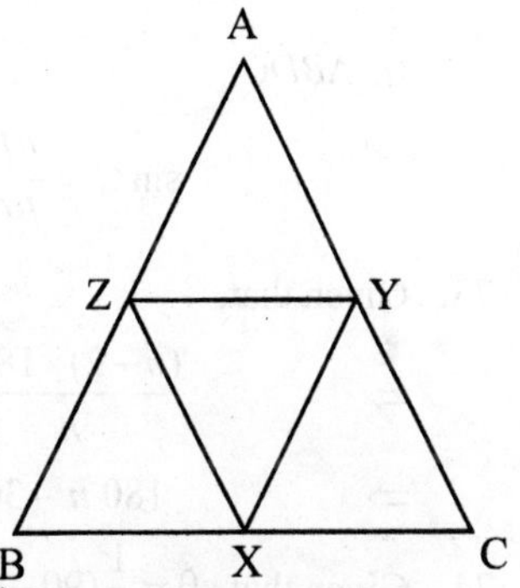

In quadrilateral AZXY,

AZ || XY & AY || XZ (By midpoint theorem)

$\Rightarrow$ AZXY is a parallelogram

II. Midpoint theorem also states that the four triangle (AZY, BXZ, CXY, XYZ) are congruent. And areas of congruent triangle are equal. (We know)

$\Rightarrow \quad ar(\Delta AZY) = ar(\Delta BXZ) = ar(\Delta CXY) = ar(\Delta XYZ)$

$\Rightarrow \quad ar(AZY) = \frac{1}{2} \times ar(\Delta ABC)$

Hence, Both I and II are correct.

70.

I. 4°

$4 = \frac{360^\circ}{n}$

$\Rightarrow n = 90$

II. 5°

$5 = \frac{360^\circ}{n}$

$\Rightarrow n = 72$

III. 6°

$6 = \frac{360}{n}$

$\Rightarrow n = 60$

IV. 8°

$8 = \frac{360^\circ}{n}$

$\Rightarrow n = 45$

All are possible.

71. $\text{Area}(\Delta ABC) = \frac{1}{2} \cdot AB \cdot AC \cdot \sin A = \frac{1}{2} \times 7 \times 12 \times \sin 30° = 42 \cdot \frac{1}{2} = 21\text{cm}^2$

72. In $\Delta\ ABC$

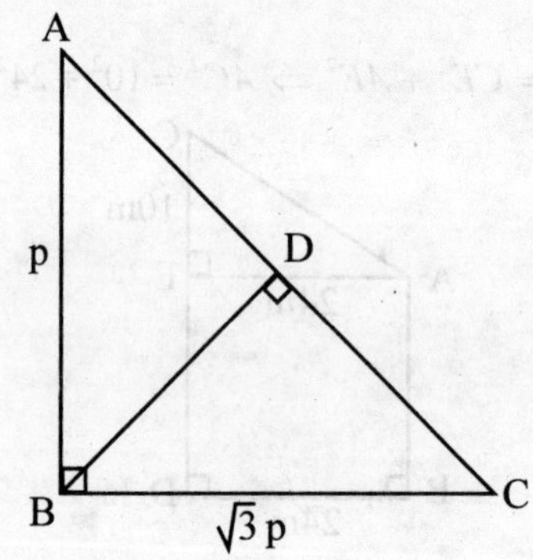

$$\tan C = \frac{AB}{BC} \Rightarrow \tan C = \frac{p}{\sqrt{3}p} \Rightarrow \tan C = \tan 30° \Rightarrow C = 30°$$

In ΔBDC,

$$\sin C = \frac{BD}{BC} \Rightarrow \sin 30^\circ = \frac{BD}{\sqrt{3}p} \Rightarrow \frac{1}{2} = \frac{BD}{\sqrt{3}p} \Rightarrow BD = \frac{\sqrt{3}}{2}p$$

73. Given that, $I - E = 120°$

$\Rightarrow \quad \frac{(n-2)\times 180^\circ}{n} - \frac{360^\circ}{n} = 120^\circ \Rightarrow (n-2)\,180 - 360 = 120\,n$

$\Rightarrow \quad 180\,n - 360 - 360 = 120\,n \Rightarrow 60\,n = 720 \Rightarrow n = 12$

74. Given that, $\theta = \frac{1}{4}(90-\theta)$ (Complementary angle means $90 - \theta$)

$\Rightarrow \quad 4\theta = 90° - \theta \Rightarrow 5\theta = 90° \Rightarrow \theta = 18°$

75. Let, AB = 11 cm, BC = 60 cm and CA = 61 cm and let X, Y, Z are mid points of BC, AC and AB respectively.

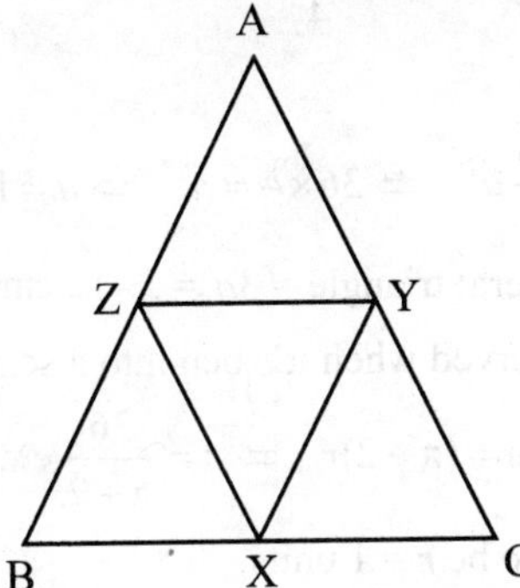

Now, By mid point theorem,

$$\text{ar}(\Delta XYZ) = \frac{1}{4} \times \text{ar}(\Delta ABC) \Rightarrow \text{ar}(\Delta XYZ) = \frac{1}{4} \times \frac{1}{2} \times 11 \times 60 = 82.5 \text{ cm}^2$$

(11, 60 and 61 are triplet such that $11^2 + 60^2 = 61^2$)

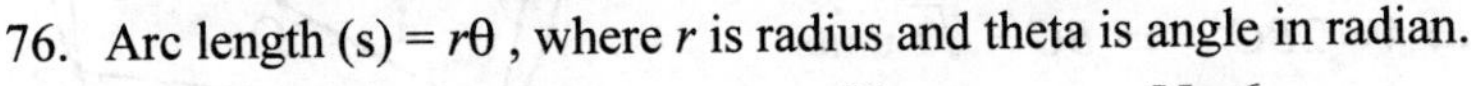
76. Arc length (s) = $r\theta$, where r is radius and theta is angle in radian.

$$\Rightarrow \quad 55 = r \times \frac{30 \times \pi}{180} \Rightarrow r = \frac{55 \times 6}{\pi}$$

$$\Rightarrow \quad r = \frac{55 \times 6 \times 7}{22} \Rightarrow r = 105 \text{ cm}$$

77. Let r be the radius of base, h be the height and l be the slant height of the conical tent. We know that, the angle between the slant height and the height is half of the vertex angle i.e., 30°.

Now, $$\sin 30^\circ = \frac{r}{l} \Rightarrow \frac{1}{2} = \frac{r}{l} \Rightarrow l = 2r$$

Now, Given that, curved surface area =100 m^2

$$\Rightarrow \quad \pi rl = 100\text{m}^2 \Rightarrow \pi \cdot r \cdot (2r) = 100\text{m}^2$$

$$\Rightarrow \quad 2r^2 = \frac{100}{\pi}\text{m}^2 \Rightarrow r = \sqrt{\frac{50}{\pi}}\text{m}$$

r
h
l
60°

By pythagorean theorem, $l^2 = r^2 + h^2 \Rightarrow (2r)^2 = r^2 + h^2$

$$\Rightarrow 4r^2 - r^2 = h^2 \Rightarrow h^2 = 3r^2 \Rightarrow h = \sqrt{3} \times \sqrt{\frac{50}{\pi}} \text{ m} \Rightarrow h = \sqrt{\frac{150}{\pi}} \text{ m}$$

Now, $$\text{Volume} = \frac{1}{3}\pi r^2 h = \frac{1}{3} \times \pi \times \frac{50}{\pi} \times \sqrt{\frac{150}{\pi}} \text{ m}^3 = \frac{50}{3} \times \frac{5\sqrt{6}}{\sqrt{\pi}} \text{ m}^3$$

$$= \frac{250\sqrt{6}}{3\sqrt{\pi}} \text{ m}^3 = \frac{250\sqrt{2}}{\sqrt{3}\sqrt{\pi}} \text{ m}^3$$

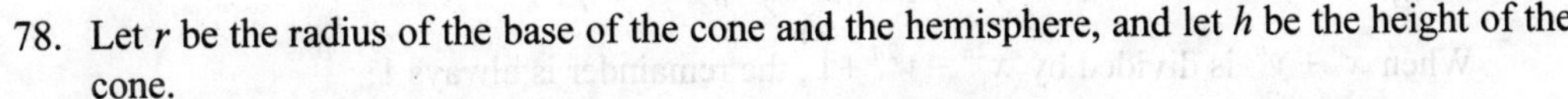
78. Let r be the radius of the base of the cone and the hemisphere, and let h be the height of the cone.

Given that, Volume of cone = Volume of hemisphere

$$\Rightarrow \frac{1}{3}\pi r^2 h = \frac{2}{3}\pi r^3 \Rightarrow \pi r^2 h = 2\pi r^3 \Rightarrow \frac{h}{r} = \frac{2}{1}$$

79. $\because$ area of an equilateral triangle (A) $= \dfrac{\sqrt{3}}{4}a^2$

where a is the side.

$\Rightarrow A = \dfrac{\sqrt{3}}{4}a^2 \quad \Rightarrow 36\sqrt{3} = \dfrac{\sqrt{3}}{4}a^2 \quad \Rightarrow 36 \times 4 = a^2 \quad \Rightarrow a = 12$ cm

Now, Perimeter of an equilateral triangle $= 3a = 3 \times 12$ cm $= 36$ cm

$\therefore$ The length of wire is conserved when it's ben into a semicircle.

$\Rightarrow 36 \text{ cm} = \pi r + 2r \quad \Rightarrow 36 \text{ cm} = (\pi + 2)r \quad \Rightarrow r = \dfrac{36}{\pi + 2}$ cm

80. Let the unit radius of the circle be $r = 1$ units.

When a square is inscribed in a circle, the diagonal of the square is equal to diameter of the circle.

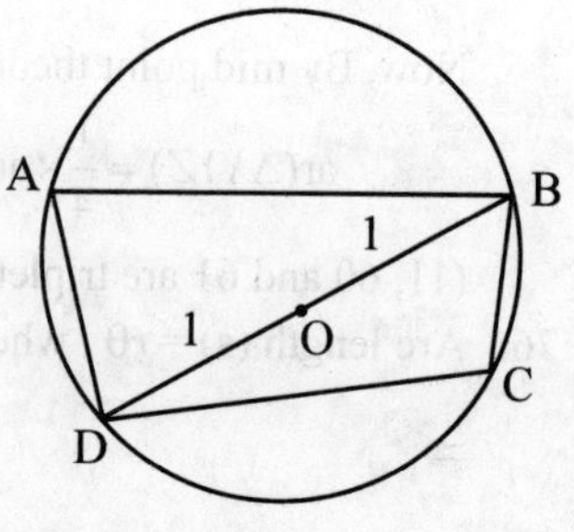

$\Rightarrow \quad BD = 2$ units

In ΔBDC, $BD^2 = BC^2 + CD^2$ (By Pythagorean theorem)

$\Rightarrow 2^2 = BC^2 + BC^2 \Rightarrow 4 = 2BC^2 \Rightarrow BC = \sqrt{2}$ units

Now, area of the square $= BC^2 \Rightarrow x = (\sqrt{2})^2 = 2$

Now, Let the unit side length of the square be S = 1 units.

When a circle is inscribed in a square, the diameter of the circle is equal to the side of the square. i.e.,

$$2r = 1 \Rightarrow r = \frac{1}{2} \text{ unit}$$

area of circle $= \pi r^2 \Rightarrow y = \pi \cdot \left(\dfrac{1}{2}\right)^2 \Rightarrow y = \dfrac{\pi}{4}$

Now,

(a) $\pi x = \pi \cdot 2 = 2\pi$ & $2y = 2 \cdot \dfrac{\pi}{4} = \dfrac{\pi}{2}$ are not equal.

(b) $2\pi x = 2 \cdot \pi \cdot 2 = 4\pi$ & $y = \dfrac{\pi}{4}$ are not equal.

(c) $\pi x = \pi \cdot 2 = 2\pi$ & $4y = 4 \cdot \dfrac{\pi}{4} = \pi$ are not equal.

(d) $\pi x = \pi \cdot 2 = 2\pi$ & $8y = 8 \cdot \dfrac{\pi}{4} = 2\pi$ are equal.

Hence, (d) is correct.

81. $x^{2n} - y^{2n} = (x^n)^2 - (y^n)^2 = (x^n + y^n)(x^n - y^n)$

Now, $\quad x^{2n} - y^{2n} + 1 = (x^n + y^n)(x^n - y^n) + 1$

When $x^n + y^n$ is divided by $x^{2n} - y^{2n} + 1$, the remainder is always 1.

82. Let N be the single digit no. Then M be also the single digit no.

We have $\quad N \times M = 252 \quad \Rightarrow N \times N = 252 \quad \Rightarrow N^2 = 252$

$\therefore$ Largest no. of single digit is 9. and $9^2 = 81 \neq 252$.

$\Rightarrow$ Single digit no. is not possible.

Now, let N be the two-digit no. i.e., $N = 10a + b$ and $M = 10b + a$

$\therefore$ $$N \times M = 252$$

$$(10a + b)(10b + a) = 252$$

We check factors of 252 in terms of two factors.

1×252, 2×126, 3×84, 4×63, 6×42, 7×36, 9×28, 12×21, 14×18

Here, 12×21 is only possible.

Either $$N = 12 \text{ and } M = 21$$

or $$N = 21 \text{ and } M = 12$$

Statement I: $N + M = 12 + 21$ or $21 + 12 = 33$ or 33

Statement II: $N > M \Rightarrow N = 21$ and $M = 12$

So, option (b) is correct.

83. The last digit in $(54D)^{100}$ depends on D.

So, the last digit in $(54D)^{100}$ is determined by the last digit of D^{100}.

$\because$ last digit of D^{100} is 1. $\Rightarrow D$ must be 1, 3, 7 or 9.

$(\because 1^{100} = 1, 3^4 = 81 \Rightarrow 3^{100} = \ldots 1, 7^4 = 2401 \Rightarrow 7^{100} = \ldots 1, 9^2 = 81 \Rightarrow 9^{100} = \ldots 1)$

Statement I: $D > 5 \Rightarrow D = 7$ or 9

Statement II: D is a multiple of 3 $\Rightarrow D = 3$ or 9

84. Given that, $$\angle A = \angle B - \angle C$$

We know that, $$\angle A + \angle B + \angle C = 180^\circ \quad (\text{In a } \triangle ABC)$$

$\Rightarrow \angle B - \angle C + \angle B + \angle C = 180^\circ \Rightarrow 2\angle B = 180^\circ \Rightarrow \angle B = 90^\circ$

$\Rightarrow \angle A + \angle C = 90^\circ \Rightarrow \angle A < 90^\circ$, $\angle C < 90^\circ \Rightarrow \angle A$ is acute.

Statement I: *ABC* is not an obtuse-angled triangle.

$\Rightarrow \angle A$ and $\angle C < 90^\circ$ and $\angle B = 90^\circ \Rightarrow$ Statement I is true.

Statement II: Angle (c) is acute.

$\Rightarrow$ $$\angle A \text{ and } \angle C < 90^\circ$$

$\Rightarrow$ Statement II is true.

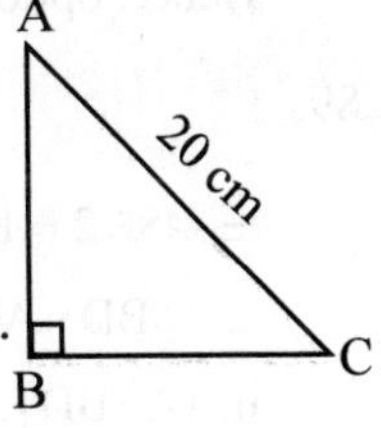

85. We know that,

In a right-angled triangle, the circumcenter is the midpoint of the hypotenuse.

So, the circum-radius is half the length of the hypotenuse. i.e., (10 cm).

86. Area of parallelogram (ABCD) $= \text{AB} \cdot \text{CB} \sin B = \text{AB} \cdot \text{BC} \sin 60^\circ$

$\Rightarrow$ $$7\sqrt{3} = \text{AB} \cdot \text{BC} \cdot \frac{\sqrt{3}}{2} \Rightarrow \text{AB} \cdot \text{BC} = 14$$

Perimeter of parallelogram $\text{ABCD} = 2(\text{AB} + \text{BC})$

Statement I: Now, we check factors of 14.

Factors of 14 are (1, 2, 7, 4)

$\therefore$ AB and BC are prime numbers $\Rightarrow$ AB = 2 or 7 and BC = 7 or 2

Now, Perimeter of parallelogram ABCD $= 2(2+7)$ or $2(7+2) = 18$

Statement II: The length of the sides are natural numbers each greater than 1 unit.

$\Rightarrow$ AB = 2 or 7 and BC = 7 or 2 again.

So, Perimeter of parallelogram ABCD = 18

87. We know that,

If two chords intersect inside a circle, then the product of the segments of one chord is equal to the product of the segments of the other chord.

i.e., $$AP \times PB = CP \times PD$$

$\Rightarrow CP \times PD = 48$ square units

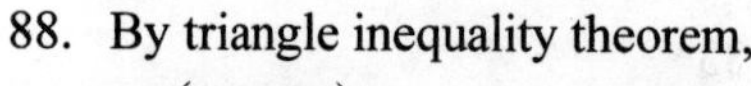

88. By triangle inequality theorem,

In (ΔABD)

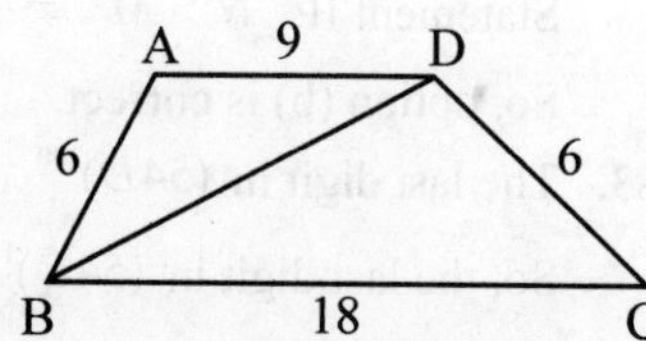

$$AB + BD > AD \Rightarrow 6 + BD > 9 \Rightarrow BD > 3$$
$$BD + AD > AB \Rightarrow BD + 9 > 6 \Rightarrow BD > -3$$
$$AD + AB > BD \Rightarrow 9 + 6 > BD \Rightarrow BD < 15$$

$\therefore \quad -3 < BD < 15 \qquad \ldots(1)$

In (ΔBCD),

$$BC + CD > BD \Rightarrow 18 + 6 > BD \Rightarrow BD < 24$$
$$CD + BD > BC \Rightarrow 6 + BD > 18 \Rightarrow BD > 12$$
$$BD + BC > CD \Rightarrow BD + 18 > 6 \Rightarrow BD > -12$$

$\Rightarrow \qquad 12 < BD < 24 \qquad \ldots(2)$

From inequalities (1) and (2), we get

$$12 < BD < 15 \ \Rightarrow BD = 13 \text{ or } 14$$

Statement I: The length of (BD) is an integer greater than 13.

$\Rightarrow \qquad BD = 14$

Statement II: The length of (BD) is an even integer.

$\Rightarrow \qquad BD = 14$

Hence, option (b) is correct.

89. $$\text{ar}(\Delta ABC) = \frac{1}{2} \times BC \times AD$$

$\Rightarrow 48 \times 2 = BC \times AD \ \Rightarrow BC \times AD = 96 \ \Rightarrow BD \times 2 \times AD = 96 \quad (\because AD \perp BC \text{ and } BD = DC)$

$\Rightarrow \quad BD \times AD = 48$

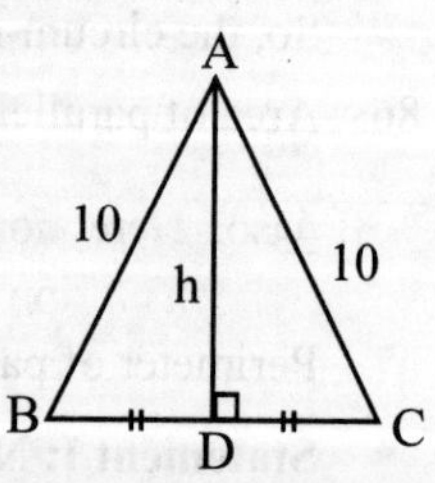

In (ΔABD), $$AB^2 = AD^2 + BD^2$$

$$\Rightarrow \qquad 10^2 = AD^2 + BD^2 \ \Rightarrow 100 = \left(\frac{48}{BD}\right)^2 + BD^2$$

$$\Rightarrow \qquad 100 = \frac{2304}{BD^2} + BD^2 \Rightarrow 100 = \frac{2304 + BD^4}{BD^2}$$

$$\Rightarrow \qquad 100BD^2 = 2304 + BD^4 \ \Rightarrow BD^4 - 100BD^2 + 2304 = 0$$

Let $(BD^2 = x)$, then $x^2 - 100x + 2304 = 0$

Using quadratic formula, we get

$$x=\frac{-(-100)\pm\sqrt{(-100)^2-4\cdot1\cdot2304}}{2\cdot1}\Rightarrow x=\frac{100\pm\sqrt{10000-9216}}{2}$$

$$\Rightarrow\quad x=\frac{100\pm\sqrt{784}}{2}\Rightarrow x=\frac{100\pm28}{2}\Rightarrow BD^2=64\text{ or }36\Rightarrow BD=8\text{ or }6$$

$$BC=2\cdot BD\Rightarrow BC=2\cdot8\text{ or }2\cdot6\Rightarrow BC=16\text{ or }12$$

Statement I: Both values of (BC) are even.

Statement II: The height of the triangle is greater than the half of the length of the base.

$\because$ $$BC\times AD=96$$

If $BC=16$, then $AD=\frac{96}{16}=6$

Here, $AD<\frac{BC}{2}$ If $BC=12$, then $AD=\frac{96}{12}=8$. Here, $AD>\frac{BC}{2}$

$\Rightarrow BC=12\Rightarrow$ Option (a) is correct.

90. Let d_1 (AC) and $d_2(BD)$ be the diagonals of the rhombus $ABCD$.

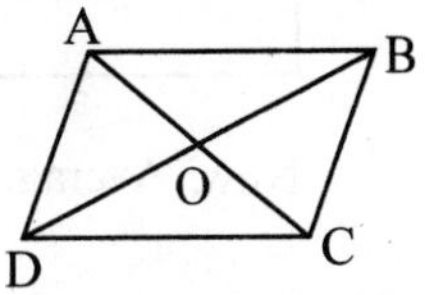

Given that, $\frac{d_1}{d_2}=\frac{5}{12}\Rightarrow d_1=5x$ and $d_2=12x$

for some constant x.

We know that, the diagonals of rhombus are bisect each other at 90°.

Now, In ΔAOB,

$$\left(\frac{d_1}{2}\right)^2+\left(\frac{d_2}{2}\right)^2=AB^2\quad\text{(By Pythagoras' theorem)}$$

$$\Rightarrow\quad\left(\frac{5x}{2}\right)^2+\left(\frac{12x}{2}\right)^2=AB^2\Rightarrow\frac{25x^2}{4}+\frac{144x^2}{4}=AB^2$$

$$\Rightarrow\frac{169x^2}{4}=AB^2\Rightarrow AB=\frac{13x}{2}$$

Statement I: $d_1+d_2=34\Rightarrow 5x+12x=34\Rightarrow 17x=34\Rightarrow x=2$

Now, $d_1=5x=10$ cm, $d_2=12x=24$ cm

$$AB=\frac{13x}{2}=\frac{13}{2}\times2=13\text{ cm}$$

Here, None of the diagonals are equal to side of the rhombus.

Statement II:

$$\Rightarrow\quad AB=\frac{13x}{2}\Rightarrow13=\frac{13x}{2}\Rightarrow x=2$$

Now, $d_1=5x=5\times2=10$ cm, $d_2=12x=12\times2=24$ cm

Also, Here, none of the diagonals are equal to the side of the rhombus.

Hence, Option (b) is correct.

91. The median is the measure of central tendency that is least affected by the presence of extreme observations in the data.

92. (a) arithmetic mean: It is used for averaging quantities that are directly additive.

(b) geometric mean: It is used for averaging quantities that are related multiplicatively, often for growth rates or ratios over time.

(c) harmonic mean: It is most suitable measure of central tendency for averaging rates, ratios, or speeds, such as price/unit, work done/hour, or km/hr. Because it gives more weight to lower values, which is appropriate when dealing with quantities that represents a rate over time or per unit.

(d) Mode: It represents the most frequent value in a dataset and is not suitable for averaging rates or ratios.

93. (d)

Marks (Class Interval)	**No. of Candidates (Frequency, f_i)**	**Midpoint (x_i)**	$f_i x_i$
10 - 20	$100-75=25$	15	$25\times15=375$
20 - 30	$75-60=15$	25	$15\times25=375$
30 - 40	$60-40=20$	35	$20\times35=700$
40 - 50	$40-0=40$	45	$40\times45=1800$
Total	$\sum f_i=100$		$\sum f_i x_i=3250$

Now, Average marks (mean) $=\frac{\sum f_i x_i}{\sum f_i}=\frac{3250}{100}=32.5$

94. New arithmetic mean $=5\times$ Original arithmetic mean $=5\times60=300$

95. Let, $f_1=45, f_2=40$ and $f_3=55$ and $x_1=2, x_2=2.5$ and $x_3=2$

Now, $\sum f_i=45+40+55=140$ and $\sum f_i x_i=45\times2+40\times2.5+55\times2=90+100+110=300$

Now, $$\text{mean}=\frac{\sum f_i x_i}{\sum f_i}=\frac{300}{140}=2.14$$

96. (a) Arithmetic mean and Geometric mean: Measures like the arithmetic mean and geometric mean are calculated using the values of all data points, rather than their position, making them mathematical average.

(b) Median: It is considered as a positional average because it is determined by its position within an ordered dataset.

(c) Mode: It is also a measure of central tendency. It is based on the frequency of values, not strictly their position in an ordered set.

97.

Distance class	**Frequency (f_i)**	**Cumulative frequency (cf)**
0-5	u	u
5-10	v	$u+v$
10-15	11	$u+v+11$
15-20	52	$u+v+11+52=u+v+63$
20-25	75	$u+v+63+75=u+v+138$
25-30	22	$u+v+138+22=u+v+160$

Now, $$\text{Median}=L+\left(\frac{\frac{N}{2}-cf}{f}\right)\times h$$

where, $L =$ lower limit of the median class = 20

$N =$ Total frequency $= u + v + 160$

$c_f =$ Cumulative frequency of the class preceding the median class $= u + v + 63$

$f =$ frequency of median class = 75, $h =$ Class width = 5

Now, Substitute these value in formula, we get

$$20.8 = 20 + \left(\frac{\frac{u+v+160}{2} - (u+v+63)}{75} \right) \times 5$$

$$\Rightarrow (20.8 - 20) \times 15 = \frac{u+v+80-u-v-63}{15} \Rightarrow 12 = \frac{17 - \frac{u}{2} - \frac{v}{2}}{15} \Rightarrow \frac{u+v}{2} = 5$$

$\Rightarrow u + v = 10$ ...(1)

Now,

Distance class	Frequency (f_i)	Midpoint (x_i)	$f_i x_i$
0–5	$3u$	2.5	$7.5u$
5–10	$2v$	7.5	$15v$
10–15	40	12.5	500
15–20	50	17.5	875
20–25	30	22.5	675
25–30	28	27.5	770
	$\sum f_i = 3u + 2v + 148$		$\sum f_i x_i = 7.5u + 15v + 2820$

Now, Mean $= \dfrac{\sum f_i x_i}{\sum f_i}$

$$\Rightarrow 17.3 = \frac{7.5u + 15v + 2820}{3u + 2v + 148} \Rightarrow 17.3(3u + 2v + 148) = 7.5u + 15v + 2820$$

$$\Rightarrow 51.9u + 34.6v + 2560.4 = 7.5u + 15v + 2820 \Rightarrow 51.9u - 7.5u + 34.6v - 15v = 2820 - 2560.4$$

$$\Rightarrow 44.4u + 19.6v = 259.6 \Rightarrow 44.4u + 19.6(10 - u) = 259.6 \quad \text{(From eqn (1))}$$

$$\Rightarrow 44.4u + 196 - 19.6u = 259.6 \Rightarrow 24.8u = 259.6 - 196 \Rightarrow 24.8u = 63.6$$

$$\Rightarrow \quad u = \frac{63.6}{24.8} \Rightarrow u = 2.56 \Rightarrow u \approx 3$$

98. $\because u + v = 10$ (from previous question) $\Rightarrow 3 + v = 10 \Rightarrow v = 7$

99.

Marks	Frequency (f_i)	Cumulative frequency (c_f)
Less than 10	5	5
Less than 20	30 – 5 = 25	30
Less than 30	$30 + k - 30 = k$	$30 + k$
Less than 40	$(48 + k) - (30 + k) = 18$	$48 + k$
Less than 50	$(55 + k) - (48 + k) = 7$	$55 + k$

Total no. of students, N = 55 + k

Now, $$\text{Median} = L + \frac{\left(\frac{N}{2} - cf\right)}{f} \times h$$

where,

L = lower limit of the median class = 20

N = Total no. of observations = $55 + k$

cf = cumulative frequency of the class preceding the median class = 30

f = frequency of the median class = k

h = class size = 10

Now, Substitute these values in formula, we get

$$\Rightarrow \quad 24 = 20 + \frac{\left(\frac{55+k}{2} - 30\right)}{k} \times 10$$

$$\Rightarrow \quad 24 - 20 = \left(\frac{55+k-60}{2k}\right) \times 10 \Rightarrow 4 = \frac{k-5}{2k} \times 10 \Rightarrow \frac{4}{10} = \frac{k-5}{2k} \Rightarrow \frac{4}{5} = \frac{k}{k} - \frac{5}{k}$$

$$\Rightarrow \frac{4}{5} = 1 - \frac{5}{k} \Rightarrow \frac{5}{k} = 1 - \frac{4}{5} \Rightarrow \frac{5}{k} = \frac{1}{5} \Rightarrow k = 25.$$

100.

Marks	Class Interval	Frequency (f_i)	Midpoint (x_i)	$f_i x_i$
Less than 10	0–10	5	5	25
Less than 20	10–20	25	15	375
Less than 30	20–30	$k = 25$	25	625
Less than 40	30–40	18	35	630
Less than 50	40–50	7	45	315
		$\sum f_i = 80$		$\sum f_i x_i = 1970$

Now,
$$\text{Mean} = \frac{\sum f_i x_i}{\sum f_i} = \frac{1970}{80} = 24.625$$

Preface

CONSUMER BEHAVIOUR is to play a significant role in 21st Century. As the emphasis is on mass-customerisation and designing marketing strategies to match the products and services to the needs, moods, noods, and pockets of the consumers. This 21st century is not going to be happy with "Consumer-satisfaction"- but is certainly with "Consumer-delight". This is possible only when the marketers understand the consumers which is not an easy task. However, consumer behaviour study has provided with good many tools and techniques to solve the problems of consumers. That is why, more importance is given to the study of CONSUMER BEHAVIOUR in all the university Departments where commerce and management subjects are taught. It is a welcome step that Bangalore University has introduced this subject at BBA level. I am happy to present this title "Consumer Behaviour" to those who are preparing for the said examination.

I do not claim any originality so far as the stuff is concerned, however, style of presentation is mine. Hence, I am totally indebted to all those foreign and Indian authors who have contributed their mite is building up this fast growing discipline.

I have tried my best to be very simple, while presenting the subject matter. Good many homely examples are given to ensure the students easy percolating and gliding. Each chapter has chapter based questions. The last chapter contains well designed skill-development cases as required by the syllabus framers of the University.

I am very much thankful to Mr. Vijay Pandey and his brothers and father who accepted to bring to light this book in record time. I am also thankful to the people who work behind the scene of publication. My special thanks to Dr. V.R. Naik and Shri S.K. Shrivastva who encouraged me to take up the work.

I welcome suggestions from both teachers and taught for refinement in the work, because I strongly believe in the concept of KAIZEN - a process of continuous improvement.

CEBMER
Mapusa
GOA 403507

Dr. C.N. Sontakki

CONTENTS

CHAPTER 1

INTRODUCTION TO CONSUMER BEHAVIOUR

BACKDROP

India is a wonderland. She is totally different from the rest of the countries of the world. India a continent not a country because of her widest and deepest diversity is getting ready to be one of the greatest powers of the world in the new millennium of 2000. Diversity is so great, yet we find a unique unity in divesity. People of India, like those of others all over the world, differ not only in the common dimensions - of age, gender, race, nationality, education, occupation, marital - status, living styles - but also in their varied activities and interests, their preferences and opinions. There is diversity in music they hear, television programme on TV channel, in dress, fashions, and what not. Diversity in people -the customers has brought in diversity in the marketers, and therefore, the manufacturers and producers. That is why, this chapter plans to discuss - Who is a consumer? How consumers behave? Why one should study consumer behaviour ? Why the field of consumer behaviour developed? Nature of consumer behaviour and allied aspects. The chapter ends with Chapter Based Questions.

WHO IS A CONSUMER ?

In common usage, we do not make any difference between a 'customer' and 'consumer.' The **customer** is someone who regularly purchases from a particular store or a company, Mr. Jimmy, thus, who shops at Bombay Dyeing Show-room dress material himself or for his family members or who uses say gasoline and oils for his car supplied by Indian Oil Company is viewed as a customer of these above cited Firms. The term 'Consumer' is more generally referred to any one engaged in activities. Therefore, a 'customer' is defined in terms of specific firm while a 'consumer' is not. This is the modern way of defining these terms. The traditional way of defining the consumers is confined to the terminology of "economic goods" and 'services'. That is, 'consumer' is a potential purchaser of product or services offered for sale, whereas 'customer' is the actual purchaser whether he consumes or not. The latter speaks of monetary aspect—as to who pays for the goods and services. For our purpose of study, we are interested in the "ultimate consumer". "Ultimate consumer" is one who purchases goods and services for the purpose of individual or household consumption. Again, we are taking of an individual making a purchase with little or no influence from others. However, in some cases a number of people can be jointly involved in a purchase decision. It may be a case of buying a new car, or a planning all India tour or all-Europe tour. In other cases, the purchaser or the buyer may be acquiring a product for some one else who has asked for certain item. These situations clearly suggest that people can take on different roles in what we are going to understand, the phrase "consumer behaviour". These roles may be-Initiator-Influencer-Buyer-User. It is equally true that in some purchase situations, each person might play one role or roles or while in other cases, a single person might take up all the roles at a time single handed. For instance, a house–wife the home ministry—an initiator and influencer—may ask her better-half—the finance minister—the buyer-to pick a tin of baby—food for their son. At another time, the husband is the finance minister-could act as initiator buyer and even user by purchasing a after-shave lotion tube. It is equally important to note that the study of consumer behaviour remains incomplete in case it treats only one consumer role

However, while emphasizing one role, it pays to devote adequate treatment to others. Therefore, we consider buyer as the individual who actually makes the purchase. Such an approach is quite useful because, even when he is told what to purchase, the buyer often makes decisions regarding the purchase timing, store of his choice, package he prefers and so on. Therefore, while allowing for the influence of others on the purchase–decision, focussing our attention on the buyer still gives good deal of flexibility while spot-lighting on one consumer role.

The term 'Consumer' is used for both personal consumers and 'organisational' consumers. *Personal consumer* buys goods and services for his or her personal use say face-cream, mouth freshner, or for household consumption say sugar, coffee–powder, or for just one number of the family say a pair of gliders for his or son or a birthday present for a friend say an imitate diamond ring. In all these cases, the goods are bought for final use, referred as "end–user" or "ultimate consumers". The other category of consumer is the *organisational consumer* that covers profit and non-profit organization. Government agencies and social institutions such as departments of local, state or central government, hospitals, colleges, churches and temples-buy products, equipments and services needed for running these organisations. Manufacturing units buy inputs to produce and sell their own products. They buy services say, advertising to communicate with their customers–present and potential. Anyone who regularly makes purchases from a store or a company is termed as 'customer' of that store or the company.

BUYERS AND USERS

Any person who buys a particular product may not be necessarily the user or perhaps only the user but not buyer. It is equally true that a person who buys the product may not be the decision–maker. For instance, a father buys a uniform for his son who is studying in secondary school–the user, or buys a pack of paste for his family members, or the mother makes a decision while buying a dress material for her grand daughter. Again, both husband and wife may buy a car together–sharing the decision to buy a car. In all these cases, the buyers are not necessarily the uses of the products. They also may not be the persons who make the product buying decisions.

In the process of acquiring and consuming a product or services, different consumer behaviour roles are played by the consumers. The following Table 1:01 gives these roles which are worthwhile to note before we go into the details of consumer behaviour.

Role	Role Description
INITIATOR	**INITIATOR** is the individual who determines that some need or want is not being fulfilled and authorizes a purchase to rectify the situation.
GATEKEEPER	**GATEKEEPER** influences the family in processing of information. He has the greatest expertise in acquiring and evaluating the information.
INFLUENCER	**INFLUENCER** is the person who influences the buying decision by some intentional or unintentional word or action actual purchase or use of product or services and both purchase and use of product or services.
DECIDER	**DECIDER** is the person or persons who actually determine as to which product or services will be chosen.
BUYER	**BUYER** is an individual who actually makes purchase or purchases.
USER/S	**USER** is a person directly involved in the use or consumption of the purchase product.

Table 1:01

For marketers – who is a target audience for sending the ad messages- a *buyer or the user?* In this regard, the marketers are divided. Some strongly believe that *buyer* of the product is the right target; some others believe that the *user* is the right person to be addressed. Still some feel that, both buyer and user is to be targetted. These things are clear from the ad messages that we come across the TV screen. Some commercials are especially designed for children. Some are meant for youngsters; some are meant for elderly people, some are meant for all. It is left to the marketers as to whom they want to target the message.

INDIVIDUAL CONSUMER BEHAVIOUR

What is consumer behaviour?

Behaviour is an activity which occurs as a result of stimuli that is aroused for which the response takes place. Behaviour is a mirror in which every one shows his or her image. Behaviour is the process of responding to stimuli. Consumer behaviour is to do with the activities of individuals in obtaining and using the goods and services; it encompasses the decision-making process that proceeds and determines purchases.

STANDARD DEFINITIONS

★ **Professor Walter C.G and Professor Paul G.W.:**

"Consumer behaviour is the process whereby individuals decide whether, what, when, where, how and from whom to purchase goods and services - Consumer behaviour is all the psychological, social and physical behaviour of potential customers as they become aware of, evaluate, purchase, consume and tell others about the products or services".

★ **Professors Leon G Schiffman and Leslie L. Kanuk:**

"It is study of how individuals make decisions to spend their available resources - time, money, effort- on consumption-related items. It includes the study of WHAT they buy, WHY they buy it, WHEN they buy it, HOW often they buy it, and how often they USE it".

★ **Engel J.F. Kollat D.T and Black Well R.D.:**

"Consumer behaviour stands for the acts of individuals directly involved in obtaining and using economic goods and services, including the decisions that precede and determine these acts. Consumer behaviour involves the study of not only of what people consume but where, how often, and under what conditions goods and services are consumed".

★ **Professor Alderson W.:**

" The study of consumer behaviour is more is the study of consumer buying than the actual consumption".

★ **James F. Engel, Roger D. Blackwilliams Paul W. Miniard:**

"Consumer behaviour refers to the actions and decision process of people who purchase goods and services for personal consumption."

★ **Bearden and Associates:**

"Consumer behaviour refers to the mental and emotional processes and physical activities of people who purchase and use the goods and services to satisfy particular needs and wants".

Consumer behaviour is the subset of human behaviour. That is, factors influencing individuals in their daily lives also influence their purchase activities. Internal influences, such as learning, motives as well as external factors such as social expectations and constraints, affect everyone in their role as consumers as well as in other capacities. In fact, it is very difficult to draw a clear line of demarcation between consumer related behaviour and other aspects of human behaviour. Consumer behaviour stands for all the acts of individuals directly involved in obtaining and using economic goods and services, including the decision–process that proceed and determine these acts. Social scientists, some times, used the word 'behaviour' to include only overt or observable actions, but it is increasingly recognised that the overt act is only one part of the decision process. Therefore, a realistic analysis of consumer behaviour must include necessarily less observable or covert processes that accompany consumption. Consumer behaviour involves the study not only of what people consume but where, how often, how much, why, when and under what conditions–goods and services are consumed. Consumer behaviour is the behaviour that consumers display in searching for purchasing, using, evaluating and disposing of, if products and services that they expect will satisfy their needs.

CONSUMERS CLASSIFIED

The term 'consumer' is used very often to describe two kinds of consuming entities. Based on this—a consumer can be **personal** or **individual** and **organizational consumer**. The 'personal' consumer buys goods and services for his own or her own use. He might buy a shaving gel, after-shave lotion or she may buy face moisturizer or cream for him or her or their members of family as the case may be; he might buy V.C.R. or V.C.D. as gift to a friend ; she might buy jewellery for her friend or a relative visiting her house. In each of these cases or situations, the goods are bought for final use by individuals, who are referred to as **End–users** or **Ultimate Consumers**. The second category of consumer is the **Organizational Consumer** who represents a profit or non–profit making business house, a government department or agency may be local, state or central and institutions such as schools, colleges, universities, hospitals, prison, a club–all of which must buy products, equipment, and variety of services in order to run their organisations. It goes without saying that manufacturing or processing companies must buy the raw materials and other components needed to manufacture and sell their own products. Similarly, service companies must buy their equipment and appliances and other service inputs to render the services that they sell. Government agencies must buy office supplies, equipment, and services to operate their departments. Other institutions must buy the materials and equipment and services to maintain their customers alongwith them. Here, we are concentrating on a personal or individual buyer and not an organizational buyer.

CONSUMER BEHAVIOUR VERSUS CONSUMPTION BEHAVIOUR

"Consumer behaviour" is the case of micro behaviour whereas "consumption behaviour" is macro–behaviour. In other words, consumer behaviour is to do with individual person while consumption behaviour, as a study, focuses on the decision–process of the individual consumer or consuming unit. It encompasses all the efforts, explain an act or acts of choice either at a given time or over a period of time. In contrast, the consumption behaviour, as a study, is to do with the explanation of the behaviour of the aggregates of consumers or the consuming units, either at a given point of time or over a period of time.

CONSUMER BEHAVIOUR VERSUS BUYER BEHAVIOUR

The sphere of consumer behaviour includes the activities of both final intermediate users. The former are called as end–users or ultimate users or consumers in true sense; while the intermediate users are the industrial units which do not consume the goods and services but add further value before they are consumed by the final users. Therefore, the phrase "buyer behaviour" means the reference made to both end–users and intermediate users. Hence, the phrase "consumer behaviour" includes only the end–users and their actions, phrase.

NATURE OF CONSUMER BEHAVIOUR

Consumer behaviour is the function of the individual involved, the product category and the present situation. This can be presented nicely as under, in the form of a pictogram :

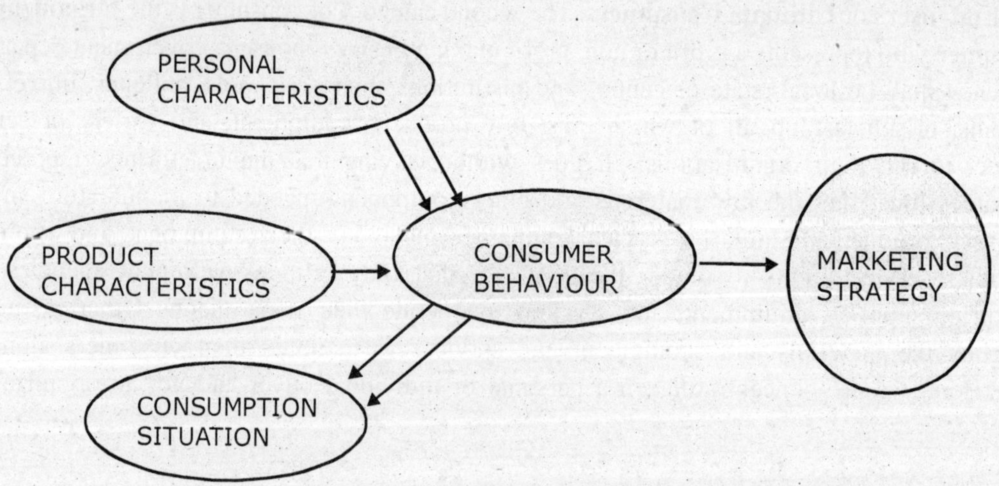

CONSUMER BEHAVIOUR IS PRODUCT-PERSON-SITUATION SPECIFIC

FIG: 1:01

What is said above is quite clear with a moments' reflection. It is our daily experience that we prefer individually, television programs telecast on alternative channels, food–items, drinks both hot and cold-soft drinks-beverages, type of literature, the shelter we live in then our closest relatives in the family and among our best friends. It is equally true that we take more or less time in evaluating a product or a service before we purchase as compared to others. That is, each person is nearly unique—or all are not similar in all respects. It is well known, that marketers have tried to group the consumers who are similar in certain respects-not identical into what are called as market segments.

What is further significant to note is that we differ not only from other customers, but the purchase behaviour of us is differing with product to product or product category to another. Take that you have selected Sansui stereo-system-PMPO by spending almost two hours, but you might spend much more time in selecting a compact disc. The time spent in case of these white goods or electronic gadgets might not coincide with that of buying a cup of coffee from a vending machine or buying a news-paper from electronic selling machine.

Still more important to note is that our purchasing and consumption behaviour for the same product will differ depending on the situation in which you are caught or operating. For instance, we in a marriage party of a close relative, where we have our close and distant relatives and friends, where our items of consumption arranged at the buffet table will be quite different than in a situation of a birthday, where age group and relations make a difference in consumption of items even if same caterer is ready with one which was at the wedding party. That is, the consumer behaviour in deeply and delicately influenced by individual characteristics, product characteristics and situational characteristics. Hence, true and practical understanding of consumer behaviour must incorporate all of these influences.

Consumer behaviour is a complex, multi-dimensional process. Consumer behaviour is an applied discipline. It is because, consumer behaviour is a sub – set of human behaviour. That is why, all behavioural sciences play decisive roles. It borrows and applies the findings from many disciplines in an attempt to solve practical problems. The disciplines that are relied upon are social psychology, individual psychology, sociology, cultural or social anthropology, economic geography, linguistics, political–science; philosophy is the study of how individuals influence and are influenced by groups; individual psychology is the study of behaviour and mental processes of an individual; sociology is the study of the collective behaviour of people in groups; cultural or social anthropology is the study of people in relation to their culture; economic geography speaks of regional differences caused by physical variations; linguistics is the study of languages of people that has its own influence on human behaviour; political–science is the study of polity variations among people; philosophy is the study of value judgements and outlooks; economics is study of people's production, consumption and hence, exchange of goods and services.

CONSUMER BEHAVIOUR AS A DISCIPLINE

The study of consumer behaviour has developed into a very significant branch of marketing discipline. Todays developments in this fast growing field, make it as an independent discipline as a specialised field of knowledge. The bigwigs of marketing found that consumers did not always behave as advocated by economists. The size of the consumer market in all the developed nations and fast developing economies of the world has been quite extensive. As many as 6 billion people all over the world represent market where they spend pretty large sums of money on variety of goods and services. What is more important was the consumer preferences were shifting and getting highly diversified. Leave aside these consumer markets, even in case of industrial–goods markets, there have been changes in consumer preferences showing a predictable purchase behaviour. The major features of homogeneity of industrial goods market remained a thing of past.

The persons engaged in studying the buying behaviour of consumers learnt clearly that there are many similarities, but all consumers were not alike. One man's goods and requirements are not the same in all cases and all the time. These consumer preferences, likes and dislikes reflected the personalities and the life-styles of these consumers. It was a clear clue for the marketing experts who came out with an idea of market segmentation. This meant a division of the total market into smaller segments, each having relatively homogenous, in respect of which a matching marketing–mix could be developed and offered for mutual gain of marketers and consumers. This segmentation led to largest markets and positioning. These are very important milestones in the development which is called as STP process. Good many techniques were developed in these areas of segmenting, targetting and positioning. Consumers were convinced of these to maximise their satisfaction with profits to the marketing men.

The other most important factors that led to the development of consumer behaviour as a discipline are–shorter life–cycles for products increased environmental concern, deeper interest in consumer protection, growth of services marketing, opening up of global markets and revolutionary growth of computers and the last but not the least, the development of highly sophisticated and refined statistical tools and techniques for indepth analyses.

CONSUMER BEHAVIOUR AND OTHER SOCIAL SCIENCES

Till the latter half of 1960s, consumer behaviour was a new field of study without a history or research of its own. In fact, consumer behaviour cannot be separated from human behaviour because, every human being is a consumer. In that sense, consumer behaviour is only a dimension of human behaviour. That is why, consumer behaviour as a discipline has borrowed heavily from the other social sciences. These social sciences that have contributed to the development of consumer behaviour are basically psychology, sociology, social–psychology, cultural–anthropology and economics. The concepts, ideas, theories found in these social sciences are found embedded in consumer behaviour. It pays the reader by acquainting himself or herself with these disciplines.

1. Psychology:

Psychology is the science that studies human mind and its functioning. It is a branch of behavioural sciences which attempts to probe and understand as to how an individual mind works while making decisions. That is, psychology is the study of an individual covering the mental aspects of motivation, perception, attitudes, personality and learning theories. All these, aspects of human mind are critical to understanding of consumer behaviour that helps one to comprehend consumption related needs of an individual, their actions and responses to different promotional communications and products and the way their experiences and personality features influence product or service choices.

2. Sociology:

Man is rational and social animal, who does not live alone. He needs a company, group to support him as a member. Sociology is the study of groups. Society is one of the most dominant and complex type of social organisations. Each society has its own set values, beliefs and customs that are reflected in culture of its own. These variables have bearing on behaviour of its members or consumers. When individuals form groups, their actions are quite different from the actions of those very individuals when they are operating all alone. The influence of group memberships, family, social class on consumer behaviour are significant for the study of consumer behaviour. The pattern of behaviour rests on the class to which a person belongs to in a society his or her status matters.

3. Social Psychology:

Social psychology is the study of an individual behaviour as a member of a group. It is the finest combination of sociology and psychology studying as to how an individual operates in a group. As a member of society, he is member of several groups or structures at the same time. Thus, he is the member of a family, belonging to a particular religion, occupation and organisation. It studies as to how the opinions of the groups they respect–say peers, reference–groups, their families, opinion–leaders influence individuals in their consumption behaviour.

4. Cultural Anthropology:

The cultural anthropology explains as to how the members or consumers behave as opposed to psychology that explains why human–beings behave. Culture is the very foundation of human behvaiour in the society. That is, cultural anthropology is the study of human–beings in society. Culture is the set of values, ideas, attitudes, habits, beliefs and other meaningful symbols that shape human behaviour in the society. It is the set of values which is subject to change. Cultural anthropology studies human–beings in society by exploring the core beliefs, values and customs that the individual inherits from his parents and grand parents and the members of the society which influence their purchase and consumption behaviour. It studies sub–cultures and helps in comparing consumers of different nationalities and cultures.

5. Economics:

Economics is the science of wealth and well-being. Humans are motivated by economic incentives and do every thing to maximise their economic returns for their efforts. With limited means having alternative uses, economics helps an individual to think and act in most rational way by choosing the best course of action open. Among other things, the study of economics is the study of consumer spending, earning, evaluation of alternatives so as to reach the best decision to have the maximum satisfaction.

Inspite of the fact that consumer behaviour, as a discipline, is only of less than 50 years of history, has grown into full–bloomed discipline having its own fund of knowledge made up of fundamental principles that can be fruitfully used in understanding consumer behaviour-the cause and effect relations. The marketing strategies can be developed and applied for the benefit of marketers and consumers. These things are clearly demonstrated by the fact that successful companies have accepted and acted upon the marketing concept. The importance of marketing research has been understood and companies are keeping aside, treasure, talent and time in understanding the unsatisfied consumer needs. In the process of learning, they have come to the conclusion that consumers are complex as individuals and vary in psychological, social needs over and above physiological needs. The consumers have needs that have priorities, distinctions that lead to development marketing programs in satisfying consumer needs and wants. Basically, marketing concept and marketing segmentation have increased importance of consumer behaviour.

WHY ONE SHOULD STUDY CONSUMER BEHAVIOUR ?

Consumer behaviour, as a sub-set of human behaviour, is today an interdisciplinary subject of keen interest and practical utility. It pays to spend our valuable time, treasure and talent. The basic reasons that justify the study of consumer behaviour are :

1. We can be Better, Wiser and Alert Consumers :

As consumers, we stand to benefit from the deep insights into our own consumption rated decisions. Consumer behaviour study .speaks of what we buy ? Why we buy ? How we buy ? When we buy ? How much we buy ? How often we buy ? What promotional components persuade us to buy? Men and women learn from the mistakes of theirs and mistakes of others. A systematic study of consumer behaviour should enable us lo be better, wiser and alert consumers to get real value for a rupee spent.

2. We can be better and Practical Marketers :

In a society, every one is a consumer, though he has a different role or roles to play. We must admit, that age, sex, caste, creed, colour, social-status and the like, have no difference in calling every one as a consumer, consumption–patterns might differ but every one remains a consumer whether he buys actually or not. Many of us are going to be marketers of future. Or already we are, then it is essential to know and recognise why and how individuals make their consumption decisions so also as marketers make more strategic–decisions. In case marketers take trouble of understanding the consumer behaviour, they are surely able to predict how consumers are likely to react to various informational and environmental cues and are able to shape and design their marketing strategies accordingly. Those marketers who know the in and outs of consumer behaviour have definitely competitive advantage over their rivals in market place.

3. We can be Better Informed Students :

As the sincere students of human behaviour, we are more deeply concerned with the clear–cut understanding of consumer behaviour. It should be noted that human behaviour, and, therefore, consumer behaviour is caused. That is, a person or a consumer does not behave this way or that way simply. There are definite reasons for his or her behaviour or reactions. The study of consumer behaviour helps us to gain an insight into why individuals act in certain consumption related ways and with learning what internal and external influences impel or propel them to act as they like. The earnest desire for understanding consumption related human behaviour has led, in fact, to a diversity of theoretical approaches to the study of consumer behaviour. Still, it is a mystery because psychologists or consumer behaviour researchers have not yet, found satisfactory solution to the question why people buy ? This has led to a wonderful world of motivation research. A student of today, remains a student though he might be a researcher or a teacher of whatever depth and width.

4. The Study makes us Multi-disciplinary Oriented :

Consumer behaviour is relatively a new field of study that was brought under spotlight in the second–half of 1960's. The reason is simple, it is basically a study of human mind. Psychology is the youngest of all social sciences. With no history or body of research of its own, this new discipline has borrowed heavily from other scientific social sciences such as social psychology, anthropology, political–science, economies, economic and commercial geography and so on. It is clear that the early theories of consumer behaviour have been based on economic theories-the essence of which hovers round the universal truth "individuals act rationally to maximise their benefits."

5. It provides Fodder for Decision–making :

Consumer is the focus of our study because, certain decisions are significantly influenced by his behaviour or expected actions and reactions that is why, consumer behaviour is an applied discipline. Such applications can be seen from two clear perspectives namely, 'micro' and 'macro'. The 'micro' perspective seeks applications of this rich fund of knowledge to problems faced by the individual firm or an organisation. That is, micro perspective involves understanding of consumers for the purpose of helping the firm or an organisation to attain its objectives. Thus, ad managers, product-designers and others in profit-driven business are interested in understanding the consumers in order to be more effective at their tasks. Not only that, even the managers of non-profit organisations such as Indian Red Cross, non-governmental organisations have been effective in applying the understanding of consumer behaviour concepts to their social activities.

On the 'macro' or aggregate level-the societal perspective- applies knowledge of consumers to aggregate level problems faced by groups or by society as a whole. That is, by societal perspective we know that consumers influence collectively economic and social conditions within the frame of society. In any market system founded on individual choice, consumers strongly influence as to what will be produced? For whom? When? Where? and with what resources? Resultantly, the collective behaviour of consumers has a significant bearing on the quality of life or standard of living. In case Indians decide to go in for privatisation, vast amounts of resources will be used to produce variety of cars, technical systems, fuel used in running these cars. It will also influence as to where many of us live-say city, sub-urban location or rural location–and how we are going to run our daily lives-What to eat ?-Where to shop? How we are entertained ? How we are meeting natural calamities ? and so on. What is more important is, that this collective desire leads not only to development of strong transportation network but also how we are adding to pollution problems and energy crisis and inter–state, regional and national problems. Thus, macro perspective analysis gives an insight into aggregate socio–economic trends based on which, the ways and means can be traced to improve the efficiency level of market system to improve the well–being of society.

APPLICATION OF KNOWLEDGE OF CONSUMER BEHAVIOUR

In the previous paragraphs, we talked about micro and macro perspective analyses of consumer behaviour. Consumer behaviour study is an independent and highly specialized study of modern marketing world. It is worth-while to note that wise marketers had noted long back that consumers did not always act or react the way the marketing theories were expounded. The reasons are many. A few of them are that consumers are the products produced by machine, they have mind, and actions or reactions, changes according to number of changing variables that create a particular situation to react accordingly. Thus, India, in 1998, gold rates went down-to those of 20 years back level-still the yellow metal did not attract the Indians thinking that prices would still come down, it did not happen. Prices shot up again. With the reopening of 21st century prices are low of gold, yet there is no sign of rise in demand. The Indian Consumer market size is mammoth of one billion. India is not a country but a continent which is constantly increasing in consumer market parameters. Billions of rupees are spent on goods and services in place of millions of rupees. Consumer income has gone up, consumer preferences have been fast changing with the opening up of Indian economy, globalization, liberalisation and privatization. Their needs are wants and wants are not only changing but diversifying. Forget about the consumer markets, even in industrial markets, where needs for goods and services were more homogeneous and concentrated than in case of consumer markets, the industrial buyers have changed their buying preference practices caused by organisational purchase behaviour.

The marketing researchers have realised that age old set of preferences, likes and dislikes, were common to social groups and units, are no more holding good. Each consumer thinks that he or she is unique and, therefore, their long–lasting tradition bond preferences died, instead they like diffentiated products reflecting their needs, personalities and life-styles. In this fast changing world of marketing which is highly competitive, dynamic and customer–driven, marketers have applied–the study of consumer behaviour for their benefit to survive, grow, flourish, continue and diversify. The major applications of consumer behaviour are :

1. Development of Integrated Marketing Concept :

A philosophy of an organisation is the dynamic sphere of marketing is referred to as a marketing concept. Marketing concept is an integration of marketing activities directed towards customer satisfaction. In other words, marketing concept is a philosophy of a business management, based upon a company-wide acceptance of the need for customer orientation, profit–orientation and recognition of important role of marketing in communication the needs of the market to all major corporate departments. Thus, the evolutionary concepts started from production. Sales, consumer and social orientation of all these integrated concept implies that the fundamental strategies of business and all aspects of both its long-term and short-term management decisions are based on a thorough understanding of needs and desires of the customers who make a market. The features of an integrated concept are-Customer Orientation-consumer satisfaction integrated managerial action and realisation of organisational goals of survival, growth, innovation, productivity, profits, social–obligations and so on.

Widespread adoption of integrated marketing concept has, all over the world, provided the impetus for the study of consumer behaviour. The companies are constantly engaged in extensive marketing research to identify unsatisfied customer needs. In doing so, the researchers found that consumers are highly complex individuals, subject to variety of psychological and social needs quite apart from their survival needs. They found that their need priorities are constantly changing, they deserve matching treatment. Thus, study of consumer behaviour brought into play integrated market concept and the development of marketing concept led to the strengthening of consumer behaviour study in its application.

2. Market Opportunity Analysis :

Consumer behaviour study intensified the market opportunity analysis. This activity involves an indepth investigation of trends and tendencies and the conditions in the market to identify consumer needs and convert them into wants that are not fully satisfied. Market analysis requires a thorough understanding of the organisation's own capabilities, the capabilities of current and future competition, the consumption patterns of potential customers, and the economic, physical and technological environment in which these elements will interact. It is worthwhile to note here, that the organisational objective is to provide superior customer value. Therefore, the major components of market analysis are **Consumer–The company–The competitors and conditions.** The analysis starts with a study of general market trends and tendencies–relating to consumers life-styles and income levels that may suggest unsatisfied wants and needs. More specific examination in focus involves assessing any unique abilities the firm might have in satisfying the unidentified consumer desires. Recent trends brought to surface have made possible new product offerings for consumer increased satisfaction. Thus, for those who are diet–conscious, we have diet–ghee, diet–sugar, diet–beer, zero–cholostrol edible oil, smokeless cigarettes, and what not ? Today, electronic devices of checking sugar count, blood pressure are a boon that waste no time, are any-time check up is possible.

3. Market Segmentation and Strategies :

All consumers are not identical in all respects. However, they have some common characteristics in terms of demographic parameters and likings and preferences. Most marketers have followed the policy of market segmentation to serve better the needs of specific groups of consumers. Market segmentation is the division of the total potential market into smaller, homogeneous segments for which they can design specific products, promotional campaigns and channels of distribution. The benefits of market segments for firms are – a more precise definition of the market, a more effective marketing programme, a better assessment of competition, better allocation of resources. As against this, segmentation involves costs-classified as product-production-promotion, inventory and management. This is, mainly based on the consumer behaviour.

4. Product Positioning and Strategies :

There is close alliance between market segmentation and product positioning. Product positioning is a creative exercise tried with every existing product. Product position is not what you do to a product, it is what you do to the mind of a prospect. Product positioning is the act of designing the company's image and value offer so that the segments' customers understand and appreciate the company's stand for in relation to competitors. It is the process of identifying the needs of market segments, product strength and weaknesses and the extent to which competing products arc perceived to meet the consumer needs. In other words, it is an attempt to project different or refined or revised product image in the market than the one that has been prevailing. The alternatives of product positioning are by making altogether different claim or highlighting the new product features-by entering in the new market segment-by introducing a new package design and so on. The success of product positioning is determined by factors–creativity in product propositions–existence of warranted competition, sizeable and profitable market segment–sensitive market segment and adequate consumer behavioural information.

5. Determining the Matching Marketing Mix :

A marketing–mix is composed of a large battery of devices which might be employed to induce consumers to buy a particular product. It is a pack of four sets of variables namely, product-price-promotion and place variables. In other words, marketing–mix is the blend or the compound of all the marketing efforts hovering round four ingredients namely, **Product–Price–Promotion–Place.** Product variables are- product line and range, product design, product package, product quality, product labelling, product branding, after–sale services and guarantees.

Price variables are, pricing policies and strategies, the terms of credit, terms of delivery, profit margin and resale price maintenance. The promotion variables are, personal selling, advertising-sales promotion, publicity, exhibitions and fairs, public–relations. The place variables are—transportation, warehousing, inventory levels and channels of distribution. Unless consumers and their behaviour is studied, it is not possible to have matching marketing mix.

6. Leads to Consumer Behaviour Research :

Consumer behaviour research is a very important aspect of marketing since consumers together form the market. To perform effectively and efficiently, the business activities that direct the flow of goods and services from producer to the final consumer or user, the needs, motives, and the desires of consumer must be understood in toto. These are the areas which form the bulk of the domain of consumer behaviour research. Consumer research is the methodology used to study consumer behaviour. There are two major theoretical perspectives concerning the study of consumer behaviour, it is but natural to find a kind of divergence in theoretical assumptions and to certain extent, in research methodology between the **positivist** and **interpretivist** approaches. Broadly speaking, **positivists** tend to be objective and empirical to seek causes of behaviour and conduct research studies that can be generalised to larger populations. The researchers of early days with their strategic management perspective were largely **positivist** in their approach. On the other hand, research done by **interpretivists** tends to be qualitative and based on small samples. They are more subjective in approach. Although they tend to view each consumption situation as unique and non-replicable, **interpretivists** seek to find common patterns of operative values, meanings and behaviour across the consumption situations.

7. Consumer Behaviour and Non-profit and Social Marketing :

There are good many non-profit organisations that they sell crime prevention, charitable services, concept of family–planning, combating aids as the profit–making units, cancer awareness, selling soaps, creams, clothes, medicines and so on. Non-profit and social organisations have ideas and services to sell or market to largest groups of consumers or constituents. These cover governmental agencies, religious organisations, colleges, universities and charitable institutions. These organisations must also appeal to the public for support in addition to an attempt of satisfying their wants or needs in society. They also need clear understanding of consumer decision–processes that assist these units. Take the finest example of the benefits of such knowledge that would have to administration of the Indian Cancer Society. Two major tasks of this organisation are: (1) to solicit public contributions for support of cancer research and (2) to encourage regular physical check up for early detection of the deadly disease. Regarding the first task, basic information, such as the characteristics of potential contributors, what motives their influence generosity, and how these motives can be most effectively appealed to, is highly useful. Regarding the second task, a sound basis for developing methods of encouraging regular physical examinations would include specific knowledge of reasons, why the examinations are avoided–the expense, the time involved, the fear of learning about the illness – for fear of death is deadlier than death. This equally applies to AIDS-which is fast spreading all over the world particularly in underdeveloped world.

8. Consumer Behaviour and governmental Decision–making :

Of late, consumer behaviour and its principles are more relevant in case of government department. Government is a big consumer. It is quite clear from two important points. Government as the custodian of general public, has to provide services to the public which has brought government into existence and has empowered. Therefore, consumer behaviour study has deeper and broader impact and spectrum on its decisions on behalf of general public. Another thing is that government has designed legislation to protect consumers or to assist them in evaluating the products and services to have maximum social advantage. Take for instance, public–transport system that provides service. It can better serve the needs of public only when the government understands the public, the variety of consumer or consumer groups. Here, local, state and national governments, decisions are involved. Further, consumer protection has been given top priority. Maximum social welfare is the crux. These actions or decisions directly affect the welfare and some indirectly. Thus, use of safety belts while driving, use of helmets for rider and pillion–rider and no driving under the influence of alcohol. Indirectly, speed limits are set. Again, the ads are regulated directly or indirectly that are not respecting the norms of truth and ethical values.

9. Consumer Behaviour and Demarketing :

Demarketing means and includes all such efforts that encourage consumers to reduce consumption of a particular product or service. Not only in advanced nations, but also in case of developing nations, though governments encourage the consumption or use of certain items may be durable or non- durable, a stage will come where, the public will have to use or reduce consumption of these items. Thus, use of petrol driven vehicles has increased with the improvement in life–styles. However, it has increased demand for petrol and this fuel is running out of supply. Hence, attempts are made to reduce the use of petrol and more of other sources of fuel namely, diesel and gas and even electricity. Similarly, today's marketing has been so easy with plastic carry bags, at the same time, plastic being non-degradable, has created garbage, pollution and health problems. Therefore, paper–bags have been slowly replacing them. Consumption of tobacco-products-particularly smoking and chewing has health hazards. Therefore, statutory warnings are given, smokeless cigarettes are developed to avoid or reduce lung–cancer, alternative hot drinks are developed in place of alcohol to reduce incidence of lever cancer or damage. Even efforts are made to shift on alternative products to chocolates as they cause teenage teeth and gum problems. Thus, demarketing efforts have atleast influenced the attitudes and motives of people if not totally stopped. It is because habits die hard.

10. Consumer Behaviour and Consumer Education :

Consumer education has a constructive role to play. Each consumer, on an average, depends on his hard earned money. Because of attractive ads, they waste their income inadvertently. Advertising is a great educator, informer, entertainer and a great cheat, at the same time. Again, human beings-being social and rational animals, in most cases act irrationally or go by emotions. For instance, they buy a brief and vest and get a coupon to get a prize of a car. Millennium offers by manufacturers have confused the people where even the things like, gold, silver, flats have lost their value. Buy a Colgate tooth brush and be eligible for Bunglow, Car, three-wheeler, sound–system and hosts of consolation prizes. These have made people act irrationally. Therefore, by studying the findings of consumer behaviour, the consumers can learn as to how best they can get better value for their hard earned rupee. Thus, they can learn from mistakes and mistakes of others. Consumer behaviour study can as well serve as the data for the development of educational prŏgrammes designed to improve consumer decision–making process in case of produce and services available on one hand and their purchasing power or status in the society on the other hand. In fact, it has become a part and parcel of formal education where children are educated at primary, secondary, higher secondary level and adults at college and university level directly or indirectly.

In a nutshell, the study of consumer behaviour has developed as a distinct and interdisciplinary subject today because of its wider applications to solve the problems of all those who are involved in making available products and services and consuming or using them in the final analysis.

CHAPTER BASED QUESTIONS

A. WRITE SHORT NOTES ON:

1. Consumer - 2. Customer 3. Consumer behaviour 4. Consumption behaviour 5. Buyer behaviour 6. Organisational buyer 7. Marketing ethics 8. unethical consumer practices 9. Unethical market practices 10. Consumer research 11. Applications of consumer behaviour 12. Importance of study of consumer behaviour.

B. SHORT ANSWER QUESTIONS :

1. What is "consumer behaviour ?" Is the same as "consumption behaviour ?"
2. What is meant by 'end-user' and 'intermediate' user.
3. Is there any difference between consumer behaviour and buyer behaviour ?
4. What is the nature of consumer behaviour ?
5. What is the significance of consumer behaviour study ?
6. In what way, consumer behaviour study can be applied to marketing ?
7. What is marketing ethics ?
8. Does consumer behaviour study give an edge over consumer to marketers ?
9. Is consumer study a weapon to exploit consumers ?
10. Are consumer unethical practices leading to counteracting unethical practices of marketers ?

C. ESSAY TYPE QUESTIONS:

1. Who is a consumer? Describe the different role a person plays as a consumer.
2. What is consumer behaviour? What is its nature?

3. What is consumer behaviour? Explain it as a discipline.
4. What is consumer behaviour? Bring home the social sciences that contributed to the development of consumer behaviour.
5. Define the phrase " Consumer behaviour". Why one should study consumer behaviour?
6. How the marketers are applying the knowledge of consumer behaviour?
7. "Consumer behaviour is an interdisciplinary subject". How for is it true?
8. Who is a consumer? How consumers are classified?
9. " A scientific study of consumer behaviour helps to solve many marketing problem". Explain.
10. What is "Consumer behaviour". Is it different from "buyer behaviour" and "Consumer behaviour"?

****

CHAPTER 2

MARKET SEGMENTATION AND CONSUMER BEHAVIOUR

BACKDROP

Taking extreme situations, there would not have been thundering success of marketing management which is one of the hot favourites of those who are panting for BBA or MBA course. The marketers would not have succeeded with one product or service to meet the requirements of every customer because each customer is different. If the marketer wants to satisfy the individual needs of every Tom Dicky and Harry because they are different and unique, each marketer has to offer products and services that exactly fit in the requirements of each consumer. It would have been possible but not feasible. The point lies in a golden way avoiding the extremes for the mutual benefit of both consumers and the marketers. Individuals have different needs and values and, therefore, marketer cannot address himself to the total population while designing a marketing-mix for a given product or a service. Hence, every marketing manager knows that there is no such thing as a single market for any given product or a service. All markets are made up of segments and these segments are made-up of sub- segments. This all happens because, consumer is unique in one respect that he is not like other person, but he is like others in some respects. Though market *is aggregate of all existing and potential buyers for a product, these markets vary widely in their nature and character making them totally heterogeneous. Hence, delimiting the given market for a product is a must for effective tapping. This delimiting implies dividing the total market into smaller units with more or less similar features. Such an attempt is market segmentation. One or more of these segments are then selected as the target market and a strategy is developed to reach this selected group. Such a strategy of the organisation is based on the three market analysis approaches namely, product differentiation, market segmentation targetting and market positioning. This chapter plans to discuss the meaning of segmentation, difference between segmentation and product differentiation, benefits of segmentation, bases for segmentation, requisites of sound segmentation and alternative strategies for market segmentation or targetting and other details of segmenting. The chapter ends with Chapter Based of Questions.*

WHAT IS MARKET SEGMENTATION ?

A **marketing segment** is a meaningful buyer group having similar wants. Hence, segmentation is a consumer oriented marketing strategy. Market segment is the portion of the market defined on the basis of the shared characteristics of people it covers. Market segmentation is the process of grouping buyers into different categories having common desires or needs. It is the process of dividing the market in order to conquer them. It is to identify the groups of buyers on the basis of difference in their desires or requirements. It is the strategy that sub–divides the target market into sub–groups of consumers with defineable, distinct ai homogeneous characteristics with a view to develop and follow a distinct and differentiated marketing programmes for each sub–group in order to enhance satisfaction to consumers and profit to the marketer.

STANDARD DEFINITIONS

★ **Mr. Dick Warrn Twedt:**

"The strategy of marketing segmentation involves the development of two or more different marketing programmes for a given product or a service with each marketing programme aimed at a different grouping of individuals whose expects reactions to the seller's marketing efforts will be similar during a specified time period."

★ **Professor William Stamton :**

"Market segmentation consists of taking the total heterogeneous market for a product and dividing it into several sub – markets or segments each of which tends to be homogeneous".

★ **Professor Philip Kotler:**

" Market segmentation is the sub–dividing of a market into homogeneous sub–sets of customers where any sub–set may conceivably be selected as a market target to be reached with a distinct marketing mix."

★ **Professor Edmond Pearce:**

" The purpose of market segmentationis to determine the differences among purchasers which may affect the choice of market area of marketing methods. The market is divided into sectors or segments in such a way that the purchases in each segment, while not being exactly identical, are of a similar nature and are influenced by similar buying motives".

★ **Mr. Schewe J.D. and Smith W.J.:**

" Market segmentation breaks the larger heterogeneous market into small homogeneous segments. The elements of each smaller segment are more similar in terms of wants, needs or behaviour than the total market. A separate marketing programme is developed to best match each segments individual needs and wants".

In fine, a **market segment** is a portion of a large market in which the individuals, groups, or organisation's share one or more characteristics that cause them to have relatively similar product needs. **A target market** is a group of customers who have similar needs and wants and for whom a firm creates and maintains a separate matching marketing mix. Thus, **market segmentation** is the process of dividing the whole heterogeneous market into relatively distinct–and–homogeneous sub–groups of consumers with common needs or characteristics and selecting one or more segments to target with distinct marketing programme. The concept of market segmentation is an extension of the marketing concept that rose to height in 1950s. This segmentation is founded on the simple but keen observation that all the existing and potential consumers are not alike; they are different as to their needs, wants, tastes, backgrounds, education, experience and income levels and these variables are subject to change of life–styles they follow.

Let us take the living example of tea-market segmentation by Hindustan Lever Limited. The company has two brands namely, **Brooke Bond** and **Lipton.** Further, **Brooke Bond** has 3 **Roses premium tea, Red Label, Green Label, Natural, Taj Mahal, Brooke Bond A-1.** In case of Lipton- the varieties available is only one Green Label. Thus, the market is segmented into seven clearly set target markets.

The market segmentation can be presented visually as under :

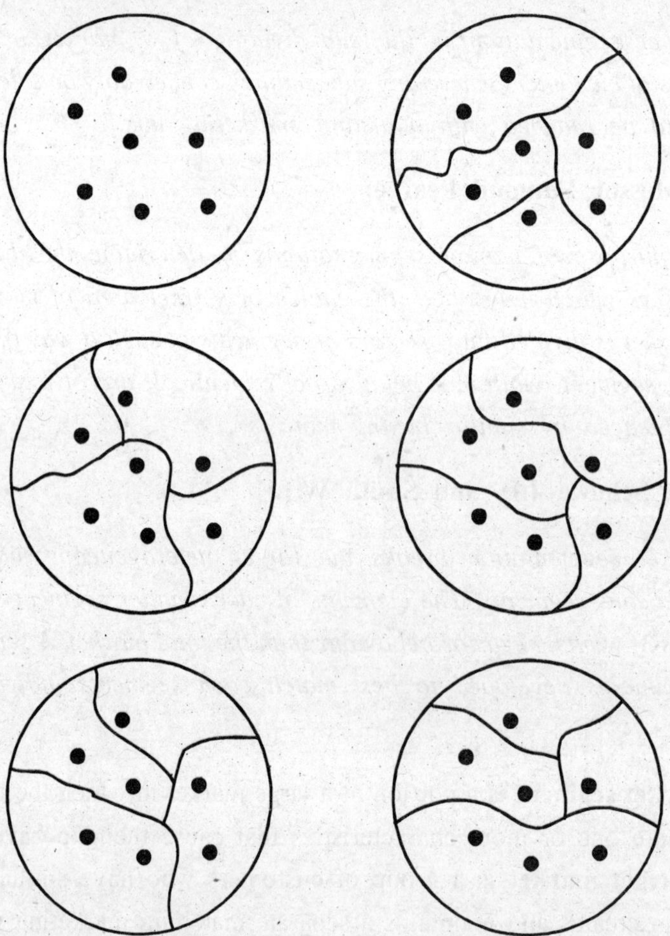

FIG: 2:01 UNSEGMENTED AND SEGMENTED MARKET

This means increased consumer satisfaction and increased profits to the company that segments its market into manageable homogeneous segments. In essence, market segmentation is a method meant for maximising market response from limited marketing resources by accepting the differences in the response characteristics of the market in question. This is the result of accepting individual differences after grouping them into broad likings and leaning.

MARKET SEGMENTATION AND PRODUCT DIFFERENTIATION

One should not equate these two distinct concepts. It is because, **"Market segmentation"** and **"product differentiation"** are not one and the same. **Product differentiation** is an attempt to bend the demands to the will of supply. The market demands is adjusted to the supply condition. That is, the manufacturers can exercise control over demand, prices and distribution through this technique. Here, the organisation has commands over demand, prices and distribution aspects. Here, the attempt is to single out the market offering from the array of competitive goods on the market If the firm succeeds in convincing that a given product is 'different' or 'better' than the one available in the market, consumers will be more inclined to buy the firm's product than its competitors. In product differentiation, colours, sizes, shapes, brand names play vital role. This approach became very popular in 1920s. However, its way is so powerful that it is practised even to-day as one of the ways to claim a piece of market.

As opposed to this, the "market segmentation" approach is bending the supply to the will of demand. It emphasizes the demand side. It is a question of adjusting product and market efforts to the needs of consumers. That is, the marketing organisation tailors its offerings to fit the carefully defined needs of specific groups of consumers, by dividing the public at large into specific sub-markets. Once the market has been clearly identified, it becomes easier for an organisation to make a product to appear different from those of competitors. Market segmentation appeared in 1950s along with new concept of marketing. It is most widely used approach today in having hold over larger share of the market.

BENEFITS OF MARKET-SEGMENTATION

Market-segmentation strategy, if used properly, can benefit both the marketing organisations and the consumers. According to **Professor Maurice I.** Mendell and **Professor Larry J. Rosenberg,** the following benefits accrue to the firms practicing such segmentation strategy.

1. A more precise definition of the market: Market segmentation improves organisation's understanding of why consumers do or do not buy certain products or services. In other words, it provides as understanding of how best to meet the changing market demands. Thus, the firm would be in a better position to spot and compare marketing opportunities.

2. A more effective marketing programme: Understanding the consumers is the first step in planning the appropriate programme to meet their needs. The information gained from market segmentation allows the organisation to plan a systematic and matching market programme that reflects the current market conditions and allows for the future modification. That is, the firm can make finer adjustments of its products and appeals.

3. Better assessment of the competition: Market-segmentation helps in assessing correctly the strengths and weaknesses of the competitors. If a competing product appears to be deeply entrenched in one segment, there is hardly any point in wasting precious resources in marketing to such a group. However, where the company product fails lo relate to a potentially strong target, management can take full advantage of the same.

4. Better allocation of resources: Once the target segment is clearly defined and identified, the limited resources at the command of the company can be channelled for best results. Thus, promotional campaigns can be co–ordinated effectively so that the consumers would recognise the promotional appeal quickly and the marketers would minimise the losses resulting from excess exposures.

THE COSTS OF MARKET-SEGMENTATION

As noted above, a market-segmentation strategy offers atleast four benefits to both the organisations and the consumers. Marketing segmentation strategy involves certain costs to arrive at these benefits which we cannot afford to forget. At times, it may so happen that these segmentation strategy costs outweigh the benefits, rendering segmentation a futile exercise. These costs are :

1. Product costs: In order to accommodate each market segment, the firm goes ahead with designing a specific product. Designing work may be in the range of mere change in label to complete rethinking of entirely new product. It involves good deal to research and development activities as a product is to be separated for each market segment.

2. Production cost: Though technological break-throughs have made possible good many firms to reduce the number of units needed achieves economies of scale, still there persists a problem of high productivity costs. The firm is expected to achieve sufficient sales volume in each market segment to justify the costs involved in separate production runs. In case the sales in each segment are insufficient to keep production lines operating continuously, the firm may be either forced to stop the production periodically or to shift the line over to the production of products for other market segments. Both of these activities are quite expensive.

3. Promotion costs: Every organisation has to develop a promotion strategy that fits each market segment. It is but natural that multiple strategies require huge expenditure on both human and financial resources to design different ads and place them in various media. Multiplicity of ads deprives the firms of quantity discounts and concessions. This results in mounting promotion costs.

4. Inventory costs: More the segments the firm wants to serve, the larger will be the inventory costs. These inventory costs work out higher because of two basic reasons namely: 1. Larger selection of products will force the firm to maintain more and more records covering location and quantity of merchandise. 2. The firm must maintain increased buffer stock for normal demand and increased safety stock for unpredicted demand levels.

5. Management costs: Market segmentation strategy consumes good deal of valuable time of management. The management is to design and implement a co–ordinated marketing strategy for each market segment. Such a co–ordinated strategy deals with product pricing, promotion and distribution. Many organisations employ specialised services of a product or market manager whose main responsibility is to develop and monitor organisation strategies.

THE BASES FOR MARKET-SEGMENTATION

Market segmentation being the key input in firm's marketing planning process, the crucial issue is to how to identify these market segments? Unfortunately, there is no one best way of segmenting the markets. That is why, marketing managers are expected to examine a variety of segmentation criterion, so as to identify those that will be most effective in defining their markets. There are two basic approaches to identify market segments. These are—"Consumer characteristics" approach and "consumer response" approach as given in the following chart.

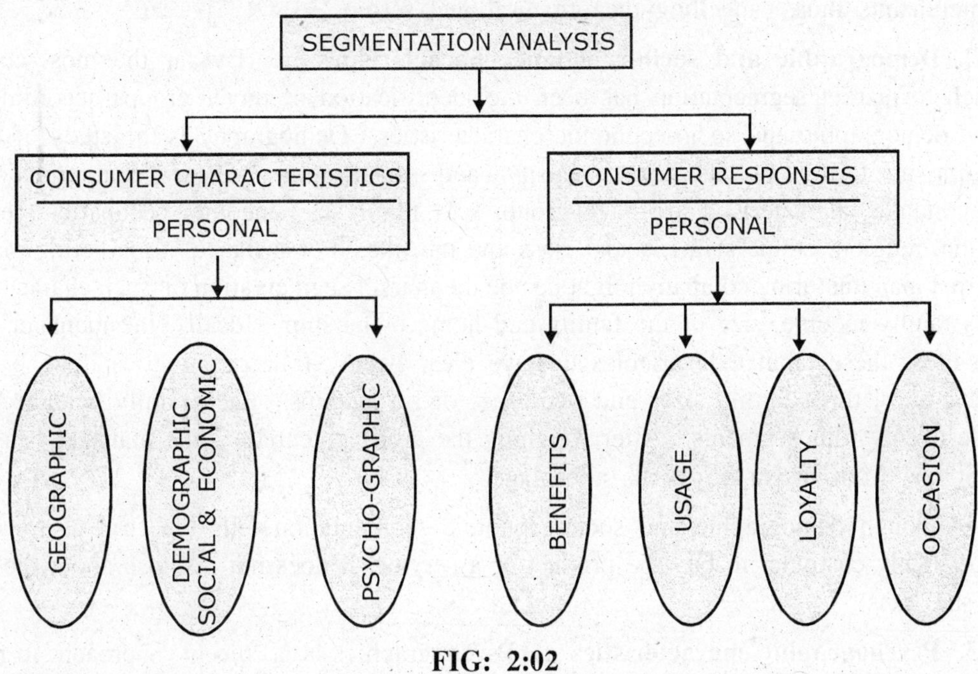

FIG: 2:02

Creative marketing manager, very often, combines these basic approaches to get as comprehensive portrayal of the segment as possible. Thus, at least seven possibilities can be exploited.

A. CONSUMER CHARACTERISTIC APPROACH

Consumer characteristic or personal segment bases measure in terms of geographic, demographic and psychographic features of consumers. These characteristics portay the consumer in terms of "Where they live ?" "Who are they ?" "How they behave ?". Following is the brief explanation of each such consumer characteristic and its variables.

1. Geographic characteristics : Historically, the first and probably the most obvious basis for segmentation of markets has been geographic characteristics of people. Regional difference in consumer tastes for products as a whole are well-known. The sellers distinguish carefully among the regions in which they operate and select where they have comparative advantage. Thus, small retailers may distinguish between nearby or surrounding customers and distant customers or far-off ones. A fertiliser dealer may have groupings as—urban and rural. A regional manufacturing company may have segments as in terms of zones say—North—South—East and West. A national manufacturer might have different sales–territories in terms of States—Divisions—Districts—Talukas—Cities and Towns.

Though geographic segmentation helps the marketers to concentrate their efforts to the exact places, organisational, promotional and distributional efforts can be fruitfully utilised. Consumers do not stick to a particular locality or a region. The geographic mobility changes consumer habits thus, cancelling the organisational set-up.

2. Demographic and socio-economic characteristics : By far the most common approach to market segmentation has been the identification of buyer groups according to a selected demographic and socio-economic characteristics. Demography is the study of human population in terms of its size, density and distribution. These demographic and socio-economic characteristic are—age–group, sex, familysize, income, occupation, level of education, religion, ethnic status, social class and the like. For instance, a marketing manager of a carpet manufacturing company might decide the market segmentation on such characteristics such as family income, size of the family and home ownership. Ideally, the manager would like to cross these tabulated variables to have clear twenty four possibilities if he has four income groups, three family sizes and two types of ownerships. Each family belongs to one of these twenty four segments. After weighing the profit potentiality, the manager can select the best one from his angle to his advantage.

Although demographic and socio-economic segmentation tells the firm as to who is the most likely customer to buy the product or a service, it does not disclose about the brand he or she likes.

3. Psychographic characteristics : Psychographics is a recent approach to market segmentation which has emerged as a major alternative to the traditional approaches. In a fundamental sense, it is not a different approach. It has been coined by **Professor E. Denby** who says that psychographics seeks to describe the human characters of consumers that have bearings on their responses to products—packings, advertising and public–relations efforts of the unit. Psychographic characteristics include variables like—personality—attitudes and life-styles.

Personality is the individual's consistent reactions to the world that surrounds him or her. The personality variables are—dominance, aggressiveness, objectivity, achievement-motivation and the like. These influence the buyer behaviour. The renowned example of this kind is the famous personality study conducted by American automobile industrialist in case of **Ford** versus **Chevrolet** cars. According to the study, **Ford** cars attracted the personalities with variables like "independent, impulsive—masculine—alert to change and self-confident". **Chevrolet** cars, on the other hand, "conservative— thrifty—prestige conscious—less masculine and seeking to avoid extremes". This applies to other consumer durable and non-durable items.

Coming to life-style, it stands for the people's activities, opinions and the sum-total of their interests and values. Thus, the customers can be grouped as SWINGERS who seek up-to-date goods—they are past posed—hedonistic style; **Status seekers** who try to buy goods that will reflect a high status in the society and the **Plain Joes** who seek ordinary, unfrilled goods that do their job.

B. CONSUMER RESPONSE APPROACH

Unlike consumer characteristic approach, the consumer response approach believes in why a consumer buys a product than asking as to who is the consumer. Consumer response to the market offerings are much more important which can be made as the bases for segmentation. These responses are benefit, usage, loyalty, and the occasion.

1. Benefit response : Under this, the consumers are sub- divided into specific groups in relation to the various benefits that the buyer is seeking from a product in particular. These benefits sought differ from product to product. These benefits are—the aspects of efficiency, prestige,durability, economy or resale value and the like. The following exhibit clears this point in case of tooth pastes in India.

Benefit sought	Major brands	Percentage of market share
1. **Cosmetic** White teeth–stop bad breath	Colgage, Close-up	65 per cent
2. **Therapeutic** Fight gum troubles	Forhans, Cibaca, Signal	25 per cent
3. **Ayurvedic** Without side effects	Vicco Vajradanti, Neem	0.5 per cent
4. **Others** White teeth–virility	Smokers, Dentobac	0.5 per cent

For instance, in case of tooth-paste, the benefits sought may be bright teeth, taste and low price. In case of an automobile, the fuel efficiency, quality, status and resale value. Segmentation based on benefits is usually most practical approach from a marketing stand point of view because of its close connection, product–planning and publicity. Thus, a company can choose the benefit it wants to emphasize, create a product and deliver it by a direct message to the group seeking that benefit or set of benefits. However, choosing a benefit to emphasize is not an easy task. The firm must be damn sure that the stated benefit or set of benefits that the buyers seek are real ones.

2. Usage response : The first departure from pure demographic segmentation is usage or volume response as a base for segmentation. Thus, the amount of usage of a particular product has its say among different consumer segments. Under this approach, the seller distinguishes the users as the **heavy, medium, light and non-user** of his product. Then he attempts to determine whether these groups differ in demographic and psychographic factors. Thus, the classification is into primary groups as **users** and **non-users** and **potential users.** Actual users are separated as to **heavy** and **light users** where an attempt is made to determine the difference that exists between them. Naturally, he concentrates on **heavy users** as they consume four to ten times the **light users.**

Non–user group is consisting of two types of people. One, those who generally do not use this product (**Non-potential Users**) and those who might use (**Potential users**). Much caution should be exercised before writing off the people as **Non-potential users** because, every rule has an exception. In fact, the seller should give thought to all the volume groups because, they present different opportunities. Here, **Potential users** are those do not use the product but who are not barred from its use for any functional, cultural or economic reason. Many a times, it may be the ignorance of the product or inertia or even a psychological resistance. However, these can be removed effectively. Thus, extensive information feed will kill ignorance, inertia will be eased by repetitive ads and psychological resistances by thematic ads.

3. Loyalty respons: Loyalty response or product-space response is the latest base used for segmentation. It works on the fact that the consumers can differentiate between the products. That is, they create a product preference scale. That is, the buyers are asked to compare the existing brands of product and rate them as they perceive them based on their liking. Such attempt helps the seller in developing an ideal brand to which a sizeable group of customers is clustering. Later, these consumers can be sub-divided on the demographic or psychographic features.

However, it is really very difficult to pin down correctly the loyalty. That way, every seller has consumers who are most loyal, moderately loyal and fickle-minded. There is no guarantee that the most loyal consumers are the heaviest users. Further, brand loyalty is very hard to measure because, it depends on the availability of competing products. Again, the reasons for faith in a particular product are too often too personal to take as the basis for effective segmentation.

4. Occasion response : The earlier noted responses of loyalty, benefit and usage are one that vary according to the immediate situation. For this reason, the shrewd marketers use occasion response to determine which situations produce optimal consumption patterns for a given product. This sense of occasion is of top significance when it comes to the question of designing the marketing mix. For instance, inexpensive local broiler chicken for daily dinner is fine but inappropriate for entertaining guests where costly 'Venky' chicken works wonders. For daily use most of them use common brand toilet soap cakes but for special occasions specific brands of toilet soap cakes. The experience has proved beyond doubt that it is possible to broaden the occasion response to product and, therefore, its demand. Thus, a traditional break-fast drink can be converted into a refreshing drink for any time of the day. Much depends on marketing creativity.

In conclusion, it can be said that the marketer may proceed to segment his market in many ways. His goal is to delimit and determine the most decisive mode of segmentation. That is, the differences among the buyers that may be the most consequential in choosing among them or marketing to them. The marketing segment strategy chosen is the key input to the marketing manager's most vital decisions. Once, the manager has the sound understanding of markets to be approached, intelligent decisions about product, price, promotion and place can be made.

REQUISITES OF SOUND MARKETING SEGMENTATION

Market segmentation has its own benefits and costs. The strength of it lies in better understanding of consumers for making intelligent marketing decisions and their implementation. The weakness of segmentation is evident from the inability of a marketer to take care of all segmentation bases and countless variables. The possibilities are so many that practically there may be one segment for each consumer as no two consumers are exactly similar. To optimise the benefits from market segmentation, every firm is to adopt a five points criteria for effective segmentation. Requisites of sound segmentation strategy are spelled out very succinctly by **Professor Martin. L.Bell** of Washington University U.S.A. These are :

1. **It is identifiable and measurable :** The segment or the group of buyers must be clearly defined. That is, who is in segment ? Who is outside the segment ? After answering these questions, it is essential to get demographic, social and cultural data about segment members. These data should permit the measurements of the size and importanc the segment as a potential project of marketing strategy. Unfortunately, obtaining segment data is seldom easy especially when the segment is defined in terms of behavioural features.

2. **It evidence of adequate market potential :** Either an actual or potential need must exist in order to segment that opens an opportunity. Actual needs are recognised needs—overt demands for existing goods and services. Potential needs can be transformed into perceived wants through education or persuasion. Potential needs are more difficult to ascertain than the actual needs. Here, marketer is to develop strategies only for substantial segments – whether actual or potential. If the segment is too small to justify a separate marketing programme, it should not be used to develop a segmentation. It is the purchasing power that represents a meaningful marketing opportunity that is likely to stem from either income or savings or credit as the case may be.

3. **It is economically accessible** : Segmentation involves a search for enough similarity among buyers, to permit the seller each search of these potential customers economically. For example, segment members could be concentrated geographically, may be shopping at the same store or may be reordering the same magazines. A segment based on motivational characteristics cannot be reached economically. If 'Closeup' tooth-paste makers attempt to reach a segment identified by the user's desire enhance sex–appeal, it may take up TV advertising. This message reaches both the intended and others. The cost per segment member is much higher in case of TV advertising. Any way, a segment should allow mass media.

4. **It reacts uniquely to marketing efforts :** A segment should make differential response to the marketing efforts put in. Different segments unless they respond in unique ways to particular marketing efforts, hardly justify the use of a separate marketing programme. Put in other words, every efficient marketer is aiming at equal marginal response from the last unit of the marketing applied in each market. That is , all the segments must not result in the same programme and levels of spending. Thus, a given segmentation, to be meaningful, should differ in their responses to marketing efforts. Differing responses will help in optimising the marketing operations by changing marketing efforts and amount involved.

5. It is relatively stable over a period of time : Marketing strategies are long-range plans that project three to five years into the future. Moreover, lead-time of upto a year often is needed to analyse market and to prepare a plan. Therefore, the segments that emerge rapidly and disappear just as quickly do not offer very good marketing opportunities for a firm that follows the generally accepted approach. Only highly innovative entrepreneurs can, at considerable amount of risk, attempt to serve these segments. It is only an exceptional case than a rule.

6. It is dynamic : A final word of caution is to be given at this juncture. One should not conclude that once a company finds its segment, its problems will be solved for ever. The marketing is changing constantly. The segments are to lose aggregations subject to modification. Technology, competition, perceptions and attitudes all are volatile. Because of such changes, marketers must monitor the market constantly to detect the changes in it to adapt the strategy accordingly. That is, nothing different than dynamic segmentation.

MARKET-SEGMENTS AND MARKETING-MIX

There is close alliance between market segments and marketing–mix. There are two possibilities of relationship between marketing–mix and marketing segmentation. In the first possibility, the direction is from market segmentation to marketing–mix. Here, the company assumes certain things about a segment and then prepares a marketing–mix based on these assumptions. For instance, it is safe for stereo manufacturers to assume that the young singles segment of the music market will need the latest in **Hi-Fi** equipment and will be willing to pay for it. In case of second possibility, the direction is from marketing mix to the market segmentation to revised mix. It implies that the company already has the product in hand and that the market is crowded with similar items. The company studies the market to see if there is any correlation between certain buyers and certain brands of products, instead of taking a blind plunge in the fray. For instance, if the company finds that college students wear mainly one brand of baggy pants, it has then a better idea of how to plan an effective marketing strategy. The company revises the product several times to meet the needs of new target market.

TARGET MARKETING

Market segmentation is the process of matching the needs of consumers with company suffering by grouping into relatively homogenous groups. Segmentation is the stepping stone to the market targetting or target marketing. It enables the marketing manager to target his market that selecting among the segments market parts bases on some definite criteria. Market targetting is the process of deciding and preparing the marketing program for market or markets. Segment profiling is a major and constructive step in this direction of market targetting. For understanding the size of segment, growth potential, segment's structural attractiveness, company objectives and resource. In fact, size and growth potential of a segment are more important.

However, deciding about size of segment is more difficult. It may so happen that small segments may not be of much interest but small companies may prefer small segments. The size and the amount of sales volume expected from a given segment helps the company in thoroughly analying the segment for successful targetting.

Naturally, all companies expect an above average increase in sales and profits. That is why they prefer segments with growth, potential. It is equally true that competitors are equally tempted by above increase sales and profits. The factors that play significant role in the process of targetting are: (1) Segments attractiveness which focusses its attention on profitability. (2) Keeping competitors at bay is more important, at times and therefore, profitability and profit get second place. (3) The company objectives and its resources. It may so happen that a particular segment may offer wide range of opportunities to the company but company may not have resources; then it should limit the selection of segmented markets. The objective of the organisation is also to decide the selection of segments. If the segments do not allow the company to realise its objectives, it has no option but to limit the targetting.

After evaluating and selecting the market segments, the company may now decides as to how to cover segments selected as target market. Thus, successful segmentation leads the way for sound and effective targetting of market or markets. Target market is one which the company serves or decides to serve. Target market implies selecting one or more market segments or consumer groups for marketing approach and design marketing programs and policies. Target market lends a definite orientation to marketing decisions which finally helps in positioning the products of its own.

ALTERNATIVE MARKET TARGETTING STRATEGIES

Although the present day theory of market segmentation seems clear, its practical application is not that simple. A successful segmentation involves a number of steps which makes consumer motivational research mandatory. A great deal of knowledge about the market is a must. The second step is the application of segmentation theory. It is to do with collection and analysis of data concerning each market segment. Third step is to apply the criteria of a good segmentation namely, its identification and measurement—its importance—economic accessibility—divergence in response to marketing effort and stability. The fourth and the last step is to develop action programmes to deliver the benefits that consumers want. This involves the development of marketing programmes for both short and long–run. With this back-ground, let us study the alternative strategies.

That way, every market can be segmented to some extent as the buyers are never all alike. However, a firm may or may not wish to shape its marketing policies around these differences. To a marketer, in fact, three alternative market targetting strategics are open. That is, in planning its marketing mix for a product or a service, a company can choose either to confront the differences among the consumers or to ignore them. If it ignores such differences, it is said to follow undifferentiated marketing in which a single marketing mix is used for the entire market. On the other hand, if it segments the market on the basis of consumer differences, it can choose to practice either differentiated marketing in which a different marketing–mix is used for each segment or concentrated marketing in which all or most of the marketing efforts are focussed on one or a few segments.

UNDIFFERENTIATED MARKETING

This strategy puts forth only one product and tries to draw in all the buyers with one marketing programme. The firm chooses not to recognise the different demand curves that make up the market. Instead, it treats market as an aggregate focussing on what is common in the needs of people rather than on what is different. It is an attempt to design a product and a marketing programme that appeals to the broadest number of buyers. It relies heavily on product differentiation to protect itself from competition, mass channels, mass–advertising and universal themes. It aims to endow the product with a superior image in people's mind whether or not this is based on any real difference.

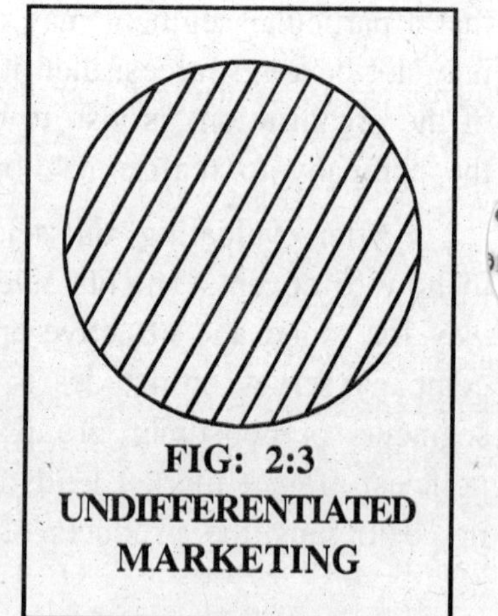

FIG: 2:3 UNDIFFERENTIATED MARKETING

The best examples can be those of **Coca-cola** and **cigarettes.** For many years, for instance, **Coca-cola** meant only one thing to consumers. It was a patented soft-drink available in a single flavour and bottle size. Its theme was "Things Go Better with **Coca-cola**" Similarly, cigarettes, regardless of brands, the size has been 2.75 inches and wrapped in a white paper and packed in a same soft container with aluminium foil. The ad theme had been "The Pause that Refreshes".

Undifferentiated marketing is primarily defended on the grounds of cost economics. A narrow product-line minimizes production— inventory—administrative—selling and distribution costs. Large-scale approach makes available discounts, research and development costs lowered, providing the largest profit in the short-run. In spite of these plus points, an increasing number of marketers have expressed strong doubts about the optimality of this strategy. The firm practising undifferentiated marketing, typically develops a product market programme aimed at meeting requirements of the entire market. When several firms in the industry do this, the result is hypercompetition for the largest segment and undersatisfaction of the buyers of smaller areas. This strategy is more vulnerable to competition than the other targetting strategies. Sometimes, appealing to the largest market results in what is known as "majority fallacy." Such an approach may provide the largest profits in the short-run; however, in the long-run, the market opportunities for smaller segments do make undifferentiated target marketing quite vulnerable. The finest example of this kind is that of American car manufacturers. American manufacturers produced only large cars while foreign firms capitalised on this smaller segment needing small cars. The successful company is one that cares for buyers of both large and small cars.

DIFFERENTIATED MARKETING

Under differentiated marketing strategy, a firm decides to operate in several or all segments of the market but designs separate products and marketing programmes for each. That is, **"differentiated strategy"** involves a different marketing mix for each segment. By tailoring their market offerings to many different segments, marketers hope to achieve additional sales and increased consumer identification with a brand or company name. Such firms work by long-standing rule of thumb. If a product is successful within a certain market segment, they move over to another product altogether to a newly created segment, leaving aside that successful product. The best example is that of **General Motors** of America that tries to produce car for every 'purse-purpose and personality'. Similarly, the cereal companies carefully differentiate among the health conscious adults, working women, fussy teen-agers and sweet-toothed children while planning new cereals. . Each group has a cereal carefully designed especially for its needs, though the company name is always in the foreground in ads and packaging. So is the case with soaps and shampoos. In India, **Hindustan Lever Limited** and **Godrej Limited** then have used this effectively. Even known Tea Companies namely, **Brooke Bond** and **Liptons** (now merged) have taken full advantage of this alternative.

FIG: 2:04
DIFFERENTIATED MARKETING

By offering product and marketing variations, the firm hopes to attain higher sales and a deeper position within each market segment. It hopes that a deep position in several segments will strengthen the customers overall identification of the company with the wide range of product field. It hopes for greater loyalty and repeat sales because, firm's offerings are bent to the desires of customers than otherwise. Good many firms have moved out of undifferentiated marketing strategy to differentiated marketing strategy. The result is multiple product offerings and channels to reach consumers. Thus, today. **Cola** is sold in different bottle sizes as well as in cans. Cigarettes are manufactured in a variety of lengths and filter types. **Chevrolet** is no longer a single style low priced car but a set of cars matched distinctly to different market segments varying in cost – size and features. The greater willingness to customerise their products, results in net increased total sales, at the same time, marketing and production costs increase. Thus, product modification costs, production costs, administration costs, inventory costs and promotion costs multiply. In view of this double – sided effect, nothing can be said about its optimality. It is definitely sales – oriented.

CONCENTRATED MARKETING

Both differentiated and undifferentiated marketing alternatives to segmentation strategies imply that the firm goes in for the whole market. However, good many firms see a third alternative namely, **concentrated marketing**. This alternative becomes a must when the firm's resources are limited. Instead of going after a small share of a large market, the firm goes after a large share of one or two or a few sub-markets. Rather than pursuing the entire market or most of its segments — as is the case with the other two alternatives — the concentrated marketing focusses on only one or a few segments. Put in another way, instead of spreading itself thin in many parts of the market, it concentrates its forces to gain good market position in a few areas.

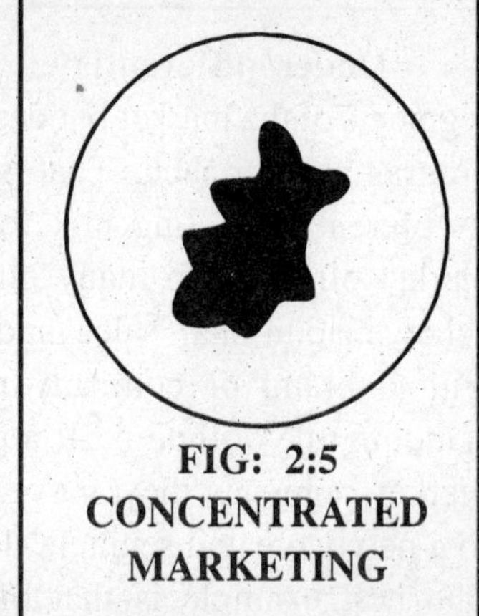

FIG: 2:5 CONCENTRATED MARKETING

The world class examples are that the **Rolls Royce Automobile Company** of United Kingdom has appealed only to the wealthiest, highest socially positioned automobile buyers. On the other hand, **Volkswagen Company** of Germany has not attempted to trade outside the low price range. This practice is more popular with publishers of textbooks. Thus, some specialise at primary level – some at high school level and others at college and university levels. Further, specialisation is also possible. For instance, **Richard.D. Irwin Incorporation** has specialised in Economics and Business titles while **John Wiley and Sons** in Mathematics and Statistics.

The major advantage of concentrated marketing is that the organisation can become a specialist in the needs of its selected market segment. Thus, firm achieves a strong market position in a particular segment or segments it serves, owing to its greater, deeper knowledge of the segment's needs and special reputation it acquires. Further, it enjoys many operating economies because of specialisation in production, distribution and promotion. This enables savings in costs through large of a small number of products; at the same time, it has positive impact on advertising and distribution. The firm can earn high rates of return on its investments provided the segment chosen is right. It is worth noting about the dangers of concentrated marketing. It involves tagging the company's future growth to one segment of the market and it means obvious risks. Putting all the eggs in a single basket restricts the growth opportunities and makes it vulnerable, particularly to a successful competitive attack or if there is a sudden dipping down and thus reducing its market share. Further, a firm that concentrates too much forgets or neglects other segments that are otherwise profitable.

SELECTING A VIABLE MARKETING STRATEGY

After having marshalled the implications of the three alternative marketing strategies, the next immediate question to be answered is that of selecting an ideal segmentation strategy. While selecting a strategy, the following points are to be given due cognisance as very ably narrated by **Professor R. William Kotraba** in his article "The Strategy Selection Chart" in Journal of Marketing. These points worth pondering are:

1. The company resources : The financial resources at the command of the company dictate the choice of strategy. Where the firm has limited financial resources, it is forced to go in for concentrated marketing as the realistic choice. On the other hand, amplicity of finance permits the firm to go in for either differentiated or undifferentiated marketing. It is the extent to which the company enjoys the financial leeway that decides the choice.

2. The product homogeneity : Most consumers do not perceive differences in such basic commodities as salt, grapes, steel, gasoline and the like. Undifferentiated marketing strategy for such products is more natural than differentiated or concentrated strategies. On the other hand, products that are capable of great variations like sound gadgets, cameras, automobiles are more suited to concentrated or differentiated marketing strategies.

3. The product-life cycle stage : When a firm introduces a new product, it releases one of the two versions. The firm's interest is to develop primary demand and, therefore, undifferentiated marketing seems more suitable strategy; or it may follow concentrated efforts to meet the requirements of a particular segment. As the product moves through its life-cycle towards the saturation stage, the firm starts to search harder for new and untapped needs in order to maintain or increase the sales level. Thus, through the mature stages of the product life-cycle, the firm tends to pursue a strategy of differentiated marketing.

4. The market homogeneity : Market homogeneity refers to the degree to which the consumers are alike in their needs, preferences, and other characteristics. In such homogeneous markets, undifferentiated marketing strategy works well. On the other hand, heterogeneous markets warrant either concentrated or differentiated marketing strategy

5. The competitor's marketing strategies : This refers to what the firm's competitors are doing. When competitors are practicing active segmentation, it is really hard for a firm to compete through undifferentiated marketing approach. If so, it loses the battle. On the other hand, when the competitors are practicing the undifferentiated marketing strategies, a firm often gains by practicing active segmentation.

6. The policy of the government: Every alert marketer gives due weightage to the impact of governmental policy on his line of activities. Government as the custodian of the society, may frame certain policies to protect and uphold the interests of the masses. For instance, it is natural that the government goes in for public distribution system in cases of essential commodities. If the products or commodities controlled by the government, the marketer has no say than to accept the guidelines. This implies that he is to change, adopt his strategy much against his will and plan.

In a nut-shell, it is clear that no particular strategy is superior to others always. That is, shrewd marketer cannot rely on a particular strategy. As marketing is a dynamic entity, it is the changing conditions that warrant adopting of a particular strategy. It is the scientific study of the set of changing circumstances that gives the way-out as to which is the best strategy under given circumstances.

MARKET–INTEGRATION AND MARKET–ORCHESTRATION

The study of market segmentation would be incomplete without understanding these two phrases namely, **"market integration"** and **"market orchestration"** the two marketing targetting techniques that attained the zenith of popularity during 1970s and 1980s.

"Market integration" is the strategy of counter segmentation of reducing the over-segmentation. For example, the need for high protein drink is found in case of teen-agers, executives, active elders of various professions. Instead of providing for each group a separate drink, a common drink for all with like needs can be developed, thus reducing costs on production and distribution. It is over-segmentation on one hand and economic problems such as inflation, shortages on the other that prompt market integration. In effect, it is counter segmentation.

"Market orchestration" is the technique of selecting the additional segments while offering the same product or service to more than one market segment. In other words, a firm is faced with the problem of market orchestration when it wishes to deliver the same product or service to more than one market segment. The firm is to decide which segments are to be included and whether it is advisable to do so. Generally, this problem is faced by hotels, restaurants, departmental stores and other business lines where customers of differing life-styles come together face to face. The common means of orchestration is price, because high and low prices have the impact of encouraging or discouraging the segments to be included in the range of service by the firm.

CHAPTER BASED QUESTIONS

A. WRITE SHORT NOTES ON:

1) Market segmentation 2) Target market 3) Undifferentiated marketing 4) Differentiated marketing 5) Concentrated marketing 6) Benefits of segmentation 7) Costs of segmenting 8) Basis of segmentation 9) Consumer characteristic approach 10) Consumer response approach 11) Essential of sound market segmentations.

B. SHORT ANSWER QUESTIONS:

1. What is market segmentation?
2. What is a target market?
3. What is undifferentiated marketing?
4. What is differentiated Marekting?
5. What is concentrated marketing?
6. What are the benefits that segmentation brings?
7. What are the cost of market segmentations?
8. What are the approaches to segmentation?

9. What is the relation between segmentations and marketing mix?
10. What do you mean by "market orchestration" and market integration?

C. ESSAY TYPE QUESTIONS:

1. What is market-segmentation? Why marketers resort to segmenting?
2. What is marketing-segmentation? Is it same as "Product-diffrentiation"?
3. Define clearly the word "market-segmentation". Explain the cases of segmentation.
4. What are the approaches to market-segmentation? Explain with examples.
5. What is targetting? What are the alternative targetting strategies?
6. What is a 'Viable' market segment? How do you select such market segment strategy?
7. What are the 'benefits' and 'costs' of market segmentation?
8. What are 'consumer response' bases of segmentation? Explain each with examples.
9. What is consumer characteristics approach to market segmentation? Explain each of these with examples.
10. What do you mean by "market-integration" and what is "market orchestration"?

****

Chapter 3

Consumer Needs and Motivation

Backdrop

Understanding consumer needs is almost half the job of meeting them on the part of a wiser marketer. Diversity in consumer behaviour—different ways of spending, the money-need not surprise us because, diversity is not a new phenomenon. It is as old as human civilisation and culture. Drinking water is a free commodity as supplied by local governments. Free in the sense, that they a litre of water costs least and whatever local government charges for it, it is only a service charge. As against this, millions of consumers are going in for bottled water-popularly called as '"ineral water". For a bottle of 1,000 ml., we are prepared to pay Rs. 10.00 for a 5 litres bottle Rs. 30.00 plus it can be refilled at Rs. 10.00. If one calculates the cost of 1,000 ml., bottled water say-'Bisleri' or 'Bailey' brands, consumers are prepared to pay at 1000 times more that what it should cost. Water has become a product of miracle. The question is why people are prepared to pay such a high price for virtually free product ? The study conducted gives three basic reasons as to why people are prepared to pay such a high price, namely, SAFETY-HEALTH-STATUS. "The marketing strategy implications of these differing motivations are:

***Safety** speaks of the danger of municipal water, stress-filteration, safe-source, low-price, home or office delivery. **Health** speaks of stress purity and taste, no additives or treatments, moderate price, home or office delivery. **Status**-speaks of stress-quality, usage-situations, exclusive taste, high-price, retail, restaurant, bar and other outlets. In, other words, the purchase motives are three. 'Health' concerns focussing on nutrition and fitness motivate some. users. These individuals want natural, untreated, pure water. 'Safety' motivates other purchases. Most of the consumers are concerned with ground water contamination and study reports of deteriorating water quality levels. The third motivating factor is that of status or-'snob-appeal'. Ordering a 'Bisleri' or 'Bailey' and serving is more chic and higher status than plain water. Human needs—consumer needs-are the basis of all modern marketing efforts. Needs are the very essence of the marketing concept. The key to an organisation's survival, profitability, and growth in a highly competitive marketing environment ; is its ability and understand and satisfy unfulfilled consumer needs better and sooner than those of rivals. This chapter, therefore, concentrates on the basic needs that operate in most people to motivate their behaviour. Hence, we shall study all aspects of motivation in so far as it has relations with human needs. The chapter ends with Chapter Based Questions.*

WHAT IS MOTIVATION?

Motivation is the driving–force within an individual generated by a state of tension caused by unfulfilled needs and wants. Motivation is a process of willingness to expand energy to achieve a goal or a reward. It is the force that activates dormant energies and sets in motion the action of people.

STANDARD DEFINITIONS

★ **Professor Dalton E McFarland:**

"Motivation is the way in which urges, drives, desires, aspirations, stirrings or needs direct, control or explain the behaviour of human beings".

★ **Professor Koontz and O' Donnell:**

"Motivation is a general term applying to entire class of drives, desires, needs, wishes and similar forces that lead to action through behaviour".

★ **Mr. Stephen P. Robbins:**

" Motivation represents an unsatisfied need, which creates a state of tension, or disequilibrium causing the individual to move in a goal directed pattern towards restoring state of equilibrium by satisfying that need"

★ **Mr. Robert Dubin:**

" Motivation is the complex of forces stating and keeping a person at work in a group motivation in something that moves the person to action and continues him in the course of action already initiated".

Thus, 'motive' is an inner state of energies that activates or moves and directs the behaviour towards a goal. In other words, motive is restlessness, a lack, a yearn, a force: Once in the grip of motive, the organism does something to reduce the restlessness to remedy the lacking to alleviate the yearn, to mitigate the force. Motivation is concerned with how behaviour gets started, is energises, sustained, directed and stopped. Motivation is the story of all those forces which are biological, social and psychological which defeat one's laziness and moves one to action either, willingly or reluctantly. In a sentence, motivation is the driving–force within an individual which impels and propells him or her to action.

MOTIVATION PROCESS

Motivation is a repetitive or circular process. If one observes these human beings keenly, they are busy in a tension reduction behaviour, consciously or subconsciously. This motivation process can be presented in the form of simple model. This model can be configured as under:

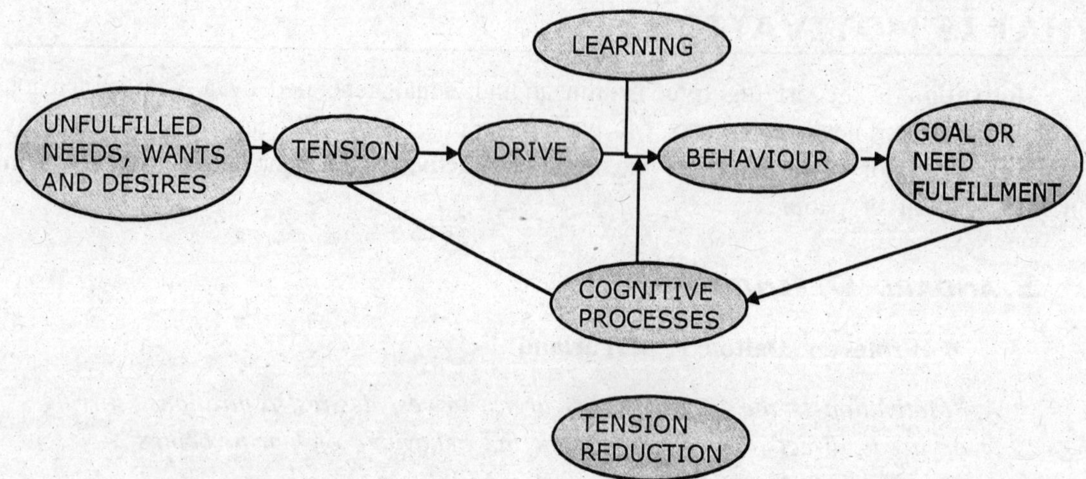

FIG: 3:01 SIMPLE MODEL OF MOTIVATION PROCESS

Each person has a stock of unfulfilled needs, wants and desires. The tension prevails so long as the need or needs are not fulfilled. The tension created leads to drive or cue making him or her to behave in one way or the other. This behaviour leads to need fulfillment thereby reducing tension. Thus, an individual who is thirsty, he feels uneasy till his thirst is quenched. This feeling of uneasiness gives him cue or desire of possible alternative soft drinks or other liquid preparations. Thus, he has plain–water, mineral–water, chilled–water, soft drinks such as **Coke, Pepsi, Thumps up, Appy Fruity, Coconut–water,** varieties of lassi, butter–milk and so on. What he is going to consume to quench his thirst is decided by learning and cognitive processes. An individual may decide to go in for chilled mineral water. Once he has the sufficient intake, his thirst is quenched and the uneasiness is replaced by satiation. This cycle of unsatisfied wants and needs → tension desire satisfaction reduced tension goes on because, his thirst will be a repeating want or need or a desire.

NATURE OF MOTIVATION

The exact nature of motivation is cleared by the special features it has. These are:

1) It is psychological: The concept or notion of motivation is purely psychological. It is so because, it is concerned with the intrinsic forces operating within an individual forces that compel him to act or not to act in a particular way. A motive is a personal and internal feeling which is invisible and intangible.

2) It is dynamic and continuous: As a mental process, it is an ever changing entity itself with every moment. We know that human needs are unlimited and go on changing continuously. Every individual needs these wants to be fulfilled and satisfaction of one need paves the way for another which is waiting for fulfillment.

3) It is complex and intericated: There is need for understanding what motivates a person to behave this way or that way. These needs are mental feelings which can hardly be measured accurately. They are vague and one has deduce or read them through external behaviour. What is important is, the needs are purely subjective. Each person follows a different approach to satisfy his needs. The same need results in different behavioural pattern. Again, one particular behaviour is the outcome of different needs. That is, no possibility of exact generalisations in case of motivation. His wants are inter–connected and are inseparable.

4) It is a circular process: Feeling of an unsatisfied need causes tension and an individual takes action to reduce this tension. It is explained that when the need is satisfied, tension is removed and the person comes to a state of equilibrium. These needs are repetitive as the day follows might and *vice versa.* Not that a person is able to quench his thirst once for all so that he can concentrate on others. There is no limit to wants and are queuing for gratification one after another.

5) Motivation and satisfaction: One should note that motivation is not the same thing as satisfaction. Motivation is the process of stimulating an individual to take a desired action. Contrary to this, satisfaction means contentment arising out of the satiating or gratification of a need. That is, motivation is the drive towards an outcome while the satisfaction is the outcome or the result experienced by an individual. These concepts of motivation and satisfaction have the cause and effect relationship.

6) A product of valence: Motivation is the product of anticipated values from a given course of action and the perceived probability that the action will lead to these values. The anticipated value is known as valence and it reflects the strength of a persons preference for an outcome in relation to others. The perceived probability is better known as expectancy and it refers to the strength of belief that a particular action will result in the desired outcome. When a person is thirsty and he has taken say chilled water, he is supposed to have reduction in tension so that he comes to nor making of behavioural pattern.

CONSUMER NEEDS AND DIVERSITY OF NEED SYSTEM

WHAT ARE CONSUMER NEEDS ?

A consumer need is one which is created whenever there is a physiological or psychological imbalance. Thus, need exists when bodily cells are deprived of food and water or when the personality is deprived of other people who serve as friends or peers or companions, though psychological needs may be based on a deficiency some–times, they are not. For example, an individual with a strong need to get ahead may have a history of consistent success. Consumer needs are desires, feelings, expectations representing their physiological and psychological world of deficiencies. Desires are more or less dreams where all cannot be met in all practicality. However, needs can be met to a greater extent. Any need or desire or expectation, feeling backed up by purchasing power is a want. If you ask a person, in terms of needs, they are countless whereas, when you ask him in terms of his 'wants', they get short listed. The basic reason is needs are many though satiable, the extent of satisfaction or meeting depends on purchasing power his or her. However, we talk of consumer needs, because needs can be converted into wants by the effective use of the power of marketing communication. Some people, experts-call drives, motives, instincts as needs. Though they are right, there is synergic difference. Let us not go by hair splitting. We use the word 'needs' as explained above.

CLASSIFICATION OF NEEDS

Most of the needs are diverse in content as well as in spanning. In fact, these needs are countless and at the same time diverse. These human or consumer needs are classified in a number of ways. A brief outline is given as under:

1. Physiological Versus Psychological Needs: Under this scheme of classification the base is a source. Physiological needs or motives are oriented directly in satisfying biological needs of an individual. These are hunger, thirst, sex, avoidance of pain, safety, rest, physical activity, protection, shelter, clothing, and so on. Conversely, psychological needs or motives focus on satisfaction of psychological needs, achievement ambition, comfort, economy, efficiency, competition, co–operation, fear, sympathy, social distinction, warmth love and affection, courtesy, style, humour, amusement, shyness, teasing, consistency, attribute, causation, categorisation, cues, novelty, independence, self-expression, ego–defence, assertion, reinforcement, affiliation, modelling and so on.

2. Conscious Versus Unconscious Needs: The basis of this classification is the degree of consumer awareness. Needs differ widely in degree to which the consumer is reaching his awareness source. Hence, conscious motives are those of which the consumers are aware, whereas unconscious when the consumer is unaware of these needs. The degree to which consumer awareness, and hence, conscious needs depend on their knowledge exposure, education and their interest to know more and more. Many a times, people are not aware of paper-napkins, picks, finger–bowls. There are many consumers who have washed their hands in finger bowls with lemon piece and drank it. Many people do not know how to use commod type toilets. I am not blaming them, but blaming their unawareness. It is also possible that people are not conscious of some of their basic wants, because they do not want to confront the true reason for their purchase. A person might see the product in terms of durability than status symbol. Forget about, durability or status, even they may not be aware of colour combinations or matching colours of clothing. The more the extent of exposure, more they will be aware and they can go for new wants which they never heard–off.

3. Positive Versus Negative Needs: The consumer wants can be either positive or negative in nature. Here, the base of classification is influence. Positive needs are those where positive influences attract them towards the desired goods. Positive attractions exert predominant influence. Say, ambition, pleasure,possession, personal appearance, cleanliness, comfort, competition, cooperation, devotion to others, social–distinction, imitation, achievement and so on. However, a few of the very important cases of negative forces do exist. Best example is the force of fear. Fear is the key that thrills. A person goes in for low-or no cholesterol edible oil because, he wants to avoid heart diseases; takes nourishing and nutrients to keep him fit and fine, he takes insurance policy because of the danger of being exposed to accident, resulting in disability, or death and so on. Negative wants direct them to be away from undesirable consequences. Want or need for unsafe sex, family–planning, non–smoking, non-gambling and so on.

Though above sets of classification of needs serves the useful purpose, they are not comprehensive where they have used only one characteristic in each case as a base for classification. In this regard a very comprehensive classification has been given by **Professor William J. Mc Gurse** in Journal of Consumer Research 2:302-319, March 1976. -He has used four two-poler motive tendencies. The following table gives the same.

A Comprehensive Classification of Needs

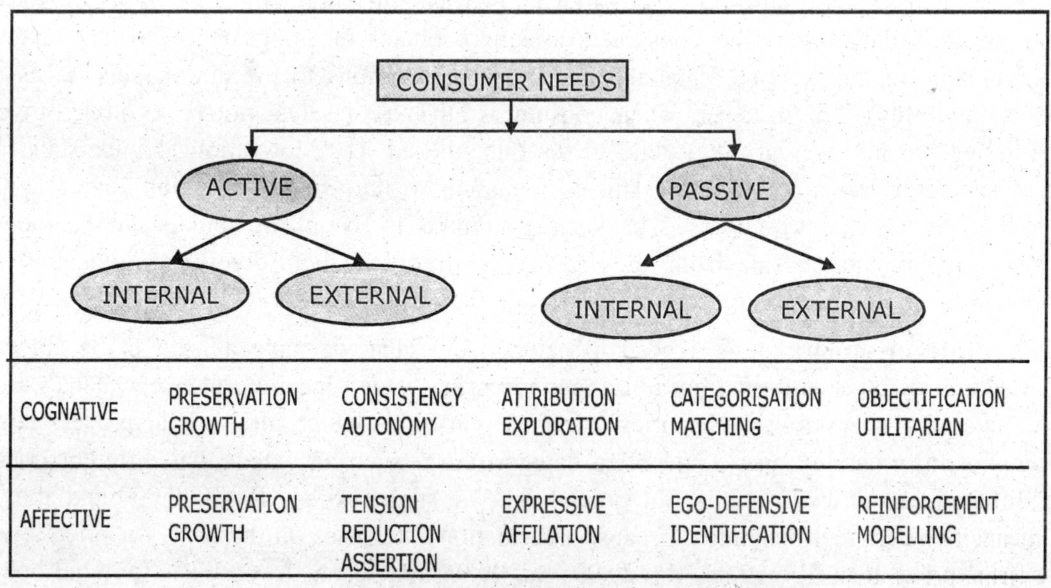

FIG: 3:02 CLASSIFICATION OF NEEDS

As shown in the above table, the relevant distinctions are cognative and affective (mental deliberation versus emotional reactions) preservation and growth (maintenance of equilibrium versus self–development), active and passive (self-initiated action versus reactive tendencies) and internal and external (achievement of new internal states versus new relationships with the environment. It is worth emphasizing here that these four means of classification are not at all intended to be mutually exclusive. Instead, this classification-fourway- used together, provide an interesting basis for identifying and appreciating sixteen major need or motivational influences on consumer behaviour.

SIXTEEN MAJOR MOTIVATIONAL INFLUENCES IDENTIFIED

Let us note these sixteen major need or motivational influences on consumer behaviour. **Professor William J. Mc Gruse** gives very nicely these sixteen motivational influences:

1. CONSISTENCY: It is need for consistency. It is a motivation to maintain coherent and organised view in the world. It is a basic desire to have all facets or parts of oneself consistent with each other. For instance, when a consumer learns that 'meal-maker' is considered nutritious it really is not very high in food value, he feels uncomfortable and attempts to find an explanation for this consistency. These facets of oneself consistent with each other include attitudes, behaviours, opinions, self–images, views of others and so on and so forth.

2. Attribution: It is need for attribute causation. This set of motives deals with our need to determine who or what causes the things that happen to us. Do we attribute the cause of a favourable or unfavourable outcome to ourselves or to some outside force. Need attribute cause has led to an area of research known as "attribution theory". This approach is to understand the reasons the consumers assign particular meanings to the behaviours of others has been used primarily for analysing consumer reactions to promotional messages in terms of credibility. For instance, when consumers attribute a sales–motive to advice given by a salesman or ad message, they tend to discount the advice. Motivation to understand or infer causes for various occurrences. This is focussed in three major directions namely, (1) inference as to the causes of various events (2) attempts by people to understand their own attitudes, values, and so on, from the behaviour they see themselves engaging in and (3) inferences about the reasons other people act the way they do.

3. Categorisation: It is a need to categorise. That we have a need to be able to categorise and organise information and experiences in some meaningful yet manageable way. Naturally, we establish categories or mental partitions which allow us to process large quantities of information. Prices are often categorised to such an extent that different prices mean different categories of goods. In case of cars we start with say Rs. 3,00,000 and above. This means two different meanings, because information categories on the basis for price level. Many firms price items Rs. 59.95, Rs. 29.99, Rs. 99.95 and so on. A reason is to avoid being categorised in the over Rs. 30.00 and Rs. 100. Motivation to categorise complex information in order to organise and deal with it more easily. In case of clothes, it is very often categorised into 'formal' 'casual' and hang–around categories.

4. Objectification: It is a need to be objective, motivation to use 'objective', external information instead of internal reflexion to draw conclusions about one's values, attitudes and the like. This objectification is similar to attribution but more passive in nature. A consumer using the amount of her indoor exercises purchased as her measure of the degree she enjoys the health improvement. An author using the amount of reference material purchased as his measure of degree he enjoys the quality writing or authorship.

5. Autonomy: It is a need for independence. It is a motivation to seek individuality and personal growth through self actualisation and development of a distinct identity. An individual's need for independence of feeling of self–government is derived from a need to establish a sense of self–worth and meaning by activing self–actualisation. For instance, reading "Do it yourself" books is done to improve oneself in a particular direction. Marketers have responded to this motive by providing products that "You do your own way" and "be your own person", “Self is given top priority”.

6. Exploration: It is a need for novelty. Motivation to seek stimulation through new events or circumstances. We people seek variety and difference simply out of a need for novelty. Perhaps, this is the prime reason for brand switching and so called impulse buying. Impulse –buying or switching from favourite to other brands to generate some excitement in one's life. The need for novelty is curvilinear and changes over time. That is, individuals experiencing rapid change generally become satiated and desire stability while individuals in stable environments become bored and desire a change. Travel industry of today has the segments of market in part promoting 'adventure' vacations or 'relaxing' vacations to groups, depending on their likely need for novelty. With the increase in competition and forceful advertising, one switches on from say 'Denim' deodorant to 'Elle-18'.

7. Matching: It is a need for comparing or matching. Motivation to develop mental images of ideal situations and regularly compare or match perceptions of actual situations to these ideals. Each has an ideal-mental image of say bicycle, mobike, car, watch, shoes, and what not. New arrivals of these are being compared to model of their mental images or ideal items. Thus, a person tries to compare actual car model to ideal car that he has or a dream–car. A strong matching motive implies that internal standards, and not characteristics of other brands are more important criteria for judging product or products.

8. Utilitarian: It is a need for utility. It is the motivation to use the external environment as a valuable resource for information and skills to solve life's problems. There cannot be better example than that of "window–shopping" to see how the available products can help around the home. That is, need or motivation for utility makes use of external environment as a rich source of information and to use skills to solve life's problems. Even the marketers use this in window display or window dressing to influence the consumer behaviour.

9. Tension Reduction: It is a need for reduction of likely tension. Motivation to reduce or avoid any tension that is generated when needs are not being satisfied, it is generally a tension of inner feeling for lacking-for having not satisfied a want or need. Many a times, people avoid new brands of products though they are good as they generate uncertainty and tension regarding the product performance. This tension is mounting with the increased competition and fast improving technology. A person who has trusted and a brand loyalty for "SONY" products will hesitate to buy **Thomson** or **Philips** products, because he has the tension or not sure of their performance.

10. Self Expression: It is a need for self-expression. It is a motivation to project one's identity to others. This motive is externally oriented and deals with the need to express ones identity to others. We human-beings have to feel the need to let others know by our actions as to who we are and what we are. This is expressed in the purchase of many products especially in case of clothing shoes, two–wheelers and four–wheelers which allows the customers to express an identity to others since these products have symbolic or expressive meanings. For instance, purchase of the latest winter sports wear is sure to reflect much more than a desire to remain warm while engaged in winter sports. I might not be knowing golf play but if I can't buy golf 'T' shirt, it speaks of what I am.

11. Ego-defensive: It is need for ego-defense. It is a motivation to protect one-self from social embarrassment and other threats to feelings of self-worth. The need to defend our identities or egos is another motive. When one's identity is threatened, one is motivated to protect one's self-concept and utilize defensive behaviour and attitudes. Many products can provide ego defence. Thus, we see good many people purchase underarm deodorant, anti–dandruff shampoos, strong bond adhesive or carpet freshener to avoid the potential of social disgrace.

12. Reinforcement: It is a need for reinforcement. It is a motivation to act in ways that have previously resulted in rewarding situations. We people are very often motivated to act in certain ways because, we are rewarded for doing so. The products designed to be used in public situations such as clothing, furniture and art work are frequently sold on the amount and type of reinforcement that will be reminded. For instance, those customers who are not really interested in diamonds, but they want to keep for the sake of keeping. To such persons the ad running the following way grants reinforcement. "Enter a room and guess you are immediately surrounded by friends sharing your excitement."

13. Assertion: It is a need for assertion. It is a motivation to strive for competition power and success. The need for assertion reflects a customer's need for engaging in those types of activities that will bring about an increase in self-esteem, as well as an esteem in the eyes of others. Individual with a strong need for assertion are more likely to complain when dissatisfied with a purchase. Consumers use some products as a means to be competitively successful. Thus, now–a–days, people are going in for pure cotton clothes and business community is taking full advantage of it. A meter of pure cotton say is available at Rs. 50, however, people who want to show that they are superior will buy Japanese cotton cloth paying Rs. 300 per metre. It is a matter of showmanship or exuberance.

14. Affiliation: It is a need for affiliation. It is a motivation to seek acceptance affection, and warm personal contact with others. In other words, affiliation is the need to develop mutually helpful and satisfying relationships with others. The need here is to share and to be accepted by others. We all know group membership, is a critical part of most of the customers lives, and many customer decisions are based on the need to maintain satisfying relationships with others. It is quite natural that, we human beings believe in the statement "birds of same feather flock together." That is why, marketers very often use such affiliation based themes in their ads. Ad may say on credit card "Don't you want to be a card–holder that is used by Bill Clinton ?"

15. Identification: It is a need for identification. It is a need to develop new identities and roles to enhance one's self concept. Acting these roles out in social settings allows expression of values and development of feelings of importance. For instance, we have different professions or lines of activities. Thus, we have associations or clubs of professionals such as doctors, wrestlers, researches, chartered–accountants, cost-accountants, bankers, architects, and so on, patronage get-together spots frequented by them in an attempt to belong or identify with that group.

16. Modelling: It is a need for modelling. It is a motivation to imitate other individuals with whom one identifies or emphasizes. In other words, the need for modelling reflects a tendency to base behaviour on that of others. Modelling is a major means by which children learn to become consumers. The tendency to model explains some of the confirmity that occurs within reference groups. Marketers utilize this motive by showing desirable types of individuals using their brands. The best example can be that of ad on wrist watches. India Today magazine used Sharukh Khan as a model and the ad states that his brand is '**OMEGA**' constellation and he owns it. Even the personality may be Action Hero of "The World is not Enough"—James Bond in Mr. Pierce Brosnan place of Sharukh Khan. Even **G.I. Joe Toys**–might have 'He–man' or 'Rambo' as the model. Thus, the **McGuire's** psychological motives developed classification is more specific than that of **A. Maslow.**

THE DYNAMIC NATURE OF MOTIVATION

Motivation is the reason for behaviour. A motive is a construct representing an un-observable inner–force that stimulates and compels a behavioural response and provides specific direction to that response. In fact, a motive is why an individual does something and not anything. Motivation is a psychological structure or construct which is subject to constant change depending on the reaction to life experiences. One thing is damn sure that 'needs' and 'goals' are constantly changing. Needs and goals are such mental concepts which are constantly growing and changing in response to an individual's physiological conditions, environment, interaction with others and their own experiences. As individuals, they attain their goals, they develop new ones. In case they do not attain their goals, they continue to strive hard for old goals or they develop substitute goals. There are three basic reasons as to why need- driven human activity never ends. These are: (1) Existing needs are never completely satisfied and, therefore, they continually impel activity designed to attain or maintain the level of satisfaction (2) As needs become satisfied, new and higher-order needs emerge that cause tension and (3) The people who "are capable of achieving their goals" set new and higher goals for themselves. The dynamism of human needs and goals can be seen under several colours or shades of mental colour scheme. These main captions under which the dynamics can be studied are (A) Ever changing needs and goals (B) Frustration and (C) Arousal of motives.

A. EVER CHANGING NEEDS AND GOALS

1. Needs are Never Fully Satisfied: Majority of human needs are never fully or permanently satisfied. The most common experience is that at regular intervals in daily–life people experience hunger needs in the form of light-tea, heavy-tea, high-tea, breakfast, lunch, dinner, and so on. Most of the people seek regularly companionship and approval from others to satisfy their social needs. Even the most complex psychological needs are rarely satisfied. For instance, a person may partially or temporarily satisfy a "power need" by working as an assistant to **CEO** of **TOP TEN OF INDIAN CORPORATE WORLD.** However, this little taste of power may not sufficiently satisfy his need and, therefore, he strives hard for his own decision-making position in the company.

2. New Needs Emerge as Old Needs are Satiated: Some of the motivational theorists believe that a hierarchy of need exists and that new, higher order needs emerge as lower order needs are fulfilled as explained by **A. Maslow**. For instance, a person who has largely satisfied his basic psychological needs may turn his efforts to achieve his friends or colleagues standard of living or life–style say by joining a social club. Having achieved the acceptance, he might seek recognition by giving lavish parties or making large charitable contributions. Here, the marketers must attune to such changing needs. Manufacturers of automobiles who stress the prestige value of their products may fail to recognise that many consumers now look elsewhere to satisfy needs for prestige, it may be through charitable gift–giving or public service. Because of this reason, manufacturers of automobiles might do better if they stress other need satisfactions say, family enjoyment, safety, as reasons for buying a new model.

3. Success and Failures Influence Goals: The researchers and scholars, after indept research, have come to the conclusion in general that the individuals who successfully achieve their goals usually set new higher goals for themselves. That is, they raise their levels of aspiration. The secret is that their success in life builds their confidence of their ability to reach higher goals. It is true opposite also. That is, conversely those who do not reach their goals, sometimes, lower their levels of aspiration. Thus, goals selection is often a function of success and failure. Thus, a college student who is aspiring for a medical course say M.B.B.S, tries for entry into dental at least, due to .family and personal hurdles. Even if he does not get dental, he is ready to accept B. Pharma-a course in pharmaceutical. The effects of success and failure on goal selection have strategy implications for marketers. It is quite natural that goals should be reasonably attainable. That is, the ads designed should not promise more than the product or service that delivers. Even a good product or service may not be repurchased or asked for it fails to line up to the expectations of consumers. A consumer is likely to regard any disappointing product or service with even lesser degree of satisfaction than its objective performance that warrants. Therefore, the advertisers who create unrealistic expectations for their products or services are likely to create or cause higher degree of dissatisfaction among consumers. The frustrations and disappointments that burst out of consumer dissatisfaction have only helped in adding fuel to the driving–force behind consumerism.

4. Substitute Goals: Whenever an individual cannot attain a specific goal or a type of goal that he anticipates will satisfy certain needs, his behaviour may be directed to a substitute goal. Such substitute goal may not be as satisfactory as the primary goal but least is sufficient to dispel the uncomfortable pinching tension that he has. It is but natural that continued deprivation of a primary goal may result in the substitute goal assuming primary goal substitute. Thus, an individual who has stopped wholesome–milk, because he is on diet might actually begin to prefer skimmed milk. A woman who cannot afford now say, 'Postman' brand cooking oil, might be happy with 'Sweekar' brand image. The examples given highlight the substitute effect going in for substitute goal may be a defensive reaction to frustration that he or she has been forced to face.

B. FRUSTRATION

Frustration is the feeling arising out of a person's failure to achieve his preset goal. It is natural that at one time or another, every one has experienced the frustration that stems from the inability to attain a goal. The barrier that comes in the way of achievement of goal may be personal to the individual, say physical or financial resources or it can be an obstacle in the physical or social environment. Irrespective of the cause, the individuals react differently to the situations of frustration. Some people are adaptive and manage to cope by finding their way around the obstacle or, if that fails, by selecting a substitute goal. Others who are less adaptive and they regard their inability to attain a goal as a personal failure and experience feelings of anxiety. For instance, a youngster who prefers a sport–car and is now happy with a minivan, instead. Under this caption, we consider three sub–points which are very relevant.

1. Defence Mechanisms: Those people who cannot cope up with frustration very often mentally try to redefine their frustration and the situations arising out of it, in order to protect their self-images and defend their self-esteem. For example, a college teacher yearning for American tour for his annual vacation cannot afford, therefore, he might select less expensive tour to Singpore or even an exclusive north Indian tour for more number of days. A person who cannot cope up may react with anger toward his head of the department or the head of the institution asking the authorities for sanction of loan out of his savings or he may persuade himself to accept either north India tour or Singpore than America. These last two possibilities are the examples, respectively of aggression and rationalisation. The other defence mechanisms are regression, withdrawal, projection, autism, identification and repression. Let us note these in brief as the students of consumer behaviour.

Aggression: Those individuals who experience frustration may restore to aggressive behaviour in an attempt to protect their self-esteem. Thus, two Indians in sea-sailing competition, burnt their yatch because of disappointing poor performance and swam to sea shore. It may so happen that consumers might boycot manufacturers in an effort to improve quality and have boycotted the retailers in a effort to have lower prices.

Rationalisation: Those individuals who experience frustration may redefine a frustrating situation by inventing plansible reasons for being unable to attain their goals. Instead, they may decide that goal is not really worth pursuing. A person instead of reaching Delhi by air, might think of Rajdhani Express, instead. Rationalisations are not deliberate lies as the individual is fully aware of the cognitive distortion that occurs as a result of the frustrating situation.

Regression: Some individuals react to frustrating situations with a kind of childish or instant and immature behaviour. A consumer buying a readymade shirt say 'Zod' brand manufactured by a wellknown **Zodiac Group** may react in such a immature manner that as he can not afford a price range of Rs. 550 Rs. 850, where other shopper is ready to buy, may go to the extent of learing away so that neither he gets nor the other shopper gets that rare piece which is the last in the stock.

Withdrawal: There are other people frustrated who resolve the matter by simply withdrawing from the situation. Mr. X who has been struggling hard to get the status of a general manager. While he finds it too difficult will leave the organisation and elsewhere his desire is fulfiled. In addition, he may rataionalise his resignation by deciding that the organisation is not true to its stated ideals and that its other members are some what shallow. In addition, he might decide to use his valuable time for more constructive activities.

Projection: A person facing a frustrating situation might react by redefining the situation by projecting blame for his own failures and inabilities over the other objects and persons. Thus, a cricketer who had target of sixer of the last ball of the last over, but gets one run, blames the bowller for changing his pace. A owner of an automobile who is driving on high way and meets with an accident because of his faults, blames the driver of opponent having no correct judgement or he may blame pot-hole on the road.

Autism: Autism or autistic way of thinking stands for a kind of thinking that is almost totally dominated by needs and emotions, with little effort made to relate it to ground reality. It is nothing but "day dreaming" or fantasizing, helps an individual to attain imaginary gratification of unfulfilled needs. For instance a person who has the strong desire to attend the third Olympic (Games of 21st century) in Beijing in China might day-dream in his bedroom being shy and honey. He might even romantisize with his dream–girl is his wonderful day dream.

Identification : Some individuals use a different defence mechanism while resolving the situation of frustration. An individual might resolve his feelings of frustration by sub-consciously identifying with other persons or situations which they consider quite relevant.

Wise marketing people have taken full advantage. Most of the commercials and advertisements are shot where a person is frustrated and by using the product or an idea or service, he overcomes frustration very successfully. For example, most of the ads on deodorants, after–shave lotions, face-creams or shampoos fall under this category. Thus, a boy who is not cared by girls, goes in for a particular brand of deodorant, or shampoo, or after shave lotions and succeeds as shown in an ad ; here, he or she identifies with the situation of frustration and the way to come out of it.

Repression: This is another way that the frustrated individuals avoid the tension of frustration. Repression is to forget about the need that is not satisfied. Forgetting a need is that forcing himself or her-self to take the need out of their conscious awareness. Many a times, the repressed needs manifest themselves indirectly. Thus, we come across a pair who has no child at all. In such cases both husband and wife might love plants or pets and engage in caring for plants and pets. To keep engage, wife might take up teaching or working in library, husband might volunteer his services in a club or collect money for orphans. As manifestation of repressed needs is socially acceptable, is very nicely called as 'sublimation'

These defence mechanisms against frustration are only illustrative. There is no limit as to how a person redefine the frustrated situations and protect their self-esteem and anxieties that are the outcome of failure to achieve.

C. MULTIPLICITY OF NEEDS

We come across hardly a man or woman who is satisfied with a single need. The consumer behaviour speaks of fulfilling more than a single need. It is our experience that we buy clothing for protection and for modesty. In addition, our clothing fulfils an unlimited range of personal and social needs. However, usually there is one overriding need that initiates behaviour. For instance, a woman may want to lose weight because, she wants to wear more stylish clothing. At the same time, she may be concerned about blood pressure. Further, she has say, noticed that her life–partner admire slimmer girls on the beach. If one takes into account the combined amount of tension produced by each of these reasons if it is strong, she will surely going in for controlled diet. However, just one of the reasons say her husbands, straying eye might serve as the triggering mechanism which we can call as prepotent need.

D. NEEDS AND GOALS VARY AMONG INDIVIDUALS

It is almost impossible to infer very accurately the motives from a behaviour of a person. Person with different needs may seek fulfillment through selection of the same goals, while people with the same needs may seek fulfillment through different goals. Let us take an example of say, five members of a consumer organisation who advocate strongly consumerism or might be that they have joined the organisation for different reason or reasons. Let us say, the first is genuinely interested in protecting consumer interests, the second may be more concerned about an increase in counterfeit merchandise, the third may seek social contacts from the meetings of the organisation, the fourth may enjoy the power of guiding or directions a large group, the last may enjoy the status provided by the membership in a strong organisation. Similarly, all five persons might be driven by the common need say 'ego' need to seek fulfillment in different ways. The first may seek advancement and recognition through a professional career. The second may become an active member in league of Women Voters. The third may join health club. The fourth may take a professional dance lessions and the last may seek attention by dominating class-room discussions.

AROUSAL OF MOTIVES

It is established fact that most of the specific needs are dormant for a longer period, is case of individuals. Arousal of motives is to do with charging the consumer behaviour. The arousal of any particular set of needs are at a specific point of time that are caused by internal stimuli found in an individuals' physiological condition, emotional or cognative processes or by external stimuli namely, external environment. Let us touch each of these for better and perfect understanding of consumer behaviour.

1. **Physiological Arousal:** Our bodily needs, at any one specific moment of time, are deeply rooted in an individuals physiological conditions at that moment of time. Thus, a drop in blood sugar level or tummy contractions will press or indicate an awareness of a hunger need. A fall in body temperature is going to induce shivering that makes an individual aware of need of warmth. Secretion of sex harmones awaken the need for sex. One thing is certain that majority of physiological cues or arousals are compulsory or involuntary. However, they arouse related needs that cause uncomfortable tensions until they are satisfied. For instance, a shivering man or woman goes in search of heat to get rid of the tension or discomfort of chill. A hungry person tries to quench his or her hunger by taking some solids or even food drinks. Consumer research has demonstrated clearly that television programs generate very often the physiological arousal in viewers that affects or influences the impact of incoming commercials.

2. **Emotional Arousal:** It is interesting and true that day dreaming or fantasizing is responsible for arousal or stimulation of dormant needs. Persons who are bored or frustrated in their attempts to achieve their goals very often engrossed in day-dreaming which we have referred earlier as autistic thinking. Day–dreaming is entering the world of imagination where they are in a back drop they wish. These situations are most desirable and filling to their aspirations. One might dream of winning a jack-pot in a horse race course, another might be in the world of "Kaun Benega Crorapati" lunched by Star Plus channel and anchored by the best actor of the century. These sweet thoughts tend to arouse dormant needs that produce uncomfort or tension and push to goal oriented behaviour. Thus, a young student who is to be an able businessman joins, to learn management, business school for that matter. A young girl student who wants to be a dancer, joins school of dance and music. Most of the ads on deodorants, perfumes, beauty-aids, are designed on emotional tints to provoke and propel an action.

3. Cognative Arousal: There are many thoughts in the kit of memory of an individual. Only they are to be cued or remended. Any random thought or personal achievement that can lead to cognative awareness of needs. Thus, an advertisement on computer course of short duration and job guarantee might trigger a young and ambitious person to talk to his parents either in person or on phone. There are good many ads on computer education, hotel management, business management, that persuade the reader to take a favourable action.

4. Environmental Arousal: Each person is a member or constituent of society being a social and rational animal. An individual cannot pull on in isolation, because, he is very much part of environment which is external to him or that surrounds him in different lives. It is his environment which reminds him of his dormant needs. Unless, the environment gives cue, these needs remain dormant. The ads on TV about pizza, fast-food, cookies, noodles all remind our school–going children, college–goers and even the office–goers remembering to have these as per their choice at internal or lunch break.

In case of environmental arousal, situation is the one which is the object in-itself. Thus, a house–wife looks at her neighbours new foam beds, she brings it to the notice of the finance minister the husband. An ad on **Bajaj scooter** for Rs. 999 with the background of bus-goer, is such instance. Not very often an advertisement or other environmental cue can produce a psychological imbalace in the mind of an ad viewer, listener or reader. Thus, a person who has colour TV with limited channels of which his proud, an ad about flat tube and 200 channels with surround sound creates a tension a sort of unhappiness until he gets that new TV.

Those people who live in an advanced and highly complex environment, experience good many opportunities for need arousal. Reverse is true in an environment which is stagnant and simple. This is the finest example as to why TV has a mixed effect on the lines of teeming million in under developed countries that they come to know so many things-new life-styles of advanced nations. This, TV. enlightens these people and their needs are aroused. This might enrich the life-style of those who can afford those new products and services and make some people who can not afford these resuling in aggressive and anti-social ways of enjoying them. Those who can not afford by their hard earnings, might resort to revolt, rob, .boycot and so an as they are unhappy. This is going to the pattern of mechanism to defend their need satisfaction.

It will not be out of place if we note the two divergent philosophies regarding arousal of human motives or needs. The **behaviourist** school considers motivation a mechanical process. That is, behaviour is seen as the response to a stimulus and elements of conscious thought are ignored. Taking an extreme example of this stimulus response theory of motivation is that the impulse buyer who reacts largely to external stimuli in a buying situation. According to this theory, the consumers cognative control is limited. That is, he or she does not **act** but **reacts** to stimuli in the market place as pin-pointed by **Professor Peter Weinberg** and **Wolfgang Gottwald,** in their research article "Impulse Consumer Buying as a Result of Emotions"—(Journal of Business Research 10, 1982 pp 43). On the other hand, COGNATIVE school believes that all behaviour is directed at goal achievement. Needs and past experiences are reasoned, categorised and transformed into attitudes and beliefs that act as pre-dispositions to consumer behaviour. These pre- dispositions are aimed at helping the individual safety needs, and they decide the direction that he or she takes to achieve this satisfaction.

TYPES AND SYSTEMS OF NEEDS

For pretty long time, the psychologists and those who had keen interest in human behaviour have tried to develop an exhaustive list of human needs or motives. Majority of lists of such human needs or motives have been diverse in content and length. What is truth that is evident from such lists is that there is good deal of unanimity or agreement so for as physiological needs and as opposed to this, there is vast disagreement of specific psychological or psychogenic needs. In 1923, **Professor Daniel Starch** of Harvard Business school of America came out with forty tour human motives which are used for appeals in ad copies. In 1938, a well know psychologist **Mr. Henry Murray** prepared a meticulously list consisting of twenty eight psychogenic needs. They have served as the basic constructs for a number of widely used personality tests. **Mr. Henry Murray** strongly believed that each one has the same basic set of' needs, But the individuals differ in their priority ranking of these needs. According to **Mr. Henry Murray,** the basic needs include many motives that are supposed to play an important role in consumer behaviour such as acquisition, achievement, recognition and exhibition as given under :

A. NEEDS ASSOCIATED WITH AN INTIMATE OBJECTS

1. Acquisition.
2. Conservancy.
3. Order.
4. Retention.
5. Construction.

B. NEEDS THAT REFLECT AMBITION, ACCOMPLISHMENT AND PRESTIGE

1. Superiority.
2. Achievement.
3. Recognition.
4. Exhibition.
5. Inviolacy (Involate attitude).
6. Infavoidance (to avoid shame, failure, humation, ridicule).
7. Defendence (Defensive attitude).
8. Counteraction (counterative attitude).

C. NEEDS CONCERNED WITH HUMAN POWER

1. Dominance
2. Deference
3. Similance (Suggestive attitude)

4. Autonomy.
5. Contrarience (to act differently from others).

D. SADO-MASOCHISTIC NEEDS

1. Aggression.
2. Abasement.

E. NEEDS CONCERNED WITH AFFECTION BETWEEN PEOPLE

1. Affiliation.
2. Rejection.
3. Nutruance (to nourish aid or protect the helpless).
4. Succorance (to seek aid, protection or sympathy).
5. Play.

F. NEEDS CONCERNED WITH SOCIAL INTERCOURSE (THE NEEDS TO ASK AND TELL)

1. Cognizance (Inquiring attitude).
2. Exposition (Expositive attitude).

Mr. HENRY MURRAY'S LIST OF PSYCHOGENIC NEEDS

SOURCE: Adapted from Henry Murry, "TYPES OF HUMAN NEEDS" in **David C. Mc Clelland** STUDIES IN MOTIVATION.

Turning our attention to structure or a system of human needs or motives, **Dr. Abraham Maslow,** a clinical psychologist, formulated a widely accepted theory of human motivation based on the idea of universal hierarchy of human needs. Maslow's theory of need or motive hierarchy assumes five level structure which rank in order of importance from lower level (biogenic) needs to higher level (psychogenic) needs. His hierarchy of needs approach is based on four basic premises namely, 1. All humans acquire a similar set of motives through genetic endowment and social interaction 2. Some motives are more basic or critical than others 3. The more basic motives must be satisfied to a minimum level before other motives are activated. 4. As the basic motives become satisfied, more advanced motives come into play. Following diagram depicts Maslow's motive and the relative predominance of motives and the number and variety of wants recognized for each motive.

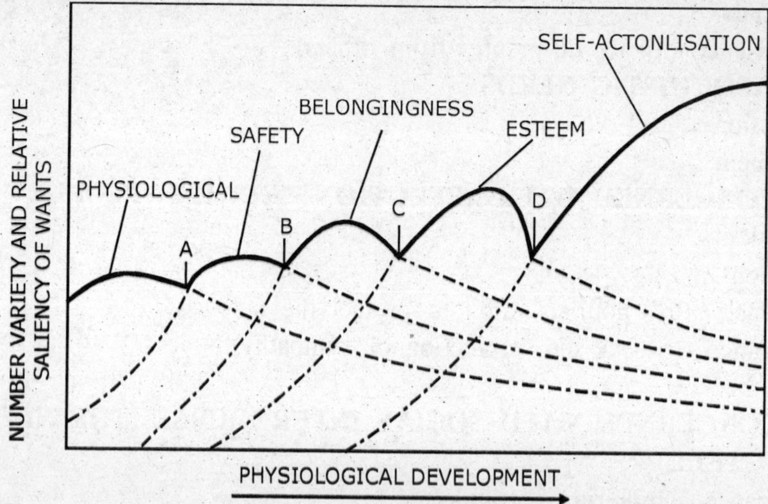

FIG: 3:03 MASLOW'S MOTIVES AND RELATIVE PREDOMINANCE

MOTIVE HIERARCHY

The concept of hierarchy underlines many schemes offered to explain and justify the structuring of motives and their influence. The most influential motive is seen as enjoying the most dominant position in the hierarchy, the second most influential holds the second most dominant position and so on running through the entire list. To be useful, the hierarchy concept must help in what factors influence the relative ordering of motives. This hierarchy can be also presented in alternative way as under :

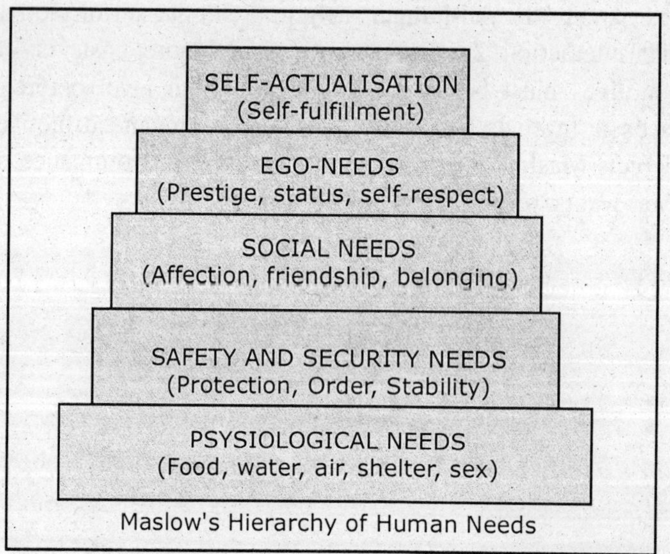

FIG: 3:04 MASLOW HIERARCHY OF HUMAN NEEDS

The presentations in figures 3:03 and 3:04 clearly partray the hierarchy needs in very clearcut way. Though as per theory, there is some overlapping between each level for the simple reason that no need is completely satisfied, for clarity each level is depicted as mutually exclusive. For this reason, though all levels of need below the dominant level continue to motivate behaviour to some extent, the prime motivator the major driving–force within the individual-is the lowest level of need that remains unsatisfied very largely. Let us note these levels from the bottom of the presentation in rectangles as noted.

1. **Physiological Needs:** The first and the most basic level of needs is physiological, in the hierarchy of needs theory. These needs are those that are required to sustain biological life such as food, water, air, shelter, clothing, sex which are biogenic in nature. According to **Dr. A Maslow** physiological needs are dominant when they are critically unsatisfied. It goes without saying that a person whose hunger is dangerously extreme, it can be cooled with food of his choice or at least those food items that make him or her tummyful. In develop ing nations and underdeveloped countries, food, water, shelter are the problems and which are getting acute as there is population explosion. In case of developed nations, the people yearn for higher level needs as they are well above these primary needs.

2. **Safety Needs:** Once the physiological needs are met, safety and security needs rank for their satisfaction. These needs or motives are to do with more than physical safety. They comprise of order, stability, routine, familiarity, control over one's life and environment and the certainty the knowledge of different dimensions of life and living. Here, fear is the key. Therefore, one saves for future, takes insurance policies against possible losses, education, vocation training that provide safety and security. Even for self-defence, one can have pepper sprays, hand–guns mace, personal body guards and so on.

3. **Social Needs:** The third level of hierarchy of needs deals with which are essential to be in a society. Man or woman for that matter is rational and social animal. These are love, affection, belonging, acceptance. That is, people seek warm and satisfying social or human relationships with others and are motivated by love for their families and best wishes or well-being of others. Social motives have given good many opportunities to the advertisers that create an awareness and arousal of social needs particularly personal care products.

4. **Egoistic Needs:** According to **Dr. A. Maslow,** a person reaches the fourth level needs once social needs are more or less met with. This fourth level is concerned with egoistic needs. Those egoistic needs are both inward and outward oriented. Inwardly oriented ego needs reflect a person's need for self-acceptance for self-esteem, for success, independence, satisfaction with whatever he does. Outwardly oriented ego needs include, the needs for prestige for reputation, status, recognition. It is very unfortunate that these ego needs, sometimes, are fulfiled in dysfunctional ways. Poverty is a curse every where. In advanced nations of the west the poor boys steal or rob the sport-shoes or cycles of others as these items are status symbols. We often hear about teenagers of American schools keep hand-guns not only to protect themselves but to command respect from other boys and girls. It also takes place in underdeveloped countries where films have deep impact on their tender minds.

5. **Self actualisation Needs: Dr. A. Maslow** is of the strong opinion that most of the people fail and get frustrated in meeting their ego needs and, hence, reaching fifth stage is a dream than a reality. Self-actualisation is self-fulfillment need that speaks of an individual's desire to fulfil his or her potential something unusual which is not easily possible for others. His or her potential is to become everything and not anything hear she is capable of becoming. **Dr. A. Maslow** puts it most aptly "What a man can be, he must be". These self-actualisation needs differ from individual to individual and age group to age group. A person might have a very strong desire to be the youngest cricketer, hockey-player, foot-baller, chess grand–master, musicians, magician, and what not ? **Mr. Maslow.** says that self-actualisation-need need not be a creative one. It is something not easier for all. These are the cases of rare and unique achievements. Most of the ads on management, computer and engineering education courses work on this.

EVALUATION OF NEED HIERARCHY

The NEED HIERARCHY as visualised and demonstrated by **Dr. A. Maslow** has both plus and minus points. It pays to browse through to know its real worth.

PLUS POINTS

1. **Need hierarchy Concept Can be of great use to Managers of Marketing : Dr. A. Maslow's** theory is based on the postulation that man is wanting animal, never all his wants are fully satisfied, all needs are not equally important. This means that where lower level needs are relatively well satisfied, they cease to be the significant motivators of behaviour but can become very important to a person when they are absent, dormant, threatened or taken away. This need hierarchy acts as a very useful tool for understanding consumer motivations and readily adaptable to marketing strategy because, basically consumer goods serve to satisfy each of the need levels. The need hierarchy provides a useful and very comprehensive frame-work for marketers trying to develop ad-appeals for their products or services. These appeals can be used basically for two marketing issues-for their solutions namely, marketing segmentation and product positioning or repositioning their products.

2. **Need-hierarchy Explains and Predicts not Only Individual but also Group Behaviour:** This is especial for those employees whose work is closely related to that of others and whose work brings them into contact with others. It explains what happens as people's desire for the "good–life" increases. When people are operating below or near substistence level, they will be motivated lo work for low wages and even be satisfied with poor working conditions. However, once, they are made aware that there is "more–life" and once they feel reasonable, well satisfied at the lowest level, they maturely want, expect and demand more. That is, so called "carrot and stick" approach that fails to work once a person moves beyond subsistence level.

3. **Need Hierarchy Limits Frequent Use of Certain Incentives:** As noted earlier, **Dr. Maslow** points out that needs can be thought of as being ranked in a hierarchy in which

one need is more important than others until its satisfied. Once that need is satisfied, the next higher need becomes predominant. However, the order can be reversed and even swing back and forth. He also pointed out that once satisfied a given need no longer motivates human behaviour. Only when, one is deprived of something, and, therefore, craves it, can be used as an incentive. It is worth noting that threat of deprivation of a need is also an incentive. This significant concept is worth appreciating as it effectively limits the frequency of use of certain incentives and restricts the use of others.

4. Versality of the Need Hierarchy: Need hierarchy is so versatile that it helps in designing promotional programmes for a single product from each level. Take the a folding home exerciser for which promotional appeals can be developed for each level of hierarchy. Take a "Bull-worker", an appeal for 'physiological' needs it speaks of its role in improving the tone of the body and individual health, a "safety–appeal" can be that how safe, it is to use against any infit trator, a "social–appeal" might speak of fun to exercise with others or being novel it can build contacts, "egoistic–appeal" can be stress on a feeling of proud of strong and sturdly body, in case of "self-actualisation appeal" its a matter of convenience and luxury of having home exercise after a full day's work every day.

THE MEASUREMENT OF MOTIVES

The interesting and curious questions before us are: How motives are identified ? How they are measured ? How do researchers know which motives are responsible for certain kinds of behaviour ? These are the questions, indeed, very difficult to answer because, the motives are hypothetical constructs—in that they can not be seen or touched, or smelled, handled or otherwise tangibly observed. This is the main reason as to why no single measurement method or measuring rod or yardstick or measure can be considered as a reliable ready index. Instead of this, researchers rely on a combination of observation and inference, self-reports and projective techniques that help in establishing the presence or strength of different motives. These tools of measurements are taken up in the very next aspect namely, motivation research.

MOTIVATIONAL RESEARCH

It is amply clear from earlier discussion, that the identification consumer motives is a complex business. Many consumers are unaware of the motives influencing their purchase behaviour that is. some motives may not reach the consumer's consciousness, and others may be repressed because, to deal with them may be uncomfortable. This presents difficulty to the marketer, who needs to understand consumers in order to design the most effective mix of marketing offers. Any direct attempts to determine such motives, say by interviewing consumers, may only yield 'surface' explanations or rationalisations that hide true strivings. The difficulties involved in discovering directly the precise motivating factors that shape buying behaviour led some marketing psychologists to devise oblique techniques for exposing hidden motives. For many years before World War II, researchers in the fields of psychology, psychiatry, sociology and social-anthropology had been conducting research into the behaviour of consumers particularly probing of sub- conscious mind. Marketing research, to-day, can be conveniently divided into two parts 'quantitative' and 'qualitative'. The former concentrates on finding the answers to the questions-'What'-'Who'-'How many'-'Where' and 'When' the consumer wants a product or service ? On the other hand, qualitative research focusses its attention mainly on 'Why' he wants a product or service ? Put in other words, consumer ability to buy is not that important; what is more important is motives behind such buying. In course time, "motivation–research" became integrated with 'traditional ' marketing research and influenced and expanded the inventory of marketing research and research techniques. It reached its peak hight of popularity in the latter half of 1950s and early 1960s.

WHAT IS MOTIVATION RESEARCH ?

Motivation research is a form of consumer research which has gained ground over the recent years. Motivation Research is the currently popular term used to describe the application of psychiatric and psychological techniques to obtain a better understanding of why people respond as they do to products, ads and various other marketing situations. It is an attempt to discover and explain why the consumer behaviour differs; why he or she behaves in a particular way ? What appeals and sales programmes will best influence his or her decision to act or buy or not to act or not to buy. It concentrates on emotional or hidden stimuli to consumer action. In the words of **Professor Edward Scriven,** it is "a set of tools borrowed from the fields of psychology and sociology, to uncover and evaluate the motives or desires that are back-bone of human behaviour in consumer market". According to **Dr. Ernest Ditcher,** it is "a research to discover the underlying motives, desires and emotions which provoke human behaviour and, in market research, true reasons why people buy or do not buy certain goods as distinct from the reasons they express or even imagine they act upon." In the words of **Professor Harry, Henry,** it signifies "the use of research techniques designed to penetrate below the surface and particularly of techniques borrowed and adopted from psychology in conjunction with the standard techniques of consumer research of the behaviour type."

Thus, motivation research is an attempt to uncover the consumer's suppressed (conscious) and repressed (unconscious) motives. In suppression, the consumer remains aware of his motives but does not care to admit their existence to others for the fear of redicule,' punishment or being ostracisd. Information about the motivating factor remains in the conscious mind, however. Repression implies a more serious rejection of knowledge about a motive because, the individual will not admit motive's existence even to himself. The concept of motivation research has been offered as a means of identifying consumer's true underlying purchase motives.

It is a careful probing beyond the surface 'why?' offered by consumers to explain their actions. As a branch of marketing research, it aims to discover the real reasons for their purchasing preferences via sample questionnaires, interviews and the like. The theme of above definitions is that motivation research is to discover underlying motives, desires, instincts and emotions which provoke human behaviour and, in market research, true reasons why people buy or do not buy certain goods as distinct from the reasons they express or even imagine they act upon. The distinguishing feature of motivation research is probing "beneath the skin", or "below the surface". To-day, the most challenging task of marketing research is to predict how people will react and why they react in a particular way in a given situation. "How they react" ? can be answered with ease and confidence. Say, how the consumers receive the new product, package advertising message and the like where surveys–cross tabulations and analysis can help us to find the answers. However, more difficult task is one of finding out "why people react in a particular way?" By merely asking consumers why they like or dislike a product or an advertisement or a package, one cannot get satisfactory answers. The answers differ widely and are misleading very often. These answers are misleading not because people are dishonest, but merely because they do not know really why ? Wrong or unbelievable answers are given because of two possible reasons: 1. Conscious or unconscious attempt to rationalise their behaviour and 2. Preference for not disclosing their real reasons for ranking. Conventional research does not answer this but motivation research does.

It is the psycho-analysis that helps in overcoming the inability or the reluctance of people to tell why they like or dislike a product or a service. This is knovvn as penetrating below the surface to reach sub-consciousness. The relations between a consumer and a product are partly conscious and partly subconscious. The 'price' of a motor car, for example, is conscious relationship. A person no doubt is worried and particular about kilometres per litre, cost per kilo-metre, boot- space—acceleration and the like but his product preference of sub-conscious type is 'sex symbol'. Car is an extension of man's personality which is known by motivation research. Precisely, motivation research is the art of finding out Why ? without asking why? The best example of understanding is one of choosing a girl as a house-wife. Ask a sober looking young man as to what type of woman he wants to marry; the answer may be: "the quiet, home-loving more concerned about the food she cooks for me than the clothes, ornaments and make-up aids she wants". Observe him at parties and get-to-gethers and you will find that he is after a woman attractively clothed with gleaming make-up appealing to his hidden motives. Thus, motivation research is something that goes beneath the line. Thus, it is an attempt to market below the line.

TECHNIQUES OF MOTIVATION RESEARCH

The techniques used in motivation research are of five types namely, Projective Techniques, Indepth–interview, Attitude–scales, observation and physiological measures.

1. PROJECTIVE TECHNIQUES

These **projective techniques** represent the test conducted to establish the personalities of the respondents and their reactions to product media-advertisement—package—product design and the like. They project or reflect the subject's thought about what he or she sees, feels, perceives—thus producing the reactions. These tests are derived from clinical psychology and work on the postulation that if an individual is placed in an ambiguous situation, he is guided by his own perception to describe the situation. They often provide an insight into the motives that lie below the level of consciousness and when the respondent is likely to rationalise his motives consciously or unconsciously; his responses tend to reflect his own attitudes and beliefs by indirection and discretion; they are his own perceptions and interpretations to the situation to which he is exposed. There are five most commonly administered tests of this kind namely;

1. Thematic Appreciation Test.
2. Sentence Completion Test.
3. Word Association Test.
4. Paired Picture Test and
5. Third Person Test.

1. Thematic Appreciation Test (TAT) : Under this test, the respondent is presented with a picture or series of pictures of a scene or scenes involving people and objects associated with goods or services in question. These are unstructured, doubtful inaction and very often neutral giving no expression or motions. The respondent is to study the picture or the pictures and construct a story. His narrations or readings are interpreted by a skilled analyst. Thus, the picture may be of a young man scribbling on a piece of a paper. Here, the respondent is to read as to whether the person in picture is writing. If so what ? for whom ? and why ? and so on.

2. Sentence Completion Test (SCT): Sentence completion tests are designed to discover emotional responses of the respondent. It is the easiest, most useful and reliable test to get the correct information in an indirect manner. The respondent is asked to complete the sentence given. For instance, the questions may be:

In case of ladies:

1. I like instant coffee because, ...
2. I use talcum powder because, ...
3. I use electric kitchen gadgets because, ...
4. I do not use pain-killers like aspirin because, ...
5. I do not like red, brown and black colours because,

In case of gents:

(a) I liked filter, tipped cigarettes because, ...

(b) I gave up smoking because, ...

(c) I love natural proteins because, ...

(d) I prefer cold–coffee because, ...

(c) I do not use foam–beds because, ...

The way the questions are asked, they do not reflect right or wrong answers. However, the emotional values and tensions are reflected in the answers so given.

3. Word Association Test (WAT): Word association test is similar to that of sentence completion test. The only difference is that instead of an incomplete sentence, a list of words ranging from twenty-five to seventy-five is given. This is the oldest and the simplest kind of test. The respondent is to match the word. That is, the word suggested by the researcher is to be associated by the respondent by the most fitting word he thinks. This is widely used to measure the effect of the brand names and advertising messages. Here, it is not possible to give all the seventy-five words. On illustrative basis, let us have fifteen words.

1. Perfume	2. Tooth–paste	3. Hair–oil ....
4. Shampoo	5. Shoes	6. Two-wheelers .....
7. Four-wheelers	8. Tyre	9. Glass wares ....
10. Ink...........	11. Pencils.........	12. Fridges ....
13. Cupboards	14. Television	15. Video–cassettes ...

Thus, a respondent may give his preferences as 'Colgate' or 'Promise' or 'Close-up' or 'Forhans' in case of tooth paste. On the basis of such answers, it is possible to determine a scale of preference.

4. Paired Picture Test (PPT): This is another very appealing and easy to administre test. Paired picture test means that the respondent is given a pair of pictures almost identical in all respects except in one. For instance, the researcher may be interested in knowing the reaction of respondents to a new brand of refrigerator. The pair of pictures may show a woman opening refrigerator which is' moderately priced with a usual brand; another picture of the same woman opening the refrigerator door of another brand. Looking to these two pictures, the respondent is to give his own feelings or reactions. Though the same pair is shown to so many, respondents, the reactions differ from person to person. Instead of using the usual figures, cartoons may be introduced. The analyst gets here the inner feelings of an individual for this analysis purpose.

5. Third Person Test (TPT): The format of this test is that the respondent is given a photograph of third person — may be a friend, a colleague, a neighbour, a star, a player, a professional and the like. The point involved is that the researcher is interested in knowing what the third person thinks of an issue as heard through the respondent. It is assumed that the respondent's answer will reveal his own inner feelings more clearly through the third person than otherwise it would have been possible. The best example of this kind the test conducted on American house-wives in connection with 'instant coffee'. Prior to the test, the attitude of house-wives was "It does not taste good". With the test being conducted, the real attitude was "A lady using instant coffee is lazy, spend- thrift and not a good house-wife". This amply clears the fact how the test revealed the naked truth. In addition to these tests of usual type, the researchers do use other qualitative techniques such as role-playing, psycho-drama, graphalogy and the like.

II. DEPTH INTERVIEWS

Depth interview technique is significantly different to the study of consumer behaviour. In contrast to the regular type of interview, a highly trained person goes on asking questions to encourage the respondent or respondents to talk freely about the subject matter in question instead of administering a readymade questionnaire. He asks the questions to draw out the emotions, feelings and opinions of the respondents and he develops the pattern of questions in such a way that he gets the maximum information. The interviewer encourages the respondent to talk freely with open mind by maintaining as encouraging, friendly unjudiced attitude. The major strength of depth interview lies in the way it extends the interviewer's grasps of the consumer's attitude and the insight into them. Such an interview lasts for an hour to three hours. As long as depth interviews do not degenerate into games of verbal wit between the interviewer and the subjects or into "free association" sessions, they can yield useful information which the familiar questionnaire session is unlikely to uncover.

This technique results in getting at the real and often hidden reasons for the consumer actions than the usual announced answers. For example, the most announced reasons for taking evening bath may be "to keep clean or because it is healthy." In contrast, the depth interview may reveal that an evening bath might symbolise "the washing away of not only the dirt but of the frustrations, worries and disappointments of the day". These depth interviews may be individual or group.

'Individual' depth interview or 'one–to–one' interview is a technique under which only one respondent is interviewed at a time, for an hour to three hours by a trained researcher — may be a copy writer, brandman, ad director and the like. Such individual interviews have been of special significance as they provide authentic feed-back for developing new products, improving the existing products and developing new ways to talk to consumer through advertising. Individual depth interviews reveal true attitudes as there is no fear of social embarrassment and free to face exposure. Individual depth interviews are appropriate in six situations : 1. Detailed probing of an individual's behaviour, attitudes, or needs is squared 2. The subject matter under discussion is likely to be of a highly confidential nature. 3. The subject matter is of an emotionally charged or embarrassing nature. 4. Certain strong, socially acceptable norms exist say baby feeding and the need to conform in a group discussion may influence responses 5. Highly detailed step by step understanding of complicated behaviour or decision making patterns is required say, planning the family holiday 6. The interviews are with professional people or with people on the subject of their jobs say, finance, personnel directors.

'Group' or 'Focussed' group interviews, on the other hand, is another variation under which eight to ten persons or eight to fifteen persons are brought together in a relaxing surroundings to discuss marketing or advertising and the subjects introduced by the trained group leader or the moderator. The group leader knows that certain defined areas are to be covered in the course of one to three hours. The proceedings are recorded for future reference. These group sessions are better substitutes for face to face contact with the consumers. These usually are employed by advertisers, mostly national advertisers, to know their impressions about new product, brands and the ad message.

Focus group interviews can be best applied to: 1. basic need studies for product ideas creation 2.new product idea or concept exploration 3. product positioning studies 4. advertising and communication research 5. background studies on consumer's frames of reference. 6. establishment of consumer vocabulary as preliminary step in questionnaire development and 7. determination of attitudes and behaviours.

One is to be aware of the draw backs of group sessions that the findings are not projectable. The participants may be too shy to open up and talk frankly about the subject and the responses may be over-shadowed by a domineering member of a group.

The standard focus group interview involves 8 to 12 individuals. Normally, the group is designed to reflect the characteristics of a particular market segment. The respondents are selected according to the relevant sampling plan to meet at a central location that generally has facilities for taping or Filming the interviews. The discussion itself led by a moderator. The competent moderator attempts to develop three clear stages in one-to-three hour interview: 1. establish rapport with the group, structure the rules of group interaction and set objectives 2. attempt to provoke intense discussion in the relevant areas and 3. attempt to summarise the groups responses to determine the extent of agreement. Generally, either the moderator or a second person prepares a summary of each session, after analysing the session's transcript.

III. ATTITUDE SCALES

Consumer attitudes are measured on specialized scales. These scales differ—as "non-comparative" and "comparative". Non-comparative rating scales require the consumer to evaluate an object or an attribute of the subject without directly comparing it to another object. It can be four point or Five point scale comparison. The following are the two Fine examples of this kind.

A. How do you like the taste of 'Mangola' ?

Like it Very much	Like it	Dislike it	Strongly dislike it
☐	☐	☐	☐

B. How do you like the taste of "Colgate Gel" to "Colgate Total"?

Like it much more	Like it more	Like it about the same	Like it less	Like it much
☐	☐	☐	☐	☐

In case of comparative rating scales such as paired comparisons involve presenting the consumer two objects say brands or packages at a time and requiring the selection of one of the to according to some criterion as overall preference, taste, or colour. Rank order scales require the consumer to rank a set of brands ads, or features in terms of overall preference, taste or importance. The constant sum scale is similar except it also requires the respondent to allocate 100 points to allocate among the objects. The allocation is to be done in a manner that reflects the relative preference or importance assigned to each object. Following is an illustrative structure of this kind.

TATA TAJA TEA			**TAJMAHAL TEA**		
1.	Colour	....	1.	Colour	....
2.	Aroma	....	2.	Aroma	....
3.	Taste	....	3.	Taste	....
4.	Density	....	4.	Density	....
5.	Package design	....	5.	Package design	....
	Points	100		Points	100

The same thing can be done by using semantic differential scale.

The best example can be presented as under:

FAST .. .. SLOW	FAST ... ... SLOW
BAD ... ... GOOD	BAD ... ... GOOD
LARGE ... ... SMALL	LARGE ... ... SMALL
HEAVY ... ... LIGHT	HEAVY ... ... LIGHT
INEXPENSIVE EXPENSIVE	INEXPENSIVE...............................EXPENSIVE

NOTE: The space left in between represent:

– Extremely – Very – Some what – neither nor

– Some what – very –extremely

In the above case, the consumer ranks **Hero Honda** as extremely fast. Bad-some what, large-neither nor, heavy-some what, Inexpensive- extremely. While in case of **Bajaj Chetak**, Slow very, Good- (extremely, Expensive-some what.)

According to Likert scales, the consumer is asked to indicate a degree of agreement or disagreement with each of a series of statements related to the attitude object as :

I. Apana Bazar is one of the most attractive stores in town:

Strongly Agree	Agree	Neither Agree nor Disagree	Disagree	Strongly Disagree
☐	☐	☐	☐	☐

II. The service at Apana Bazar is not satisfactory:

Strongly Agree	Agree	Neither Agree nor Disagree	Disagree	Strongly Disagree
☐	☐	☐	☐	☐

III. The service at Retail shop is very important to me.

Strongly Agree	Agree	Neither Agree nor Disagree	Disagree	Strongly Disagree
☐	☐	☐	☐	☐

In order to analyse responses to Likert's scale, each response category is assigned a numerical value. In this example each expected category has == Strongly Agree = 5; Agree = 4; Neither Agree nor Disagree 3; Disagree = 2 and Strongly disagree 1. These can be in reverse order also.

IV. OBSERVATION

Observation can be 'personal' or 'mechanical' of the consumer reactions at a retail out let, or social get together and so on. Observation can be fruitfully used when (1) the behaviour of interest are public (2) they are repetitive, frequent or predictable and 3. they cover relatively brief time of span. An observation study needs five decisions on the part of researcher: (A) Natural versus contrived situation: The question is do we wait for a behaviour to occur in its natural environment or do we create an artificial situation in which it will occur? (B) Open versus Disguised observation. That is to what extent the consumers are aware that some one is observing their behaviour ? (C) structured versus unstructured observation: Are we limiting our observations to predetermined behaviour or are we to note whatever that occurs ? (D) Direct or Indirect observation: Are we cmna to observe the behaviours themselves or merely the outcomes of the behaviour? (E) Human or mechanical observation: outcomes of the observations be made mechanically or by selected and experienced persons ?

V. PHYSIOLOGICAL MEASURES

Physiological measures are **direct observations** of physical responses to a stimulus such as an advertisement. These responses may be controllable like eye movements or uncontrollable like the galvanic skin response. One such case of *direct measurement* is that of attention.

1. Eye pupil dialation: Changes in the size of pupil of the eye appear to be related to the amount of attention that a person is giving a message. A pupilometer can measure these changes very accurately.

2. Eye Tracking: An eye camera can track movements of the eyes relative to the advertisement red and watched. The paths of the eyes can then be mapped to determine (a) What parts of the message were attended to, (b) What sequence was used in viewing the message and (c) how much time was spent on each part.

3. Tachistoscopic Test: A tachistoscope is a slide projector with adjustable projector speeds and levels of illumination. Thus, the advertisement can be shown very rapidly and/ or dimly. Ads are tested to determine at what speeds the elements, such as product, brand and headline, are recognised. Speed of recognisation of various elements in the ads and readership that is - attention are highly correlated.

4. Theatre Tests: Theatre tests involve showing commercials along with television shows in a theatre : Some, such as the one maintained by ASI Market Research, have dials at each seat which viewers use constantly indicate their interest – attention—in the show or a commercial.

5. Brain Wave Analysis: There is some evidence that electro–encephalographs can indicate the amount and type of attention given to an advertisement or package.

Of all these techniques, the eye–tracking appears to offer greatest potential as opined by experts.

The **indirect tests** of attention include theatre tests, day-after call, recognition tests and starch scores. Let us touch upon these in brief.

1. Theatre Tests: These tests involve showing commercials along with television programmes in theatre. Viewers complete questionnaires designed to measure which commercial and what aspects of those commercials attracted their attention.

2. Day-after Call: This is popularly called as DAR. It is the most popular method of measuring the attention getting power of television commercials. The individuals are interviewed the day after a commercial is telecasted on a programme they watched. Racall of the commercial and recall of specific aspects of the commercial are interpreted as a reflexion of the amount of attention.

3. Recognition Tests: Recognition tests are tests in which the commercial of interest, or key part of it along with other commercials are shown to target market members. Recognition of commercial, or key parts of the commercial is the measure. This technique appears to work better than standard recall measures.

4. Starch Scores: These are the most popular technique for evaluating the attention attracting power of print advertisements. The respondents are shown advertisements from magzine issues they have recently read. For each ad., they indicate which parts-say headlines, illustrations, copy blocks, slogans-they recall reading. Three main 'scores' are computed namely: (A) Noted: The per cent who recall seeing the ad in that issue (B) Seen associated: The per cent who recall reading a part of the ad that clearly identifies the brand or advertiser and (C) Read most : The per cent who recall reading 50 per cent or more of the copy.

Starch scores allow an indirect measure of attention to the overall ad and to key components of advertisement.

STRENGTHS AND WEAKNESSES OF MOTIVATION RESEARCH

Strengths: Motivation research had made certain positive impressions in the field of marketing. In particular, it has encouraged marketers to move away from demographic analysis only to go beyond income, occupation and education. The question of why consumers behave as they do, once raised, was not be dismissed, and a richer conceptual frame-work came into use as researchers and businessmen discovered the utility of other theoretical approaches. It has become obvious that in order to understand the 'why' of consumer actions, the question of 'how' consumers behave had to be answered first and that motivation theory had limited explanatory scope. Consumer behaviour as a major marketing topic grew out of the demise of motivation research. Unfortunately, motivation research has not received much attention in recent marketing literature. However, there is considerable existing opportunity for the application of motivation concepts to research in consumer behaviour. These concepts take us further far beyond towards understanding and predicting consumer response.

Weaknesses: Motivation, research, though provides a valuable starting place for more traditional consumer research, it suffers from some serious limitations. **First** motivation research studies, because of the intensive methods used, require longer time duration ranging from one hour to three hours. As a result, the sample sizes were most often small—frequently less than hundred people—and the statistical analysis could not be conducted with any confidence. The use of depth-interviews and psycho-analytical techniques was also open to question—because, the explanatory and predictive power of such techniques is still a major issues of contention among the psychologists. **Secondly,** another specific difficulty of motivation research is that results are to be analysed only by trained psychologists and these psychologists do not agree in interpretation. The businessman who is riven the research findings is at the mercy of the researcher, in a sense, because, he can not interpret the raw data himself.

In conclusion, it can be said that though it is costly in terms of expertise and time, it pays to invest. It is a must in case where marketing problems are wrapped more in emotional involvement, where conventional techniques fail. Due price is to be given for better results on which the marketing manager is to work. Motivation research is comparatively a new tool. Whether it is just a fad or a valuable new and sharp tool of marketing management, only time factor is to decide. With the growth in this youngest branch of science, this motivation research is sure to make a permanent settlement in the realm of marketing research.

CHAPTER BASED QUESTIONS

A. WRITE SHORT NOTES ON:

1. Consumer needs 2. Classification of needs 3. Motivational influences given by Me. Gurie 4. Nature of motivation 5.Motivation arousal 6. Frustration?. Systems of needs 8. Motivational research 9. Techniques of motivational research 10. Measurement of motivation.

B. SHORT ANSWER QUESTIONS:

1. What are needs?
2. How do you classify human needs?
3. What is the nature of motivation?
4. What is frustration?
5. What is measurement of motivation?
6. What is motivation research?
7. What are projective techniques?
8. How do you evaluate need hierarchy?
9. What are attitude scales?
10. How do you measure attitude?

C. ESSAY TYPE QUESTIONS :

1. What do you mean by human needs? Are they different from motives?
2. What are human needs? How do you classify them?
3. What is motivation? How it can be measured?
4. What is need of hierarchy system? What are its merits and limits?
5. What is frustration? What defences are open to fight frustration?
6. What is arousal of motives? How one can arouse human motives?
7. Explain clearly the five level hierarchy needs with example and diagram.
8. What are the strengths and weaknesses of motivational research?
9. Explain the concept of ever changing needs.
10. Explain the motivational influences identified by Mc. Gurie.

******

CHAPTER 4

PERSONALITY AND CONSUMER BEHAVIOUR

BACKDROP

Wise marketing men have long understood the importance of personality cult in individuals which is quite strong among men and women. Individual uniqueness is the heart of the study of consumer behaviour and personality cannot be separated from individual consumers. Personality and self-concepts are so vibrant that marketers can really appeal to their strong sense and proud feeling of their personality and self-concept. In view of this, this chapter plans to discuss meaning and nature of personality - theories of personality, personality and market segmentation deeper implications, meaning and application of psychographic analysis. Meaning of self-concept self-concept and consumer behaviour - marketing applications -Chapter ends with Chapter Based Questions.

WHAT IS PERSONALTY?

We people very often talk of the word **personality;** we use the word very commonly. The term 'personality' is very elusive and difficult to define and the concept differs from individual to individual, situation to situation and time to time. Personality is the supreme realisation of the innate individuality of a particular living-being. Personality of a person is what a fragance is to a flower. Personality is the sum–total of all those elements which attract others towards that individual or person.

STANDARD DEFINITIONS

★ **Professors Leon G. Schiffman** and **Leslie Lazar Kaunk:**

"Personality stands for "those inner psychological characteristics that both determine and reflect how a person responds to his or her environments".

★ **Professors David L. Loudan and Albert J. Della Bitta:**

" behaviour of an individual will be organised into a coherent pattern which has consistency throughout various situations".

★ **Professors Del I.Hawkins, Roger J. Best and Kenneth A. Coney:**

" Personality is individual's characteristic response tendencies across similar situations."

★ **Professor Gordon W. Allport:**

" Personality is the dynamic organisation within an individual of those psychological systems that determine his unique adjustment to environment."

★ **Mr. Morgan and King:**

" Personality is the characteristics and distinctive trait of an individual, the stable and shifting pattern of relationship between these traits and the way the traits interact to help or hinder the adjustment of a person to other people and situations."

Very often we say that a friend of mine has a personality which is very aggressive, competitive, out–spoken, witty and ambititious. These are, in–fact, characteristics, art how they respond to a wide range of situations which definitely differ from person to person basically. Let us take a simple case; there is a variety of fish which lays thousands of eggs at a time yet it keeps very cool; conversely hen lays one egg at a time and makes big noise as if it is an achievement of unique type. This speaks of response to situation of laying eggs. One is cool and calm and another is overt and publicizing.

In that sense every person has a personality made up of psychological and physical characteristic features or traits. By personality, one should not go only by internal or external but both integrated into one. Though behavioural scientists have given wide variety of definitions and as such are not unanimous of what is a personality. That is, there cannot be one standard definition for the word personality. Though experts differ as to unanimity, they do agree on three major aspects of similarity among the divergent definitions. These three similarities are: 1. All focus on unique characteristics that account for difference between individuals rather than how people are alike. 2. All of them stress the consistency of an individual's dispositions rather than changes in his or her actual behaviour across different situations and 3. All definitions do not exclude behavioural tendency to reflect as to how an individual's personality will tend to influence his or her actions or reactions to environmental situations.

In clear and simple words, personality is something that represents the reactions of an individual's moods, values, attitudes, motives and habitual ways of responding to a situation or situations. The thrust is on inner characteristics — describable in terms of qualities, attributes, traits, factors, innate gifts, mannerisms which clearly distinguish, one individual from another individual's which help him or her to influence others or have a sway on others or attract others towards him or her.

THE NATURE OF PERSONALITY

Personality, as we have conceived for the purpose of consumer behaviour study has at least three properties which spell out its exact nature. These are : 1. Individual differences 2. Consistency and endurance and 3. Possibility of change.

1. Individual Differences: Personality reflects individual differences. It is because, the inner characteristics that make up an individual's personality are unique and they represent highly divergent combination of factors. That is why, no two individuals are exactly alike in all respects, even if they are twins or quadruplets. Nevertheless, many individuals tend to be similar in terms of a single personality characteristics. For example, many people can be described as 'high' in sociability. That is, these persons show higher degree of interest in social or group activities. It is equally true that there are some people who want to be away from social or group activities whom we can call as low in sociability. Some experts have thought of broadly four types of personalities—such as 'Sanguine', 'caustic', 'melancholic' and 'cool'.

Personality is a very useful concept because, it enables one to categorise consumers into different groups on the basis of a single trait or a few traits together. If each person were different in all respects, it would be almost impossible to group these persons or consumers into segments, and there would be hardly any reason to develop standardized products and, therefore, promotional campaigns.

2. Personality is Consistent and Enduring: An individual's personality is commonly fund oth consistent and enduring. Many a times, we hear people taking about their children of these features. A mother may say "my son has been stubborn from the day he was born." Another father might comment "my daughter has been cool and smiling since her birth." These statements support that an individual's personality has both consistency and endurance. The stable nature of personality projects that it is unreasonable for anybody to convince them. What is important is an individual's personality may be consistent, but his behaviour changes because of psychological, socio-cultural and environmental factors. Coming to endurance, it speaks of the taking or everlasting features. Though a person's personality is enduring, he has to bend than to break.

Both the qualities are essential for marketers to explain or predict consumer behaviour in terms of personality. The stable nature of personality suggests that its not reasonable for marketers to try to change the consumer's personalities to conform to certain products. At the best, they might learn which personality characteristics influence specific consumer responses and attempt to relevant traits in their target group of consumers. In the light of the fact that personality is though consistent, the consumption behaviour very often varies due to psychological, socio-cultural and environmental forcus, that affect behaviour, an individual is specific needs or motives, attitudes, reactions to group preserves, and even responses to newly available brands may cause a change in person's behaviour. It is worth noting here that personality is only one of the combination of factors that influence how a consumer behaves.

3. **Personality can change:** There are no two opinions about the nature of personality that its consistent and enduring. However, it may change under certain circumstances. The reasons for such change may be major events in life, say birth of a child, the death of a dear and near one's a divorce, a major career promotion, compulsory retirement, accident, chronic disease and so on. An individuals's personality changes not only is response to abrupt events, but also as a part of gradual maturity or aging process. Thus, a person enjoying 'Sanguine' type personality might turn "melancholic'. This is also a lesson for the marketers to turn and try to appeal to his or her personality change – over for the mutual benefit.

THEORIES OF PERSONALITY

There is no shortage of theories explaining the concept of personality. Each has his own way of profounding the basic concept and its mechanism. However, at some selected theories can not escape our attention as they are important and widely accepted all over the world. These are: 1. Freudian 2. Jungian 3. Neo-Freudian and 4.Trait. Let us know the implications of these selected theories for discussion that explain concept of personality.

1. FREUDIAN THEORY OF PERSONALITY

Sigmond Freud's Psychoanalytic theory of personality is the main-stay of modern psychology. This psychologist built the theory on the premise that unconscious needs or drives, especially sexual and other biological drives, are at the heart of human motivation and personality. **Mr. S. Freud** developed his theory on the basis of patients' recollections of early childhood experiences, analysis of their dreams, and the specific nature of their mental and physical adjustment problems.

Interacting systems of Personality

Mr. S. Freud proposed, on the basis of analysis, that the human personality consists of three interacting systems namely, 'id'—'the superego'—and the 'ego'. In other words, every individual's personality the product of a struggle among the three interacting forces as said above. According to **S. Freud** the 'id' is the source of strong inborn drives and urges such as aggression and sex. In other words, 'id' was conceptualised as a 'warehouse' of primitive and impulsive drives-basic physiological needs such as thirst, hunger, sex and shelter — for which the individual seeks immediate satisfaction without concern for the specific means of satisfaction. The 'id' operates on what is called as the "pleasure principle". That is, it acts to avoid tension and seeks immediate pleasure. However, it tends to operate at a very subjective and unconscious level and is not fully capable of dealing with objective reality. What is more important is that many of its impulses are not acceptable to the value system operating in the society. Thus, a person who is very thirsty because of excessive heat, his 'id' would urge him to grab at any rate by any means to have a cold drink. He is not bothered as to what he is doing is right or wrong. He may, beg, borrow or steal and in extreme cases, he might be so aggressive and angry that his behaviour might lead to criminal activities.

In contrast to 'id', the 'super–ego' is conceptualised as the individual's internal expression of society's moral and ethical codes of conduct. The role of 'super-ego' is to see that the individual satisfies his needs in a socially acceptable fashion. In other works, it words as a 'brake' that restraints or inhibits the impulsive forces behind the 'id'. It contributes the moral part of the individual's psychic structure through internalising the values of society. When a friend of yours buys a brand new car that too a luxury car, 'id' actually fosters feelings of jealousy. Because of 'super–ego' this 'ill-felling' is cut and makes persons to compliment or congratulate on his buying a new car, as it is within the social norms.

Coming to 'ego' it represents individual's conscious control. It works as interval monitor that attempts to balance the impulsive demands of the 'id' and socio-cultural contraits of the 'super ego'. The following rectangle explains these three forces as to how they work in determining human behaviour as part of personality.

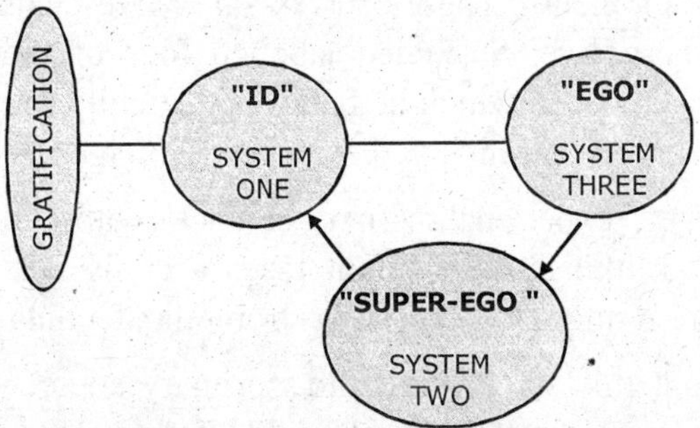

FIG: 4:02 INTER-RELATIONSHIPS BETWEEN ID ; EGO AND SUPER-EGO

The 'ego' comes into being because of the limitations of the 'id' in dealing with the real world. The 'ego' develops the individual's capabilities of realistic thinking and ability to deal appropriately with his environment by learning and experience. It operates on what is known as the "Reality Principle", which is capable of post–poning the release of tension until that time when it will be effectively directed at coping with the external environment. In the above case of a thirsty person, who grabs a soft-drink 'id' did not worry of values or morality ; here, the 'ego' reasons as to why not ask for, instead of grabing. That is why, 'ego' is called as the executive of the personality, as it serves as a focal point for effective action in an environment.

According to **S. Freud,** the individual's total personality develops and is defined by the relationships among these three forces of 'id'-'ego' and 'super-ego'. The 'ego' serves as an administrator of the interaction between moral standards of super-ego and the socially acceptable desires and attempted expressions of the 'id'. This generally results in realistic compromises between very basic strivings and accepted behaviour. Most of these compromises are said to be accomplished at the unconscious level. Freudian Psychology, in fact, argues that a vast portion of our behaviour is unconsciously motivated or affected by conscious factors that only occasionally reach the individual's conscious level of awareness. Therefore, one must appreciate what factors are influencing the consumer at unconscious and sub–conscious levels, to understand fully the causes of behaviour and the interactions of personality.

Though 'ego' is quite capable of resolving many of the conflicts that arise between the three personality forces, on some occasions no resolution is reached, and the individual is placed under good deal of tension. It is here that defence mechanisms are used to fight tension or frustration. These defence mechanisms can be repression, projection, identification and lastly reaction formation is which unconscious feelings held towards others are consciously expressed as opposites. For instance, you may have hostile feelings about me but they are turned to sweet feelings.

STAGES OF PERSONALITY DEVELOPMENT

In addition to explaining the three interacting forces or systems, **S. Freud** emphasized that an individual's personality is formed as he or she passes through a number of stages of infancy and childhood development. These stages are five namely, **Oral, anal, phallic, latency and genital**. Of these five, **S. Frued** labelled four of these stages of development of conform to the area of the body on which he believes that child's sexual instincts are focussed at the time. Let us see what he meant by these specific stages of personality development.

1. Oral Stage: An infant experiences first contact with the outside world through mouth—eating, sucking, drinking. The infant faces a crisis at the end of the stage when he or she is kept away from mother's breast or from the feeding–bottle.

2. Anal Stage: Throughout this stage, the child's primary source of pleasure is the process of elimination. A second crisis develops at the end of this stage when parents try to train the child to attempt toilet activities.

3. Phallic Stage: The child experiences self–oriented. Sexual pleasure during this phase with the discovers of sex-organs. A third crisis occurs when the child experiences sexual desire for the parent of opposite sex. The question is how the child resolves this crisis because it affects later relationships with persons of the opposite–sex and with authority figures.

4. Latency Stage: Mr. S. Freud believed that the sexual instincts of child lie dormant from the age of five until the beginning of adolescence and that no important personality changes occur during this period.

5. Genital Stage: At adolescence, the individual develops a sexual interest in persons of opposite–sex, beyond. Self- oriented love and love for parents.

An adult's personality is determined, according to **Mr. S. Freud,** by how well he or she deals with the crises that are experienced while passing through each of these stages especially the first three. For example, in case a child's oral needs are not satisfied adequately during the First stage of development, the person may become capable of fixing or arranging at this stage and as an adult display a personality that comprises of such traits as dependence or excessive oral activity like gum-chewing and smoking. When an individual is fixated or capable of fixing at the anal stage, the adult personality may display other traits such as excessive need for neatness and cleanliness.

1. APPLICATION OF FREUDIAN THEORY

Marketers have used successfully the psychoanalytic theory in a number of ways. These are described as under :

1. **Product Personality and Freudian Theory** The researcher streses the idea that human drives are largely unconscious and that the consumers primarily anaware of their reasons, particularly, true, for buying what they want to buy. The researchers and, therefore, marketers tend to focus on consumer purchases and/or consumption situations, treating them as a reflection and extension of consumer's own personality. Put in other words, ones appearance and possessions—how well is one groomed ? What one wears, carries and displays are taken to reflect the individual's personality.

2. **Brand Personality and Freudian Theory:** It is found that consumers tend to ascribe various descriptive "personality–like" traits or characteristics, to different brands in a wider variety of product categories, the ingredients of brand personalities. Consumers not only ascribe personality traits to products and services, but also they tend to associate personality factors with specific colours. In some cases, various products, even brands, associate specific colour with personality such as connotation. For example, 'Coca Cola' is associated with red, which connotes 'excitement' ; in case of 'Mangola' yellow is associated with 'novelty' and in case of 'kali-cola' black connotes 'sophistication'. For this reason, brands wishing to create a sophisticated personal or an up-scale premium image, use labelling or packaging that are primarily black. To go still further, a combination of black and white has the effective power of communicating that a product is carefully engineered, high–technology, and sophisticated

in design. The case of **IBM** has used all black cases with a few selected red–buttons and bars to pack its products. Each odour and colour combinations speak of masculine and femine products and services. These differ from country to country.

Thus, Chinese consumers to possess a definite gender, products and services, they perceive as under. 'Masculine' products and services—beer, car, cigarette, coffees, credit card, hair–cut, legal services, long–distance tele–phone, restaurant, scotch, sneakers, tennis–raquet, took paste and wine, while "Feminine" products and services are bathing–soap, clothes–dryer, day care, dish-washing liquid, electric–iron, facial-tissue, hair-spray, shampoo, washer and so on. Talking of colours associated with personality link having definite marketing implications thus, blue commands respect and authority and marketing insights are most favoured American colours, **IBM** holds the title to blue ; associated with Club soda ; men seek products packed in blue houses painted blue are avoided ; low–calorie, skimmed–milk, coffee in blue can, perceived as 'milo'. Coming to "yellow colour', it marks caution; novelty; temporary and warmth. The marketing implication are— eyes register it fastes, coffee in yellow can taste 'weak'; stops traffic, sells a house. In case of "green colours" it stands for secure; natural; relaxed or easy going; living things. The marketing implications are; good boost, environments; associated with vegetables and chewing gum; 'Canada Dry' Soft drink sales increased when it changed sugar free package from red to green and white. Coming to "red colour", it speaks of human, exciting, hot, passionate and strong. The marketing insights are: makes food 'smell' better; coffee or a red can perceive as 'rich' ; women have preference for bluish red ; men have a preference for yellowish red ; **Coca cola** 'owns' red. In case of "Orange Colour", it projects being powerful affordable and informal. The marketing insight is that it draws attention quickly. In case of "brown colour", it speaks of informal and relaxed, masculine, nature. The marketing insights are coffee in a dark brown can is perceived as "too strong" ; men seek the products packaged in brown. Turning to "white colour", it represents goodness, purity, chastity, cleanliness, delicacy, refinement and formality. The marketing insights are : Suggests reduced calories : Pure and wholesome food : Clean, bath products, femine. In case of "black colour", it speaks of sophistication, power, authority, mystery. The marketing insights are: Powerful, clothing and hightech electronics. Coming to silver, gold and platinum colours, they speak of regal, wealthy, stately. The marketing insight is that each colour suggests premium price.

3. Influencing Consumer's appeals: Freudian psychologic or psychoanalytic theory as a basis for attempts is influencing consumers. One such application is the appeal to fantasy which plays an important role in the operation of the pleasure principle namely, the 'id'. Fantasy has been used in promotions of products such as perfumes, shoes and jeans. These products have definite brands say 'Channel-5' or 'Fantacy', Nike, or Puma ; Jeans-say 'Levis', 'Wrangler', respectively.

4. Resolving the Conflicts: Freudian psycho proponents suggest that various appeals assist in resolving conflict which can develop between 'id', 'ego' and 'super-ego' is some purchase situations. Appeals directed to the 'id', but disguished by a veiled appeal to the super ego, are said to result in a situation which can be satisfactorily resolved by 'ego'. That is, 'ego' directs behaviour that is also acceptable to the 'id' and 'super-ego'. The idea has been translated into ad themes such as those of inner wears of women like bras, panties, corrects, tights, swim wear and the like–say 'Lovable' brand. The ad illustration may be a young woman shown above the waist wearing only the bra in the intimate company of a young man. However, to avoid wrong interpretations, wedding rings that appear the wedding bands are prominently displayed on the left hand of both young woman and man to satisfy moral concerns of 'super-ego'. The 'ego' then reasons that sex is socially acceptable behaviour under conditions of marriage, and, therefore, the sexual allusive hint of the ad is acceptable. This approach is called as "triple appeal" because 'id' is sexually stimulated and allowed to engage in fantasy while the moral requirements of the 'super ego' are conciliated or placketed. The 'ego' attributes the acceptability of the two forms of expression between both 'id' and 'super-ego'. Though most people take for granted that Freudian applications to marketing are restricter sex, it is wrong. Therefore, good many themes which deal with sex appeal and fantacy. These are to illustrated to deal with wish fulfilment, aggression, escape from life pressures. Thus, in case of city-dewellers, the ad may appeal to them to be free from the tensions of city–life and settle away from city–areas. This is done by builders and contractors. In case of sport events, the aggression is rejuvenated such as car-racing, bike-racing, hockey, boxing and wrestling. The tourism industry speaks of a break—a grand relief—by using 'home away from home' pinpointing places like Simla, Darjeeling, Kulu-Mrainali, Kodai-Canal, Ooty in India.

5. Defence Mechanism–the base for promotional strategies : An understanding of different defence mechanism by the marketing people, will go a long way in designing marketing and promotional strategies. In each case of defence, mechanisms used to fight tensions and, sometimes, frustrations, the ad appeal, heals and projects a reconciliation of conflicts. Taking identification as a defence mechanism, many men and women experience conflict arising from hostile feeling they hold towards their spouses for whom they believe they should have strong affection. In case a well-known personality who has had martial problems and has publicity reconciled with his or her spouse, can be a very strong identification for these people.

2. JUNGIAN PERSONALITY THEORY

Mr. Carl Jung was a colleague and contemporary of **Mr. S. Freud.** The modern psychology consists of good deal of contributions and his theories and insights pertaining to personality types have been made. Particularly very useful for marketers by the **Myers-Briggs** Type indicators—a personality inventory—that measures the following pairs of Jungian- inspired psychological dimensions. These dimensions are four namely, 1. Sensing-intuting 2. Thinking-feeling 3. Extroversion- introversion and 4.Indulging – perceiving. Each of these four pairs of dimensions reflect two distinctly different personality features or characteristics that project a picture as to how consumers respond to the surrounding world.

The attempt to match each psychological dimension from all pairs with each psychological dimension in all other pairs would result in too many different combinations of personality types of productive study. In order to avoid this extremely complex situation, a consumer researcher recently investigated the consumption relevance of just two pairs of dimensions namely, sensing-intuting and thinking-feeling.

The following square gives the summary characteristics of selected Jungian personality types.

SENSING- THINKING (ST): ★ Rational in decision–making ★ Logical and empirical in view points ★ makes decisions following an 'objective' orientation ★ Heavily weighs economic consideration—most price sensitive ★ Will extend considerable effort to search for decision making information ★ Risk avoider ★ Materialism reflects personal or private motives ★ Short-term horizon in making decisions.

SENSING-FEELING (SF) : ★ Emperical view point ★ Propelled by personal values rather than logic ★ Makes decisions following a 'subjective' orientations ★ Likely to consider others when making a decision ★ Shares risks with others ★ Materialism reflects how objective will impact on others ★ Short-time horizon in making decisions.

INTUTING-THINKING (NT) : ★Takes a broad view of personal situation or world. ★ Relies heavily in imagination, yet uses logic in approaching decision. ★ Imagines a wider range of options in making a decision ★ Weighs options mentally ★ Willing to take risk or be speculative in decisions ★ Long time horizon in making decisions.

INTUTING-FEEZING (NF) : Takes a broad view of personal situation or world. ★ Imagines a wide range of options in making a decision ★ Highly "people oriented-likely to consider others views. ★ Makes decisions following a subjective orientation. ★ Least price sensitive ★ Risk seeking-venture and novelty seeking. ★ Indefinite time horizon in making decisions.

SOURCE: Adapted from **Stephen J. Gould, "Jungian Analysis and** Psychological Types : An Interpretative Approach to Consumer Behaviour" Edited by **R.H. Holman and M.R. Soloman.**

The sensing (S) intuting (N) dimensions capture how consumers find out-about 'things'—obtaining and processing the information—and the thinking (T) and feeling (F) dimension are opposite ways of making decisions or decision making styles. By cross-tabulating the specific characteristics of the two pairs of psychological dimension; four possible types are derived as under :

The rectangle showing selected Jungian personality types :

	THINKING (G)	FEELING(F)
SENSING (S)	Sensing-thinking (ST)	Seeing-Feeling(SF)
INTUTING (N)	Intuting-thinking (NT)	Intuting-feeling (NF)

Using psychological inventory, such as the **Myers-Briggs Type** Indicators or by creating an independent consumer specific personality inventory, it is possible to learn how these four personality types impact on consumer informal processing, consumers decision making, and other consumption related issues—personality—related insights that would enable marketers to better satisfy consumer needs. Let us take a solid case of four individuals facing the investment decision to purchase shares of reputed companies—corresponding to the four Jungian personality types. The ST and SF types—that is, sensing types—would tend to continuously study balance sheets and financial performance to secure their facts. However, the ST— sensing thinking type-consumers would arrive at their own decision and might eventually use discount broker. In contrast, the SF-sensing feeling type—consumers would want to include the recommendations of others as part of the decision and would be more likely to use a full-service brokerage house. The NT-intuting thinking—and NF—Intuting- feeling-types of consumers would be more likely to make their stock selections intuitively or in terms of a lunch. The NT-intuting-thinking types consumers would rely heavily on their own imaginations, yet would carefully think of all options open. Although NF- intuting and feeling type consumers would also rely on their imagination while considering the options open; the final decision might be expected to reflect inputs from others in any way say, comments overheard in some social get together say a cocktail party or high tea party.

It is most likely that shrewd advertising and marketing executives are sure to use some of these Jungian personality types intuitively in creating consumer targeted messages in marketing of investment proposals—in shares, bonds, real–estates and even other forms of investments.

3. NEO-FREUDIAN PERSONALITY THEORY

Those persons who did not believe in 'id-based' theory of personality argued that an individual develops a personality through countless attempts to deal with others in social setting. These social theorists, are collectively called as the "Neo-Freudian School", viewed individuals as striving to overcome feelings of inferiority and searching for ways to obtain love, security and brotherhood. Their argument minimised the roles of 'id-based' instincts that **Freud** emphasized instead, they stressed that childhood experiences is relation to others produce the feelings of inferiority, insecurity and lack of love. These feelings motivate individuals to perfect themselves and also to develop methods to cope up with anxieties produced by such feelings of inferiority. In other words, several Freud's colleagues disagreed with his contention that the personality is primarily instinctual and sexual in nature. Instead, these neo- Freudians believed that social relationships are fundamental to the formation and development of personality. **Mr. Alferd Adler** viewed human-beings as seeking to attain various rational goals which he called 'life–style'. He also placed much emphasis on the individual's efforts to overcome feelings of inferiority. That is, they strive for superiority. Another noted neo-Freudian **Mr. Harry S. Sullivan** stressed that people attempt continuously to establish significant and rewarding relationships with others. He was more concerned, unlike others, with the individuals efforts to reduce tensions such as anxiety. The first major consumer behaviour study using a neo- Freudian approach was based on the theoretical scheme of **Karen Horney.** Like **Sullivan**, she focussed on the input of child- parent relationship, especially the individual's desire to conquer feelings of anxiety, she identifies the major needs which are acquired as a consequence of individuals attempting to find solutions to their problems in developing a personality and dealing with others in a social environment. These ten needs were then classified into three major orientations which describe general strategies for relating others. These orientation groups were nothing but personality groups. These groups have been :

(1) Compliant (2) Aggressive (3) Detached.

1. Compliant oriented personalities are those persons who move towards people and stress the need for love, approval, modesty and affection. These individuals tend to exhibit large amounts of empathy and humility, and are unselfish.

2. Aggressive oriented personalities are those persons who move against people and stress the need for power, strength and the ability to manipulate others. They desire to excel and win admiration.

3. Detached oriented persons are those persons who move away from people. These individuals stress the need for independence, freedom and self-reliance in their dealings with others. The significant consideration is that no strong emotional ties develop between themselves and others. These are the persons who want to be free from obligations.

A CAD-Compliant, Aggressive, Detached–instrument was developed on Horney's theory is used to measure people's interpersonal orientations or personalities within a consumer context. The outcome of the study indicated that different products and brands were used by individuals having different personality types. In case of 'Complaint' types they preferred known brand names and used more mouth wash and toilet soaps. 'Aggressive' types, they preferred to use razors instead of electric, shavers, used more Cologne and after–shave lotions and purchased 'Old Spice' brands and 'Van Heusen' brand of shirts. The 'detached' types appearant have the least aware of these brands. It was also found that these 'detached type of personalities appear to be less involved in purchasing than those of the earlier two types of personalities.

It is likely that many marketers have used some of these neo-Freudian theories intimately. For instance, marketers who position their products or services as providing an opportunity to belong or be appreciated by others in a group or social setting would seem to be guided by **K. Horney's** characteristics of a compliant individual and that of 'aggressive' type and attempts were made to touch 'detached' type where lies the greater potential. Inspite of this social theories have not been found of greater application to consumer behaviour area. Experts are of the opinion that additional research is very essential to refine the scales to generate a wider base of findings for which it is possible for the marketers to design marketing strategies.

4. TRAIT THEORY

The **Trait Theory** constitutes a major departure from the basically qualitative measures that typify the Freudian and Neo-Freudian movements namely, personal observation, self-reported experiences, dream analysis, projective techniques and so on. The orientation of trait theory is primarily quantitative or empirical which focusses the measurement of personality in terms of specific psychological characteristics, called traits.

WHAT IS TRAIT ?

A 'trait' is defined by **professor J.P. Guilford** in his book 'Personality' as "any distinguishing, relatively enduring way in which one individual differs from another." The concept of 'trait' is based upon three basic assumptions or propositions: (1) Individual possesses relatively stable behavioural tendencies (2) People differ in the degree to which they possess these tendencies and (3) When identified and measured, these relative differences between individuals are useful in characterising their personalities. Therefore, it is crystal clear that traits are general and are relatively stable personality characteristics which influences tendencies to behave.

The most popular personality concepts which are used to explain the behaviour of consumers have been trait and factor theories. Trait theories are concerned with the construction of personality tests or inventories that pinpoint the individual differences in terms of specific traits. The factor theories are based on the quantitative techniques of factor analysis, which explores the inter-relationship between various personality measures across a large number of individuals.

Selected single trait personality tests are increasingly being developed specifically for use in consumer behaviour studies. These tests meansure just one trait, such as say self-confidence-or tact. These tailor-made personality tests measure such traits as consumer innovativeness–that is, how receptive a person is -to new experiences, consumer susceptibility to inter-personal influence (SUSCEP) guages how consumers respond to social influence, consumer materialism that is assesses the degree of consumer's attachment to 'World' possessions and consumer ethnocentrism that is identifies consumer's likelihood to accept or reject foreign products.

APPLICATION OF TRAIT AND FACTOR THEORIES

Researchers have learnt, based on experience using personality inventories to understand consumption-related behaviour, that, it is generally more realistic to expect personality to be linked to how consumers make their choices and to the purchase or consumption of a broad product category than a specific brand. For example, there is more likely relationship between personality and whether or not an individual owns a convertible sports car, than between personality and the brand of convertible sports car purchased.

In terms of their use in studying consumers, the advantage of trait and factor theories is that they are based on a number of readily available and standardized personality inventories and evaluative techniques. Using these techniques, a large number of researchers have tried to find out a relationship between personality and the consumer behaviour. It is gratifying to note that these efforts have been met with greater degree of success. Let us take some selected cases. Ford and Chevrolet are owners personalities.

Researchers used in this case what is known as EPPS (Edwards Personal Preference Schedule) to know personalities associated with brand or products. According to the findings those who own 'Ford' cars attracted the personalities with variables like—"independent-impulsive—masculine—alert to change and self-confident." 'Chevrolet', on the other hand, showed variables like "Conservative–thrifty—prestige conscious—less masculine and seeking to avoid extremes." And other famous example is that of coffee—particularly 'instant'. The research study was conducted by interviewing house-wives. It was interesting to find that ladies given to answer that those "ladies who use instant coffee are very busy"; another answer was "those who use instant coffee are lazy." Though controversial, the real answers came with busy personalities are lazy personalities.

. Though personality studies did help in tracing the traits and their behaviour, the experts in the field are of the opinion that : (1) The personality tests have been inappropriately employed in consumer studies (2) Personality tests have not always been carefully administered whenever used in consumer studies. (3) Many studies have searched for relationship between personality and specific aspects of consumer behaviour like brand choice, brand loyality over a period of time, amount of product use (4) A sizeable number of studies have focussed on specific personality traits such as tolerance for ambiguity, rigidity, self-actualisation, need for affiliation and their relationship to certain types of behaviour. In view of these limitations, future research should be more carefully designed and must employ more relevant tests of consumer's personalities. In fact, one development in this regard is the use of personality concept as a part of a larger research package to understand consumers better as each consumer has a personality.

CONSUMER PSYCHOGRAPHICS

Psychographics is a recent approach to market segmentations which has emerged as a major alternative to the traditional approaches. In fundamental sense, it is not a very different approach. The word 'psychographics' has been coined by **professor E. Denby** who says that psychographics seeks responses to products—packaging, advertising and public relations efforts of a marketing unit. The psychographic characteristics include variables like personality, attitudes and life styles.

Already we all know about the concept of personality which stands for individual's consistent relations to the words that surrounds him or her. The personality variables, among other things, are dominance, aggressiveness, objectivity, achievement-motivation and the like. These influence the consumer behaviour.

When one talks of ‘attitude’–it stands for expressions of inner feelings that reflect whether the person is favourably or unfavourably pre-disposed to some 'object' say a brand, a service, or a retail shop. They are the outcome of psychological processes, and therefore, are not observable but must be inferred from what people say or what they do. Attitude in the context of consumer behaviour is a learnt predisposition to behave as consistently favourable or unfavourable way with respect to a given object. The practical implications for marketing strategies that result from understanding these factors.

Turning our attention to ‘life–styles' the term was coined for the first time by **Alfred Adier** some 60 years ago as "the goal a person shapes for himself and the ways he uses to reach it”. It is a unique pattern of living which influences and is reflected by one's consumption behaviour. The experts have classed these life–styles as 'Swingers', ‘Status–seekers’ and 'Plain Joes'. Swingers are those who seek upto date-latest goods and services ; they are hedomistic and are fast posed. 'Status-seekers' are after those goods and services that reflect and refine their status in the society or circles. 'Plain Joes' are those who seek ordinary, unfrilled goods and services that do their job. Marketers can design their marketing strategies accordingly by studying the types of life–styles.

SELF CONCEPT

Another closely related concept is 'self–or 'self–image' to that of personality that we have already studied. Consumer the Rex of marketing world has a number of long lasting images of his own or her own. As closely related concepts to personality, these 'self–images' or perceptions of ‘self’, speak of the individuals tendencies to buy products and services, patronising a selling outlet. 'Self-concept' or 'self-image' has become a popular approach in recent years to investigate the possible relationship between how individuals perceive themselves and what behaviour they exhibit as consumers.

WHAT IS SELF–IMAGE OR SELF–CONCEPT ?

The first person to come out with new concept is **Professor William James** who writes in his book THE PRINCIPLES OF PSYCHOLOGY-VOL. 1 as "The sum-total of what a man can call ‘his’—not only his body and his psychic powers, but his clothes and his house, his wife and children, his ancestors and friends, his reputations and workers, his lands and horses, and yatch and bank-account." In the words of **Theodore. M. Newcombe,** self–concept is "the individual as perceived by the individual in a socially determined frame of reference." **Mr. M.J. Sirgy** defines in his book, "Self concept is Consumer Behaviour" a Journal of Consumer Research, December 1982 as "Self–concept" can be defined "as the totality of the individual’s thoughts and feelings having reference to him or herself as an object." In alternative words, self-concept is composed of an individual he holds about himself or herself.

THE SELF IMAGE-ITS MAKE-UP

It is already noted that each individual has an image of him-self or her-self as a certain kind of person, with certain traits, habits, possessions, relationships and always behaving like other images and personality, the individual's self-image is unique, the out come of that person's background and experience. Individual's develop their self-images through interactions, with other people. To start, with, their parents and then other individuals or groups with whom they relate over the years.

The literature on consumer behaviour lacks no variety so far as self-images are related as developed by behaviorists. One such model developed and advocated by **Professor Elizabeth. C. Hirschman,** in her article" Toward a General Theory of Compulsive Consumption" —in Journal of Consumer. Research—19th September, 1992., projects four specific kinds of self-images; (1) Actual self-image—how consumers in fact themselves ? (2) Ideal self-image—how consumer would like to see themselves? (3) Social self-image—how consumers feel other see them? and (4) Ideal social self–image how consumers would like others to see them. Some others researchers have taken up one more type namely: Expected self-image—how consumers expect to see themselves at some specified future time ? In fact, the expected self-concept or image is somewhere between the actual and ideal self–images. It is some what like a future oriented combine of "what is" (the actual self-image) and what consumers would like "to be" (the ideal self-image). Further, expected self–image is likely to be more valuable to marketers than the 'actual' or 'ideal' self–image as a guide for designing and promoting the products because, the exputed self image provides the consumers with a realistic 'opportunity' to change the 'self.'

HOW THE CONCEPT OF SELF-IMAGE DEVELOPS ?

The behaviourists have designed variety of theories as to how people develop their self-images or concepts. The basis for most of these theories as noted earlier is social interaction. Here, an attempt is made to touch four such theories.

1. **Self-Appraisal:** Some of the theorists strongly believe that a person fashions a self-concept by labelling his own dominant behaviour patterns according to what is socially acceptable and unacceptable behaviour. For instance, certain behaviour of individuals are classified as 'social' and 'anti-social'. By observing his own behaviour, a person might start to develop an awareness that his behaviour falls into general category 'anti- social', with repeated confirmation of his label, a position of a person's self-concept or image emerges, playing a dominant role, in how he views himself or her-self.

2. **Reflected Appraisal:** This is another theory of self- concept development which is also called as "looking glass self" theory. This theory basically holds that appraisals a person receives from others would be the self-concept or self-image. The extent of this influence depends upon characteristics of the appraiser and his or her appraisal.

4. Biased Scanning: This theory of biased scanning views self-concept development in terms of identity aspirations and biased scanning of the environment for information to confirm how well a person is meeting his or her aspirations. Accordingly, it suggests that a person who aspires or is motivated to be a good pleader, will seek out information that helps to conform his aspiration and filter out information that contradicts. Thus, perceptual scanning is biased toward seeing one self as one likes to be ; that is biased toward self gratification.

The above discussion makes it amply clear that these theories of self-image development take different views as to how people look at themselves. In reality, these theories are working to a certain extent, if not in toto. One's self-concept is very likely to be shaped to varying degrees according to how one perceives oneself to others, one's levels of aspirations and biased selection of information about one self, the labelling of oneself according to how one perceives society categorized one-self, and the reflected appraisals of significant others.

CONSISTENCY OF THE SELF

Psychologists agree that a person's conception of 'self' displays a high degree of consistency, particularly in the short-run, in spite of the fact that these theories vary on the development of a self-image. This relatively fixed structure of 'silk' is due to two conditions namely,

(1) 'Self' has an inertial tendency like other many systems. That is, it tends to resist change.

(2) Once the 'self' becomes established, change becomes less likely because of selective perception of environmental information.

That is, the 'self' tends to interpret concepts in terms of the 'self'. Thus, the ideas formed from a new experience are easily absorbed into the existing organisation of 'self' when the experience is perceived as consistent with the existing structure. In contrast, ideas perceived as inconsistent with the present structure, are either rejected or altered to fit into the 'self', since they pose a threat to the individual.

MEASURING SELF–CONCEPT

In order to use the concept of 'self', there is need to measure it. The most common measurement approaches the Semantic Differential as used in attitude scales. **Mr. N.K. Malhotra** has developed a set of 15 pairs of adjectives that offer promise of being applicable across a variety of settings. These terms used in **square** given were found effective in describing ideal, actual and social self-concepts, automobiles and celebrities. They could be, thus, used to ensure a match between the self-concept actual or ideal of a target market, the product concept and the characteristics of an advertising spokesman.

MEASUREMENT SCALES FOR SELF-CONCEPTS, PERSON-CONCEPTS AND PRODUCT CONCEPTS.

1. Rugged	Delicate
2. Excitable	Calm
3. Uncomfortable	Comfortable
4. Dominating	Submissive
5. Thrifty	Indulgent
6. Pleasant	Unpleasant
7. Contemporary	Non-contemporary
8. Organised	Unorganised
9. Rational	Emotional
10. Youthful	Mature
11. Formal	Informal
12. Orthodox	Liberal
13. Complex	Simple
14. Colourless	Colourful
15. Modest	Vain

SOURCE: **N.K. Malhotra,** " A scale measure self–concepts, person concepts and product concepts" Journal or Marketing Research, November 1981- p. 462

SELF–CONCEPT AND CONSUMER BEHAVIOUR

In the market place, consumers' self-perceptions can have a strong influence on their behaviour. The following figure makes it amply clear:

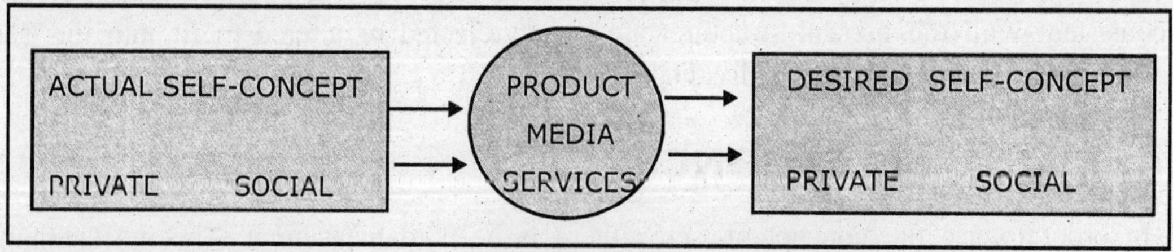

FIG: 4:02 THE SELF–CONCEPT AND CONSUMER BEHAVIOUR

Take the example of the way an individual perceives various products could be affected by the image he has of him–self. Preferences might actually develop, in fact, for certain brands because the consumer perceives them as reflecting his own self-image. It is a also true that certain other brands may be desired because the consumer views them as imprinting an image that he presently does not possess but aspires to have. Consumer researchers and marketing people have been able to develop strong interest in concept of self-image because of these and other possible influences.

MARKETING APPLICATIONS OF SELF–IMAGE CONCEPT

The concept of **self–image** is a very powerful marketing concept with its deeper implication and applications in the field of consumer behaviour. The concept has been a boon to the marketers who harp on consumer as an unique individual and there are fair chances of better results in the marketing field. Let us take some cases :

1. **Positioning the product or products:** The use of self–concept by an alert marketing manager is explained by several logical steps-sequence that leads to relationship between the self-concept and product purchase. These sequential steps are : (1) An individual has a self-concept. The self-concept is formed through interaction with parents, peers, teachers and significant others. (2) One's self–concept is of value the individual. (3) Because to self-concept is valued, individuals strive to enhance their self concept. (4) Certain products serve as social symbols and communicate social measuring about those own or use such products (5) The use of products as symbols communicates measuring to one's self and others causing an impact on the individual's private and social self-concept. (6) As a result, individuals often purchase or consume products, services and media to maintain or enhance a desired self–concept.

2. **Creates Brand Image :** There is close relationship between self-image and brand image. Brand has many dimensions. For individuals a particular brand is value for money brand is personality, brand is symbol of status, brand is a quality, brand is a mark of life-style. The brand image creates impact purchase and consumption. One expects both expects and actual satisfaction. That is, the band purchased must be instrumental in delivering the individuals the desired self-concept. This is nicely depicted in the figure: 4:03 and advertising copy 4.6

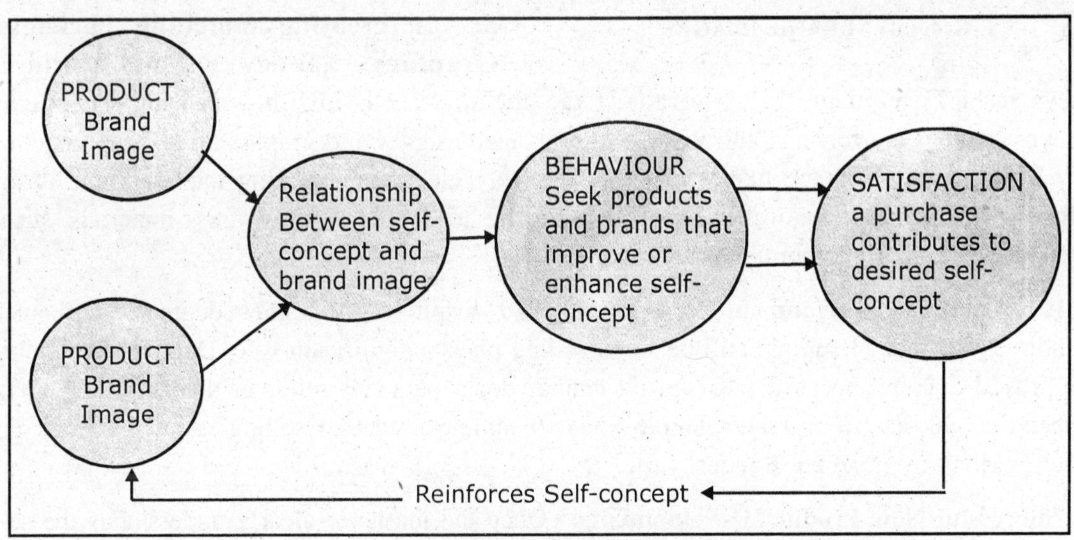

FIG: 4:03 THE RELATIONSHIP BETWEEN SELF–CONCEPT AND *BRAND IMAGE INFLUENCE.*

FIG 4:4:04 BRAND IMAGE

3. A Base for Market Segmentation: As against demographics in the base for market segmentations, marketers can develop into psychographics such as personality type life-styles and self concept. Though each individual is entirely different if one takes all his features—both physical and more 20 mental, these persons of having some similar characteristics—especially the self-image, a market can be segmented and it works for the mutual benefit of both the consumers and producers and therefore, marketers.

4. In Designing an Advertising Copy : The ad copies that can be created to understand and create self-image and it can be maintained and improved. These ads cover products, services and ideas that we associate with self-image. Image conscious consumers go in for these ads, as they are meant for them is in ad medium — may be journals, newspapers, television or even outdoor ads.

5. In creating matching Packaging Graphics: Today, in this world of showmanship, everything that is to attract the attention needs big showmanship. Packaging are what clothes are for man and women. A package designer and graphic artist come together, they can appeal self-image like anything. It is very clear with packing materials and design with graphics in case of jewelleries, glass products, household articles, dress materials, shoes, perfumes particularly premium brands.

6. It Strengthens the Hands of Personal Selling : Inspite of the highest degree of refinement in indirect methods of selling, still personal selling plays a significant role. Door to door selling has proved a boon, because what an ad cannot do, a personal saleman does it. Self-image concept, he can use to a greater degree. Finest example is that of selling vacuum cleaner and water purifier by **Eureka Forbes Ltd.**

7. Successful New Product Development : Once the marketer clearly understands the self-image the inner self of target audience, he will be able to develop a product or service or an ideas that matches the inner self of consumers where he or she gets the best value for his or her rupee spent. New product development is not that easy as hardly 2 or 3 product ideas reach the stage of commercialisation out of 100 ideas.

In spite of these merits of application, the notion of 'self–concept' suffers from serious limitations which are likely to reduce the usefulness of the concept of marketing applications. These limitations are : (1) Difficulty in measuring these self–concept : (2) There is no unanimity as to the concept of 'self', itself, as different interpretations create uncertainty regarding its use in understanding consumers. (3) The self-image concept stresses consumers' self-awareness at the conscious level and tends to minimise the importance of subconscious or unconscious level of influence. Such deeper main-springs of behaviour can have far-reaching influence on consumer behaviour.

CHAPTER BASED QUESTIONS

A. WRITE SHORT NOTES ON:

1. Personality 2. Nature of personality 3. Freudian theory 4. Neo- freudian theory 5. Jungian theory 6. Trait theory 7. Self–concept 8. Stages of personality development 9. Personality traits 10. Application of psychographics in marketing

B. SHORT ANSWER QUESTIONS:

1. What is personality?
2. What is the nature of personality?
3. Explain the substance of Freud's Theory of personality.
4. Discuss Jungian theory of personality.
5. Explain Neo-Freudian theory of personality.
6. Explain Trait theory of personality.
7. What are the stages of personality development?
8. How does personality develops?
9. What is self-image?
10. What is the make-up of self–image?

C. ESSAY TYPE QUESTIONS

1. What is personality? What is its nature?
2. What is self-image? How it develops?
3. How can you measure self-concept? In what, way can you apply in marketing?
4. What is psychographics? Explain the variants of psychographics and its application.
5. Explain Freudian theory of personality. How is it applied in marketing?
6. What is Jungian theory of personality? To what extent is it applied in market places?
7. Explain Neo-Fruedian theory of personality and its application to marketing situations.
8. What is trait theory? What is its use influencing consumer behaviour?
9. What are the psychographics used in marketing? Are they superior to demographics?
10. Explain as to how 'id', 'ego' and 'super–ego' work as a system.

*****

CHAPTER 5

PERCEPTION AND CONSUMER BEHAVIOUR

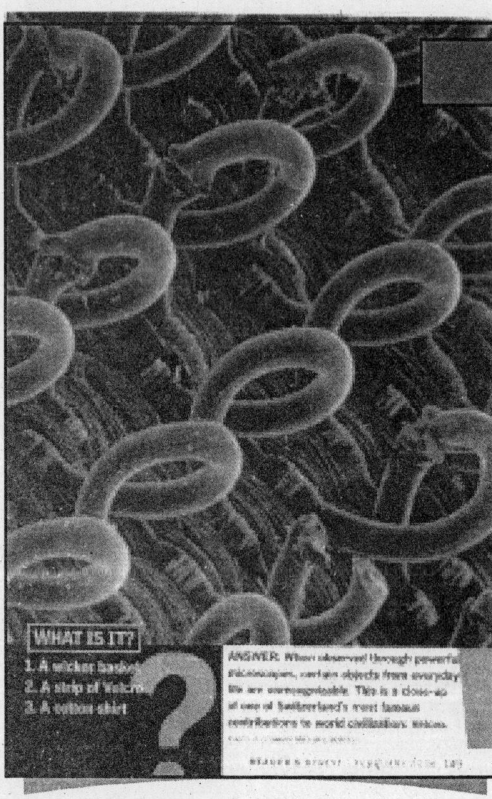

BACKDROP

A given situation or an event, phenomenon, though similarly exposed to individuals, they react or interpret the same in different way. This kind of difference in human behaviour is something which speaks of perception . These impressions are not necessarily based on actuality or reality. Thus, we experience on beach side, at the time of sunset or sunrise that sky is kissing the sea and that setting sun is getting immersed in water. Again, when we look at railway track, the two parallel lines, we feel that meeting each other at a distance. Again, the concept of miraj - a sensation of water-flow at a distance during hot summer day. In reality, these are not true. Perception is based on sensations from the outside world which is influnced by past experience or learning expectations, fantacies, beliefs, values personality and so on. Marketers have understood the importance of perception in consumer behaviour. Accordingly, they have been able develop and apply marketing strategies to enhance the consumer satisfaction and self-profit maxisation. Hence, this chapter plans to discuss the various aspects and implications of perception in the area of consumer behaviour. The chapter ends with Chapter Based Questions.

WHAT IS PERCEPTION?

Though different individuals are exposed to the same stimuli under same conditions, each individual identifies the stimuli, selects them, organizes them and interprets them in an unique way. How unique will be the interpretation rests on his needs, wants, values, beliefs, personal experiences, moods and expectations among other things. **Perception** is the way a person sees the world around him. **Perception** is a personal phenomenon on the basis of which an individual acts or reacts. Perception is being influenced by the attributes of stimuli like size, colour, intensity and the context in which it is seen or heard or both. Like beauty, perception is one that lies in the eyes of the beholder.

STANDARD DEFINITIONS:

★ **Professors Schiffman and Kanuk:**

"Perception is the process by which an individual selects, organizes and interprets stimuli into meaningful and coherent picture of the world."

★ **Mr. Stephen P. Robbins:**

" A process by which individuals organize and interpret their sensory impressions in order to give meaning to their environment."

★ **Mr. Fred Luthans:**

"Perception is an important mediating cognitive process which persons make interpretations of the stimulus or situation they are forced with."

★ **Professors Del. I. Hawkings, Roger J. Best and Kenneth A. Coney:**

"Perception consists of those activities by which an individual acquires and assigns meaning to stimuli."

★ **Mr. Joseph Reitz:**

"Perception includes all those process by which on individual receives information about his environment seeing, hearing, feeling, tasting and smelling. The study of these perpetual process shows their functioning is affected by three classes of variables - the objects or events being perceived, the environment in which perception occurs and the individual doing the perceiving."

Perception is that process which involves seeing, receiving, selecting organising, interpreting and giving meaning to the things that surround us. In other words, perception is the process of selecting, organising and interpreting on attracting measuring to the events that are happening in the environment that surrounds us. The essence of perception is that it is on mental process. Perception is fundamentally a cognative or thinking process. As perception is an intellective process it is purely subjective in nature.

'PERCEPTION' VERSUS 'SENSATION'

Very often, one is confused with these two words namely, 'perception' and 'sensation' as both are cognative or thinking activities of mental faculty. The behaviour scientists maintain strongly that all the knowledge of the world depends on the senses and their stimulation. However, a normal sensory dates are inadequate to produce or to explain the total picture of the world as a person experiences. By this time, we are told that earth is round. However, the question still arises as how round is the round world? Very few people of us know that there are 27 million cells in our eyes each. That human body is a makeup or complex of cells. That is why? one should know as to what these terms mean.

'Sensation' is the response of physical sensory organ. The physique senses are vision, hearing, touch, smell and taste. These five senses are attached by stimuli continuously - both internal and external to the human body and the reaction in particular sense organ take place because of these. The human sense organs or the sensory receptors are - eyes, ears, nose, mouth and skin. The instances of sensation can be a reaction of eye to a colour, ear to sound, nose to smell, mouth to taste and skin to touch feeling.

'Perception' is much more than mere **'sensation'** because it correlates integrates and comprehends diverse sensations and information from many organs of the body by means of which a person identifies the things or objectives. **'Perception'** is determined by both physiological and psychological factors because **'Perception'** is developed out of previous experience or learning, feeling and motives. Thus, activisation of the ears to hear what another person is saying is **sensation** and the inference of what is heard is perception.

Stated very succinctly, **sensation** is the mental process as exposed **perception** is a complex mental phenomena. Perception is more complex as it is influenced by more variables. A person becomes just cautions of the quality stimulus by **sensation.** That is, he is just aware of stimulus like colour, form, shape, smell, feel or touch and smell. By **perception,** on the otherhand, he arrives at the concrete measuring. The degree of activity differs in case of sensation and **perception.** Comparatively, a person is more or less inactive in case of sensation while in case of perception he is active. In perception, he understands perfectly sensation as he tries to make interpretation. Last but not the least, **sensation** is something a part of perception because, sensation takes place first and paves the way for perception. In case of sensation, only sensory organs and nerves are active. While in case of perception the entire human body is activised.

Sensation is the immediate and direct response of sensory organs to simple stimuli say ad, a brand name or say a package. Sensitivity to stimuli varies form person to person and the quality of sensory receptors. It is well-known that some smell well over long–distance, while some clearly see the object is with naked eyes those which are far–off. Further, sensation for a stimulus rests on the differentiation if input. Even when one has a very strong sensory input, and static and unchanging environment may provide little or no sensation at all. For instance, a person residing by the side of national highway or railway track has no sensation from car or train producing traffic noises. Increase or decrease of one or two sounding horns never be noticed. The senses fail to detect small differences in sensory inputs when the quantum of sensory inputs is pretty high.

THE PERCEPTION PROCESS

Perception is not one short activity. It is a process having logical steps and stages one leading to another. Any information processing model explains this best. It has the logical stages of exposure, attention, interpretation and memory. Of these four logical stages, the first three are the part and parcel of perception process. The following figure, gives a model of information processing model which we can be used as perception process model.

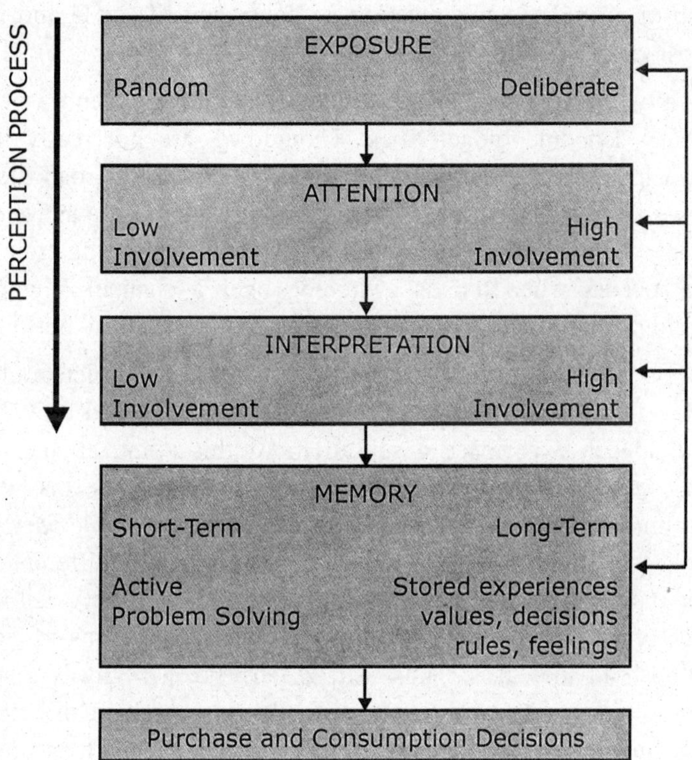

FIG: 5:01 INFORMATION PROCESSING FOR CONSUMER DECISIONs

1. EXPOSURE:

Exposure occurs when a stimulus comes within the range of an individuals sensory nerves. For an individual to be exposed to a stimulus, there is need for that stimulus to be placed within the persons relevant environment. That is, the individual need not receive the stimulus for exposure to have occurred. It is worthwhile to note that both **perception** and

memory are extremely selective. One of the massive amounts of the information available an individual can be exposed to only a limited amount. Again, of the information to which the individual exposed, only a small percentage is alloweded to and passed on to the central processing part of the brain for interpretation. That is, an individual is generally exposed to not more than a small fraction of the available stimuli. Thus, one normally watches only one television channel at a time, reads one magazine, a newspaper or the book at a time and so on. the reasons for this selectivity is self-prompted. Generally, the answer to which stimuli will impress us most-will depend on our goals which we want to attain. These goals may be immediate or long–range. Thus, immediate goals could involve seeking stimuli such as TV for amusement, an ad for purchase–decision or a complement of enhance ones self-concept or self-image. On the other hand, long–term goals might involve studying a particular text to pass the coming examination getting a degree, to make himself a better marketing manager or all of the three. An individual's goals and the types of information needed to attain those goals are a function of an individual's existing and expected life-style and short–term motives as hunger or curiosity.

We are generally exposed to a large number of stimuli on a more or less random basis during our daily busy schedule of activities. Though we are not really interested, we hear messages on radio while driving our vehicles, bulletin–boards on road side display ads and the like. However, we zip and zap the excessive or unwanted exposure as we apply self selecting on tension. 'Zipping' occurs when one fast forwards through a commercial on a prerecorded program. 'Zapping' occurs when one is switching over a channel whenever a commercial appears. Zipping and zapping increased because of remote control button.

The explanation at this stage of exposure will not be complete unless one has clear-cut idea as to the level experiencing a sensation. **Absolute threshold** refers to the honest level at which an individual can experience a sensation. At this point, an individual can detect a difference between 'something' and 'nothing' and this point will be that individuals absolute threshold for that stimulus. Thus, Mr. X senses 6 decibel sound while Mr Y 10 decibels. That is, absolute threshold for sound in case of these two individuals is different. Many individuals' ability to differentiate sensory characteristics such as taste, smell, feel, visibility is small. These senses are more likely to become increasingly dull under conditions of constant stimulation and the absolute threshold increases. Thus, Mr. Z derives on highway for say, half an hour crossing bill–boards on one side ways, it is difficult to say which bill board has gone into his head making an impression. This is known as 'adaptation' which means "getting used to" certain sensations. This sensory adoption is a problem for majority of TV advertisers because of advertising clutter. This makes them to change their ad campaign from time to time to keeping spell–bound and tied the largest audience.

Turning to **differential threshold,** it stand for the smallest detectable difference between two values of the same stimulus. This is better knows as "Just Noticeable Difference" (J.N.D). A German scientist of erstwhile 19th century **Mr. Ernst Weber** discovered that the just noticeable difference between two stimuli was an amount relative to the intensity of initial stimulus. To measure the differential threshold for a stimulus, one generally changes its intensity in very small amounts. An individual's threshold exists when he or she first notices that the stimulus has changed. The difference between this value and the starting is the **justice noticeable difference, Weber** law states that stranger the initial stimulus, greater the additional intensity needed for the second stimulus to be perceived as different. For instance, an automobile manufacturer increases the price of a particular model say by Rs. 500, quite possible that it is not noticed because the increase falls below the J.N.D. the difference in price may become noticeable if the increase is say of Rs. 2,000 or more. To be noticed, an additional level of stimulus equal to the J.N.D. to be added to make the difference noticeable. Similarly, if the reduction in price of the same model is say Rs. 500, it is again quite unlikely to be noticed as it falls below J.N.D.

The marketers use Weber's law to predict as to how the consumers will respond to differences between marketing variables or changes in these variables. In some cases, the marketers aim is to have consumers detect differences such as when product improvements are made or price is reduced. In other cases, the aim is to have the differences skipping the attention such as reductions in product size or quality, increases in present price or changes in packaging when package design is significant for easy recognition by the consumer.

Coming to **subliminal perception,** it is a situation when the stimulus is below the threshold of awareness and is perceived. That is, people can also perceive stimuli that are below the level of conscious awareness. In such situation, the stimuli which are otherwise too weak or brief to be consciously seen or heard prove strong enough to be perceived. This means that the threshold of conscious awareness is higher than the absolute threshold for effective perception. This concept of subliminal perception caught the imagination of people in the late 1950s when it was told that advertisers could expose consumers to subliminal messages without them being consciously aware of this. It was firmly believed that such messages could motivate people to buy products or act in the ways beneficial to advertiser without really being aware of why they did so. The findings of researches were that individuals could perceive below their level of conscious awareness but their purchase behaviour was not influenced by the subliminal perception. Again, in 1970s there was revival of interest in subliminal perception when the advertisers were charged of using embeds in their print advertisements to motivate the consumers to buy their advertised products. "Embeds" are the disguised stimuli that are not readily recognised by consumers. These embeds are believed to be planted in print ads to stimulate or influence consumers purchase behaviour.

2. ATTENTION:

Attention occurs when the stimulus activities one or more sensory nerves and the resulting sensations go to the brain for processing. We are continuously exposed to thousands of time more stimuli than we can process. Experts in marketing say that on average a super market has 18,000 individual items for sale. Normally, it takes hours to attend each of these items. Hence, the consumers are selective in attending to marketing as well as other messages.

This selectivity has major implications for marketing managers and others concerned with communicating to consumers. The implication is that any one wishing to communicate effectively with consumers must understand as to how to get attention after getting exposure. The question is what determines or influences attentions. Though attention occurs always within the context of a situations the same individual may denote different levels of attention to the same stimulus in different situations. Therefore, attention is determined by three factors namely – the stimuli the individual and the situation.

A. Stimulus Factors

Stimulus factors are the physical characteristics of the stimulus. Independent of our individual characteristics, a number of stimulus characteristics are sure to attract ones attention. These are:

(1) Size and Intensity: The size of the stimulus influences the probability of paying attention. Larger stimuli are more likely to be noticed than smaller ones for instance, a full page ad is more likely to be noticed than a quarter page ad. Like size, the intensity has its own impact. The insertion frequency - the number of times the ad appears in the same issue of a magazine, has an impact similar to the ad size. The study conducted clearly demonstrated that multiple insertions were found to crease recall by 20 per cent and by 200 per cent in another. The intensity–the loudness or brightness–of a stimulus operates much the same way as size.

(2) Colour and movement: Both colour and movement serve to attract attention with brightly coloured and moving items being more noticeable. Thus, a brightly coloured package is more likely receive attention than a dull package. A four coloured ad is likely to have more quick attention than black and white. Again, some thing unusual is likely to draw the attention. In a magazine, where all ads are coloured and only in black and white with emphasis is on contrast, its black and white to sure to be preferred to colour ads.

(3) Position: Position stands for the placement of an object in a person's visual field. That is, the objects placed near the centre of the visual field are more likely to be notices than those near the edge of the field. This is the basic reason as to why the consumer goods manufacturers compete keenly for eyeland space in grocery stores. Similarly, in advertisements on the right hand side pages receives more attention than those on the left.

(4) Isolation: Isolation is something separating a stimulus object from other objects. The use of 'white space' is based on this principle in case of ads in print media. We have ads that hardly any message or the text, but yet the blank ad becomes more a unique thing in attracting the attention.

(5) Format: Format is the manner in which the message is presented. It may be simple or complex. It is found generally, that simple and straight forward presentation receive more attention than those which are complex and intricate. The elements in message that increase the effort needed for processing the message tend to reduce the attention. Ads that lack clear visual point of reference or one which have in appropriate movement – too fast or too slow or jumpy increase the processing efforts and decrease the attention. In the same say, audio-messages which are difficult to understand due to say foreign accents, inadequate volume, deliberate distortins, sound background noises and the hike likely to reduce the attention. What is important to note is that format interacts with individual differing characteristics. That is, one may like very complex presentation as he finds if interesting or challenging. That is, why, like other stimulus elements format should be developed with reference to a specific target market in question.

(6) Compressed messages: The findings of research differ in this regard. Some say that compressed messages increase the attention. Again, some say that a spread and complex increases the attention. One can say in general, that compressed commercials do not distract from attentions and may increase the same. However, attention varies with the type of message, the product and the nature of the audience. Further, the interpretation assigned to the context of compressed message varies and is not always favourable.

(7) Quantity of information: Information quantity refers to the total stimulus field than to any particular item in that field. Though there is good deal of variation among the individuals, all consumers or individuals have limited capacities to process the information. The information overload takes place when the individuals are congronted with so much information that they cannot or will not attend to all of it. They become frustrated and either postpone or giveup the decision, instead, making a random choice or use a suboptimal portion of the total information to available.

Infact, there cannot be standardised rules or guidelines as to how much information the consumers can or will use what is important is that it should be sufficient enough to make a purchase decision. It is upto the markers to findout what will be the sufficient information. However, most important information should be provided first and it should be highlighted as a part of the literature or body of the ad or the text matter. More detailed and less important data can be provided in special form of brochures, video tapes and informercials.

B. Individual Factors

Individual factors are the *characteristics* of an individual. The basic factor is:

(1) Interest or need: Interest or need is the primary individual characteristic which influences attention. Interest is the reflection of overall life-style as well as the result of both long-term goals and plans and short–term needs. Short–term goals and plans are heavily influenced by the situation. Further, individuals differ in their ability to attend the information available as noted earlier. Individuals seek out-exposure- and examine–attend to–information relevant to their current needs. For instance, an individual planning a holiday vacation will be definitely interested in vacation-related advertisement. Individuals attending to specialised medium say Business Week or India Today will be receptive of ads on related products and services. Parents with young children are more likely to notice and read warning labels on products like food supplements than the individuals having no children.

C. Situational Factors

Situational factors cover those stimuli in the environment other than focal stimulus - the ad or package - and/or temporary characteristics of the individual which are induced by the environment. These are time pressure or crowded store, programme involvement and contrast.

(1) Time pressure or Over–crowding: The individuals in a hurry are less likely to attend to the available stimuli than those who have extra time. Individuals in an unpleasant environment like an over crowded store or a store which is too noisy or too warm or too cold will not attend to many of the available stimuli as they try to minimize their time in such an environment.

(2) Programme Involvement: The print, radio and TV advertisements occur in the context of a given program, a magazine or a news paper. On an average, the audience is responding to the medium because of the program or editorial content and not necessarily the ad. It is already noted that people resort to zapping because of wise use of time. A high level of programme involvement greating increases both the viewership or readership or the listenership and the impact of the print message or audio–visual commercials on radio and televisions.

(3) Contrast: Contrast stands for the tendency on the part of audience to attend more closely to stimuli that contrast with their background than to stimuli that blend with it. Contrast is a primary component of aware wining headlines. It is the colour, contrast along with the design because of which many are interested in an ad.

3. INTERPRETATION

Interpretation is the finest task of assigning of measuring to sensations. It is the function of Gesatalt or the pattern formed by the characteristics of the stimulus, the individual and the situation. This can be configured as under:

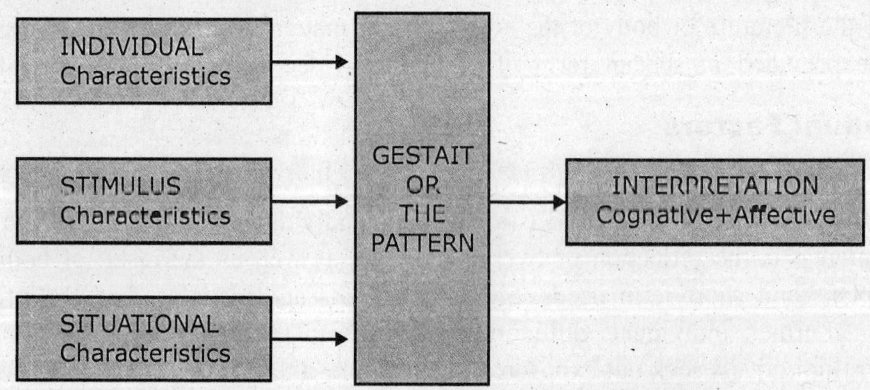

FIG: 5:02 - DETERMINANTS OF INTERPRETATION

The INTERPRETATION stage is the final stage in process of perception and involves both cognative and affective interpretation.

Cognative interpretation is the process whereby stimuli are placed into the existing categories of meaning cognative component is the formal component of interpretation. Cognative interpretation is an interactive process. Any addition of new information to the existing categories is bound to alter those categorieses and their relationships with the other categories. For instance, when the compact disc player was first introduces to the consumers, they grouped it in the general category of record players in order to evaluate it. With further learning or experienced and information, the customers gained the detailed and perfect knowledge to categories it giving it an independent status. It is the individual's interpretation and not the reality that influences the behaviour. For instance, a company may introduce a high quality new brand at lower price than existing brands, say in case of soap, because the company has a more efficient production and marketing process. In case the consumers this "lower price" to "lower quality, the new brand will not be a success though objectively a reality. This only emphasizes the distinction to be made between semantic measuring and psychological measuring. For instance, the semantic meaning of the word "on sale" is "a price reduction from the normal level". However, when it is applied to fashion clothes the psychological meaning is these clothes are or soon will be out of style. That is, it subjective interpretation than objective that influences the behaviour.

Coming to affective interpretation is the emotional response or feeling response triggered by a stimulus like on advertisement. There are normal responses and varied responses too as in case of cognative interpretation. Thus, when we see an ad showing peace and tranquility, a pair of white pigeons are shown flying. The same ad may be looked negatively by a person who hates white flying pigeons. This also suggests the subjectivity.

FACTORS INFLUENCING INTERPRETATION

In the world of marketing, these marketing stimuli have meaning only when interpret them individually. Further, situational and stimulus features that influence interpretation.

A. Individual Features:

Individual features or characteristics have impact on interpretation. The most important one are two namely, learning and expectations.

(1) Learning: Learning is any change in the content or organisation of long-term memory learning is the result of information processing. We have learnt also the natural and artificial or man–made things. Thus, we were told as to what is time, space, columns and habits of eating, drinking, resting and so on. Each term has a specific meaning in a given culture. Thus, 'dinner' is an evening meal and is supposed to be lighter. Again, we have learnt about "fried chicken" of a particular in the context to a nation. Thus, we Indians have adopted the slogan "I am losing it" - of the ads given by McDonalds of America - in case of Pizza. Basically, it is a communication process where advertisers intend to get back the response as to what they intend. The target audience has learnt when the intended message is interpreted in the same way as that of a sender of message.

(2) Expectations: The individuals interpret the given stimulus consistently with their expectations. Thus, one expects the dark brown pudding to taste like chocolate and not like vanilla, because dark pudding is generally chocolate flavoured and not vanilla pudding is cream coloured. Consumers evaluate the performance of a well known brand or more expensive brand as higher than that of an identified product with an unknown brand name or a lower price. Further, consumers attribute ads of new or unknown brands to well known brands.

B. Situational Features:

There are situational features that influence the interpretation the common situational feature are:

(1) Temporary features: Temporary situational features are temporary in that they do not last longer. It may be that they reoccur again with the lapse of time. The examples may be thirst, hunger, restlessness that influence the interpretation of a given stimulus. like moods. When a person is thirsty any stimulus has positive impact on interpretation say an ad on soft drink, a bottle of soft drink on table or a mineral water given at working table.

(2) The amount of Time available: The meaning that is assigned depends on the availability of time for interpretation. Time is very important input at the command of the individual and he or she wants to use it in the best possible manner. The stimulus say the wonderful information relating his product requirements may be useless if he doesn't have a sufficient time to interpret. The meaning attached gets distorted.

(3) Physical Features of situation: the physical characteristics of the situations like temperature the surround disturbances external distractions and even the way the information is processed and the like influence the interpretation. Thus, a persons reading or assuring measuring changes with unduly hot or cold climate or temperature, the rush in the situation say, a retail unit outside distractions will influence badly.

(4) Proximity: Proximity is a tendency to perceive objects or events which are closer to one another as being related. Thus, the ads on soft–drinks are not generally broadcasted or telecasted during the new bulletin. However, we see these ads very often repeated when running commentary is on. As these soft drinks are upbeat and fun products, their sale may be affected badly with bad news.

C. Stimulus Features:

It is the stimulus that sets the basic structure to which an individual responds. The structure and the nature of the product, package, advertisement or sales presentation have the deep impact on the nature of the mental processes which are activated and on the final meaning attached to the message. In view of the critical significance of the measuring associated with stimuli the shrewd marketers started using 'Semiotics'. **Semeiotics** is the science that explains as to how meaning is created maintained and alterer. It focuses on signs which may be anything conveying meaning and include word, pictures, music, colours, forms, smells, gestures, providing prices and the like.

Colours can be used to illustrate the importance of semiotics. The red and blue hues of Pepsi logo, the golden arches of McDonald's the pink flavour to the telecom brand Hutch blue colours have made marketing colourful. Again, the source of message affects its interpretation. Thus, price, presentation format, the type of message and the nature of the product influences as it how promotional claims can be interpreted. Thus, consumers interpret quality claims for a brand in the light of price of a given brand against competing one. In a nutshell, all aspects of message have influence on one's interpretation.

DISTORTION OF PERCEPTION

By this time, we all know that individuals interpret the stimuli in their own unique way. Interpretation of stimuli by individuals is dependent on their earlier experiences, motives, beliefs and interest at the time of perception. A number of factors influence individuals that are likely to distort their perceptions. These are outlined below:

1. Physical Appearance:

Individuals may or may not consciously recognise what they tend to attribute the quantities which in their opinion are associated with certain individuals, to others who may resemble those persons. The research findings have proved beyond doubt that people enjoying in good look or appearance are perceived or more successful in business than those who have average look. Analogically, the attractive models in ads and commercials prove more persuasions and have more positive impact on consumer attitudes and behaviour than average looking models.

2. Stereotyping Stimuli:

People tend to form 'pictures'or 'projections' in their minds of the meanings of different types of stimuli. It was **Mr. Walter Lippman** who used the word 'stereotype' to describe the 'biases' involved in perceives the people. If signifies the fact that a bias is involved when the perceives aces or perceives the other person or an individual involved on the basis of a single class or a category to which he or she belongs. The other meaning of stereotypes is general concurrence on the attributed traits and also the possibility of a discrepancy between attributed traits and the actual traits. Thus, Mr. X may have distorted perception in case of sales people where he opines: "All salesmen are fast speakers and bluff a lot". This may not be true cent percent.

3. Irrelevant Stimuli:

Individuals form perceptional judgement based on their weightage given to irrelevant cues or stimuli. That is, the individuals are faced with difficult perceptual judgement; they respond to some irrelevant stimuli in a given circumstance or a situation. We come across the people who five undue importance to the shape of headlamps and tail-lights, colour of car instead of considering technical specifications of fuel efficiency, resale value, maintenance costs and so on.

4. Maiden Impression:

It is rightly said "First impression is the last impression". That is, first impressions are often lasting even when the perceiver is not exposed to sufficient relevant or predictive information. In case of under developed countries that are fast developing, people do not have faith in their our products and services. Thus, "anything imported" is good or the best. This happened to Indians till the end of last century in case of items like cameras, shoes, clothes, watches, including nail cutters. Today, "Made in India" feeling is having top–priority as other people have shown in Indian products.

5. Snappy Judgements:

We have people who jump to the conclusion rather than arriving at conclusion. These judgements are snappy based on semi-baked feedback or insufficient information. The person is impatient so much so that he has no waiting capacity to take all facts in to consideration that are vital to his conclusion. Thus, when a person remarks: "Short men are shrewd but dangerous," it may be true but in not all short persons. Contrary to this, shortmen may be dull and sober. That is why, the advertisers make a strong claims about their products or services to win over the confidence of audience.

6. Halo Effect:

The phrase, 'halo effect' was used for the first line in 1920 to describer process in which judges used the general impression - favourable or unfavourable - to evaluate certain specific traits. In this context, 'halo' is a barrier or a screen that keeps the perceiver away from actually seeing the trait he is judging. It is a tendency to evaluate one attribute or aspect of stimulus to distort reactions to its other attributes or properties. This is quite common in case of brand or line extensions where the producer or the marketer may take advantage of the reputation of the brand. We know that Ponds is a well known brand for face powder or body talc, but has come out with soaps. People are made to think that soap made by Ponds is better because it is well known for face and body talc. The companies want to go in for extreme diversification. Thus, 'Anchor' brand switch gears are well known. Here, the company has recently introduced 'dental–cream' to encash on brand reputation.

CONSUMER IMAGE AND CONSUMER BEHAVIOUR

Consumer image or enduring perception has its own impact on consumer behaviour or consumer decision–making. An image is the total picture or perception of something that individuals form by processing all the information they are exposed over a period of time. The consumer behaviour research findings say vehemently that consumers develop enduring perspectives or images with regard to product brands, prices, stores the company making and the risk involved if they purchase and use the product in question. Let us touch these important perceptions and their implications.

1. Perceived Product Quality:

Consumers are likely to assess the quality of a product on the basis of variety of informations they relate to a product. These cues or the stimuli are both **intrinsic** and **extrinsic.**

Intrinsic cues relate to the physical attributes to the product like size, colour, smell, taste which are used to measure the quality level of a product. Thus, the quality of a perform is judged on the bases of fragrance, colour, physical apperance of the container and the packaging. It is usual to find the washing powders and detergents and even cakes are blue coloured because bluing is associated with brightening and whitening. Though physical attributes are used to judge the quality, there is no intrinsic relationship with the product quality. This has been proved beyond doubts in case of blind tests differentiating the soft–drinks one from another in taste. The **extrinsic** cues or stimuli are brand image, price or even the country of origin, not excluding ads, packaging or group pressure. The consumers lacing actual experience with the product lend to judge the quality on the basis of extrinsic cues they relate high quality to high price on the strength "cheap is costly" or "higher the price - better the quality". In the past some countries were known for quality products. Thus, Italian leather products, Egyptian cotton, German glass wares, French wine and perfumes and so on. However, sometimes associating cheaper products as of lower quality is wrong. Thus, China has been able to produce computer (PC) at a cost of Rs. 4000, even best quality laptops are the cheapest. Thus, Lenov brand electronics are the best entring global markets of west and other Asian countries.

2. Perceived Service Quality:

It is really difficult to evaluate the quality of services as opposed to products. This is so because of special features of services. Services are intangible, variable, perishable and are produced and consumed simultaneously. That is, services cannot be stored. A product is something which is tangible which can be seen touched and can even tried before purchase. However, service cannot be seen, touched, handled or experienced in advance by the intending consumer. In order to compare and judge the quality of services the consumers use extrinsic cues or stimuli. Thus, a doctors services can be judged on the bases of the educational qualifications, infrastructural facilities of his consulting section, hospital, staff, patient resting facilities, the way the patients are consulted and treated through hear–say reports.

Unlike products, services are first sold and then produced and consumed simultaneously. Products can be checked making them subject to quality control setup. However, production is not possible in case of services. Thus, a wrong operation of a patient, is a permanent loss leading consumer dissatisfaction leading fall in the service image. Consumers evaluate service quality on the basis of magnitude and direction of the gap between the expectation of service and consumer's perceived value of the service actually delivered. Thus, a checking in of hotel and comparing the quality of services will be certain expectations - in terms of behaviour of receptionist, roomboy, timely serving of tea, coffee, breakfast, lunch or dinner, attending the guests at the bells–point or a telephone call and so on. His expression is the base for evaluating the quality of service. Some renowned researchers have indicated certain dimensions of service quality which consumer would use. These are:

(1) Apperance of the physical facilities, equipment, personnel and communication materials

(2) Ability to perform the promised service accurately and dependably

(3) Willingness to help customers and provide prompt service

(4) Knowledge and courtesy of employees and their ability to convey trust and confidence

(5) Caring, individualised attention that the firm provides to the customers.

Other reseacher in this regard deals with two dimensions namely, 'outcome' and 'process'. The outcome aspect focuses on the reliability of delivering of core service. The process aspect focuses on how the core service is delivered. Thus, a pharmacy in the town is known for reliable services of providing original medicines at genuine rates at the counter. The process aspect is that of door–delivery at no extra cost within the radius of four kilo-meters. It is found that there is positive relationship between service. Quality, consumer satisfaction and purchase intension. Higher the service quality and higher service–satisfaction result in high level of consumers purchase intentions.

Image is an important factor in service differentiation as service are intangible. The format or the design of physical environment is significant in creating an image of services. That is why, the offices of consultants of all kinds, retail–stores are quite impressive. The image of service provider is measured by the physical environmental communication cues. Here, the physical objects and the behaviour of service personnel create an impression boosting the consumer confidence.

3. Price Perceptions:

Consumer perception of prices is very significant even from the angle of marketer so that he can modify his product prices to the requirements of consumers. Whether a consumer perceives the price of a product or service as 'high' or 'low' of 'fair' is having deep impact on his or her buying intentions and post-purchase satisfaction. The price perception is quite complex. The living example is that of explosive growth of cheap ball-point pens, the makers of Parker Pen repositioned its pens based on price during 1980's and offered the low price pens. The results were not only discouraging but disastorous because Parkers image was not consistent with its price. It was because Parker was known for high price and quality. The company had to revert back in 1989 high priced pens. This helped the company to have upswing in earnings. The implications are: Consumers have certain expectation of what the price is or shortage of a product or service. Such expectations may or may not reflect the actual price of the product or service and the consumers very often associate price of a product or service to its quality.

According to one research study consumers react to price variable as under:

(1) Consumers seemingly use price as an indicator of product quality and as an indicator of purchase cost.

(2) Consumers develop reference, prices as standards for evaluating prices they see in the market place. A reference or standard price is any price that consumer uses as the bases for comparing another price.

(3) Consumer's reference prices are not constant and get modified by shopping experiences. That is, exposure to higher or lower than the reference. Price is likely to result in upwards or downward movement of the reference price.

(4) Consumers tend to develop a range of acceptable prices around the reference or standard price. Prices above or below the reference price are likely to be judged as unsuitable and may lead to decreased willingness to purchase the product.

(5) Factors like brand–image or store–image can often the strength of the perceived price quality relationship.

(6) In case the prices for different alternatives are perceived as similar, then price is unlikely to influence the choice between these alternatives.

4. Store Perception:

Consumers have store perception. There are five major components of stores image namely, location, design, product–assortment, services and the personnel each of which contributes to consumer perception of the place or store from which he or she buys. Mere physical attributes do not talk of a stores image. There are intangible factors which too influence consumer perception of stores image like advertising, interpersonal communication and experience. Further, consumer perceptions of stores are greatly influenced by consumers own self-perception and motives. Added to that, consumer's self–images influence the places in which they shop.

5. Perceived Risk:

Whenever consumers make decisions to buy any new brand, there is an element of uncertainty of the consequences and perception of risk is involved in most of such buyings. Risk perception is the consumers' perception of uncertainty that they face when they are unable to foresee different consequence of their purchase–decision. The relevant risk dimensions are the uncertainty and the consequences. It is worth emphasizing here that the influence of risk depends on the individuals perception. That is, the risk actually may or may not exit and even if a real risk exists but is not perceived, it will not influence consumer behaviour.

Among several types of risks, the most important ones are:

(1) **Financial or monetary risk:** It is the risk of not matching cost with value. That is, the price paid is not really compensated by the product quality and quantity.

(2) **Performance risk:** The risk of not performing of what it is expected to perform. The person may spend a lot on its repairs or bring it to the order.

(3) **Physical risk:** Possible harm caused to user and others by actual usage.

(4) **Social risk.** It is a kind of risk of not meeting expectations of reference group or groups causing social embarrassment.

(5) **Psychological risk:** It is risk caused by loss of self–esteem or self-image as a result of poor choice and making a person feel stupid.

The degree of risk perception varies and depends upon the person, product, situation and the culture, among the consumers. Some consumers who are high-risk perceive's or risk avoides limit the product choices to a limited number of safe alternatives to avoid risk of poor selection. Consumers who are low-risk perceivers or risk takers tend to consider their choices from a wider range of available product alternatives. Product category is also an element of perceived risk. Consumers are more likely to perceive risk is the selection of high-involment product categories. The situations say, shopping situations also contribute to perceived risk. The shopping situations maybe–face to face, telephone, mail, direct mail, internet or online shopping. Except in case of door to door or face to face selling situations, consumers face higher risks.

Risk Reduction Strategies

Consumers employ certain strategies to reduce risk involved in purchases.These strategies help them to reduce perceived risk because they build his or her confidence in making purchase decisions. These strategies are designed either to increase certain associated with the product purchase or to minimise negative purchase consequences. These are:

(1) **Acquisition of Additional Information:** This helps the consumers to assess the risk involved in much better way.

(2) **Practising Brand Loyalty:** Consumers play safe by remaining loyal to a brand which already tried by them. There is greater risk in trying on untried brand.

(3) **Trying only Trusted Brands:** Wise customers learn from the mistake of others. A tried and trusted brand is definitely a popular one.

(4) **Relating Price with Quality:** Consumers go ahead with their purchases on the quality price realising higher the price better is the quality.

(5) **Banking on Stores Image:** While selecting the products, they select the retail outlet which is trust worthy.

(6) **Insisting on Guarantees and Warranties:** Some customers are quite frank and blunt and are wise. They insist of all sorts of refund in case product fails to meet this expectations.

(7) **Buying Smallest Sized Packs:** In most of the consumer non-durable products, the consumer buy smallest sized pack - to avoid or reduce risk instead of buying large pack. Even manufacturers and marketers have realised this point.

(8) **Reducing the Level of Expectations:** They have also learnt to reduce the level of expectations. It is worth while not to except too much before purchasing. This strategy with certainly reduce psychological effects.

CHAPTER BASED QUESTIONS

A. WRITE SHORT NOTES ON:

1. Perception 2. Sensations 3 Perception process 4. Exposure 5. Attentions 6. Interpretation 7. Distortion of perceptions 8. Consumer image 9. Perceived quality 10. Perceived risk 11. Absolute threshold 12. Diffential threshold

B. SHORT ANSWER QUESTIONS:

1. What is perception?
2. What is sensation?
3. What is perceived service quality?
4. What is price-perception?
5. What is stores-perception?
6. What are risk reduction strategies?
7. What are absolute and differential thresholds?
8. What are stereotying stimuli?
9. What is consumer image?
10. What is perceived risk?

C. ESSAY TYPE QUESTIONS:

1. Define clearly the term 'perception'. Is it same as 'sensation'?
2. What is perception? Explain the stages of perception-process.
3. What is exposure? Explain its implications.
4. What is attention? What sets of factors influence attention?
5. What is 'interpretation'? What factors influence this stage of perception?
6. What is 'perception'? How it gets distorted?
7. What is consumer 'image'? How consumers perceive quality and price?
8. What is perceived-risk? How consumers reduce the risks involved?
9. Explain the concept of 'perception' with regard to consumer behaviour.
10. "Perception is only the part of processing information". Explain.

*****

CHAPTER 6

LEARNING, MEMORY AND CONSUMER BEHAVIOUR

BACKDROP

Learning is one of the basic foundations of human behaviour. Humans and infra humans learning in either simple or complex way, from the very maiden opportunity to react to sensory stimuli. Learning takes place every where and throughout the very existence of an individual. Learning is a broad behavioural area that encompasses training, acquisition of skills, education, management development or other similar change or modification process. How individuals learn is a matter of great interest and significant to academicians, psychologists, consumer researchers and marketers. Marketers are deeply concerned with it, in that they are interested in teaching them about product, products attributes and potential consumer benefits; they want to tell as to where to buy products? How to use them? How to maintain them including their disposal. The closely allied concepts of learning are the memory and the involvement. Hence, this chapter plans to discuss in detail these concepts as relevant to consumer behaviour .The Chapter ends with the Chapter Based Questions.

WHAT IS LEARNING?

Learning is a relatively long lasting change in behaviour due to experience. Learning is change in behaviour, organisation of behaviour and reinforcement of any new activity. Learning is a permanent change in behaviour through education, training, practice and experience.

STANDARD DEFINITIONS

★ **Professor E.R. Hilgard:**

" Learning is a relatively permanent change in behaviour that occurs as a result of prior experience"

★ **Mr. W. Mc Gehee:**

" Learning has taken place if an individual behaves, reacts, responds as a result of experience in a manner different from the way he formerly behaved."

★ **Mr. Schiffman and Mr. Kanuk:**

" The process by which individuals acquire the purchase and consumption knowledge and experience that they apply to future related behaviour."

★ **Mr. A.A Mitchell:**

" Learning is any change in the content or organisation of long term memory."

★ **Mr Stephen P. Robbin:**

"Learning is any relatively permanent change in behaviour that occurs as a result of experience."

The above definitions make certain points clear as, what is learning? **First**, learning involves change–may be good or bad.

Learning takes place when a person links the stimuli and response to the environment. For instance, an individual is attracted by the fragrance of a perfume (stimuli) and he intends to purchase (drive), and if he is satisfied with the price and other factors (positive–reinforcement) he purchases it. **Secondly**, such a change in the behaviour must be relatively permanent because, learning is said to be there only when permanent change occurs. **Thirdly**, such change in behaviour is acquired through experience. Learning is said to be there only when behaviour change in there. That is, mere change in thought process or attitude does mean learning. **Fourthly** for learning, there is need for experience in one form or the other. Such an experience may be acquired either directly through practice or through observation or reading. **Fifthly**, learning is a continuous process. It is a life–long process. Through constant learning, consumer improves his decision making capacity.

In a nutshell, learning is a concept and none has seen learning. It is a continuous process and gets modified and refined as a result of exposure to new information and personal experience and becomes bases for the future observable behaviour. Hence, one infers that learning has taken place if an individual behaviour reacts or responds as a result experience, in a manner different from the way the individual behaved formerly. Thus, with the exposure to new ads in case of say soft–drinks, he may change his purchase behaviour from one brand to another or one product to another in the category.

COMPONENTS OF LEARNING

Certain components which are fundamental to learning process are present. These are, drives, cues, response, reinforcement and retention. To have perfect understanding of learning process, it is but essential to learn about these components.

1. Drive : Motivation is the driving force which impels an individual to action and is the outcome of unfulfilled need or needs. In case an individual has strong motivation to learn something there is increased likelihood of learning taking place. For instance, a person who is interested in acquiring MBA degree in marketing is definitely involved to get all the detailed information about the institutes that offer best services for his money and time other, not interested will hardly pay any importance to such details. That is, learning occurs because of the presence of drive; drive is any thing that impels action. It is a strong stimuli. It is the drive that arouses in an individual an urge to respond to the stimuli forming the basis of behaviour. One should know the difference between a 'motive' and a 'drive.' A motive is based on the needs or goals. That is, motive is a purposeful or directed towards specific goal while devine is an increased probability of activity without actually specifying the nature of activity.

2. Cues: Cues are also stimuli but are not as strong as drives. That is, cues are relatively weak stimuli not strong enough to arouse consumers but have all the potential of providing directions to motivated activity. Cues are the objects existing, in the environment as provided by the individual. If motives or drives stimulate the learning process, it is the cues that serve the purpose for stimuli and give direction to these motives or drives. Thus, all ads, displays, packings, price, information that are cues, will help the consumers in making purchase decision.

That is why, marketers have fully explained this situation. That is, the cues are used by marketers to stimulate the consumer in congruence their expectations and aspirations. Marketers are to exercise due care so as to not to fall beyond consumer expectations. Thus, designer clothes and jewelleries need exclusive shopping centres and ads to be given in cases of magazines which these elite class read.

3. Response: The stimuli result in responses the way an individual reacts to a cue or a stimulus to response. This response could be physical or mental in nature leading to learning. For instance, a marketer of washing machines keeps on providing cues to target consumers through promotional activities and may not be successful in getting the final purchase behaviour because of good many reasons, though consumers are motivated to buy. It is also possible that target consumer forms favourable image about a particular model supported by resources of purchase washing machines. Consumer are exposed to many cues providing direction at the same time and each cue competes for attention. It is important to note that cues or stimuli may be significantly affected by earlier learning as a result of response- reinforcement. The non–physical form of response may be attitudes, familiarity, perception and the life. That is, learning can takes place even when the responses are not over.

4. Reinforcement: Reinforcement is fundamental to the process of learning. Without reinforcement, one cannot observe any measurable change in behaviour. Reinforcement of a specific response increases the likelihood for the response to occur. Reinforcement can be anything that both increases the strength of response and tends to induce repetitions of the behaviour that precedes the reinforcement. Reinforcements the set of environmental events that increases the likelihood of specific response occurring in the future as a result of particular cues or stimuli. If a college girl wins a title " Miss Commerce" at inter–collegiate level by using a particular brands of make–up–aids, she is likely to buy and use the advertised brands of cosmetics again and again. In this case, learning has take place through reinforcement as she was amply rewarded by getting title "Miss Commerce."

5. Retention: Retention is nothing but the stability of learnt behaviour maintained by an individual over a period of time. That is, retention is well learnt response which is achieved vith the passage of time. These is a tendency for the conditioned response of continuous for long period of time, under repeated conditions of positive reinforcement. Whenever a consumer is very much satisfied with a particular brand of a product, he or she will learn to use the product through positive reinforcement. For instance, when a house–wife gets a very warm treatment at a newly opened Big Bazar, she is forced of visit the **Big Bazar** in the wake of her receiving good treatment. Therefore, she patronizes, the store. It becomes her stores later it becomes her companys stores. Thus, this learnt behaviours is seen when she patronizes and recommends it to others. This is retention. It is quite possible that her patronage may be broken by competition if they are able to extend continued better services than the present store.

BEHAVIOURAL LEARNING THEORIES

Behaviour psychologists believe firmly in observing changes in an individual's responses that result due to exposure to specific external, environmental stimuli. That is why, behavioural learning theories are also referred to as "Stimulus response" theories. Behavioural theories are founded on stimulus- response (SR) orientation and it is believed that learning occurs through connection between the stimulus and the response. It is already noted that an individual is said to have, learnt when he or she responds in a particular or predictable manner to a known stimulus. Of the several theories of learning two are of much relevance from marketing point of view. These are "classical conditioning" and "instrumental conditioning."

I Classical Conditioning: Generally, we think of conditioning as a kind of automatic response to something as an outcome of repeated exposure to it. For example, if a young man or woman visits say **Pizza Hut's** outlet, the reaction may be conditioned from repeated pleasant visits to the unit. This classical conditioning matches one stimulus with another that already gets a given response and over a period of repeated visits, the new stimulus will start causing the same or quite similar response. It was **Ivan Pavlov** who pioneered the study of classical conditioning.

In his study, he noticed that since his hungry dogs salivated (unconditioned response) at the sight of food (Unconditioned stimuli), the connection between food and salivation is not taught and is just a reflex reaction. **Mr.Pavlov** reasoned that a mental stimulus like as the sound of a ringing bell could also cause the dogs to salivate if it was closely associated with the unconditioning stimulus (food). To test this reasoning, **Mr. Pavlov** rang a bell while giving food to the dogs. After a good number of repetitions the dogs learnt the connection between bell and the food. When they heard the bell (conditioned stimulus) even in the absence of food, they salivated (conditioned response).

Classical conditioning is a process in which a previously neutral stimulus acquires the ability to elicit a response by repeated association with a stimulus that naturally produces a similar response. This concept or theory of learning can be configured as under:

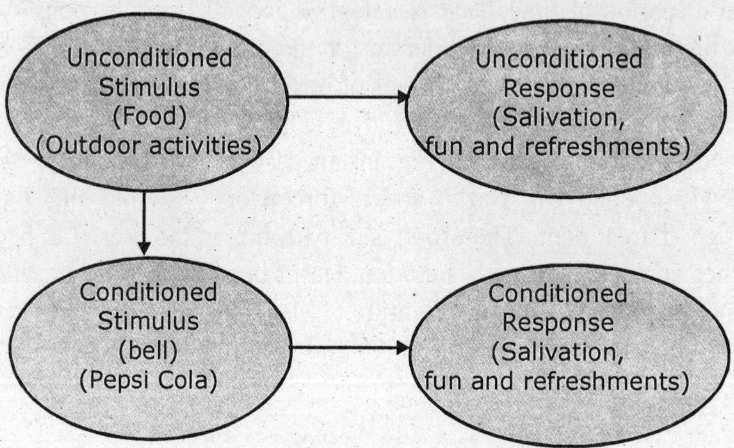

FIG: 6:01 CLASSICAL CONDITIONING

Formally, classical conditioning was viewed as being a reflexive action. However now, it is viewed as cognative associative learning leading to the acquisition of new knowledge pertaining to the world. The association of conditioning stimulus (bell) and unconditioned stimulus (bell and food) influenced the "expectations" of dogs and this in turn, influenced their behaviour) Salivation.

According to **Mr. Chris Janiszewski** and **Luk Warlop**. conditioning encourages attention to the advertised brand in the promotions that are released. It is important to note that two factors are signified for learning to occur through the associative process. The first is contiguity-conditioned stimulus and unconditioned stimulus must be close in time and spare. The second factor is repetition- the frequency of association. The more the frequency of unconditioned and conditioned stimuli occurring together, the stronger the association between them will develop. Here, consumer can be conditioned to develop positive imprints and images of brands through associative process.

CORE CONCEPTS OF CLASSICAL CONDITIONING

The study of classical conditioning culminates into three core concepts which are very important for understanding consumer behaviour. These core concepts are repetition stimulus generalization, and stimulus discrimination. It pays to know each in brief with examples.

4. Repetition: People forget the things which they are supposed to remember. However, one way of increasing the retention of learning in repetition. It is strongly believed that repetition works strengthening the bond of association and thus showing the process of forgetting. Learning follows a pattern which is known as "learning curve". The concept of learning curve is configured as under:

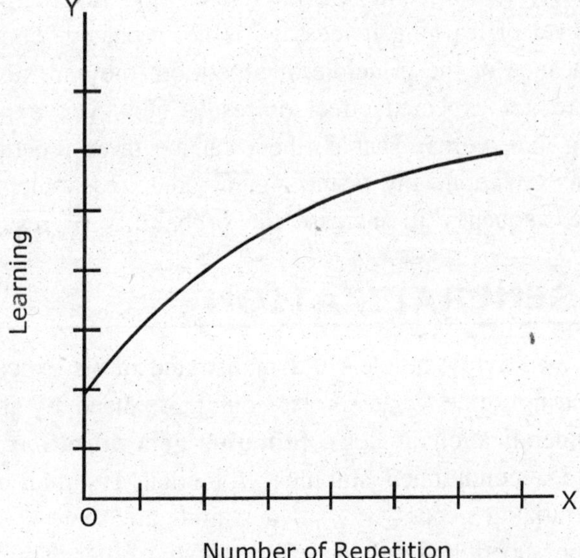

FIG.: 6:02 STANDARD LEARNING CURVE

From the above diagram it is quite evident that along 'X' axis are the member of repetitions and along 'Y' axis the amount of learning is measured. In a standard case, the rate of learners is quite rapid in the early stages. At later stage, the rate of learning per repetition decreases as the amount of learning accumulates. This means that there is limit to the amount of repetition that helps learn and beyond a limit, the attention and the rate of learning decline.

In the same fashion the ad repetition curve takes the shape of learning curve. Advertisers must ascertain whether the general pattern shown by the learning curve actually fits their particular, product or products. For example in case of low involvement buying situations fewer exposures are needed to register simple facts such as brand names or brand recognition. In situations where the standard learning curve is appropriate, there can be significant implications. For instance, an advertiser may be willing to repeat a given ad message several times. It is because of this reason that brand name is very often repeated a number of times is a single ad. The advertiser gives only for fractional increases in consumer learning for the repetition beyond a point. Still another implication is that of 'advertising wear out' It is the result of over exposure because of much repetitive advertising leading to individual's boredom, disinterest and decreased attention and retention of the ad message. According to experts this wear out effect can be reduced by changing or varying the ad message. This is possible by repeating the same message or the theme in a variety of formats. Thus, in case of **Coca Cola,** the slogan "Thanda yane **Coca Cola**" has been changed over to "Thanda ka tadaka". When the advertising context is changed in different versions of the same ad without changes in cosmetic features is called as substantive variation approach. Experts know that consumers exposed to ads with substantive variation process more information about product attributes and the formed attitudes are more resistant to change compared to ads with cosmetic variation.

Consumers are likely to forget most of it rapidly and repetitive becomes a must, just to maintain consumer level of learning if message is not repeated. Though the advertisers are convinced of the significance of the principle of repetition, they are not sure as to how many repetitions are a must to get expected effect or result. However, experts opine that "three-hit theory" is something that works. That is, there can be three repetitions. More than three, will be waste of resources without any result. Again, other researchers are of the opinion of having twelve exposure frequency to increase the probability of result.

B. STIMULUS GENERALIZATION

Going back to **Ivan Pavlov**, he clearly demonstrated in his experiments that dogs could learn to salivate on hearing some what similar sound produced by jingling keys. This was the case of stimulus generalization in dogs. **Stimulus generalization** is the phenomenon in which stimuli similar to a conditioned stimulus will elicit a similar response without prior learning. Stimulus generalizing occurs when two stimuli are seen as similar and the effects of one, therefore, can be substituted for the effects of the other. This principle states that a new but similar stimulus or stimulus situation will produce a response that is the same or similar as that produced by the original stimulus. The more the new stimulus is like the conditioned stimulus, the more probable it is that the new stimulus will produce the same conditioned response. The process of stimulus generalization seems to occur automatically unless stopped by discriminative learning. In fact, stimulus generalization makes consumer life much easier and allows them to simplify the process of evaluation because they need not make separate judgements for each and every stimulus.

Consumers use stimulus generalization from the past experience to put them into categories, as increasing number of new products are brought in the market place. That is why some local or regional marketers make use of this principle by using nearly identical packaging for their products so that they resemble some well known brands in appearance. This is quite common in case of products in edible oils, iodized salt, mineral water and so on. In case of companies of repute, they may or may not follow the policy of stimulus generalizations. The companies that follow this stimulus generalization are **Bajaj, Sony, Pepsi, Coke, Philips Lakme** using family branding. The companies that do not use stimulus generalizing are **HLL, P & G , ITC** and the like. Instead, they believe in stimulus discrimination.

The various forms of stimulus generalization can be product line extension, product form extension and product category extension. Another thing is that of family branding.

Product line extension is the strategy of introducing variations of the same product. The differentiation is seen in terms of colour, packaging, size, flavour and the like. The core product value remains the same. Thus, **Lux soap** is available in white, pink, yellow colours. **Maggi soups** are available in different flavours. **Product form extension** is the strategy whereby the same product is made available in different physical forms. Thus, **Dettol Soap** is available in cake and liquid form. **Drugs** are made available in form of tablets, syrups, injections, inhalers and so on.

Product category extension is the strategy of diversifying the products into different categories using the same established brand name. Thus, **Maggi** noodles, Maggi tomato sauce, Maggi soups, **Ponds** soap, powder, cream, **Lux** soap, shampoo and so on. The success of product category extension strategy depends on the quality image of the parent brand because only then consumers are more likely to bring positive association to the non–category extensions.

Coming to **family branding** it is the practice of marketing the whole product mix of a company under the same family brand name. The aim is to take advantage of consumers' tendency to generalize favourable brand associations from one successful product to the next. Thus, **Sony, Nikon, Philips, Canon, Epson, Tata, Lakme, Ponds, Pepsi, Coca cola** do use this.

C. STIMULUS DISCRIMINATION

In simple term, **stimulus discrimination** is quite opposite of stimulus generalization. Discrimination is a reaction to differences among similar stimuli, unlike reaction to similarity of stimuli. Stimulurs discrimination is the process in which the organisation responds only to the exact, original conditioned stimulus and not to similar stimuli. The ability to discriminate among the stimuli is learnt. That is, the consumer learns to discriminate and select a particular stimuli from the other similar stimuli the stimuli which has established a unique brand image in his or her mind. With good many competing brands available in the market the marketing firms look out for such stimulus by which consumers are able to discriminate the brands in the product category. On the contrary, the consumer finds it difficult to select the most satisfactory products due to the small difference between the products and brands. Hence, they feel that by paying close attention to these differences, stimulus discriminates can be achieved. The concept of "Product or brand positioning" is based on stimulus discriminations that strives to create a brand's unique image in the minds of consumers. The marketers who offer 'me too' type of products attempt to encourage stimulus generalization among consumers while innovators and market leaders strive hard to convince consumers to discriminate and consider their brands as different from general type of products and other brands in the same product category. The famous example is that of "Hot and sweet tomato chilli sauce" of **Maggi** that convinced the consumers by using the USP namely, "Its different".

It is rather difficult under most situations, to unsettle a leader brand from its position after stimulus discrimination has taken place. Thus, everyone knows of Vicks Vaporub or Rasna concentrates. They are leaders, because they were the first to introduce these products and get longer period to teach consumer through variety of marketing communications to associate the brand name with a product. In general, it is more probable that consumers learn to discriminate if the period of learning is longer and associate a brand name with a product. It is our usual experience that consumers are exposed to numerous marketing stimuli, some encouraging stimulus generalizing and other stimulus discrimination. The key to attain stimulus discrimination is effective product or brand positioning.

HOW TO MAKE CLASSICAL CONDITIONING MORE EFFECTIVE?

According to **Frances K. Mc Sweeney** and **Calvin Bierley,** four conditions make classical conditioning more effective. These are:

1. Concentrate on core stimuli:

There should be no other stimuli that could overshadow the unconditioned stimulus.

2. No prior associations to other brands or product categories:

Unconditioned stimuli should have no previous associations to other brands or product categories. For example, if a company manufacturing alcohol decides to use Sharukh Khan in its ad to convey macro image to its target group, the campaign is bound to be ineffective because of his earlier association with Pepsi Cola.

3. Unfamiliar and single presentation:

The unconditioned stimulus should be familiar and should be presented above. That is if a spokes–person or ambassador appears is too many different brand ads, the effect is reduced . Thus, Big-B Amitabh Bachan is seen practically in most of the ads- say Reid and Taylor, Everyday, Boroplus, Navrathna Oil, Hajmola, Dabur Honey and so on. Same is the case which Sharuk Khan

4. Fresh conditioned stimulus:

Classical conditioning is more effective when conditions stimulus is needs. Generally consumers have established associates for well known products and brands. The best example are **Gattu of Asian Paints, Muscleman of MRF. Roaring Lion of Metro Goldwin Mayors**

(MGM). In such cases, it is difficult for them to link their products with new unconditioned stimulus.

II INSTRUMENTAL CONDITIONING

Instrumental conditioning or operant conditioning is a type of learning in which the consequences of behaviour lead to changes in the probability of that behaviour's occurrence. Instrumental conditioning involves developing association between stimulus and response. However, it requires the subject to discover a correct response that will be reinforced. Mr. **Burrhus Skinner** an American psychologist discovered this operant conditioning. Accordingly, the process of operant or instrumental conditioning rests on the unconditioned stimulus after certain behaviour is shown. The unconditioned response is said to be the positive or negative reinforcement of the performed behaviour. The effect of operant conditioning is based on the strengthening or weakening of the performed behaviour. A positive, stimulus (reward) increases the chances of the given behaviour being performed again (strengthening). On the contrary a negative stimulus (punishment) reduces the chances of given behaviour being repeated again.Let us take a case to learn much about his concept. In a Pavlov–like experiment situation, dogs or rats are provided with two levels instead of just one. One level–on–push produced food (reward) and the other an electric shock (punishment). The rats or dogs would quickly learn to press the lever that produced food and avoid the lever that produced an electric shock, when they felt hungry. Here, learning occurs because the consequence of a repeated behaviour is rewarding. This concept or theory of operant conditioning can be configured as under:

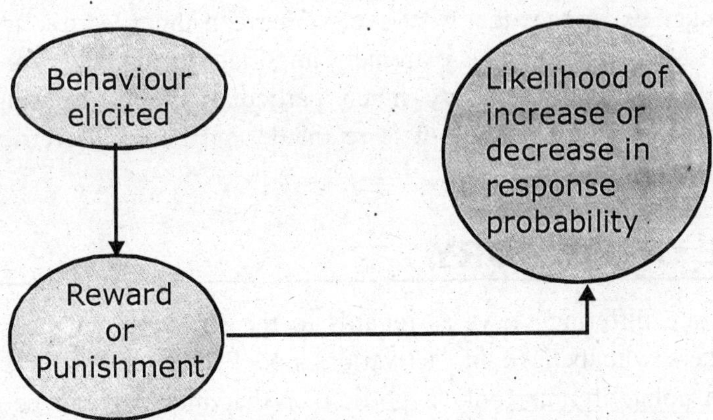

FIG: 6:03 INSTRUMENTAL CONDITIONING

In the context of consumer behaviour, instrumental conditioning points out that learning takes place by means of trial and error processes and consumer's experience more agreeable results (outcomes or rewards) is case of some purchases than others. Favourable consequences reinforce the behaviour and increase the likelihood of its repetition. This is, the consumer will purchase the product again, unfavourable outcome will decrease that likelihood. For instance, a house wife having a problem in finding a satisfactory washing detergent.

Looking to ads on TV, she tries a brand of washing powder say, 'Tide' and she feels succeeded. What is true of washing detergents is true of other consumer non–durable products. That is, advertised brand helps them to avoid uneasy feeling and embarrassment.

DISTINCTION BETWEEN 'CLASSICAL' AND 'INSTRUMENTAL' CONDITIONING

Professor **David Krech** and his associates bringout distinction between the two in atleast three respects.

1) "Classical conditioning" involves an already established response to another stimulus. However, in case of "instrumental conditioning," no previous stimulus response necessary. That is, the learner is expected to discover 'correct' response.

2. In "classical conditioning" the outcome is independent of learners action. In case of "instrumental conditioning" the outcomes dependent on learner's action

3) "Classical conditioning" affects development and changes in opinion, tastes and goals. However, in case of "operant conditioning", it influences changes in goal oriented behaviour.

MEMORY

Memory is the total accumulation of prior learning experiences. Memory represents the information that an individual retains or stores and that he or she can recall for future use. It is known through experience that what ever matter consumer learn, often, they are unable to retrieve it readily. Memory process is extremely important to the understanding of consumer behaviour. Cognative theorists are very much particular about learning as to how the information gets stores in memory? how it is retained? and how it is retrieved during brand evaluation.

STRUCTURE OF MEMORY

In fact, there are differing views as regards to the structure and operation of memory. One group of thinkers who believe in 'activation model', consumers have a large memory store. At any given point of time, only a portion of that memory can be activated for use and remaining inactive portions of memory are not available to recall information stored in memory. Another line of thinkers think of three separate store houses of memory namely, 'sensory', 'short term' and long term memory. One should take these as separate physical entities but as distinct processes of memory functioning having distincts features. The information is retained temporarily in case of 'sensory' and 'short–term' stores before being finally stored in the long–term memory. This memory system can be configured as under:

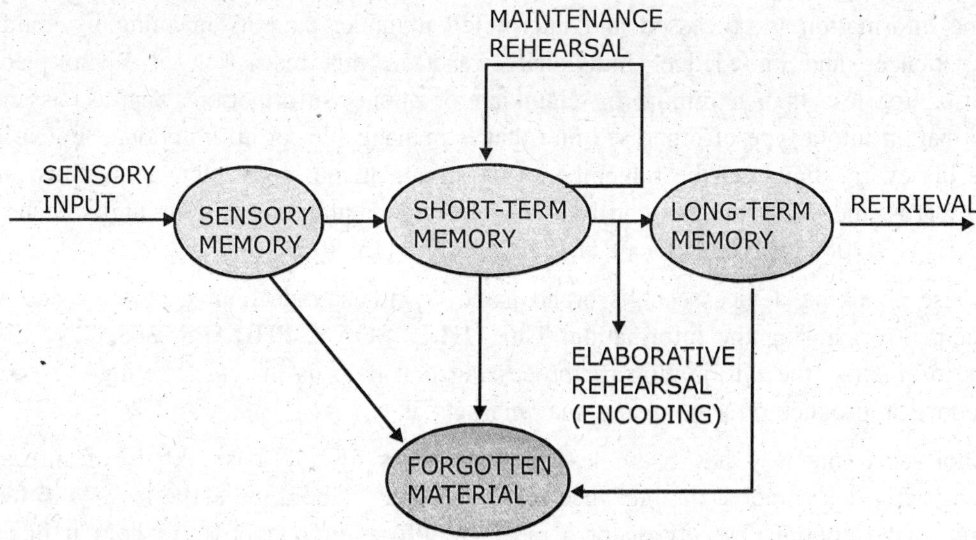

FIG: 6:04 THE MEMORY SYSTEM

Sensory Memory

Input in the form of **sensation** is produced by senses to start with. According to experts, the senses do not relay the complete images like camera. Instead, each sense organ receives a fragment portion of information such as say, flower's shape, colour, smell, feel, taste and so on. Each sense transmits its bit of sensation to the brain at the same time,where all sensation are coo-rdinated and perceived as a single image, in a single moment of time. It is believed generally that these exist memory stores for sensations which accumulate all the sensory inputs and appear to produce information in a form that closely resembles the actual stimuli. For instance, if we close our eyes after observing an object, in natural setting or on TV monitor, we can see the "after–image" in our minds eye for just a fraction of a second. Different stores do not actually or physically exist but only appear to exist because of different levels of processing in the memory. For instance, some stimuli may receive only a cursory processing say brown, medium sized ceiling fan. In this case, the consumer might also process the stimuli more elaborately.

In case the image of sensory input is not processed, it is lost immediately. In day–to –day life, all of us are exposed to good deal of information and much of it fails to make an impression or appealing to our mind. According to **Mr. Daniel Coleman**, the brain 'tags' automatically and unconsciously all perceptions with a positive or negative value. This, tagging, of value to perceptions takes place within the first fraction of a second of perception and tends to remain unless further information is processed. This means that first impressions are more lasting often.

Short–term Memory

Short–term memory is real and is more like a workspace for information processing. It has short capacity. That is, less than a minute and an average seven pieces of information.

Here, the information is processed and interpreted to understand its meaning by combining past experiences and knowledge. Individual's react to this restriction of seven pieces of information for less than a minute by chunking or slicing information. That is, assembling the information into a type of organised unit that is manageable for an individual. For instance, a figure of several digits can be remembered by slicing it into convenient lots. Thus, in case of credit cards, the digits are, sixteen but can be divided into four of four digits each. Thus, a number of 5420860011168083 can be 5420 8600 1116 8083.

These chunking devices can be brand name, symbols trade marks, colours, and so on. These help in organising the information. Thus, **HLL, SONY, PHILIPS, 555, 777,** all these helps us to organise the information for processing that help us in remembering the company that produce a product or a company that markets it.

Short–term memory has been described in terms of two basic kinds of information processing activities namely, 'maintenance rehearsal' and 'elaborative activities'. **Maintenance rehearsal** is the continual repetition of a piece of information in order to hold it in current memory for use in problem solving or transferral to long term memory. If the information in the short term memory is not rehearsed and is transferred to long term memory it will be forgotten through the process of decay within less than a minute. **Elaborative activities** are the previously stored experiences, values, attitudes, beliefs and feelings to interpret and evaluate information is working memory as well as add relevant previously stored information. Elaborative activities serve to redefine or add new elements to memory.

Short–term memory is closely analoguous to what we normally call as thinking. It is an active, dynamic processes, not a static structure.

Long–term Memory

Long–term memory is an unlimited permanent storage. That is, the system of long–term memory can be viewed as relatively permanent place of storage for information that has undergone elaborative rehearsal. It can store numerous types of information such as concepts, decision, rulers, processes, affective or emotional states and the like. An almost unlimited amount of information can be retained here for several years or just for a few minutes. This information is constantly organised and updated as new links between the chunks of information are formed. According to some experts, people never forget any thing that has reached long–term memory. What is really forgotten is the key that unlocks the material located in the memory. Another group of experts hold a view that information stored in long–term memory does not just decay over a period time but appears to be forgotten because of interference with retention.

THE MEMORY PROCESS

Processing of information requires that first of all it should be filtered through short–term memory, then stored in long term memory and then finally retrieved when needed. The information available for storage in the long memory rests on the amount of rehearsal it gets by an individual. In case there is no rehearsal for an input, it leads to fading and finally the material is lost. There are two views about rehearsal. **View one** says that rehearsal is needed to amplify the weak signals by sensory memory store. The **view two** is that rehearsal involves relating new data to old to make the information meaningful. For instance, a person may remember by repetition 'Head and shoulder' as more effective anti dandruff shampoo including its price.

The capacity of short–term memory is very limited and only one to three bits of information can be conveyed in a 15 second commercial, as noted earlier only sufficiently simple information stands a better choice of being absorbed with least loss. The encoding process selects and assigns a word or visual image to perceived material. Rehearsal serve the purpose of holding the material long enough for encoding to occur. Learning a picture or visual takes lesser time than verbal inputs.However, both visual and verbal inputs are important informing a complete mental image. An individual doesn't perform encoding by directly representing an object in memory. The information is stored in a generalised form which has meaning for the individual. It is because of this reason that for different individuals the same object may get represented differently in memory. For instance, if one looks at a well grown tree and he is asked to think of an animals X may think of monkey, Y may think of bear and Z may think of cow. It is also found that pattern of encoding differs between men and women. Women are comparatively faster in recalling TV commercials. It is also brought to light that highly involving portions of programme may require viewers to allocate good amount of their cognative resources to such scenes. This results in reduced encoding and storing or commercial related information. Due to this, commercial shown within or adjacent to high involving settings are likely to be more effective with relatively lesser elaboration.

Whenever there is an information load or overload – over exposure of information at line–individuals experience the difficulty is encoding and storing all of it that causes confusion and consumers make less than optimum purchase decisions. That is, consumers are known to make less effective choices when exposed to too much high quality information or too much information. Then how much information should be considered as overload? The research findings say that consumers get confused when faced with 10 to 25 choices and with information about 15 to 25 product attributed.

The process of storage in long–term memory represents the organising and reorganising of information received after rehearsal from short term memory. It is reported that when consumers have little knowledge about brands, episodic exposure creates learnings as per the order in which the information is received. The product features of later introductions in the same product category, though quite novel, seem to be superfluous and uninteresting. 'Episodic' memory represents the information organised in a sequence as each event occurred after the preceding one. Thus, a person remembers the dates of first meeting of his 'would be', the date and time of marriage, the first issue and so on.

Finally, we will sumup the discussion on memory with some useful memory concepts that are relevant to advertising world. These are : 1) Unique and unexpected ad messages have greater potential for being remembered as these are not likely to be forgotten. 2) The beginning and the ending messages stand out most and interfere with remembering the material in between because of retroactive and proactive inhibition. 3) Ad messages that encourage immediate elaborative rehearsal of the information help in its retention 4) Chunking help processing and retention of more information in the limited time or space 5) The material that is meaningful and relevant to the individual is learnt more quickly and stands better chance of being retained in the long term memory 6) Visual information is more memorable than verbal 7) Incomplete or open ended messages encourage involvement and increase information retention.

CHAPTER BASED QUESTIONS

A. WRITE SHORT NOTE ON:

1. Learning 2. Components of learning 3. Drives 4. Cues 5. Response 6. Reinforcement 7. Retention 8. Classical conditioning 9.Learning Curve 10. Stimulus generalizations 11. Instrumental Conditionings 12. Memory.

B. SHORT ANSWER QUESTIONS:

1. What is stimulus discrimination?
2. What is instrumental conditioning?
3. What is memory?
4. What is classical-conditioning?
5. What is sensory-memory?
6. What is short-term memory?
7. What is long-term memory?
8. What is memory process?
9. What is learning?
10. What is learning curve?

C. ESSAY TYPE QUESTIONS:

1. What is learning? What are the components of learning?
2. Explain the classical conditioning theory.
3. Explain in full instrumental conditioning theory.
4. What is classical conditioning theory? What are its core concepts?
5. What is classical conditioning? How one can make classical conditioning more effective?
6. What is instrumental conditioning? Distinguish clearly between 'classical' and 'instrumental' conditioning.
7. What is memory? Explain the structure of memory
8. What is memory? What are different concepts of memory?
9. What is memory? Explain the memory process
10. What is memory? What are the advertising implications of memory in consumer behaviour?

******

CHAPTER 7

CONSUMER ATTITUDES AND CONSUMER BEHAVIOUR

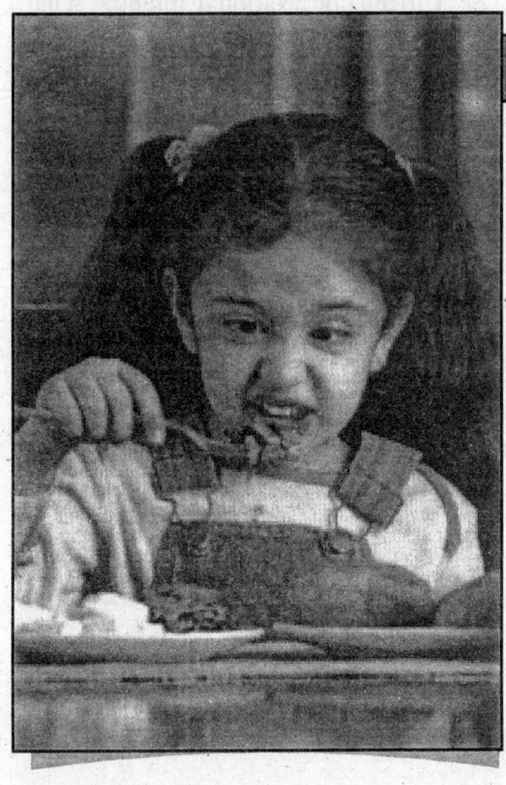

BACKDROP

The concepts of attitude and life-style occupy a focal position in both social psychology and consumer behaviour. Applied researchers are keenly and deeply interested in their own authority but as guides to behaviour. Same is the case with life-style concept which is gaining ground of late as quality of life has improved all over the world in general and developed nations in particular. Consumer behaviourist must master these two components to use them in helping the consumers to get best for what they pay and marketers best for their value adding. Here, I plan to touch upon the meaning, characteristics, functions and attitude development. Structure and structural models — theories and attitude change. The chapter ends with Chapter Based Questions.

ATTITUDE

There is no unanimity over the definition of word 'attitude' among the galaxy of social psychologists. In fact, **Mr. Martin Fishbein** points that there are not less than one hundred, definitions of attitude. We can not afford to go through all hundred definitions for time and spatial restrictions and reasons. Before one goes to atleast most widely accepted definitions, one can attempt what the attitude means ? An average person knows the significance of values, beliefs and attitudes on the behaviour of human-beings. Very often beliefs and attitudes are conflicting. That is why, there is nothing wrong in knowing these conflicting words.

WHAT IS BELIEF ?

'Belief' or an 'opinion' is an emotionally neutral cognition or a 'knowledge' that the individual holds about some aspect or object in the environment. That is, the things a person 'knows' to be true from his own point of view, consisting of the facts about something as he sees them at a given point of time. A 'belief' is a subjective concept of truth. An individual's beliefs are based more on emotion than reason and may or may not agree with facts. Even a false thing, for some, becomes a fact. A student using a **'Casio'** brand calculator has a feeling that he may go wrong it he uses say 'Orpat' or 'Citizen' or 'National' brand. That is why, good many students after calculating a given matter, they try to confirm their answers with others. Take another example, that Mr. India may believe, that 'Black Panthers' is a radical revolutionary organisation, which, to him, is a fact—a belief—and, therefore, is neutral in feelings or affect.

WHAT IS AN ATTITUDE ?

An 'Attitude' is a sister concept of belief. **An attitude,** by definition is not neutral, but is insistent, dynamic and stirred up. In above, examples, his beliefs are neutral. However, his attitudes are not neutral ; they may be positive or negative. Naturally, one can have a belief without an associated attitude about some object. For instance, a brand of pencil eraser. Here, the product exists which may be necessary but he neither hates it nor loves it. He is not willing to argue or join a conspiracy to destroy a particular brand of eraser and he holds no attitude towards the object. However, it would be possible to hold an attitude without also holding a belief. Attitudes cannot exist in vacuum. They must be directed toward something–either as concrete as a brand of an automobile or as abstract as the idea of freedom or responsibility.

STANDARD DEFINITIONS

★ **Professors David Krech and Richard Crutchifield:**

An attitude is "an enduring organisation of motivational, emotional, perceptual, and cognative processes with respect to some aspect of the individual's world."

★ **Professors Paul F. Second and Carl W. Backman:**

"It is a set of regularities of an individuals feelings, thoughts and pre-dispositions to behave towards some aspect of the world."

★ **Mr. Smith and Mackie:**

"An attitude is any cognative representation that summarizes our evaluation of an object, which may be the self, other people, things, action, events or ideas."

★ **Mr. Gordon W. Allport:**

" Attitudes are learned dispositions to responds to an object or class of objects in a consistently favourable or unfavourable way."

★ **Mr Richard E. Petty D.T. Wegener and I.R. Fabriger:**

"An attitude is the way we think, feel and act towards some aspect of our environment such as retail store, television programme or product."

Thus, it is a learned pre-disposition to respond in a consistently favourable or unfavourable manner with respect to the given object. An attitude is an individual's favourable or unfavourable inclination towards an attribute of an object. Attitudes are learned or acquired rather than inborn in an individual. They are formed as a result of the person's experimental and observational behaviour. An attitude speaks how far it is positive or negative, favourable or unfavourable or pro or con a person facts toward an object. That is, attitude as a feeling or evaluative reaction to objects. Attitudes are an expression of inner feelings that reflect whether a person is favourably on unfavourably predisposed to some 'object' - say a brand, or a retail outlet. Because they are an outcome of psychological processes, attitudes are not directly observable, but must be inferred from what people say or what they do. It is a trigger to action, as toward the purchase, of a brand of hair cream or an automobile, a hate of communists that can lead to enlisting in the military or negative feelings toward the colour of one's hair–black or brown or silver white or snow-white.

Attitudes immediately define one's position with the characteristics of dynamism and stirring. That is, the individual may not always be fully conscious of his attitudes. Opinions can be expressed, but attitudes must be inferred from the feeling–tone of verbal expressions or perhaps from the individual's overt behaviour. That is why, consumer behaviourists assess attitudes by asking questions or making inferences from behaviour. For instance, if a researcher determines from questioning a consumer that the individual in question consistently buys Hindustan Lever products and recommends them to friends, the researcher in likely to infer that the consumer has a positive attitude or favourable attitude towards Hindustan Lever products.

An attitude is much like a trigger of a rifle. If the conditions are right within individuals psychological field, if enough pressure is applied, it will lead to action, either overt behaviour, or anger, hostility, love or compassion toward the attitude object. Thus, an American individual may have mildly negative attitude towards rice as it is fattening ; he does not like its relatively bland taste. Again American cooks hardly cook rice is a nice way. However, an American seeing Japanese in motion pictures depicting cultural habits, he may develop a liking for rice as a cheap supply of calorie intake and it can be prepared in different or variant forms and might appreciate to such an extent that no meal is complete without rice. That is, he does not get hostile, worked up or emotional, nor does he decide to take violence against the next oriental movie on eastern cultural eating habits. However, if the psychological field is changed to such state that his better-half serves him pancakes with a side order of mushy rice for breakfast, he may push his plate aside, not eat at all and thrust out his lips and look sily. This patent illustration demonstrates clearly that the conditions are right for his attitude to emerge as on overt behavioural act, although his attitudes toward, rice were not action-prone previously.

CHARACTERISTICS OF AN ATTITUDE

The exact nature of an attitude can be figured out by its characteristic features. These are:

1. Attitudes are Learned Predispositions: What has been seen earlier is that there is general agreement over this point that attitudes are acquired or learned. That is, attitudes relevant and purchase behaviour are formed as a result of direct experience with the product, information acquired from others, or exposure to mass media, advertisers and other forms of direct marketing such as a catalog of a retail house or mail order house. A word of caution is hinted here that while attitudes may result from behaviour, they are not the same as behaviour. Instead, they reflect either a favourable or an unfavourable evaluation of the altitude object. Attitudes have a motivational quality as learned pre–dispositions. That is, they might propel a consumer toward a particular behaviour or repel the consumer away from a particular behaviour.

2. Attitudes have Consistency: Attitudes are relatively consistent with the behaviour they reflect. Despite their consistency, however, attitudes are not necessarily permanent and they do change. What consistency users, here in this context, is to be understood clearly. Generally, one expects that consumer behaviour to correspond with their attitudes. For instance, when Indias, is flooded with so many foreign cars, might prefer Italian car than Japanese and naturally one expects to buy Italian car than Japanese. Put alternatively, when consumers are free to act as they wish, we anticipate that their actions will be consistent with their attitudes. However, there are cases: where there is gap instead of consistency between attitude and behaviour. For instance, Indian consumer, on account of affordibility, might go us for Japanese car, instead of Italian model. Therefore, one should take into account the situational influence on consumer behaviour and attitudes.

3. Attitudes Occur within the Situation: The attitudes occur within and are affected by the situation. Here, situation means events or circumstances that influence the relationship between an attitude and behaviour at a particular point of time. A specific situation can cause consumers to behave in ways seemingly consistent with their attitudes. Let us take a case of a house–wife Mrs. Ducks who purchases shampoo each time a different brand. Though her brand switching behaviour may seem to reflect a negative attitude or dis-satisfaction with the brands she tried, it may be that she is actually influenced by a specific situation—may be her decision to go economical. Though she uses 'Head and Shoulder' Shampoo, she might go for an ordinary brand which is cheaper say, 'Gleem' brand. Opposite is also true. Let us take Mr. Johny who rents a car always from Mr. Albert Pinto each time he goes on holidays, we assume that he will change as some other, Mr. Paul is renting it at cheaper rate and has brand new cars for rental. However, contrary to our belief Mr. Johny says Mr. Albert Pinto is 'Okay' though he charges little more as he has long–standing relations.

In fact, individuals have variety of attitudes toward a particular behaviour, each corresponding to a particular situation. For instance, Mr. Gopi considers 'Hotel Aram' opt for lunch but not for dinner for he comes home usually in time back from his college. However, this Friday, he came late from college, fully hungry, not prepared to have dinner at home, prefers to go in for 'Hotel Aram' because, odd hours and his tiredness and extreme, uncontrollable hunger. Can we say that he has changed his attitude ? Not certainly. Here, it is significant to note that consumer attitudes vary from situation to situation and his way of approach is that decides the linkage between attitude and consumer behaviour.

4. Attitudes have An Object: By very nature, attitudes are having an object. The very concept of attitude is based on object which forms the heart of attitude and consumer behaviour. The word 'object' as interpreted in consumer behaviour literature, means specific consumption or marketing, related concepts such as products, product categories, brand, service, possessions, product, use, causes or issues, people, ads, price, medium or a dealer. It can be a physical or abstract thing. The base or foundation for all attitude study and research is object. Therefore, the researchers tend to be object-specific. For instance, if we are interested in learning consumers' attitude towards three major apparel retailers, our object might include-"The Ultra" "The Trend" and "The Readymade"- say in Panjim City of Goa State.' Similarly, if we are examining the consumer attitudes toward major brands of 'Note-book' computers, our object must be say **IBM'-Dell, Compac, Toshiba, Gateway–2000** and so on. These "note book" PCs' are also called as 'laptop' PCs.

5. Attitudes have Direction Degree and Intensity: Any attitude expresses how a person feels about an object. It clearly expresses three things namely direction, degree and intensity.

The expression of 'direction' means a person not neutral but either positive or negative or favourable or unfavourable or for or against. The object in question when one turns to the 'degree'—it speaks of the extent to which he likes or dislikes the object. Terminally, when we come to the 'intensity' it speaks of the level of sureness or confidence of expression about the object or how strongly a person feels about his on her conviction. One may think that 'degree' and 'intensity' are same or seem the same. They are closely related but not same or synonymous. For instance, Mr. Jackson thinks that 'Tornado' brand vacuum cleaner made by **Eureka Forbes Ltd.** is poorly designed. This indicates that his attitude is negative and the degree is negative feeling of quite large. However, the individual may have very little conviction or feeling of sureness which we call 'intensity' that he is right. His attitude, thus, could be more easily changed is a favourable direction than a person who feels a strong conviction that 'Tornado' vaccum cleaner is poorly made. These three aspects namely, 'direction', 'degree' and 'intensity' of an individual's attitude toward a product or service provide the marketing people an estimate of his or her readiness to act toward the product or service. However, a shrewd marketer understands how important the consumer attitude is as against other attitudes and the situational blockades like his ability to pay, or ethical considerations that come in the way of purchase decision.

6. Attitudes have Structure : Human attitudes represent a well-organised structure, meaning thereby they have internal consistency and possess inter- attitudinal centrality and they tend to be stable so that they are generalizable, having varying degrees of salience Attitudes stand themselves not in isolation. On the contrary, they are associated or tied in with each other to form a complex unit. It means that there should be certain degree of consistency because they are related that warrants fitment between them failing which conflict will set in. They should also show greater degree of consistency than more peripheral attitude do, because the more central attitudes are related to a larger number of other attitudes. Because attitudes cluster or cling into a structure, they tend to show stability over a period of time. Further, attitudes tend to be generalisable. In other words, a person's attitude towards a specific object tends to generalise toward a class of objects. Thus, a person who purchases Philips, dry-cell for torch and for his walkman, finds them less effective and, therefore, he generalizes that all Philips products are not designed and manufactured properly. It is also true that a person uses over a period Anchor brand electrical products and finds them quite reliable, goes in for newly introduced product by 'Anchor' say toothpaste, he says whatever, 'Anchor' brand products are there, they are good, though it may not be actually so. Another aspect of structure is that some attitudes are more important or salient to him or her than others. Thus, an Indian lady might feel that "buying Indian" is more important to her than "saving money." Such attitudes increase salience.

FUNCTIONS OF ATTITUDES

Human attitudes perform certain functions in furthering consumer behaviour. After all, an attitude is the way we think, feel and act toward some aspect of our environment say a retail outlet, a television serial or a product or a service. The functions the attitudes perform is that they attempt to enhance the image they hold of themselves. In other words, functions of attitudes are the motivational bases which shape and reinforce positive attitudes toward goal objects, perceived as need satisfying and negative attitudes toward objects perceived as purchasing or threatening. In fact, these functions help us in understanding why people hold the attitudes toward psychological objects. The following figure makes this amply clear.

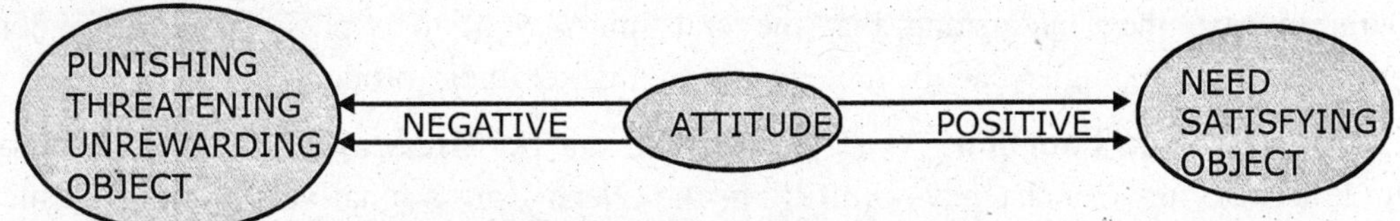

FIG: 7:01 ATTITUDE DEVELOPMENT AND FUNCTION BASED ON PERCEIVED NEED SATISFACTION OR HARM AVOIDANCE

Source: Adopted from- Consumer Behaviour, Written by **D.L. Loudin and A.J. Della Bitta.**

These functions of attitude arc four as discussed by **Mr. Daniel Katz** in his article "The Functional Approach to the Study of Attitudes" in journal entitled PUBLIC OPINION-Quarterly 24 published in I960. These arc : Adjustment function — the ego- defensive function, the value expressive function and the knowledge function. Let us note these in brief.

1. Adjustment Function: Mr. Daniel Katz found that adjustment function directs the people towards pleasurable or rewarding objects and away from unpleasant and undesirable ones. That is, it serves the utilitarian concept of maximizing reward and minimizing punishment. The attitudes, thus, of consumers depend to a large degree on their perceptions of what is need satisfying and what is punishing or penalising. Because the consumers perceive products, services or stores as providing need-satisfying or unsatisfying experiences, one may expect their attitudes towards these objects to vary in relation to the experiences that have occurred in consumers life. For instance, Mr. Toffee is a heart patient and is a learned person. He has read that, the latest research findings are controversial. One research finding says " 'Kissing' in normal sense is soothing for heart–troubles", another research finding say" 'Kissing' is harmful to heart patients" In such a situation, it is Mr. Toffee's own reading or attitude as to what is right and accordingly he might go in for or keep away.

2. Ego-Defensive Function: The attitudes that are formed to protect the 'ego' or 'self–image' from threats that help in fulfilling ego defensive function. In practical world, many overt or outward expressions of such attitudes reflect the opposite of what the person perceives himself or herself to be. For instance, a house lady who decides to buy a household flour–mill say 'NAVDEEP' Safari Auto Mill fully automatic, made by **M/s Navdeep Products Ltd.** of Daman, costing Rs. 10,500, may defend herself very strongly as the decision taken by her as very correct or at the same time might pass on the blame for the advice of her neighbour. Such ego-defence attitudes help us in protecting one's self-image, and very often one is unaware of them.

3. Value Expressive Function: If ego-defensive attitudes are formed to protect one's self-image, the value expressive attitudes help in expressing a person's focally held values. That is why, most consumers adopt certain attitudes in an attempt to translate their values into something more tangible and easily expressed. It happens very often with traditional or conservative persons. That is, those who do not want to show of their riches, might develop the attitude of being open in expressing their hidden purchasing power. Thus, Mr. Stingy might buy costly and bright clothes, above average automobile and so on. Here, marketer can gain in case they understand that the consumers want to express their value concept. Accordingly ad compaigns can be developed to market their products.

4. Knowledge Function: As rightly pointed out by **Mr. Mark P. Zauna** in his article "Attitude-Behaviour Consistency—Fulfilling the Need for Cognative Structure"-edited by **Mr. Thomas K. Srull** in title CONSUMER RESEARCH-VOLUME 16, Association for Consumer Research, Provo, UT-1989, humans have a need for a structured and orderly world and therefore, they seek, consistency, stability, definition and understanding. Out of this need, develops attitude towards acquiring knowledge. Further, the need to know tends to be specific. That is why an individual, who does not play chess nor he wishes to learn about such an indoor game, is most unlikely to seek knowledge or an understanding of the said game. This is surely going to influence the amount of information search devoted to this particular topic. In essence, out of our need to know develop attitudes about what we believe that we need or do not need to understand. In addition, attitudes enable consumers to simplify the complexity of the real world. One knows fully well that real world is too complex for us to cope up with easily. Therefore, we develop good number of mechanisms to simplify the complex situations. This process of simplification warrants sensory thresholds and selective attention and, therefore, attitudes. Attitudes allow us to categorise or group objects as a way of knowing about them. Thus, when a new detergent powder say 'Tide' is introduced to the housewives, try to categorise it into a group of other brands of detergent powders about which they know something. In case the house–wife develops a positive attitude towards 'Tide' than others, she must have taken into account plus points of 'Tide' as compared to other alternative detergent brands made by the same company or different companies.

DEVELOPMENT OF CONSUMER ATTITUDES

Development and formation of attitudes and beliefs is basically a learning process. In reality, acquisition of attitudes supports the contention that learning occurs in a process of the interaction of an individual with others. That is, attitudes are formed as the result of all the influences that have on an individual namely, perception, learning, memory, product positioning, motivation, personality, emotion, self-concept, life-styles and so on. By nature, attitude is an enduring organisation of motivational, emotional, perceptual and cognative processes with respect to some aspect of our environment. It is a learnt predisposition to respond in a consistently favourable or unfavourable manner with respect to a given object, that is, as noted earlier, our attitude is the way we think, feel and act towards some aspect of our environment, may be a store, a product or service. All attitudes ultimately develop from human needs and the values people place upon the objects that satisfy those perceived needs.

SOURCES OF ATTITUDE DEVELOPMENT

There are sources of attitude development. Many of the human needs and, therefore, attitudes, hence there is need for creating an awareness of needs first and attitudes. These sources are :

1. Personal Experiences: We as consumers or buyers come in contact with objects in everyday living environment. All of us are not fully aware of all objects. These objects are both existing and therefore old and science and technology and other elements of environment brings in new objects. As consumers, we evaluate these new and revaluate the old and this process of evaluation helps in developing attitudes towards objects. Let us take a case of house–wife who was using 'Singer' food processor and it worked very well for eight months. Being unsatisfied, she bought Kenwood processor that she expects would work at least two years as per guarantee and warranty conditions. By evaluating the new brand revaluating the old, she develops an attitude towards the food–processor.

Direct experience of ours with sales representatives, products, services and the dealers, helps us to create and shape our attitudes toward those market objects. Several factors, however, influence how we are going to evaluate such direct contacts. These factors are 'needs', 'selective perception' and 'personality'. Individual needs differ and change over a period of time. As a result, people can develop different attitudes toward the same object at different points in their lives. It is equally important that personal perception that people operate on their own personal interpretation of reality. That is why, the way people interpret the information about objects say-product, store, service and so on affects their attitude towards them. Personality is another factor that influences the way the people process their direct experiences with the objects. The nature of personality that is more important some are aggressive, some are passive, still some are introverts and some are extroverts. These differences affect the attitudes they form.

2. Group Associations: A most all of us are influenced to one degree or the other by other members in the groups to which they belong. Attitudes become target for the influence of these groups. Our attitudes towards products, services, stores, ads, ethics, welfare and a complex of such subjects that form objects are influenced deeply and strongly by the groups we value and honour with which we are associated. These groups are family, work and peer groups, cultural and sub-cultural groups all influence our attitude development. Children in family form their attitudes toward objects influenced by role behaviour of different members of family. Work and peer groups that work on"birds of same feather flock together" represent groups of like-minded persons. The like mindedness may be, age group, work group, play group, affinity group having definite objectives in common. Though children come from all types of families, they meet as friends, players, and associates, by which group values affect their individual attitudes. Thus, a group of college goers might decide to have 'Nike' shoes or 'Puma' shoes and 'Nike' T-shirts or 'Adidas' T-shirts and 'Lee' Jean pants, they form that all college goers are identified as college students than primary school children. These group have definite concepts, group programmes which speak of their needs and attitude. Again, cultural and sub-cultural groups do have their value system and accordingly they develop their attitudes.

3. Influence Groups: These are called as opinion leaders. A consumer's attitude is formed and shaped and changed through personal contact with influential people, may be highly respected friends, relatives experts, and stars. These are opinion–leaders who are respected by their followers where individual follower's attitudes are influenced by opinion leaders. It is our usual experience, that we hear a speech of an expert in very flowery language and diction and accent. We also try to copy. If your favourite hero has parted his hair left side, you do so, your dresses, inner wears will be of his brand and so on.

Advertisers have taken full advantage of this. As Amitabh Bachhan is the anchor of "Kaun Banega Crorepati" goes on sharing his ideas with participants. When the Zodaic signs of both are Libra, coming from U.P. State, suffering from asthma, the participants and audience think that they arc more or less equal to Amitabh. Attitudes are influenced by commercial ads of "Slice of life" type. These ads show 'typical' people fore–fronting a 'typical' problem or problems and finding similarly a 'typical' solution to the problem or problems. The most common examples of .these kind are, use of "Head and Shoulder" shampoo, "Colgate total", after shave lotion "Old Spice" or 'Denim', hair cream "Brylcream" or "Clinic-Active" and so on. The following figure gives a comprehensive picture of inter-personal relationships of an attitude toward a product or a brand and other related psychological processes.

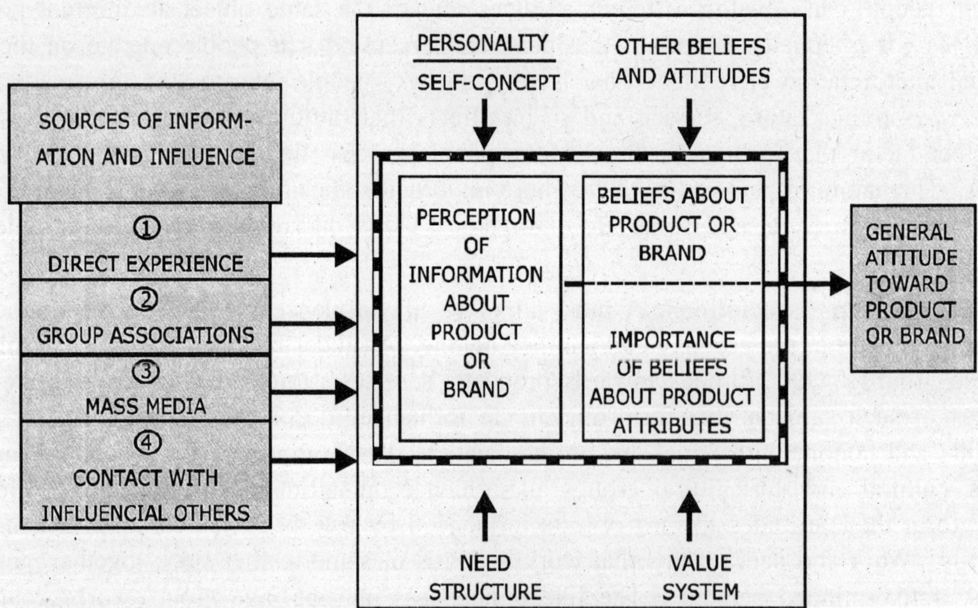

FIG: 7:02 INTER-PERSONAL RELATIONSHIPS OF AN ATTITUDE TOWARD A PRODUCT OR BRAND AND OTHER

RELATED PSYCHOLOGICAL PROCESSES

The summary of above figure is that several sources provide consumers with information and influence about products, services, stores and other objects. An individual consumer or prospect gets the information selectively and distorts the information according to his or her individual needs, values and personality and in order how well the information 'fits' with the existing beliefs and beliefs he or she has. This processed information initiates either development, change or conformation in the consumer's beliefs about the product and the importance of each of the products; attributes to her and her current needs. Out of this entire process the outcome is a synthesis of a general attitude.

STRUCTURE OF ATTITUDES AND STRUCTURAL MODELS

Each attitude is a structure or make up of several components. In order to understand the relationship between attitudes and the behaviour, psychologists have tried their best to design models that help in capturing the underlying dimensions of an attitude. In this regard the researcher's focus has been on specifying the composition of an attitude to explain or predict the consumer behaviour in a perfect way. Though, there are good many theories or structural models, an attempt is made to four major models namely, Tri-component Attitude Model, Multi-attribute Attitude Models, trying to consume Model and the Attitude toward the Ad-Model. Naturally, each structural model attempts to provide us with differing perspective on the member of component parts of an attitude and how these parts are arranged or inter-related.

1. TRI-COMPONENT ATTITUDE MODEL

According to the **tri-component attitude model** attitudes consists of three major components namely, COGNATIVE, AFFECTIVE and CONOTIVE components**.** The conotive component is also called as BEHAVIOURAL component. This model can be presented as under in two distinct ways in the form of configurations which are given.

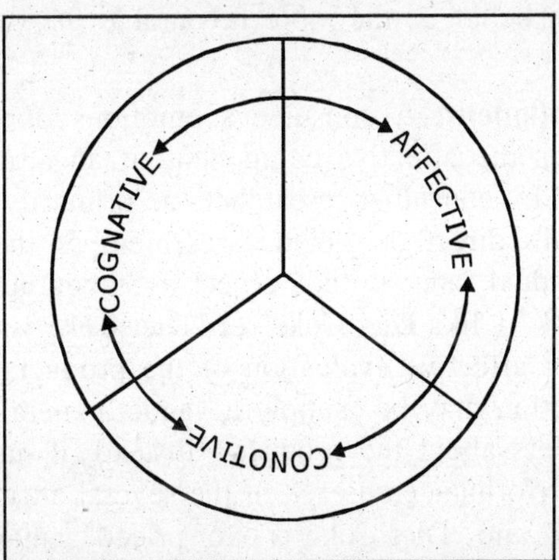

FIG: 7:03 SIMPLE MAKE UP OF TRI-COMPONENT ATTITUDE MODEL ANOTHER CONFIGURATION CAN BE GIVEN AS UNDER

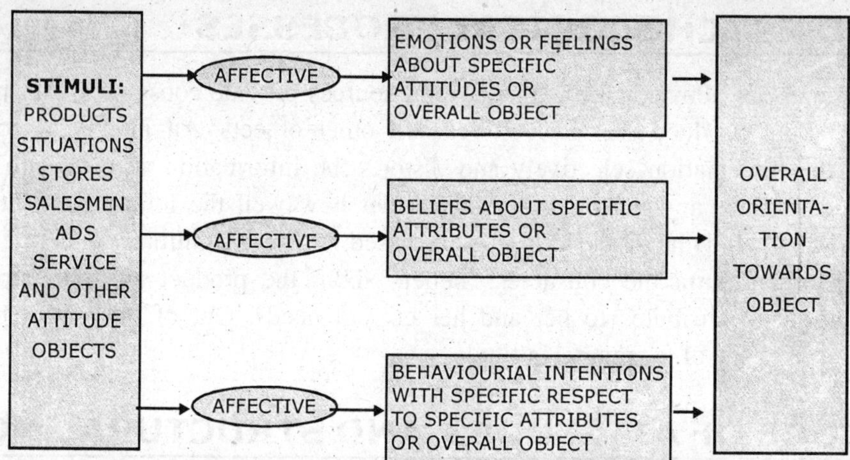

FIG: 7:04 - ATTITUDE COMPONENTS AND THEIR MANIFESTATIONS

According to Tri–component Attitude model, attitudes comprise of three major components namely, cognative, affective and conotive. Let us know about each component.

1. The Cognative Component: This component is made up of a person's cognitions. That is, the knowledge and perceptions that are acquired by combination of direct experience with the 'attitude object' and related information from different sources. This implies consumer beliefs about an object. This knowledge and resulting perceptions commonly take the form of beliefs. That is, consumer believes that the attitude object has various attributes and that specific behaviour will lead to specific outcomes. Thus, a person has strong belief that a Diet Coke, has almost no calories, caffaine free competitive price and has been made by a big company. The total configuration of beliefs about this brand of soda represents the cognative component of an attitude toward **Diet Coke.** It is significant to bear in mind that beliefs need not be correct or true. They only exist. Most of the beliefs about attributes are evaluative in nature. For instance, high gas mileage, attractive styling and reliable performance are generally viewed as positive beliefs. Therefore, the more positive beliefs associated with a brand and more positive each belief is, the more favourable the overall cognative component is presumed to be.

2. The Affective Component: A consumer's emotions or feelings about a particular product or a brand constitutes the affective component of an attitude. These emotions and feelings are frequently treated by consumer researchers as primarily evaluative in nature. That is, they capture an individual's direct or global assessment of the attitude object. That is, the extent to which the individual rates attitude object as 'favourable' or 'unfavourable', 'good' or 'bad'. A consumer who says "1 like Diet Coke" or "Diet Coke is terrible soda" is expressing the results of an emotional or affective evaluation of the product. It is worth while to note here that this overall evaluation may be simply a vague, general feeling developed without cognative information or beliefs about the product. Instead, it may be the result of several evaluations of the products' performance on each of the several attributes. Thus, the statements such as "Diet Coke tastes bad" and "Diet Coke is over-priced", may mean negative affective reaction to specific aspects of the product which will determine the overall reaction to this brand of soft-drink, in combinations with feelings about other attributes.

Affect laden experiences manifest as emotionally charged states like happiness, sadness, shame, disgust, anger, distress, guilt or say surprise. Such emotional states may enhance or amplify positive or negative experiences that later recollections of such experiences' may impact what comes to mind and how the individual acts. Since products, like other objects one reacts, are evaluated in the context of a particular situation, a consumer's affective reaction to product as well as beliefs about the product may change as the situation changes. For instance, a consumer may believe that (1) **Diet Coke** has caffaine and (2) Caffaine keeps one awake. These beliefs may cause a positive response when consumer needs to stay awake to prepare for an examination and a negative response when he wants to drink something late in the evening that will not keep him awake later. While feelings are often the outcome of evaluating specific attributes of a product, they can proceed and influence cognitions.

3. The Conative Component: Conation is concerned with likelihood or tendency that an individual will undertake a specific action or behave in a particular way with regard to the attitude object. This behavioural component of an attitude is one's tendency to respond in a certain manner toward an object or activity. Thus, a series of decisions to purchase or not to purchase 'Diet Coke' or to recommend it or not to his or her relatives and friends is a reflection of the behavioural component of an attitude.

Since, behaviour is generally directed towards an entire object, it is less likely to be attribute specific than either beliefs or affect. However, this is not always the case especially with respect to retail outlets. For instance, good many consumers buy canned products at discount or grocery outlets but purchase meats and fresh vegetables at regular markets. This instance makes it clear that there can be common reaction behaviourally to specific beliefs so far as outlets are concerned, but same is not the case with the products, because one has to buy or not to buy the complete product.

What is important to note in case of all three components of attitude tend to be consistent as indicated by the following figure:

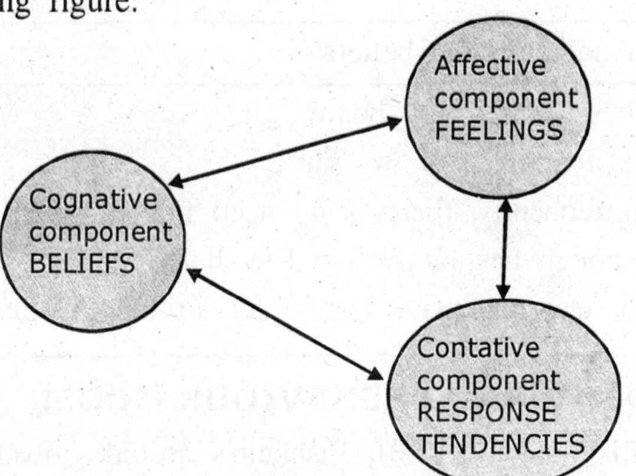

FIG: 7:05 ATTITUDE COMPONENT CONSISTENCY

2. MULTI–ATTRIBUTE ATTITUDE MODELS

These models project consumer attitudes with regard to an attitude 'object'-may be a product, service, a direct mail catalog, an ad, brand a cause or an issue–as a function of consumers' perception and assessment of the key attribute or beliefs held with regard to the particular attitude 'object'. In fact, there is no lack of such models. However, those proposed by **Mr. Martin Fishbein** and his associates have stimulated the greater amount of research interest. Therefore, here only three models of this version are considered without going into the deeper technicalities. These models considered are-the – **Attitude toward object model, attitude toward behavioural model and the theory of Reasoned action model.**

A. THE ATTITUDE-TOWARD OBJECT MODEL

According to this model, (ATO), the consumer's attitude toward a product or specific brands of a product is a function of the presence or absence and evaluation of certain product specific beliefs and or attributes or both. Put in other words, consumers generally have a favourable attitudes towards those brands that they believe have an adequate level of attributes that they have evaluated as positive and they have unfavourable attitudes towards those brands about which they feel do not have an adequate level of desired attributes or have too many negative attributes.

The following is the case constructed is case of Laptop or Notebook PC of a **HELEVETTE PLACARD**, popularly known as HP based on Fishbein Attitude Toward object model of two consumer belief system :

Mr. Jonny—having favourable beliefs

★ 'HP—Lap-top' is worth the extra price charged.
★ 'HP—Lap-top' is small size having black case giving really a professional look.
★ 'HP—Lap-top' has truely innovative Key-board.
★ 'HP—Lap-top' has been highly evaluated by computer magazines in India.
★ 'HP—Lap-Top' is most suited for a person like me who is always on the move.

Mr. Dodo—having an unfavourable beliefs.

★ 'HP—Lap-Top' has normal Key–board.
★ 'HP—Lap-Top' is too high for my purse.
★ Since I travel infrequently, there is no need for 'HP—Lap- top' PC.
★ 'HP- Lap-Top' though has innovative Key board might break down soon.
★ I am looking for less expensive and larger lap-top PC than HP computer company.

B. THE ATTITUDE TOWARD BEHAVIOUR MODEL

According to this model, **(ATB),** Fishbein's attitude toward behaviour model is the individual's attitude toward behaving or acting with respect lo an object, rather than the altitude toward the object itself. The appeal of the altitude toward behaviour model is that it merely corresponds closely to an actual behaviour than the attitude toward object model as we have seen. This model can be presented as under in the form of a figure.

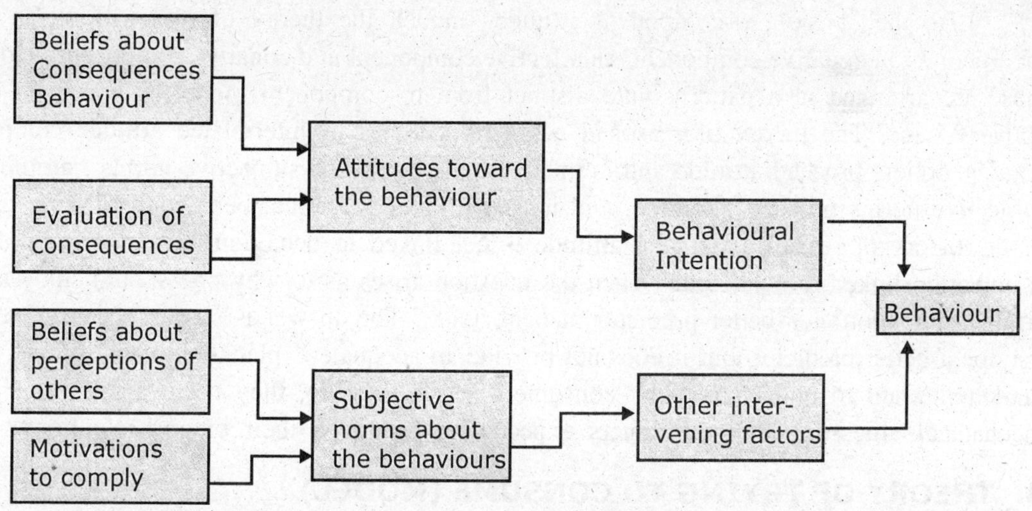

FIG:7:06 THE RELATIONSHIP OF COMPONENTS IN FISHBIEN'S BEHAVIOURAL MODEL

To take an example of a consumer who is comparing two brands of automobiles say 'Opel Astra Glimca' costing 8,95,000 against 'Opel Astra Club' costing 10,09,000 rupees, his act of purchasing—that his altitude toward the behaviour reveals more about the potential act of purchasing than does simply knowing his attitude toward the brand. He might have positive attitude towards 'Glimca' and unfavourable altitude about 'Club' for his own reasons.

C. THE THEORY OF REASONED ACTION (MODEL)

The theory of reasoned action (TRA) builds on other research conducted by **Fishbein** and his associates. It represents a comprehensive integration of components into a structure that is designed to lend both better explanation and prediction of behaviour. This can be configured as under:

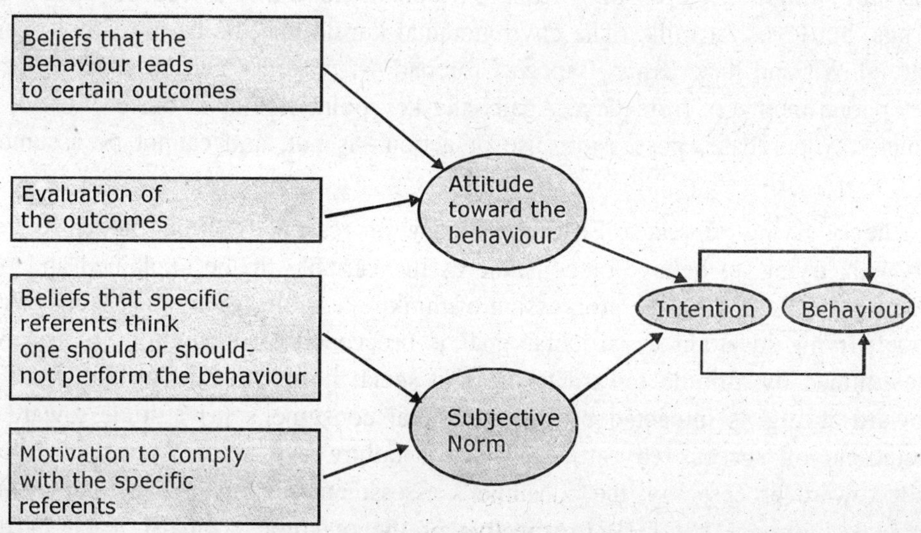

FIG: 7:07 A SIMPLIFIED VERSION OF THE THEORY OF REASONED ACTION

Like the basic tri-component attitude model, the theory of reasoned action model incorporates a cognative component, an affective component and conative component. However, these are arranged in a pattern quite distinct from tri-component model as portrayed by the above figure. The theory of reasoned action is a series of interrelated attitude components namely, beliefs proceed attitude and normative beliefs proceed subjective norms : attitudes and subjective norms proceed intention, and intention proceeds actual behaviour. Taking the side of the theory of reason action, an attitude is not linked to behaviour as strongly or directly as intention linked to behaviour. Then the question arises as to why one should study attitude at all if 'intention' is a better predictor of behaviour ? The answer is simple because 'intention' may be a better predictor, but it does not provide an adequate explanation of behaviour. When marketers want to understand why consumers act as they do, they need more than basic mechanical measure of consumers expectations—namely, their buying intentions.

3. THEORY OF TRYING TO CONSUME (MODEL)

The theory or model of reasoned action operates pretty well, when an outcome, say the actual purchase of a product, is closely proceeded by deliberation leading up to the purchase and there is nothing to block the consumer behaviour say buying a cable digital T.V, set or a camera. However, **Fishbein** model in this connection does not specifically account for those circumstances when a particular action or behaviour is directed towards a goal, as pointed out by **Richard P. Bagozzi** and **Palm R. Warsaw**. Therefore, attempts were made to extend. Fishbein theory of reasoned action in order to make it accommodating, the consumer goals as expressed by their "trying" to consume. The theory of **trying to consume (TTC)** is designed to account for good many cases where the action or outcome is not certain but reflects consumer's attempts to consume or purchase instead. In such cases, there are personal impediments and environmental impediments or both, that might prevent the desired action or outcome from occurring. Here, a consumer may face personal hurdle, may be he wants to buy **Rayban Sunglasses** upto only Rupees 1,000 or he wants to reduce weight from 90 kgs to 85 kgs, but loves fat milk. The environmental hurdle may be he is ready to participate in programe of Wit and luck "Kaun Banega Crorepati"—but is not getting contact line to four megopolies phone members from Goa. Again, the key point is that in these cases of "trying, the outcome—say purchase, possession, use or action—is not, and cannot be assumed to be certain.

The theory trying to recast **Fishbein theory** of reasoned action model by replacing 'behaviour' with 'trying' to behave or consume as the variable tlo be explained and predicted or any one of them. Following the progression of thinking established by the theory of reasoned action model, trying to attain a particular goal is proceeded by "intention" to try, which in form is determined by attitude toward trying and social norms toward trying also. Further, attitude toward 'trying' is impacted by the individual consumer's (a) attitude toward success and expectations of success (b) attitude toward failure and expectations of failure and (c) attitude toward process say, the consumer's assessment of how the 'act' of trying to do something makes him or her feel, irrespective of the outcome. Finally, each of the three attitudes—towards success, towards failure and towards process—are determined by the summation of the 'product' of the consequences likelihood—say the chance "that an individual anticipates" Working better from strict dieting—and consequences evaluations—say, how pleasant it would be for an individual to "look better."

For the benefit of readers, two things are given—a box of selected instances of potential impediments that might impact on trying and the figure of theory of trying lo consume.

BOX showing **SELECTED INSTANCES OF POTENTIAL IMPEDIMENTS THAT MIGHT IMPACT ON TRYING TO CONSUME :**

POTENTIAL PERSONAL IMPEDIMENTS

1. "I wonder whether my finger nails will be longer by the time of my wedding".
2. "I want to try to lose fifteen pounds by next summer."
3. "I am going to try to get tickets for a Broad-way Show for your birthday".
4. "I am going to attempt to give up smoking by my birthday."
5. "I am going to increase how often I go to the gym from two to four times a week."
6. "Tonight, I am not going to have dessert at the restaurant."

POTENTIAL ENVIRONMENTAL IMPEDIMENTS

1. "The first ten people to call in will receive a free "T" shirt."
2. "Sorry, the shoes did not come in this shipment from Italy"
3. "There are only three bottles of champagne in our store- room, You came in today than tomorrow."
4. "I am sorry, we cannot serve you, we are closing the restaurant because of a problem with the oven".
5. "Please come other than Sunday as bridge remains closed."

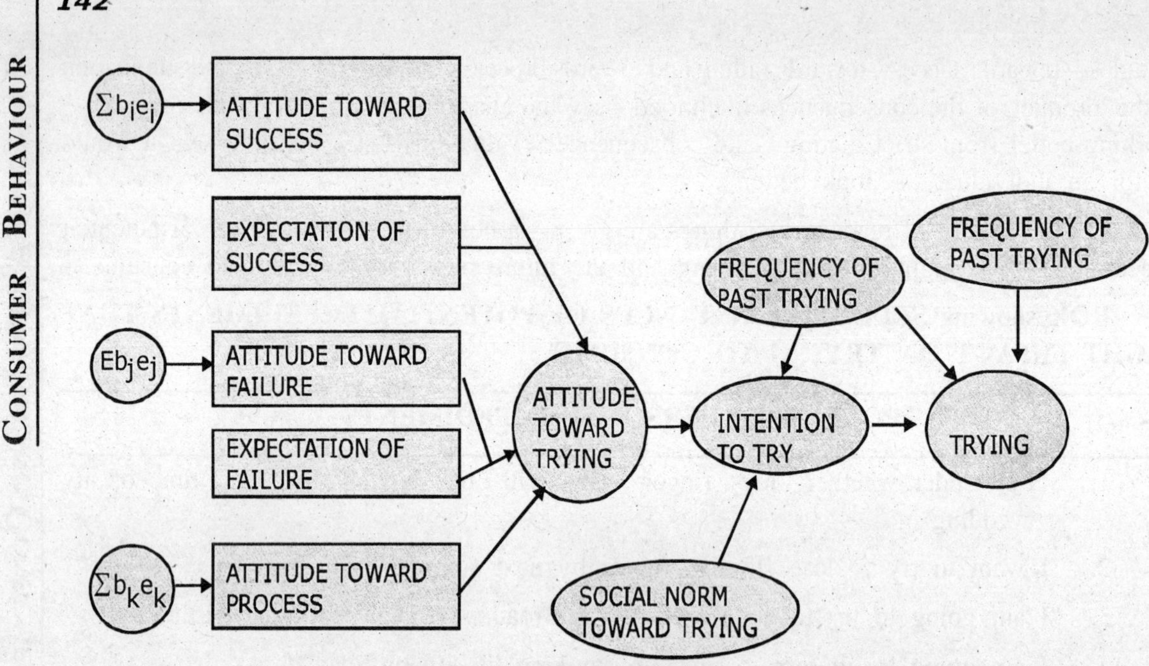

FIG: 7:08 THE THEORY OF TRYING TO CONSUME

NOTE: Regarding the "E be", the b's are consequence likelihoods, the "es" are the consequence evaluations; subscript 'i' refers to consequences contingent on success; subscript 'j' refers to consequences contingent on failure and subscript 'k' refers to consequences associated with the process of trying, independent of success or failure considerations.

The above model proposes unique additions that the frequency of past trying—the consumer's prior experience with trying—impacts on both intention to try and the act of trying, and that frequency of past trying—the consumer's most recent experience with trying—impacts solely on trying. The results of initial research support the importance of these two factors in providing an understanding and predictive accuracy of trying to consume.

4. THE ATTITUDE TOWARD THE ADVERTISING MODEL

This attitude toward the Ad model is an attempt to understand the impact of advertising or some other promotional vehicle say, catalog on consumer attitudes toward a particular product or brand. This model projects the consumer forms various feelings or affects and judgements or cognitions as the result of exposure to an ad. These feelings and judgements in turn affect the consumer's attitude towards the ad and beliefs about the brand acquired from exposure to the ad. Finally, the consumer's attitude toward the ad and beliefs about the brand influence his or her attitude toward the brand. The researchers maintain that it is critical to clearly distinguish between and separately measure cognitive evaluations of the ad-that is, judgements about ad and affective responses toward the ad that is, feelings experienced from exposure to the ad, in assessing consumer attitude toward an ad. This model is presented in the following configuration.

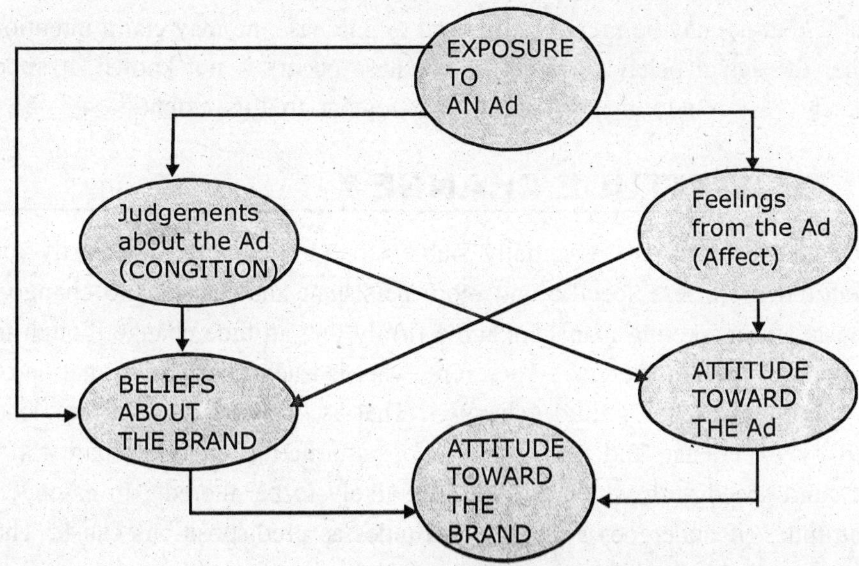

FIG : 7:09 A CONCEPTION OF THE RELATIONSHIP AMONG ELEMENTS IN AN ATTITUDE—TOWARD THE AD MODEL

SOURCE : Journal of Consumer Research 14-Dec. 87P. 431-in the article "The Power of Feelings in Understanding Advertising Effects"—by **Julie. A. Edell** and **Marian Chapman Bruke.**

In the final analysis, research reveals that attitudes toward ads in general seem to have little impact on the attitude toward a specific advertisement. However, the attitude toward a specific type of advertising—say "comparative advertising"—may have some impact on the attitude toward a specific advertisement. These findings, if supported or corroborated, would establish that individuals who profess to 'hate' ads in general or to 'love' advertising, would still be likely to form a unique attitude toward a specific advertisement. In case of responses to a specific type of advertising, the implications are different. It appears that if a consumer dislikes comparative advertising, a specific comparative ad would be negatively influenced by the attitude toward comparative advertising in general.

ATTITUDE CHANGE

Earlier we have learnt about formation or development of consumer attitudes. Social scientist and marketing scholars and practiceners have assumed since very long that there is necessarily a relationship between attitude change and behavioural change. Interestingly, experimental evidence has indicated that behaviour change will often lead to attitude change. However, whether attitude change leads to behaviour change has not been clearly demonstrated. Thus, an advertiser is able to inculcate positive attitudes and a positive image within a consumer does not necessarily mean that the consumer will actually buy his product. In fact, there are data to show that he may be more predisposed to purchase, he may claim intentions to purchase but whether the actual behavioural act of purchase occurs is not known. It appears that there is insufficient research to make a definite statement to this extent.

WHAT IS ATTITUDE CHANGE ?

As noted attitudes are essentially stable structures and are not easily modified unlike motives, attitudes are less specific and more persistent and resistant to change. Thus, social psychologists and marketing managers agree firmly that attitude change though not impossible, is difficult and expensive exercise. Therefore, what is true of attitude formation or development is equally true in case of attitude change. That is, attitude changes are learned, they are influenced by experience and other sources of information and personality affects, both the receptivity and speed with which attitudes are likely to be altered. In essence, the structural state of attitudes, if undergoes a change, attitudes as predispositions can be changed towards an object, a product or service or anything for that matter which is the object of attitudes. However, one thing is sure that changing attitudes is not an easy job that calls for good deal of talent, treasure and time on the part of those who wish to encash on such attitude and, therefore, behaviour change in consumers.

STRATEGIES OF ATTITUDE CHANGE

How one can change effectively the consumer attitudes to influence their behaviour favourably to one's advantage? It is a matter of strategy. There are certain attitude change strategies that promise good performance in this regard. Different authors and researchers have given these strategies in the way they thought most fitting. Notwithstanding these differences, one can go in for a compendium of five strategies. These are :

(1) Changing the consumer's basic motivational function

(2) Associating the product with an admired group or event.

(3) Resolving two conflicting attitudes

(4) Altering components of the multi-attribute model and

(5) Changing consumer beliefs about competitor's brands.

1. CHANGING THE BASIC MOTIVATIONAL FUNCTION :

One of the effective strategies for changing consumer attitudes toward a product or brand is to make needs prominent. One method for changing motivation is known as the functional approach as profounded by **Professor Daniel Katz and associates.** According to this approach, altitudes can be classified in terms of four functions namely, **UTILITARIAN, EGO-DEFENSIVE, VALUE EXPRESSIVE and KNOWLEDGE.** These are discussed earlier in this chapter.

However, here is an attempt to use them as solutions to the issue of attitude change. Let us touch these to have perfect understanding.

A. THE UTILITARIAN FUNCTION:

Generally, we find that we as consumers hold certain brand attitudes partly because of a brand utility. It is but natural that our attitude towards that brand or product tends to be favourable because it has helped us in the past, may be in a small way. In such situations, one way of changing attitudes in favour of a product is by showing the people that the brand or product can serve an utilitarian purpose that they may not have considered. For instance, a particular brand of detergent soap powder that provides the user stain remover, colour conditioner and even providing more brightness and smoothness to washed clothes. That way, all producers say like **Hindustan Lever, Proctor and Gamble, Godrej** and so on. However, a company which guarantees all, at–least cost comparatively, appeals to an utilitarian benefit. The side given ad is one of the finest ads.

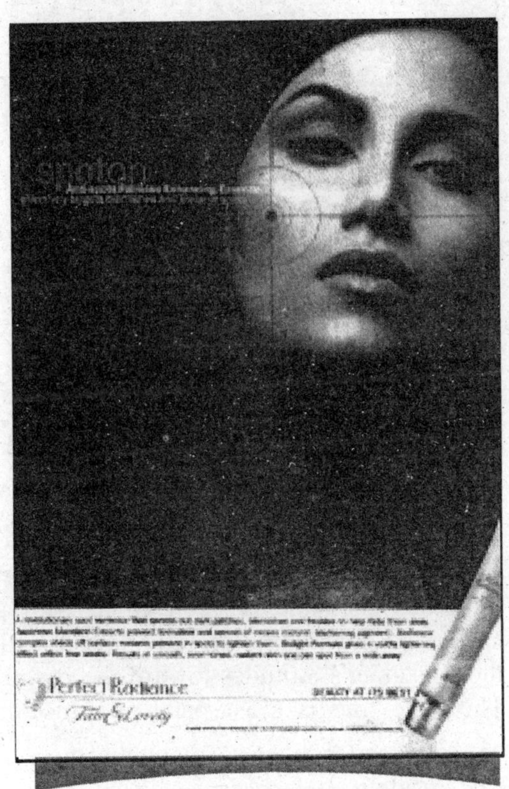

FIG: 7:10 AD APPEALING TO UTILITARIAN BENEFIT

B. THE EGO-DEFENSIVE FUNCTION:

It goes without saying that majority of the people want to protect their self-images from their inner feelings of uncertainly or doubt. Most of the advertisements for cosmetics and personal care products, by acknowledging this need, increase both their relevance to the consumer and the likelihood of a favourable attitude change by offering or guaranteeing reassurance to the customer's self-concept. The side given advertisement on pimples-speaks of 'skin–care'.

C. THE VALUE EXPRESSION FUNCTION:

As noted earlier, attitudes are an expression or reflexion of the consumer's general values, life-style and the outlook. In case a consumer segment holds a positive attitude toward owning say latest electronic gadget, then their attitudes toward new electronic devices are likely to reflect that orientation most generally. Similarly, if a segment of consumers has a positive attitude toward being "in fashion" then their attitudes towards high fashion clothing are likely to reflect this view point. By knowing target consumer's attitudes marketing can better anticipate their values, life-styles or outlooks and reflect these

FIG:7.11 AD SHOWING ATTITUDE EXPRESSION FUNCTION

FIG:7.12 AD SHOWING ATTITUDE EXPRESSION FUNCTION

characteristics in their advertisements and direct marketing efforts. Following are the two advertisements that explain this function.

D. THE KNOWLEDGE FUNCTION:

Persons have a strong need to know and understand the people and the things with whom they come in contact. The consumer's "need to know" a cognative need, is very important to marketers concerned with product positioning. In–fact, most of the product and brand positioning are attempts to satisfy the consumer's "need to know" and to improve the consumer's attitudes toward the brand by emphasizing its merits over competitive brands. The following is the typical advertisement that demonstrates this fact.

Instead of using separately, these functions, they may be combined or dovetailed for best results. It is because different consumers may like or dislike the same product or a service for different reasons. Here, functional frame-work for examining the attitudes can be very useful. Take a case of some three consumers who have all positive attitudes toward say hair care products. However, one may be responding mainly to the fact that the products work well speaking of utilitarian function. The second consumer may have the inner confidence to agree with the point "When you know beautiful hair does not have to cost a fortune"—Speaking of an ego-defence function. The third consumer's favourable attitudes might reflect the realisation that has for many years stressed 'value' either equal or better products for lesser value—emphasizing the knowledge function.

2. ASSOCIATING THE PRODUCT WITH A SPECIAL GROUP, EVENT OR CAUSE:

Attitudes are related to certain groups, social events or causes atleast in part if not in full. It is possible to alter attitudes toward products services and brands by pointing out their relationship to a particular social group, events or causes. The following are the three ads which demonstrate how one's attitude can be changed by associating the product with a social group, event or cause.

FIG:7:13 AD SHOWING ASSOCIATING PRODUCT WITH SOCIAL GROUP

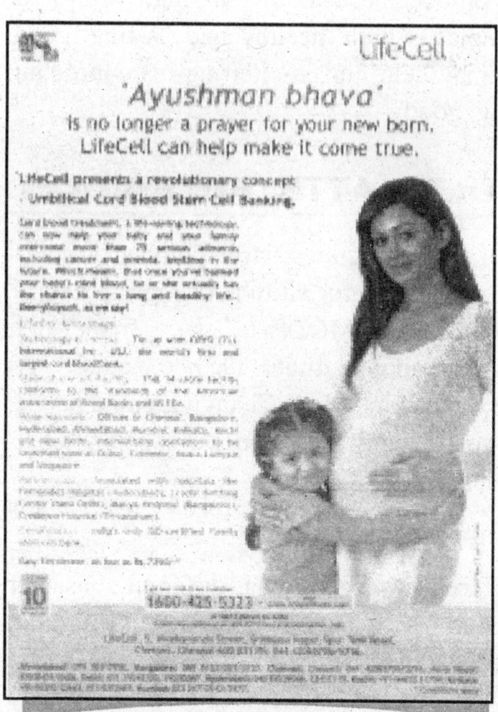

FIG:7:14 AD SHOWING ASSOCIATES OF PRODUCT WITH EVENT

FIG:7:15 AD SHOWING ASSOCIATES OF PRODUCT WITH CAUSE

3. RESOLVING TWO CONFLICTING ATTITUDES

Shrewd marketers can take full advantage of attitude change strategies of actual or potential conflict between attitudes. It is specially true if consumers can be made to view that their attitude toward a brand is in conflict with another attitude. In–fact, they may be induced to change their evaluation of the brand. Let us take a case of Dr. Williams Snow, a young lecturer in Business Management who likes cooking too as a hobby. His belief is that Branola cooking medium is advertised as "healthy" oil taken from rice bran, is not particularly good tasting when used in food preparation. His view is consistent with the more general belief that foods that are good for you taste 'bad', and foods that are bad for you taste 'good'. When Dr. William Snow is introduced the brand namely 'Carnola' extracted edible oil by one of his closest friends namely, Dr. Alien Smith, who happens to be an associate in his department, he becomes eager to try the new product because of his high regard for Corns Ltd. products and the friend's claim that 'Carnola' is both 'healthy and 'tasting' good. In case Dr. William Snow likes the product, then this may help him in alterting his conflicting beliefs about a good being both 'healthy' and 'tasting good'.

4. ALTERING COMPONENTS OF MULTI–ATTRIBUTE MODEL

Earlier, in this chapter, we had the opportunity of studying quite a large number of multi-attribute attitude models. These models have deeper implications for attitude change strategies. Using the standard Fishbein ATTITUDE-TOWARD-OBJECT MODEL, as a framework, it pays to consider the following strategies for bringing about attitude change namely (A) Changing relative evaluation of attributes represented by subscript e_i (B) Changing brand beliefs represented by subscript b_i (C) Adding an attribute combined represented by sub-script $b_i e_i$ and (D) Changing the overall brand rating represented by sub–script A.

A. Changing Relative Evaluation of attributes (e_i)

The market for many product categories is structured so much so that different consumer segments are attracted to brands that offer different features or beliefs. For instance, within a product category such as headache remedies, there are product brands like 'Crocin', 'Microfined Aspro', 'Dispirin' and 'Anacin'. Here, people believe that each is different and by taking only a particular brand will give them relief. Similarly, is the case of chewing gum, the market can be divided into 'regular' and 'sugar–free' or 'regular' and 'bubble–gum'.

When a product category is divided naturally according to distinct product features, in general, marketers have usual opportunity to persuade consumers to 'Crossover'. That is, to persuade consumers lo prefer one version of the product. It may be say with 35 mm single lens to reflex Camera switching over from a highly automatic point and shoot camera say made by **Kodak or Yashica, or Cannon** ;—it may be *vice-a-versa* too. This strategy is equal to altering the relative evaluation of conflicting product attributes. Put alternatively, it serves to upgrade consumer beliefs about one product attribute, and either go ungrade someother attribute which parallel or convince consumers that it is not in conflict with the upgraded attribute. In case the consumer's evaluation of one attribute can be upgraded, then there may well be a shift in the consumer's overall attitude or even in the intention to buy because, the attributes to be changed usually are significant and distinctive. If Mrs. Urmi switches to a decaffained coffee which helps her to sleep at night, she may switch over to decaffained tea because she associates 'decaffaination' with relaxation.

B. Changing Brand Beliefs (b_i)

This is a second cognative—oriented strategy for changing attitudes which concentrates on changing beliefs or perceptions about the brand itself. This is the most common form of an ad appeal. We all experience daily that advertisers are constantly reminding us that their product has 'more' or is 'better' or 'best' in terms of some important product attribute. Thus, 'Surf excell' claims superiority over other brands by having appeal "Surf Excell Hai Na". Indigo whitening agents have tried to establish that whitening ability, while—'Ujala' has upper hand than 'blue bird' and other similar brands with the appeal"Aaya Naya Ujala—Char Bundwala". Such an attempt is going to change the altitude of a consumer.

CONDITIONS TO BE FULFILLED TO CHANGE BRAND BELIEFS

To change the beliefs about brands a special atmosphere is to be created which is possible through certain conditions that warm the background. These conditions are also called as caveats. These are : (1) The attempts to change consumer perceptions about brand attribute will not work if the brand does not actually have the attribute in question in the long run. (2) The marketers can encourage consumers to expand the use of a brand to situations in which it would normally be used if consumer's attitude toward the brand are either too similar or too dissimilar from the brands usually used in the situation. For instance, it can be done by promotional advertising say Rs. nine off on a bottle of **Bru Coffee** originally costing say Rs. 49-00 of say 125 grams. Or issue of discount coupon of Rs. 5 each on tetrapack edible oil of **Dhara** of one litre. To attain this, a marketer will have to stress shared or common attributes between the two brands in case they are really quite dissimilar or promote unique attributes between the brands if they are really 'quite similar (3) The changes in the relative evaluations of both the attribute and the brand attribute beliefs must not be two drastic because, too extreme an advertising position is likely to result in rejection of the entire message. This caution is based on the theory of "assimilation—contrast" as pointed out by **Mr. Carl 1. Hovland, O.J. Harvey** and **Malafer Sherif** in an article "Assimilation and contrast effects in Reactions to consumer inclinations and Attitude change"-in JOURNAL OF ABNORMAL AND SOCIAL PSYCHOLOGY-55, July 1957-pp 244-52.

This theory of "assimilation contrast gives a warning to the marketers who are trying to change attitudes by alterning the relative evaluations of either attributes or brand attribute beliefs must be careful to avoid "over-sell" or "over-selling" their case. This theory states that consumers will assimilate or accept only moderate changes and not drastic. That is, when the change suggested by a message, is too extreme, the contrast likely will result in distortion of the whole message and finally rejection of the message because it is too extreme. For instance, many people reject the notion of high chlolestral contributing to heart diseases and strokes, because they cannot envision denying themselves butter, red meat or cheese or the rich desserts.

C. ADDING AN ATTRIBUTE (COMBINED b_ie_i)

This is another cognative strategy which consists of adding an attribute. This is attained by either adding an attribute that previously has been ignored or one that represents an improvement or technological innovation. The following ad is one which deals with the first situation. That is, adding the attribute which was forgotten. Another way is adding an attribute which means an actual product change or technological innovation which is much easier than stressing the previously ignored attribute. The two ads given on the below explain the situation. One and situation dealing with the adding of an attribute to change the attitude.

FIG:7:16 AD SHOWING ADDING AN ATTRIBUTE

FIG:7:17 AD SHOWING ADDING AN ATTRIBUTE

D. CHANGING THE OVERALL BRAND RATING (A_o) :

It is yet another cognative oriented strategy that consists of attempts to alter the consumers' overall assessment of the brand directly. Such strategy depends or banks very often on some form of global statement, as indicated in the ad given below such as "this is the largest selling brand" or "the one all others try to imitate". Or a similar claim that sets the brand apart from all its competitors. This is a very common case with two wheelers or other consumer durables.

5. CHANGING THE BELIEFS ABOUT COMPETITORS' BRANDS

Changing consumer beliefs about the attributes of competitive brands or product categories is another approach to attitude change strategy. Thus, an ad on **'MOOV'** a pain balm speaks of its superiority over **'Iodex'** which has the inherent black colour and greasy that spoils the clothes, unlike **'Moov'.** Thus, is a comparative advertising which must be done with extra care and caution because it may boomrang by giving visibility to competing brands and the claims made in ads.

Comparative advertising is a strategy which is increasingly used by marketers in an advertising that claims product superiority over one or more explicitly named or implicitly identified competitions, either on overall basis or on selected product attributes. Comparative advertising is very much useful for target market selection and for brand positioning strategies that emphasize the diffenitial advantage of the so called "under dog" product over leading brands. Though, it is used widely, some criticis maintain strongly that comparative ads may assist recall of the competitor's brand at the expense of the advertised product or brand.

6. THE ELABORATION LIKELIHOOD MODEL (ELM)

The **Elaborate Likelihood Model** (ELM) proposes more of global view that consumer attitudes are changed by two very clearly different "routes of persuasion", as compared to the different specific strategies of attitude that we have touched so far. These are "Central route" and "Peripheral route".

The **central route** is relevant particularly to attitude change when a consumer's motivation or ability to assess the attitude object is high. That is, attitude change occurs because the consumer seeks information relevant to the attitude object itself. When consumers are ready and willing to the effort to comprehend, learn or evaluate the available information about the attitude object, learning and attitude change occur through the central route. For instance, a person having consized of quality standards — he might shift to national to international product. Thus, in case of cordless phones, Mr. Paul might go in for **'Sony'** brand-made in Japan model SPP 206 rather than **'Orpat' OTT126** of India. In contrast, when a consumer motivation assessment skills are low, learning and attitude change occur through '**peripheral route**' without the consumer focussing on information relevant to the attitude object itself. Attitude change is very often an outcome of secondary inducements in such cases. For instance, ads or appeals such as rupees Five–off coupon, free samples, encouraging an endorsement made by a celebrity in any field, attractive packing and so on.

COGNATIVE DISSONANCE THEORY

WHAT IS IT ?

The credit of developing this theory of dissonance goes to **professor Leon Festinger** who made in 1957. This theory states that a kind of discomfort or dissonance occurs when a consumer holds conflicting thought about a belief or an attitude object. According to **professor Leon Festinger, "Cogantive dissonance"** is a psychological state which results when a person perceives that two cognitions (thoughts) both of which he believes to be true do not fit together. That is, they seem inconsistent. The resulting dissonance produces a tension which serves to motivate the individual to bring harmony to inconsistent elements and thereby reduce psychological tension. For example, you have made a commitment-down payment or placed an order for a product which is an expensive one, say an automobile or say personal computer with high tech- configuration- You begin to feel cognative dissonance when you think of unique positive qualities of the brands not selected.

Dissonance occurs because making a relatively permanent commitment to a chosen alternative warrants one to give up the attractive features of the unchosen alternatives. This is something inconsistent with the desire for those features. Thus, habitual and most limited decision–making will not produce post purchase dissonance, since one does not consider any attractive features in an unchosen brand that do not exist also in the chosen brand. For instance, a consumer who has an evoked set of say four brands of coffee could consider them equivalent on relevant attributes except price and, therefore, always purchases the least expensive brand. Such purchase would not produce post purchase dissonance. However, because purchase decisions often require some amount of compromise, post-purchase dissonance is quite normal. Nonetheless, it is likely to leaves. consumers with an uneasy feeling about their prior beliefs or actions a feeling that they tend to resolve by changing their attitude to conform into their behaviour. Thus, in case of post-purchase dissonance, attitude change is frequently an outcome of an action or behaviour. The conflicting thoughts or dissonant information that follow a purchase are prime factors that induce consumers to change their attitudes so that they will be in consonance with their actual purchase behaviour. The figure post-purchase behaviour.

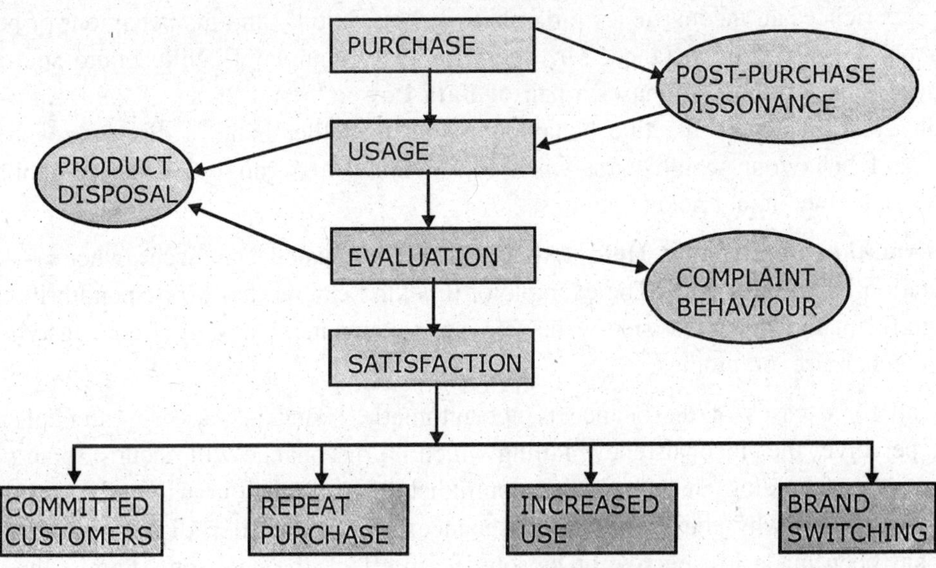

FIG: 7:18 POST-PURCHASE CONSUMER BEHAVIOUR

WHAT CAUSES DISSONANCE ?

It is natural that once you are committed to one alternative or the other, you are wondering as to: 1. Did I make the right decision? (2) Should I have done something else ? This is a very common reaction after making a difficult and relatively permanent decision. Doubt or anxiety of this type is referred to as post purchase dissonance. The probability of consumer experiencing post–purchase dissonance, as well as the magnitude of dissonance is the function of : (1) **The degree of commitment or irrevocability of the decision :** The easier it is to alter the decision, the less likely the consumer is to experience dissonance. (2) **The importance of the decision to the consumer :** The more important the decision, the more likely dissonance will result. (3) **The difficulty of choosing among the alternatives :** The more difficult it is to select from among the alternatives, the more likely the experience and magnitude of dissonance : The decision difficulty is the function of the number of alternatives considered, the number of relevant attributes associated with each alternative, and the extent to which each alternative offers attributes not available with other alternatives. (4) **The individual's tendencies to experience anxiety :** Some persons have higher tendency to experience anxiety than do others. The higher the tendency to experience anxiety, the more likely the person will experience post purchase dissonance.

This dissonance can arise is three basic ways :

1. Any Logical Inconsistency: Any logical inconsistency can create dissonance for instance, "all toffees are sweet while toffee is sour".

2. Any Experience of Inconsistency: Dissonance can occur or can be created when an individual experiences an inconsistency either between his attitude and his behaviour or between two of his behaviours. For instance, Mr. Sporty actively complaints **Nike** brand sports wear on many occasions and then purchases a pair of **Bata Power** Sports Shoes. This is an example of an inconsistency between the two behaviours. On the other hand, a discrepency between an attitude and behaviour would exist when Mr. Drinkwater strongly dislikes gambling but bets on the outcome of a cricket game.

3. Disconfirmation of Strongly Held Expectation: Dissonance can occur when a strongly held expectation is disconfirmed. The example of this kind can be that Mrs. Spendthrift expects to find significant savings at a side-walk sale on pavements, but only finds unstylish and damaged merchandise of clothes.

In case of all three cases of dissonance is not automatic. That is, it is essential for a person to perceive the inconsistency failing which no dissonance will occur. Some persons are very capable of holding an attitude that contradicts their behaviour without perceiving the contradiction. That is why, they suffer no dissonance. As noted earlier, dissonance arises after a decision has been made irrespective of its source. In effect, the decision commits the person to certain positions or attitudes when prior to that time that person was capable of adjusting his or her attitudes or behaviour to avoid dissonance.

HOW TO REDUCE COGNATIVE DISSONANCE

Any person who is experiencing cognative dissonance has three ways of reducing it. These ways are (1) Rationalisation (2) Seeking additional informations that is supportive or consistent with his or her behaviour and (3) Either eliminating or alterning some of the dissonant elements which can be accomplished either by forgetting or by suppressing dissonant elements or by changing his attitude, so that it can no longer be dissonant with another attitude or behaviour. Each of these strategies may be used unmixed or in any combination of his or her choice.

These three cases can be best illustrated by the following examples. Take the case of Mr. Photoson who has purchased a **Nikon** 35 mm, single lens reflex camera for Rs. 32,500 after considering seriously other brands say **Canon, Pantax, Minolta, Olympus,** more or less in the same price range. In addition to his investment of hard-earned money of Rs, 32,500 he has also invested invisibly in terms of much thinking and searching time and considerable amount of 'ego' in his purchase–decision. Therefore, the amount of stake visible and invisible— in this purchase is really significant. After evaluating the plus and minus points of each brand of camera, he has selected the brand **Nikon**. Subsequent to purchase Mr. Photoson finds his camera hard to focus and the lenses too difficult to change. In addition, he experiences a strap broken on the carrying case. He now begins to doubt the wisdom of his purchase. Here, a tension arises from his two beliefs that (a) "**Nikons** are well constructed precision cameras" and (h) "My **Nikon** is difficult to focus for clear pictures, it takes too such time and effort to change the lenses, and the strap on carrying case has broken".

Mr. Photoson can reduce the tension arising froms these two cognitions by rationalising any fine camera can have its faults and that the retailer probably treated the carrying case with abuse which caused the eventual strap break. Instead, he might seek information that reinforces his belief that **Nikon** cameras are among the very best in the world, thereby amplifying the strong points of camera such as rapid film advancement, nice- styling, and a solid shutter click, indicating durability. Finally, a third option is to change his opinion toward Nikon cameras say: "They are not good cameras ; I should have purchased **Minolta.**"

This solid example clearly demonstrates a very common type of marketing phenomenon namely, post purchase dissonance. This occurs when a person makes a decision to buy a brand from among several alternative brands available within a product category. This discomfort or dissonance becomes very strong when the consumer makes large commitment in the purchase. As seen, such commitment involves both money and searching time, effort and ego. A purchase decision involving choice among brands, therefore, of say chewing–gum at a super–market check out counter is unlikely to produce much of such perceptible dissonance. Goods requires the consumer to commit much of himself or his money or both, however, are sure to generate considerable post–purchase dissonance. Therefore, in general, durable and luxury goods are more likely to produce dissonance than convenience goods, for they usually require larger consumer invest in time, money and ego.

During purchase–decisions, dissonance can result when the consumer recognises alternative brands have both positive and negative characteristics. Hence, after making a decision, he realizes that he has acquired some relatively undesirable traits of the selected brand while foregoing some relatively desirable traits of the alternative brands. At this juncture, the consumer might even rate the unchosen alternative higher than the brand purchased. In the mind of customer, positive attributes of unchosen brands and negative features of the chosen brand are emphasized. This period of post purchase process is called as the 'regret' phase, and it is usually very brief. The following phase is termed as dissonance reduction period. At this stage, the consumer is very likely to evaluate the chosen brand more positively than at the time of purchase and he may evaluate the unselected brands less positively.

ATTRIBUTION THEORY

WHAT IT MEANS?

Attribution theory, is a group of loosely interrelated social psychological principles, attempts to explain as to how people assign causality—blame or credit—to events on the bases of either their own behaviour or behaviour of others. It focusses its attention on how people assign causality to events and how they form or alter attitudes as an outcome of assessment their own behaviour or the behaviour of the other people or things. Thus, an individual might say, "I contributed to **CRY CARD** Unit because it really helps the helpless or people in need; take another expression say, "He tried to persuade me to buy that unknown auto-focus, auto–winding camera because he would get higher rate of consumers."

In attribution theory, the heart of the body is why ? "Why did I do this ?" "Why did he get me to switch on to other brand ?" That is, the process of making inferences about one's own or another's behaviour is an inseparable component of attitude formation and changes.

Attribution theory describes attitude formation and change as an outgrowth of people's speculations as to their own behaviour and experiences, attribution towards others and attributions towards things. Let us touch these three aspects.

1. **Own Behaviour/Self–perception:** In this context "Self-perception" theory is to do with self- behaviour. Self–perception theory deals with inferences or judgements as to the causes of their own behaviour. In terms of consumer behaviour, self perception theory suggests that attitudes develop as consumers look at and make judgements about their own behaviour. Stated in an attenative and simple way, if Miss Beautiful purchases a journal 'Femina' on her way to office, she is apt to conclude that she likes very much 'Femina'—that is—she has a positive attitude towards this periodical. In fact, drawing inferences from one's own behaviour is not always as simple or as clear cut as the case of 'Femina' journal. To appreciate the real complexity of self–perception theory, it is essential to distinguish between the 'internal' and 'external' attributions. Let us take a case of Mr. Johny who has just competed using a popular computer graphics program say **Microsoft** for the first time and that his overhead projector transparency has come out exceptionally good. After review his work he expresses to himself "a master–piece" ; he says "I am really a perfect professional making great presentations". This statement is a case of "internal attribution", because, he is giving himself credit for the outcome. That is he speaks of his exceptional ability, his skill and his extra ordinary efforts.- To put in his own words "This graphic presentation is good because of me". On the other hand, if Mr. Johny comes to the conclusion that his successful graphic presentation transparency was due to factors beyond his control say because of user-friendly program, the assistance by a friend, or just good luck', this would be the case of an external attribution. In such a case, he might express "My graphic presentation is great because of beginner's luck".

The distinction between internal and external attributions can be of strategic marketing importance. For example, it is in the best interest of the firm, generally that producers **Microsoft** Graphs program if Personal Computer users, especially in experienced personal computer users, internalized their successful use of the graphics package. If they - internalized such a positive experience, it is more likely that they will repeat the behaviour and become a 'Satisfied' regular user. In case, however, they externalise their success, it would be preferable that they attribute it to the **Microsoft Graphics** programme, rather than to an incidental environmental factor such as **"beginner's luck"** or "friend's help" or "instructions".

Consumers are likely to accept credit personally for success- internal attribution-and to credit failure to others or to outside events-external attribution, according to the principle of "DEFENSIVE ATTRIBUTION" For this reason, it is most crucial that marketers offer uniformly high quality products that allow consumers to perceive themselves as the reason for success, that is, "I am competent". Further, a company's ad should serve to reassure consumers, particularly inexperienced ones, that the products will not let them down but will make them 'Victors' instead.

It is very important to note in this regard that self-perception theorists have explored the situation in which consumer compliance with a minor request affects subsequent compliance with a more substantial request. This strategy is popularly known as "foot in the door technique". It is based on the premise that individuals look at their prior behaviour and conclude that they are the kind of persons who say 'yes' to such requests. Such self-attribution serves to increase the possibility that they will agree to a similar, or more substantial request. In other words, if someone is ready to donate Rs. 25 to the **Cancer Society of India,** they may be persuaded to donate much more than Rs. 25 when they are convincingly approached. Here, the initial donation, in effect, is 'foot in the door.'

The latest research, into the 'foot in the door' technique, has concentrated on understanding how specific incentives say–rupees of coupons of differing amounts—finally influence consumer attitudes and subsequent purchase behaviour. It follows that different size incentives create different degrees of internal attribution which, in turn lead to different amounts of attitude change. For instance, those persons who try a brand without inducements or individuals who buy a brand repeatedly are more likely to infer increasingly positive attributes toward the brand from their respective behaviours. They may say individually, "I purchase this brand because I like it". Contrary to this, individuals who try a free-sample are less committed to changing their attitudes toward the brand say, "I tried this brand because it was free."

Stating succinctly, it is not the biggest incentive that is most likely to lead to positive attitude change, contrary to what might be expected. Marketers, therefore, run the risk if an incentive is too big, that consumer's might externalize the cause of their behaviour to the incentive and be less likely to change their attitudes and less likely to make future purchases of a given brand. The better policy is to go by moderate dose of incentive which is just big enough to stimulate initial purchase of the brand but still small enough to encourage consumers to internalize their positive usage experience and allow a positive attitude change to occur.

2. **Attributions Toward Others:** Understanding of attributions toward others is also important because of the variety of potential applications to consumer behaviour and marketing. It is established fact that every time a person asks 'Why?' about any statement or action or reaction of another person—may be a family member, a friend, or even a sales person or third party where attribution theory is applicable. Take for example if Manav and Mansee are in an electronic shop thinking of purchasing a digital cellular phone, a sales person's recommendation, that they buy an **Erricson** phone for Rs. 12,000 more than the one they were initially considering would logically be the question 'Why?' If the couple Manav and Mansee concluded that the sales person suggested the **Erricson** phone because of its superior features—say, size, battery life, clarity, range weight, they would be likely to judge sales-person's motives as 'sincere' and would possibly purchase the more expensive model. However, if they conclude that the person who is selling was interested only in the greater amount of commission he would get

by selling more expensive model, then they might judge sales person as 'insincere' and not buy the more expensive model. They might leave, indeed, and go elsewhere because, they no longer trust the sales person or the store. They would be asking themselves, "Is sales person trying to sell the more expensive model because of superiority of model or because of the higher commission?"

This particular illustration clearly suggests that in evaluating the words or deeds of others, the consumer tries to determine if the other person's motives or skills are consistent with the consumer's best interests. If these motives or skills are judged as congruent, the consumer is likely to respond favourably. Otherwise, the consumer is likely to reject the other person's words and make purchases elsewhere.

3. Attributions Toward Things: Consumer researchers have not left even a things because they are interested in consumer attributions toward things because, products or services can readily he thought as 'things'. In this sphere of judging product performance that consumers are most likely to form product attributions. Apparently, they want to find out why a product meets or does not meet their expectations. In this connection they could attribute the product's successful performance or failure to the product itself, to themselves, to other people or situations or to some combination of these variables. To recapitalise an earlier example of Mr. Johny who developed an excellent overhead projector transparency, he could attribute his thundering success to **Microsoft** graphic package—which means product attribution—to his own skill or to his friend who helped him in getting good transparency.

How One Tests One's Attributions?

Once the attributions about product's performance or a person's words or actions, we often attempt to determine whether the inference we have made is correct? According to a leading attribution theorist **Mr. Harold H. Kelly,** individual's acquire conviction about observations by acting like "naive scientists", that is, by collecting additional information in an attempt to confirm or disconfirm prior inferences. In collecting such information, consumers often use the criteria which is made up of four points : (1) Distinctiveness—The consumer attributes an action to a particular product or persons if the action occurs when the product or person is present and does not occur in its absence. (2) Consistency over time—Whenever the person or product is present, the consumer's inference or reaction must be the same or nearly same. (3) Consistency over modality—The inference or reaction must be the same, even when the situation in which it occurs, varies. (4) Consensus—The action is perceived in the same way by other customers.

To validate the criteria an example can be given of a lawn mower of a particular brand say **'Lawn Man'.** Mr. Green Lover a home owner who is proud of his lawn observes that his lawn grass seems to be cut more evenly with his new **Lawn Man** mower than his old brand Lawn Boy. He is likely to credit the Lawn Man with improved appearance of his lawn—distinctiveness. Further, if Green Lover finds that his new **Lawn Man** produces the same high quality results each time he uses it, he will tend to be more confident about his initial observation—consistency overtime. Similarly, he will also be more confident if he finds that his satisfaction with **Lawn Man** extends across a wide range of other related tasks such as nulching grass and picking up leaves—Consistency over modality. Finally, Mr. Green Lover will have still more confidence in his inference to the extent that his friends who own **Lawn Man** also have similar experience—consensus.

Almost like Mr. Green Lover, one should go about gathering additional information from one's experiences the people and things and use this information to test one's own inferences.

CHAPTER BASED QUESTIONS

A. WRITE SHORT NOTES ON:

1. Attitude 2. Belief 3. Features of attitude 4. Functions of attitude 5. Sources of attitude development 6. Structure of attitudes 7. Cognative component 8. The affective component 9. Conative component 10. Theory of reasoned action

B. SHORT ANSWER QUESTIONS:

1. What is Attitude?
2. What is the theory of trying to consume?
3. What is attitude change?
4. What are the strategies of attitude change?
5. What is ELM?
6. What is cognative dissonance?
7. What is attribution theory?
8. What are the function of attributes?
9. What is structure of attitude?
10. What is theory of reasoned action?

C. ESSAY TYPE QUESTIONS:

1. What is an attitude? What are the characteristics of an attitude?
2. Define the word 'attitude' explain the function of attitudes.
3. What is development of consumer attitudes? What are the sources of attitude development?
4. Explain the structure of attitudes in detail.
5. What is multi-attribute attitude? Explain the models of this kind.
6. What is attitude change? What are the strategies of attitude change?
7. What is cognative dissonance theory? How one reduces this cognative dissonance?
8. What is attribution theory? Explain its implication.
9. What is the "theory of trying to consume?" What are its main stages?
10. What is motivational function? What strategies are employed in it?

*****

CHAPTER 8

FAMILY AND CONSUMER BEHAVIOUR

BACKDROP

The concept of family is really very complex and ever-evolving core institution in many of the world's societies or nations. Family has deep impact on the consumer behaviour or its members. , Family is the seat of learning. Thus, a child learns how to enjoy candy by observing any elder brother or sister and learns the use and value of money by listening to and watching his or her parents. Group-decisions are made—say buying a new sound systems, a two-wheeler, a new car, a vacation trip or to join a local or out of town college and the like. It is the family that provides the best opportunity for product exposure and trial and imparts consumption values to its members. Therefore, an attempt is made to give clear cut meaning of family, households, functions of the family, family decision-making, family roles-dynamics of husband and life decision-making. Children as buyers, the family life cycle—traditional and modern consumer socialization and other allied aspects. The chapter ends with a Chapter Based Questions.

WHAT IS THE FAMILY ?

Though **family** is a familiar and commonly used word, it is not that easy to define the term in precise terms. The main reason for this is, family composition and structure and the roles played by the members of family are always subject to change. However, traditionally the word 'family' is defined as two or more persons related by blood, marriage or adoption who reside together. In a dynamic sense, the individuals who constitute a family might be described as the members of the basic social group who live together and interact to satisfy their personal and mutual needs.

Presented in another way family is both primary and reference group ; primary as characterized by intimate, face–to–face interaction and reference–group because with members referring to certain family values, norms and standards in their behaviour. These two factors however, are not the sole reasons accounting for the strength of the family's influence. It is, rather, **first**, the fact that bonds within the family are likely to be much more powerful than those in other small groups. **Second,** contrary to the other groups to which the consumer belongs, the family functions directly in the role of final consumption. That is, family operates as an economic unit, earning and spending money. In doing this, family members must establish individual and collective consumption priorities, decide on products and services and brand that fulfil their needs, and also decide where these items are to be bought and how they are to be used in furthering family members' goals. Further, consumer's attitudes toward spending and saving and even the brands and products purchased have been moulded by the families they grew up often quite indelibly.

'FAMILIES' VERSUS 'HOUSEHOLDS'

One should note clearly the difference between the two words namely, 'families' and 'households'. The word 'family' is different from 'household'. A 'household' includes the related family members and all the unrelated persons who occupy a housing unit—may be a house, apartment, group of rooms and so on. Thus, households may be of two main types—families and non-families. In US, most Americans live in households. Those who do not, live in group quarters such as military barracks, prisons, nursing–homes and college dormitories. However, the word 'family' is more limited and refers to a group of two or more persons related by blood, marriage, or adoption and residing together as household. 'Families' comprise seven out of ten households approximately. However, it should be noted that marketers are interested not only in the concept of families but also households. Since, both may form the basis or frame- work of much consumer decision–making and buying behaviour. In Western world three types of families dominate namely, the 'married', the 'nuclear' and the 'extended'.

The simplest type of family is 'married–couple' which consists of a husband and wife its terms of number of members and relations. As a household, the 'married–couple' generally is representative of newly married who have not yet started family and the older couples who have already raised their children.

A 'nuclear–family' is one where husband and wife and one or more children are existing. This type of nuclear family, together with at least one grand-parent living within the household, is called as an extended family. The concept of two generation or extended three generation family which was quite common is fast disappearing after 1970 and 1980s because of mobility. Contrary of this, because of divorce, separation and out of wedlock births, there has been a rapid increase in the number of 'single-parent family' households consisting of one parent and at least one child. What type of family is 'typical' is very difficult to say which varies from culture to culture. In a country like Canada, the nuclear family is most common. While is countries like India and Thailand a family consists of a head of household, married adult children and grand children.

FUNCTIONS OF THE FAMILY

Family as a social and economic unit, has to perform certain basic functions which is most relevant to consumer behaviour. These are : economic-well being, emotional support, suitable family life–styles and socialisation of family members. Let us acquaint with these family functions.

1. **Economic Well-being:** Though families in affluent nations of North America, Europe and Asia are no longer formed basically for economic security, providing financial means to its dependents is unquestionably a basic family function. That is, how the family divides its responsibilities for providing economic well-being has changed beyond one's imagination during the last 35 years. The traditional roles of husband and wife as bread winner and house keeper are no longer valid. It is quite common to see that married women with children in industrialized nations are employed outside the home and their husbands are sharing household responsibilities. The economic role of children is also changed. Today, inspite of the fact that many teenage children work but they rarely assist the family financially. Instead, many teenagers are expected to pay for their own assessments, education and prepare themselves financially independent as a mark of self-sufficiency.

2. **Emotional Support:** Provision of emotional nourishment—love, affection and intimacy—to its member is significant function of the contemporary family. In fulfilling this function, the family provides support and encouragement and helps its members in coping with decision–making and personal and social problems. In order to make it easier for working parents to show their love, affection and support to their children. This is evident from the fact that greeting card companies are having a busy schedule of making available all types of cards to be given by parents to children and vice–a–versa. In case the family cannot provide much of assistance when it is needed, it may turn to a professional counsellor or psychologist as an alternative. It is quite evident, in many societies and communities, educational and psychological centres are available that are designed to assist parents who need help to improve their children's learning and communication skills so that better adjustment can be brought about.

3. Matching Family Life-styles : Establishment of a matching or suitable life-style for the family is yet another significant function in terms of consumer behaviour. Up-bringing, experience and personal and jointly held goals of the spouses determine the importance placed on education or career, on reading, on TV viewing, on the learning of computer skills, on the frequency and quality of dining out, and on the selection of other entertainment and recreational activities. Researchers have clearly identified a shift in the nature of family 'togetherness'. Whereas a family being together once meant doing things together, today it means being the in the same household and each person doing his or her own thing. Family life-style; commitments, including the allocation of time is greatly influencing consumption patterns. For instance, the increase in the number of married women working outside the home has reduced the time they have for household chores and has created a market for convenience productions and fast food restaurants. Again, with both parents working, an increased emphasis is being placed on the idea of 'quality–time' rather than 'quantity–time' spent with children and other family members. Realising the scarcity of quality family time. Hotels and tourism units have planned a variety of vacation and weekend packages targeted at couples and their children.

4. Socialisation of Family Members : Family becomes an important unit providing basic framework in which consumer socialisation occurs. The socialisation of family members, ranging from young children to adults, is a central family function. In case of young children, this process includes imparting to children the basic values and modes of behaviour consistent with the culture. These include generally moral and religious principles, inter-personal skills, dress and grooming standards, appropriate manners and speed, and the selection of suitable educational and occupational career-goals. To illustrate how this socialisation responsibility is expanding, parents are increasingly anxious to see their young children possess adequate computer skills, almost before they are able to walk or talk as early as twelve months after birth. Because, the parents seem to be so intensively interested in their young children learning about using a computer, hardware-software developers are rapidly, creating products targeted at parents seeking to buy such items for their very young children. Socialisation of skills manners, values and qualities are imparted to a child directly through instructions and indirectly through observation of behaviour of parents and older siblings. The birthday party for a young child is an instance of a social event designed to help children acquiring skills. According to mothers of 2 to 5 years old children, encouraging the child to assist in planning and conducting of a birthday party is the part of the overall process of socialisation and is a means of teaching important social skills. Thus, consumer socialisation is the process by which young people acquire skills, knowledge, and attitudes relevant to their functioning as consumers in the market place. Learning including acquiring consumption related knowledge is a life-long process. However, the quantity and nature of learning that take place before early adulthood- say around 18 years as well as its impact on subsequent learning, are sufficiently unique to justify focussing on this time period.

Marketers frequently target parents looking for assistance in the task of socialising their children. To this end, marketers are sensitive to the fact that socialisation of young children provides an opportunity to establish a foundation on which later experiences continue to build throughout life. These experiences are reinforced and modified or modified as the child grows into adolescence, the teenage years, and eventually adulthood. Let us note about socialisation of children adults and intergenerations.

Consumer Socialisation of Children

This aspect of childhood socialisation that is particularly relevant to the study of consumer behaviour in consumer socialisation, which is the process by which children acquire the skills, knowledge and attitude necessary to function as consumers. A variety of studies have focussed on how children develops consumption skills. These alternative ways are observation of parents who act as models, shared experiences, assisting busy working mothers. Consumer socialisation also serves as a tool by which parents influence other aspects of socialisation process. Consumer socialisation has two clear components, (a) Socialisation directly related to consumption, such as the acquisition of skills and knowledge concerned with budgeting prices, brand attitudes, and actual product usage (b) Socialisation indirectly related to consumption, such as the underlying motivations that spur a young man to purchase his first razor or a young girl to want her first bra. Both the types of socialisation are significant. The indirect component of consumer socialisation is often of most interest to marketers, who want to understand as to why people buy their products. The direct component of consumer socialisation is often of greatest interest to academic consumer researchers who have broader goods of understanding all aspects of consumer behaviour.

Adult Consumer Socialisation

It goes without saying that the socialisation process is not confined to child-hood alone; on the contrary, it is an on-going process. It is now accepted that socialisation begins in early childhood and extends throughout a person's entire life. For instance, when a newly married couple establishes a separate household their adjustment to times and consuming together is part of this continuing process. Similarly, the adjustment of a retired couple who decide to move from Goa to Jaipur is also part of ongoing socialisation process. Even a family that is welcoming a pet into their home, as a new family member, must face challenge of socialising the pet so that it fits into the family environment.

Inter-generational Socialisation

As we experience, it is quite common for selected product loyalty or brand preference to be transferred from one generation to another—inter-generational brand transfer may be even three or four generations within the same family. For instance, specific brand preferences for products like ketchup, sauce, coffee, dental cream, dress materials that are frequently passed on from one generation to another. The best 'example is that of "Complete man"—ad of **Raymonds,** "Woodword's Gripe Water" for kids. Another type of transfer between family members is the giving of loaning of 'things'. When a family member gives or loans to another family member some hand the down clothing, an old car or colour TV, a set of dishes or flat ware, a sofa, or a ten year old base-ball glove, these transfers serve as substitutes for the purchase of such items in the market place.

Following is a very simple model of socialisation process. It focusses on the socialisation of young children but that can be extended to family members of all ages.

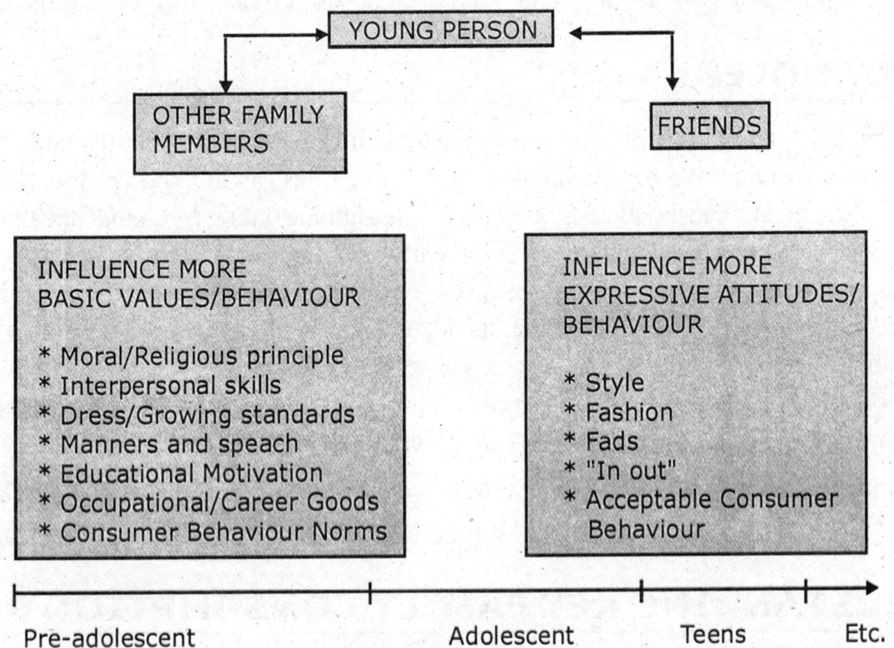

FIG: 8:01 A SIMPLE MODEL OF THE SOCIALISATION PROCESS

It is important to note in the above figure that the arrows run both ways between the young person and other family members and between young person and his or her friends. This two- directional arrow signifies that socialisation is really a two way lane, in which the young person is both socialized and influences those who are doing socialising. Supporting this view is the reality that children of all age often influence the opinions and behaviour of their parents.

FAMILY DECISION–MAKING

Shrewd marketers examine frequently and thoroughly the attitudes and the behaviour of the one family member whom they believe to be the major decision-making, though many marketers recognise the family as the basic decision-making unit. In certain cases, they do examine the attitudes and behaviour of the person most likely to be the primary user of the product or service. For example, in case of men's underwears or vests, it is frequently purchased by women for their husbands and unmarried sons in Western world unlike in India. It becomes a common place to seek the views of both the men who wear these underwears and vests and woman who buys it. By considering both the likely user and likely purchaser, the marketer gets a clear picture of the consumption process.

Decision–making by a family differs in many ways from individual decision–making. For a family to function as a cohesive unit, roles or tasks of the family are played by different members as individuals or as in group. This takes us to the study of family roles.

FAMILY ROLES

Roles or tasks are distributed and played by members of the family and its members as it is a cohesive unit. We have a number of tasks or roles as to Who is doing the laundry? preparing the meals and breakfasts ? Setting the dinning table ? taking out the household daily garbage? Who is working or running with a dog ? or a family bird (pet) to care ? In a dynamic society, family–related roles are constantly changing. It is quite natural that when good many married women working outside the house and greater assumption of household tasks by men, marketers should be very sensitive as to how shifting family roles may affect composition of their target markets. Further, they must he very careful to design their ads in the ways that are most appropriate and acceptable, to their target markets.

Marketers are more interested in family consumption roles than the household chores though they cannot be ignored. The key family consumption roles are, therefore very important.

UNDERSTANDING KEY FAMILY CONSUMPTION ROLES

The experts differ in number of key roles of consumption. Some have given eight roles which are significant from the point of view of a family as a consumption unit. There are others who are happy with live roles. It pays to take up both set of EIGHT roles and FIVE roles.

Eight Family Consumption Roles : One group of experts profounds eight family roles in decision–making process. A look at these roles provides deeper insight into how family members interact in their various consumption related roles. These are:

1. **Influencers:** Influencers are the members of the family who provide information to other members about a product or service.
2. **Gate-keepers:** These are the members of a family who control the flow of information about a product or service into the family.
3. **Deciders:** These are those family members with the power to determine unilaterally or jointly whether to shop for, purchase, use, consume, or dispose of specific product or service.
4. **Buyers:** These are family members who make the actual purchase of a particular product or service.

5. **Preparers:** These are the family members who transform the product into a form most suitable for consumption by other family members.
6. **Users:** These are the members who use or consume a particular product or service.
7. **Maintainers:** These are the members who service or repair the product so that it will provide continued satisfaction.
8. **Disposers:** These family members are the ones who initiate or carry out the disposal or discontinuation of a particular product or service. It may be noted that the word 'members' is repeated in each case. It may be even a 'member' only depending on the situation and family traditions and customs.

Five Family Consumption roles: Another group of experts which is happy with five consumption roles and these are:

1. **Information gatherer(s) :** The individual who has expertise and interest in a particular purchase. Different individuals may seek information at different times or on different aspects of purchase.
2. **Influencer(s):** The person who influences the alternatives evaluated, the criteria considered, and the final choice.
3. **Decision-maker(s) :** The individual who makes the final decision. Of course, joint decisions also occur
4. **Purchaser(s):** The household or family member who actually purchases the product.. This is typically an adult or a teenager.
5. **User(s):** The user of the product. For many products there are multiple users.

The following figure gives a model of family decision–making process in case of products needed by children.

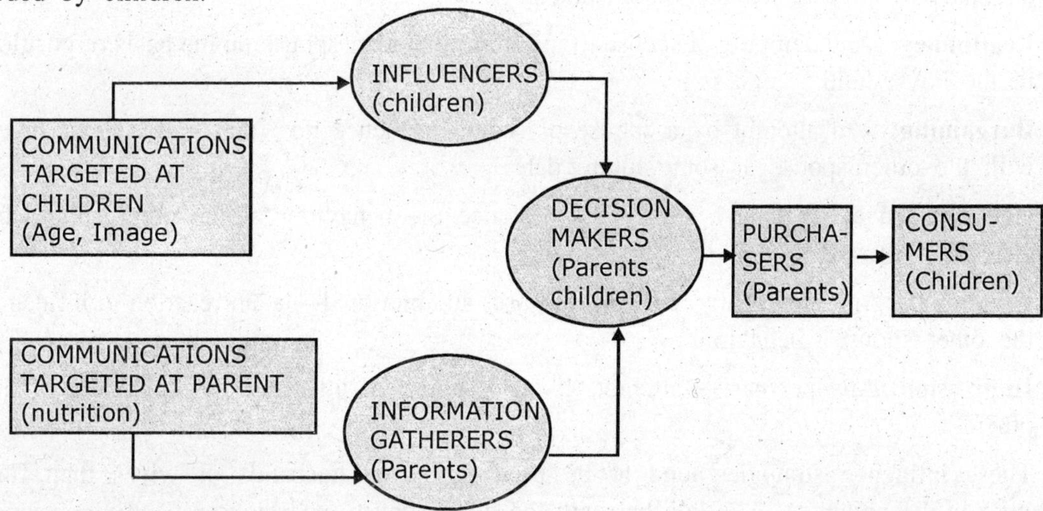

FIG: 8:02 FAMILY DECISION–MAKING PROCESS FOR CHILDREN

The number and identity of the family members who play these roles vary from family to family and product to product. In some cases a single family member will independently assume a number of roles; in other cases, a single role will be performed jointly say by two or more members. In other cases, still one or more of these basic roles may not be required. For instance, a family member may be walking down the snack food at a local super market when he picks out an interesting new chocolate candy. His selection does not directly involve the influence of other family members. Therefore, he is the decider, buyer, and in a sense, the gate-keeper; however, he may or may not be the sole consumer or user. Products may be consumed by a single member of a family say (beer, hot–drink, medicine, lipstick), consumed or used directly by two or more family members (vegetables, shampoo, soap, bread, canned –juices, fish), or consumed indirectly by the entire family (central air-conditioning, a home TV set, car, mobile, burgler–alarm system, wall–paintings, weekend trip.)

INFLUENCING SPOUSES AND RESOLVING CONSUMER CONFLICTS

Husbands and wives frequently find themselves in disagreement about when to save, or spend ? How much to spend on some item, what colour fits best, or as to where to buy something ? When it comes to making of purchase decisions in the family the above stated cases are just a few of the many consumer behaviour–decisions where there must be disagreement. To avoid or resolve the potential disagreements, husbands and wives commonly attempt to influence each other to arrive at what they feel to be the best outcome. Some six influence strategies for resolving husband and wife consumptions related conflicts have been identified by **Mr. William J. Quals** These are:

1. **Expert:** An attempt by a spouse to use his or her superior information about decision alternatives to influence the other spouse.
2. **Legitimacy:** An attempt by a spouse to influence the other spouse on the basis of position in the house-hold.
3. **Bargaining:** An attempt by a spouse of secure influence now that will be exchanged with the other spouse at some future date.
4. **Reward:** An attempt by a spouse to influence the behaviour of the other spouse by offering a reward.
5. **Emotional:** An attempt by a spouse to use an emotional—laden reaction to influence the other spouse's behaviour.
6. **Impression:** Any persuasive attempts by one spouse to influence the behaviour of the other.

These influence strategies tend to be used by either husbands or wives then find themselves in disagreement or in conflict with the other spouse regarding a specific consumer decision. For instance, it is our common experience that different family members want to eat at different types of restaurants, see different films or go a different type of family vacation: These are only illustrative and there can be countless possibilities of potential family consumption conflicts that might be warranting resolution. Following is a list of some common issues on which married couples disagree.

Reasons for diagreement	P	E	R	C	E	N	T	A	G	E
1. How to spend money?				22 per cent						
2. Which TV programme to watch?				20 per cent						
3.Hiscipling the children? (How)				19 per cent						
4. How to spend weekends?				15 per cent						
5. House keeping (how to)?				10 per cent						
6. Which friends to see ?				08 per cent						
7. Spending money for kids (How-How much)?				06 per cent						

Scale = lcm = .10 percent

Source: **'Roper's America** : "Money is the Root of Most Arguments" : Adweek's Marketing Week, April 2, 1990.

Dynamics of Husband-Wife Decision-making

Naturally, marketer's interest in the relative amount of influence that a husband and wife when it comes to family consumption choices. In this regard, **Mr. Kim. P. Corfman** says in his article, "Perceptions of Relative Influence : Formation and Measurement"—in Journal of Marketing Research May 28, 1991 pp. 125-136. Most husband-wife influence studies classify family consumption decisions as "Husband-dominated." "Wife-dominated", "joint" (equal or syncratic) and 'automatic'—solitary or unilaterally. The studies that have examined both the extent and nature of husband-wife influence in family decisions have found that such influence is fluid and likely to shift, depending on the specific product service, the family role structure orientations, and the specific stage in the decision making process. These factors are also influenced by changing life-styles, particularly the changes in family life-style options associated with women working outside the home and increased occurrence of dual income households.

CHILDREN

It is the experience of good many parents that young children attempt to influence family decisions as soon as they possess the basic communication skills needed to interact with other family members. Older children are likely to participate most directly in family consumption activities. Research has made it clear that children between 5 and 14 of age some 165 billion US dollars in 'direct' purchasing power and that they influence about 400 billion US dollar of their household's overall purchase decisions. In India the corresponding figures are 15,000 crores and 3,500 crores. These children influence a wide range of family decisions such as choice of vacations, stereo-gadgets and home computers. The parent-child relationship, as it relates to consumer behaviour can be viewed as an 'influence-versus-yield' situation. Specifically, children attempt to influence their parents to make a purchase to yield. It is quite evident that children not only attempt to influence their parents to make purchases of special interest to them say candy, food–drinks but also products of remote interest say laundry detergents and bathing soaps as they see TV—which is clear from observation of shoppers in a super market.

The impact of TV on children in so deep, though as children grow older, the less they are interested TV watching. The impact of T.V. ads appears to vary among children of different age groups. Research has clearly demonstrated that pre-schoolers as young, one to three years of age understand TV commercials and are able to discern the programme content and the commercial. As it is said the TV ads have great influence on children. For instance, the TV commercials targeted to adults have great influence especially on lipstick — in the age group of 9 and 10 as girls perception of the product and brand because they associate product with being an adult. Older children recognise more symbolism in advertising than youngers do.

Good many consumer goods companies target special marketing efforts at preteen "Latechkey" children (those who are home alone for atleast part of each school day while their parents work). There children receive special attention from such companies such as **General Foods, Whirlpool, Lipton, American Home Products.** One such programme is 'Kidsmarts' a magazine that is being distributed by Lipton covering over 1.5 lakh households to help working parents and their children make after- school hours more productive and secure until their parents come home. This magazine contains good many ads and cents off coupons for Lipton products that are consumed by children in this age group.

Marketers have identified a separate market segment of 'teen–agers'. There are six reasons as to why teenagers need special attention. These reasons are: 1. Teens spend a lot of money 2. Teens also spend family money 3. Teens influence what the family buys 4. Teens are trend-setters 5. Teens are a growing market and 6. Teens are future consumers. They deserve special attention in addition to above reasons because teens enjoy shopping. The income of teenage boys and girls is mostly earned by working after school hours. A significant number of teenagers work at least part time and most of their income is discretionary. In terms of spending patterns, high-school students studying in 7th through 12th standards are most interested in sports and fitness. Boys between ages of 16 and 19 spend much of their money on films, dating, entertainment, clothing, car expenses, while girls of the same age group spend on clothing cosmetics and fragrances. It is very interesting to note that while communicating with teens, the 'younger–teens'—13 and 15 years prefer to be called as 'teens' where as older teens—between 17 and 19 dislike the word 'teens' or 'teenagers' as they feel as if they are adults.

Coming to college goers or college age children, it represents another significant family sub-group. As per one estimate 12.5 million college students represent 33 to 60 billion US dollar market with US dollars 13 billion discretionary income. The corresponding figures in India are 15 million representing 16,000 crores with 10,000 discretionary income. This market consumes wide range of necessities such as books, personal clothing, petrol as well as selective purchases of goods and services like rock-concerts and spring vacations. Like pre–teen and teen-agers, college students are still in the process of establishing many of their brand preferences and shopping habits. In many cases, college students exert greater influence on the purchase decisions of their families. Marketers cannot ignore these college goers as they are present and future prospects as consumers.

THE FAMILY LIFE–CYCLE

The concept of **family life–cycle** has attracted the attention of sociologists and consumer researchers as a means of projecting what was once a rather steady and predictable series of stages that more families progressed through. The concept of family life–cycle has proven very valuable for the marketer, particularly for segmentation purposes. With the advent of many diverse family and life-style arrangements, the age old concept of FLC has not remained as a rule. These factors have been increasing divorce rates, the explosive number of out of wedlock births and the 30 years (1980s to 2000s) decline in the number of extended families that transpired, as many young families moved to advance their jobs and career opportunities. Inspite of the decline in its predictive precision, FLC yet remains a useful marketing tool when one keeps in mind that there are family and life-style arrangements that are not accounted for the traditional representation. In fact, FLC analysis enables marketers to segment families in terms of a series of stages spanning the life course of a family unit. The FLC is a composite variable created by combining systematically such commonly used demographic variables as "marital–status", "size of family", "age of family members"—focussing on the age of the oldest or youngest child—and 'employment status' of the head of the family or household. In view of its importance the concepts of FLC is divided into 'Traditional' and 'Modern' FLC studies.

TRADITIONAL FAMILY LIFE–CYCLE

The term 'life-cycle' refers to the progression of stages through which individuals and families proceed over time, starting with bachelorhood and ending with dissolution of the basic unit. Different researchers have expressed various preferences in terms of the number of FLC stages, ranging between five to almost nine. The five stage traditional FLC model serves our needs. These five basic stages are : 1. Bachelorhood 2. Honey–mooners 3. Parenthood 4. Post- parenthood and 5. Dissolution. Let us go into the details and implications of these FLC stages.

1. STAGE 1-BACHELOR–HOOD

This is the first stage of FLC. It consists of young single men and women who have established households apart from their parents. Though most of the members of this FLC stage are fully employed, many are college, or graduate students who have left their parents' homes. Young single adults are apt to spend their incomes on rent, basic home furnishings, the purchase and maintenance of automobiles, travel and entertainment, and clothing and accessaries. The members of the 'bachelorhood' stage frequently have sufficiently large disposable income to indulge themselves. Marketers target these singles for a wide variety of products and services. It can be anything that is consumer durable or non-durables that matches to the bachelorhood.

In most larger cities, there are travel agents, housing developments, health clubs, sports clubs, and other service product marketers that find this FLC stage a very lucrative target niche. Meeting, dating, and mating are very prominent occurrences or concerns of many young adults, typically are beginning their working lives after recently completing college or some other form of career or job training. It is relatively easy to reach this segment, because many special interest publications target singles. For instance, 'GQ' and "Play boy" are meant for young sophisticated single male audience, whereas 'Cosmopolitan' and 'Glamour' are directed to young single females. Marriage marks transition from the bachelorhood stage to the honeymooner stage. Engaged and soon to be married couples have combined income of more than 50,000 US dollars which works out 30 percent greater than the average US households. Therefore, they are the target for many products and services. The bridal industry is of the order of 35 US billion dollars a year.

2. STAGE 2: HONEY–MOONERS

In India bridal market is picking up and is of the size of 22,000 crores. The honey moon stage starts immediately after the wedding vows are taken and generally continues until the arrival of the couple's first child. This stage of FLC serves as a period of adjustment to life. These couples have available a combined income that often permits a life-style that provides them with opportunities for more indulgent purchasing of possessions or allow them to save or invest their extra income because, many young husbands and wives both work.

Honey–mooners have a considerable startup expenses while establishing a new home ; they need major and minor appliances, bed-room and living-room furniture, carpeting, drapes, dishes and a host of utensils and accessory items of utensils and necessary items. During this stage FLC, the advice and experience of other married couples are likely to be important to newly married. Even they search for new product information which are normally made available by magazines like, "Sweethome"—"House–Keeping", "Interior", "Design" and so on.

3. STAGE 3 : PARENT–HOOD

When a couple has its first child, the honeymoon is considered to be over. The stage of parent–hood—sometimes called as full-nest stage—usually extends over more than 20 year period. Because of its long duration, this stage can be divided into shorter phases namely, pre-school phase, the elementary school phase, the high school phase and the college phase. Throughout these parent–hood phases, the interrelationships of family members and the structure of the family gradually change. Furthermore, the finance resources of the family change significantly as one or both parents progress in a career and child rearing and educational responsibilities gradually increase and finally decrease as children become self-supporting.

As increasing in the number of births among baby boomers that is those born between 1946 and 1964 has resulted in a "baby boomlet" in America. These parents are older—34 and 52, better educated, more affluent, and more socially aware. As many as 44 per cent also feel that they are better parents to their children than their parents were to them. Their children often become the focus of their lives and they spend money accordingly. "Boomer" parents have become an important target group for companies that serve baby market. The best example can be that of Johnson and Johnson limited. They are also important market for many insurance services and investment activities.

At the same time, good many magazines cater to the information and entertainment needs and children. For instance, there are special interest publications such as "Humpty Dumpty" specially designed for the young child just learning to read. 'Child–care', 'Doctor', 'Teen–Ager' and so on.

4. STAGE 4 : POST-PARENT HOOD

It is but natural that post-parent-hood is traumatic for some parents and liberating for others because, parent–hood extends over a long period of time and when all the children have left home this stage is rightly called as 'empty nest' stage that signifies for many parents almost a rebirth, as time for doing all things, they could not do while children were at home and they had to worry about soaring educational expenses. For a mother, it is a time to further her education to enter or re–enter the job-market, to seek new interests. For the father, it is the time to indudge in new hobbies. For both husband and wife, it is the time to travel, to entertain, perhaps to refurnish their home or to sell it in favour of a new home or condominium.

It is during this stage of FLC that married couples tend to be most comfortable financially. Today's **'empty–nesters'** have more leisure time. They travel more frequently, take extended vacations, are more likely to purchase a second house in a warmer climate than the earlier period. They have higher disposable incomes because of savings and investments and they have fewer expenses. For this reason, families in post parenthood stage are really an important market for luxury goods, new automobiles, expensive furniture, vacations, tour and travel over long distances for longer period.

It is found that many empty nesters retire while they are still in good health. It is because, voluntary retirement provides the opportunity to pursue new interests, to travel, and to fulfill unsatisfied needs. It is hotels, airlines car-leasing companies that quickly respond to this market with discount to customers above 45 to 50 years. Some unfortunate families have to take compulsory retirement and they care much for savings for better health and minimum standard of decent living.

The older consumers tend to use TV as an important source of information and entertainment. They prefer programmes that provide opportunity to "keep up with what is happening", especially news and public affairs programmes. In addition, a number of special interest magazines cater exclusively to this market such as, 'Old Age' 'Modern Maturity', 'Active Life'. They deal with the relevant and specific issues of such consumers.

5. STAGE 5 : DISSOLUTION

Dissolution–the basic family unit occurs with the death of one spouse. When the surving spouse is in good health, is working or has adequate savings and has a supportive family and friends, the adjustment is easier. The surving spouse-usually the wife-often tends to follow a more economic life-style. Many surving spouses seek each other out for companion ships; others into second or even third or fourth marriages in Western world. Indian conditions, with cultural values, it is the rare case of second or third marriage. Those families who do not have supportive family households and friends prefer to take shelter in "Homes for the Aged". In Western world the sons and daughters try to support and meet their parents on festive and special occasions unlike in Indian back-ground. Today, in India too, parents of empty nests are going for well managed "Homes for Aged" and spend their life in the company of others who enjoy the same or similar status.

This is the last phase of FLC which might be an ordeal and agony for some and those who are lucky, it is tolerably well-till death.

MODERNISED FAMILY LIFE–CYCLE

During recent years, many changes in the family have occurred particularly in smaller family size say, postponement of marriage, rising divorce rates and the like. To compensate these limitations, consumer researchers have been trying to search out expanded FLC models that reflect better diversity of family and life-style arrangements. Here is an attempt to give two such modernized FLCs. One is given by **Professor-Patrick E. Murphy** and **William A. Staples** in their article. "A Modernized Family Life Cycle"-Journal of Consumer Research 6-(June, 1979. Which is given as under) :

The above model depicts along the main horizontal the stages of the traditional FLC and above and below the man horizontal selected FLC stages that account for some significant non- traditional family households that the marketers are increasingly targeting. The heart of socio-demographic forces that drive this expanded FLC model include divorce and later marriages, with or without the presence of children. This model only recognises that started in marriage, ignoring such single parent house-holds as unwed mothers and families favoured because a single person or single person's adopt a child, though it gives greater reality.

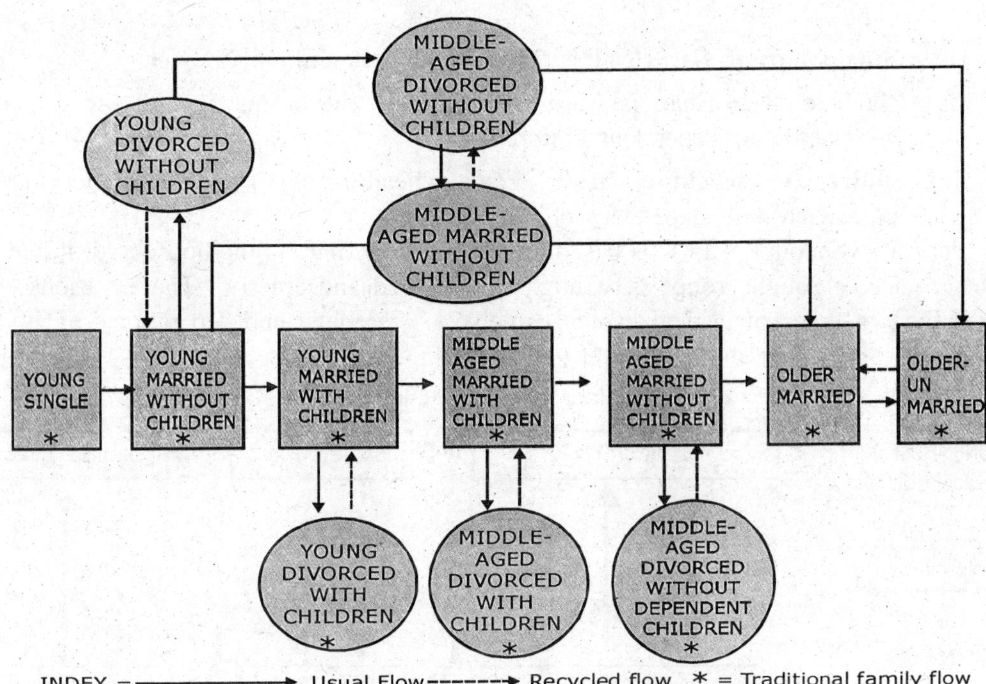

FIG: 8:03 AN EXTENDED FAMILY LIFE-CYCLE SCHEME ACCOUNTS FOR ALTERNATIVE CONSUMER LIFE-STYLE REALITIES

Let us take another model. This would another conception of life– cycle, which includes such stages as divorced single parents and middle aged marriage without children. This modernised version is described as consisting of the following groups. These groups are

1. **Bachelor 1:** Head is 18—34 single (never married, divorced, separated involved), no dependent children
2. **Young Couple:** Female head is 18-34, couple (married or unmarried, no children).
3. **Full–Nest-I:** Female head is 18-34, couple (married or unmarried) youngest child under 6.
4. **Full–Nest—11:** Female head is 18—36, couple (married or unmarried), youngest child 6 or over.
5. **Single Parent I:** Head is 18-34, single (never married,divorced, separated, widowed) youngest child under 6.
6. **Single Parent II:** Head is 18—34, single (never married,divorced, separated, widowed), youngest child 6 or over.

7. **Bachelor II:** Head is 35—64, single (never married, divorced, separated, widowed, no dependent children.
8. **Children's Couple:** Female is 35-64, Couple (married or unmarried) no dependent children.
9. **Delayed Full–Nest:** Female head is 35—64, couple (married or unmarried) youngest child under 6.
10. **Full–Nest III:** Female head is 35—64, couple (married or unmarried), youngest child 6 over.
11. **Single Parent III:** Head is 35-64, youngest child 6 or over.
12. **Bachelor III:** Head is 65 or older, single (never married, divorced, separated, widowed), no dependent children.
13. **Older Couple/Empty–Nest:** Female head is 65 or older, couple (married or unmarried), no dependent children.

. The above model of FLC is based on age, of the female in the household, if appropriate which, is traced through groups of young, middle aged and elderly. These various ages are affected by two types of critical events namely, 1. marriage and separation—by divorce or death and 2. arrival of the first child and departure of the last child. Thus, the modernised FLC accounts for four household types over three age groups through a person's life-time.

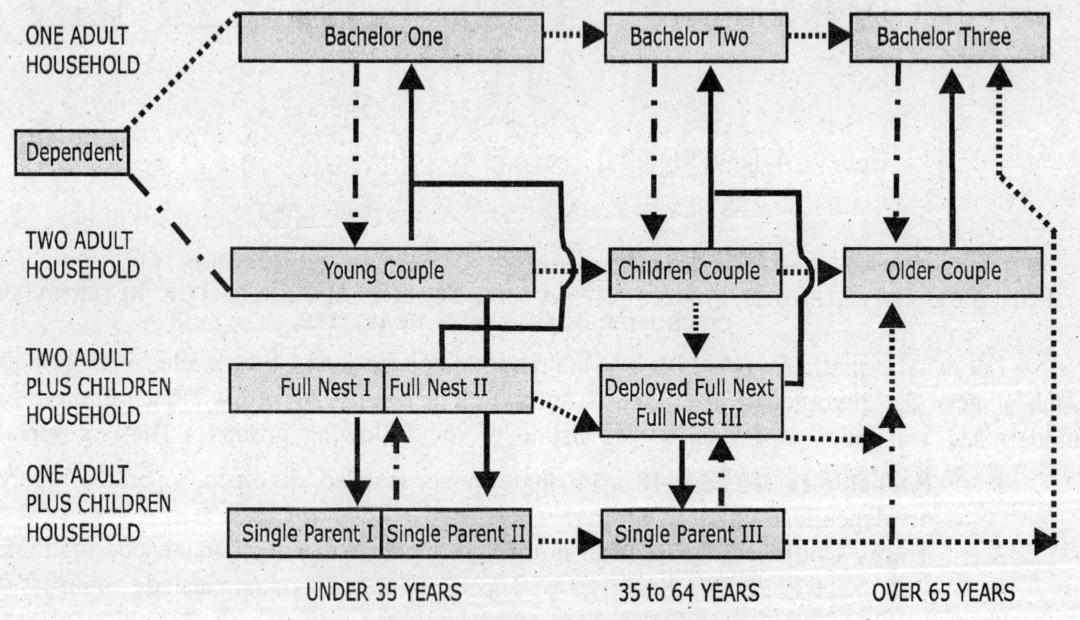

FIG: 8:04 THE REDEFINED FAMILY LIFE CYCLE

Source: **Mary C. Gills** and **Ben Mr. Enis,** "Recyclings the FLC: A Proposal for Redefinition" in Andrew Mitchel, ed., Advances in Consumer Research: Vol. 9. 1982-p.274.

RELATIONSHIP BETWEEN FLC AND CONSUMER BEHAVIOUR

Despite the fact that in modified version of FLC as many as their teen stages have been suggested there is no unaminity among researchers and research studies as to the most appropriate categorisation of FLC. In spite of different age groups, researchers have agreed that there is relationship between FLC and the consumer behaviour. One of the largest Departmental Stores of America namely **Sears and Roeburk** Incorporation found that people leaving the "Young Single" stage and entering the "Young family" stage were far more likely to buy and own all types of appliances. Such information is of help to buyers, marketers, advertisers and sales staff. For instance, sales people are encouraged to identify the customer's position in the life cycle as to how many kids? What are their ages? and so on, so forth so as to sell products to appropriate prospects.

Research based on FLC has revealed many consumption differences across household life cycles stages, indicating that the model is a useful segmentation tool. It is a good prediction of individual attitudes and leisure activities. The model is strongly and significantly related to food and beverage consumption, major and minor appliance ownership, dollar value of major household acquisitions and dollar value of home entertainment devices. It is very important to note that age as opposed to life cycle plays significant role in segmenting the market that age is more sensitive for products and services classified mainly as luxurious and for a diverse catch all category. However, for most products, life cycle analysis allows the marketer to achieve a richer understanding of market.

CHAPTER BASED QUESTIONS

A. WRITE SHORT NOTES ON :

1. Family 2. FLC 3. Stages of FLC 4. Functions of family 5. Socialisation of family members. 6. Family decision-making 7. Family roles 8. Dynamics of husband and wife decision making 9. Children 10. Teenagers II. Modified FLC 12. Consumer behaviour and FLC.

B. SHORT ANSWER QUESTIONS:

1. What is a family ?
2. Explain the functions of family.
3. What is family decision-making ?
4. Explain the dynamics of husband and wife decision-making.
5. Explain children of different age groups on family decision
6. What is traditional FLC ?
7. What stages are involved in FLC ?
8. What is modernized FLC ?
9. What is the relationship between FLC and consumer behaviours?
10. What is a family role? Explain briefly all possible family rules.

C. ESSAY TYPE QUESTIONS:

1. Define the term 'family' as a student of consumer behaviour Explain the functions of a family as a unit.
2. What do you mean by family ? Explain family roles in key area of consumption.
3. Explain the dynamics of husband and wife decision–making How spouses influence and resolve conflict ?
4. "Dynamics of spouse decision making is influenced by several factors". Explain.
5. What is the place of 'Children' of different age group in family decision-making?
6. Who are 'teenagers'? Why marketers have given too much importance to this segment?
7. What is FLC? What stages are those in a traditional FLC? Explain each.
8. What are modified FLCs? or models? Why they have been developed?
9. What is the relationship between family, family life–cycle and consumer behaviour?
10. Do you think the terms 'family' and 'household' are the same? If not, how do you distinguish between them and why?

*****

CHAPTER 9

CULTURE AND CONSUMER BEHAVIOUR

BACKDROP

Culture as a man-made environment has deep impact on his or her behaviour that too consumption, over there fore consumer behaviour. That is why, this chapter, is devoted to the various aspects of culture namely its meaning, its characteristics — its measurement value system-sub culture—its aspects and other relevant aspects and interaction. The chapter ends with a Chapter Based Questions.

WHAT IS CULTURE ?

The word **culture** is so all pervasive; that it, is difficult to define in a few words. The word 'culture' is so compact that its impact and richness can hardly be projected in concrete terms. It is difficult to fix the boundaries of culture. Culture stands for man-made environment in·which he lives along with his companions as a social and rational animal. In that sense, it stands for the whole of that complex which impounds in its compound every thing about his knowledge, beliefs, art, morals, laws, customs, habits and other capabilities that he has acquired by human as a member of society. Culture is a distinctive way of a group of people and their complete design for living. As we are the students of consumer behaviour, culture can be defined as the sum total of all learnt beliefs, values, habits, attitudes, customs and taboos that serve to direct consumer behaviour of members of a particular society. Therefore, there cannot be any culture without a society and society without culture. A culture is a distinctive way of life of. a group of people — their design for living in it. "It is a way of life that determines our conduct and behaviour towards every thing that surrounds us visible and the invisible. It encompasses all aspects of the environment—both tangible and the values of a whole society having bearing as what we do, what we see, what we use, what we judge as nicely put by **Professor Philip Kotler**. The pictures: 9:01 and 9:02 explain it clearly.

STANDARD DEFINITIONS

★ **Mr. D.J. McCort and Naresh K. Malhotra:**

"Cultured is complex whole that including knowledge, belief, art, law, morals, customs, and any other capabilities and habits acquired by humans as members of the society."

★ **Professors Howard and Seth:**

" Culture is a selective, man-made way of responding to experience, a set of behaviour pattern."

★ **Mr. John Bull:**

" Culture is a distinct way of life of a group of people ad their complete design of living."

★ **Mr Philip R. Cateora and J.M. Hess:**

" It is the man-made part of the mans environment- the sum total of his knowledge, beliefs, art, morals, laws, customs and any other capabilities and habits he acquired as the member of society."

★ **Professor Philip Kotler:**

" It is the way of life that determines our conduct and behaviour towards everything that surrounds us visible and invisible. It encompasses all the aspects of the environment – both tangible and intangible-that have been created by man. It cover the attitudes and values of the whole society having bearing on what we do, what we see, what we use, what are judge."

Habits, customs, attitudes, beliefs, values and taboos are the part of culture. It is worth while to understand each component. A 'habit' is an act of repetition and that turns automatically. It may be buying habit, an eating habit, a drinking habit, or even wearing a pair of socks, shoes and dress. A 'custom' is a habit followed by many people belonging to a specific group. It may be a custom of two meals a day, painting lips and nails, having a morning or evening bath, visiting churches, temples, mosques a week. An 'attitude' is a state of readiness to make value judgements in support of or against people, product, an idea or service. Thus,

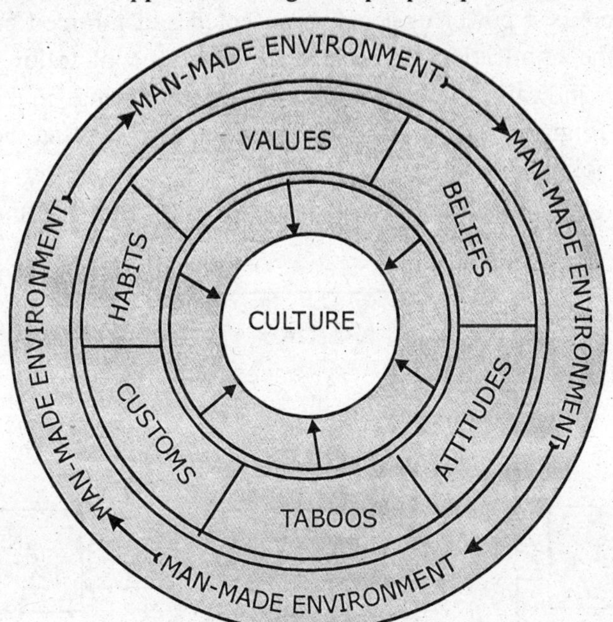

FIG: 9:01 CULTURE A CREATION OF HUMAN EFFORTS

each one is having his or her attitude towards life, people, politics, events, nation, religion— practically for everything that affects his life and living. These attitudes are ever changing. That is, an attitude can be positive or negative, permanent or temporary and *vice–a–versa.* A 'belief is a subjective concept of truth. An individual beliefs are based more on emotion than reason and that may or may not agree with facts. Beliefs consists of very large number of mental and verbal statements that reflect a person's particular knowledge and assessment of something say a person, a stores, a product, a brand, a service and so on. For instance, even a false thing, for him, becomes a fact. Thus, a student uses **Casio** calculator has a feeling that he may go wrong if he uses a **'Citizen' or 'National' or 'Orpat'.** A 'value' is a desirable condition or state of affairs. Group members have values in common. It may be possessing of a wealth, an University degree, a good health, a charming beauty and so on. These values differ from group to group and time to time. Values are also beliefs. Values differ from other beliefs, however, because they meet the following criteria : (1) They are relatively few in number. (2) They serve as guide for culturally appropriate behaviour. (3) They are enduring or difficult to change. (4) They are not tied to specific objects or situations and (5) They are widely accepted by the members of society. Therefore, in broad sense, both values and beliefs are mental images that affect a wide range of specific attitudes which, in turn influence the ways a person is likely to respond to a specific situation. For instance, the criteria a person uses to evaluate alternative brands in a product category say **Kodak—Fuji—Olumpys-Minolta, Yashika, Richo** and other brands, say of 35 mm with similar functions. Or his or her eventual preference for one of these brands over the other are influenced by both person's general values say—perceptions as to what constitutes quality and the measures of country of origin — and specific beliefs say, particular perceptions about the quality of Indian made versus Japanese made camera. A 'taboo' is a powerful negative accepted and inforced custom by the members of a group. It has the implication that evil consequences will follow in case it is violated. Thus, colour of cloth indicating mourning is blue in Iran, white in India and Japan, purple in America, black in England and so on. A Hindu woman shall salute her husband every night and morning before joing him in bed and leaving bed in the latter part, failing which her husbands life-span shall vanish. Thus, Judo-Christian group cannot violate the "Ten Commandments."

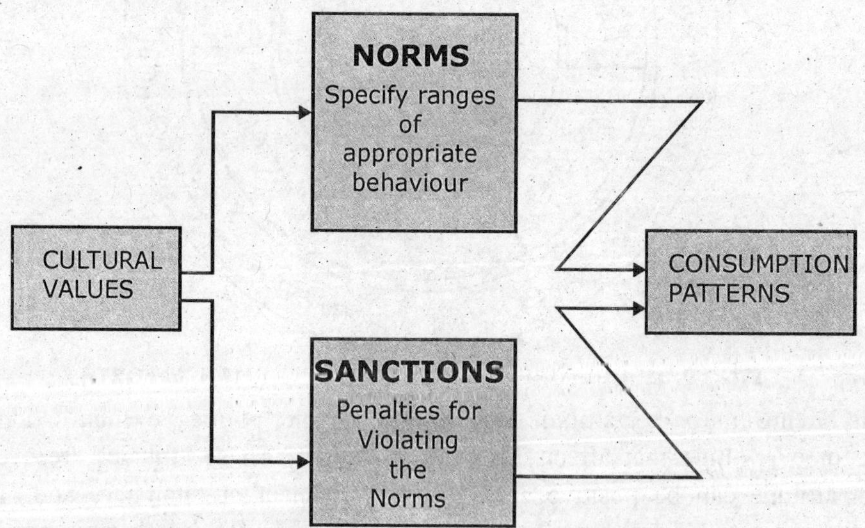

FIG: 9:02 VALUES, NORMS, SANCTIONS AND CONSUMPTION PATTERNS

CHARACTERISTICS OF CULTURE

Some great writer once said, "Culture is not an exotic motion studied by a select group of anthropologists in the South Seas. It is a mold in which are all cast, and it controls our daily lives in many unsuspected ways." With this note, let us attempt to know the basic characteristics of culture that helps us to spell out its nature in terms of scope and complexity of it.

1. Culture is Invented:

Culture is not a gift of God. Though we say "God made the world and man made the garden," God made the wonderful earth and man made the environment of his own for his way of living which we call it culture. That is why, we have highly cultured people on one side, and uncultured people on the other. That is, culture does not exist on its own. Some where waiting for being discovered. People were responsible for inventing the culture. This greatest invention consists of three interdependent systems or elements namely, (1) An ideological system, or mental component, that consists of the ideas, beliefs, values, the ways of reasoning that human beings learnt accept in defining what is desirable and undesirable. (2) A technological system that consists of the skills, crafts, and arts that enable humans to produce material goods derived from the natural environment and (3) An organisation system that makes it possible for humans to co–ordinate their behaviour effectively with the actions of others like family system and social class.

2. Culture is Learnt:

Culture is not innate or instinctive. Unlike innate biological characteristics say sex, skin, hair colour or intelligence, culture is learnt. At an early age, we begin to acquire from our social environment a set of beliefs, values, habits, attitudes, customs and taboos that make culture. In case of children, the learning of these acceptable cultural values and customs is reinforced by the process of playing with their toys and persons. As children play, they act out and rehearse important cultural lessons and situations. The early life of learning is charged with full of emotions. This cultural learning prepares them for later real life circumstances. Let us see as to how culture is learnt and other aspects of learning.

How culture is learnt ?

Anthropologists have clearly identified three clear-cut forms of cultural learning namely, 'formal', 'informal' and 'technical'. 'Formal' learning is one in which adults and older siblings teach a young family member "how to behave". "Informal learning" is one which allows the child to learn primarily by imitating the behaviour of selected others such as family members, friends or television heroes and "Technical learning" is one in which the teacher instructs the child in an educational environment about what should be done ? how it should be done ? and why it should be done ? Enculturation and Acculteration:

Anthropologists think of two ways of acquiring the culture. According to them, the learing of one's own culture is known as "enculturation." The learning of a new a foreign culture is known as "acculturation". This distinction is more important for marketers because acculteration is what helps in planning to sell their products in foreign or multi-national markets. In such cases, marketers must study the specific culture or customs of their potential target markets to determine whether their products will be acceptable to its members and, if so, how they can best communicate the characteristics of their products to persuade the target market to buy.

Language and Symbols:

The members of society must be able to communicate with each other through a common language to acquire a common culture. Shared meaning will not exist and true communication will not take place in absence of a common language. To communicate effectively with their audiences, marketers must use appropriate symbols to convey desired product images or features. These symbols can be verbal or non-verbal. Verbal symbols include TV announcement or an ad in magazine, or a newspaper or any outdoor vehicles. Non-verbal communication symbols include items like figures, colours, shapens and even textures to lend additional meaning to print or broadcast ads, to trade–marks and packaging or product designs. Basically, the symbolic nature of human language sets it apart from all other animal communication. A symbol can be anything that stands for something else. Any word is a symbol. Thus, the word 'razor' or 'car' or 'hurricane' 'susuki' all stand for a specific meaning to convey. Thus, "Mercides Benz" stands for a luxury car. It is possible for a person to experience cognatively a visualisation for a product because the human mind can process symbols. Clearly, the ability of humans, to understand symbolically how a product, service, or idea can satisfy their needs makes it easier for marketers to sell the features and benefits of their offerings.

Rituals:

Culture includes various experiences and behaviour, in addition to language and symbols. A ritual is a type of symbolic activity consisting of a series of steps occurring in a fixed sequence and repeated over a time. Rituals extend over human life-cycle from birth to death, in practice, including a host of intermediate events say confirmation, graduation, wedding. These rituals can be very public, elaborate, religious or civil ceremonies or they can be as mundane as an individual's grooming behaviour or flossing. Ritualized behaviour is typically formal and is often scripted behaviour. Most significant from the angle of marketers is the fact that rituals tend to be repleted with ritual artifacts or products. That are associated with or some how enhance performance of the ritual: For instance, tree ornaments, stockings and various food–items are linked to the ritual of Christmas celebration: So is the case of rituals of Hindus say that of naming a baby, starting of education, house–warming, marriage, death and so on. In addition, there is ritual behaviour not only that a ritual is performed traditionally. Ritual behaviour is any behaviour that is made into a ritual.

3. Culture is Shared:

Culture is a group phenomenon shared by human-beings living in organized societies and kept relatively uniform by the social pressure. A particular belief, values or practice must be shared by a significant portion of the society to be considered as a cultural characteristic. That is, culture is viewed as a group customs that link together the members of society. Common language, of course, is the critical cultural component that makes it possible for people to share values, experiences and customs. Different social institutions within a society transmit the elements of culture and make the sharing of culture a reality. The main one, are family educational institutions and houses of worship. Another very important overlooked institution is mass-media. Thus, advertising, promotional–efforts and public–relation play significant role.

4. Cultures are Similar yet Different:

All cultures portray certain similarities. For instance, some elements are commonly found in all societies say athelic sports, body–adornment, a calendar, cooking, courtship, dancing, education, family, gestures, government, housing, language, law, music, art, religious ritual and host of other items. These however, differ deeply and greatly from society to society or community to community that are the root cause of differences in consumer behaviour. Thus, a prayer as a ritual is different in different religions though they appear similar and a must.

5. Culture is Gratifying and Persistent:

Culture as invented, satisfies basic biological needs as well as learned needs. It consists of habits that will be maintained and supported and reinforced as long as those practice them are gratified. Cultural elements are handed down from generation to generation because of this gratification. That is why, we people are comfortable doing the things in a customary way. For a given thing, we have common habits or customs and conventions. Thus, Christians do not eat meat on Fridays while Hindus on say, Mondays and Thursdays. Muslims pray five times a day, while Hindus once or twice a day. Normally, people from coastal areas prefer evening bath than morning. Lip-painting and nail painting is common among advanced nations than under developed — as we all as high society and low society people.

6. Culture is Dynamic:

Culture has to evolve continually, if it is to function in the best interests of society to fulfit its need gratifying role. Good many factors produce cultural changes within a given society—say, change in technology, shifts in population, resource crunches, wars, changing values and customs borrowed from other cultures. India is, like most of the other countries, has been male dominated where generally husband is a finance minister and wife is minister for home affairs. However, this has undergone a change. Ladies are working, reaching new heights in sports, and dominating at home where both finance ministers have to strike balance to take up home ministry. Thus, in the past we never heard of "paternity leave" but to day it is there with "maternity leave". To start with ladies went in for part–time jobs, but are now competing with males in full–time offers and over–time. This has made marketer's job really difficult where he has to carefully and closely monitor the socio-cultural environment in order to market the existing products effectively or to develop new products. That is, marketers are to carefully analyse as to who are the users, buyers — ladies only or gents only or both together. In addition, they will have to know where they shop ? When they shop How they shop ? constant monitoring of cultural changes, the marketers are benefitted.

7. Culture is Adoptive, Adaptive, Organized and Integrated:

By nature, culture is adoptive, because any one can copy any element of it to suit his requirements or his group, it adaptive in the sense culture gradually and continuously keeps changing inspite of our strong resistance. Some societies are quite static and with a very slow rate of change, while others are more dynamic and sensitive, resulting in a very rapid changes in the folds. What is more important is culture is organised and integrated. That its, components are fitting well. Although every culture has some inconsistency in its elements, it tends to form a consistent and well-integrated whole.

8. Culture Fixes Norms:

A given culture operates primarily by setting rather loose boundaries for individual behaviour and by influencing the functioning of such institutions as the family structure and mass-media. That is, culture provides the frame-work within which individual and household life-styles evolve. The boundaries that culture sets on behaviour are named as 'norms'. Norms are the rules that specify or prohibit certain behaviours in specific situations. Norms are derived from "cultural values". Cultural values are widely held beliefs that affirm what is desirable. It is not necessary for a cultural values to be logically consistent. Some tension or strain between conflicting cultural values is characteristic of most advanced societies. This arrangement of abstract, and some times conflicting, cultural values that characterise industrialized societies leaves scope for variety of distinct life-styles to evolve within each society. Violation of cultural norms results in 'sanctions' or penalties ranging from mild social disapproval to banishment from the group. Conformity to norms usually gives implicit or obvious rewards only when a child is learning the culture—which we call as socialisation—or an individual is learning a new culture—called usually as acculturation. In other situations, conformity is expected without reward. For instance, generally we expect people to arrive on time for business and social appointments. On their arrival on time, no one openly compliments them, however, if they come late one is getting angry for their late arrival.

MEASUREMENT OF CULTURE

In measuring the culture as a valid concept, good many techniques are used. The techniques which we have already covered earlier relate to projective tests used by psychologists to study motivation and personality and the attitude measurement techniques used by social psychologists and sociologists are equally used in measuring culture. Not repeating them again, we shall consider some more significant techniques namely, (1) Content Analysis (2) Consumer Field Work and (3) Value Measurement Instruments.

1.Content Analysis:

Generally, conclusions about a society or specific aspects of a society can be drawn from the examination of the content of its messages. That is "Content Analysis" concentrates on the content of verbal, written and pictorial communications. Content Analysis can be used as a relatively objective means of determining as to what social and cultural changes have occurred in a particular society. For instance, consumer researchers use content analysis to examine how African Indians or Indian Africans, women, the elderly and animated spoke-characters have been depicted in mass media. It may be a content analysis to study trends in the use of animated spoke- characters-say Micky—Mouse, Tom and Jerry, and so on. Content Analysis is used by both marketers and public–policy makers interested in comparing the ad claims of competitions within a specific industry. It is equally used in evaluating the nature of ad claims targetted to specific audiences-say-women, children elderly or the youth.

2. Consumer Field Work

Anthropologists very often go deep in the environment under study while examining a specific society. They select a small sample of people from a particular society and observe their behaviour carefully. Based on these observations, researchers draw definite conclusions about the values, beliefs and custom of the society under investigation. For example, if a researcher is interested in how people select CDs (Compact Discs), they might post trained observers in record stores and note how specific types of CDs are selected, say— 'rap' versus 'country', 'jazz' versus 'classical', 'rock' versus 'sound tracks, 'instrumental' versus 'vocal' or say "Hindustan" versus "Kernatic" so and on. They may go beyond this. That is—how often and long read the description, How many times take off the CD from display and the like?

The outstanding features of Field observation are: (1) It takes place within a natural environment (2) It is performed without the respondent's awareness (3) It focuses on observations of behaviour. Instead of field observation, the researchers become "participant observers". That is, they place themselves in the position of salesman and not the reactions of customers. Both these above methods need highly skilled researchers to be overt and impartial in their judgements. In addition to field work methods, "indepth interview" and "focus-group" sessions can be used to have a just look at culture.

3. Value Measurement Survey Instruments:

. As against the conclusions drawn on observations, of late, researchers came with questionnaire or survey method. For this, researchers use data collection instruments called "Value Instruments" to ask the people how they feel about such basic personal and social concepts like freedom, comfort, national security and peace. Research into the relationship between people's values and their actions as consumers still is in its infancy. It is an area that is sure to recieve increased attention for it taps a broad dimension of human behaviour, however, that could not be explored effectively before the availability of standardized value instruments.

Luckily a variety of popular "value instruments" have been used in consumer behaviour studies. These include : (1) the Rokeach value Survey (2) the list of values (LOV) and the (3) Values and life styles (VALS 2).

1. The Rokeach Value Survey:

This is most widely used instrument, RVS is a self-administered value inventory that is divided into two parts, each part measuring different but complementary types of personal values. The first part consists of 18 terminal value–items which are designed to measure the relative importance of 'end' states of existence personal goals. The second part consists of 18 instrumental value items, which measure basic approaches an individual might take to reach end-state values. Thus, the first half of the measurement instrument deals with ends and the second half considers means.

2. The List of Values (LOV):

This LOV is a related measurement instrument that is designed to be used in surveying consumer's personal values. The LOV scale asks consumers to identify their two most important values from NINE value list based on the terminal value of the Rokeach value survey. The terminal value list consists of 18 such values say : (1) A comfortable life (2) An exciting life (3) A world at peace (4) Equality (5) Freedom (6) Happiness (7) National security (8) Pleasure (9) Salvation (10) Social recognition (II) True friendship (12) Wisdom (13) A world of beauty (14) Family security (15) Mature love (16) Self-respect (17) A sense of accomplishment (18) Inner harmony. The LOV scale uses nine –point scale, asks the respondent to choose two most important ones say-warm relationship with others and sense of belonging, or sense of accomplishment and inner harmony or social recognition and a comfortable life and so on.

3. Value and Life–Style (VALS–2)

Perhaps the most widely used of life-style and psychographic research by marketing mangers is 'VALS' programme developed by SRI International – Value and Life–style. It was introduced in 1978 which provided a systematic classification of American adults into nine distinct value and life-style patterns. Because of laxities was replaced by VALS -2- in 1969. Which has more of psychological base them the original and in more activity and interest based. SRI International has identified three basic SELF ORIENTATIONS namely. (1) Principle Oriented 2) Status Oriented and (3) Action Oriented.

VALS-2 is linked with good many dates bases. Again, VALS-2 system is linked tall the major geodemographic systems like PRIZM. VALS-1 suffers from shortcoming of the originals. These arc : (1) VALS-1 are individual measures 2) a few individuals are 'pure' in term of 'Self orientation 3) The values and demographics measured by VALS-2 are inappropriate.

Inspite of these problems VALS-2 is the most complete general segmentation system ever made available. Hence, it is widely used by managers for different types of segments.

The Yankelovich Monitor

The **Yankelovich monitor** is a commercial research that systematically tracks social trends designed to reflect a variety of cultural values. It was first conducted in 1976 and has been updated annually since then. At present, the monitor tracks about sixty social trends. Thus

Trend No 3-speaks of physical fitness and well being Trend No. 12 deals with novelty and change Trend No. 32—speaks of concern about environment Trend No. 46 deal with s non –commitment to buy American if applied Indian Trend No. 47 deals with responsiveness to fantacies Trend 52 speaks of accommodation to technology and Trend 55 deals with need for control and so on.

By carefully interpreting social trends and determining which demographic segments are most affected by a particular group of trends, the 'monitor' is capable of providing advance warning of likely shifts in demand for various product categories. Armed with these insights, corporate executives are in a better position to make more enlightened decisions. Accurate and timely decisions pay generally rich dividends.

DYG SCAN-A Scanning Programme

In America, this DYG SCAN has been conducted as yearly DYG SCAN study since 1986. This is a research programme that tracks 38 social values among different segments of American population, as well as business, government and academic leaders. Further, on optional service called 'Brandlink' extends specific trends to the actual image assessment of a subscribing firm's brands. The sample can be selected from among the general public, users of specific products without regard to brand, users of the brand itself or users of competitor's brands. The brand's image as well as of its close competitors, can then be compared.

These illustrations of "standardized value", and social trend measurement instruments" are promising techniques for meaningful consumer behaviour research, especially segmentation analysis. These values can be used with profit by combining other variables to predict shifts in consumption patterns—particularly is cases of new product, concepts, repositioning of products and adjusting a firm's general marketing efforts.

VALUE SYSTEM CORE VALUES

Cultural values are significant to the well–knit and organised nature of any country. A cultural value can be expressed as a widely held belief or sentiment that some activities, relationships, feelings or goals are important for the community's identity or well being. In the words of **Professor Milton Rockeach,** values as centrally held and enduring beliefs that guide actions and judgements across specific situations and beyond immediate goals more ultimate end states of existence. Values, therefore, produce inclinations to respond to specific stimuli in standard ways. That is, cultural values are widely held beliefs that affirm what is desirable. These values affect behaviour through norms which specify an acceptable range of responses to specific situations. Put in other words, a specific behaviour is expected to either help or hinder the attainment of some values or group of values. Then the consumers are motivated to engage in behaviours designed to enhance the achievement of certain values and to avoid those behaviours perceived to hinder the attainment of certain value states or system.

These values are culturally determined. That is, they are learnt from social–integration say' families and friends, in settings such as schools and temples and churches, mosques and gurudwaras. Value system strongly influences the consumer behaviour. Even though specific situations might dictate slightly different actions, overall there is much similarity in consumer behaviour within a given culture like tastes, methods of shopping and so on. It is because of these that every marketer has to understand society's basic or core–structure of values that the strategy decisions do not apply in the tract of ingrained cultural patterns. It is much easier to synchronise with the culture than to attempt to change fundamental cultural–values.

CORE VALUES OF CULTURE

Though values of culture differ from nation to nation and from time to time, the process of socialisation that takes place all over the world, there are certain basic values which can hardly be ignored. The most widely held core values of culture are :

1. Individualism:

This value is highly complex and intricately interrelated with such things as freedom, democracy, nationalism and patriotism. Modern generations place strong value on being themselves. Self-reliance, self-interest, self-confidence, self-esteem, and self-fulfilment are all most popular expressions of 'individualism'. Striving for individualism seems to be linked to the rejection of dependency. That is, better to rely on one-self than others. The stress on individualism has been very strong during the last three decades all over the world. One talks of "Age of Me". The "Age of Me" emphasizes self-fulfilment, self-actualisation, inner motivation and a focus on self and is clearly demonstrated by slogan. "I want to be me", "I do it my way", "I am looking out for number one."

In terms of consumer behaviour, an appeal to individualism frequently takes the form of reinforcement of consumer's sense of identity with products or services that both reflect and emphasize that identity. For instance, ads for high-style clothing and cosmetics usually promise that their products will enhance the consumer's exclusive or distinctive character and set him or her apart from others.

2. Freedom:

Freedom is another very strong cultural value. In India and America and other countries which are wedded to the concept of democracy, this is a very significant value. These countries have historical roots in such democratic ideals as "freedom of speech", "freedom of the press" and "freedom of worship" and other similar values. As an outgrowth of these democratic beliefs in freedom, Indians have a strong preference for freedom choice, the opportunity to choose from a variety and under a wide range of alternatives. This individual preference is reflected in the large number of competitive brands and product variations that are quite visible on the show-cases and shelves of the modern super–market or departmental store. Any individual has a freedom to choose from wide variety and range in terms of sizes, colours, flavours, features, styles, and even special ingredients, say in case of tooth paste all natural ingredients free of sugar. That is why, all industrial or manufacturing units have wide ranging products be it a biscuit, a candy, a cookie, or a TV, camera, calculator, fridge and so on. There are more than 125 Sony walkmans, 900 **Philips** colour TV sets, 3500 **Sieko** watches, 2000 calculators of **Casio.**

3. Equality:

All democratic countries and socialistic pattern countries believe in the intrinsic equality of people. That is, each person is a worthwhile individual. To be very apt, "all are equal before God". That is, everyone has an equal right to life, liberty and the persuit of happiness and equal opportunity to strive for social and economic rewards. Recently, this value has caught special push for social and economic equality in the country. One of the most profound impacts during this decade has been womens, demand for equality that has brought about changes in marital and matrimonial relationships and the average Indian family. Women's lib movement represents a value structure that has resulted in greater independence and equality of women in social and economic spheres. It is clear that more and more women are outside the house occupying higher positions, changing sex roles within the house, so much so that they have their own life-styles.

4.Activity:

Human life is full of activities by its very nature. Generally, culture stresses activity particularly work as a predominant value. This has a religious background. Every religion has said, "idleness is an evil." That is, each individual is expected to work and work hard, earn money and be thrifty. This work has a origin in religious discipline as a mean of livelihood. Today Indians, like others, believe in not only working but working hard and smart. Of late, we see increased productivity and affluence have resulted in replacing long hours of hard work with longer vacation periods, shorter work weeks and more paid holidays. Resultantly, it has increased leisure time in which activity in still stressed, but it is manifested in innumerable leisure pursuits as clearly demonstrated by increased participation in sports, trips, tours and outings. Thus, readymade–garment makers, shoe–makers have encashed on it. All shoe companies say that their shoes are the best for sports. It is equally true of garment–maker, travel–agencies, hotel and tourism units. As everyone has sometime given the wants to utilize the maximum for work and select those lines of activities where there a possibility of leisure time. Leisure time and “pressure time” are the status symbols of life- style. Thus, marketers have been popularising, ‘instant coffee’, fast–food outlets, which fitwell into the life-styles of these categories. Lack of time adds to life-style stresses. Thus, compulsory shopping eats away leisure time adding to the consumers' pressure. Though people want the things, they hate shopping as it is a drugery and eats good deal of leisure time. This has led to tele–shopping, ready gourmet-mixes, ready–yogurt, and so on. Thus, a woman might prefer Pillsbury **Atta** than buying wheat and get floured.

5. Progress and Achievement:

Progress is passport to the prosperity and being a positive way of looking at things leads to achievement or individual success. The promise of progress is another core–cultural value. This is related to achievement and success, efficiency and practicality. For any positive thinker past is past but future is more important. Progress and achievement are possible in future though one has failed in the past. That is why, achievement is heavily stressed. The success stories of the achievers show the way for future achievement. In achievement self–reliance and initiative have an upper hand. From marketing points of view, the achievement orientation has placed higher value on symbols or brands. Thus, buying a car is an achievement but buying a **Fiat Siennaa, Mercedes Benz,** a mark of serious and rare achievement. Even the persons' residential area is a mark of an achievement and progress.

6. Efficiency and Practicability:

Efficiency and practicability are other core–cultural values. Those who are moved by efficiency, productivity and practicality they appreciate technical excellence and constant search for better ways of doing the things. A country's economic system is structured on these that support one another. People are interested in performance features or functional features of products such as speed, economy, safety and durability. These are all rational motives. This appreciates of the practical as opposed to the intellectual and emotional helps in explaining as to why certain ad messages are very effective. Those who are practical oriented are oriented toward informality. Formal things are age–old and only adoption of informal things set a trend and pace for change and improvement. Thus, formal habits in manners, dress, speech and social relationships have been playing important role. Thus, 'Hollo' is 'Hieh' or 'Hai'. ‘Pop’ for father, Mom for mother, ‘B-I-L’, for brother in law. 'Cal' for calculator, 'mot' or 'mobile phone' and what not? Formality, brings in rigidity and less of comfort. Informality looks more real and practical.

7. Materialism or Material Comfort:

There are two clashing values namely 'materialism' and ‘spiritualism.’ Materialism or material comfort signifies the attainment of 'good life' and living,—a life that provides a decent living, an automobile an independent bunglow, washing machine, home–theatre, micro-oven, an air–conditioner, a hot–tub, vacuum cleaner, lawn–movers, colour TV set and convenience oriented and pleasure providing goods and services. This concept of materialism is a relative concept changing from individual to individuals, time to time and area to are. Closely related to materialism is the growing trend towards hedonism or devotion to pleasure. It is but natural that we seek maximum pleasurable sensation with minimum effort. This hedonism has led to production and marketing of products ranging from luxury cars and homes to many foods and such pass times as electronic video games. One thing is sure that this material comfort, or materialism brings in money, and hefty spending yet not happiness, for materialism is not everything.

8. External Conformity:

One might embrace freedom of choice and individualism as we have seen earlier, but these accepted subject to reality of confirmity. External conformity is very necessary process by which the individual adopts to society. Experienced have said "-no social organisation, no culture. No form of institutionalized relationship whatever could exist without the process of interaction we call confirmity. In terms of consumer behaviour, confirmity or uniformity takes the form of standarized goods and services. Standardized products have been made possible by mass productions. The mass–production gives wide choice of standardized products enabling the individual to maintain his individuality or conforming. In this context, individualism and conformity coexist side by side. This is called as "ping-pong" relationship that has given much lee-way. Thus, male and female executives are conforming less to work–place dress codes. That is, these executives have wider options as against the formal codes of the time. The reasons they give is that uniformity increases their efficiency by 50 per cent.

9. Control over Environment:

Though man is slave of nature or environment inspite of big chunk strides in science and technology, yet he wants to control his environment. This kind of control is not only confined to control of weather conditions and harnessing of sun and tides as sources of energy, but also extends to areas like genetic–engineering, astrophysics and so on. We hear regularly and on festive occasions that gadgets have come that stitch our clothes, knit our clothes, attach buttons, spin, chop, slice and dice our vegetables. The magazines give latest gadgets of desire speaking highly about variety of micro-ovens VCR, VCP, TV with remote control, sense detector, home computer, automatic soda, tea coffee maker-which make life more comfortable. When we hear of the car-phone, CD player, video camera, pulsating shower massage and auto–lights, we perceive them as modern frills. The other dimensions this value is countervailing trend. That is, people move away from these complex host of products to simplicity as a reaction to all unavoidable complexity of modern hedonistic life. They are victims of traffic jams, delayed take–of planes, junk mail and other unavoidable hassles of contemporary life. They are found to make their living simple to derive more satisfaction and contentment.

10. Youthfulness:

Generally, men and women attempt to look younger than what they are. This value places present price on youthfulness. Youthfulness should not be confused with youth-which describes an age grouping. It is a state of mind and state being sometimes expressed as "young at heart", "young in spirit", or "young is appearance." As a result, marketers have got designed ads creating a sense of urgency about retairing youth and fearing aging which one cannot avoid. Thus, hand cream speaks of "young hands" skin-or face cream speaks of "teenage skin and complexion"; fragrance and make up-aids speak of "young skin" or "wrinkle free skin"—appealing ad appeal "defy aging". The detergent ad asks the reader "can you match their hands with their ages"? These ad themes promise the customer the benefits of youthfulness.

That is why people spend good deal of money on deodorants, shampoos, colognes, and other cosmetics in order to improve or hide their true bodily features. We spend on detergents, cleansers, waxes, rug–shampoos, room–deodorizers and electric bug killers to conquer the dirt, germs, pests and odours that surround us.

11. Fitness and Health:

Health is wealth and fitness is means to wealth. Fitness and health have emerged as other core values of culture. This value has manifested in a number of ways—in terms of indoor and outdoor–games, jogging, taking dog for morning walk (run.). Use of vitamins. "Fitness" trend is an outgrowth of the "my generation". Younger and older generations or different age groups of same generation are engaged in one or more athletic activities that have boosted market for fitnesses related products and services. Fitness and health are becoming life–style choices for many consumers. These health awareness have been instrumental in opening up of well–known chain sports class or gyms. Traditional food manufacturers have been modifying their ingredients to keep pace with "stay slim". Therefore, diet beer, oil–less roties or chapaties, white–meat in place of red–meat. The concept of "no-cholesterol" also "no- preservative" snack products. "Sugar free sugar" and such other products have come.

12. Religious and Moral Orientation:

People of the country do believe in God and some not. This is the case with every country and every religion. India a continent itself, we have many religions and cults. Roughly 85 per cent of people believe in God directly or indirectly and about 74 per cent believe past birth or life after death. Religion and commandments of religion decide their behaviour. Religion affects the birth, growth and death of a person and depending on the religion, the rituals are performed at each stage of life–cycle. Like Western World, Indians are also God fearing and believe in divine power. They do believe that God is almighty and decides the fate of an individual. As against divine powers, we have evil powers that make them accepting superstitions and good deal of human lives and material wealth is being wasted. We come across richest and poorest gods and godesses in India like richest and poorest person. Each religion has sub-cults and groups and the ethical standards differ. Among Christians, we have judo–Christians who have very high and strong moral and ethical quality. Resultantly, these people tend to view the world in absolute terms and tolerate few gray areas; that is, they judge the things in terms of good or bad, right or wrong, ethical or unethical. In the name of religion, we have conventions or customs as to the days of fasting, days of visiting the temples, mosques, number of times they are to pray and offerings to be made. Prayer among Hindus is incomplete without flowers, certain preparations as 'abishek' and last 'naivedy' with breaking of coconuts and so on. This has given big push floriculture, forest product, horticultural products and dairy products. Even at each event, birth, naming, schooling, 'upnayan' wedding, conception, longevity celebrations and death, we have special performances. Even we have post–death rituals.

13. Humanitarianism:

Each country has well-to-do and ill-to-do people. There are people who have lot of wealth and income. They have a problem as how to spend judiciously? On the contrary, there is a vast majority which is faced with how to have two square meals a day. India is known for poverty which is our own creation because whatever development has taken place is eaten away by population explosion. However, rich people will have to help poor directly or indirectly because those who live in glass houses cannot afford to quarrel with surrounding people or their neighbours. That is why, religion also preaches to help the poor and down–trodden. In India, in like other countries, we have generous people and always ready to help the needy. We have regional, national and international organisations that care for the forlorn, down–trodden, under–privileged in one form or the other. Poverty is a curse and has to be eradicated at any cost; however, it is impossible to do so, for reasons best known to us. However, these charitable institutions such as **United Fund, Red Cross, CARE** and individuals on compassionate grounds do help. Even when natural calamities occur, all other nations help those nations which are hit hard. There are organisation which discourage begging and rehabilitate these people.

SUB-CULTURE AND INTEGRATION

WHAT IS A SUB-CULTURE ?

Culture has a potent influence on all consumer behaviour. Individuals are brought up to follow the set of beliefs, values, customs and taboos of their society. That is, they are made to accept behaviour that is acceptable as per norms of society that as good, right and ethical and attempted to eschew from those which are considered as taboos. Culture having bigger societies have segmentation possibility as it is measurable like other variable. However, marketers segment larger societies into smaller sub-groups or sub-cultures that consist of people who are similar in terms of their ethnic origin, their customs and the ways or the ways they behave. These sub-cultures provide significant marketing opportunities for an astute marketing executive.

The members of a specific sub-culture have beliefs, attitudes, customs that set them apart from other members of the same society. Further, they adhere to most of the dominant beliefs, values and behavioural patterns of larger-society. Therefore, one can define "sub-culture" as a distinct cultural group that exists as an identifiable segment within a larger and more complex society.

Thus, the cultural profile of a society or nation is a composite of two distinct elements namely. (1) The unique beliefs, values and customs subscribed to by members of specific sub- cultures and (2) The central or core cultural themes that are shared by most of the population irrespective of specific sub-cultural memberships. These sub-culture categories can be based on nationality, religion, geographic region, race, age, gender, occupation and social class. In case we take America as a base country, by nationality there may be French, Puert-Rican, Korean, Chinese, English, Indian, Pakistani, Sri Lankan, Japanese and so on; by Religion-Catholic, Jews, Hindus, Mohamedans, Buddhist and so on; by geographic region—may be Southern, Northern Eastern, Western, South–Eastern Mid-eastern, North–Eastern, mid-northern, North Western; by Race, African, American, Caucasian, Asian- American; by Age, Teens, Xers (adolescents), Middle aged and Elderly; by gender, Male and Female; by occupation- Engineers, Professors, Cooks, Plumbers, Carpenters, Doctors, Tailors, Painters and so on; by social class—lower, middle and Upper–Upper, Upper–Middle. Middle–Middle, Lower–Middle and Lower. This applies to all countries taking any country as a base. Here, for better understanding, as a model we take America as a base country where, like India we have people from all walks of life from an angle. America is a confluence of all nations, all religions all age–groups, all occupations, all social classes, all races and all genders.

The following is the figure which clearly and succinctly makes clear the concept of sub-culture. That is, it depicts the relationship between culture and sub-culture.

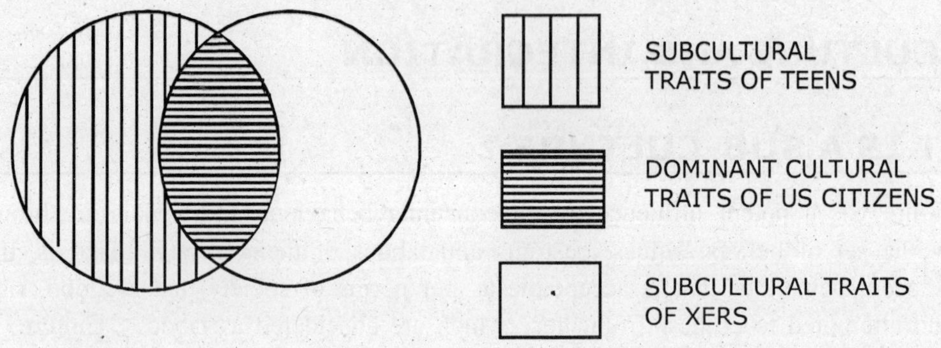

FIG:9:03 RELATIONSHIP BETWEEN CULTURE AND SUB-CULTURE

In order to explain the concept of sub-culture and its dimensions and implications, a case of USA is taken as a true symbol of democracy, confluence of immigrants from maximum countries, its achievements in science and technology, population control, results of sub-cultures is the form of manifestations of success in each field, say sports, economic, social, environmental control and what not. A simple example is the latest medal tally at 27th Olympic Games at Sydney where USA ranked top and India was struggling for bronze as was the 26th Olympic Games of 1996. USA provides a model for study of sub-cultures.

SUB–CULTURE CATEGORIES

The standard ways in which sub-cultures are categoried can be based on Nationality, Religion, Geographic and regional, Race, Age, Sex, Occupation and Social Classes : Here the first six are touched as the last two arc already discussed in earlier chapters.

1. NATIONALITY AND SUB-CULTURES:

The citizens of America those who were born in that country see themselves as American yet they retain a strong sense of identification and pride in the language and customs of their ancestors. This ancestrol pride is manifested most deeply in consumption of ethnic–foods, is travel to the "home land", and in the purchase of countless cultural artifacts ethnic clothing, art, music, foreign language news papers when it comes to consumer behaviour. Interest in these goods and services has spread rapidly as younger Americans attempts to better understand and more closely associate with their ethnic roots. To highlight the significance of ethnic origin as a sub–cultural market segment, the following pages are devoted to the study of Hispanic American Sub-culture.

Hispanic Sub-Cultures

Hispanic-American means those Americans who are originally from Spain or Portugal speaking Spanish or Portuguese languages. As per Advertising Age October 2nd 1995, these Hispanic Americans represent about 9 per cent of the US population exceeding 25 million with a purchasing power of US dollars 205 million. In contrast to other American population segments, Hispanic-Americans are younger-several years younger than the non-Hispanic; they represent members of larger families and they are more likely to live in an 'extended family' household, with members consisting of several generations. With 53 percent population increase during the decade from 1980 to 1990 and an expected growth rate of 27 per cent for the decade 1990 to 2000, Hispanic-Americans are projected to supplant African-American minority group. Hispanics have become dominant minority in cities of New York, Los-Angeles, San Diego, Phoenix, San-Francisco and Denver and they represent the majority in San Antonio and Texas. This sub-culture group is considered as a single market based on a common language and culture or as separate sub-cultural markets that correspond to different Hispanic countries of origin. There are 12 Hispanic sub-groups now identified in America. The three largest groups are—Mexican-Americans, Puerto-Ricans and Cubans which share the total Hispanic Americans namely 60 per cent, 12 per cent and 5 per cent. These sub–cultures are highly concentrated is California, Texas, New-York and Florida with more than 10 per cent of their members. Two things we must note about Hispanic American culture group—Consumer behaviour and defining Hispanic market.

So far as the former part is concerned, existing evidences indicate that Hispanic and Anglo-consumers differ in terms of a variety of significant buyer behaviour variables. Thus, Hispanic consumer's prefer well-established brands and those who are traditional prefer to shop at smaller shops or stores. In short, they prefer well-known or familiar brands, buy brands perceived to be more prestigious, are fashion conscious, prefer to shop at smaller personal stores, they buy brands advertised by their ethnic-group stores, tend not to be impulse buyers, they clip and use cent–off coupons, they are likely to buy what their parents bought, prefer fresh to frozen or prepared items, tend to be negative about marketing practices and government intervention in business.

Coming to the defining and segment, the Hispanic market, marketers have used six variables to define as to who is Hispanic. There are : Spanish Surnames, Country of Origin, country of family ancestry, Spanish spoken at home, self- identifications and the degree of identifications. What is important to note is that each of the major Hispanic sub-cultural groups have some distinct beliefs, values and customs. That is why, a marketing strategy that may be very successful with Puerto-Ricans in New-York might be an utter failure with Cubans in Miami. It is because of this reason that some marketers segment the Hispanic market by appealing to the distinct cultural values of a specific nationality. Others segment the Hispanic market in terms of the degree of acculturation to the dominant American cultural values, customes, artifacts and rituals.

The impact of nationality sub-cultures is really significant. Americans right in the midest of its largest immigration waves in the decades. The Census of 1990 found almost 20 million foreign born persons living in U.S. of which about 25 per cent of them arrived within five years. Over the last 40 years, 8 per cent was the highest population that were foreign borns. These people are so many countries of which top rankers are taken as per their share; they rank from Mexico from Philippines, Canada, Cuba, Germany, England, Italy, Korea, Vietnam, China, El Salvador and India. These foreign born residents with largest populations live in decsending order in Los Angeles County California, Roda County Florida, Cook County Illinois, Queens County New York, Kings County New York, Organge County, California, San Diego County California, Harris County Texas, New York County New York, Santa Clara County California, Bronx County New York and San-Francisco County California. Their impant ranges from 21.7% to 8.3 per cent respectively of twelve states of foreign born residents—namely, California, New York, Hawai, Florida, New Jersey, D.C, Massachusetts, Rode Islands, Texas, Nevada, Connecticut and Illinois.

2. RELIGIOUS-SUB-CULTURES:

It is reported that there are more than 200 different well organized religious groups. Of these predominant ones are the Protestant denominations, Roman Catholicism, and Judaism. The members of all these religious groups at times are likely to make purchase decisions that are influenced by their religious identity. Generally and commonly, consumer behaviour is directly affected by religion in terms of products that are symbolically and ritualistically associated with the celebration of various religious holidays. For instance, Christmas has become the major gift-buying season of the year. Whatever research findings are there, one comes to know that leisure activities of individuals with different religious backgrounds found several interesting clarifications. Thus, Protestant and Catholic consumers possessed a stronger leaning towards solitary leisure pursuits, while Jewish consumers preferred activities that provided companionship and stimulation (sensory) or companionship or sensory stimulation. In religious matters both husband and wife decision–making has religious tint. It is found that husbands in pro–religions and Catholic families were the major influencers in making specific purchases or purchase–decisions : as against this, in case of Jewish families, husbands and wives shared equally in most of the decisions.

These religious needs and practices go beyond their original meaning and purpose. For instance, the dietory laws or norms for an observant Jewish family represent an obligation. For example, the marks "U" "K". on food packages are the symbols that the food meets Jewish dietary norms or laws. For non- observant Jews and non-Jews, however, these marks often signify that the food is pure and wholesome.

3. GEOGRAPHIC AND REGIONAL SUB-CULTURES

America is a vast country enjoying wide range of climatic and geographic conditions. It is natural that most of the Americans have a sense of regional identifications and use this identification as a way of describing others. Thus, one can address a person as "Southern gentleman", 'Western', or 'Eastern', 'Mid West', "West Coast-gentleman" and so on. Any person who has travelled across the United States might have noticed good many differences in consumption behaviour especially when it comes to foods and drinks. For instance, "a mug of black coffee" with milk and sugar is preferred in the East. We do have differences in consumption of' staple food such as bread. In South, and Mid–west, soft white bread is preferred, while on East and West Coasts, firmer breads made of whole wheat, rye and French and Italian breads are favoured.

Regional differences in consumption patterns in segment product purchase, ownership or usage levels occur between major metropolises. Further, each metropolitan area can be segmented as down towns and edge cities. This kind of difference by redefining local markets speak of varying life-styles. This helps in planning marketing efforts and implementing them say promotional and distinction policies.

4. RACIAL-SUB-CULTURES:

Major American racial sub-cultures are Caucasian, African- American, Asian–American and American-India. Inspite of the differences in life-styles and consumer spending patterns existing among these groups, vast majority of racial oriented consumer research has spot lighted that consumer differ between African-Americans, Caucasian and Asian–American Consumers. These are touched in brief for the benefit of readers.

THE AFRICAN-AMERICAN CONSUMER

African-American consumers are numbering more than 30 million constituting largest racial minority in America, that means 12 per cent of total population. They have a purchasing power ranging between 250 to 275 US billion dollars. African- Americans are sought after by marketers who are to target about 845 US dollars (million) in advertising and promotions. This, important sub-cultural grouping represented a single 'undifferentiated' African American market consisting of consumers who have a uniform set of consumer needs. However, the real thing is that they are a diverse group consisting good many sub-groups, each having distinct backgrounds, needs, interests, and options. What is applicable to white majority, has been divided into variety of market segments, each with its own distinctive needs and tastes, is applicable to African-American market.

IT'S SEGMENTATION

The basis of segmentation of African-American market has largely been approached in terms of economics or social class- standing. That is, by decisions based on income, education and occupation. Care is taken to give top priority to the values of life-styles of the African-American middle-class. Most of the researchers have found that value orientations of middle-class African Americans are quite similar to those of middle class Caucasian. These African-Americans are committed to middle- class-values than their white counterparts. They care, for instance, security of loved ones and need for accomplishment. These are involved in Internet and Frequent Black-oriented on-line sites are particularly up-scale group. A survey of subscribers to BOBC-Black On Black Communication,"an Afrocentric" newsletter delivered by e-mail, revealed that 77 per cent have a four year college degree and 45 per cent have household incomes in excess of 50,000 US dollars; ten per cent of them have incomes more than 1,00,000 US dollars. There has been a rapid rise of African-Americans to higher ranks in middle class. As "Claritas a Research Company that produces PRIZM demo-graphic clusters, has found 'Black Enterprise' Cluster-where widespread number of middle-class and affluent clusters, each made up of a mix of different racial and ethnic sub-cultures.

CONSUMER BEHAVIOUR FEATURES OF AFRICAN AMERICAN CONSUMERS

There are some meaningful differences in terms of product preferences and brand purchase patterns, though there are many similarities between African-Americans and the rest of the America is terms of consumer behaviour. That is, African-American Consumers tend to prefer popular or leading brands are loyal and are unlikely to purchase private label and generic products. Marketers have cared for herbal values that they give in treating their hair and skin. Some meaningful differences exist among White, African-American and Hispanic –American consumers in purchase, ownership and use of a diverse group of products.

REACHING THE AFRICAN-AMERICAN AUDIENCE

The basic problem of American marketers is "how to reach African- American Consumers?" Traditionally, marketers have subscribed to one or two distinct marketing strategies. Some have followed the policy of running all the advertising in general mass media on the belief that African-Americans have the same media habits as whites. Others have followed a policy of running additional advertising in selected media directed exclusively to African- Americans. Alongwith general advertising, news papers, magazines and other media are directed specifically to African- Americans, of late, marketers have sought the help of African- American ad agencies.

ASIAN AMERICAN CONSUMERS

The Asian-American population, consisting of Chinese, Philippines, Japanese, Asian Indians, Korean and Vietnamese, is currently more than 7.5 million in size and is the fastest growing American minority. Between 1980-1990, the whites, African-American and Hispanic populations is US, grew at the rate of 6, 13, and 53 per cent respectively; during this same time period, the Asian population grew by 108 per cent.

The latest Census reports say that Asian-American market consists of 3 per cent of US population consisting of 23 per cent Chinese, 19 per cent Filipin, 12 per cent Japanese, 11 per cent Asian Indians and 11 per cent Koreans, with the remainder being drawn from a variety of Asian Countries. Today, nearly 40 per cent of all new immigrants to this attractive country are from Asia. These are largely family oriented, highly industrious, and strongly driven to achieve middle class life–style. For marketers, they are an attractive market for increasing the number of marketers.

WHERE ARE THE ASIAN-AMERICANS IN AMERICA ?

Asian-Americans are largely urban people who are presently concentrated are smaller number of large American cities. About 58 per cent live in Los Angeles, San-Francisco and Hawai, while 18 per cent reside in New York, Philadelphia and Washington D.C. The present, Asian American population is 10 percent of total California's population. Which is an increase from 1.20 million is 1980 to 2.80 million in 1990. By the year 2010, there will be approximately 8 million Asians is California. It would be the greatest folly if the marketers treat the Asian-American market as one single market. Most of the Chinese as well as other Asian-Americans do not live in down-town urban areas; they live in the suburbs.

HOW TO UNDERSTAND THE ASIAN-AMERICAN CONSUMER ?

Local news-papers and weekly news magazines frequently portray the accomplishments of Asian Americans, by nature and hard–working, very family oriented, and strivers for excellence in educational pursuits. Throughout, Asian-American students consistently won a substantial share in academic awards and scholarships. The American Statistical Bureau classifies them as the basis of their per capita income, their own business than non-Asian American minorities. Those who do not own their own business are largely in professional, technical or managerial occupations. They tend to be better educated and more computer-literate than the general population.

As consumers, the Asian Americans spend about 38 billion US dollars on consumer goods and services every year. They value quality and willing to pay for it. This segment tends to enjoy status of loyal customers, frequently more-male oriented when it comes to consumption decisions and attracted to retailers who make it known that they welcome Asian-

American patronage. It is important to remember that Asian-Americans are really drawn from diverse cultural backgrounds. That is why, although Asian- Americans have many similarities, marketers should approach this overall group with much caution as they are not totally homogeneous. Asian-American models are used in effective reaching of a segment.

5. Age sub–culture:

Each major age sub–grouping of the population can be thought as a separate sub–culture because important shifts occur in the demand for specific types of goods and services say " teen agers-13,19", "20-30 years" and "65 plus". Since children below 13 years will be discussed later we will concentrate on age groups called "generation X", "baby boomers" and "mature" and "elderly". These age segments have been singled out because of their distinctive life-styles that qualify them for consideration as sub-cultural groups.

The Generation X Market

This age grouping is often referred to as 'Xers', 'busters' or 'slackers' as opposed to boomers—and 'twentyfive somethings'— which consists of approximately 46 million—between 18-29 years olds who spend about 125 billion US dollars per annum. They do not like labels and do not want to be singled out and marketed too-ironically enough. Unlike their parents, who are frequently baby–boomers, they are in no rush to marry, start a family, or work excessive hours to earn high-salaries. For 'Generation X' Consumers, job-satisfaction is typically much more important than salary. 'Xers' reject the values of older co-workers who may neglect their families while striving lo secure higher salaries and career advancements. For generation X, it is more significant to enjoy life and to have a life–style that gives freedom and flexibility. Good many Xers are much more interested in say tennis shoes, furniture for their apartments and camping equipment than ocian–front condos. Owing one's own home is very often considered as negative that reduces an individual's flexibility. Xers find often good jobs either difficult or impossible to find, though upward mobility has traditionally been the dream of Americans.

How to Appeal to Generation X ?

Members of generation X are proud of their sophistication. They do purchase good brand names, say Sony, Hitachi but not necessarlly designer labels although they are not necessarily materialistic. They want to be recognised by marketers as a group in their own right and not as mini baby- boomers. Therefore, the ads targetted at them must focus on their style of music, fashions and the language. One important key for marketers appears to be their sincerity. Xers are not against advertising but only opposed insincerity. The baby boomer's media does not work with generation X. For instance, the age group between 18 and 24 years has the lowest per cent of daily news-papers readership of all the age groups. Xers arc the MTV generation and while three major US TV net-works attract an average of only 18 per cent of the 18-29 group, the Fox network claims 38 per cent of its viewers are in this group. The success of Fox has had with Xers may be due to such programm's as "Married Children", "The Sumpsons", "In Living Colour", and" Beverly Hills, 90210". Further, more cable TV such as MTV, Comedy Central, and E ! have been very successfully impacting this audience is reaching. It means that a low key approach works best in reaching this market segment.

The Baby Boomer Market

"Baby boomers" have been identified as a desirable target by marketers because (a) they are the single largest distinctive age category alive today, (b) they frequently make important purchase decisions and (c) they contain a small sub-segment of trend-selling consumers-very well called as "yuppies" or young upwardly mobile professionals—who have influence on the consumer tastes of the other age segments of society. The question is who are baby-boomers? Baby-boomers is the age segment of the population born between 1946 and 1964. Thus, baby boomers are the broad age category that extends from about mid-30's and early 50's. Baby boomers represent more than 40 per cent of the adult population. In addition to this magnitude, they are valuable because they comprise about 50 percent of. all those in professional and managerial occupations and more than half of those with atleast college degree in hand.

As consumers, baby boomers, they enjoy buying for themselves, for their homes or apartments, and for others–they are consumption oriented. The nature of the products and services they most need or desire change as baby-boomers age. For instances, because of the aging of this market segment, **Levis** is featuring "relaxed fit" jeans, sales of "lines–less" bifocal glasses to new customers who are up substantially and sales of walking shoes have grown rapidly. The market for baby-boomers can be further sub-divided in sub-segments as 1) younger-boomers-in the range of mid thirties and early 40s; 2) older boomers, those who are ranging from mid forties and early fiftys. For instance, older boomers came of age in a period when it was much easier to start with a career and afford at home. Thus, though many younger boomers are well educated and "aspire to the good life", they have found much more difficult to attain their dreams.

Among baby–boomers yuppies are most sought after sub-group of baby-boomers as they are affluent. Though they constitute only 5 per cent of the population, they are well–off financially, well- educated, and enviable professional or managerial careers. Yuppies are associated with status or premium brand names such as **BMWs or Volvo Stalon Wagons, Rolex watches. Cable TV, Kenwood food** processors. Today, Yuppies are maturing and they are shifting their attention away from expensive status type possessions to travel, physical fitness, planning for second careers or some other form of new directions for their lives.

Mature and Elderly Consumers:

Like other people, Americans are aging. Baby–boomers, are stating to hit 50-whom we call "mature adults", there arc plenty of preboomers-45to 65 years-and the number of elderly consumers is growing twice as far as the overall U.S population.

Who are 50-plus and Elderly consumers ?

There has been increased focus on the group Americans who have reached their 50th birthdays as the population of America has grown older. As per Census Bureau between 1990 and 2020, as the number of individuals less than 50 years old decreases, the 50 and older will grow by 76 percent. Not only do people over 50 have money, but also they tend to spend it. For instance, they purchase 43 per cent of all new cars including 57 per cent of all new Cadillacs. Typically, they dine out frequently and spend more than any group on travel and recreation. As compared to the general populations, members of the 50 plus segment watch more TV, read more news- papers and listen less to radio. More specifically, the 50 plus market prefers both TV and radio news programmes, game shows and mature market magazines.

More than 30 million Americans are within 65 plus age category representing 12 per cent of the country's population. This number is to be expected to be more than double in the next 50 years, with the elderly people segment eventually accounting for 20 per cent of the American population. This expected growth in the elderly population can be explained by the declining birth rate, the aging for the huge baby boomers segment and improved medical diagnosis and treatment. This is rightly called by a wise man as "Age of Wave".

The question is who are elderly ? In the US 'Old-age' is officially assumed to begin with a person's 65th birthday that is when the individual qualifies for full social security and medicare. However, people who are 70 years old still tend to view themselves as "middle aged". Research clearly demonstrated that people's perception of their ages are more important in determining behaviour than their chronological aged. That is, the number of years lived. People may, in fact, at the same time have a number of different perceived or "Cognative ages". Specially, 'elderly' consumers perceive themselves to be younger than their chronological ages on four perceived age dimensions. These are : feel age how old they feel ? 'the look age' how old they look ? 'do age' how involved they are in activities favoured by members of a specific age group ? and 'interest age' how similar their interests are to those of members of a specific age group ? For marketers, these findings underscore the significance of looking beyond chronological age to perceived or cognative age while appealing to mature consumers and to the possibility that cognative age might be used to segment the mature market.

The next two aspects are how one is to understand elderly market ? and how to segment this market ? Though American marketers ignored this older market, now they are quite anxious to know more and more about elderly customers. They are interested especially in the very attractive sub-segment of affluent elderly who have more discritionary income than any other adult-age segment. With more and more elderly consumers receiving the benefits of private pension and retirement programs, today, in addition to social–security and Medicare, and with an increasing percentage working longer, this age segment is even more likely to be able to afford new products and services in the future.

The elderly are by no means a homogeneous sub-cultural group. There are those who, as a matter of choice, do not have colour TVs or touch tone telephone services, whereas there are others who have the latest desk-top computers and spend their time surfing the internet with an increased appreciation that the elderly constitute a diverse age segment, more attention is now being given to identifying ways to segment the elderly into meaningful groupings. One such simple segmentation scheme is consisting of three age categories as (a) 'the young'— 65 to 74 years, 'the old'-75 to 84 and the "old old" those 85 years and above. This market segmentation approach provides useful consumer-relevant insight. For example, the consumers of 'the young age' group tend to have both health- and many, therefore, comprise of profitable market for all types of discretionary and luxury items. In contract, the 'old old' group they look for various specialized housing and medical services.

The marketers found these customers very rational and wise and they prefer to do daily marketing—in mall markets and hardly use .credit cards. For communicating with these, visual message are targetted to elderly avoiding gender and life style stereo types. The marketers should avoid high frequency sounds, reduce clutter, use larger type sizes and introduce fewer ideas per unit of time.

6. SEX AS A SUB-CULTURE

Sex-roles are important cultural component and hence it can be a sub-culture. It is because there is close relationship between sex roles and consumer behaviour. It is a fact that all societies tend to assign certain traits and roles to males and others to females. It is in American Society, masculine traits were—aggressiveness, competitiveness, independence and self-confidence right from the beginning. The traditional feminine traits were neatness, tactfulness, gentleness and talkativeness. In terms of role differences women have historically been cast as 'house–keepers' with the major responsibility for child care and men as the 'bread winners'. Today, times have changed still, advertisers do consider these roles and traits in appeals. In every society, it is quite common to find products that are either exclusively or strongly associated with the members of one sex. In the U.S., shaving–equipments, cigars, pants, ties, and working clothes were historically male products; bracelets, hair-spray, hair–dyers and sweet-smelling colognes were feminine products. It is not hard and fast. Thus, we see today males going for fragrances and hair–dyes; same is the case with vitamins. That 'male only' or 'female only' products are fast dying. That is why, the marketers should think of not only 'sex' but "perceived sex" for developing ad campaigns.

WORKING WOMAN AND THEIR DIMENSIONS

It is natural that today woman is not only a decision maker as a member of family but a bread winner. Marketers are more interested in not only woman but married woman that to'o working. Marketers have recognized that married women who work outside are a large grouping market segment because her needs are quite different from mere-house wife. The size in so big that approximately 72 per cent of American woman work full time and 3.5 percent of American families consist of the traditional arrangement of bread winning husband and a marketing wife. Young women with young children have been the last of grouping segment of female work–force. In this connection, marketers are to think of segmentation, knowing shopping patterns and finally reaching them.

In connection with segmentation, marketers thought of it to provide richer framework and developed the categories that differentiate the motives of working and non-working women. For instance, good many studies have divided female population into four segments—as "stay home house wives", "plan-to work house wives", "just-a job working woman" and "career oriented working women." The last two categories are important. "Just a job" working women seem to be motivated to work primarily by a sense that family requires the additional money whereas, "career oriented" working women, who tend to be in a managerial or professional position, are driven by a need to achieve and succeed in their chosen careers.

So far as patterns of shopping are concerned, working women spend less time on shopping than non-working women. They accomplish this "live economy" by shopping less often and by being brand and store loyal. Working women also arc likely to shop during evening hours and on the week ends as well as to direct through direct-mail catalogs.

The question is how to reach them and approaching working women is of greatest significance. Today's working women and mothers do not like them to be shown in ads in bathing suits or sweedish Bikini team. She should not be used as sex object while presenting, or positioning the products.

SUB-CULTURAL INTERACTION

Simultaneously, all consumers are the members of more than one sub-culture segment. Thus, African-American Bapist working–wife living in the Pacific North–west is an example. For this reason, marketers should strive to understand as to how multiple sub-cultural memberships interact to influence target consumers relevant consumption behaviour. The promotional strategy should not be limited to target a single sub–cultural membership.

CHAPTER BASED QUESTIONS

A. WRITE SHORT NOTES ON :

1. Culture 2. Sub–culture 3. Features of cultrue 4. Measurement of culture 5. Bases of culture. 6. Age and sub-culture 7.Nationality and sub-culture 8. Sec as sub culture 9. Race and sub-culture 10. Baby boomers 11. Working women 12. Parenthood

B. SHORT ANSWER QUESTIONS:

1. What is culture ?
2. What are the features of a culture ?
3. What is sub-culture ?
4. How can one measure culture ?
5. Discuss Nationality as a base for sub–culture.
6. Discuss Race as a base for sub-culture.
7. Explain geographically as a base for sub-culture.

8. Explain Age as a base for sub-culture.
9. Discuss Sex/gender as a base for sub-culture.
10. Explain integration and sub-cultures

C. ESSAY TYPE OF QUESTIONS :

1. Define culture — from the angle of consumer behaviour and its impact.
2. Explain the term culture and pin point its nature through its features.
3. What is measurement of culture ? How one measures it ?
4. What is a sub-culture ? What are its bases ?
5. Discuss 'nationality' as a base of sub-culture.
6. Explain 'geography' and 'region' as base of sub-culture.
7. Discuss Age as a base of sub-culture.
8. Explain Sex as a base of sub-culture.
9. Discuss Race as a base of sub-culture.
10. What lessons marketers learn from central values and segmenting them?

*****

CHAPTER 10

GROUP DYNAMICS, REFERENCE GROUPS AND CONSUMER BEHAVIOUR

BACKDROP

Though individual behaviour is the result of both internal and external environment of an individual, his or her behaviour is deeply influenced by the company he keeps or the group of which he belongs. Though you do not like to wear formal dress at annual social at college, you have to accept it because you are part and parcel of group where you cannot swim against the current. Similarly, the group values are to be respected. It is evident from our daily routine activities. Whenever you meet your friends and relatives you wish them, greet them whether they do or not. That is, group dynamics and group influences have impact on consumer behaviour. That is why, this chapter is designed to explain the concepts of group, group dynamics and reference groups from the angle of consumer behaviour. The Chapter ends with Chapter Based Questions.

WHAT IS GROUP?

In a broad sense, 'group' refers to two or more persons who share a set of norms, values or beliefs and have certain implicitly or explicitly defined relationships to one another such that their behaviours are interdependent. But in other words, group is the combination of two or more people with a purpose of achieving their common and shared goals through their integrated efforts.

STANDARD DEFINITIONS

★ **Mr. G. C. Homans:**

"Any number of people who share goals, often communicate with each other over a period of time, and are few enough so that each individual may communicate with all the others, person to persons."

★ **Mr. Marwin E. Shaw:**

" Group is two or more persons who are interacting with one another in such a manner that each person influences and is influenced by each other."

★ **Prof. Clours R. Shepherd:**

" A group may be defined as the aggregation of small number of persons who work for common goals, develop a shared attitude and also aware that they are part of a group and perceive themselves as such."

★ **Mr. David H. Smith:**

" A group is set of two or more individuals who are jointly characterised by a network of relevant communication, a shared sense of collective identity and one or more shared disposition with associated normative strength."

To state in simple words, a 'group' is a collection of two or more interacting individuals with a stable pattern of relationships between them who share common goals and who perceive themselves as being a group. 'Group' stands for coming together of small number of persons who work for common goals, develop the shared attitude and aware that they are part of a group.

'GROUP' VERSUS 'REFERENCE–GROUP'

'Group' and 'reference–group' are to be clearly understood because they are not one and the same. A 'Group' is two or more persons who share a set of norms, values, or beliefs and have certain implicitly or explicitly defined relationships to one another such that their behaviours are inter–dependent. A reference–group is a group whose presumed perspectives or values are being used by an individual as the basis for his or her current behaviour. That is, reference–group is simply a group that an individual uses as a guide for behaviour in specific situation. A reference–group is a group of people whom you refer to while making purchase decisions or any other decisions. Reference–groups influence consumer behaviour buy building aspirations for the individual and helping him or her to choose the product or service for a chosen life style. These reference–groups are small and associations of relatives or non–relatives or both.

We all belong to a number of different groups and go on adding to many more groups. Whenever we are actively involved with a particular group, it functions as a reference group. As the situation changes, people base their behaviour on entirely different group which then becomes this reference–group. It is quite possible that an individual may be member of several reference–groups at a time and use only one group as a point of reference in a given situation.

FEATURES OF REFERENCE–GROUP

The above definitions make clear the exact nature of group or reference–group. These are:

1. Two or more persons: Since interaction is the heart of group, there is need for at least two persons to form a group. This group has no upper limit of membership. However, a group is a group that permits meaningful and purposeful interaction among the members.

2. Group identity: A member of group has collective identity and he takes pride in being the member of such group. He is to take active interest in participations of group activities. Thus, a hostel for working employees gives an opportunity to share the benefits of comfortably staying irrespective of their employment status, religion and so on. It is because they are hostel –mates or room–mates in a hostel.

3. Voluntary association: Group membership is not compulsory. He or she can join or leave a group membership at any time. Further, one group doesn't come in the way of its members to join other groups at the same time. It is left to the sweet will and discretion of the individual as to join or not join and which group or how many groups at a time.

4. Perfect interaction: The members of the group are to interact with one another and share their ideas and communicate with one another. It does not mean that all the members should interact with every one at a time. neither it is possible nor it is worthwhile. What is important is to keep well informed of the latest developments.

5. Sharing of goal interest: Members of the group are there to attain the set goals of the group which helps them to share their problems mutually. It is the shield for protection and promotion of legitimate interests. That is, what is good for the group is good for its members. He has the greatest sense of belongingness and sharing of group concerns.

WHY GROUPS?

Individuals form the groups. Why they do so? Three must be certain reasons for the same. The possible reasons are:

1. Off from the regular work environment: It is very rare that an individual gets a job to his satisfaction. That is, he finds either that his capabilities are falling short or not perfectly or fully utilized. In most cases, we come across the second category people who feel that they much better qualified for the job they are placed in, such persons have greater hunger for using their superior capabilities in some other way. Hence, they find group as an outlet to let out their tatents. Thus, an employee joins a **Lion's Club** or **Rotary club** or other associations where their tatents get some edge

2. Group as a rich source of socialisation: Everybody agrees that man is a rational and social animal. He has both technical and social imperatives. He wants to belong, to associate with others than working in isolation. This is, one of the basic drives being a associative animal. We find the people have a strong desire to socialise, to drive away psychological fatigue and boredom of the daily routine. Thus, the business executives and businessmen are the members of good many social clubs where they want to reduce the uncertainties of social environment by sharing their feelings, concern and perspectives with one another.

3. Group meets affiliative needs: It is well known that the birds of the same feather flock together. People like the people who are more like them is several respects. Groups help is meeting the affiliative needs of people by allowing them to involve in different kinds of social activities of their choice. Thus, in a school or college students come out for a cup of tea or coffee, a movie, or a musical concert, even though they are studying in different standards or classes. Members of a group interact towards attaining common goal or need. The affiliative needs like friendship, support, love and affection are taken upmost care. Thus, a student studying at post graduate level may develop a practice of meeting the members of the family of their friends on each Sunday or different families on certain time interval to share the concerns and well–being of the family members,

4. Group makes possible group decisions: Though many decisions are made single handed which can be quick, because no one is to be consulted. However, all the quick decisions are not sound decisions. Hence, group–decisions will help in arriving at sound decisions as more and more people participate in decision–making process. These decisions are more rational because "two heads are better than one". That is why, group or team or combined decisions are sounder, yielding better results. It is because the group members think in similar direction at different attitudes

TYPES OF REFERENCE–GROUPS

Reference–group provides points of comparison by which one can evaluate attitudes and behaviour. A reference–group is a person on group that serves as a point of comparison for an individual informing either general or specific values, attitudes or behaviour. Each one of us like to evaluate our abilities and opinion based on comparable abilities and opinions of others as we are socialising. That is why, it is said that reference group is that group that an individual uses as a point of reference in determining his own judgements, preferences, beliefs and behaviour. The reference–group for an individual may be single person or it can also be a very large aggregate of persons, such as social institution or class or even a sub culture. The individual can refer a reference–group either in part or full. The individual may or may not be member of any particular reference–group or it is quite possible that he may even negatively make use of the reference group. An individual reference groups are those groups which may have direct or indirect influence on the person's attitudes or behaviour.

An individual can be a member of reference–group such as family and would be said to be part of membership group. The same individual may aspire to belong to a chess club and would be said to be part of an aspiration group. A disclaimant group is one to which an individual may belong to or joint and then reject groups, values. An individual may also regard the membership in a specific group as something undesirable to be avoided. Such a group is a dissociative group. Membership groups and aspiration groups are viewed positively, disclaimant and dissociative groups are viewed negatively. Hence, marketers advertise to appeal to the desire, to be part of a group and very rarely appeal to the desire to avoid or disclaim a group. Marketers prefer to focuses on membership and aspiration groups.

Membership groups

Positive reference–groups are significant and are classified as primary and secondary and formal and informal. Primary group is one where a person maintains regular contact with family members, friends and business associates. These persons together constitute primary group. People who meet less frequently such as those who meet during morning walk, or club members, constitute a person's secondary groups. Seen from the marketer's point of view, primary groups are more important because they influence consumers product beliefs, tastes and perfumes and have a more direct effect on purchase behaviour. Research has proved that the members of the primary groups are more likely to buy the same brands.

Groups can also be divided on the basis whether they have a **formal** structure such as president, executive and secretary and the like in hierchical order with specific roles. The structure of an **informal** group is loosely defined. For instance, when three individuals become computer applications and on every last Sunday of each month meet for dinner, then it would be considered as an informal group. The ramifications of membership groups can be:

1. Primary informal groups: The family and peer groups are primary informal groups. They are the most important because is the closences and contact frequencies between the individual and other group members. This provides for a more encouraging environment for the exchange of information and influence about consumption related topics. As a result of this, the admen frequently show consumption among friends and family.

2. Primary formal groups: Primary formal groups are those which have a more formal structure and with which the consumer comes in contact less frequently than primary informal groups. Business groups that come together of work or daily basis are the finest examples of primary formal groups. Advertising use such groups as a means of winning product approval.

3. Secondary informal groups: These groups do not have formal structure and meet once in a while. The examples of these kinds are women kitty–parties, sport groups that get together infrequently. When an individual goes on shipping in a group of three or more consumers, there is more likelihood that he or she will purchase more than initially planned because such groups often directly influence his or her purchases.

4. Secondary formal groups: These groups are not really important to marketers because they are structured, meet only infrequently and are not cohesive. The instances of this kind are associations of teachers, railers and the like.

ASPIRATION GROUPS

Non–membership group with positive attraction are called as aspiration groups and known for exerting a strong influence. One can think of two types of such aspiration groups namely, anticipatory and symbolic.

1. Anticipatory aspiration groups: These are the groups that an individual anticipates to join at some future time. Generally, the individual has some direct contact with such group or groups. For example, the individual may wish to join a group higher in the organisation hierarchy. The individual aspiration is more likely to be an outcome of anticipated rewards that go with higher position in an organisation such as power, status, prestige, money and other perks. Marketers appeal to the desire of individuals to increase this position by moving to a higher aspiration group and frequently advertise clothing, automobiles, liquor and other products within the context of business success and prestige. In case of liquors we come across the ads on **Signature** brand and **Johnnie Walker** brand. These products project the highest advancement of business leaders.

2. Symbolic aspiration groups: Here, the individual admires these groups but is unlikely to join them despite acceptance of the groups beliefs and attitudes. Experts found that individuals establish a vicarious connection with such groups by purchasing a product associated to the aspiration group. Thus, a tennis fan may buy **Nike** sports jacket and shoes because many tennis stars wear these. It is significant for an influence of this kind that product is visually obvious. Marketers use certain celebrities to advertise the product and thereby appeal to the symbolic aspirations of consumers. Thus, Dhoni-the Indian wicket keeper-batsman is taken as a brand ambassador of many consumer durable goods.

NATURE OF REFERENCE GROUPS

Reference–groups enjoy certain unique features which clearly state the nature of these reference groups these features are:

1) Norms: Reference–groups establish certain norms. 'Norms' are the rules and standards of behaviour that group establishes. Group members are expected to conform to these which may be with regard to the appropriateness of clothes, shoes, eating habits or brands of cosmetics and so on. For formal groups, norms are generally unwritten but are quite well understood. Behaviour deviation from the norms can result in sanctions.

2) Values: Reference–groups set values. 'Values' are shared beliefs among the group members regarding what behaviours are appropriate or inappropriate. Cultures and subcultures define largely the values. However, they vary significantly by family and peer groups. For example, one family may give more weightage to material possessions and another spiritual attainment.

3) Roles: Reference–groups define roles 'Roles' refer to functions that an individual assumers or that the group assigns to her him or her to accomplish group objectives. For example in a group buying behaviour, such a family, marketers can identify specific roles that individuals assume. These roles may be of initiator influences- decider–buyer and user . In case of an organisation, the role of gatekeeper is very important because this individual has the most control over the flow of information.

4) Status: Reference–group projects status. 'Status' is the achieved or ascribed position that the individual occupies within the groups hierarchy. As one may expect, greater power and influence goes with higher status. For example, the executive vice president of an organisation has a higher status than sales executive. The Vice–president will most likely to have a large well furnished office symbolising his or her status. In an Indian, family, the head of the family has higher status than any other member.

5) Socialisation: Reference–groups is an incarnation of socialisation 'Socialisation' refers to the process by which new members learn the group system of values, norms and expected behaviour patterns. When an individual vacates one job and joins another organisations he or she is expected to learn the informal rules and expectations from the work groups, in addition to formal rules and expectations.

6) Power: Reference–group establishes power. 'Power' is the potential ability to influence behaviour to change the course of events, to overcome. resistence and to get people to do things that they would not otherwise do. It is the ability to influence people or things usually obtained through the control of important resources. A group's influence on its members, behaviour is closely associated to its power. Though there are different sources of power, four are more relevant to consumer behaviour context. These are:

* **Reward Power:** 'Reward power' is the groups, ability to reward the individual. The strength of reward power increases with the size of the rewards as perceived by the individual. The family can reward the child with approval and praise. A business unit can reward its employees with promotion, money and status. The purchase of say clothes or mobile by an individual can be rewarded by compliments from relevant group member and can reinforce the consumers choice.

* **Co-enrcive power:** 'Coercive power' refers to the power of the group to use disapproval, withholding rewards or even punishing an individual. The greater the significance of the group the greater is its power. Parents can punish the child. An organisation has the power to terminate the services of an individual or social group can exclude the individual for their services deviant behaviour.

* **Expert power:** 'Expert power' is one that influences the outcomes from the experience expertise and the knowledge of the individual or group. Consumers regularly accept influences from friends they perceive as being more knowledgeable and experienced and make purchases on their recommendation. Similarly, a sales persons may be taken as an expert source on the company's product.

* **Referent power:** 'Referent powers' is one that flows from the feelings of identification that the individual has with the members of the group. The greater the similarity of beliefs and attitudes between the individual and the group members, the greater the referent power of the group.

GROUP DYNAMICS

Simply stated, 'group–dynamics' is the study of forces operating within a group. Group –dynamics is concerned with the interaction of the individuals in a face to face relationship. This group–dynamics is the social process by which people interact face to face in small groups. Group–dynamics is to do with determining the interactions and forces between group members in a social situation.

If applied to "Consumer behaviour", the study of group–dynamics it is an aid to the marketers to know and understand the impact of different types of groups in individuals, buying behaviour process. Each type of group with which individual is associated some important needs and functions for the individual. As no single group satisfies all the needs of an individual member, one tries to belong to different groups. A family unit helps in biological maintenance, secondary and a feeling of belongingness. As to the contrary, sorolity is purely social in nature and takes care of the other needs of the individual like helps us providing more educational information, in developing contacts for future employment, get associated through membership to other social organisations, provide information on buy of speciality or shopping products. What is more important is each group exerts a variety of influence on the individuals by making him adhering to group norms. It also guides his behaviour in his consumption behaviour.

WHEN GROUPS DO EXERT INFLUENCE?

The findings of the research show very clearly that in adopting to group norms, individuals not only accept the values established by family, peer–groups or organisation, out also use, group norms to define themselves. That is the individual's develop their self–concept on how they think others see them. It means that reference groups provide the points of comparison by which individuals evaluate their own attitudes and behaviour.

FACTORS OF GROUP INFLUENCE

The group influence on an individuals buying behaviour rests an three important factors. These are outlined below;

1. Attitude towards the group: An individuals succeptibly to group influence varies widely. The buying behaviour of a consumer is more likely to be influenced by the group, if the individual : (i) views the reference group as a credible source of information about the product or services. (ii) values the views and reactions of group members with regard to buying decisions (iii) accepts the rewards and sanctions allotted out by the group for proper or improper behaviour.

2. Nature of the group: A given reference group is more likely to influence a group member's behaviour if they are : (i) cohesive, that is having similar values and norms (ii) frequently interacting and thus creating more opportunities to influence members (iii) distinctive and exclusive; that is, the membership in the group is highly regarded.

3. Nature of the product: The nature of the product also determines the degree of influence a group has on an individual. Groups are more likely to be influential for products which are : (i) visible such as clothing and ii) exclusive that might speak of status such as **BMW** automobile.

REFERENCE–GROUP INFLUENCES ON CONSUMERS

It is clearly demonstrated that reference groups do exert influence on a wide range of consumption behaviour. Research projects that groups exert informational comparative and normative influences. It pays to understand these three.

(1) Informational influence: Reference–groups along with other influence sources can exert informational influence by offering information to assist in making decisions. For instance, chat groups on the internet provide good deal of information on subjects like internet and intranet sites. This type of influence occurs when a consumer accepts information as reliable form a reference group member and beleives that the information is going to enhance his or her knowledge about product or service choice. Informational influence is significant as it affects the time and effort the consumer devotes in information search and decision making. The consumers who can get reliable information from others may be easily reluctant to engage in time consuming information search when making purchase decisions. It is sometimes important for the marketers to increase the likelihood that consumers engage in information search, though informational influence can reduce the time devoted to information search. In case the product or service is new and superior, few consumers are likely to know about its benefits. Thus, an ad that enhancers product awareness and encourage consumers to compare products may be essential.

Informational influence is founded on the similarity of the group's members to the individual or the expertise of the influencing group members. For instance, an individual may encounter several members of a given group using a particular brand of tooth–paste. He or she then takes it as an evidence that it is a good brand and decides to buy it or , one may decide to buy a particular brand and model of dot matrics printer because a friend who is very knowledgeable about printers owns or recommends it. Here, conformity is the result of information shared by the group that works. Consumers are more likely to seek expert advice from personal sources such as friends, relatives or neighbours rather than than market–controlled sources like ads or salespersons, because they believe personal sources are more reliable. Consumers may view ad claims with doubt because the ads convey market controlled information about the product or service.

Informational influence is likely to be more significant when the consumer perceive financial, social or performance risk in buying a product. Thus, a consumer who is buying an automobile seeks information from knowledgeable relatives, friends, neighbours or even sales–people due to considerations of costs, social visibility and mechanical failures. He seeks expert advice in case of high–tec and complex products like computer, where informational influence is more important. Informational influence is significant when the consumer's objective is to seek knowledge, the source creditably is acceptable, the source is viewed as expert and finally, the behaviour is acceptance of influence. This has implications for marketers. As source expertise and credibility affect informational influence, marketers can use sources viewed as credible and export for a given product category. For instance, marketers of say sports shoes will use well known sports, figures and stars.

2. Comparative influence: Consumers have a tendency of comparing constantly their attitudes with reference to those of members of important groups. They, serve as a base and the individuals urge to seek support to his or her attitudes and behaviour. Individuals are inclined to associate with groups with which they agree to stay away from group with which they disagree to accomplish this. Resultantly, the process of comparing oneself to other members of the group and evaluating as to whether the group turns out to be supportive becomes the basis of comparative influence. For instance, when a new student in a college meets other students, he or she is likely to compare their attitudes towards education, students union, sports, movies, dress and the like. In the process, the student also identifies brands of jeans, shoes, two–wheeler, cosmetics they use. The student is attracted to those students who appear to be similar to him or her because they reinforce existing attitudes and behaviour.

The consumer's objective is self-maintenance and enrichment in accepting the comparative influence. To increase his or her self concept, the individual associates with groups that have similar attitudes and behaviour. This grants reinforcement and ego gratification. The source of power is referent power and the individual identifies himself or herself with the group. Individuals who are influenced by comparative influences should have similar features to those whose influence is being accepted. Consumers are more likely to seek information from those friends that they believe are similar to them and regard as reliable. The implications for admen is that they should use spokesmans or spokespersons perceive by consumers as being similar to them. For instance, the famous ad of **HLL** on Surf using Lalitaji - representing middle class. Even the salesman who consumers believe to be having similar tastes, attitudes, or even religion are likely to be more, successful in influencing purchase decisions.

3. Normative influence: Normative influence also called as "utilitarian influence' means a social pressure designed to encourage confirmity to the expectations of others to gain a direct reward or to avoid any sanctions. Take for instance, the type of clothes, music, shoes and the like you buy or for that matter hair style you follow and compare with your friends. Chances are you and your friends have made similar choices. The finest example is that of **Banasthali, Vidyapeeth** of Rajasthan for women where teachers, students and other staff members are to wear only khadi and remain vegetarian though many students and teachers may have non –vegetarian dishes and prefer wearing other than khadi, they have to conform to expected code of conduct on the campus at any rate.

Normative influence affects confirmity which is the tendency for an individual to behave as the group behaves. It is the finding of the research that conformity and brand choice seem to be related, though brand choice congruity may not be the only way to express conformity. An individual may also conform by performing the activities that the group wants the members to perform. For instance, the individual's actions at a party might be influenced by whether the comparisons are parents or his or her friends. In each case individual is conforming to different set of expectations with regard to appropriate behaviour. What is very much interesting is that conformity varies by culture. For instance, Indian and other Eastern cultures are more group oriented and individuals are likely to go with the group aspirations and desires.

The conditions for conformity are : (1) "The individual" is more committed to group and values membership in it (2) "The group" provides significant rewards for compliance and punishment for lack of compliance (3) "The individuals behaviour in conforming is visible to members of the group. On the other hand, **rejection of conformity** depends on (1) The strength of value–system of the individual (2) The intensity of group pressure for conformity (3) The degree of individual commitment to the group (4) The value placed on individuality.

RELATIVE IMPORTANCE OF INFORMATIONAL, COMPARATIVE AND NORMATIVE INFLUENCE

The three components of influence consumer behaviour, the relative importance is the function of product and consumer characteristics.

1) Product characteristics: Consumers have a tendency to yield readily to informational influence when the products are technologically complex like computers, automobiles, airconditioners, washing machines, sound gadgets or the products that require objective evaluation for selection like insurance policies, tax–savings schemes of investment. Susceptibility of informational influence is pretty high when the product or service purchases perceived as risky or when the brands are very different from one another. The products that are viewed as means of self–expression and identity, or consumed in public like paintings, music, jewellery, clothing watches, shoes are most likely candidates for comparative as well as normative influence, According to researchers, reference–groups can influence two types of decisions: (1) Whether we all buy a product within a given product catgory and 2) What brand we are likely to buy. The influence would also depend on whether the product is typically consumed in private or public and whether is a necessity or luxury.

One prediction can be that because we must buy items of necessity, reference–groups are likely to have little influence on what we buy. However, they might exert some influence in case of luxury items. Another prediction is that products consumed in public give others the opportunity to observe which brand we have bought. On the contrary, few people see which brands we buy and consume in private. Different brand images communicate different meanings to people and due to this, reference–groups are likely to exert considerable influence on the brand we buy, when the product is consumed in public but not when it is privately consumed. Another dimension is that it is difficult for groups to develop norms and use sanctions for violating them when the product is consumed in private. However, publicly consumed product provides opportunities for sanctions. Hence, the conclusion that follows in that reference–groups influence product category choice for luxuries but not necessities and influence brand choice for products consumed in public but not those consumed in private.

2) Consumer characteristics: By nature, some consumer personalities are such that they make them readily yielding to influence by others. The consumer researchers have developed the scale of "susceptibility to interpersonal influence". Those susceptible to interpersonal influence seek to enhance their self image by possessing products that they believe others will approve. These are the individuals who are also willing to conform to expectations of others regarding which products and brands to buy. A personality characteristic referred to as " attention to social comparison information" is also related to normative influence. Those individuals who are high on this personality trait are likely to pay a great deal of attention to what others do and use this information as a guide to their own behaviour. Strong ties between individuals also affect the degree of influence. As a result of strong ties, individuals, presumably, want to maintain their relationship with group members and are more inclined to accept group norms. In general, informal influence producer more shifts in beliefs than does normative influence.

APPLICATIONS OF REFERENCE–GROUP INFLUENCES

Shrewd marketers employ informational, comparative and normative group influences in developing communication strategies.

Advertising makes use of informational influence through expert spokespersons who communicate the information about product features and performance. One way is to employ a character posing as an expert like doctor for commonly used medicines or the engineer for technical products. Another way is to use real celebrity who has expertise in product area, such as Andre Agassi or Ronalde in **Nike** tennis apparel or shoes for foot ballers. Consumers are likely to believe a testimony from Agassi or Ronalde for the product category as reliable.

Advertising applies comparative influence by using either an actual referent in the form of a 'typical consumer' or use a celebrity as a symbolic referent with whom consumer identify because he or she is likeable or attractive. The typical consumer, say Lalitaji in Surf ads, persuades consumers that people like themselves have used the advertised product. The typical consumer is a referent because he or she mentions common needs and problems faced by the target group members. In case of symbolic referent the ad takes advantage of his or her fame in some area of activity and not his or her expertise in the product category. In case of **Santro Zinx** automobile, Mr. Sharukh Khan and **Fiat Palio Sachin Tendulkar** are symbolic referents.

Marketers use very frequently normative influence by showing up group approval in ads for a particular brand. Ads on **Orient PSPO, Cocacola, Pepsi, Taj Mahal tea,** brands of paints, clothing and personal care products and the like, are the examples of advertisings simulation of social approval. Typically, the ad shows an individual who is important to the consumer such as spouse, friend, neighbour or business associate and express approval or praise of the consumers, choice. Normative influence is also exerted through fear appeals that attempt to show the potential result of not using the product. Take the case of ads of **Colgate** Suraksha Chakra, ads of simple remedies, anti–dandruff shampoos, deodorants that speak of this. Such ads emphasise that use of product changes disapproval to approval or may even enhance the users attractiveness.

In case of personal selling, informational influence operates when the salesman is viewed as knowledgeable about the product category and is a legitimate source of information. Such influence reflects expert power. When a consumer believes that the salesman is similar in needs and characteristics as the consumer himself or herself, comparative influence operates and the salesman is likely to exert influence through referent power. Normative influence is exerted during bargaining between buyer and the salesman to achieve favourable terms of exchange and the influence reflects reward or coersive power. Bargaining is likely to occur when price is negotiable, as may happen in case of real estate, automobiles and home appliances.

CHAPTER BASED QUESTIONS

A. WRITE SHORT NOTES ON:

1) Group (2) Reference–group (3) types of reference–groups (4) Membership groups (5) Aspiration–groups (6) Nature of groups (7) Power (8) Group dynamics (9) Factors of group influence (10) Informational influence (11) Comparative influence (12) Alternative influence.

B. SHORT ANSWER QUESTIONS:

1. What is a group?
2. What is a reference–groups?
3. What are the features of reference–groups?
4. Why are groups formed?
5. What are the types of groups?
6. What are membership groups?
7. What are aspiration groups?
8. What is the nature of reference–groups?

9. What is group-dynamics?
10. What is informational influence?
11. What is comparative influence?
12. What is normative influence?

C. ESSAY TYPE QUESTIONS:

1. What is group? Can you call any group a reference-group?
2. What is a reference-group? What are its features?
3. What is a reference-group? Why reference-groups are formed?
4. What is a group? Explain different types of reference-groups?
5. What is group-dynamics? When groups to exert influence?
6. What is group influence? Explain different types of influence
7. What is an influence (group)? What is relative importance of such influences?
8. Explain the applications of group dynamics by the marketers.
9. What is power? Explain different types of power as an essential feature of reference-group?
10. What are aspiration-groups? Explain the types of aspiration-groups.

*****

CHAPTER 11

SOCIAL CLASS AND CONSUMER BEHAVIOUR

BACKDROP

Social class is another factor that has deep impact on consumer behaviour. Therefore, an attempt is made to give the clear-cut meaning of social class, its characteristics-the measurement of social class-life style, profiles of social classes, social class mobility and the relevant concepts of affluent consumer and the non-affluent consumer and selected consumer behaviour applications of social class and other allied aspects. The chapter ends with a Chapter Based Questions.

WHAT IS A SOCIAL CLASS ?

Though **social class** can be taken as a range of social positions or a continuum on which each member of society can be placed, the consumer behaviour researchers have divided this continuum into a small comparatively small number of specific social classes, or strata. Within this framework, the concept of social class is used to assign individuals or families to a social class category. Founded on this practice, 'Social Class' is defined as "the division of members of a society into a hierarchy of distinct status classes, so that members of each class have relatively the same status and members of all other classes have either more or less status". In other words, 'Social Class' is a group consisting of a number of people who have approximately equal positions in a society. In this context, it is better to be aware that these positions may be achieved rather than ascribed, with some opportunity existing for upward or down–ward movement to other classes.

Each country in the world has its own social class system which represents a hierarchical division of a society into relatively distinct and homogeneous groups with respect to attitudes, values and life-styles. Thus, every one is aware of the concept of social class but find it difficult to explain in concrete words. Thus, China has five groups as 'Super-rich'— 'Rich', 'Well-to do' 'Decent–living' and 'Poor'. Americans and Canadians use the words social class and social standing synonymously where they mean social rank. Thus, Americans have three major classes each having two sub-categories except in case upper class which his three sub-ranks. Thus, upper- Americans can be "upper–upper class", Lower–upper class, and Upper– Middle class Americans are classed as 'Middle middle' and 'Working class'. Lower Americans as 'Upper lower' and 'Lower lower' class. In most of the under developed, we have basically three way classification where middle–class has upper and lower. That is, upper–class, upper Middle class. Middle middle class, Lower Middle class and 'Poor' or lower class.

For a social class system to exist in a society, the individual classes must meet five basic criteria. They are: They must be

(1) Bounded

(2) Ordered

(3) Mutually exclusive

(4) Exhaustive and

(5) Influential. 'Bounded' means that there are clear breaks between each social class.

That is, it is necessary that a rule be devised for each class that will include or exclude any particular individual. 'Ordered' means that classes can be arranged or spread out in terms of some measure of prestige or status from the highest to the lowest "Mutually Exclusive" means that an individual can only belong to one social class, though movement from one to another is something unavoidable 'Exhaustive' means that every member of a social system must fit in some class. 'Influential' means that they vary in behaviour pattern between the classes. The following Figure gives factors — social ranking and unique behaviours.

SOCIO ECONOMIC FACTORS		SOCIAL RANKING	UNIQUE BEHAVIOUR
Occupation		Upper Class	Preferences
Education	→	Upper Middle	Purchases
Ownership		Middle Middle	Consumption
Income		Lower Middle	Consumption
Heritage		Lower Class	Communication

FIG: 11:01 SOCIAL STANDING DERIVED AND BEHAVIOUR INFLUENCES

NATURE OF SOCIAL CLASS

Before we go to know about nature of social class, let us know about "Social class" and "Social Status" because, it clears the way for our perfect understanding. Researchers very often measure "Social class" in terms of "Social Status". That is, they have attempted to define each social class by the amount of Social Status the members of that class have in comparison with members of other social classes. In terms of Social class research also called as "Social stratification", status is frequently thought as the relative rankings of members of each social class in terms of specific status factors. For instance, wealth, power and the prestige are the three factors frequently used while estimating social class. While considering consumer behaviour and marketing research, status is very often defined in terms of one or more of the following convenient demographic variables or Socio-economic variables such as family income, occupational status, and educational attainment. These socio-economic variables, as expressions of status, are used by marketing managers and other practitioners on a daily basis to measure social class. The nature of social class is featured by following characteristics:

CHARACTERISTIC FEATURES

1. A Social Class Exhibits Status:

As noted earlier, social class and status are not interchangeable words, though they are closely related to each other. Status refers generally to one's rank in the social system as perceived by the members of society. Therefore, an individual's status is the function of .not only of social class to which he belongs but also of his personal characteristics. For instance, eminent lawyers have high rank in social system like scientists and researchers. However, every lawyer has no equal status. Thus, Mr. Nani-Palkhiwala might charge to serve a notice Rs. 3,000 but another lawyer may do it for mere Rs. 30. Moreover, an individual personal contribution to society helps in determining his or her status. Thus, a lawyer makes a classic and historic judgement, has higher status than any other lawyer who fights a case what is important is each society establishes certain set of values. These values are reflected in the ideal types of people in that society. That is, .those who conform more to the ideal accorded they enjoy more respect and prestige, while others ranked lower. It is possible that is one nation members of armed forces might have been accorded highest prestige while another politicians or educators, or business community. The criteria used and the weights given in relative terms determined by the society's value system is respected. However, the factors determining status are—authority over others, power—political, economic, musde, ownership of property, income, consumption patterns and life-style, occupation, education, public service, ancestry and association with others.

The status symbols have been changing. In the past, a status symbol of middle class was a "Sewing machine" but today it is “Washing machine” or personal 'Computer' or “Sound System.” However, status is exhibited by a social class.

2. Social Class is Hierarchical:

Social class categories are generally ranked in a pattern of hierarchy, ranging from low to high status. That is, members of specific social class members of other social classes as having either more or less status than they do. To many people social class categories suggest that others are either equal to them or superior to them or inferior to them. This hierarchy of social class is important lo marketers. It is because their purchasing power and consumption pattern or style of life differs widely. Thus, upper–class people do not buy lower–class products and any class juxtaposed between upper and lower will buy neither upper–class nor lower –class products. Thus, taking a pair of shoes, upper–class might go for 'Adidas', ‘Puma’, ‘Woodland’, ‘Phonix’, but middle–class might go for normal brands of ‘Bata’ or ‘Carona’, ‘Action’, While lower–class might go for unbranded products. Taking a cup of tea, upper class might buy a premium brand tea in the range of say Rs. 3,000 to 6,000 per kg. middle class, might go in for Rs. 300-Rs. 600 per kg. while poor class might go in for Rs. 80 to Rs. 125 per kilo-gram.

3. Social Class is Multi–dimensional:

Social class is multi-dimensional as it is founded on a number of components. These are not equivalent to, or determined solely by, occupation or income or any one criterion. However, they may be indicated by or be related to, one or more of these measures, it is significant for the marketers to realize that some of these variables are more reliable, substitutes or ‘Proxies’ than others. The variable income is a misleading as an indicator of social class position. Yet money is what an individual associates with the idea of social class, for more than anything else. Occupation generally provides a good clue to one's social class. Housing is another social class ingredient.

4. Social class as a Frame Reference:

Social class membership serves consumers as a frame of reference a reference –group-for the development of their attitudes and behaviour. In the context of reference–groups members of a specific social class may be expected to turn most often to other members of the same class for cues or dues regarding an appropriate behaviour. Members of a particular social class say upper–class may aspire to advance their social standing by emulating the behaviour of the middle–class. To attain this goal, they might read middle–class magazines, do middle–class things-say visit museums and advance their education and hang out at middle class restaurants so that they can observe middle–class behaviour by hanging out at middle class restaurants.

5. Social Class Restricts Behaviour:

The interaction between classes is limited because the class members are more comfortable and find reinforcement with those of "themselves like" in term of values and behaviour patterns. Resultantly, members of the same social class tend to associate with each other and not to any large extent with members from another social class because they share similar educational backgrounds occupations, income levels or life-styles. This factor of limited interaction impedes interpersonal communication between different classes about ads, products and other marketing elements. Classification of society into social classes has enabled the researchers to note the existence of shared values, attitudes and behaviour patterns among members within each social class and differing values, attitudes and behaviour between social classes. Consumer researchers have been able to relate social class standing to consumer attitudes relating specific products and to examine social-class influences on the actual consumption of products.

6. A Social Class is a Homogeneous Unit:

Social class is viewed as homogeneous division of society in which people within a class have similar attitudes, activities, interests and other behaviour patterns. This means for a marketer that group of people are exposed to similar media, purchase similar products and service and shop in similar stores. This kind of homogeneity permits the marketer in many cases to effectively segment the market by social class and to develop really appealing marketing mixes.

7. A Social Class is Dynamic:

A social stratification systems in which people have some opportunity for upward or downward movement are known as open systems. People in closed systems have inherited or ascribed status. That is, they are born into one social level and are unable to leave it. The difference between a system based on earned or achieved status against one based on inherited status is significant with regard to mobility. Advanced nations have open systems than traditional bound developing countries. Even in case of open systems the extent of mobility is within a short range of time. Sometimes, as entire occupational category seeks to raise its status through public–relations efforts, as evidenced by attempts in many fields to attach the word professional to their job activities-say law-enforcement officers, accountants, and even say truck drivers. As said, rural part of India vibrantly caste based representing a closed system that has existed for thousands of years characterised by high degree of social inequality forcing a person by birth to stick to a permanent social position. Thus, a son of Carpenter cannot be a teacher or a trader. This degree of discipline is fast fading as rural youngsters move over to urban areas. In this sense, the social class is dynamic – particularly where such class systems are open. With the changing times, the rigidity of closed or curtained is fast vanishing.

8. A Social Class Depicts a Market Segment:

The strata of social classes provide ready market segments for products and services. The consumer researchers have been able to relate product usage to social class membership. Thus, certain magazines are meant for a particular social class, say "Business India" for rich class and upper middle–class while "India Today" might be meant for middle-middle class. It does not mean that, a reader who is used to read "Business India" will not read "India Today". Those who read say 'Fortune' "Business Today" are not likely to ready say other general magazines. Thus, 'Fashion', 'Design', 'Inside out' are meant for rich groups than lower. These magazines provide a valuable service by providing practical information on a variety of topics and issues which pertain high, low, and middle income groups. These 'up–scale' and 'down-scale' publications give the requirements of these subscribers or readers. That is, all types of marketing organisations can tap this information for their marketing strategies.

SOCIAL CLASS CATEGORIES

A sociologist agrees over the possible number of distinct class divisions that are a must to desirable effectively the class, structure of a country. They range from TWO categories to NINE class categories. The following square gives these seven district categories.

TWO-CATEGORY SOCIAL CLASS SCHEME

Blue-Collar	**White-Collar**
* Upper	* Upper
* Lower	* Lower

THREE CATEGORY SOCIAL CLASS SCHEME

Blue Collar	**Grey Collar**	**White Collar**
* Lower	* Lower	* Lower
* Middle	* Middle	* Middle
* Upper	* Upper	* Upper

FOUR-CATEGORY SOCIAL SCHEME :

* Lower * Lower–middle* Upper–middle * Upper

FIVE-CATEGORY SOCIAL CLASS SCHEME

* Lower * Working Class * Lower middle * Upper middle
* Upper–Lower * Lower–middle * Middle * Upper–middle * Upper.

SIX-CATEGORY SOCIAL CLASS SCHEME

* Lower-lower * Upper-lower * Lower–middle * Upper–middle
* Lower–upper, * Upper–upper.

SEVEN-CATEGORY SOCIAL CLASS SCHEME

* Real lower-lower * A lower group of people but not the lowest * Working class
* Middle class * Upper-middle * Lower-upper * Upper-upper.

NINE-CATEGORY SOCIAL CLASS SCHEME

*Lower-lower * Middle–lower * Upper-lower * Lower-middle * Middle-middle
* Upper–middle, * Lower-upper, * Middle-upper * Upper-upper.

Each class category has a definite percentage of population at a given movement of time: The following is the Three, Six and Five category social class giving percentages in America, Japan and India on the eve of year 2000.

COUNTRY:	**America**	**Japan**	**India**
CLASS CATEGORY			
THREE STAGE			
Upper	21.1%	15%	12%
Middle	30.3%	28%	23%
Lower	48.5%	57%	65%
	100.00%	100.00%	100.00%
SIX STAGE:			
Upper–upper:	0.9%	0.8%	0.6%
Lower–upper:	12.7%	13.8%	12.4%
Upper–middle :	7.2%	8.3%	9.8%
Lower–middle:	28.4%	25.5%	21.3%
Upper–lower:	44.0%	29.8%	24.7%
Lower–lower:	19.5%	21.8%	31.2
	100.00%	100.00%	100.00%
FIVE STAGE:			
Upper	13.6%	14.8%	9.0%
Upper–middle	7.2%	18.2%	11.1%
Middle	28.6%	24.0%	20%
Lower–middle	18.9%	13.9%	14.3%
Lower	31.7%	29.1%	45.6%
	100.00%	100.00%	100.00%

THE MEASUREMENT OF SOCIAL CLASS

Despite the fact that behavioural scientists agree that "Social- class" is a valid concept but disagree on its measurement. Researchers, it appears that, are uncertain about the underlying dimensions of social class structure. That is why to get the best, researchers have attempted to give several measures which are broadly classified as-**SUBJECTIVE, REPUTATIONAL** and **OBJECTIVE** measures. Let us note these in brief :

THE SUBJECTIVE MEASURES

The **Subjective** approach to measure the social class, the individuals arc asked to estimate their own class positions. A typical example of this approach is the following QUESTION: "Which one of the following five categories best describes your social class : Upper, Upper-middle, Middle, Lower–middle and Lower Class ?

Upper ☐
Upper–middle ☐
Middle ☐
Lower–middle ☐
Lower ☐
Do not Know/Refuse to Answer ☐ / ☐

The result of classification of social–class membership is based on the participants self perception or self images. That is social–class is treated as a 'personal' phenomenon, one that reflects an individual's sense of belonging or identification with others. This feeling of social-group membership is called by experts as class-consciousness' because most people are reluctant to categorise themselves as either 'lower' or 'upper' class, the middle class ends with an unrealistically large share. That is, subjective measures of social–class membership tend to an under-abundance of people who classify themselves as 'middle' class-understanding the 'upper' and 'lower' class. Further, it is more likely that subjective perceptions of one's social class membership is related to product usage and consumption preferences as it is a matter of reflection of one's "self- image".

REPUTATIONAL MEASURES

The **reputational** approach for measuring social class requires selected community informants to make initial judgement concerning the social-class membership of others within a community. The final task of assigning community members to social class positions, however, belongs to the trained researcher. That is, sociologists have widely used this approach to get a better understanding of the specific class structures of communities under study, consumer Researchers are, however, more concerned with the measurement of social class to understand market and consumption behaviour much more than mere social structure. With this accent in focus, these measure is prove more impractical. That is, this approach demands that members of a community to rank each other in the status system. This approach is limited to small communities because the citizens must know each other in order to rank each other and marketers cannot use with effectiveness.

THE OBJECTIVE MEASURES

As opposed to the purely subjective and reputational methods, which require people to think of and project their own class standing or that of community members, **OBJECTIVE** measures comprise of selected demographic or socio-economic variables concerning the individual's under study. These variables are measured through questionnaires that ask respondents several factual questions about themselves, their families or their places of residence. When the selecting objective measures of social class, the most researchers favour one or more of the following variables namely-occupation, amount of income, and education. To these socio-economic factors, they add 'geo- demographic' clustering data in the form of zip-code and residence-neighbourhood information. These socio-economic indicators are significant especially as the means of locating concentrations of consumers with specific social class membership.

The objective measures of class are of two categories namely, "Single–variable indexes" and "Composite-variable indexes".

1. SINGLE VARIABLE INDEXES

A **single variable** index makes use of one socio-economic variable to evaluate social class membership. These can be, occupation, education, income possessions and other variables. Let us touch these :

A. OCCUPATION:

Occupation is widely accepted and, perhaps, the best documented measure of social class because, it implies occupational status. The significance of occupation as a social class indicator is demonstrated by the frequency with which people ask others they meet for the first time, "What you are?" or "What do you do for a living?" The answer to this question serves as a guide for the person to size up others. The occupation is a main source of purchasing power. The following chart gives the ranking of occupations in terms of honesty and ethical

Occuation	Ranking in1999	Percentage
Medical	01	62%
University and College teachers	02	60%
Bankers	03	61%
Journalists	04	53%
Executives	05	50%

Lawyers	06	46%
Police–force	07	45%
Engineers	08	40%
Religious people (priests/clergies)	09	38%
Computer people	10	37%
Contractors	11	35%
Real estate agents	12	30%
State Officials	13	25%
Advertising people	14	20%
Politicians	15	19%
Insurance Agents	16	17%
Local Officials	17	16%
Sales personnel	18	15%
Tailors/outfitters	19	14%
Carpenters	20	13%

Note: Per cent is based on weightage given on opinion poll.

Source: Times of India Sunday Edition.

B. EDUCATION:

The level of a person's formal education is another most commonly accepted approximation of social class standing. Generally speaks, the more the education a person has, the more likely it is that the person is well paid-correlated to higher income earnings—and has an admired or respected position. Brouse through the following table.

All House	Total	Primary level	High school	Higher Secondary level	College level	Post college level
Under Rs. 5,000	1922%	31.11%	28.80%	21.08%	10.91%	10.91%
Rs. 5,000-9,999	17.25%	18.07%	17.10%	18.21%	17.20%	15.67%
Rs. 1,000-14,999	15.00%	13.02%	12.02%	15.00%	16.70%	18.28%
Rs. 15,000-24,999	17.87%	15.67%	16.80%	16.85%	18.21%	21.81%
Rs. 25,000-49,999	15.98%	12.07%	13.08%	14.07%	20.02%	20.74%
Rs. 45,000-64,999	8.40%	6.05%	7.07%	8.90%	10.04%	10.11%
Rs. 65,000-74,999	4.52%	3.50%	3.40%	3.40%	4.10%	5.09%
Rs. 75,000 & above	1.76%	1.50%	1.73%	1.79%	1.83%	2.00%
Number of Households in thousands	85,680	65,670	69,981	75,699	83,893	85,680

DIFFERENTIAL	-	20010	4311	5718	8194	1787
Average Income monthly	Rs.	Rs.	Rs.	Rs.	Rs.	Rs.

DSO Department of Statistics Goa State

C. INCOME:

Individual or family income is yet another socio-economic variable often used to approximate social class standing. The researchers who prefer income as a measure of social class use either amount or source of income. The following table gives both amount of income and the source of income in case of typical categories.

Amount of Income	Source of Income
Under Rs. 25,000 pa.	Social Welfare
Rs. 25,000-Rs. 34,999	Private finanacial help
Rs. 35,000-Rs. 44,999	Wages
Rs. 45,000-Rs. 54,999	Salary
Rs. 55,000-Rs. 64,999	Profits/royalties
Rs. 65,000-Rs. 74,999	Consultency fees
Rs. 75,000-Rs. 84,999	Earned Wealth
Rs. 85,000-Rs. 99,999	Inherited wealth
Rs.1,00,000-Rs.1,49,999	Inherited income
Rs. 1,50,000 and above	Interest dividends royalties

Though income is a popular estimate of social class standing, not all. Consumer researchers agree that it is an appropriate index of social class. It is because a blue collar automobile mechanic and a white collar assistant bank manager may both earnan yearly income of say Rs. 75,000, yet because of social class differences; each will spend that income earned in a different way. How they decided to spend their incomes reflect different values. Within this context, it is the difference in values that is an important discriminant of social class between people and not the amount of income they earn.

In case this view is supported, it is a matter of acceptance of market behaviour of "Under privileged" upper middle class and "Over privileged lower middle class formities with the same basic annual incomes. In fact, "over privileged" lower class consumers can be differentiated from underprivileged upper middle–class by their ownership of certain products, say motor boats pick-up traders, lawn-mowers and other in–house products. We also see that upper middle class with the same income spend relatively greater amounts on private club memberships, extra expenditure on education and cultural activities and so on. In addition, it is personal values than income that act as the function of attitude or behaviour than of income level.

Thus, these "attitudinally affluent" consumers may represent a broad segment who do not have the income needed to be considered affluent in todays' society, yet to have the best. They buy less but buy better quality-giving priorities and gradually working their way toward having as to what they want.

D. POSSESSIONS:

Good many sociologists have used 'possessions' as an index of social–class. One such scheme of evaluation is that of Chaplin's social status scale. It's focal point is on the preference of certain items of furniture and assessories in living rooms such as floor, floor-covering, drapes, fire–place in cold regions, library table, telephone, books cases etc., and the condition of the room such as its cleanness, organisation or arrangement or general atmosphere. Experts conclude of social class based on the bases of these observations. For instance, extent of home decorations reflect a social standing. It is generally found that lower middle–class families are likely to place their TV sets in living room, while middle and upper middle–class usually place their TV sets in the bed-room or family room. In case of upper class, it might be placed in special room TV room or Home theatre-room.

E. OTHER VARIABLES:

The other variables are quality of neighbourhood and rupee value of their residences. Though, they are not used very oftens, they are frequently used informally to support or varify a social class membership assigned on the basis of occupational status or income.

II. COMPOSITE VARIABLE INDEXES

Composite indexes combine a number of socio-economic factors systematically to form an overall measure of social class standing. These indexes reflect the indepth complexity of social class than single variable indexes. Thus, the research exploring consumers perceptions of mail and phone order shopping reveals that the higher the socio-economic status—in terms of a composite income, occupational status, and education—the more positive the consumers ratings of mail and phone order being, relative to in-store shopping. The research has also demonstrated that 'down-scale' consumers—a low composite of these three variables were less positive toward a magazine and catalog shopping and more-positive toward in-store shopping than more 'up-scale' socio-economic grouping. Possessed with such information, the retailers target their stores merchandise to working class more on-down-scale consumers would have a real challenge using direct marketing catalogs and telephone selling approaches. Contrary to this, retailers, concentrating on up-scale consumers, have been especially effective in developing catalog programs targeted to specific segments of affluent or up-scale customers.

Among the widely available composite indexes, two are more vibrant and effectively used namely, "Index of Status Characteristics." and the "Socio economic Status Score."

INDEX OF STATUS CHARACTERISTICS

A classic composite measure of social class is one which is developed by Warner Known as **Warner's Index** Status Characteristics. WISC is a weighted measure of following socio-economic variables such as occupation, source of income and not amount of income, house type and residing area—quality of neighbourhood.

SOCIO-ECONOMIC STATUS SCORES

This is being developed by Bureau of the Census of America which combines three basic socio-economic variables namely, occupation, family income and educational attainment. This socio-economic status scores is popularly called as SES.

The following two squares give the two popular composite social class measures. These are duly altered to meet the Indian situation taking into account the special conditions prevailing in India. Similar measure can be developed based on these models.

Index of Status Characteristics:

Composite Index	Variables	Weight
Index of Status Characteristics (ISC)	Occupation	4
	Source of Income	3
	Type of House	2
	Residing area:	1

Composite	Index	Variable
Socio-economi (SES)	Status Score	Occupations
		Family Income
		Education Level

PROBLEMS OF SOCIAL CLASS MEASUREMENT

Though variety of single and composite measures to categorise the social classes, quite a serious problems are associated with this delicate task. The experts those, who have brought this topic under focus, have brought to light at least four problems. These are :

1. Ranking is based on averages:

The ranking of a social class is based simply upon an average of the person's position on several status dimensions. This ignores the inconsistencies which are sure to arise from an individual rating high on one dimension–say income—but low on the other say–education.

2. Assumption of Stability:

A person's social class is assumed to be stable and thus the effects of mobility are ignored.

3. Reference-group effects are ignored:

An individual identifies only with the social class in which she or he is categorised, thus ignoring reference group effects from other classes.

4. Measurement of an entire family:

The social class of an entire family may be measured by examination of characteristics of only the adult male wage earner, thus ignoring characteristics of other family members particularly the employment and education of the adult female in the family.

In a nut shell–resolving these and other problems would make the concept of social class more useful especially from the view point of marketers. Though the size of different classes varies from country to country depending on the method of classification used and may shift over a time, yet it is quite significant to realize that bulk of the market for most of the products exists in the broad upper and middle class groups. The highest class is made up of negligible part of the total population. It is because of its wealth, it is important to the marketing of certain luxury items. However, this group is too small in number to provide the focal point for the majority of marketers. On the other hand, at the rock–bottom the class spectrum is one that finds a market that, even though sizeable, doesn't have sufficient income for good many products. That is why, most consumer goods marketers concentrate on the remaining group.

LIFE-STYLE PROFILES OF THE SOCIAL CLASSES

The consumer researchers have found based on evidences that within each class of the social classes, there is a constellation of specific "life-style factors" that tend to distinguish the member of each class from the members of all other social classes. That is, life style factors stand for shared beliefs, attitudes, activities, and the behaviours. The significance of social stratification for the marketer is these differences in values, attitudes and behaviour of each classes. These differences provide a basis on which to segment markets and obtain an enhanced understanding of the behaviour of consumers. That is, within each class, we find different life–styles. What researchers say is that there is a diversity of family situation and a nearly unbelievable range in income totals are comprised within each class and there is considerable variation in the way individual members of a class realize these class goals and express these values. What is more important is that it is almost impossible to point to a clear line of demarcation where one class changes to another. That is why, there is bound to be sort of overlapping of social strata. Inspite of these, some major distinctions can be made between each of the social class. In order to identify life–styles in each class, the total society is divided into five way classification as upper class, upper-middle, middle-middle lower-middle and lower class, unlike six way classification in the Western world. The following Table gives five way classification based on income as the base from all sources.

SOCIAL CLASS	Annual Income	Percentage
Upper–class :	Rs. 3,50,000 and above	2%
Upper–middle class :	Between 3,49,000 and 1,99,999	20%
Middle–middle class :	Between 2,00,000 and 1,49,999	08%
Lower–middle class :	Between 1,50,000 and 74,999	25%
Lower–class:	Below 75,000 rupees	45%
		100%

From the table it is clear that hardly 2 per cent of total social classes belong to upper class, 53% middle class and 45% lower class. Let us identify the life–styles of each class and sub- life styles that each social group has :

UPPER CLASS AND LIFE–STYLES

The upper class has annual income over 3,50,000 rupees. This is consisting of the wealthiest aristocrats and people having new wealth than inherited. This group is a nice blend of many life-styles. These consists of landlords, politicians, business people, merchants, financers, and professionals. These represent, established, old money and new money groups. The features of the life–styles are that, is money that multiplies; pay attention to status, therefore, premium brands having luxurious life having international residence, friendship, relations. Life is upholder family reputation, reflecting the excellence of one's breeding and displaying a sense of community responsibility. 'New rich' or 'new money' is that group which consists of executive elite—founders of large business wealthy doctors, lawyers. Their problem is how to spend, therefore, very liberal in spending on each occasions like birthday, engagement, marriage and even general. They spend lavishly as if money has no value. Their dwellings may be, palatial, independent luxurious bungalow having thousands of square metres with facilities, consisting of separate main building, guest house, servant quarters and so on. What is important is they find an excuse to spend as they an trillionnaires, billionaires and millionaires. One can imagine of the' style of eating, food needs, bathing, and so on. Their life–styles an more of showmanship and exuberance. Thus, main at functions like wedding, they might present the invite as they attended the function instead of invitees presenting something to the hosts.

MIDDLE CLASS AND THEIR LIFE–STYLE

The word middle in general accounts for 53 per cent of which upper rest 20 per cent middle middle 8 per cent and lower middle 25 percent. These are again business people, self-employed, doctors, lawyers, authors, professors, engineers, all types of professionals including consultants. Depending on three sub–groups life–style changes. Taking only a means of transport, upper middle class might go in for say Fiat Unò group-say Sierra, jubilees and Fia Uno EXA/ CPSMET. In addition, they might have two wheelers, may be motorcycles or scooter of various makes : They might take their food and stay at 5 to 3 star establishments. They have national level friends, relations and even NRI status. Middle–class in general has a standard life–style which is maintained at any rate in various aspects of food, shelter, clothing and other belongings. In case of functions of importance, middle class may request for "no presents please". Or it may be "presentation of bouquets only."

LOWER SOCIAL CLASS AND LIFE–STYLES

Lower class accounts for almost 45 per cent consisting of poor and honest and hard working to make both the ends to meet. Their life–style, if speak in terms of vehicles, may be restricted to two wheelers of various normal brands even scooters, scooties and bicycles. They depend on public transport system. They go to low–grade restaurants, shops and so on. Their life is struggle yet they attempt to improve. Inspite of these, their income falls short of expenses on occasions of birth, marriage death of a person in extreme cases. Their life–style is not very permanent. On the occasions of social functions normally, they expect the presents. They take it is as a matter of pride that one who receives more presents is more accepted by guests.

SOCIAL CLASS MOBILITY

The study reveals that the social class membership is not very hard and fixed in India as in some of the countries whether developed or developing. The main reason is spread of education social–contacts, the strong urge to improve. While individuals can move either up or down in social–class standing from the class position held by their parents. There is mostly upward mobility due to education and exposure. Recognising that individuals often aspire to the life–style and possessions enjoyed by members of a higher social class, the marketers very often incorporate symbols of higher–class membership-both as products and props in all targetted lower social class audience. Another feature of social class mobility is that products and services traditionally within the reach of ones social class may filter down to lower social classes. Thus, plastic surgery was mainly for very rich and bollywood stars. Now, plastic surgery has come close to middle class and if necessary in case of poor too in exceptional case. Just as we have upward mobility, we come across the downward mobility too. Due to economic reasons particularly the high rise inflation, downward mobility is seen. This is more so in case of middle–class ranges as upper, middle and lower as these groups of victim of the circumstances.

CLASSIFICATION OF CONSUMERS FROM MARKETING ANGLE

From the marketing point of view, the consumers are classed as "affluent" and 'non-affluent'.

THE AFFLUENT CONSUMER:

For marketers, **affluent** households constitute a special category and a special segment because its members have incomes that provide them with a disproportionately larger share of all discretionary income. This additional discretioning income can be tapped for purchase luxuries of all kinds in terms of property, vehicles, music–system, travel, stay, and other services. The media exposure is also very specific. That is, the media habits of these affluent consumer are different than general population. These submits to costly magazines on lines- fashion designs films, health food, and sex magazines like Cosmopolitan. On an average, spend 14% of their daily time on watching TV programs, 80% of them spends to CTV, listen to 47% of time to radio and so on. This affluent consumer class is urban and rural. At present, much of marketing are housing their attention for urban to rural -a virgin area which is yet to be tapped.

THE NON-AFFLUENT CONSUMER

For marketers, consumers are more important whether they are affluent or **non-affluent** because they are interested in total profits than mere profits. In case they forget non-affluent consumer, they are real losers as this class constitutes the majority. If we take the poor class it deposits 45 per cent- higher than any other class calculated individually class wise. In such cases all the copies are specially designed for these downscale consumers — which are meant for "average Joes" and average Janes—reflecting the modest life–styles, as they cannot efford premium brands—products and services.

SELECTED CONSUMER BEHAVIOUR APPLICATIONS OF SOCIAL CLASS

Definitely social–class profiles provide a broad but clear picture of the values, attitudes and behaviour that distinguish the members of various social–classes. The study of this level will help the marketers to design appropriate marketing strategies to get the best from consumers and give the best to consumers too. These selected areas are, clothing and shopping, pursuit of leisure, saving, spending and credit.

1. Clothing—Fashion and Shopping:

When we come to clothing, we are to remember what Greek philosopher Epictetus once said. "Know, first, who you are, and then adorn yourself accordingly." The wisdom that lies in this statement is more relevant to clothing marketers today because, it is the clothes that speak of social class or self–image. It is well-known that most of the people dress to fit their self-images that include their perceptions of their own social class and class membership. However, modern ads appeal not only to adhere to their social class but they try to change their behaviour. Thus, an ad reading "One need not be a millionaire to look a millionaire". However, depending on social class membership, the people spend on clothing.

Fashions are often repeated, with the trends. Members of specific social classes differ in terms of what they consider fashionable or in good taste. For instance, upper middle class consumers have a strong preference for T Shirt, caps and other clothing that offer an external point of identification where they may go in for **Nike, Peter England, Van Hausen.** In case of dress material, **Raymonds, Vimal, S. Kumar** and so on. In case of Jeans, **Lee,** Levis and so on. While in case middle middle class might go in for, **Ruf and Tuff, Flying Machine, Lee** in case of jeans and T shirts, other brands say **Lacoste, Alligator, Exemburg** and so on. In case of dress materials may be **BSL, Gwailior** suiting and so on.

Social–class is an important variable in determining where a consumer shops. These, rich and upper middle–class have departmental stores, star super markets, show–rooms while middle–class have specific stores and poor class people go to retailers of their choice.

2. The Pursuit of Leisure:

Social–class membership is also closely related to the choice of recreational and leisure time activities. For instance, upper class consumers are likely to attend theatre and concerts, clubs to play bridge or attend college football or baseball games. Lower class consumers tend to avoid television watcheres and fishing enthusiasts and they enjoy drive in movies and baseball games. The lower class consumers spends more time on commercial types of activities say bowling plays, pools or billiards or visiting taverns and craft activities like model–building, painting and wood working projects rather than cerebral activities such as reading visiting museums and so on.

3. Saving, Spending and Credit:

Upper class consumers are more fortunate to have future-oriented and confident of their financial acumen. That is why they are more insting invest in insurance, stock and real estate. The lower classes are generally more concerned with immediate gratification, as and when they save they are basically interested in safety and security. Therefore, it is not surprising that when it comes to bank credit card usage, members of the lower social classes tend to use their bank credit cards for instalment purchases, while members of the upper social classes pay their credit card bills in full each month. Put in other words, lower class purchases tend to use their credit cards to buy now and pay later" for things they might not otherwise be able to afford, while upper class purchases use their credit cards as a convenient substitute for hard cash.

CHPATER BASED QUESTIONS

A.. WRITE SHORT NOTES ON:

(1) Social class. (2) Nature of social class. (3) Social–class categories. (4) Measurement of social class. (5) Subjective measures. (6) Reputational measures. (7) Objective measures. (8) Problems of social class measurements. (9) Classification of consumers from marketing angle. (10) Selected consumer behaviour applications. (11) Social class mobility 12) Single variable indexes.

B. SHORT ANSWER QUESTIONS:

1. What is a social-class category ?
2. Explain the nature of social-class.
3. Explain different social-class categories.
4. What are the different measures of social-class ?
5. What problems one faces in measuring a social-class ?
6. What is social-class mobility ?
7. Discuss the social-class life-style profiles.
8. How consumers are classified by marketers for effective marketing efforts and results?
9. Do marketers encourage upward or downward class movement ?
10. "In each class, life-styles differ." Do you agree ? Give reasons.

C. ESSAY TYPE QUESTIONS:

1. What is a 'social class' ? What is its nature ?
2. Define 'Social-class'. Explain the various characteristics.
3. "Existence of a social-class should fulfil certain conditions". Explain these conditions.
4. What is social-class category ? How many class categories are accepted ?
5. What is measurement of social class ? What are the measures of social class category?
6. What are variable indexes ? Explain single variables used.
7. What are composite variable indexes ? How many types of indexes are there ?
8. What is social-class measurement ? What problems one faces in measuring social-class?
9. Explain the idea of social-class life-style profiles.
10. What is social-class mobility ? What is the role of marketers in this regard?

*****

CHAPTER 12

CONSUMER DECISION PROCESS

BACKDROP

We are well placed to understand as to how consumers make purchase-decisions because, the earlier chapters have provided a unique framework comprising of psychological, social and cultural aspects or variables that have impact on consumer decisions and decision-making process. Mr. Napoleon I expressed very aptly when he said: "Nothing is more difficult, and therefore, more precious than to be able to decide." In the light of this, a consumer-decision should bring him real satisfaction a real value for money. Hence, this chapter plans to discuss the vital aspects such as what is a consumer decision-making ? Views on consumer decision-making. A model of consumer decision-making dealt at length and touching some standard consumer decision-models and allied aspects. The chapter ends with a Chapter Based Questions.

WHAT IS A DECISION ?

Each day, each one of us makes countless **decisions** regarding every aspect of our daily lives. We generally, however, make these decisions without stopping to think about how we make them and what is involved in the particular decision-making process itself. In simple terms, a "decision is the selection of an option from two or more alternative choices." In other words for a person to make a decision, a choice of alternatives must be available. When a person has choice between making a purchase and not making a purchase, a choice between brand 'A' and brand 'B', or a choice of spending time doing 'X' or 'Y', that person is in a position to make a decision. On the contrary, if the consumer has no alternatives from which to choose and is literally forced to make a particular purchase or take a particular action, say use a prescribed medication, then this single "no choice" does not constitute a decision, such a no-choice decision is commonly referred to as a "Hobson's choice". Coming to the ground realities, no-choice purchase or consumption situations are fairly very rare. From our earlier reading, in **chapter nine on culture**, 'freedom' is one of the core values expressed in terms of wide range of product and service choices. That is, if there is almost always a choice, then there is almost, always an opportunity for consumers to make decisions. Furthermore, experimental consumer research reveals that providing consumers with a choice when there were originally none can be a very good business strategy. One that can substantially increase sales. Take for example, when a direct mail electrical appliance catalogue displayed two soda makers instead of just one-say, the original soda–maker at Rs. 1,500 and 'new' only slightly larger for Rs. 2,500, the addition of the second comparison soda–maker seemed to stimulate consumer evaluation that significantly increased the sales of the original soda-maker.

TYPES OF CONSUMER DECISIONS :

There are good many options possible for a consumer in today's world of free market economy. However, these options may be boiled down into Five main types of decisions. Infact, these are the five dimensions of decision. These are: (1) What to buy? (2) How much to buy? (3) Where to buy? (4) When to buy? and (5)How to buy? Let us touch these dimensions.

1. What to Buy ?

Deciding 'what' to buy is one of the consumer's most basic tasks. No buying activity may take place unless this fundamental decision is made. A consumer's product or service decision encompasses not only the general category of products desired say appliances, but more specifically, the narrower range of items say 'kitchen' appliances. Consumers are to make even decisions on brands, prices and product features. Thus, a house–wife is going for say micro-oven where almost all leading companies such as **IFB, L.G., Kenstar, Philips, BPL, Panosonic, National** and so on. She is to select not only a maker, make, range of watts-say between 700 —1000, number of digital controls, levels of power and the price say ranging from Rs. 8,999, Rs. 13,333 or even Rs. 15,999. This is a specific decision as to what will be purchased and with this particular decision finalised, the consumer moves closer to the completion of overall purchase decision process.

2. How much to Buy ?

A second basic decision by the consumer relates to "how much" of the item will be purchased. For instance, when a shopper is buying green vegetables he determines how much he is going to buy, say quarter kilo, or half kilo, three quarters or full one kilo. Much depends on family households, consumption pattern, facilities for cold storage and so on. However, there are certain items which are not available in broken sizes say, you cannot buy a one third car, half a fridge. At the most, the components can be bought say in case of car-wind screen,\ tyre, tube, rear glass or compression bulbs and so on in case of latter. Even in case of packed consumer non-durables and durables certain manner quantity is to be brought as they are packed is case of standard products. In case of edible oil say 'Dhara' we have tetra–pack of one litre or pouch of one litre and five litres containers. In case of tea. power we have package ranging from 125 grams to 1,000 grams.

3. Where to Buy ?

Yet another determination is to be made by the consumer involves 'where' the selected product or service is to be purchased. This is really very significant decision which interacts with the earlier decision thoroughly on what to buy ? To products, although physically the same, are likely to be perceived differentially because of other facets associated with them. For example, consider a case of air–conditioner sold with delivery, installation and is home servicing guaranteed by a full service departmental store compared with the same model priced lower but sold at no–frill basis in a discount house, with none of the above services. It is certain that consumers are likely to perceive the same based, well known company's product in different ways, based on the nature of the prices and the services extended. A rational buyer prefer a departmental store and an emotional buyer a discount house. It may not be necessarily true as perception plays significant role. Consequently, what one buys is closely related to decisions of where one decides to buy. Certainly, not all sales outlets are similar and consumers have many option regarding the outlet, services offered, merchandise lines, prices charged and so on.

4. When to Buy ?

Purchase or buying has time "value of time" dimension. This decision is influenced by such factors as urgency of the need and availability of the chosen item. The other elements such as store opening times, periods of sales and clearances, availability of transportation, and freedom of family members to shop all have bearing on when one purchases. We see most of the consumers wait for special offers Diwali, Christmas, where fabulous discounts are given, exchange offers are made available, zero-percent interest financing is granted; on the other hand some consumers are lured by such 'quick–sale' or aggressive selling gimmicks. They are cool and buy whenever they think the product is wanted. Much depends on number of factors, say, quality considerations, store royalty, after sale services and so on.

5. How to Buy ?

This is the final dimension of decision of 'how' to buy which is really a complex issue good many factors affect as to how the consumer buys. Some of the key factors are — consider some of the alternative strategies consumers use ; shop extensively or buy from the first outlet, pay cash or buy on credit, have it delivered or take it home ; pay cash down and buy it on hire purchase on instalment basis. It may be bought directly or indirectly as the case may be. Countless purchasing patterns occur in the market place, with each consumer relying in whatever 'strategy seems to work well for himself or herself, the crux of problems that consumer must solve should benefit from the old logic of a computer than the 'hit and miss' decision approach taken by some consumers.

In fact, these types of decisions are dimensions of a single purchase. That is purchase processes involves taking into account, what, ? how much, ? when, ? where ? and how ? to buy.

LEVELS OF CONSUMER DECISION–MAKING

It is crystal clear that not all consumer decision-making situations receive or require the same degree of information search. In case all the purchase–decisions required extensive effort, then the consumer decision–making would be an exhausting process that left little time for any thing else. If all the purchases were routine, on the other hand, they would tend to be monotonous and would provide little pleasure or novelty. On a continuum of effort ranging from very high to very low, one can clearly distinguish three levels of consumer decision–making. These are : (1) Extensive problem solving (2) Limited problems solving and (3) Routinized response behaviour. The following figure makes this continuum very clear.

CHARACTERISTICS OF CONSUMER PROBLEM SOLVING APPROACHES

Variables	Routine Problem Solving	Limited Problem Solving	Extensive Problem Solving
1) Purchase involvemnent level	Low Medium	High	
2) Problem Recognition	Automatic	Semi-automatic	Complex
3) Information Search and Evaluation	Minimal	Limited	Extensive

4) Purchase Orientation	Convenience	Mixed	Shopping
5) Post-purchase Processes	Very Limited	Limited	Complex
	Habit	Inertia to purchase	Loyalty if satisfied
	Brand Loyalty	Brand Switching if dissatisfied	Complaint if dissatisfied

FIG: 12:01 ALTERNATION PROBLEM SOLVING APPROACHES AND THEIR FEATURES

As we have seen the range of consumer problem solving approaches can be placed of a spectrum from routine to extensive patterns, where we earlier mentioned a continuum starting with extensive problem solving and ending with routined response behaviour. These are explained as under:

1. Routine Problem Solving : This is called as R.P.S. or R.R.B. namely, routinised response behaviour. At this level, consumers have some experience with the product category and a well established set of criteria with which to evaluate the brands they are considering. Only in some cases, they may search for small amount of additional information ; in other cases they simply review what they already know. Put in other words, when consumers buy a brand they have purchased before, it is usually involving little or no information seeking and is performed quickly and easily. Consumers are brand loyal and tend to buy in a habitual, automatic and unthinking way.

2. Limited Problem Solving : This is known as L.P.S. At this level of problem solving, consumers already have established the basic criteria for evaluating the product category and the various brands in the category. They, however, have not fully established performances concerning a select group of brands. Their search for addition information is more like "fine-tuning", they must gather additional brand information to discriminate among the various brands. That is, when consumers buy a new brand in a familiar product category, it usually involves a moderate amount of information seeking and time in choosing.

3. Extensive Problem Solving : It is the extreme case with is also known as E.P.S. When the consumers have no established criteria for evaluating a product category or specific brands in that category or have not narrowed the number of brands they will consider to a small manageable subset — their decision–making efforts can be classified as an extensive problem solving. At this level, the consumer needs a great deal of information to establish a set of criteria, on which to judge specific brands and a correspondingly large amount of information concerning each of the brands to be considered. That is, when consumers buy in an unfamiliar product category, it is usually involving the need for obtaining substantial information and a longer time to choose. They must form a concept of the new product category and determine the criteria to be used in choosing a brand.

In this regard Figure 12.01 highlights the major characteristics of each of these decision approaches. The above explained three types of consumer decision making may be generalized toward a typical customer problem solving model consisting of four basic types of activities in the process of purchasing. The consumer's four steps are: 1. Problem recognition 2. Information search and evaluation 3. Purchase–decision and 4. Post-purchase behaviour. The assumptions underlying this and other decision process approaches to consumer behaviour appear to the following. These are: 1. Two or more alternative exist, so that a choice must be made by the consumer 2. Consumer evaluative criteria facilitate the forecasting of each alternative's consequences for the consumer's goals or objectives 3. The consumer uses a decision rule or evaluative procedure to determine the chosen alternative and 4. Information obtained from external sources or memory or both sources and memory are used in the application of the decision rule or evaluative procedure.

It should be noted, however, that for certain purchase situations, some consumers do not engage in a pre–purchase decision process. For instance, they may not have stored information, it may not be retrieved or retrievable and they may not search externally. Thus, some purchases may occur as a result of approaches other than a decision process. They can occur out of necessity; they can be derived from certain culturally man dated life–styles ; they can be interlocked with other purchases ; they can reflect purchase preferences acquired in early childhood ; they can result from confirmity to group norms or imitation of others, they can be made in the basis of various surrogates; they can even occur on a more superficial basis.

MODELS OF CONSUMERS

Before we try to understand the views of consumer decision–making, or decision–making process models, it pays to consider several schools of thought that depict consumer decision–making in distinct and clear different ways. The phrase 'Models of Consumers' refers to a general 'view' or perspective as to how and why individuals behave as they do. Here is an attempt to examine models of consumers in terms of the following views, namely, 1. an economic view 2. a passive view 3. a cognative view and 4. an emotional view.

1. An Economic View:

In the field of theory of Economics, which portrays a world of perfect competition, the consumer has often been characterised as making rational decisions. This model is called "the economic man" theory. It has been criticized by consumer researchers for a number of reasons. To behave rationally, in economic sense, a consumer will have to 1. be aware of all available product alternatives 2. be capable of correctly ranking each alternative in terms of its benefits and disadvantages and 3. be able to identify the one best alternative. However, realistically, consumers rarely have all of the information, or sufficiently accurate information, or even an adequate degree of involvement or motivation, to make the so called 'perfect' decision.

According to **Mr. Herbert. A. Simon,** a leading social scientist, the classical economic model of an all-rational consumer is unrealistic for the following reasons : (a) people are limited by their existing skills, habits and reflexes (b) people are limited by their existing values and goals and (c) people are limited by the extent of their knowledge. Consumers operate in an imperfect world in which they do not maximise their decisions in terms of economic consideration, such as price-quantity relationships, marginal utility, or indifference curves. In fact, the consumer generally is unwilling to engage in extensive decision-making activities and will settle, instead, for a 'satisfactory' decision, one that is "good enough". For this reason, the economic model is often rejected as too idealistic and simplistic in its approach.

2. A Passive View

As opposed to "rational economic view" of consumers, it is the "passive view" that depicts consumer as basically submissive to the self-serving interests and promotional efforts of marketers. Consumers are perceived as impulsive and irrational purchases, in passive view, ready to yield to the aims and arms of marketers. At least to some degree, the passive model of consumer was subscribed by the hard-driving super-salesmen of old who were trained to regard the consumer as an object to be manipulated. The second decade of 20th century salesmanship text highlights the long held belief in the dominance of the salesman over the unresisting, somewhat passive manner. The essence of the text is the developing a selling processes calls for development of four stages that salesman should get first individual attention of consumer developing attention into interest and converting this interest into desire and finally desire into action to buy.

The principal limitation of this passive model is that it fails to recognise that the consumer plays an equal, if not dominant role in many buying situations—sometimes by seeking information about product alternatives and selecting the product that appears to offer the greatest satisfaction and at other times impulsively selecting a product that satisfies the mood or emotion of the moment. All that we have referred to motivation, selection perception, learning, attitudes, communication and opinion leadership serves to support the proposition that consumers are rarely objects of manipulation. Therefore, this simple and single minded view must be rejected as unrealislic.

3. A Cognative View :

This third model depicts the consumer as a "thinking problem solver". Within this framework, consumers frequently are pictured as either receptive to or actively searching for products and services that fulfil their needs and enrich their lives. The 'cognative' model focusses on the processes by which consumers seek and evaluate information about selected brands and retail outlets. The ad on Master Card is basically a cognative appeal which is given on the next page. Within the context of the cognative model, consumers are viewed as information processors. Information processing leads to the formation of preferences and, finally, to purchase intentions. Consumers also may use a preference formation strategy that is "other based", in which they allow another person to make the selection for them.

In contrast to the "economic view", the "cognative view" recognizes that the consumer is unlikely to even attempt to obtain all available information about every choice. Instead, consumers are likely to cease their information seeking efforts when they perceive that they have sufficient information about some of the alternatives to make a 'satisfactory' decision. As this information-processing view point suggests, consumers often develop short-cut decision rules which are called as 'heuristics' to facilitate the decision making process. They also use decision rules to cope up with exposure to too much information which is called as "information overload".

This cognative problem solving view describes a consumer who falls somewhere between the extremes of the economic and passive views, who does not or cannot have total knowledge about available product alternatives and, therefore, cannot make perfect decisions, but who nevertheless actively seeks information and attempts to make satisfactory decisions. The cognative model appears to capture the essence of a well- educated and involved consumer who seeks information on which to base consumption decisions.

4. An Emotional View:

Marketers frequently prefer to think of consumers in terms of either economic or passive models, although they are long aware of the 'emotional' or 'impulsive' model of consumer decision making. However, in reality, each of us is likely to associate deep feelings or emotions such as joy, fear, love, hope, sexuality, fantasy and even little 'magic' with certain purchases or possessions. These feelings or emotions are likely to be highly involving. For instance, a person who misplaces a favourite fountain pen might go to great length to look for it, despite the fact that he or she has six other's at hand. Possessions also may serve to pressure a sense of the past and act as familiar transitional objects when one is confronted with an uncertain future. If we were to reflect on the nature of our recent purchases, we might be surprised to realize just how impulsive some of them were. Rather than carefully searching, deliberating and evaluating alternatives before buying, we are just as likely to have made many of these purchases on impulse, on a whim or because we were 'emotionally driven'. When a consumer makes what is basically emotional purchase, less emphasis is placed on the search for pre-purchase informations.

Instead, more emphasis is placed on current mood and feelings "go for it". This is not to say that emotional decisions are rational. We know that buying products that afford emotional satisfaction is perfectly rational consumer decision. Further, in case of a good number of products the choice of one brand over another has little to do with rationality. There are good many instances of this kind. For example, good many consumers buy designer lable clothing, not because they look any better in them, but because the status labels make them feel better. When a woman chooses between two dresses, it is quite natural that she decides in favour of the one that makes her feel better. This is a 'rational' decision. If a man with a wife and three children purchases a four seater Maruti 800 for hismelf, of course, the neighbours might consider about his level of rationality. No question would arise if the same man selected **Fiat UNO** or **Fiat Sienna** as each might be impulsive emotional purchase decision. Advertisers have fully recognised this and are renewing interest thereby increasing importance of 'emotional' or 'feeling oriented' advertising.

Last, but not the least, consumer 'mods' are also important is decision making. 'Mood' is a 'feeling state' or 'state of mind' unlike an emotion, which is a responsive to a particular environment, a mood is more typically an unfocussed, pre-existing state—already present at the time a consumer 'experiences' an advetisement, a retail environment, a brand or a product.

Mood appears to be important for consumer decision–making because, it influences the aspects as to when consumers shop, where they shop and whether they shop alone or with others. It is likely to influence how the consumer responds to actual shopping environments—point of purchase. That is why, some retailers attempt to create a mood for shoppers, even though shoppers enter the store with pre–existing mood. Research demonstrates that a store's image or atmosphere can affect shopper's moods. In turn, shoppers' mood can influence howlong they stay in the store, as well as other behaviour that retailers wish to encourage. In general, individuals in a positive mood recall more information about a product than those in a negative mood. However, as the results of one study suggest introducing a positive mood at pop decision say background music pop display is unlikely to have a meaningful impact on specific brand choice unless a previously stored brand evaluation already exists.

A MODEL OF CONSUMER DECISION–MAKING :

Before we go to different standard consumer buyer models, let us be conversant with simple model which presents consumer decision–making that reflects cognative problem solving consumer and the emotional consumer to some degree. This model is the simplest consisting of three major components namely **input process and output**. The following is the pictorial presentation of the same.

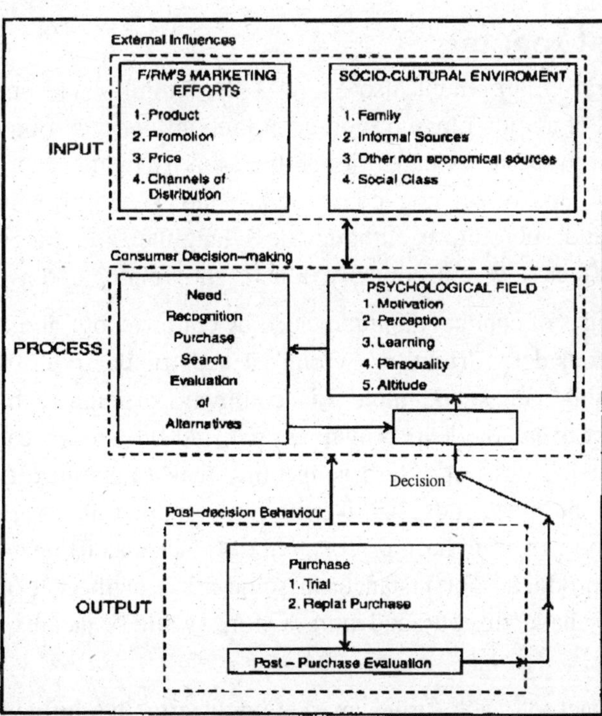

FIG: 12:02 A SIMPLE MODEL OF CONSUMER DECISION–MAKING

INPUT

The **INPUT** component of consumer decision making model draws on external influences that serve as sources of information about a particular product and influence a consumer's product related values, attitudes and the behaviour. The main among these input factors are "marketing mix" activities of a unit that attempts to communicate the benefits of products and services to potential consumers, and the non-marketing 'socio-cultural' influences, which, when internalised, affect the consumer's purchase decisions. Let us note these marketing inputs and socio-cultural inputs.

1. Marketing Inputs:

An unit's marketing activities are the direct attempts to reach, inform, educate and persuade the consumers to buy and use its products and services. These inputs to the consumer's decision–making process take the form of specific marketing-mix strategies that consist of product (size, package, guarantees and warranties) promotion (mass media advertising), direct marketing, personal selling other promotional efforts) Price (pricing policies and strategies) and Place (distribution channels of distribution to move the products from the producers to the consumers). Finally, the impact of the firm's marketing efforts in large measure is governed by the consumer's perception of these efforts. Thus, marketers are sure to do well to remain diligently alert to consumer perceptions by sponsoring consumer research than to rely on the\ intended impact of their marketing messages.

2. Socio-cultural Inputs

The socio-cultural environment also exerts a deeper influence on the consumer. Social-cultural inputs consists of a wide range of non-economic influences. For instance, the comments of a friend, on editorial in the news-paper, usage by a family member, or an article is Consumer Reports are all specific and direct non–commercial sources of information. The influences of social class, culture, and sub-culture, although less tangible, are important input factors that are internalised and effect how consumers evaluate and finally adopt or reject the products.

The unwritten codes of conduct communicated by culture subtly indicate which consumption behaviour should be considered 'right' or 'wrong' at a particular point of time. For example, colour hair–dying among ladies and men was common particularly black ; but today's ads given by **Godrej** and **Loreal** of deep brown, brown, red brown are catching the imagination of both men and women. It is not a surprising thing that a person might think of black head hair and brown beards and mustaches. Unlike the firm's marketing efforts, socio-cultural inputs do not necessarily support the purchase or consumption of a specific product but may influence consumers to avoid a product. For instance, in some areas of the nation and in social circles, animal rights activities have discouraged purses, furs, wallets, jackets, caps made of animal skins and furs.

The overall impact of each firm's marketing efforts the influence of family, friends, neighbours and society's existing code of behaviour are all inputs that are likely to affect what consumer purchases and how he buys and what he buys.Because these influences may be; directed to the individual or actively sough by the individual a two headed arrow is used to link the **INPUT** and **PROCESS** segments is the Model given earlier.

PROCESS

The 'process' component of the model is concerned with how consumers make decisions. In order to understand this process, one should consider the influence of the psychological concepts. The psychological field represents the internal influences motivation, perception, learning, personally and attitudes that affect consumer's decision-making processes what they need or want ? their awareness of various product choices, their information-gathering activities and their evaluation alternatives. The act of making a consumer decision consists of three stages, as given in the model, namely, 1. Need recognition 2. Pre-purchase search and 3. Evaluation of alternatives. Let us touch these in brief.

1. Need Recognition:

The recognition of a need is likely to occur when a consumer is faced with a 'problem'. For instance, take a case of a newly retired couple Mr. John and Mrs. Joan who have recently sold their home in a suburb of Belgaum District and relocated to Pune city in Pune district. Although Mr. John is looking forward to spend time in the golf–course, Mrs. Joan has been promising herself for the past two years that she was going to learn how to send and receive e-mail and access the Internet, using their home computer. Some of her friends who are already retired have told her how convenient and inexpensive it is to keep is touch via e-mail, and they have also spoken to her about the wealth of free information available on internet-everything from apple pie receipes to zoo visiting hours. Now, turning to Mrs. Joan, she is also interested in being able to check on the performance of the couple's current investments is several stocks and mutual funds. Further, her friends have told her that the easiest way to gain access to e-mail and the Internet is by subscribing to an on-line service. Because Mrs. Joan had been talking about subscribing to an on-line service for a while, for Mother's Day, her children and grand children presented her with a new, external 28,800 bps Modern and they even attached it to her computer and telephone line for her.

Among consumers, there seem to be two different need or problem recognition styles. Some consumer's are 'actual state' types who perceive that they have a problem when a product fails to perform satisfactorily. In contrast, other consumers are "desired state types" for whom the desire for some thing new may trigger the decision process. Need or problem recognition can also be viewed as either 'simple' or 'complex'. 'Simple' problem recognition refers to needs that occur frequently and that can be dealt with almost automatically, such as becomes hungry and puchasing a candy bar from a vending machine. Complex problem recognition, however, is characterised as a state in which a problem develops over a time, as the "actual state" and the "desired state" gradually move apart. For instance, after several years of driving a car, the owner may begin to consider trading in for a new one to avoid growing repair bills.

2. Pre-purchase Search:

Pre-purchase search begins when a consumer perceives a need that might be satisfied by the purchase and consumption of a product. The recollection of the past experiences drawn from long-term memory storage—might provide the consumer with adequate information to make the present choice. On the other hand, when the consumer has had no prior experience, he or she may have to engage in extensive search of the outside environment for useful information on which to base a choice of his or her.

The consumer generally searches his or her memory—the psychological field of model before seeking external sources of information regarding a given consumption related need. Past experience is considered an internal source of information. The greater the relevant past experience, the less external information the consumer is likely to need to reach a decision. Many consumer decisions are based on a combination of past experience internal sources and marketing and non- commercial information-external sources. The degree of the perceived risk can also influence this stage of the decision process. In high–risk situations, consumers are likely to engage in complex information search and evaluation ; in low-risk situations, they are likely to use very simple search and evaluation tactics.

An examination of the external search effort associated with the purchase of different product categories TV sets, VCRs or PCs found that, as the amount of total search effort increased, the consumer attitudes toward shopping became more positive and more time was made available for shopping. Not surprising, the external search effort was greatest for consumers who had the least amount of product category knowledge. From this, it follows that less consumers know about a product category and the more important the purchase is to them, the more time they will make available and 'the more extensive their prepurchase search actively is likely to be. Additionally, by conducting focus–groups and other consumer research on how expectations of a product influence consumers' subsequent interactions and evaluation of product information, firms are assisted in their efforts to device better products and to promote their products' strongest points.

How much information a consumer will gather also depends on various situational factors such as experience, social acceptability value related considerations. In addition, on-line service sources can be tapped may be 'personal' and 'impersonal'. Personal on line service sources are friends, neighbours, relatives, co-workers, computer store sales people calling on the line vendor. The impersonal on the service sources are newspaper articles, magazine articles, consumer reports, direct mail brochures and trial offer from on line vendors, information from product ads.

3. Evaluation of Alternatives

While evaluating potential alternatives, consumers tend to use two types of information. 1. a list' of brands from which they plan to make their selection the evoked set and 2. the criteria they will use to evaluate each brand. Making a selection from a sample of possible brands is a human feature that helps is simplifying the decision making process.

A. EVOKED–SET

The 'evoked set' refers to the specific brands a consumer considers in making a purchase within a particular product category, within the context of consumer decision–making. The evoked set is also called the 'consideration' set. A consumer's evoked–set is distinguished from his or her 'inapt set' which consists of brands the consumer excludes from purchase consideration because they are felt to be unacceptable say seem as inferior and from the 'inert–set' which consists of brands the consumer is indifferent toward, because they are perceived as not having any particular advantage. Regardless of total number of brands in a product category, a consumer's evoked–set tend to be quite small or an average, often consisting of only three to five brands. The evoked–set consists of small number of brands the consumer is familiar with, remembers, and finds acceptable one. This evoked set can be given as under :

As the following figure (12:03) indicates, it is essential that a product be part of a consumer's evoked set if it is to be considered at all. The final terminal positions in the model that do not end in purchase would appear to have perceptual problems. For instance: 1. Brands may be unknown because of the consumer's selective exposure to advertising media and selective perception of advertising stimuli 2. Brands may be unacceptable because of poor quantities or attributes or inappropriate positioning in either advertising or inappropriate positioning in either advertising or product characteristics 3. Brands may be perceived as not having any special benefits are regarded indifferently by the consumer 4. Brands may be overlooked because they have not been clearly positioned or sharply targeted at the consumer market segment understudy and 5. Brands may not be selected because they are perceived by consumers as unable to satisfy perceived needs as fully as the brand that is chosen. In each of these instances, the implication for marketers is that promotional techniques should be designed to impart a more favourable, perhaps more relevant product image to the target consumer. This may also warrant a change in product features or attributes—more or better features.

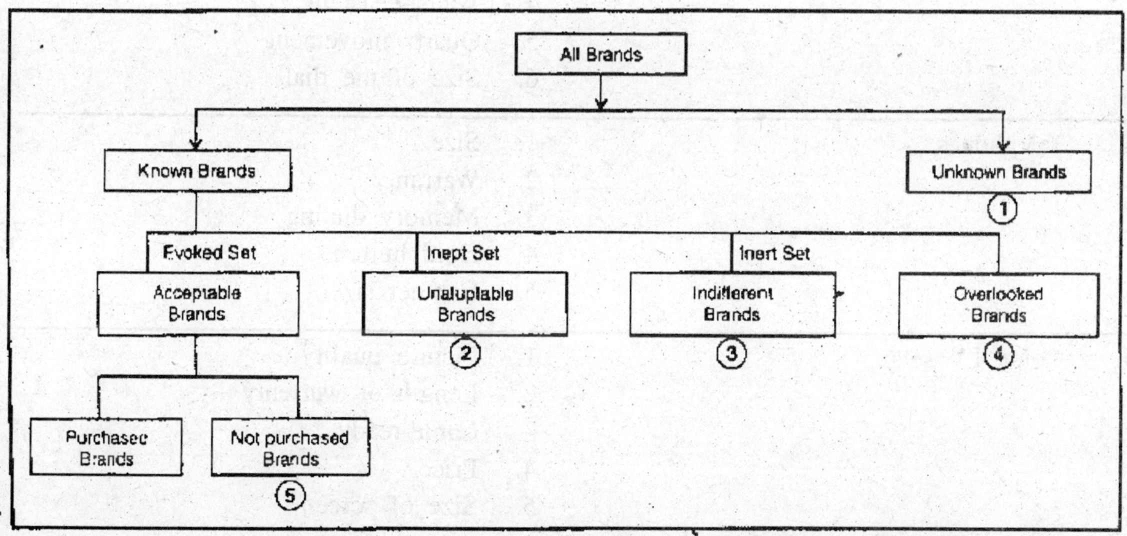

FIG: 12:03 THE EVOKED SET AS A SUBSET OF ALL BRANDS IN PRODUCT CLASS

B. CRITERIA USED FOR EVALUATING BRANDS

The criteria that the consumers use to evaluate the brands constitute their evoked sets usually are expected in terms of important product attributes. Examples of product attributes that consumers have used as a criteria in evaluating nine product categories are given in the following figure. These nine selected products are PCs, shoes, wrist watches, telephones, colour TVs, Frozen dinners, 35 MM cameras, Fountain pens and Men's dress/shirts.

Product	Product Attributes used as criteria
1. Personal Computers:	1. Processing Speed 2. Price 3. Type of display 4. Hard-disk size 5. Amount of memory 6. Laptop or desk-top
2. Shoes/Sneakers:	1. Styling 2. Colour 3. Price 4. Air-sole 5. High–top/Low–top 6. Purpose-tennis, basket ball
3. Wrist-watches:	1. Watch-band 2. Alarm feature 3. Price 4. Water-resistent 5. Quartz movement 6. Size of the dial
4. Telephones:	1. Size 2. Warranty 3. Memory dialing 4. Hold button 5. Cabinet size
5. Color TV sets:	1. Picture quality 2. Length of warrenty 3. Cable-ready 4. Price 5. Size of screen

6.	Frozen Dinners:	1. Taste 2. Type of main course 3. Type of side–dishes 4. Freshness 5. Preparation requirements
7.	35 MM Cameras:	1. Auto–focus 2. Bill-in-flash 3. Automatic film–loading 4. Lens type 5. Size and weight
8.	Fountain Pens:	1. Balance 2. Price 3. Gold nib 4. Smoothness 5. Ink-reserve
9.	Men's Dress/shirts:	1. Size 2. Fabric 3. Colour 4. Collar–style 5. Cuff-style

FIG:12:04 POSSIBLE PRODUCT ATTRIBUTES USED AS PURCHASE CRITERIA FOR NINE PRODUCT CATEGORIES.

After having gone through the experience of comparing evaluating different brands or models a product and finding the one that just feels, looks and forms right or forms right or both. Surprisingly and interestingly, research shows that when consumers discuss such "right products", there is little or no mention of price, brand names are not often on top of mind ; items often reflect personality characteristics or childhood experience and it is very often "Love at first sight."

Let us go back to the case of Mrs. Joan, the retiree who is contemplating of subscribing to a computer on line services. After calling the major on-line service companies for literature on this services, reading the direct mail she received at home and a number of magazine articles, and talking to a few knowledgeable friends, Mrs. Joan's evoked set might consist of "India on Line" Computer Age, 'Computer Today' and Prodigy. On the basis for her recently acquired information, Mrs. Joan has tentatively decided that on the line source she subscribes to should have the following three features : (a) local telephone number access (b) the ability to "surf the net" and the (c) availability of stock market and mutual fund information that she would need to monitor her investments. Evaluating the information supplied by each of the on-line sources, Mrs. Joan might mentally or on paper construct a table that compares the availability of specific features of the three services in her evoked set.

As part of her search process, Mrs. Joan might have also acquired information about other, relevant issues or attributes that could influence her final choice. For instance, she has learnt that both India on line and prodigy offer 24 hours, 7 days-a week technical, support; computer today offers more business and reference features ; and India on line has the most user-friendly interface. Additionally, whereas an India on line and 'Computer today' will allow her to use her modern at its highest speed of 28000 bps Prodigy's maximum user speed was at the time of her decision only 14400 bps.

CONSUMER DECISION RULES

Decision rules are to be non-compensatory when good performance on one evaluative criterion does not offer or offset or compensate for poor performance on another evaluation criterion of the brand. These consumer rules are often referred to as heuristics decision criterion strategies and information processing strategies which are procedures used by consumers to facilitate brand or other consumption related choices. These rules reduce the burden of making complex decisions by providing guidelines or routines that make the process less taxing and cumbersome. These, consumer decision making rules are broadly classed into two major categories namely 1. Compensatory and 2. Non-compensatory.

Compensatory Decision Rules:

In following compensatory decision rules, a consumer evaluates brand options in terms of each relevant attribute and computes a weighted or summate score for each brand. The computed score reflects the brands relative merit as a potential purchase choice. The assumption is that the consumer will select the brand that scores highest among the alternatives evaluated. A unique feature of a compensatory decision rule is that it allows a positive evaluation of a brand on one attribute to balance out a negative evaluation on some other attribute. For instance, a positive assessment of the energy savings made possible by a particular brand or type of high bulb may offset on unexceptable assessment in terms of the bulb's diminished light output.

Non-compensatory Decision Rules:

In contrast, non-compensatory rules do not allow consumers to balance positive evaluations of a brand on one attribute against a negative evaluation on some other attribute. For instance, in case of an energy source–light bulb, the products negative—unacceptable- rating on its light output would not be off-set by a positive evaluation of its energy savings. Instead, this particular light bulb would be disqualified from further consideration. If Mrs Joan's choice of an on line service was based on maximum connection speed, a non-compensatory decision rule would have eliminated 'Prodigy'. For the benefit of readers, three non-compensatory rules are considered at this level.

1. Conjunctive Decision Rule:

The consumer establishes a separate, in following a conjunctive decision rule, minimally acceptable level as a cut–off point for each attribute. If any particular brand falls below the cut of point on any one attribute, the brand is eliminated from further consideration. It is because, the conjunctive rule can result in several acceptable alternates, it becomes essential in such cases for the consumer to apply an additional decision rule to arrive at a final selection. For instance: to accept a satisfactory brand. The conjunctive rule is particularly useful in quickly reducing the number of alternatives to be considered. The consumer can then apply another more refined decision to arrive at a final choice.

2. Disjunctive Decision Rule:

This disjunctive decision rule is the "mirror image" of the conjunctive decision rule. In applying this decision rule, the consumer also establishes a separate, minimally acceptable level as the cut–off point for each attribute which may be higher than the one normally established for a conjunctive rule. In case, if a brand alternative meets or exceeds the cut –off established for any one attribute, it is accepted. Here, again a number of brands might exceed the cut off point, producing a situation in which another decision rule is required. When this occurs, the consumer may accept the first satisfactory brand as the final choice or apply some other decision rule, perhaps more suitable.

3. Lexicographic Decision Rule:

The consumer first ranks the attributes in terms of perceived relevance or importance in following lexicographic decision rule. The consumer then compares the various brand alternatives in terms of the single attribute that is considered most important. If one brand scores sufficiently high on this top- ranked attribute regarding the score on any of the other attributes, it is selected and the process ends. When there are two or more surviving brand alternatives, the process is repeated with the second highest-ranked attributed and so on, until reaching the point of that one of the brands is selected because, it exceeds the others on a particular attribute.

The highest ranked attribute, the one applied first, may reveal something about the individual's basic consumer or shopping orientation, with the lexicographic rule. For instance, a 'buy best' rule might indicate that the consumer is quality'oriented ; a "buy a most-prestigious brand" rule might indicate that the consumer is status oriented to 'buy least expensive' rule might reveal that the consumer is economy minded. The ad on **Fiat Sienna** appeals to consumers who appreciate safety and practicability, yet do not assume that such car also cannot be exciting and fun to drive.

OUTPUT

The **output** portion of the consumer decision–making model concerns two closely associated kinds of post-decision action namely, "purchase behaviour" and "post purchase evaluation". The prime aim of both the activities is to increase the consumer's satisfaction with his or her purchase.

1. Purchase Behaviour:

Generally, consumers make three types of purchases namely, trial–purchases, repeat purchases and long-term commitment purchases. When a consumer purchases a product or a brand for the first time and buys a smaller quantity than usual, this purchase is considered as a trial. Thus, a trial is the exploratory phase of purchase behaviour in which consumer attempts to evaluate a product through direct use. For instance, when consumers purchase a new brand of say, mouth wash about which they may be uncertain, they are likely to purchase –smaller trial quantities than if it were a familiar brand. Consumers can also be encouraged to try a new product or brand through such promotional tactics as free samples, coupons and sale prices or sale prices or both.

When a new brand in established product category-say, tooth–paste, washing detergents, shampoos, soft–drink and the like is found by trial to be more satisfactory or better than other brands, consumers are likely to repeat the purchase. Repeat purchase behaviour in closely related to the concept of brand loyalty which most firms try to encourage, because it contributes to greater stability in the market place. Unlike trial, in which the consumer uses the product on a small–scale and without any commitment, a repeat purchase usually signifies that the product meets with the consumer's approval and that he or she is willing to use it again in larger quantities.

Of course, trial is not always feasible though possible. For instance, with most of the consumer durables such as washing machines, clothes dryers, freezers or microwave ovens—a consumer usually moves directly from evaluation to a long-term commitment through purchase without the opportunity for an actual trial.

Let us go back to Mrs. Joan and her decisions on selecting an on-line service. If she wanted to personally try each of the services in her evoked set "India On-line", Prodigy and "Computer-Today" she could receive a free trial disc and several free hours of usage from each of these on-line vendors. However, she feels that setting up software for each service on her home computer would be just too much of a hassle for her. Instead, she asks around her retirement community and finds individuals who subscribe to each. She discovers that users of on-line services delight in showing others their computer systems and what their on-line services have to offer. Mrs. Joan is especially impressed with how easy to "Surf the net" using version of browser software that India on–line users can download and use for free. So she decides that she will try India on–line, using the disc that she received in the mail that offers her ten free hours of trial usage.

2. Post-Purchase Evaluation:

As the consumers use a product they evaluate its performance in the light of their own expectation particularly during a trial–purchase. There are three possible outcomes of these evaluations namely, (a) Actual performance matches expectations leading to a neutral feeling (b) Performance exceeds expectations, causing what is known as "positive disconformities of expectations" that leads to satisfaction and (c) Performance is below expectations, causing "negative disconformities of expectations" and resulting is dissatisfaction. Consumer's expectations and satisfaction are closely linked for each of these outcomes. That is, consumers tend to judge their experience against their expectations when performing a post-purchase evaluation. The two go hand in hand.

A very significant component of post–purchase evaluation is the reduction of uncertainty or doubt that the consumer might have had about the selection. As a part of their post-purchase analysis, consumers try to treasure themselves that their choice was a wide one. That is, they attempt to reduce 'post-purchase cognative dissonance'. They do this by adoption of the following strategies as noted in earlier chapter else-where. They may rationalise the decision as being wise ; they may seek ads that support their choice and avoid those of competitive brands; they may attempt to persuade friends or neighbours to buy the same brand and thus, confirm their choice or they may turn to other satisfied owners for reassurance. The degree of post-purchase analysis that consumers undertake depends on the importance of the product decision and the experience acquired is using the product. When the product lives up to expectations, they probably will buy it again. When the product's performance is disappointing or does not meet their expectations, however, they will search for more suitable alternatives. Thus, consumer's post-purchase evaluation 'feeds–back' as experience to the consumer's psychological field and serves to influence future related decisions.

Let us conclude the discussion with the case of Mrs. Joan as her post-purchase evaluation of India on line. Mrs. Joan was pleasantly surprised at how easy it was to install 'India on–line, on her computer. She just followed the directions on the disc and then followed the directions on the screen, and within about 10 minutes, she was on line and had her own e-mail address and pass-word. In not more than 30 minutes from the time she first put the India Online install disc in her disc-drive, she had corporated e-mail message to several of her friends and had sent them. In fact, she could not wait to start receiving their replies, now that they would have her e-mail address. She also had even more fun in "surfing the net" than she ever thought would. Mrs. Joan also found the experience of being able to send e-mail to her many friends and relatives anywhere in the world to be a great time and cost saver. Additionally, she was very much impressed with the amount of investment information she could obtain from India online information that would help her both select and keep tract of her stocks, bonds, and mutual funds. Although India online did not offer her specialized, extra cost investment news and services, she found that the investment services it did offer were more than adequate for her needs. For instance, she found that she could enter up to 100 investments into her India online 'port-folio' and each time she opened her 'port folio' screen she would find updated quotes for each one. In a nutshell, Mrs. Joan is thus tool thrilled with her experience as an India. Online user, and she wonders as to how she ever got along without it.

SOME STANDARD CONSUMER BEHAVIOUR MODELS

Broadly, one can classify **buyer behaviour models** into four broad categories namely **Economic, Learning, Psycho–Analytical** and **sociological.** For the benefit of beginners, an attempt is made to present these broad buying behaviour models with the possible ramifications.

1. The economic model: According to this Economic models of buying behaviour, the buyer is a rational animal and his buying decisions are totally governed by the concept of utility. Stated briefly, if he has certain amount of purchasing power, a set of needs to be met and a set of products to choose from, he will allocate the amount over the set of products in a very rational manner or a very clear intention of maximising the utility of the benefits he is going to derive.

This purchasing–power or economic model of consumer behaviour is certainly undimensional. It leads us to the four important predictions about the buyer behaviour. 1. Lower the price of the product, the bigger the quantity that will be bought which is generally called 'price effect'. 2. Higher the purchasing power, the higher the quantity that will be bought which is known as 'income effect'. 3. Lower the price of a substitute product, the lower quantity will be bought of the original product which is known as "substitute effect". 4. Higher the provisional expenditure, the higher the sales which is most commonly known as "communication effect".

Though these predictions are highly acceptable, the assumption of **Rationality** in behaviour is challenged by behavioural scientists for this **Economic Model** fails to give satisfactory explanation of how consumers behave. The model tries to explain as how a consumer ought to behave rather than how he behaves. That is why, it cannot be applied to real life situation as one cannot measure utility in precise terms. The model centres round the 'product' and 'income' and 'consumer' and closely related psychological and sociological factors.

2. Learning model: All theories of buyer behaviour have been basically based on a learning model namely, Stimulation Response or more popularly known as SR model. SR learning theory is very useful to modern marketing and marketers. Learning, as noted earlier, refers to a change in the behaviour which occurs as a result of practice. It is a change in the behaviour that results from previous experience and behaviour in similar situations. What is important, learning is the product of reasoning, thinking, information processing and, of course, perception. Therefore, behaviour is deeply affected by the learning experiences of the buyers.

Of all the psychologists, **Pavlovian** stimulus or learning model of buyer behaviour is widely accepted. He says that buyer behaviour is capable of being manipulated by human drives, stimuli, and responses of the buyer. This model banks on man's ability to learn forget and discriminate. Learning process involves there steps namely, **drive** a strong internal stimulus which impels action. When it is directed towards a drive- reducing object, it becomes a motive. A drive—need—thus motivates a person for action to satisfy the need. Here, the objectives are the stimuli which satisfy our drives. **cures** are weak stimuli. **cures** determine when the buyer will respond. Say, we have cues such as a product, an advertisement relevant to the situation and the existing in our environment. Response is the final stage which is needed to fulfill the drive or a need which was acting as a strong stimulus. Thus, the thirst, can be quenched by a soft-drink supported by an ad. These sequential components of learning link stimulus cue and response finally resulting in a habit. In marketing, it is better known as a learning brand loyalty brand images and store patronage. Repeated reinforcement leads to habit formation and the decision process for an individual, becomes a matter of routine.

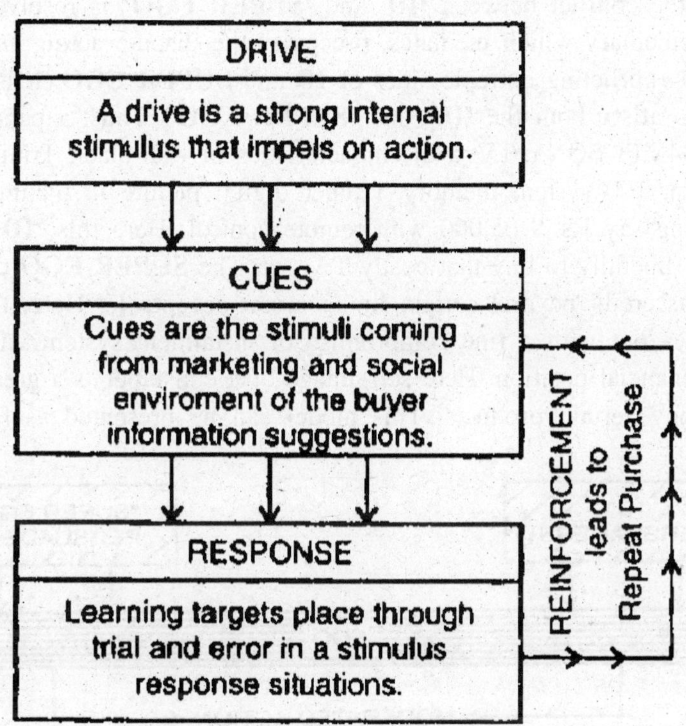

FIG: 12:05 S.R. MODEL—A LEARNING MODEL

It is worth emphasizing here that we learn through trial and error and changes-in our behaviour are brought about by practice as experience. The SR model of **Pavlovian** learning is made clear by the following figure.

3. The psychoanalytical model: The psychoanalytical model draws from Freudian psychology. According to this model, the individual consumer has a complex set of deep seated motives which drive him towards certain buying decisions. The buyer has a private world with all his hidden fears, suppressed desires and totally subjective longings. His buying action can be influenced by appealing to these desires and longings.

According to **Mr. Freud,** human personality has three parts namely,

1. the **'ID'**, the source of all mental energy which drives one to an action.

2. the 'SUPER EGO', the internal representation of what is socially approved—one's conscience.

3. the **'EGO'**, the conscious director of ID' impulses for finding satisfaction in socially acceptable manner.

In other words, ID' represents one's animal or basic impulses, 'instincts' and cravings for immediate and total satifaction. These instints might be even anti-social. The **'SUPER EGO'** or conscience reflects one's idealised or mended behaviour pattern via media between the extremes, that is, the conflict between **'ID'** and **'SUPER EGO'** is resolved by **'EGO'**. The **'EGO'** is the intermediary which mediates, processes the dispute acting as a rational control centre between the conflicting extreme sides of **ID** and **SUPER EGO**. It is **EGO** that directs one's behaviour to satisfy both the **'ID'** and **'SUPER EGO'**. Thus, a person is interested in buying say KV-L34MFI SONY TV with characteristics of—Hi-black Trinitron screen—super drum sound system, 100 system memory, I tuner digital picture in picture, A/V Stereo, LD compatibility costing say Rs. 1,05,000 with remote control. Here this **'ID'** demands the use of consumer credit liberally to buy that costly T.V. set. The **SUPER EGO** dissuades him from heavy borrowing as credit beyond certain limits is not acceptable. Here, the **EGO** acts as a mediator and comes out with a fine compromise of instalment system without away strains and drain on his financial position. Here self–image of a consumer is a great motivating force inducing him to buy certain products. This model can be presented as follows:

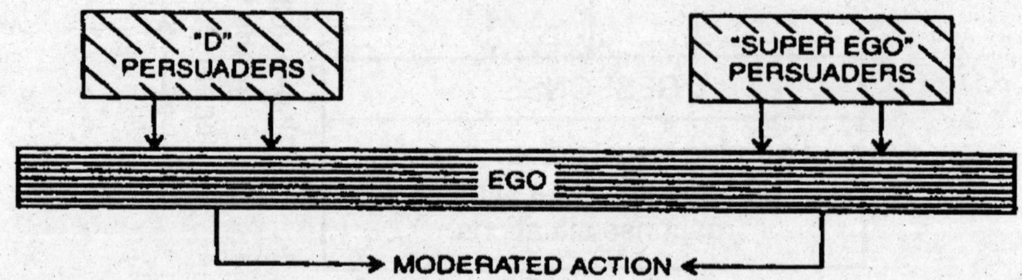

FIG: 12:06 PSYCHOANALYTICAL MODEL

4.Sociological model: According to the sociological model, the individual buyer behaviour is influenced by society—by intimate groups as well as social classes. That is, his buying decisions are not totally determined by the concept of utility. That is his buying decisions are governed by social compulsions. As a part of sociological model—two important variations can be considered namely, one that of **Nicosia** and another **Howard Sheth.**

The marketing scholars have tried to build buyers-behaviour models purely from stand point view of marketing man. Here, **Francisco Nicosia** of 1966 and **Howard Sheth** model of 1969 are of this category. These models are systems models where human being is analysed as a system with stimuli as an INPUT and behaviour as an **OUTPUT**. Before we touch up these two outstanding systems, models, let us understand as at what is systems model.

The systems approach one can think of three elements namely 'inputs'—'processing' and 'output'. One more element one can add that is "feed back loop". Let this be presented as under :

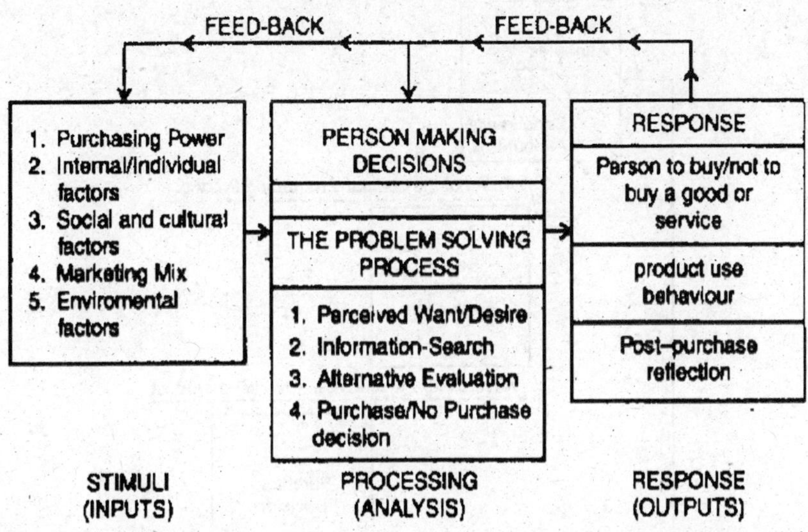

FIG: 12:07 BUYER BEHAVIOUR MODEL UNDER SYSTEMS APPROACH

Here **'inputs'** cover purchasing power, marketing mix, intra- personal influences inter-personal influences and other environment factors. Intra-personal influences are motivation—perception- learning attitudes and the buyer's personality. Inter-personality influences are family social class reference groups and culture. Other environmental factors are economic conditions, pending legislation, fashion trends and technological advancements.

The **'outputs'** cover the buyer's attitudes, opinions, feelings, and the preferences as influenced by buying process and buyer's actions such as patronage, brand and store loyalty, positive or negative influences upon other potential buyers. The buying responses are choice of product brand dealer quantities and the like.

A. FRANCISCO NICOSIA MODEL

As well known consumer motivation and behaviour expert **Mr. Franscioso** presented his buyer model in 1966 which attempts to establish linkages between the marketing firm and its consumer. The essence is how the activities of the firm influence the consumer and result in his decision to buy. According to his model, the messages from the firm first influence the predisposition of the consumer towards the product. He developed a certain attitude towards the product depending on the situation. It leads to a search for the product or on evaluation of the product. In case, these steps have a positive impact on him, it may result in decision to buy. This is the sum and substance of the "activity explanation" in the **Franscioso** Nicosia's model. His model lumps these activities into four basic fields.

Field ONE has two sub–fields namely, firm's attributes and the consumer attributes. An ad message from the firm reaches consumer's attributes. Depending on the way the message.

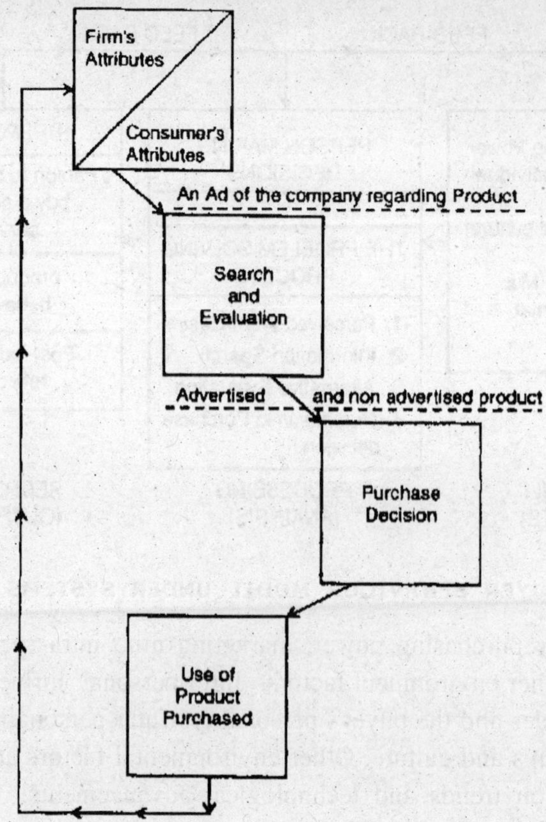

FIG: 12:08 F. NICOSIA MODEL

is received by the consumer, a certain attribute may develop and this becomes the input for the Field **two**, Field **two** is the area of search and evaluation of the advertised product and other alternatives. If this process results in a motivation to buy, it becomes the input for field three. Field **three** consists of the act of purchase. The Field **four** consists of use of the purchased item. There is an output from Field four — feed back of sales results to the firm. His model can be presented in Fig. 12:08.

B. THE HOWARD SHETH MODEL

John Howard and **Jagdish Sheth** presented their buyer model in 1969. It is an integrated model. It assumes problem solving approach in buying and adopts input-output or systems approach in buying. **Howard** introduced learning process in buying. Satisfaction leads to brand loyalty. Discontentment creates brand switching buy the buyers. In other words, the logic of this model is that there are inputs in the form of stimuli. There are outputs beginning with attention to a given stimulus and ending with purchase. In between these inputs and the outputs, there are variables affecting perception and learning. These variables are 'hypothetical' as they cannot be directly measured at the time occurrence.

This model can be presented in both simple and more sophisticated forms as under :

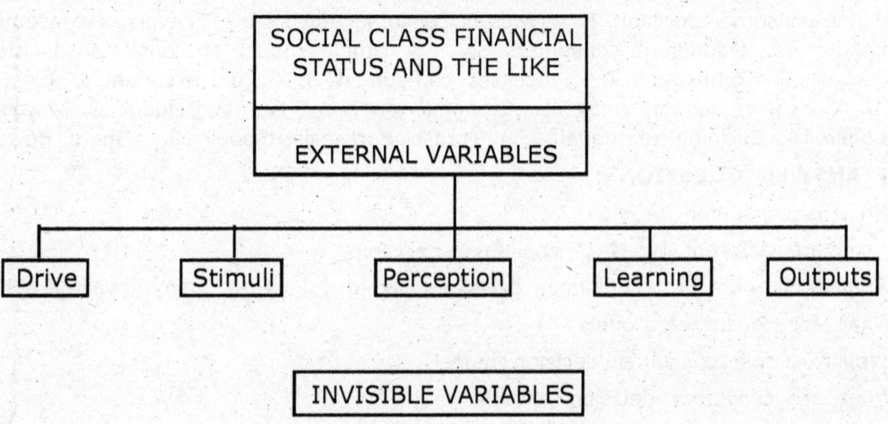

FIG: 12:09SIMPLE MODEL

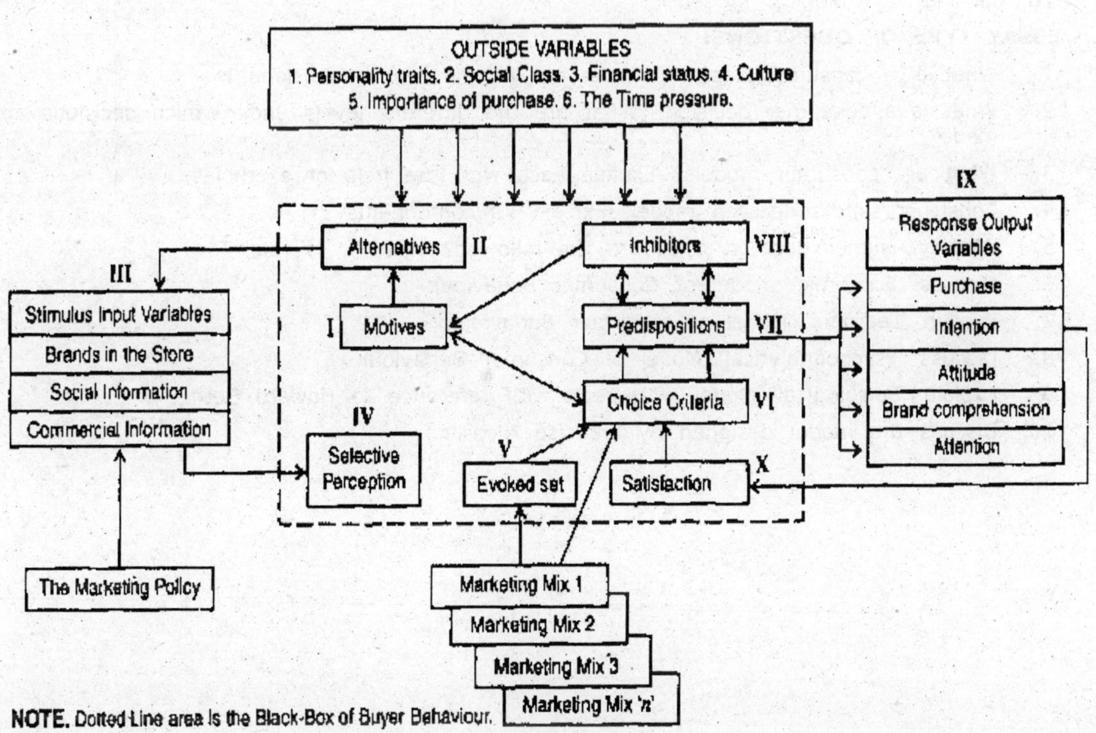

FIG: 12:10 HOWARD AND SHETH BUYER BEHAVIOUR MODEL

CHAPTER BASED QUESTIONS

A. WRITE SHORT NOTES ON :

1. A consumer decision 2. Types of consumer decisions 3. Levels of consumer decision making 4. Models of consumers 5. A simple model of consumer decision making 6. Input component 7. Processes component 8. Output component 9. Evoked set 10. Consumer decision rules 11. Economic model 12. Learning Model 13. Psychoanalytical model 14. Sociological model 15. Fransciso Nicosia model 16. Sheth Howard Model.

B. SHORT ANSWER QUESTIONS:

1. What is consumer decision ?
2. What are different types of consumer decisions ?
3. What is the level of consumer decision-making ? How many levels are there ?
4. What are consumer models ?
5. Explain simple consumer decision model.
6. What are consumer decision rules ?
7. Explain 'Economic' Model.
8. Discuss Learning Model.
9. Explain psychoanalytical model.
10. Elucidate sociological model.

C. ESSAY TYPE OF QUESTIONS:

1. What is a consumer decision? What are its different dimensions ?
2. What is a consumer decision? What are the different levels under which decisions are made?
3. What are consumer models? Explain each with the help of examples.
4. Construct simple decision model. Explain its components.
5. What consumer decision rules are the followed? What is is role ?
6. Discuss 'Economic' model of Consumer Behaviour.
7. Explain 'Learning' Model of consumer Behaviour.
8. Discuss 'Psychoanalytical' Model of Consumer Behaviour.
9. Explain 'Sociological' Model in general with reference to Howard Seth.
10. Discuss the model designed by Franciso Nicosia.

*****

CHAPTER 13

INNOVATIONS AND DIFFUSION PROCESS

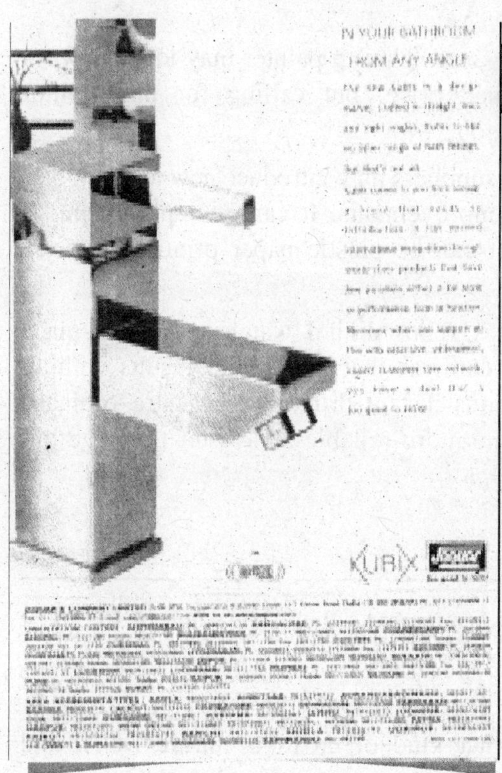

BACKDROP

Today, innovation is about much more than new products. It is about reinventing business processes and building entirely new markets that meet untapped customer needs. Most important as the internet and globalisation widen the door of new ideas, its about selling and excuting the light ideas and bringing them to market in record time. In the 1990s, innovation was about technology and control of quality and cost. Today, it is about taking corporate organisations built for efficiency and rewiring them for creativity and growth. Innovation is fundamental to both growth and change, and most marketing firms and industrial units will be unable to survive in the era of massive marketing competition if they fail to innovate. The business conditions are so fast changing making them to innovate or perish. The question is not only making available a innovative product or service, but to see that consumers accept these at faster rate to get the best both for the producer, marketer and the consumer. Hence, this chapter plans to discuss about innovation and its diffusion in all its format as necessary from the angle of consumer behaviour. The Chapter ends with Chapter Based Questions.

WHAT IS INNOVATION?

Innovation is an idea, practice or product perceived to be new by an individual or a group. Innovation is the adoption of a new idea, product or process which is prospectively useful. Innovation is a product, service, attribute or idea that consumers within a market segment perceive as new and that has an effect on existing consumption patterns. Creativity and imagination applied in a business context is innovation. Though innovation means change, every change is not innovation. Innovation is present only when the innovator is required to engage in productive, problem solving. It is the design and the development of something new as yet unknown and not in existence. Which will establish a new economic configuration out of its and existing elements innovations induce demand and seek an opportunity to same the present and potential market at a profit. It is a means to translate ideas into products and products into profits through undivided consumer satisfaction.

Innovation as a, process has four basic directions namely:

1) Improvement and development of existing forms: Thus a manufacturer of earth moving equipment can introduce features *geared* to make operate more quickly or efficiently an ad agency may offer new direct services.

2) Improvement and development of existing processes: Thus, a painter may identify ways of printing a specific items improve finish or reduce price, but calling for only minor modifications, if any, is a clients layout or presentation.

3) Introduction of novel production process: Thus, suppliers may introduce new technologies to the production resulting in costs cuts and quality improvements. It can be replacement of spinning Jenny by Silicon chip in spinning process. It can be plastic paper printing in place of individually painted bulletins.

4) The introduction of new products and services: Thus digital–watches came in place of usual winding watches. So is the case with digital–cameras that give better results without films. The marketers of today are able to give fast–foods instant drinks to the extra–ordinary time. Conscious customer, courier services have become more reliable and faster than age old traditional postal channels.

SOME INNOVATION MYTHS

There are many mistaken view about how innovation works there are some of the most pervasive ones.

Firstly, people think innovation is to do about creating hot new products. It is a myth. The really is not always so. Some companies have that kind of creative culture DNA, and others do not. Innovation audits after reveal hidden strengths in areas such as brand recognition and sound delivery.

Secondly, innovation is about crazy creativity. It is a myth. The reality is not at all. Creativity is an essential part of the innovation process, but matrics, discipline, focus, and leadership are just as crucial.

Thirdly, innovation is expensive and takes time. It is a myth. The reality is that fasting to innovative is far more expensive. This is especially true at a time. When technology and information are being commoditised and globalised.

Fourthly, you need hundreds of ideas, because the failure rate for innovation is high. It is a myth. The reality is may be so. Companies that have innovation architecture in place can handle many concepts at one time. Most do not and would do better to focus on fewer, bolder ideas that shift product strategies reframe brands, and meet consumer needs.

Fifthly, metrics guarantee higher innovation success rates is a myth. The reality is you can measure your way to mediocrity. The right financial bundle rates and other metrics are essential but they are only part of a larger innovation process. Design strategies such as early prototyping and customer observation are as important as measurement for innovation productivity.

Innovations bring about changes in consumer's consumption pattern. Some innovation influence various dimensions of acquiring the products namely- how? When? Where? Why?. Thus, we are exposed of 'internet shopping' which has changed the way one buys certain types of products. The way are use the products and services is also changed by innovations. Use of micro–wave, e-mail or voice-mail have changed our ways of coding food and communicating. Disposable items, diapers, cameras, syringes, have revolutioned the way we live. Innovations have also influenced consumers' disposition behaviour–say ideas about recycling have brought about new product innovations in recyclable packaging materials and reusable containers. Music needed on computers hard–disk to disposed of differently from albums, CDs and cassettes. What is more important is the list of innovation goes on increasing with the passage of time. The key is the thirst and hunger for more effective and better quality products and services. As long as this thirst and hunger is there, there is no escape from innovations. Further. rapid strides in the fields of science and technology are responsible for designing and offering ultra radically new and highly complex products that promise ease, comfort and convince to consumers.

'Knowledge economy' as we know it is being eclipsed by something new call it the "creative economy" Increasingly new core competency is creativity the right brain stuff that smart companies are now harnessing to generate top line growth. Now, it is not just about maths and science anymore. It is about creativity, imagination and above all innovation what is unfolding is the commoditization of knowledge. In case of creative companies of the world, the clear facts of innovation success are As reported by Business week in "The world's most innovative companies" dated April 24, 2006.

1. Overall revenue growth: 56 per cent
2. Sales from new products and services 50 per cent
3. Customer satisfaction 47 per cent
4. Return on investment in innovation 30 per cent
5. Number of new products and services 30 per cent

6. New product success ratio 20 per cent
7. Higher prices: 11 per cent

What is more important to note is the innovators or users of innovations are tended by higher prism for their products and services as the companies more up is experience chain. Thus, per cup coffee, seeds are priced at 1 to 2, ground coffee, in the range of 5 ¢ to 25 ¢ per cup, service coffee per cup- ranging from 75 ¢ to 150 ¢ and experience coffee, ranges from $ 2 to $ 5. As reported by Business Week in an article "Get creative" dated August, 8–15 2005.

TYPES OF INNOVATIONS

Experts in the field, classify the innovations based on the degree of novelty namely, **continuous, dynamically continuous and discontinuous.** The innovation continued can be configured as under:

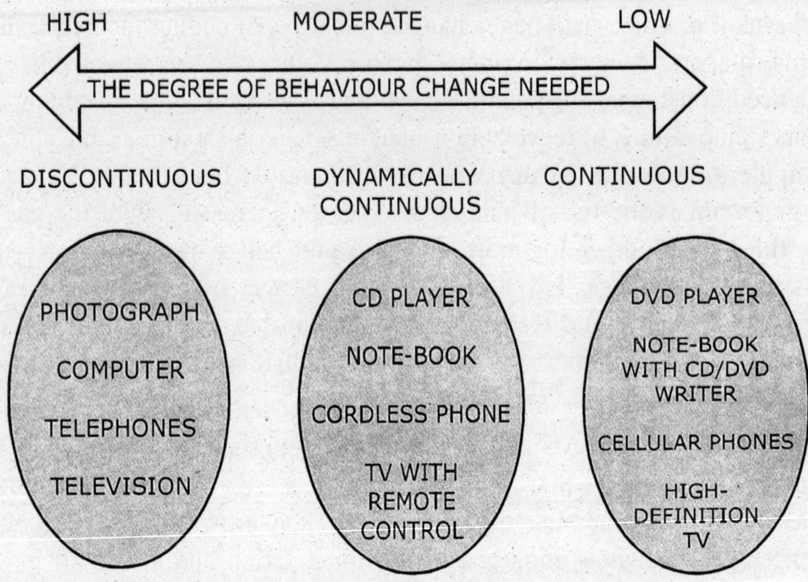

FIG: 13:01 – TYPES OF INNOVATIONS

Innovations vary in how much behavioural change they require on the part of consumer. Discontinuous innovation readically new products when they are first introduced and require good deal of change in consumption patterns. Continuous innovations are in most cases, extensions of existing products and require very little change.

A **continuous innovations** is one which has a limited influence on the consumption behaviour of consumers. Consumer use a product representing continuous innovation in the same way as they use those products that came before to their notice. Product alteration is on a continuous basis. Adoption of such products requires minor changes in behaviour that are important to consumers. Most of the new products that are introduced in the market represent continuous innovations they are newer models whether they are fans, computers, tooth brushes or automobiles.

A **dynamically continuous innovation** is one that affects consumers' consumption behaviour in a deeper or pronounced way. Adoption requires a moderate change in important behaviour or major change in an area of behaviour that is of low or moderate significance to the individual. The examples can be those of internet shopping, digital–cameras, notebook computers, electric–cars, cordless–phones. Perhaps the best example can be real, juke–box as it requires changes in the way one acquires, uses and disposes of music and may use other technologies like CD and DVD writers.

A **discontinuous innovation** represents a product which is entirely new to the consumer where he hardly knows about it. The best example can be that of sky car is an automobile which flies, floats and runs on road. It has a speed of 480 kms per hour and takes off and lands vertically like a helicopter. Once up on time, heart transplant, lever transplant MRI scanning, TV, Components, inkjet and laser printers – and the host of such cases were not known to us. That is, they were the cases of discontinuous innovation. Today, they are dynamically continuous innovations with the lapse of time, they will be continuous too.

THE DIFFUSION PROCESS

Diffusion is a process by which the acceptance of an innovation or new product, a new idea, a new service, is spread by communication to the members of a special system over a period of time. The diffusion process is the manner in which innovations spread over time to other consumers through communication across a market. The research in the field traces the penetration and acceptance of an innovation across its life–cycle. A new product's phases of life cycle begin from introduction and end in decline in a standard case. These are associated categories of adoptors by the time of adoption, with this progession.

The diffusion process identifies innovation in the introductory phase of life cycle; there we find **early adopters** during the growth period, the **early majority** and the **late** majority adopts the product in its maturity period and **laggards** or late adopters are the last to adopt the product. These life–cycle phases are significant because they are limited to different marketing strategies during the product life–cycle. During the introduction phase, the marketers objectives are related to establishing distribution, building brand awareness among members of the target market and encouraging trial to begin the diffusion process. As the product gains acceptance, the marketer defines its early adopters. It now tends to strengthen its foodhold in the market by shifting from the objective of creating brand awareness to one of broadening product appeals and increasing product availability by increasing its distribution.

The profile of adopter groups and the product life–cycle phases and profile of adopter groups can be configured as under:

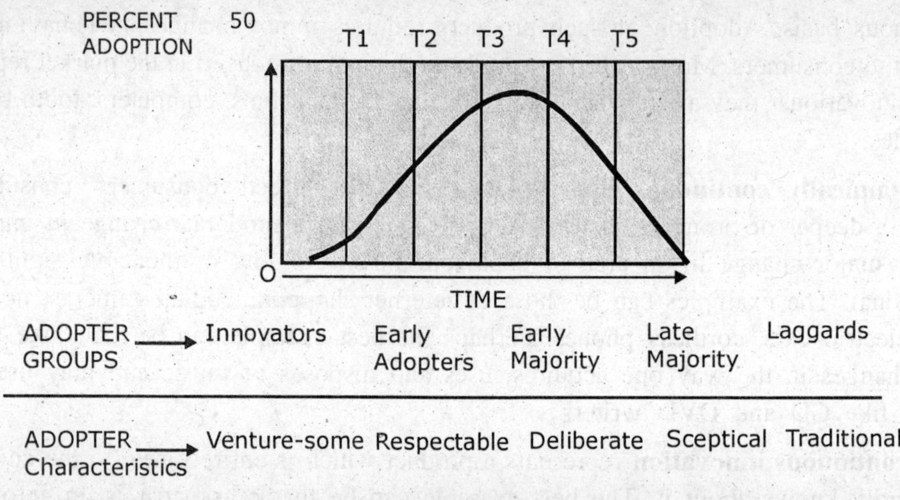

FIG : 13:02 PROFILE OF ADOPTER GROUPS

P < C STATES	INTRODUCTION	GROWTH	MATURITY		DECLINE
Adopter groups	Innovators	Early Adopters	Early Majority	Late Majority	Laggards
Percentages	2.50%	13.50%	34.00%	34.00%	16.00%
Characteristics	Venturesome	Respectable	Deliberate	Sceptical	Traditional

FIG: 13:03 P < C PHASES AND PROFILE OF ADOPTER GROUPS

As the brand matures, competition intensity increases gradually and sales begin to level off. The marketer starts emphasising price appeals, starts sales promotions and thinks of modifying the product to gain competitive advantage. Majority of the adopters enter the market at this stages mainly because of the influences of early adoptions. The majority that has already gone through the process of product adoption does not rely much on media any more for information. When the brand is viewed to have entered in its decline phase, lower prices become more relevant and the marketer considers revitialising the brand, or adopt the strategy of harvesting or divesting. It is during the decaying maturity and the decline phase of product life cycle that laggards enter the market.

ADOPTION DECISION

Innovation adoption calls for an individual or a group of consumers to decide on buying a new product. The process of diffusion commences when early adopters influence then reference group members and other acquaintances. This adoption is the first step in the diffusion process.

The adoption of an innovation is likely to be reasonably involving decision for most of those. Who are among the first to buy the product and can be represented by a hierarchy of effects model. That is, "adoption process" is fundamentally a term used to describe extended decision making by consumers when a new products, service or the idea is involved high involvement in product or purchase situation is likely for discontinuous innovations. For instance, a decision to buy a DVD–writer or having a laser eye–surgery is a case of high involvement decision. Most of continuous innovations trigger limited decision making. In case of low–cost, low–risk innovations, involvement level will be much lower involving. The adoption process and the extended decision–making will configure as under:

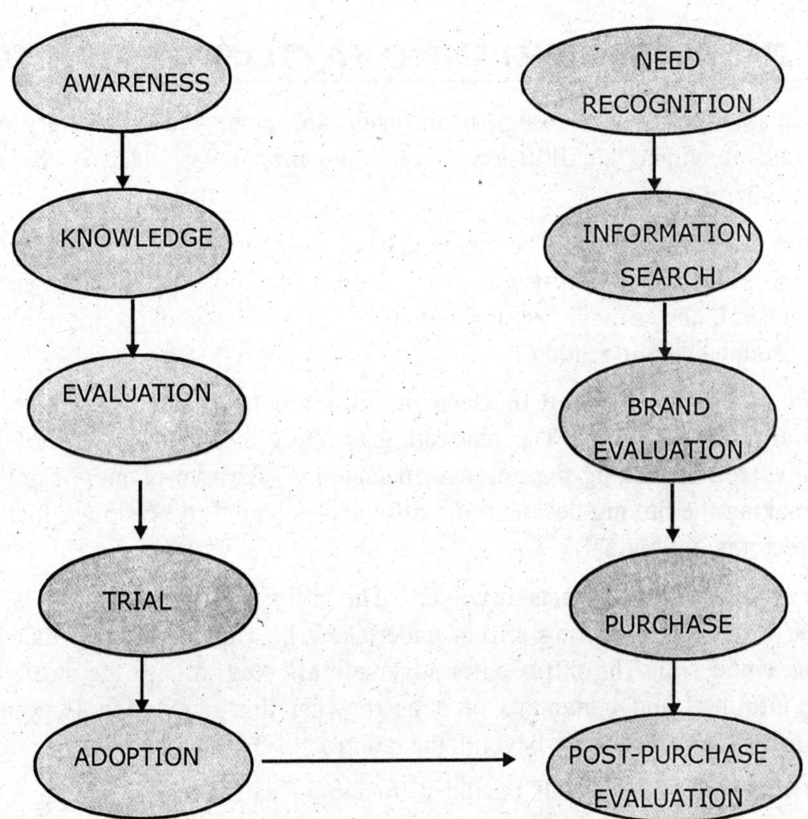

FIG: 13:04 ADOPTION PROCESS AND EXTENDED DECISION–MAKING

It is evident from the above process that consumer becomes aware and recognises the need for the product, the consumer gets involved in information search. The next step is brand evaluation. Next is trial of the product before making a purchase decision and finally the consumer decides whether to adopt the product. As a result of using the products, the consumer makes post purchase evaluation.

In this process of adoption, product trial is more significant than in most other decision because the perceived risks are likely to be much higher with the use of a new product. Certain types of products or services are difficult, or even impossible to try on sufficiently acceptable bases such as electric cars or laser eye surgery. Post–purchase evaluation is having important implications because of the expenses involved or the complexity of many products and the fast changes in technology. For instance, result of unfavourable post purchase evaluation the market for personal computers may be shrinking instead of increasing rapidly. As a result of favourable evaluation, many of the personal computer adopters have either upgraded their old sets or discarded them by purchasing the latest say fourth–generation computers.

The result of consumers' decision–process can be adoption or recycle of the product. Many consumers' evaluation of cellular–phones has led them to reject this really wonderful product because they consider the convenience of keeping a cellular phone is not worth the expenses involved.

FACTORS INFLUENCING INNOVATION–DIFFUSION

The product's adoption and subsequent diffusion are dependent on its very nature. That is, the tune or rate at which the diffusion of an innovation takes place is the result of at least ten factors. There are:

1) The type of target group: The target market for new products is a very significant factor that influences the rate of diffusion. Some groups are more elevated to accept change than others. In general, the affluent, young and the highly educated groups tend to try and accept the new products more readily.

2) Number of people involved in Decision–making: It means ask whether a decision is made by an individual or group. The innovation is likely to spread more rapidly, if fewer individuals are involved in making the purchase decision. When two or more family members are involved in making the buying decision, the diffusion is bound to be slower than innovation the basically affect the individual.

3) Extent of marketing efforts involved: The diffusion of innovation is very much influenced by the extent of marketing efforts undertakes. It is quite possible that innovations may be the most wonderful. The diffusion is adversely affected in case the sufficient number of members are informed and convinced of it as to what this innovation is going do. That is, the rate of diffusion is not well beyond the control of marketer.

4) Need–fulfilment: Faster will be the diffusion in case more involving and obvious is the need that innovation satisfies. The best example is that of anti–dandruff shampoo. The rate of diffusion of anti–dandruff shampoos has been faster as they gained faster trial among those who were uncomfortable with dandruff. The other examples of this kind are 'Viagra' the male impotency drug and 'Rogaine' cure for hair loss or baldness.

5) Compatibility: Compatibility is the degree to which the innovation is consistent with an individual's and groups–needs, attitudes–beliefs and past experiences. As a result, the more its consistency the faster is its diffusion. Microwave oven was introduced in 1970s in India, the diffusion was slow because it was not compatible with Indian family values. Even interest bankers or shopping is not consistent with the established habits of majority of Indians. The result is show diffusion of innovation.

6. Relative Advantage: The rate of diffusion is higher and rapid in case consumers perceive any innovation as something better in matching their relevant need as compared to the existing ones. The consumers take into account both the cost and the performance while considering the relative product advantage. An innovation must have either the performance or the cost advantage over the existing products or services, to be successful. Thus, in case of newer computer processors, they have performance advantage over the earlier one because of which diffusion has been faster.

7. Complexity: Much depends on product complexity or simplicity. In case the innovation is much different to understand and difficult to use, its diffusion is bound to be slower. On the contrary, product simplicity and ease of use that speeden the diffusion process. Majority of the computer makers have overcome the initial complexity that they had by telling their customers that their computers are user friendly.

8. Observability: It is the ease with which consumers can observe positive effects of adopting innovation. The diffusion is more rapid if the positive effects are easily observable. Thus, take the case of cellular–phones, colour–printers, TV–sets, automobiles where the positive effects are easily visible.

9. Triability: Trial of a product before adoption has its own impact. Trial is not a problem in case of low cost low-risk items like daily used consumer non–durable and some durable items. There are some item which are demonstrated but tried in lesser scale. Thus, no marketer sells the items of computers celluarphone, fax–machines, printers, cameras and so on trial basis. The only way is buy and try and not try and buy. This reduces the speed of diffusion.

10. Perceived risk: There are certain risks associated with buying a new product. These we call as perceived risks. The risk of adopting an innovation can be financial, physical, performance or social as we have noted earlier whatever the nature of perceived risks. What is important to note is, the more to risk, associated with trying a now product, slower will be the diffusion process.

DIFFUSION PROCESS AND TIME–FACTOR

Time is a significant element of diffusion theory and relates to the 'time–adoption' of an innovation by consumer considering whether consumers are earlier or later adopters and the rate of diffusion. That is, the speed and extent with which individuals and groups adopt the innovation. Let us concentrate as these two aspects namely, **Time of Adoption** and **Rate of Diffusion.**

TIME OF ADOPTION

After examining more than 500 studies on diffusion, **Mr. Everelt-H-Rogers** concluded that there are fine categories of adoption based on time of adoption. These are : Innovators early adopters- The early majority- The late majority and laggards. The following box gives two versions of innovation diffusion.

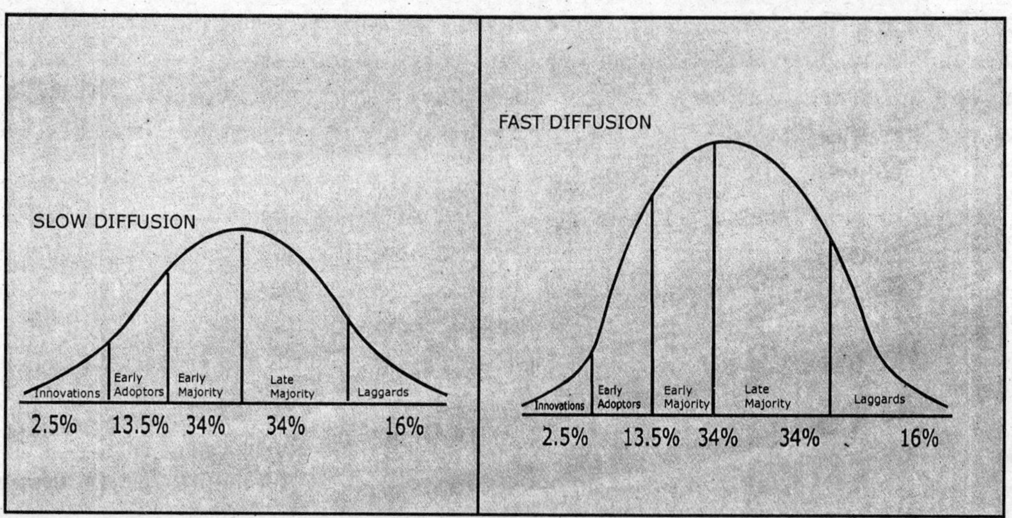

FIG: 13:05 ADOPTER GROUPS AND THE SPEED OF DIFFUSION

1. Innovators: On an average the first 2.5 per cent of all those consumers who adopt the new product and the technology enthusiasts, is the share of innovators. They have an eagerness bordering almost on obsession to try new products sources and ideas. Innovators are generally younger, better–educated having higher incomes, are mostly cosmopolitans and active outside their community than those of non–innovators. They are less reliant on group norms, use other innovators as reference–group other than local peers. They are more self confident using extensively the commercial media, sales–people and professional sources in learning about new products and ideas.

2. Early Adopters: The early adopters represent on average not 13.5 per cent of those who adopt new product and they are likely to be the opinio–leaders in local reference–groups. These are the individuals who admire a technologically new product not so much for its features as for its abilities to create a revolutionary break through in the way things are normally attained. Though, thcy arc not among the earliest individuals to adopt the product, yet they adopt the product in the early stage of its life–cycle. They are successful, we educated and younger, than their press. They are likely to be more reliant on group norms and values than innovations and are more oriented to the local community than having cosmopolitan look. They are the persons who take calculated risk on an innovation but are more concerned with the possible failure. These early adopter are willing to use commercial, professional and interpersonal information sources. As they are opinion–leader, they are likely to transmit word of mouth influence and, because of this, they are the most important group in determining the success or failure of the new product.

3) The early majority: The early majority represent 34 per cent and are deliberate and cautious with regard to innovations. They are looking for innovations that offer incremental predictable improvements of an existing technology. They adopt innovations earlier them most of then social group but only after the innovation is viewed successful with others. They collect more and more information and evaluate more brands than is the case with early adopters and hence, the process of adoption takes longer. They are price–sensitive and like to see that competitors enter the market. They are socially–active, older, less well–educated and less socially mobile than early adopters and are very rarely leaders. They depend heavily on interpersonal sources of information and are very significant link is the process of diffusing new ideas because of their position between earlier and later adopters.

4) The late majority: The late majority are some what sceptical about innovations who represent 34 per cent of adopters. They are conservative, wary of progress, rely on tradition and generally adopt innovations in response to group–norms and social pressure or due to decreased availability of the persons product rather than positive evaluation of the innovation. They are older, with less than average, income and education and have less social status and mobility than those who adopt earlier. They are the individuals who place high–value on bundled products that include everything they need to connect to the device. It may be a sound–gadget or an internet.

5) Laggards: Like innovators, the laggards are the least likely to rely on the group's norms. They represent 16 per cent of the adopters of the total. Laggards are tradition bound, dogmatic and make decisions in terms of the past values. By the time, they adopt an innovations it is old and has been superseded by something new. They are suspicious of new products and alienated from a technologically progressing society and adopt innovations with great reluctance. There are people yet they have not bought personal computers where the sets barely cost Rs. 5,000 to Rs. 10,000. These can be called as laggards because, they must enjoy this as a necessity than a luxury.

RATE OF DIFFUSION

The rate of diffusion is to do with the cumulative level of adoption of an innovation over the time among groups. For instance, it has taken a little over three decades since TV war introduced in India and not even 75 per cent of households own TV set. Even adoptions of some other low cost innovation are much slower in India especially in rural markets. Most of the households in rural India do not own telephone connection. Even in cities, so far, the use of automatic teller machines (ATM) is growing very slowly. In developed and developing countries the speed of adoption of innovations is fast increasing. It is a research finding that consumers are adopting innovations more readily than used to. The reasons that explain why innovations are adopted more quickly are: (1) With the increase in household income, new products are likely to be more affordable (2) Rapid technological advances require quicker adoption cycles (3) As technology is becoming more standardised, it reduces consumer's risk perception associated with the adoption of a new product (4) Information regarding innovations is communicated rapidly and is accessible to the consumer costly.

ROLE OF COMMUNICATION IN DIFFUSION–PROCESS

By very definition, the communication in the key element that influences the diffusion across the markets. Consumers depend on market controlled mass–media namely, advertising for information about new products introductions. The ad campaigns are designed to create an awareness among potential consumers and communicate relevant information about features and benefits. The researchers have come to the conclusions that early–adopters are more dependent on mass media for information. In due course, they show greater reliance in friends and relatives to help them in evaluating new products and word of mouth influence increases in importance as early–adopters progress from awareness to knowledge, evaluation, trial and adoption stages. This is very well configured as under:

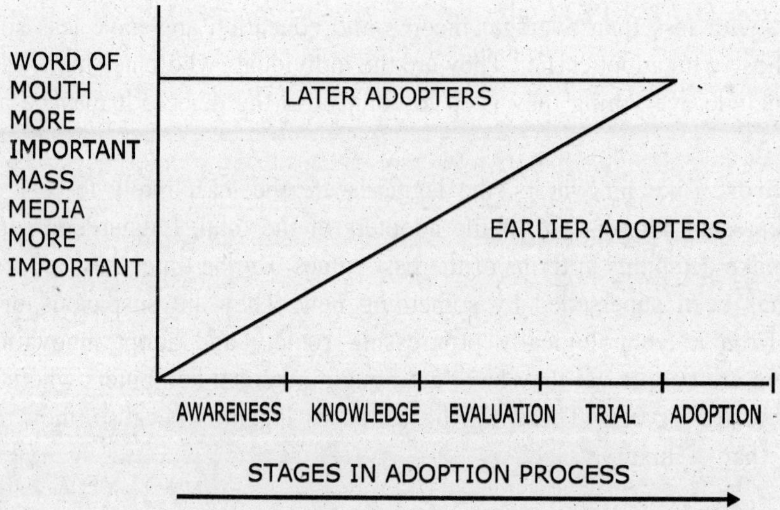

FIG: 13:06 RELATIVE IMPORTANCE OF MASS–MEDIA AND WORD OF MOUTH IN THE ADOPTION OF INNOVATION

In case of later–adopters, word of mouth is more likely to influence throughout the adoption process. They learn about innovations from relative and friends and neighbours rather than the mass media. In order to create awareness among early adopters, the role of mass media, advertising providing informational content is more important in the process of diffusion. For later–adopters, advertisers can try to encourage favourable word of mouth by using influential spokespersons.

For diffusion to occur across groups, positive word of mouth must first start within the groups. However, favourable word of mouth within just groups is not enough for diffusion to occur but needs the spread of information across different groups. This does happen because consumers spread information by interacting with individuals outside their own groups.

CHAPTER BASED QUESTIONS

A. WRITE SHORT NOTES ON:

1. Innovation 2) Diffusion 3) Continuous innovation 4) Dynamically continuous innovations 5) Discontinuous innovations 6) Adopters categories 7) Innovators 8) Early-adopters 9) The early-majority 10) The late majority 11) Laggards 12) Word of mouth influence.

B. SHORT ANSWER QUESTIONS:

1) What is an innovation ?
2) What is diffusion-process?
3) What is continuous innovation?
4) What is dynamically continuous innovation?
5) What is discontinuous innovations?
6) What are the factors of diffusion of innovations?
7) What is perceived risk?
8) Who is innovator? Who is early adopter?
9) Who is early majority? Late-majority and laggard?
10) Word of mouth influence- What is it?

C. ESSAY TYPE OF QUESTIONS:

1) What is innovation? How experts have classify the innovations?
2) What is diffusion-process? What factors affect diffusion-process?
3) What is diffusion-process and what is its relation with time factor?
4) What are adopter categories? How experts have classified these adopters ? What are their features?
5) What is "time of adoption" and "rate diffusion" ? Explain the role of communication in diffusion process.

*****

CHAPTER

14

CONSUMER SATISFACTION AND CONSUMER BEHAVIOUR

BACKDROP

Customer satisfaction is the focal point of most of the marketing strategies in the present context. In a buyers' market where the customer is the Rex, business begins and ends with the customer. The customer is the kingpin of any economy that permits survival, growth, profitability, liquidity and the image of an organisation. When so much emphasis is laid on customer, the foundation of this Rex is his or her satisfaction and delight. Every marketer and hence, manufacturer has to meet the expected level of satisfaction. It is not only maintaining but expanding or enhancing it at any rate. If the king is happy, every subject will be happy irrespective of his or her role. Hence, this chapter is designed to learn the details of various dimensions of consumer satisfaction which is a positive mark of consumer behaviour. The Chapter ends with Chapter Based Questions.

WHAT IS SATISFACTION?

Satisfaction is the evaluation of an emotion. Satisfaction is the outcome of the interaction between the expectations of a consumer and the performance of the product or service. Satisfaction is the reward a consumer gets for the sacrifice he or she makes. Customer satisfaction is an emotional and sentimental issue. It is a feeling of satisfaction over the job well done in the past. Consumer satisfaction is the customer's perception that a vendor has met his expectations fully, efficiently and promptly. That is, customer satisfaction is the summary of the opinions of the customer about the vendor. Right from the pre-marketing to post–installation service and complaint management, the customer comes in contact with different. departments and functions of vendor–organisation and the opinion he forms about the supplier is the sum total of his experience and perception he has while coming in contact with different sections of vendor unit.

A customer finally chooses a vendor who provides him more value for his money and with whose product or service is quite satiated or happy. Higher the level of satisfaction, better is the emotional and sentimental attachment to that brand and more intense is the relationship between the customer and the vendor. It is not only a matter of product or service. What is more is that consumer expects good deal of guidance in solving his problems that crop up at any time. In case his problem or problems are solved by the vendor, it is the vendor who creates a committed and loyal customer who is instrumental in creating new customers.

FACTORS INFLUENCING CUSTOMER SATISFACTION

The factors that influence the consumer satisfaction can be lumped into broad categories. There are human and product factors. Instead, we can also call them as vendor behaviour and vendors product and service performance.

For the sake of convence these two sets of factors can be configured very well.

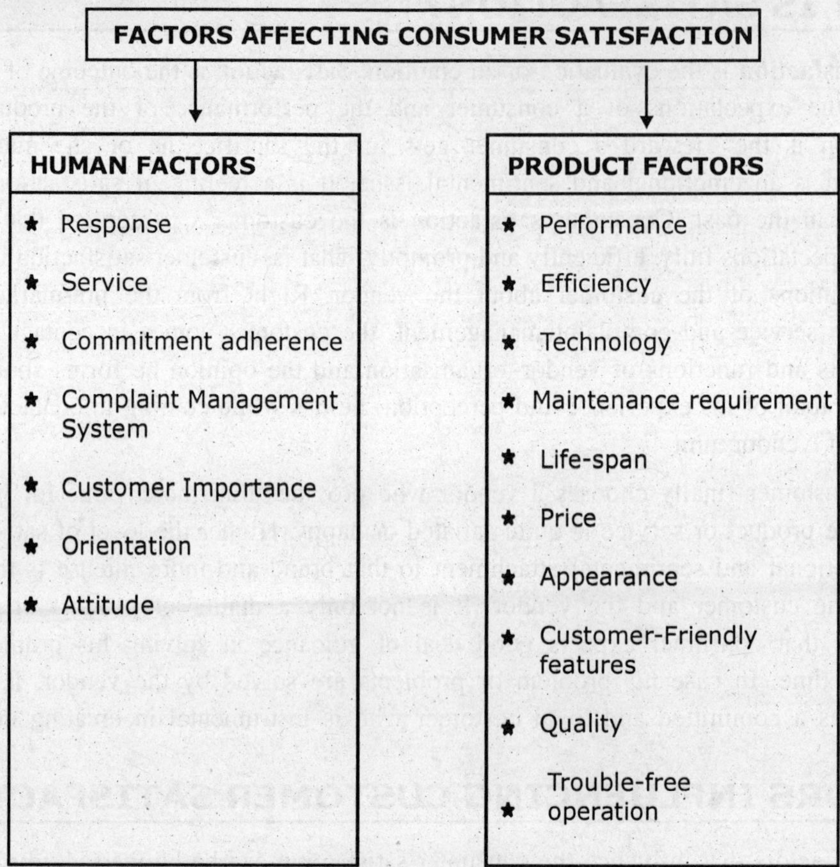

FIG : 14:01 FACTORS INFLUENCING CONSUMER SATISFACTION

The Vendors product and service performance

The Vendor Behaviour

The vendor behaviour deals with the behaviours of management and employees of the vendor. The vendors response, service and complaint handling are the factors more dependent on the attitude and skill of the human resources available with the vendor.

This is dependent on the capabilities of the vendor and the skill of his employees. It speaks of engineering and technology developed or acquired by the vendor. The quality system being followed by the vendor and its effectiveness has a bearing on the trouble–free operation and lesser maintenance requirement of the product.

There are two aspects of the product that worry a customer after the product has come into operations namely, efficiency and performance and its maintenance requirements. If a particular product is having frequent breakdowns or is requiring expert attention more frequently, it becomes a source of irritation for the customer in addition to financial loss. Again, if the equipment is not operating efficiently and consuming higher resources, it is a source of worry and headache for the customer. It is observed that in a customers segment, many customers, are able to tolerate slight inefficiency but will not beyond a limit and get totally detached with the vendor if the product needs high level of maintenance. In core segment, equipment efficiency is as important as its availability. Too much maintenance may offer availability of the equipment and it may derail the economic calculations of the customer. This speaks of the importance of effective complaint management system and service.

MEASUREMENT OF CUSTOMER SATISFACTION

Every customer has the options and there are competition waiting to spoil and exploit the situation in today's globalised economy featured by highest degree of competition. The interested parties can leverage small amount of dissatisfaction by feeding exaggerated information about a vendor to the prospective unsuspecting customer. That is, it is necessary that the supplier has the information and feedback on the image he enjoys in the minds of customers. To evaluate the image the supplier enjoys, he needs information. He needs to know what comes to the mind of customer on remembering the supplier. Is supplier efficient? innovator? modern? quality supplier? reliable ? or untrustworthy? Is he keeping commitments? Is he different from his competitors and so on ? It is, therefore, essential for the supplier to measure customer satisfaction and take appropriate actions to maintain and improve it. A successful business house needs quantitative measures. One such measurement is that of consumer satisfaction. The measurement of consumer satisfaction provides a supplies with the quantitative feed back. In case such measurement is done at regular intervals as is done in case of other business parameters like turnover and profits, the supplier gets feedback as to whether his efforts have produced positive results or there is need for corrective action.

TYPES OF INFORMATION AND FEEDBACK

Measurement of customer satisfaction can provide the detailed information and feed–back in respect of business, customers, supplier, competition and performance related.

1. Business Related: The business related information and feed back is in respect of:

Measurement of customer satisfacts allows the company to judge the effectiveness of its business from and provide information on how much customer centric it is?

★ I Provides quantitative information on : (i) How many customers have been lost (ii): How much business has been lost? (iii) How much business volume and profit, customer decay has been caused?

★ Satisfied customer acts as an extended marketing arm of the supplier.

★ A customer is unsatisfied when the supplier has either not come up to his expectation or honoured his commitment.

★ Due to human psychology, an unsatisfied customer is more likely to speak often on the subject than a satisfied customer. It is rightly said that a single unsatisfied customer can used the effect of five satisfied customers.

★ Only a portion of unsatisfied customers complain while the rest simply switch to other competitors. They are lost simply without the knowledge of supplier.

★ Loss of customer is loss of an opportunity and profitability.

★ Customer satisfaction increases the business and profitability of the supplier by increasing his realisation there by getting premium, for his products and services.

★ Customers satisfaction channelises the organisation resources to bridge the gap between the supplier and customer.

2. Customer Related: The customer related information and feed back relates to the variables like:

★ The number of customers lost.

★ The type of customers lost.

★ The reasons for the loss of customers.

★ The possible preventive measure to avoid customers loss.

★ The value assignment by the customers to supplied product and service rendered.

★ The customer decision–making variables and weightage given to each variable.

★ The customer's needs and requirements.

★ To make the supplier organisation more customer centric organisation .

3. Supplier Related: The supplier related information and feedback is related to:

★ Supplier's strengths and weakness as compared to his competitors.

★ Suppliers core competitors as perceived by the market for future growth and prosperity of the supplier.

★ To regain the position lost by knowing the exact position.

★ Supplier's market standing and image.

★ The customer's perception of supplier

★ Success rate of the supplier in keeping customers satisfied.

4) Competitor Related: Measurement of customer satisfaction provides competitor related information and feedback on:

★ The business lost to the competitors its extent and impact.

★ The strengths and weaknesses of different players or competition .

★ The studying of competitors- the gap between the actuals and bench marking.

5) Performance Related: The information and feedback related performance is to do with:

★ Opportunities open for improvement with the existing programme, product and services.

★ The details of actual cost of customer–turnover.

★ The level of performance as perceived by the customers.

★ The sharpenes of company's competitive edge and its improvement for better future.

CUSTOMER SATISFACTION MEASUREMENT METHODS

The methods of measurement of consumer satisfaction should cover the basic areas namely, identification of customer needs, customer value fixing on factors, suppliers and competitors, bench–mark, variations on deviations between the customers expectation and supplier's achievement and the variation between supplier's achievement and the bench mark. There are number of direct and indirect methods of measure customer satisfaction having their own sets of merits and demerits.

The direct Methods:

The most commonly used direct methods are- front line agencies, one–to–one meeting, complaint or appreciation letter, customer meet and customer survey.

The Indirect Methods:

The indirect methods that are most important ones, are two namely, customers complaints and customer loyalty.

MEASURES OF CUSTOMER SATISFACTION

There are basically three measures of consumer satisfaction which are outlined as under:

1) Customer satisfaction index: Customer Satisfaction Index (CSI) is the measure of overall satisfaction the customer is getting from the supplier in meeting his expectations. This measure is expressed in the ratio of two variables and can be–

CSI: Total marks obtained by the supplier based on customer feedback: Total marks expected by the customer for supplier to obtain CSI helps the supplier to trace how much he is able to come closer to meet the customer's needs and expectations. It is a performance measuring parameter by the organisations for measuring the workforce effectiveness. Companies can also use this index for the overall monitoring of their progress in becoming customer–centric.

2) Excellence index/Competitive index: This is the measure of standing of supplier in comparison to individual best targets- bench marks- against different activities. This also indicates the gap, which a supplier has to cover to excellence. It is defined as:

EI/CI total marks obtained by the supplier based on customer feedback: Total marks based on benchmark for excellence.

Like **CSI,** the companies can use **EI** for overall monitoring of its progress in actioning excellence. This factor has more relevance than **CSI** as it indicates competitive edge the company has or lacks as the case may be. The business growth or differentiality and, hence, customers preference that company can have in the market depend on this factor. **CSI** is after all an expectation of the customer, but **EI/CI** is a reality and it can have a definite bearing on **CSI**. It can actually drive the business direction and growth of the company in the future.

3. Bench marking: Bench–marking refers to the practice of looking to corporations worldwide for the best practice methods followed by them and incorporating the same suitably at the time of formulating the action plan for course of correction based on the result of the survey. These are the companies most preferred by the customers. A company becomes most preferred because of an attribute or bunch of attributes. For instance, company X may be known for 'engineering skill', while company Y for the service it provides. The importance of this is that there are the highest standards achieved by different companies in different segments but if the company wants to become globally competitive and excel then it most set its large on exceeding these parameters. Bench–marking also helps in assessing the gap between the company and the best in the fied. This acts mirror to the supplier to judge his own performance in comprises to that of others in the field.

SOURCES OF CUSTOMER DISSATISFACTION

Every organisation wants that its class of customers is to be one which is satisfied. That is, consumer expectations are marked by product and service quality. However, all customers can not be satisfied in toto. However, it is important to know as to why the customer dissatisfaction prevails. It is obviously on account of discrepant expectations and realisations. It is equally true that the nature of expectations is ambiguous. These situations of ambiguity occur in different ways. These can be:

1) Lack of experience with the existing product: Thus, when a customer is going in for CD writer, he may not be having exact picture in mind as to what to expect. This happens because it is his first experience or encounter in buying CD writer.

2) The product in question may be new or an innovation: The customer may be marketing for the first time on on-line say on–line banking or undergoing laser eye–surgery.

3) Variation in quality of service: In case of tailors all don't cut and stitch the same style or way nor the same tailor performs the same way at different times.

4) Absence of service while making purchase–decision: For instance while placing orders over the telephone for in home teleshopping, the service may not be physically present.

5) Expectations built up due to comparisons: It is usual that a product is judged in comparison with some standard or criterian, say discription on a leaflet or use of a similar product by one's friend or neighbour or one's own experience with a similar product.

LEVEL OF CONSUMER DISSATISFACTION

The levels of dissatisfaction are four as explained by experts.

1) General level dissatisfaction: This kind of dissatisfaction is associated with consumption system itself. Thus, dissatisfaction may be with the demand and supply or use of goods and services. Here, dissatisfaction occurs due to the non–availability of goods and services at all places– shopping system or say dissatisfied with the choice possibilities-purchase system.

2) Category level dissatisfaction: In such case, certain product classes may be the cause of dissatisfaction for the customer. Say, Mr. X may be dissatisfied with washing machines, packaged tours microwave ovens.

3) Brand level dissatisfaction: The customer has dissatisfy over a certain brands of manufacturer, retailers and so on. Thus, a person has dissatisfaction with say, **'Shudda Salt'** of **Nirma** company. Another may not be happy with **Nike** shoes.

4) Specific level dissatisfaction: This can happen with reference to services and offers of units selling products. Mr Z may be dissatisfied with the repair service extended to him in case of his home appliance.

INHIBITING AND FACILITATING FACTORS OF CONSUMER SATISFACTION

It pays to know and understand the dimensions of **satisfaction** and **dissatisfaction.** A distinction between **facilitators** and **inhibitions** is quite essential because satisfaction may have more than one dimension. Thus, too high a price may deter consumers (inhibition) whereas a reasonable price may not stimulate them enough to purchase or accept the product. In other case, it is unfriendly salesman who may put–off prospective buyer while a friendly salesperson may not act as a sufficient reason for buying. However, the manufacturers or the producers have to dispose of inhibitions and encourage facilitating factors, for the good of all.

Similarly, the presence of inhibiting and facilitating factors can influence consumer satisfaction. Likewise, inhibiting factors cause dissatisfaction and removal of the same may prevent dissatisfaction but may not create satisfaction. Facilitating factors cause satisfaction only when inhibition is not there. The role of inhibition and facilitation is very clear from the following diagram.

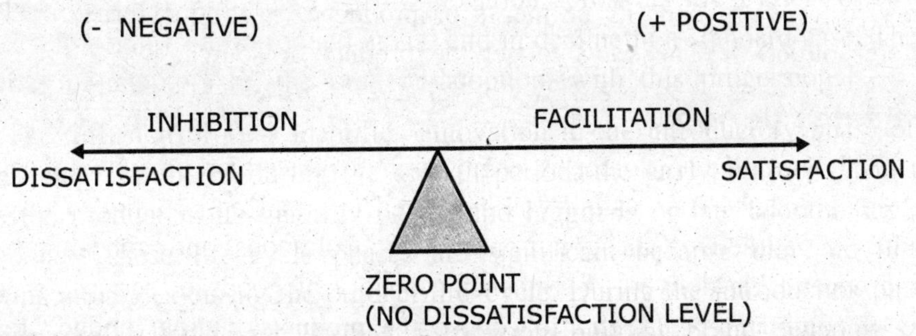

(- NEGATIVE)

(+ POSITIVE)

INHIBITION

FACILITATION

DISSATISFACTION

SATISFACTION

ZERO POINT

(NO DISSATISFACTION LEVEL)

FIG: 14:02 ROLE OF INHIBITION AND FACILITATION

A good example of dissatisfaction caused by inhibition can be that of public transport system in any metros of India. Generally, the daily commuters are dissatisfied if their expected bus on a particular route is late or if no vacant seats are available. However, if this is taken for granted if the right on time and manage to get the seats. Hence, the priority of public –transport management is to remove or do away with inhibition and not the creation of facilitation. Take another example of air–passengers. The inconvenience and discomfort is caused to these passengers when the staff resorts to flash strikes or go in for go–slow–tactics. These inhibition factors are causes of dissatisfaction for the passengers. Here, facilitating factors like providing snacks or lunch at the airport is not compensating the inhibition caused by the delay in flights. Therefore, the management in above cases is to make all efforts to remove inhibiting factors so that facilitating factors may result into positive effect.

WAYS OF ENHANCING CUSTOMER SATISFACTION

In the present–day business world, markets are changing their colours very fast whereby consumer needs and expectations undergo a sudden change. This volatility of market conditions makes the producers and marketers to keep their valuable customers by adding value by offering products and services the way they want it. The intelligence lies is offering a unique product to every individual customer goes in a big way in adding value to the customer decision–making process in a situation. Where the customer is to select and choose from both large and astonishing number of options, features, price structures and delivery modes.

It goes without saying that consumer satisfaction is a continuous process which does not begin or end with a purchase. On the contrary, it encompasses the entire ownership experience from selecting a product, to purchase, through after care to repeat purchase. Very definitely, there are three phases in the customer satisfaction process namely, **pre sales, during sales and after sales periods.** The customer expectations and their experience will in combine determine the level of satisfaction. Following is the period wise expectations.

I Presale Period:

During this period, the customer's expectations are developed through various information sources like advertising word of mouth and the like. Hence, the availability of clear and useful information on the following points is very much expected. ★ The product or service. ★ Quality of product or service. ★ Core benefits or merits of products. ★ Product price. ★ Product availability or stores outlet. ★ The ways to get the product or source.

II During the sales period:

It is the period when the customer is engaged in experiencing on how to deal with enquiries and sale of products. The customer is interested in ★ an opportunity to inspect the products ★ having an attractive sales environment ★ Courteous and appealing service. ★ Reasonable and dependable delivery ★ Enhancing quality of goods or services ★ Prompt redressal in case of compliant ★ Liberty to choose the products without undue pressure.

III The after sales period:

This is the period when the customer has started using the product. This expectations during this period are: ★ Support service or advice pronsion if called for ★ Prompt replacement or refund, if necessary ★ A simple and smooth complaint forward procedure ★ Efficient repair and maintenance service ★ Efficient and effective consumer follow up process.

It is evident from the above elucidation that customer satisfaction stretches beyond core product or service offering. Hence, the marketers are to start by attempting to understand the elements that together decide the consumer satisfaction levels only then, list the elements in proper and logical sequence so as to identify as to what is to be done to increase customer satisfaction level. This is possible only when the marketers design a system that facilitates interaction with their customers. These interactions will prove equally important as the quality of the core product or service offered by the company in the long run. The **steps that are to be taken to enhance consumer satisfaction** can be:

1) Conduct Regular Research: Regular research is a must to keep track of the changing levels of customer satisfaction. It is market research that gives excellent clues to the firm about what the present and prospective customers think of marketers product, product competition. History is full of evidences where the biggest companies were thrown to the winds as they totally forgotten the customers. Thus, **Colgate Pamolive** having engaged highest market share upto 1970s, did not bother about consumers. Suddenly in 1980s and 1990, the competitors made their way. Again, in 2000's has come back by remembering its customers. What is true of **Colgate** is true of **Ford Motor company**. Regular researches enable the marketers to keep at best of changing consumer levels. That is why, automobile makes are mailing customer satisfaction surveys form to their new buyers once after two months they bought the car and after eighteen months again, they can get similar like levels of satisfaction from dealers at definite time interval. The point is, don't sell to forget the customer.

2) Customer Feedback: Most of the companies are endowing their customers to give feed back and using this data as the means of maintaining their contact and dialogue. This feed back is about the company products and the extended services. This goes on records nicely analysed and helps in solving the problems faced by the consumers right in time. The feedback forms are sent to the customers asking relevant questions as to the products/s bought, the quality of service rendered and so on. Once the feedback is analysed, it helps in framing marketing strategies.

3) Add Value through Sound pricing: Perhaps pricing is very sensitive area of marketing mix. This 'p' has to be handled with utmost care and caution. It is the price and pricing which will establish short–term and long–term relation with the customer. Customers compare product quality *vis a vis* the price. They patronize those producers and marketers at which the high quality is maintained at competitive prices. That is, they stay with you, in case your quality and price standards are a perfect match as compared to your competition. What strategy or method you follow is not important. What is more important is, that it adds value to the customer.

4) Create Distribution Equity: Shrewd marketers have realised that with greater availability of goods which are close substitutes to competing brands, having good distribution supply–chain can actually prove to be the differentiating factor in deciding customer value satisfaction and loyalty too. While it is easier for companies to match product technologies and prices, it will not be easier to duplicate an effective distribution system offering a value–chain of quality services at the same time. Today, the companies are out to enhance customer satisfaction and value by increasing their distribution equity by monitoring and sustaining the quality of their product till it reaches the door step of a customer. Thus, **Videocon International** has built its own fleet of carriers to move electronic products dust free. **HLL** conducts two transit tests to ensure that products reach the outlet infact and is stored securely in warehouse. **Reliance** petrol bunks on highways of India provide not only diesel, petrol and parts but also supply purified and cooled water with backed up bathroom and toilet facilities. Of late, they are having restaurant facilities too. As the market gets tougher by brands, it is the strong distribution equity that will determine the consumer satisfaction in the long run.

5. Plan Consumer Benefitting Promotions: The promotions that deliver value to customers well help in building customer support and hence, satisfaction. The promotions must match the needs of different types of customers. The overall impact should be to attack and cover maximum number of customers. For those who are brand switchers, bargain deals lure them. In case of regular buyers lower prices in general and seasons off will work wonders. Offer lesser prices to win their confidence that in the days of inflation you are really helping them. The prospective customers who are postponing the purchases as is the case with consumer durables, grant them different types of discounts and relates and also zero interest instalment purchases. For encouraging new brand or product, give it as a sample. Whenever it is possible in case of consumer non durables- along with established brand. This serves two purposes of communication and the introduction of new product. Sales–promotion modes will help a lot. Since we too are to woo consumers, consumer promotions play significant role.

6. Develop Customer Satisfaction Oriented Culture: Every organisation should strive to make customer satisfaction its work culture. Development of such a culture should percolate from the top management through its initiative, thinking, direction and action. The development of this work–culture should not remain only a lofty objective. The management has to train the employees, develop skills, change their level of knowledge and attitude. The organisation is to ensure that its systems on quality, performance, services and complaint redressal are designed to achieve excellence and consumer satisfaction. System have as much role to play is development of organisational culture as human being and their habits. In fact, systems are to organisation what habits are to human beings.

7) Effective Complaint Management: In spite of best efforts of the supplier in ensuring that he supplies quality goods and provides efficient service and puts the customer on top of his priority list, he may frustrate the customer by not coming up to his expectations, resulting in complaints. A compliant is a positive indication that though the customer is unsatisfied, he may continue to remain with the supplier. The vendor must welcome complaints. Complaints are like early warning systems and indicate customer dissatisfaction. If not handled properly or given due attention or discouraged, it may lead to loss of customer himself or herself. If handled properly, they result in enhanced customer satisfaction and loyalty. It may also and to the mutual understanding existing and may further strengthen the relationship bond between the vendor and the customer. Customer complaints bring benefits or merits to the supplier in the forms of reliable feedback, deficiency identification, source of identifying generic defects, bring home customer needs and preceptions and above all a source of market trend and behaviour. Fast and comprehensive resolution of complaints helps in growth of business and customer retention. It adds value to the image of the company. The company should develop and use good complaint management system. Good complaint management system is one which is single–window system, responsive, maintains spare sorting and classifying complexes, follow up of complaint redressed has a time schedule, analytical base.

CHAPTER BASED QUESTIONS

A. WRITE SHORT NOTE ON:

(1) Consumer satisfaction (2) Vendor behaviour (3) Product performance (4) Types of information and feedback (5) Measurement of customer satisfactions (6) Direct methods (7) Indirect methods (8) Bench marking (9) Customer dissatisfaction (10) Levels of dissatisfaction

B. SHORT ANSWER QUESTIONS:

1. What is consumer satisfactions ?
2. What are factors of consumer satisfaction?
3. What is vendor behaviour?
4. What is product performances?
5. How one measures consumer satisfaction?
6. What are different types of information and feedback?
7. What is customer dissatisfaction?
8. Trace the source of customer dissatisfaction?
9. What are inhibiting and facilitating factors?
10. How to enhance consumer satisfaction?

C. ESSAY TYPE OF QUESTIONS:

1. What is customer satisfaction ? What are the factors of customer satisfacttion?
2. What is customer satisfaction? How is it measured?
3. What is feedback? What are different types of feedback and information?
4. Explain in detail direct and indirect methods of customer for satisfaction measurement.
5. What is customer dissatisfaction? What are the sources of the same?
6. What is level of Customer dissatisfaction? How many levels are there?
7. What are inhibiting and facilitating factors of customer satisfaction?
8. How can one enhance customer satisfaction?

******

CHAPTER 15

CONSUMERISM AND CONSUMER BEHAVIOUR

BACKDROP

Modern business is an integral part of current day society. This business has far reaching impact on social and economic life of we people. As a socio-economic institution, each marketing company is to deliver the goods and services or deliver the standard of living or a life-style as per the aspirations of the members of this society. It has a great social responsibility which means an intelligent and objective concern for the well-being of society. The major aspect of social responsibility of modern marketing is consumerism and protection of consumers against the ill-treatment meted out by it. This chapter plans to discuss at length the concept of consumerism, issues and solutions with special reference to Indian conditions. The chapter ends with Chapter Based Questions.

WHAT IS CONSUMERISM ?

Though **the consumer** occupies the central position in the entire scheme of free economy, consumer has not always received the attention that he deserves. In the practical world, consumer is the 'slave' and not the 'sovereign' as described. The consumer's well-being or the welfare lies in the fulfillment of their normal, legitimate, interest, expectations and aspirations as to goods and services they need. Each and every consumer is worried, hurried till buried about the quality, quantity, price and the timeliness of the supply of goods and services. The term 'consumerism' has come into existence in the early 1960s when it was coined by the business community in the western world particularly in America. It is the social force designed to protect the interests of consumers by organising the consumers to bring pressure on business community to heed to their say. **Professor David W.Cravens** and **Gerald E. Hills** have defined it as "an environmental force intended to aid and protect the consumer through the exertion of legal , moral and economic pressure on business." It can be defined as simply "let the seller beware". It is the social force protecting the consumer and aiding the consumer; it is engaging in actions intended to stir public–opinion and generate public–pressure. Put in other words, it is an organised efforts to fight against the unfair marketing practices with a view to secure consumer protection. Consumerism is the society's attempt to bring back the balance in the exchange between the buyers and sellers as the strength of power is normally in favour of sellers rather than buyers.

This concept of consumerism challenges the very foundations of so called marketing concept or the philosophy as studied in the early chapters. According to **Professor Peter Drucker** the 'modern guru' on management, consumerism dispute and defy the four premises of marketing concept. These premises are:

1. "Consumers know their needs".

2. "Business knows exactly their needs".

3. "Business provides necessary and useful information." and

4. "The products and services readily fulfil consumer aspirations on one hand and business premises on the other."

Consumerism as a protection against unfair business practices and injustices, aims at removing or uprooting them or reducing them if not eliminating them totally. There are countless injustices and unfair practices on which the business community stuffs its profits and guffs the consumer satisfaction. To illustrate a few of this kind, these can be — supply of unsafe products, adulteration, fictitious pricing, defective and deceptive packing and packaging, false and misleading advertisements and sales literature, hoarding, profiteering, black-marketing, price collusion, misuse of brands, planned obsolescence, short measures and weights, release of those goods which are not really wanted by the consumers nor can they afford, face-lifting of products and the like.

THE SCOPE OF CONSUMERISM

The consumer movement stands for the organised efforts of the consumers seeking to redress, restitute and remedy for the dissatisfaction of the standard of living. It is a social movement seeking to challenge the rights and powers of the buyer in relation to those of sellers. Consumerism, like democracy, is a movement by the people, of the people and for the people where the role of common man is uncommon. Consumerism encompases two basic areas namely,

1. Removal of reduction of the discontent and dissatisfaction created in the business relations between the buyers and sellers in market. That is, the marketing activities of the sellers should guarantee consumer satisfaction at any rate as envisaged as a major plank of modern marketing concept.

2. Protecting the consumers against the prejudices caused in the exchange relations. Though these are two major aspects of consumerism, the modern concept of consumerism encompasses one more very important aspect namely, consumer interests in environmental forces influencing their quality of life. Thus, consumerism is the social demand for the refinement in marketing practices by revamping them to be more truthful, efficient, informative, responsive, sincere and making the business house more concerned about the quality of life. The quality of life of a common man depends on making available more and more public goods and services at reasonable costs than the private goods and services. The real benefits of development are to percolate to the commonest of the common man in the country. Thus, better high-ways, safe–dams and bridges, state roads, approach roads, better education, better means of transport, better health services, clean and crimefree cities and towns, better communication system, sound preservation of scarce resources, national heritage, elimination of environmental pollution, up holding of cultural values and the like. Thus, the social responsibility of marketing is not only delivery of standard of living but improved quality of life standards for one and all.

CONSUMER PROTECTION

Consumer protection is a form of social action which is hatched to attain the well-being of the society namely–consumers. Since an individual consumer is considered more vulnerable, in the modern world, to exploitation and harassment by the manufacturers and distributors or sellers because of environmental complexities of business operations, technological changes, application of mass production techniques, vast resources, man-power and acrimonious advertising, it is necessary that the various groups of society namely, government, judiciary, business units, traders, voluntary associations of consumers are to play their due role to protect and promote the consumer interests - economic, social and environmental all rolled into one.

THE NEED FOR CONSUMER PROTECTION :

The fundamental cause of consumer movement in any country has been the consumer dissonance. 'Dissonances' means after-purchase doubts, disillusionment, disappointment and dissatisfaction. **Dr.Peter Drucker** has rightly put 'consumerism' as 'the shame of modern marketing'. There will be no scope for consumerism in case all the marketers practice the marketing concept in its real sense and entirity because, true marketing is 'consumer oriented'. **Professor Philip Kotler,** an authority on marketing management, looks at consumerism as the final expression of the marketing concept because, it is that philosophy which force the product managers and marketing managers to look at the goods and services purely from the eyes and brains of consumers. As 'satisfied–consumer' is the real asset that an alert marketer can develop and build, he should welcome the pressure of consumer protection as he can capitalise on that for his own benefit. The real need for consumer protection arises because of atleast five points. There are:

1. Consumer needs physical protection against products and services that are spurious, unsafe and endanger health and property and, therefore, his welfare it may be bottle of aerated water or gas–cylinder or even a bottle of medicine or a jar of wine.

2. Consumer needs protection against deceptive and unfair trade practices followed by unscrupulous business community. Today, the consumer is the victim of business mal-practices and frauds. Thus, a contractor may cheat a humble and needy flat-buyer. A village rustic may be supplied with an out-dated medicine by a pharmacy.

3. Consumer needs protection against the abuse of monopoly and restrictive trade practices. Every one knows that monopoly is a prolific source of consumer exploitation in terms of higher prices and lowering the quality standards because, the monopolist can control quality, and therefore, price and quality.

4. Consumer needs protection against deaf-ears on non-commitment of sellers. In most cases the receipt as a documentary proof makes it clear that the goods once sold will not be taken back. If the goods are not up to the specification, they are to be accepted back and due adjustment is to be made in the receipt. Unfortunately, many greedy sellers pass on the defective goods under this clause and thus deceive the consumers for faults of theirs.

5. Consumer needs protection against pollution of all kinds. The producers make use of inputs for making available the much needed output of goods and this process results in ecological and environmental imbalances badly affecting the health and property of consumers through air, water, food, noise and relation pollution. Thus, chemical factories, fertilizer plants collienies have been known for air and water pollution endangering the health of flora and fauna of a particular region.

CONSUMER RIGHTS

When voluntary efforts fail, compulsory impositions will work as the last resort. Government of the nation does it through legislation. The American experience became the foundation for other nations to think of consumer rights. It was **John F.Kennedy,** the Ex-President of America who spelled out the consumer rights in the year 1962. These six rights have been outlined below:

1. **Right to Protection of health and safety :** There are good many products that are likely to cause physical danger to consumer health, lives and property; they contain potentially harmful substances that are dangerous from the consumer welfare point of view. The best examples of this kind are - food–additives, colours, emulsifiers and preservatives. The health hazards that are likely to arise are to be eradicated or reduced considerably. Again, in case of food items and drugs-both life saving and life sustaining - safety is to be guaranteed. We have toy for the innocent kids, appliances to the house–wife, containers of plastics and other metals in which we store our daily needs, man-made fabrics so close to our bodies, the chemicals and drugs we take in for the functioning of the human body all need safety. One thing is encouraging to-day that recent legislation have shifted the responsibility for such unsafe items on the shoulders of sellers rather than on buyers. We cannot afford to forget the misleading and arinonious advertisements that make tall claims about the product and product performances. The actual experience, however, is far from the truth. Thus, a manufacturer of textiles claims that his cloth is wash and wear but is not so actually; a detergent soap cake is sufficient to wash a heap of clothes; a mixer-grinder is supposed to run continuously for 45 minutes; the cloth that does not shrink; the fact colours that will not fade and so on. The consumers who rely on the ads and buy the products and eventually find themselves cheated systematically are to be protected.

2. **Right to be Informed :** This implies that the manufacturer and the dealer is expected to disclose all the material facts that are going to affect the economic and social interests of consumers. The supply of information means that it should be adequate, accurate, up-to-date, verifiable and legal as to quality, quantity, performance and other product characteristics. Such an information gives wide choice to the consumers in making sound decision on buying. That is why, the product, package or the package is to give the full details about the name of the product, composition, dosages, date of manufacturing, date of expiry, batch number, warnings, antidote and the like. In addition, it should clarify as to the name of the manufacturer, price with or without tax. Such information saves the consumer against the possible deceit. Again, when whatever the seller states he should mean what he says.

3. **Right to Choose :** Consumer satisfaction is the final aim of modern marketing and is the philosophy of marketing concept. A wise trader or dealer or manufacturer is one who maximises his profits by maximising the consumer satisfaction. Consumer satisfaction can be increased by giving the consumer the widest choice. Choice means the widest range of products in quality and brand varieties at reasonable prices. Right to choose implies that monopoly is brushed aside by the consumers. It is they who are to decide as to what product they want of what quality-in what quality and within what price range. It is not the producer to decide as to the quality, price and quantity. In other words, the manufacturer has no control over quality and price which are the mainstay of monopoly conditions. It means that consumers encourage competition among the producers and dealers so that there will be improvement in quality standards, reduction in costs and increase in quality produced so that maximum number of consumers can be benefitted.

4. Right to be Heard: Consumer has the right to ventilate and register his dissatisfaction, disagreement and get his complaint heard and weighed. This right is of vital importance because, all the earlier three rights are meaningless if this right is not granted. In absence of this right, the very purpose of granting other rights is defeated. This gives scope for the government to intervene between the buyers and sellers and thus grant the justice to the aggrieved party. In the practical world of business, the business houses turn their dealf ears to the genuine complaints by the consumers. The complaints and protests go to cold storage. This right makes these firms to accept these complaints and protests as feed-back in future policy formulations.

5. Right to be Redressed: This right is one step ahead of the previous right. The complaints and protests are not just to be heard, the aggrieved party is to be granted compensation by bringing due adjustment in the transaction that has taken place. There should be fair and just settlement of deserving claims. In to-day's world, most of us buy the goods in good faith on the strength of advertisements. We know the truth only when we use the product; we realise that the product falls short of its performance standards, or it is below the quality standards and so on. In such cases, the dissatisfied customer must get back his value for money. As per the terms and conditions of sales, the customer is to be compensated. This is done through warranties and guarantees. Such adjustments are sure to reduce cognitive dissonance or the post-purchase doubts and laxities. Further, the right to be heard means existence of an acceptable mechanism for asserting the rights.

6. Right to Quality of life : Quality of life refers to the perceived well-being of people, in groups and individually, and the well-being of the environment in which these people live. Of late, consumerism has been defined as an improved 'quality of life'. It means that the environmental problems affect the very life of consumers. Air pollution, water pollution, food pollution noise pollution and relation pollution, and the legacies of on-going industrialisation. The social costs of these pollutions are at the cost of social benefits. In other words, each kind of pollution is eating away the social benefits existing. The consumers are asking of social responsibility of manufacturers and dealers. That is, when resources are used for the output, these pollutions are bound to be there and sure to affect the quality of life. That is why whenever a decision is taken to start a factory say, fertiliser, the company should be prepared to recycle the polluted water and air for the benefit of mankind.

CONSUMER PROTECTION IN INDIA

In fact, consumer movement had its birth in 1960s through the ramblings of its formation had started as early as 1920s. It has been delayed and there is dire need for consumer protection in India at least for six solid reasons outlined below :

1. Consumer has no say: In India at least, the consumers are the 'slaves' in the market condition than the 'kings' India presents a case of 'sellers' market' even to-day though we are daring to say that the things have changed. The consumer is to be satisfied with whatever is available in the market He has no real choice and hence, no enhanced satisfaction. This is why the Indian consumer is hurried, worried till buried about the fair-play of fair-trade practices so that he can get real value for his hard earned money.

2. Lack of get-to-gether: It is really very difficult to organise the consumers in India as India is a very vast country and the consumers are widely scattered speaking different languages, belonging to different religions, sects and sub-sects, cultures and sub-cultures, different political affiliations. Added to this, economic and social backwardness is acting as a formidable barrier in coming together of consumers in the form of concrete organisation or organisations.

3. Ignorance and illiteracy: Ignorance, illiteracy, economic and social backwardness have given ample scope for the organised business community to exploit these consumers. This merciless exploitation should stop; it can be stopped by forming strong organisations of consumers on regional basis with federal relations at the centre. The consumers need and should not depend on government legislation alone. To wage a war against profiteering, black-marketing, adulteration, hoarding, fraudulent packing, deceitful weighing and measuring and misleading advertising, there is need for such organisations.

4. To get the latest goods and services: To-day's science and technology is making available new ideas, products and services. The producers and dealers must make these available to the consumers. The latest products are comparatively lower in costs, better in quality and are available in sufficient quantities. For instance, the service of glucose count in case of diabetic patients is possible within two minutes in just 20 rupees is against the time consuming and costly conventional method. The calculators are available at much lower prices with increased functions and digits. On one side, wide variety of goods are available ; on the other, it is getting complicated to select right kind of product for use as they are highly sophisticated. This advance of science and technology has made marketing much more professional or impersonal warranting the services of technicians for proper selection for right use.

5. Acrimonious advertising : Consumer sovereignty is greatly influenced by advertising—the great educator, fine entertainer and the worst cheat. Though advertisements are supposed to provide maximum possible information for consumer decision making, majority of the ads are false and misleading making the consumers to take them at fairly high rate of discount. The truth and ethics\ have reached the rock- bottom. The advertisers of to-day, make all tall claims, mix truth with lies thus losing the consumer confidence. Under the shield of advertising, producers are bringing out those products which people really do not need nor they can afford. Such false persuaders are to be hooked and booked which is possible only when the consumers act unitedly through an organisation of their own.

6. Indiscriminate environmental pollution : India is a rich nation so far as flora and fauna are concerned. However, if the things are left to themselves and unbridled, very soon the beauty of this country will be sapped very badly by the pollution of water, air, noise, food and relations caused in the name of industrialisation. Ecological imbalances caused have already shown their fury in the most crude forms. Thus, indiscriminate cutting of forests has reduced the annual rain-fall and increased the chances of floods in river basins. Historical mounments like Taj-mahal and Radhaswami temple are getting\ enmasked by the thick smoke gushed out by the chimneys of industrial units. The lives and property in around chemical complexes are in danger resulting in health hazards. To raise voice against these factors that work against the quality of life, there is need for strong consumer organisations.

THE WAYS TOWARDS CONSUMER PROTECTION

When one thinks of ways and means of providing much warranted consumer protection, two alternatives are open namely, self- regulation and government intervention.

1. Self-regulation:

When we talk of self-regulation, two types of associations come to the rescue of consumers and business community namely, consumer associations and business associations.

A. Consumer Associations : It is heartening to note that there are consumer associations coming up in a big way in every nook and corner of the country. At national level, four organisations are worth mentioning. These are : Consumer Guidance Society of India of Mumbai, Consumer Education and Research Centre of Ahmedabad, Consumer Action Forum of Kolkata and Karnataka Consumer Service Society of Bangalore. However, these regional organisations suffer from lack of persistent and motivated members, inadequate finance and are over-shadowed by the influence of politicians and big business shots. It means that such organisations succeed only when they are guaranteed to timely and adequate finance, whole hearted membership and effective leadership.

B. Business Associations : Self-regulation by the business community through fair-deal to resellers and consumers does wonders. Just as road accidents can be reduced by respecting the traffic rules and regulations, so also the itch for confrontation can be eradicated if the business community behaves through the chambers of commerce and trade associations. If the business morality is high and the ethical and moral values are at the top, there is no reason why business community should go fut. In fact, the business community should welcome the complaints and the suggestions made by the consumers as the golden opportunities for further and future refinement. Oppositions, complaints, criticisms are to be accepted sportively and democratically as the part of the game for their own betterment and benefit.

II. Government Intervention:

When the efforts of self-regulation fail, that pave the way for governmental intervention as the last resort. It means that well organised business community has failed to regulate itself; the government as the custodian of the nation, takes the side of weaker side of the battle namely, the consumers. This results in consumer legislation for protecting the legitimate interests of the consumers.

During post-war period, particularly after Independence, Government of India has come out with good piece of legislation containing the unethical practices and abuses in the arena of marketing. These enactments are : Industries Development and Regulation Act of 1951, M.R.T.P. Act of 1969, Weights and Measures Act of 1956. The Prevention of Adulteration Act of 1954, The Drugs and Cosmetics Act of 1954. The Trade and Merchandise Act of 1958, The Essential Commodities Act of 1955 and Consumer Protection Act of 1986.

Of all these Acts, the last one is most specific and relevant to our discussion. There is, now, a greater recognition that consumers need to be protected not only from the effects or restrictive and unfair trade practices but also from defective goods or deficient services, over–charging of prices and unscrupulous exploitation. Further, ignorance of the consumer of his rights, lack of consumer movement and lack of due recognition to the consumer organisations have also been some important factors which led to the enactment of Consumer Protection Act of 1986.

The Consumer Protection Act of 1986 has envisaged the setting up of the councils at the centre and state levels called as Central Consumer Protection Council and the State Consumer Protection Council. The Central Government is empowered to constitute the Central Consumer Protection Council which may consist of 150 members. Its term is three years. The Council may meet as and when necessary but not less than three meetings of the Council shall be held every year. The Council is empowered to constitute among its members, "working groups". Every working group so constituted shall perform such functions as are assigned by the Central Council.

The objectives of the Central Council are to protect and promote the rights of consumers such as:

1. The right to be protected against marketing of goods which are hazardous to life and property.
2. The right to be informed about the quality, quantity, potency, purity, standard and prices of goods so as to protect the consumer against unfair trade practices.
3. The right to be assured, wherever, possible access to a variety of goods at competitive prices.
4. The right to be heard and to be assured that consumer's interests will receive due consideration at appropriate forums.
5. The right to seek redressal against unfair trade practices or unscrupulous exploitation of consumers and
6. The right to consumer education.

The State Governments are also empowered to establish State Consumer Protection Councils for their respective states. This Council shall consist of such members as may be notified by the State Government by Notification from time to time. The objectives of State Council are to protect and promote the rights of consumers within the state as referred to in case of Central Council.

The Consumer Protection Act of 1986, provides for the establishment of three–tier quasi judicial machinery for redressing the consumer grievances as under: 1. District forums. 2. State Commissions and 3. National Commission.

1. **District Forums:** A Consumer Disputes Redressal Forum to be known as "District Forum" is required to be established by the State Government with the prior approval of the Central Government. It consists of a president and two members appointed by the State Government. The president of the Forums is required to be a person who is, or has been or is qualified to be a District Judge. Of the two members, one must be a person of eminence in the fields of education, trade or commerce and another must be a lady social worker. The jurisdiction of the Forum is to entertain complaints where the value of goods or services and the compensation claimed is less than rupees one lakh. The complaint may be lodged by affected consumer himself or the registered consumer association to which he belongs.

2. **State Commission:** A Consumer Disputes Redressal Commission to be known as "State Commission" is required to be established by the State Government with the prior approval of the Central Government is the State. It shall consist of a president and two other member's to be appointed by the State Government. The president should be a person who is or has been a Judge of a High Court, appointed by the State Government. The two other members should be persons of ability, integrity and standing and should have adequate knowledge or experience of, or have shown capacity in dealing with problems relating to economics, law, commerce, accountancy, industry, public–affairs or administration, one of whom should be a woman. The jurisdiction of the State Commission is original as well as appellate. The original juridiction is vested in the Commission in respect of complaints where the value of goods or services and the compensation so claimed exceed rupees one lakh but not more than rupees ten lakhs. The appellate jurisdiction is confined to appeals from the orders of any District Forum within the State. Further, the State Commission may call for the records and pass orders in any consumer dispute where it appears that a District Forum has acted illegally or with material irregularity, or has exceeded its jurisdiction or has failed to exercise its jurisdiction.

3. **National Commission** : The highest authority to settle the consumer disputes under the Act is an independent statutory body called National Commission. This "National Commission" shall consist of a president and four other members to be appointed by the Central Government. The president should be a person who is or has been a Judge of the Supreme Court. The persons to be appointed as members should be the persons of ability, integrity, and standing with adequate knowledge or experience or capacity in dealing with the problems relating to economics, law, commerce, accountancy, industry, public–affairs or administration. One of the members should be a woman. The jurisdiction of the National Commission is original as well as appellate. The original jurisdiction is limited to the complaints where the value of goods or services and compensation exceeds rupees ten lakhs. The appellate jurisdiction is confined to appeals against the order of any State Commission. Thus, no appeal can veto National Commission against the orders of the District Forum. It is the State Commission of each state which can entertain and appeal from the District forums of the state. Further, the National Commission is empowered to call for records and pass appropriate orders in any consumer dispute where it appears that the State Commission has acted illegally or with material irregularity or exceeded its jurisdiction or has failed to exercise its jurisdiction.

AMENDMENTS TO CONSUMER PROTECTION ACT 1986

Some progressive amendments to the Consumer Protection Act 1986 were introduced by Consumer Protection (Amendment) Ordinance 1993. The salient features of the Amendment are:

1. **Rise in Pecuniary Jurisdiction:** The pecuniary jurisdiction of the District Forum was raised from Rs. I Lakh to Rs. 5 lakhs and that of State Commission from Rs. 10 lakhs to 20 lakhs.

2. **Empowerment of State Governments:** The State Governments are empowered to establish more than one District Forum in a District if it deems fit in large cities. The raising of the pecuniary jurisdiction will increase the work load of the District Forums as most of consumer complaints do not exceed Rs. 5 lakhs in value. Hence, it would be in the consumer' interest that the State Governments establish more than one Forum in Districts with large cities.

3. **Institution of Complaint:** Now an aggrieved party can institute a complaint at any place where branch office of the opposite party of an establishment is situated.

4. **Right to Institute in Group:** Right is granted to one or more consumers to institute a complaint where there are several consumers having the same interest.

5. **Incorporation of UTP:** The definition of Unfair Trade Practice has been incorporated at length in the Act thereby dispensing with the need to refer to MRTP Act for definition of UTP.

6. **Change in Definition of Consumer Goods:** Use by consumer of goods bought and used by him exclusively for the purpose of earning livelihood in case of self-employment is excluded from the definition of commercial purpose.

7. **Inclusion of Housing Construction:** Housing Construction has been included within the definition of services, which was misused by developers.

8. **Grant of Additional Powers to Fora:** The Fora have been given additional authority and power. The Fora can issue an order to the opposite party directing him to remove defect a deficiency in the services in question, to discontinue the UTP or RTP or not to repeat them not to offer hazardous goods for sale and to provide for adequate costs to parties.

OTHER ACTS REGULATING MARKETING ACTIVITIES :

In the interest of consumers, the Government of India has passed good many Laws and enactments. The total legislation covers variety of laws and Acts. However, we are concerned here with those that directly related to marketing of goods and services. These are :

(1) Dangerous Drug Act 1930

(2) Agricultural Produce, Grading and Marketing Act 1937

(3) Drugs and Cosmetics Act 1940

(4) Drug Control Act 1950

(5) Emblems and Name (Prevention of Improper Use) Act 1950

(6) Indian Standards Institution (Certification Marks) Act, 1952

(7) Standard Weights and Measures Act, 1952

(8) Drugs and Magic Remedies (Objectionable Advertisements) Act 1954.

(9) Prevention of Food (Adulteration) Act of 1954 and 63

(10) Essential Commodities Act 1954

(11) Trade and Merchandise Mark Act 1958

(12) Essential Services Maintenance Act 1968

(13) Monopolies and Restrictive Trade Practices Act, 1969

(14) Drugs and Cosmetics Control Act, 1976

(15) Standard Weights and Measures (Packaged Commodities) Rules 1957

(16) Prevention of Black Marketing and Maintenance of Supplies of Essential Commodities Act, 1980

(17) MRTP Act, 1984

(18) Consumer Protection Act. 1986

(19) Indian Sale of Goods Act, 1930.

Indirectly good many Acts apply to marketing are namely Shop Establishment Act of 1872; Partnership Act of 1932, Indian Companies Act of 1956, Indian Contract Act of 1872 and so on. It is not possible to touch all Acts. It is already noted at length Consumer Protection Act of 1986. Therefore, we have brief reference to some of the Acts which more closely related to marketing area. These are ;

1. Prevention of Food Adulteration Act, 1954 :

This Act was passed with a view to protect public health prohibiting adulteration of food. An adulterated food article is one which is injurious to public health. It is injurious when the product quality is not as demanded or claimed, it contains an injurious substance, any constituent of the article has been taken away, it has been prepared, packed or kept under insanitary conditions, it is unfit for human consumption, it is obtained from deceased animals, it is poisonous or deleterious, it contains a prohibited preservative, its quality or purity falls below the prescribed standards.

2. The Essential Commodities Act, 1955 :

Perhaps, it is one of the major public interest protection where it was passed with the object of controlling the interests of general public, the production, supply and the distribution of trade and commerce in certain commodities declared essential. A number of products are included under it. Whenever a company markets these products, the provisions of this Act apply to it and influence its product, distribution and pricing decisions. In 1974, it was further amended with provision against hoarders, black marketers and profiteers. It is made compulsory to display the prices of essential commodities.

3. Weights and Measures Act of 1956 :

This Act was passed to safeguard the consumers against the exploitation of weighing measures of commodities by traders. Under this Act, standard weights and measures were introduced. What is more important is they switched over to metric system of weights and measures which simplified the procedure of calculation.

4. The Monopolies and Restrictive Trade Practices Act of 1969 :

This Act was passed in order to regulate the monopolistic and restrictive trade practices followed by the companies. It applies to all such undertakings, except government and public undertakings. MRTP Commission has been set up to enquire into monopolistic and restrictive trade practices upon a complaint by any trade or consumer association. The main objectives, in brief, are checking the concentration of economic power which is detrimental to the society, checking such monopolistic and restrictive trade practices which are injurious to public well being and welfare and vesting powers with the Central Government to investigate and control mergers and expansions.

5. Drugs and Cosmetics Act 1940 :

Like food products, drugs of sub-standard quality are manufactured and marketed. The prices charged for medicines are not proportional to the cost of production. The high prices charged for medicines are high enough to exploit the needy consumers by greedy druggists. There is great danger of dumping spurious and duplicated medicines which are not life saving but life taking medicines.

6. Packaged Commodities Order—1975 :

This is a significant measure in favour of consumers. The order provides that producers of several packaged products should print on the packages the contents, weights, price, date and month of manufacture date of expiry name and address of the manufacturer meant for retail trade.

7. ISI Certification Act, 1952 :

Any product manufactured as per the standard laid down by ISI may be certified under the ISI Certification Act of 1952. Under the Act ISI has been authorised, to issue licences for the use of ISI marks to firms producing goods, by the Central Government, which conform to the ISI specifications. ISI marks indicate that (1) the goods are subjected to quality control (2) the goods are well safeguarded against sub-standard goods (3) they have more than a claim by the manufacturers (4) they earn the buyers confidence (5) they will act as a third party guarantee. For proper implementation ISI Officers inspect samples by paying surprise visits. The samples are seated and sent for testing.

8. Sale of Goods Act 1932:

Sales of Goods Act was passed in 1932 with the intention of safeguarding the interests of both the buyers and sellers as a deal involves exchange. In case the deals go well as per normal practices, there is nothing wrong. However, both the buyers, and sellers, do fall prey to the unexpected, unjustified acts of others and they are to suffer for no fault of theirs. This deals with sale, an agreement to sales goods dealt in and their classification, conditions and warrantees, rules regarding delivery payment, rights of unpaid seller and so on.

9. Other Acts:

There are other Acts namely, Drug Control Act 1950, Dangerous Drug Act 1930, The Drugs and Magic Remedies or Objectionable Advertisement Act of 1954, Household Electrical Appliances (Quality Control) Order 1976, that are equally important.

In a nutshell, we can say that these Acts protect the legitimate interest of consumers in a number of ways. With a separate Act namely, Consumer Protection Act of 1986 as amended up to date, these enactments make the consumer protection much more meaningful and purposive and effective. What is more important is that more passing of Acts has no meaning. The persons for whom it is passed should take full advantage of these. That is there should be public awareness among organisations and associations, so that the group cry is louder than individual and more demonstrative.

CHAPTER BASED QUESTIONS

A. WRITE SHORT NOTES ON:

1. Consumerism 2.The scope of consumerism 3. Consumer rights 4. Consumer protection in India. 5. Ways towards consumer protection 6. Consumer Protection Act of 1986. 7. Latest amendments. 8. District Forum 9. State Commission 10 National Commission 11. Acts supporting Consumer Protection 12. Right to Protection of health and safety. 13. Right to be informed 14. Right to choose 15. Right to be heard 16. Right to be redressed 17. Consumer movement in India. 18. Self-regulation 19. Consumer association 20. Business association intervention.

B. SHORT ANSWER QUESTIONS:

1. What is consumerism ? Is it a revolution or evolution ?
2. What are consumer rights?
3 What is the scope of consumerism ?
4. What is self regulation ?
5. How self regulation is possible ?
6. What is State intervention ?
7. What factors led to the passing of consume protection Act of 1986 ?
8. What are the latest amendments to it ?

9. Are consumer happy with the three tier justice ?
10. What other Acts support the interests of consumers ?
11. How can one make the consumers movement strong ?
12. Explain three rights of consumers.
13 Explain any five Acts that support consumers.
14. What is the mechanism of consumer complaints redressals ?
15. In intervention self regulation is better or government intervension is better ?

C. ESSAY TYPE QUESTIONS:

1. What is consumer movement ? Why it has been slow in India ?
2. What is consumerism ? What is its Scope ?
3. What are the rights of consumers ? Explain each at length.
4. What is consumerism ? What is the need for it ?
5. What is the position of consumer protection in India ?
6. What are the ways towards consumerism ?
7. What is Consumer Protection Act of India ? What are its highlights ?
8. What latest amendments have taken place in this connection ?
9. What all other Acts support in protecting consumer interests ?
10. Are consumers happy with the functioning of the mechanism of consumer protection ? Give reasons.

*****

CHAPTER 16

SKILL DEVELOPMENT IN CONSUMER BEHAVIOUR

BACKDROP

It is our general experience that in a real sense everyone is a manager because he or she performs all the functions of marketing management namely, planning, organising, directing, coordinating, communicating-motivating and controlling. Only thing is that he or she is not aware that he or she is performing these functions. That is why, he or she does not call himself or herself a manager. According to such thinkers, only company executives who get the work done through others, are managers in real sense. One need not be management expert; to give the example of house wife as the best example of general manager, marketing manager, products manager and personnel manager and even in some cases financial manager. However, these people of whom we talk are not standardised in that, their performance is not the same because, they might have acquired successfully the educational qualifications in a formalised way of learning. What is appreciated is that acquisition of knowledge is one thing and application of the same to actual situation is not one and the same. One may hit because, he knows to hit . However, where to hit to get results is more important with time and cost dimensions. This speaks of skill development. This chapter is devoted to actual cases of skill development in the dynamic area of consumer behaviour.

SKILL DEVELOPMENT CASE ONE

Required:

Conduct an informal interview of a local retail store owner and determine what demographic and some economic segments the store appears to satisfy. How did the owner select this segment or segments?

Solution:

I had the opportunity of visiting the retail store owner in Belgaum city who owns a retail store dealing in consumer durables. I took prior appointment and Mr. Aloknath, the owner manager, was kind enough to spend his valuable time on Sunday the 2nd April 2006. I did spend time with him and collected good deal of informating about his business at retail store "Sweet Home." On the next two days, I prepared a profile of his consumers that would help in market segmentation.

HOUSEHOLDS IN INCOME BRACKETS

Income Range	Households	Percentage
Less than 6,000 per year	8,500	34%
Between 6,001-25,000 per year	5,000	20%
Between 25,001-50,000 per year	4,000	16%
Between 5,000 - 1,00,000 per year	2,500	10%
Between 1,00,001 - 2,00,000 per year	1,500	6%
Between 2,00,001 - 3,00,000 per year	1,250	5%
Between 3,00,001-5,00,000 per year	1,000	4%
Between 5,00,001-7,50,000 per year	750	3%
More than 7,50,001-per year	500	02.%
	25,000	**100%**

HOUSEHOLD BY EDUCATION LEVEL

Level of Education	No of Household	Percentage
Primary education level	5,000	20%
Secondary education level	5,500	22%
Higher secondary level	7,500	30%
Graduation Level	5,000	20%
Post–graduation level	2,000	08%
	25,000	**100%**

HOUSEHOLDS BY AREA

Area	Households	percentage
City proper	8,750	35%
City extension	6,250	25%
Outside Corporate units	5,000	20%
Suburbs	3,750	15%
Rural	1,250	05%
	25,000	**100%**

HOUSEHOLDS WITH ADDITIONAL EARNERS

Number (Additional)	Households	Percentage
None	6,250	25%
One member	7,500	30%
Two members	5,000	20%
Three members	3,000	12%
More than three members	1,750	07%
	25,000	**100%**

HOUSEHOLD BY FAMILY SIZE

Number of Members	Households	Percentage
Two members	6,750	27%
Three members	7,500	30%
Four member	5,000	20%
Five members	3,750	15%
More than Five	2,000	08%
	25,000	**100%**

HOUSEHOLD BY FAMILY LIFE–CYCLE

Stage	No.of household	Percentage
Bachelorhood	6,250	25%
Honeymooners	5,000	20%
Parenthood	3,750	15%
Post–parenthood	8,750	35%
Dissolution	1,250	05%
	25,000	**100%**

Thus, the demographic and socio-economic features identified for market segments are -income level level of education, geographic spread, additional income earners, family size, family life cycle.

Mr. Aloknath picked up these demographic and socio -economic features because, the demand potential for household gadgets differs as per these features. Any segment which is large enough and profit potential is one that has the given top priority. Thus, size of TV would depend on family income and family size of course the size of the drawing room. The demand for second TV will depend on family life cycle income and members of the family.

SKILL DEVELOPMENT CASE TWO

Required:

Conduct formal interview to the managers of three retail clothing stores. Determine the degree to which they believe consumer personality and self image are important to the marketing activities of the store.

Solution:

The three retail stores I contacted in Bangalore city were Pantaloon, A little larger (ALL) and life–style. I spent each day three hours with each of the retail clothing managers and drawn the information about the importance of consumer personality and self image. The extent and degree of these two variables as to marketing were:

Variables	Storespantaloon	ALL	Life–style
1. Consumer personality	5 points	8 points	10 points
2. Self image	8 points	7 points	10 points

NOTE: Points as given by the managers of three stores out of 12 points each.

For our analysis purpose? I prefer to give two points and three points weightage. Then the results will be:

Retial stores	Variables Personality	Self Image	Total Points
Pantaloon	5x2=10	8x3=24	=34
aLL	8x2=16	7x3=21	=37
Life style	10x2=20	10x3=30	=50

From the above table it is clear that highest value is placed on personality by life style, followed by aLL and Pantaloon. Coming of self image again the highest value is placed by Life–style, followed by aLL and Pantaloon. Taking above personality the ranking is that life style Life style ranks Ist, followed by aLL and the third place is that of Pantaloon. In case of self image the first plane goes to Lifestyle, followed by Pantaloon and the last place is occupied by aLL. taking both the variables, again, Life–style stands first, aLL second and Pantaloon third.

Whatever may be the cause, it means that all three firms lay much emphasis on personality and self-image. The reason is the product they are dealing in are basically adding to the personality of a consumer. Where he tries to find self–image.

SKILL DEVELOPMENT CASE THREE

Required:

Visit three local restaurants and assess how each attracts clientele in different stages of the family life cycle.

Solution:

The term 'lifecycle' represents the progression of stages through which individuals and families proceed over time, starting with bachelorhood and ending with the dissolution of the basic unit. The experts have classified these family life–cycle into two categories as traditional and modern dividing them into five to thirteen stages. Here, tradition FLC is taken as a base for our analysis namely the stage of bachelorhood honeymooners parenthood, post parenthood and dissolution.

The three local restaurants I visited in Goa . Where Annapurna, Shivsagar and Ruchira. The post–visit findings are given in case of each restaurant.

State of FLC	Annapurna	Shivsagar	Ruchira
1. Bachelor hood	1. Ad in local paper 2. Ads on Local cable 3. Word of mouth	1. Ads in Local paper 2. Ads on local cable 3. Word of mouth	1 Ads in local paper 2. Ads on local cable 3. Word of mouth
II. Honeymooners	1. Word of mouth 2. Ads in regional papers	1. Ads in localpaper 2. Ads in regional paper	1. Ads on local cable 2. Ads in regional paper
III Parenthood	1. Word of mouth ads 2. Local news paper ads	1. Word of mouth ads 2. Local news paper ads	1. Word of mouth ads 2. Local news paper ads
IV Post parenthood	1. Local cable ads 2. Local print media	1. Local cable ads 2. Local print media	1.Local cable ads 2. Local print media
V. Dissolutions	1. Word of mouth 2. Personal contact	1. Word of mouth 2. Personal contact	1.Word of mouth 2. Personal contact

Thus, the different ways of attracting customers to the restaurants are – ads in local and regional paper local cable line; and personal contact. As there is competition and local consumers each is forced to compete with one another.

SKILL DEVELOPMENT CASE FOUR

Required:

You are the owner of two furniture stores, one catering to upper middle–class consumers and the other to lower middle–class consumer. How do social class differences influence each store's

a) Product lines and styles

b) Advertising media selection

c) The copy of communication styles used in advertisements

d) Payment policies

Solution

The concept of 'social class' is rather difficult to define as the concepts of riches and poverty are relative to place, time and person. Again, classification can be sample three classes and extending up to nine way classifications. However, theme is the same, though the degree is differing. Thus, I have used **Fourway social class category- "Lower"- "Lower–Middle, "Upper–Middle and "Upper class".** On the basis of annual income each class can be:

Social Class	Income bracket (Annual)
1. Upper Class	Rs. Above 5 crores
II Upper Middle	Rs. between 1.5 cores 4.999 crores
III Lower Middle	Rs. between 5,00,000 149.99 lakhs
IV Lower–Class	Rs. below 1,00,000

INFLUENCE OF SOCIAL CLASS DIFFERENCE ON:

A. Product lines and styles: When we move to upper middle–class having an income of 1.5 crores to 5 crores, they can afford to spend freely on stylish and elegant household furniture. The problem is how to spend. Instead of paying income tax, they prefer to spend lavishly on furniture items. They go in for latest, westernized furniture items. The style is one which is modern. On the contrary the lower middle class having an income group in the range of 5 lakh to 149.99 lakhs. They prefer to go gain for Indian style, but sophisticated furniture.

B. Advertising Media Selection: In case of upper middle–class, the costliest magazines- namely magazine like Architect, Life and style, Prism Interior and Exterior- and so on are suitable for print media. Again, on TV medium ads can be given. Incase of lower middle class, same magazines, and other local print media works. Of course TV ads have an impact. Other ways are organising the shows and exhibitions.

C. The copy and communication styles: In case of upper class, the copy and communication styles are more personal speaking life–style self image to a greater extent.

In case lower middle–class, self–image and life–style have bearing but not so sharply. The celebrities and the special serials "Rich and famous" - "Bold and beautiful" are the themes.

D. Payment Policies: In case of upper middle–class, the members are divided into two categories. One group goes in for lumpsum payment asking for fabulous discount. Others, ask for easy instalment payments. Infact, there is no need from them to avail the credit facilities.

In case of lower middle–class, also, these two types of customers come. The payment policies should be flexible- left to the parties. What is important is to strike the deal. It is so because, payment is not a problem. It can be outright purchase or instalment basis.

SKILL DEVELOPMENT CASE FIVE

Required

For each of the following Products and services indicate whom you would go to for information and advice.

a) The latest fashion is clothes

b) Banking

c) Air–travel

d) Vacation destination

e) A personal computer

For each situation, indicate person's relationship to you and your reasons for selecting him/her the source of information and advice.

Solution

Information is the base for making decisions. As a consumer I have to make sound and timely decision where I need reliable, adequate, latest information. What information I seek from which source, the extent all will depend on the decision–situation. I dependent on those sources which I think are most dependable and help in making decision to buy or not to buy.

A. The Latest Fashion Clothes: The world of fashion is really wonderful as it is colourful and ever changing. When I am to get informations with regard to fashion clothes the sources My sources of information are: (1) I Collect information by going through the fashion magazines, the ads that appear on TV Monitor, and will write to fashion designers, visit fashion clothes stores and out lets. Further, I will depend on the word of mouth advertising to listening to those people who are fashion–lovers.

Here, there is no question of personal relationship except in case of Wood of mouth informations who are my friends and relatives who like fashions in clothing and tell me what design suit most to my personality and self–image.

B. Banking: Banking is part and parcel of modern living. We need credits, security of funds, best sources without and presure. In case of banking sources I first got different banks and meet customer relationship officer who gives good deal of informations about bank and banking products that the bank has to offer as compared to its competition. Further, I also ask my peers, friends and relatives about the bank and its services as to how they evaluate. I would ask for the explanations on each point I need information before I come to conclusion.

I have my own relatives and friends working in banks whom I rely on as they are my well wishers.

C. Air Travel : Of late because of keen competition and increase in the consumer disposable income, each one is trying to fly by air than travel by train and road transport say by special taxi. I prefer air travel as it save time, and it is cost effective too. To travel by air, I collect informats from travel agents, air–ways officials and also depend on advertisements regarding tariffs and special sources they offer. After comparing, these, asking for the real experience of relatives and friends I decide to go by air.

I rely on ads in news–paper, information supplied by travel agents of course, relatives and friends. As it is highly competitive, each one wants to give facts and facts only to woo customers which is to my advantage.

D. Vacation Destination: To enjoy vacation with family, there is real need for a viable travel plan and its execution. Fortunately, there are travels and tours companies that organise tours and travels both inside. The country as well as outside the country depending on the size of the pocket, the size of the family the time that is available for tour.

The tour companies have standardised plans quoting per head expenditure, for the duration of tour covering longer distances, taking to many sites–wonders of the world, including stay and food and security. It is possible to organise such tours onour own. However, one doesnot have expertise. It is an art and needs good deal of contacts and alternatives to the changing situations.

Hence, I will visit local offices of travels and tours of fame like, Thomas Cook, TCI, COX and Kings, SOTC and so on. The variety of plans to choose. Again, I will collect information from friends and relatives who have travelled through these tour companies. I rely on the information given by these companies, their ads and my relatives and friends.

E. Personal Computer: Computer, especially personal computer was a luxury in 1990s where one was to pay atleast Rs. 1 lakh for the unit with dotmatrix printer and the pretty slow moving gadget. Today, a standard computer is available for Rs. 25,000 to 30,000 with fastest and latest confirguration with CD–ROM facilities. PC has become as necessary as TV set. With the lapse to time, further improvements and reduction in cost is possible due to technological advancement.

When I am to buy a computer, I will seek information from competitive computer dealers in city or cosmopolies. Again, daily newspapers and magazines give latest arrivals with all the details of cost, improvement in performance and so on. Again, I will write the makers of PCs who are designers and assemblers. Further, I cannot lose sight of my relatives and friends who have gone in for latest PC. I rely on the information made available by the specialised magzines and news papers. Again, the competitive dealers can give competitive bid rates and services guarantees and so on. This information helps in getting latest PC- real value for my money.

******